# High Energy Physics-1980
(XX International Conference, Madison, Wisconsin)

**AIP Conference Proceedings**
Series Editor: Hugh C. Wolfe
**Number 68**
Particles and Fields Subseries No. 22

# High Energy Physics-1980
(XX International Conference, Madison, Wisconsin)

Part 2

Editors
**Loyal Durand and Lee G. Pondrom**
University of Wisconsin

**American Institute of Physics**
New York                                    1981

Copying fees: The code at the bottom of the first page of each article in this volume gives the fee for each copy of the article made beyond the free copying permitted under the 1978 US Copyright Law. (See also the statement following "Copyright" below). This fee can be paid to the American Institute of Physics through the Copyright Clearance Center, Inc., Box 765, Schenectady, N.Y. 12301.

Copyright © 1981 American Institute of Physics

Individual readers of this volume and non-profit libraries, acting for them, are permitted to make fair use of the material in it, such as copying an article for use in teaching or research. Permission is granted to quote from this volume in scientific work with the customary acknowledgment of the source. To reprint a figure, table or other excerpt requires the consent of one of the original authors and notification to AIP. Republication or systematic or multiple reproduction of any material in this volume is permitted only under license from AIP. Address inquiries to Series Editor, AIP Conference Proceedings, AIP.

L.C. Catalog Card No.   81-65032
ISBN 0-88318-167-3
DOE CONF- 800724

## TABLE OF CONTENTS

Foreword . . . . . . . . . . . . . . . . . . . .

### HADRONIC INTERACTIONS, THEORY AND EXPERIMENT

<u>Hadron Spectroscopy, Experiment</u>: B. French, CERN, Organizer    1

    Evidence for Neutral $A_1$ and H Resonances in Charge Exchange, J.A. Dankowych, et.al. (presented by N.R. Stanton) . . . .    2

    General Features of the Reaction $\pi^+ p \to \Delta^{++} \pi^0 \pi^0$ AT 8 GeV/c, N.M. Cason, et.al. . . . . . .    5

    Unitary Description of Scalar Mesons in $\pi\pi$ and $\bar{k}k$, A.B. Wicklund . . . . . . . . . . . . . .    #

    E Meson Observation in $\pi^- p$ Interactions at 3.95 GeV/c, C. Dionisi, et.al. (presented by P.F. Loverre) . . . . .    8

    Search for Narrow $B\bar{B}(\pi)$ States in 16 GeV/c $\pi^- p$ and 5 GeV/c pp Interactions, S.U. Chung . . . . . . . . . . .    11

    Search for Baryonium States in the Reaction $pp \to pp\bar{p}p$ at 11.75 GeV/c, M.W. Arenton . . . . . . . . . . . . .    #

    Search for Narrow Baryonium States in $K^+ P$ Reactions, D. Frame, et.al. (presented by I.S. Hughes) . . . . . .    15

    Review of N-N Scattering, A. Yokosawa . . . . . . . .    18

<u>Hadron Spectroscopy, Theory</u>: R. Jaffe, MIT, Organizer    21

    Multiquark States and Exotics, A.J.G. Hey . . . . . . .    22

    Baryons With Chromodynamics, N. Isgur . . . . . . . .    30

    Difficult States in the Quark Model: Glueball and the Pion, J.F. Donoghue . . . . . . . . . . . . . .    35

<u>High Energy Hadronic Interactions, Exclusive Processes, Experiment and Theory</u>: L.W. Jones, Michigan, Organizer    43

    Small Angle Hadron Proton Elastic Scattering at Fermilab Energies, J. Lach . . . . . . . . . . . . . .    44

    Forward $\pi^- p$, $\pi^\pm$ He, and pHe Elastic Scattering at the SPS Energies, A.A. Vorobyov . . . . . . . . . . .    46

NOTE: The symbol # in place of a page number means that no manuscript was available.

First Observation of A Dip at $t=-1.4$ $(GeV/c)^2$ in $\bar{p}p$
Differential Elastic Cross Section at 50 GeV/c, M. Poulet .   48

Hadron Elastic Scattering and Diffraction Dissociation at
High Energies and Small Momentum Transfer, R.L. Cool, et.al.
(presented by H. Sticker) . . . . . . . . . . . . .   49

Measurement of the Total Cross Sections of $\Sigma^-$ and $\Xi^-$ on
Protons and Deuterons Between 75 and 135 GeV/c, Bristol,
Cambridge, Geneve, Heidelberg, Lausanne, Queen Mary College,
Rutherford Collaboration, (presented by A.A. Carter) . . .   51

Elastic Scattering Theory, R. Henzi . . . . . . . . .   52

Spin Effects in Nucleon-Nucleon Elastic Scattering,
K.M. Terwilliger, et.al. . . . . . . . . . . . . .   54

pp Spin Correlations at High $p_t$, I.P. Auer, et.al. (presented
by H. Spinka) . . . . . . . . . . . . . . . . .   56

Measurement of the Polarization Parameter in pp Elastic
Scattering at 200 GeV/c, W. Bartl, et.al. (presented by
F. Bradamante) . . . . . . . . . . . . . . . .   57

Polarization in $\pi^- p \to \pi^0 n$ Charge Exchange Reaction in the Low
Momentum Transfer Region at 40 GeV/c, A. Derevtchikov . . .   #

Proton Diffraction Excitation Function, J. Lee-Franzini . .   58

$(3\pi)$- Nucleon Collision in Coherent Production on Nuclei at
40 GeV/c, Dubna/Milan Collaboration, (presented by G. Bellini)   #

Dip and Kink Structures in High Energy Diffraction
Dissociation Processes, T.T. Chou . . . . . . . . . .   59

Measurement of Exclusive Hypercharge-Exchange Reactions at
35 to 140 GeV/c, E.N. May, et.al. . . . . . . . . .   60

Hyperon Polarization at Fermilab, K. Heller . . . . . .   61

Color, Flavor, and the Pomeron, J.W. Dash . . . . . .   62

<u>High Energy Hadronic Interactions</u>: R.C. Hwa, Oregon, Organizer   63

Description of Inclusive $\pi^+/\pi^-$ Production Ratios, $\bar{p}p$
Annihilation and Proton Diffraction Dissociation at Medium
Energies With the Quark Model, M. Markytan . . . . . .   64

Small $p_T$ Hadron Reactions by Monte Carlo Simulation,
Y. Ishihara and C. Iso (presented by C. Iso) . . . . . 68

Massive Meson Production, Correlations, Primordial $p_T$ etc.
K. Takahashi . . . . . . . . . . . . . . . 70

Meson Structure Functions and A-Dependence in Single Particle
Inclusive Hadron Fragmentation, Bari-Brown-CERN-Fermilab-
MIT-Warsaw Collaboration (presented by P.H. Garbincius) . 74

$\pi^-$-Neon Collisions at High Energy, W.D. Walker . . . . 77

A Review on DTU-Parton Model for hh and hA Collisions,
C.B. Chiu . . . . . . . . . . . . . . . 80

Multiparticle Jets From $\pi^+/K^+/p$ p Collisions at 147 GeV/C
Compared to $e^+e^-$ Annihilations, S.P. Ratti . . . . . 84

Study of the Basic Properties of Low-$p_T$ Multiparticle Systems
Produced in pp Collisions Removing Leading Protons, M. Basile,
(presented by B. Esposito) . . . . . . . . . . . 86

<u>High $P_T$ Hadronic Interactions, Experiment</u>: R. Cool, Rockefeller,
Organizer                                                              91

Preliminary Large Transverse Energy Cross Sections Measured
With A 2π Calorimeter Trigger, C. Favuzzi, et.al. (presented
by K. Pretzl) . . . . . . . . . . . . . . . 92

Results From Fermilab Experiment E236 on High -$P_t$ Events
From Proton-Proton Collisions, R.W. Williams . . . . . 98

Angular Dependence of High-$p_T$ $\pi^\circ$ Production, D.L. Owen et.al.
(presented by P.D. Grannis) . . . . . . . . . . 101

A Measurement of the Transverse Momenta of Partons and of
Jet Fragmentation as a Function of $\sqrt{s}$ in p-p Collisions,
H.J. Besch et.al. (presented by A.F. Rothenberg) . . . . 102

Diquark Fragmentation Studies at the CERN ISR, Aachen -CERN-
Harvard-Munich-Northwestern-Riverside Collaboration,
(presented by G.J. VanDalen) . . . . . . . . . . 108

Production of High $P_T \pi^\circ$'s at the CERN ISR, Athens-Brookhaven,
CERN Collaboration (presented by T. Fields) . . . . . 111

<u>High $p_t$ Hadronic Interactions, Theory</u>: J. Owens, Florida State,
Organizer                                                              113

QCD Corrections to Hadronic Large $p_T$ Scattering,
I. Hinchliffe . . . . . . . . . . . . . . . 114

Higher Twist QCD Terms in High-$P_T$ Pion Production, E.L. Berger
et.al. (presented by T. Gottschalk) . . . . . . . 118

Direct Evidence for the Three-Gluon Coupling, E. Reya . . . 123

Properties of Meson Structure Functions, R.L. Thews . . . 128

Direct Leptons and Photons in Hadronic Interactions, Experiment:
L. Resvanis, Athens, Organizer . . . . . . . . . . . . . 133

A Study of Direct Single Photons and Correlated Particles in
Proton-Proton Collisions at $\sqrt{s}$ = 62.4 GeV, H.J. Besch, et.al.
(presented by H.J. Besch) . . . . . . . . . . . 134

Direct Single Photon Production and A Search for Diphoton
Resonance or Continuum Production at Fermilab Energies,
B. Cox . . . . . . . . . . . . . . . . . . 137

Observed Differences in Event Structures of High-$p_T$ $\pi^\circ$ and
Single Photon Events Produced in pp Collisions in the CERN
ISR, Athens-BNL-CERN-Copenhagen-Lund-Rutherford-Tel Aviv
Collaboration (presented by D.C. Rahm) . . . . . . . 140

Massive Electron Pair Production at the ISR, Athens-Athens-
Brookhaven-CERN-Syracuse-Yale Collaboration (presented by
C. Kourkoumelis) . . . . . . . . . . . . . . 144

Muon Pair Production (M > 4.1 GeV/c$^2$) By $\pi^-$ at 150, 200 and
280 GeV/c in the CERN - NA 3 experiment, CEA, CERN, COLLEGE
DE FRANCE, Ecole Polytechnique, LAL Collaboration (presented
by Ph. Mine) . . . . . . . . . . . . . 147

Experimental Determination of Nucleon, $\pi$ and K Structure
Functions from Massive Dimuons Produced at 150 and 200 GeV/c.
NA3 Collaboration (presented by D. Decamp) . . . . . 149

Drell-Yan Muon Pairs at 40GeV/c From Tungsten Using $\pi^\pm$, $K^\pm$,
$\bar{P}$ and P Beams, M.J. Corden et. al. (presented by R.J. Homer) 154

A High-Statistics Study of Dimuon Production by 400 GeV/c
Protons, T.J. Roberts, et. al. (presented by T.J. Roberts) 156

Mu Pair Production at the ISR, P.L. Braccini . . . . . . #

Anomalous Low Mass $e^+e^-$ Pair Production in 17 GeV/c $\pi^-$p
Collisions, G. Abshire, et. al. (presented by P.D. Grannis) 158

Observation of a Direct Low-Mass $e^+e^-$ Continuum in $\pi^-p$ Interactions at 16 GeV/c, SLAC -Johns Hopkins University-Caltech Collaboration (presented by R. Stroynowski) . . . 161

<u>Direct Leptons and Photons in Hadronic Interactions, Theory</u>: A. Contogouris, McGill University, Organizer . . . 162

Large-$p_T$ Direct Photon Production and Photon-Hadron Correlations in QCD, A.P. Contogouris . . . . . . 163

Direct Photons, F. Halzen and D.M. Scott (presented by F. Halzen) . . . . . . . . . . . . 172

Perturbative QCD and Large $P_T$ Photoproduction, M. Fontannaz, et. al. (presented by D. Schiff) . . . . . . . . 180

Transverse Momentum Distribution of Dimuons in the Drell-Yan Process, D.E. Soper and J.C. Collins (presented by D. Soper) 185

## NEW PARTICLE PRODUCTION

<u>New Particle Production I</u>: C. Brown, Fermilab, Organizer 189

Diffractive Hadroproduction of Charmed D Mesons, L.J. Koester et. al. (presented by L.J. Koester) . . . . . . . . 190

Production of Charm in the SFM at the ISR, CCHKK and ACCDHW Collaborations (presented by G. Sajot) . . . . . . . 192

Production of Charmed Particles at the CERN ISR, Aachen-CERN-Harvard-Munich-Northwestern-U.C. Riverside Collaboration (presented by F. Muller) . . . . . . . . . . . . 193

$\Lambda_c^+$ Production in $\nu$-D Interactions, IIT-Maryland-Stony Brook-Tohoku-Tufts Collaboration (presented by T. Kitagaki) . . 194

Observation of Charmed Baryon Production in $\nu p$ Interactions Aachen-Bonn-CERN-Munich(MPI)-Oxford Collaboration (presented by P.C. Bosetti) . . . . . . . . . . . . . . . . 196

Production of J/$\psi$ and Other Particles in $\pi^-$N Collisions at $\pi^-$ Momenta 140-197 GeV/c, R. Barate et. al. (presented by J.G. McEwen) . . . . . . . . . . . . . . . . . 197

High Statistics Study ($\sim 10^6$ Events) of J/$\psi$ Production and $\Upsilon$ Production in the Energy Range 150 to 280 GeV by $\pi^{\pm}$, $p^{\pm}$ Incident Particle, G. Burgun et. al. (presented by P. Delpierre) . . . . . . . . . . . . . . . . 201

Multimuon Production in 280 GeV $\mu^+$ Iron Interactions, European Muon Collaboration (presented by R.P. Mount) 205

Open and Hidden Charm Muoproduction, A.R. Clark, et. al. (presented by A.R. Clark) . . . . . . . . . . . . 212

Photoproduction, Experiment: T. Nash, Fermilab, Organizer   219

  Measurement of the J/ψ Photoproduction Cross Section, J.P. Cumalat (presented by I. Gaines) . . . . . . . . . 220

  Results of the WA4 Photoproduction Experiment at the CERN SPS, B.D'Almagne . . . . . . . . . . . . . . 221

  Evidence for $D\bar{D}$ Diffractive Photoproduction Off Silicon, E. Albini, et. al. (presented by E. Bertolucci) . . . . . 227

  Shadowing of Virtual Photons, W.A. Loomis et. al. (presented by W.A. Loomis) . . . . . . . . . . . . . . 230

  Compton Scattering and Quasi-Elastic Omega Photoproduction at 50-130 GeV, A.M. Breakstone, et. al. (presented by D.B. Smith) . . . . . . . . . . . . . . . . . 233

New Particle Production II, Experiment: C. Baltay, Columbia, and J. Fry, Wisconsin, Organizers   236

  Measurement of Prompt Neutrino Fluxes in a Beam Dump Experiment, CERN-Dortmund-Heidelberg-Saclay Collaboration (presented by K. Kleinknecht) . . . . . . . . . . 237

  Experimental Study of Prompt Neutrino Production in 400 GeV Proton-Nucleus Collisions, CHARM Collaboration (presented by F. Niebergall) . . . . . . . . . . . . . . . 242

  Study of Prompt Neutrino Production in Proton-Cu Interactions Using BEBC in Connection With Narrowband $\nu_e$ Results, Aachen-Bonn-CERN-IC London-Oxford-Saclay Collaboration (presented by P.O. Hulth) . . . . . . . . . . . 247

  28 GeV/c Beam Dump Experiments, J.M. LoSecco . . . . 252

  Hadronic Production of Prompt Single Muons, B.C. Barish et.al. (presented by K.W. Merritt) . . . . . . . . . . . 257

Phenomenology of New Particle Production: E. Reya, Dortmund, Organizer   262

  Leptoproduction of Heavy Quarks, J.P. Leveille and T. Weiler, (presented by T. Weiler) . . . . . . . . . . . 263

  QCD Fusion Models for Heavy Quark Production, M. Gluck . . 268

  Gluon Fusion, V. Barger, et. al. (presented by W.Y. Keung) . 271

New Quark and Weak Boson Signatures at $\bar{p}p$ Colliders,
F. Halzen and D.M. Scott (presented by D.M. Scott) . . . 273

The Intrinsic Charm of the Proton, P. Hoyer . . . . . 277

Heavy Flavor Production With Final State Interaction,
B. Margolis . . . . . . . . . . . . . . . . . . . 279

Diffractive Photoproduction of Strangeness and Charm-Overlap
of QCD and Regge Models, D.P. Roy . . . . . . . . 282

Higher Order Corrections for $J/\psi$ Photoproduction, J.F. Owens 284

QCD Corrections to Neutrino Charm Production, T. Gottschalk 286

Charm and Heavy Leptons in Neutrino Interactions:
Production and Decay, R.V. Konoplich, et. al. . . . . . 288

Suppressed Production of F, F*, and $\Sigma c$, P. Mukhopadhyay . 290

<u>Quark Searches, Experiment and Theory</u>: A. Zichichi, CERN and
G. Morpurgo, INFN-Genoa, Organizers  292

    Quark Search in PETRA, J. Von Krogh . . . . . . . . #

    Quark Search in High Energy Experiments, G. Valenti . . . 293

    Quark Confinement, Michael Creutz . . . . . . . . . 296

    Additional Evidence for Fractional Charge of 1/3 e on Matter,
    G.S. LaRue, et. al. (presented by G.S. LaRue) . . . . . 302

    Search for Quarks in Matter, M. Marinelli and G. Morpurgo,
    (presented by G. Morpurgo) . . . . . . . . . . 308

## LOW ENERGY WEAK INTERACTIONS AND WEAK DECAYS

<u>Low Energy Weak Interactions, Experiment</u>: F. Reines and
J. Schultz, UC-Irvine, Organizers  317

    Parity Violation in Proton-Nucleus Scattering at 6 GeV/c,
    N. Lockyer, et. al. (presented by T.A. Romanowski) . . . 318

    Experimental Constraints on Models of CP Violation,
    M.P. Schmidt, et. al. (presented by M.P. Schmidt) . . . 322

    Neutrino Instability, H.W. Sobel, et. al. (presented by
    H.W. Sobel) . . . . . . . . . . . . . . . . . 326

    Neutrino Oscillation Phenomenology, V. Barger . . . . . 334

Emulsion Experiments on Short-Lived Particles, Experiment : J. Prentice, Toronto, Organizer ... 341

    Observation of Charmed Particles Produced by High-Energy Photons in Nuclear Emulsions Coupled With a Magnetic Spectrometer, Photon-Emulsion and Omega-Photon Collaborations (presented by G. Diambrini-Palazzi) ... 342

    Observation of Charmed $F^+$ Meson Produced by Neutrino Interaction in Emulsion, R. Ammar et. al. ... 348

    Charmed Particle Production and Decay Lifetimes and a Neutrino Oscillation Test, K. Niu ... 352

    Production of Charm Particles in Proton-Emulsion Interactions at 400 GeV/c, Bombay-Chandigarh-Delhi-Jammu Collaboration (presented by P.K. Malhotra) ... 361

Weak Decays, Experiment: G. Goldhaber and G. Abrams, UC-Berkeley Organizers ... 363

    Measurement of the Vector and Axial Vector Coupling Constants in $\Lambda^\circ \to p + e^- + \bar{\nu}$, D.A. Jensen et. al. ... 364

    Measurement of the Decays $\tau^- \to \rho^- \nu_\tau$ and $\tau^- \to K^{*-}(892) \nu_\tau$ Using the Mark II Detector at SPEAR, J. Dorfan ... 368

    Recent Results on Decays of D Mesons From Mark II, SLAC-LBL Mark II Collaboration (presented by A.J. Lankford) ... 373

    Measurement of Inclusive $\eta$ Production in $e^+e^-$ Interactions Near Charm Threshold, Crystal Ball Collaboration (presented by F.C. Porter) ... 380

Weak Decays, Theory: M. Suzuki, U.C.-Berkeley, Organizer ... 385

    Decay of Charm, N. Cabibbo ... #

    Penguins and the $\Delta I=1/2$ Rule, C.T. Hill ... 386

    The Role of Non-Spectator Interactions in Charm and Bottom Decays, V. Barger, et. al. (presented by J.P. Leveille) ... 390

    Comment on W-Exchange in B-Decay, S.P. Rosen ... 396

        QUANTUM FLAVOR DYNAMICS AND UNIFIED THEORIES

QFD I: R. Peccei, Max-Planck Institute, Organizer ... 397

    CP Violation in the Six-Quark Model, M.B. Wise ... 398

Parity Violation in Nuclei, D. Tadić . . . . . . . . . 404

Horizontal Interactions, R.D. Peccei . . . . . . . . 411

Operator Analysis of New Physics, H.A. Weldon . . . . . 415

QFD II : E. Ma, Hawaii, Organizer         421

Radiative Corrections in the Standard Model, E. Ma . . . 422

Multi-W and Z Bosons, V. Barger, et. al. (presented by
W.Y. Keung) . . . . . . . . . . . . . . . . . . 427

Departure From Weinberg-Salam Model and Grand Unification,
N.G. Deshpande . . . . . . . . . . . . . . . . . 431

Higgs Meson and Radiative Corrections, K.T. Mahanthappa . . 436

Possible Spin-One Resonances in the Strongly Coupled Higgs
Sector, M. Kobayashi and T. Matsuki (presented by
M. Kobayashi) . . . . . . . . . . . . . . . . . . 440

QFD III : R. Slansky, Los Alomos, Organizer         444

Decoupling Theorems and Effective Field Theories, B.A. Ovrut
and H.J. Schnitzer (presented by H.J. Schnitzer) . . . . 445

Neutral Lepton Mass Matrix, J.A. Harvey, et. al. (presented
by P. Ramond) . . . . . . . . . . . . . . . . . . 451

Calculability of the N-P Mass Difference in Gauge Theories,
J. Kiskis . . . . . . . . . . . . . . . . . . . 455

Sequential Internal Supersymmetry, Y. Ne'eman and
S. Sternberg (presented by Y. Ne'eman) . . . . . . . . 460

QFD IV : J. Pati, Maryland, Organizer         463

Probing the Hierarchy of Grand Unification Through
Conservation Laws, J. Pati . . . . . . . . . . . . 464

Majorana and Dirac Masses for Neutrinos, E. Witten , , , #

Aspects of Unified Gauge Theories, Q. Shafi . . . . . . 470

Majorana Masses for Neutrinos and Neutron Oscillations (N↔N̄)
as Tests of Unification Models With Intermediate Mass Scales,
R.N. Mohapatra . . . . . . . . . . . . . . . . . 478

Neutrino Oscillations of the Second Class, V. Barger, et. al.
(presented by P. Langacker) . . . . . . . . . . . . 483

**Dynamical Symmetry Breaking, Theory**: M.A.B. Bég, Rockefeller, Organizer ... 488

    Dynamical Symmetry Breaking and Hypercolor, M.A.B. Bég ... 489

    Testing Technicolor Theories, G.L. Kane ... 493

    Fermion Masses and Weak Isospin in Technicolour Models, P. Sikivie ... 496

    Dynamically Broken Gauge Theories, H.R. Pagels ... 501

    Suppression of Superheavy Magnetic Monopoles in Grand Unified Theories, So-Young Pi ... 505

**Weak Decays, Theory**: D. Nanopoulos, CERN, Organizer ... 509

    Quark Flavor Mixing and its Physical Implications, Ling-Lie Chau Wang ... 510

    CP Violation at High Energies, M.K. Gaillard ... 517

    Soft CP Violation: Present Status, G. Senjanović ... 524

    A Quantum Structuredynamic Model of Quarks, Leptons, Weak Vector Bosons, and Higgs Mesons, O.W. Greenberg and J. Sucher (presented by O.W. Greenberg) ... 531

    A Solution to the Problem of the Fermion Masses, D.V. Nanopoulos ... 533

## $e^+e^-$ PHYSICS AND ELECTROMAGNETIC PROPERTIES, EXPERIMENT AND THEORY

**QED and Electromagnetic Properties of Particles**: Magnetic Moments, Form Factors, Low Energy $e^+e^-$ Interactions, Experiment and Theory, T. Devlin, Rutgers University, Organizer ... 536

    Radiative Widths of K* and ρ Mesons, D. Berg, et al., (presented by P.A. Thompson) ... 537

    Measurement of the $\Xi^0$ and $\Xi^-$ Magnetic Moments, R. Handler, et. al. (presented by R. Handler) ... 539

    Magnetic Moments of Quarks in Baryons and Mesons, J.L. Rosner ... 540

    Observation of a $\phi'(1.65)$ Vector Meson in $e^+e^-$ Annihilation at DCI, J.C. Bizot, et. al. (presented by J.C. Bizot) ... 546

    Electromagnetic Form Factors of Hadrons, B.T. Chertok ... 547

Measurement of the Elastic Electron-Neutron Cross Section at High $Q^2$, S. Rock, et. al. (presented by B.T. Chertok) . . . 550

Test of Electro-Weak Theories at Petra, A. Böhm . . . . . 551

$e^+e^-$ Physics I: W. Frazer, U.C.-San Diego, Organizer . . . 562

    Two-Photon Reactions in $e^+e^-$ Colliding Beams, W.R. Frazer. . 563

    Structure Functions and High Twist Contributions in Perturbative Quantum Chromodynamics, S.J. Brodsky and G.P. Lepage (presented by S.J. Brodsky) . . . . . . . 568

    Two-Photon Results From SPEAR, A. Roussarie . . . . . . 573

    Two Photon Processes at PETRA, W. Wagner . . . . . . . 576

    Rho Rho Production by Two Photon Scattering, TASSO Collaboration (presented by E. Hilger) . . . . . . . 586

High Energy $e^+e^-$ Reactions (R, Inclusive Distributions, Jets) Experiment: H. Spitzer, DESY, Organizer . . . 589

    Measurements of R and Search for New Thresholds at PETRA, D. Cords . . . . . . . . . . . . . . . . . . . 590

    Features of Inclusive Hadron Production in $e^+e^-$ Annihilation at PETRA, D. Pandoulas . . . . . . . . . . . . 596

    TASSO Results on Jets and QCD, TASSO Collaboration, (presented by S.L. Wu) . . . . . . . . . . . . . . . . 604

    First Results from CELLO, H. Oberlack . . . . . . . . #

High Energy $e^+e^-$ Interactions: S. Orito, DESY, Organizer . . . 615

    Recent Results From the JADE Collaboration (presented by S. Yamada) . . . . . . . . . . . . . . . . . 616

    PLUTO Results on Jets and QCD, PLUTO Collaboration (presented by V. Hepp) . . . . . . . . . . . . . . . . 622

    Results on Jets, QCD and Lepton Production From the MARK J, H.B. Newman . . . . . . . . . . . . . . . . 627

Interpretation of $e^+e^-$ Reactions, Theory: P. Hoyer, Nordita, Organizer . . . 638

    Using $e^+e^-$ Cross-Sections to Test QCD and to Search for New Particles, R.M. Barnett . . . . . . . . . . . . . 639

The Pertubative Calculation of Event Shapes in $e^+e^-$ Annihilation, R.K. Ellis, et.al. (presented by A.E. Terrano)    644

A QCD Analysis of Jets in $e^+e^-$ Annihilation, A. Ali . . .   648

Jet Acollinearity and Quark Form Factors, W.J. Stirling . .   654

Monte Carlo Jet Generation, R. Odorico . . . . . .   659

$e^+e^-$ Physics IV, G. Hanson, SLAC, Organizer     663

    Crystal Ball Studies of the Reaction $\psi' \to \gamma\gamma\psi$, T.H. Burnett    664

    Radiative Transitions From the $\psi(3095)$ and $\psi'(3684)$ to Ordinary Hadrons, D.L. Scharre . . . . . . . . . .   668

    Hadronic Decays of the $\eta_c$, K. Königsmann . . . . . .   675

    Observation of $\eta_c$ (2980) in $\psi'$ (3684) Radiative Decay, SLAC-LBL Mark II Collaboration (presented by G. Trilling) .   679

    Production and Decays of D* Mesons, H.F.W. Sadrozinski . .   681

$e^+e^-$ Physics V, B. Gittelman, Cornell, Organizer     686

    Parallel Session on Upsilon Spectroscopy, B. Gittelman. . .   687

    Experimental Evidence for the $\Upsilon$-Decay Into 3 Gluons, C. Grupen . . . . . . . . . . . . . . .   689

    Resonance Parameters of $\Upsilon$ and $\Upsilon'$ and Inclusive Spectra Measured at DORIS, W. Schmidt-Parzefall . . . . .   692

    New Results on $\Upsilon'$ (10.01) Hadronic Decay, F. Messing . .   696

    Results on the $\Upsilon$ Resonances From the CUSB Group at CESR, J.K. Yoh . . . . . . . . . . . . . . .   700

    First Results on Bare b Physics, E.H. Thorndike . . . .   705

**Onium Theory and Spectroscopy, Theory,** C. Quigg, Fermilab, Organizer     712

    (Quark) Onium Theory and Spectroscopy, C. Quigg . . . .   713

    A Fit of Heavy Quarkonia, A. Martin . . . . . . .   715

    Inverse Scattering and the $\Upsilon$ Family, C. Quigg and J.L. Rosner (presented by J.L. Rosner) . . . . . .   719

    QCD Corrections to Quarkonia Decays, W. Celmaster . . .   725

QCD Inspired Bag Model of Quarkonium, P. Hasenfratz, et. al. (presented by J. Kuti) . . . . . . . . . . . . 728

E(1440): Glueball or Quarkonium?, S. Meshkov . . . . . 732

## LEPTON-NUCLEON INTERACTIONS

eN, μN, νN Interactions, 1: Cross Sections, Scaling, Experiment, A. Benvenuti, CERN, Organizer  734

New Results on Inclusive νFe Charged Current Interactions, CERN-Dortmund-Heidelberg-Saclay Collaboration (presented by J.G.H. de Groot) . . . . . . . . . . . . . 735

Experimental Study of Neutral and Charged Current Cross Sections and y-Distributions for (Anti) Neutrinos, CHARM Collaboration (presented by K.H. Mess) . . . . . . . 738

Recent Results From the CFRR Neutrino Experiment at Fermilab, B. Barish, et. al. (presented by M. Shaevitz) . . . . . 741

Neutral Current Interactions in νN and νP Collisions and Study of the Produced Hadrons, H. Yuta . . . . . . . 746

Evidence for Gluon Radiation in High Energy Neutrino Interactions, Berkeley-Fermilab-Hawaii-Seattle-Wisconsin Collaboration (presented by V.J. Stenger) . . . . . . 752

Energy Distribution and Average Transverse Momentum of Produced Hadrons in Deep Inelastic Muon Scattering, European Muon Collaboration (presented by F.W. Brasse) . . 755

eN, μN, νN Interactions, II, U. Amaldi, CERN, Organizer  760

BEBC and Gargamelle Data on Hadronic Final State in ν Interactions, C. Matteuzzi . . . . . . . . . . . 761

Measurement and Analysis of $F_2^{\nu,\bar{\nu}}$ and $xF_3^{\nu,\bar{\nu}}$ in the $Q^2$-Region 0.5 - 40 GeV$^2$, Gargamelle SPS-Collaboration (presented by H. Weerts) . . . . . . . . . . . . . . . 766

The Structure and the Amount of the $q\bar{q}$ Sea and Broadening of Charm Jets, CERN-Dortmund-Heidelberg-Saclay Collaboration (presented by J. Knobloch) . . . . . . . . . . . 769

Deep Inelastic Muon-Nucleon Scattering at High $Q^2$, Bologna-CERN-Dubna-Munich-Saclay Collaboration (presented by M. Klein) . . . . . . . . . . . . . . . . 773

Structure Function Measurements in Muon-Iron and Muon-Proton Scattering, and a QCD Analysis, European Muon Collaboration (presented by P.R. Norton) . . . . . . . . . . . . 777

New Results on Polarized Electron-Proton Scattering at SLAC, BERN-Bielefeld-KEK-Kyoto-Peking-Saclay-SLAC-Tsukuba-Yale Collaboration (presented by K.P. Schüler) . . . . . . 781

<u>Interpretations of eN, µN, νN Reactions, Theory,</u> R. Petronzio, CERN, Organizer      784

    Quantum Chromodynamics and Deep-Inelastic Scattering, A.J. Buras . . . . . . . . . . . . . 785

    On the Shape of Hadron Structure Functions, F. Martin . . 797

    Normalizing the Renormalization Group in Deep Inelastic Leptoproduction, R.L. Jaffe . . . . . . . . . . 801

    Structure Functions and High Twist Contributions in Perturbative Quantum Chromodynamics, S.J. Brodsky and G.P. Lepage (presented by S.J. Brodsky) . . . . . . . 805

## COSMIC RAYS

<u>Cosmic Rays: Experiment and Theory,</u> G. Yodh, NSF/Maryland and G. Cassiday, Utah, Organizers      811

    Total Cross Section and Scaling at 100 TeV from Cosmic Rays, G.B. Yodh . . . . . . . . . . . . . . #

    Hadronic Interactions Around 50 TEV, R.W. Ellsworth, et. al. (presented by R.W. Ellsworth) . . . . . . . . . 812

    Simulation of Centauros, R.W. Ellsworth, et. al. (presented by T.K. Gaisser) . . . . . . . . . . . . 816

    First Operation of Fly's Eye ($E>10^{17}$ eV), G. Cassiday . . #

    Next Generation of Cosmic Ray Experiments--Ideas, D. Cline #

    $\pi/p$ and N/CH Ratios at 300 -3000 GeV at Mountain Altitudes, A. Amatuni . . . . . . . . . . . . . #

    Data Photographs From the Auckland Cosmic Ray Telescope, P.C.M. Yock . . . . . . . . . . . . . 820

    A Cosmic Ray Interaction of Energy Greater Than 130 TeV T.A. Koss, et. al. (presented by R.J. Wilkes) . . . . . 824

Non Linear Effects in Nuclear Matter and Self-Induced Transparency (SIT), G.N. Fowler and R.M. Weiner (presented by R.M. Weiner) . . . . . . . . . . . . . . . . 826

## NEW ACCELERATORS AND NEW EXPERIMENTAL TECHNIQUES

<u>New Accelerators and Collider Developments in the 80's,</u> D. Cline, Wisconsin, Organizer . . . . . . . . . . . . 828

    Cornell 50 x 50 GeV Storage Ring, B.D. McDaniel . . . . 829

    LEP at CERN, W. Schnell . . . . . . . . . . . . . 833

    CERN SPS Antiproton-Proton Collider, C. Rubbia . . . . . #

    Tevatron Phase I Antiproton-Proton Collider at Fermilab, D. Young . . . . . . . . . . . . . . . . . . . #

    Isabelle at BNL, J. Sanford . . . . . . . . . . . . #

    HERA, B.H. Wiik . . . . . . . . . . . . . . . . 837

    CHEER, Canadian High Energy Electron Ring (presented by R.J. Hemingway) . . . . . . . . . . . . . . . . 839

    Survey of Other ep Machine Designs Around the World, W. Lee   #

    Status of the Fermilab Energy Saver/Doubler Dipole, J.R. Orr   846

    Performance, Construction and Installation of Superconducting Magnets at Isabelle, E. Bleser . . . . . . . . . . . #

    Development of RF Superconducting Cavities for Large Storage Rings, E. Picasso . . . . . . . . . . . . 849

    A Muon Storage Ring for Neutrino Oscillations Experiments, D. Cline and D. Neuffer (presented by D. Neuffer) . . . . 856

    The IHEP Accelerator Storage Complex Status Report, L.D. Soloviev . . . . . . . . . . . . . . . . . 858

    The Tristan-KEK Future Project, T. Nishikawa . . . . . . 859

<u>New Detectors and Experimental Techniques,</u> T. Ludlam, Brookhaven Organizer   862

    Status and Prospects for High Resolution Streamer Chambers for Heavy Quark Studies, J. Sandweiss . . . . . . . . #

    Holographic Photography of Bubble Chamber Tracks : A Feasibility Test, L. Montanet . . . . . . . . . . . 863

High-$p_T$ Event Trigger and Processor for the ISR Axial Field Spectrometer, BNL-CERN-Copenhagen-Lund-Rutherford-Tel Aviv Collaboration (presented by C.W. Fabjan) . . . . . . . 867

Particle Detection Techniques Under Development at Brookhaven, V. Radeka . . . . . . . . . . . . . . . #

First Experiences with a Fastbus System at Brookhaven, L.B. Leipuner, et. al. (presented by L.B. Leipuner). . . . 873

## GAUGE THEORIES AND MATHEMATICAL PHYSICS

<u>Gauge Theory I,</u> E. Witten, Harvard, Organizer      876

    Semiclassical Approach to Large N Expansion, A. Jevicki . . 877

    Lattice Field Theory for Strings, D. Weingarten . . . . . 882

    QCD From the Effective Lagrangian Point of View, J. Schechter 886

    The Importance of Being Topologically Excited, D.G. Caldi 891

<u>Gauge Theory II:</u> Lattice Theories, Loop Spaces, Model Theories, R. Jackiw, MIT, Organizer      895

    Monte Carlo Studies of Lattice Gauge Theories, C. Rebbi . . 896

    The Fluctuating String in QCD, R.B. Pearson . . . . . . 902

    Monte Carlo Study of SU(2) Gauge Theory at Finite Temperature J. Kuti, et. al. (presented by J. Kuti) . . . . . . . 906

    Hidden Symmetry of the Point Magnetic Monopole, R. Jackiw . 911

    Induced Gravitation, S.L. Adler . . . . . . . . . . 915

<u>Mathematical Physics,</u> N. Christ, Columbia, Organizer      916

    Anomalous Ward Identities and Path Integration, K. Fujikawa 917

    Hidden Symmetry in Yang Mills, L. Dolan . . . . . . . 923

    Comments on the Integrability of the Loop-Space Chiral Equations, C. Gu and L.C. Wang (presented by L.C. Wang) . . 929

    General Classical Solutions of the Euclidean $CP^{n-1}$ Model, A.M. Din and W.J. Zakrzewski (presented by A.M. Din) . . . 936

<u>New Theoretical Developments,</u> P. Ramond, Caltech, Organizer      941

    Ternary Algebras as the Basis of a Dynamical Theory of Subconstituents, I. Bars . . . . . . . . . . . . 942

Composite μ and τ Families in an $E_6$ Unified Model, G.L. Shaw    948

Infrared Properties of the Coupling Constant in Non-Abelian Gauge Theories, J.S. Ball . . . . . . . . . . . . 954

A Relaxation Method for the Euclidean Yang-Mills Action Functional and its Application to n=2,3 Multimonopole Solutions, S.L. Adler and T. Piran (presented by S.L. Adler)    958

<u>Supersymmetry Supergravity,</u>    R. Arnowitt, Northeastern, Organizer    963

    Prospects for Supergravity, B. Zumino . . . . . . . . 964

    Is Flavor Proliferation Explicable by Supergravity?, P. Frampton . . . . . . . . . . . . . . . . 970

    Applications of Superfield Feynman Diagrams, M.T. Grisaru    . 971

    Riemannian Superspace Reduction and Supergravity Geometry in Superspace, R. Arnowitt and P. Nath (presented by both) . . 975

    Geometric Ghosts and Unitarity, Y. Ne'eman . . . . . . 981

    High Spin Fields, T.L. Curtright . . . . . . . . . . 985

<u>QCD I,</u> G. Sterman, SUNY-Stony Brook, Organizer    989

    Infrared Divergences in Quantum Chromodynamics, A. Andrasi, et. al. (presented by J. Frenkel) . . . . . . . . . 990

    Intrinsic Transverse Momentum in Gauge Theories, J.C. Collins    996

    Hadron Wavefunctions and Structure Functions in QCD, T. Huang 1000

    Summing QCD Corrections in Infrared Sensitive Quantities, M. Ciafaloni . . . . . . . . . . . . . . . . 1006

<u>QCD II,</u> W.A. Bardeen, Fermilab, Organizer    1012

    Condensation of $(G^a_{\mu\nu})^2$ in Quantum Chromodynamics, Y. Kazama    1013

    Local Covariant Operator Formalism of Non-Abelian Gauge Theories-a non-perturbative approach to quark confinement in the Heisenberg picture-, I. Ojima . . . . . . . . . 1017

    Operator Ordering and Feynman Rules in Gauge Theories, N.H. Christ . . . . . . . . . . . . . . . 1022

    Chiral Symmetry Breakdown in Large-N QCD, E. Witten . . . 1026

Glueballs in QCD Sum Rules, V. Zakharov . . . . . . . 1027

QCD III, E. Golowich, Massachusetts, Organizer       1031

   Glueballs and Oddballs: Their Experimental Signature,
   C.E. Carlson . . . . . . . . . . . . . . . . 1032

   Hadronization Problem in Quark and Gluon Jets, R.C. Hwa . . 1038

   Scattering Quarks Off the Vacuum, J.F. Donoghue . . . . 1042

   Nonperturbative Vacuum and Hard Scattering Processes,
   N. Sakai . . . . . . . . . . . . . . . . . . 1045

   Goldberger-Treiman Constants of Dynamical Near-Goldstone
   Modes, M.A.B. Bég . . . . . . . . . . . . . . 1048

   A Goldstone Pion With Bag Confinement, R.W. Haymaker and
   T. Goldman (presented by R.W. Haymaker) . . . . . . 1051

## PLENARY SESSIONS

### ACCELERATORS AND EXPERIMENTAL TECHNIQUES       1055

Accelerator and Instrumentation Prospects of Elementary Particle
Physics, A.N. Skrinsky, Novosibirsk . . . . . . . . 1056

### WEAK INTERACTIONS       1094

QFD and Unification, H. Sugawara, KEK . . . . . . . . 1095

Low Energy Weak Interactions and Decays, G.H. Trilling,
UC-Berkeley . . . . . . . . . . . . . . . . . 1139

Low-Energy Weak Interactions: Theory, S. Pakvasa, Hawaii . 1164

### HADRON DYNAMICS, EXPERIMENT AND THEORY       1195

Light Quark-Hadron Spectroscopy, L. Montanet, CERN . . . . 1196

Hadron Dynamics, V. Zakharov, ITEP, Moscow . . . . . . 1234

Characteristics of Inclusive and Exclusive Final States at Low
$p_T$, all Beams, Experiment: Multiplicities, Correlations,
Polarization, A-dependence, Exclusive Processes, Cosmic Rays,
H. Miettinen, Helsinki . . . . . . . . . . . . . 1277

Hadron Structure From Lepton Beams, F. Sciulli, Caltech . . 1278

Experimental Results in Deep Inelastic Hadron Interactions,
J. Lefrancois, France . . . . . . . . . . . . . . . 1318

Perturbative QCD, C.H. Llewellyn Smith, Oxford . . . . . . 1345

## $e^+e^-$ PHYSICS AND NEW PARTICLE PRODUCTION  1378

New $e^+e^-$ Physics, B.H. Wiik, DESY . . . . . . . . . . 1379

New Flavor Production in $\gamma$, $\mu$, $\nu$, and Hadron Beams, S. Wojcicki,
Stanford . . . . . . . . . . . . . . . . . . . . 1430

Phenomenology of New Particle Production, R.J.N. Phillips,
Rutherford Lab . . . . . . . . . . . . . . . . . . 1470

New Flavor Spectroscopy, K. Berkelman, Cornell . . . . . . 1499

## NEW THEORETICAL DEVELOPMENTS  1530

Recent Progresses in Gauge Theories, G. Parisi, INFN, Frascati  1531

New Ideas and Speculations, L. Susskind, Stanford . . . . . 1569

Special Session: The future of High Energy Physics, Organized by
B. Durand . . . . . . . . . . . . . . . . . . . . 1570

Local Organizers . . . . . . . . . . . . . . . . . 1571

Advisory Committees . . . . . . . . . . . . . . . . 1572

Assistant Scientific Secretaries . . . . . . . . . . . 1573

Conference Staff . . . . . . . . . . . . . . . . . 1574

Conference Delegates . . . . . . . . . . . . . . . . 1575

Participation by Country . . . . . . . . . . . . . . 1593

Contributed Papers . . . . . . . . . . . . . . . . . 1594

Chapter 7

Cosmic Rays

# COSMIC RAYS: EXPERIMENT AND THEORY

G. Yodh, NSF/Maryland
G. Cassiday, Utah, Organizers

## HADRONIC INTERACTIONS AROUND 50 TEV

R. W. Ellsworth and G. B. Yodh
Department of Physics
University of Maryland, College Park, Md. 20742

T. K. Gaisser
Bartol Research Foundation of The Franklin Institute
University of Delaware, Newark, De. 19711

## INTRODUCTION

Large emulsion chambers[1] have by now accumulated a significant number of cosmic ray interactions with visible energies of twenty to several hundred TeV, corresponding to interaction energies of 50-1000 TeV. The interactions fall into two classes: (1) local interactions in the detector and (2) interactions in the atmosphere above detector. In this paper we concentrate on interpretation of local interactions in the detector, called C-jets by the Brazil-Japan group.[2] The existence of complicated scanning selection criteria, together with large fluctuations and the steep energy spectrum of the cosmic ray beam, make it essential to carry out a detailed Monte Carlo simulation in order to interpret the data. In particular, the most important limitation of the technique is the fact that only energetic secondary $\gamma$-rays (e.g. from decay of $\pi^o$'s produced in the interaction) are detected. Charged hadrons typically escape the detector without further interaction.

We carried out such a simulation some time ago[3] in which we showed that the main features of the energy dependence and distributions of pseudo-rapidity of secondary photons could be understood as a consequence of the various selection effects superimposed on a straight forward (scaling) extrapolation of accelerator data. The large observed fraction of events with big masses ($M_\gamma$) of clusters of secondaries and with high $P_T$ did not, however, emerge from the original simulation, in which all secondaries were chosen from ordinary $P_T$-distributions without a high-$P_T$ component.[4] A major result of the simulation that we report here is that inclusion of a hard scattering component as suggested by parton models of hadronic processes may also account for this aspect of the data also. Indeed we find that the $M_\gamma$ - distribution is rather sensitive to details of the hard scattering model.

## SIMULATION

In order to make use of the large amount of existing experimental data on inclusive single particle production at high energies, we have adopted an independent emission scaling model. Secondary particle center-of-mass (CM) momenta are drawn at random from probability distributions which result from radial scaling[5] of the invariant inclusive cross sections. Thus the momentum of each secondary is uncorrelated with that of any other--there is no dynamical clustering. Secondary particles are created until the available CM energy is exhausted. Details of the interaction model and simulation methods have been given elsewhere.[4]

The transverse momentum distribution we use has the two-component form

$$\frac{dn}{P_T dP_T} \propto A e^{-B_i(x) P_T} + \frac{C}{(P_T^2+1)^2} e^{-(48x' + 38x'^3)},$$

where $B(x) = \begin{cases} 2/(.25 + |x|), & |x| \leq .2 \\ 4.44, & |x| > .2 \end{cases}$

for secondary mesons and $B = 6, 4, 4$, for "fragment" $\pi$, P, K. The second term is the high $P_T$ component, the form of which is taken from the work of Halzen and Luthe,[6] which gives an explicit parametrization of the energy dependence of the high $P_T$ component. The constant C/A is related to the fraction F of secondary pions which are chosen from a distribution proportional to the second term in Eq. (1), in which $x' \equiv 2(P_T + 1)/\sqrt{s + 1156}$.

The energy dependence of F reflects the threshold behavior of the high $P_T$ process, which arises from the low probability of a parton-parton subenergy with a large fraction of $\sqrt{s}$. An energy increases a given subenergy, and hence a given $P_T$, corresponds to an ever smaller fraction of the CM energy. Here we take

$$F = \begin{cases} (3.4 \times 10^{-4}) \, s^{.605} & S < 30,000 \text{ GeV}^2 \\ (1.4 \times 10^{-2}) \, s^{.254} & 30,000 < S < 400,000 \\ 0.40 & S > 400,000. \end{cases}$$

T. Shibata[7] has recently made an extensive calculation of high $P_T$ processes at cosmic ray energies which incorporates the full machinery of the modern parton model, including gluon scattering, scale breaking and parton transverse momentum. He extended the hard scattering calculations, which are normally done for near 90° in the center of mass, to the forward fragmentation region, which is relevant for the cosmic ray calculation in which only fast secondaries are visible. The inclusive distributions obtained from the simple parametrization described above are roughly in agreement with the parton model calculation. The absolute magnitude of the high $P_T$ component we use is in good agreement with that of Shibata, but the shape is somewhat flatter. Accordingly, we have also tried a distribution with the exponent -2 in equation (2) replaced by -2.75. The simulation also took into account the following features of the experiment:
(1) The falling energy spectrum of the incident hadrons ($E^{-2.7}$); (2) multiple interactions in the target; (3) energy resolution (25%); and angular resolution.

## EFFECTS OF HIGH $P_T$

A feature of the Chacaltaya data that is often emphasized is the $M_\gamma$-distribution. This is a histogram of events classified by the invariant mass of $\gamma$-rays within a certain angle of the energy-weighted center of the event. The cone is defined (event by event) so that if the $\gamma$-rays come from isotropic decay of a cluster of particles and if selection effects can be ignored, then $M_\gamma$ is the cluster mass times the fraction of $\pi^0$'s in the cluster. Specfically, $M_\gamma$ is obtained by the following algorithm: define the invariant mass of $\gamma$'s inside a cone $\theta$ as $M_\gamma(\theta)$.

Then

$$M_\gamma(\theta) = (\Sigma_\theta p^\gamma_\mu)^2 = |\Sigma_\theta E_\gamma \Sigma_\theta E_\gamma \theta^2|^{\frac{1}{2}}$$

Also define the total transverse momentum inside $\theta$,

$$P_T(\theta) = \int^\theta P_T \, d\theta = \Sigma E_\gamma \theta_\gamma$$

For small $\theta$, $M_\gamma(\theta) < \frac{4}{\pi} P_T(\theta)$, and for large $\theta$, $M_\gamma(\theta) > \frac{4}{\pi} P_T(\theta)$. Define $M_\gamma \equiv M_\gamma(\theta_0)$, where $\theta_0$ is the solution of $M_\gamma(\theta_0) = \frac{4}{\pi} P_T(\theta_0)$. In practice this involves an interpolation.

Nearly half the events among the 68 reported have $M_\gamma > 3$ GeV, a much larger fraction than found in the simulation of Ref. 3. Events with large $M_\gamma$ are also seen among the lower energy data, but with a lower frequency. Moreover, the events with large $M_\gamma$ also have high multiplicity. On the basis of these features of the data, the Japan-Brazil group argue that there is threshold (around 50 TeV) for production of events with large clusters or fireballs.

It is clear from the definition $M_\gamma$ that there is a correlation between transverse momentum and $M_\gamma$. Halzen[6b] suggested some time ago that the events with large $M_\gamma$ might be due to hard scattering. Although this is probably not true on an event by event basis, our simulation shows that a large fraction of events with large $M_\gamma$ can arise as a consequence of the hard scattering component. Figure 1A and B shows the Brazil-Japan $M_\gamma$ distribution,[8] normalized to the simulation results for the case of no high $P_T$ tail (C = 0 in equation (2)). Much better agreement is obtained when the hard scattering component is included, as shown in Fig. 4B and 4C. However, the softened $P_T$ distribution, as for the parton model calculation of Ref. 7 does not produce as good agreement as the original $(P_\perp^2 + 1)^{2.0}$ distribution.

## CONCLUSIONS

Simulations are capable of producing an $M_\gamma$ distribution that is consistent with the data, provided a sufficiently large component of hard scattering is assumed. We must emphasize that the experiment is sensitive only to the small $P_T$ part of hard-scattering, a region in which calculations do not have a firm theoretical foundation. A test of the basic idea that events with large $M_\gamma$ are correlated with hard scattering would be to look for asymmetries corresponding to jet production, as suggested by Gaisser and Sidhu.[9] Arata[10] has made a study of azimuthal structure in terms of the fireball model. He finds that azimuthal structure is present in C-jets with $M_\gamma > 3$ GeV but not in those with $M_\gamma > 3$ GeV. He also shows that a model similar to that used here but without the hard scattering does not reproduce the azimuthal structure. It remains to be investigated whether the hard scattering model will reproduce the azimuthal structure in the data.

## ACKNOWLEDGEMENTS

This work has been supported by the National Science Foundation, the Department of Energy, and the University of Maryland Computer Science Center.

## REFERENCES

1. See T. K. Gaisser and G. B. Yodh, Annual Reviews of Nucl. and Particle Science 1980 (to be published) and Y. Fujimoto, Proc. 16th Int. Cosmic Ray Conf. (Koyto) 14, 308 (1979), for reviews and for the references on emulsion chamber terminology, techniques and results.

2. Y. Fujimoto et al., Brazil-Japan Emulsion Chamber Collaboration, in Cosmic Rays and Particle Physics, 1978 (Proc. Bartol Conf., ed. T. K. Gaisser) AIP Conf. Proceedings 49 p. 94.

3. R. W. Ellsworth, G. B. Yodh and T. K. Gaisser, Ibid., p. 111.

4. An error was discovered in the simulation of Ref. 4 which had the effect of incorrectly giving extra transverse momentum to secondaries in those events with multiple interactions in the thick target. When this error was corrected, the simulation without high $P_T$ gave only a small number of events with $M_\gamma > 3$ GeV. This error had no effect on results involving only longitudinal quantities, such as distributions of $E_\gamma/\Sigma E_\gamma$. This was the situation reported by R. W. Ellsworth, G. B. Yodh and T. K. Gaisser at the 1979 Cosmic Ray Conference at Kyoto, paper HE 3-37 (unpublished).

5. E. Yen, Phys. Rev. D10, 839 (1974). See also K. Kinoshita and H. Noda, Prog. Theo. Phys. 49, 896 (1973). See F. E. Taylor et al., Phys. Rev. D14, 1217 (1976) for further data on radial scaling.

6. F. Halzen and J. Luthe, Phys. Letters 48B, 440 (1978) and F. Halzen, Nucl. Phys. B92, 404 (1975).

7. T. Shibata, Proc. 16th Int. Cosmic Ray Conf., Kyoto, 7, 3462 (1979).

8. Presented in a Rapporteau talk by Y. Fujimoto, Proc. 16th In. Cosmic Ray Conf. (1979). A subset of events, those for which the gammas are exhausted before determining an $M_\gamma$, have been removed from both the data and the simulation.

9. T. K. Gaisser and D. P. Sidhu, Proc. 15th Int. Cosmic Ray Conf. (Plovdiv) 7, 18 (1979).

10. N. Arata, Proc. 16th Int. Cosmic Ray Conf. (Kyoto) 6, 283 (1979).

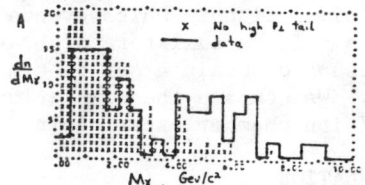
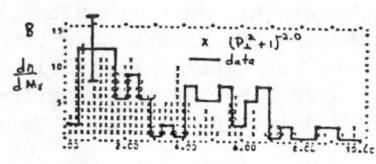
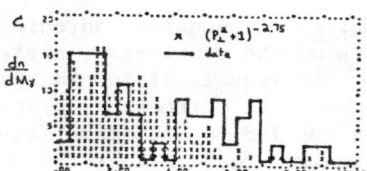

Figure 1. Actual and simulated $M_\gamma$ distributions for (A) no high $P_T$ tail; (B) a tail of the form $(P_\perp^2 + 1)^{-2.0}$ (C) $(P_\perp^2 + 1)^{-2.75}$.

# SIMULATION OF CENTAUROS

R. W. Ellsworth,[a] Jordan Goodman[a] and G. B. Yodh
Department of Physics, University of Maryland,
College Park, Maryland 20742

and

T. K. Gaisser[b] and Todor Stanev[c]
Bartol Research Foundation of The Franklin Institute
University of Delaware, Newark, Delaware 19711

## INTRODUCTION

The Centauro events[1] found in the Japan-Brazil emulsion chamber have attracted a great deal of attention because of the anomalous feature that they appear to be due to interactions of about 1000 TeV in which few if any neutral pions are produced even though many secondary hadrons are produced. The multiplicities involved are such that the events cannot be due to statistical fluctuations in the relative number of neutral and charged pions in a single interaction. Because of the complexity of the experimental selection criteria it is desirable to compare the data to simulations in which these are taken into account as far as possible.

Preliminary results of several such calculations were reported at the Kyoto Conference.[2,3,4] Only one of these[2] considered effects of heavy nuclei, and it was not clear what primary spectrum and composition were assumed in generating the events. The simulation of Ref. 3 contained only proton events of very high energy, with the result that most of the simulated showers were larger than the observed events. Only average values of measured parameters were reported in Ref. 4. We report here preliminary results of a simulation in which events were generated from a primary spectrum with a large admixture of heavy primaries and with energy thresholds chosen to obtain a set of events in the same size range as the experiment. We compare the simulations with data from all three of the large emulsion chamber experiments.[5]

## THE CALCULATION

The model used for hadronic interactions, as well as the general features of the simulation, are similar to those used in the accompanying paper on local interactions in the detector.[6] The modifications made here correspond to the special features of emulsion chambers as applied to studies of interactions in the overlying atmosphere. The essential fact is that the detectors are sensitive only to energetic photons. Jets produced in the emulsion chamber within 4 radia-

a. Work supported in part by the National Science Foundation.
b. Work supported in part by the U. S. Department of Energy.
c. On leave from the Institute for Nuclear Research and Nuclear Energy, Sofia, Bulgaria. Joint research supported in part by the U.S. National Science Foundation and by the Bulgarian State Committee for the Promotion of Science and Technical Progress under the U.S.-Bulgaria Cooperative Science Program.

tion lengths are classified as photons; those with deeper points of initiation are classified as hadrons. Hadrons and photons from the primary and subsequent interactions of a cosmic ray nucleus in the atmosphere above the detector appear in the detector as a family of parallel jets. The morphology of the events is described more fully in Ref. 7.

The simulation procedure consists of the following steps: (1) select the mass and energy of a primary cosmic ray nucleus from a spectrum[8] including protons, helium and iron; (2) compute its cascade in the atmosphere, including nuclear fragmentation and secondary hadronic and electromagnetic interactions; (3) record on magnetic tape the position and energy of each hadron ($\pi^{\pm}$, K, p or n) and each photon above an energy threshold at the detector; and (4) for each hadron decide whether it interacts in the detector, and if so compute the fraction, $K_\gamma$, of its energy deposited as visible electromagnetic energy. Each hadron which deposits $E_h > 2$ TeV in the detector is kept, as is each electromagnetic jet with $E_\gamma > 1.5$ TeV. We then find the energy-weighted center of each event and add up all visible energy inside a circle of radius 20 cm. We choose this radius in light of the size of individual modules in the Chacaltaya emulsion chamber, which are 40 x 50 cm.[1] All events with $\Sigma E_{vis} > 100$ TeV inside the circle are included in the simulated sample for comparison with the data.

## COMPARISON TO DATA

Centauro events are characterized by an anomalously large fraction of secondary energy in the hadronic component (hence few $\pi^\circ$'s in the primary interaction). To assess the anomalous events properly it is necessary to compare them to ordinary events. This was done at the Kyoto conference with results as summarized in the rapporteur talk by Fujimoto.[9] Figure 1 shows the data of the three experiments compared to our simulations. We conclude by inspection of the Figure that five events show an excess of hadronic energy well beyond the range of conventional explanation. Two of these are Centauro events including the original event with virtually no electromagnetic energy) and two were classified as mini-Centauros ($n_h \sim 10$) by the Brazil-Japan group. The fifth also has small $n_h$ and is from the Pamir experiment. Only two of these (the Centauros with $n_h \sim 50$) have anomalous multiplicity.

The rate of events with $\Sigma E_{vis} > 100$ TeV calculated in our simulation from the primary spectrum given in Section II is 2-3 x $10^{-7}$ $m^{-2}$ $sec^{-1}$ $sr^{-1}$, whereas the reported rate of such events from the emulsion chamber experiments is $\sim 10^{-8}$ $m^{-2}$ $sec^{-1}$. The cause of this discrepancy is at present not clear. We note, however, that the experimental scanning efficiency appears to be lower for events near threshold. Some of the difference may also be accounted for if events near the edges of modules in the emulsion chamber are not counted. If, for example, events within 5 cm of the edge of a 40 x 50 cm rectangle are rejected the collection area is reduced by 0.6. It is also possible that some further selection criteria may have been used to obtain preferentially those events due to especially high energy interactions near the detector. We have not included such effects. The

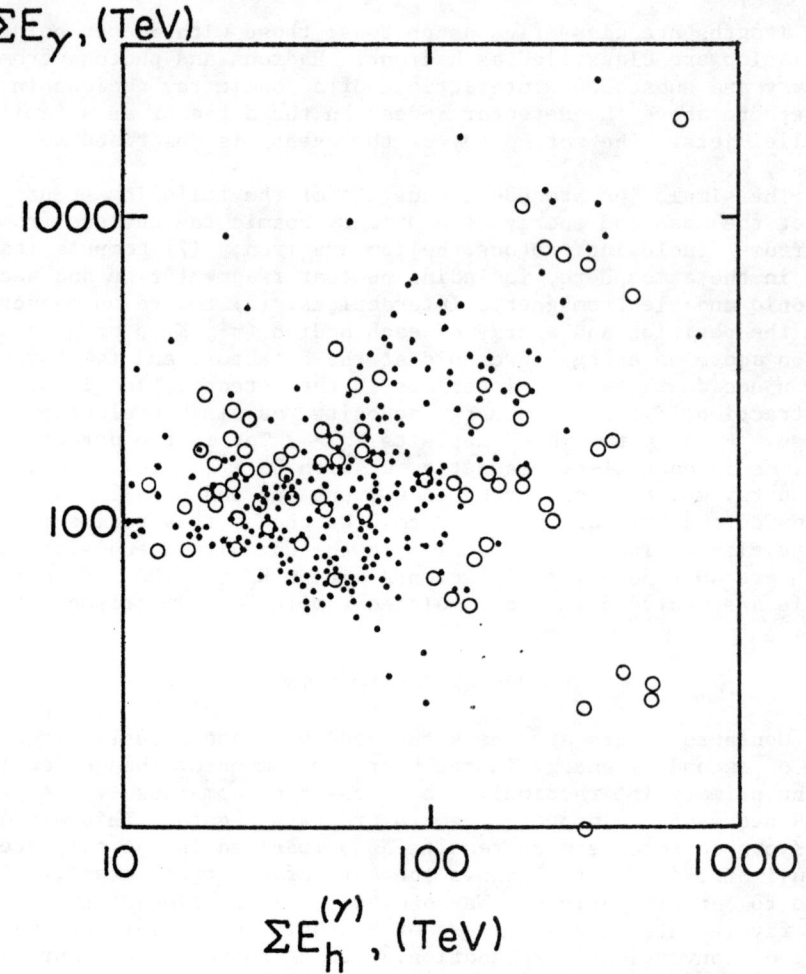

Fig. 1 Scatter plot of total electromagnetic energy vs. total hadronic energy. Open circles-experiment; points-simulation.

relative importance of events with large values of $\Sigma E^{(\gamma)}/\Sigma E_\gamma$ could conceivably be increased by application to the simulation of unknown selection effects. It is, however, difficult to see how this could change our basic conclusion, which is based on 381 simulated events with $\Sigma E_{vis} > 100$ TeV, as compared to 73 observed events. Further study is required to resolve the discrepancy between calculated and observed overall rate of events.

Because of the low statistics there is no contradiction between the rate of anomalous events among events with $\Sigma E_{vis} > 100$ TeV reported by the Chacaltaya group (4/50), by the Pamir group (1/30 or 1/100) and by the Fuji group (0/15). Further comparison of results of the three experiments as well as further data collection is clearly desirable to confirm the existence of the effect.

Two classes of explanations for Centauro events can be imagined: (a) those involving a new kind of interaction of ordinary hadrons beyond some threshold energy and (b) those involving exotic components of the primary beam. In case (a) Centauros, if they exist, would be detectable at $\bar{p}p$ colliders and at ISABELLE (unless the threshold is unreasonably sharp and just beyond the reach of the machine).

In proposing possible interpretations of Centauros, various authors have considered the high $p_T$ of the events as an important clue to the nature of the process. A value of $<p_T> = 1.7 \pm 0.7$ GeV/c for hadrons produced in Centauro interactions is claimed.[1] It should be emphasized (as stated in Ref. 1) that the height of interaction could only be obtained for the original Centauro which occurred only about 50 m from the detector. We have found[7] that the mean $p_T$ for this event is roughly 1 GeV/c rather than 1.7. Nothing is known experimentally about the $p_T$ distribution in the other cases.

A complete version of this work will be published elsewhere.

## REFERENCES

1. C. M. G. Lattes, et al., Proc. 13th Int. Cosmic Ray Conf. (Denver) 3, 2227 and 4, 2671 (1973); M. Tamada, Nuovo Cimento 41B. 245 (1977) and Y. Fujimoto and S. Hasegawa in Cosmic Rays and Particle Physics-1978 (Bartol Conf. ed. T. K. Gaisser) A.I.P. Proceedings 49, 317 (1978); C. M. G. Lattes, Y. Fujimoto, and S. Masegawa, Physics Reports (in press).
2. A. M. Dunaevskii, et al., Proc. 16th Int. Cosmic Ray Conf. (Kyoto) 7, 154 (1979).
3. B. S. Acharya, M. V. S. Rao, K. Sivaprasad and Srikantha, Rao, Proc. 16th Int. Cosmic Ray Conf. (Kyoto) 6, 289 (1979).
4. M. Shibata, Proc. 16th Int. Cosmic Ray Conf. (Kyoto) 7, 196 (1979).
5. The three large emulsion chamber experiments are the Brazil-Japan collaboration at Mt. Chacaltaya [Ref. 1], the Pamir collaboration [S. G. Bayburina et al., Proc. 16th Int. Cosmic Ray Conf. (Kyoto) 7, 75 and 241 (1979)]; and the Mt. Fuji collaboration [M. Akashi et al. Proc. 16th Int. Cosmic Ray Conf. (Kyoto) 13, 87 and 98]. Further bibliography on this subject may be found in Ref. 7.
6. R. W. Ellsworth, T. K. Gaisser and G. B. Yodh, these proceedings.
7. T. K. Gaisser and G. B. Yodh, Ann. Revs. of Nucl. and Particle Science, 30, 475 (1980).
8. J. A. Goodman et al., Phys. Rev. Letters 42, 854 (1979).
9. Y. Fujimoto, Proc. 16th Int. Cosmic Ray Conf. (Kyoto) 14, 308 (1979). See also Ref. 1d.
10. D. Sutherland in Cosmic Rays and Particle Physics-1978 op. cit. p. 503; J. D. Bjorken and L. D. McLerran, Phys. Rev. D20, 2353 (1979). See also S. A. Chin and A. K. Kerman, Phys. Rev. Letters 43, 1292 (1979) and A. K. Mann and H. Primakoff, Phys. Rev. D (in press).

DATA PHOTOGRAPHS FROM THE AUCKLAND COSMIC RAY TELESCOPE

P.C.M. Yock

Department of Physics, University of Auckland, Auckland, New Zealand

ABSTRACT

Raw data photographs from the cosmic ray telescope at the University of Auckland with which evidence for heavy particles was obtained [1-3] are presented and discussed.

INTRODUCTION

The apparatus that was used in this work is described in detail in Refs 1-3. Essentially it comprised a telescope with six scintillator planes, two wide-gap spark chambers, and absorbers. The scintillators provided time-of-flight data over a 2m path length and also ionization measurements before and after traversal of a 68.4 g/cm$^2$ steel absorber. Masses and charges of particles were determined from these measurements. The pulses from the scintillator planes were recorded by adding and displaying them sequentially on single sweeps at 10 ns/cm (nominal) of a Hewlett Packard 183A oscilloscope. The oscilloscope was viewed by a camera and the pulse height and timing data obtained from photographs of the traces. A small sample of these photographs is reproduced here to provide a direct record of the accuracies with which such measurements may be made. Four types of traces were photographed at intervals throughout the experiment: (1) muon traces, (2) proton traces, (3) 50 MHz oscillator traces, and (4) heavy particle traces. They are considered here in turn.

MUON TRACES

Many photographs of muon traces were taken during the course of the experiment and used to measure the relative cable and tube delays involved in the timing measurements and to monitor any drifts. Two examples taken nine months apart with several muon traces on each are shown in Fig.1. Times-of-flight of individual particles are given by the average of the spacings of the first and fifth, and second and sixth pulses. For muons this quantity was found to drift by $\leq$ 0.1 ns during the course of the experiment [2]. Note that

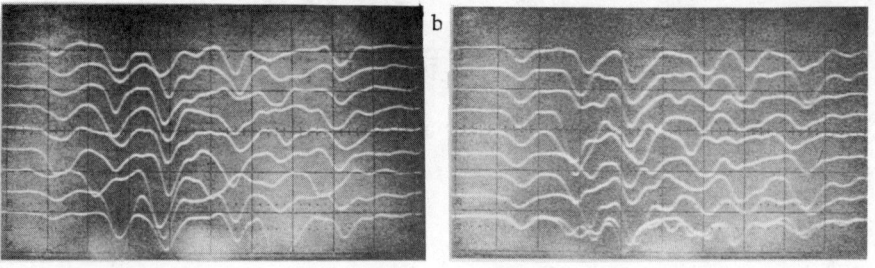

Fig.1 Muon calibration photographs taken July 1979 and April 1980.

the pulses are negative, clipped, anode pulses. Pulse separations were measured using the constant fraction technique (50%) with a low-power travelling microscope on the original half-size Polaroid photographs. For each trace the measurements were made relative to the (internal) graticule at the location of the trace [4]. Traces with small, badly shaped pulses were excluded. This was not expected to produce any systematic timing error when averaged over several events, and this was indeed found to be the case when stopping protons were used as particles of known speeds to check the timing calibrations [2].

## PROTON TRACES

Two examples of proton photographs which were used to calibrate the pulse height measurements are shown in Fig.2. These photographs were obtained with a delayed coincidence trigger and an absorber of reduced thickness (15.5 g/cm$^2$) to enable slow protons to penetrate

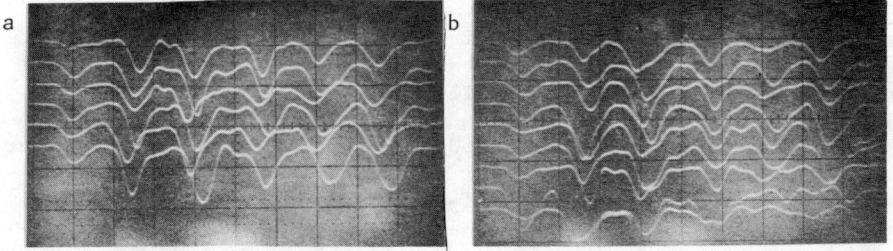

Fig.2 Proton calibration photographs taken August 1979 (2 hour exposure) and February 1980 (5 hour exposure).

it, whilst at the same time stopping slow muons. The greater spacing between the pulses in the proton photographs as compared to the muon photographs is readily apparent [5]. For the proton runs (and also the heavy particle runs discussed below) the oscilloscope gain was reduced by a factor of two relative to the muon runs. Not all triggers in the proton runs were caused by protons of course. The bottom two traces in Figure 2b are obviously multiparticle events, and the bottom trace in Figure 2a was produced by a deuteron [6]. The proton photographs may be used to deduce the accuracy of the TOF measurements for slow particles. With cable and tube delays appropriately taken account of, the third pulse should be midway between the first and the fifth for any trace, and likewise the fourth pulse midway between the second and sixth [7]. As may be checked from Fig.2, the accuracy with which the proton traces satisfy these tests corresponds to a time-of-flight accuracy of $\simeq$ 400 psec. Two independent checks of this TOF accuracy, reported in Ref 2, were consistent with this result.

## 50 MHz OSCILLATOR TRACES

The accuracy of the timing measurements obviously depends on the linearity, jitter, fluctuations and long-term constancy of the

time base of the oscilloscope. The geometry and spot-size of the
CRT also affect the timing accuracy. The effects of all of these
factors are already included in the timing tests mentioned above
using muons and protons and which were all carried out assuming
a constant (i.e. drift-free) value of the time base throughout the

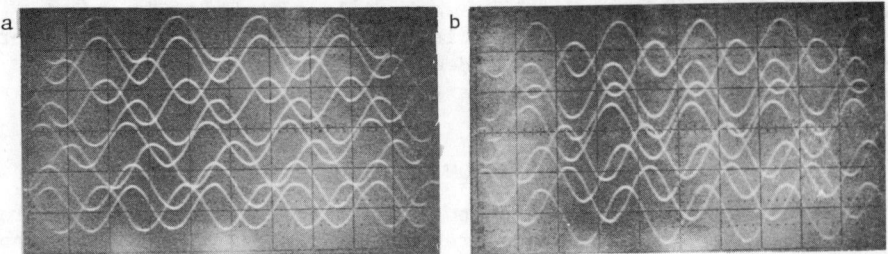

Fig.3 50 MHz oscillator photographs taken September 1979 and
April 1980.

course of the experiment. However, as a further check, many 50 MHz
oscillator photographs such as those shown in Figure 3 were taken
during the experiment. The oscillator used in these tests was a
Tektronix 180A crystal controlled oscillator. For each sweep of
the oscillator two overlapping 60 ns intervals in the eight middle
divisions were averaged in order to mimic the averaging process
which is involved in the time-of-flight measurements [5,8]. Amongst
400 oscillator sweeps photographed at various intervals throughout
the experiment no fluctuation of the time base greater than 0.7%
away from its overall average was observed. This figure includes
the effects of any imperfections of the oscillator itself, of course.
The trigger for the oscilloscope was essentially random in the
oscillator runs (actually derived from cosmic rays).

## HEAVY PARTICLE TRACES

Six photographs spanning a ten month period are shown in Fig.4.
These were obtained with the same trigger as that used in the proton
runs but with the full absorber thickness of 68.4 g/cm$^2$. Most
triggers were clearly caused by more than one particle [7]. Of the
arrowed traces two correspond to known particles and the remainder
apparently to heavy particles. These events are discussed in detail
in Refs 1 and 2. There is however sufficient information contained
in the photographs reproduced here to check fairly accurately the
measured masses and charges reported in Refs 1 and 2 for these
particular events. Figs 4a - 4f are, respectively, 25, 6, 4, 7, 24
and 7 hour exposures taken in June, July, August and December 1979,
and February and April 1980. The arrowed traces are, respectively,
event b (Ref.1), event c (Ref.1), a triton (Ref.1), event e (Ref.2),
a deuteron (Ref.2), and possibly a very heavy preceded particle
(Ref.2).

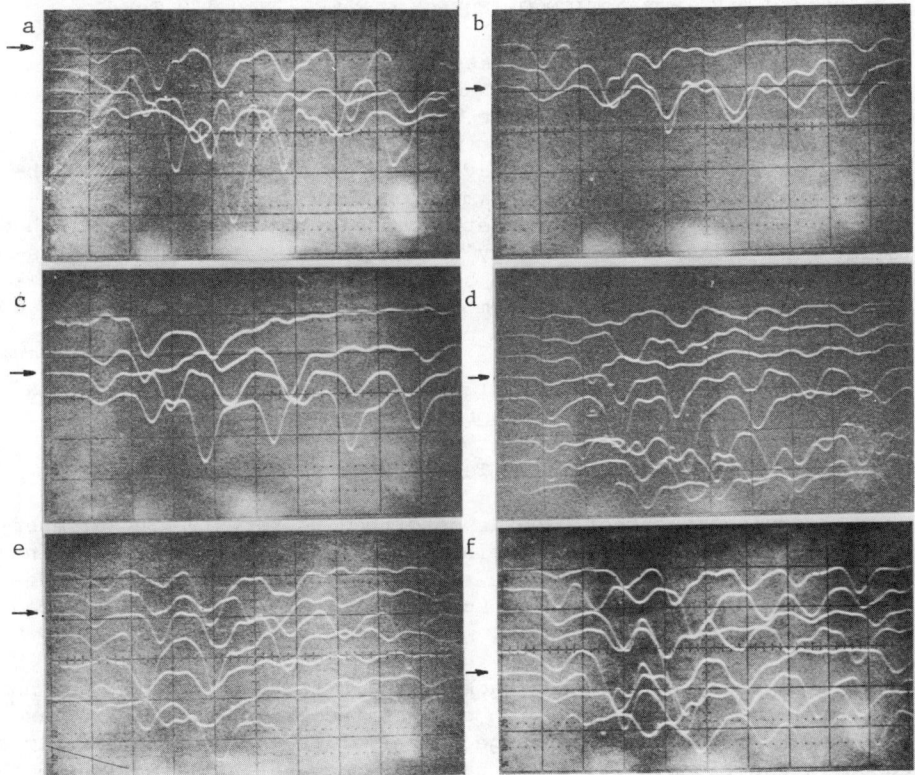

Fig. 4 Some heavy particle photographs.

REFERENCES AND FOOTNOTES

1. P.C.M. Yock, Phys. Rev. D (July 1st issue, 1980).
2. P.C.M. Yock, "Further evidence for heavy particles in the cosmic radiation", submitted to Phys. Rev. Lett. (May, 1980).
3. P.C.M. Yock, Phys. Rev. D <u>18</u>, 641 (1978).
4. This is a particularly important consideration for measurements being made from the photographic reproductions shown here which are not particularly free of distortion.
5. Proton speeds may be deduced from the photographs through use of the formula $12.72/\beta = \tau_{15} + \tau_{26} - \bar{\tau}_{15,\mu} - \bar{\tau}_{26,\mu} + 12.84$ where the notation of Ref. 3 has been used and the pulse separations $\tau_{ij}$ are in nanoseconds.
6. Masses and charges may be deduced from the oscilloscope photograph using the methods detailed in Ref.3. The bottom particle of Fig. 2a would have been too slow to penetrate the absorber had it been a proton, and its last two pulse heights imply it was a deuteron.
7. These actual timing requirements for single particles are $\tau_{15} - 2 \cdot \tau_{13} = \bar{\tau}_{15,\mu} - 2 \cdot \bar{\tau}_{13,\mu}$ and $\tau_{26} - 2 \cdot \tau_{24} = \bar{\tau}_{26,\mu} - 2 \cdot \bar{\tau}_{24,\mu}$.
8. This averaging process also effectively removes the effect of a small but visible 10 MHz component in the oscillator's output.

# A COSMIC RAY INTERACTION OF ENERGY GREATER THAN 130 TeV

T.A. Koss, R.J. Wilkes and J.J. Lord
Visual Techniques Laboratory, FM-15
University of Washington, Seattle, WA 98195

ABSTRACT: A very high energy family of particles has been observed in a large balloon-borne emulsion chamber. The family consists of at least 31 electromagnetic cascades of energy greater than 0.5 TeV; the total visible energy is 130 TeV. From studies of cascade development, 27 showers can be identified as γ-initiated, 2 as definite neutral hadrons, and 2 more as possible neutral hadrons. No cascades associated with charged particles have been found.

A very high energy family of particles has been observed in a large balloon-borne emulsion chamber. The family consists of at least 31 electromagnetic cascades of energy greater than 0.5 TeV; the total visible energy is 130 TeV.

The emulsion chamber used is described in ref. 1. Shower energies for 10 cascades were determined by performing track counts in several emulsion layers. In the remaining cases, energies were determined by comparing the x-ray spot densities, using the 10 cascades with track counts as benchmarks.

All cascades were traced back to their starting points, and none were found to be associated with charged incoming tracks. Two cascades began abruptly with a large number of tracks, characteristic of hadron interactions. Two other cascades begin at an effective depth of 7.6 r.l., and are thus also possible hadron interactions. All other cascades have starting points and structures typical of photon-initiated showers.

The particle family entered the chamber at a large zenith angle (approximately $67°$), and thus the effective thickness of the chamber was greatly increased. Most of the showers are very nearly parallel. From geometric measurements, we have been able to determine that the primary vertex is located at a distance of 150 (+150,-75) m. At the balloon's float altitude, this corresponds to approximately 0.07 $g/cm^2$ of air. In addition, most showers passed through about 2 $g/cm^2$ of aluminum and wood before entering the chamber's sensitive volume.

The event axis passes through a corner of the stack, so more than half of the large angle ($\theta \gtrsim 3 \times 10^{-4}$ rad) tracks cannot be observed. A fiducial area $6 \times 7$ $cm^2$ was defined, and this area, containing the central cluster of cascades, was scanned by microscope to ensure that no cascades of energy $\geq 0.5$ TeV were missed. The 19 cascades in the fiducial area remain in-geometry at least down to an effective depth of 10 r.l. In the following discussion, only these 19 cascades are considered.

0094-243X/81/680824-02$1.50 Copyright 1981 American Institute of Physics

For each cascade, angles and $p_T$ values were calculated relative to the energy-weighted center of the event. In Fig. 1, the fractional energy spectrum ($x' = E_\gamma/\Sigma E_\gamma$) is compared with equivalent data from the Chacaltaya experiment [2]. In Fig. 2a we show the rapidity distribution and in Fig. 2b the "normalized pseudorapidity" ($E_0 \cdot \theta$) distribution is compared with that of Centauro III [2], which has some similar features: $E_{vis}$ = 150 TeV, R = 230 m, n = 37.

Data presented here are preliminary, and we are continuing our analysis of this event.

## REFERENCES

1. The data presented here were obtained from the Japan-US Galactic Electron Collaboration; for names of participants, and experimental details, see J. Nishimura et al., Astrophysical Journal **238**, 394 (1980).
2. Brasil-Japan Emulsion Chamber Collaboration: Proc. 16th Int. Cosmic Ray Conf. **6**, 362 (1979); Proc. 15th Int. Cosmic Ray Conf. **7**, 208 (1975); "Cosmic Rays and Particle Physics 1978," T. Gaisser, ed., p. 94 (AIP Conference Proceedings No. 49), AIP, New York, 1979.

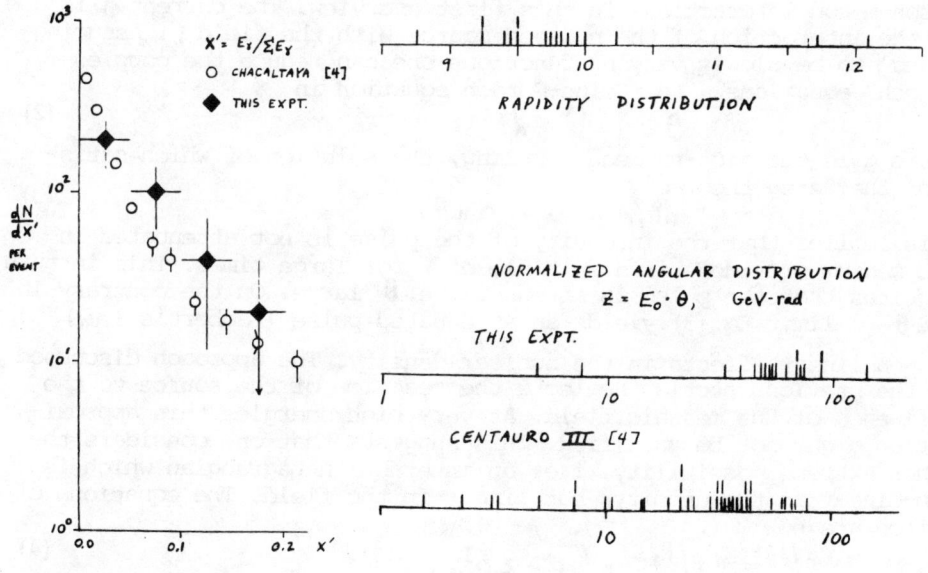

Fig 1          Fig 2

## NON LINEAR EFFECTS IN NUCLEAR MATTER AND SELF-INDUCED TRANSPARENCY (SIT)*

G.N. FOWLER, Physics Department, University of Exeter
R.M. WEINER, Physics Department, University of Marburg

In hadron collisions with nuclei at high energies the supression of cascading (transparency effect) which is observed is commonly interpreted in terms of a coherent interaction between the projectile and the nuclear matter traversed. In the first encounter of the projectile within the nucleus an intense coherent mesonic field is expected to be created. It thus seems plausible that the response of the nuclear medium will be non-linear in the mesonic field. Furthermore and particularly in the case of heavy ion collisions at high energy one has the added complication that quite apart from possible high meson field intensities the nuclear density is also expected to undergo large fluctuations implying a significant interaction in the meson field source.

A. <u>Non-linear effects in the mesonic fields</u>. The pion field satisfies the Klein Gordon equation.

$$(m^2 + \Box)\varphi = j \qquad (1)$$

where j is the current induced by $\varphi$ and m the pion mass; we neglect meson-meson interactions in this first overview. The current j is due to the interaction of the nucleon source with the field $\varphi$. Assuming j and $\varphi$ to be slowly varying functions one can reduce the coupled "Bloch" equations to a sine-Gordon equation in

$$\theta = g \int^t \varphi \, dt \qquad (2)$$

where g is the pion-nucleon coupling, the solution of which satisfies the "area theorem"

$$d\theta/dz \sim -\sin\theta \qquad (3)$$

This implies that the intensity of the pulse is not attenuated in the medium provided $\theta$ is a multiple of $\pi$ for large times. This in turn requires that the pulse is intense, i.e. $\theta$ large. On the contrary in the $\theta \to 0$ limit Eq. (3) yields an attenuated pulse (Lambert's law)

B. <u>Non-linear effects in the nuclear density</u>. The approach discussed in the previous section neglects the reaction of the source to the influence of the mesonic field. At very high energies this approximation might not be justified. This suggests that one considers the other extreme possibility, i.e. an interaction Lagrangian which is non-linear in the density $\rho$ but linear in the field. The equations of motion are now

$$\partial^2\rho/\partial t^2 - \partial^2\rho/\partial z^2 = -b\rho - c\rho^3 + g\varphi \qquad (4)$$

$$(m^2 + \Box)\varphi = g\rho \qquad (5)$$

where b and c are constants.

---

*Work supported in part by NATO and GSI, Darmstadt. A more detailed account will be published in Proc."Orbis Sci."1980, ed. by A. Perlmutter and L. Scott, Plenum Press.

We now assume that the fields may be treated as c numbers (coherence property) and that the intensities are high enough for classical behaviour to be a reasonable approximation. The solution of eqs.(4),(5) is the usual soliton pulse

$$\rho \sim \operatorname{sech}\left[\frac{1}{\tau}\left(t-\frac{z}{v}\right)\right] \tag{6}$$

Possible physical consequences:
1. If a coherent meson field is indeed responsible for the transparency effect, then we should expect it to manifest itself in the Bose-Einstein correlations of secondary pions as described in reference[1] particularly for large positive pseudo-rapidity where a one dimensional approximation is expected to hold.

2. There should exist a threshold in the coherent pion field intensity above which transparency should occur.

3. The suppression of cascading, if due to SIT, should be a periodic function of the intensity of the coherent meson field.

4. Early observations in cosmic ray physics and more convincing recent work at LBL[2] with emulsions has provided evidence for the existence among the fragments arising in heavy ion collisions of "anomalous" objects with a lifetime $\geq 10^{-13}$ sec. and a mean free path several times smaller than that to be expected from normal nuclear interactions. It is conceivable that in heavy ion collisions the non-linear excitations produced could be metastable (soliton property) and could be carried off by an outgoing nuclear fragment. The interaction of such an excited fragment with ordinary nuclei might lead to a transverse instability (growth) which could increase every time a collision takes place until eventually the excitation disintegrates. This would lead to a decrease of the mean free path compared with what one would expect from "normal" events.

References

1. G.N.Fowler and R.M.Weiner - Physical Review D17, 3118 (1978)
    G.N.Fowler, N.Stelte and R.M.Weiner - Nucl.Phys.A319,349 (1979)

2. E.M.Friedländer et al. - Phys.Rev.Lett. 45, 1084 (1980)

Chapter 8

New Accelerators and New Experimental Techniques

# NEW ACCELERATORS AND COLLIDER DEVELOPMENTS IN THE 80's

## D. CLINE, WISCONSIN, ORGANIZER

# CORNELL 50 x 50 GeV STORAGE RING[*]

B. D. McDaniel
Cornell University, Ithaca, N. Y. 14853

## ABSTRACT

A 50 x 50 GeV electron-positron storage ring facility is described. The design of this colliding beam facility employs the use of superconducting r.f. accelerating cavities to make very significant savings in capital and operating costs. Results of cost optimization procedures are provided for several cases. A parameter list is given for the chosen configuration, together with a cost estimate and time schedule.

## INTRODUCTION

We are proposing for consideration, the construction of an $e^+e^-$ storage ring colliding beam facility of 50 x 50 GeV which will operate over the range from 30 to 100 GeV in the center of mass. The physics capabilities of such a machine have been widely discussed and will not be described here. Obviously, at the present time the most exciting goal is the production of the $Z°$ and the study of its decay modes.

The field of electron-positron physics in the United States is well covered to about 36 GeV in the center of mass. CESR nicely bridges the energy region from SPEAR to PEP with excellent running conditions at the Upsilon. PEP is competing with PETRA for the physics in the range from 16 to 36 GeV. However, as we look into the future of electron-positron physics of the next decade we see that the U.S. physics has a problem. At the present time it seems clear that CERN is about to embark on the construction of LEP. That machine, in its first phase is expected to have initial operation at an energy of 50 x 50 GeV by 1986. In its next phase it is to go to an energy of 86 x 86 GeV. The LEP machine, which is designed to achieve ultimately very high energies, is a very expensive machine, costing a major part of one billion dollars to construct. It seems difficult for the United States, with its present construction commitments, to compete with that machine. We should not commit our efforts to building an equivalent machine since that would be wasteful duplication and it would likely come into operation at a later date. If we are to build a larger machine than LEP, the cost will be so excessive that no start can be made in this country for many years because of the large burden of present facilities and operating costs. It therefore appears unreasonable to project at this time a machine which would have a design peak energy equal to or greater than that of LEP.

[*]Supported in part by the National Science Foundation under Contract NSF-C537.

0094-243X/81/680829-04$1.50 Copyright 1981 American Institute of Physics

However, by constructing a machine which is dedicated to somewhat more restricted goals we believe it is possible to provide, at a much lower cost, a machine which will compete successfully for the Z° physics against LEP. We list the requirements for a Z° factory: It must compete well with LEP for the Z° physics, have high luminosity, be timely, must have multiple simultaneous interaction regions to accommodate a significant body of the U.S. user community, and it must be financially feasible in the appropriate time frame. The ring we propose satisfies these criteria. It would have a luminosity of $3 \times 10^{31}$ cm$^{-2}$sec$^{-1}$ at the energy of the Z°, a luminosity that is the same as that proposed for LEP at the same energy. We also believe it can be constructed on a time scale to equal or surpass that of LEP.

## SUPERCONDUCTING R.F. ACCELERATING CAVITIES

In order to meet the criterion of modest cost, a basic ingredient of the proposal is that superconductivity technology be applied to the r.f. acceleration of the stored beams. In 1974 we had developed a superconducting accelerating cavity at S-band frequency and installed it in the synchrotron where it was tested and operated for several months. This cavity demonstrated the real utility of such a device. However, with the advent of CESR, in order to focus our efforts on the early completion of the CESR construction this program was greatly attenuated. However, even so, we have carried on a program to study the various effects which limit the ability to obtain high field, how to reliably produce and process the cavities, and how to utilize mass production methods to produce accelerating structures. We have made much progress in all of these areas and feel that the technology for the production of superconducting structures is now under good control and that reliable structures can be reproducibly and cheaply made. We feel that the questions which remain are not inherently concerned with the technology of superconductivity, but with the interaction of the accelerated particles with the superconducting device. We now believe that 1500 MHz is the appropriate frequency for a superconducting system for the storage ring we have in mind. We have made several prototype cavity structures and now have had several two-cell cavities under test which provide an accelerating gradient of 4.2 MV per meter and a Q in excess of $3 \times 10^9$. We believe that we are now ready to produce a full scale prototype of a complete cryostat and accelerating section as shown in Fig. 1.

Fig. 1

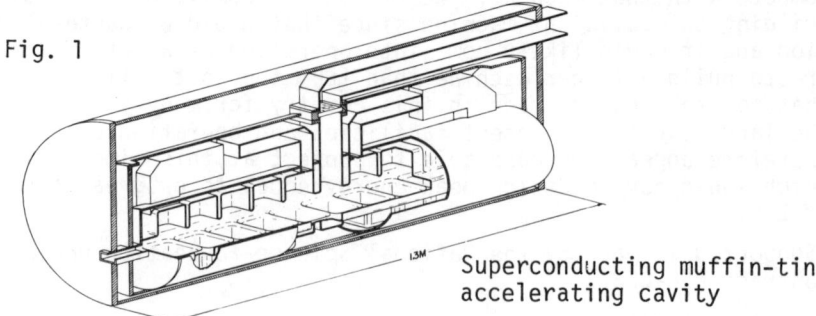

Superconducting muffin-tin accelerating cavity

## STORAGE RING DESIGN

In preparation for the recent Woods Hole HEPAP sub-panel meeting, we prepared a design study of a 50 x 50 GeV electron-positron storage ring which would satisfy the criteria given above. We incorporated the use of the superconductivity technology and applied optimization equations to the specific design energy, namely 50 GeV beams. We chose a design luminosity of a moderate value. We studied the problems of experimental apparatus in the straight sections and concluded that detectors of full capability can be installed in a straight section having a free space as short as 4 to 6 meters between storage ring elements. Thus austere choices were made in order to achieve the performance and cost goals of the project. It is especially important to note that this machine, designed for peak luminosity at 50 x 50 GeV, can provide the same luminosity as LEP at that energy, and at a much lower capital and operating cost.

The principal characteristics are displayed in Table I. There are four interaction areas, two of 6 meters and two of 4 meters length. In making this design we have assumed that we can operate without compensating solenoids as has been done at PETRA. The 4 and 6 meter interaction points will have luminosities of 3 and 2 times $10^{31}$/cm$^2$/sec respectively. Two suitable construction sites have been found on the campus of Cornell University.

TABLE I
BASIC PARAMETERS

| | | | |
|---|---|---|---|
| Nominal Top Energy per Beam | 50 GeV | R.F. Frequency | 1.5 GHz |
| Number of Interaction Points | 4 | Energy Loss per Turn | 1.1 GV |
| Free Space for Experiments | 4 and 6m, 2 each | R.F. Voltage per Turn | 1.5 GV |
| Luminosity at 50 GeV | $3 \times 10^{31}$/cm$^2$/sec | Cavity Gradient | 3.0 MV/m |
| Tune Shift | 0.03 | Length of Superconducting R.F. Structure | 500 m |
| Beam Current per Beam | 3.7 mA | Total R.F. Power | 10 MW |
| Circumference | 5.5 km | Total Power for the Facility | 25 MW |
| Bending Radius | 500 m | | |

## COMPARATIVE COSTS

We have made calculations of the costs for various system configurations based on a minimization of capital cost plus 5 years of operational cost. These are shown in Table II in 1980 dollars, and include contingency. Case (a) is the design which is proposed and provides a beam energy of 50 GeV at a peak luminosity of $0.3 \times 10^{32}$/cm$^2$/sec. If the fixed costs, $58M, are excluded, the system cost is $107M for the storage ring proper. If instead of using superconducting cavities, normal cavities are used, the corresponding figure is $220M. Configurations for higher luminosity and energy are also shown.

In all cases we see that the use of superconducting cavities reduces the capital cost by about a factor of two.

TABLE II
CAPITAL COSTS OF VARIOUS STORAGE RINGS
(Fixed costs, $58M, excluded)

| | Beam Energy GeV | Peak Luminosity ($10^{32}$/cm$^2$/sec) | Supercond. Cavities $M | Normal Cavities $M |
|---|---|---|---|---|
| a. | 50 | 0.3 | 107 | 220 |
| b. | 50 | 1.0 | 155 | 265 |
| c. | 86 | 0.3 | 283 | 634 |
| d. | 86 | 1.0 | 383 | 698 |

All costs in 1980 dollar values.

TABLE III
50 X 50 GeV STORAGE RING FACILITY COST

| | |
|---|---|
| Total Storage Ring Cost | 106 |
| Fixed Costs | |
| Injector | 29 |
| IR Halls and Cranes | 8 |
| Site Work and Utilities | 9 |
| Auxiliary Lab and Office Space | 12 |
| Total Fixed Costs | 58 |
| TOTAL FACILITY COST | $164 M |

(All costs in $M, 1980 values, 20% contingency included)

## SUMMARY

Using the superconducting parameters of case (a), we find the costs as given in Table II. We see that the total cost including contingency is $164M (1980 dollars). Taking an overview, we find that the superconducting components of the system, which are so important in reducing the overall cost, amount to only $22M, i.e. 15% of the total cost for the laboratory complex.

We feel that about 18 to 24 months of R and D work will be required before we are completely ready to specify all the components and obtain funding. By scheduling a year to get approval for funding and two and one-half years for construction we would plan to turn on in early 1986.

We feel that we are in the following position. We have demonstrated our ability to make cavities cheaply, reliably and in mass production. It is now necessary to make important studies of the interaction of the beam with the cavities to study resonance effects such as higher mode loss damping, and coupling of synchrotron betatron instabilities. In addition, there may be some non-resonant microwave instabilities which must be examined, that is, effects internal to the bunch. We plan to study these effects by constructing two prototype cavity sections which we will put into CESR next year to test and study. In the meantime we will refine our lattice design, system layout, and site planning. We would also start prototype work on other elements of the system. We would expect that by the time funding approval is obtained we would be ready to begin construction.

1. J. Kirchgessner et al, IEEE, NS 22-3, 1141 (1975).
2. Design Study for a 50 x 50 GeV Collider, CLNS 80/456, Internal Report, Floyd Newman Laboratory of Nuclear Studies, Ithaca, NY 14853.

# LEP AT CERN

W. Schnell
CERN, Geneva, Switzerland

## ABSTRACT

LEP is a large electron positron collider in which energies of up to 130 GeV per beam can be reached in several stages of construction. The circumference of the storage ring, which will be built adjacent to the present CERN site, is 30 km.

## INTRODUCTION

This project for a Large Electron Positron collider, called LEP, has been under study at CERN since 1976. The present design has been described in a detailed Design Report[1]. It has been optimized for a nominal beam energy of about 90 GeV at $10^{32}$ cm$^{-2}$ s$^{-1}$ maximum luminosity, to be obtained with room-temperature RF cavities of the highest shunt impedance considered feasible. The design permits extension to 130 GeV by means of superconducting cavities. The LEP circumference is about 30 km. Of the eight interaction regions foreseen, four are designed for the maximum luminosity and a free space of ± 5 m between the nearest quadrupoles, the other four for twice the free space and half the luminosity.

Much emphasis is being put on starting colliding-beam physics at the earliest possible date, albeit at reduced energy and luminosity. With $1/6$ of the nominal RF system installed a beam energy of more than 50 GeV can be obtained, which is expected to be very satisfactory for $Z^0$ physics. This stage of RF installation has, therefore, been given special attention and, combined with other measures of austerity, it will form "Phase One" of LEP operation. In Phase One only four experimental areas will be equipped and the RF system will be placed adjacent to two diametrically opposite beam crossings.

## THE GENERAL LAYOUT

The layout of LEP, near the present CERN site, has been chosen so as to permit e-p collisions in an SPS bypass as well as injection of protons into the LEP tunnel at a later stage. In addition, it is now proposed to use the PS-SPS complex as the LEP injector.

The LEP ring and its experimental areas are situated underground. About thirty test borings have been carried out so far to determine the exact depth of the upper rock boundary. Roughly one third of the LEP tunnel will lead through the Jura limestone and it is planned to start drilling a reconnaissance tunnel in the near future, in order to explore the quality of this type of rock.

0094-243X/81/680833-04$1.50 Copyright 1981 American Institute of Physics

The experimental areas will also be under ground, five of them under the flat part of the terrain and accessible via vertical shafts, the remaining three under the Jura mountains, with roughly horizontal access tunnels.

## LATTICE, COLLECTIVE PHENOMENA

The main arcs of LEP are formed by a regular separate-function FODO lattice. Before the beam reaches an RF straight section and the subsequent collision point its dispersion is suppressed, i.e. particles of different energy within the beam are superimposed on the same orbit. The necessary optics are formed from essentially the same elements as the main arcs.

Single beam collective phenomena determine the current which can be stored at the injection energy. The actual injection energy, 22 GeV, has been chosen as the minimum permissible value. The dominant two-beam phenomenon is the incoherent beam-beam limit. These effects are usually described by the maximum permissible beam-beam tune shift $\Delta Q$ and the design has been made assuming $\Delta Q = 0.06$ as an attainable limit. From the empirical data available this seems possible and half the value appears as a safe lower limit. There is enough flexibility in the LEP lattice to keep the luminosity proportional to $\Delta Q$ down to that lower limit, should the need arise.

## MAGNET SYSTEM

The dipole magnets are C-shaped, made from precision stamped laminations. The magnets will be excited by simple aluminium bar conductors instead of the usual multiturn coils.

The unusually low field, 0.123 T at 130 GeV, permits a reduction of the steel filling factor to less than 0.3 without leading to saturation. To this end, spacers are pressed into the laminations by the punching die, so that the laminations of 1.55 mm thickness will be spaced at 5.5 mm pitch once they are stacked in a jig (Figure 1). Then the assembly is placed in a mould and the space between laminations is filled with "concrete" i.e. a low shrinkage, corrosion-resistant mortar composed of cement and silica sand. So far, two full-size dipoles have been completed at CERN.

The strengths of the lattice quadrupoles and sextupoles form one of the potential limitations of machine energy; they have been designed to allow operation up to 130 GeV. The excitation coils will be fabricated from anodized aluminium strip of graded thickness, permitting the entire coil, of stepped cross-section, to be wound as a homogeneous unit.

## VACUUM SYSTEM

The high critical energies of the synchrotron radiation hitting the dipole chambers and the low injection field pose new problems for LEP although the basic design, a water-cooled chamber made of extruded aluminium and a linear sputter ion pump using the dipole field, is similar to that of existing machines.

A lead shield, surrounding the chamber, will prevent an excessive amount of radiation in the tunnel where corrosive and toxic chemicals are generated from air and humidity.

The distributed outgassing load will be absorbed by a linear distributed sputter ion pump with cells of 50 mm diameter. The pump anodes will be formed by five superimposed layers of thin stainless steel strips inserted between the titanium cathodes. A prototype has been tested.

## INJECTION

The preinjector contains a linear accelerator up to 600 MeV. An accumulation ring acts as a buffer between the linear accelerator and the existing PS and SPS machines, in which 22 GeV will be reached in successive acceleration. Modest additions to the existing PS RF systems will permit acceleration to about 3.5 MeV at which energy the beam will be transferred to the SPS via the existing complex of transfer systems. After acceleration in the SPS by means of an additional RF system the beam will be transferred to LEP at 22 GeV.

The vacuum chambers and the magnets of both existing machines can tolerate the synchrotron radiation without modification. The LEP injection cycles can be interlaced with the normal proton acceleration cycles in such a way that fixed-target proton physics at CERN is not noticeably affected by the additional task of LEP injection.

## RF SYSTEM

At the nominal parameters, the accelerating structure will consist of 768 five-cell slot-coupled cavities fed by high-power CW klystrons. To each of these cavities a low-loss, H-mode, storage resonator will be coupled and the coupled system will be excited at both its resonant frequencies. This makes the stored energy oscillate between the two resonators, spending on average half the time in the low-loss environment. In this way the dissipated power is decreased by a factor 1.5. The design, employing a spherical storage resonator, is shown in Figure 2.

In Phase One only 128 cavities and storage cavities, powered by 16 klystrons, will be installed.

The performance of room-temperature accelerating structures might be further improved by more intense modulation or pulsing. However, only superconducting accelerating cavities will permit the achievement of the full performance potential of LEP. This is discussed in a separate contribution to this conference[2].

References
1. The LEP Study Group, CERN Int. rep. ISR-LEP/79-33 (1979).
2. E. Picasso, contribution to this Conference.

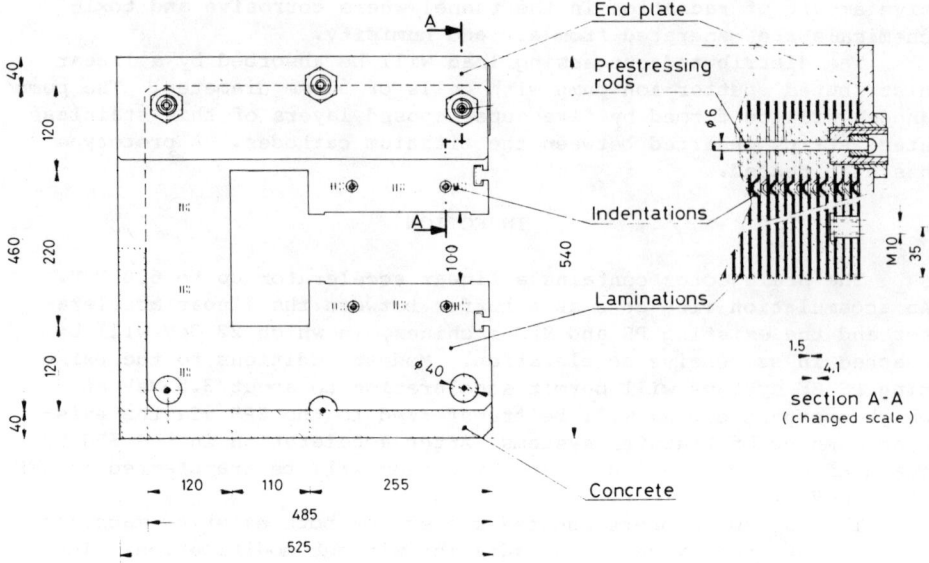

Fig. 1 : Concrete-filled dipole magnet.

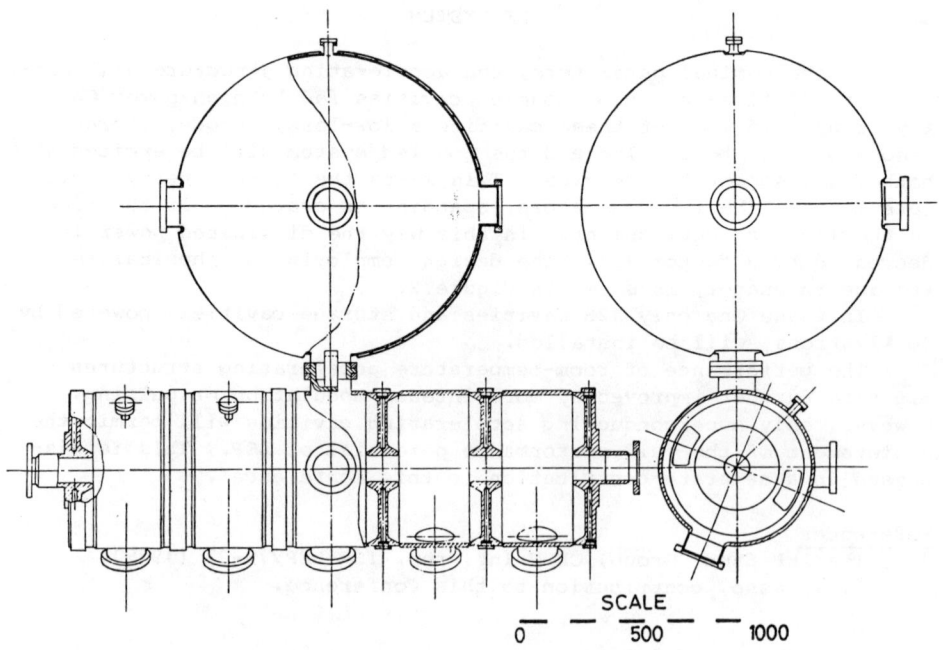

Fig. 2 : RF cavity with spherical storage cavity.

# HERA

B.H.Wiik - Deutsches Elektronen-Synchrotron DESY, Hamburg, Germany.

DESY has proposed to construct an electron-proton colliding ring HERA designed to collide 820 GeV protons with 30 GeV electrons yielding 314 $GeV_2$ in the center of mass and a maximum momentum transfer of 98400 $GeV^2$. The kinematical region available with HERA is shown in Fig. 1, the black dot at the left hand corner represents the region covered by a 1 TeV fixed target machine. Note that HERA can explore the region above 100 GeV in the center of mass, the presumed mass scale of weak interactions.

A feasibility study[1]) of the HERA project done in collaboration with ECFA has been completed and a discription of the project can be found in that report.

The general machine parameters are listed in table 1. The machine, to be constructed on a site ajoining DESY, has a fourfold symmetry: four 360 m long straight sections are joined by four arcs with a geometric radius of 797.6 m yielding a total circumference of 6451.2 m. The counter rotating beams cross at an angle of 10 mrad in the middle of the long straight sections.

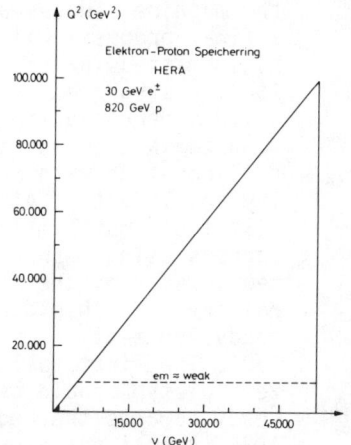

Fig. 1 - Kinematical region in $Q^2$ and $\nu$ which can be explored with HERA

A system of vertical bending magnets turns the transverse polarization of the electrons in the arcs into a longitudinal polarization in the interaction point. The machine and the four experimental halls will be burried some 10-20 m below the surface avoiding any disturbance to the site.

To guide the protons around the ring a total of 716 5.69 m long superconducting dipoles with an induction of 4.725 Tesla and 236 1.8 m long quadrupoles with a nominal gradient of 74.4 T/m are needed. The conductor is a niobium-titanium superconductor imbedded in a copper matrix. Both the magnet yoke and the bore will be at room temperatur. A warm bore is needed to avoid any excessive heat load on the refrigerator system from higher order mode losses caused by the short, very intense proton bunches.

The design of the electron ring is based on the experience gained with the PETRA ring. Although the PETRA components in principle could be used directly some changes are made to simplify the design and to reduce the cost. Notably the magnets and the vacuum system are considered as a single system where the magnets are used to shield the environment against the intense synchrotron radiation produced by the circulating electron beam. This leads to a design using a 4 mm thick Cu beam pipe with discrete pumps spaced 2 m apart and a simple arrangement of the coil.

|  | p-ring | e-ring |
|---|---|---|
| Nominal energy (GeV) | 820 | 30 |
| $s = Q^2_{max}$ (GeV$^2$) | 98400 | |
| Luminosity (cm$^{-2}$sec$^{-1}$) | $0.35 \times 10^{32}$ | |
| Polarisation time (min) | | 19.5 |
| Number of interaction points | 4 | |
| Length of straight sections (m) | 360 | |
| Free space for experiments (m) | 10 | |
| Circumference (m) | 6451.2 | |
| Bending radius (m) | 579.436 | 550.0395 |
| Magnetic field (Tesla) | 4.725 | 0.1819 |
| Total number of particles | $6.72 \cdot 10^{13}$ | $0.78 \cdot 10^{13}$ |
| Circulating current (mA) | 500 | 58 |
| Energy range (GeV) | 100 → 820 | 10 → 33 |
| Emittance ($\varepsilon_x/\varepsilon_z$) (x$10^{-8}$m) | 0.47/0.23 | 3.2/0.8 |
| Beta function $\beta^*_x/\beta^*_z$ (m) | 7/0.6 | 5/0.15 |
| Dispersion function $D^*_x/D^*_z$ (m) | 0.5 | 0 |
| Beam-beam tune shift $\Delta Q_x/\Delta Q_z$ | 0.0007/0.0007 | 0.014/0.01 |
| Beam size at crossing $\sigma^*_x$(mm) | 0.18(0.97)** | 0.40 |
| Beam size at crossing $\sigma^*_z$(mm) | 0.038 | 0.035 |
| Number of bunches | 224 | |
| Bunch length (cm) | 9.5 | 1.1 |
| RF frequency (Mhz) | 208.189 | 499.665 |
| Maximum circumferential voltage (MV) | 100*** | 288 |
| Total RF power (MWatt) | 2 | 13.2 |
| Filling time (min) | 8.5 | 5 |
| Injection energy (GeV) | 40.0 | 14.0 |
| Energy loss / turn (MeV) | | 140.2 |
| Critical energy (keV) | | 109 |

\* At the interaction point
\** Including the bunch length
\*** 25 MW is foreseen initially

Table 1
Basic parameters

Note that a luminosity of $0.35 \times 10^{32}$ cm$^{-2}$s$^{-1}$ is achieved with a linear tune shift of $7 \times 10^{-4}$ for the protons and 0.014 for the electrons.

DESY is continuing the machine studies and a final proposal will be submitted in April of 1981. A strong effort to build superconducting magnets is underway in collaboration with Saclay and Industry. At DESY the first 1 m long magnets using a shortened version of the final cryostat should be ready for test in March 1981. The first full size prototyp should be ready towards the end of 1981. A full scale prototype of the quadrupol magnet is being constructed at Saclay and also this magnet should be ready for test towards the end of 1981. The preparation for the civil engineering work is progressing well and the necessary construction could start in 1982. Once the project has been approved the construction work can be completed in four years such that the electron ring can be operating in the fifth year and the proton ring in the seventh year of construction. The two years between the completion of the electron ring and the turn on of the proton ring will be used to install the proton magnets and to commission the electron ring - i.e. learn how to operate a high current multibunch electron machine and how to produce polarized electrons. However, some time will also be available for e$^+$e$^-$ physics using the central part of the ep-detectors.

The HERA program is very ambitious, it is based on new technologies and it tries to collide electrons and protons which has never been done before. However, the scientific reward should be great and HERA should ensure DESY of an exciting future.

1) Study of the Proton-Electron Storage Ring Project HERA
   ECFA 80/42 - DESY HERA 80/01, 17 March 1980.

CHEER

CANADIAN HIGH ENERGY ELECTRON RING

Talk presented by

R.J. Hemingway (Canadian Institute of Particle Physics)

## ABSTRACT

The Institute of Particle Physics (IPP) in Canada have received funds from the Natural Sciences and Engineering Research Council (NSERC) to pursue a study which looks at the feasibility of adding an external electron storage ring at one of the long straight sections of the Tevatron. The machine, as currently configured, has a 300 MeV Linac injector, a 300 MeV accumulator ring, a 2 GeV booster synchrotron, and a 10 GeV storage ring holding 120 mA of either electrons or positrons. Particular attention has been paid to beam polarisation and the design of the interaction region.

## INTRODUCTION

The contributions of deep inelastic lepton nuclear scattering to the discovery of the internal structure of nucleons during the 70's are well known. The possibility of confirming many of our current theoretical ideas or even of discovering entirely new phenomena, by extending deep inelastic scattering to entirely new ranges of $Q^2$ is one of the obvious experimental challenges of the 1980's. The technique that clearly promises access to the largest area of the $Q^2-\upsilon$ plane is the collision of high energy electron and proton beams. CHEER, in its current configuration will be located at straight section D$\phi$ of the Tevatron at Fermilab (see Fig. 1). The Feasibility Study [1], of which only machine details are given here, started in the spring of 1979 when IPP selected E-P collisions as a most promising major project for H.E. Physics in Canada. The following characteristics were deemed to be important.

- The machine should provide an increase of useful $Q^2$ by at least a factor of 10 above fixed target experiments (present and future)
- To exploit fully the ELECTRO-WEAK interaction the machine should provide electrons and positrons in both helicity states
- Due to the complex on-going US construction program, the project should attempt to minimise interference
- The machine should be operational at least 3 years before TRISTAN at KEK or HERA at DESY.
- The project should be modest in cost.

Since the Canadian government has stated its intention to increase the fraction of the GNP presently awarded to R & D over the next few years it is our intention to design and construct a large fraction of the machine and detector in Canada.

The Feasibility Study, which will be finished in September 1980, has selected the Tevatron as the prime target since a) it will have the highest energy protons and 2) it is expected to have an earlier completion date than other machines.

Table I compares some of the features of CHEER with other Lepton-Quark experiments. For an estimated machine cost of 30 million dollars (1980 USA), electron-proton collisions could be investigated at Fermilab in 1985 with comparable Luminosity and Kinematics to those of TRISTAN and HERA. Both $Q^2_{max}$ and $\nu_{max}$ are a factor of 25 higher than the best muon beam the Tevatron will provide.

Table I Lepton - Quark Experiments

|  | $E_e$ (GeV) | $E_p$ (GeV) | $Q^2_{max}$ (GeV$^2$) | $\nu_{max}$ (TeV) | L (cm$^{-2}$sec$^{-1}$) | Completion Date | Cost ($US) |
|---|---|---|---|---|---|---|---|
| TRISTAN | 20 | 300 | 24,000 | 13 | $0.3 \times 10^{32}$ | 1988-89 | 240M[a] |
| HERA | 30 | 820 | 98,400 | 52 | $0.35 \times 10^{32}$ | 1988-89 | 330M |
| CHEER | 10 | 1000[b] | 40,000 | 21 | $0.4 \times 10^{32}$ | 1985 | 30M |
| TEVATRON | 750 | 1 | 1,500 | 0.75 | $0.7 \times 10^{30}$[c] | 1982-3 | 10M |

a) $e^+e^-$ option only

b) Now under construction

c) 1m $H_2$ target

## THE MACHINE

A schematic outline of the machine is shown in Fig. 2 and a list of parameters is given in Table II.

Electrons (or positrons) are produced by a 300 MeV Linac and injected into a 300 MeV accumulator ring. This ring, with a circumference of 19.75m, is a smaller version of PIA at DESY - it accumulates and damps $10^{11}$ particles into a single bunch. The bunch is then transferred to a booster synchrotron of 98.75m circumference, which accelerates to 2GeV before transfering to CHEER.

Many of the basic machine parameters are dictated by those of the Tevatron which has been designed and is under construction, viz- the RF frequency, the bunch separation, the emittance and the low β straight section. We have assumed rebunching of the Tevatron beam by a factor 7 to give 159 total bunches, each with $1.4 \times 10^{11}$ particles. CHEER, with a circumference of 1343m, will contain 34 bunches giving 120mA of stored current. The RF system is designed at 804 MHz and must compensate the 1.04 MW of radiated power. Total filling times are estimated at 2 seconds for electrons and 10 minutes for positrons.

### Table II   Machine Parameters of CHEER

| | | |
|---|---|---|
| Pre-Injector | 300 | MeV LINAC |
| Accumulator | 300 | MeV |
| Injector | 2 | GeV SYNCHROTRON |
| Storage Ring | 10 | GeV |
| Filling Time | \multicolumn{2}{l}{2 seconds for $e^-$, 10 minutes for $e^+$} |
| # of Electrons | \multicolumn{2}{l}{$10^{11}$ electrons/bunch in 34 bunches} |
| Total Current | 120 | mA |

Lattice Characteristics at 10GeV

- Circumference     1343.5 m
- Bend Radius     100. m
- Polarization Time     35. min
- Gross Radius     157. m
- 2 straight sections     180. m each
- 60 cells, FODO, 90° phase advance
- Momentum Compaction $\alpha = .00293$
- Energy width $\sigma_E = .00085$
- Damping times $\tau_x = .0105$ sec., $\tau_E = .0053$ sec.
- Emittance $\varepsilon_x/\varepsilon_z = .0472/.0094$ πmm.mrad. (coupling = 0.5)
- $\beta_{max} = 24$ m, $\eta_{max} = .93$ m

### RF Characteristics

| | |
|---|---|
| Installed RF power | 2.1 MW |
| Radiated RF power | 1.02 MW |
| RF frequency | 804 MHz |
| Radiation loss/turn | 8.49 MeV |
| Critical energy | 22.2 keV |

### Intersection Region

| | |
|---|---|
| Free space for detector | 13 m |
| $\beta^*_x/\beta^*_z$ | .55/2.055 m |
| $\sigma_x = \sigma_z$ | .16 mm |
| Crossing Angle | 0° |
| Beam-beam tune shift | $\Delta\nu p = 0.0018$, $\Delta\nu e = 0.0345$ |
| Dispersion $\eta^*$ | 0. |
| Luminosity | $.4 \times 10^{32}$ cm$^{-2}$ sec$^{-1}$ |

(Assume TEVATRON $\varepsilon$ (95%) = .026 mm.mrad.
$\beta^*_p$ = 5m, 1.4 $10^{11}$ protons/bunch, 159 bunches).

## THE INSERTION GEOMETRY AND POLARISATION

Two major constraints have to be recognised - a) the limited free space between the low $\beta$ superconducting quadrupoles of the Tevatron and b) the importance of both helicity states for electrons and positrons. FIG. 3 shows a possible lay-out. Transverse polarisation, which should build up naturally within ~15 minutes, is rotated into the longitudinal direction using a scheme proposed by Christ, Farley and Hereward [2]. This scheme, which keeps CHEER coplanar with the Tevatron, should provide both helicity states for both charges. The total horizontal bend of 69 mrad is provided by four bending magnets of decreasing strength as the interaction point is approached. With this arrangement only a few watts of radiated power go down the cold bore of the Tevatron beam pipe.

## THE LUMINOSITY

The low $\beta$ insertion of the CHEER lattice has been chosen to match the round proton beam. With $\beta^*_p$ of 5m, the transverse beam size is 0.16 mm and leads to a luminosity for zero degree crossing angle of $0.4 \ 10^{32}$ cm$^{-2}$ sec$^{-1}$.

## TIMETABLE

Given a receptive attitude by our funding authorities, we estimate 1.5 years to complete a detailed design and then 3 years for construction. Operation could commence in mid 1985.

## CONCLUSION

We believe that the CHEER project represents an excellent opportunity for Canada to contribute more fully to international high energy physics, and specifically represents a very real possibility for an E-P collider in N. America by mid 1985.

## REFERENCES

1. CHEER WORKSHOP PARTICIPANTS

    A.N. Kamal (University of Alberta), D. Beder (University of British Columbia), R.K. Carnegie, K.W. Edwards, A. McPherson, M.K. Sundaresan, P.J.S. Watson (Carleton University), J. Fraser, M. Harvey, H.C. Lee, J. McKeown, H. Schneider, S. Schriber (Chalk River Nuclear Laboratories), C.S. Kalman (Concordia University), G. Karl (University of Guelph), S. Conetti, P.G. Estabrooks, R.J. Hemingway (Institute of Particle Physics), H.K. Quang (Laval University), J. Jovanovich (University of Manitoba), P. Patel, D.G. Stairs, J. Trischuk (McGill University), C.K. Hargrove, P. Hobson (National Research Council), B. Campbell, V. Elias, N. Isgur, J.F. Martin, J.D. Prentice, T.S. Yoon (University of Toronto), R. Servranckx, Y.M. Shin (University of Saskatchewan), E. Blackmore, D. Gurd, P. Jackson, J. Ng (Triumf), W. R. Frisken, P. Padley, L. Turnbull (York University), G. McKeon, R. Migneron, J.B. Mitchell (University of Western Ontario), G. Danby (Brookhaven National Laboratory), L. Hand (Cornell University), K. Anderson, F. Merritt, J. Pilcher (Enrico Fermi Institute), M. Harrison, D. Johnson, F. Mills (Fermilab), S. Wipf (Los Alamos), A. Skuja (University of Maryland), L. Krauss (Massachusetts Institute of Technology), D. Laughton (Princeton), E. Williams (Vanderbilt University), B. Norum (University of Virginia).

2. N. Christ, F.J.M. Farley, H.G. Hereward  NIM $\underline{115}$, 227 (1974)

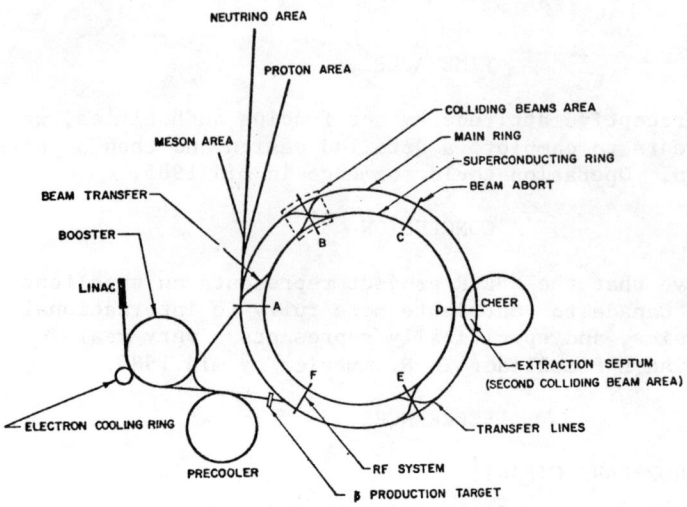

FIGURE 1. SCHEMATIC PLAN OF THE FERMILAB TeV PROGRAM

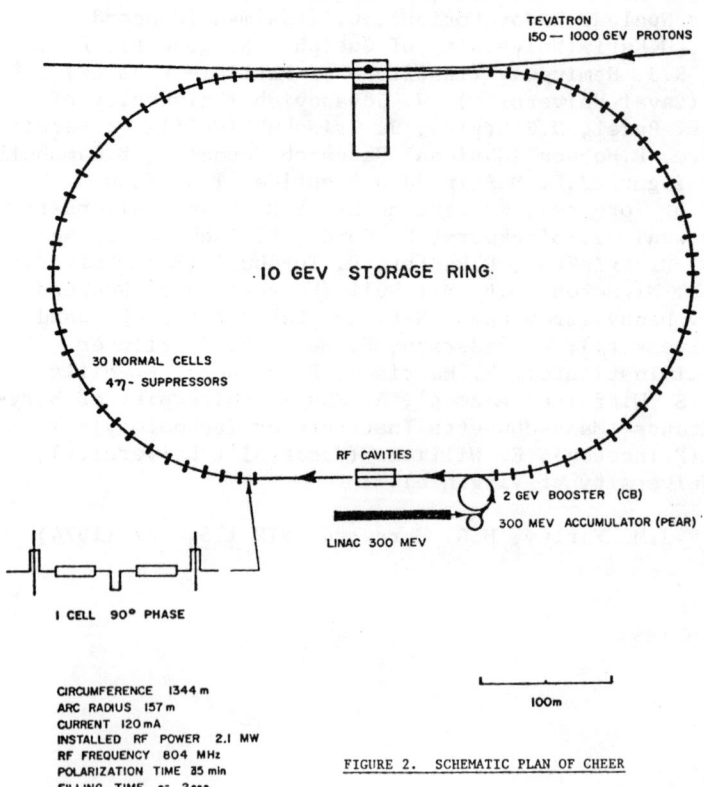

FIGURE 2. SCHEMATIC PLAN OF CHEER

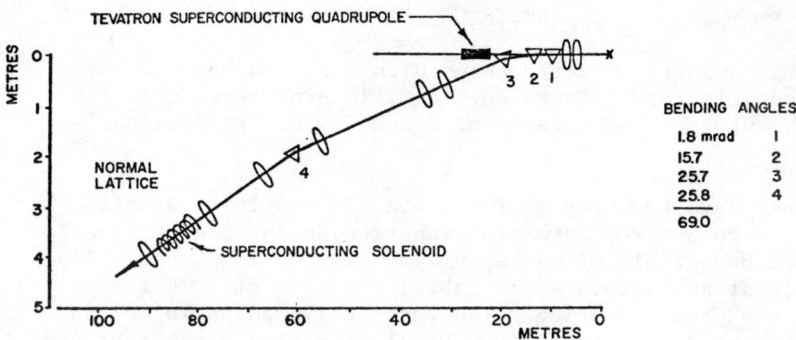

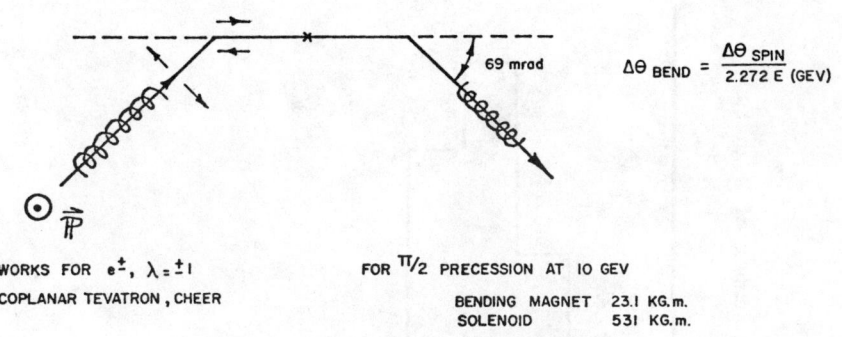

FIGURE 3. SCHEMATIC OF INSERTION REGION

## STATUS OF THE FERMILAB
## ENERGY SAVER/DOUBLER DIPOLE

J. R. Orr
Fermi National Accelerator Laboratory*
P. O. Box 500, Batavia, Illinois 60510

### ABSTRACT

The Fermilab Energy Saver/Doubler dipole design has now been shown to be satisfactory. There were some internal problems with the cryostat, but these have been found and fixed. Production is underway.

The Energy Saver/Doubler will contain 774 dipoles. Briefly, the dipole is a cold-bore, warm-iron superconducting magnet capable of producing a bend field of 44 kG at 4400 amp for 1 TeV operation. The coil, made of Rutherford style cable, consists of 2 shells clamped in a $\cos\theta$ configuration. The inner coil radius is 1.5 in. The overall magnet length is 21 ft. Detailed design parameters and specifications are given in the Fermilab Superconducting Accelerator Design Report.[1]

An indication of the overall uniformity and quality of these magnets can be seen in Figures 1 and 2. Figure 1 is a histogram of the quench currents, at the required Doubler ramp rate of 200 amp/sec, for the first 50 magnets. Figure 2 is a set of histograms of the first 2 normal ($b_n$) and skew ($a_n$) multipoles for the same set of magnets. The multipoles are measured in units of $10^4$ (in.)$^n$.

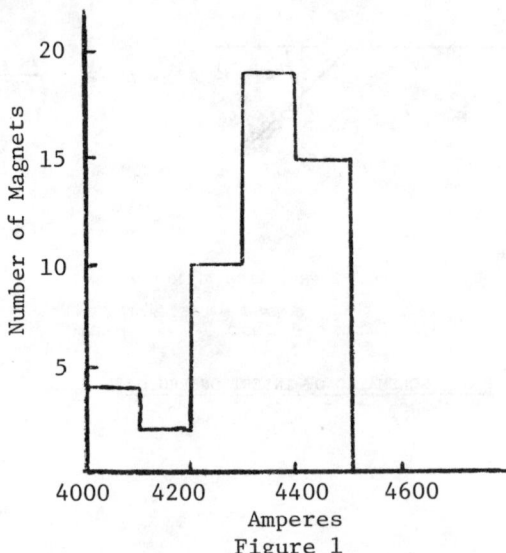

Figure 1

---

*Operated by Universities Research Association, Inc., under contract with the U.S. Department of Energy.

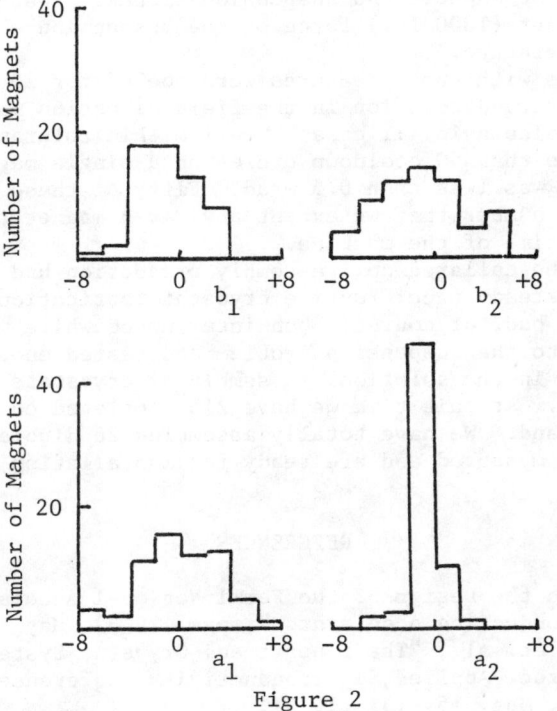

Figure 2

Shortly after the onset of production a change was made in the cryostat design in order to simplify fabrication.[2] During the routine magnetic testing of completed magnets a defect in the suspension system within this new cryostat began to manifest itself. The most obvious symptom was a change in the direction of the dipole field from one cooldown cycle (between 300°K and 4°K) to another. Full field ramping and quenching had no effect on the field direction.

The observed field rotation was sometimes as large as several milliradians per cooldown. The specified tolerance in the uncertainty of this parameter is $\pm 0.5$ mrad.[3]

The problem was traced to the breakage of the "anchor", the only constraint (against rotation) between the cold coil assembly and the warm outer cryostat wall. The anchor failure was due to stress caused by the large thermal gradient which appears along the magnet during a rapid cooldown.

The problem was solved by increasing the number of anchors from 1 to 4 and by returning to an older cryostat design which allowed better positional control of the remaining 32 G-10 suspension blocks which are arrayed in 4 rows 90° apart between the coil and the warm outer tube of the cryostat.

An additional improvement was incorporated into the iron yoke at the same time. Spring loaded bolt assemblies were threaded through the yoke above the anchors and suspension blocks. These assemblies provide a constant (1000 lb.) force on the suspension elements independent of temperature.[4]

Ten magnets with these features were cooled for 20 to 50 cycles each and the observed rotation in the field direction was less than 0.1 mrad per cooldown in all cases - well within tolerance. In fact, after more than 50 cooldown cycles on a single magnet the total excursion was less than 0.5 mrad. Fifty of these cycles are about twice the number that we expect any given magnet to undergo in the entire lifetime of the machine.

Although the collared-coil assembly production had continued at a more or less steady pace, routine cryostat fabrication and final magnet assembly had, of course, been interrupted while we searched for a solution to the suspension problem and tested enough magnets to be confident in the solution. Assembly of cryostats and magnets has now resumed. At this time we have 215 completed collared-coil assemblies on hand. We have totally assembled 26 dipoles; 21 of these have been measured and are ready for installation in the accelerator tunnel.

## REFERENCES

1. "A Report on the Design of the Fermi National Accelerator Laboratory Superconducting Accelerator", pgs. 29-32, May 1979.
2. G. Biallas, et. al., "The Support and Cryostat System for Doubler Magnets", Proc. Applied Superconductivity Conference, 1978, IEEE Trans. Mag., Mag. 15, 131 (1979).
3. "A Report on the Design of the Fermi National Accelerator Laboratory Superconducting Accelerator", pg. 30, May 1979.
4. R. A. Lundy, "State of the Energy Doubler", to be published in the proceeding of the 1980 Applied Superconductivity Conference held in Santa Fe, September 1979.

# DEVELOPMENT OF RF SUPERCONDUCTING CAVITIES FOR LARGE STORAGE RINGS

E. Picasso
CERN, Geneva, Switzerland

## ABSTRACT

The present status of RF superconducting cavities is given. Electric fields of 3 MV/m are achieved in single cell cavities, in a reproducible manner.

## INTRODUCTION

Since the early 1960's[1] we have known that it is possible to sustain an RF electric field of many millions of volts per meter in large volume by the expenditure of a few watts of RF power. The hope is to achieve maximum values of RF electric field of 40 to 50 MV/m. For the time being, this hope is a dream; lower values have been achieved up to now, typically of a few MV/m, in the range of frequencies used to accelerate $e^+e^-$ beams in storage rings[2]). However, it is by now recognized that substantial savings in the cost of high energy $e^+e^-$ machine can be realized if superconducting cavities capable of sustaining several MV/m can be used[3]). In an $e^+e^-$ storage ring of the dimension of LEP[4]) at 86 GeV, where synchrotron radiation power for the two beams is about 25 MW, the power dissipation at the cavity walls is about 60 MW if normal conducting cavities are used. For a gradient of 1 MV/m the use of superconducting cavities, where there is practically no power dissipation at the cavity walls, requires about 4 MW of power consumption at the refrigerator to operate the cavities at 4.2 K. If an accelerating gradient ($E_{acc}$) of about 3 MV/m is obtained the power consumption at the refrigerator is about 16 to 20 MW depending on the quality factor (Q) and the working temperature of the cavities (typically for Nb cavities $Q = 1.3 \times 10^9$, $T \sim 4.2$ K at a frequency $f = 500$ MHz)[5]).

Superconducting RF cavities are characterized by two quantities: the Q-value and the limiting field.

## 1. The factor of merit Q

The "unloaded" Q-value is a measure of the merit of an RF cavity as resonator and is defined as the ratio of the energy stored in the cavity to the energy lost due to RF dissipation in the cavity per radian of the RF cycle. The Q-value gives the economy of a superconducting accelerating section. Q-values of $2 \times 10^9$ are obtained in a cavity resonating at 500 MHz. RF losses are not considered to be the main concern. Above 1 GHz a cavity, whose walls are made of Niobium, must work at 1.8 K which is very costly. Below 1 GHz one can work at 4.2 K, mainly due to the fact that below 1 GHz the residual resistance, which is temperature independent, dominates (see Fig. 1). If $Nb_3Sn$, which has a critical temperature $T_c = 18$ K, could be used, then also high frequency resonators could be operated at 4.2 K.

## 2. Limiting Field

Let us consider the ultrarelativistic range and let us consider only cavities of either cylindrical or spherical shape, or muffin-tin[6] structure.

Three important quantities must be considered: peak magnetic field ($H_p$), peak electric field ($E_p$) and the accelerating fields ($E_{acc}$). Figure 2 gives the present state of the art of $E_p$ versus frequency. A few points to note on the peak magnetic (electric) fields are: i) peak fields usually decrease with increasing cell numbers; ii) $H_p$ (and $E_p$) increases with the frequency. All values obtained for $H_p$ are much smaller than the critical magnetic field $H_c \sim 200$ mT.

Figure 3 gives the present state of the art for $E_{acc}$ versus frequency. At present, accelerating electric fields of 3 MV/m can be reached at almost all frequencies, though C-band seems favoured for higher fields.

It is worthwhile discussing in a little more detail the nature of the limitation. It seems that three sources of limitation could be responsible for the present situation: i) electron loading due to multipacting, ii) electron loading due to field emission or thermoionic emission, and iii) breakdown defects due to thermal instabilities, with possible additional complications due to magnetic instabilities.

i) <u>Electron loading</u> can be due either to resonant or non-resonant phenomena. Resonant electron loading is a low-pressure high-frequency electron conduction phenomenon in which the time-of-flight of an emitted electron from one surface to another is in synchronism with the frequency of the field and on impact releases more than one electron (multipacting). The resonant electron loading occurs in two varieties: one- and two-side multipacting. The one-side multipacting is at present the limitation in HE PL accelerators at Stanford University. One way to avoid multipacting is to choose the shape of the cavity in such a way as to disturb one of the conditions necessary for it. There are many ways to avoid multipacting but I will mention but a few of them.

a) To decrease the radial component of the electric field in such a way that the impinging energy of a resonant electron is less than 40 eV. If this condition is satisfied then the probability of emitting a secondary electron is less than unity. A cavity shape which satisfies this condition is the cylindrical one with sharp corners.

b) Another condition is to design the resonator in such a way that the axial component of the electric field $E_z > E_r$. In a spherical resonator, where $E_z$ is greater than the radial component $E_r$, the electron trajectory is displaced, i.e. the electron "walk away". At Genoa University spherical cavities have been developed and promising results have been obtained at 4.5 GHz[7]. At Wuppertal and CERN[8] cavities of this shape have been constructed at 3 GHz and at 0.5 GHz.

c) To depress multipacting one must reduce the secondary electron emission coefficient $\delta$ to be less than unity over the whole range of energies. This requires surface preparation of Nb walls, thermal treatment, RF processing, etc.

Spherical cavities and sharp cornered cylindrical cavities or grooves in muffin-tin cavities do not show evidence of multipacting[2].

ii) <u>Non-resonant electron loading</u> is a problem as yet unsolved in cavities with a frequency below 1 GHz. The electron current increases with voltage and from the "Fowler-Nordheim" plot and using work function 4 eV a field enhancement factor can be derived. This coefficient is the ratio between the effective electric field responsible for pulling out the electrons from the metal surface divided by the macroscopic electric fields on the metal surface. The value of this ratio up to 2000 has been observed in electron loading cavities. These values are too high to be understood in terms of geometric factors, such as surface roughness. I feel that the electron loading phenomenon has not yet been sufficiently understood and more powerful diagnostic methods are needed in order to clarify and interpret the complex physical processes.

A chain of carbon resistors and solid state X-ray detectors is used at CERN[9] for temperature and X-ray mapping of the cavity surface. Together with a visual inspection of the cavity interior during operation, a better insight into field breakdown mechanisms and electron loading is obtained. Figure 4 shows a temperature map of a spherical cavity of 500 MHz in operation at CERN. The surface of the body is projected onto a plane, temperatures are plotted along the $\Delta T$ axis.

iii) <u>Thermal limitations</u> are associated with the existence of "bad spots" where either the temperature is close to the critical temperature of the superconducting material or normal conducting spots (regions), initial in a superconducting state at zero magnetic field, soon become normal as the field is applied by proximity effect[2]. Better cooling conditions or purer Niobium surface (lower tantalium impurities) must be used. The last point is a difficult metallurgical one because the large quantities of Nb required to construct a low-frequency resonator (500 MHz or lower) are normally contaminated by tantalium at levels of several hundred parts per million.

I am confident that at 500 MHz an accelerating field of 3 MV/m can be reached in a single-cell cavity without encountering too many difficulties. Higher values of $E_{acc}$, let us say about 5 MV/m or higher, are harder to obtain without dealing with high electron loading for the cavity that damage the Niobium surface of the resonator. Another effect of electron loading is to excite higher order modes in the resonator. One of the major problems in applying

RF superconducting cavities to storage rings is the extraction of the power dissipated at the walls by the higher order modes excited by the beam when it goes through the cavity[10]).

Two very important questions for large storage rings which are currently investigated are:

      i) what results can be obtained in a single cell at frequencies lower than 500 MHz?

and

      ii) what results can be obtained in a multicell structure at frequencies of about 500 MHz?

On the last point, four coupled spherical cavities have been constructed at CERN and at the end of the year we hope to have the first results.

Feeding high power (typically 100 kW) must be investigated and the effect of higher order modes excited by the bunched beam eliminated. In order to expose a superconducting cavity to an accelerator environment the KfK Karlsruhe Laboratory is constructing two cylindrical cavities at 500 MHz to be tested at DORIS. Cheap methods of construction of the cavities and cryogenic system are essential before taking any decision to introduce this technique in such a big project as LEP.

In my opinion, which is perhaps over optimistic, if we confine our consideration to up to 3 MV/m, one or two years are still needed in order to understand the behaviour of a complex structure in the Laboratory, plus at least another two years to operate the multicell structure in DORIS. I do not think we will be in a position to take any decision on the construction of a storage ring operating at 350 to 500 MHz using RF superconducting techniques before at least five years from now. At higher frequencies, where the dimensions of the resonators are smaller and the electron loading is probably less important, it may be possible to be in a position to use the superconducting cavity in a storage ring earlier. For accelerating fields above 3 MV/m, let us say 5 MV/m, more development is needed and more understanding of the surface physics of superconducting material is required[11]). I strongly feel that an effort has to be made to develop methods of manufacture that do not require the use of bulk Niobium.

To summarize, the first phase of LEP will be built with normal conducting cavities; later on, superconducting cavities could be installed, depending on the state of the art, in order to obtain higher energies.

## References

1. Banford, A.P. and Stafford, G.H., 1961 Plasma Phys. $\underline{3}$, 287. Wilson, P.B., Nucl. Instr. Methods, 20, 336 (1963).
2. Citron, A. Report on the Workshop on RF Superconductivity, held at Karlsruhe, July 2-4, 1980. XIth International Conference on High Energy Accelerators, July 7-11, 1980, CERN, Geneva.
3. Design Study Proposal for a High Energy, High Luminosity Electron-Positron Collider based on Superconducting RF Cavities, The Floyd R. Newman Laboratory of Nuclear Studies, Cornell University, Ithaca, New York 14853, May 1980.
4. W. Schnell, LEP at CERN, Madison Conference, 1980.
5. Bernard, Ph., Cavallari, G., Chiaveri, E., Haebel, E., Heinrichs, H., Lengeler, H., Picasso, E., Picciarelli, V. and Piel, H. First results on a Superconducting RF Test Cavity for LEP. XIth International Conference on High Energy Accelerators, July 7-11, 1980, CERN, Geneva.
6. Lengeler, H., Design of Superconducting Structures, Talk given at the Workshop on RF Superconductivity, KfK Karlsruhe, 2-4 July 1980, CERN/EF 4/8/80.
7. Lagomarsino, V., Manuzio, G., Parodi, R. and Vaccarone, R. 1978 ASG, IEEE Trans. Mag. 15.
8. Klein, V., Proch, D., Lengeler, H. First Results with Superconducting 3 GHz Cavities of Spherical Shape. XIth Int. Conf. on High Energy Acc., July 7-11, 1980.
9. Piel, H. and Romijn, R., Temperature Mapping on a Superconducting RF Cavity in Sub-cooled Helium, CERN/EF 80-3, June 1980.
10. Hutton, A., Present Status of the LEP Project, XIth Int. Conf. on High Energy Accelerators, CERN, Geneva, July 7-11, 1980 CERN-ISR-TH/80-37.
11. Halbritter, J., Theoretical Aspects in RF Superconductivity, KfK Report 3019, September 1980, Karlsruhe.

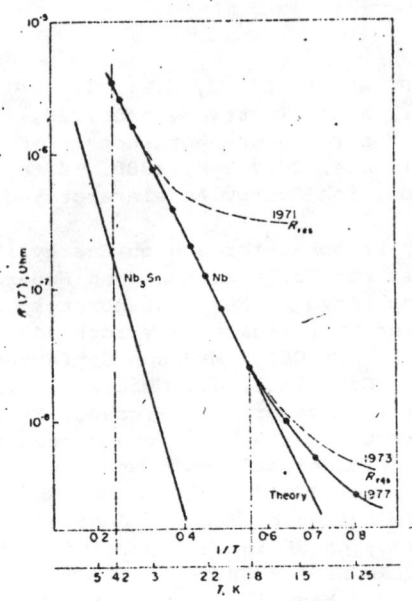

Fig. 1 : Surface resistance as a function of temperature obtained in a superconducting Nb cavity at f = 3.7 GHz. The straight lines correspond to the value calculated with (1) for Nb and $Nb_3Sn$. By improved treatments $R_{res}$ could be lowered to values ~ $10^{-9}$ ohm (courtesy P. Kneisel).

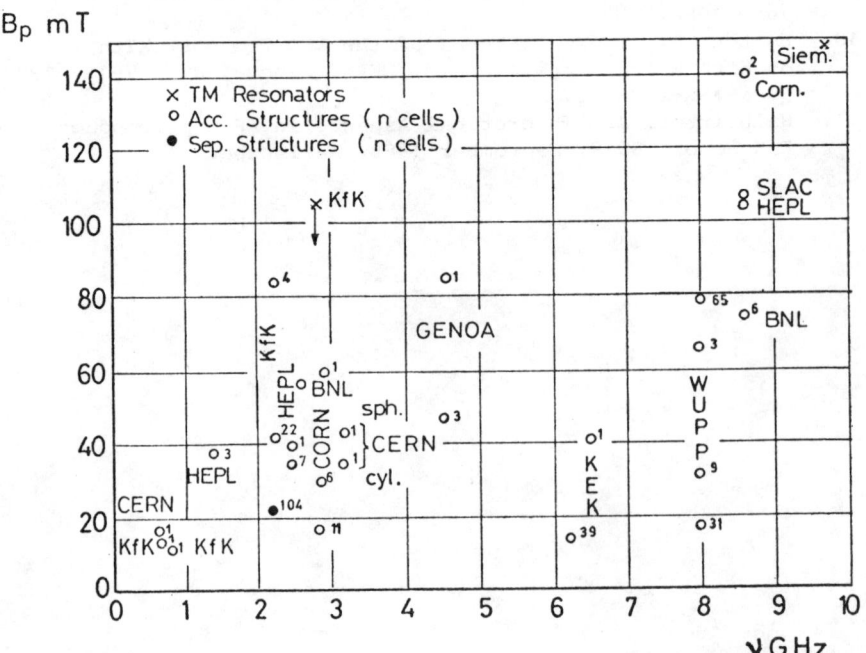

Fig. 2 : Peak magnetic surface field reached versus frequency. (courtesy A. Citron)

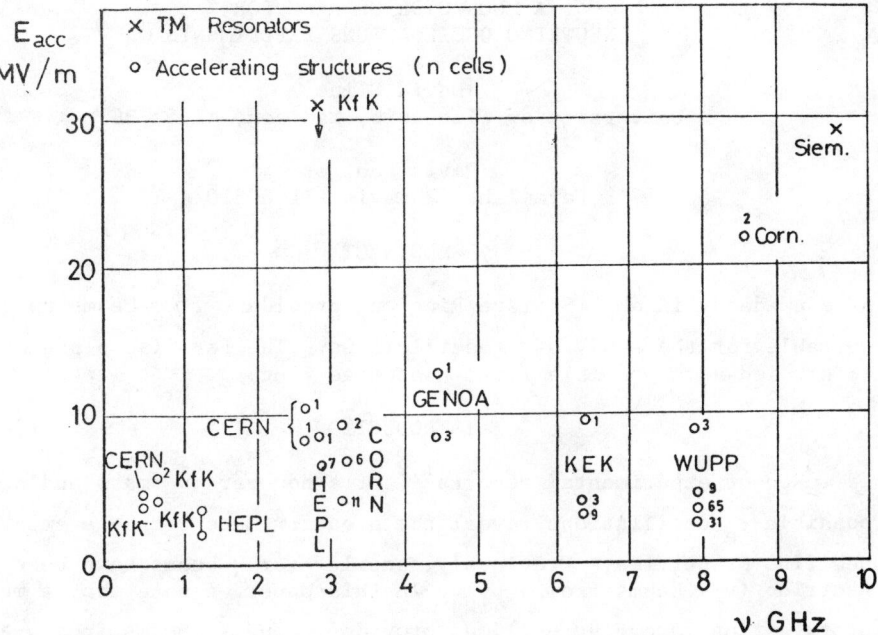

Fig. 3 : Peak accelerating field reached versus frequency. (courtesy A. Citron)

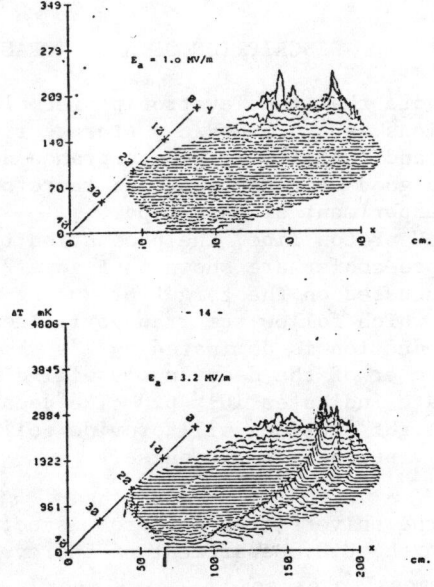

Fig. 4 : Temperature map of the cavity. The surface of the cavity body is projected on a plane, temperatures are plotted along the $\Delta T$ axis.

# A MUON STORAGE RING FOR NEUTRINO OSCILLATIONS EXPERIMENTS

David Cline
University of Wisconsin, Madison, WI 53706

David Neuffer
Fermilab,* Batavia, IL 60510

## ABSTRACT

$\mu^{\pm}$ decay in a $\mu^{\pm}$ Storage Ring can provide $\nu_e$, $\nu_\mu$ beams uniquely suitable for the study of $\nu$ oscillations. The Fermilab $\bar{p}$ precooler is studied as a possible first $\mu$ storage ring.

## INTRODUCTION

Recent experimental reports[1,2] of a non-zero $\nu_e$ mass and of possible $\bar{\nu}_e$ oscillations reveal the need for more complete study of neutrino properties. Previously, accelerator $\nu$ beams have been muon neutrino ($\nu_\mu$) beams from $\pi \to \mu \nu_\mu$. In this paper we note that a muon storage ring (see Figure 1) can provide $\nu_e$ and $\bar{\nu}_\mu$ beams from $\mu \to e \nu_e \bar{\nu}_\mu$ as earlier suggested by Wojicicki and Collins.[3] We further note that a $\mu$ storage ring provides clean $\nu$ beams of precisely knowable flux, and therefore an excellent tool for the study of $\nu_e$ and $\nu_\mu$ oscillations.

## DESCRIPTION OF A $\mu$ STORAGE RING

We also note that the Tevatron $\bar{p}$ precooler (see Figure 2) inescapably functions as a 4.5 GeV/c $\mu$ storage ring during the first ms of its cycle, and that its large acceptance designed for $\bar{p}$ acceptance make it a very good storage ring, and therefore a candidate for use in the first experiment of this type.

The 80 GeV proton line, the production target, the transport line and the pre-cooler are shown in Figure 2. Pulses of $1.8 \times 10^{13}$ protons are focussed on the target producing many secondary particles ($\pi$, k, $\bar{p}$, etc.) which follow the transport line to insertion in the ring. The production is dominated by $\pi$'s which decay ($\pi \to \mu \nu$) and a substantial number of the decay muons will circulate in the ring, a first estimate indicates $10^{10}$ $\mu$.[4] The decay of these muons in precooler straight sections will provide collimated $\nu_e$ and $\bar{\nu}_\mu$ beams with $\sim 8 \times 10^8$ $\nu$ per beam per p pulse.

---

*Operated by the Universities Research Association, Inc., under contract with the U.S. Department of Energy.

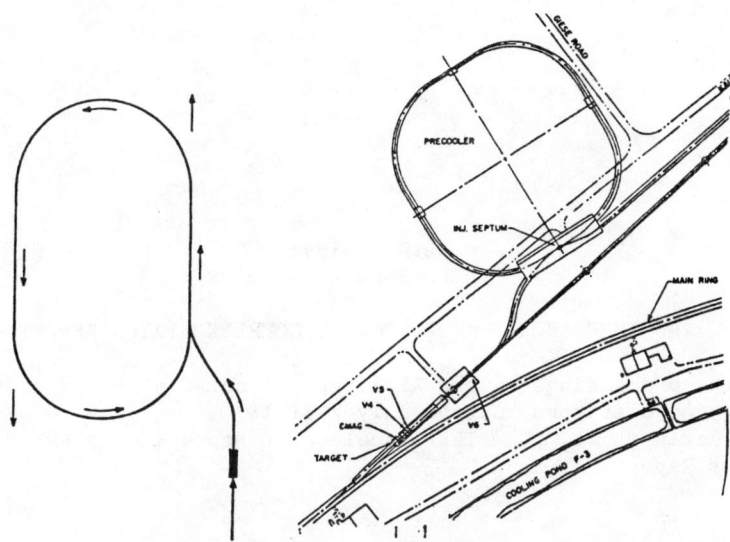

Figure 1: A μ Storage Ring.   Figure 2: The $\bar{p}$ Precooler/μ Storage Ring.

Modifications of the precooler to increase and acceptance and to increase the decay straight section length could increase this flux by a factor of ~10 and the proton pulse period of 10 seconds can be reduced from 10 seconds with $\bar{p}$ cooling (parisitic ν beam) to two seconds (dedicated mode). These intensities and designs are discussed in Reference 4, and will be improved in future work.

## EXPERIMENTAL COMMENTS

The precooler μ storage ring can provide adequate event rate for a variety of experiments. A 100 ton detector, 0.5 km away will receive ~4-400 events/day with $5 \times 10^8 - 5 \times 10^9$ $\nu_e \bar{\nu}_\mu$/pulse, $10^4$-$10^5$ pulses/day. The Fermilab 15' bubble chamber could also observe events. A suitable compromise between detector size, sensitivity, and cost is left as a challenge to interested experimenters. Since the ν flux can be precisely known from monitoring the decaying muon current, the μ storage ring can provide a unique tool for future ν experiments.

## REFERENCES

1. F.H. Reines, et al., this proceedings.
2. V.A. Lyubimov, et al., ITEP-62, submitted to Yadernaya Fizika (1980).
3. S. Wojcicki, unpublished (1974), T. Collins, unpublished (1975).
4. D. Cline and D. Neuffer, paper contributed to XX International Conference on High Energy Physics (1980).

L.D. Soloviev
IHEP, Serpukhov, USSR

THE IHEP ACCELERATOR STORAGE COMPLEX STATUS REPORT

See the proceedings of the Xl International Conference on High
Energy Accelerators, Geneve July 7-11 1980.
A.I. Ageev et al.  The IHEP Accelerator Storage Complex
Status Report.

# THE TRISTAN - KEK FUTURE PROJECT

Tetsuji Nishikawa
KEK, National Laboratory for High Energy Physics

## ABSTRACT

Since 1973, a storage ring complex, TRISTAN(Transposable Ring Intersecting Storage Accelerators in Nippon) has been proposed at the KEK, the National Laboratory for High Energy Physics in Japan.[1] The original TRISTAN plan was a colliding beam scheme with electrons (or positrons) at 17 GeV and protons at 180 GeV. Recently, however, the TRISTAN was extended to higher energies for it is ascertained that achieving a momentum transfer larger than 100 GeV/c is highly desirable to investigate strong and weak intereactions.[2]

I. Outline

The new version of the TRISTAN is designed so as to have the largest size that can be accommodated on the KEK site(Fig 1). The total circumference of the rings installed in the same tunnel is raised up to 3016 m from 2036 m of the previous version. This makes it possible to accelerate electrons up to 25 ∿ 30 GeV and protons more than 300 GeV leading to the center-of-mass energy for e-p collisions as high as 170 GeV.

II. e-p Colliding System

The principal parameter for e-p collisions is given in Table 1.

The exisiting KEK facilities can be used as the injectors for the e-p colliders, i.e., the proton and electron beams in the TRISTAN rings are provided from the 12 GeV KEK proton synchrotron and the 2.5 GeV electron linac for the synchrotron radiation facility (Photon Factory) respectively.

The maximum proton energy in the TRISTAN is about 300 GeV with the superconducting magnet ring of 4.5 T. To reduce a large superconductor magnetization, a conventional magnet ring proton booster will be installed above or below the superconducting ring.

The maximum electron energy is determined by the available rf power to compensate the energy loss due to synchrotron radiation. If a 20 MW rf power is fed into the ordinary rf-cavity system of 200 m long, then an electron beam of 70 mA will be obtained at about 25 GeV. The electron linac for the Photon Factory will be completed by the end of 1981. However, its energy of 2.5 GeV is not enough to inject the beam directly into the TRISTAN ring since the damping time is too long as compared with the repetition time of the linac. Therefore an intermediate accumulater ring of about 6 GeV will be used between the 2.5 GeV linac and the TRISTAN ring.

The proposed lattice of Tristan rings has four-fold symmetry with four long straight sections for rf accelerations and experimental insertions. A schematic diagram of the colliding system is shown in Fig 2. The lattice structure and orbit functions have been carefully studied both for electron and proton rings, including the typical design of experimental insersions. In particular, utilization of polarized electrons for studies of the weak interaction physics has been investigated.

The luminosity for e-p collision caluculated with optimized parameters is given in Fig 3.

## III. $e^+$-$e^-$ Colliding System

The TRISTAN entire Project will be divided into phase I and phase II. At phase I, the electron ring will be constructed with the accelerator enclosure for the whole project. The electron-positron colliding beam experiments will be carried out in the electron ring.

For this purpose the electron ring will be arranged as in the dotted line in Fig 2. The rather long straight section prepared for e-p collision can be used to install a long rf cavity system of about 400 m in length. This makes it possible to attain higher energies with smaller rf power. The optimum luminority for $e^+$-$e^-$ collision vs beam energy is shown in Fig 4. The development of superconducting rf cavity may lead the maximum attainable energy as high as 40 GeV.

## IV. Status

In addition to the design study, technical research and development of superconducting magnet and cavity system, ultra-high vacuum system, high power rf system and computer control system are in progress. The construction of TRISTAN phase I is hoped to be authorized from Fy 1981 for a construction period of five years.

Table 1

General parameters of the TRISTAN electron-proton collider

| | |
|---|---|
| Circumference (m) | 3016 |
| Number of intersections | 4 |
| Average radius of curved section (m) | 333 |
| Length of straight section (m) | 230 |
| Free space for experiments (m) | 20 |

| Beam | Electron | Proton |
|---|---|---|
| Energy Range (GeV) | 6-27 | 90-300 |
| Range of $Q^2$ (GeV$^2 \cdot c^{-2}$) | $2 \times 10^3 - 3 \times 10^4$ | |
| Number of Bunches | 40 | 40 |
| RF Frequency (MHz) | 500 | 80 |
| Injection energy (GeV) | 6 | $\geq 30$ |

REFERENCES

1. T. Nishikawa, Proc. of the U.S.-Japan Seminar on High Energy Accelerators, KEK(1973) p. 209, Proc. the IX th Int. Cont. on High Energy Accelerators, SLAC(1974) p. 584 etc.
2. Y. Kimura, to be published in Proc. of the XI th Int. Conf. on High Energy Accelerators, CERN(1980), etc.

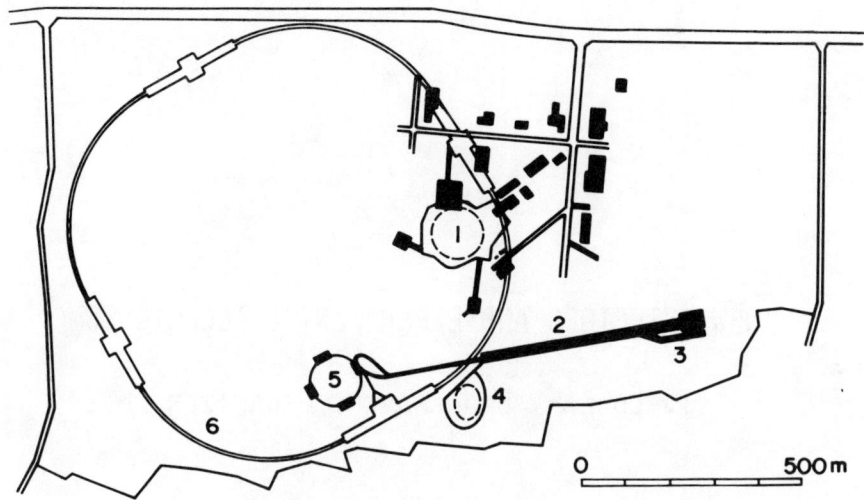

Fig. 1 Plan view of TRISTAN.
1. 12 GeV P.S. 2. 2.5 GeV electron linac. 3. 0.2 GeV linac for $e^+$ generation.
4. 2.5 GeV electron storage ring for synchrotron radiation research. 5. 6 GeV $e^{\pm}$ accumulator ring. 6. TRISTAN ring.

Fig. 2 Layout of the TRISTAN e-p and $e^+-e^-$ collider.

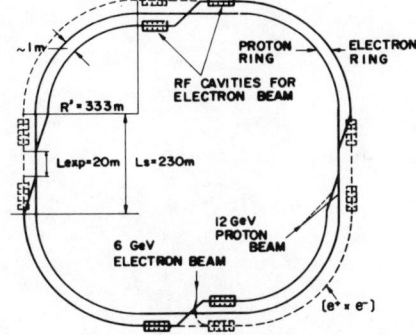

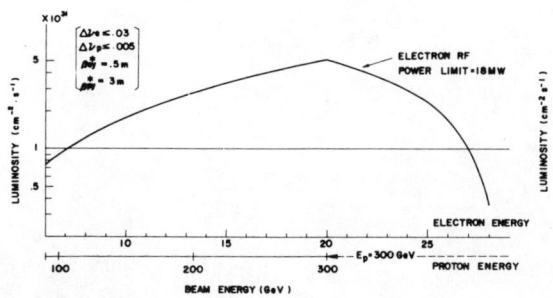

Fig. 3 Luminosity of the TRISTAN e-p collider.

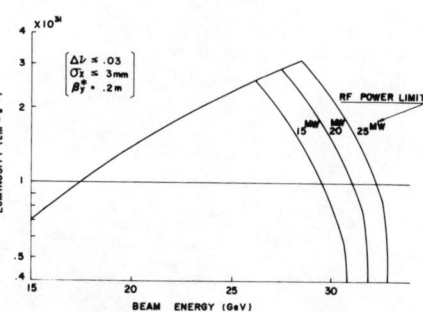

Fig. 4 Luminosity of the TRISTAN $e^+-e^-$ collider.

# NEW DETECTORS AND EXPERIMENTAL TECHNIQUES

T. Ludlam, Brookhaven, Organizer

HOLOGRAPHIC PHOTOGRAPHY OF BUBBLE CHAMBER TRACKS:
A FEASIBILITY TEST

L. Montanet
CERN, Geneva, Switzerland

Since the discovery of the $J/\psi$, high-resolution techniques have regained interest around particle accelerators to observe directly particles with new flavour such as charm and beauty.

At CERN, we have built[1] a small rapid-cycling hydrogen bubble chamber, using the technique developed a few years ago for the construction of track-sensitive targets. With this lexan bubble chamber (LEBC), it has been possible to observe directly, for the first time, charm particles produced in hadron interactions on a hydrogen target. The spatial resolution achieved, of the order of 40 µm, allows the detection of particles with a lifetime $\sim 10^{-12}$ s with good efficiency. The results of a test experiment, NA13, done at CERN in a 320 GeV/c $\pi^-$ beam, have been published[2]. The cross-section for charm production at this energy is found to be $\sim 40$ µb.

A resolution of $\sim 40$ µm is not enough, at SPS energies, to detect with good efficiency particles with a lifetime of $10^{-13}$, i.e. that of most "stable" charm particles. Moreover, even a rapid-cycling bubble chamber remains rather limited in terms of sensitivity: it is difficult to achieve better than a few events/µb in a hadronic beam. This is linked, to some extent, to the limited depth of field with which it is possible to achieve a good resolution in classical optics.

An alternative to these limitations is to use holography for recording the information. In principle, the holography technique opens the possibility of reaching a resolution of the order of a few microns (hence the observation of charm particles) and relaxes the limitation on the depth of field, allowing to increase the sensitivity by $\sim 20$ over classical optics.

To test these ideas, the holography technique has been applied to a small heavy-liquid bubble chamber in the conditions of a particle physics experiment[3]. A resolution of 8 µm over a depth of field of 10 cm was obtained.

A small heavy-liquid bubble chamber, BIBC [4], was used for the tests. It is a small cylinder of 65 mm diameter, 35 mm of depth, equipped with two windows made of optical quality BK7 glass.

Figure 1 gives a schematic view of the experimental set-up. A monomode, Q-switched ruby laser delivers 10 ns single-light pulses of 10 mJ and of wavelength 694 nm. Light pulse, beam and bubble chamber expansion system are synchronized. The beam is a 140 GeV $\pi^-$ coming from a CERN-SPS fast extraction (2.4 µs).

Holograms were made both on glass plates and on polyester-based film, both coated with Agfa-Gevaert 10E75 emulsion, at 10 to 18 cm from the beam line.

The mean value of bubble density was 300 to 400 bubbles/cm for minimum-ionizing particles. Bubbles of 8 µm diameter can be reconstructed with very good contrast.

Since the reference beam travels through the object field, the image quality is sensitive to turbulence in the bubble-chamber liquid as well as to the total number of bubbles present in the chamber. Still, it seems possible to "freeze" 200 beam tracks in a hologram without significant deterioration of the quality of the reconstructed picture in a chamber of the size of BIBC.

Figure 2 shows a segment of track of 220 µm, made of 8 µm bubbles, 1000 bubbles/cm, with a ∿ 100 µm delta ray. Figure 3 shows a $\pi$ interaction at 140 GeV/c, the full length corresponding to 1.2 mm of real space.

I would like to thank Paul Lecoq for helpful discussions and assistance in the preparation of this communication.

\* \* \*

## REFERENCES

1. H. Leutz et al., to be published in NIM.
2. W. Allison et al., Phys. Lett. 93B, 509 (1980).
3. M. Dykes, P. Lecoq, D. Güsewell, A. Hervé, H. Wenninger, H. Roger, B. Hahn, E. Hugentobler, E. Ramseyer and M. Boratav, CERN/EF 80-2, submitted to NIM.
4. B. Hahn, E. Hugentobler, E. Ramseyer, to be published.

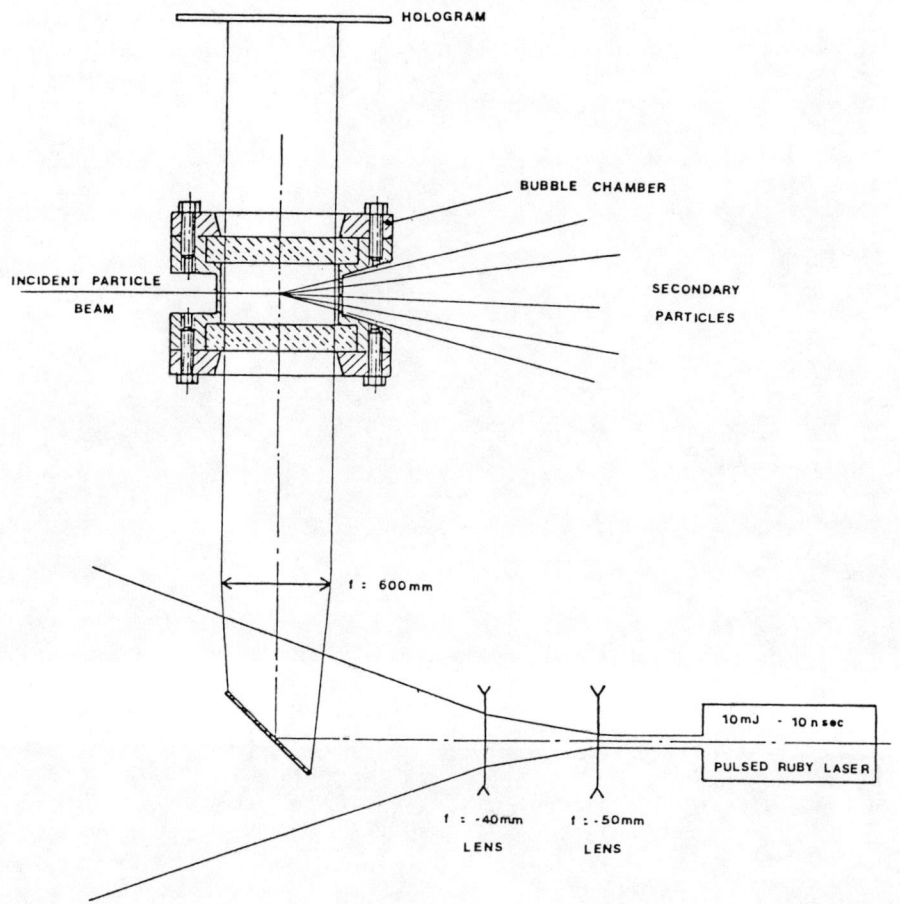

Fig. 1

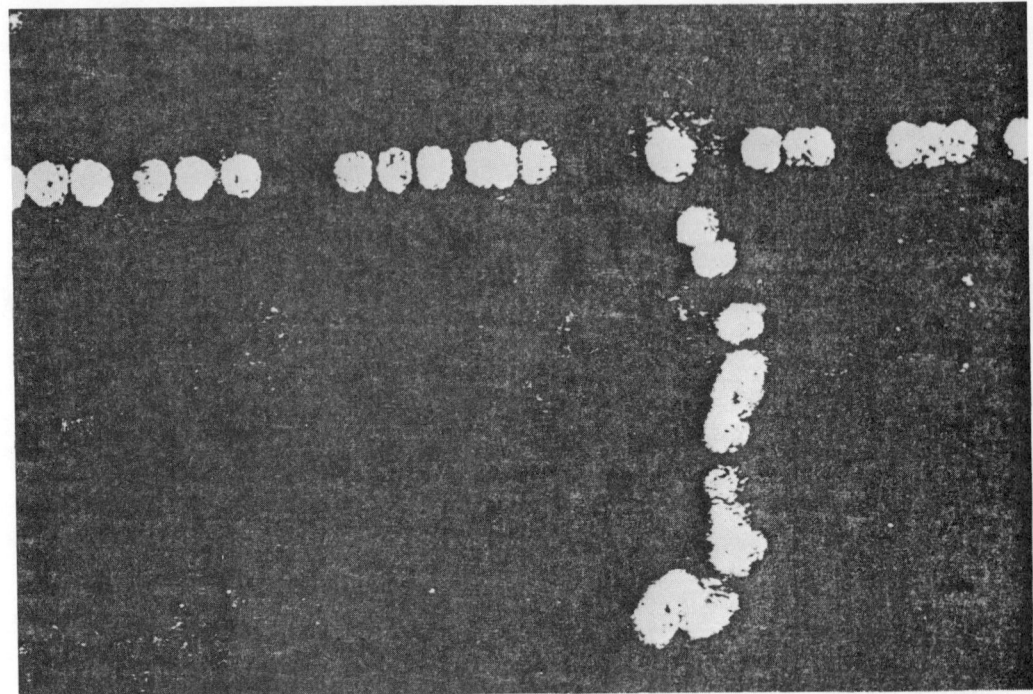

Fig. 2

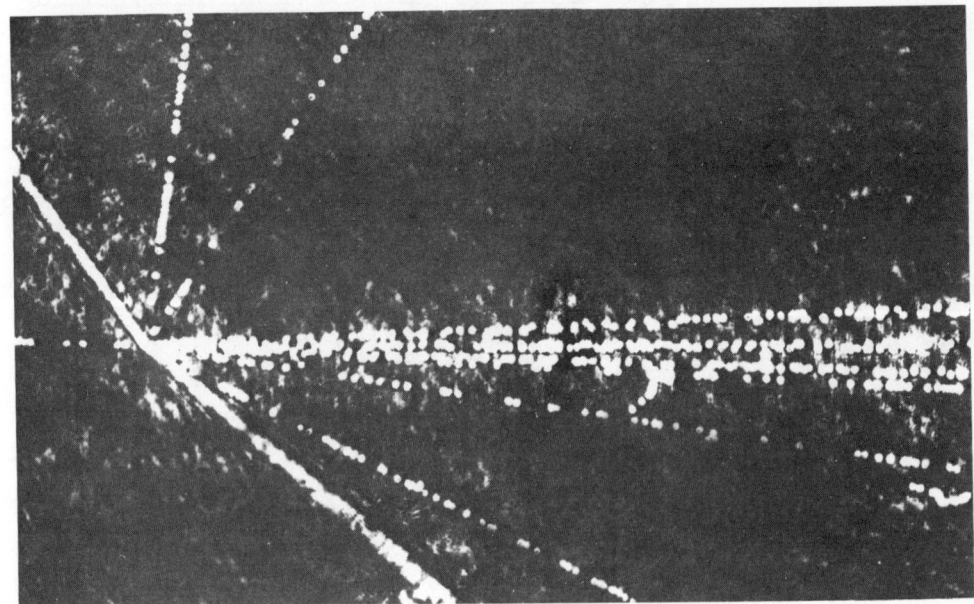

Fig. 3

## HIGH-$p_T$ EVENT TRIGGER AND PROCESSOR
## FOR THE ISR AXIAL FIELD SPECTROMETER

BNL-CERN-Copenhagen-Lund-Rutherford-Tel Aviv Collaboration
CERN, CH-1211 Geneva 23, Switzerland
presented by C.W. Fabjan

### ABSTRACT

Drift-time information from the central detector and energy deposit from a surrounding hadron calorimeter are used for complex trigger and on-line selection procedures. High-$p_T$ ($p_T \gtrsim 5$ GeV/c) charged particles and special event topologies can be selected with high efficiency. The hardware approach and performance is discussed.

### INTRODUCTION

Recently the Axial Field Spectrometer (AFS) was brought into operation at the CERN ISR. It is at present being exploited by the BNL-CERN-Copenhagen-Lund-Rutherford-Tel Aviv Collaboration [1] in a broad physics programme to investigate the hard scattering of nucleon constituents. Such events occur with cross-sections in the μb to nb range, or even smaller, may be produced by different incident particles (pp, p$\bar{\text{p}}$, αp, αα collisions), frequently in the environment of very high interaction rates (present luminosity is typically $> 2 \times 10^{31}$ cm$^{-2}$ s$^{-1}$, which corresponds to a particle flux of $> 10^7$ s$^{-1}$ into a 4π detector). Central to efficient study of these relatively "rare" events are trigger and filter schemes, matched in processing power to the primary interaction rate and of sufficient discrimination to permit practical off-line analysis methods. Acceptance criteria, based for example on the momentum of a charged track, are typical for high-$p_T$ studies; energy deposit in a hadron calorimeter provides information on event topologies, e.g. on "jets". Information combined from tracking devices, calorimeters and particle identifiers suggests sensitive detection of or searches for states containing heavy quarks. Their production rates at hadron machines are enormous: at the ISR about $10^9$ c$\bar{\text{c}}$ systems and $10^6$ b$\bar{\text{b}}$ systems are being produced daily!

### HIGH-$p_T$ SINGLE PARTICLE TRIGGER

Charged particles are momentum-analysed in a "pictorial" drift chamber [2] (cylinder divided into 4° cells with up to 42 space points per track) immersed in a 0.5 T magnetic field, coaxial with the ISR beams. The poles of the magnet were specially shaped for completely unobstructed acceptance over the full azimuth in a polar range of $0 < \theta < 15°$, $40° < \theta < 140°$, $165° < \theta < 180°$. A three-layer Čerenkov arm (thresholds at γ = 4, 10, 20), interleaved with proportional chambers (PCs) provide particle identification over 1 sr. Liquid-argon shower detectors are installed, but will be replaced by an 8 sr hadron calorimeter at present under construction. The physics programme for 1980 emphasizes two topics: the study of events contaning a high-$p_T$ ($p_T > 5$ GeV) $\pi^0$ or direct photon, using the liquid-argon shower counters, and secondly the inclusive spectra of identified high $p_T$ charged particles and their associated event topology. Of particular interest is the $p_T$ range 5 GeV/c $\lesssim p_T \lesssim$ 12 GeV/c, requiring highly selective, yet highly efficient triggering and filtering methods. This is accomplished with the *High $p_T$ Single Particle Trigger*, a sequence of increasingly more selective conditions as schematically summarized in Table I. A series of pretriggers achieves a reduction factor of a few hundred, reducing the rates of candidate events to a few kHz. At this rate it becomes possible to address the heart of the trigger chain, ESOP [3,4]. ESOP is a microprogrammable processor for high-speed data treatment. It is built with TTL logic in NIM/CAMAC mechanics, and consists of dedicated autonomous sub-units which operate concurrently under control of a common command module. This configuration permits one 48-bit microcoded instruction to be executed every 125 ns. Its high degree of parallelism gives an over-all speed of one order of magnitude faster than typical mini-computers.

Table I High-$p_T$ particle trigger (typically: $L = 2 \times 10^{31}$ cm$^{-2}$ s$^{-1}$)

| DETECTOR | TRIGGER | RATE | DECISION TIME | |
|---|---|---|---|---|
| Forward scint. hodoscope | Interaction trigger | $6 \times 10^5$ s$^{-1}$ | 10 ns | (1) |
| PC groups barrel hodoscope | Single particle pretrigger (SPT) | $6 \times 10^4$ s$^{-1}$ | 60 ns | (2) |
| Wire hits in PC1 and PC2 | RAM threshold, $p_T$ from 2 to 5 GeV | $\sim 10^4$ s$^{-1}$ | 700 ns | (3) |
| Pulse height on radial groups of 12 d.c. wires | d.c. fast or processor | $\sim 2 \times 10^3$ s$^{-1}$ | 800 ns | (4) |
| Drift time measurement on 16 wires/sector | ESOP program. processor $p_T$: 2.5 to 5 GeV | few s$^{-1}$ | 200-300 µs | |

Comments

(1) Scintillators provide "unbiased" detection and time of interaction

(2) Hardwired coincidence between barrel scintillators and wire groups from two PC planes. "Roads" accept particles with p > 1 GeV/c.

(3) Individual wire hits in PCs are compared with map of possible hit combinations for a high-$p_T$ particle. Hit combinations are stored in a computer-loadable memory and allow change of threshold.

(4) Analogue signals from radial groups of wires in a sector are summed; discriminator logic checks for track candidates.

Detectors external to the drift chamber (PCs through RAM logic, calorimeter modules) point to sectors in the drift chamber containing a high-$p_T$ track candidate. Tracks crossing from one $4°$ sector into the neighbouring must be allowed, as well as up to two tracks/sector. For each sector the drift time from four, equally spaced groups of wires are read out. Each group consists of four wires, on which several checks are executed (solution of right-left ambiguity, $t_0$ corrections, max. drift time, etc.) and then the drift times are combined into track segments, represented by master points. Track finding then proceeds on the totality of master points from each pair of sectors; a sagitta is computed for each track and compared with a preset cut-off value.

The global efficiency of this trigger chain, which includes effects of the extended source of interactions, edge effects and misalignments of detectors, is found to be very high: at a threshold setting of e.g. 5 GeV/c we find a 95% acceptance for the SPT condition, ∼ 80% for the RAM trigger and 65% for the full processing chain, which increases to 80% at $p_T$ = 6 GeV/c. Figure 1 is indicative of the performance achieved so far. It shows the $p_T$ spectrum of <u>all</u> charged particles; the data were taken at the indicated thresholds and subsequently processed through the off-line analysis chain. The data which are corrected only for the previously mentioned geometrical acceptance, are normalized to the "expected" spectrum, defined as Expected = 3.3 × Measured $\pi^0$ spectrum. The quality of the trigger events is further emphasized by the fact that even at the highest thresholds of 5 GeV/c the purity of the sample is very high: almost 10% of all analysed events contain a particle with $p_T \gtrsim$ 5 GeV/c, as evaluated by the full off-line programme. The remaining events have resulted in a valid trigger principally for two reasons: Due to the reduced resolution of the momentum measurement at the trigger stage

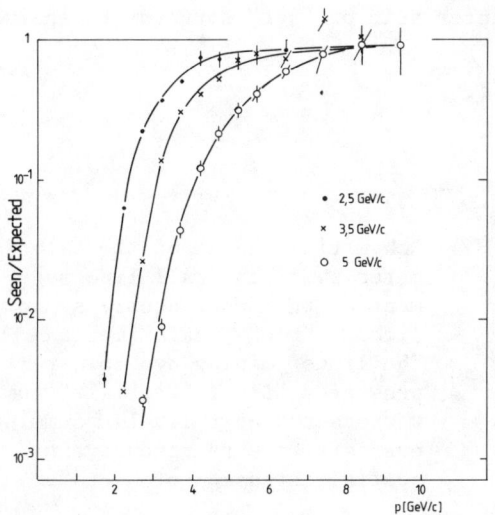

Fig. 1

The measured $p_T$ spectrum of charged particles, normalized to 3.3 times the $\pi^0$ flux, taken with three different trigger thresholds. In the $p_T$ range from 5 GeV/c to 10 GeV/c the particle flux drops by approximately four orders of magnitude. Only one particle in $10^9$ will have a $p_T \gtrsim$ 10 GeV/c.

a track below threshold may simulate a larger $p_T$ track. This is a dominant effect as lower $p_T$ particles are very much more abundant (E $d^3\sigma/dp^3 \sim p_T^{-8}$). These tracks are recognized off-line, where the complete drift chamber information provides improved momentum resolution. A second important source of false triggers is caused by two low momentum tracks, which overlap in a sector and simulate a high $p_T$ particle at the ESOP decision stage.

## THE CALORIMETER TRIGGER

The hadron calorimeter, at present under construction, uses uranium plates as absorber, sampled with 2.5 mm scintillator plates, which are read out by 2 mm thick wavelength shifters [5]. High granularity has been achieved by reading cells of 20 × 20 cm$^2$ cross-section with two wavelength shifters for each of the two longitudinal subdivisions (the first 6 radiation lengths and the subsequent 3.7 absorption lengths). The calorimeter modules surround the drift chamber in a box-like structure, such that four identical "walls" provide $2\pi$ azimuthal coverage. Energy leaking through the four corners created by the walls is measured in copper-scintillator calorimeters. The completed calorimeter will have 3200 individual signal channels. For trigger purposes, however, the dynode signals of the electromagnetic (e.m.) and hadronic segments at constant azimuth ("rows") are summed, resulting in 48 primary e.m. signals and 48 plus 4 (from Cu modules) hadronic signals. These 100 signals are transmitted via air core cable (v = 0.95c) to the counting room for further processing. As indicated in Fig. 2, the hadronic sums of adjacent rows are combined, with an overlap of one. Triggering on energy deposit in such a narrow slice will preferentially select configurations, where most of the energy is given to a single hadron. Likewise, six rows of cells ($\Delta\Omega \approx 1$ sr) are combined to "jet sums" to achieve sensitivity for topologies expected to be characteristic of "jet" structure. Again,

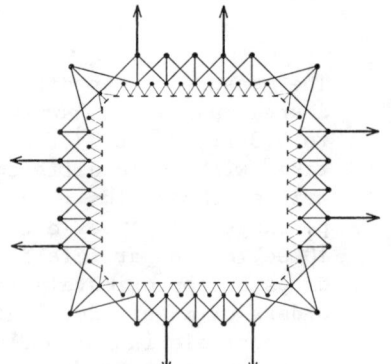

Fig. 2

Schematic diagram of the calorimeter sums. A small line segment on the inner square symbolizes a "row" of calorimeter cells. The lines joining two rows represent a "Single Particle" sum, whereas the next level of summing over six rows represents the spatial extension of "jets".

geometrical overlap of 67% in the jet sums minimizes edge effects in the azimuthal trigger efficiency. The energy sums provide also a measure of the total deposited energy, $E_{tot}$, which is used as a pre-trigger and strobe for ADCs. A total of 123 fast, linear sums are generated in this way (48 e.m. single particle (E.M.), 48 hadronic single particle (S.P.), 24 jet sums (J), $E_{tot}$ (E), $E_{tot}$ e.m. ($E_{e.m.}$), $P_{T,tot}$ ($P_T$)). These sums are interrogated by multilevel discriminators (3 levels for E.M., 5 S.P., 4 J, 6 E, 4 $E_{e.m.}$, 3 $P_T$) resulting in 493 signals. Subsequently, the definition of the e.m. single particle trigger is refined by requiring a spatial coincidence of an E.M. signal with a consistent value of an S.P. All the information thus obtained is "multiplicity"-encoded, and transmitted in encoded form onto 33 signal lines. Other detectors (PCs, DCs, etc.) may also provide information to be correlated with the calorimeter signals for which 17 encoded lines are provided. Logic decisions on the information pattern present on these 50 signal lines are carried out by an array of high-speed logic units. Each distinct trigger pattern, represented by a unique signal pattern on the 50 lines, is stored in a 50-bit memory. This arrangement allows parallel decision processing, downscaling of abundant triggers, and priority decisions. A list of various trigger combinations is provided in Table II.

Table II Selection criteria for some special final states

| EXPERIMENT | TRIGGER |
|---|---|
| 1. Identified single particle inclusive and correlations | BB $\cdot$ ESOP $\cdot$ $2^{-n}$; JET $\cdot$ ESOP; SP $\cdot$ ESOP |
| 2. Identified pairs ($\bar{p}p$, $\bar{K}K$ ..) | BB $\cdot$ ESOP(1) $\cdot$ ESOP(2) |
| 3. Jet studies and S.P. in calorimeter | BB $\cdot$ $2^{-n}$; E(1) $\cdot$ $2^{-n(1)}$ E(2) $\cdot$ JET(1) $\cdot$ $2^{-n(2)}$; E(3) $\cdot$ JET(2);... |
| 4. Flavour jets | E $\cdot$ E.M.(1) $\cdot$ ESOP; E $\cdot$ E.M.(2) $\cdot$ ESOP(1) $\cdot$ ESOP(2) |
| 5. Multileptons ($\geq 2$) | $E_{e.m.}$ $\cdot$ E.M.(1) $\cdot$ ESOP(1); E.M.(2) $\cdot$ ESOP(2) |
| 6. Inclusive electrons | E $\cdot$ E.M. $\cdot$ ESOP $\cdot$ $2^{-n}$ |
| 7. Correlations | E $\cdot$ SP(1) $\cdot$ SP(2) .. $2^{-n}$; E $\cdot$ JET(1) $\cdot$ JET(2) .. $2^{-n}$ |

Comment

The indices in parenthesis refer to a) different ESOP processors with different decision criteria or to b) one or more calorimeter sums satisfying different thresholds.

## ACKNOWLEDGEMENTS

The ESOP project owes its success to many people; we are particularly indebted to its designers, T. Lingjaerde and A. Fucci, who also contributed crucially in the running-in phase. Throughout the development phase the work and advice of D. Jacobs and D. Marland were essential. S. Cairanti and G. di Tore made invaluable technical contributions. We are grateful to P. Zanella and D. Williams for constant support.

## REFERENCES

1. Members of the collaboration are: H. Gordon, R. Hogue, T. Killian, T. Ludlam (BNL); J.C. Berset, O. Botner, M. Burns, D. Cockerill, G. Delavallade, C.W. Fabjan, P. Frandsen, A. Hallgren, B. Heck, J.H. Hilke, P. Jeffreys, G. Kesseler, M. Kreisler, J. Lindsay, W. Molzon, B.S. Nielsen, Y. Oren, P. Quéru, L. Rosselet, E. Rosso, A. Rudge, M. Sciré, J. v.d. Lans, D.W. Wang, Ch.J. Wang, W.J. Willis, W. Witzeling (CERN); H. Bøggild, E. Dahl-Jensen, I. Dahl-Jensen, Ph. Dam, G. Damgaard, K.H. Hansen, J. Hooper, R. Møller, S.Ø. Nielsen, B. Schistad (Copenhagen); T. Akesson, S. Almehed, G. von Dardel, G. Jarlskog, B. Lörstad, A. Melin, U. Mjörnmark, A. Nilsson (Lund); M.G. Albrow, N.A. McCubbin, M. Evans (Rutherford); O. Benary, S. Dagan, D. Lissauer (Tel Aviv).
2. D. Cockerill et al.; Performance of the AFS-Vertex Detector at the CERN ISR, to be published in Proc. Int. Conf. on Experimentation at LEP, Uppsala, 1980.
3. T. Lingjaerde; A Fast Microprogrammable Processor, CERN-DD/75/17 (1975).
4. B. Heck et al.; A Trigger on Charged Particles in the High Transverse Momentum for Experiment R807, presented at the Int. Conf. on Computing in High Energy and Nuclear Physics, Bologna, 1980.
5. O. Botner et al.; A Hadron Calorimeter with Wavelength Shifter Read-out, submitted to Nucl. Instr. and Meth. (1980).

# FIRST EXPERIENCES WITH A FASTBUS SYSTEM
## AT BROOKHAVEN

L.B. Leipuner, R.C. Larsen, D.S. Makowiecki,
W.M. Morse, T.K. Rudolf and W.P. Sims
Brookhaven National Laboratory, Upton, NY 11973

R.K. Adair, J.K. Black, S.R. Blatt
M.K. Campbell, H. Kasha and M.P. Schmidt
Yale University, New Haven, CT 06520

## ABSTRACT

A new concept in high energy data acquisition systems called Fastbus has been developed and implemented at Brookhaven. The system which is capable of sub-gigabit/sec speeds has been operating for some time now. A number of modules including an on-bus processor, a PDP11 interface, 32 channel coincidence latches, a 16 channel scaler, a 32 channel µ-clock device, a 60 nsec memory and a predetermined time module have been developed and built. Features of the system include extensive use of ECL logic and a water cooled crate with conduction heat transfer within a module. The system is used in an on-line experiment at the AGS.

Experiment #735, now being executed at the A.G.S. is implemented in FASTBUS. This is, to our knowledge, the first experiment to be done using this system. Briefly this data acquisition system is designed to operate and gather data from high energy physics experiments at sub-microsecond speeds. It is a 32 bit bus system capable of operating at near-gigabit rates. The bus can be segmented with each segment operating asynchronously when not communicating with each other. The system is based upon a very fast integrated circuit family, emitter coupled logic or ECL.

The experiment is described in another paper in these proceedings.[1]

All of the detectors are scintillation counters. There are more than 200 counters used and most are monitored individually. Each counter is connected to a NIM discriminator and most of the logic is done in a special trigger circuit built for the purpose. Most of the trigger logic is built with ECL but, having been built for a previous experiment, not in FASTBUS format.

Most counters are connected through standard discriminators to a set of 7, 32 channel coincidence latches. The latches are triggered twice, one each event to record both the real data and accidental counts. These coincidence latches, capable of ∼ 3 nsec resolutions are the main input paths for data at the event level. The other path is through the µ clock device which contains 32, 100 nsec/tick, clocks. These clocks are all reset and started by a trigger and stopped by a pulse from a set of counters which record the positron from the µ decay. They time out at 6.4 µsec. The counters, besides being connected to the latches are connected to trigger logic and to a "BOX" finder which determines which segment of the detector was hit. The trigger logic looks for one of 6 types of triggers. The type is passed to the computer by setting one of the latches and the "OR" of all types triggers further data processing. The box finder determines which clocks to read and again passes the data on through the latches. The trigger logic is monitored by scalers that keep track of the experiment on an accelerator pulse by pulse basis.

Data from the latches and clocks is passed to a processor which rejects events which are patently impossible and edits the data. Latch bits are translated to latch addresses. Only the clocks adjacent to the stopping block are read. All the encoded data for good events, is passed to a fast ($\sim$ 60 nsec) memory. The 4K by 16 bit memory can store $\sim$ 400 events, much more than a single accelerator pulse produces. The controller inhibits the trigger while it is taking data. A PDP11 is connected to the memory through an interface. The memory accepts data at FASTBUS rates and sends it to the PDP11, on demand, at its own rate. Synchronization with the AGS is provided by a predetermined time module. This module also contains an I/O register used to sense and control other parts of the experiment including the trigger logic.

With the exception of the discriminators which are in NIM format and the "BOX" and trigger logic, built in ECL for a previous experiment all modules are in FASTBUS format. The "BOX" logic will be redone in FASTBUS for our next run later this year. This move will improve our experiment somewhat.

The on-bus processor is based upon the ECL 10800 bit slice processor chip set. It contains special registers which encode the latch bits to latch addresses. It is a programmable device but does not yet contain an arithmetic logic unit. The processor determines which clocks to read and the information from them is tagged with their addresses and passed on to memory. In the next run the processor will present data to a special device, based upon FPLA's (for field programmable logic arrays) which will determine the sector. The processor also aborts events which contain either too little or too much information (shower events).

The PDP11 interface allows full access to the computer from the bus and full access to the bus via memory mapping registers from the 11. There are 8 - 32 bit registers whose contents are 'OR'ed with the lower bits of PDP11 addresses to form FASTBUS addresses if the 11's address is in the correct range. An interrupt facility for the 11 from the bus is implemented. Broadcasts are made from this device and the bus arbitration logically resides here. It thus serves as a segment interconnect when viewed from the bus.

Latches are coincidence types with $\sim$ 3 nsec resolving times. They are double buffered to allow the recording of accidental as well as real events. The scaler module has 16 scalers of 28 bits each. The 4000 word × 16 bit memory operates in $\sim$ 60 nsec and can be used in either normal, first-in first-out mode or first-in last-out mode. This experiment uses first-in first-out mode. The $\mu$ clock module operates with 100 nsec resolutions and has 32 channels. There is a computer controlled pre-determined time module to synchronize the data taking with the AGS. This module also contains an input and a output register for reading equipment status and driving annunciators.

The system uses a rather novel and quite convenient cooling system. The crate is water cooled with a 1/4 inch water line supplying more than enough cooling for a 1500 watts crate. The cooling lines would not be noticeable in the usual melee of coaxial cables if they were not white. Heat is transferred from the chips to the crate by several conduction paths. For modules which dissipate less than 30 watts simple conduction to the cover plates via a conducting rubber pad suffices. For modules with up to 75 watts for dissipation an aluminum lattus under the chips is also used. A simple cam mechanism provides the necessary thermal contact between the crate and modules. The system is quite clean, and effective.

For our next experiment we are building a multi-segment system with computer controlled discriminators, A $\rightarrow$ D converters, and an improved on-bus processor added to the system. Commercially built scalers will be available this fall and other units are being planned. Some debugging facilities are available now and others are being designed.

This work was supported in part by the U.S. Department of Energy under Contract Nos. DE-AC02-76CH00016 and DE-AC02-76ER03075.

## REFERENCES

1. See the paper presented by M.P. Schmidt, et al., in these proceedings.

Chapter 9

Gauge Theories and Mathematical Physics

# GAUGE THEORY I

## E. Witten, Harvard, Organizer

## SEMICLASSICAL APPROACH TO LARGE N EXPANSION

Antal Jevicki
Brown University, Providence, RI 02912

### ABSTRACT

We review some recent work done in collaboration with B. Sakita, H. Levine and N. Papanicolaou on developing new techniques for studying the planar limit of QCD.

### EFFECTIVE HAMILTONIANS FOR N = ∞

Quantum chromodynamics at large N provides phenomenologically an appealing picture of strong interactions.[1] In contrast, however, the actual problem of performing the large N expansion represents a rather formidable task. One would certainly like to avoid a direct summation of large N Feynman diagrams (which is obviously hopeless); as a possible procedure we have been formulating a nonperturbative semiclassical approach to this problem.

The first stage of this approach deals with summarizing the large N graphs in terms of an effective Hamiltonian.[2] This effective Hamiltonian is based on the collective field formalism, which represents a quantum change of variables to appropriate collective degrees of freedom. The usefulness of this procedure lies in the fact that in this new representation the large N limit becomes particularly simple: it is determined by classical stationary points of the effective Hamiltonian.

Specifically, in an SU(N) Yang-Mills theory one starts with the standard $A_0 = 0$ gauge canonical formalism and the change is made to the gauge invariant loop-space variables

$$\phi(c) = \text{Tr}\left(e^{ig\oint \vec{A}\cdot d\vec{x}}\right) \quad ; \quad \pi(c) = \frac{\hbar}{i}\frac{\partial}{\partial\phi(c)} \quad (1)$$

Here $\phi(c)$ are arbitrary space-like loops. To exemplify how one goes about deriving the new effective Hamiltonian let us describe for notational simplicity the analog example of a single U(N) matrix degree of freedom:

$$H = -\frac{\hbar^2}{2}\frac{\partial^2}{\partial M^2} + V(M) \quad (2)$$

Corresponding to the Gauss law here we have the restriction to SU(N) singlet states:

$$\hat{J}|\rangle = 0 \quad ; \quad \hat{J} = [M, \frac{\partial}{\partial M}] \quad (3)$$

The U(N) invariant collective variables are $\phi_k = \text{Tr} \exp(ikM)$ and the conjugate $\pi_k$. The change of variables involves the chain rule of differentiation and the evaluation of the Jacobian which is found in a self-consistent way. This leads to the effective Hamiltonian which essentially has the form

$$H_{eff} = \sum \left\{ -\frac{kk'}{2} \pi_k \phi_{k+k'} \pi_{k'} + \frac{\hbar^2}{2} \omega\Omega^{-1}(k,k')\omega' \right\} + V[\phi] \quad (4)$$

Here $\omega(k)$ and $\Omega(k,k')$ are functionals of the field $\phi_k$ and read:

$$\Omega(k,k') = kk' \phi_{k+k'} \quad ; \quad \omega = k \sum \phi_k \phi_{k-k'} \quad (5)$$

Of special importance in the above form is the appearance of the quantum $\hbar^2$ term which is an analog of the centrifugal barrier in quantum mechanics. The $N \to \infty$ limit is determined by stationary points of this effective Hamiltonian. The $\hbar^2$-term is of the same order as the potential term and, therefore, it plays a crucial role in this limit. For example, the vacuum is given by the static equation which is obtained by varying $V_{eff}$:

$$\hbar^2 \left\{ \frac{1}{2} \left( \int \frac{\phi(y)}{x-y} \right)^2 - \int \frac{\phi}{x-y} \int \frac{\phi}{y-y'} \right\} + V(x) - c = 0 \quad (6)$$

One can explicitly solve this nonlinear, singular integral equation and the result agrees with the original work of Ref. 3; on the other hand, this example also exhibits the complexity of the full Yang-Mills problem. Namely, in that case one has a nonlinear equation for a field defined on loops since essentially $\phi(k)$ gets replaced with $\phi(c)$. For the explicit form see the paper of Sakita.

Now that one has succeeded in formulating the summation of planar diagrams in terms of an effective equation the basic difficulty is that this still represents a rather complex nonlinear problem. Due to a nonstandard nature of this equation a direct approach to solving it does not seem to offer much hope. However, we have also introduced an alternative line of development which leads to the possibility that the problem may be reduced to ordinary space, in fact to the original classical equation. This will be described in the following section.

As a final comment we would now like to show that the collective field procedure can also be used to derive an effective action for Yang-Mills theory which then provides nonequal time correlation functions at large N. Namely, in the Euclidean formalism one would like to change integration variables to the general Wilson loop variables W(c) so that

$$Z = \int dA_\mu \, e^{-S(A)} = \int_c \pi dW(c) \, J \, e^{-S} \quad (7)$$

The essential problem now is how to determine the Jacobian J, however, in the Hamiltonian approach one had to do the same and consequently, we just generalize the Jacobian used there to four dimensions. We therefore obtain an effective action for large N: $S_{eff} = S - \log J$ with the corresponding equations of motion:

$$\frac{\partial S}{\partial W(c)} = -\sum \Omega^{-1}(c,c')\, \omega(c') \tag{8}$$

where

$$\Omega(c,c') = \frac{\partial W(c)}{\partial A_\mu} \frac{\partial W(c')}{\partial A_\mu} \quad;\quad \omega = \frac{\partial^2 W(c)}{\partial A_\mu^2} \tag{9}$$

These are certain functionals of Wilson loop variable and for example explicitly:

$$\omega(c) = \oint dx_\mu \oint dx'_\mu\, \delta(x-x')\, W(c_{xx'})W(c_{x'x}) \tag{10}$$

where the loops on the right hand side represent splitting of the loop c at point of intersection. We mention that Eq. 8 agrees with the Migdal-Makkeenko equation.[4] However, for this effective action formalism we obviously cannot have a direct connection with the original classical equations, and in this respect the Hamiltonian approach seems to be advantageous.

## CLASSICAL DYNAMICS AT LARGE N

The semiclassical quantization method is an extremely useful technique for studying nonlinear Hamiltonian systems. We have established based on several different models that there is a direct relevance of this method to the 1/N expansion.[5] Essentially what we show is that the large N effective Hamiltonians are nothing but the original classical Hamiltonians with certain boundary conditions. Consequently, some special real-time classical solutions directly determine the sum of large N diagrams.

For example, in the matrix model problem the effective equations obtained from the effective Hamiltonian of Eq. (4) read:

$$\frac{d}{dt}(\Omega^{-1}\phi) = \frac{\partial V_{eff}}{\partial \phi} \tag{11}$$

where $V_{eff}$ contains the $\hbar^2$ term. On the other hand, the classical equations of motion are

$$\ddot{M} = \frac{\partial V}{\partial M} \tag{12}$$

Obviously if M(t) is a classical solution with vanishing angular momentum the associated field $\phi_k = \text{Tr}\exp(ikM)$ obeys Eq. (11) but without the $\hbar^2$ term. However, one can show that a classical solution with

$$J \equiv i[M,\dot{M}] = \hbar(1-\delta_{ij}) \tag{13}$$

precisely reproduces the effective Eq. (11). A simple way to demonstrate this is to consider the constraint (Eq. (13)) writing it in the adjoint representation:

$$\tilde{M}_{\alpha\beta} P_\beta = \hbar C_\alpha \qquad (14)$$

where $\alpha \equiv (ij)$ ; $P_\alpha = \dot{M}_{ij}$. Imposing a gauge condition $M = \text{diag}(\lambda)$ we see that the constraint can be solved by

$$P_\alpha = \delta_{ij}\dot{\lambda}_j + \hbar \tilde{M}^{-1}_{\alpha\beta} C_\beta \qquad (15)$$

This is a canonical transformation and now the kinetic term of the Hamiltonian becomes

$$\text{Tr}(\dot{M}^2) = \sum_i \dot{\lambda}_i^2 + \hbar^2 C_\alpha (\tilde{M}^{-2})_{\alpha\beta} C_\beta \qquad (16)$$

Comparing this with the first two terms in the effective Hamiltonian one finds that with the special choice for C (given in Eq. (13)) the two coincide. What happens is that the nonzero angular momentum contribution exactly simulates the $\hbar^2$ centrifugal barrier term. Therefore, we have reached the following simple conclusion: classical solutions with special quantum boundary conditions (Eq. (13)) provide a sum of planar diagrams.

The above phenomena appears to be of quite general nature, for instance we have established that it holds for arbitrary U(N) invariant vector type models. Here one has a complex scalar field multiplet ($\phi_a(x)$, $a = 1,2,\ldots N$) and the U(N) charge is given by

$$Q_{ab} = \frac{i}{2} \int \left( \dot{\phi}_a^* \phi_b - \dot{\phi}_a \phi_b^* \right) \qquad (17)$$

We find that the classical equations again reproduce the large N quantum theory provided that the classical solutions obey the constraints

$$\begin{aligned} \phi^*(x) \cdot Q \cdot \phi(y) &= -2iN\hbar\, \phi^*(x)\cdot\phi(y) \\ \dot{\phi}(x) \cdot Q \cdot \dot{\phi}(y) &= -2iN\hbar\, \dot{\phi}^*(x)\cdot\dot{\phi}(y) \end{aligned} \qquad (18)$$

plus a third one with mixed time derivatives. Now one still has to show that the above relations do not contradict the equations of motion: this works out basically because $Q_{ab}$ is a conserved quantity. In fact, for the case of infinitely many degrees of freedom the constraints are simply solved by $Q_{ab} = \hbar N \delta_{ab}$. This simple constraint then provides the quantum boundary conditions, and the corresponding classical solutions determine the large N limit. For example, for $(\phi^* \cdot \phi)^2$ theory the lowest energy solution reads:

$$\phi_n^{cl}(x,t) = A_n \exp\{i(k_n x - \omega_n t)\} \qquad (19)$$

with $\omega_n^2 = k_n^2 + m^2 + g^2/2N |A|^2$. Imposing the constraint we obtain the well known gap equation, and the above solution obviously represents the large N vacuum. The quantity $\phi^*(x,t) \cdot \phi(y,t)$ is nothing but the equal time correlation function.

For O(N) invariant vector models one has a similar situation. For example, in the case of Helium-like Hamiltonians with two degrees of freedom $\vec{x}, \vec{y}$ the constraints are found to be:[6]

$$L^2 = (\hbar N)^2 \quad ; \quad L_{ij} x_i \dot{x}_j = L_{ij} y_i \dot{y}_j$$
$$L_{ij} x_i y_j = L_{ij} \dot{x}_i \dot{y}_j = 0 \tag{20}$$

where $L_{ij}$ is the conserved O(N) generator.

For Yang-Mills theory we also see a possibility for this phenomena. Namely, in anology with the discussion following Eq. (14) we may have:

$$J_o^\alpha \left(\frac{1}{D^2}\right)_{\alpha\beta} J_o^\beta = \hbar^2 \sum_{c,c'} \omega(c) \, \Omega^{-1}(c,c') \, \omega(c') \tag{21}$$

where $J_o^\alpha$ is an external density and $D^2 = D_i D_i$. However, due to the complexity of the right hand side we have not yet managed to work out the constraints which need to be imposed on the classical field $A_i(x,t)$ in order to make the above identification possible. Furthermore, as we have done in the above examples it is important to show the consistency of constraints with the equations of motion, this would then assure that the solution exists.

The equations presented in this section remind one of the Bohr-Sommerfeld method where certain quantization conditions have to be imposed on the classical dynamics. In effect, what we have obtained are some type of Bohr-Sommerfeld conditions for these more complex systems. The correspondence with the effective large N equations essentially means that this semiclassical quantization method becomes exact in the $N \to \infty$ limit.

## REFERENCES

1. G. 't Hooft, Nucl. Phys. B72, 461 (1974); G. Veneziano, Nucl. Phys. B117, 519 (1976); E. Witten, Nucl. Phys. B160, 57 (1979).
2. B. Sakita, Phys. Rev. D21, 1067 (1980); A. Jevicki and B. Sakita, Nucl. Phys. B165, 511 (1980); Phys. Rev. D22, 467 (1980); A. Guha and B. Sakita, City College preprint 1980.
3. E. Brezin, C. Itzykson, G. Parisi and J. B. Zuber, Comm. Math. Phys. 59, 35 (1980); For comparison of the spectrum see: M. Mondello and E. Onofri, Parma preprint 1980.
4. Yu. M. Makkeenko and A. A. Migdal, Phys. Lett. 88B, 135 (1979); D. Forster, Phys. Lett. 87B, 87 (1979); T. Eguchi, ibid pp. 91.
5. A. Jevicki and N. Papanicolaou, Nucl. Phys. B169 (1980); A. Jevicki and H. Levine, Phys. Rev. Lett. 44, 1443 (1980) and Ann. of Phys. (to appear).
6. L. D. Mlodinow and N. Papanicolaou, Berkeley preprint 1980; F. Berezin, Comm. Math. Phys. 63, 131 (1978).

# LATTICE FIELD THEORY FOR STRINGS

Don Weingarten*
Indiana University, Bloomington, IN 47405

## ABSTRACT

By constructing an equivalent euclidean lattice field theory, we show that a path integral for closed Nambu-Goto strings interacting by breaking and joining is pathological.

Some time ago Goddard, Goldstone, Rebbi and Thorn[1] showed that the spectrum of states of a free quantized string governed by the Nambu-Goto action is the same as the spectrum of states of the dual model without loop corrections. In particular, it contains a tachyon. Now tachyons have space-like momenta so they can occur with arbitrarily negative energies, and using this fact you can form two particle states with arbitrarily negative energies in their center-of-mass system. So if an interaction is turned on permitting strings to break and join and the interacting states are found by perturbation theory, it appears that all states become unstable and decay instantaneously into a divergent population of negative energy tachyons. A similar problem arises for the Dirac hydrogen atom when it is coupled to photons if nothing is done about the negative energy states.

In 1974, Bardakci and Bardakci and Halpern[2] suggested that this appearance of instability might be a figment of perturbation theory similar to the fictitious instability which occurs in a conventional quantum field theory if a perturbation expansion is made around an unstable classical solution. An instability of this sort shows up, for example, in a $\phi^4$ scalar field theory with the sign of the mass term flipped in the lagrangian

$$\partial_\mu \phi \partial^\mu \phi + m^2 \phi^2 - \lambda \phi^4 .$$

To zeroth order in $\lambda$ the theory has tachyons. If you perturb in $\lambda$, every state becomes unstable and decays into an infinite population of tachyons. But it is all a result of misusing perturbation theory. In reality the theory has a spectrum of stable states. They just can not be found by $\lambda$ perturbation. Bardakci and Halpern's conjecture was that the tachyons and related instability of string theory are similarly fictitious.

What I would like to do here is outline a proof[3] that at least for a euclidean lattice version of closed Nambu-Goto strings interacting by breaking and joining, Bardakci and Halpern's conjecture is wrong.

* Work supported in part by the U.S. Department of Energy

The instability is real and not just a figment of perturbation theory. In the process of demonstrating this result, a lattice field theory will be constructed which is equivalent to a path integral for interacting Nambu-Goto strings. This construction is perhaps interesting by itself. Finally, I'll show that by adding additional interactions beyond breaking and joining the theory can be stabilized and Bardakci and Halpern's conjecture comes true after all.

Let me begin by setting up a euclidean lattice path integral for interacting Nambu-Goto strings. Let $\Lambda$ be some finite segment of a d-dimensional hypercubic lattice $Z^d$. The dimension d can be any positive integer. Let $q_1, \ldots q_n$ be some collection of closed, oriented loops in $\Lambda$. Then the euclidean Green's function for this set of loops is defined as

$$\langle q_1, \ldots q_n \rangle = \sum_s \exp[-\alpha \, a(s) - \beta h(s)] . \qquad (1)$$

The sum is over all oriented surfaces s with boundaries given by the $q_i$ and connected to at least one $q_i$; $a(s)$ is the area of s, and $h(s)$ is the number of handles in s. $\alpha$ determines the string tension and $\exp(-\beta/2)$ is the bare dual coupling constant, $g_o$. Since every handle added to a surface has two ends and gets a factor of $\exp(-\beta)$, you can assign a factor of $\exp(-\beta/2)$ to each end and take $\exp(-\beta/2)$ as the coupling strength for a string to emit a handle.

Now I am going to set up something like a lattice gauge theory. The Green's functions it gives for strings will exactly reproduce those I have just defined. For simplicity, only the case of $\beta = 0$ will be considered. A generalization of the following discussion to $\beta > 0$ is straightforward and discussed in Ref. [3].

On the same $\Lambda$ I had before, let each nearest neighbor link $(x,y)$ be assigned a complex number $U(x,y)$ with $U(x,y) = U(y,x)^*$. For each closed oriented loop $\ell$ on the lattice, let $U(\ell)$ be the product of $U(x,y)$ along the loop. For each link $(x,y)$ let the measure $d\mu_{(x,y)}$ be

$$d\mu_{(x,y)} = \pi^{-1} \, d\text{Re} \, U(x,y) \, d\text{Im} \, U(x,y) \, \exp - |U(x,y)|^2 .$$

Our Green's function is then

$$\langle q_1, \ldots q_n \rangle = Z^{-1} \, \Pi \, [\int d\mu_{(x,y)}] \, U(q_1) \ldots U(q_n) \, \exp[\gamma \sum_p U(p)]$$

$$(2)$$

$$Z = \Pi \left[ \int d\mu_{(x,y)} \right] \exp\left[ \gamma \sum_p U(p) \right]$$

The action which enters this path integral is the same as for a U(1) lattice gauge theory but in place of Haar measure on U(1) we now have a gaussian measure.

If you expand this expression as a power series in $\gamma$, then by a slight adaptation of Wilson's construction of the strong coupling expansion for lattice gauge theories[4] you can show that path integral (1) for Nambu-Goto strings is exactly reproduced. When you expand in $\gamma$, one plaquette term U(p) comes from the exponential for each power of $\gamma$. Then the crucial step is to convince yourself that the gaussian measure gives an integral of 1 for each possible way of sewing plaquettes together to give one of the surfaces summed over in the Nambu-Goto case considered before, and a contribution of 0 for each set of plaquettes which can not be sewed together in one of these ways. For a surface of area a, a factor of $\gamma$ appears a times, thus its weight is $\gamma^a = \exp(-\alpha a)$, with $\alpha = -\ln\gamma$. this result is independent of the number of handles, we have $\beta = 0$.

So the lattice field theory reproduces the interacting string theory. But is easy to see that the lattice field theory is pathological. The vacuum-to-vacuum amplitude and a large set of Green's functions all diverge. Consider the vacuum to vacuum amplitude:

$$Z = \Pi[\pi^{-1} \int d\text{Re } U(x,y) \, d\text{Im } U(x,y) \exp(-|U(x,y)|^2)] \exp[\gamma \sum U(p)].$$

Each of the terms U(p) is a product of four U(x,y). Since the U(x,y) which enter are distinct, each U(p) can be made arbitrarily negative or positive. There is no constraint on its sign. Thus one can show that in some sectors of the region of integration $\Sigma U(p)$ blows up rapidly. The only term which can produce convergence, $\exp[-\Sigma|U(x,y)|^2]$ in the measure, falls only as the exponential of a quadratic. It can be shown that the rising fourth order term $\Sigma U(p)$ always wins in some region of phase space and Z blows up. By similar arguments[3] a large set of Green's functions can be shown to blow up in addition.

Since this theory is equivalent to the theory of interacting strings, it follows that the interacting string theory itself is pathological. Thus by allowing strings to break and join one does not obtain a theory free of pathologies as Bardakci and Halpern suggested.

But the difficulties can easily be corrected. One merely has to replace $|U(x,y)|^2$ by $|U(x,y)|^2 + f[U(x,y)]$, where $f[U(x,y)]$ is some function with real part rising faster than a fourth power of $|U(x,y)|$ as $|U(x,y)|$ becomes large. If you now go back and redo the expansion of (2) in powers of $\gamma$ to find the corresponding string theory, it turns out that the effect of $f[U(x,y)]$ is to introduce additional string-string interactions where strings cross on their interiors. $f[U(x,y)] = |U(x,y)|^6$, for example, introduces extra interactings wherever three strings cross.

Actually, lattice gauge theories with Haar measure can be considered the result of a certain singular choice of $f[U(x,y)]$.

Now it follows from a result of Lüscher[5] that lattice gauge theories have a self-adjoint hamiltonian bounded from below and are free of tachyons. Does a similar statement hold for regularized lattice string theories with other choices of $f[U(x,y)]$? By an adaptation of Lüscher's proof for lattice gauge theories, it can be shown that a large class of $f[U(x,y)]$, including $|U(x,y)|^6$, also yield satisfactory mass spectra free of tachyons. Corresponding results can be proved for the case of $\beta > 0$.

The moral of the story is that Bardakci and Halpern were wrong for strings interacting only by breaking and joining but correct if appropriate additional interactions are introduced on the string's interior.

References

1. P. Goddard, J. Goldstone, C. Rebbi, and C. B. Thorn, Nucl. Phys. B56, 109 (1973).

2. K. Bardakci, Nucl. Phys. B68, 331 (1974); B70, 397 (1974); K. Bardakci and M. B. Halpern, Nucl. Phys. B73, 295 (1974); Phys. Rev. D10, 4230 (1974).

3. D. Weingarten, Phys. Lett. 90B, 280 (1980).

4. K. G. Wilson, Phys. Rev. D10, 2445 (1974).

5. M. Lüscher, Commun. Math. Phys. 54, 283 (1977).

## QCD FROM THE EFFECTIVE LAGRANGIAN POINT OF VIEW

Joseph Schechter*
Physics Department, Syracuse University, Syracuse, N.Y. 13210

### ABSTRACT

A low energy effective Lagrangian for QCD is discussed.

The recent developments in QCD present a beautiful picture of high energy hadronic interactions but leave a "low energy hole" where effects like confinement come into play. On the other hand, the work of the pre-QCD decade secured a framework-that of effective low energy Lagrangians (or "current algebra")-which seems to fill this gap in a sense. I illustrate this with a complimentarity chart:

| Low energy QCD | Effective $\mathcal{L}$ |
|---|---|
| i) Difficult to explain why color-singlet, "confined" particles appear | i) One works (by assumption) only with "confined" particles. |
| ii) Quark mass effects are difficult to treat | ii) Quark mass differences easy to treat |
| iii) "$\frac{1}{N_c}$" approximation instructs to work in tree approximation | iii) One works in tree approximation |
| iv) Associated factor $\propto N_c^{-1/2}$ for emission of an extra meson | iv) factor $\propto F_\pi^{-1}$ for emission of an extra meson. |

Here I shall look at QCD from the point of view of an effective Lagrangian rather than vice versa. While of course the long-term goal is to derive the low energy Lagrangian from the field equations of QCD it may be useful as an intermediate stage to try to turn things around and start from an "old fashioned" effective Lagrangian, $\mathcal{L}$ and see what QCD has to say about constraining it.

There are two stages in constructing $\mathcal{L}$.

i) Isolate a <u>few fields</u> which are assumed to be most relevant for approximating the dynamics. Spin zero fields permit easy treatment of the possibility of spontaneous breakdown.

ii) Find an $\mathcal{L}$ which has the same symmetry structure as the basic one. This guarantees the consistent saturation of Ward identities.

---

*Supported in part by a grant from the U.S. Department of Energy, under contract number DE-AS02-76ER03533.

First let me review the situation in the pre-QCD days.
i) The relevant fields form a 3x3 matrix of spin 0 mesons:

$$M_a^b = S_a^b + i\phi_a^b \sim \bar{q}_b(1+\gamma_5)q_a \qquad (1)$$

For deriving current-algebra theorems it is convenient to use a non-linear approximation

$$M \approx \langle S \rangle e^{i\phi/\langle S \rangle}; \quad \langle S \rangle \approx \frac{F_\pi}{2}\mathbb{1}; \quad \eta' = \frac{\text{Tr}\phi}{\sqrt{3}} \qquad (2)$$

ii) The basic quark model has the structure (Gell-Mann)

$$\mathcal{L}_{basic} = -\sum_{a=1}^{3}(\bar{q}_a\gamma_\mu\partial_\mu q_a + m_a\bar{q}_a q_a) + (\text{unknown strong interaction})$$

$$\underbrace{\phantom{XXXXXX}}_{U(3)\times U(3) \text{ inv't}} \quad \underbrace{\phantom{XXXX}}_{(3,3^*)+(3^*,3)} \quad \underbrace{\phantom{XXXXXXXX}}_{\text{assumed} \approx U(3)\times U(3) \text{ inv't}} \qquad (3)$$

We have available the chiral invariants

$$I_m = \text{Tr}(MM^\dagger)^m \qquad (U(3)\times U(3) \text{ inv't})$$
$$J = \det M + h.c. \qquad (SU(3)\times SU(3)\times U(1) \text{ inv't}) \qquad (4)$$

A low energy effective $\mathcal{L}$ with the same structure as (3) is

$$\mathcal{L} = -\frac{1}{2}\text{Tr}(\partial_\mu M \partial_\mu M^\dagger) + \sum_a A_a(M_a^a + M_a^{a\dagger}) - V(I_m, J), \qquad (5)$$

where the only real question is whether the arbitrary function V should contain J or not. If J is not present one has $m(\pi)=m$ (isoscalar particle), which is a ridiculous spectrum. However if J is present one finds reasonable predictions[1] for the $\eta'$ mass, $\eta-\eta'$ mixing angle, $\eta \to 3\pi$ decay, "EM mass differences", etc. Thus nine years ago there existed a reasonable effective $\mathcal{L}$ "searching" for an $SU(3)\times SU(3)\times U(1)$ theory of strong interactions from which it could be derived.

At first glance QCD looks like a bad candidate since its interaction term (color indices deleted),

$$i\sum_a \bar{q}_a \gamma_\mu A_\mu q_a \qquad (6)$$

manifestly has the too large symmetry, $U(N_F=3)\times U(3)$. However the heroic work of many people showed that the Adler-Bell-Jackiw anomaly for the elusive $U(1)$ current, $J_\mu^5$

$$\partial_\mu J_\mu^5 \equiv G, \quad G = \frac{g^2}{16\pi^2} N_F F\tilde{F} = \partial_\mu K_\mu \quad (m_a = 0 \text{ limit}) \qquad (7)$$

had a physical effect even though equal to a total divergence. This arises from topologically inequivalent vacuums labelled by $\theta$ and necessitates adding to $\mathcal{L}_{basic}$

$$\Delta \mathcal{L} = -\theta G/2N_F. \qquad (8)$$

Back to the old effective $\mathcal{L}$. It satisfies $\partial_\mu J_\mu^5 \propto \frac{\partial V}{\partial J}(\det M - \det M^\dagger)$, which is purely a function of "matter" fields, rather than $\partial_\mu J_\mu^5 = G$. There is clearly no way to mock up the QCD symmetry structure without introducing at least the gluonic "degree of freedom" G in the effective $\mathcal{L}$. (We were persuaded to do this by ref.2). Accepting this, how can we get $\partial_\mu J_\mu^5 = G$ to follow automatically? First note that Noether's theorem gives

$$\partial_\mu J_\mu^5 = -i\mathrm{Tr}(M^\dagger \Box M - M \Box M^\dagger).$$

Here we evaluate $\Box M$ from the equation of motion. Clearly a scalar term in $\mathcal{L}$ proportional to G which breaks U(1) is needed. Our first try was to add a term proportional to $G(\det M - \det M^\dagger)$. However on differentiating with respect to $M^\dagger$ we get $-G(\det M^\dagger)(M^\dagger)^{-1}$ rather than just $-G(M^\dagger)^{-1}$. The solution is to replace $\det M$ by $\ln(\det M)$. So far our effective Lagrangian is

$$-\frac{1}{2}\mathrm{Tr}(\partial_\mu M \partial_\mu M^\dagger) + \mathrm{Tr}(A(M+M^\dagger)) - V(I_m) + \frac{i}{4N_F} G(\ln \det M - \ln \det M^\dagger)$$

Note the effect of a U(1) transformation $M \to e^{i\theta/N_F} M$ on the last term is to add a term $-\theta G/2N_F$ to $\mathcal{L}$. This has exactly the desired behavior (see (8)). But we are still not done. Noting from (2) that $\eta'$ is approximately proportional to $(\ln \det M - \ln \det M^\dagger)$ we see that there is no mass term for $\eta'$ present, only a $G\eta'$ mixing term. We can try to treat G as a "real particle" by adding the terms $cG^2/2 - \frac{\lambda^2}{2}(\partial_\mu G)^2$, ($\lambda$=real) and hope to get an $\eta'$ mass by diagonalizing the (mass)$^2$ matrix in $\eta'G$ space:

$$\begin{pmatrix} 0 & \sqrt{3}/2F_\pi \lambda \\ \sqrt{3}/2F_\pi \lambda & -\frac{c}{\lambda^2} \end{pmatrix}$$

But this yields $m^2(\eta')m^2(G) = -3/(4F_\pi^2 \lambda^2)$ so G must be a <u>tachyon</u>. We must try again. <u>Don't</u> add a kinetic term for G so its equation of motion will lead to its elimination in terms of $\eta'$. Substituting back in will give an $(\eta')^2$ term. This does it and we arrive at the prototype effective $\mathcal{L}$:

$$\mathcal{L} = -\frac{1}{2}\mathrm{Tr}(\partial_\mu M \partial_\mu M^\dagger) - V(I_m) + \mathrm{Tr}(A(M+M^\dagger)) + \frac{c}{2}G^2 + \frac{i}{4N_F}G(\ln \det M - \ln \det M^\dagger + 2i\theta) \qquad (9)$$

Eliminating G replaces the last two terms by

$$\frac{1}{32cN_F^2}(\ln \det M - \ln \det M^\dagger + 2i\theta)^2 \qquad (10)$$

Using (10), c is identified from $m^2(\eta') \approx 3/cF_\pi^2$. Some brief comments:

i) Setting $\theta=0$ in the above (with G eliminated) gives an effective $\mathcal{L}$ of the same form as the pre-QCD one since $(\ln \det M - \ln \det M^\dagger)^2$ is a (complicated) function of J. The previous results[1], which include SU(3) symmetry breaking effects, still hold. $\mathcal{L}$ can be upgraded to be of the "hard pion" type by including vector and axial vector fields ($V_\mu \equiv \ell_\mu + r_\mu$; $A_\mu \equiv \ell_\mu - r_\mu$) for example by replacing

$$\partial_\mu M \to \partial_\mu M - i\tilde{g}\ell_\mu M + i\tilde{g}Mr_\mu \qquad (11)$$

and adding spin 1 kinetic and mass terms. $\tilde{g}$ is Sakurai's coupling constant.

ii) By minimizing the total potential, the model can be used as a "theoretical laboratory" for discussing[4] the dependence of the vacuum energy density on the angle $\theta$.

iii) Connection with "$\frac{1}{N_c}$" approximation[2]: Eq.(10) behaves as $1/N_c$ for large $N_c$ and $\eta'$ can be thought of as a Nambu-Goldstone boson in this limit. This gives two mass scales in the problem. One is due to (10) and is expected to be of $\mathcal{O}(\Lambda=\text{QCD scale})$. The other is due to the "quark mass" terms and is of $\mathcal{O}(m_a)$. Thus

$$m(\eta') = \mathcal{O}(\Lambda) > m(8 \text{ pseudoscalars}) = \mathcal{O}(m_a). \qquad (12)$$

iv) The model can be formulated[5] as a kind of four dimensional Schwinger model using the "topological gauge field" $A_{\mu\nu\delta}$. This may lead to further insights.

An important point in the present formulation is that it's not correct to think that (symbolically)

$$|\eta'> = \alpha|\text{glue}> + \beta|\text{quark matter}>.$$

Rather it is more like a kind of duality

$$|\text{glue}> = |\eta'> = |\text{quark matter}>.$$

The mathematical $\eta'$ (SU(3) singlet) is thus expected to mix with $\eta$ in the usual way.

Finally I will briefly sketch a recent extension[6] of this work (still in an exploratory stage) to include a scalar ($0^{++}$) glueball, H. This seems to give a connection with confinement and predicts the decay pattern for a scalar glueball. To get the same symmetry structure as QCD we demand (assuming $m_a=0$) that both

$$\partial_\mu J_\mu^5 = G$$
$$\theta_{\mu\mu} = H, \qquad H = -\frac{\beta(g)}{2g}F^2 \qquad (13)$$

follow from $\mathcal{L}$ automatically. This enlarged $\mathcal{L}$ has more freedom so I break up its analysis into two stages. Stage 1: only H present. Stage 2: H, G, and M all present. It is amusing that the stage 1 Lagrangian[6] is unique:

$$\mathcal{L}_1 = -\frac{a}{2}H^{-3/2}(\partial_\mu H)^2 - \frac{1}{4}H\ell n\frac{H}{\Lambda^4}, \qquad (14)$$

where a is a constant related to the mass of the h excitation ($H \equiv$ <H>+Zh) and $\Lambda$ is a kind of QCD scale. The potential in (14) has a non-trivial minimum giving a vacuum energy density

$$-\text{<H>}/4 = -\Lambda^4/(4e) \equiv -B/4, \quad (e=2.7\ldots) \qquad (15)$$

This fits in nicely with the bag picture[7] in which the inside of the bag has a vacuum energy=0 while the outside has a <u>negative</u> energy density $\equiv -B \approx -(135 \text{ MeV})^4$. Note that the effective $\mathcal{L}$ should describe the <u>outside</u> world. From (15) we find $\Lambda \approx 250$ MeV. This work is in the same spirit as that of some other authors.[8] Our simple derivation shows that the key ingredient is the trace (i.e. scale) anomaly in (13), rather than any detailed models. The stage 2 $\mathcal{L}$ is discussed in ref.6; there are the interesting features. i)$\Lambda$ now depends on F and is increased due to the presence of matter, ii) $0^{++}$ gluonium $h^\pi$ has, if sufficiently massive, a favored decay channel $\eta\eta'$ which may be a useful experimental signature.

## REFERENCES

1. J. Schechter and Y. Ueda, Phys. Rev. <u>D3</u>, 176 (1971); <u>3</u>, 287 4 (1971); <u>4</u>, 733 (1971); <u>8</u>, 987 (E) (1973).
2. E. Witten, Nucl. Phys. <u>B156</u>, 269 (1979); G. Veneziano, Nucl. Phys. <u>B159</u>, 213 (1979); P. DiVecchia, Phys. Lett. <u>85B</u>, 357 (1979).
3. C. Rosenzweig, J. Schechter, and G. Trahern, Phys. Rev. <u>D21</u>, 3388 (1980); P. DiVecchia and G. Veneziano, CERN preprint TH2814 (1980); P. Nath and R. Arnowitt, Northeastern preprint NUB-2417 (1979); E. Witten, Harvard preprint HUTP-80/A005 (1980).
4. See P. DiVecchia and G. Veneziano and E. Witten in ref 3 above.
5. A. Aurilia, Y. Takahashi, and P. Townsend, CERN preprint TH-2815 (1980).
6. J. Schechter, Phys. Rev. <u>D21</u>, 3343 (1980); A. Salomone, J. Schechter, and T. Tudron, Syracuse preprint in preparation.
7. For a review see P. Hasenfratz and J. Kuti, Phys. Rep. <u>40</u>, 75 (1978).
8. For example, I.A. Batalin, S.G. Matinyan, and G.K. Savvidi, Sov. J. Nucl. Phys. <u>26</u>, 214 (1977); H. Pagels and E. Tomboulis, Nucl. Phys. <u>B143</u>, 485 (1978); N.K. Nielsen and P. Olesen, Nucl. Phys. <u>B144</u>, 376 (1978); H.B. Nielsen, Phys. Lett. <u>80B</u>, 133 (1978); R. Fukuda and Y. Kazama, Fermilab preprint 80/55 (1980).

# THE IMPORTANCE OF BEING TOPOLOGICALLY EXCITED

D.G. Caldi*
Lawrence Berkeley Laboratory, Berkeley, California 94720

## ABSTRACT

We identify a class of Euclidean configurations which appear to be dominant in the functional integral of the $CP^{N-1}$ models. These configurations are point-like topological excitations, and they may be viewed as constituents of instantons, although they are defined independently of instantons through a continuum duality transformation. We show not only that these configurations survive as $N \to \infty$, but that in the plasma phase they are responsible for the effects encountered within the 1/N expansion — confinement, $\theta$-dependence, and dynamical mass generation.

## INTRODUCTION

This is a report of work done with K. Bardakci and H. Neuberger, much of which is contained in a paper entitled "Dominant Euclidean Configurations for All N".[1]

We want to re-emphasize the usefulness and importance of quasi-classical methods in the study of quantum field theories. The essential feature of this approach is, of course, to identify field configurations which dominate the Euclidean functional integral. On the other hand, the 1/N expansion has received much attention recently, and its relation to the quasiclassical approximation needs clarification.[2] We will show that in the two-dimensional $CP^{N-1}$ models, point-like topological excitations (vortices, merons, instanton quarks) are important in the functional integral; that they survive as $N \to \infty$; and that in the plasma phase they are responsible for the results seen in the 1/N expansion, namely confinement, $\theta$-dependence, and dynamical mass generation.

The $CP^{N-1}$ model is defined by the following action for N complex fields $z_\alpha$,

$$S = \frac{1}{g^2} \int d^2x \, (D_\mu z_\alpha)^*(D_\mu z_\alpha) \, , \quad z^*_\alpha z_\alpha = 1 \, , \quad \alpha = 1, \ldots N \, , \qquad (1)$$

where $D_\mu = \partial_\mu + iA_\mu$, and $A_\mu = iz^*_\alpha \partial_\mu z_\alpha$. The models are considered a good laboratory for QCD since they are asymptotically free, conformally invariant, and topologically non-trivial. They are also 1/N expandable (but note N here is the number of flavors, not color),

---

*Research supported by the High Energy Physics Division of the U.S. Department of Energy under contract No. W-7405-ENG-48.

and the results are those mentioned above.

Classical solutions, i.e. instantons, are a very restrictive set of configurations and so by themselves are not important statistically; they also violate cluster decomposition. One needs more entropy. So far, one has been able to mix up instantons and anti-instantons only in the dilute gas approximation. However, recent calculations of the Gaussian fluctuations around the exact multi-instanton solutions have shown that the instanton quarks (merons) are liberated from their bound state, the instanton, and form a plasma, which is very different from the dilute gas. We shall identify a class of topological excitations which have sufficient entropy to be statistically significant and which satisfy cluster decomposition. Furthermore, although they are defined independently of instantons, they may be viewed as constituents of instantons, and indeed, the instanton configurations form a subset of these configurations.

## TOPOLOGICAL EXCITATIONS IN $CP^{N-1}$

To find configurations with sufficient entropy to be statistically significant, a change of variables is useful.

Let $z_\alpha = e^{i\theta_\alpha}\rho_\alpha$; $\sum_{\alpha=1}^{N} \rho_\alpha^2 = 1$. Then

$$S = \sum_\alpha \int (\partial_\mu \rho_\alpha)^2 + \sum_\alpha \int (\partial_\mu \theta_\alpha)^2 \rho_\alpha^2 - \sum_{\alpha\beta} \int \partial_\mu \theta_\alpha \partial_\mu \theta_\beta \rho_\alpha^2 \rho_\beta^2 . \quad (2)$$

Topological excitations are seen to exist since $\varepsilon_{\mu\nu}\partial_\mu\partial_\nu\theta_\alpha$ does not have to be zero but may equal $2\pi \sum_i q_i^\alpha \delta^2(x-x_i^\alpha) \equiv J_\alpha(x)$. We now proceed by a continuum duality transformation to constrain the functional integral to a well-defined configuration and then sum over all possible configurations. The resulting field theory is

$$\mathcal{Z}_\theta = \int \left[\frac{\Pi d\rho_\alpha}{\rho_\alpha}\right]\left[\delta\left(\sum_\alpha \rho_\alpha^2 - 1\right)\right][\Pi dB_\alpha]$$

$$\left[\delta\left(\sum_\alpha B_\alpha - \frac{1}{2\pi}g^2\theta\right)\right] \exp\left\{-\frac{1}{g^2}\left[\sum_\alpha \int (\partial_\mu \rho_\alpha)^2 + \right.\right. \quad (3)$$

$$\left.\left. + \frac{1}{4}\sum_\alpha \int \frac{1}{\rho_\alpha^2}(\partial_\mu B_\alpha)^2\right] + \lambda \sum_\alpha \int \cos\left(\frac{2\pi}{g^2}B_\alpha - \theta\rho_\alpha^2\right)\right\},$$

where we have incorporated the constraints $\varepsilon_{\mu\nu}\partial_\mu\partial_\nu\theta_\alpha = J_\alpha$ via

Lagrange multipliers $B_\alpha$, and then integrated out the $\theta_\alpha$ and summed over all $J_\alpha$'s. The term $\lambda \cos \frac{2\pi}{g^2} B_\alpha$ describes a Coulomb gas of the topological excitations or vortices, and since $\lambda$ is the fugacity and has dimensions of mass$^2$, dynamical mass generation has occurred. This system of Sine-Gordon fields is complicated by being coupled to the $\rho_\alpha$ fields which describe the effects of the "dialectric medium". But it is just this complication that provides the large amount of entropy in these configurations.

It has been argued that instantons cannot be important in the $N \to \infty$ limit since their contribution goes like $e^{-cN}$ and hence disappear. Without necessarily agreeing with this argument, it is plausible that if instanton quarks are liberated to form a plasma, then the action per quark is reduced by $N$, so that their contribution survives in the $N \to \infty$ limit. This is confirmed by an examination of (3), since if we make the following rescalings: $g^2 N = \kappa$, $\theta = \bar{\theta} N$, $\rho_\alpha = g \bar{\rho}_\alpha$, $B_\alpha = g^2 (\bar{B}_\alpha + \frac{\bar{\theta}}{2\pi})$; and exponentiate the $\delta$-function constraints we find

$$\mathcal{Z}_\theta = \int [du][dv] \exp \left\{ N \left[ \ln I(u,v,\bar{\theta}) - \frac{i}{\kappa} \int d^2 x \, u \right] \right\},$$

$$I = \int \left[ \frac{d\bar{\rho}}{\bar{\rho}} \right] [d\bar{B}] \exp \left\{ \int d^2 x \left[ -(\partial_\mu \bar{\rho})^2 - \frac{1}{4} \frac{1}{\bar{\rho}^2} (\partial_\mu \bar{B})^2 \right. \right. \quad (4)$$
$$\left. \left. + \lambda \cos [2\pi \bar{B} - \bar{\theta}(\kappa \bar{\rho}^2 - 1)] + iu\bar{\rho}^2 + iv\bar{B} \right] \right\}.$$

Hence we see that factorization has taken place so that the topological excitations survive in the large N limit.

If we consider the criterion for confinement

$\mathcal{Z}_\theta \frac{d^2 \mathcal{Z}_\theta}{d\theta^2} \neq 0$, and isolate the following term in the large N limit,

$$\frac{1}{V} \left( \frac{\partial^2}{\partial \bar{\theta}^2} \ln I \right)_{\bar{\theta}=0} = -\lambda \langle (\kappa \bar{\rho}^2 - 1) \cos [2\pi \bar{B}(0)] \rangle, \quad (5)$$

we see that the topological charge density (and so the Wilson loop) is proportional to $\lambda$. Hence if topological excitations are projected out of the functional integral by setting $\lambda = 0$, the Wilson loop vanishes and there would be no $\theta$-dependence and no confinement. Furthermore, the gas of topological excitations must be in the plasma phase since otherwise $\langle \cos(2\pi \bar{B}) \rangle = 0$ due to long-range correlations in the dipole phase.

Whether the approach taken here in the $CP^{N-1}$ model can be generalized to QCD depends on finding a parametrization of the degrees of freedom for which the techniques used here would be appropriate. This is currently under investigation.

## REFERENCES

1. K. Bardakci, D.G. Caldi and H. Neuberger, Nucl. Phys. B, to be published.
2. See ref. 1 for extensive references.

# GAUGE THEORY II: LATTICE THEORIES, LOOP SPACES, MODEL THEORIES

R. Jackiw, MIT, Organizer

# MONTE CARLO STUDIES OF LATTICE GAUGE THEORIES

Claudio Rebbi
Brookhaven National Laboratory, Upton, N. Y. 11973

## ABSTRACT

Numerical analyses of lattice gauge theories done with the Monte Carlo method are reviewed and the physical implications of the results are briefly discussed.

. . .

Quantum gauge field theories describe very successfully a variety of particle interactions. Theoretical results usually derived in the framework of a standard perturbative expansion compare very well with experiments whenever the effective coupling constant is weak. More problematic is to formulate theoretical predictions for strong values of the coupling constant. A very interesting approach consists in defining the theory on a space-time lattice[1], usually a hypercubical lattice with spacing a. This formulation provides of course a regularization of ultraviolet divergences but also, which is more relevant, allows for a strong coupling expansion.

Field variables, which are now finite group elements $U_{ij}(=U_{ji}^{-1})$, are associated to all the links of the lattice. After rotation of time to imaginary time the quantum mechanical averages are defined, in the standard way, via a sum over all configurations of the system (i.e. all the possible values of $U_{ij}$'s), weighted by the factor $\exp(-\frac{1}{g^2} S)$ ($\hbar = 1$), where the action is defined as $S = \Sigma S_\square$, $S_\square$ being the action of an elementary square, or plaquette, of the lattice. $S_\square$ in turn is defined as follows. If $i_1 i_2 i_3 i_4$ are the four vertices of the plaquette, one forms first the Wilson loop factor

$$W_\square = \mathrm{Tr}(U_{i_1 i_2} U_{i_2 i_3} U_{i_3 i_4} U_{i_4 i_1}) \qquad (1)$$

and then one takes for $S_\square$ a suitable function of $W_\square$ (usually a positive definite function that will vanish if all $U_{ij}$'s equal the identity).

If the group elements $U_{ij}$ are regarded as spin variables the sum defining the quantum averages can be regarded as a statistical sum of the kind used to study the thermodynamical behavior of a system, the dimensionality of the lattice being the only reminder that one is actually considering a quantum field theory. In this correspondence, $g^2$ plays the role of the temperature, so that weak coupling and low temperature, or, respectively, strong coupling and high temperature domains coincide.

The study of the "thermodynamical" structure of the system is crucial for the understanding of its quantum mechanical properties: indeed, the final goal is to proceed to a continuum limit where the lattice spacing a goes to zero. But to keep physical correlation lengths $\ell_i$ fixed, the dimensionless correlation lengths $n_i = \frac{\ell_i}{a}$ expressed in number of lattice sites must tend to infinity. In other

0094-243X/81/680896-06$1.50 Copyright 1981 American Institute of Physics

words the continuum limit can be recovered only renormalizing $g = g(a)$ so that it approaches a critical value $g_c$ (possibly $g_c=0$) as $a \to 0$. A study of the phase structure of the system becomes then of paramount importance.

For a while Monte Carlo simulations have been an extremely important tool in the analysis of standard thermodynamical systems. Very recently it has been realized that such computations can provide results also for four-dimensional systems and are then applicable to a study of lattice gauge theories. Pioneered by Wilson[2], this approach has been actively pursued by G. Bhanot, M. Creutz, B. Freedman, L. Jacobs and myself at the Brookhaven National Laboratory and, in the very limited space allowed, I shall briefly illustrate some of the results.[3,4]

I shall concentrate on systems where the gauge group is a finite group.[5] A very interesting feature of lattice gauge theories is indeed that the "spins" $U_{ij}$ may belong to a finite group as well. If this group, call it $\mathscr{F}$, is a subgroup of a continuous group $\mathscr{G}$ and if it is sufficiently dense in $\mathscr{G}$, a study of the system with gauge group $\mathscr{F}$, which offers various technical advantages, may provide very relevant information on the system with gauge group $\mathscr{G}$.

Figs.1-6 illustrate the results of Monte Carlo simulations of $Z_N$-models (systems with gauge groups $Z_N$, i.e. the subgroup of $U(1)$ with elements $e^{\frac{2\pi i n}{N}}$, $n = 0, 1, \ldots N-1$.[3] What are displayed are thermal cycles, that is simulations in which the temperature is not kept fixed, but is slightly varied after one or a few upgrades of all the spins are completed. Plotted is E, the expectation value of $S_\square$, versus $\beta = \frac{1}{g^2}$. Hysteresis loops signal phase transitions and we see that the $Z_2, Z_3$ and $Z_4$ models exhibit one phase transition (which further analysis shows to be of the first order). With $Z_5$ the pattern changes. The $Z_6$ and $Z_8$ models have two phase transitions (of higher order), one which stays at $\beta \approx 1$, irrespective of the order of the group, the other one moving toward $\beta = \infty$ (i.e. $g^2=0$) as N increases. This anticipates the phase structure of the system with gauge group $U(1)$. The phase for small $\beta$ is the strong coupling phase, in which through strong coupling expansions one can show that the charges of the group are confined. The lowest critical value, $\beta'_c$, indicates a transition to a spin wave phase; this is the phase which evolves into the quantum photon field as the continuum limit $a \to 0$ is taken. The second, at $\beta''_c$, is a transition to a very ordered state, characteristic of having a finite gauge group and therefore an action gap. It moves toward $\beta = \infty$ as $N \to \infty$ and is not present in the U(1) system. A direct numerical analysis of the U(1) model can also be done,[3,6] but approximating this system with a $Z_N$-model with N sufficiently large[7] reduces remarkably the computer time required and allows for higher statistics. Also, by spin compression techniques,[8] very large systems, suitable for renormalization group analysis, could be considered.

The simplest representative of a non-Abelian gauge theory is the model with SU(2) as gauge group. Finite subgroups of SU(2) are related to the rotation groups of the regular polyhedra: if $\tilde{\mathscr{F}} \in SU(2)$, factoring out its center $Z_2$ produces a symmetry group $\mathscr{F} = \tilde{\mathscr{F}}/Z_2$ of a regular polyhedron (with the exceptions of the dihedral groups, which

however do not properly reflect the non-Abelian nature of the system.)
In Figs. 7 - 9 thermal cycles obtained with Monte Carlo simulations of
models with gauge groups $Q$, $\tilde{T}$ and $\tilde{O}$ are presented.[4] $Q$ is the 8-element
group of quaternions, $\tilde{T}$ and $\tilde{O}$ are the 24- and 48-element subgroups of
SU(2) which reduce to the rotation groups of the tetrahedron and octa-
hedron when the center is factored out. The diagrams show that all the
systems have a single phase transition which moves toward $g = 0$ as the
order of the group increases, in agreement with the analysis of Ref.9
done for the SU(2) model directly. There is strong evidence then that,
contrary to the case of the U(1) Abelian model, the non-Abelian system
with gauge group SU(2) has just one confining phase. The comparison
between the results for the $Z_8$ and $\tilde{O}$ models (Figs. 6 and 9) is particularly illuminating: both systems have the same action gap, which sets
the scale for the low temperature transition to the highly ordered phase.
But the thermal cycle of the $Z_8$-model indicates a higher temperature
phase transition, which is conspicuously absent in the corresponding
cycle for the $\tilde{O}$ model. Fig. 10 presents a comparison between the curves
$E = E(\beta)$ for the $\tilde{O}$ model and the SU(2) model itself.

These results on non-Abelian systems with finite gauge groups have
been confirmed in Ref. 10, where the model with the 120-element subgroup of SU(2), $\tilde{I}$, related to the rotation group of the icosahedron is
also considered. The $\tilde{I}$ system seems extremely appropriate for simulations of the SU(2) lattice gauge theory.[11]

The numerical studies I have described represent only a very small
fraction of the work which has been done by Monte Carlo simulations.
The method appears very powerful, it has already provided us with a
wealth of relevant information and will certainly allow many further
interesting investigations.

## REFERENCES

1. K. G. Wilson, Phys. Rev. D10, 2445 (1974).
2. K. G. Wilson, Cornell University preprint, 1979.
3. M. Creutz, L. Jacobs and C. Rebbi, Phys. Rev. D20, 1915 (1979).
4. C. Rebbi, to appear in Phys. Rev. D21, 1980.
5. Results for systems with continuous gauge groups are presented in M. Creutz's contribution to this Conference.
6. B. Lautrup and M. Nauenberg, CERN preprint, 1980.
7. $Z_N$ models with large N have been used to approximate the U(1) system by T. A. DeGrand and D. Toussaint in their very interesting numerical analysis of monopole excitations, University of California at Santa Barbara preprint, 1980.
8. L. Jacobs and C. Rebbi, BNL preprint, 1980.
9. M. Creutz, Phys. Rev. D21, 2308 (1980).
10. D. Petcher and D. H. Weingarten, Indiana University preprint,1980.
11. An analysis of this model on a lattice extending for 16 sites in each dimension, done in collaboration with G.Bhanot,is in progress.

Caption for all figures: the average value E of the plaquette action is plotted versus $\beta = \frac{1}{g^2}$, as measured in thermal cycles done for the systems with gauge group $Z_2$(Fig. 1), $Z_3$(Fig. 2), $Z_4$(Fig. 3), $Z_5$(Fig.4), $Z_6$(Fig. 5), $Z_8$(Fig. 6), $Q$(Fig. 7), $\tilde{T}$(Fig. 8), $\tilde{O}$(Fig. 9) and SU(2) and $\tilde{O}$ compared (Fig. 10).

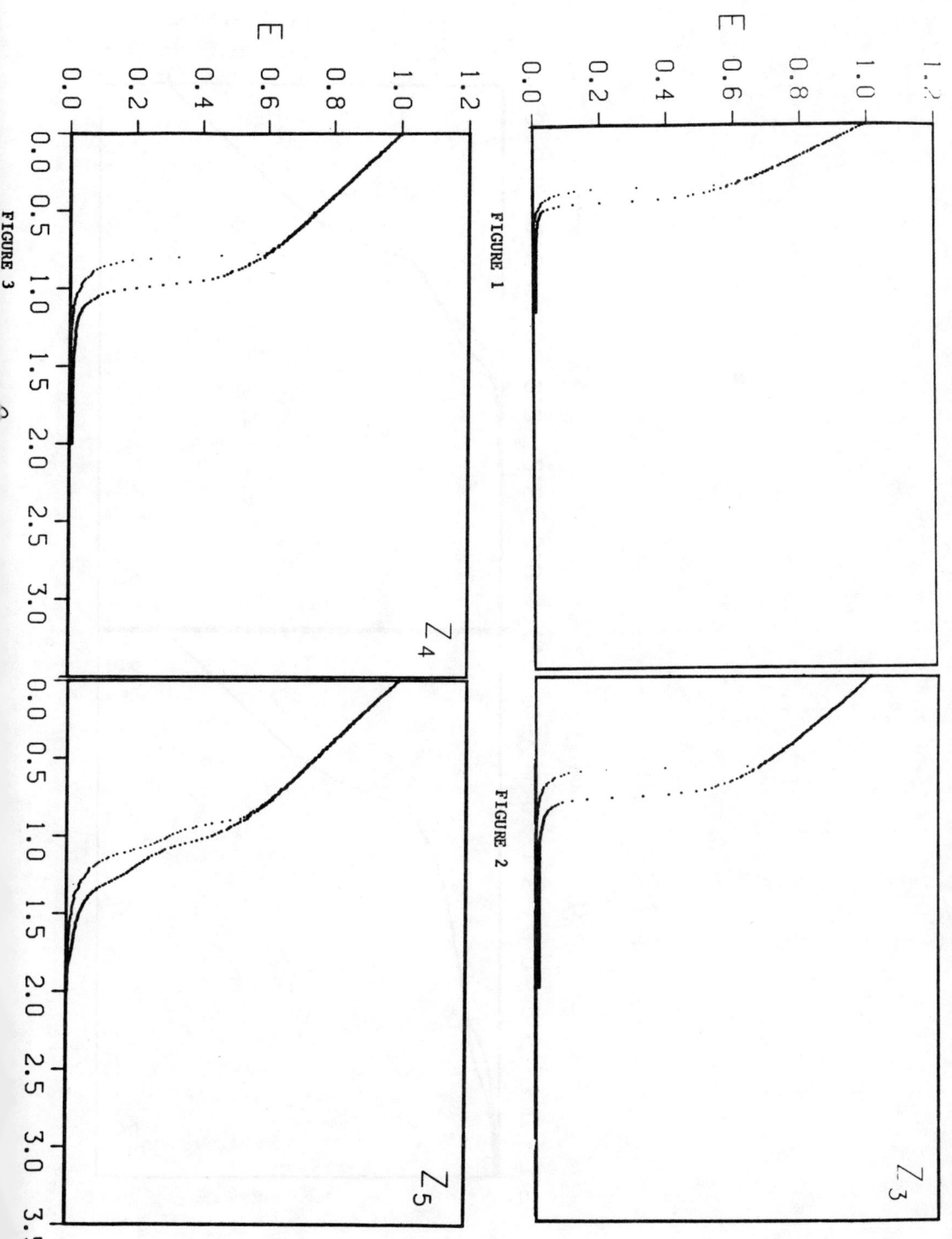

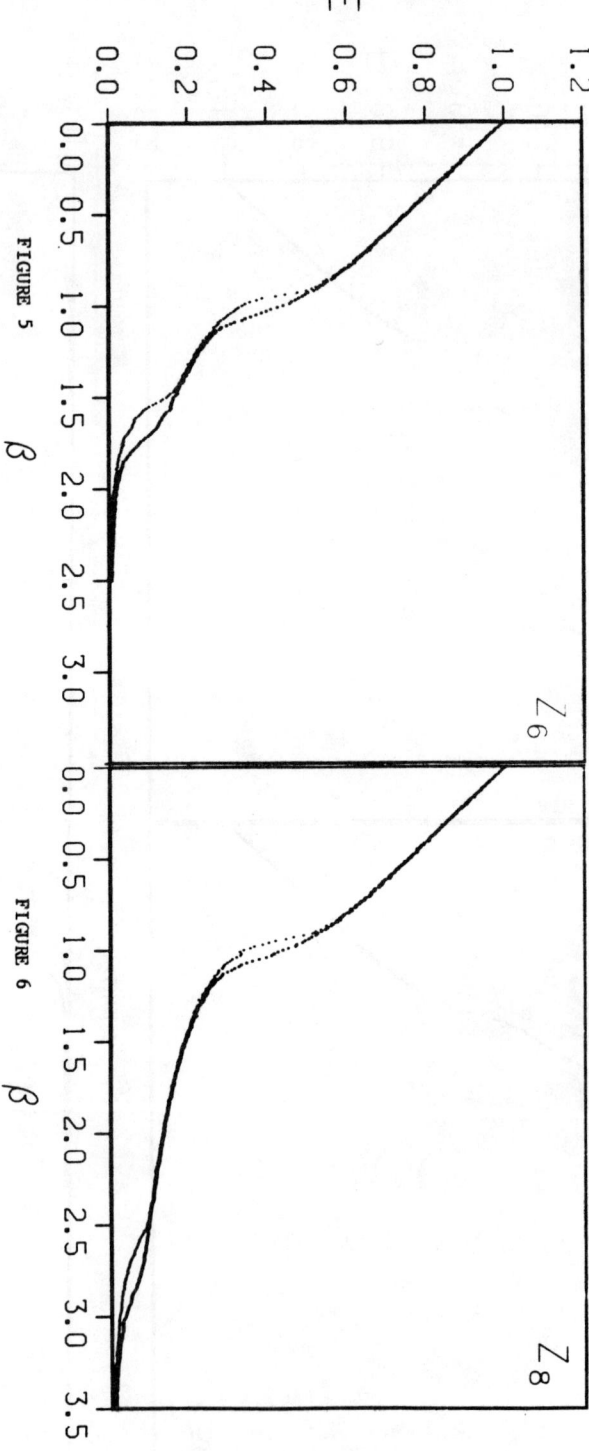

FIGURE 5

FIGURE 6

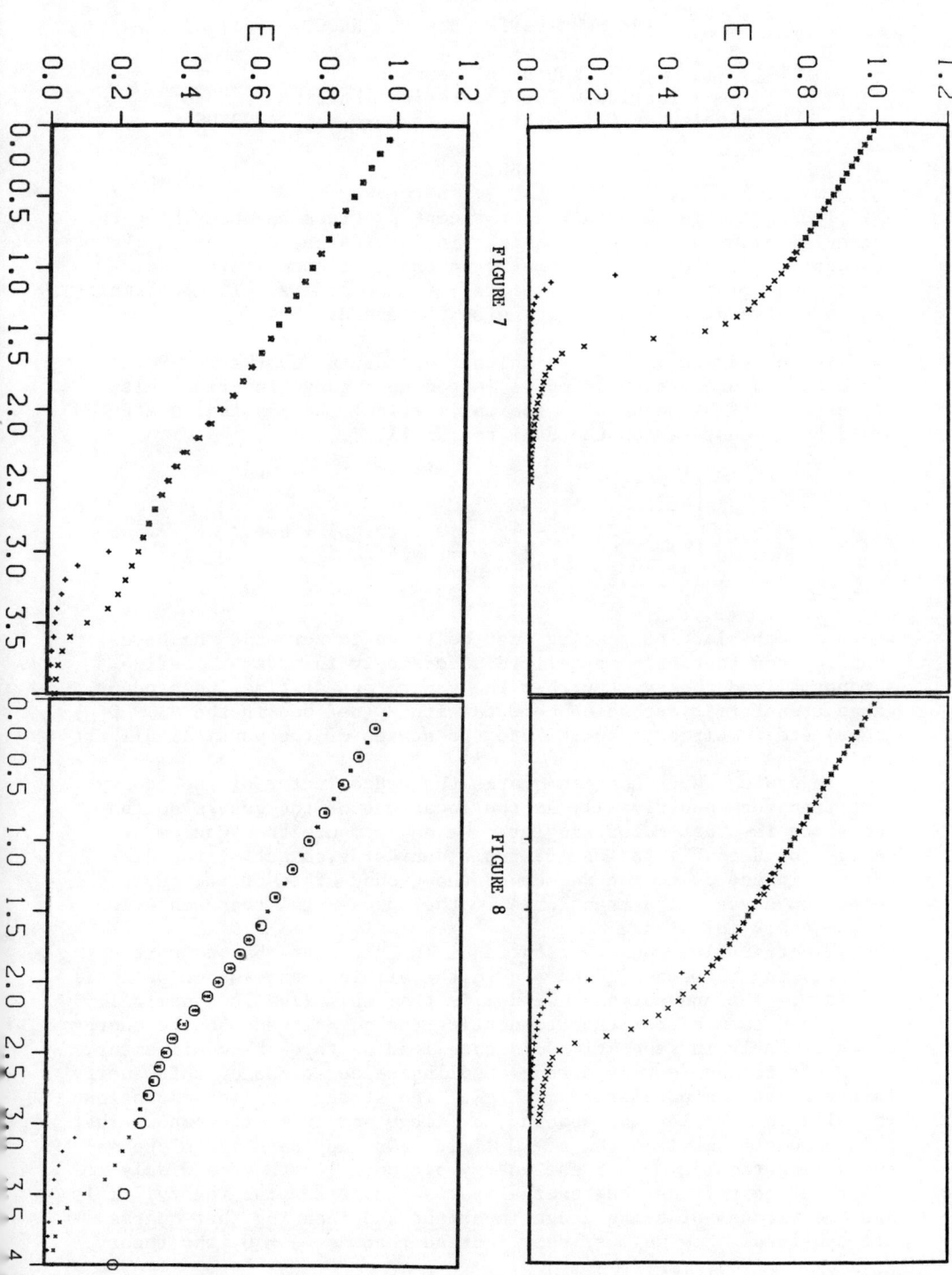

FIGURE 7

FIGURE 8

# THE FLUCTUATING STRING IN QCD

Robert B. Pearson
Institute for Theoretical Physics
University of California, Santa Barbara, CA 93106

## ABSTRACT

I describe in this talk some recent progress in describing the potential between two static quarks in lattice gauge theory. The restoration of rotational symmetry in the continuum limit is discussed and demonstrated approximately in low orders. The renormalization scales of the theory are also discussed.

In an imaginary world in which there are no quarks but SU(3) gauge fields are present, and we introduce a gauge invariant ultraviolet cut-off by means of a spatial lattice, the physical states would be eigenstates of the Hamiltonian [1,2].

$$H = \frac{g_0^2}{2a} \left\{ \sum_{\substack{\text{links} \\ \ell}} C_2(\ell) - \frac{2}{g_0^4} \sum_{\substack{\text{plaq.} \\ p}} (\text{tr}(U_p) + \text{h.c.}) \right\}$$

where a the lattice spacing must be taken to zero and the bare coupling constant $g_0$ renormalized accordingly in order to define a renormalized theory. Further the generators of time-independent gauge transformations which commute with H (defined in the $A_0 = 0$ gauge) are required to annihilate the states of the physical Hilbert space.

In states which contain quarks the gauge sector of the theory must transform nontrivially at the locations of the quarks so that the state is a net color singlet. We may compute the "Coulomb energy" of a configuration of static quarks by computing the difference in energy to the vacuum of the ground state of the super selection space which transforms in the fundamental representation at the quark locations.

Electric flux conservation requires that there be one unit of flux flowing between a q and a $\bar{q}$ in the simplest case which we shall consider. The conventional wisdom is that this flux is dynamically restricted to a narrow tube connecting the $q\bar{q}$ pair and so the energy grows linearly in separation distance leading to quark confinement.

This is born out by strong coupling calculations of this energy in the expansion parameter $x = 2/g_0^4$. The strong coupling expansion is valid for the lattice spacing a large and it must be shown that this property survives the $a \to 0$ limit. One approach is to choose the string tension T or the energy per unit length of a widely separated $q\bar{q}$ pair as the renormalization scale for the theory. This has the virtues of being gauge invariant and insuring that quarks are confined. Now we must show instead that as $a \to 0$ the theory maps onto the usual perturbative regime of QCD at small distances. One way to show this is to compute the renormalization group β-function and show that for $g_0$ small it agrees with the perturbative one. There is evidence that this is the case [3], but that the joining takes place fairly abruptly at a value $g_0 \cong 1$. This allows a zero parameter connection between the usual mass scale of perturbative QCD and the string tension. Since both quantities are in principle measurable this allows the theory to make a nontrivial prediction.

Various renormalization schemes give values for the mass scale which differ by multiplicative constants which may be determined from a one-loop weak coupling calculation [4]. In particular for the lattice one defines

$$a\Lambda_{SL} = e^{-8\pi^2/11g^2} \left(\frac{16\pi^2}{11g^2}\right)^{51/121}$$

where $g = g_0(a,T)$ and take the limit $a \to 0$. Extrapolating a sixth-order strong coupling series for the string tension we find the numerical relation

$$\sqrt{T} \cong 80 \, \Lambda_{SL}$$

This definition of $\Lambda$ differs by a constant from the euclidean lattice gauge theory [5], which differs in turn from the continuum momentum space definition by another constant [6]. Combining these results we obtain the relation

$$\sqrt{T} \cong 3 \, \Lambda_{MS}$$

This result is subject to fairly large uncertainties of several types, but is reasonable. The status of this calculation will improve in time.

The above results depend on being able to compute the string tension which was for a pair of quarks which lie on a principal axis of the lattice. It has recently been suggested [7] that this particular quantity may be anomalous undergoing at a nonzero value of the coupling constant a phase transition, called a roughening transition, where the r.m.s. transverse size of the string, which is finite for strong coupling, diverges causing a weak singularity in the string tension. This would prevent one from analytically continuing from strong to weak coupling.

It turns out for any lattice gauge theory the string is in the roughened phase even in the strong coupling limit if one excludes the special directions where the quarks are on principal axes. (To avoid any roughening none of the components of the relative distance should vanish.) The problem of computing the tension is complicated in general directions because the number of degenerate states in zeroth order is infinite. To find the zeroth order wave function of the string one must diagonalize the first-order potential exactly. In the plane this problem reduces to a degenerate Fermi gas of free fermions (the Y-links) in a box of length X+Y. The ground state has a single-particle band filled up to some fermi edge. In more than two dimensions the problem is more complex but is describable in terms of a Fermion theory with quartic interactions.

In the solvable planar case the string tension through first order has the form

$$T = \frac{g_0^2}{2a^2} \frac{4}{3} (\cos\theta + \sin\theta) \left( 1 - \frac{x}{2\pi} \sin\left(\frac{\pi}{1+\cot\theta}\right) + \ldots \right)$$

Near the value mentioned before as the joining region in the β-function this expression is almost θ independent. This is a value of the strong coupling parameter where first-order effects are important and higher order effects are just becoming important. One expects in weak coupling that all of the higher order terms contribute to a rotationally invariant potential. The third-order strong coupling corrections to the potential show signs of this already.

We may compute in the zeroth order ground state the r.m.s. thickness σ of the string at the midpoint of a string of length L. One finds the behavior seen in continuum string models

$$\sigma \sim \sqrt{\ln L}$$

This implies that even at x=0 the off axis string is in the roughened phase.

The physical differences between the on-axis and off-axis cases are due to a mass gap in the spectrum of the on-axis string and its absence for the off-axis case. The roughening transition is associated with the vanishing of this gap in the on-axis string. Since the off-axis string has a partially filled band and thus no gap they are "rough" even in the strong coupling limit. The massless case is more physical and so off-axis calculations should give better insight into continuum physics. As one example the massless string has long-range interactions which are attractive with, in the planar case at 45 degrees

$$V(r) = Tr + \text{constant} - 0.06 \, x/r + \ldots$$

Preliminary estimates give about twice the coefficient of 1/r in the (1, 1, 1) direction. This seems quite plausible phenomenologically.

Studying the angular dependence of the tension at the point where X, Y, or Z vanishes one discovers that the tension is not analytic but has discontinuous derivatives. In zeroth order for example

$$V(r) = \frac{g_0^2}{2a} \frac{4}{3} (|X| + |Y| + |Z|)$$

This discontinuity is again traceable to the mass gap in the string spectrum. It may be seen that the mass gap is just twice the energy of a single step or "kink" in the string away from a principal axis.

$$m_{kink} = \lim_{x \to \infty} (V(x,1,0) - V(x,0,0))$$

This mass gap is the most easily computed and most sensitive test in the Hamiltonian lattice gauge theory for the presence of roughening. A calculation of the kink mass through fourth order in strong coupling for various systems confirms the previously reported results [7] for roughening the 2+1 dimensions for abelian systems (Z(2) and U(1)). In 3+1 dimensions for the abelian systems the observed points where the kink mass vanishes are not discernably different from the points where the theory undergoes a transition to a non-confining phase. For the non-abelian systems (SU(3)) the place where the kink mass vanishes seems to be far into the weak coupling side of the crossover region and so our evidence is not conclusive as to the presence or location of such a transition.

In summary there seem to be new and interesting phenomena to be uncovered in the lattice gauge theory of QCD when one looks carefully at the physics of the crossover region and when one looks in new directions on the lattice.

## ACKNOWLEDGEMENTS

This talk describes work performed with J. Kogut, J. Richardson, J. Shigemitsu, and D. Sinclair. We are all grateful to the Institute for Theoretical Physics for its hospitality during the time this work was performed. This work was supported by the following grant: NSF-PHY77-27084. We would also like to thank P. Hasenfratz for communicating his work to us prior to publication as well as useful discussions with D. Gross and M. Lüscher.

## REFERENCES

1. K. G. Wilson, Phys. Rev. D 10, 2445 (1974).
2. J. B. Kogut, L. Susskind, Phys. Rev. D 11, 395 (1975).
3. J. B. Kogut, R. B. Pearson, J. Shigemitsu, Phys. Rev. Lett. 43, 484 (1979).
4. W. Cellmaster, R. J. Gonsalves, Phys. Rev. D 20, 1420 (1979).
5. D. Gross (unpublished).
6. A. Hasenfratz, P. Hasenfratz, CERN Report No. TH.2727-CERN, 1980 (unpublished).
7. A. Hasenfratz, E. Hasenfratz, P. Hasenfratz, CERN Report No. TH.2890-CERN, 1980 (unpublished).

## MONTE CARLO STUDY OF SU(2) GAUGE THEORY AT FINITE TEMPERATURE

J. Kuti[+], J. Polónyi and K. Szlachányi

Central Research Institute for Physics
H-1525 Budapest, P.O.Box 49.

### ABSTRACT

We find numerical evidence for the phase transition between the confinement phase and free Coulomb phase of SU(2)Yang-Mills theory with lattice cut-off. The search for the critical temperature is based on a Monte Carlo study of the string tension between a heavy $Q\bar{Q}$-pair in heat bath. The arbitrary normalization $0.2$ GeV$^2$ is used for the string tension at zero temperature when a smooth extrapolation of the lattice theory to the continuum limit is carried out. Our numerical estimate for the critical temperature is $T_c \sim 160 \pm 30$ MeV in the absence of quark degrees of freedom. It is suggestive that the phase transition is of second-order.

### ORDER PARAMETER AND CORRELATION FUNCTION

In order to study the string tension in a heat bath we have to introduce an external color source Q at location R and a color sink $\bar{Q}$ at the origin. The free energy $V(\beta,R)$ of this heavy $Q\bar{Q}$-pair is related to the correlation function of thermal Wilson loops by the formula

$$\langle tr W(0)\, tr W^+(\vec{R}) \rangle_{thermal\ average} = e^{-\beta V(\beta,R)} \qquad (1)$$

$\beta = 1/T$ and $k_B = 1$. The thermal Wilson loop $W(\vec{R})$ is defined as a closed path in the fictive imaginary time direction,

$$W(\vec{R}) = \mathcal{P} \exp\left[i \int_0^\beta dt\, A_0(t,\vec{R})\right] \qquad (2)$$

where $\mathcal{P}$ denotes path ordering in the standard fashion.

The free energy $V(\beta,R)$ is a measure of the potential energy between the heavy $Q\bar{Q}$-pair at finite temperature. In the confinement phase we expect the behavior

$$\langle tr W(0)\, tr W^+(\vec{R}) \rangle \sim const\ e^{-\beta \sigma(\beta) R} \qquad (3)$$

at large distances. The free Coulomb phase is characterized by a screened Coulomb potential, and accordingly,

$$\langle tr W(0)\, tr\, W^+(\vec{R}) \rangle \underset{R\to\infty}{\sim} const\left(1 + \beta \frac{g^2}{4\pi R} e^{-\varkappa R}\right) \tag{4}$$

The effective string tension $\sigma(\beta)$ at finite temperature is defined by Eq. (3). The Debye screening length $\varkappa^{-1}$ in Eq. 4 is known to be $\varkappa^2 = \frac{2}{3} g^2 T^2$ in the lowest order of perturbation theory.

The trace of the gauge invariant thermal Wilson loop may be regarded as an order parameter in SU(2) gauge theory at finite temperature. The thermal average of $tr W(\vec{R})$ measures the free energy of an isolated quark with respect to the vacuum. It vanishes in the confinement phase, since the infinite free energy of the isolated quark is in the exponent of Eq.(1). The order parameter $tr W(\vec{R})$ is some non-vanishing constant in the Coulomb phase where it is related to the finite self-energy of an isolated free quark on the lattice.

## MONTE CARLO METHOD

The Monte Carlo method[1,2] was applied for the calculation of the order parameter $tr W(R)$ and for the evaluation of the correlation function $tr W(0) tr W(R)$. In our program the heat bath method of Creutz[2] was implemented for sweeping through all lattice sites in each step of the iteration towards thermal equilibrium.

The behavior of the correlation function is shown in Fig. 1 at $\beta = 4a$ (where $a$ is the lattice spacing) and $g^2 = 2$ in the confinement phase. We find the numerical value $\sigma(\beta) a^2 = 0.54$ for the tension as extracted from the exponential shape of the correlation function. Creutz measures $\sigma(0) a^2 = 0.6$ at $g^2 = 2$. His result corresponds to zero temperature within some technical limitations.

Fig. 2 depicts the correlation function in the free Coulomb phase at $\beta = 4a$ and $4/g^2 = 2.3$. The arrow marks the value of $\langle tr W \rangle^2$ which is the asymptotic limit of the correlation function.

In our search for the phase transition point the order parameter never exhibits a discontinuous jump at $T_c$ and a second-order transition is suggested. It is also supported by the observation of large fluctuations near the critical point.

The critical coupling constant is shown in Fig. 3 at various temperatures. The interpretation of these results requires a smooth extrapolation to the continuum limit of the theory.

Our calculated Monte Carlo points in Fig.3 follow the renomalization group relation

$$T_c a = \text{const} \, [g^2(a)]^{-\frac{51}{121}} \cdot \exp\left[-\frac{12\pi^2}{11 g^2(a)}\right] \tag{5}$$

for $4/g^2 \geq 2$. The othervise arbitrary constant in Eq.(5) is determined by the Monte Carlo points. The best estimate of the critical temperature with the presented extrapolation to the continuum limit is

$$T_c = (0.35 \pm .05)\,\sigma(0), \quad \text{or} \quad T_c = 160 \pm 30 \text{ MeV} \tag{6}$$

with the arbitrary normalization $\sigma(0) = 0.2 \text{ GeV}^2$.

The dotted line in Fig.3 is an estimate of the critical temperature in the strong coupling limit:

$$T_c \cong a^{-1}\,\sigma(0)/\ln 5$$

There are finite temperature corrections to the relation between the lattice spacing $a$ and coupling constant $g$ which we calculated in Coulomb gauge on the one-loop level. These corrections are small for $4/g^2 > 2$ where $T_c \ll a^{-1}$. The dashed line in Fig.3 includes this finite temperature correction.

After the work reported here was complete and presented similar work [3] came to our attention.

REFERENCES

1. K.G. Wilson, Cornell preprint /1979/.
2. M. Creutz, Brookhaven preprint /1979/.
3. L.D. McLerran and B. Svetitsky, SLAC-PUB-2572 /1980/.

+ Speaker at the conference

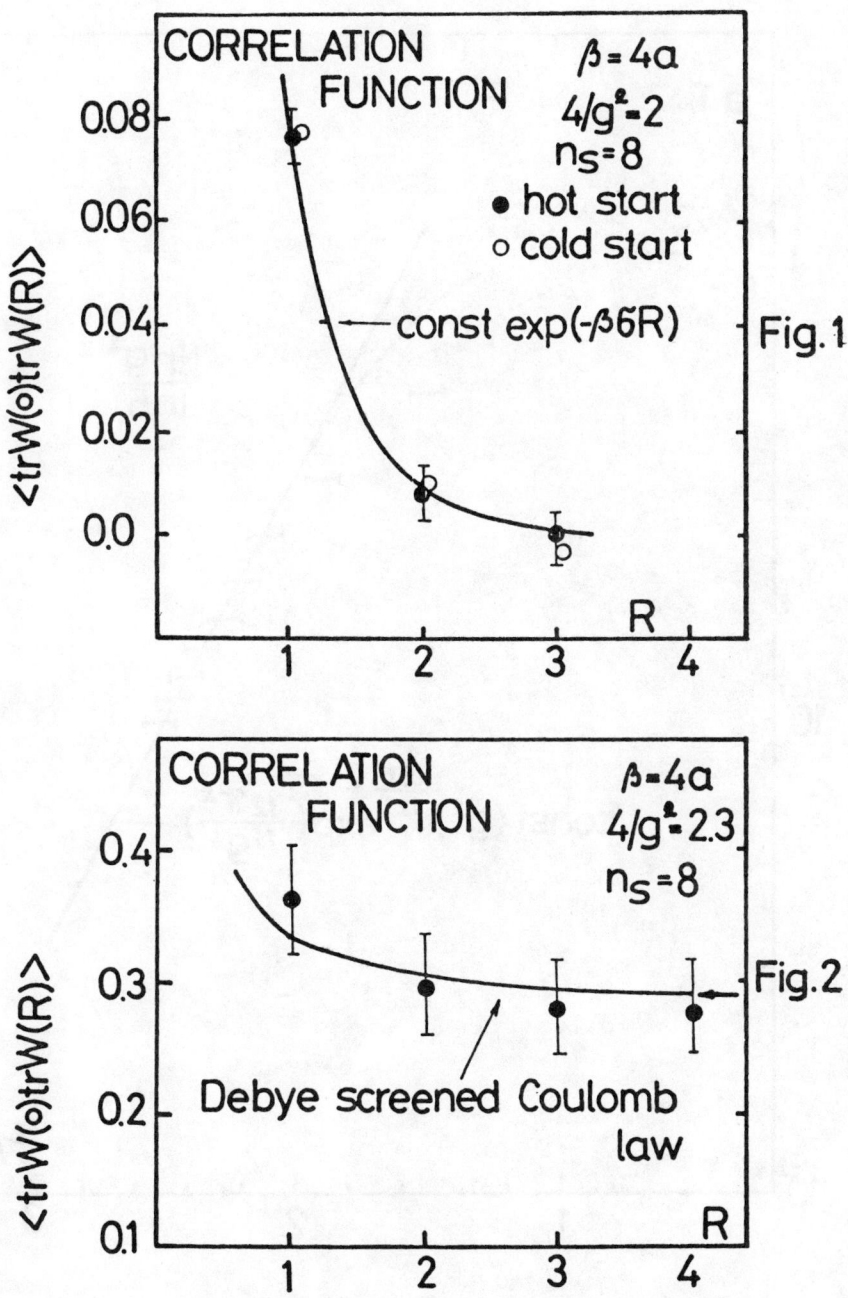

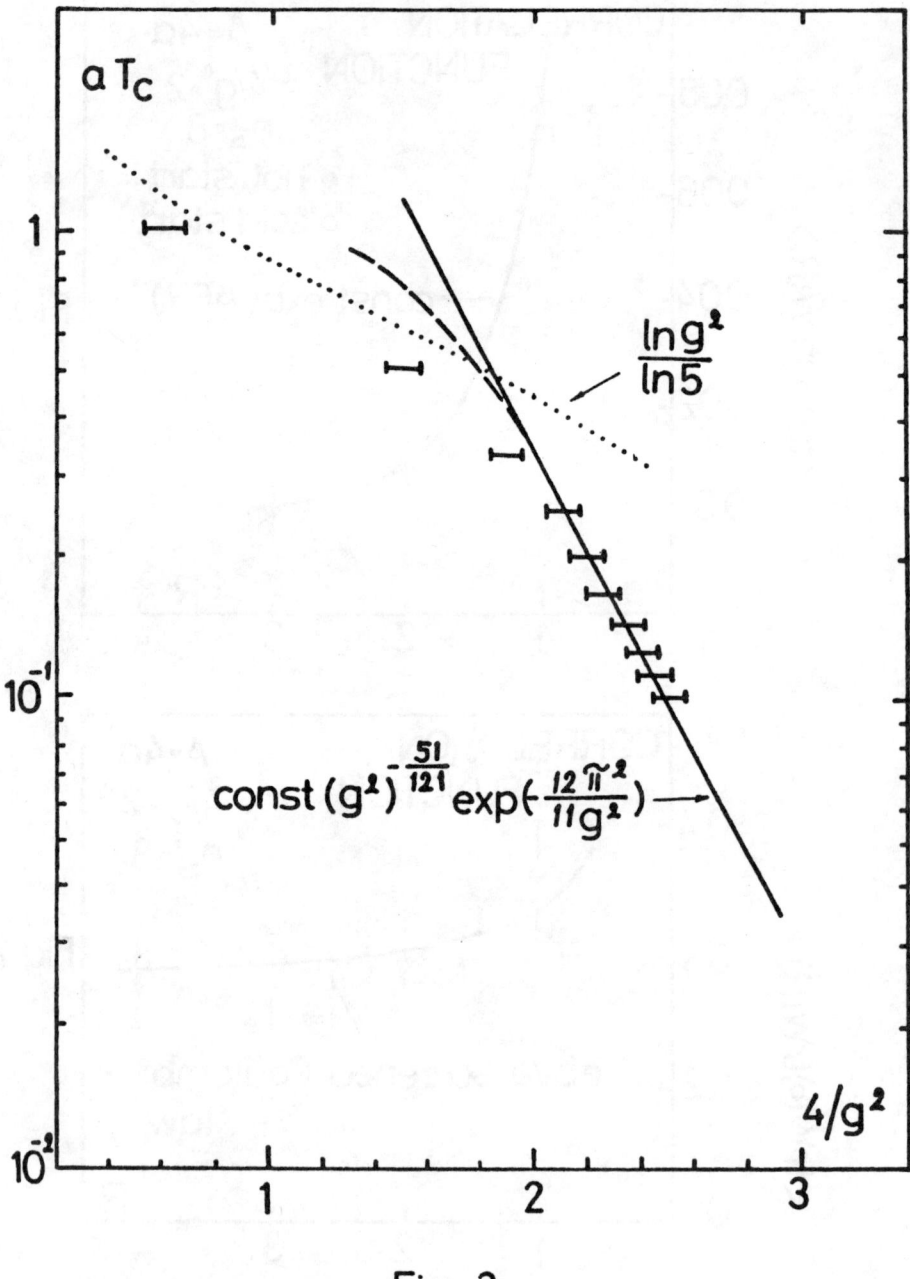

Fig. 3

# HIDDEN SYMMETRY OF THE POINT MAGNETIC MONOPOLE

R. Jackiw[*]
Institute of Theoretical Physics, UCSB
Santa Barbara, CA 93106

## ABSTRACT

A previously unnoticed $O(2,1)$ symmetry of the point magnetic monopole is described.

The next talk will be about numerical determination of multiple magnetic monopole solutions to Yang-Mills-Higgs theories. But first I shall take a few minutes to acquaint you with an elementary property of the point magnetic monopole which I recently found and which had not previously appeared in the literature that I thoroughly searched.

I am here concerned with the dynamics of a point particle with mass M and charge e, moving in the field of a strength-g point magnetic monopole, which is taken to be infinitely heavy and placed at the origin of the coordinate system. The equations for the dynamical variable $\vec{r}(t)$, the coordinate of our particle, are governed by the Lagrangian

$$L = L_{kin} + L_{int}, \qquad L_{kin} = \tfrac{1}{2}Mv^2, \qquad L_{int} = \frac{e}{c}\vec{v}\cdot\vec{A} \tag{1}$$

$$\vec{v} = \frac{d}{dt}\vec{r}, \qquad \vec{B} = \vec{\nabla}\times\vec{A} = g\frac{\vec{r}}{r^3} \tag{2}$$

which gives rise to the familiar Lorentz force law.

$$M\frac{d^2}{dt^2}\vec{r} = \frac{e}{c}\vec{v}\times\vec{B} = -\frac{eg}{c}\frac{\vec{r}\times\vec{v}}{r^3} \tag{3}$$

The symmetries that characterize this dynamics are the subject of my presentation.

Before describing the symmetries, let me remind you that there is a long story concerning the vector potential $\vec{A}$ which specifies the magnetic monopole. In order to accomodate (2), $\vec{A}$ must necessarily possess some singularities, and these give rise either to the "Dirac string,"[1] or to the "Wu-Yang sections."[2] Related to this is the fact that the angular momentum, which is a conserved quantity by virtue of the rotational invariance of the dynamics, has an unconventional form. In addition to the familiar kinematic contribution $\vec{r}\times M\vec{v}$, there is a further term, first given by Poincaré, which arises from the electromagnetic fields; $\vec{J} = \vec{r}\times M\vec{v} - (eg/c)(\vec{r}/r)$.

The rotational invariance of our problem, and the corresponding conserved quantity $\vec{J}$, are of course familiar. Also familiar is the symmetry of time-translation invariance. Under a time-translation, $t \to t + t_o$, the dynamical variable changes according to $\vec{r}(t) \to \vec{r}(t+t_o)$, which in infinitesmal form reads $\delta\vec{r} = \vec{v}$. The consequent conserved quantity is the Hamiltonian, which measures the energy; $H = \tfrac{1}{2}Mv^2$. Note

---
[*]Permanent address: Center for Theoretical Physics, Massachusetts Institute of Technology, Cambridge, MA 02139

0094-243X/81/680911-04$1.50 Copyright 1981 American Institute of Physics

that the Lagrangian is not invariant under a time-translation, but the action $I = \int dt L$, is.

The two further symmetries, to which I call your attention, are similar to the time-translation. They too describe invariance against further redefinitions of the time variable; they leave invariant not the Lagrangian but the action; and they lead to constants of motion.

Consider first time-dilatation.

$$t \to \alpha t \qquad (4)$$

One verifies that transforming the dynamical variable as a denstiy of weight $-\frac{1}{2}$

$$\vec{r}(t) \to \alpha^{-\frac{1}{2}} \vec{r}(\alpha t), \quad \delta \vec{r} = t\vec{v} - \tfrac{1}{2}\vec{r} \qquad (5)$$

leaves the action and the equations of motion invariant. The conserved quantity

$$D = tH - \frac{M}{4}(\vec{r}\cdot\vec{v} + \vec{v}\cdot\vec{r}) \qquad (6)$$

involves the virial.

The second symmetry transforms the time by a special conformal transformation [translation of the reciprical time].

$$\frac{1}{t} \to \frac{1}{t} - \frac{1}{t_o} \qquad (7)$$

The transformation law for $\vec{r}(t)$, which leaves the action and the equations of motion invariant, is

$$\vec{r}(t) \to (1 - \frac{t}{t_o}) \vec{r} (\frac{t}{1 - t/t_o}), \quad \delta \vec{r} = t^2 \vec{v} - t\vec{r} \qquad (8)$$

and the corresponding constant of motion involves H and D.

$$K = -t^2 H + 2tD + \tfrac{1}{2}Mr^2 \qquad (9)$$

The above invariances are realized both classically and quantum mechanically. In the latter context $\vec{r}$ and $\vec{v}$ are operators that satisfy the equal time algebra

$$[r^i, r^j] = 0, \quad [r^i, v^j] = \frac{i\hbar}{M} \delta^{ij}, \quad [v^i, v^j] = \frac{i\hbar e}{Mc^2} \varepsilon^{ijk} B^k \qquad (10)$$

and D is written in the symmetrized form already used in (6).

Even though D and K are constants of motion, their commutator [Poisson bracket] with H does not vanish; rather one finds, using (10) that

$$[H,D] = i\hbar H, \quad [H,K] = 2i\hbar D, \quad [D,K] = i\hbar K \qquad (11)$$

These allow verifying the time-independence of D and K by use of the formulas

$$\frac{dD}{dt} = \frac{i}{\hbar}[H,D] + \frac{\partial D}{\partial t} = 0, \quad \frac{dK}{dt} = \frac{i}{\hbar}[H,K] + \frac{\partial K}{\partial t} = 0 \tag{12}$$

The three-parameter set of transformations of the time: translations, dilatations, and conformal transformations, form a three-parameter Lie group, with generator algebra (11). This group is recognized to be $O(2,1)$. Hence when it is combined with the $O(3)$ group of rotations, we see that the problem admits an $O(3) \times O(2,1) = O(3,1)$ group of invariances.

The generators however are not all independent. The Casimir operator of the $O(2,1)$ group is related to that of the rotation group.

$$C = \tfrac{1}{2}[KH + HK] - D^2 = \tfrac{1}{4}[J^2 - (\tfrac{eg}{c})^2 - \tfrac{3}{4}\hbar^2] \tag{13}$$

[The last term is a quantal re-ordering contribution, absent in the classical mechanical context.] (13) shows that at fixed angular momentum, the entire dynamics of the magnetic monopole-electric monopole interaction is characterized by a single, irreducible, unitary, hence infinite-dimensional, representation of $O(2,1)$, labelled by the monopole strength. Within quantum mechanics, $J^2$ is quantized in units of $\hbar^2 j(j+1)$, and so is $eg/c$ in units of $\hbar\mu$ [$2j, 2\mu$ = integers].[1] Consequently C is also quantized.

We recognize that the situation here is quite analogous to that of the Coulomb/Kepler problem, which describes the dynamics of an electric monopole in the field of another infinitely heavy electric monopole. There too, the obvious rotational invariance, when combined with the hidden Runge-Lenz invariance, gives rise to an $O(3) \times O(2,1) = O(3,1)$ symmetry of the unbounded motion, and an $O(3) \times O(3) = O(4)$ symmetry of the bounded motion. [In the monopole problem there is no bounded motion.][3]

One may use the symmetry to give an algebraic treatment for monopole dynamics, either classically or quantum-mechanically. Since the answers have been available for many years, the algebraic approach does not produce any new results; rather it supplements the analytic treatment and, in a sense, explains why a simple, closed-form answer is attainable.[4]

One may also do something else: the $O(2,1)$ invariance allows for quantization procedures, other than the conventional equal-time one. Rather one may quantize on other characteristic curves of the $O(2,1)$ group. This is analogous to the situation in relativistic field theories where the $O(3,1)$ Lorentz invariance [or in massless theories, the $O(4,2)$ conformal invariance] allows for quantization on surfaces other than the fixed-time ones. That this should be an interesting alternative, which organizes the same physical content in novel ways, was first stressed by Dirac.[5] In our simple example the idea can be worked out completely and explicitly.

Here there is no time to give details of the above calculations. You may find them by spending 80 cents on a copy of the submitted paper or you may await its publication in <u>Annals of Physics</u>.[6] One obvious calculation does not however exist yet. In the Coulomb/Kepler problem, the $O(3,1)/O(4)$ invariance allows for the introduction of a coordinate system which makes the hidden symmetry manifest and facilitates computation -- this is the Fock sphere. An analogous formalism for the magnetic monopole, which would expose its hidden symmetry, has thus far not been indentified.

In conclusion let me remind that all the above symmetries are a property of the non-relativistic point monopole. How they may appear when the dynamics is relativistic, or when the monopole has additional structure is not known. However one may expect that some of our results may extend beyond the context here emphasized. The reason for this hope is that the <u>interaction</u> Lagrangian in (1) allows for an even larger set of symmetry transformations than the O(2,1) group discussed here. One may verify that <u>any</u> reparametrization of the time, $t \to f(t)$, leaves $\int dt\, L_{int}$ unchanged. This general invariance is reduced to O(2,1) by the fact that $\int dt\, L_{kin}$ is invariant only under the restricted transformations. Other kinetic terms may very well allow for other symmetries.

## REFERENCES

1. P.A.M. Dirac, Proc. Roy. Soc. (London) <u>A133</u>, 60 (1931).
2. T.T. Wu and C.N. Yang, Nucl. Phys. <u>B107</u>, 365 (1976) and Phys. Rev. <u>D16</u>, 1018 (1977).
3. For a review of the hidden symmetries of the Kepler problem see M. Bander and C. Itzykson, Rev. Mod. Phys. <u>38</u>, 330, 346 (1966).
4. The O(2,1) group has been previously utilized to analyze various dynamical problems; for a review see B. Wybourne, <u>Classical Groups for Physicists</u> (Wiley, New York, NY, 1974). An application to the $1/r^2$ potential, very similar to the one I give here for the magnetic monopole, is by V. de Alfaro, S. Fubini, and G. Furlan, Nuovo Cimento <u>A34</u>, 569 (1976).
5. P.A.M. Dirac, Rev. Mod. Phys. <u>21</u>, 392 (1949).
6. R. Jackiw, Annals of Physics [in press].

# INDUCED GRAVITATION

Stephen L. Adler
The Institute for Advanced Study
Princeton, New Jersey 08540

This talk (in an expanded version) will be included in A. Zichichi, ed., The High Energy Limit, Proceedings of the 18th International School of Subnuclear Physics "Ettore Majorana" (the 1980 Erice Lectures), to be published by Plenum Press. Brief accounts of the principal results have appeared in S. L. Adler, Phys. Rev. Lett. 44, 1567 (1980) and S. L. Adler, Phys. Lett. B (in press).

# MATHEMATICAL PHYSICS

N. Christ, Columbia, Organizer

# ANOMALOUS WARD IDENTITIES AND PATH INTEGRATION

Kazuo Fujikawa*
Institute for Nuclear Study, University of Tokyo
Tanashi, Tokyo 188, Japan

## ABSTRACT

The path integral formulation of anomalous Ward-Takahashi (WT) identities is briefly summarized.

## INTRODUCTION

There exist two well known anomalous WT identities in quantum field theory, the chiral (Adler-Bell-Jackiw[1]) anomaly and the trace anomaly. It has been recently pointed out that the derivation of those WT identities can be considerably simplified in the path integral formalism[2] and that those anomalies are characterized by the appearance of two non-commuting operators.[3] In the following we exemplify the path integral formulation by considering the chiral anomaly for non-Abelian gauge theories.[4]

## PATH INTEGRATION

We start with the Yang-Mills field coupled to fermions in Euclidean space $R^4$

$$\mathcal{L} = \bar{\psi}[i\not{D}-m]\psi + (1/2g^2)\mathrm{Tr}F^{\mu\nu}F_{\mu\nu} \tag{1}$$

where

$$\not{D} \equiv \gamma^\mu(\partial_\mu - A_\mu), \quad \gamma_5 \equiv \gamma^4\gamma^1\gamma^2\gamma^3$$
$$A_\mu \equiv iA_\mu^a T^a, \quad [T^a, T^b] = if^{abc}T^c \tag{2}$$

In Euclidean space $\not{D}$ becomes a Hermitian operator, and we consider the eigenvalue equation

$$\not{D}\varphi_n(x) = \lambda_n \varphi_n(x), \quad \int dx\, \varphi_n(x)^\dagger \varphi_m(x) = \delta_{n,m} \tag{3}$$

The classical fields $\psi(x)$ and $\bar{\psi}(x)$ are expanded as

---

* Visitor at State University of New York at Stony Brook from February 1 to July 15, 1980.

$$\psi(x) = \Sigma_n a_n \varphi_n(x), \qquad \bar{\psi}(x) = \Sigma_n \bar{b}_n \varphi_n(x)^+ \qquad (4)$$

with $a_n$ and $\bar{b}_n$ the elements of the Grassmann algebra. The path integral measure is then defined by[5]

$$d\mu = \prod_x [\mathcal{D}A_\mu(x)] \mathcal{D}\psi(x) \mathcal{D}\bar{\psi}(x)$$
$$= \prod_x [\mathcal{D}A_\mu(x)] \prod_n da_n d\bar{b}_n \qquad (5)$$

where $[\mathcal{D}A_\mu(x)]$ includes the gauge fixing and Faddeev-Popov factors. Under the local chiral transformation

$$\psi(x) \to \exp[i\alpha(x)\gamma_5]\psi(x), \qquad \bar{\psi}(x) \to \bar{\psi}(x)\exp[i\alpha(x)\gamma_5] \qquad (6)$$

the Lagrangian changes as

$$\mathcal{L} \to \mathcal{L} - \partial_\mu \alpha(x) \bar{\psi}\gamma^\mu \gamma_5 \psi - 2mi\alpha(x)\bar{\psi}\gamma_5\psi \qquad (7)$$

which defines the currents $j_5^\mu(x) = \bar{\psi}\gamma^\mu\gamma_5\psi$, $j_5(x) = \bar{\psi}\gamma_5\psi$. Under (6), the coefficients in (4) are transformed as

$$\psi(x)' = \exp[i\alpha(x)\gamma_5]\psi(x) \equiv \Sigma a_n' \varphi_n(x) \qquad (8)$$

i.e.,

$$a_n' = \Sigma_m \int dx\, \varphi_n(x)^+ \exp[i\alpha(x)\gamma_5]\varphi_m(x) a_m \qquad (9)$$
$$\equiv \Sigma C_{n,m} a_m$$

By noting $da_n$ = left-derivative,[5] the Jacobian for this (infinitesimal) transformation is given by

$$\det[C_{n,m}]^{-1} = \det[\delta_{n,m} + i\int \alpha(x)\varphi_n(x)^+ \gamma_5 \varphi_m(x) dx]^{-1}$$
$$= \exp[-i\Sigma_n \int \alpha(x) \varphi_n(x)^+ \gamma_5 \varphi_n(x) dx] \qquad (10)$$

We evaluate the quantity $A(x) \equiv \Sigma \varphi_n(x)^+ \gamma_5 \varphi_n(x)$ in (10) by summing the series starting from small eigenvalues in (3) (i.e., $|\lambda_n| \lesssim M$)

$$A(x) = \lim_{M\to\infty} \Sigma\, \varphi_n(x)^+ \gamma_5 f[(\lambda_n/M)^2] \varphi_n(x)$$

$$= \lim_{M\to\infty} \mathrm{Tr} \int \frac{d^4k}{(2\pi)^4}\, e^{-ikx} \gamma_5 f[(\not{D}/M)^2] e^{ikx}$$

$$= -\mathrm{Tr}\{*F^{\mu\nu} F_{\mu\nu}\} \int \frac{d^4k}{(2\pi)^4}\, f''(k^2)$$

$$= -1/(16\pi^2)\,\mathrm{Tr}\, *F^{\mu\nu} F_{\mu\nu} \qquad (11)$$

with $*F^{\mu\nu} \equiv \frac{1}{2} \epsilon^{\mu\nu\alpha\beta} F_{\alpha\beta}$. Here we changed the basis from $\varphi_n(x)$ to plane waves; the result (11) is independent of any smooth function[2] $f(z)$ with $f(0) = 1$ and $f(+\infty) = f'(+\infty) = \cdots = 0$. The chiral transformation law of the measure (5) becomes

$$d\mu \to d\mu\, \exp[i/(8\pi^2) \int \alpha(x) \mathrm{Tr}*F^{\mu\nu} F_{\mu\nu} dx] \qquad (12)$$

The generating functional of Green's functions is defined by

$$Z(J,\eta,\bar{\eta}) \equiv \frac{1}{N} \int d\mu\, \exp\{\int dx[\mathcal{L} - J_\mu A^\mu + \bar{\psi}\eta + \bar{\eta}\psi] \qquad (13)$$

The WT identities are summarized by the variational derivative (the change of integration variables does not change the integral itself)

$$\frac{\delta}{\delta\alpha(x)} Z(J,\eta,\bar{\eta}) \equiv 0 \qquad (14)$$

From (6), (7) and (12), we obtain, for example,

$$\partial_\mu \langle [j_5^\mu(x)\psi(y)\bar{\psi}(z)]_+ \rangle = 2mi \langle [j_5(x)\psi(y)\bar{\psi}(z)]_+ \rangle$$

$$-i\delta(x-y)\langle[\gamma_5\psi(y)\bar{\psi}(z)]_+\rangle - i\delta(x-z)\langle[\psi(y)\bar{\psi}(z)\gamma_5]_+\rangle$$

$$-i/(8\pi^2)\langle[\mathrm{Tr}*F^{\mu\nu}F_{\mu\nu}\psi(y)\bar{\psi}(z)]_+\rangle \qquad (15)$$

The Minkowski version is obtained by a suitable Wick rotation.

## DISCUSSION

We add several comments on the above formulation.
(i) By assuming term-by-term integration in (11), one obtains

$$n_+ - n_- = \nu \qquad (16)$$

with $n_\pm$ = number of zero eigenvalue solutions in (3) with $\gamma_5 \varphi_n(x) = \pm \varphi_n(x)$, and $\nu$ the Pontryagin index. For well-behaved gauge fields, the relation (11) is thus regarded as a local version of the Atiyah-Singer index theorem[6] (16). Eq. (16) indicates that a unitary transformation to plane waves $\not{D} \xi_n(x) = \lambda_n \xi_n(x)$ (i.e., interaction picture) should be treated carefully in gauge theory, as the Jacobian factor in (10) is now replaced by $\Sigma \, \xi_n^+(x) \gamma_5 \xi_n(x) = 0$; a naive unitary transformation between basis vectors belonging to different (local) indices fails.
(ii) The appearance of the anomaly and its association with divergent Feynman diagrams is related to the fact that

$$[\not{D}, \gamma_5] \neq 0 \qquad (17)$$

Namely, $\not{D}$ and $\gamma_5$ cannot be simultaneously diagonalized (Cf. Heisenberg uncertainty relation). When one diagonalizes $\not{D}$ by (3), $\gamma_5$ becomes ill-defined and $\text{Tr} \, \gamma_5 \sim \Sigma \, \varphi_n^+ \gamma_5 \varphi_n \neq 0$ in (11). This characterization is also shared by the trace anomaly.[3] In path integration, (17) suggests that the Jacobian becomes non-trivial, as the invariance of $\mathcal{L}$ and non-invariance of $\not{D}$ leads to the non-trivial behavior of $D\psi D\bar{\psi}$ under the $\gamma_5$ transformation.
(iii) A characteristic property of the renormalizable theory is that the <u>form</u> of the Lagrangian is not altered by higher order effects. Our derivation of WT identities relies solely on this invariant form of $\mathcal{L}$. It is tempting to regard (15) as a <u>bare</u> form of WT identity. This, however, does not necessarily suggest that all the terms in (15) separately represent finite quantities.[1] For the vanishing momentum transfer the left-hand side of (15) vanishes, and the right-hand side is expected to be made finite by the ordinary renormalization factors. This provides an intuitive way of seeing the Adler-Bardeen theorem.
(iv) In the Pauli-Villars or dimensional regularization method, the path integral measure becomes invariant under $\gamma_5$ transformation: For the Pauli-Villars regulator the Bose-like regulator field cancels the Jacobian, and for the dimensional regulator the integral (11) vanishes for $n = 4 - \epsilon$. The variation of $\mathcal{L}$ in (7) should thus contain extra terms to obtain the ordinary WT identity.

This is indeed the case: The regulator field contributes to currents in the Pauli-Villars regulator, and $\{\not{D}, \gamma_5\}_+ \neq 0$ in the dimensional regulator if one chooses $\gamma_5$ suitably.

(v) If one assumes a smooth <u>global</u> limit of the local transformation law (12) specified by the parameter $\alpha(x)$, one obtains

$$d\mu \to \sum_\nu d\mu_{(\nu)} \exp[-2i\nu\alpha] \quad (18)$$

where $\alpha$ is a constant and $\nu$ the Pontryagin index. This procedure gives rise to the chirality selection rule specified by the index theorem (16) and to the $\theta$-vacuum prescription.[7] We, however, note that these global properties are of more restricted validity than the local relations (11) and (12), as the topological property such as the index theorem is valid for a rather restricted class of field configurations. Besides, a naive global limit of (12) to obtain (18) generally fails if the spontaneous breakdown of the chiral symmetry takes place. On the contrary, the local relation (11) is a statement on the unitary transformation of basis vectors for a well-behaved operator, and it is relatively insensitive to the precise behavior of field variables at space-time infinity.

## CONCLUSION

The present formulation shows that the anomalous WT identities are not really "anomalous" but rather they represent one of the most fundamental and intrinsic properties of quantum field theory.

## REFERENCES

1. S. Adler, Phys. Rev. <u>177</u>, 2426 (1969); J. Bell and R. Jackiw, Nuovo Cim. <u>60A</u>, 47 (1969); W. Bardeen, Phys. Rev. <u>184</u>, 1848 (1969).
2. K. Fujikawa, Phys. Rev. Lett. <u>42</u>, 1195 (1979); Phys. Rev. <u>D21</u>, 2848 (1980). In this second reference, $[\gamma_\mu, \gamma_\nu]$ in eqs. (2.15) and (2.24a) should be replaced by $(1/2)[\gamma_\mu, \gamma_\nu]$.
3. K. Fujikawa, Phys. Rev. Lett. <u>44</u>, 1733 (1980) Cf. S. Hawking, Comm. Math. Phys. <u>55</u>, 133 (1977).
4. Some of the earlier works on the chiral anomaly are H. Fukuda and Y. Miyamoto, Prog. Theor. Phys. <u>4</u>, 347 (1949); J. Steinberger, Phys. Rev. <u>76</u>, 1180 (1949); J. Schwinger, *ibid.* <u>82</u>, 664 (1951). See also C. Hagen, Phys. Rev. <u>177</u>, 2622 (1969); R. Jackiw and K. Johnson, *ibid.* 182, 1459 (1969): S. Adler and W. Bardeen, *ibid.* <u>182</u>, 1517 (1969).
The gravitational chiral anomaly has been calculated by T. Kimura, Prog. Theor. Phys. <u>42</u>, 1191 (1969); R. Delbourgo and A. Salam, Phys. Lett. <u>40B</u>, 381 (1972); T. Eguchi and P. Freund, Phys. Rev. Lett. <u>37</u>, 1251 (1976); N. K. Nielsen et al., Nucl. Phys. <u>B140</u>, 477 (1978).
The trace anomaly as a dilatation anomaly has been considered

by S. Coleman and R. Jackiw, Ann. Phys. 67, 552 (1971);
R. Crewther, Phys. Rev. Lett. 28, 1421 (1972); M. Chanowitz
and J. Ellis, Phys. Lett. 40B, 397 (1972) and Phys. Rev. D7,
2490 (1973).
The trace anomaly as a conformal anomaly has been discussed by
D. Capper and M. Duff, Nuovo Cimento 23A, 173 (1974); M. Duff,
S. Deser and C. Isham, Nucl. Phys. 111B, 45 (1976); J. Dowker
and R. Critchley, Phys. Rev. D13, 3224 (1976); L. Brown and
J. Cassidy, *ibid.* D15, 2810 (1977); H. Tsao, Phys. Lett. 68B,
79 (1977); T. Yoneya, Phys. Rev. D17, 2567 (1978).
The renormalization properties of anomalous WT identities have
been investigated by S. Adler and W. Bardeen, Phys. Rev. 182,
1517 (1969); A. Zee, Phys. Rev. Lett. 29, 1198 (1972);
J. Lowenstein and B. Schroer, Phys. Rev. D7, 1929 (1973);
K. Nishijima, Prog. Theor. Phys. 57, 1409 (1977); S. Adler,
J. Collins and A. Duncan, Phys. Rev. D15, 1712 (1977) and
references therein.

5. F. Berezin, The Method of Second Quantization (Academic, N. Y.,
1969); S. Coleman, Lectures at 1977 Erice Summer School.
6. M. Atiyah and I. Singer, Ann. Math. 87, 484 (1968); M. Atiyah,
R. Bott and V. Patodi, Invent. Math. 19, 279 (1973). See also
references quoted in Ref. 2 above.
7. G. 't Hooft, Phys. Rev. Lett. 37, 8 (1976); Phys. Rev. D14,
3432 (1976); R. Jackiw and C. Rebbi, Phys. Rev. Lett. 37, 172
(1976); C. Callan, R. Dashen and D. Gross, Phys. Lett. 63B, 334
(1976).

# HIDDEN SYMMETRY IN YANG MILLS

L. Dolan
The Rockefeller University, New York, N. Y. 10021

## ABSTRACT

The nature of the new symmetry is discussed.

## INTRODUCTION

In lieu of solving the quark gluon system exactly, what is needed now to make further progress on the strong interactions is a systematic nonperturbative approximation of the color gauge field theory. Various schemes, namely large N, semiclassical approximations, and lattice gauge theories have been suggested to this end. Although large N was an immediate and surprising success for relativistic scalar field theories[1], its calculation for gauge theories has remained elusive. Semiclassical approximations, although inherently nonperturbative and thus useful to probe a whole range of phenomena inaccessible to standard perturbation theory such as the U(1) problem[2] and tunneling in finite temperature[3] field theories, are in general still valid only in the weak coupling regime. To probe the strong coupling region the lattice theory has produced a calculational technique. Yet the recent observation of the existence of the roughening phenomenon[4] has cast doubt on the relevance of these calculations to the continuum world.

Since quantum color SU(3) contains no dimensionless parameter, it may in fact be necessary to consider Yang Mills coupled to fermions in order to introduce an expansion parameter for some solvable systematic nonperturbative approximation. But clearly any new symmetry of even the pure gauge theory will be an extremely useful guide in this attempt.

It has long been understood that symmetries of a theory often give valuable information about the solution without knowing it exactly. The subject of the material presented here is the search for a new symmetry in Yang Mills.[5]

We begin with a description of Yang Mills reformulated as a functional field theory whose fundamental variable is the path dependent field $\psi = Pe^{\oint A \cdot d\xi}$ defined on a closed path in four dimensional Euclidean space. These are the variables in which we expect to see the symmetry most naturally. Since it is unlikely that there exist further local continuous symmetry currents expressed in terms of the local gauge fields $A_\mu^a(x)$, this new symmetry appearing in the functional formulation probably gives rise to nonlocal conserved currents in the original variables.

A reparameterization invariant Lagrangian of the functional fields is given and is used to construct path dependent Noether currents associated with the normal space time symmetries. In the analogous two dimensional chiral models, it is demonstrated that the nonlocal hidden symmetry currents can be derived as Noether currents of an infinite parameter hidden symmetry group.[6]  The group is exhibited.  Due to the similarity of the chiral equations and the gauge theory functional formulation, it is likely that the group responsible for the new symmetry in Yang Mills is the same as that found for the chiral theory.  The qualitative nature of this symmetry in the gauge theory is that it is a particular path dependent gauge transformation.

## SPACE TIME SYMMETRIES

The fundamental field of the functional formulation of Yang Mills is defined by

$$\psi[\xi] = P\, e^{\oint d\xi \cdot A}$$

$$= \lim_{N \to \infty} e^{(\xi^1 - x)\cdot A(x)} e^{(\xi^2 - \xi^1)\cdot A(\xi^1)} \ldots e^{(x - \xi^N)\cdot A(\xi^N)}$$

$$= P\, e^{\int_0^1 ds\, \dot\xi_\mu(s) A_\mu(\xi(s))} \tag{1}$$

The closed path in four dimensional Euclidean space is parameterized by four functions $\xi_\mu(s)$ where $0 \le s \le 1$ and $\xi_\mu(0) = \xi_\mu(1) \equiv x_\mu$

A reparameterization invariant Lagrangian is

$$\mathcal{L}[\xi] = \int_0^1 \frac{ds}{\sqrt{\dot\xi^2(s)}} \, \text{tr}\, \frac{\delta \psi}{\delta \xi_\mu(s)} \frac{\delta \psi^{-1}}{\delta \xi_\mu(s)} . \tag{2}$$

Here $\psi^{-1} = \lim_{N\to\infty} e^{-(x-\xi^N)\cdot A(\xi^N)} \ldots e^{-(\xi^1 - x)\cdot A(x)}$

and the gauge covariant derivative $\frac{\delta}{\delta \xi_\mu(s)}$ is given by

$$\psi[\xi + \delta\xi] - \psi[\xi] = -\delta x_\mu [A_\mu(x), \psi]$$

$$+ \int_0^1 ds\, \delta\xi_\mu(s) \frac{\delta \psi}{\delta \xi_\mu(s)} . \tag{3}$$

If we now vary $\psi \to \psi + \Delta\psi$ then the change in the Lagrangian $\mathcal{L}[\xi]$ is

$$\Delta\mathcal{L} = \int_0^1 \frac{ds}{\sqrt{\dot\xi^2(s)}} 2\, \text{tr}\left\{ \frac{\delta}{\delta \xi_\mu(s)}(\Delta\psi \frac{\delta\psi^{-1}}{\delta\xi_\mu(s)}) \right.$$

$$\left. - \Delta\psi\, \psi^{-1} \frac{\delta}{\delta\xi_\mu(s)}(\psi \frac{\delta\psi^{-1}}{\delta\xi_\mu(s)}) \right\} . \tag{4}$$

The connection with Yang Mills theory is made in the following way.  Impose the equations of motion

$$\frac{\delta}{\delta\xi_\mu(s)}\left( \psi^{-1} \frac{\delta \psi}{\delta\xi_\mu(s)} \right) = 0 . \tag{5}$$

Since $\frac{\delta}{\delta\xi_\mu(s)}(\psi^{-1}\frac{\delta}{\delta\xi_\mu(s)}\psi) \sim \psi^{-1}_{x;\xi(s)} D_\mu F_{\mu\nu}(\xi)\psi_{\xi(s):x} \dot{\xi}_\nu(s) = 0$,

if $A^a_\mu(\xi)$ is a solution to the Yang Mills equations $D_\mu F_{\mu\nu}(\xi) = 0$, then $\psi$ satisfies Eq. (5).

If $\Delta \mathcal{L}[\xi]$ given by Eq. (4) can be written as $\int_0^1 ds \frac{\delta}{\delta\xi_\mu(s)}\Lambda_\mu(s,[\xi])$ without use of the equations of motion, then a functionally conserved quantity is $J_\mu(s,[\xi]) = 2\,\text{tr}\Delta\psi\frac{1}{\sqrt{\dot{\xi}^2(s)}}\frac{\delta}{\delta\xi_\mu(s)}\psi^{-1} - \Lambda_\mu(s,[\xi])$.

For a space time symmetry transformation on the path $\xi_\mu(s) \to \xi_\mu(s) + \delta\xi_\mu(s)$, let $\psi[\xi]$ transform as a scalar: $\psi[\xi] \to \psi[\xi+\delta\xi] \equiv \psi[\xi]+\Delta\psi$ infinitesimally. For translations, $\xi_\mu(s)\to\xi_\mu(s)+c_\mu$ (where $c_\mu$ is constant) and $\Delta\psi = \int_0^1 dt\, c_\sigma \frac{\delta}{\delta\xi_\sigma(t)}\psi$. The associated conserved quantity is

$$J_\mu(s,[\xi]) = c_\sigma \{2\text{tr}\int_0^1 dt\, \frac{\delta\psi}{\delta\xi_\sigma(t)}\frac{1}{\sqrt{\dot{\xi}^2(s)}}\frac{\delta\psi^{-1}}{\delta\xi_\sigma(s)} - \delta_{\mu\sigma}\mathcal{L}[\xi] \}. \quad (6)$$

Note that the change $\Delta\psi = \int_0^1 dt\, c_\sigma \frac{\delta\psi}{\delta\xi_\sigma(t)}$ induced by a full translation of the path corresponds to a change in the local field variable

$$A^a_\mu(x) \to A^a_\mu(x+c) \equiv A^{a'}_\mu(x) \sim A^a_\mu(x) + c_\sigma\partial_\sigma A^a_\mu(x). \quad (7)$$

That is to say $P e^{\oint d\xi\cdot A'(\xi)} = \psi[\xi + c]$. Functionally conserved currents corresponding to SO(4) transformations and dilations have also been derived.[5]

## HIDDEN SYMMETRY GROUP

In order to investigate new symmetries in Yang Mills theory we now discuss the local two dimensional chiral models. The Lagrangian density and equations of motion are

$$\mathcal{L}(x) = \frac{1}{16}\text{tr}\,\partial_\mu g(x)\,\partial_\mu g^{-1}(x) \quad (8)$$

and

$$\partial_\mu(g^{-1}\partial_\mu g) = 0 \quad (9)$$

where the matrix field $g(x)$ is an element of some group G. Note the similarity of Eq.'s (8) and (9) with Eq.'s (2) and (5). The infinitesimal transformations giving rise to the first two nonlocal hidden symmetry currents as Noether currents in this model are for $n = 1$ and $\bar{2}$:

$$\Delta^{(n)}g = -g\lambda^{(n)} \quad (10)$$

$$\lambda^{(1)} = 4[\chi^{(1)}, T]$$

$$\lambda^{(\bar{2})} = 4 \{[\chi^{(\bar{2})}, T] + 1/2 [\chi^{(1)}, [\chi^{(1)}, T]] \}$$

$$\chi^{(1)}(y,t) = 1/2 \int_{-\infty}^{\infty} dx\, \varepsilon(y-x)\, A_o(x,t)$$

$$\chi^{(\bar{2})}(y,t) = 1/2 \int_{-\infty}^{\infty} dx\, \varepsilon(y-x)\, \partial_o \chi^{(1)}(x,t)$$

$$+ 1/4 \int_{-\infty}^{\infty} dx\, \varepsilon(y-x)\, [A_o(x,t), \chi^{(1)}(x,t)] \qquad (11)$$

Here $T = T^e \rho^e$ for $\rho^e$ constant and $T^e$ are the matrix generators of the group G.

The transformations (10) shift the Lagrangian density without using the equations of motion respectively by

$$\Delta^{(1)} \mathcal{L} = \partial_\mu 1/2\, \mathrm{tr}\, \{(1/2\, \varepsilon_{\mu\nu}[\partial_\nu \chi^{(1)}, \chi^{(1)}] + \varepsilon_{\mu\nu} A_\nu) T\}$$

and $\Delta^{(2)} \mathcal{L} = \partial_\mu 1/2\, \mathrm{tr}\, \{(\varepsilon_{\mu\nu}[\partial_\nu \chi^{(1)}, \chi^{(\bar{2})}]$

$$+ \varepsilon_{\mu\nu} \frac{1}{6} [\,[\partial_\nu \chi^{(1)}, \chi^{(1)}], \chi^{(1)}\,]$$

$$+ \varepsilon_{\mu\nu} [A_\nu, \chi^{(1)}] \,) T \} . \qquad (12)$$

The associated conserved Noether currents are

$$\mathcal{J}_\mu^{(1)} = [A_\mu, \chi^{(1)}] - \varepsilon_{\mu\nu} A_\nu - 1/2\, \varepsilon_{\mu\nu}[\partial_\nu \chi^{(1)}, \chi^{(1)}] \qquad (13a)$$

and

$$\mathcal{J}_\mu^{(2)} = [A_\mu, \chi^{(\bar{2})}] + 1/2\, [\,[A_\mu, \chi^{(1)}], \chi^{(1)}\,]$$

$$- \varepsilon_{\mu\nu}[\partial_\nu \chi^{(1)}, \chi^{(\bar{2})}] - \frac{1}{6} \varepsilon_{\mu\nu}[\,[\partial_\nu \chi^{(1)}, \chi^{(1)}], \chi^{(1)}\,]$$

$$- \varepsilon_{\mu\nu}[A_\nu, \chi^{(1)}]. \qquad (13b)$$

Furthermore, the infinitesimal transformations (10) can be used to construct the generators $M_e^{(n)}$ of a group:

$$M_e^{(n)} \equiv -\int d^2y\, \Delta_e^{(n)}(g(y))\, \frac{\delta}{\delta g(y)} . \qquad (14)$$

For the first two transformations given in (10) we find

$$[M_e^{(1)}, M_d^{(1)}] = C_{eda} M_a^{(\bar{2})} \qquad (15)$$

where $C_{eda}$ are the structure constants of the field group G. In general, we will need to keep the entire set of hidden symmetry currents to close the algebra

$$[M_e^{(n)}, M_d^{(m)}] = f_{eda}^{(n)(m)(j)} M_a^{(j)} . \qquad (16)$$

where $f_{eda}^{(n)(m)(j)}$ are the structure constants of the infinite parameter hidden symmetry group.

Return now to the functional Yang Mills theory. Local gauge transformations on $A_\mu(\xi) \to U(\xi)A_\mu(\xi)U^{-1}(\xi) + \partial_\mu U(\xi) U^{-1}(\xi)$ result in $\psi[\xi] \to U(x)\psi[\xi]U^{-1}(x) = \psi + [S(x), \psi]$. Where $x_\mu$ is the origin of the closed path $\xi_\mu(0) = x_\mu = \xi_\mu(1)$ and $U(x) = e^{S^a(x)\frac{J^a}{2i}} \equiv e^{S(x)}$ for SU(2). We note that $\Delta\psi = -\psi S(x)$ also leaves the functional Lagrangian Eq. (2) invariant and the associated path dependent Noether current for this "internal" symmetry transformation is

$$J_\mu(s,[\xi]) = S^a(x) \, 2 \, \text{tr} \frac{1}{\sqrt{\dot\xi^2(s)}} \, \psi^{-1} \frac{\delta\psi}{\delta\xi_\mu(s)} \, \frac{\sigma^a}{2i} \tag{17}$$

Arguing in analogy with the chiral models, we now look for a transformation

$$\Delta\psi = -\psi S = -\psi[X, T] \tag{18}$$

where the gauge function S depends on the entire path instead of only on the origin $x_\mu$, and $T = \rho^a \frac{\sigma^a}{2i}$ for $\rho^a$ constant. For Eq. (18) to be a symmetry of the equations of motion i.e., if $\psi$ is a solution to (5) so is $\psi + \Delta\psi$, it must be that

$$1/\sqrt{\dot\xi^2(s)} \, [\frac{\delta^2 X}{\delta\xi_\mu^2(s)}, T] + [F_\mu(s), [\frac{\delta X}{\delta\xi_\mu(s)}, T]] = 0 \tag{19}$$

Here $F_\mu(s) \equiv \frac{1}{\sqrt{\dot\xi^2(s)}} \, \psi^{-1} \frac{\delta\psi}{\delta\xi_\mu(s)}$.

Let $F_\mu \frac{\delta X}{\delta\xi_\mu} = -\frac{\delta X}{\delta\xi_\mu} F_\mu$ and $F_\mu T \frac{\delta X}{\delta\xi_\mu} = -\frac{\delta X}{\delta\xi_\mu} TF_\mu \tag{20}$

From (19) and (20) construct the functionally conserved quantity:

$$J_\mu(s,[\xi]) = 1/2 \, [F_\mu(s), X] + \frac{1}{\sqrt{\dot\xi^2(s)}} \frac{\delta X}{\delta\xi_\mu(s)} . \tag{21}$$

Note that Eq. (21) has the same form as Eq. (13a) when the chiral fields are solutions. Similarly, Eq. (13b) reduces to $\mathcal{J}_\mu^{(2)} = [\partial_\mu X^{(1)}, X^{(1)}] + 1/3 \, [[A_\mu, X^{(1)}], X^{(1)}]$ for solutions. Thus the next current in functional space is

$$J_\mu^{(2)}(s,[\xi]) = \frac{1}{\sqrt{\dot\xi^2(s)}} \, [\frac{\delta X}{\delta\xi_\mu(s)}, X] + \frac{1}{3}[[F_\mu(s), X], X]. \tag{22}$$

## CONCLUSION

In conclusion we remark that a reparameterization invariant Lagrangian density has been given which is used to construct path dependent Noether currents of space time symmetries. By working in the chiral theory, we exhibited the underlying symmetry group responsible for the previously hidden symmetry currents. Since the functional formulation of the gauge theory is the chiral theory with ordinary derivatives replaced by functional ones, it is reasonable that hidden symmetry exists in Yang Mills as well.

Although we have not yet been able to explicitly construct these currents as path dependent Noether currents of an internal symmetry transformation, our equations are consistent with Polyakov's program[7] for complete integrability when the fields are solutions to the equations of motion. We note here that although Polyakov has claimed to prove that conserved currents exist in 3 dimensions he has not actually constructed them. Owing to the need to define the functional derivative by a rigorous regularization procedure, this remains an open problem.

The approach of path dependent Noether currents given in this paper is thus a complement to Polyakov's suggestion and may prove a useful way to find the hidden conserved quantities. Lastly, we note that functional field variables have now appeared in two contexts associated with color gauge theory: 1) complete integrability or the existence of new conserved quantities and 2) duality transformations.[8] It may thus be useful to investigate if there is any connection between these two properties of a system (for example, can the existence of the duality transformation be used to find an unsuspected symmetry) and how they are related to its nonperturbative description.

## REFERENCES

1. R. Jackiw and L. Dolan, Phys. Rev. $\underline{D9}$, 3320 (1974).
2. G. 't Hooft, Phys. Rev. $\underline{D14}$, 3422 (1976).
3. L. Dolan and J. Kiskis, Phys. Rev. $\underline{D20}$, 505 (1979).
4. A. Hasenfratz, E. Hasenfratz, and P. Hasenfratz, CERN preprint TH 2890.
5. L. Dolan, Rockefeller preprint DOE/EY/2232B-202.
6. L. Dolan and A. Roos, to be published Phys. Rev. $\underline{D22}$, (October, 1980).
7. A. M. Polyakov, Phys. Lett. $\underline{82B}$, 247 (1979) and Nuc. Phys. $\underline{B164}$, 1971 (1980).
8. G. 't Hooft, Nuc. Phys. $\underline{B138}$, 1 (1978) and $\underline{B153}$, 141 (1979).

# COMMENTS ON THE INTEGRABILITY OF THE LOOP-SPACE CHIRAL EQUATIONS

Chaohao Gu
Department of Mathematics, Fudan University, Shangai, China

Ling-Lie Chau Wang[*]
Brookhaven National Laboratory, Upton, N.Y. 11973 USA

## INTRODUCTION

The path ordered phase factor $\Phi_{ab} \equiv P \exp\int_a^b dx_\mu A_\mu(x)$ along a loop (or part of a loop) has many desirable features. It is the minimal necessary object[1] describing a gauge theory, has simple gauge transformation properties[2] and is the spin-like quantity for lattice gauge theory.[3] Thus it is natural, may even be essential, that we formulate gauge theory in the loop space. Last year (1979), it has been realized that Yang-Mills equations give chiral-like equations in the loop space.[4,5] This beautiful realization immediately raised the hope that the loop-space chiral equations may also be a total integrable system, just like the ordinary-space chiral equations,[6] and thus lead to the full solutions of the Yang-Mills equations. However here we want to discuss some intricate properties of the integrability conditions of the loop space chiral equations, which do not have their correspondence in the ordinary chiral equations.

In part I we shall briefly demonstrate how the ordinary space chiral equations provide the existence conditions for the infinite number of conserved currents and how these currents are related to the so-called inverse-scattering equations, whose integrability is provided by the original chiral equations. In part II we briefly introduce the loop-space chiral equations and then discuss the integrability conditions of the non-local currents in two possible different situations. In the first case, the "generating" functions are functionals of the loop alone. We show that the integrability conditions are not satisfied and higher order conserved non-local currents do not exist. In the second case, the "generating" functions are functionals of the loop as well as a parameter the integrability conditions at a restricted point of the parameter are satisfied, however there is an infinite fold of arbitrariness.[9] It indicates that additional guiding principles are needed in addition to the original loop-space chiral equation in order to uniquely determine the infinite conserved non-local currents as functionals of the loop and the parameter.

## I. CHIRAL FIELDS IN TWO DIMENSIONS

The chiral fields are simply described by the following equations,

$$A_\mu(x) \equiv g^{-1}(x)\partial_\mu g(x), \qquad \text{curvatureless}, \qquad (1.1)$$

$$\partial_\mu A_\mu(x) = 0, \qquad \text{continuity-like}, \qquad (1.2)$$

which follows from a Lagrangian density $\mathscr{L} = \text{Tr}\partial_\mu g^{-1}(x)\partial_\mu g(x)$. Following the procedure of Brezin et al.,[6] we consider $A_\mu(x)$ being the first conserved current.

$$V_\mu^{(1)}(x) \equiv A_\mu(x) = \varepsilon_{\mu\alpha}\partial_\alpha\chi^{(1)}(x). \tag{1.3}$$

We shall see that $\chi^{(1)}(x)$ is needed to construct higher current. So we re-express eq. (1.3) as

$$\partial_\mu\chi^{(1)}(x) = -\varepsilon_{\mu\beta}A_\beta(x). \tag{1.3}'$$

Then integrability of $\chi^{(1)}(x)$ is

$$\partial_\nu\partial_\mu\chi^{(1)}(x) - \partial_\mu\partial_\nu\chi^{(1)}(x) = 0. \tag{1.4}$$

From eq. (1.3), we have

$$-\varepsilon_{\mu\beta}\partial_\nu A_\beta(x) + \varepsilon_{\nu\alpha}\partial_\mu A_\alpha(x) = 0, \tag{1.5}$$

for $\mu = \nu = 1$, $-\partial_1 A_2(x) + \partial_1 A_2(x) = 0,$ \tag{1.5a}

for $\mu = 1$, $\nu = 2$, $-\partial_2 A_2 - \partial_1 A_1 = 0.$ \tag{1.5b}

Equation (1.5a) is automatically true and eq. (1.5b) is just eq. (1.2). Therefore $\partial_\mu A_\mu = 0$ is the necessary and sufficient condition for the integrability of $\chi^{(1)}(x)$.

Now we construct the second current,

$$V_\mu^{(2)} \equiv \mathcal{D}_\mu\chi^{(1)}, \qquad \text{where } \mathcal{D}_\mu \equiv \partial_\mu + A_\mu(x). \tag{1.6}$$

Using eqs. (1.1) and (1.2), we can easily show $\partial_\mu V_\mu^{(2)} = 0$. Similarly we can construct higher currents

$$V_\mu^{(n)} = \varepsilon_{\mu\nu}\partial_\nu\chi^{(n)} = \mathcal{D}_\mu\chi^{(n-1)}. \tag{1.7}$$

Again eqs. (1.1) and (1.2) give $\partial_\mu V_\mu^{(n)} = 0$. We shall call these $\chi^{(n)}$'s the "generating" function.

Next we shall demonstrate the connection of these infinite conserved currents with the inverse-scattering (or the linear) equations. Multiplying eq. (1.7) by $L^n$ and sum over all n,

$$\sum_{n=1}^\infty \varepsilon_{\mu\nu}\partial_\nu L^n\chi^{(n)} = L\sum_{n=1}^\infty \mathcal{D}_\mu L^{n-1}\chi^{(n-1)}, \tag{1.8}$$

where L is an arbitrary constant, eq. (1.8) can be rewritten as, due to $\chi^{(0)} = 1$,

$$\varepsilon_{\mu\nu}\partial_\nu \sum_{n=0}^\infty L^n\chi^{(n)}(x) = L\mathcal{D}_\mu \sum_{n=0}^\infty L^n\chi^{(n)}(x).$$

Defining $\phi(x,L) \equiv \sum_{n=0}^\infty L^n\chi^{(n)}(x), \tag{1.9}$

we obtain $\varepsilon_{\mu\nu}\partial_\nu\phi = L\mathcal{D}_\mu\phi$, or $[\partial_\mu - A_\mu - L^{-1}\varepsilon_{\mu\nu}\partial_\nu]\phi(x,L) = 0,$ \tag{1.10} which is the inverse-scattering equations[6] for the chiral equations.

Now we discuss the integrability conditions of $\phi$. From eq. (1.10) it is easy to obtain

$$\partial_\mu \phi(x,L) = -\frac{1}{1+L^2}[A_\mu(x) + L\varepsilon_{\mu\alpha}A_\alpha]\phi(x,L). \quad (1.11)$$

Differentiating eq. (1.11) once more, and using eq. (1.11), it follows

$$\partial_\nu \partial_\mu \phi(x,L) = -\frac{1}{1+L^2}[\partial_\nu A_\mu(x) + L\varepsilon_{\mu\alpha}\partial_\nu A_\alpha(x)]\phi(x,L) +$$

$$\frac{1}{(1+L^2)^2}[A_\mu(x) + L\varepsilon_{\mu\alpha}A_\alpha(x)][A_\nu(x) + L\varepsilon_{\nu\beta}A_\beta(x)]\phi(x,L). \quad (1.12)$$

The integrability conditions of $\phi(x,L)$, $\partial_\mu \partial_\nu \phi(x,L) - \partial_\nu \partial_\mu \phi(x,L) = 0$, for arbitrary L, gives the $L^0$ term, $\partial_\nu A_\mu - \partial_\mu A_\nu + [A_\mu, A_\nu] = 0$; the $L^1$ term, $-\varepsilon_{\mu\alpha}\partial_\nu A_\alpha + \varepsilon_{\nu\beta}\partial_\mu A_\beta + [\varepsilon_{\mu\alpha}A_\alpha, A_\nu] + [A_\mu, \varepsilon_{\nu\beta}A_\beta] = 0$; the $L^2$ term, $-\partial_\nu A_\mu + \partial_\mu A_\nu + [\varepsilon_{\mu\alpha}A_\alpha, \varepsilon_{\nu\beta}A_\beta] = 0$; the $L^3$ term, $\varepsilon_{\mu\alpha}\partial_\nu A_\alpha - \varepsilon_{\nu\beta}\partial_\mu A_\alpha = 0$. They imply simply $\partial_\nu A_\mu - \partial_\mu A_\nu + [A_\mu, A_\nu] = 0$, $\partial_\mu A_\mu = 0$. Therefore we see that the original equations of motion provide the integrability of $\phi(x,L)$ and also provide the conditions for constructing infinite number of conserved currents.

## II. CHIRAL FIELDS IN LOOP SPACE

Let us consider the phase factor along the loop $\ell = x^\mu(s)$ as shown in Fig. 1a,

$$\Phi_{02s10} = \psi(\ell) = P \exp(i\oint A_\mu dx_\mu). \quad (2.1)$$

The functional differentiation of the loop phase factor is defined as the change in $\psi(\ell)$, as $\ell$ changes to $\ell'$, which is infinitesimally deformed from $\ell$ at s (Fig. 1b),

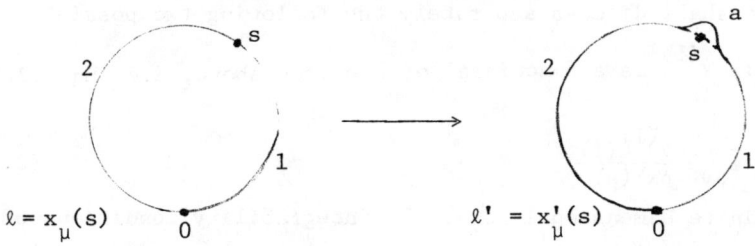

Figure 1a                                      Figure 1b

$$\frac{\delta\psi(\ell)}{\delta x^\mu(s)} = \frac{\psi(\ell')-\psi(\ell)}{dsdx^\mu(x)} = \Phi_{02s}f_{\lambda\mu}[x(s)]\frac{dx_\lambda(x)}{ds}\Phi_{s10}. \quad (2.2)$$

It is just the parallel transported "normal flux" per unit area that went through the little area a shown in Fig. 1b. Define the loop space gauge potential

$$\mathscr{F}_\mu(\ell,s) \equiv \psi(\ell)^{-1} \frac{\delta\psi(\ell)}{\delta x^\mu(s)} = \Phi_{01s} f_{\lambda\mu}[x(s)]\dot{x}_\lambda(s) \Phi_{s10}. \tag{2.3}$$

Functionally differentiating it again, and after some work, one can show[4,5]

$$\frac{\delta\mathscr{F}_\mu(\ell,s)}{\delta x^\nu(s')} - \frac{\delta\mathscr{F}_\nu(\ell,s')}{\delta x^\mu(s)} + [\mathscr{F}_\mu(\ell,s), \mathscr{F}_\nu(\ell,s')] = 0, \tag{2.4}$$

which just gives the projected Bianchi identity:

$$(\mathscr{D}_\mu f_{\nu\lambda}[x(s)])\dot{x}_\lambda + (\mathscr{D}_\nu f_{\lambda\mu}[x(s)])\dot{x}_\lambda + (\mathscr{D}_\lambda f_{\mu\nu}[x(s)])\dot{x}_\lambda = 0, \tag{2.4a}$$

and

$$\frac{\delta\mathscr{F}_\mu(\ell,s)}{\delta x_\mu(s)} = 0, \tag{2.5}$$

which gives the projected Yang-Mills equation

$$\mathscr{D}_\mu f_{\mu\nu}[x(s)]\dot{x}_\nu(s) = 0. \tag{2.5a}$$

The geometric meaning of the loop space equations (2.4), (2.5) is that the loop phase factor arrived from an initial loop to a given final loop is independent of the different volumes swapped out by the intermediate loops if the Bianchi identity and the sourceless Yang-Mills equations are satisfied.

Again eq. (2.5) is like a continuity equation so we try to follow the same procedure as for the chiral fields and identify the first current, here we specify in two dimension though the conclusion is general,

$$V_\mu^{(1)}(\ell,s) \equiv \mathscr{F}_\mu(\ell,s) = \varepsilon_{\mu\nu} \frac{\delta\chi^{(1)}}{\delta x^\nu(s)}. \tag{2.6}$$

This satisfies eq. (2.5), but the question is whether eq. (2.5) provides the sufficient conditions for the integration of $\chi$ from eq. (2.6). We shall discuss separately the following two possible cases.

Case (1): $\chi^{(1)}$ is a functional of the loop above, i.e. Eq. (2.6) reads

$$\mathscr{F}_\mu(\ell,s) = \varepsilon_{\mu\nu} \frac{\delta\chi^{(1)}(\ell)}{\delta x^\nu(s)}. \tag{2.6}'$$

Just as the finite dimensional case, the integrability conditions of $\chi^{(1)}(\ell)$ is

$$\frac{\delta\chi^{(1)}(\ell)}{\delta x^\nu(s')\delta x^\mu(s)} - \frac{\delta\chi^{(1)}(\ell)}{\delta x^\mu(s)\delta x^\nu(s')} = 0. \tag{2.7}$$

From $\frac{\delta\chi^{(1)}(\ell)}{\delta x^\nu(s)} = -\varepsilon_{\nu\mu}\mathscr{F}_\mu(\ell,s)$, eq. (2.7) gives

$$\varepsilon_{\mu\alpha} \frac{\delta\mathscr{F}_\alpha(\ell,s)}{\delta x^\nu(s')} = \varepsilon_{\nu\alpha} \frac{\delta\mathscr{F}_\alpha(\ell,s')}{\delta x^\mu(s)}, \tag{2.8}$$

for $\mu = \nu = 1$, $\dfrac{\delta\mathcal{F}_2(\ell,s)}{\delta x^1(s')} = \dfrac{\delta\mathcal{F}_2(\ell,s')}{\delta x^1(s)}$, (2.8a)

which is false, unless $s' \to s$; for $\mu = 1$, $\nu = 2$,

$$\dfrac{\delta\mathcal{F}_2(\ell,s)}{\delta x^2(s')} = -\dfrac{\delta\mathcal{F}_1(\ell,s')}{\delta x^1(s)}, \qquad (2.8b)$$

which become eq. (2.5) only in the limit $s' \to s$. Therefore we see that higher conserved currents cannot be constructed by this procedure. In ref. (7) we demonstrate this point by solving the 2-dimension Yang-Mills equations explicitly.

Since the higher conserved currents do not exist, we cannot construct the inverse-scattering equation following the procedure given in Section I; however, by analog to eq. (1.10), we can construct one,

$$\left[\dfrac{\delta}{\delta x^\mu(s)} + \mathcal{F}_\mu(\ell,s) - \gamma\varepsilon_{\mu\nu}\dfrac{\delta}{\delta x^\nu(s)}\right]\Phi(\ell,\gamma) = 0. \qquad (2.9)$$

What are the conditions for the integration of $\Phi(\ell,\gamma)$? Equation (2.9) can be rewritten as

$$\dfrac{\delta}{\delta x_\mu(s)}\Phi(\ell,\gamma) = -\dfrac{1}{1+\gamma^2}[\mathcal{F}_\mu(\ell,s) + \gamma\varepsilon_{\mu\alpha}\mathcal{F}_\alpha(\ell,s)]\Phi(\ell,\gamma). \qquad (2.9a)$$

Requiring, for arbitrary $\gamma$,

$$\dfrac{\delta^2\Phi(\ell,\gamma)}{\delta x^\mu(s)\delta x^\nu(s')} - \dfrac{\delta^2\Phi(\ell,\gamma)}{\delta x^\nu(s')\delta x^\mu(s)} = 0, \qquad (2.10)$$

one obtains the following conditions for arbitrary $s$, and $s'$

$\dfrac{1}{\gamma^3}$ term: $\varepsilon_{\mu\alpha}\dfrac{\delta\mathcal{F}_\alpha(\ell,s)}{\delta x^\nu(s')} - \varepsilon_{\nu\beta}\dfrac{\delta\mathcal{F}_\beta(\ell,s')}{\delta x^\mu(s)} = 0$ (2.10)

$\mu = 1$, $\nu = 1$, $\dfrac{\delta\mathcal{F}_2(\ell,s)}{\delta x^1(s')} - \dfrac{\delta\mathcal{F}_2(\ell,s')}{\delta x^1(s)} = 0$, (2.10a)

$\mu = 1$, $\nu = 2$, $\dfrac{\delta\mathcal{F}_2(\ell,s)}{\delta x^2(s')} + \dfrac{\delta\mathcal{F}_1(\ell,s')}{\delta x^1(s)} = 0$; (2.10b)

$\dfrac{1}{\gamma^2}$ term: $-\dfrac{\delta\mathcal{F}_\mu(\ell,s')}{\delta x^\nu(s)} + \dfrac{\delta\mathcal{F}_\nu(\ell,s)}{\delta x^\mu(s')} + [\varepsilon_{\mu\alpha}\mathcal{F}_\alpha(\ell,s'), \varepsilon_{\nu\beta}\mathcal{F}_\beta(\ell,s)] = 0$, (2.11)

$\dfrac{1}{\gamma}$ term: $-\varepsilon_{\mu\alpha}\dfrac{\delta\mathcal{F}_\alpha(\ell,s')}{\delta x^\nu(s)} + \varepsilon_{\nu\beta}\dfrac{\delta\mathcal{F}_\beta(\ell,s)}{\delta x^\mu(s')} +$

$[\varepsilon_{\mu\alpha}\mathcal{F}_\alpha(\ell,s'), \mathcal{F}_\nu(\ell,s)] + [\mathcal{F}_\mu(\ell,s'), \varepsilon_{\nu\beta}\mathcal{F}_\beta(\ell,s)] = 0$, (2.12)

$\gamma^0$ term: $\dfrac{\delta\mathcal{F}_\mu(\ell,s')}{\delta x^\nu(s)} - \dfrac{\delta\mathcal{F}_\nu(\ell,s)}{\delta x^\mu(s')} + [\mathcal{F}_\mu(\ell,s'), \mathcal{F}_\nu(\ell,s)] = 0$. (2.13)

We see that they require much more than the loop space chiral equations (2.4) and (2.5) for integrability.

**Case (2).** $\chi^{(1)}$ is not only a functional of the loop but also a parameter s. Now Eq. (2.6) becomes

$$\mathscr{F}_\mu(\ell,s) = \lim_{s'\to s} \varepsilon_{\mu\nu} \frac{\delta \chi^{(1)}(\ell,s)}{\delta x_\mu(s')} = \varepsilon_{\mu\nu} \frac{\delta \chi^{(1)}(\ell,s)}{\partial x_\mu(s)} \qquad (2.6)''$$

Thus the integrability condition of Eq. (2.6) becomes

$$\frac{\delta \chi^{(1)}(\ell,s)}{\delta x^\nu(s)\delta x^\mu(s)} - \frac{\delta \chi^{(1)}(\ell,s)}{\delta x^\mu(s)\delta x^\nu(s)} = 0. \qquad (2.7)'$$

Notice that all parameters coincide at a point at s. Then from $\frac{\delta \chi^{(1)}(\ell,s)}{\delta x^\nu(s)} = \varepsilon_{\nu\mu}\mathscr{F}_\mu(\ell,s)$ the integrability condition becomes Eq.(2.8) with $s'\to s$. Thus the equation of motion Eq. (2.5) does provide integrability of $\chi^{(1)}(\ell,s)$ from (2.6)'. However the peculiar situation here is that Eq. (2.6)' constraints $\chi(\ell,s)$ only when the parameter s' of $\delta x_\mu(s')$ coincide with s of $\chi(\ell,s)$, thus does not constraint $\chi(\ell,s)$ enough and there are infinite many $\chi(\ell,s)$'s that can satisfy Eq. (2.6)''. This is another manifestation that additional informations are needed in order to integrate the system from one point of the loop to the other uniquely.

Since $\chi^{(1)}(\ell,s)$ can be constructed, now we can follow the same procedure as Eq. (1.6) and (1.7) in the ordinary chiral field to construct the higher currents.

$$V_\mu^{(n)}(\ell,s) = \lim_{s'\to s} \left[ \frac{\delta}{\delta x_\mu(s')} + \mathscr{F}_\mu(\ell,s) \right] \chi^{(n-1)}(\ell,s).$$

Notice here the arbitrariness in $\chi^{(n-1)}(\ell,s)$ reflects directly in the next current $V_\mu^{(n)}(\ell,s)$. Similarly following the same procedure as in Eqs. (1.8) to (1.10), one obtains the linearized equation for $\Phi(\ell,s,L)$, which is in the form of Eq. (2.9) with $\gamma = L^{-1}$,

$$\lim_{s'\to s}\left[\frac{\delta}{\delta x^\mu(s')} + \mathscr{F}_\mu(\ell,s) - L^{-1}\varepsilon_{\mu\nu}\frac{\delta}{\delta x_\nu(s)}\right] \Phi(\ell,s,L) = 0. \qquad (2.9)'$$

The integrability conditions in this limit of $s'\to s$ are just Eqs. (2.10) to (2.13) with $s'\to s$, which imply eqs. (2.4) and (2.5), the equations of motion.

In conclusion, the above discussions indicate that the loop-space chiral equations are not a totally integrable system in the ordinary sense. The loop-space chiral equations do not provide enough information for the integration of loop space currents from one point of the loop to another in a unique way. However, in spite of such difficulties, the observation that the Yang-Mills equations give the loop-space chiral equations is such a beautiful one, with further insight it is bound to lead to new understanding of the gauge theories.

Acknowledgement: We would like to thank A. M. Polyakov and I. Ya. Aref&#234;va for their responsive correspondences after receiving our manuscript and the warm discussions afterwards. They obviously were aware of the points elaborated here though they did not spell them out in detail in their papers. One of us (LLCW) would like to thank Prof. L. D. Soloviev for inviting her to attend the International Workshop on High Energy Physics at Protvino (Serpukov) Sept. 22-28, which made such direct discussions possible.

## REFERENCES

*This talk was delivered by L.-L. Chau Wang.

1. S. Mandelstam, Phys. Rev. $\underline{175}$, 1580 (1968); C. N. Yang, Phys. Rev. Lett. $\underline{33}$, 445 (1974).
2. C.-H. Gu and C. N. Yang, Scientia Sinica $\underline{18}$, 483 (1975); T. T. Wu and C. N. Yang, Phys. Rev. $\underline{D12}$, 3840 (1975); C.-H. Gu, Fudan Jour. No. 2, 51 (1976); Phys. energ. fort et phys. nucl. 2, 98 (1978).
3. K. Wilson, Phys. Rev. $\underline{D10}$, 2445 (1979); A. Polyakov (unpublished). M. Creutz, Phys. Rev. $\underline{21D}$, 2308 (1980), and the references therein.
4. A. M. Polyakov, Phys. Lett. $\underline{B82}$, 247 (1979).
   A. M. Polyakov, Gauge Fields as Rings of Glue, Landau Institute preprint 1979 (to be published in Nuclear Physics)
5. Y. S. Wu, Physica Energiae Fortis et Physica Nuclearis $\underline{3}$, 382 (1979);
   I. Ya. Aref'eva, Lett. Math. Phys. $\underline{3}$, 241 (1979).
6. V. E. Zakharov and A. V. Mikhailov, Sov. Phys. JETP $\underline{47}$, 1017 (1978); M. Lüscher and K. Pohlmeyer, Nucl. Phys. $\underline{B137}$, 46 (1978);
   E. Brezin, C. Itzykson, J. Zinn-Justin and J. Zuber, Phys. Lett. $\underline{82B}$, 442 (1979);
   A. Ogielski, M. Prasad, A. Sinha and L.-L. Chau Wang, Phys. Lett. (1980).
   For a review see L.-L. Chau Wang, "Bäcklund Transformations, Conservation Laws and Linearization of the Self-duel Yang-Mills and Chiral Fields," Talk presented at the 1980 Guangzhou Particle Theoretical Physics Conference, January 1980.
7. Chaohao Gu and Ling-Lie Chau Wang, "On the Loop-space Formulation of Gauge Theories," Fudan University and Brookhaven National Laboratory, preprint BNL-28051 (1980).
8. Tohru Eguchi and Yataka Hosotani, "Integrability Condition in Loop Space," University of Chicago preprint EFI 80/27 (1980).
9. In reference 7 we did not discuss this case because we felt that the equations were not constraining enough. We thank L. Dolan, A.M. Polyakov and I. Ya. Aref&#233;va for stressing the importance of this alternative.

# GENERAL CLASSICAL SOLUTIONS OF THE EUCLIDEAN $CP^{n-1}$ MODEL*

A.M. Din
LAPP, Annecy-le-Vieux, France

W.J. Zakrzewski**
University of Durham, Durham, U.K.

## INTRODUCTION

It is generally believed that nonabelian gauge theories govern dynamics of elementary particles. This belief is supported by impressive successes of perturbative calculations performed in the regions of large momenta which is possible due to the property of the asymptotic freedom exhibited by these theories. However, the infrared properties of these theories are not well understood yet and the problem of proving or disproving their confining property remains one of the most important challenges at present. New hopes for progress in this area are associated by some in the existence of a nontrivial topological structure of these theories. Instantons, specific classical solutions to Euclidean equations of motion, have been used to estimate various nonperterbative properties of the theory and are the basis of the Princeton group approach[1] to the theory of hadrons. However, several technical complications necessitate approximations and prevent the results from being completely reliable. At the same time Jevicki and Levine[2] pointed out the relevance of the classical solutions in Minkowski space to the 1/N limit of the corresponding quantum theory.

Some of these questions can be studied in simple models. Recently such a two-dimensional field theoretical model was developed.[3] The model, being a generalization of the O(3) σ model, is described by a field which takes values in the n-dim. complex projective space ($CP^{n-1}$). It has several properties in common with non-abelian gauge theories - it is asymptotically free and conformally invariant at a classical level, it allows 1/N expansion (leading to dynamical generation of mass and a confining long range field), it has nontrivial topology which allows for the existence of nontrivial solutions to the classical equations of motion.[4] In this talk we discuss a general way of obtaining all finite action solutions to the classical euclidean equations of motion. The

---

* Presented at the XX International Conference on High Energy Physics, Madison, Wisconsin, 17-23 July 1980.

** Participating Guest, Lawrence Berkeley Laboratory, Berkeley, California 94720.

discussion is primarily based on two papers of ours.[5]

## $CP^{n-1}$ MODEL IN TWO-DIMENSIONAL EUCLIDEAN SPACE

It is convenient to parameterize the field of this model in terms of

$$z = \left[ z_1(x), z_2(x), \ldots z_n(x) \right] \quad (1)$$

where $x = (x_1, x_2)$, $\bar{z} \cdot z = 1$, and we consider only equivalence classes of 1. ie. $z' \sim e^{i\theta(x)} z$. The Lagrangian density is given by

where
$$\mathcal{L}(z) = \overline{D_\mu z} \cdot D_\mu z$$
$$D_\mu f = \partial_\mu f - (\bar{z} \cdot \partial_\mu z) f. \quad (2)$$

The model exhibits a global $U(n)$ and a local $U(1)$ invariance. We shall be interested in the solutions of the equations of motion of the model; i.e. in the solutions of

$$D^2 z_\alpha + \mathcal{L}(z) z_\alpha = 0. \quad (3)$$

A class of such solutions is provided by the instantons and anti-instantons. These are the solutions of

where
$$D_\pm z_\alpha = 0$$
$$x_\pm = x_1 \pm i x_2. \quad (4)$$

The finiteness of the action gives $z_i = \dfrac{p_i(x_+)}{|p(x_+)|}$ or $z_i = \dfrac{p_i(x_-)}{|p(x_-)|}$

where $p_i$ is a polynomial. We shall show that for $n > 2$ there are further finite action solutions of 3. The method used by us was adapted from Borchers and Garber,[6] and Barbosa[7] who obtained all solutions of the equations of motion of the $O(N)$ $\sigma$ model (which can be derived from the above model by restricting $z_i(x)$ to be real). The method considers a solution $z$ and two spaces $H_\ell$, and $H'_{\ell'}$ defined as

$$H_\ell = \{ D_- z, D_-^2 z, \ldots D_-^\ell z \}$$
$$H'_{\ell'} = \{ D_+ z, D_+^2 z, \ldots D_+^{\ell'} z \} \quad (5)$$
$$\ell + \ell' = n - 1$$

Then one shows that $z$, $H_\ell$, $H'_{\ell'}$ are mutually orthogonal. For the case of $CP'$ (ie. $n = 2$) the orthogonality of $z$, $D_+ z$, and $D_- z$ coupled with the linear dependence of these vectors implies that either $D_+z = 0$ or $D_-z = 0$ i.e. all finite energy solutions of equations of motion correspond to instantons or antiinstantons. Next step in the construction consists of proving the existence of an analytic vector $f$ ($\partial_- f = 0$), belonging to $\{z, H_\ell\}$ which satisfies $\bar{f} \cdot D_-^i z = \omega \delta^{i\ell}$ $i = 0, \ldots \ell$ where $\partial_+ \omega - (\bar{z}\partial_+ z)\omega = 0$. Thus one shows that $\{f, H_\ell\}$ can be spanned out by $\{f, \partial_+ f, \ldots \partial_+^\ell f\}$ and that $z$ can be expressed in terms of $f$ and its derivatives. The explicit expression is given by

$$z^{(\ell)} = \frac{\hat{z}^{(\ell)}}{|\hat{z}^{(\ell)}|}$$

$$\hat{z}^{(\ell)} = \partial_+^\ell f - \sum_{j,i=0}^{\ell=1} \partial_+^i f \, M_{ij}^{(\ell)-1} \, \partial_+ M_{j\ell-1}^{(\ell)}$$

$$M_{ij}^{(\ell)} = \overline{\partial_+^i f} \cdot \partial_+^j f \quad i,j = 0, \ldots \ell-1$$

(6)

Then it is a matter of algebra to show that for any rational $f(x_+)$, $z^{(\ell)}$ defined by 6 fulfills the equations of motion 3. As there exists a one-to-one correspondence between $e^{i\theta}z$ and $\lambda(x_+)f(x_+)$, where $\lambda(x_+)$ is an overall factor it is easy to see that for $\ell = 0$ $z^{(0)} = \frac{f}{|f|}$ giving instanton solutions and that for $\ell = n-1$ $D_+ z^{n-1} = 0$ thus showing that $z^{(n-1)}$ represents antiinstanton solutions. For $0 < \ell < n-1$, $D_+ z^{(\ell)} \neq 0 \neq D_- z^{(\ell)}$ and we have generally new solutions, which as we shall argue below, can be represented as mixtures of instantons and antiinstantons.

Let us briefly mention that there is another way of expressing the general solution.[6] It involves a lowering-like operator $P_+$ whose action is defined by

$$P_+ g = \partial_+ g - \frac{g(\bar{g} \cdot \partial_+ g)}{|g|^2} \quad . \tag{7}$$

It is easy to check that $\hat{z}^{(\ell)} = P_+ \hat{z}^{(\ell-1)} = P_+^\ell \hat{z}^{(0)}$. In an analogous way one can define a raising like operator $P_-$, and then show that $[P_+, P_-] \hat{z}^{(\ell)} \sim \hat{z}^{(\ell)}$. The algebraic nature of these operators is currently under consideration.

All solutions of the equations of motion have a finite action and a finite topological charge. Moreover both come as multiples of $2\pi$. In fact it is easy to show that

$$S_\ell = 2(I_{\ell+1} + I_\ell)$$
$$Q_\ell = 2(I_{\ell+1} - I_\ell) \tag{8}$$
$$I_\ell = \int d^2x \, \partial_+ \partial_- \log \det M^{(\ell)}$$

(for more details etc. see ref. (5)). It is also easy to show that all new solutions (i.e. for which $D_\pm z \neq 0$) are unstable under small fluctuations. For if we take a solution $z$ and consider a small orthogonal fluctuation $\phi$ ($\overline{z} \cdot \phi = 0$) resulting in a field configuration $z' = z\sqrt{1 - |\phi|^2} + \phi$ then it is easy to show that $S(z') < S(z)$ for $\phi \sim D_+ z$. The fluctuation operator has thus negative modes indicating that all new solutions are not minima of the action but only its saddle points. The appearance of negative modes prevent a straightforward calculation of the fluctuation operator determinant. We are currently trying to determine the number of negative and zero modes for each solution. Preliminary results indicate that these numbers are such as to justify an instanton-antiinstanton mixture interpretation of these new solutions.[8] We have also looked at the inclusion of fermions. We have been able to show that new solutions provide us with examples in which the index theorem is satisfied in a nontrivial way (i.e. there exists both helicity positive and helicity negative fermionic zero modes normalizable on a sphere). We are also looking at the generalization of the approach and its application to other models.

The general outlook is reasonably encouraging. Our technique for finding classical solutions may generalize to further models. In particular it may lead to further solutions of the $HP^{n-1}$ models (coupled to gravity) and may even shed some light on ways of finding solutions to nonabelian gauge theories. The new solutions of the $CP^{n-1}$ models may improve instanton gas calculations and provide a new insight into comparing them with the results of 1/N expansions. Moreover one should seriously look for classical solutions in Minkowski space and determine their relation to the 1/N expansion.

## REFERENCES

1. C. Callan, R. Dashen and D. Gross, Phys. Rev. Lett. **44**, 435 (1980) and references there in. D. Gross, R. Pisarski and L. Yaffe - Princeton preprint (1980).
2. A. Jevicki and H. Levine, Harvard preprint HUTP-80/A017 (1980), Brown preprint BROWN-HET-418 (1980).

3. H. Eichenherr, Nucl. Phys. B146, 215 (1978); E. Cremmer and J. Sherk, Phys. Lett. 74B, 341 (1978); V. Golo and A. Perelomov, Phys. Lett. '79B, 112 (1978).
4. A. D'Adda, P. di Vecchia and M. Lüscher, Nucl. Phys. B146, 63 (1978) and ibid B152, 125 (1978).
5. A.M. Din and W.J. Zakrzewski, Lapp preprints LAPP-TH-17 and TH 21 (1980).
6. M.J. Borchers and W.D. Garber - Comm. Math. Phys. 72, 77 (1980).
7. J. Barbosa, Trans. Am. Math. Soc. vol. 210, 75 (1975).
8. A.M. Din and W.J. Zakrzewski - in preparation.

# NEW THEORETICAL DEVELOPMENTS

P. Ramond, Caltech, Organizer

TERNARY ALGEBRAS AS THE BASIS OF
A DYNAMICAL THEORY OF SUBCONSTITUENTS [*,†]

Itzhak Bars [**]
Yale University, Physics Department
J.W. Gibbs Laboratory
New Haven, Connecticut 06520

The $SU(3) \times SU(2) \times U(1)$ gauge model works well at low energies ($10^{-15}$ - $10^{-16}$ cm). Because it contains large numbers of degrees of freedom and parameters, it is considered by many theorists as a good phenomenological model which should be derivable from a more complete and fundamental theory. Thus, we ask the question: What is the correct theoretical description of physics at much shorter distances than the present accelerator energies? Two possibilities have emerged (i) Grand unification schemes in which a large number of fields are taken as elementary (ii) Composite schemes of quarks, leptons, gauge bosons and higgs bosons whose goal is to make a viable theory in terms of few subconstituents.

Grand unification is based on the idea that the gauge principle which is successful at low energies can be extrapolated all the way to $10^{16}$ - $10^{19}$ GeV. Using symmetry schemes one could unify the many degrees of freedom within few representations of a unifying group. In a model based on the exceptional group $E_8$, recently proposed in collaboration with M. Günaydin[1], the maximum such unification has been achieved: Only one and the smallest possible representation for each spin are used, namely, the 248 adjoint representation for gauge bosons and fermions and the 3875 for higgs bosons. Note that the "successful" grand unification groups, including $E_6$, are all members of the E-series in the classification of Lie algebras[2] (i.e. $SU(5)$ is isomorphic to $E_4$, $SO(10)$ is isomorphic to $E_5$). The exceptional group $E_8$ is the last member of the series and has very special properties. The major prediction of the model, independent of detailed calculations, is that the next three $SU(5)$ families to be discovered below 1 TeV are of V+A type with respect to weak interactions. If $E_8$ or some other special scheme of this type proves to be successful we may believe that it is fundamental. Otherwise, in my opinion, grand unification remains as a useful but probably not fundamental approach.

[*] Research supported by DOE Contract No. EY-76-C-02-3075
[†] Talk delivered at (i) IXth International Colloq. on Group Theoretical Methods in Physics, Cocoyoc, Mexico, June 1980; (ii) XXth International Conference on High Energy Physics, Madison, Wisconsin, July 1980.
[**] Alfred P. Sloan Foundation Fellow

0094-243X/81/680942-06$1.50 Copyright 1981 American Institute of Physics

The idea of subconstituents is at its infancy. Several schemes have been proposed and the field is growing. The common factor in all the schemes is the goal of constructing a theory with few degrees of freedom. But what kind of dynamics should the subconstituents satisfy in order to produce the dynamics of a gauge theory (SU(3) x SU(2) x U(1)) at low energies, including the phenomenologically successful quarks, leptons, gauge bosons and higgs bosons as effective degrees of freedom? I will describe some ideas developed since 1978 in collaboration with M. Günaydin[3,4] which are based on ternary algebras[3,5].

In the ternary algebraic approach our basic idea is to try to give physical meaning to the mathematical fact that ternary algebras are building blocks of all Lie algebras and Lie superalgebras. Given that gauge theories are based on Lie algebras, it seems natural to explore ternary algebras as a basis for the dynamics of the fundamental subconstituents.

Ternary algebras close under triple products (abc). An example of a triple product is

$$(abc) = a.(\bar{b}.c) + c.(\bar{b}.a) - b.(\bar{a}.c)$$

where a,b,c belong to some vector space and the product (.) could be associative as well as nonassociative. The relation of the ternary algebra to the Lie algebra can be seen by a grading of the form

$$\begin{array}{ccccccc} \cdots & -1 & 0 & +1 & \cdots \\ \cdots & \tilde{U}_b & S_{ab} & U_a & \cdots \end{array}$$

where the Lie algebra generators $U_a$, $S_{ab}$, $\tilde{U}_b$ (which are multiplied by infinitesimal parameters) obey the commutation rules

$$[U_a, \tilde{U}_b] = S_{ab}$$

$$[S_{ab}, U_c] = U_{(abc)}$$

$$[S_{ab}, S_{cd}] = S_{(abc)d} - S_{c(bad)}$$
$$\vdots$$

This is explained fully in refs. (3,5).

There is a one-to-one correspondence between ternary algebras and (symmetric) coset spaces. Thus any physical application based on ternary algebras can be reproduced via coset spaces and vice versa. But ternary algebras provide an unusual way of looking at coset spaces and suggest mathematical structures useful for physical applications which are not available with the usual methods. For example a ternary algebra formed by the direct product of n quarternions $a = H_1 \otimes H_2 \otimes \ldots \otimes H_n$ leads to $SO(3 \times 2^n)$ for n = even and $sp(3 \times 2^n)$ for n = odd. Thus, it can be related to the coset spaces $SO(3 \times 2^n)/SO(2^n) \times SO(2^{n+1})$ and $Sp(3 \times 2^n)/Sp(2^n) \times Sp(2^{n+1})$. From the usual coset space methods the above quaternionic structure of this space is not at all obvious while in contrast, it is an explicit input in the ternary algebraic method. Similarly, octonionic ternary algebras associated with the coset spaces $E_6/SO(10) \times U(1)$, $E_7/SU(8)$, $E_8/SO(16)$, etc., provide the octonionic properties as an input. These structures do not emerge in standard coset space approaches. Vast classes of associative (non-octonionic) ternary algebras with such properties can be found in refs. (3,5). You may construct your own brand new ternary algebra, since no complete classification is available. Thus, ternary algebras provide a really novel tool for physical applications.

In our approach the fundamental fields are associated with the elements of the ternary algebra. We have called them ternons. Gauge bosons, quarks, leptons, etc., are taken as composites of ternons. As described below, at this stage it appears quite likely, as shown in ref. (4), that the dynamics of a low energy gauge theory will emerge from a ternon theory in four dimensions. Furthermore, a convergence of ideas seems to be developing between the ternary algebra point of view and Harari's rishon scheme[6], if one uses the ternary algebra of a complex octonion[7] from which the SO(10) and $E_6$ groups can be constructed. Also, the recent ideas of Ellis, Gaillard and Zumino[8] which developed from SO(8) supergravity[9] are in agreement with our approach since their SU(8) composite gauge potential, just like our gauge potentials, are constructed from a coset space. Their coset space $E_7/SU(8)$ is associated with the ternary algebra formed by the direct product of two octonions $O_1 \otimes O_2$, where the first octonion is purely imaginary while the second one contains only the five directions ($e_0$, $e_7$, $e_4$, $e_5$, $e_6$). Both of the ternary algebras mentioned above leading to $E_6$ (Rishons) and $E_7$ (supergravity) are special cases of the ternary algebra of two arbitrary octonions[3,5] $O_1 \otimes O_2$ which leads to $E_8$.

The general procedure for constructing a composite gauge potential from any ternary algebra was given in ref. (4), where supersymmetric schemes including composite fermions were also described. The simplest case of a rectangular M x N complex ternon field $\phi(x)$, including the quantum theory, has been investigated more thoroughly in collaboration with M. Günaydin. In this case our approach is related to the Grassmannian generalization of the $CP_N$ type[10] composite gauge potential, except that in 4 dimensions we consider a more general model[11].

The U(M) gauge potential is given by $A_\mu(\phi) = iW^\dagger \partial_\mu W$, where the (M+N) × M matrix W is constructed from the N × M matrix $\phi(x)$ as follows

$$W(\phi) = \begin{pmatrix} \phi(1 + \phi^\dagger \phi)^{-1/2} \\ (1 + \phi^\dagger \phi)^{-1/2} \end{pmatrix}$$

and it satisfies automatically $W^\dagger W = 1_M$. If $\phi(x)$ is subjected to a <u>global</u> U(M+N) transformation

$$\phi \to \phi' = (\alpha\phi + \beta)(\gamma\phi + \delta)^{-1} \quad ; \quad U_{M+N} = \begin{pmatrix} \alpha & \beta \\ \gamma & \delta \end{pmatrix} \quad ,$$

then $A_\mu(\phi)$ transforms like a <u>local</u> U(M) gauge field

$$A_\mu(\phi') = U_M^\dagger(x)(A_\mu(\phi) + i\partial_\mu)U_M(x)$$

where $U_M(x)$ depends on $\alpha,\beta,\gamma,\delta$ as well as $\phi(x)$.

Recall that $\phi(x)$ belongs to the coset space U(N+M)/U(N) × U(M). The above construction of the gauge field A reminds us of the gauge formulation of gravity, where the elementary field, the vierbein, belongs to the coset space of the (Poincaré group)/(Lorentz group), and the connection $\omega_\mu^{ab}$ is a composite gauge potential for the Lorentz subgroup.

The ternon dynamics will be described by the Lagrangians

$$L_4 = -\frac{1}{4g^2} \text{Tr}[F_{\mu\nu}(A(\phi))]^2$$

$$L_2 = \frac{1}{2\lambda^2} \text{Tr}[(D_\mu W)^\dagger (D_\mu W)]$$

in 4 and 2 dimensions respectively. It is interesting to note that according to a theorem due to Narasimhan and Ramanan[12] all the information of a <u>classical</u> U(M) gauge field is contained in the construction above, provided N is larger than a certain number. Our aim is, of course, to show that the <u>quantum</u> ternon theory leads to the dynamics of a quantum gauge theory at low energies <u>only</u>. Thus, we will not be bound, a priori, by the classical limits on the minimum dimension of $\phi(x)$.

The composite gauge potential has the same cubic and quartic couplings of a gauge field. If its propagator behaves at low energies like the propagator of an elementary gauge field then the low energy physics will reduce to a gauge theory. This is already known to occur in the $CP_N$ model[10] and in its Grassmannian generalizations[4,13]. To investigate this problem we need to calculate the generating function in the quantum theory of ternons

$$Z(J) = \int d\mu(\phi) \, e^{-S(\phi)} + \int Tr(A_\mu(\phi)J^\mu)$$

The measure $d\mu(\phi)$ is given in ref. (4). The path integral can be manipulated by introducing auxiliary fields and integrating out the original ternon fields $\phi(x)$. Then, we arrive at a new form[4b]

$$Z(J) = \int [dA_\mu] \, e^{-\int \frac{1}{4g^2} Tr(F_{\mu\nu}(A))^2 \, + \, \text{corrections}} + \int A \cdot J$$

where the path integral is done over an <u>elementary</u> gauge field. Thus, we learn that the effective action is the Yang-Mills action plus corrections which are a function of only the covariant derivative $D_\mu$. These corrections can depend only on $F_{\mu\nu}$ and its derivatives as well as possible non-local terms associated with $(D_\mu)^{-2}$ etc. Thus, provided we can show that the corrections are absent or neglible at low energies we will obtain a low energy gauge theory. This point remains to be proven. The large N limit could be used as a tool to investigate this problem as discussed in ref. (4b).

Fermions can be included easily in the theory of ternons. A supersymmetric scheme based on superfield techniques was included in our original paper[4], but supersymmetry is not a requirement for the inclusion of fermions. More intricate supersymmetric models can probably be constructed via superternary algebras[3].

We started out with two goals: (i) To obtain the dynamics of a gauge theory at low energies from the dynamics of ternons (ii) To arrive at the correct phenomenological variables, namely, quarks, leptons, etc., as composites of ternons. We have shown that the first goal is likely to be achieved in our framework and we are now ready to investigate the second one. We are encouraged that other attempts such as the Rishon scheme[6] and the supergravity scheme based on $E_7/Su(8)$[8] are compatible with our ternary algebra approach as already mentioned above. Given the rich structures provided by ternary algebras we have no doubt that the model building stage will be rewarding.

References

(1) I. Bars and M. Günaydin, Yale preprint YTP80-09, to be published in Phys. Rev. Lett. and manuscript in preparation.

(2) This remark is due to F. Gürsey.

(3) I. Bars and M. Günaydin, J. of Math. Phys. 20, 1977 (1979) and lectures by I. Bars and M. Günaydin in the Proc. of the 8th Int. Colloq. on Group Theor. Methods in Physics March 1979, Kiriat Anavim, Israel.

(4a) I. Bars and M. Günaydin, Yale preprint YTP79-05, to be published in the Phys. Rev. D.

(4b) I. Bars and M. Günaydin, Yale preprint YTP80-14, to be published Phys. Lett.

(5) J. L. Kantor, Sov. Math. Dokl. 44, 254 (1973) and Trudy Sem. Vector Anal. 16, 407 (1972) (Russian); B. N. Allison, Am. J. Math 98, 285 (1976) and Trans. Am. Math. Soc. 114, 75 (1976).

(6) H. Harari, Phys. Lett. 86B, 83 (1979).

(7) This observation followed from a conversation with H. Harari. This ternary algebra is described by Kantor in ref. (5) and also in ref. (3).

(8) J. Ellis, M. K. Gaillard, B. Zumino, CERN preprint, April 1980.

(9) E. Cremmer and B. Julia, Phys. Lett. 80B, 48 (1978).

(10) A. D'Adda, M. Luscher, P. DiVecchia, Nucl. Phys. B146, 63 (1978).

(11) This model was independently constructed for different physical applications by a number of authors:
C.K. Chao, T.T. Sheng, Y.T. Nan, Scientiae Sinica 22, 34 (1979);
A.P. Balachandran, A. Stern, G. Trahern, Syracuse preprint SU-4213-126 (1978);
I. Bars in ref. (3); F. Gürsey and C. Tze, Yale preprint, Aug. (1979);
L. Lukierski, Summer Inst. Elem. Part. Phys., Kaiserlautern, Germany, Aug. 1979.

(12) M. S. Narasimhan and S. Ramanan, Am. J. Math. 83, 563 (1961) and 85, 223 (1963).

(13) E. Brezin, S. Hikami and J. Zinn-Justin, Saclay preprint 1980.

(14) J. Lukierski and B. Milewski, Phys. Lett. B93, 91 (1980).

# COMPOSITE μ AND τ FAMILIES IN AN $E_6$ UNIFIED MODEL

Gordon L. Shaw[*]
University of California, Irvine, Ca. 92717

In this talk, I summarize a recent unified model developed[1] with R. Slansky based on the exceptional group $E_6$ in which the electron family (e, $\nu_e$, u and d) is elementary and the μ and τ families are composite. The model predicts a rich (including charge 2 leptons and color 8 quarks) composite fermion spectrum (an $SO_{10}$ 144) not far above the τ family. The presence of these composite fermions is crucial for the self-consistency of the dynamics of the model. Composite scalars occur and these may play a role in the gauge hierarchy problem.[2] Features of the symmetry breaking are discussed using the Dynkin representation group theoretic formalism[3] which has quite general interest.

As we have heard in this Conference, unification presents problems of substantial difficulty: 1) The large number of "fundamental" fermions has not as yet been understood. The six quark flavors (including the expected t quark) together with the e, μ and τ leptons and their associated neutrinos now fill three generations of (left-handed) $SU_5$ families[4] ($\bar{5}$ + 10). In the electron family $f_e$, the $e_L^-$, $\nu_{eL}$ and $\bar{d}_L$ are assigned to the $\bar{5}$ and the $e_L^+$, $u_L$, $\bar{u}_L$ and $d_L$ to the 10. This pattern repeated for the muon family $f_\mu$ and $f_\tau$ does not appear very orderly. 2) The heirarchy problem assiciated with the enormous mass ratio $\sim 10^{13}$ between the unification mass $\sim 10^{15}$ GeV and the weak vector boson mass $\sim 10^2$ GeV is still not understood. 3) The huge number of Higgs scalars and arbitrary associated couplings suggests that perhaps these are not all fundamental.

Included in possible solutions to these problems have been models which a) embed $SU_5$ together with a local family group into a yet larger unified group,[5] b) introduce hypercolor[6] to produce bound scalars and intermediate mass scales, c) obtain all the known fermions, gauge bosons and Higgs scalars as composites of more elementary constituents.[7] Since the electroweak and color gauge theories appear so elegant, we do not wish to consider them as "effective" low-energy interactions as presumbly they would be in these latter composite models. Our unified $E_6$ model does combine some of the features of the above approaches and has some fermions (and all the gauge bosons), such as those in one of the families, as elementary and the remaining families (as well as some scalars) are tightly bound composites.

We start in the usual manner with a simple Lie group G' that unifies the electroweak forces from $SU_2^W \times U_1^W$ and QCD based on $SU_3^C$, and introduce a new binding force responsible for the deeply bound composite mediated by a single vector boson coupled to a current derived from our new (broken) symmetry $U_1^t$. We unify G' × $U_1^t$ into a simple group G. Requiring the fermion assignments, including the

---

[*] Supported in part by the National Science Foundation.

composite ones, to be a) vectorlike under color and electric charge to conserve parity in the strong and electromagnetic interactions and b) flavor chiral, we were led to the familiar $G' = SO_{10}$ which contains the electron family in the $\underline{16}$ dimensional spinor representation. The $\underline{16}$ branches into the $\underline{1} + \underline{5} + \underline{10}$ of $SU_5$ where the $\underline{1}$ is a left-handed antineutrino (there have been a number of suggestions[8] on how to dispose of the left-handed antineutrino and provide the left-handed neutrino with an acceptably small mass). The new binding charge and $G'$ are then unified in $E_6$:

$$E_6 \supset SO_{10} \times U_1^t \quad . \tag{1}$$

The <u>fundamental</u> fields in this $E_6$ theory include the 1) $\underline{78}$ gauge vector bosons transforming as the adjoint of $E_6$, 2) $\underline{27}$ of spin 1/2 left-handed fermions (and the CPT-conjugated $\overline{\underline{27}}$ of right handed fermions) where the $\underline{27}$ is complex and the lowest-dimensional nontrivial representation of $E_6$, 3) various scalars including an elementary $\underline{78}$ of scalars that can form bound states with the elementary $\underline{27}$ of fermions. The branching rules for $E_6$ irreps into those of $SO_{10} \times U_1^t$ (with the $Q^t$ charge determined up to an overall normalization) are easily computed. For the $\underline{27}$ and $\underline{78}$ we have

$$\underline{27} = \underline{1} \ (4) + \underline{10} \ (-2) + \underline{16} \ (1)$$
$$\underline{78} = \underline{1} \ (0) + \underline{45} \ (0) + \underline{16} \ (-3) + \overline{\underline{16}} \ (3) \tag{2}$$

where the $Q^t$ charges are given in brackets. We assume that neither the $\underline{27}$ nor the $\underline{78}$ of scalars get superheavy masses (these scalars do not vanish by the Higgs mechanism), but are heavy enough and the $U_1^t$ binding strong enough that the composites appear pointlike to all probes made so far, such as in lepton-pair production or g-2 experiments. Furthermore the vector boson mediating the $U_1^t$ force must be very heavy (perhaps $\geqslant 10$ TeV) and the coupling constant $\alpha_t(Q^2)$ must be very large at smaller $Q^2$. Naively, this does not appear possible since $U_1^t$ is supposed to be unified in $E_6$ at large $Q^2$ and the elementary particles in our $E_6$ Lagrangian suggest that the theory is asymptotically free. However, as we note below, their are enough tightly bound fermions predicted in the model to <u>temporarily</u> drive all the couplings $\alpha_i$ (including the electroweak and color forces) <u>strong</u>.

We assume that the fermion-scalar combinations in (2) with the largest negative product of $Q^t$ charges bind deeply as shown in Table I. The spectrum of left-handed fermions classified by $SO_{10}$ representations is

$$f_L = (\underline{1} + \underline{10} + \underline{16}) + (\underline{16}) + (\underline{16} + \underline{144}) \tag{3}$$

The first set in (3) consists of the elementary fermions in the $\underline{27}$, the second set arises from the binding of the $SO_{10}$ $\underline{1}$ fermion to the $\underline{16}$ of scalars and is the most tightly bound composite; the third set arises from the binding of the $\underline{10}$ of fermions to the $\underline{16}$ of scalars.

TABLE I.  $\underline{27} \times \underline{78}$ fermion-scalar bound state patterns.

| Scalar ($Q_1^t$) | Fermion ($Q_2^t$) | $SO_{10}$ bound states | $Q_1^t \cdot Q_2^t$ |
|---|---|---|---|
| $\underline{16}$ (-3) | $\underline{1}$ (4) | $\underline{16}$ | -12 |
| $\overline{\underline{16}}$ (3) | $\underline{10}$ (-2) | $\underline{16} + \underline{144}$ | -6 |
| $\underline{16}$ (-3) | $\underline{16}$ (1) | $\underline{10} + \underline{120} + \underline{126}$ | -3 |

A fourth set omitted from (3) would be due to the binding of the $\underline{16}$ and $\overline{\underline{16}}$ which gives a $\underline{10} + \underline{120} + \underline{126}$. We emphasize that the composite fermions in (3) are not bound states involving the elementary fermion $\underline{16}$, whereas this fourth set is. This last set is the least strongly bound and we shall neglect it.

A very nice feature of (3) is the occurrence of three families of $\underline{16}$'s. The electron and muon families can be assigned to the first two $\underline{16}$'s in (3); without further analysis, it is not possible to decide which is elementary. The $\tau$ family should either be in the third $\underline{16}$ or in the $\underline{144}$ which also contains a $\overline{\underline{5}} + \underline{10}$ of $SU_5$. The $\underline{16}$ and $\underline{144}$ are bound with the same overall binding strength, so that a priori it is not possible to select which one has the lowest-mass states. What is significant is the prediction of many more fermions not too far above the $\tau$ and $b$ masses, some having "exotic" features: the $\underline{144}$ contains quarks in color $\underline{6}$'s and $\underline{8}$'s and leptons with charge 2 as we see in Table II.  (Recall the charge $Q^{em} = I_3^W + Y^W/2$.)

These predicted $\underline{144}$ of deeply bound fermions are crucial to our unified model since they provide the self-consistency for $\alpha_t$ to be strong enough to bind and yet the electroweak and color forces are perturbative at $Q^2 \sim 10^2$ GeV$^2$. The scenario is as follows. If the elementary $\underline{78}$ of vector bosons, $\underline{78}$ of scalars and $\underline{27}$ of fermions are the only contributions to the one-loop approximation to the running coupling constant equations, then the $E_6$ theory is asymptotically free. However, if the composite fermions in (3) are tightly bound enough so that they also contribute to the one-loop approximation over, e.g., the range of 100 GeV to 10 TeV, they can quickly destroy the asymptotic freedom[9] of QCD and push $\alpha_c(Q^2)$ into a strong coupling regime. We readily estimate (see Eq. (4) in Ref. (1)) that the $\underline{144}$ can drive $\alpha_c(Q^2) = .1$ at $(Q^2)^{1/2} = 100$ GeV to $\alpha_c(\bar{Q}^2) \gg 1$ at $(\bar{Q}^2)^{1/2} \gtrsim 5$ TeV. As seen in Fig. 1, we also speculate that at some larger $Q^2$ (at energies above which the bound-state fermions are no longer point like) the $\alpha_i(Q^2)$ all decrease again. Note that due to the nonperturbative region in Fig. 1 we can no longer calculate[10] quantitatively $\sin^2\theta_w$ or the proton lifetime (the leptoquark vector bosons in the $SU_5$ subgroup of $E_6$ will mediate proton decay as usual and thus must still be given a superheavy mass).

In addition to the composite fermions, composite scalars are predicted in our model. These composite scalars might play a role in

TABLE II.  Branching Rules

$$SO_{10} \supset SU_5$$

$\underline{10} = \underline{5} + \underline{\overline{5}}$

$\underline{16} = \underline{1} + \underline{\overline{5}} + \underline{10}$

$\underline{45} = \underline{1} + \underline{10} + \underline{\overline{10}} + \underline{24}$

$\underline{120} = \underline{5} + \underline{\overline{5}} + \underline{10} + \underline{\overline{10}} + \underline{45} + \underline{\overline{45}}$

$\underline{126} = \underline{1} + \underline{\overline{5}} + \underline{10} + \underline{\overline{15}} + \underline{45} + \underline{50}$

$\underline{\overline{144}} = \underline{5} + \underline{\overline{5}} + \underline{10} + \underline{\overline{15}} + \underline{24} + \underline{40} + \underline{\overline{45}}$

$$SU_5 \supset SU_2^W \times SU_3^c \times U_1 \qquad U_1 \text{ Generated by } 3Y^W$$

$\underline{5} = (\underline{2},\underline{1})(3) + (\underline{1},\underline{3})(-2)$

$\underline{10} = (\underline{1},\underline{1})(6) + (\underline{1},\underline{\overline{3}})(-4) + (\underline{2},\underline{3})(1)$

$\underline{15} = (\underline{3},\underline{1})(6) + (\underline{2},\underline{3})(1) + (\underline{1},\underline{6})(-4)$

$\underline{24} = (\underline{1},\underline{1})(0) + (\underline{3},\underline{1})(0) + (\underline{1},\underline{8})(0) + (\underline{2},\underline{3})(-5) + (\underline{2},\underline{\overline{3}})(5)$

$\underline{40} = (\underline{2},\underline{1})(-9) + (\underline{1},\underline{\overline{3}})(-4) + (\underline{2},\underline{3})(1) + (\underline{1},\underline{8})(6) + (\underline{3},\underline{\overline{3}})(-4) + (\underline{2},\underline{\overline{6}})(1)$

$\underline{45} = (\underline{2},\underline{1})(3) + (\underline{1},\underline{3})(-2) + (\underline{1},\underline{\overline{3}})(8) + (\underline{2},\underline{\overline{3}})(-7) + (\underline{1},\underline{6})(-2) + (\underline{3},\underline{3})(-2) + (\underline{2},\underline{8})(3)$

$\underline{50} = (\underline{1},\underline{1})(-12) + (\underline{1},\underline{3})(-2) + (\underline{2},\underline{\overline{3}})(-7) + (\underline{1},\underline{6})(8) + (\underline{2},\underline{8})(3) + (\underline{3},\underline{\overline{6}})(-2)$

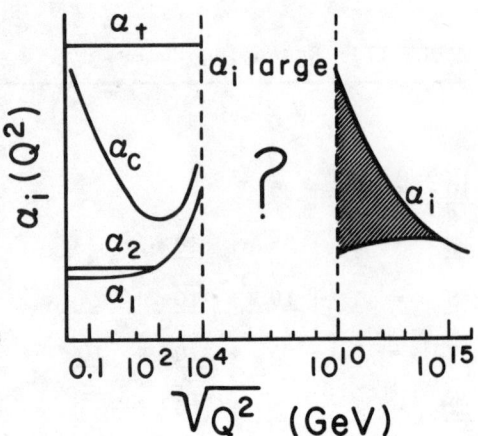

FIG. 1. Speculated behavior of coupling constants $\alpha_i$ versus energy in our $E_6$ model. $\alpha_t$ is the coupling of the new tight binding force resulting from the $U_1^t$ in the decomposition $E_6 \supset SO_{10} \times U_1^t$. The steep rise at 100 GeV is due mainly to the contribution of the composite 144 in (3) to the renormalization of the running coupling constants. At energies large compared to the binding of these states, the $\alpha_i(Q^2)$ decrease.

understanding the gauge hierarchy problem. Suppose the super strong breaking is due to an explicit Higgs mechanism. At that large $Q^2$, the scalars that do the weak breaking need not exist. At lower $Q^2$ where $\alpha_t(Q^2)$ is large, these composite scalars are formed and may be available, to do the weak breaking.

We emphasize a general remark about <u>any</u> model of composite leptons. The extraordinarily accurate measurements of g-2 for the electron and muon pose, at present, the strongest restriction on the constituent masses. In, particular, there is agreement[11] for the muon g-2 between experiment and the "usual" theory (QED + hadronic + weak) to a part in $10^8$. The corrections to g-2 due to the composite structure can be estimated[12] (in a renormalizable theory) by using a sidewise dispersion relation. These corrections come from energies above the bound-state threshold $m_c$ and are of order $m/m_c$ where m is the composite lepton mass. The muon measurements would then require $m_c \gtrsim 10^7$ GeV. In certain cases, smaller, constituent masses can be used. The g-2 corrections from perturbation theory are estimated to be of order $mm_F/m_B^2$ for $m_B \gg m_F (> m)$, here $m_F$ and $m_B$ are the masses of the constituent fermion and boson.

We note that some general group-theoretical features are discussed using Dynkin's representation theory. In particular, the weight analysis of the irreps and the discussion of the symmetry breaking are presented in this powerful, yet simple language. These results are of quite general interest. In certain cases, the vector-boson mass matrix of a Yang-Mills gauge theory with local symmetry G can be obtained in a very simple way without constructing an explicit Higgs potential. If c defines the symmetry breaking direction in weight space, the mass eigenvalue of the vector boson associated with weight $\alpha$ is related to the scalar product $(c,\alpha)$. Suppose the spin-zero fields (with nonvanishing vacuum expectation value) $\phi(\underline{r},\lambda)$ transform as the $\underline{r}$ irrep of G where $\lambda$ denotes the weight. Then this result holds either for a) $\underline{r}$ = adjoint irrep and $\lambda = 0$, or b) $\lambda$ is an "extreme" weight defined so that if $\lambda - \alpha$ is a weight in $\underline{r}$, $\lambda + \alpha$ is not (note that all the weights of the $\underline{27}$ of $E_6$ are extreme as are the non-zero weights of the $\underline{78}$). Constraints, e.g. that $SU_3^c$ is not broken, are easily imposed. A brief summary of the Dynkin formalism and these results is given in the Appendices of Ref. 1. For an in-depth discourse see the review by Slansky.[2]

Finally we mention that the crucial problems of calculating very deep bound states from very heavy constituents, and the breaking of the chiral symmetry of our theory remain, of course, to be investigated.

## REFERENCES

1. G. L. Shaw and R. Slansky, Phys. Rev. D, in press.
2. E. Gildener, Phys. Rev. D$\underline{14}$, 1667 (1976); S. Weinberg, Phys. Lett. $\underline{82B}$, 387 (1979).
3. E. B. Dynkin, Amer. Math. Soc. Trans. Ser. 2 $\underline{6}$, 111 (1957); $\underline{6}$ 245 (1957); W. Mc Kay, J. Patera and D. Sankoff, Computers in Non Associative Rings and Algebras ed. by R. Beck and B. Kolman (Academic, N.Y. 1977) p. 235; R. Slansky, Phys. Rep. C (to be published).
4. H. Georgi and S. L. Glashow, Phys. Rev. Lett. $\underline{32}$, 438 (1974).
5. H. Georgi, Nuc. Phys. B$\underline{156}$, 126 (1979); P. Frampton and S. Nandi, Phys. Rev. Lett. $\underline{43}$, 1460 (1979); M. Gell-Mann, P. Ramond and R. Slansky, Supergravity, ed. P. van Nieuwenhuizen and D. Z. Freedman (North-Holland, Amsterdam, 1979) p. 315.
6. S. Weinberg, Phys. Rev. D$\underline{13}$, 974 (1976); L. Susskind, ibid. $\underline{20}$, 2619 (1979); G't Hooft, Cargese lectures 1979 (unpublished).
7. There are many papers on composite fermions. See, e.g., footnote 4 of Ref. 1. Also see, J. Ellis, M. K. Gaillard, L. Maiani and B. Zumino, CERN Th-2841, 1980.
8. M. Gell-Mann, P. Ramond and R. Slansky, Ref. 5; E. Witten, Phys. Lett. $\underline{91B}$, 81 (1980).
9. D. Gross and F. Wilczek, Phys. Rev. Lett. $\underline{30}$, 1343 (1973); H. D. Politzer, ibid. $\underline{30}$, 1346 (1973).
10. H. Georgi, H. Quinn and S. Weinberg, Phys. Rev. Lett. $\underline{33}$, 451 (1974).
11. J. Bailey et al., Nuc. Phys. B$\underline{150}$, 1 (1979).
12. G. L. Shaw, D. Silverman and R. Slansky, Phys. Lett $\underline{94B}$, 57 (1980); S. J. Brodsky and S. D. Drell, SLAC-2534 (1980).

# INFRARED PROPERTIES OF THE COUPLING CONSTANT IN NON-ABELIAN GAUGE THEORIES

James S. Ball
University of Utah, Salt Lake City, Utah 84112

This work was done in collaboration with F. Zachariasen based on earlier work with with P. Lucht, University of Utah and R. Anishetty, M. Baker, and S. Kim at University of Washington. This non-perturbative investigation is based on using the Dyson eq. together with the Ward identity to produce an integral equation for the gluon propagator. This approach is most easily illustrated in the case of scalar QED. The Dyson eq. for the electron (spinless) propagator is

$$S^{-1}(P) = S_0^{-1}(P) + e^2 \int d^4k \, \Gamma_\mu^0 \, D_{\mu\nu}(k) \, S(P+k) \, \Gamma_\nu$$

where the bare quantities $\Gamma_\mu^0$ and $S_0^{-1}$ as well as the photon propagator $D_{\mu\nu}$ are assumed known. The Ward identity

$$k_\mu \Gamma_\mu = S^{-1}(P+k) - S^{-1}(k)$$

can now be "solved" to obtain the longitudinal vertex as follows:

$$\Gamma_\mu(\ell) = \frac{S^{-1}(P') - S^{-1}(P)}{P'^2 - P^2} (P+P')_\mu \ .$$

If the longitudinal vertex is a good approximation for the full vertex in the IR region (which is the case in QED) we obtain a closed integral eq. for $S(P)$ which has the solution

$$S(P) \sim \left[\frac{1}{P^2 - m^2}\right]^{1+\frac{\alpha}{\pi}},$$

which is the correct IR result for QED.

To maximize the similarity to QED, we choose to work with QCD in the axial gauge. In this gauge the Ward identity is the simple generalization of QED and no ghosts are present. The disadvantages of this gauge are as follows: 1.) Because of the gauge four-vector n the tensor forms are more complicated; 2.) A second scalar variable $1/\gamma = (n \cdot q)^2/n^2 q^2$ exists; and 3.) The treatment of $n \cdot k$ denominators in loop integrations is difficult.

For the case of pure glue QCD the Dyson eq. is represented by the following diagrams:

This becomes a closed integral eq. with our basic assumption that the full vertex is well approximated in the IR by the longitudinal vertex which is determined by the Ward identity. In addition to the basic assumption, several further assumptions are used to simplify the eq. We assume $\Delta_{\mu\nu} = Z(q^2) \Delta°_{\mu\nu}$ where $\Delta°_{\mu\nu}$ is the bare propagator. The scalar function is assumed to be independent of $\gamma$, which is true in UV and should be true in IR if our results have anything to do with confinement. We can now obtain the following closed integral eq. for Z without any 2 loop terms by forming $n \cdot \Pi \cdot n$:

$$-q^2 (1 - \frac{1}{\gamma}) Z^{-1}(q) = -q^2 (1 - \frac{1}{\gamma}) + \int d^4k \, K(k,q,n) \, Z(k)$$

$$+ \int d^4k \, L(q,k,n) \, \frac{Z(k) \, Z(k')}{Z(q)} .$$

## RENORMALIZATION PROCEEDURE

First, all terms proportional to $\Lambda^2$ (UV cutoff) are removed so that $\Pi_{\mu\nu}(0) = 0$. Note that if $Z = \Lambda^2/q^2$ is used for all $q^2$ the integral produces a constant term and this term is also removed in this process. A mass scale is introduced by defining $Z(q^2) = Z(M) \, Z_R(q)$ which then provides a definition of $g^2(M)$.

Renormalization group arguments can then be used to determine the following large $q^2$ behavior of Z and $g^2(q)$:

$$Z => \frac{1}{(1 + \frac{16}{11} b \, g^2(M) \, \ln q^2/M^2)} \, 11/16$$

$$g^2(q) => \frac{1}{1 + \frac{16}{11} b \, g^2(M) \, \ln q^2/M^2}$$

where b is the usual b of non-Abelian gauge theories: $b = 11/48 \, C_A/\pi^2$.

## NUMERICAL RESULTS

The final integral equation is a non-linear eq. with two non-trivial integrals. Because of numerical difficulties the integrals can only be calculated accurately for two orders of magnitude for $q^2$. However, because of the scaling properties of the equation, the full range can be covered simply by varying $g^2(M)$, and requiring requiring that the Z's in overlapping regions satisfy

$$Z\left(g^2(M_2), \frac{q^2}{M_2^2}\right) = \frac{Z\left(g^2(M_1), \frac{q^2}{M_1^2}\right)}{Z\left(g^2(M_1), \frac{M_2^2}{M_1^2}\right)}.$$

The general feature of the solution is that it has a $1/q^2$ singularity as $q^2 \to 0$. A trial form that worked well was

$$Z(q) = \frac{.375 M^2}{q^2} + .625 \frac{q^2(1.5 M^2)}{M^2 (q^2 + .5M^2)} \frac{1}{\left(1 + \frac{16}{11} b g^2(M) \ln \frac{q^2 + .6M^2}{1.6 M^2}\right)}$$

for $g^2(M) = 2\pi$. The resulting form for $g^2(q^2)$ is approximately

$$\frac{g^2(q)}{2\pi} = \frac{.215 M^2}{q^2} + \left(1 - \frac{.378 M^2}{.8M^2 + q^2}\right) \frac{1}{1 + .424 \ln\left(\frac{.35 M^2 + q^2}{1.35 M^2}\right)}.$$

Shown below are: Fig. 1 shows the input curve versus output values (points), dashed curve is asymtotic the form; Fig. 2 is $g^2$ vs $q^2$ for various fixed values $\sqrt{1/\gamma} = n \cdot q/nq$, which shows that Z, and hence $g^2$ are approximately independent of $\gamma$ and Fig. 3 is $-\beta(g)/g$ vs. $\sqrt{C_A/4\pi}$ g.

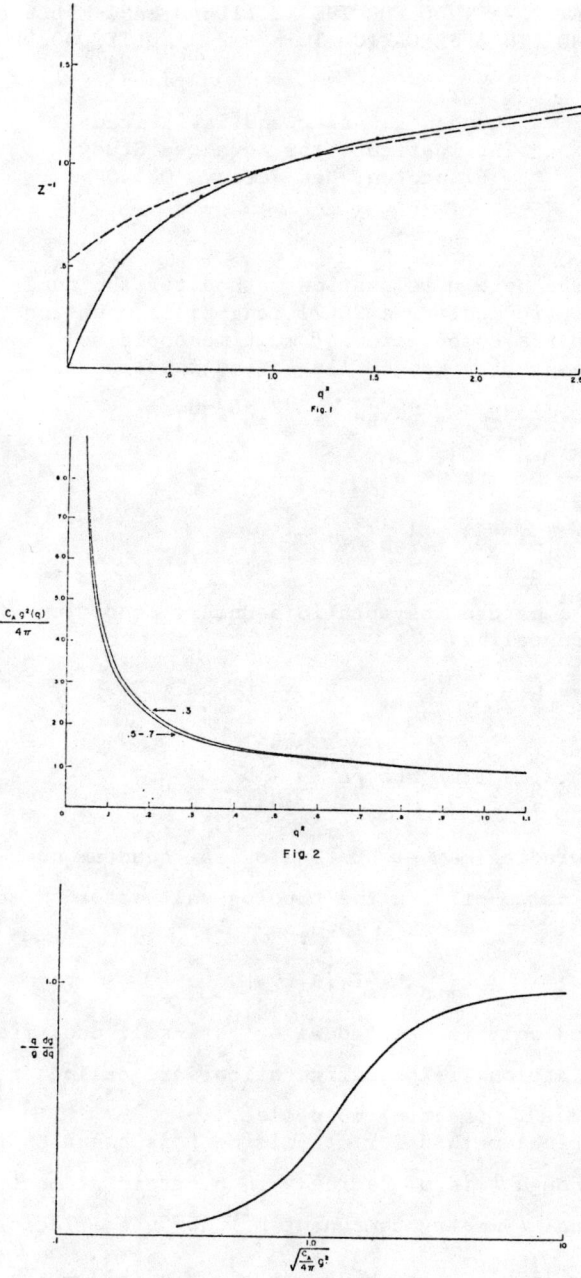

Fig. 1

Fig. 2

Fig. 3

# A RELAXATION METHOD FOR THE EUCLIDEAN YANG-MILLS ACTION FUNCTIONAL AND ITS APPLICATION TO n = 2,3 MULTIMONOPOLE SOLUTIONS

Stephen L. Adler and Tsvi Piran
The Institute for Advanced Study
Princeton, New Jersey  08540

We describe here a relaxation method for the numerical solution of the static Euclidean SU(2) Yang-Mills equations, and its application to the computation of multimonopole solutions. The static Euclidean SU(2) Yang-Mills action functional is

$$F = \int d^3x [\tfrac{1}{2}(\vec{E}^j \cdot \vec{E}^j + \vec{B}^j \cdot \vec{B}^j) - g^2 \vec{b}^0 \cdot \vec{J}^0]$$

$$\vec{E}^j = -\frac{\partial}{\partial x^j} \vec{b}^0 - \vec{b}^j \times \vec{b}^0 \qquad (1)$$

$$\vec{B}^j = \epsilon^{jk\ell}\left(\frac{\partial}{\partial x^k}\vec{b}^\ell + \tfrac{1}{2}\vec{b}^k \times \vec{b}^\ell\right) ,$$

and carries the natural asymptotic boundary condition[1] (modulo a coordinate rescaling)

$$\lim_{|\vec{x}|\to\infty} \vec{b}^0 = \hat{b}(\hat{x}) \quad , \quad \hat{b}\cdot\hat{b} = 1 \quad , \qquad (2)$$

$$n = \frac{1}{8\pi} \int d^2S^i \, \epsilon^{ijk} \epsilon^{abc} \hat{b}^a \frac{\partial}{\partial x^j}\hat{b}^b \frac{\partial}{\partial x^k}\hat{b}^c ,$$

with n the winding number or topological quantum number. When $\vec{J}^0 = 0$, the minimum of F in the topological sector n can be shown to be[2]

$$F_{min} = 4\pi |n| , \qquad (3)$$

and is attained only for self-dual or anti-self-dual field strengths; these minimal action field configurations are called 't Hooft-Polyakov[3] Prasad-Sommerfield[4] (multi-) monopoles.

Our numerical method for minimizing F is based on the observation that although F is of degree 4 in $\vec{b}^\mu$, it is at most quadratic in each internal symmetry component $b^{1\mu}$, $b^{2\mu}$, $b^{3\mu}$ individually,

$$F = \int d^3x \, [\tfrac{1}{2} b^{1\mu} Q_{\mu\nu}(b^2,b^3) b^{1\nu} + b^{1\mu} R_\mu(b^2,b^3) + S(b^2,b^3)]$$

$$= \text{cyclic permutations } (b^{1\mu} \to b^{2\mu} \to b^{3\mu} \to b^{1\mu}) . \qquad (4)$$

The quadratic form $b^{1\mu} Q_{\mu\nu} b^{1\nu}$ is positive semidefinite and symmetric, and can be made positive definite when a suitable gauge-fixing term is included in $F$. Hence we can use relaxation methods designed for linear problems (quadratic actions), provided we relax the internal 1, 2 and 3 components in succession:

$$\begin{array}{l} \text{relax } b^{1\mu} \ (b^{2\mu}, b^{3\mu} \text{ fixed}) \\ \text{relax } b^{2\mu} \ (b^{1\mu}, b^{3\mu} \text{ fixed}) \\ \text{relax } b^{3\mu} \ (b^{1\mu}, b^{2\mu} \text{ fixed}) \end{array} \quad . \quad (5)$$

The method which we use is the overrelaxed Gauss-Seidel algorithm.[5] After discretization, $F$ takes the general form

$$F = \frac{1}{2} \sum_{i,j=1}^{K} \alpha_{ij} U(i) U(j) - \sum_{i=1}^{K} V_i U(i) + \gamma \quad , \quad (6)$$

and minimization requires solving the equations

$$\frac{\partial F}{\partial U(i)} = 0 \Rightarrow \sum_j \alpha_{ij} U(j) - V_i = 0 \quad . \quad (7)$$

The iteration procedure is defined by

$$U^{(n-1)}(i) = n-1^{\text{th}} \text{ approximation}$$
$$U^{(n)}(i) = n^{\text{th}} \text{ approximation}$$
$$U^{(n)}(i) = U^{(n-1)}(i) + \delta(i) \quad , \quad (8)$$
$$\delta(i) = \frac{\omega}{\alpha_{ii}} [V_i - \sum_{j=1}^{i-1} \alpha_{ij} U^{(n)}(j) - \sum_{j=i}^{K} \alpha_{ij} U^{(n-1)}(j)] \ .$$

By direct substitution and some algebra, one finds

$$F^{(n)} - F^{(n-1)} = (\tfrac{1}{2} - \tfrac{1}{\omega}) \sum_i \alpha_{ii} \delta(i)^2 < 0 \quad \text{for} \quad 0 < \omega < 2 \ , \quad (9)$$

since the positive definiteness of $b^\mu Q_{\mu\nu} b^\nu$ implies that all the $\alpha_{ii}$ are positive. In practice, one adjusts the overrelaxation parameter $\omega$ to a value between 1 and 2 which maximizes the rate of convergence. Despite the complexity of $F$, it is easy[6] to find a discretization, analogous to the usual 5-point formula for the Laplacian, in which each node point is coupled only to itself and its nearest neighbors.

In applying this method to the monopole system, we look for the degenerate limit of n = 2,3 multimonopoles in which $\vec{b}^0$ has a single zero. These can be realized within the minimal 6 function axi-symmetric Ansatz,[7]

$$\vec{b}^0 = h_1 \hat{z} + h_2 \hat{\rho}_n \quad,$$

$$\vec{b}^j = \hat{\phi}^j \left( \frac{f_1 - n}{\rho} \hat{z} + \frac{f_2}{\rho} \hat{\rho}_n \right) - (\hat{z}^j a_1 + \hat{\rho}^j a_2) \hat{\phi}_n \tag{10}$$

$$\hat{z} = (0, 0, 1)$$
$$\hat{\rho}_n = (\cos n\phi, \sin n\phi, 0) \qquad \hat{\rho} = \hat{\rho}_1$$
$$\hat{\phi}_n = (-\sin n\phi, \cos n\phi, 0) \qquad \hat{\phi} = \hat{\phi}_1$$
$$\rho = (x^2 + y^2)^{1/2} \quad,$$

for which a convenient choice of gauge-fixing is

$$\Delta F = \int d^3x \, \frac{1}{2} (\partial_z a_1 + \partial_\rho a_2)^2 \quad. \tag{11}$$

Using the iterative method described above, with the Dirichlet boundary conditions

$$h_1 = f_2 = a_2 = 0 \qquad \text{on} \quad z = 0$$
$$a_1 = f_2 = h_2 = 0, \quad f_1 = n \qquad \text{on} \quad \rho = 0$$

$$\left.\begin{array}{l} h_1 = h \cos\theta \\ h_2 = h \sin\theta \\ f_1 = f \cos\theta \\ f_2 = f \sin\theta \\ a_1 = -\sin\theta/r \\ a_2 = \cos\theta/r \\ h = 1 - n/r \\ f = n \cos\theta \end{array}\right\} \begin{array}{l} \text{on outer boundary } \rho = 10, \; 0 \leq z \leq 10 \\ \qquad\qquad\qquad\quad z = 10, \; 0 \leq \rho \leq 10 \\ \\ (r, \theta = \text{usual polar coordinates}) \quad, \end{array}$$

and making an analytic correction for the contribution to $F$ from outside the outer boundary, gives the following results on a $50 \times 50$ mesh,

| topological quantum number n | number of iterations | CPU time on Virtual Machine 370/IBM3033 | Final F value | Bogomol'nyi bound |
|---|---|---|---|---|
| 2 | 185 | 75 sec | 25.170 | $8\pi = 25.132$ |
| 3 | 165 | 48 sec | 37.879 | $12\pi = 37.698$ |

Plots of the action density $\propto \vec{E}^j \cdot \vec{E}^j + \vec{B}^j \cdot \vec{B}^j$ are given in the figures, together with a comparison plot for the known[4] analytic n=1 solution. Our results for the n=2 case are consistent with (but considerably more accurate than) the recent variational calculations by Rossi and Rebbi.[8] Further numerical studies are planned with non-vanishing source density $\vec{J}^0$.

REFERENCES

1. See e.g., S. L. Adler, Phys. Rev. D<u>19</u>, 1168 (1979).
2. E. B. Bogomol'nyi, Yad. Fiz. <u>24</u>, 861 (1976)[Sov. J. Nucl. Phys. <u>24</u>, 449 (1976)]; S. Coleman et al., Phys. Rev. D<u>15</u>, 544 (1977)
3. G. 't Hooft, Nucl. Phys. B<u>79</u>, 276 (1974); A. M. Polyakov, Zh. Eksp. Teor. Fiz. Pis'ma Red. <u>20</u>, 430 (1974)[JETP Lett. <u>20</u>, 194 (1974)].
4. M. K. Prasad and C. M. Sommerfield, Phys. Rev. Lett. <u>35</u>, 760 (1965).
5. W. F. Ames, <u>Numerical Methods for Partial Differential Equations</u> (Academic Press, N.Y., 1977), p. 113.
6. F. Bauer, O. Betancourt and P. Garabedian, <u>A Computational Method in Plasma Physics</u> (Springer-Verlag, N.Y., 1978), Ch. 3.
7. P. S. Jang, S. Y. Park and K. C. Wali, Phys. Rev. D<u>17</u>, 1641 (1978); N. S. Manton, Nucl. Phys. B<u>135</u>, 319 (1978).
8. C. Rebbi and P. Rossi (unpublished).

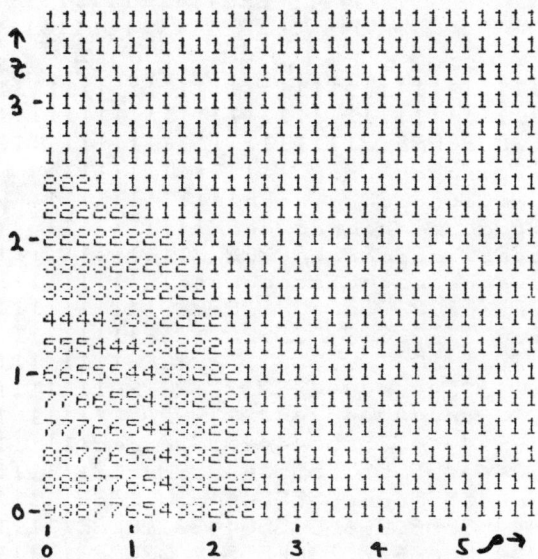

Fig. 1. Action density for n=1 solution

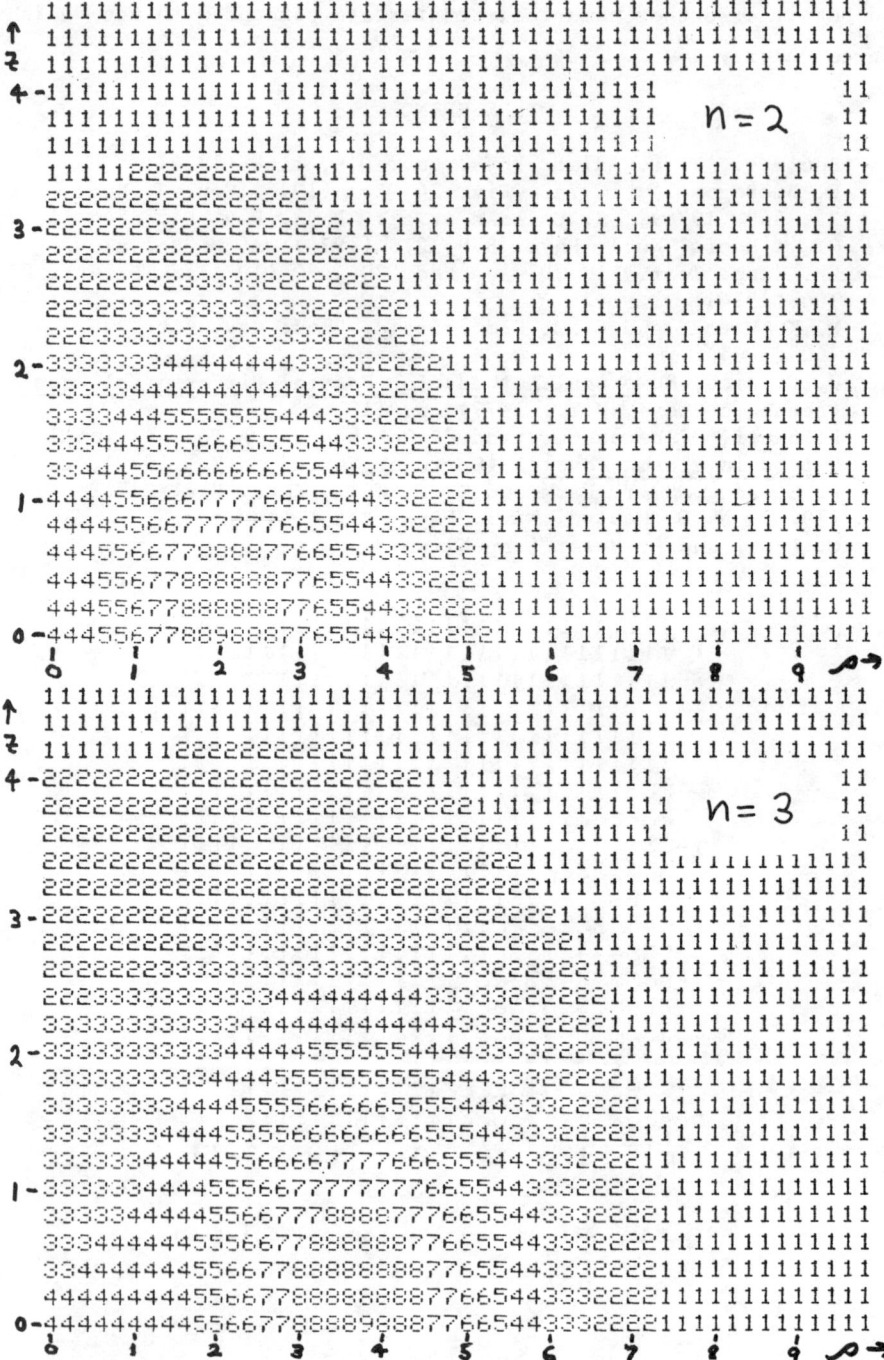

Fig. 2. Action density for n=2,3 solutions.

# SUPERSYMMETRY, SUPERGRAVITY

R. Arnowitt, Northeastern, Organizer

## PROSPECTS FOR SUPERGRAVITY

B. Zumino
Fermi National Accelerator Laboratory, Batavia, Illinois 60510
and
CERN, Geneva, Switzerland

Present particle phenomenology appears to be successfully described by a renormalizable gauge theory based on the low energy group

$$SU(3)_{colour} \times SU(2)_{left} \times U(1) \quad . \tag{1}$$

The three effective coupling constants of this theory vary with energy according to the renormalization group equations and become approximately equal at an energy of about $10^{14}$–$10^{15}$ GeV, where one can attempt a grand unified theory (GUT) based on a gauge group with only one coupling constant and containing the low energy group (1). As discussed by Georgi and Glashow,[1] the minimal GUT is that based on the group SU(5). All other viable GUTs with larger groups proposed till now go through SU(5). The GUT picture has had some successes, notably the determination,[2,3,4] in reasonable agreement with experiment, of the low energy weak angle[2,3,4] which parametrizes nuclear current couplings and certain relations,[3,4,5] between quark and lepton masses.

The grand unification energy is very large and not much smaller than the Planck mass $m_p \simeq 10^{19}$ GeV, where gravity becomes important. A natural way to unify gravity with lower spin fields is provided by supergravity,[6] the supersymmetric extension of Einstein's gravity. In working out this appealing idea one is faced with the difficulty that the largest supergravity theory (largest theory with a supermultiplet with maximum spin not exceeding two) is the N=8 theory (N is the number of supersymmetries) which has an SO(8) internal symmetry. Now, the group SO(8) is too small, it does not contain the low energy group (1), and the fundamental sypermultiplet of the N=8 theory (which consists of 1 spin 2, 8 spin 3/2, 28 vectors, 56 spin 1/2 and 70 scalars) is not rich enough to accomodate all known quarks, leptons and spin-one gauge fields.[7]

Perhaps these difficulties are only apparent and what is really important is that SO(8) can contain $SU(3)_{colour} \times U(1)_{e.m.}$, which in our present picture is the exact gauge group of nature. It is a general feature of supergravity theories that the vector fields are exactly the right number and in the right representation to become the gauge fields of SO(N). For $N \leq 4$ it has been possible to introduce a gauge coupling g for the vector fields.[8] Supersymmetry requires it to be accompanied by a mass term $m_2 \sim g/\kappa$ for the spin 3/2 fields and by a cosmological term $\sim (g/\kappa^2)^2$. Perhaps this gauging can be achieved also for the N=8 theory. As suggested by Hawking,[9] such an

0094-243X/81/680964-06$1.50 Copyright 1981 American Institute of Physics

enormous cosmological constant could give rise to a foamy structure of space time. Over distances large compared to the Planck length the foamy structure would not be visible and space time could look smooth and approximately Minkowskian. At the same time the SO(8) internal symmetry would be masked and a larger approximate effective symmetry could result. However interesting this point of view, it is hard to see how one would proceed to implement it. In the following we shall assume that the cosmological constant vanishes, due to cancellation among different contributions to it.

An approach which seems more promising at this time is that based on the observation that N=8 supergravity (<u>without</u> the gauge coupling g) has not only an SO(8) symmetry but also a less obvious SU(8) internal symmetry. This symmetry was observed first[10] as a true global symmetry of the equations of motion. The complex transformations of the SU(8) group are realized on the spin-one fields as duality rotations which transform the electric into the magnetic field strengths. More recently, Cremmer and Julia[11] have formulated the N=8 theory in such a way that the SU(8) in a certain sense becomes local. They achieved this by adding redundant scalar fields (which could actually be eliminated by going to a particular gauge) and by introducing SU(8) gauge vectors as auxiliary non-propagating fields, without a kinetic term of their own. Nevertheless, Cremmer and Julia suggested that the SU(8) gauge vectors could become dynamical, i.e. their propagator develop a zero mass pole, in analogy with a similar phenomenon known to occur in the $CP^{n-1}$ non-linear model in two space-time dimensions when[12] treated in the large n approximation. In effect one is assuming[13,14] that the fields of the fundamental supermultiplet of N=8 supergravity bind together to form another supermultiplet containing spin-one fields in the adjoint representation of SU(8). Just as the group SU(8) is large enough to contain the group of grand unification, the supermultplet of bound states can be sufficiently rich to contain all known quarks, leptons, vectors and scalars. All fields considered elementary at the present time are taken instead to be composite.

One then could have the following picture. At an energy comparable with (or higher than) the Planck mass $m_p$, supersymmetry is valid and also the larger internal symmetry SU(8). The dynamics is specified by the gravitational constant $\kappa \sim 1/m_p$. Coming down to the GU mass, all supersymmetries are broken and SU(8) is broken down to the GU group, say SU(5). This symmetry breaking is spontaneous and characterized by the vacuum expectation value $<\phi>$ of some scalar field. At the GU mass the interactions are determined by a dimensionless coupling constant $g \sim \kappa<\phi>$ which emerges as the GU gauge coupling constant. The GUT appears as a low energy effective theory derived from the dynamics which is valid at the Planck mass. That it is a renormalizable gauge theory can be perhaps understood from the so called "zero mass decoupling theorem."

This theorem states that if a large mass (say $m_p$ for definiteness) occurs in a field theory, and if a subset of fields or states remains <u>massless</u> as $m_p \to \infty$, then these states decouple from the rest in the limit and their effective interactions at low energy are renormalizable (with additional non-renormalizable interactions inversely proportional to some power of $m_p$). The zero mass decoupling theorem is physically intuitive and can actually be proven, if the original field theory is a well defined renormalizable theory (for instance a gauge theory) and the zero mass states are those corresponding to some of the fields of the original theory.[15] Presumably, it can also be proven for zero-mass bound states. Indeed, if their effective interactions were not renormalizable, divergences would arise in the computation of vertex functions, for which the only cut-off can be the inverse size of the bound states, which we expect to be $m_p$. This would result in masses of order $m_p$, or a breakdown of perturbation theory for the effective interactions of the bound states.* We wish to apply the theorem to the case where the original theory is $N=8$ supergravity and the zero mass states are composites of the fundamental fields. A consequence of the zero mass decoupling theorem is that the zero mass states cannot have spin higher than one, since their interaction would be necessarily non-renormalizable. For the same reason, the spin-one fields must be gauge fields and the zero mass spin one-half fields must form an anomaly free set. Scalar fields can survive in any number. All fields or states which do not belong to this subset with renormalizable interactions must disappear as $m_p \to \infty$, by acquiring very large masses (or perhaps by becoming unbound).

According to Cremmer and Julia, the composite fields of the $SU(8)$ gauge vectors are expressions (bilinear plus higher order terms) in the fundamental scalar fields. These expressions are not identical with the currents of the global $SU(8)$ transformations mentioned above, which differ from them by the presence of terms containing the other fundamental fields (spin 1/2, 1 and 3/2) and which furthermore are non-local for the spin 1, on which they generate duality rotations. Nevertheless, the gauge $SU(8)$ and the global $SU(8)$ are clearly related and could be identified by going to a particular fixed gauge, where only the global $SU(8)$ remains. Ellis, Gaillard, Maiami and I[13,14] have reinterpreted the suggestion of Cremmer and Julia and have tried to determine the supermultiplet containing the conserved currents of the global $SU(8)$. For a massless supersymmetric system with $N \leq 4$ and maximum spin 1, all currents are local and this supermultiplet would be the supercurrent supermultiplet,[16] which contains also the energy momentum tensor and the N spinor currents. One can put the fields of this supermultiplet on the mass shell and one obtains in this way a massive supermultiplet of supersymmetry with $N \leq 4$ spinor charges. The maximum spin is 2 and is a singlet. States with lower spins are in representations of $Sp(2N)$. As the mass tends to zero the supermultiplet breaks up into a finite number of massless supermultiplets in which the various helicities are in representations of $SU(N)$. It is not difficult to identify which of these massless supermutliplets contains the $SU(8)$ currents in the adjoint representation and one finds that it is of the type given below in (2), (3). This argument is reliable for $N \leq 4$; we assume that the multiplet has the same form for $N=8$,

*This argument is due to M. Veltman.

We come thus to the conclusion that the relevant supermultiplet of states is given by (A,B,...=1,2...8)

$$\left(\frac{3}{2}\right)^A, \quad (1)^A_B, \quad \left(\frac{1}{2}\right)^A_{[BC]}, \quad (0)^A_{[BCD]}, \cdots \left(-\frac{5}{2}\right)^A, \tag{2}$$

to which one must add the TCP conjugate states

$$\left(\frac{5}{2}\right)_A, \quad (2)_{AB}, \quad \left(\frac{3}{2}\right)_{A[BC]}, \cdots \left(-\frac{3}{2}\right)_A, \tag{3}$$

(a set of k antisymmetrized lower indices is equivalent to 8-k antisymmetrized upper indices). It contains spin-one states in the adjoint representation but also other spin-one states. Ideally one would like to give masses by super-Higgs effect to all unwanted states, those which do not belong to the surviving zero mass subset with effective renormalizable interactions. However, it is easy to see that this cannot be done in an SU(5) invariant way, and not even is an SU(3)×SU(2)×U(1) invariant way. For instance, in order to give an invariant mass to a spin 5/2 state, one needs all the helicities 5/2, 3/2, 1/2, -1/2, -3/2, -5/2 and they must be in the same representation of the invariance group. It is clear that the supermultplet (2), (3) does not contain all the necessary helicities. It is also easy to convince oneself that no irreducible supermultiplet and no finite set of irreducible supermultiplets will do.* Clearly, this is due to our desire to be left with complex (chiral) massless representations; it is very easy to construct finite sets of massless supermultiplets which combine to massive ones, in a vector-like way, simply by starting from a finite mass supermultiplet and going to the limit as the mass tends to zero when it breaks up into massless ones.

The above discussion suggests that one may need an infinite dimensional set of irreducible supermultiplets. This would mean that the dynamics at the Planck mass involves states of arbitrarily high spin most of which become infinitely massive after the symmetry breaking, as $m_p \sim \langle\phi\rangle \to \infty$. Those higher spin states which, like the graviton, remains massless, will be left with non-renormalizable interactions proportional to reciprocal powers of $m_p$.

*A formal proof of this impossibility has been worked out by M. Gell-Mann, private communication. In a recent Harvard preprint, HUTP-80/A050, P. Frampton considers finite sets of supermultplets and states that his solutions have enough helicity states to give masses to all higher spin states. However these masses are not $SU(3)_{colour}$ invariant and would give rise to breaking of this group already at energies comparable with the Planck mass.

In an attempt to extract as many properties of the GUT as possible from the dynamics at the Planck mass, without actually knowing this dynamics (nor the$_{14}$ possibly infinite-dimensional multiplet), Ellis, Gaillard and I$^{14}$ have assumed that all surviving low energy states are already contained in the single supermultiplet (2), (3). We have then looked for the maximal set of states of spin 0, 1/2 and 1 contained in it which can have renormalizable interactions, without however keeping those SU(8) representations which are obtained by saturating in (2), (3) upper with lower indices (traces). With these (admittedly very drastic) simplifications, we found that the breaking of SU(8) to SU(5) (rather than SU(6) or SU(7)) is preferred. For the maximal set of left-handed states we found two solutions, of which only one is vector-like for $SU(3)_{colour} \times U(1)_{e.m.}$. This solution has three generations of $10+\bar{5}$ of SU(5).

Irrespective of the validity of this particular set of states, one may ask by what techniques one can obtain more information about the GUT. Now, in our picture, dynamics at the Planck mass involves higher spin states. It appears at this moment, that no consistent classical local Lagrangian describing the interaction of higher spin fields exists. Therefore the only available technique for describing these interactions is that of on-shell scattering amplitudes (S-matrix). It is not difficult to find the constraints imposed on these amplitudes by supersymmetry$^{17}$ and it seems also possible to formulate the spontaneous breaking process. In this way one can obtain, in principle, not only the group representations of the GUT but also the details of the interaction, the Higgs potential and the Yukawa coupling. Perhaps this approach$_{18}$ can give some understanding of the socalled hierarchy problem.

Incidentally, having taken the point of view that higher spins are entering anyway the dynamics at (or above) the Planck mass one may even try to relax the restriction to supersymmetries with N<8 and admit spins higher than 2 in the fundamental supermultiplet. Our present point of view that local Lagrangians are low energy approximations to a theory formulated in terms of on shell amplitudes seems to give us this freedom. Still, there seems to be some virtue in keeping with N=8 supergravity, a theory remarkable for its convergence and its symmetries.

## ACKNOWLEDGMENT

This lecture is based on work done in collaboration with J. Ellis, M.K. Gaillard and L. Maiani.

## REFERENCES

1. H. Georgi and S.L. Glashow, Phys. Rev. Lett. <u>32</u>, 438 (1974).
2. H. Georgi, H.R. Quinn and S. Weinberg, Phys. Rev. Lett. <u>33</u>, 451 (1974).
3. M.S. Chanowitz, J. Ellis and M.K. Gaillard, Nucl. Phys. <u>B128</u>, 506 (1977).
4. A.J. Buras, J. Ellis, M.K. Gaillard and D.V. Nanopoulos, Nucl. Phys. <u>B135</u>, 66 (1978).

5. D.V. Nanopoulos and D.A. Ross, Nucl. Phys. B157, 273 (1979).
6. D.Z. Freedman, P. van Nieuwenhuizer and S. Ferrara, Phys. Rev. D13, 3214 (1976); S. Deser and B. Zumino, Phys. Lett. 62B, 335 (1976); S. Ferrara and P. van Nieuwenhuizen, Phys. Rev. Lett. 37, 1669 (1976); S. Ferrara, J. Scherk and B. Zumino, Phys. Lett. 66B, 35 (1977); D.Z. Freedman, Phys. Rev. Lett. 38, 105 (1977); A. Das, Phys. Rev. D15, 2805 (1977); E. Cremmer, J. Scherk and S. Ferrara, Phys. Lett. 68B, 234 (1977); E. Cremmer and J. Scherk, Nucl. Phys. B127, 259 (1977); B. de Wit and D.Z. Freedman, Nucl. Phys. B130, 105 (1977).
7. M. Gell-Mann, Talk at the 1977 Washington Meeting of The American Physical Society.
8. D.Z. Freedman and A. Das, Nucl. Phys. B120, 221 (1977); D.Z. Freedman and J.H. Schwarz, Nucl. Phys. B137, 333 (1978); E. Cremmer and J. Scherk, unpublished.
9. S.W. Hawking, private communication.
10. S. Ferrara, J. Scherk and B. Zumino, Nucl. Phys. B121, 393 (1977).
11. E. Cremmer and B. Julia, Phys. Lett. 80B, 48 (1978); Nucl. Phys. B159, 141 (1979). The first attempt to relate this work to phenomenology was made by T. Curtright and P. Freund, in "Supergravity," Ed. P. van Nieuwenhuizen and D.Z. Freedman, (North Holland, Amsterdam, 1979), p. 197.
12. A. d'Adda, P. Di Vecchia and M. Lüscher, Nucl. Phys. B146, 63 (1978), B152, 125 (1979); E. Witten, Nucl. Phys. B149, 285 (1979).
13. J. Ellis, M.K. Gaillard, L. Maiani and B. Zumino, LAPP preprint TH-15/CERN preprint TH-2481 (1980) to appear in the Proc. of the Europhysics Study Conf. on "Unifications of Fundamental Interactions," Erice, 1980; J. Ellis, Proc. of the First Workshop on Grand Unification, Durham, New Hampshire, 1980.
14. J. Ellis, M.K. Gaillard and B. Zumino, CERN preprint TH-2842/LAPP preprint TH-16 (1980), to appear in Phys. Lett.
15. See H. Schnitzer's lecture in these proceedings for numerous references.
16. S. Ferrara and B. Zumino, Nucl. Phys. B87, 207 (1975); M. Sohnius, Phys. Lett. 81B, 8 (1979); for $N=4$ the supercurrent has been worked out by E. Bergshoeff and B. de Wit.
17. M.T. Grisaru, P. van Nieuwenhuizen and H.N. Pendleton, Phys. Rev. D15, 996 (1977); M.T. Grisaru and H.N. Pendleton, Nucl. Phys. B124, 81 (1977).
18. E. Gildener, Phys. Rev. D14, 1667 (1976).

## IS FLAVOR PROLIFERATION EXPLICABLE BY SUPERGRAVITY?

Paul Frampton
Lymen Laboratory of Physics
Harvard University, Cambridge, MA 02138

The observed recurrence of quark-lepton families $(u,d;e,\nu_e)$, $(c,s;\mu,\nu_\mu)$, $(?,b;\tau,\nu_\tau)$ presents a deep mystery since almost everything happening around us would seem to be unaffected if all but the first family were thrown away. Also, the simplest unification scheme for color and electroweak forces based on SU(5) works nicely for one family, unless we take seriously the predictions for quark masses arising from details of the symmetry breaking. Only partial success in explaining the multiple families is obtained by embedding SU(5) in a larger SU(N) using principles such as anomaly freedom, asymptotic freedom, absence of color exotics, and inclusion of all possible superheavy mass terms consistent with the symmetry. This applies at least to the use of fundamental representations (i.e. single-column Young tableaus) of SU(N).

Supergravity suggests using SU(8) with two-column tableaus generated within SU(8) supersymmetric supermultiplets as explained by Zumino in an accompanying talk[1]. Embedding SU(3) x SU(2) x U(1) of color and flavor naturally in SU(8) and requiring that this subgroup be exact at high energy necessitates that all nonrenormalizable states be real under SU(3) x SU(2) x U(1). This appears impossible with any finite number of supermultiplets – certainly impossible from bilinears of the basic supergravity multiplet without additional very ad-hoc assumptions (e.g. Veltman's theorem[1]).

Instead, focusing on spin 1/2 and anomaly-freedom, and combining higher helicities 3/2 and 5/2 into appropriate massive higher-spin states, gives a result with maximum spin 3 and four light families[2]. Because of the finite number of SU(8) supersymmetric supermultiplets, one conclusion is that color or electromagnetic gauge symmetry is broken. Hopefully, the resultant gluon or photon effective mass is infinitesimal due to inverse powers of the Planck mass. Breaking of color may well be related to the reported observations of free fractional charges. Or else, it may signify an unanswerable objection to this supergravity approach.

1. Bruno Zumino, these proceedings, and references therein.

2. P. H. Frampton, Harvard preprint HUTP-80/A050(July 1980).

# APPLICATIONS OF SUPERFIELD FEYNMAN DIAGRAMS

M.T. Grisaru*
Brandeis University, Waltham, MA 02254

## ABSTRACT

This report describes a calculation using superfield Feynman diagrams, to establish that the three-loop $\beta$-function vanishes in N=4 super Yang-Mills theory. A similar calculation shows that the trace and chiral anomalies in supergravity are members of a supersymmetric multiplet.

I will describe in this report two results that can be obtained using as a main tool superfield Feynman diagrams[1]. The first result is the fact that the $\beta$-function vanishes at the three-loop level in N=4 supersymmetric Yang-Mills theory[2] (it had been established already that it is zero at the one- and two-loop level[3]). The second result is an understanding of the supersymmetric nature of one-loop anomalies in supergravity. The work was done in collaboration with M. Rocek and W. Siegel. We understand that the vanishing of the three-loop $\beta$-function was also established at Dubna in a computer calculation using ordinary field theory (not superfield) methods[4].

The N=4 theory is a conventional model with the following field content[5]: One Yang-Mills vector, four Majorana fermions, three scalars and three pseudoscalars. All the fields are massless and (matrices) in the adjoint representation of the gauge group. The Lagrangian is:

$$\mathcal{L} = \text{Tr}\left\{ -\frac{1}{4} F_{\mu\nu}^2 - \frac{1}{2}(D_\mu A_i)^2 - \frac{1}{2}(D_\mu B_i)^2 - \frac{1}{2}\bar{\lambda}_k \not{\partial} \lambda_k \right.$$
$$\left. - \frac{1}{2} g \bar{\lambda}_k [\alpha_{k\ell}^j A_j + \gamma_5 \beta_{k\ell}^j B_j, \lambda_\ell] \right.$$ (1)
$$\left. + \frac{1}{4} g^2 \left( [A_i, A_j]^2 + [B_i, B_j]^2 + 2[A_i, B_j]^2 \right) \right\}$$

where the $\alpha$'s and $\beta$'s are 4x4 SU(2) x SU(2) matrices, i,j = 1,2,3, k,l = 1,2,3,4. The action is invariant under supersymmetry transformations and under SU(4) rotations of the fermion and spin zero fields. It is scale invariant at the classical level. Remarkably enough, quantum corrections do not break the scale invariance, at least to the three-loop level.

The model is renormalizable and supersymmetry (and the SU(4) symmetry) restricts the number of independent renormalization constants. These are in general gauge dependent. We find that in a particular, supersymmetric gauge they are finite; the theory has no divergences whatsoever (through two and probably three loops).

The calculation of the three-loop $\beta$-function is manageable if done in terms of superfields. We use so-called N=1 superfields

---

*Work supported in part by NSF Grant PHY79-20901

$$V = C + \theta^\alpha \chi_\alpha + \cdots + \theta \sigma^\mu \bar{\theta} A_\mu + \bar{\theta}^2 \theta^\alpha \lambda_\alpha + \cdots$$
$$\phi_i = \tfrac{1}{\sqrt{2}}(A_i + i B_i) + \theta^\alpha \lambda_{i\alpha} + \tfrac{1}{\sqrt{2}} \theta^2 (F_i - i G_i) \tag{2}$$

where V is a real superfield containing the Yang-Mills vector and one of the fermions. The three $\phi_i$ are "chiral" superfields (in the "chiral representation") containing the remaining three fermions and the spin zero fields. We use two-component spinors. The action is:

$$S = \mathrm{Tr}\left[\int d^4x\, d^4\theta\, e^{-gV} \bar{\phi}_i e^{gV} \phi_i + \tfrac{1}{64 g^2} \int d^4x\, d^2\theta\, W^\alpha W_\alpha \right.$$
$$\left. + \tfrac{ig}{3!} \int d^4x\, d^2\theta\, \varepsilon^{ijk} \phi_i [\phi_j, \phi_k] + \text{h.c.} \right] \tag{3}$$

$$W_\alpha = \bar{D}^2 (e^{-gV} D_\alpha e^{gV}) \tag{4}$$

and the D's are covariant spinor derivatives (in chiral representation)

$$D_\alpha = \tfrac{\partial}{\partial \theta^\alpha} + 2i\, \sigma^\mu_{\alpha\dot{\alpha}} \bar{\theta}^{\dot{\alpha}} \partial_\mu \qquad \bar{D}_{\dot{\alpha}} = \tfrac{\partial}{\partial \bar{\theta}^{\dot{\alpha}}} \tag{5a}$$

$$\{D_\alpha, \bar{D}_{\dot{\alpha}}\} = 2i\, \sigma^\mu_{\alpha\dot{\alpha}} \partial_\mu \tag{5b}$$

Its gauge invariance makes it possible to gauge away the unphysical fields appearing in eq. (2). In the resulting ("Wess-Zumino") gauge the Lagrangian is that of eq. (1). However it is preferable to work in a manifestly supersymmetric gauge.

By superfield power counting[1] the three-point function $\langle T\phi\phi\phi\rangle$ for the chiral superfields is finite. Therefore to establish that the $\beta$-function vanishes it is sufficient to show that the two point function $\langle T\phi\phi\rangle$ is finite. After quantization (adding gauge fixing terms and ghosts) Feynman rules can be derived[1]. We work in a supersymmetric Feynman gauge where the free propagators have the form

$$\underset{\theta_1 \qquad \theta_2}{\longrightarrow} \quad \sim \quad \frac{\delta^4(\theta_1 - \theta_2)}{p^2} \tag{6}$$

The $\delta$-function is

$$\delta^4(\theta_1 - \theta_2) = (\theta_1 - \theta_2)^2 (\bar{\theta}_1 - \bar{\theta}_2)^2 \tag{7}$$

and satisfies the relations

$$\delta^4(\theta_1 - \theta_2)\, \delta^4(\theta_1 - \theta_2) = 0$$
$$\delta^4(\theta_1 - \theta_2)\, (D)^m (\bar{D})^n\, \delta^4(\theta_1 - \theta_2) = 0 \qquad m, n < 2 \tag{8}$$

Therefore a loop such as shown here

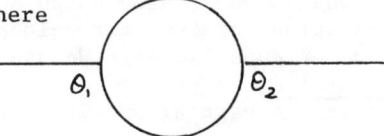

would give zero were it not for the fact that the vertices have associated with them spinor derivatives $D_\alpha$, $\bar{D}_{\dot\alpha}$ which act on the $\delta$-functions. For example

$$\delta^4(\theta_1-\theta_2) D^2 \bar{D}^2 \delta^4(\theta_1-\theta_2) = 16\, \delta^4(\theta_1-\theta_2) \qquad (9)$$

A general supergraph contains a certain number of lines, with propagators as above, and integration over the $\theta$'s associated with each vertex. The general strategy[1,6] is to work on a graph by performing integration by parts of the various D's across vertices, making use of eqs. (5b) and (9), and use left over $\delta$-functions to do $\theta$-integrations. If less than two D's and two $\bar{D}$'s act in a loop one gets zero by repeated use of (8). If too many D's act on a given propagator some of them get replaced by momentum factors (as a consequence of eq. (5b)) which often cancel denominator factors in the propagators. Eventually one is reduced to evaluating an ordinary integral associated with a diagram composed of scalar lines. For example the superfield diagram below (solid lines = chiral fields, wavy lines = real (vector) fields) leads to the evaluation of the integral associated with the corresponding reduced diagram in which two of the lines have been contracted. For examples of this procedure see Refs. (1) and (6).

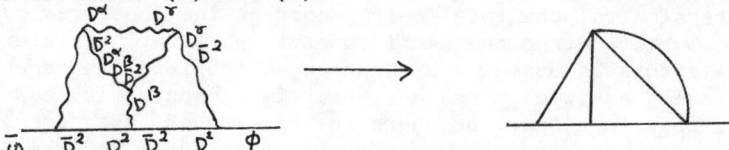

We have used these methods to calculate the divergent part of the three-loop chiral field propagator $\langle T\phi\phi\rangle$. We have found that, even before doing momentum integrations, all the divergences cancel and have therefore concluded that $\beta(3\text{ loop})=0$. The fact that the theory has only one $\beta$-function can be established using gauge invariance and invariance under SU(4) transformations[2]. As mentioned above there are in fact no divergences whatsoever and it seems very likely that we are dealing with a nontrivial (barring infrared problems the S-matrix is not unity) <u>finite and conformally invariant four-dimensional field theory.</u> The result also encourages one to think that N=8 supergravity, which has some features in common with N=4 super Yang-Mills theory might also be more finite than one would expect. At the one loop level there are some indications that this may be the case.

I turn now to a brief discussion of the one-loop anomaly question. It has been known for sometime that the trace, chiral and supersymmetry current anomalies for globally supersymmetric models (scalar and vector multiplets) are themselves members of a supersymmetric multiplet[7]. By contrast in supergravity the contributions to the trace anomaly from the graviton and gravitino (and their ghosts) and to the chiral anomaly from the gravitino and its ghosts are, when suitably normalized

$$T^\mu_\mu = \frac{41}{768\pi^2} {}^*R\,{}^*R \,,\quad \partial^\mu j^5_\mu = \frac{21}{768\pi^2} R\,{}^*R \qquad (10)$$

and it is difficult to see how the numbers 41 and 21 can be associated with members of a multiplet. I should emphasize that the presence of these anomalies has no bearing on the issue of finiteness of the S-matrix in supergravity.

Recently we have developed methods for doing perturbation theory with supergraphs in supergravity[8]. These methods can be used to calculate the one-loop anomalies of the supergravity multiplet in a background supergravity field. Since the calculations are manifestly supersymmetric they are bound to give anomalies which are in a multiplet. We use background field methods and a choice of gauge fixing term which has the curious consequence (see ref. 1 for a similar result in the supersymmetric Yang-Mills case) that the supergravity superfield gives no contributions to the anomaly. Instead the full contribution comes from certain chiral ghost superfields. The reason is simply that with our gauge choice the contributions from the supergravity superfield to the relevant diagrams do not have enough spinor derivative at vertices to overcome the presence of two many $\Theta$'s as in eq. (8). The final result indicates that the anomalies do form a multiplet, and that the trace anomaly in eq. (10) is correct while the chiral anomaly is wrong. This is not to say that the calculations of the anomaly for a spin 3/2 gravitino chiral current[9] are incorrect. Rather, it appears that the correct expression for the total chiral current in supergravity includes contributions from the antisymmetric part of the vierbein field. While this is normally a nonpropagating field and does not affect physical quantities such as S-matrix elements, it can make its presence felt in situations where the background has nontrivial topology. This is consistent with certain "superindex theorems" established in Ref. 9. The conclusion is that just like in global supersymmetry, the anomalies in supergravity do form a multiplet.

## REFERENCES

1. M.T. Grisaru, M. Rocek and W. Siegel, Nucl. Phys. B159, 429 (1979).
2. M.T. Grasaru, M. Rocek and W. Siegel, "Zero 3-loop $\beta$-function in N=4 super Yang-Mills Theory" (to be published).
3. E.C. Poggio and H.N. Pendelton, Phys. Lett. 72B, 200 (1977); D.R.T. Jones, Phys. Lett. 72B, 199 (1977).
4. We thank A. Slavnov for informing us of this result.
5. L. Brink, J.H. Schwarz and J. Scherk, Nucl Phys. B121, 77 (1977), F. Gliozzi, J. Scherk and D. Olive, Nucl. Phys. B122, 253 (1977).
6. L.F. Abbott and M.T. Grisaru "The 3 loop $\beta$-function for the Wess-Zumino Model" Nucl. Phys. (to be published).
7. For a review see M.T. Grasaru "Anomalies in Supersymmetric Theories" in "Gravitation" (Cargese 1978), M. Levy and S. Deser, Eds. (Plenum Press 1979).
8. M.T. Grisaru and W. Siegel "Supergraphity" (in preparation). See also the Proceedings of the 1980 Nuffield Supergravity Workshop (Cambridge University Press, to be published).
9. S.M. Christensen and M. Duff, Nucl Phys. B154, 301 (1979).

# RIEMANNIAN SUPERSPACE REDUCTION AND SUPERGRAVITY GEOMETRY IN SUPERSPACE

R. Arnowitt and Pran Nath
Department of Physics, Northeastern University, Boston, MA  02115

## I. INTRODUCTION

Riemannian superspace is interesting for two reasons. The first is that the Riemannian superspace provides us with a framework for the study of local supersymmetry and thus generates a possible scenario for the unification of fundamental interactions including gravitation. This is possible because the gauge group of the theory is large enough to accommodate the Einstein, the Yang-Mills and the supersymmetry gauge groups. It also turns out that the <u>dynamics</u> of the Riemannian superspace produces some unique properties for this theory at the quantum level such as the ultraviolet finiteness of the S-Matrix to all orders in the quantum loops. The second reason for interest in the Riemannian superspace geometry is that the tangent space group of the Riemannian theory is $OSp(3,1|4N)$ and contains as a subgroup $O(3,1) \times O(N)$ which characterizes the tangent space of supergravity. Thus the non-Riemannian geometry of supergravity may be viewed as embedded in the larger Riemannian geometry. In this sense the Riemannian geometry represents the more general case.

Perhaps the simplest way to illustrate why a larger geometrical framework such as the Riemannian superspace may provide a useful device for geometrical constructions of supergravity is to make some observations regarding the usual approaches. The first feature of the usual constructions of extended supergravity arises due to the lack of unification between the Lorentzian and the Yang-Mills parts of the geometrical quantities such as connections, curvatures and field strengths of the theory. The reason for this, of course, is that supergravity is not a unified theory since its tangent space group is a product group $O(3,1) \times O(N)$ rather than a unified group such as $SU(5)$. Thus what one would like is an embedding of the product group in a larger group which can unify these two aspects. The second feature which arises in the usual superspace formulations of supergravity concern the fact that the vacuum expectation values of some of the components of torsion and curvature are non-vanishing. This represents a novel assumption since in conventional gauge theory one does not start out by postulating non-vanishing vacuum expectation values for the dynamical quantities. Such vacuum expectation values are expected to arise through the phenomenon of spontaneous or dynamical symmetry breaking. Again one may ask if this can be achieved in a larger geometrical framework. These observations have led us to examine supergravity geometry in the larger framework of the Riemannian superspace.

The reduction procedure which extracts the supergravity geometrical quantities from the Riemannian geometry is somewhat intricate and we shall discuss this in some detail later. Briefly it consists of two steps. (i) A contraction to a specific $O(3,1) \times O(N)$ part of the Riemannian tangent space group $OSp(3,1|4N)$ is made. (ii) The limit $k \to 0$ is taken where k is a parameter that appears in the Riemannian geometry.[1] Some of the results that emerge from this approach are as follows:

1. One finds that the $O(3,1)$ Lorentzian and the $O(N)$ Yang-Mills connections of supergravity arise as parts of a single Riemannian object $\omega_{MN\Lambda}$ according to[2]

$$h_{rs\Lambda} = \omega_{rs\Lambda})_{k \to 0} \; ; \quad h_{ij\Lambda} = \frac{1}{4k} \omega_{aibj\Lambda} \eta^{ab}]_{k \to 0} \; , \qquad (1.1)$$

where the parameter k enters in the tangent space metric diag $\eta_{AB} = (\eta_{mn}, k\, C_{ab})$. Here $\eta_{mn}$ is the Lorentzian metric and C is the charge conjugation matrix. The reduction thus leads to closed form solutions for the supergravity connections valid for any N. This is a nontrivial result as torsion constraints have not been employed.

2. The mechanism of spontaneous symmetry breaking is seen to exist in the dynamics of Riemannian geometry which produces the supergravity vacuum state in the $k \to 0$ limit.

3. The condition that a smooth (i.e., non-singular) $k \to 0$ contraction exists, is sufficient to show that once the above vacuum part of the geometry is separated out, the remainder of the Riemannian geometry automatically contracts to supergravity geometry. Thus supergravity geometry uniquely arises as the smooth contraction of Riemannian geometry and the <u>kinematic</u> torsion constraints are deduced for all N.

4. When a gauge complete vierbein[3] is inserted, the condition that the contraction be smooth uniquely determines the dynamics of the fields to be that of supergravity, at least for N = 1, 2.

## II. THE k-DEPENDENCE OF RIEMANNIAN GEOMETRY

First we point out the obvious k dependences. We define the tangent space covariant derivative $\mathcal{D}_\Lambda$ by

$$\varphi \mathcal{D}_\Lambda = \varphi [\overleftarrow{\partial}_\Lambda - \tfrac{1}{2} X^{MN} \omega_{MN\Lambda}] \; , \qquad (2.1)$$

where $X^{MN}$ are the matrix representations of the generators of $OSp(3,1|4N)$ which obey the algebra

$$[X^{MN}, X^{PQ}\} = f_{RS}^{MNPQ} X^{RS} \; . \qquad (2.2)$$

In Eq. (2.2) $f_{RS}^{MNPQ}$ are the $OSp(3,1|4N)$ structure constants which depend on the inverse tangent space metric $\eta^{PQ}$ and thus have terms of order $1/k$.[4] The explicit k dependence of the Riemannian quantities can then be easily seen if one examines the factors of $X^{MN}$ and the structure constants as for example in the torsion

$$\mathcal{T}_{AB}^{\;\;C} = V^C_{[\Sigma} \mathcal{D}_{\Lambda\}} (-1)^{\Sigma A} V_A^\Lambda V_B^\Sigma \; . \qquad (2.3)$$

In addition to the above k dependence there is also a hidden k dependence of the theory appearing through the $\omega_{MN\Lambda}$, etc. Explicitly one finds

$$\omega_{ABC} = \tfrac{1}{2}[\Omega_{ABC} + (-1)^{BC}\Omega_{ACB} - (-1)^{A(B+C)}\Omega_{BCA}] \quad , \qquad (2.4)$$

$$\Omega_{ABC} = V_A{}^\Lambda (-1)^{BC+C}[V_{\Lambda,\Sigma}{}^D - (-1)^{C(\Lambda+\Sigma)+\Lambda+\Sigma+\Lambda\Sigma}V_{\Sigma C,\Lambda}{}^D](-1)^{B\Sigma}V_B{}^\Sigma \eta_{DB} .$$
$$(2.5)$$

### III. THE REDUCTION AND THE CONNECTION FORMULAE

Having identified all the k dependence in the theory, the reduction of the Riemannian connections to the supergravity connections proceeds in a straightforward fashion. As stated earlier we first contract the tangent space to a specific $O(3,1) \times O(N)$ part. This means that $\mathcal{D}_\Lambda$ is decomposed so that

$$\mathcal{D}_\Lambda = \tilde{D}_\Lambda - [X^{mai}\omega_{mai\Lambda} + \tfrac{1}{2}\bar{X}^{aibj}\omega_{aibj\Lambda}] \quad , \qquad (3.1)$$

$$\tilde{D}_\Lambda \equiv \partial_\Lambda - \tfrac{1}{2}[X^{mn}\omega_{mn\Lambda} + Y^{ij}\omega_{ij\Lambda}] \quad ,$$

$$Y^{ij} = X^{aibj}k\eta_{ba} \quad , \quad \omega_{ij\Lambda} = \tfrac{1}{4k}\omega_{aibj\Lambda}\eta^{ab} \quad . \qquad (3.2)$$

The supergravity covariant derivative $D_\Lambda$ is then proven to be given by the limit $k \to 0$ of $\tilde{D}_\Lambda$, i.e., $D_\Lambda = \tilde{D}_\Lambda)_{k\to 0}$. The results that emerge from the contraction are given in Ref. (4). When a supergravity gauge complete vielbein is inserted in the connection formulae, one finds on expansion in orders in $\theta^\alpha$ that the results agree on shell with those obtained by Brink et al. in Ref. (3).

### IV. SMOOTHNESS CONDITION ON GEOMETRIC CONTRACTION

The problem of how to characterize supergravity in superspace has been an ongoing one since the fundamental work of Wess and Zumino[3] on the N = 1 theory. The technique adopted by Wess and Zumino was to impose the so-called torsion constraints on specific components of the torsion tensor, i.e.,

$$T_{ab}^m = 2i(\eta\gamma^m)_{ab} \quad ; \quad T_{rs}^m = 0 = T_{na}^m \quad , \qquad (4.1a)$$

$$T_{ab}^c = 0 = T_{an}^c \quad . \qquad (4.1b)$$

Subsequent authors have extended these ideas to higher N. However, the nature of the torsion constraints change with N, and no systematic geometrical insights seem to have emerged.

What we should like to discuss for the remainder of this note is an alternate approach to superspace dynamics. This procedure is based on the technique discussed above of carrying out the geometrical contraction of Riemannian geometry by taking the $k \to 0$ limit. What we will see is the following: If one separates out the vacuum part of the geometry, the condition merely that a smooth (i.e., non-singular) $k \to 0$ contraction of the remainder of Riemannian geometry exists, implies (i) the geometry of the contracted space is that of supergravity and (ii) the __dynamics__ of supergravity must hold. This result has been explicitly verified for N = 1, 2, and we will give arguments that it may also hold for arbitrary N. From this picture then, one has the remarkable result that starting with a gauge complete off shell vierbein __geometrical contraction is equivalent to supergravity dynamics__, i.e., only one theory, supergravity, can result in such a contraction.

To examine the conditions that a smooth contraction exists, we recall the results of Sec. 3 where it was seen in Eq. (3.1) that the Riemannian covariant derivative $\mathcal{D}_\Lambda$ could be broken into its $O(3,1) \times O(N)$ part $\tilde{D}_\Lambda$ and its coset part $K_\Lambda$, i.e., $\mathcal{D}_\Lambda = \tilde{D}_\Lambda + K_\Lambda$. In Sec. 3 it was shown that on shell $\tilde{D}_\Lambda$ smoothly limits to the supergravity covariant derivative $D_\Lambda$ as $k \to 0$. We now assume that this is also true off-shell, i.e., we impose the boundary conditions at $\theta^{\alpha i} = 0$ that $\omega_{mn\mu}$ and $\omega_{ij\mu}$ reduce to the supergravity Lorentz and Yang-Mills connections.

The remaining part $K_\Lambda$ [which can be read off Eq. (3.1)] is not so simple. There are $1/k$ factors arising from the inverse Riemannian tangent space metric $\eta^{AB}$ appearing in matrix elements of the coset generators $X^{mai}$, $\bar{X}^{aibj}$ and $k$ factors from lowering the middle index on the connections. We require therefore that a non-singular limit exist for $K_\Lambda$, i.e., that the coefficients of any residual $1/k$ terms vanish. This is easily seen to lead to the following (off-shell) superspace "smoothness" conditions:

$$P_{main}(z) \equiv \Omega_{main}(z) + \Omega_{naim}(z) = 0 , \qquad (4.2)$$

$$\Omega'_{aibjm}(z) \equiv \Omega_{aibjm}(z) - \langle 0|\Omega_{aibjm}|0\rangle = 0 , \qquad (4.3)$$

where $\Omega_{ABC}(z)$ is given in Eq. (2.5).

## V. DETERMINATION OF $K_\Lambda$ AND KINEMATICAL TORSION CONSTRAINTS

Before examining the dynamical aspects of the contraction, we show that indeed a smooth $k \to 0$ contraction of Riemannian geometry yields precisely the supergravity geometry. We have already seen that the $O(3,1) \times O(N)$ part $\tilde{D}_\Lambda$ correctly limits to the supergravity covariant derivative. The Riemannian geometry, however, possesses an additional coset piece $K_\Lambda$ not a priori found in supergravity. We will now show however, that $K_\Lambda$ can be uniquely determined in terms of the supergravity torsion $T^C_{AB}$ in the $k \to 0$ limit, and thus in the contraction limit is not a new quantity.

The condition determining $K_\Lambda$ arises from the fact that the total Riemannian torsion $\mathcal{T}^C_{AB}$, defined in Eq. (2.3), must of course vanish. This yields then the condition that the supergravity torsion must be the negative of the coset torsion (formed from $K_\Lambda$) in the $k \to 0$ limit, i.e.,

$$[V^C_{[\Sigma} K_{\Lambda\}}(-1)^{\Sigma A} V^\Lambda_A V^\Sigma_B]_{k \to 0} = - T^C_{AB} . \qquad (5.1)$$

One may indeed show from Eq. (5.1) that all non-zero components of $K_\Lambda$ are determined in terms of $T^C_{AB}$ in the $k \to 0$ limit.

Thus Riemannian geometry possesses a smooth contraction only to supergravity geometry. Note that these results hold for arbitrary N.

Having shown that the Riemannian geometry does indeed contract smoothly to the supergravity geometry, one may enquire as to where the torsion constraints fit into this picture. One may in fact <u>deduce</u> the kinematical torsion constraints directly from the smoothness condition Eq. (4.2). To see this, recall from Sec. 2, that the supergravity torsion $T_{AB}^C$ is the $k \to 0$ limit of

$$\tilde{T}_{AB}^C \equiv V_{[\Sigma \Lambda\}}^C \tilde{D}_\Lambda (-1)^{\Sigma A} V_A^\Lambda V_B^\Sigma , \qquad (5.2)$$

where $\tilde{D}_\Lambda$ is given in terms of the Riemannian connections in Eqs. (3.2), (2.4) and (2.5). Consider, for example $\tilde{T}_{n\;ai}^{\;m}$. One finds

$$\tilde{T}_{nmai} = \tfrac{1}{2}[\Omega_{naim} + \Omega_{main}] . \qquad (5.3)$$

But in the $k \to 0$ limit the r.h.s. must vanish as this is just the smoothness condition Eq. (4.2). We have therefore $T_{nai}^m = 0$. In a similar fashion one may show from the smoothness condition that

$$T_{rs}^m = 0 ; \quad T_{ai\;bj}^{\;m} = 2i(\eta\gamma^m)_{ab}\delta_{ij} . \qquad (5.4)$$

Note again that these results hold for arbitrary N.

## VI. DYNAMICAL EQUATIONS

Not only does the requirement of a smooth contraction imply the kinematical constraints, but they also appear to contain the <u>dynamics</u> of supergravity as well. In order to obtain further information from Eq. (4.2) it is necessary to insert in a gauge complete off-shell vierbein. Such a vierbein has been explicitly calculated by Brink et al.[3] for N = 1 supergravity using Breitenlohner auxiliary fields[5] and we consider this case here. Brink et al. find, to linear order in $\theta^\alpha$

$$V_\mu^m(z) = e_\mu^m(x) + i\bar{\psi}_\mu \gamma^m \theta - i\bar{x}^m \gamma_\mu \theta + \ldots$$
$$V_\mu^a(z) = \tfrac{1}{2}\psi_\mu^a(x) + [(\Gamma_\mu + iM)\theta]^a + \ldots$$
$$V_\alpha^m(z) = i(\bar{\theta}\gamma^m)_\alpha + \ldots ; \quad V_\alpha^a(z) = \delta_\alpha^a + \ldots \qquad (6.1)$$

where $e_\mu^m(x)$ and $\psi_\mu^a(x)$ are the N = 1 supergravity fields and $\chi_m^a(x)$, $M(x)$ two of the Breitenlohner fields. By direct substitution, we find that the boundary condition is automatically obeyed (and indeed could have been omitted from our definition of the geometrical contraction). The smoothness condition, Eq. (4.2) to zero'th order in $\theta^\alpha$ yields

$$P_{man}(x,\theta^\alpha=0) = i(\bar{\chi}_n\gamma_m + \bar{\chi}_m\gamma_n) = 0 \qquad (6.2)$$

and hence $\chi_m^a(x) = 0$. However, by the Breitenlohner gauge transformation laws, the vanishing of $\chi_m^a$ implies that all Breitenlohner fields take on their mass shell values and that the gravitino field equation holds. From this it is easy to deduce the Einstein equations and hence the entire N = 1 supergravity dynamics is a consequence of the smoothness condition Eq. (6.2). Of course, one may now go back and verify that the dynamical torsion constraints, Eqs. (4.1b) also hold. However, these equations are now supernumeraries, since we have already deduced all the supergravity dynamics.

Do these results generalize to arbitrary $N \leq 8$? If in fact Breitenlohner type auxiliary fields generalize to higher N, the answer is yes.[6] We speculate therefore that indeed, in a Breitenlohner formalism, the dynamics of supergravity for $N \leq 8$ is universally stated in Eq. (4.2).

## ACKNOWLEDGMENTS

We should like to express our indebtedness to P. van Nieuwenhuizen for suggesting that we examine the $k \to 0$ limit in the coset space and to B. Zumino for useful discussions. Research is supported in part by the National Science Foundation under Grant No. PHY80-08333. One of us (PN) gratefully acknowledges the hospitality of CERN, where part of this work was done.

## REFERENCES

1. The earliest discussion regarding the meaning of the parameter k in superspace geometries as well as first discussion of non-Riemannian geometries is given in B. Zumino, Proc. Conf. on Gauge Theories and Modern Field Theory, Boston, 1975, MIT Press, Cambridge, Mass. (1976). (Eds. R. Arnowitt and P. Nath)
2. The notation of Brink et al. of Ref. (3) is used throughout this paper.
3. P. Nath and R. Arnowitt, Phys. Lett. <u>65B</u>, 73 (1976). J. Wess and B. Zumino, Phys. Lett. <u>66B</u>, 361 (1977); L. Brink, M. Gell-Mann, P. Ramond and J. H. Schwarz, Phys. Lett. <u>74B</u>, 336 (1978); <u>76B</u>, 417 (1978); W. Siegel and S. J. Gates, Nucl. Phys. <u>B147</u>, 77 (1979); Y. Ne'eman and T. Regge, Riv. Nuovo Cimento Ser. III 1 (No. 5) (1978); L. Brink and P. Howe, Phys. Lett. <u>88B</u>, 268 (1979); L. Castellani, S. J. Gates and P. van Nieuwenhuizen, Stony Brook Preprint.
4. R. Arnowitt and P. Nath, Phys. Rev. Lett. <u>44</u>, 223 (1980). P. Nath and R. Arnowitt, Nucl. Phys. <u>B165</u>, 462 (1980).
5. P. Breitenlohner, Phys. Lett. <u>67B</u>, 49 (1977).
6. The N = 2 gauge complete vierbein using minimal auxiliary fields of L. Castellani, P. van Nieuwenhuizen and S. J. Gates (SUNY at Stony Brook preprint ITP-SB-80-9) automatically satisfies Eq. (4.2) while the boundary conditions at $\theta^\alpha = 0$ allow one to deduce all the N = 2 supergravity dynamics.

# GEOMETRIC GHOSTS AND UNITARITY

Y. Ne'eman[*,†,+]

Tel Aviv University, Israel and University of Texas, Austin

## ABSTRACT

We review the geometrical identification of the renormalization ghosts and the resulting derivation of Unitarity equations (BRST) for various gauges: Yang-Mills, Kalb-Ramond and Soft-Group-Manifold.

## THE CONVENTIONAL TREATMENT

The (geometric) Yang-Mills Lagrangian $L_{INV} = -\frac{1}{4} F^a_{\mu\nu} F^{\mu\nu}_a$ has no inverse Fourier-transform needed to write a propagator. It thus had to be supplemented by a gauge-fixing term $L_{FIX} = -\frac{1}{2} C^a C_a$ (with $C^a = \partial_\mu A^a_\mu$ for example) or more formally, the linear $L_{FIX} = -\sigma_a \Sigma^a$, with $\sigma_a$ a (boson) scalar Lagrange multiplier, ensuring the appearance of $a\delta(\Sigma)$ in the functional integration. To ensure Unitarity (i.e. cancellation of contributions due to redundant components of A or its F ) Feynman introduced ghosts[1]. The ghost Lagrangian $L_{GH} = \bar{X} \hat{m} X^a$ involves scalar ghost $X^a$ and antighost $\bar{X}_a$ fields, $\hat{m}$ is given by the gauge variations of $\Sigma^a$ (or $C^a$).

$$\delta A^a_\mu = D_\mu \epsilon \rightarrow \delta \Sigma^a = \hat{m} \epsilon^a \qquad (1)$$

$X^a$ abd $\bar{X}_a$ have to be assigned Fermi statistics, to produce a minus sign guaranteeing cancellations between their closed loops and the redundant contributions. The quantum Lagrangian

$$L = -\frac{1}{4} F^a_{\mu\nu} F^{\mu\nu}_a - \sigma_a \Sigma^a + \bar{X}_a \hat{m} X^a \qquad (2)$$

obeys Slavnov-Taylor invariance and provides generalized Ward-Takahashi identities. 't Hooft and Veltman then completed the proof of renormalizability, adding regularization. The conditions for unitarity have since been reduced to invariance under BRST (discrete) supertransformations [2]. Let $\Lambda$ be a (constant) odd element in a Grassmann manifold, and take ($\delta$ is the gauge infinitesimal variation of (1)),

$$\epsilon^a = X^a \Lambda \; , \; \Lambda^2 = 0 \; , \; \{\Lambda, X\} = \{\Lambda, \bar{X}\} = 0 \; ; \; s = -\frac{\partial}{\partial \Lambda} \delta \qquad (3)$$

$$\therefore sA^a_\mu = D_\mu X^a; \; sX^a = \frac{1}{2} f^a_{bc} X^b X^c = \frac{1}{2}[X,X]^a \; ; \; s\Sigma^a = \hat{m} X^a \qquad (4)$$

Add the formal prescriptions $s\bar{X}_a = \sigma_a$ , $s\sigma_a = 0$ (5)

$\therefore sL_{INV} = 0$ ; $sL_{FIX} = -\sigma_a \hat{m} X^a$; $sL_{GH} = \sigma_a \hat{m} X^a - \bar{X}_a s^2 \Sigma^a$; so that provided $s^2 A^a_\mu = 0$ (to make $s^2 \Sigma^a$ vanish), which also implies $s^2 X^a = 0$ (using Jacobi's identity or algebraic closure of the $\delta$ variations), one has $sL = 0$. Note that in the more conventional quadratic $L_{FIX}$, $sC^a = \hat{m} X^a$, $s\bar{X}_a = C_a$, so that $s^2 \bar{X}_a = \hat{m} X^a \neq 0$. The transformation s, with $s^2$ vanishing on $A^a_\mu$, $X^a$ and (in the linear treatment only) on $\bar{X}_a$, has the features of an exterior differential.

## THE GEOMETRICAL IDENTIFICATION

The geometrical identification of the ghosts, as vertical components of the connection $P$, a principal fiber bundle $(P, M, G\bullet)$ has been recently demonstrated [3]. Using $x^\mu$ as co-ordinates on the base manifold $M$ and $\alpha^i (i = 1..n)$ on the group fiber $G$, a section $\Sigma$ may be

---

\* Wolfson Chair Extraordinary of Theoret. Physics
† Supported in part by the U.S.-Israel Binational Science Foundation
+ "      "     "  "   "    "  U.S.DOE, Grant EY-76-S-05-3992

0094-243X/81/680981-04$1.50 Copyright 1981 American Institute of Physics

given by $\alpha^i = 0$, lifting the $x^\mu$ onto it. The dot (•) represents right-multiplication $P \times G \to P$ so that $(p.g) \bullet g^1 = p \bullet (gg^1)$, $\forall\, p \in P$, $\forall g, g^1 \in G$. To produce that action, the dot maps the n-dimensional Lie algebra $A_G$ of $G$ onto the differential operators of $P_*$, the tangent manifold. The (quasi) inverse map (since $P_*$ has 4 + n dimensions) is given (through contraction $dz^M \lrcorner \frac{\partial}{\partial z^N} = \delta^M_N$) by the connection $w^a$. Let $z^M(x^\mu, \alpha^i)$ represent local co-ordinates on $P$. On $\Sigma$,

$$w^a = \Xi^a_i(\alpha, x)\, d\alpha^i + A^a_\mu(\alpha, x)\, dx^\mu \tag{6}$$

We remind the reader that although he may be used to $A^a_\mu$ being only $x^\mu$ dependent, any gauge transformation will introduce $\alpha^i$ dependence, through the homogeneous term $e^{i\alpha^i \lambda_i} A^a_\mu(x)\, e^{-i\alpha^i \lambda_i}$ ($\lambda$ is a basis $A_G$). For $\Sigma$ a global (trivial) horizontal section, $\Xi^a = \Xi^a_i\, d\alpha^i$ is the left-invariant Cartan 1-form. More generally, $\Xi^a = X^a$, the ghost field. To prove it, one writes the Cartan-Maurer structural equation on $P$. This corresponds to the statement that the curvature

$$F^a = \tfrac{1}{2} F^a_{MN}\, dz^{[M}\, dz^{N]} = (\tfrac{\partial}{\partial z^M} w^a_N - \tfrac{\partial}{\partial z^N} w^a_M + \tfrac{1}{2} f^a_{bc} w^b_M w^c_N)\, dz^{[M}\, dz^{N]} \tag{7}$$

has nothing but horizontal components
$F^a = \tfrac{1}{2} F^a_{\mu\nu}\, dx^{[\mu} dx^{\nu]}$. Denoting by $\bar{d}$ the full differential over $z^M$ (more generally the exterior derivative, i.e. a curl when acting on a one-form, such as $X^a$) and by $\ell$ and $d$ the $\alpha^i$ and $x^\mu$ components respectively, we have

$$\ell \oint = d\alpha^i \tfrac{\partial}{\partial \alpha^i} \oint,\quad d\oint = dx^\mu \tfrac{\partial}{\partial x^\mu} \oint,\quad \bar{d} \oint = dz^M \tfrac{\partial}{\partial z^M} \oint = \ell \oint + d \oint \tag{8}$$

$d\alpha^i\, F^a_{i\mu} = 0 = \ell A^a_\mu - D_\mu X^a$, $d\alpha^{[i}\, d\alpha^{j]}\, \tfrac{1}{2} F^a_{ij} = 0 = \ell X^a - \tfrac{1}{2}[X, X]^a$
i.e. the first two BRST equations (4), provided

$$\Xi^a \equiv X^a,\quad \ell \equiv s \tag{9}$$

Indeed, $\bar{d}^2 = d^2 = s^2 = sd + ds = 0$ from cohomology.

We now have to verify whether indeed the fermi statistics of $X^a$ derive indeed from its being a 1-form: in non exterior-calculus words, a differential, anticommuting because of the antisymmetrization of differentials in a measure. Perturbation physics is described by the generating functional ($G$ is the group volume)

$$e^{i\Gamma} = G \int D(\phi)\, e^{iS_{INV}(\phi)}, \text{ with } \phi = A^a_\mu \lambda_a dx^\mu \tag{10}$$

with $S_{INV}(\phi) = \int L_{INV} d^\mu x$, the Yang-Mills action. We fix the gauge by inserting $1 = \int \delta(\Sigma)\, d\Sigma$; however, $d\Sigma = d^i \partial_i \Sigma = s\Sigma$ with no $\partial_\mu$ contribution through our choice of co-ordinates for $\Sigma$. Using the representation $\delta(\Sigma) = \int d\sigma e^{i\sigma\Sigma}$ and the Berezin intregral $s\Sigma = \int d\bar{X} e^{i\bar{X} s\Sigma}$ (with $\int \hat{d}\theta \oint(\theta) \equiv - i\tfrac{\partial}{\partial \theta} \oint(\theta)\big|_{\theta=0}$)

$$e^{i\Gamma} = \int D(\phi, X, \bar{X}, \sigma) e^{iL} \tag{11}$$

where we have replaced the group volume $G$ by an integral over $X$, the measure on the fiber. The functional integral is indeed geometrical, and the fermi statistics of $X, \bar{X}$ derive from their nature as differentials or 1-forms.

Ojima has recently suggested a geometric derivation for (5). Complexifying the fiber $G$, one writes for (8) $\bar{d} = s + r + d$, where $r = d\bar\alpha \tfrac{\partial}{\partial \bar\alpha}$ and $rs + sr = 0$, $r^2 = 0$. Note that since spin and

statistics do not co-relate, the $r$ equations may violate CPT and differ from the $s$; so does $\bar{X}$ differ from $X$ in its BRST variations. The resulting new (rr), (rs) equations are

$$r\overline{X} = \tfrac{1}{2}[\overline{X},\overline{X}] \; ; \quad rX + s\overline{X} + [X,\overline{X}] = 0 \qquad (12)$$

Fixing $s\overline{X}_a = \sigma_a$, the second equation reads $rX = -\sigma-[X,\overline{X}]$. Applying s to the (rr) equation, we have $r\sigma = -[\sigma,\overline{X}]$. We now apply s to both sides of the (rs) equation. The left hand side is $-rsX = -\tfrac{1}{2}r[X,X] = +[\sigma,X] + [[X,\overline{X}],X]$ where we have used the value $rX$ again. For the right hand side, we have $-s\sigma-\tfrac{1}{2}[[X,X],\overline{X}] + [X,\sigma]$. The triple brackets cancel on both sides by Jacobi's identity, and we find $s\sigma = s^2 \overline{X} = 0$. Note that $L_{FIX} + L_{GH} = -s\;(X\Sigma)$.

Since $sL = 0$, and $L$ is a 4-form over $x^\mu$ so that $dL = 0$ by saturation, we have geometric closure of $L$ over $P$,

$$dL = 0 \qquad (13)$$

## SUPERGROUPS AND SPONTANEOUS BREAKDOWN

In ref.[4] we have shown that for G a supergroup, H⊂G its even subgroup (indices $a,b...\in H$; $i,j...\in G/H$)

$$\left. \begin{array}{l} \omega^a = \chi^a + dx^\mu A^a_\mu \\ \omega^i = \eta^i + dx^\mu \xi^i_\mu \end{array} \right\} \qquad (14)$$

Due to the odd nature of G/H, the $\eta^i$ are scalar boson fields, corresponding to a Goldstone-Higgs set. The $\xi^i_\mu$ is a vector-ghost set. This picture was used[4] to reproduce chiral spontaneously-broken $SU(3)_L \otimes SU(3)_R$, applying the supergroup Q(3). It may also provide the interpretation of SU(2/1) as a unified weak-electromagnetic theory (see our paper in session QFDIII of these Proceedings).

## THE KALB-RAMOND FIELD

The usefulness of this geometric formalism was displayed in recent work on the $B^u_{\mu\nu}$ antisymmetric tensor gauge field[4]. Both the original Abelian model [4] and the more recent non-Abelian[5] have been shown to represent scalar fields, the latter a non-linear $\sigma$-model. The Lagrangian is here

$$L = \tfrac{1}{4}\epsilon^{\mu\nu\rho\sigma} B^a_{\mu\nu} F_{a\rho\sigma}(A+H) - \tfrac{1}{2}m^2 H^a_\mu H^\mu_a - \tfrac{1}{4}F^i_{\mu\nu}(A) F^{\mu\nu}_i(A) \qquad (15)$$

$$B^a = \tfrac{1}{2} B^a_{\mu\nu} dx^\mu \wedge dx^\nu \; ; \; H^a = H^a_\mu dx^\mu \; ; \; A^i = A^i_\mu dx^\mu \, , \, A^m = 0 \qquad (16)$$

a denotes the adjoint representation of G; i the subgroup G´ and m the quotient G/G´.

The first attempts to supply ghosts and BRST equations was done with no geometric guidance and failed. Applying geometry, Thierry-Mieg produced straight-forward results[6]. For instance, the full B "connection" is

$$\tilde{B} = \tfrac{1}{2}B_{\mu\nu}dx^\mu\wedge dx^\nu + \tfrac{1}{2}\tilde{B}_{ij}d\alpha^i\wedge d\alpha^j + \tilde{B}_{\mu i}dx^\mu\wedge d\alpha^i \qquad (17)$$

and we see that we have as ghosts a scalar boson $\phi$ and a vector fermion $\xi_\mu$ as in (14),

$$\phi = \tfrac{1}{2}\tilde{B}_{ij}d\alpha^i\wedge d\alpha^j \; ; \; \xi_\mu = \tilde{B}_{\mu i} d\alpha^i \qquad (18)$$

and the BRST equations represent the vanishing of all components of $\tilde{D}_{A+H}\tilde{B}$ other than the fully horizontal[6] (in $dx^\mu\wedge dx^\nu\wedge dx^\rho$).

## THE SOFT GROUP MANIFOLD

To treat non-internal gauges (such as gravity and supergravity) we have introduced[7] the Soft Group Manifold (SGM). The SGM becomes a principal bundle upon application of the equations of motion (spontaneous fibration), with a subgroup F⊂G as fiber[8]. We thus get pseudo-closure $d_{(F)} L = 0$.

More recently, we have checked[9] that the geometric identification of ghosts and of the BRST operator s holds over $G$ (the SGM). We have used the algebra of Lie derivatives $L_\varepsilon$ which holds off-mass-shell. However, the conventional quantum Lagrangian is not invariant under this algebra $L_\varepsilon L = \xi \lrcorner dL + d (\xi \lrcorner L)$. The second term is a divergence and does not matter, but $dL \neq 0$ as long as $L$ is not reduced by fibration to space-time only. Invariance has generally been obtained by adding a quartic ghost term in $L$, cancelling its vertical components[10].

The geometric derivation also fixes the "natural" commutation relations among the various ghosts and fields[11].

## REFERENCES

1. R. P. Feynman, Acta Phys. Polon. **26**. 697 (1963); B. S. DeWitt, "Dynamical Theory of Groups and Fields", Gordon & Breach, N.Y. (1965); L. D. Faddeev and V. N. Popov, Phys. Lett. **B25**, 29 (1967)
2. C. Becchi, A. Rouet and R. Stora, Commun. Math. Phys. **42**, 127 (1975); I. V. Tyutin, FIAN 39 (1975).
3. J. Thierry-Mieg, These de Doctorat d'Etat (Paris Sud, 1979) and J. Math. Phys. (to be pub.); also NUOVO Cim. A (to be pub.); Y. Ne'eman, Proc. XIX Int. Conf. on HEP (Tokyo, 1978), eds S. Homma et al, Phys. Soc. Jap. Pub. (1979), 552. See also ref 8.
4. M. Kalb and P. Ramond, Phys. Rev. **D9**, 2273 (1974). E. Cremmer and J. Scherk, Nucl. Phys. **B72**, 117 (1974).
5. D. Z. Freedman, report CALT 68-624 (1977) unpub. P. K. Townsend, CERN report TH. 2753 (1979) unpub.
6. J. Thierry-Mieg, report HUTMP 79/B86.
7. Y. Ne'eman and T. Regge, Rivista del Nuovo Cimento I, N5 (1978).
8. J.Thierry-Mieg and Y. Ne'eman, Ann. of Phys. (NY) **123**, 246 (1979)
9. Y. Ne'eman, E. Takasugi and J. Thierry-Mieg, Phys. Rev D (to be pub.); also report TAUP 798-79 (revised version).
10. R. E. Kallosh, JETP Lett. **26**, 575 (1977). E. S. Fradkin and M. A. Vasiliev, Phys. Lett. **72B**, 70 (1977).
11. Y. Ne'eman and J. Thierry-Mieg, in Recent Developments in General Relativity, Proc. II Marcel Grossmann Symposium (Trieste 1979) A. Ruffini, ed.

# HIGH SPIN FIELDS

Thomas L. Curtright
The Enrico Fermi Institute, University of Chicago,
Chicago, IL 60637 and Physics Department, University of Florida,
Gainesville, FL 32611*

This is a brief qualitative review of some recent attempts to formulate local field theories of fundamental, structureless particles with arbitrary spin. Despite these attempts there are no known consistent theories, based on a local lagrangian, describing higher (s>2) spin particles which interact either mutually, with Einstein gravity, or with lower spins (i.e. s<2). We outline here the technical reasons why such local formulations are nonexistent and we indicate areas possibly deserving further study.

We are motivated to study higher spin field theories because of their intrinsic mathematical interest (as was Dirac[1]), and because linear supermultiplets with manifest invariance groups large enough to contain $SU(3)_c \times (SU(2) \times U(1))_{ew}$ require spins $s \geq 5/2$. Conceivably, such high spin linear representations of extended supersymmetries may not be present in Nature, even if supersymmetry is. Nevertheless, the higher spin formalism deserves consideration for possible physical applications, even if such applications are only to approximate higher spin composite particles, or to describe higher spin auxiliary fields in extended supergravities.

Historically, Fierz and Pauli[2] first emphasized the technical benefits of describing relativistic particles with any spin by using a local lagrangian framework. They discussed all spins s<5/2 in detail, including all gauge invariances present for s<5/2 when m=0. They also pointed out that higher spins require a large number of auxiliary fields, for massive particles, in order to obtain wave equations defining a unique on-shell mass and spin. The necessary structure involving such auxiliary fields delayed the explicit construction of even free field lagrangians for arbitrary spin and nonzero mass for 35 years[3]. Only within the last two years was it finally realized that the m=0 case is conceptually quite simple for arbitrary spin[4-8].

All higher spin massless fields may be viewed as gauge fields. For example, consider a free spin 5/2 particle described by the symmetric spinor-tensor $\psi_{\mu\nu}$. Under the Lorentz group, $\psi_{\mu\nu}$ transforms as $(3/2 + 1/2, 1) + (1/2, 0)$, and carries spins $5/2 + 2(3/2) + 3(1/2)$. By analogy with known lower spin cases, one expects the gauge invariance under

$$\delta\psi_{\mu\nu} = \partial_\mu \eta_\nu + \partial_\nu \eta_\mu \tag{1}$$

to reduce the number of physical propagating massless modes to only helicities +5/2 and -5/2. This is not quite correct. There is no local lagrangian of the form $\bar\psi \partial \psi$ which yields an invariant action

---

*Permanent address.

0094-243X/81/680985-04$1.50 Copyright 1981 American Institute of Physics

under (1) unless constraints are imposed on $\eta_\mu(x)$, the spinor-vector gauge parameter. A sufficient and simple constraint is

$$\gamma^\mu \eta_\mu(x) = 0, \qquad (2)$$

which leads to the unique gauge invariant action

$$A_{5/2} = \int dx (\bar{\psi}_{\mu\nu} \not{\partial} \psi^{\mu\nu} + 2\bar{\psi}_\mu \not{\partial} \psi^\mu - \tfrac{1}{2} \bar{\psi} \not{\partial} \psi - 4\bar{\psi}_\mu \partial_\nu \psi^{\mu\nu} + 2\bar{\psi} \partial_\mu \psi^\mu) \qquad (3)$$

where $\psi_\mu = \gamma^\nu \psi_{\mu\nu}$, $\psi = \psi_\mu{}^\mu$, and $\psi_{\mu\nu}$ is Majorana for simplicity.

The restricted gauge invariance of (1) and (2) is still sufficiently strong to eliminate all spurious intermediate helicity modes. This was established by explicit examination of propagator pole residues[5,6], by reducing the action to (Hamiltonian) canonical form[8], and by explicity constructing supermultiplets combining such s=5/2 gauge fields with known lower spins[7].

These same considerations apply to all higher spins described in terms of totally symmetric tensors and spinor-tensors, with additional constraints on the fields for s>3. Namely,

$$\gamma^\mu \gamma^\nu \gamma^\lambda \psi_{\mu\nu\lambda\cdots\rho} = 0 \quad \text{for} \quad s \geq 7/2$$

$$\phi^\mu{}_\mu{}^\nu{}_{\nu\lambda\cdots\rho} = 0 \quad \text{for} \quad s \geq 4. \qquad (4)$$

The gauge transformations are the now obvious extensions

$$\delta \phi_{\mu\nu\lambda\rho\cdots} = \partial_{(\mu} \xi_{\nu\lambda\rho\cdots)} \quad \text{with} \quad \xi^\nu{}_{\nu\rho\cdots} = 0$$

$$\delta \psi_{\mu\nu\lambda\cdots} = \partial_{(\mu} \eta_{\nu\lambda\cdots)} \quad \text{with} \quad \gamma^\nu \eta_{\nu\lambda\cdots} = 0 \qquad (5)$$

Unique actions follow from postulating invariance under (5) with (4).

Three other formal developments for free fields with arbitrary spin are noteworthy. The BRS formulations of the gauge invariances above are known[6,9] and the corresponding covariant quantization of the free fields has been carried out. A more "geometrical" foundation for free higher spins has been initiated through the construction of generalized Christoffel symbols[10] (which include gauge invariant objects) formed by taking higher derivatives of the fields. Finally, "vierbein" (i.e. non-symmetric tensor) formulations have been given which are equivalent to the totally symmetric tensor formalism[11]. This last approach avoids the above constraint discontinuity at s=7/2 (cf. eqn. (4)).

Free higher spin fields are thus pleasingly described as local gauge fields. Unfortunately, the introduction of local interactions leads to difficulties in at least three areas: 1) Higher spin charges; 2) Consistency conditions; 3) Causality violations. We briefly summarize results on these difficulties.

First, there are no higher spin (>1) charges[12] to which higher

spin (>2) massless fields could couple at zero momentum. Thus all
such fields must decouple as $q \to 0$. (Perhaps more dramatic, a related
analysis shows $m=0$ Born amplitudes for $s=5/2$ coupled to gravity must
vanish identically[13].)

Second, the equations of motion for massless higher spin fields
are inconsistent when minimal coupling to gravity is attempted[14]
unless unrealistic conditions are imposed on the curvature of space-
time (e.g. Weyl flat spaces). This may be seen by differentiating
the field equations. These inconsistencies are directly related to
the loss of gauge invariance for the minimally coupled action. More
physically, unless the necessary curvature conditions are met, energy-
momentum does not seem to be conserved[15]. For the massless case,
nonminimal couplings apparently do not alleviate these inconsistencies.

Third, if masses are introduced, the wave equations can be made
consistent, but a new problem arises: Acausal propagation[16]. In
general, spacelike characteristic surfaces pass through points with
nonvanishing curvature, even in the weak field limit.

We now indicate some open questions inspired by the above
difficulties. a) What constraints apply to $m>0$, $s>2$ Born amplitudes
in the presence of gravitational interactions? b) Can both local
mass terms and local nonminimal couplings be introduced to yield
consistent, causal higher spin theories in such a way that the $m \to 0$
limit is unavoidably singular? c) Interactions with gravity aside,
can higher spins be consistently and causally self-coupled?

One naturally guesses analogies with Yang-Mills theory or
Einstein gravity as self-coupling candidates, but a more promising
approach for answering the third question may be by imitating the
nonlinear Born-Infeld model of self-coupled electrodynamics, where
$L = [1 + KF_{\mu\nu}F^{\mu\nu} - \frac{1}{4} K^2 (*F_{\mu\nu}F^{\mu\nu})^2]^{\frac{1}{2}}$. For higher spins, a nonlinear
lagrangian might be similarly composed of gauge invariant general-
ized Christoffel symbols. (Recent work suggests that supersymmetry
may be essential in constructing such nonlinear theories[17].)

The third question above also brings up the crucial notion of
"manifestly chargeless" sources, which vanish identically as $q \to 0$,
as may indeed be found in the Born-Infeld model. Such sources
appear natural from another point of view: "Massive dual formulations" for higher spin fields[18]. Massive dual fields describe spin
s when $m>0$, but collapse to give no propagating physical degrees of
freedom when $m=0$ (except for the special cases $s=1$ and $s=3/2$). Thus
the massless limit is quite singular, even for the free field case.
It is an open question, related to b) above, whether such massive
higher spin dual fields can interact locally in a fully consistent,
causal manner in the presence of gravitation.

Finally, we note the allowed possibility that higher spins can
be made to interact only if all such spins are present (cf. conventional dual models). A possible novel approach here is based on
representations of the de Sitter group[19].

In summary, although consistent local interacting field theories
involving higher spin fields are not known, the kinematic foundations
upon which such models must be constructed have been considerably

clarified in the last two years by extending the concept of gauge invariance to include all spins. Massless fields with arbitrary spin appear to have insurmountable difficulties when coupled individually to gravity. Massive fields have not been similarly investigated to the point of an impasse, and in our view represent the most promising possibilities for local higher spin field theories with gravitational coupling.

Coincident with many of the higher spin investigations described above has been the discovery that the N=8 supergravity theory has a nonlinearly realized "local" SU(8) invariance[20] which may permit supersymmetric theories to incorporate $SU(3)_c \times (SU(2) \times U(1))_{ew}$ without the need for fundamental higher spin (>2) fields. This is still conjecture. There is a distinct possibility, however, that some higher spin formalism may play a useful role in the phenomenological application[21] of the N=8 theory.

## REFERENCES

1. P. A. M. Dirac, Proc. R. Soc. A155, 447 (1936).
2. M. Fierz and W. Pauli, Proc. R. Soc. A173, 211 (1939).
3. L. P. S. Singh and C. R. Hagen, Phys. Rev. D9, 898; 910 (1974).
4. C. Fronsdal, Phys. Rev. D18, 3624 (1978).
5. J. Fang and C. Fronsdal, Phys. Rev. D18, 3630 (1978).
6. F. A. Berends, J. W. van Holten, P. van Nieuwenhuizen, and B. de Wit, Phys. Lett. 83B, 188 (1979); Nucl. Phys. B154, 261 (1979); J. Phys. A:13, 1643 (1980).
7. T. L. Curtright, Phys. Lett. 85B, 219 (1979).
8. C. Aragone and S. Deser, Phys. Rev. D21, 352 (1980).
9. C. Fronsdal and H. Hata, Nucl. Phys. B162, 487 (1980).
10. B. de Wit and D. Z. Freedman, Phys. Rev. D21, 358 (1980).
11. C. Aragone and S. Deser, Brandeis preprint, 1980; M. A. Vasiliev, Lebedev Institute preprint N14, 1980.
12. S. Weinberg, Phys. Rev. 135, B1049 (1964); S. Coleman and J. Mandula, Phys. Rev. 159, 1251 (1967); R. Haag, J. T. Lopuszanski, and M. Sohnius, Nucl. Phys. B88, 257 (1975).
13. M. T. Grisaru, H. N. Pendleton, and P. van Nieuwenhuizen, Phys. Rev. D15, 996 (1977).
14. H. A. Buchdahl, Nuovo Cimento 10, 96 (1958); 25, 486 (1962); S. M. Christensen and M. J. Duff, Nucl. Phys. B154, 301 (1979); C. Aragone and S. Deser, Phys. Lett. 86B, 161 (1979).
15. C. Aragone and S. Deser, Brandeis preprint, 1979.
16. G. Velo and D. Zwanziger, Phys. Rev. 186, 1337 (1969).
17. S. Deser and R. Puzalowski, Brandeis preprint, 1980.
18. T. L. Curtright, Chicago preprint, EFI 80/04; T. L. Curtright and P. G. O. Freund, Chicago preprint, EFI 80/05.
19. C. Fronsdal, Phys. Rev. D20, 848 (1979); see also, J. Fang and C. Fronsdal, UCLA preprint, UCLA/79/TEP/23.
20. E. Cremmer and B. Julia, Nucl. Phys. B159, 141 (1979).
21. T. L. Curtright and P. G. O. Freund, in "Supergravity," P. van Nieuwenhuizen and D. Z. Freedman (eds.), North-Holland (1979); J. Ellis, M. K. Gaillard, and B. Zumino, CERN preprints TH-2842 and, with L. Maiani, TH-2841, 1980.

# QCD I

G. Sterman, SUNY-Stony Brook, Organizer

# INFRARED DIVERGENCES IN QUANTUM CHROMODYNAMICS

A. Andraši, M. Day, R. Doria and J. C. Taylor
Department of Theoretical Physics, Oxford University,
Oxford, England.

C. Carneiro, J. Frenkel and M. Thomaz
Instituto de Fisica, Universidade de São Paulo,
São Paulo, Brasil.

## ABSTRACT

We show that, in general, Bloch-Nordsieck inclusive cross-sections are not infrared finite in perturbative QCD.

Our purpose will be to demonstrate in a simple way, that there are processes in which non-leading infrared (IR) divergences coming from two soft gluons do not cancel between real and virtual diagrams (ref. 1). Since a necessary condition for non-cancellation is an initial state containing at least two coloured particles, the simplest reaction to look at is :

$$q + \bar{q} \rightarrow \gamma^* + \text{soft gluons} \qquad (1)$$

where q and $\bar{q}$ are quark and anti-quark, $\gamma^*$ is a virtual photon and the soft gluons are required in a physical cross-section of finite energy resolution. We take the initial (massive) quark and anti-quark to be each colour and spin averaged.

To order $g^4$, where $g$ denotes the QCD coupling constant, we find uncancelled IR divergences, which are non-leading. In order to get an understanding of this behaviour, let us compare reaction (1) with the process :

$$\gamma^* \rightarrow q + \bar{q} + \text{soft gluons} \qquad (2)$$

In this case, the IR divergences cancel simply in consequence of unitarity (ref. 2). To see this, consider the photon self-energy for $E_{\gamma^*} > 2m$, an example of which is shown in Figure 1, where the continuous lines denote the massive quark and antiquark. The broken lines represent massless gluons with soft momenta. As is well known, this diagram is IR finite. Therefore, in particular, its imaginary part, which can be related

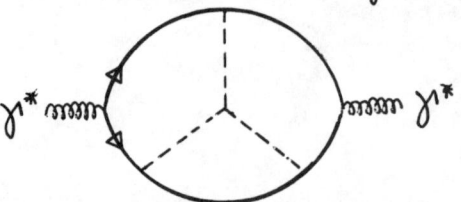

Fig. 1 - Diagram representing the photon self-energy.

to process (2) via unitarity, will be finite. An example of diagrams contributing to the cross-section of this reaction is shown in Figure 2.

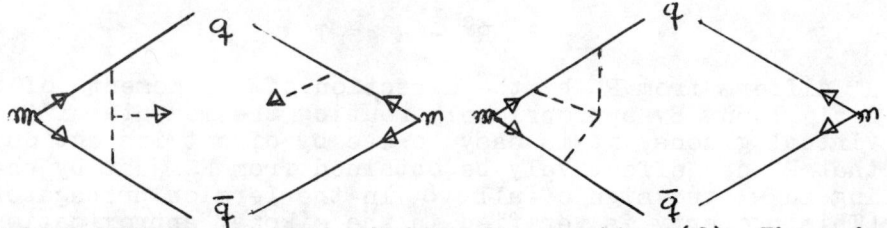

Fig. 2 - Diagrams contributing to reaction (2). The r.h.s. of the graphs denote the complex conjugate of the l.h. ones.

We will now make use of the time-reversal invariance of our theory, in order to transform (2) in a process more directly related to our reaction (1), namely :

$$q + \bar{q} + \text{soft gluons} \rightarrow \gamma^* \qquad (3)$$

Examples of graphs contributing to this process are shown in Figure 3.

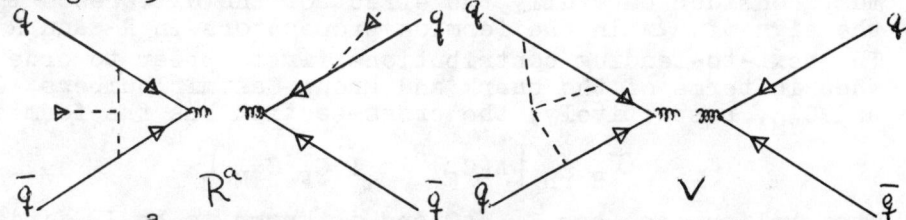

Fig. 3 - $R^a$ denotes diagrams where initial soft gluons are absorbed and V represents a purely virtual graph required in Bloch-Nordsieck inclusive cross-section.

Here, the soft gluons are <u>absorbed</u> and, in consequence of the time-reversal invariance, the IR divergences will also cancel. In contrast, in process (1) the soft gluons are <u>emitted</u> from the initial state. As we will see, this difference is crucial for the existence of uncancelled IR divergences in this reaction. Examples of diagrams contributing to this process are shown in Figure 4.

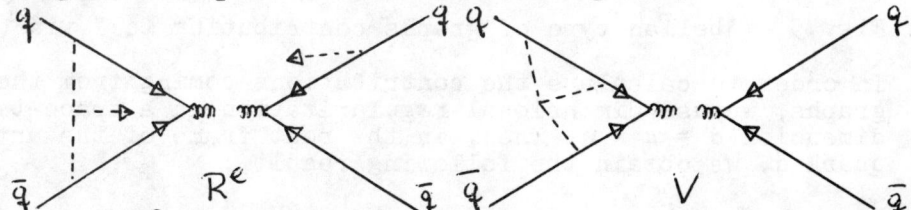

Fig. 4 - $R^e$ denotes diagrams where soft gluons are emitted from the initial state.

We will denote by I the uncancelled IR divergences arising from these diagrams. Comparing Figures 3 and 4, we see that we can obtain I as follows :

$$R^e - R^a = I \qquad (4)$$

$R^a$ differs from $R^e$ by the direction of the momenta of the real gluons. By appropriately routing the momenta of the virtual gluons, it is easy to see by direct inspection, that $R^a$ can effectively be obtained from $R^e$ just by changing in $R^e$ the sign of all $i\varepsilon$ in the fermion propagators. (This property is verified in the eikonal approximation, which is sufficient for our purpose). Equation (4) leads imediately to the conclusion that the leading IR divergences do cancel in all orders in perturbation theory. To see this, we recall that for the leading IR divergences we can neglect the $i\varepsilon$ in the fermion denominators (ref. 3). Then $R^a$ and $R^e$ become equal, so that we obtain :

$$(R^e - R^a)^{lead} = I^{lead} = 0 \qquad (5)$$

However, for the next-to-leading IR divergences, we must consider carefully the effect of the difference in the sign of $i\varepsilon$ in the fermion propagators in $R^e$ and $R^a$. The next-to-leading contributions first appear to order $g^4$, when in terms of the quark and group Casimir numbers, $C_F$ and $C_{YM}$, respectively, the cross-section has the form :

$$\sigma_{Born} \left[ A(C_F)^2 + I' C_F C_{YM} \right] \qquad (6)$$

The abelian case has $C_{YM} = 0$ and is known to be IR finite, so A must be IR finite. Therefore, it is sufficient to restrict ourselves to graphs which contribute to $I'$ in equation (6), some of which are shown in Figure 5.

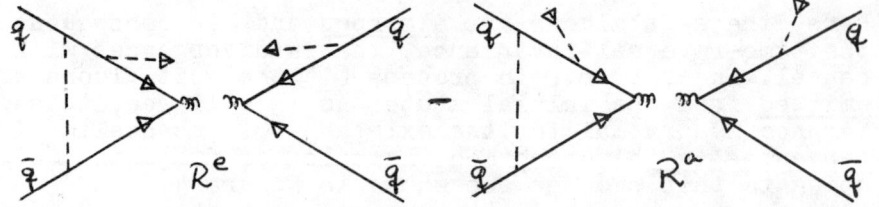

Fig. 5 - Abelian type of graphs contributing to I via (4).

In order to calculate the contributions coming from these graphs, we use dimensional regularization in a space-time dimension $d = 4 + \eta$. Then, in the rest frame of the antiquark $\bar{q}$, we obtain the following result :

$$I'_{a.} = \frac{1}{16\pi^2} \frac{1}{\beta} B(\beta) \frac{1}{2\eta} g^4 \qquad (7)$$

Here, $\beta$ denotes the speed of the quark $q$: $\beta = |\vec{q}| q_0^{-1}$ and $B(\beta)$ represents the bremsstrahlung probability function:

$$B(\beta) = \frac{1}{\beta} \ln\left(\frac{1+\beta}{1-\beta}\right) - 2 \qquad (8)$$

The graphs shown in Figure 5 do also contribute to A in equation (6). However, there are also other abelian graphs contributing to A only. These are obtained by interchanging the order in which the soft gluon vertices appear in Figure (5). It is easy to show that the class of graphs thus obtained, gives a contribution equal to $-I'_a$. Therefore, as is well known (ref. 4), all the IR divergences do cancel in the abelian case.

We consider now the contribution coming from intrinsically non-abelian graphs, represented schematically in Figure 6.

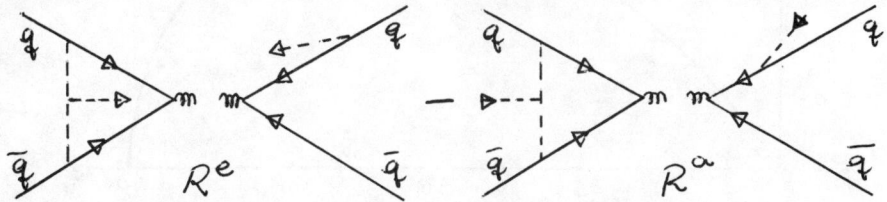

Fig. 6 - Non-abelian graphs contributing to I.

For the contributions arising from these diagrams, we obtain the result:

$$I'_{n.a.} = \frac{1}{16\pi^2} (-1) B(\beta) \frac{1}{2\eta} g^4 \qquad (9)$$

We remark that there are also other graphs contributing equally to $R^e$ and $R^a$, so that their contribution cancels in equation (4). For instance, consider the fourth-order diagrams containing self-energy insertions on the gluon line. In this case, due to the fact that the self-energy is an even function of the external momenta, it is easy to see that $R^e = R^a$, and, therefore, these graphs will not contribute to this order.

Adding equations (7) and (9), and using (8), we find the following simple result for the uncancelled IR divergences present to fourth-order in process (1):

$$I = C_F C_{YM} \frac{1}{16\pi^2} \left(\frac{1}{\beta} - 1\right) B(\beta) \frac{1}{2\eta} g^4 \qquad (10)$$

We have obtained this result in the Feynman and Coulomb gauges, thereby verifying explicitly the gauge invariance of equation (10).

A similar result is obtained for quark-quark scattering, except for an overall minus sign, which arises because the antiquark in (1) is replaced, in this case, by

a quark which has opposite colour charge. Physically, the
presence of uncancelled IR divergences in non-abelian
gauge theories arises from the new possibility of colour
being exchanged between the initial particles participating in the reaction. Clearly, this possibility is absent
in the abelian case, or even in the non-abelian case
where we have only single coloured particle in the initial state, in which case the IR divergences do cancel
(ref. 5).

We represent the $\beta$ dependent part of equation (10)
in Figure 7.

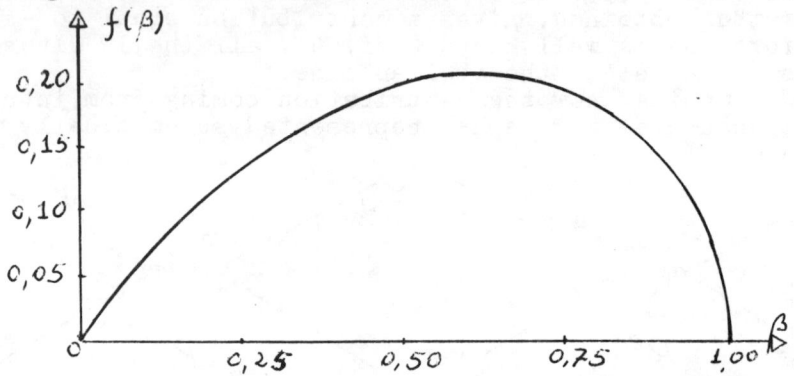

Fig. 7 - Graphical representation of $f(\beta) \equiv (\frac{1}{\beta} - 1) B(\beta)$

In the region $\beta \ll 1$, $f(\beta) \simeq \frac{2}{3}\beta$ and the whole contribution comes from the abelian graphs shown in Figure 5. On
the other hand, as $\beta \to 1$, there is a strong interference
between the contributions coming from both abelian and
non-abelian graphs, making $f(\beta) \simeq (1-\beta) \ln(1-\beta)$. It is
worthwhile to note that in this case $f(\beta)$ is inversely
proportional to the square of the quark energy in the laboratory system, and, hence, it tends to zero rapidly for
high energy processes.

Of course, since (10) is infinite, it is clear that
the cross-section for process (1) is not a physically sensible quantity to calculate, at least in perturbation theory. Therefore, we conclude that the impulse approximation,
where the quarks are treated as assymptotically free particles and the nature of the bound state is neglected,
cannot be entirely correct. In a very simplified model of
bound quarks, we have found that the IR divergence is replaced by a logarithm of the size of the bound-state.

We have also shown, that equation (10) is consistent
with the Kinoshita-Lee-Nauenberg theorem (ref. 6). Higher
order non-leading contributions modify the fourth-order
result expressed by this equation in two ways. First, the
square of the coupling constant $g^2$ is replaced by the
square of the running coupling constant $g^2(k)$ given by

the renormalization group equations (ref. 7). Furthermore, there will appear, in addition, an exponential of the bremsstrahlung probability function. More precisely, to all orders in perturbation theory, we obtain for the next-to-leading IR divergences the following result :

$$I = \frac{C_F C_{YM}}{16 \pi^2} \left(\frac{1}{\beta} - 1\right) B(\beta) \int_0^\Delta \frac{dk}{k} k^{2\eta} g^4(k) \times \exp-\left[\frac{C_{YM}}{8\pi^2} B(\beta) \int_k^\Delta \frac{dk'}{k'} k'^\eta g^2(k')\right] \quad (11)$$

Here, $\Delta$ denotes an arbitrary upper limit, which defines the soft gluons as being those satisfying the condition $k, k' < \Delta$. The derivation and the physical implications of this equation will be discussed in more detail in a future communication.

Acknowledgements : J.F. thanks the Conselho Nacional de Pesquisas (Brasil) for financial support and M. L. Frenkel for helpful discussions.

## REFERENCES

1- R. Doria, J. Frenkel and J. C. Taylor, Nucl. Phys. B168 93 (1980); A. Andrasi, M. Day, R. Doria, J. Frenkel and J. C. Taylor, Oxford Preprint 37/80.
2- T. Applequist, J. Carrazone, J. Kluberg-Stern and M. Roth, Phys. Rev. Lett. 36, 768 (1976).
3- E. Poggio and H. Quinn, Phys. Rev. D14, 578 (1976); C. P. Korthals Altes and E. de Rafael, Phys. Lett. 62B, 320 (1976); T. Kinoshita and A. Ukawa, Phys. Rev. D16, 332 (1977); J. Frenkel, R. Meuldermans, I. Mohammad and J. C. Taylor, Nucl. Phys. B121, 78 (1977).
4- F. Bloch and H. Nordsieck, Phys. Rev. 52, 54 (1937); D. R. Yennie, S. C. Frautschi and H. Suura, Ann. Phys. (N. Y.) 13, 379 (1961).
5- S. B. Libby and G. Sterman, Phys. Rev. D19, 2468 (1979).
6- T. D. Lee and M. Nauenberg, Phys. Rev. 133, 649 (1964); T. Kinoshita, J. Math. Phys. 3, 1950 (1962).
7- D. J. Gross and F. Wilczek, Phys. Rev. Lett. 30, 1343 (1973); N. D. Politzer, Phys. Rev. Lett. 30 1346 (1973).

# INTRINSIC TRANSVERSE MOMENTUM IN GAUGE THEORIES

J. C. Collins[*]
Joseph Henry Laboratories, Princeton University, Princeton, N.J. 08544

## ABSTRACT

The Drell-Yan process can be computed reliably at low transverse momentum.

## INTRODUCTION

This talk summarized some work with D. Soper.[1] Our purpose is to compute the effect of gluon bremsstrahlung on high-energy hadronic processes. Although our methods are general, we explain them by treating the Drell-Yan process as an example:

$$H_1 + H_2 \to \mu^+ + \mu^- + \text{anything}.$$

If $p_1^\mu$ and $p_2^\mu$ are the momenta of the hadrons then we write the dimuon momentum as

$$q^\mu = x_1 p_1^\mu + x_2 p_2^\mu + q_T^\mu, \tag{1}$$

where $q_T \cdot p_1 = q_T \cdot p_2 = 0$. The Drell-Yan limit is $\sqrt{s} \to \infty$ with $x_1$ and $x_2$ fixed.

The parton model for this process[2] is that the dimuon is formed by annihilation of a quark from one hadron against an antiquark from the other, with the same distributions as in deep inelastic scattering. Initial and final state interactions are ignored. Experiments[3] are in at least qualitative agreement with the model, and it can be proved[4] that the model is correct in QCD (modulo the usual logarithmic corrections). However the proofs only apply when either $q_T \sim \sqrt{s}$ or $q_T$ is integrated over. The work[1] summarized below shows how to compute the $q_T$-dependence of the cross-section when $q_T \ll \sqrt{s}$. We can compute in a systematic way beyond the leading-logarithm approximation even at $q_T = 0$. There are striking power-law effects.

## SKETCH OF PROOF

Our proof starts from the power-counting arguments of Ref. 4, which show that the dominant cut graphs for the cross-section have in axial or Coulomb gauge the form of Fig. 1.

[*]Supported in part by the NSF under Grant No. PHY78-01221.

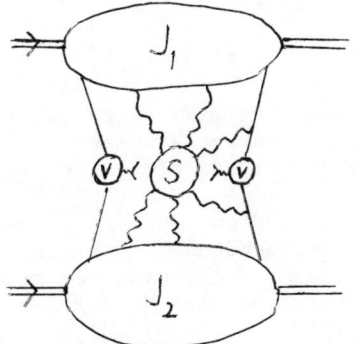

Fig. 1. Dominant graphs for cross-section have this form

There, the momenta in $J_1$ and $J_2$ are collinear, i.e., almost on the light-cone and parallel to the parent hadron. The momenta in the vertices V are ultraviolet, i.e. all their components are order $\sqrt{s}$. The momenta in S are soft, i.e. all components much less than $\sqrt{s}$. The soft region S attaches by gluons only to any part of the jets $J_1$ and $J_2$. It is crucial that the gluons have spin -1. (Contrast Ref. 5 for the spin-0 case.) When $q_T \sim \sqrt{s}$ there are extra graphs present.[4,5]

Now, if $q_T$ is of order $\sqrt{s}$ or is integrated over, then the total effect of the soft glue on the inclusive cross-section must vanish. But it is clear from one-loop calculations that the cancellation is incomplete at low $q_T$. Our arguments to compute its effect use the methods of Grammer and Yennie[6], as developed in Refs. 7 and 8.

We first construct a factorization of soft exchanges between the two jets. When a soft gluon couples to a line collinear to $p_1$ we have a contribution $V^\mu D_{\mu\nu}$, where $V^\mu$ is the vertex and $D_{\mu\nu}$ the soft gluon propagator.[6-8] It is dominated by $\mu = +$ (in light-cone coordinates) so that

$$V^\mu D_{\mu\nu} \sim V^\mu k_\mu D_{+\nu}/k_+. \tag{2}$$

Application of Ward identities then shows that the soft exchanges are the same as from incoming, massless, eikonal quarks (Fig. 2). We used Ref. 9 to find the correct $i\varepsilon$ for the $k_+$ in eq. (2), and we assumed the cancellation of IR divergences in Fig. 2 when the eikonal quarks are massless (cf. Ref.10).

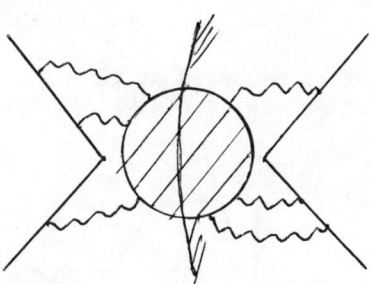

Fig. 2. Soft exchanges after use of Ward identities

There are some complications since the gauge group is non-abelian. An important problem is that a given line's momentum may range all the way from soft to collinear, given different decompositions of the form of Fig. 1. for a single graph; we have a subtraction procedure to eliminate contributions to Fig. 2 when some lines in the shaded bubble are collinear.

The result involves a convolution in transverse momentum, so it is convenient to work in impact-parameter space

$$\frac{s d\tilde\sigma}{dx_1 dx_2} = \int d^2 q_T \, e^{i q_T \cdot b} \frac{s d\sigma}{dx_1 dx_2 d^2 q_T} \tag{3}$$

Then we proved that for $b \gg 1/Q$

$$\frac{s d\tilde{\sigma}}{dx_1 dx_2} \alpha |V(Q/\mu, g(\mu))|^2 \tilde{P}_1(x_1, b, s) \tilde{P}_2(x_2, b, s) E(b\mu, g(\mu)), \qquad (4)$$

where $\mu$ is the renormalization point, V is the U.V. vertex of Fig. 1 and $\tilde{P}$ is the Fourier transform of a transverse momentum distribution[5] $P(x, k_T, s)$. E contains the effect of soft exchanges and could be absorbed into the $\tilde{P}s$.

Now P and $\tilde{P}$ are gauge-dependent and depend on $|4p \cdot n^2/n^2| \sim s$, where n is the gauge-fixing vector. The s-dependence comes only from soft and UV momenta. We use Ward identities for the gauge-dependence and then the Grammer-Yennie method to obtain (for $1/b \ll \sqrt{s}$)

$$s \frac{\partial}{\partial s} \tilde{P}(x, b, s) = \tilde{P}(x, b, s) K(x\sqrt{s}, b), \qquad (5)$$

where K is computable if $b \ll 1/\Lambda$. By extracting the s-dependence of $\tilde{P}$ and making some redefinitions, we obtain ($b \ll 1/Q$)

$$\frac{s d\tilde{\sigma}}{dx_1 dx_2} \alpha \hat{P}_1(x_1, b) \hat{P}_2(x_2, b) |\hat{V}(Q/\mu, g(\mu))|^2 e^{-2F}, \qquad (6)$$

where the new quark distributions are s-independent, and

$$2F = \int_{1/b}^{\sqrt{x_1 x_2 s}} \frac{d\mu'}{\mu'} \gamma(g(\mu')) \ln \frac{\sqrt{x_1 x_2 s}}{\mu'}, \qquad (7)$$

with $\gamma(g)$ a computable function.

The renormalization group is used to set $\mu = Q$ in $\hat{V}$, so that it can be computed. Now the generalized Sudakov form-factor $e^{-2F}$ provides a cut-off for all large b with the cut-off becoming stronger as s increases. So we mainly need $\hat{P}$ when $b \ll 1/\Lambda$. For this we prove an expansion (cf. Ref. 5)

$$\hat{P}(x, b) \sim \int_x^1 \frac{d\xi}{\xi} C(\xi/x, b\mu, g(\mu)) f(\xi, \mu)$$
$$\sim f(x, 1/b) + O(g^2(1/b)), \qquad (8)$$

where $f(x, \mu)$ is the parton distribution measured at $Q^2 = \mu^2$.

The qualitative consequences of eq. (6) are given in Fig. 3.

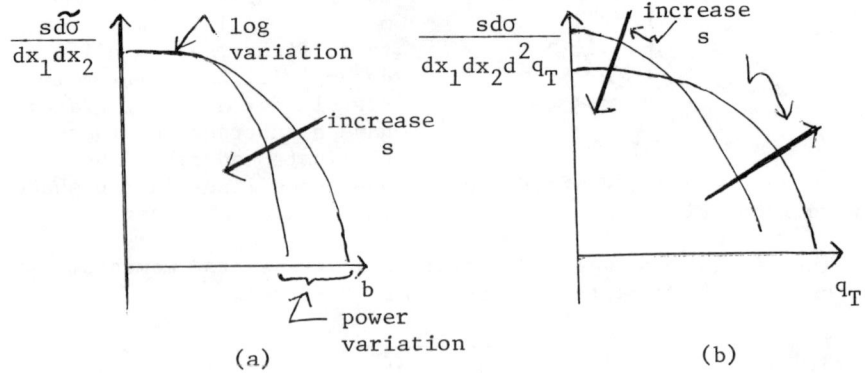

Fig. 3. Drell-Yan cross-section from eq. (6).

Quantitative calculations can be made. One result is

$$\frac{s d\sigma}{dx_1 dx_2 d^2 q_T}(q_T=0) = \frac{N(x_1,x_2,s)}{s^{a/2} \Lambda^{1-a}}, \tag{9}$$

where a is a number close to 1 and N only depends logarithmically on s. Such effects are seen in the data of Ref. 12.

## CONCLUSIONS

1. We are able to compute the Drell-Yan cross-section for all $q_T$. If s is not too big a well-defined but perturbatively uncomputable $P(x,k_T)$ appears. At large s only the usual $f(x)$'s appear.

2. The impact-parameter picture is natural.

3. There are important power-law effects.

4. The methods can be applied to many other processes (e.g. $e^+e^- \to h_1 h_2 x$, elastic scattering, nucleon form factor).

## REFERENCES

1. J. C. Collins and D. E. Soper, "Intrinsic Transverse Momentum II: Gauge Theories" (in preparation).
2. S. D. Drell and T. M. Yan, Ann. Phys. 66, 578 (1971).
3. J. LeFrancois, this conference.
4. D. Amati, R. Petronzio and G. Veneziano, Nucl. Phys. B146, 29 (1978); R. K. Ellis, H. Georgi, M. Machacek and H. D. Politzer, ibid, B152, 285 (1979); S. Libby and G. Sterman, Phys. Rev. D18, 3252 (1978); A. Mueller, Phys. Rev. D18, 3705 (1978).
5. J. C. Collins, Phys. Rev. D21, 2962 (1980).
6. G. Grammer and D. R. Yennie, Phys. Rev. D8, 4352 (1973).
7. J. C. Collins and G. Sterman, "Wee Partons in QCD" (in preparation).
8. J. C. Collins, "Algorithm to compute corrections to the Sukakov Form Factor," Phys. Rev. D.
9. C. DeTar, S. D. Ellis, P. V. Landshoff, Nucl. Phys. B87, 176 (1975).
10. R. Doria, J. Frenkel and J. C. Taylor, Nucl. Phys. B168, 93 (1980); J. Frenkel, this conference.
11. G. Parisi and R. Petronzio, Nucl. Phys. B154, 427 (1979).
12. A. S. Ito et al., FERMILAB-Pub-80/19-EXP.

# HADRON WAVEFUNCTIONS AND STRUCTURE FUNCTIONS IN QCD*

Tao Huang†
Stanford Linear Accelerator Center
Stanford University, Stanford, California 94305
and
Institute of High Energy Physics, Beijing, China

## ABSTRACT

We present[1] the theoretical and empirical constraints on the hadronic wavefunction and hadronic structure functions. In particular, we obtain a new type of low energy theorem for the pion wavefunction from the $\pi^0 \to \gamma\gamma$. Thus we can get the probability of finding the valence $|q\bar{q}\rangle$ state. All these constraints allow us to construct a possible model which describes hadronic wavefunctions, probability amplitudes, and distributions.

The underlying link between hadronic phenomena in quantum chromodynamics at large and small distance is the hadronic wavefunction. By studying the wavefunctions themselves, one could in principle understand not only the origin of the standard structure functions, but also the nature of multiparticle longitudinal and transverse momentum distribution and helicity dependence, as well as the effects of coherence. In this talk, we will discuss the theoretical and experimental constraints on the hadronic wavefunction and structure functions and construct a simple model to implement these constraints.

We define the states at equal $\tau = t + z$ on the light-cone using the light-cone gauge $A^+ = A^0 + A^3 = 0$. The amplitude to find $n$ (on-mass-shell) quarks and gluons in a hadron with 4-momentum P directed along the Z direction and spin projection $S_z$ is defined as $\psi_{S_z}^{(n)}(x_i, k_{\perp i}, s_i)$ [see Fig. 1] $[x_i \equiv (k_i^+/P^+)]$, where by momentum conservation $\sum_{i=1}^n x_i = 1$ and $\sum_{i=1}^n k_{\perp i} = 0$. The $s_i$ specify the spin projection of the constituents. The state is off the light-cone energy shell. For each fermion or antifermion constituent $\psi_{S_z}^{(n)}(x_i, k_{\perp i}, s_i)$ multiplies the spin factor $u(\vec{k}_i)/\sqrt{k_i^+}$ or $v(\vec{k}_i)/\sqrt{k_i^+}$. The wavefunction normalization condition is

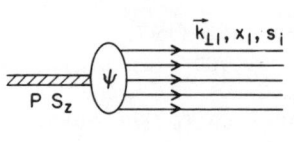

Fig. 1. The amplitude to find n(on-mass-shell) quarks and gluons in a hadron.

$$\sum_{(n)(s_i)} \int \left| \psi_{S_z}^{(n)}(x_i, k_{\perp i}, s_i) \right|^2 [d^2 k_\perp][dx] = 1 \quad . \tag{1}$$

---

* Work supported in part by the Department of Energy, under contract DE-AC03-76SF00515, and by Academia Sinica of China.
† This work was done in collaboration with S. J. Brodsky (SLAC) and G. P. Lepage (Cornell University, LNS).

0094-243X/81/681000 -06$1.50 Copyright 1981 American Institute of Physics

We will discuss the following theoretical and experimental constraints on the wavefunctions and the structure functions.
(a) The predictions of perturbative QCD for the large transverse momentum tail of the Fock state $\psi(x_i, k_{\perp i})$. For the case of inclusive reactions, the standard quark and gluon structure functions, which control large momentum transfer inclusive reactions at the large scale $Q^2$ can be found

$$G_{a/H}(x_a, Q^2) \equiv d_a^{-1}(Q^2) \overline{\sum_{n, s_i, S_z}} \int^{k_{\perp a}^2 < Q^2} [d^2 k_\perp][dx]$$
$$\times \left| \psi_{S_z}^{(n)}(x_i, k_{\perp i}, s_i) \right|^2 \delta(x - x_a) \quad , \qquad (2)$$

where $d_a^{-1}(Q^2)$ is due to the wavefunction renormalization of the constituent a. Note that only terms which fall-off $|\psi|^2 \sim (k_{\perp a}^2)^{-1}$ (modulo logs) contribute to the $Q^2$ dependence of the integral; in general, unless x is close to 1, all Fock states in the hadron contribute to $G_{a/H}$. These contributions are analyzable by the renormalization group and correspond in perturbative QCD to quark or gluon pair production or fragmentation processes associated with the struck constituent a. Multiparticle probability distributions are simple generalizations of Eq. (2).

Recently, it has been shown that exclusive processes such as form factors and large angle elastic scattering can be systematically analyzed in perturbative QCD.[2] For example, the (helicity conserving) hadronic form factors to leading order in $m^2/Q^2$ and to all orders in $\alpha_s(Q^2)$ take the form

$$F(Q^2) = \int [dx][dy] \, \phi^\dagger(x, \tilde{Q}_x, s') \, T_H(x, y, Q) \, \phi(y, \tilde{Q}_y, s) \, , \qquad (3)$$

where $\tilde{Q}_x = \left(\min_i x_i\right) Q$, $T_H$ is the hard scattering amplitude for the virtual photon to scatter the valence quarks from p to p+q; it can be expanded in powers of $\alpha_s(Q^2)$. The quantity $\phi(x, Q)$ is the "distribution amplitude" for finding the valence quark with light-cone fraction $x_i$ in the hadron at relative separation $b_\perp^2 \sim O(1/Q^2)$.

$$\phi(x_i, Q, s_i) \equiv \prod_{i=1}^{n} \left[ d_i^{-1}(Q^2) \right]^{\frac{1}{2}} \int^{k_{\perp i}^2 < Q^2} [d^2 k_\perp] \, \psi^{(n)}(x_i, k_{\perp i}, s_i) \quad , \qquad (4)$$

The large $Q^2$ dependence of $\phi$ (i.e., the large $k_\perp$ tail of $\psi$) is in fact completely determined by the operator product expansion near the light-cone,[3] and in QCD can be calculated from the perturbative expansion in the irreducible kernel for the quark constituents.[1] To order $\alpha_s(Q^2)$ one only requires single gluon exchange, and we find

$$\phi(x_i, Q^2) = \phi(x_i, Q_0^2) + \frac{C_F}{\beta} \int_{Q_0^2}^{Q^2} \frac{d\ell_\perp^2}{\ell_\perp^2} [dy] \, \alpha_s\left(\frac{\ell_\perp^2}{y_i(1-y_i)}\right)$$
$$\times \left[ V(x_i, y_i) - \delta(x - y) \right] \phi(y_i, \ell_\perp^2) \quad , \qquad (5)$$

where

$$V(x,y) = 2\left\{x_1 y_2 \theta(y_1-x_1)\left(\delta_{h_1\bar{h}_2} + \frac{\Delta}{y_1-x_1}\right) + (1 \leftrightarrow 2)\right\} = V(y,x) \quad . \quad (6)$$

This result[4] is derived in the region where $\ell_\perp^2/[y_1(1-y_1)]$ is large compared to the off-shell energy $\langle \mathscr{E} \rangle$ in the wavefunction.
(b) Exact boundary conditions for the valence Fock state meson wavefunctions from the meson decay amplitudes. The leptonic decays of mesons give an important constraint on the valence $|q\bar{q}\rangle$ wavefunction at the origin,[2]

$$\lim_{Q\to\infty} \phi_M(x_i,Q^2) = a_0 x_1 x_2 = \begin{cases} \dfrac{3}{\sqrt{n_c}} f_\pi x_1 x_2 & \text{for } \pi \\[2ex] \dfrac{3\sqrt{2}}{\sqrt{n_c}} f_\rho x_1 x_2 & \text{for } \rho_L \end{cases}, \quad (7)$$

where $f_\pi \simeq 93$ MeV is the pion decay constant for $\pi^+ \to \mu\nu$ and $f_\rho \simeq 107$ MeV is the leptonic decay constant from $\rho^0 \to e^+ e^-$. The analogous result holds for all zero helicity mesons. Because the $Q \to \infty$ distribution amplitude has zero anomalous dimension, this constraint is independent of gluon radiative correction and can be applied directly to the non-perturbative wavefunction

$$a_0 = 6 \int_0^1 [dx][d^2 k_\perp] \psi_M^{\text{non-pert}}(x_i, k_\perp) \quad . \quad (8)$$

On the other hand we can also obtain an exact low-energy constraint on $\psi(x_i, k_\perp = 0)$ for the pion in the chiral limit $m_q \to 0$. The $\gamma^*\pi \to \gamma$ vertex defines the $\pi^0 - \gamma$ transition form factor $F_{\pi\gamma}(Q^2)$ (Fig. 2a)

$$\Gamma_\mu = -ie^2 F_{\pi\gamma}(Q^2) \varepsilon_{\mu\nu\rho\sigma} P_\pi^\nu \epsilon^\rho q^\sigma \quad . \quad (9)$$

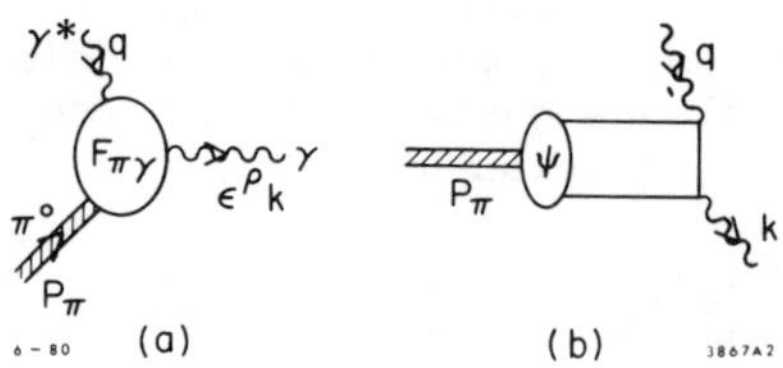

If $m_q \to 0$, then the valence $|q\bar{q}\rangle$ contribution to $F_{\pi\gamma}(Q^2)$ is (Fig. 2b)

$$F_{\pi\gamma}(Q^2) = 2\sqrt{n_c}\,(e_u^2 - e_d^2)\left\{\int_0^1 dx_1 \int_0^{x_1^2 Q^2} \frac{dk_\perp^2}{16\pi^2}\,\frac{\psi(x_i, k_\perp)}{Q^2 x_1} + (x_1 \leftrightarrow x_2)\right\}. \quad (10)$$

Fig. 2. (a) The $\pi$-$\gamma$ transition form factor $F_{\pi\gamma}(Q^2)$; (b) the lowest order diagram which contributes to $F_{\pi\gamma}(Q^2)$.

In fact, gauge-invariance requires that the valence $|q\bar{q}\rangle$ state should give exactly ½ of the total decay amplitudes at $q^2 \to 0$.[5] Thus from the $\pi \to \gamma\gamma$ decay rate and Eq. (10), we find

$$\int_0^1 [dx]\,\psi(x_1, k_\perp = 0) = \int_0^1 dx_1\,\psi(x_1, k_\perp = 0) = \frac{\sqrt{n_c}}{f_\pi} \quad . \quad (11)$$

This is a new type of low-energy theorem for the pion wavefunction which is consistent with chiral symmetry and the triangle anomaly for the axial vector current. This large-distance result, together with the constraint on the valence wavefunction at short distance from the $\pi \to \mu\nu$ leptonic decay amplitude, leads to a number of new results for the parametrization of the pion wavefunction, which we discuss below.

(c) We can show that the evolution equations which specify the large $Q^2$ behavior of the distribution amplitudes and incoherent distribution functions G are correctly applied for $Q^2 \gtrsim \langle \mathscr{E} \rangle$, where $\langle \mathscr{E} \rangle$ is the mean value of the off-shell energy in the Fock state wavefunction, $\mathscr{E} \equiv \sum_i \mathscr{E}_i \equiv \sum_{i=1}^n \left((\vec{k}_\perp^2 + m^2)/x\right)_i$ i.e., $\langle \mathscr{E} \rangle$ is a measure of the "starting point" for evolution due to perturbative effects in QCD*

In order to organize the predictions for hadronic matrix elements and all of the distribution functions and amplitudes, we shall make the following prescription:

(i) We assume the Fock state wavefunction $\psi^{(2)}$ for the 2-quark state in the non-perturbative domain depends only on the off-shell energy variable $\mathscr{E}$. [This ansatz, which is true for non-relativistic theories, can be justified, if we use the Bethe-Salpeter equation with an instantaneous energy independent kernel.[6]] For the n-particle case, we shall assume the Fock state wavefunction $\psi^{(n)}$ is a symmetric function of the $\mathscr{E}_i$, i.e., $\psi^{(n)} = \psi^{(n)}(\mathscr{E}_i)$. Although we have no strong argument for this form, we shall use it as an illustration of the effect of the non-perturbative wavefunction. Thus we find

$$G_{a/H}^{non-pert}(x_a) \xrightarrow[x_a \to 1]{} (1-x_a)^{2n_s - 1} g\left(\mathscr{E}_{min}^i\right) \quad , \qquad (12)$$

where $n_s = \min(n_H - n_a)$ is the minimum number of spectator constituents in the hadron H after removing the particle (or subcomposite) a, and $\mathscr{E}_{min}^i = m_\perp^2/x_i$ is the minimum value of $\mathscr{E}_i$.[7] Notice that if we can neglect the quark masses [i.e., for $(1-x_a) \gg m^2/\langle k_\perp^2 \rangle$] we obtain the spectator rule[8]

$$G_{a/H}^{non-pert}(x_a) = C_{a/H}(1-x_a)^{2n_s - 1} \quad . \qquad (13)$$

For example, $n_s = 1$, for the meson case; $n_s = 2$ for the baryon case.†
We can see that the non-perturbative contribution can dominate the perturbative predication in the $x \sim 1$ domain.[1]

(ii) An (approximate) connection between the equal-time wavefunction in the rest frame and in the infinite momentum frame wavefunction can be established by equating the energy propagator $M^2 - \mathscr{E} = M^2 - \left(\sum_{i=1}^n k_i^\mu\right)^2$ in the two frames.

---
* The actual limit of the $k_\perp^2$ integration is $k_\perp^2/[z(1-z)] \gtrsim \langle \mathscr{E} \rangle$, where z is the fraction of the radiated gluon. The correct argument of $\alpha_s$ is $\alpha_s(k_\perp^2/(y-z))$. [y is the fraction of the quark.]
† In addition, QCD evolution increases the exponent of Eq. (13).[1]

If we kinematically identify

$$\frac{(q^0+q^3)_i}{\sum_j^n q_j^0} \leftrightarrow x_i \equiv \frac{k_i^+}{P^+}, \quad \vec{q}_{\perp i} \leftrightarrow \vec{k}_{\perp i}, \tag{14}$$

Then the rest frame wavefunction $\psi_{CM}(\vec{q}_i)$ which controls binding and hadronic spectroscopy implies a form for the IMF wavefunction $\psi_{IMF}(x_i, k_{\perp i})$. For two particle state there is a possible connection[9]

$$\psi_{CM}(\vec{q}^2) \leftrightarrow \psi_{IMF}\left(\frac{k_\perp^2 + m^2}{4 x_1 x_2} - m^2\right). \tag{15}$$

In order to implement these constraints it is convenient to construct a simple model of the hadronic wavefunction. By using the connection (15) from the harmonic oscillator model[6] we can get the wavefunction at the infinite momentum frame

$$\psi^{(2)}(x_i, k_\perp, s_i) = A \exp\left[-R^2\left(\frac{k_\perp^2 + m^2}{x_1} + \frac{k_\perp^2 + m^2}{x_2}\right)\right]. \tag{16}$$

It is certainly a function of $\mathcal{E}$. From Eqs (7), (11) and (16) we can obtain ($m_q^2 R^2 \ll 1$)

$$R = \frac{1}{4\pi f_\pi} \simeq 0.17 \text{ fm}, \quad A \simeq \frac{\sqrt{3}}{f_\pi}. \tag{17}$$

The probability of finding the valence $|q\bar{q}\rangle$ state in the pion is thus

$$P(q\bar{q}) = \int [dx][d^2 k_\perp] |\psi(x_i, k_\perp)|^2 = 1/4. \tag{18}$$

Alternatively, if we use a power law form

$$\psi(x_i, k_\perp) = \frac{A_\alpha}{\left(\frac{k_\perp^2 + m_q^2}{x(1-x)} + \mu^2\right)^\alpha}, \tag{19}$$

we find ($m_q^2 \ll \mu^2$)

$$P(q\bar{q}) = \frac{1}{2} \frac{\alpha - 1}{2\alpha - 1}, \tag{20}$$

which again leads to 1/4 for large α. For the linear potential case, where α = 3, we have $P(q\bar{q}) = 1/5$. The distribution amplitude for the Gaussian form depends only upon the quark mass. In Fig. 3 we give the prediction of the perturbative QCD[10] for the pion form factor. Note that $\langle \mathcal{E} \rangle \sim 0.7$ GeV$^2$ is reasonable compared to

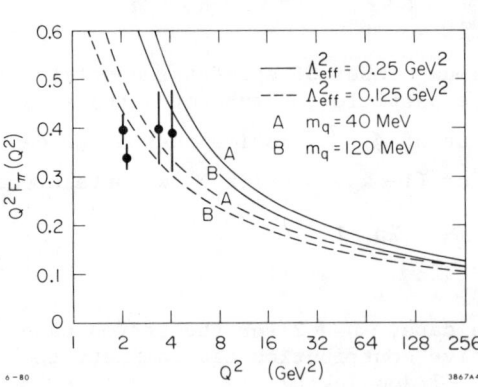

Fig. 3. QCD prediction for the meson form factor for the distribution function $\phi(x_i)$.[11]

the QCD $\Lambda^2$. For the multiparticle case, perhaps the simplest generation for the Fock state wavefunctions in the non-perturbative domain is the Gaussian form

$$\psi_{S_z}^{(n)}(x_i, k_{\perp i}, s_i) = A_n \exp(-R_n^2 \mathcal{E}) = A_n \exp\left[-R_n^2 \sum_{i=1}^n \left(\frac{\vec{k}_\perp^2 + m^2}{x}\right)_i\right]. \tag{21}$$

The parametrization is taken to be independent of spin. This ansatz for the wavefunction has the additional analytic simplicity of (a) factorizing in the kinematics of each constituent and (b) satisfying a "cluster" property.

A similar analysis has been applied to the baryon wavefunction; we find

$$\phi(x_i, Q_0^2) \propto x_1 x_2 x_3 \exp\left[-R^2 \left(\frac{m_1^2}{x_1} + \frac{m_2^2}{x_2} + \frac{m_3^2}{x_3}\right)\right] \quad . \tag{22}$$

In addition, we can consider the sea quark effects and the high twist effects. These results will be given elsewhere.

We conclude that theoretical and empirical constraints on wavefunctions and structure functions have been presented. In particular we obtain a new type of low-energy theorem and the probability of finding the valence $|q\bar{q}\rangle$ state in the total pion wavefunction is ~0.2 to 0.25, for a broad range of confining potentials. This work represents a first attempt to construct a model of hadronic wavefunction and hadronic structure which is consistent with data and QCD at large and small distances.

I would like to thank S. J. Brodsky, G. P. Lepage and M. Barnett for their help in preparing this talk. I also enjoyed the hospitality of the Aspen Center for Physics where I prepared the talk.

## REFERENCES

1. S. J. Brodsky, T. Huang and G. P. Lepage, SLAC-PUB-2540.
2. G. P. Lepage and S. J. Brodsky, Phys. Rev. D (to be published); Phys. Lett. 87B, 359 (1979); Phys. Rev. Lett, 43 545 (1979); Erratum ibid., 43, 1625 (1979); SLAC-PUB-2447.
3. S. J. Brodsky, Y. Frishman, G. P. Lepage and C. Sachradja, Phys. Lett. 91B, 239 (1980); A. V. Efremov and A. V. Radyushkin, Dubna preprints JINR-E2-11535,-11983 and -12384; A. Duncan and A. H. Mueller, Phys. Lett. 90B, 159 (1980); Phys. Rev. D21, 1636 (1980).
4. Higher order kernels entering the evolution equation include all two-particle irreducible amplitudes for $q\bar{q} \to q\bar{q}$. However, these corrections to $V(x_i, y_i)$ are all suppressed by powers of $\alpha_s(Q^2)$, because they are irreducible.
5. J. Schwinger, Phys. Rev. 82, 664 (1951); S. L. Adler, Phys. Rev. 177, 2426 (1969); J. S. Bell and R. Jackiw, Nuovo Cimento 60A, 47 (1969); R. Jackiw, Lectures on a Current Algebra and its Applications, Princeton University Press, 1972, pp. 97-230.
6. For example, see Elementary Particle Theory Group, Beijing University, Acta Physics Sinica 25, 415 (1976).
7. Examples of this result for $G_{q/M}$ and $G_{q/B}$ have recently been given by A. De Rujula and F. Martin, MIT preprint CTP 851 (1980).
8. S. J. Brodsky and R. Blankenbecler, Phys. Rev. D10, 2973 (1974).
9. An equivalent result was also obtained by V. A. Karmanov, ITEP-8, Moscow (1980).
10. The $Q^2 \to \infty$ asymptotic result was first obtained by G. R. Farrar and D. R. Jackson, Phys. Rev. Lett. 43, 246 (1979).
11. The data are from C. Bebek et al., Phys. Rev. D13, 25 (1976).

# SUMMING QCD CORRECTIONS IN INFRARED SENSITIVE QUANTITIES

Marcello Ciafaloni
Scuola Normale Superiore, I-56100 Pisa
INFN, Sezione di Pisa

## ABSTRACT

On the basis of some eikonalization properties of QCD it is argued that most of the large perturbative corrections are summed by modified evolution equations which include a rescaling in the argument of the running coupling constant.

## INTRODUCTION

In this talk I shall deal with some closely related subjects, motivated by the existence of large QCD corrections in structure and fragmentation functions for $x \to 1$, Drell-Yan process etc.[1] First I will give a resummation prescription[2,3] to all orders of a class of these corrections and then I will show that the basis of such prescription lies[4] in some eikonalization properties of soft gluon emission. Finally I will comment on the $e_+ e_-$ multiplicity problem, which is also related to soft gluon emission, and I will argue that recent calculations[5,6] sum in fact the leading QCD corrections to all orders.

## RESUMMATION PRESCRIPTION FOR LARGE N

Important first order QCD corrections were found in structure functions[7] for large N and in the Drell-Yan formula[8]. In the latter case one has

$$D_N^{DY} = (D_N^{DIS})^2 \left[ 1 + \frac{\alpha_s}{\pi} C_F \left( \frac{1}{2}\pi^2 + (\log N)^2 + O(1) \right) + \ldots \right] \quad (1)$$

where $N$ denotes the moment index with respect to the $\tau = Q^2/s$ or $x = Q^2/\nu$ variables and $C_F = 4/3$ for 3 colors. Due to the large factors $\pi^2$ and $(\log N)^2$ the corrections' size is $\sim 100\%$ and it is obviously important to control them to all orders.

In the Drell-Yan case the idea is that large corrections are due to two effects:

(i) <u>Dependence of momentum scales on kinematics</u>. The scale which controls QCD corrections is taken to be (see below) the maximum transverse momentum of emitted gluons. One has $q_\perp^2 \lesssim Q^2(1-\tau)^2$ for Drell-Yan and $q_\perp^2 \lesssim Q^2(1-x)$ for structure functions[8,2]. The

(1-x)-dependence provides, upon q integration, factors of log(1-x) in front of the bremmstrahlung spectrum dx/(1-x). Therefore, $\alpha_s (\log N)^2$ terms arise in the coefficient function[*] and a fraction of them survives in (1) because of the different phase space boundary in DY compared to DIS. This remark can be generalized to higher orders in $\alpha_s$ because leading log corrections come from iteration of the basic gluon emission process in a physical gauge (Fig.1).

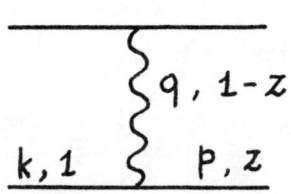

Fig.1
Basic rung of the ladder

The logN dependent terms then arise because $\alpha_s(k^2)$ is to be replaced by $\alpha_s(q^2_{MAX}) = \alpha_s((1-z)k^2)$, due to the inequality

$$k^2 \geq \frac{q^2}{1-z} + \frac{p^2}{z} \quad . \qquad (2)$$

(ii) <u>Continuation from spacelike to timelike Q</u>. Both DIS and DY structure functions are proportional to the quark e.m. form factor $F_q(Q^2)$, evaluated at spacelike and timelike Q values respectively. In the process of analytic continuation $\log Q^2$ acquires a phase $i\pi$, and since $-\log F_q \sim \alpha_s \log^2 Q^2$, this explains the $\pi^2$ coefficient in (1).

In more detail, the resummation prescription amounts for structure functions to the following modified evolution equation

$$\frac{d \log D_N}{d \log k^2} = C_F \int_0^{1-Q_0^2/k^2} \frac{dz}{1-z} \alpha_s(k^2(1-z))(z^N - 1) =$$
$$= - C_F \frac{\alpha_s}{\pi} \log N (1 + b \alpha_s \log N + \ldots) \qquad (3)$$

where the maximum gluon mass is $k^2(1-z)$ (Eq.2). Analogously for the Sudakov form factor one has

$$-2 \log F \cong - \log D \Big|_{N \cong \frac{Q^2}{Q_0^2}} = \frac{C_F}{\pi} \alpha_s (\log Q^2)^2 [1 - \frac{1}{3} b \alpha_s \log Q^2 + \ldots] \quad (4)$$

The statement is that when $\alpha_s \log N \sim \alpha_s \log Q^2 \sim O(1)$, (3) and (4) sum all contributions $\log Q^2 (\alpha_s \log Q^2)^k$ to logD which in this limit are of the same order. As a consequence, large corrections of type

---

[*] When I speak separately of coefficient function and anomalous dimension contribution, I refer to the minimal subtraction scheme in dimensional regularization, or any other scheme which differs by a change of scale which is logN independent.

(i) and (ii) exponentiate. For large N one has[3]

$$D_N \cong F_q^2(Q^2) / F_q^2(Q^2/N) \quad \text{(DIS)}$$
$$\cong F_q^2(-Q^2) / F_q^2(Q^2/N^2) \quad \text{(DY)} \tag{5}$$

where the $N^2$ and minus sign in the 2nd equation account for the different DY kinematics. The factor $\exp\{\frac{\alpha_s}{\pi}[\frac{1}{2}\pi^2 + (\log N)^2] + \ldots\}$ resulting in Eq.(1) is actually welcome, because the parton model formula is off by a factor $2 \div 3$ with respect to experimental data.

## EIKONALIZATION PROPERTIES

Let me now stress the point that the proof of Eq.(3) on the basis of the kinematical argument (i) is actually incomplete, because higher order irreducible diagrams contain comparable terms which are due to infrared singularities rather than to rescaling of the running coupling constant. For instance, if the anomalous dimension had terms $(\alpha_s \log N)^k$, $(k \geqslant 2)$

$$\gamma_N = -\frac{\alpha_s C_F}{\pi} \log N + \alpha_s^2 [c_2 (\log N)^2 + \ldots] + \alpha_s^3 [c_3 (\log N)^3 + \ldots], \tag{6}$$

they should be <u>added</u> to those in Eq.(3) (which all vanish in the $b \to 0$ limit). It is known, on the other hand that $c_2 = 0$ [9], due to a cancellation of the 'infrared singular' $(\alpha_s \log N)^2$ contributions to $\gamma_N$. I want to argue that similarly $c_3 = 0$ etc., and that the origin of such cancellations lies in some eikonalization properties of QCD. In fact, eikonalization in QED is equivalent to independent emission of wee photons. The coefficients $c_i$ should then vanish because they roughly measure the integrated correlations.

To be more precise, let me consider the flavor non singlet fragmentation function $D_{NS}$. In a physical gauge all mass singularities come from diagrams with a two quark state in the t-channel (Fig.2).

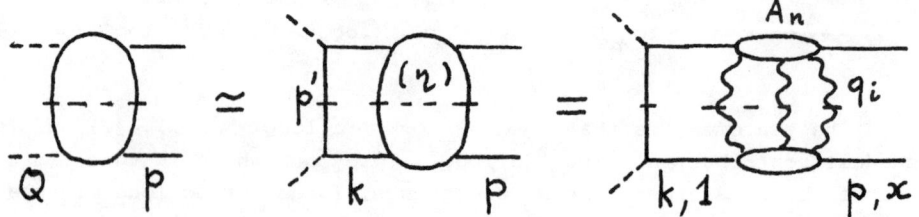

Fig.2
Quark fragmentation function for light-like gauge vector $\eta$.

Moreover, in the light-cone gauge ($\eta \sim Q-p/x \sim p'$) soft gluon emission from the recoiling quark $p'$ is suppressed. It is then sufficient to evaluate the bremmstrahlung of wee gluons $q_i$ ($m \ll q_{"_i} \ll Q$) from a fast quark of momentum $Q \gg m$ (quark mass), with final colors summed over.

Due to the explicit $\eta$-dependence of $A_\eta$ there are two variables for the emitted gluons, $z_i = (q_i \eta)/(p \cdot \eta)$ and $2p \cdot q_i = q_{\perp i}^2 / z_i$, so that $q_{i\mu} = z_i p_\mu + (p \cdot q_i) \eta_\mu + q_{\perp \mu}$. The basic emission probability (Fig.1) is then

$$\frac{dw}{dz\, d(p \cdot q)} = w(q) = \frac{\alpha_s}{\pi} \frac{C_F}{z(p \cdot q)} (1 + O(z)). \tag{7}$$

Upon q-integration the factors $(p \cdot q)^{-1} [z^{-1}]_{NS}$ give rise to collinear, $\log Q^2$ [infrared, $\log N$] singularities of $D$.

Consider now many gluon emission: it has been shown[4] that, if $A_n$ is a sum of tree diagrams,

$$w(1,\ldots,n) = \prod_i w(i) \left[ 1 + O\left(\frac{z_i}{z_{i+1}}\right) \right], \tag{8}$$

i.e., it factorizes provided $0 < z_1 \ll \ldots z_n \ll 1$ (strong ordering region in $z_i$). This statement is non trivial - and in fact it needs the cooperation of all contributing diagrams - because it holds irrespective of strong ordering assumptions on $p \cdot q_i$. Therefore, upon $q_i$ integrations, we pick up an infrared leading term [$(\alpha_s \log N)^k$ factor] but both leading and subleading collinear logs.

The factorization property (8) leads to exponentiation of $D_N^{NS}$ and therefore to vanishing of $c_i$ in (6) provided the region $z_i \cong z_j$ is neglected. Is this justified? In the two gluon case it is easy to see that, unlike QED, the triple gluon coupling gives rise to strong correlations in this region ($z_1 \cong z_2$). However, the <u>integrated</u> correlations partly cancel against radiative corrections to single gluon emission, except for the gluon self-energy diagrams which give rise to the b-dependent terms in Eq.(3).[10] The extension of this analysis to all orders is now under study.

## THE MULTIPLICITY PROBLEM

In $e_+ e_-$ annihilation, extra logs coming from the kinematical boundary are very important also when the soft gluon is the one observed ($z \to 0$) rather than the emitted one ($z \to 1$). In fact in the singlet case the analog of (3) is roughly

$$\frac{d\, D_N^S(k^2)}{d \log k^2} = \int_{Q_0^2/k^2}^{1-Q_0^2/k^2} dz\, P(z)\, \alpha_s(k^2 z(1-z))\, z^N\, D_N^S(zk^2), \tag{9}$$

where now $P(z)$ is singular also for $z \to 0$ and therefore the rescaling $k^2 \to zk^2$ in the jet mass $(p^2 \leq zk^2$, Eq.(2)) is important. Eq.(9) modifies the anomalous dimension even for frozen $\alpha_s$ ($b=0$) and gives

$$\gamma_N^{e_+e_-} = -\frac{(N-1)}{2} + \sqrt{(\frac{N-1}{2})^2 + C\alpha_s} = \frac{C\alpha_s}{N-1} - \frac{C^2\alpha_s^2}{(N-1)^3} + \cdots \quad (10)$$

$$\xrightarrow[N \to 1]{} \sqrt{C\alpha_s} + O(\alpha_s) \quad , \quad C = N_c/\pi$$

Otherwise stated, the $N=1$ limit of $\gamma_N^{e_+e_-}$ relevant to the multiplicity is not infinite, but acquires a $\sqrt{\alpha_s}$ singularity whose perturbative remainder is the variable $\alpha_s/(N-1)^2$. As a consequence, the renormalization group implies for the multiplicity

$$<n> \sim \exp \int^{\log Q^2} [\sqrt{C\alpha_s(t)} + O(\alpha_s)] dt =$$

$$= (\log Q^2)^{\text{power}} \exp[2(\frac{N_c}{\pi b} \log Q^2)^{\frac{1}{2}}] \quad . \quad (11)$$

The situation here is to be contrasted with the DIS case where $x \to 0$ (Regge limit) goes in the direction of enlarging the $k^2$ phase space instead of restricting it. Therefore no maximally singular contributions $\sim \alpha_s^2/(N-1)^3$ of type in (10) are expected in DIS. The ones found are $\sim \alpha_s^2/(N-1)^2$ and, since $N-1 \sim \sqrt{\alpha_s}$, they contribute to the $O(\alpha_s)$ term in (10) and to the power of logarithm in (11). In conclusion, the exponential term should not be affected by higher order irreducible diagrams. Note finally that the $\alpha_s^2/(N-1)^3$ term of kinematical origin should instead come out with the correct coefficient in the 2nd order $e_+e_-$ calculation now in progress[11].

To sum up, due to the fact that soft gluon emission partly eikonalizes, most large perturbative corrections are calculable to all orders. In a physical gauge it is sufficient to plug in the correct kinematics into the dressed ladders, thus leading to the modified evolution equations (3) and (9).

## REFERENCES

1. C. Llewellyn-Smith, rapporteur's talk at this Conference.
2. G. Parisi, Phys. Lett. 90B, 295 (1980); G. Curci and M. Greco, Phys. Lett. 92B, 175 (1980).
3. D. Amati, A. Bassetto, M. Ciafaloni, G. Marchesini and G. Veneziano, Nucl. Phys. (to be published).
4. M. Ciafaloni, Phys. Lett. (to be published).
5. W. Furmanski, R. Petronzio and S. Pokorski, Nucl. Phys. B155, 253 (1979).

6. A. Bassetto, M. Ciafaloni and G. Marchesini, Nucl. Phys. B163, 477 (1980).
7. See, e.g., A. Buras' talk at this Conference.
8. See, e.g., G. Altarelli, R. K. Ellis and G. Martinelli, Nucl. Phys. B143, 521 (1978) and B157, 461 (1979).
9. E. G. Floratos, D. A. Ross and C. T. Sachrajda, Nucl. Phys. B129, 66 (1977) and B152, 493 (1979); G. Curci, W. Furmanski and R. Petronzio, CERN preprint TH 2815 (1980).
10. M. Ciafaloni and G. Curci, in preparation.
11. R. Petronzio, private communication.

## ACKNOWLEDGEMENTS

This talk is mostly based on work done in Refs. 3,4 and 6. I wish to thank my colleagues D. Amati, A. Bassetto, G. Marchesini and G. Veneziano for innumerable conversations on the subject. I wish also to thank G. Curci, G. Parisi and G. Sterman for interesting discussions and suggestions.

# QCD II

W.A. Bardeen, Fermilab, Organizer

# CONDENSATION OF $(G_{\mu\nu}^a)^2$ IN QUANTUM CHROMODYNAMICS

Yoichi Kazama
Fermi National Accelerator Laboratory, Batavia, Illinois 60510

This talk is based on the results obtained in collaboration with R. Fukuda.[1]

In the following, I wish to discuss the property of the QCD vacuum by focusing on the expectation value of the color-singlet composite operator $(G_{\mu\nu}^a)^2$, the square of the gluon field strength tensor. Just as the $\bar{q}q$ operator plays the primary role in studying the "quark content" of the vacuum, especially its chiral property, $(G_{\mu\nu}^a)^2$, measuring the "gluon content" of the vacuum, gives us a valuable bit of information on its complex structure. Theoretically, if one assumes importantce of certain field configurations, such as dilute gas of instantons,[2] one finds $<0|(\alpha_s/\pi)(G_{\mu\nu}^a)^2|0>$* to be non-zero and positive. On the other hand, through the QCD sum rules[3] this quantity is directly related to the properties of physical resonances and has been determined experimentally to be $\sim 0.012$ GeV$^4$.

Now a natural and an important question is whether we can deduce from QCD that $(G_{\mu\nu}^a)^2$ condenses quite generally without assuming dominance of certain configurations. We have addressed ourselves to this question and have the answer in the affirmative.[1] In the true non-perturbative vacuum of QCD, $(G_{\mu\nu}^a)^2$ condenses with positive (magnetic) sign.

The standard method to study condensation of an operator, call it $\hat{\phi}$, is to construct the effective potential for it by first computing the generating functional in the presence of a source term $J\hat{\phi}$ and then Legendre-transforming the result. This procedure, although straight forwardly implementable in such theories as Gross-Neneu model[4] or O(N)-symmetric $\lambda\phi^4$ model[5] in the large N limit, is next to impossible to carry out for QCD at present. This calls for a trick. A trick is provided by the trace anomaly equation[6,7] in the presence of the constant source J coupled to the operator $\hat{\phi} \equiv \int d^4x\, 1/4(G_{\mu\nu}^a)^2$. Let us recall that in QCD with massless quarks, the trace anomaly equation, in the vacuum state, takes the form $<0|\theta_\mu^\mu|0> = (2\beta(g)/g)<0|\,1/4(G_{\mu\nu}^a)^2|0>$. Because of the Poincaré invariance of the vacuum, the left hand side is nothing but 4 times the energy density $\varepsilon = <0|\theta_{00}|0>$. Thus if we can derive a similar equation in the presence of $J$, we will obtain $\varepsilon(J)$, from which the effective potential can be constructed.

Our ability to derive such an equation rests on the fact that the operator $\hat{\phi}$ appears already in the action. In the generating functional $Z = \exp iW$, the addition of the above source term amounts to a simple change $\int d^4x\, 1/4(G_{\mu\nu 0}^a)^2 \rightarrow (1+J_0)\int d^4x\, 1/4(G_{\mu\nu 0}^a)^2$ (where 0 signifies bare quantities). In the axial gauge specified by the constraint $\delta(\eta_\mu A_0^\mu)$, this $(1+J_0)$ factor in turn may be eliminated by

---

*It is to be emphasized that, throughout, $<0|(G_{\mu\nu}^a)^2|0>$ means the difference between $(G_{\mu\nu}^a)^2$ in the true vacuum and in the perturbative vacuum.

the rescaling $g_{J0}^2 \equiv g_0^2/(1+J_0)$, $A_{J0}^\mu \equiv (1+J_0)^{1/2} A_0^\mu$, and we are left with the generating functional identical <u>in form</u> to the one without the source. This we know how to renormalize and we can derive the anomaly equation from it. There is however one complication, which can be readily seen by expanding $W(J)$ in powers of $J$;

$W(J_0) = W(0) + J_0 \delta W/\delta J_0 + \frac{1}{2} J_0^2 \delta^2 W/\delta J_0^2 + \ldots$. As $(-\delta/\delta J_0)$ effects an insertion of a "hard" operator, we need to remove the extra infinities caused by the multiple insertions of $\phi_0$. This turns out to lead to mixing of infinite number of operators $\{\phi_0^n\}_{n=1,2,\ldots,\infty}$ under renormalization. Without giving the details let me just assert that this is handled by renormalizing $J$ by a Z-factor which itself depends on the source: $J_0 = J Z_J[J] = J(Z_J^{(0)} + J Z_J^{(1)} + J^2 Z_J^{(2)} + \ldots)$. Here the renormalized source $J$ is defined such that $(-\delta/\delta J)^n W(J)|_{J=0}$ gives the renormalized n-fold insertion of $\hat{\phi}$. This evidently is quite analougous to the case of the usual coupling constant renormalization and is of no mysterious nature.

Now that we have done away with technical complications, we can follow the analysis of Ref. 7 and get the desired anomaly equation. For the vacuum state it reads

$$\frac{1}{4} \langle 0 | \theta_{\mu J}^\mu | 0 \rangle = \varepsilon(J) = (\beta(g_J)/2g_J)(1+J)\phi , \qquad (1)$$

where $g_J^2 = g^2/(1+J)$ and $\phi \equiv (-\delta/\delta J)W(J)$. This is an exact result, including, in particular, long wave length excitations. With $\varepsilon(J)$ at hand, we can perform the Legendre transform and obtain the effective potential

$$V(\phi) = (\beta(g_J)/2g_J)(1+J)\phi - J\phi . \qquad (2)$$

It seem at first that we still do not know how to compute $\varepsilon(J)$ or $V(\phi)$. The important observation is that $\phi$ is the independent variable so that we must substitute $J = -\partial V/\partial \phi$ everywhere $J$ appears in the above equation. We then, realize that it is <u>a non-linear differential equation for $V(\phi)$</u>, which may be solved at least for small g. With $\beta(g_J) = -b_0 g_J^3 - b_1 g_J^5 - \ldots$, we get the solutions sketched in Fig. 1. (For $\phi < 0$, the energy will only go up.) Referring the details to the original paper, we see from this figure that we have a stable non-perturbative vacuum for which $\phi > 0$. (Stationary value for the upper curve, e.g., is $g^2\phi = \mu^4 \exp(-c-1-2/b_0 g^2)$, c is a constant, which clearly shows the non-perturbative nature of the vacuum.) This is the result announced in the beginning. It is worth emphasizing that the negative character of $\beta$ function is crucial for this result.

A brief remark would be helpful here: a similar potential obtained in the past by various people[8] should not be confused with the one we have discussed. Their potential is not for the composite singlet operator $(G^a_{\mu\nu})^2$, but for $A^a_\mu$, the octet vector potential. Vacuum expectation value of the composite operator is what is relevant for QCD sum rules.

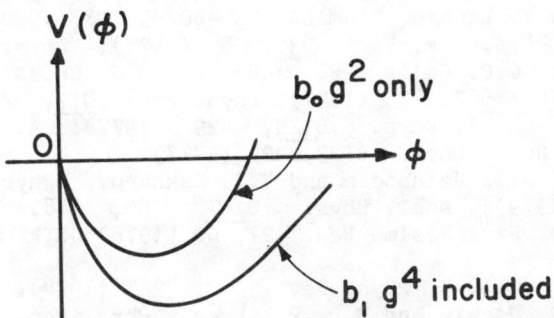

Fig.1. Effective potential corresponding to two different truncations of β function

In the remainder, I would like to describe how the condensation of $(G^a_{\mu\nu})^2$ is related to confinement. For the confinement problem a useful object is the vacuum average of the Wilson loop operator

$\Psi(C) \equiv \langle\Psi(C)\rangle \equiv \langle 1/N \mathrm{Tr} P \exp ig\oint_C A_\mu dx^\mu\rangle$, where C denotes a closed loop. Recently it has been shown[9] that this object can be made finite by the usual renormalization procedure once one isolates the self-energy of the test particle. Then, in the axial gauge the renormalized $\Psi(C)$ satisfies the renormalization group equation, $(\mu\partial/\partial\mu + \beta(g)\partial/\partial g)\Psi(C) = 0$. Let us for simplicity take the loop to be a circle of radius r. Noting that $g^2\partial/\partial g^2$ effects an insertion of $\hat{\phi}$, after Wick rotation, the above equation can be readily be transformed into

$$(\partial/\partial\sigma + M^2(\sigma))\Psi(C) = 0 , \quad (3)$$

where σ is the area of the loop and $M^2(\sigma) = (\beta(g)/g\sigma)\Delta\Phi$, $\Delta\Phi \equiv \int d^4x(\phi_c(x) - \phi)$. Here $\phi_c(x)(\phi)$ is the vacuum expectation value of $\phi(x)$ in the presence (absence) of the loop, i.e. $\phi_c(x) = \langle\phi(x)\Psi(C)\rangle/\langle\Psi(C)\rangle$. This equation tells us that the area dependence of the Wilson loop is governed by the integrated difference of condensation with and without the loop. If the local difference tends to zero rapidly away from the loop, ΔΦ would be proportional to σ (i.e. $M^2(\sigma) \sim$ constant) and one obtains so called the area law. The fact that the difference enters is of great importance for the correct sign. The Wilson loop being a source of chromo-electric flux, $\phi_c(x)$ near the loop is less magnetic than φ i.e., ΔΦ<0. This together with the negative character of β function gives the desired sign for the area dependence.

## ACKNOWLEDGMENT

We are grateful to Bill Bardeen for his interest in our work and useful discussions.

## REFERENCES

1. R. Fukuda and Y. Kazama, FERMILAB-Pub-80/55-THY (1980).
2. G. 'tHooft, Phys. Rev. Lett. $\underline{37}$, 8 (1976); Phys. Rev. $\underline{D14}$, 3432 (1976); C.G. Callan, R. Dashen and D.J. Gross, Phys. Lett. $\underline{63B}$, 334 (1976); $\underline{66B}$, 375 (1977); Phys. Rev. $\underline{D17}$, 2717 (1978); A. Polyakov, Nucl. Phys. $\underline{B120}$, 429 (1977); A. Belavin and A. Polyakov, Nucl. Phys. $\underline{B123}$, 429 (1977).
3. M.A. Shifman, A.I. Vainstein and V.I. Zakharov, Phys. Rev. Lett. $\underline{42}$, 297 (1979); Nucl. Phys. $\underline{B147}$, 385, 448, 519 (1979); Zh. Eksp. Teo. Fiz. Pis'ma Red. $\underline{27}$, 60 (1978) JETP Lett. $\underline{27}$, 55 (1978).
4. D. Gross and A. Neveu, Phys. Rev. $\underline{D10}$, 3235 (1974).
5. S. Coleman, R. Jackiw and H.D. Politzer, Phys. Rev. $\underline{D10}$, 2491 (1974); M. Kobayashi and T. Kugo, Prog. Theo. Phys. $\underline{54}$, 1537 (1975); L. Abbott, J. Kang, and H. Schnitzer, Phys. Rev. $\underline{D13}$, 2212 (1976).
6. R.J. Crewther, Phys. Rev. Lett. $\underline{28}$, 1421 (1972); M.S. Chanowitz and J. Ellis, Phys. Lett. $\underline{40B}$, 397 (1972); Phys. Rev. $\underline{D7}$, 2490 (1973).
7. J.C. Collins, A. Duncan, and S.D. Joglekar, Phys. Rev. $\underline{D16}$, 438 (1977).
8. G.K. Savvidy, Phys. Lett. $\underline{71B}$ 133 (1977); S.G. Matinyan and G.K. Savvidy, Nucl. Phys. $\underline{B134}$, 539 (1978), N.K. Nielsen and P. Olesen, Nucl. Phys. $\underline{B144}$, 376 (1978); H. Pagels and E. Tomboulis, Nucl. Phys. $\underline{B143}$, 485 (1978); R. Weder, Phys. Lett. $\underline{85B}$, 249 (1979); P. Minkowski, University of Berne Preprint (1980).
9. V.S. Dotsenko, S.N. Vergeles, Landau Institute preprint (1979); W.A. Bardeen and C.T. Hill, private communication.

LOCAL COVARIANT OPERATOR FORMALISM OF NON-ABELIAN GAUGE THEORIES
- A non-perturbative approach to quark confinement
in the Heisenberg picture -

Izumi OJIMA[*]
Research Inst. for Math. Sci., Kyoto Univ., Kyoto 606, Japan[**]

ABSTRACT

A manifestly covariant and local operator formalism of non-Abelian gauge theories is successfully formulated on the basis of BRS invariance. By this invariance as a quantum version of local gauge invariance, criteria for physical states and observables are dictated so as to preserve the principles of quantum theory: elimination of negative norms, physical S-matrix unitarity, etc. In this framework of relativistic QFT, the logical structure of quark confinement problem is clarified together with its criteria.

## 1. INTRODUCTION

All the four types of interactions ruling the nature are nowadays believed to be intermediated universally by gauge fields. Apart from (the traditional) QED, all these are non-Abelian gauge fields, which have been so far treated only by the path-integral method for lack of a consistent operator formalism. In spite of the powerfulness of the path-integral formalism as a calculational method, the absence of the notions of the state vector space and the Heisenberg operators in it obstructs us to get an insight into the general and fundamental aspects of the logical structure of the theory in a non-perturbative fashion. The understanding of these aspects seems quite necessary not only at such abstract level as the problems to assure consistency of the theory and unitarity of the physical S-matrix, etc., but also for the resolution of the outstanding problem of quark confinement, where the question "what are physically observable objects of the theory?" should be answered.

We discuss, in this article, these problems on the basis of the manifestly covariant and local Heisenberg-operator formalism of non-Abelian gauge theories obtained by T. Kugo and myself[1,2], the essence of which is explained in §2. Its application to the quark confinement is discussed in §3 and §4, where the general consequences derived from such basic ingredients of relativistic QFT as Lorentz covariance, locality, etc., work quite effectively as technical tools, in combination with the subsidiary condition specifying physical states, the notion of observables and with the "Maxwell" equation.

---

[*] Fellow of the Japan Society for the Promotion of Science (until August 1980).
[**] Address after September 1980: The Institute for Advanced Study, Princeton, NJ, 08540.

0094-243X/81/681017/-05$1.5 Copyright 1981 American Institute of Physics

## 2. INDEFINITE METRIC AND SUBSIDIARY CONDITION
### - Kinematical "confinement" by quartet mechanism -

The major difficulty encountered in constructing the covariant local operator formalism of gauge theories is the necessity to introduce an <u>indefinite metric</u> in the theory, which means the presence of negative probabilities. In order to achieve a physically meaningful interpretation, we should exclude unphysical negative norms from the "physical world" of the theory which should be defined suitably by setting up a <u>subsidiary condition</u> to specify a <u>physical state</u>. By this condition, the <u>physical subspace</u> $\mathcal{V}_{phys}$ consisting of physical states is picked out in the total state space $\mathcal{V}$ with indefinite metric.

<u>Theorem 1</u>[2]   If the following three conditions are satisfied,
- (0) hermiticity of the Hamiltonian H
  ($\to$ pseudo-unitarity of the total S-matrix S),
- (i) time invariance of $\mathcal{V}_{phys}$: $H\mathcal{V}_{phys} \subset \mathcal{V}_{phys}$
  ($\to S\mathcal{V}_{phys} = S^{-1}\mathcal{V}_{phys} = \mathcal{V}_{phys}$),
- (ii) positive-semidefiniteness of $\mathcal{V}_{phys}$:
  $|\Phi\rangle \in \mathcal{V}_{phys} \to \langle\Phi|\Phi\rangle \geq 0$,

then physical scattering processes are described consistently in the physical Hilbert space $H_{phys} \equiv \overline{\mathcal{V}_{phys}/\mathcal{V}_0}$ ($\mathcal{V}_0 \equiv \mathcal{V}_{phys}^\perp \cap \mathcal{V}_{phys}$: zero-norm subspace of $\mathcal{V}_{phys}$) in terms of the physical S-matrix $S_{phys}$ defined by
$$S_{phys}|\hat{\Phi}\rangle \equiv \widehat{S|\Phi\rangle} \quad \text{for} \quad |\Phi\rangle \in \mathcal{V}_{phys}, \quad |\hat{\Phi}\rangle \equiv |\Phi\rangle + \mathcal{V}_0 \in \mathcal{V}_{phys}/\mathcal{V}_0.$$
These $H_{phys}$ and $S_{phys}$ represent the ordinary state space and S-matrix in the theory without indefinite metric and the latter satisfies the usual <u>unitarity</u>, $S_{phys}^\dagger S_{phys} = S_{phys} S_{phys}^\dagger = 1$, without the contributions from unphysical negative norms.

These conditions (0)~(ii) are shown as follows to be satisfied in our formalism with the Lagrangian density
$$\mathcal{L} \equiv \mathcal{L}_S(A,\varphi) - A_\mu^a \partial^\mu B^a + \alpha B^a B^a/2 - i\partial^\mu \bar{c}^a \cdot (D_\mu c)^a, \tag{2.1a}$$
$$\mathcal{L}_S(A,\varphi) = -F_{\mu\nu}^a F^{a,\mu\nu}/4 + \mathcal{L}_{matter}(\varphi, \mathcal{D}_\mu \varphi). \tag{2.1b}$$
By the hermiticity assignment of Faddeev-Popov (FP) ghosts, $c^\dagger = c$, $\bar{c}^\dagger = \bar{c}$, the condition (0) is satisfied: $\mathcal{L}^\dagger = \mathcal{L}$, $H^\dagger = H$. Since the system (2.1) is invariant under the <u>BRS transformation</u>
$$\delta A_\mu^a = (D_\mu c)^a \equiv \partial_\mu c^a + g(A_\mu \times c)^a, \quad \delta\varphi_i = igc^a T_{ij}^a \varphi_j,$$
$$\delta B^a = 0, \quad \delta c^a = -g(c\times c)^a/2, \quad \delta\bar{c}^a = iB^a, \tag{2.2}$$
we have its corresponding <u>conserved</u> Noether charge $Q_B$
$$Q_B \equiv \int d^3x (B^a(D_0 c)^a - \dot{B}^a c^a + ig\dot{\bar{c}}^a(c\times c)^a/2) = Q_B^\dagger, \tag{2.3}$$
satisfying the remarkable <u>nilpotency property</u>
$$Q_B^2 = \{Q_B, Q_B\}/2 = \delta Q_B/2i = 0. \tag{2.4}$$
Using this BRS charge $Q_B$, we adopt as the <u>subsidiary condition</u>
$$Q_B|\text{phys}\rangle = 0, \tag{2.5}$$
by which the condition (i) is automatically satisfied owing to the conservation of $Q_B$. Although the condition (2.5) looks quite different from the Gupta-Bleuler condition $B^{(+)}(x)|\text{phys}\rangle = 0$ in QED, it is really shown to reproduce the latter one in the Abelian cases owing to their speciality, which means that (2.5) is a natural and non-trivial extension of the Gupta-Bleuler condition to non-Abelian cases.

In order to verify the condition (ii), we consider the algebra consisting of the BRS charge $Q_B$ and the FP ghost charge $Q_C$ defined by
$$Q_C \equiv i\int d^3x[\bar{c}^a \overset{\leftrightarrow}{\partial}_o c^a + g\bar{c}^a(A_o \times c)^a] = Q_C^\dagger \;, \tag{2.6}$$
$$[iQ_C,c] = c \;, \qquad [iQ_C,\bar{c}] = -\bar{c} \;, \tag{2.7}$$
satisfying the relations
$$[iQ_C, Q_B] = Q_B \;, \tag{2.8}$$
$$iQ_C|*,N> = N|*,N> \;; \quad <*,M|**,N> \propto \delta_{M,-N} (N \in \mathbb{Z}). \tag{2.9}$$
Since all the possible representations[3,4] of this algebra with the properties (2.3), (2.4), (2.6), (2.8) and (2.9) are only <u>BRS-singlets</u> characterized by $Q_B|\alpha>=Q_c|\alpha>=0$ without any $|*>$ satisfying $Q_B|*> = |\alpha>$ and <u>quartets</u> consisting of pairs of BRS-doublets $\{|N>, Q_B|N> = |N+1>;$ $|-(N+1)>, Q_B|-(N+1)> = |-N>\}$ with $<N|N'> = \delta_{N,-N'}$, we obtain the following theorem on the assumption of asymptotic completeness:

<u>Theorem 2</u>[2]   $P^{(n)} = \{Q_B, \exists R^{(n)}\}$, $n \geq 1$,
where $P^{(n)}$ is the projector onto "n-unphysical-particle sector" containing n-members of quartets besides arbitrary number of BRS-singlet particles.

By this theorem, the condition (iii) is verified,
$$Q_B|f> = 0 \Rightarrow <f|f> = <f|\sum_{n\geq 0}|P^{(n)}|f> = <f|P^{(0)} + \{Q_B, \exists R\}|f>$$
$$= <P^{(0)}f|P^{(0)}f> \geq 0 \;,$$
on the inevitable assumption that all the BRS-singlet-particles have positive norms. Note that from Thm. 2 we can easily show
$$\sum_{n\geq 0} P^{(n)} \mathcal{V}_{phys} = (P^{(0)}\mathcal{V})^1 \cap \mathcal{V}_{phys} = Q_B \mathcal{V} = \mathcal{V}_0 \;,$$
which asserts that "unphysical" particles belonging to quartets appear in $\mathcal{V}_{phys}$ but only in zero-norm combinations and that they are invisible physically. By this <u>quartet mechanism</u>, any member of quartets is "confined" into unphysical world, and, in the visible physical world, there remain BRS-singlets only.

In view of the quartet mechanism, quarks and gluons will be confined <u>if</u> they have asymptotic 1-particle states belonging to some quartets, namely <u>if</u> $[iQ_B,q] = ig\lambda^a c^a q/2$ and $[iQ_B,A_\mu] = D_\mu c$ have discrete poles due to the bound states in channels of c-q and $A_\mu$-c [3].

The application of this formalism to Einstein gravity[5,6,7], vierbien formalism[5], supergravity[8] and to free massless higher spin fields[8] have been extensively and successfully made.

## 3. OBSERVABLES AND "MAXWELL" EQUATION
- Confinement vs. superselection rule -

In order to assure the consistency of the probabilistic interpretations not only in the scattering theoretical aspects so far discussed but also in the measurements of a <u>physical quantity</u> A, we should at least have the condition
$$<\Phi|A|\chi> = <\chi|A|\Phi> = 0 \quad \text{for} \quad |\Phi> \in \mathcal{V}_{phys}, \; |\chi> \in \mathcal{V}_0 \;, \tag{3.1}$$
namely, no effect of zero-norms $|\chi> \in \mathcal{V}_0$ should be observed through measurement of A. We call any operator A satisfying (3.1) an <u>observable</u>. Although this requirement seems quite modest, it leads really in this formalism to a clear result that an observable should essentially be BRS or gauge invariant:

<u>Theorem 3</u>[9,2]   The condition (3.1) is equivalent to
$$[Q_B,A]\mathcal{V}_{phys} = [Q_B,A^\dagger]\mathcal{V}_{phys} = 0 \tag{3.2}$$

or $A\mathcal{V}_{phys} \subset \mathcal{V}_{phys}$ and $A^\dagger \mathcal{V}_{phys} \subset \mathcal{V}_{phys}$ (3.3)
and defines an operator $\hat{A}$ in $H_{phys}$ by
$$\hat{A}|\hat{\Phi}\rangle = \widehat{A|\Phi\rangle} \quad \text{for } |\Phi\rangle \in \mathcal{V}_{phys}, |\hat{\Phi}\rangle \in \mathcal{V}_{phys}/\mathcal{V}_0. \quad (3.4)$$
Combining this theorem with the Reeh-Schlieder theorem[10,9,2] valid in a relativistic QFT which assets $\mathcal{V} = \overline{\mathcal{F}(\mathcal{O})|0\rangle}^w$ (w: weak closure) for a polynomial algebra $\mathcal{F}(\mathcal{O})$ of local operators in any finite space-time region $\mathcal{O}$, we obtain

Theorem 4[9,2]  The following three conditions for a <u>local</u> operator $\phi \in \mathcal{F}(\mathcal{O})$ are all equivalent:
   (i) $\phi$ is a local <u>observable</u>,
   (ii) $[Q_B, \phi] = 0$, (3.5)
   (iii) $Q_B \phi |0\rangle = 0$. (3.6)

Now at this stage, we need dynamical information brought by the "Maxwell" equation of the Yang-Mills field[9]:
$$\partial^\mu F^a_{\mu\nu} + gJ^a_\nu = \{Q_B, (D_\nu \bar{c})^a\}, \quad (3.7)$$
where $J^a_\nu$ is the conserved Noether current of the gauge transformation of the first kind, containing the FP ghost contributions also. Combining this and Thm. 4 with <u>locality</u>, we can prove the <u>color singletness of local observables and of local physical states</u>:

Theorem 5[9,2]   $[\hat{Q}^a, \hat{A}] = 0$ for <u>local</u> observable A,
Theorem 6[9,2]   $\hat{Q}^a |\hat{\Phi}\rangle = 0$ for <u>local</u> physical state $|\Phi\rangle$
                 $(|\Phi\rangle \in \mathcal{V}_{phys} \cap \mathcal{F}(\mathcal{O})|0\rangle)$,
where $\hat{Q}^a$ is an <u>unbroken</u> global "color" charge. So, the color confinement, $\hat{Q}^a H_{phys} = 0$, will have been achieved, for example, if the Reeh-Schlieder property holds restricted to physical subspace: $\overline{\mathcal{V}_{phys} \cap \mathcal{F}(\mathcal{O})|0\rangle}^w = \mathcal{V}_{phys}$. This condition does not hold automatically in contrast to the one for the total space $\mathcal{V}$, as can be seen from the example of QED where <u>charge superselection rule</u> holds allowing such charged states as electrons. There are, however, some indications in favor of color confinement that charge superselection rule of QED is due to the speciality of <u>Abelian</u> gauge theory. Note that <u>confinement of the global color charge</u> leads without any gap to <u>quark confinement</u>, because the "behind-the-moon" argument usually placed against a confinement is not applicable to this case[9,2].

Closer analysis of the "Maxwell" equation in the next section reveals the crucial role of massless spectra in determining the fate of the color degree of freedom.

## 4. "MAXWELL" EQUATION AND GOLDSTONE THEOREM
### - Confinement vs. Higgs phenomenon -

Rewriting the "Maxwell" equation (3.7) in the form
$$gJ^a_\mu = \partial^\nu F^a_{\mu\nu} + \{Q_B, (D_\mu \bar{c})^a\},$$
we notice that the color current $gJ^a_\mu$ consists of two conserved currents, $\partial^\nu F^a_{\mu\nu}$ and $\{Q_B, (D_\mu \bar{c})^a\}$. Since the color charge $gQ^a \equiv \int d^3x\, gJ^a_0$ is <u>almost</u> equal to the anticommutator, $N^a \equiv \int d^3x \{Q_B, (D_0 \bar{c})^a\}$, of $Q_B$, <u>vanishing</u> in the physical world, up to the term $G^a \equiv \int d^3x\, \partial^i F^a_{0i}$ which <u>vanishes</u> on the local states owing to locality, the "Maxwell" equation can be naturally viewed to tend to confine its own charge. To make clear the meaning of "almost", we refer to the <u>Goldstone theorem</u>[11], according to which the charge $G^a$ vanishes <u>unless</u> $\partial^\nu F^a_{\mu\nu}$

contains discrete <u>massless</u> spectrum. The same theorem, however, tells us that the charge $\overline{N^a}$ is <u>almost</u> always broken spontaneously owing to massless pole appearing in $F.T.<0|TA_\mu B|0> = iF.T.<0|TD_\mu c\bar{c}|0> = -p_\mu/p^2$. In this case, "almost" means, "unless $\det(\mathbb{1}+u) = 0$", with the parameter u defined as the pole residue of $A_\mu \times \bar{c}$ at $p^2 = 0$.

Gathering together these sets of information, we obtain

<u>Theorem 7</u>[2]    On the assumption $\det(\mathbb{1}+u) \neq 0$, the charge $Q^a$ is broken spontaneously iff $\partial^\nu F^a_{\mu\nu}$ has no discrete massless spectrum.

<u>Corollary</u>[2]    If a gauge boson becomes massive, then the charge in the corresponding channel is broken spontaneously. (Converse Higgs theorem)

So, if we do not want the color symmetry to be broken, there remain only two possibilities:

a) $\partial^\nu F^a_{\mu\nu}$ contains a massless pole ($\det(\mathbb{1}+u) \neq 0$),

or  b) $u = -\mathbb{1}$.

While gluons will not be confined in the case a), the seemingly rare case b) asserts the color confinement straightforwardly: $u = -\mathbb{1} \Rightarrow G^a = 0$, $gQ^a = N^a \Rightarrow \hat{Q}^a = 0$ in $H_{phys}$. There may be several hurdles on the way to prove $u = -\mathbb{1}$, but once it is done, the quark confinement will have been achieved with whole consistency in this formalism of relativistic quantum theory of gauge fields.

I would like to thank Yamada Science Foundation and the Japan Society for the Promotion of Science for financial supports.

## REFERENCES

1. T. Kugo and I. Ojima, Phys. Lett. <u>73B</u>, 459 (1978); Prog. Theor. Phys. <u>60</u>, 1869 (1978); <u>61</u>, 294, 644 (1979).
2. T. Kugo and I. Ojima, Suppl. Prog. Theor. Phys. No. 66 (1979).
3. T. Kugo, Phys. Lett. <u>83B</u>, 93 (1979); Nucl. Phys. <u>B155</u>, 368 (1979).
4. N. Nakanishi, Prog. Theor. Phys. <u>62</u>, 1396 (1979).
5. N. Nakanishi, Prog. Theor. Phys. <u>59</u>, 972, 2175 (1978); <u>60</u>, 1190, 1890 (1978); <u>62</u>, 779, 1101, 1385 (1979); <u>63</u>, 656, 2078 (1980); <u>64</u>, No. 2 (1980).
6. T. Kugo and I. Ojima, Nucl. Phys. <u>B144</u>, 234 (1978); K. Nishijima and M. Okawa, Prog. Theor. Phys. <u>60</u>, 272 (1978).
7. N. Nakanishi and I. Ojima, Phys. Rev. Lett. <u>43</u>, 91 (1979).
8. H. Hata and T. Kugo, Nucl. Phys. <u>B158</u>, 357 (1979); C. Fronsdal and H. Hata, Nucl. Phys. <u>B162</u>, 487 (1980).
9. I. Ojima, Nucl. Phys. <u>B143</u>, 340 (1978); I. Ojima and H. Hata, Z. Phys. C, Particles and Fields <u>1</u>, 405 (1979); I. Ojima, ibid. <u>5</u>, No.**227**(1980).
10. H. Reeh and S. Schlieder, Nuovo Cim. <u>22</u>, 1051 (1961).
11. J. Goldstone, A. Salam and S. Weinberg, Phys.Rev. <u>127</u>, 965 (1962); H. Reeh, Forts. Phys. <u>16</u>, 687 (1968); F. Strocchi, Comm. Math. Phys. <u>56</u>, 57 (1977).

## OPERATOR ORDERING AND FEYNMAN RULES
## IN GAUGE THEORIES

N. H. Christ[*]
Columbia University, New York, N. Y. 10027

The considerable importance of Yang-Mills theories in particle physics certainly justifies a careful investigation of the formulation of the quantum theory in a variety of gauges. In this talk I would like to discuss both the Hamiltonian operator and the corresponding Feynman rules in a class of gauges specified by imposing conditions on only the spatial components of the vector potential. The most familiar example (to which I will limit this discussion) is the Coulomb gauge

$$\partial_i V_i = 0 \ . \tag{1}$$

The work that I will describe was done in collaboration with T. D. Lee.[1]

Perhaps the simplest gauge in which to construct a canonically quantized Yang-Mills theory is the $V_0 = 0$ gauge. The conventional Hamiltonian operator is

$$H = \tfrac{1}{2} \int d^3r \ \text{tr} \{ \pi_i^2(r) + B_i^2(r) \} \tag{2}$$

with

$$B_i(\vec{r}) = \epsilon_{ijk} \{ \partial_j V_k - i \tfrac{g}{2} [V_j, V_k] \} \ . \tag{3}$$

For the SU(N) gauge theory the matrix $V_i(\vec{r}) = \sum_\ell V_i^\ell(\vec{r}) T^\ell$ where $T^\ell$, $1 = \lambda \leqslant N^2 - 1$ is an N x N traceless, Hermitian matrix. The unconstrained fields $V_i^\ell(\vec{r})$ and their conjugate momenta obey the usual commutation relations

$$[V_i^\ell(\vec{r}), \pi_j^{\ell'}(\vec{r}')] = i \delta^{\ell\ell'} \delta_{ij} \delta^3(\vec{r} - \vec{r}') \ . \tag{4}$$

The Hamiltonian equations of motion determined by H of Eq. (2) do not include Gauss' law which instead is imposed as a condition on the physical states

$$(D_i \pi_i)^\ell |\psi\rangle = (\partial_i \pi_i^\ell + g f^{\ell ab} V_i^a \pi_i^b) |\psi\rangle = 0 \ . \tag{5}$$

---

[*] This research was supported in part by the U.S. Department of Energy.

Let us assume for the moment that this conventional formulation of the $V_0 = 0$ gauge quantum theory is correct.

We can now change to the Coulomb gauge by a straightforward transformation of variables. Given $V_i(\vec{r})$, let $U(r)$ be the gauge function transforming $V_i(r)$ into a Coulomb gauge vector potential $A_i$:

$$V_i(\vec{r}) = U(\vec{r}) A_i(\vec{r}) U^{-1}(\vec{r}) + \frac{i}{g} U(\vec{r}) \partial_i U^{-1}(\vec{r}),$$

$$\partial_i A_i = 0.$$
(6)

Of course, when $V$ and $A$ are of order $\frac{1}{g}$ this transformation possesses Gribov ambiguities.[2] However, the questions which I wish to discuss arise even when $V$ and $A$ are small so that Eq. (6) can be interpreted perturbatively. We may parametrize the $SU(N)$ matrix $U$ by $N^2 - 1$ generalized Euler angles $\phi_1, \cdots \phi_{N^2-1}$, $U = U(\phi_\alpha)$ so that $U(r)$ in Eq. (6) determines $N^2 - 1$ functions $\phi_\alpha(\vec{r})$, $U(r) = U(\phi_\alpha(r))$. Thus Eq. (6) relates two possible sets of variables $V_i(\vec{r})$ and $\phi_\alpha(\vec{r})$, $A_i(\vec{r})$. If we view the $V_0 = 0$ gauge Hamiltonian as a Schrödinger operator

$$H = \tfrac{1}{2} \int d^3 r \left\{ \left[ -i \frac{\delta}{\delta V_i^\ell(r)} \right]^2 + \left[ B_i^\ell(\vec{r}) \right]^2 \right\}$$
(7)

we may use the chain rule to change variables to $A_i$ and $\phi_\alpha$. Precisely this technique for transforming to Coulomb gauge has been used for other purposes by Gribov[2], Gervais and Sakita[3], Creutz, Muzinich and Tudron.[3]

We are making a change of variables in a quantum theory quite analogous to the change from Cartesian to polar coordinates in single-particle quantum mechanics. For a particle moving in two dimensions we have

$$-\tfrac{1}{2}\left(\frac{\partial^2}{\partial x^2} + \frac{\partial^2}{\partial y^2}\right) = -\tfrac{1}{2}\tfrac{1}{r}\frac{\partial}{\partial r} r \frac{\partial}{\partial r} - \tfrac{1}{2}\tfrac{1}{r^2}\frac{\partial^2}{\partial \phi^2},$$
(8)

or if we extract a factor of $r^{-\tfrac{1}{2}}$ from the states so the Jacobian disappears from the inner product, $H$ becomes

$$\sqrt{r}\, H \,\frac{1}{\sqrt{r}} = -\tfrac{1}{2}\frac{\partial^2}{\partial r^2} - \tfrac{1}{2}\tfrac{1}{r^2}\frac{\partial^2}{\partial \phi^2} - \tfrac{1}{8}\tfrac{1}{r^2}$$
(9)

acquiring an extra potential-like term $-\tfrac{1}{8}\tfrac{1}{r^2}$ which distinguishes the form of the Cartesian and curvilinear Hamiltonia.

We can now do precisely the same thing for the Yang-Mills case. The transformed Hamiltonian depends on $A_i^\ell$, $\phi_\alpha$ and their conjugate momenta

$P_i^\ell$, $P_\alpha$. However, since the Hamiltonian in $V_0 = 0$ gauge is invariant under time-independent gauge transformations, all the $\phi_\alpha$-dependent terms are proportional to $p_\alpha$. However, when expressed in our new variables, the Gauss-law condition (5) becomes

$$p_\alpha(\vec{r}) | \psi \rangle = 0 . \qquad (10)$$

Thus the gauge variables do not appear at all when the Hamiltonian is restricted to the physical states, obeying Eq. (10). We find

$$H = \tfrac{1}{2} \int d^3 r \{ \mathcal{J}^{-1} E^\ell(\vec{r})^\dagger \mathcal{J} E^\ell(\vec{r}) + [B_i^\ell(\vec{r})]^2 \} \qquad (11)$$

where

$$-E_i^\ell(\vec{r}) = P_i^\ell(\vec{r}) + g \int d^3 r' \langle \vec{r}, \ell | \partial_j (\partial_j \mathcal{D}_j)^{-1} | \vec{r}', k \rangle f^{kab} A_i^a(\vec{r}') P_i^b(\vec{r}') , \qquad (12)$$

and $\mathcal{J} = \det(\partial_i \mathcal{D}_i)$. This is not the usual Coulomb gauge Hamiltonian! There are extra ordering pieces. In fact, this result is not new but was proposed in a quite different form by Schwinger[4] in 1962. This problem has also been addressed by R. Utiyama and J. Sakamoto[5] using a similar technique, but their result appears to be different from ours.

If we again extract a factor of $\mathcal{J}^{-\tfrac{1}{2}}$ from the states, $H$ can be written as

$$\tilde{H} = \sqrt{\mathcal{J}} \, H \, \frac{1}{\sqrt{\mathcal{J}}} = \tfrac{1}{2} \int d^3 r \{ [E_i^\ell(\vec{r})]^2 + [B_i^\ell(\vec{r})]^2 \}_W + V_1 + V_2 \qquad (13)$$

with

$$V_1 = -\frac{g^2}{8} \int d^3 r \langle \vec{r}, \ell | (\partial \cdot \mathcal{D})^{-1} \partial_i t^{\ell'} | \vec{r}, n \rangle \cdot \langle \vec{r}, \ell' | (\partial \cdot \mathcal{D})^{-1} \partial_i t^\ell | \vec{r}, n \rangle ,$$

and

$$V_2 = -\frac{g^2}{8} \int d^3 r \int d^3 r' \langle \vec{r}, \ell | \delta_{ii'} - \mathcal{D}_i (\partial \cdot \mathcal{D})^{-1} \partial_{i'} | \vec{r}', n' \rangle \cdot \langle \vec{r}', \ell' | \delta_{i'i} - \mathcal{D}_{i'} (\partial \cdot \mathcal{D})^{-1} \partial_i | \vec{r}, n \rangle \cdot \langle \vec{r}, n | t^\ell (\partial \cdot \mathcal{D})^{-1} \partial_j \partial_i (\partial \cdot \mathcal{D})^{-1} t^{\ell'} | \vec{r}', n' \rangle \qquad (14)$$

where the $t^\ell$ are adjoint representation generators and $\mathcal{D}_i$ the covariant derivative in Coulomb gauge. The subscript $W$ in Eq. (13) implies a Weyl ordering or complete symmetrization of the non-commuting factors, e.g.

$$\{f(q) p^2\}_W = \tfrac{1}{4} p^2 f(q) + \tfrac{1}{2} p f(q) p + \tfrac{1}{4} f(q) p^2 . \qquad (15)$$

Finally, to obtain the Feynman rules with propagators given by the covariant T* product evaluated by symmetrical integration we must use the Hamiltonian in this Weyl-ordered form. Thus the "covariant" Feynman rules follow from

$$\int d[P_i] d[A_i] e^{i \int d^4 x \, P_i^\ell \dot{A}_i^\ell - i \int \tilde{H} dt}$$

$$= \int d[A_\mu] \delta(\partial_i A_i) \det(\partial_i \mathcal{D}_i) e^{-\frac{1}{4} i \int d^4 x \, F_{\mu\nu} F_{\mu\nu} - i \int dt (V_1 + V_2)}. \quad (16)$$

The new terms implied by $V_1$ and $V_2$ contribute first in the two-loop order.

The conclusions described above are of course only correct if the original $V_0 = 0$ gauge quantum Hamiltonian was correct. The validity of our $V_0 = 0$ gauge quantum theory can be demonstrated by showing its equivalence to that defined by the usual covariant-gauge Feynman rules.[1]

### References

1. N. H. Christ and T. D. Lee, Phys.Rev. D (in press).
2. V. N. Gribov, lecture at the 12th Winter School of the Leningrad Nuclear Physics Institute, 1977.
3. J.-L. Gervais and B. Sakita, Phys.Rev. **D18**, 453 (1978).
   M. Creutz, I. Muzinich and T. Tudron, Phys.Rev. **D19**, 531 (1979).
4. J. Schwinger, Phys.Rev. **127**, 324 (1962), ibid. **130**, 406 (1963).
   R. P. Treat, Phys.Rev. **D12**, 3145 (1975).
5. R. Utiyama and J. Sakamoto, Prog.Theor.Phys. **55**, 1631 (1976).

# CHIRAL SYMMETRY BREAKDOWN IN LARGE-N QCD

Edward Witten
Department of Physics, Harvard University
Cambridge, MA 82138

## ABSTRACT

Chromodynamics with n flavors of massless quarks is invariant under chiral $U(n) \times U(n)$. We show that in the limit of large number of colors, under reasonable assumptions, this symmetry group must spontaneously break down to diagonal $U(n)$.

# GLUEBALLS IN QCD SUM RULES

V. Zakharov
ITEP, Moscow

I would like to make a few observations on glueballs based on the original paper worked out by V. Novikov, M. Shifman, A. Vainshtein and myself.[1] The original is quite lengthy, more than one hundred pages, so that in no way could I pretend to be complete.

1. The general framework remains the same that we are pursuing now for a few years, that is the QCD sum rules. The picture is that asymptotic freedom, valid at short distances, gets violated at a larger scale by interaction of the quarks and gluons with the vacuum fields. Formally, we proceed from the general operator expansion and introduce various vacuum expectation values such as $<0|\alpha_s G^a_{\mu\nu} G^a_{\mu\nu}|0>$ where $G^a_{\mu\nu}$ is the gluon field strength tensor. These matrix elements set the scale for the resonance masses.

We probe glueballs with corresponding currents constructed from $G^a_{\mu\nu}$. Thus, for the scalar glueballs the relevant current is

$$j_s = \alpha_s G^a_{\mu\nu} G^a_{\mu\nu}$$

Furthermore, define the two-point function $M(q)$

$$M(q) = <0|: \int d^4x T\{j_s(x) j_s(0)\}|0> e^{-iqx}$$

The sum rules then look as

$$\frac{1}{\pi M^2} \int \text{Im}\, \Pi'(s) \exp(-s/M^2) ds = \frac{2M^2}{\pi^2} \alpha_s^2(M)$$

$$+ \frac{<0|\alpha_s G^2|0>}{M^2} \left( \frac{32\pi}{b} - 4\alpha_s \right)$$

$$- \frac{2\alpha_s}{\pi M^4} <0|g_s^3 f_{abc} G^a_{\mu\nu} G^b_{\nu\alpha} G^c_{\alpha\mu}|0> + O(1/M^6) \quad . \tag{1}$$

Note that in the limit of very large $M^2$ only the first term in the r.h.s. survives. The others become important at some point determined by the numerical value of $<G^2>$. As is mentioned above, this $M^2$ sets the natural scale for the resonance mass.

2. Certainly, we cannot go into quantitative analysis here. The first qualitative observation, however, is that the sum rules indicate a new mass scale for the glueballs.

Indeed, from experience with the $\rho$-meson sum rules we learn that at $M^2 = m_\rho^2$ the power correction is about 10 percent of the leading, bare graph. In the gluonic channel the same happens at

$$M^2_{crit} \sim 6 \text{ GeV}^2 \qquad (2)$$

which is about 10 times larger than $m_\rho^2$.

The simplest explanation as of where the difference comes from is that the power correction is associated now with the Born type graph while the bare graph contains a loop (see Fig. 1). In the case of $\rho$ channel both the power correction and bare graphs are one-loop ones. This gives a factor $(16\pi^2)$ of enhancement for the power correction. Since $<G^2>$ is of dimension 4, the proper factor to compare the masses squared is $\sqrt{16\pi^2}$

$$(\mu^2)_{glueball} \sim 4\pi m_\rho^2 \quad .$$

3. But this is actually an oversimplification. The real leading power correction comes from a different source. The point is that we can derive the low-energy theorem:

$$M(0) = i \int dx <0|T\{F^2(x)F^2(0)\}|0> = \frac{32\pi^2}{b}<0|\frac{\alpha_s}{\pi}G^2|0>$$

$$F^a_{\mu\nu} \equiv g_s G^a_{\mu\nu}$$

The knowledge of $M(0)$ can be translated into the language of power corrections since the high energy behavior of the total amplitude is controlled by the perturbation theory and is known.

The derivation of the low-energy theorem deserves a further comment since it is quite general and applies in some other cases as well. Namely, $M(0)$ can be represented as a derivative of $<F^2>$ with respect to the bare strong interaction coupling constant

$$M(0) = \frac{d}{d\left(-\frac{1}{g_0^2}\right)} <0|F^2|0> \qquad (3)$$

On the other hand, the dependence on the bare coupling constant is fixed by the renormalization-group argument, $<F^2> \sim M_0^4 \exp(-8\pi^2/bg_0^2(M))$, and we come to (3).

As another application of the technique, let me mention that

$$\frac{d}{dm_q}<F^2> = -\frac{24}{b}<\bar{q}q> \qquad (4)$$

This equation tells us, among other things, how the vacuum expectation value $<F^2>$ evolves with switching on of a small quark mass. From this one can estimate that

$$\langle F^2\rangle_{\text{no light quark}} \simeq (2 \sim 3) \times \langle F^2\rangle_{\text{real world}}$$

4. Turning back to the glueball business, let me emphasize that the new mass scale is welcome by the phenomenology. Indeed, the $\eta'$ mass vanishes (as shown by Witten) in the large $N_c$ limit and still it is larger than the "typical" hadronic mass, that is $m_\rho^2$. Moreover, $\Gamma(\psi \to \eta\gamma)$ is dual to the $\sim 2$ GeV$^2$ of the gluonic continuum and still is considered to be a respectable quark-made meson. All the puzzles are resolved once one appreciates the fact that the so-called hadronic mass scale is not unique, as is usually assumed, but differs from channel to channel as is indicated by the sum rules.

In the paper we elaborate the point and introduce the corresponding parameter which depends on the quantum numbers and substitues for the standard $1/N_c$ counting. At this point I should admit that, although qualitative results are encouraging, the quantitative conclusions are less definite. The reason is that the $\rho$-meson disturbs QCD vacuum only slightly. Then the unperturbed vacuum is a good zero approximation, and we have succeeded to relate, say, $m_\rho^2$ to $\langle\bar{q}q\rangle$. Now we need a real knowledge of the vacuum wave function.

By counting the powers of $\alpha_s$ one gets convinces that the leading power correction $\sim\langle F^2\rangle$ corresponds to the graph when both gluons from the current are absorbed (and returned back afterwards) by the vacuum. There are no such graphs in the $\rho$ case. Many theoreticians would propose then to use the instanton calculus to find explicit answers for the graphs discussed. Unfortunately, it would not be reasonable to follow such advice, in our mind. Indeed, the instanton density itself is distorted by the vacuum condensate:

$$d_{\text{eff}} = d_0(\rho)\exp\left[\frac{\pi^4 \rho^4}{8\alpha_s^2(\rho)}\langle 0|\frac{\alpha_s}{\pi}G^2|0\rangle\right] \quad (5)$$

One can find, using this equation, how far the dilute instanton gas approximation extends and find that it is bounded to $Q^2 \gtrsim 30$ GeV$^2$ or so, that is pushed into an uninteresting domain.

Another guess still can be true: although instantons are not applicable literally they might give a correct message on quantum numbers for which "mixing with the vacuum" is important.

In conclusion, there is an example of what we understand by the "quantum numbers of instantons." Consider pure gluedynamics, with no light quarks. Then we would believe that the leading vacuum fluctuation shares the instanton property of being a self-dual (anti self-dual) field. In particular,

$$\int \langle 0|T\{G^2(x), G^2(0)\}|0\rangle d^4x \simeq \int \langle 0|T\{G\tilde{G}(x), G\tilde{G}(0)\}|0\rangle d^4x$$

(no light quarks)

Switching in light quarks we, on one hand, introduce relatively light η' and, on the other hand, force the G$\tilde{G}$-correlator to vanish. Therefore

$$f_{\eta'}^2 m_{\eta'}^2 = \frac{7g}{b} \langle \frac{\alpha_s}{\pi} G^2 \rangle$$

the relation which does hold well numerically.

### References

1. V. Novikov, M. Shifman, A. Vainshtein, V. Zakharov, preprint ITEP-87, 88 (1980). See also several recent papers by the same authors. Further references are given therein.

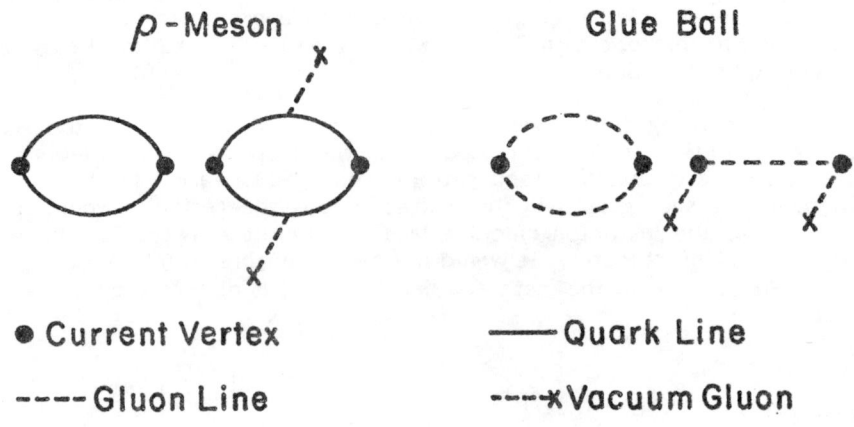

Fig. 1

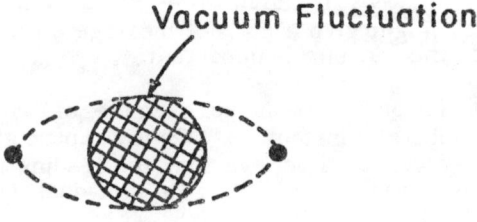

Fig. 2

# QCD III

E. Golowich, Massachusetts, Organizer

GLUEBALLS AND ODDBALLS: THEIR EXPERIMENTAL SIGNATURE

C. E. Carlson
College of William and Mary, Williamsburg, VA 23185

ABSTRACT

We study hadrons with a 2-gluon valence component, emphasizing states not found in the quark model. We predict that spin 0 and 1 glueballs will be narrow, with widths ranging from a few MeV to less than one MeV, and that spin 0 glueballs will mimic s$\bar{s}$ quarkonium states in their decay. We also comment on photonic couplings.

This talk will report on work done with J. Coyne, P. Fishbane, Franz Gross, and S. Meshkov. J. Donoghue has already given a general talk at this Conference about glueballs, which will allow me to omit certain topics including glueball masses, and S. Meshkov has reported to this Conference other aspects of the present work. This talk will concentrate on certain properties of glueballs made from two gluons with emphasis on particular states that we will call "oddballs."

It seems that the confining potential of quantum chromodynamics leads to a discrete spectrum of flavorless bosonic states whose valence component contains no quarks. These states[1,2] of the pure glue field are called glueballs and are classifiable according to how few gluons can form them. The 2-gluon (2g) section contains[2] $J^{PC}=J^{++}$ and $J^{-+}$ states for all J. The (odd)$^{-+}$ states do not couple to the $q\bar{q}$ quark model sector; we refer to these glueballs of exotic quantum numbers as "oddballs."

We will show that glueballs are narrower, sometimes much narrower, than ordinary hadrons. We distinguish a decreasing sequence of widths. The broadest is a geometric mean between an ordinary hadronic width - O(100 MeV) - and an OZI-violating[3] decay width - O(1 MeV). This follows because glueball decay involves the formation of a quark pair whereas an OZI-violating decay involves both annihilation of one pair and formation of another. We take 30 MeV for this value. Various mechanisms that we discuss suppress different channels. We expect the J=0 and $1^{++}$ states to have total widths of O(3 MeV). For the $J^{PC}=\text{odd}^{-+}$ oddball states, we expect total widths less than 1 MeV; the interesting exclusive channel $1^{-+} \to 2(0^{-+})$ is for practical purposes closed. The oddball will also have dynamically suppressed photonic couplings, and all of their production modes are small.

Another testable property is the flavor content[4] of the decay products. Heavy quarks are favored in the J=0 glueball decay, with a ratio of kaons to pions of 0.9 to 1.4 to 2 as the glueball mass runs from 1 to 2.5 GeV to infinity.

We now expand these remarks, beginning with a discussion of the dominant diagram for non-oddballs, which we call the "U-graph," Fig. 1(a).

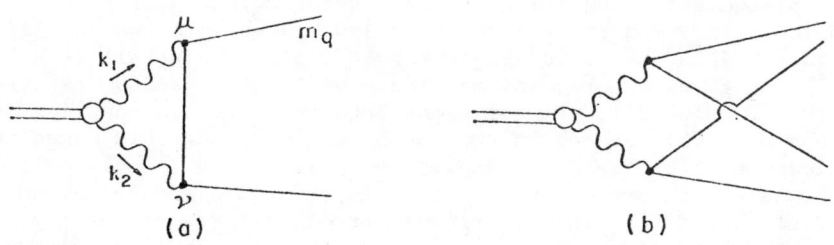

Fig. 1. Lowest order graphs contributing to hadronic glueball decay. (a) The U-graph, decay into a $q\bar{q}$ pair. (b) The crow's foot, decay into 4 quarks.

It is familiar[4] in the two lepton decay of the $\pi^0$. (Not shown in this $q\bar{q}$ production graph, which is a short distance process, is the long range component which polarizes an additional quark pair from the vacuum to make color-neutral hadrons in the final state, e.g., a 2-meson state.) The vertex $T_{J^{PC}}(k_1,k_2)$ is an unknown piece of such a graph. In accordance with quark model ideas about composites we take[5]

$$T_{0^{++}} = f\,(k^2-\Lambda^2+i\epsilon)^{-1}\tfrac{1}{2}F^{\mu\nu}F_{\mu\nu} \to f(k^2-\Lambda^2+i\epsilon)^{-1}(k_1\cdot k_2\,\epsilon_1\cdot\epsilon_2 - k_1\cdot\epsilon_2\,k_2\cdot\epsilon_1) \quad (1)$$

$$T_{0^{-+}} = f\,(k^2-\Lambda^2+i\epsilon)^{-1} F^{\mu\nu}F^{\lambda\delta}\epsilon_{\mu\nu\lambda\delta} \to f(k^2-\Lambda^2+i\epsilon)^{-1}\epsilon_{\mu\nu\lambda\delta}\,k_1^{\mu}\epsilon_1^{\nu}k_2^{\lambda}\epsilon_2^{\delta} \quad (2)$$

where $k=\tfrac{1}{2}(k_1-k_2)$ is the relative momentum at the vertex, $\epsilon_i$ is the polarization vector of the gluon $i$, $\Lambda^2$ is a scale factor, and $f$ is a coupling strength. Calculation of the U-graph is then straightforward. The result, to $O(m_q/M_G)$, contains Spence functions in the constant terms; for $\Lambda=\tfrac{1}{2}M_G$ these Spence functions can be analytically evaluated, and we quote the resulting amplitude in this case. Omitting color factors and normalizing to $\bar{u}u = 2m_q$,

$$\mathcal{M}_{0^{++}} = \frac{fg^2}{2\sqrt{2}\,\pi^2}\frac{m_q}{M_G}\left\{\ln^2\frac{M_G}{m_q} - 2\ln\frac{M_G}{m_q} + \frac{\pi^2}{12} + 2\ln 2 - i\pi\ln\frac{M_G}{m_q}\right\} \quad (3)$$

$$\mathcal{M}_{0^{-+}} = \frac{fg^2}{2\sqrt{2}\,\pi^2}\frac{m_q}{M_G}\left\{\ln^2\frac{M_G}{m_q} - 2\ln\frac{M_G}{m_q} + \frac{\pi^2}{12} - i\pi\ln\frac{M_G}{m_q}\right\} \quad (4)$$

For both amplitudes the imaginary part dominates the real part; when $M_G/m_q=6(3)$ the ratios of the magnitudes are 3.06(2.8) and 12.5(20.5) for $0^{++}$ and $0^{-+}$ respectively. We conjecture that this dominance of the imaginary part, in which the intermediate gluons are on-shell, is true for any spin, dominating by a factor ten or more in amplitude squared.

Amplitudes (3) and (4) contain a factor $m_q/M_G$; vector quark-gluon couplings don't allow decay of J=0 states into two massless fermions. This factor is absent for a "typical" glueball (e.g., a $2^{++}$) and gives an order of magnitude width suppression. Regarding the $1^{++}$ glueball, symmetry considerations[6] forbid coupling of J=1 and of (odd)$^-$ states to two on-shell gluons and this makes the imaginary part of the decay amplitude vanish. The $1^{++}$ glueball width, as determined by the U-graph, is thus also reduced by an order of magnitude. The logarithmic terms are present for any spin and enhance light quark production, but for J=0 the linear mass factor enhances heavy quark production. Keeping only the imaginary term in (3) or (4), the strange quark production relative to up quarks is in the ratio

$$[(m_s/m_u)(\ln M_G/m_s)/(\ln M_G/m_u)]^2 \tag{5}$$

The other lowest order diagram which contributes to glueball decay is given in Fig. 1(b); we call it the "crow's foot." This mode is available to decay of all $J^{PC}$ states, unlike Fig. 1(a), and for the oddballs odd$^{-+}$ the crow's foot is the only diagram. To compare with the U-graph, we take the unitarity-inspired viewpoint that integration over phase space describes the total decay width. For the U-graph two-body phase space is trivial; for the four-body phase space of the crow's foot the integration is more complicated and gives logarithms. We find for the crow's foot, to $O(m_q/M_G)$,

$$\Gamma'_{CF}(0^{++}) = \frac{g^4 f^2}{288 \pi^5 M_G} \cdot \frac{1}{4} \left\{ \ln^2 \frac{M_G}{m_q} - \frac{8}{3} \ln \frac{M_G}{m_q} + 1.32 \right\} \tag{6}$$

$$\Gamma'_{CF}(0^{-+}) = \frac{g^4 f^2}{288 \pi^5 M_G} \cdot \frac{1}{4} \left\{ \ln^2 \frac{M_G}{m_q} - \frac{8}{3} \ln \frac{M_G}{m_q} + 0.97 \right\} \tag{7}$$

We have again used $\Lambda = (1/2)M_G$ and assumed equal mass quarks at each production vertex.

The ratio of widths for the two graphs gives us a measure of the total width due to the crow's foot. We find, keeping only the imaginary part of the U-graph and inserting the color factors,

$$\frac{\Gamma'_U(0^{++})}{\Gamma'_{CF}(0^{++})} = \frac{8.9}{3.4} \left(\frac{m_q}{M_G}\right)^2 \frac{\pi^2 \ln^2 \frac{M_G}{m_q}}{\ln^2 \frac{M_G}{m_q} - \frac{8}{3} \ln \frac{M_G}{m_q} + 1.32} \tag{8}$$

and the same for the $0^{-+}$ state with $1.32 \to 0.97$. There is large cancellation in the denominator, and one needs the $O(m_q/M_G)$ terms for $M_G/m_q \lesssim 6$. We therefore give the ratios (8) for $M_G/m_q=10$; for the $0^{++}$ and $0^{-+}$ cases these are 6.5 and 23.8 respectively. These numbers will increase for smaller values of $M_G/m_q$. We take a factor of 10 to reqpresent the relative contribution to the width of the U-graph and the crow's foot.

This factor leads us to two conclusions. First, the U-graph is in general dominant over the crow's foot. The properties of the U-graph therefore generally determine the properties of the non-oddball states, e.g., the flavor-dependence of the decay as in Eq. (5). The $1^{++}$ state is more complicated, because here the imaginary part of the U-graph vanishes[6]; however, the crow's foot is similarly suppressed for light quark pairs, where the intermediate gluons can be more nearly on-shell. The relative contribution of U-graph and crow's foot for the $1^{++}$ requires further study, although the total width of the $1^{++}$ will be at least ten times decreased. The second conclusion is that the total widths of all 2g oddballs are doubly suppressed because they can decay only through the crow's foot graph. First, the crow's foot for non-oddballs is already small compared to the normal U-graph, and second, the Yang theorem[6] applies to all odd-+ states and further suppresses the crow's foot amplitude. We may conservatively expect the width of a 2g oddball to be an MeV or less.

Two body decay channels are experimentally interesting. Fig. 2 shows how we realize such final states for oddballs. This graph can in fact be calculated[7], but for the two pseudoscalar decay of the oddball, symmetry considerations obviate the need for this. Suppose the mass of the $J^{PC} = 1^{-+}$ oddball is below the $\eta_c\eta'$ threshold. The $(\pi\pi)$, $(KK)$, $(\eta\eta)$, and $(\eta'\eta')$ channels all have CP even and are forbidden; $(\eta\pi)$ is forbidden by isospin. A small amount of $(\eta\eta')$ is allowed because SU(3) flavor symmetry is not exact. In addition, some $(\eta\eta')$ is allowed to the extent that the $\eta$ and $\eta'$ contain admixtures of pseudoscalar glueballs. However, a strong two pseudoscalar signal is not an experimental indication of the decay of the $1^{-+}$ oddball.

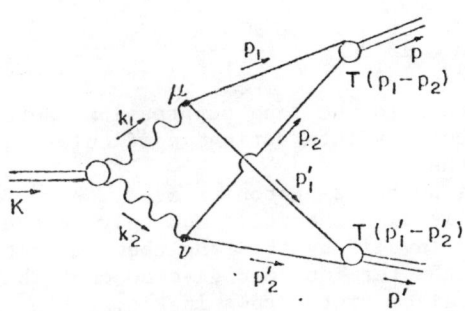

Fig. 2. An exclusive crow's foot, decay into 2 mesons.

Another area where the oddballs are distinguished is photonic production and decay. Fig. 3(a) shows the photonic decay of a quark model state into an oddball. (The same graph in another channel describes the photonic decay of an oddball.) These processes are suppressed by two mechanisms; the vanishing of the coupling to two

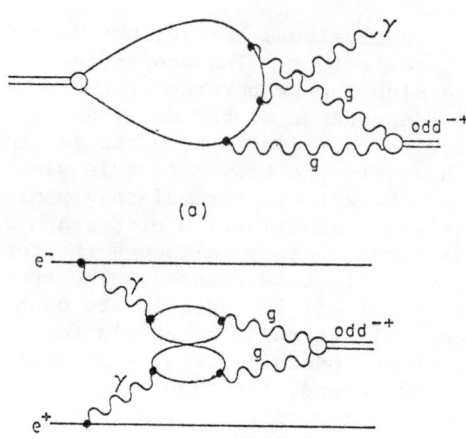

Fig. 3. Photonic couplings of glueballs. (a) photonic production of oddballs by hadron decay. (b) 2 photon production of oddballs.

on-shell gluons[6], and the fact that for oddballs bremsstrahlung comes only from between the emitted gluons. We expect the latter effect to be important because physically bremsstrahlung is preferentially a long distance process and the interior line from which emission comes is hard. For all spinless particles the suppression in the amplitude is

$$(1-M_G^2/M^2)/(2-M_G^2/M^2) \qquad (9)$$

where M is the mass of the decaying state. Eq. (9) gives at least a factor four in the width, and increases as $M_G^2/M^2$ decreases from unity. Assuming a conservative factor of 4, we estimate

$$\frac{\Gamma(\psi \to \gamma + 1^{-+})}{\Gamma(\psi \to \text{all}=3g)} = \frac{12}{5} \frac{\alpha}{\alpha_c} R \frac{1}{4} \approx 0.2\% \qquad (10)$$

where 12/5 is a color factor and R is the Yang suppression factor of 10. This number is experimentally interesting, particularly considering the narrowness of the $1^{-+}$.

Production of the oddballs by the 2-photon process, shown in Fig. 3(b), or by gluon fusion, is also special. Not only is the equivalent photon approximation invalid by the Yang theorem, but 2-quark intermediate states in the fermion loop give no contribution. Only the 4-quark intermediate state drawn in Fig. 3(b) contributes.

While the approximations and assumptions we have made in reaching our conclusions are undoubtedly crude, they provide a useful guide for the experimental quest for glueballs. Most surprising of all, it seems to us, would be their continued non-observation.

We thank A. Soni and J. Rosner for very helpful discussions. We thank the Aspen Center for Physics for hospitality during part of this work. C. E. C. and F. G. are supported in part by NSF PHY-79 08240, and P. M. F. is supported in part by NSF PHY-79 01757.

## REFERENCES

1. H. Fritzsch and P. Minkowski, Nuovo Cim. $\underline{30A}$, 393 (1975); P. Freund and Y. Nambu, Phys. Rev. Lett. $\underline{34}$, 1645 (1975); R. Jaffe and K. Johnson, Phys. Lett. $\underline{60B}$, 201 (1976); J. Kogut, D. Sinclair, and L. Susskind, Nucl. Phys. B $\underline{114}$, 199 (1976); J. Bolzan, W. Palmer, and S. Pinsky, Phys. Rev. D$\underline{14}$, 3202 (1976); D. Robson, Nucl. Phys. B$\underline{130}$, 328 (1977); P. Roy and T. Walsh, Phys. Lett. $\underline{78B}$, 62 (1978); K. Koller and T. Walsh, Nucl. Phys. B. $\underline{140}$, 449 (1978); K. Ishikawa, Phys. Rev. D $\underline{20}$, 731 (1979); H. Suura, Phys. Rev. Lett. $\underline{44}$, 1319 (1980); J. Bjorken, SLAC Summer Institute on Particle Physics, SLAC-PUB-2372.
2. J. Coyne, P. Fishbane, and S. Meshkov, Phys. Lett. $\underline{91B}$, 259 (1980).
3. See I. Muzinich and F. Paige, Phys. Rev. D $\underline{21}$, 1151 (1980).
4. S. Drell, Nuovo Cim. $\underline{11}$, 693 (1959); S. Berman and D. Geffen, Nuovo Cim. $\underline{18}$, 1192 (1960); K. Ishikawa, Phys. Rev. D$\underline{20}$, 2903 (1979).
5. The $0^{++}$ coupling in Eq. (1) is S-wave. A $_2$ D-wave coupling is also possible, with tensor structure $k_1^2 k_2^2 \varepsilon_1 \cdot \varepsilon_2 + k_1 \cdot k_2 \; k_1 \cdot \varepsilon_1 \; k_2 \cdot \varepsilon_2 - k_2^2 \; k_1 \cdot \varepsilon_1 \; k_1 \cdot \varepsilon_2 - k_1^2 \; k_2 \cdot \varepsilon_2 \; k_2 \cdot \varepsilon_1$. We ignore this coupling, which requires more convergence than the monopole. We thank L. Trueman and L. N. Chang here.
6. C. N. Yang, Phys. Rev. $\underline{77}$, 242 (1950).
7. Such calculations can be simplified by using the zero-binding approximation for the dynamical part T of the ordinary hadron vertices (S. Brodsky, private communication).
8. C. A. Levinson, H. J. Lipkin, and S. Meshkov, Nuovo Cim. $\underline{32}$, 1376 (1964).

## HADRONIZATION PROBLEM IN QUARK AND GLUON JETS

Rudolph C. Hwa
University of Oregon, Eugene, Oregon 97403

### ABSTRACT

The hadronization problem is treated in the recombination model. The recombination function is determined by extracting the dressed-quark distribution from the hadron structure functions measured in deep inelastic scattering and lepton-pair production processes at high $Q^2$. The quark-antiquark distributions in quark and gluon jets are calculated in leading order QCD. The convolution of the two in the recombination model yields fragmentation functions that are in agreement with available data in both normalization and shape.

### INTRODUCTION

The hadronization problem is important in bridging the gap between perturbative QCD calculations and experimentally measured hadron distributions. In a quark jet, for example, the quark, antiquark and gluon distributions at some low $Q^2$ due to gluon-bremsstrahlung and pair creation from an initial quark produced at some high $Q^2$ in an $e^+e^-$ annihilation process, say, can be calculated in perturbative QCD. However, the formation of pions from those partons requires a knowledge of the pion wave function, which is not calculable from first principles. That is, of course, the hadronization problem. In the absence of a definitive solution to the confinement problem one could either study jet properties that are independent of hadronization (such as energy flow),[1] or make Monte-Carlo analysis of the data,[2] or do what can best be done with the help of some sensible phenomenology. We shall follow the last course: the phenomenology involves the determination of the pion wave function at low $Q^2$ from the pion structure function at high $Q^2$.

### VALON REPRESENTATION

The wave function of a hadron at low $Q^2$ can best be described in the valon representation.[3,4] A valon is a valence quark cluster, which may be thought of as a dressed quark or a constituent quark.[5] The parton distribution in a valon when probed at high $Q^2$ can be calculated in QCD. Thus whereas the hadron wave function in terms of the partons is complicated, there being many Fock states that are important, in terms of valons it is simple since to a good approximation a nucleon has only three valons and a pion two. The valon distribution in the hadron, $G_{v/h}(y)$, is related to the hadron structure function $F_h(x,Q^2)$ by a convolution

$$F_h(x,Q^2) = \sum_v \int_x^1 dy\, G_{v/h}(y) F_v(x/y, Q^2) \tag{1}$$

where $F_v$ is the valon structure function which can be determined in leading-order QCD calculation (for large $Q^2$). Using data for $F_h$, we have extracted $G_{v/h}(y)$ for both nucleons and pion.[3,4] Knowing that, which for pion is $|<U\bar{D}|\pi>|^2$ we immediately have also the recombination function $R^\pi(x_1,x_2,x)$ which is $|<\pi|U\bar{D}>|^2$. The result turns out to be[4]

$$R^\pi(x_1,x_2,x) = x_1 x_2 x^{-2} \delta(\frac{x_1}{x} + \frac{x_2}{x} - 1) \tag{2}$$

where $x_1$ and $x_2$ are momentum fractions of the two recombining valons, while $x$ is that of the final pions.

In the analysis it is necessary to adjust the two parameters $\Lambda$ and $Q_0$ in order to fit the deep inelastic scattering data. They are found to be 0.65 and 0.8 GeV, respectively. The value of $Q_0$ gives an estimate of the inverse-size of a valon and is numerically reasonable: greater than 0.7 GeV, the rough inverse-size of a hadron, but less than 1 GeV, the on-set of precocious scaling. Moreover, it is roughly the value of primordial $<k_T>$.

## FRAGMENTATION FUNCTION

The quark and gluon fragmentation functions $D(x,Q^2)$ can be calculated in the recombination model according to[6]

$$xD(x,Q^2) = \int \frac{dx_1}{x_1} \frac{dx_2}{x_2} F(x_1,x_2,Q^2) R^\pi(x_1,x_2,x) \tag{3}$$

for the decay into a pion. The two valon distribution $F(x_1,x_2,Q^2)$ in the jet can be calculated in perturbative QCD. In this calculation there is a convolution of three evolution functions connected to one bifurcation point (see Fig. 1)

$$F_{i \to k\ell}(x_1,x_2,Q^2) = \sum_{jj'j''} \int \frac{dk^2}{k^2} \frac{\alpha(k^2)}{2\pi} \int_{x_1+x_2}^{1} dy G_{i \to j}(y,Q^2,k^2)$$

$$\cdot \int_{x_1}^{y-x_2} \frac{dz}{y} P_{j \to j'}(\frac{z}{y}) \frac{x_1}{z} G_{j' \to k}(\frac{x_1}{z},k^2,Q_0^2) \frac{x_2}{y-z} G_{j'' \to \ell}(\frac{x_2}{y-z},k^2,Q_0^2) \tag{4}$$

This function is in agreement with the jet calculus of Konishi, Ukawa and Veneziano.[7] The actual calculation involves taking the moments of both (3) and (4), and using the solution of the renormalization group equation for the moments of the G functions. It is arithmatically complicated but conceptually straightforward.[6]

There is one conceptual point that needs to be made. Eq. (4) describes the distribution of quark $q_k$ at $x_1$ and antiquark $\bar{q}_\ell$ at $x_2$ in a jet initiated by a quark or gluon of type i. The $q_k$ and $\bar{q}_\ell$ are at $Q_0$ and correspond to the valons, suitable for recombination. If we substitute (4) and (2) into (3) we would, however, have accounted for only part of the pions in the final state. The reason is that gluons at $Q_0$ carry a sizable fraction of the jet momentum and they must hadronize also. We account for their hadronization by first converting all gluons to $q\bar{q}$ pairs, then adding those $q\bar{q}$ pairs to the distribution $F(x_1,x_2,Q^2)$ and finally using (3) for the calculation of $D(x,Q^2)$.

Note that there are no adjustable parameters in the calculation. The resultant $D(x,Q^2)$ for quark and gluon jets are absolute predictions. The only data with definite pion identification in the jet are for $e^+e^-$ annihilation at about 5 GeV.[8] They are shown in Fig. 2. The solid line is our prediction at Q = 5 GeV. The agreement in both shape and normalization is remarkably good. It gives support to the recombination model for hadronization. The gluon fragmentation function can also be calculated, but there is not data for comparison.

## ACKNOWLEDGMENT

The collaboration of Dr. V. Chang has been crucial in this work. I am grateful for his untiring efforts. This work is supported in part by the Department of Energy under contract EY-76-S-06-2230, TA 4, Mod. A008. By acceptance of this article, the publisher and/or recipient acknowledges the U.S. Government's right to retain a non-exclusive, royalty-free license in and to any copyright covering this paper.

## REFERENCES

1. See, for example, B. Wiik in these Proceedings.
2. R.D. Field and R.P. Feynman, Nucl. Phys. B136, 1 (1978).
3. R.C. Hwa, Phys. Rev. D22, 759 (1980).
4. R.C. Hwa, OITS-122, to be published in Phys. Rev.
5. N. Cabibbo and R. Petronzio, Nucl. Phys. B137, 395 (1978).
6. V. Chang and R.C. Hwa, Phys. Rev. Lett. 44, 439 (1979); OITS-138 Phys. Rev. (to be published).
7. K. Konishi, A. Ukawa, and G. Veneziano, Phys. Lett. 78B, 243 (1978); 80B, 259 (1978); Nucl. Phys. B157, 45 (1979).
8. R. Brandelik et al., Nucl. Phys. B148, 189 (1979).

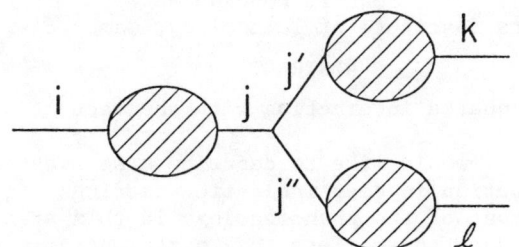

Fig. 1

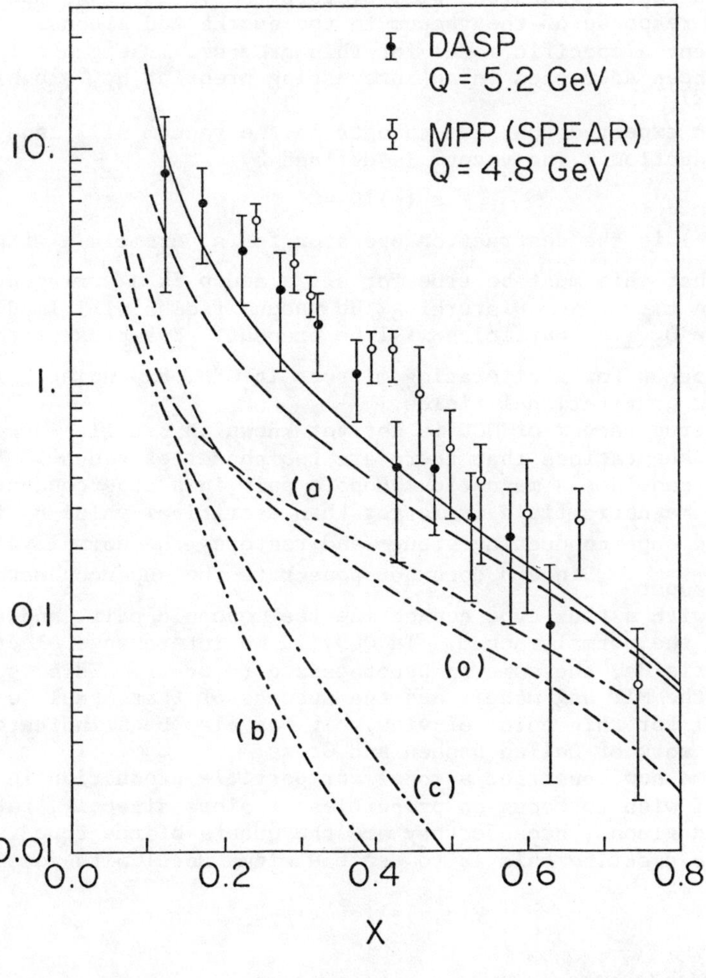

Fig. 2

## SCATTERING QUARKS OFF THE VACUUM

John F. Donoghue
Massachusetts Institute of Technology, Cambridge, Ma. 02139

### ABSTRACT

A model for quarks interacting with the vacuum is described.

In this talk I would like to describe a somewhat unusual method of particle production in deep inelastic reactions - quark vacuum scattering[1]. Most of the phenomenology in this area is done using field theoretic ideas appropriate to a perturbative vacuum such as we are used to in QED (where we may work in a power series in $\alpha$ about $\psi = 0$). However the vacuum of QCD is much more complicated and is assumed to be a condensate of quark and gluon fields. Quarks produced in deep inelastic reactions will disturb the vacuum, and one must ask about the response of the vacuum to the quarks and gluons. Below I will present a specific model for this process. The general problem has also been addressed in an interesting preprint by Voloshin and Zakharov[2].

It is expected that the changes in the vacuum will lead to particle production. The vacuum is defined by

$$a_\alpha(p)|0\rangle = 0$$

where $a_\alpha(p)$ is the destruction operator for a particle $\alpha$ with momentum $p$. That this must be true for all $\alpha$ and $p$ is a stringent constraint on the state. Disturbing the vacuum fields will lead to $a_\alpha(p)|0'\rangle \neq 0$, <u>i.e.</u> particles will be present. Examples of this are known to occur for accelerating mirrors in QED, expanding universes, and strong gravitational fields.

The true vacuum of QCD is not yet known in detail. However, there are suggestions that there are two phases of vacuum. To motivate this consider a magnetic monopole pair in a superconductor. Where the magnetic field is larger than a critical value $H_c$ it destroys the superconducting vacuum and restores the normal vacuum, $E_{normal} > E_{super}$. Since $\vec{H}$ does not penetrate the superconductor, we are left with a flux tube connecting the monopole pair, and this tube is all of the normal vacuum. In QCD, if we interchange electric and magnetic fields, the same is hypothesized to occur. This is the basis of the MIT bag model, and the success of that model is a possible signal for this point of view. It has also been indicated by the instanton work of Callen Dashen and Gross.

Let me now construct a model for particle production in such a vacuum. I wish to focus on properties of pions directly, rather than quarks and gluons, because they are the quanta of the true vacuum. One way to describe this is to use the sigma model with potential

$$V(\sigma, \vec{\pi}) = \lambda(\sigma^2 + \vec{\pi} \cdot \vec{\pi} - F_\pi^2)^2 \tag{1}$$

Here, σ plays the role of the vacuum field with σ=0 being the perturbitive vacuum and σ=-$F_\pi$ the true vacuum. The difference between these two cases is

$$V(0,\vec{\pi}) - V(-F_\pi, \vec{\pi}) = \lambda F_\pi^4 - 2\lambda F_\pi^2 \vec{\pi}\cdot\vec{\pi}$$

$$= B - \frac{2B}{F_\pi^2} \vec{\pi}\cdot\vec{\pi} \qquad (2)$$

In the second line, I've relabeled the Energy/Volume difference $\lambda F_\pi^4 = B$, where B is the bag constant and is known to have the value $B = (130 \text{ MeV})^4$. Note the pion coupling in Eq. 2. When the vacuum changes phase we can excite pion pairs with the coupling

$$L_{int} = \sigma^2(x) \vec{\pi}\cdot\vec{\pi} \qquad (3)$$

with

$$\sigma^2(x) = \frac{2B}{F_\pi^2} \theta(V) \qquad (4)$$

where V is the space of the perturbative vacuum.

Essentially the same interaction can be derived in a bag model framework. (Let me neglect the pion's mass in what follows). In the true vacuum $m_\pi^2 \approx 0$, but this means that in the perturbative vacuum $m_\pi^2 < 0$. This signals the breakdown of chiral symmetry. Then

$$L_{int}(x) = m_\pi^2(x) \vec{\pi}\cdot\vec{\pi} \qquad (5)$$

where $m_\pi^2(x)$ is nonzero only the perturbative vacuum. Its mass can be estimated by the techniques of Donoghue and Johnson by removing the volume energy from the contribution to the pion's mass

$$m_\pi^2 \Big|_{Pert} = \frac{-1}{<\frac{1}{2}p>} \left\{ \frac{4\pi}{3} BR^3 \right\} = -12R^2 B \approx -.7 \left(\frac{2B}{F_\pi^2}\right) \qquad (6)$$

The two estimates agree remarkably well on the size and form of the coupling.

To produce pions we will consider σ(x) as an external field governed by scattered quarks and gluons, and study the behavior of the vacuum using

$$|out> = U|0> \qquad (7)$$

where

$$U = T \exp\left\{-i\int d^4x \, H_{int}(x)\right\} \qquad (8)$$

We can count pions by using the number operator

$$N = \sum_i \int d^3k \; a_i^+(k) a_i(k) \tag{9}$$

This leads to an expression for the number of pions produced

$$N_{out} = \langle out | N | out \rangle$$

$$= 12 \int \frac{d^3k}{(2\pi)^3 2k_o} \frac{d^3k'}{(2\pi)^3 2k_o'} \left| \int d^4x \sigma^2(x) \, e^{i(k+k')\cdot x} \right|^2 \tag{10}$$

For a discussion of which diagrams have been included in this, see Ref. 1.

As an example let me consider $e^+e^- \to q\bar{q}$. The $q\bar{q}$ separate and form a flux tube of gluonic field between them. Quark pairs will be produced from the flux tube and a cascade process will result. This is the basis of an interesting model advocated by Casher, Nussinov and Neueberger. I will simplify it by neglecting breakup, since we are focusing only on the vacuum. The minimum volume which can be produced is $V_{tot} = n_F V_o$, where $n_F$ is the number of particles produced from conventional fragmentation processes and $V_o$ is the perturbative volume of each particle. The radius is estimated using Gauss's law to be

$$\rho^4 = \frac{2\alpha}{\pi} \frac{4}{3} \frac{1}{B}$$

Solving Eq. 10 then leads to[1]

$$N_{out} = \frac{3}{16\pi} (1 - 1/\sqrt{2}) \rho V_o \left( \frac{2B}{\frac{F^2}{\pi}} \right)^2 n_F$$

$$\approx .16 \, n_F$$

This estimate is a significant fraction of the total number of particles produced.

In summary, the above is a new mechanism for particle production due to quark and gluon fields rearranging the vacuum. A simple estimate indicates that it can be competitive with other mechanisms. However, before it can become acceptable for phenomenology, a more realistic picture of jet development must be studied.

REFERENCES

1. J.F. Donoghue, MIT preprint CTP#829 - scheduled for publication in Phys. Rev. D22. Because of space limitations I have omitted most references throughout this talk. All references which would have been included can be found in the above paper.
2. M. Voloshin and V. Zakharov, DESY preprint 80/41.

# NONPERTURBATIVE VACUUM AND HARD SCATTERING PROCESSES

N. Sakai*
Fermi National Accelerator Laboratory, Batavia, Illinois 60510

## ABSTRACT

A number of interesting suggestions for the QCD nonperturbative vacuum have been advocated in recent years by a group of people in Copenhagen. I review briefly some of the main ideas. I also describe an attempt to obtain the physical effects of the nonperturbative vacuum by studying hard scattering processes such as $e^+e^- \to$ hadrons.

## NONPERTURBATIVE VACUUM

To explore the nontrivial structure of the QCD vacuum, many people have calculated the effective potential (=vacuum energy density) V for the gluon field $A_\mu^a$ by using a simplifying Ansatz of constant chromomagnetic field ($a$ = color index, $\mu$ = Lorentz index)[1,2]

$$A_y^{a=3} = Hx, \quad A_\mu^a = 0 \text{ otherwise.} \tag{1}$$

In the one-loop approximation Savvidy has found that this effective potential has a minimum for $H \neq 0$ and the perturbative vacuum (H=0) is unstable[1] as shown in Fig. 1. The nonperturbative vacuum at the minimum has a constant chromomagnetic field and may be called "color-ferromagnetic vacuum." However, N.K. Nielsen and P. Olesen pointed out that the effective potential actually has an imaginary part and the color-ferromagnetic state is unstable.[3] They found that the large anomalous color magnetic moment of non-Abelian gluons causes the unstable mode when the gluon spin is parellel to the chromomagnetic field H. By giving nonvanishing vacuum expectation values for the unstable mode,[6] one can further lower the vacuum energy. To lower the energy <u>density</u> by a finite amount, H.B. Nielsen and M. Ninomiya noticed that the lowering of the vacuum energy <u>density</u> requires an unstable mode field configuration which is extended all over the space.[7]

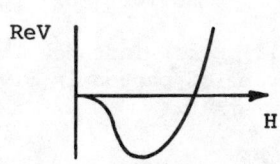

Fig. 1 Real part of the effective potential in the one-loop approximation.

---

*On leave of absence from Department of Physics, Tohoku University, Sendai, 980, Japan.

Thus they were naturally led to consider a periodic field configuration which exhibits domains of chromomagnetic flux. They also obtained an estimate of bag constant in agreement with the phenomenological value. Closer correspondence with the superconductor and with the flux concept of 'tHooft has also been discussed recently.[8]

One of the serious problems of the proposed nonperturbative vacuum is the lack of rotational or Lorentz invariance. H.B. Nielsen and P. Olesen estimated the quantum fluctuations of the chromomagnetic flux and proposed a "quantum liquid picture" to restore the invariance: quantum fluctuations are so large that the QCD vacuum looks like a "liquid."[9] The resulting picture is a random distribution of chromomagnetic flux tubes like spaghetti, although it is not yet well-formulated in detail.

## HARD SCATTERING PHENOMENA

In extracting physical effects, we make two basic observations: 1) Hard scattering processes probe a small space-time region which is likely to be well-approximated by constant chromomagnetic field. ii) By considering the total cross section we circumvent the difficult task of constructing hadrons explicitly.[10]

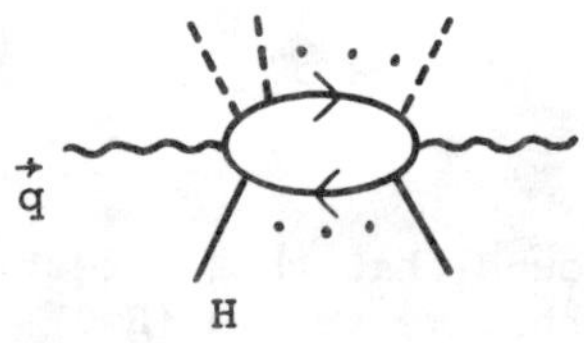

The effect of the nonperturbative vacuum in $e^+e^- \to$ hadrons can be obtained by calculating the vacuum polarization $\pi^{\mu\nu}$ due to quarks in the constant chromomagnetic field H (see Fig. 2). If we separate $\pi^{\mu\nu}$ into perturbative part $\pi^{\mu\nu}$ (H=0) and nonperturbative part $\Delta\pi^{\mu\nu}$, we find no ultraviolet divergence in $\Delta\pi^{\mu\nu}$ in the one-loop order. We evaluate the asymptotic behavior of the nonperturbative effect $\Delta\pi^{\mu\nu}$ for $-q^2 \gg gH \gg m^2$ (m:quark mass). For instance the longitudinal vacuum polarization in the rest frame of photon is given by (H is along the z-direction)

Fig. 2 Vacuum polarization in the field H.

$$\Delta\pi^{zz} \simeq \frac{(gH)^2}{3\pi^2 q^2} \ln \frac{-q^2}{2gH} , \qquad (2)$$

If one naively calculates $\Delta\pi^{zz}$ with only two H field insertions attached to the quark loop instead of summing all H field insertions, one obtains a mass singularity (2gH in the logarithm is replaced by $m^2$). The correct asymptotic behavior (2) can be interpreted in terms of parton language: If a quark is produced in the nonperturbative vacuum, the virtuality of the quark extends from order $q^2$ down only to the nonperturbative mass scale gH (quarks tend to be off-shell by an amount gH).

To restore the rotational and Lorentz invariance we use the Nielson-Olesen proprosal[9] and average over the rotated and boosted color field configuration. If we average over spacial directons of H only (rotational averaging), we obtain the nonperturbative correction $\Delta R$ for $R=\sigma_{had}/\sigma_{\mu\bar{\mu}}$

$$\Delta R/R_{parton} \simeq -\frac{2}{9}\left(\frac{gH}{q^2}\right)^2, \qquad (3)$$

which is about the same order of magnitude as the perturbative $g^4$ order corrections around $q^2=1\sqrt{3}$ GeV$^2$ ($gH\approx 3.42\Lambda^2$, $\Lambda\approx 0.5$ GeV). If we take the Lorentz invariant averaging, we obtain only real part for $(gH)^2$ order of $\Delta\pi^{\mu\nu}$. This result agrees with that of the ITEP group.[11] In this case the nonperturbative effect to R can behave at most as $(gH/q^2)^4$.

## REFERENCES

1. S.G. Matinyan and G.K. Savvidy, Nucl. Phys. $\underline{B156}$, 1 (1979); G.K. Savvidy, Phys. Lett. $\underline{71B}$, 133 (1977).
2. H. Pagels and E. Tomboulis, Nucl. Phys. $\underline{B143}$, 485 (1978).
3. N.K. Nielsen and P. Olesen, Nucl. Phys. $\underline{B144}$, 376 (1978); Phys. Lett. $\underline{74B}$, 304 (1978).
4. L.S. Brown and W.I. Weisberger, Nucl. Phys. $\underline{B157}$, 285 (1979).
5. A. Yildiz and P.H. Cox, Phys. Rev. $\underline{D21}$, 1095 (1980).
6. J. Ambjørn, N.K. Nielsen, and P. Olesen, Nucl. Phys. $\underline{B152}$, 75 (1979).
7. H.B. Nielsen, and M. Ninomiya, Nucl. Phys. $\underline{B156}$, 1 (1979); H.B. Nielsen, Phys. Lett. $\underline{80B}$, 133 (1978).
8. J. Ambjørn and P. Olesen, NBI-HE-80-14; J. Ambjørn, B. Felsager, and P. Olesen, NBI-HE-80-22.
9. H.B. Nielsen and P. Olesen, Nucl. Phys. $\underline{B160}$, 380 (1979).
10. A. Patkós and N. Sakai, Nucl. Phys. B $\underline{168}$, 521 (1980).
11. M.A. Shifman, A.I. Vainstein, and V.I. Zakharov, Nucl. Phys. $\underline{B147}$, 385 (1979).

# GOLDBERGER-TREIMAN CONSTANTS OF DYNAMICAL NEAR-GOLDSTONE MODES*

M.A.B. Bég
Rockefeller University, New York, N.Y. 10021

Within the framework of a model for the QCD vacuum, an argument due to Landau is used to assess the magnitudes of the Goldberger-Treiman constants of dynamical near-Goldstone modes. The values so obtained for $f_\pi$ and $f_K$ are in fair agreement with observation.

1. The motivation for the work[1] which I shall describe is twofold: (a) the puzzle of pseudo-scalar meson states in conventional (QCD-based) spectroscopy and (b) the challenge offered by the theoretical calculability of the Goldberger-Treiman constants of these states in QCD.

2. That pseudoscalar meson states are something of an enigma has been recognized for a long time. The usual quark-model picture would have us view them as non-relativistic $^1S_o$ bound states of the $q\bar{q}$ system. A different perspective is afforded by the interpretation of these states as near-Goldstone modes[2]; in the <u>exact</u> Nambu-Goldstone limit (all quarks devoid of current mass), the pseudoscalar mesons are massless Goldstone bosons and can not be viewed as non-relativistic bound states.

We first show that for near-Goldstone -- as opposed to exactly massless Goldstone -- modes the clash with the naive quark model is less serious. To this end, we consider a meson P, composed of quarks $q_1$ and $\bar{q}_2$ and denote by m their average current mass and by M their common dynamical or Nambu-Goldstone mass. If the principle known as PCAC, to wit

$$m << M \qquad (1)$$

is satisfied, standard chiral-dynamics tells us that $\mu_P$, the mass of the pseudoscalar state, is given by

$$\mu_P^2 = m <\bar{q}\ q> /f_P^2 \qquad (2)$$

$f_P$ being the Goldberger-Treiman constant of the state P.

Note that on the right hand side of Eq. (2), there occurs a rather large factor with the dimension of mass:

$$<\bar{q}\ q>/f_P^2 \sim 4000 - 6000 \text{ MeV} \qquad (3)$$

It is the magnitude of this number that permits us to envisage a reconciliation between PCAC and the non-relativistic limit:

$$2m+2M-\mu_p \ll 2m+2M+\mu_p \tag{4}$$

[To illustrate our argument, we have assumed in Eq. (4) that the constituent quark mass is the linear superposition of current and dynamical masses; in Eq. (8) below, constituent masses will be inferred from experiment and this linearity assumption will not be used.]

One may indeed verify that there exists a region in the (m,M) plane in which Eqs. (1), (2) and (4) are mutually compatible i.e. the non-relativistic limit and PCAC can peacefully co-exist. The rest of our discussion is predicated on the assumption that results derived in this mass region can be continued into the real world.

3. To calculate the $f_p$'s we shall take a slightly "unorthodox" view of the QCD vacuum[3]. We assume that the vacuum may be regarded as a $q\bar{q}$ plasma and that confinement is to be interpreted as a consequence of almost total screening of the color-charge a la Debye and Hückel. In flavor space such a picture implies a condensate, $\langle\bar{q}q\rangle \neq 0$, and non-vanishing values for the $f_p$'s; in color space it may be associated with an Anderson-Higgs mechanism with concomitant generation of an effective mass for the color-gluons. We thus have spontaneous breakdown of chiral symmetry in flavor space and very short-range $qq$ and $q\bar{q}$ forces.

With short range forces, $q_1\bar{q}_2$ scattering may be parametrized in the neighborhood of the threshold as

$$k \cot \delta = -1/|a| \tag{5}$$

where k is the c.m. momentum and $\delta$ is the phase shift. The negative sign of the scattering length in Eq. (5) ensures that there is a bound state in the neighborhood of the threshold. We identify this state with P and, following Landau and others[4], continue[5] Eq. (5) to the P-pole to determine the $Pq_1\bar{q}_2$ coupling constant:

$$\frac{G^2_{Pq_1\bar{q}_2}}{4\pi} = \frac{(M_1+M_2)^3}{2(M_1 M_2)^{3/2}} \cdot [1 - (\frac{\mu_p}{M_1+M_2})^2]^{1/2} \tag{6}$$

Now Eq. (1) implies the Goldberger-Treiman formula:

$$f_p \approx \frac{(M_1+M_2)g_A}{G_{Pq_1\bar{q}_2}} \tag{7}$$

Within the framework of the quark-parton description of weak currents, we may set $g_A=1$ and thus obtain

$$f_P^2 \approx \frac{(M_1 M_2)^{3/2}}{2\pi(M_1+M_2)} \cdot [1 - (\frac{\mu_P}{M_1+M_2})^2]^{-\frac{1}{2}} \qquad (8)$$

With[6] $M_u \approx M_d \sim 350$ MeV, $M_s \sim 475$ MeV we find the theoretical values of the Goldberger-Treiman constants for pions and kaons to be:

$$f_{\pi^+} \approx 100 \text{ MeV} [130 \text{ MeV}] \qquad (9)$$

$$f_{K^+} \approx 128 \text{ MeV} [155 \text{ MeV}] \qquad (10)$$

the numbers in square brackets being the experimental values corresponding to a value of the Cabibbo angle such that $\tan\theta_c = 0.232$.

4. Considering the drastic nature of the extrapolations involved, in applying Eq. (8) to pions and kaons, the agreement between theory and experiment is surprisingly good. The fact that Eq. (8) affords a somewhat better estimate for $f_K$ than for $f_\pi$ may be indicative that this agreement is not entirely fortuitous; the underlying assumptions happen to be more reasonable for kaons than for pions.

*Work supported in part by the U.S. Department of Energy under Contract Grant Number DE-AC02-76ER02232B.

1. M.A.B. Bég, Rockefeller University Report No. COO-2232B-198 (1980).
2. J. Goldstone, A. Salam and S. Weinberg, Phys. Rev. 127, 965 (1962).
3. Cf. J. Finger, D. Horn and J. Mandula, Phys. Rev. D20, 3253 (1979).
4. L. Landau, JETP 12, 1294 (1961); M.A.B. Bég, BNL Internal Rep. No. 5569 (1961); S. Weinberg, Phys. Rev. 130, 776 (1963).
5. The continuation would be impossible in the presence of CDD poles; such poles would herald the elementarity of P.
6. The constituent masses are averages of values derived from the meson and baryon sectors respectively [Input: $\omega, K^*$, N and $\Lambda$ masses].

# A GOLDSTONE PION WITH BAG CONFINEMENT

Presented by Richard W. Haymaker
Louisiana State University, Baton Rouge, LA  70803

T. Goldman[†]
California Institute of Technology, Pasadena, CA  91125

## ABSTRACT

A model of the pion is presented in which chiral symmetry is dynamically broken prior to confinement and in which a bag like boundary condition is imposed on the relative q q̄ coordinate.

## INTRODUCTION

Chiral symmetry has not fit comfortably in bag models[1] for reasons that were understood from the beginning; the bag boundary condition breaks chiral symmetry explicitly and hence one can not treat properly the goldstone nature of the pion. Attempts to handle this problem include (i) the introduction of new degrees of freedom outside the bag which couple to the surface in order to conserve the axial current,[2] and (ii) requiring $m_\pi = 0$ as a constraint.[3] In our model,[4,5,6] we take the pion to be a qq̄ state in the bag like all other mesons, but we change the order in which the various interactions are treated. If the dynamical symmetry breaking is treated properly first, then we find the explicit chiral breaking is much less severe a problem and perhaps no problem at all.

Our calculation proceeds in three steps: (I) We break chiral symmetry dynamically and redefine the vacuum employing the Nambu Jona-Lasinio mechanism (NJL)[7] prior to imposing a bag boundary condition. (II) We remove the center of mass motion by confining only in the relative coordinate. We found new boundary conditions for this relative coordinate bag. (III) We then impose our boundary condition on the pion wave function $\phi_\pi$ and give some preliminary results on the pion mass, $M_\pi$, the pion decay constant, $f_\pi$, and the diameter of the pion bag.

## DYNAMICAL SYMMETRY BREAKING (DSB)

Starting from QCD with no mass terms, we suppose that a chiral multiplet containing the pion is the most attractive channel and that it is driven to negative mass-square. We remove this state and treat it as an NJL driving term parametrized by a strength $G^2$ and a physical cut-off $\Lambda$ giving the scale of this bound state. After DSB, the parameters $(G^2, \Lambda^2)$ go naturally over to $(M, f_\pi)$ where M is the quark mass. One can make a connection between these parameters and QCD: See Pagels and Stokar[8] and Ref. (5). We are assuming here that QCD has sufficient strength to break chiral symmetry:  See

[†]Permanent address LASL, Los Alamos, New Mexico  87545

Caldi[9] and Ref. (5) for estimates of the strength of various driving terms.

I would like to make one important observation in this approach: before confinement the pion is a zero mass bound state with binding energy 2M and hence its wave function is exponentially damped, $\phi_\pi(r) \sim e^{-Mr}$. The other mesons that arise in the NJL Mechanism occur near and above the $q\bar{q}$ threshold and hence their wave functions are either flat or oscillatory. The pion has a doubly special status: besides being a goldstone mode, the subsequent application of a boundary condition can be expected to give an exponentially small mass shift.

## RELATIVE COORDINATE CONFINEMENT

The static cavity approximation to the MIT bag model treats mesons as an uncorrelated $q\bar{q}$ pair. Since the $q\bar{q}$ correlations are responsible for DSB we depart from the static cavity and remove the center of mass motion, leaving a correlated $q\bar{q}$ wave function. The following table contrasts the relative confinement with the static cavity.

### TABLE 1: STATIC CAVITY, vs RELATIVE CAVITY

| STATIC (3 space) | | RELATIVE (6 space) |
|---|---|---|
| | SURFACE | |
| $f \equiv \vec{r}^{\,2} - R^2 = 0$ | | $f \equiv (\vec{r}_1 - \vec{r}_2)^2 - R^2 = 0$ |
| | NORMAL | |
| $\hat{n} \propto \vec{\nabla} f$ | | $\hat{n}_6 \propto (\vec{\nabla}_1, \vec{\nabla}_2) f \propto (\hat{n}, -\hat{n})$ |
| | CURRENT | (for separable $\phi \sim q_1 \bar{q}_2$) |
| $\dfrac{\partial \rho}{\partial t} + \vec{\nabla} \cdot \vec{J} = 0$ | | $0 = \left(\dfrac{\partial}{\partial t_1} + \dfrac{\partial}{\partial t_2}\right)\rho_1\rho_2 + (\vec{\nabla}_1, \vec{\nabla}_2)\begin{pmatrix}\rho_2 \vec{J}_1 \\ \rho_1 \vec{J}_2\end{pmatrix}$ |
| | | (generalize to non-separable $\phi$) |
| | LINEAR BOUNDARY CONDITION | |
| $i\hat{n} \cdot \vec{\gamma} \psi = \psi$ | | $\gamma^0 \phi \vec{\gamma} \cdot \hat{n} - \vec{\gamma} \cdot \hat{n} \phi \gamma^0 = i(\phi + \gamma^0 \hat{n} \cdot \vec{\gamma} \phi \hat{n} \cdot \vec{\gamma} \gamma^0)$ |
| | LEADING TO ZERO FLUX | |
| $\hat{n} \cdot (\bar{\psi}\vec{\gamma}\psi) = 0$ | | $(\hat{n}, -\hat{n}) \cdot \begin{pmatrix} \text{Tr } \bar{\phi}\gamma^0 \phi \vec{\gamma} \\ \text{Tr } \bar{\phi}\vec{\gamma}\phi\gamma^0 \end{pmatrix} = 0$ |

This development is for the special case of equal mass, center of mass frame, spherical bag and non-interacting particles. See Ref. (6) for the generalization of the quadratic boundary condition.

## CONFINED $\phi_\pi$

Prior to confinement, $\phi_\pi = \phi_\pi^{(0)} + \phi_\pi^{(1)}$ where

$$\phi^0(x,P) = \begin{pmatrix} \dfrac{\vec{F}\cdot\vec{\sigma}}{2M} & \dfrac{E+2M}{4M}F \\ \dfrac{E-2M}{4M}F & \dfrac{\vec{F}\cdot\vec{\sigma}}{2M} \end{pmatrix} \quad (1)$$

$$\phi^1(x,P) = \frac{3EM}{\pi^2 f_\pi^2 D_\pi^{-1}(s)} \int_0^\infty \frac{k^2 dk}{(k^2+M^2)^{1/2}} \frac{\begin{pmatrix} E\vec{G}\cdot\vec{\sigma} & (2(k^2+M^2)+EM)G \\ (2(k^2+M^2)-EM)G & E\vec{G}\cdot\vec{\sigma} \end{pmatrix}}{s-4(k^2+M^2)} \quad (2)$$

where

$$F = j_0(qr), \quad \vec{F} = i\,\hat{r}\,q\,j_1(qr) \quad (3)$$

and $G$, $\vec{G}$ similarly with $q \to k$. Since we are relaxing the large $r$ boundary condition and replacing it with a condition at finite $r$, we have considered the inhomogeneous Bethe-Salpeter equation allowing the free wave term $\phi_\pi^0$ to be present. $D_\pi$ is this pion propagator with residue $-1$.

Applying the linear boundary condition gives:

$$\left(\frac{E}{2M} j_0(qR) - \frac{q}{M} j_1(qR)\right) \frac{\pi^2 f_\pi^2 D_\pi^{-1}(s)}{M^2} + \frac{3E}{4M}\left\{\frac{2}{R}\int_{-\infty}^\infty \frac{kdk\,(k^2+M^2)^{1/2}\sin kR}{(k^2-q^2)}\right.$$
$$\left. + E\frac{d}{dR}\frac{1}{R}\int_{-\infty}^\infty \frac{kdk\,\sin kR}{(k^2+M^2)^{1/2}(k^2-q^2)}\right\} = 0 \quad (4)$$

The smallest root of Eq. (4) can be well represented by an exponential form as follows:

$$M_\pi = \frac{8.0 e^{-2MR}}{(R)^{1/2}} \text{ GeV} + \frac{4\pi}{3} B \left(\frac{R}{2}\right)^3 \quad (5)$$

We have also added a volume energy for a bag of diameter $R$ ($R$ is $q\bar{q}$ separation). If we take $M = 300$ MeV, $\Lambda = 1.0$ GeV, we obtain $f_\pi = 92$ MeV (93 MeV: exptl). Further taking MIT value of $B$, $B^{1/4} = .14$ GeV gives $m_\pi = 120$ MeV, $(R_{\pi/2})$ (bag radius = 3.5 GeV$^{-1}$). This bag size is consistent with other mesons in the bag model.[1]

## CONCLUSION

The pion gets mass only at the last step. It is small because it is already a deeply bound state. An $M_\pi$ value of 120 MeV reflects a <u>small</u> chiral breaking since it is comparable to $M_\pi$ induced by current quark masses $M$, of 5—10 MeV. If we attribute confinement as being due to QCD growth of $M_q$ with scale, then this picture is consistent with the standard view that all explicit chiral breaking is due to quark mass terms. Any other confining mechanisms that are chiral invariant do not effect the value of $M_\pi$.

## REFERENCES

1. A. Chodos, et al., Phys. Rev. D9, 3471 (1974).
2. See R. L. Jaffe, Erice Lectures, 1979, MIT preprint CTP-814 for a review of hybrid chiral bags.
3. J. F. Donoghue and K. Johnson, Phys. Rev. D21, 1975 (1980).
4. T. Goldman, R. Haymaker, "A Goldstone Pion with Bag Confinement" Cal Tech preprint CALT-68-782.
5. T. Goldman, R. W. Haymaker, "Dynamically Broken Chiral Symmetry with Bag Confinement", in preparation.
6. T. Goldman, R. W. Haymaker, "Bag Boundary Conditions for Confinement in the $q\bar{q}$ Relative Coordinate", in preparation.
7. Y. Nambu, G. Jona-Lasinio, Phys. Rev. 122, 345 (1961).
8. H. Pagels and S. Stokar, Phys. Rev. D20, 2947, (1979).
9. D. Caldi, Phys. Rev. Letters 39, 121 (1977).

Chapter 10

Accelerators and Experimental Techniques

# ACCELERATOR AND INSTRUMENTATION PROSPECTS OF ELEMENTARY PARTICLE PHYSICS

A. Skrinsky, INP Novosibirsk, Speaker

H. Schopper, DESY, Chairman

S. Aronson, Brookhaven,

P. McIntyre, Fermilab,

Scientific Secretaries

# ACCELERATOR AND INSTRUMENTATION PROSPECTS OF ELEMENTARY PARTICLE PHYSICS

## A.N. Skrinsky
### Institute of Nuclear Physics, Novosibirsk, USSR

I will try consider briefly the shifts in the accelerator field and in the field of instrumentation, which were most significant in facilitating in recent years the progress of elementary particle physics and will be in the near future. The choice of questions under consideration will inevitably be quite subjective, and discussion on detector problems may be effected by my insufficient information. For these sins I ask to excuse me beforehand.

1. First of all, a few general remarks - quite obvious ones. The solution of problems of elementary particle physics, which become more and more complicated, forces to shift to larger scale for all the systems, both for accelerators and detectors. This, in its turn, becomes accessible because of a wide use of the module principle for construction of systems with maximum homogeneity of modules. This permits making systems significantly cheaper and prolonging their life time.

Of the same fundamental importance is wide and complete computerization. In the modern systems of high energy physics the computing devices became as intrinsic as magnets or counters; they are responsible for all the functions of total and continuous pick up of information on installations and processes, control of the systems and data processing.

2. The greatest event in the area under consideration is the exploration of the colliding beam method. Colliding beam experiments starting with electron-electron beams in Stanford and Novosibirsk, electron-positron in Novosibirsk, Orsay and Frascati and proton-proton at CERN became one of the main sources of fundamental information in elementary particle physics, and their significance will only increase in future. We shall have a special detailed discussion of colliding beams /1/.

3. It is well-known how important for implementation of electron-positron colliding beams was the existence of <u>radiation cooling</u> for light particles even at low energies. Radiation cooling enabled one to stack intense positron beams, to compress transverse dimensions of $e^+e^-$ colliding beams down to small sizes (to a few microns even now) and to maintain the beams compressed despite strong perturbations of particle motion caused by the field of encountered beam, that, in its turn, permits

achieving high luminosity.

Cooling will be of the same fundamental importance also for implementation of the proton-antiproton colliding beam experiments, which became accessible after development of the <u>electron cooling</u> in Novosibirsk and the <u>stochastic cooling</u> at CERN /2-4/.

These methods complement each other substantially in their possibilities. Stochastic cooling is especially effective for beams of low density with large emittance (i.e. at small 6 - dimensional phase density). Electron cooling is the most effective particularly for getting low-temperature ("narrow") beams of heavy charged particles (protons, antiprotons, ions). It is not excluded that cooling with the circulating electron beam will turn out to be useful for suppressing the diffusional beam cross-section growing with time of proton-antiproton colliding beams of high energies /3/. Let me note that at energies $\gtrsim$ 10 TeV an important and positive role will be played by radiation cooling for increasing luminosity of proton-antiproton colliding beams.

The use of <u>ionization cooling</u> can open up very interesting possibilities in getting intense muon beams of high energy including implementation of muon colliding beams of sufficiently high luminosity (see 10.14; 12.1).

3.1. Continuous cooling of the particle beam in a storage ring gives an important possibility for carrying out experiments with a superthin internal targets /5/ wherein diffusional growing of the beam size (because of multiple scattering on the target substance and due to fluctuations of ionization losses) is suppressed by intensive cooling. Thus, fine "spectrometric" experiments become possible with <u>ultimate high luminosity</u> which is only determined by the injector productivity and the cross-section for single-scattered particle loss on the target substance, that is impossible to achieve in the ordinary set-up of experiments. A very important advantage of such a set-up is that all events are continuously distributed in time. Experiments of this kind - electro-excitation of nuclei - for a few years are being performed on electron storage ring VEPP-2 /6/.

Another application of the superthin target mode of operation is generation of secondary beams with good tagging (of kind of the generated particle and its momentum) using registration of accompanying particles. Such a mode gives 100% duty cycle, the relative intensity of the secondary beam is determined by the ratio of the interaction cross-section in use and the total cross-section, beam emittance is determined by interaction properties and by the size of the primary circulating beam (cooled) at the section of interaction.

3.2. Similar set up of experiments with continuous cooling is reasonable even under conditions when the target cannot practically be made so dense that the life-time of a particle in the storage ring would be determined by the collision with target and not by residual gas in the vacuum chamber (another restriction in this case can be the difficulty of achieving of high enough stored currents). But in any case, the luminosity is much, much higher, than that in the one flight mode with the same target and accelerator, and beam qualities are the best.

Such a situation is characteristic for experiments with the **polarized gas-targets** which nowadays permit one to have even for hydrogen or deuterium up to $10^{12}$ atom/cm$^2$ only, which corresponds to average vacuum in a storage ring better than $10^{-9}$ Tor. Naturally, the most interesting work with a polarized target is in a storage ring with polarized beams.

Another interesting example of this kind is the target of free neutrons which is especially promising for detailed study of $\bar{p}n$ interaction at low and medium energies / 3 /.

4. Nowadays we are at the stage when important improvements for accelerator field are under implementation.

First of all, a wide use of superconductivity has started. The use of superconducting magnetic systems even now permits increasing the maximum guiding field from 20 kG to 45 kG (using Ni, Ti alloys) and correspondingly gaining energy for proton and antiproton beams (at a given scale of accelerator facility). There is a possibility to reach 100 kG in the near future (the use of Ni, St alloys). An important fact is the significant reduction in energy consumption which is especially high in the storage ring case.

I would like to draw attention to the fact that at small fields up to 20 kG (with ferromagnetic formation of a magnetic field in the storage ring or slow accelerator) the use of superconducting coils permits construction of extremely miniature magnet systems (design works of the High Energy Laboratory, JINR, Dubna).

However, superconducting magnet systems should still demonstrate long operation with the intense beams which are planned for most projects; that longevity requires a special care.

While the superconducting magnet systems are at the beginning of use for accelerator facilities, in the detector systems they have already permitted an important advance in producing magnetic spectrometers. It is reasonable to use superconductive systems when it is required either to attain the maximum stationary magnetic field or it is necessary to have a particularly large volume occupied by the magnetic field of the detector.

5. The use of high magnetic guide field enables one to increase the energy of heavy particle beams, but this way is closed for electrons and positrons because of excessive increase of synchrotron radiation losses. However, the use of superconductive magnetic structures turns out to be efficient for reaching higher luminosity for electron-positron colliding beams at low and average energies / 1 / and also for producing irradiating structures for various applications of synchrontron radiation.

6. The use of superconductive resonators in RF accelerating structures, which have already been used in RF separators for beams of secondary particles, will be essential for accelerator progress. Up to now it is not clear whether an increase of accelerating gradient of these systems higher than 5 or 10 MeV/m will be achieved, but, at any rate, such systems will permit increasing noticeably (by 1.5-2 times) the energy of cyclic electron-positron storage rings /7,8 /.

7. A sharp increase in acceleration rate in linear accelerating structures (up to 100 MeV/m and maybe somewhat higher) one can achieve in a pulsed mode (with normal conductivity of resonators). Such accelerators could be called superlinacs. The problem of achieving an appropriate surface strength with respect to high voltage break-down as well as the problem of developing accelerating structures for relativistic particles with minimum overvoltage, one can consider as solved in principle /9,10 /. The basis of possible progress in this field is the development of pulsed short-wave generators of a fundamentally new level of pulsed power (order of gigawatt). Two directions in the development of pumping systems seem to be most promising.

One of these directions is connected with the fast progress in the technology of high power pulsed relativistic electron beams / 9 /. Already now when solving the controlled fusion problems the pulsed power of electron beams of a few gigawatts is achieved for durations of the order of a microsecond, with transformation of a substantial part of the beam energy into the energy of the RF electromagnetic field. The present day task is to make these generators more efficient, more sensitive in control over the amplitude and phase and to develop them for the regime of comparatively high repetition rates.

Another direction / 11 / is connected with the fact that modern big proton accelerators (not even mentioning future accelerators) have an energy stored in the beam of millions joules, good properties of high energy proton beams (small energy spread-only tens of MeV at an energy 500 GeV and a small emittance) permit rather easily (with the help of a bending modulator) deep bunching along the beam with the required wave length of the order of one centimeter. With the beam passing through the corresponding diaphragmed waveguide one can transmit an energy of a proton beam in the electromagnetic field of this linear accelerating structure with an accelerating rate up to 100 GeV/km. Let us call such a mode of operation the proton klystron mode. By injecting particles to be accelerated after the exciting proton bunch one can obtain a wide range of particles of high energies (see 10.).

So, it is possible to transfer nearly full energy E of the basic proton accelerator to accelerating particles with the beam intensities up to 10% of the initial beam intensity. By lengthening accelerating structure and exciting consecutive sections with various proton superbunches one can proportionally raise an energy of accelerating particles with corresponding loss in their intensity.

8. I am rather embarassed with the necessity to say only a few words on detectors. This area is almost limitless.

8.1 The progress in detector systems is strongly connected with permanently proceeding revolution in electronics. Namely the revolution enables one to create modern fast track devices and to handle the large flows of information.

The very rough upper estimation of information on an individual event in a big detector ($10^7$ resolvable elements of space x $10^2$ resolvable time moments x $10^2$ resolvable values of amplitudes = $10^{11}$ resolvable elements ) shows that the number of elements is large, so that computer image of an event is quite informative - -or as it is said sometimes now - is quite pictorial. As a rule, thousands of these events should be registered in a second i.e. the full information flows are very large.

Therefore, development of faster processors is very important. Apparently the "Fast-bus system" developed at Brookhaven is the record one which provides the processing rate up to Gigabit per second / 12 /. But even this rate is insufficient for the purpose of processing the full information flow if the information is considered as totally uncorrelated and of equal value.

The wide use of parallel taking and processing of information is of great significance, as well as the use for this purpose of the more perfect programmable microprocessors, that enables to record and further to use for analysis only potentially interesting information. There could be several levels of decision on the further more detailed recording and processing the information and even several levels of triggering of the detector devices.

8.2 Detector systems using now can be huge in their size. Especially large are neutrino detectors /13/ and the multikiloton detectors for the study of proton stability /14/.*

But of extreme importance is also the line of the microdetectors development when for achieving necessary information the ultimately high spatial, time and amplitude resolutions are used (either one kind or combined). The International Symposium in Italy (September, 1980) will specially be devoted to microdetectors.

8.3 Let us consider now the progress and prospects in some certain detecting methods.

The discharge track devices are improving greatly. Revolution in electronics enables one to use the finer properties of electric discharge in various media. Already now the spatial resolution in a liquid-argon chamber is $\sigma_x$ = 8 mkm /15/, in a gas chamber - 20 mkm /16/ and the time resolution achieved $\sigma_t$ = 20 pikoseconds /17/.

One can confidently predict that further improvement and miniaturization of electronic components (as well as their lower cost) and, may be, the use of integrated sensitive and electronics processing components will further facilitate the progress in track detectors.

8.4 It is quite promising the direction of "active targets" with a fast (electronic) information taking which is direct outgrowth of the bubble chambers and high pressure gas chambers /18/. One of the versions of such a target is a set of fine semiconductor counters /19/ with a longitudinal resolution 10 mkm, designed, in particular, for measuring the life-time of D-mesons generated in the substance of the target itself. Possibilities of this device are expanded especially with the add of the transverse resolution for each counter (the prototype of the device is already manufactured providing transverse resolution $\sigma_x$ = 10 mkm).

* I try to cite this Conference contributed papers.

But with the use of technological means of modern microelectronics — thin silicon plates production, ion implantation, molecular epitaxy, laser and in not too distant future the X - ray lithography with the use of synchrotron radiation of electron storage rings, the use of integrated circuits in production of the whole channel with up to transport of information into processor — the prospects open up for the real revolution in the whole this field.

The latter note is valid also for the system of information read-out, optical, in particular, for the detectors of any kind.

8.5 Quite interesting possibilities appear when using the thin-wire scintillation hodoscopes / 20 /. Good results obtained for information read-out when using the avalanche photodiodes and microchannel electron multipliers. It looks realistic today to have hodoscopes with the spatial resolution up to 100 mkm, the length along filaments of 1 m and event rate up to $10^7$ Hz.

8.6 Interesting prospects open up for small bubble chambers at operation with a very small bubbles / 21 / - - the resolution already achieved is of 10 mkm. Especially attractive in this case is the use of holographic information taking which enables one (maintaining the same resolution) to increase sharply the image optical depth (a 10 cm image depth is already achieved). The main efforts in this case are put on the further information processing. Note, the holographic way of taking information is apparently feasible for streamer chambers /22/. Actually, for holographical detection in the real detectors it is reasonable to tend for using the filmless way of taking information, i.e. microchannel multiplying plates, the large area semiconductor counters with the necessary spatial resolution and perhaps some other methods.

8.7 Note, the hybrid emulsion and rapid-cycle buble chambers with the counter aiming the interesting events and adding high time resolution are still of interest, especially for operation with very high multiplicities and complex unknown events.

In particular, the hybrid bubble chambers can be adequate to the work with linear electron-positron colliding beams at super high energies / 9 / when at an average repetition rate of tens Hz the luminosity at a single interaction should be very high.

8.8 In conclusion of the detector section let me note the quite extensively developing methods of direct measurements (or at least estimates) of relativistic $\gamma$ - factors of particles under study. With the energy growth this problem becomes more and more complicated and important. Among these methods I would mark the gas Čerenkov counters (especially those with measurements of the Čerenkov radiation circle /23/), detectors of transition radiation /24/, the use of relativistic dependence of ionization loss (at high energies in gases), radiation at channeling in monocrystals, which is most successfully applied for positive particles, and the synchrotron radiation. For various cases the optimal methods can be different and sometimes optimal may be their combination. Some methods, for instance, registration during channeling, is mostly applicable for tagging the secondary particles falling down the target when directions of their motion are sufficiently collinear. Total absorption calorimeters are under rapid development, too (see, in particular, /59/).

9. Let us consider now the possibilities for generation high quality beams of a possibly wide set of particles both the primarily accelerated and secondary. The progress in this direction determines to a significant extent the development of elementary particle physics. When we talk on a particle beam we have in mind in this case the situation when particles live so long, that at reasonable intensities the number of collisions between these particles and atoms (other than the nucleus which caused the generation of these particles) is sufficient for the study.

Among the characteristics of beams significant from the point of view of elementary particle physics, energy and intensity have obvious importance. An increase in the energy of projectile particles leads to an increase in reaction energy for fundamental processes under study. In ultrarelativistic case this energy increases as $\sqrt{E}$ in the experiments with stationary target and as E in colliding beams. An increase in intensity makes possible both the observation of more rare processes and higher accuracy of experimental data, which frequently supplies qualitatively new information of fundamental importance. As a bright illustration of the latter may serve the discovery in laser experiments of the parity violation in atomic transitions and, consequently, the discovery of electron-nucleon weak interaction due to neutral currents / 25 /.

In addition to energy and intensity, the following qualities of beams are of a very great importance: smallness of their emittance, monochromaticity and optimum of their time structure. The smallness of emittance permits minimizing the transversal size of interaction region between particles of a beam and the substance of the target, which improves, say, momentum analysis of reaction product. Concerning the time structure of the beams it's worth mentioning that sometimes it's beneficial to have the shortest intense bunches separated by long vacant sections for helping to avoid, for example, homogeneous cosmic background, for the use of primarily triggered detectors like bubble chambers, and for separation over the velocities; in other cases it is beneficial to have the beams, continuously distributed in time, loading optimally the detecting electronics and getting the possibility to "tag" each interesting particle by the products accompanying its production.

Note, that in the most general case the main space-time characteristics of the beam is the 6 - dimensional phase volume occupied by the beam. Its smallness permits, in principle, getting a beam structure required for given experiment by appropriate transformations.

9.1. In recent years, obtaining polarized beams has become more and more important. The opinion accepted earlier that spin effects, for strong interactions at any rate, become weaker and weaker at higher energy turned out to be absolutely incorrect. More than that, one can say that it is impossible to develop the quantitative theory of elementary particles without experimental study of the spin properties.

9.2. Of the same importance for polarization experiments, as producing polarized beams, is the progress in producing polarized targets. The best condensed targets with polarized protons and polarized or aligned deuterons, having the polarization level up to 60%, are the complex carbon-hydrogen compositions with a fraction of protons up to 10% of their weight (the recognition of a reaction with protons is carried out by kinematics) and operate at liquid helium temperatures. An important disadvantage of these targets is their insufficient irradiation damage resistance.

Gas targets with polarized protons and deuterons have effective thickness up to $10^{12}$ $p/cm^2$ only, but polarization degree is nearly 100%. These properties make them especially convenient for experiments with internal targets in storage rings. Targets with polarized electrons (ferromagnetic and gaseous) have rather limited value ($e^+e^-$, $e^-e^-$ reaction energies are too small with a stationary target); so, polarized colliding beams are required (see 11.).

10. Let us consider now, very schematically, the possibilities of generation of the beams (we use the word beam in the same meaning as we used it earlier, see 9.) of all known stable enough particles. The optimal secondary

beam generation is often multi-step and complex process. And at many stages the use of cooling and super-thin target mode is effective.

10.1. Protons. The proton accelerators continue to grow in energy and intensity being the basis for the great class of experiments with beams of secondary particles, also.

Even now energies up to 500 GeV are accessible; in the not so far future the DOUBLER at I TeV will be put into operation; the UNK project at 3 TeV is under way. The subject of consideration of ICFA (International Committee on Future Accelerators) was an accelerator at energy 20 TeV. Construction of such an accelerator may become the first all over the world project and will contribute both to solution of problems of elementary particle physics and, let us hope, to solution of other problems of our unquiet world.

Modern intensity of proton beams of the highest energy is $10^{13}$ p/sec; further increase of their intensity is connected with a solution of a problem of further sharp improvement in the "beam hygiene", which is of particular importance for accelerators using superconductivity, that it is assumed to be used in every project for proton accelerators at super high energy. The use of superlinacs with the proton klystrons opens up interesting possibilities for getting protons of higher energies using existing facilities.

Naturally, the record intensities for medium energies belong to meson factories (up to $10^{16}$ p/sec). Further increase in intensity will be permitted by the growth in power of RF generators and by solution of radiation problems.

In the field of lower energies the electrostatic tandem generators ensure excellent beam properties. The biggest tandem generator, for 60 MeV protons, is nearing completion in Daresbury. However, many corresponding experiments, e.g. spectrometric appear to be feasible (and without sharp energy limit) with the help of storage rings with electron cooling in the super-thin target operation mode /3/.

Obtaining polarized proton beams is connected with the design of intense sources of polarized protons and, in the case of cyclic accelerators at high energies, with overcoming the depolarizing effects of spin resonances. The experience of Argonne laboratory has shown experimentally the possibility (and usefulness) of acceleration of polarized protons up to rather high energies.

New possibilities are already seen now for filling

cyclic proton accelerators with polarized particles up to the total intensity of the given accelerator. The main way is the use of the proton polarized $H^-$-beams, which may have the same intensity as polarized $H^+$-beams, and the use of charge exchange injection into the accelerator, that permits one to increase by several orders of magnitude current circulating in the accelerator compared to the current of the $H^-$ - source /26/. Additional increase in injection multiplicity and improvement in the stored beam emittance one can achieve by introducing electron cooling during the injection process. Only for meson factories are there yet no possibilities for the intensity of polarized proton beams to approach the intensities of ordinary beams.

Acceleration up to very high energies in cyclic accelerators is accompanied by numerous spin resonances. This question was thoroughly studied theoretically and ways were found for overcoming detrimental effect of resonances, including producing magnetic structures which eliminate these resonances completely / 27 /.

The problem of obtaining polarized protons of high energies after initial stacking in a booster is especially simplified with use of superlinacs, in particular, with the use of proton klystrons.

Since presently there are no pure polarized targets of condensed substance, an especially important role could be played by the experiments in storage rings with internal gas target which enable one to operate with nearly pure initial spin states. One should pay attention that even longitudinal polarization of circulating beam near the target can be made stable / 28 / for achieving states with the given helicities.

10.2. Nuclei. "Relativistic nuclear physics" turned out to be more interesting than it was expected earlier ("porridge on porridge"). Such experiments give both ideas on supercompressed nuclear substance and supply data on fundamental interaction (study of inclusive processes). Already nowadays accelerated uranium nuclei are obtained with energy up to 10 MeV/nucleon and $10^9$ U/s and nuclei to carbon with 5 GeV/nucleon energy and up to $10^7$ C/s intensity. An implementation of projects is under way which sharply raise the ceiling of available energies and intensities. In some cases coherent methods of acceleration could be used, including "smoketron" devices.

This table represents the expected maximum parameters (energy and intensity) of the nuclear beams for one of the biggest projects VENUS (Berkeley):

Table 1.

|    | 1 GeV/nucleon | 20 GeV/nucleon |
|----|---------------|----------------|
| Ne | $0.8 \cdot 10^{12}$ | $1.2 \cdot 10^{11}$ |
| Kr | $2 \cdot 10^{11}$ | $3 \cdot 10^{10}$ |
| U  | $0.7 \cdot 10^{11}$ | $1 \cdot 10^{10}$ |

This project also envisages an operation in colliding beam mode. Let us note that ISR is already operated in the mode colliding deuteron beams and operation is planned in the near future with colliding beams of $\alpha$ -- particles.

Obtaining beams of polarized deuterons of high energies is even simpler than for the case of protons (because of smallness of the anomalous magnetic moment).

10.3. **Electrons.** Electron accelerators and storage rings play a very essential role both in experiments on elementary particle physics and in various applications (in particular, for generation of synchrotron radiation).

The record in electron accelerators belongs to SLAC; the available energy there is in excess of 30 GeV and in the near future will attain 50 GeV at intensity up to $10^{14}$ e⁻/s.

Both electrons and positrons of higher energies are obtained presently on proton accelerators due to the process $pZ \to \pi^0 X$; $\pi^0 \to 2\gamma$; $\gamma Z \to e^+ e^- Z$. Nowadays it is possible to obtain electron beams of quite good quality with energy up to 300 GeV at intensity up to $10^8$ e±/s (separation with synchrotron radiation, for example, /27/).

A sharp increase in intensity (up to $10^{13}$ e±/s) of electron beams with energy of hundreds GeV will feasible after design of superlinacs for linear electron-positron colliding beams (see 11.3).

Intensities of polarized electron beams have reached $10^{11}$ e⁻/s at SLAC. Intense polarized circulating beams are obtained due to radiative polarization in storage rings. Using intense circularly-polarized radiation (e.g. laser or spiral-undulator beam) travelling against electron beam it is possible to achieve much higher polarization rate of circulating electrons (and positrons) /30,31/. A sharp extension of possibilities in obtaining beams of polarized electrons of high energies will become accessible upon development of superlinacs (see 11.3).

10.4. **Positrons.** In the field of electron accelerators the presently available intensity of positron beams of entirely full energy achieves 1% intensity of electron intensity at worse quality of the beam. The use of intermediate storage rings with radiation cooling can essentially improve the quality of positron beams and increase their intensity.

Obtaining beams of polarized positrons in experiment still was necessarily connected with radiative polarization in storage rings.

At energies higher than 100 GeV, as mentioned above, possibilities for positron beams, including polarized beams, are the same as those for electrons.

10.5. Photons. Intensities and energies of beams of quanta obtained as bremsstrahlung at electron accelerators and also as a result of decay of neutral pions at proton ones are quite high. However, an important problem is beam separation and energy tagging for quanta hitting the targets. The latter is especially complicated for proton accelerators, and even so complicated that first of all one has to obtain $e^{\pm}$ beams of known energy and only after that following the ordinary procedure of measuring the energy of remaining $e^{\pm}$ an energy of the bremsstrahlung quantum can be tagged. The same technique of tagging energies of photons obtained on internal (superthin) targets is also convenient for obtaining intense fluxes of gamma-quanta in electron accelerators (storage rings).

Interesting prospects in obtaining intense, monochromatic and, at the same time, appropriately polarized beams of gamma-quanta of high energies is a backward Compton-effect on electrons travelling in cyclic storage rings at high energies.

For obtaining such quanta with energy E one should have electrons with energy E and polarized photons with energy higher than $(m_e c^2)^2/E$. Under these conditions, zero-angle scattered photons will have full energy E (almost independently on the initial photon energy). At the scattering angle $m_e c^2/E$ the photon energy will be much less. So, for effective monochromatization one needs to measure scattering angle (e.g. the position of photon event on a long-distant located target), and electron beam should have as small angular spread as possible. It is useful, additionaly, to measure an energy of the simultaneously scattered electron. At energies up to 50 GeV it is reasonable to employ synchrotron radiation in spiral undulators. In this case, it is necessary to ensure the electron interaction exclusively with photons, emitted inside the angle $m_e c^2/E_{tot}$. Irradiating particles can travel either in the same storage ring ($e^+e^-$ colliding beams) or in a special storage ring at a substantially lower energy. Some interesting possibilities can arise if the photon beams of short-wave intense electron beam lasers (without mirrors) could be used / 32 /. At energies higher than 50 GeV one can use photons of short-wave lasers of usual type.

Intensity of such beams of gamma-quanta corresponds to transfer of all stored electrons to these quanta with the life-time due to this process of thousands seconds (up to $10^8$ $\gamma$ /s).

10.6. Neutrons. Neutron fluxes with an energy up to tens of MeV are obtained mainly with nuclear reactors (including pulsed reactors) and at deuteron and proton accelerators. For monochromatization of reaction energy the fast separators and the time-of-flight methods of detection are used. I cannot help but drawing attention to the fact that it is a very attractive possibility for the energy range from tens eV to hundreds keV to use very powerful and highly collimated synchrotron radiation (with the quantum energy higher than 1.6 MeV) from electron storage rings at an energy $\gtrsim$ 10 GeV irradiating a berrilium target. Small transverse dimensions of the effective neutron source (achievable dimensions are down to 10 mkm×1 mm), short pulse (fractions of nanosecond) and a very low duty factor ($\lesssim 10^{-5}$) at high average intensity (up to $10^{14}$ n/s) ensure by many orders of magnitude better conditions for the study of neutron reactions using the time-of-flight method. In the lower part of the mentioned energy range the small transverse dimensions of the source make very effective the use of Bragg's monochromatization with the use of bent crystals, and also make effective obtaining polarized neutrons with the help of magnetic mirrors.

At higher energies an interesting pulsed source of neutrons can be obtained at meson factories with the use of charge-exchange ($H^- \rightarrow H^+$, $D^- \rightarrow D^+$) stacking of accelerated protons or deuterons in a cyclic storage ring, and using fast extraction onto the target.

At energies > 100 MeV an optimum method for obtaining quite monochromatic and well directed neutrons is the use of the decay reaction for accelerated deuterons with required energy per nucleon. In the superthin target mode an intensity for well collimated and quite monochromatic neutron flux can be achieved close to that for the deuterons and also good tagging with the remaining proton of the same energy. The use of polarized deuterons enables one to have neutrons with a good degree of polarization.

The use of charge-exchange reaction $pZ \rightarrow n(Z+1)$ permits doubling energy for neutrons obtained at a given cyclic accelerator but the beam quality in this case is worse. The cross-section of elastic charge-exchange falls down rapidly with proton energy growth ($\sigma_{ex} \approx 2/E^2_{GeV}$ mb) and at energies higher than tens of GeV one has to use the reaction pp $\to n\pi p$ with the useful cross-section of 0.2 mb having with proton accelerators up to 0.5% efficiency of transforming of protons to neutrons.

10.7 Antiprotons. Development of electron and stochastic cooling gives the possibility of obtaining the high intensity, absolutely pure, monochromatic and small-emittance antiproton beams. The first projects of antiproton storage rings under implementation and under preparation /33-36/ will give $(1 \div 5) \cdot 10^7$ $\bar{p}$/s but the ways are visible for increase by two orders of magnitude in production efficiency /3/.

The stacking will be performed at an energy 0.5÷5 GeV. The antiprotons can be decelerated to very low energies or be accelerated up to energies available for proton accelerators (or even higher when using proton klystrons). Of special interest are the studies with antiprotons at low energies with continuous electron cooling in obtaining intense and long-life protonium fluxes -p$\bar{p}$-electromagnetically-bound states /3,58/.

When using continuously cooled with electrons antiproton beams, which interact with a longitudinally polarized gas target at the storage ring section with stable longitudinal polarization of the circulating beam, one can achieve polarized antiproton beams with intensity up to 10% of the intensity of the initial antiprotons / 3 / with their subsequent acceleration (or deceleration) up (or down) to the energy required.

10.8. Antideuterons. With the same storage rings being designed for obtaining antiprotons one can get absolitely pure beams of antideuterons with intensity only by 3-4 orders of magnitude lower than that for antiprotons / 3 /. Such beams can turn out to be interesting for the study of nuclear states consisting of nucleons and two antinucleons.

10.9. Antineutrons. At energies up to tens of GeV the most profitable is to obtain antineutrons due to elastic charge-exchange reaction $\bar{p}p \to \bar{n}n$ (the cross-section at high energies is about $\sigma_{ex} = 15/E^2$(GeV) mb) with tagging, if possible, by the remaining neutron of low energy. The intensity of antineutrons will be up to $\sigma_{ex}/\sigma_{tot}$ of the system efficiency for antiprotons. The use of polarized antiprotons will enable obtaining the beams of polarized antineutrons with an intensity one more order of magnitude lower additionally (because of losses during antiproton polarization).

At still higher energies one has to get antineutrons in the reaction $\bar{p}p \to \bar{n}\pi^- p$ with cross-section of fractions of mb having worsened quality of the resulting beam (even with tagging). The antineutron intensity can reach a fraction of a percent of the antiproton intensity.

The antineutron beam with excellent quality, intensity up to $10^{-4}$ that of the antiprotons and with ideal tagging by the remaining $\bar{p}$ can be obtained by stored and accelerated antideuterons: $\bar{d}p \to \bar{n}pp$.

10.10 Pions. Obtaining beams of charged pions is the most explored way among the secondary particle beams production at high energies. Here, I would like to draw attention only to the tempting prospects for obtaining pure, rather monochromatic and well-collimated pion beams by their acceleration in superlinacs with acceleration rate higher than $2m_\pi c/\tau_\pi = 0.4 \, MeV/cm$; in this case, the most natural is the use of a proton klystron /11/. When using optimal conversion systems, for each ten protons with energy $\gtrsim 100$ GeV one can have one either positive or (and) negative pion with energy of a few GeV which is fit for further acceleration. In order to decrease the number of muons accompanying the beam of accelerated pions one should tend maximum acceleration rate.

Let me note here that at energies higher than hundreds GeV the number of events with full cross-section induced by neutral pions in a condensed target becomes substantial. So, at initial proton energy of 1 TeV with intensity $10^{13}$ p/s more than $10^5$ events will be caused by neutral pions having average flight length of 20 mkm. But, of course, the problem of identification of these events is extremely difficult.

10.11. Kaons. Unfortunately, for acceleration of charged kaons the accelerating gradients higher than 3 MeV/cm are required; that is still out of reality. There is some hope to achieve such gradients using special modification of proton klystron: all the protons, occupying the whole circumference of a big proton accelerator should be compressed into one (or several, with long distances in-between) bunch about 1 cm long and injected into a special linear wave-guide structure /37/. In this case, inside the bunch a very strong longitudinal electric field shall appear, breaking (decelerating) the protons of the bunch. Consequently, negative particles travelling with protons together inside the bunch shall be accelerated. So, the scheme gives possibility, in very principle, to accelerate $K^-$. Neutral kaons could be produced using charge exchange or charge loss reaction of accelerated K⁻ with a target. The development and design of enough damage-resistant systems of such kind is, of course, a task for the future. But up-to-now at high energies an optimum method for setting up kaon--beams production may turn out to be the use of the thin target mode (and at energies and intensities enabling effective cooling - the superthin target mode) at proton storage ring with the best available tagging (correspondingly with very complicated trigger). Since the total cross-section for generation of every kind of kaons in p-p reactions is large (fractions of mb), there are many kaons generated on this target. Naturally, for making more pure experiment one will have to use the whole set of the charge, momentum, velocity and gamma factor selection techniques, and, while recording products of KN reaction, one should most carefully take into account the quantum numbers of particles produced.

10.12. **Hyperons.** A new circumstance at superhigh energies is the long life-time for hyperons. Even at 100 GeV the long-lived hyperons live for tens of meters distances. Nevertheless, for separation of initial beam from the beam of produced negative and neutral hyperons (or positive of significantly deviated momentum) one should use strong magnetic fields, but this problem becomes easier linearly with energy growth. All the rest said on carrying out experiments with kaons remains valid even in this case (inclusive cross-sections, in particular, are of the same order).

10.13. **Antihyperons.** At not very high energies (rather to hundreds of GeV) the use of elastic charge-exchange reactions $\bar{p}p \to \bar{Y}Y$ ($\sigma_{ex}/\sigma_{tot} \approx 10^{-2}/E^2_{GeV}$) with tagging using by-product hyperons (being nearly at rest) in the (super) thin target mode in antiproton storage ring seems to be the optimum for obtaining antihyperon beams. Apparently, antihyperons produced in such a process by polarized antiprotons will preserve a noticeable polarization level.

At higher energies one will have to proceed in the same way as in the case of hyperons; the inclusive cross-section for antihyperon production in pp collisions is by only one order of magnitude lower than that for hyperons.

10.14. **Muons.** In order to have very pure, high energy and most intense muon beams with a very small emittance and good monochromaticity it is reasonable to proceed as follows /11/:

a) to obtain as many as possible pions with energy of 1 GeV on the target, with strong focusing, in nuclear cascade using the proton beams of energy $\gtrsim$ 100 GeV;

to let pions decay in the possibly stronger focusing channel;

c) to cool muons (with ionization cooling) in a special ring with targets placed at the sections with a very strong focusing;

d) to accelerate muons up to required energy in the short-pulse cyclic accelerator or (better) in a superlinac. The intensity of the muon beam can reach up to 10% of the intensity of the basic proton synchrotron (with use of the proton klystron mode).

In order to get polarized muon beams of high energy it seems to be most profitable to use monochromatic pion beams accelerated in a superlinac by injecting them into a special ring with strong magnetic field pulsed or superconducting). The structure of the ring should be designed in the way to have dynamically stable longitudinal polarization / 28 / of circulating muons (at injection energy, at any rate) equal in both long straight sections, which occupy, say, 3/4 of the circumference of the ring. The muons produced in the forward hemisphere will have momentum quite close to the pion momentum; muons of inverse helicity (moving backward in rest frame of the pions) deviate strongly over the momentum and can easily be removed from the ring. Polarization of the produced muon beams can be quite high (approaching the ratio of straight sections length to the circumference).

10.15. Neutrinos. The beams of muon neutrinos of high energies, well-directed and of useful intensity of a few percent of the intensity of the basic proton synchrotron, can be obtained with beams of accelerated pions. In order to decrease the neutrino beam diameter near the detecting facilities, which are located behind shielding of required thickness, it is profitable to perform the pion decay in a special storage ring with relatively long straight sections /11/.

Both muon and electron neutrinos of the same intensity can be obtained in the track of this kind by injecting into the track the accelerated cooled muons /11/.

Thus, a combination: "superlinac - special race-track" can be a multipurpose installation.

As to the beams of $\tau$ - neutrinos connected with a heavy lepton, it might turn out that their main source will be decay of $\tau$ - leptons of $\tau^{\pm}$ - pairs produced by $\gamma$ - quanta on the target nuclei / 37 /. At high enough energies nuclear form-factor does not decrease the $\tau$ - pair production. $\gamma$ - quanta can be obtained both with the help of proton and electron beams of high energy. More specifically one can evaluate the flux of $\tau$ - neutrinos with electrons. In a thick target the number of produced $\tau^{\pm}$ pairs will be of about $(m_e/m_\tau)^2 = 10^{-7}$ with respect to the number of incident electrons. So, the neutrino flux from a superlinac can be of the order of $10^6$ $\nu_\tau$/sec inside the angle $m_\tau c^2/E_e$ with average energy about $E_e/4$. It is hard to expect that in the case of protons the neutrino beam quality will be higher.

It is not excluded, that in a 100% duty cycle mode it would be possible to design the trigger system on $\tau$ - lepton and similar events production, for facilitating selection of $\nu_\tau$ events in neutrino target.

11. Colliding beam experiments became the main supplier of fundamental data in physics of elementary particles. Many electron-positron storage rings are in operation now (see Table 2 and Figure 1). Certainly, the colliding beam experiments are essentially needed at the highest energies.

11.1 However, the $e^+e^-$ colliding beams shall necessarily be developed and advanced not only at the highest energies. In particular, this necessity is connected with the fact that detailed study of quark-gluon systems in the field of low and average energies is of primary importance at present since it permits quantitative study of quantum chromodynamic effects, in particular, connected with the asymptotic freedom to confinement transition. Such experiments are especially suitable for electron-positron colliding beams, but to this end a sharp increase in luminosity of installations is required. The possibility and usefulness of this were proved by experience of VEPP-2M designed specifically for increased luminosity and, correspondingly, yielding increased accuracy of experimental data in the energy range up to $\sqrt{s} = 1.5$ GeV. Even now the possibilities are seen for making installations with luminosity up to $10^{33}$ cm$^{-2}$s$^{-1}$ at full energy of 4-5 GeV.

Other directions in improvement of electron-positron installations also promise to give important result. The possibility to work with polarized beams is very useful. In addition to sharp increase in absolute accuracy of measurements of produced particle masses /38,39/ (see Table 3), even the work with transversely polarized colliding beams helps in understanding the spins of produced final and intermediate formations. Implementation of experiments with longitudinally polarized beams permits obtaining qualitatively new information on the spin dependent strong interactions, and to study weak interactions, for instance, of b-quarks in the region of $\Upsilon$ -mesons.

Table 2.

| Storage ring laboratory | Particles | $\sqrt{S}$ (GeV) | $L_{max}$ (cm$^{-2}$ sec$^{-1}$) | Start | |
|---|---|---|---|---|---|
| VEP-1 (Novosibirsk) | e⁻ e⁻ | 0,32 | $5 \cdot 10^{27}$ | 1965 | Stop |
| Stanford storage rings | e⁻ e⁻ | 1 | $2 \cdot 10^{28}$ | 1965 | Stop |
| VEPP-2 (Novosibirsk) | e⁺ e⁻ | 1,4 | $3 \cdot 10^{28}$ | 1966 | Stop |
| ACO (ORSAY) | e⁺ e⁻ | 1,1 | $1 \cdot 10^{29}$ | 1967 | |
| ADONE (Frascati) | e⁺ e⁻ | 3 | $6 \cdot 10^{29}$ | 1970 | |
| CEA (Cambridge) | e⁺ e⁻ | 4 | $3 \cdot 10^{28}$ | 1971 | Stop |
| SPEAR (Stanford) | e⁺ e⁻ | 8,2 | $2 \cdot 10^{31}$ | 1972 | |
| VEPP-2M (Novosibirsk) | e⁺ e⁻ | 1,4 | $3 \cdot 10^{30}$ | 1974 | |
| DORIS (Hamburg) | e⁺ e⁻ | 11 | $10^{30}$ | 1974 | |
| DCI (ORSAY) | e⁺ e⁻ | 4 | $10^{30}$ | 1976 | |
| VEPP-4 (Novosibirsk) | e⁺ e⁻ | 4(11) | $3 \cdot 10^{28} ( \quad 10^{31})$ | 1979 (1981) | |
| PETRA (Hamburg) | e⁺ e⁻ | 38 | $5 \cdot 10^{30} (10^{32})$ | 1979 (1980) | |
| CESR (Cornell) | e⁺ e⁻ | 11(16) | $2 \cdot 10^{30} (10^{32})$ | 1979 (1980) | |
| PEP (Stanford) | e⁺ e⁻ | 28(36) | $2 \cdot 10^{30} (10^{32})$ | 1980 (1981) | |

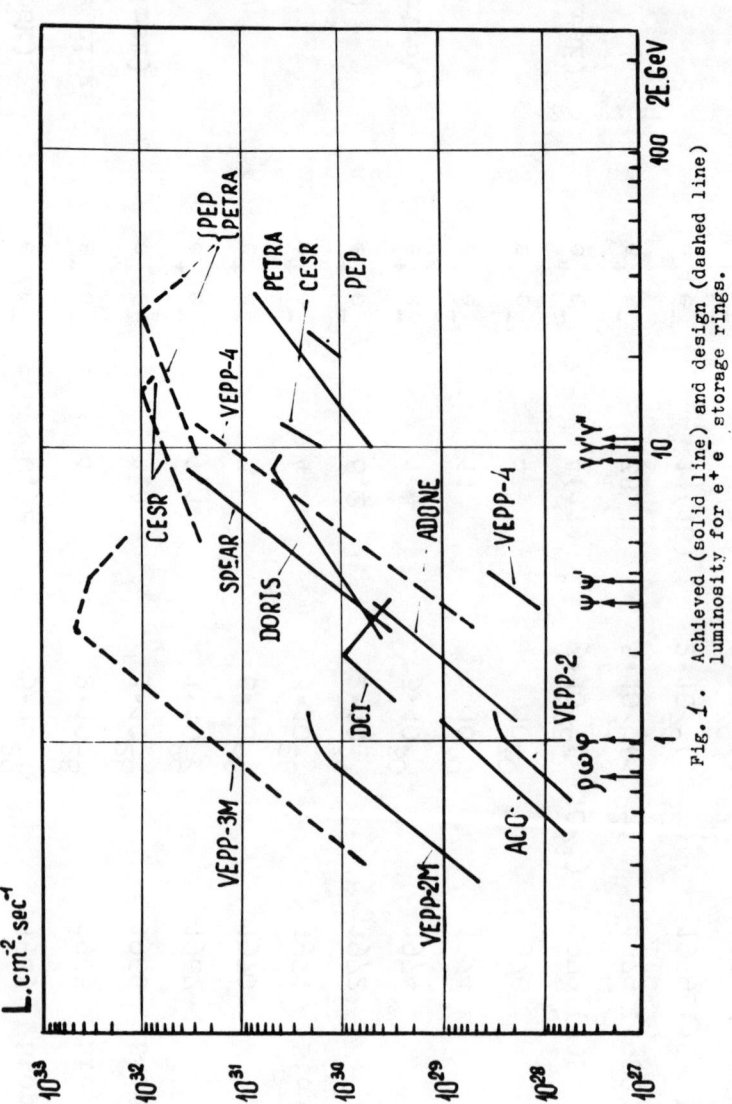

Fig. 1. Achieved (solid line) and design (dashed line) luminosity for $e^+e^-$ storage rings.

Table 3

| Particle | Mass, MeV | |
|---|---|---|
| | High Precision Measurement* | Old World Average |
| $K^{\pm}$ | 493.670±0.029 | 493.668±0.018 |
| $\Phi$ | 1019.54±0.12<br>1019.52±0.13 | 1019.62±0.24 |
| $\Psi$ | 3096.93±0.09 | 3097.1±0.9 |
| $\Psi'$ | 3686.00±0.10 | 3685.3±1.2 |

\* High precision measurements have been performed at VEPP-2M and VEPP-4 using the resonance depolarization method of the absolute energy calibration.

The possible sharp increase (higher than one order of magnitude) in monochromaticity of electron-positron reactions opens up interesting possibilities /40/. So, one can proportionally raise the fraction of resonance reactions, that is of special importance for $\Upsilon$ - mesons, and study the inner structure of $\Psi$ - mesons (even for the purpose of proving that it does not exist). Note, that even higher monochromaticity can be achieved with $p\bar{p}$ - colliding beams under continuous electron cooling /3/.

11.2. But the main trend in the field of electron--positron colliding beams remains the tendency to higher energies.

Already now the total energies up to $\sqrt{s}$ = 40 GeV become accessible (PETRA, PEP). An intensive development of the LEP project is under way (first stage-up to $\sqrt{s}$ = 100 GeV, second - up to 250 GeV), the project of the new storage ring at Cornell and also the HERA project enables, in principle, obtaining $e^+e^-$ energies up to 100 GeV (see Table 4). Note, that at these high energies in cyclic storage rings (despite the overlapping of spin resonances) implementation of $e^+e^-$ polarized colliding beams is feasible /41/. The new and interesting is the project of quasilinear single-pass $e^+e^-$ colliding beams at SLAC /42/ (Fig. 2) at an energy up to $\sqrt{s}$ = = 100 - 140 GeV.

Further increase in energy of electron-positron colliding beams in cyclic storage rings (now conventional) is almost unrealistic because of the catastrophic rise in loss by synchrotron radiation that forces to enlarge the installation both in dimensions and power consumptions as the square of energy. Therefore, the main direction in development becomes linear colliding beams /9/.

In the plans for linear colliders at super high energies even from the initial stage the possibilities are considered of using long superconducting structures with recuperation of accelerated particle energy and of using pulsed superlinacs /43/. Several projects of linear $e^+e^-$ superconducting colliders are being developed now - Cornell, CERN, Hamburg /44-46/. The collider project VLEPP based on superlinacs is being developed in Novosibirsk and adjustment of its most important components is being performed /9,10,47,48/. Let me outline briefly the schematic of the Novosibirsk project and its possibilities.

11.3. The general layout of the facility may be represented as follows (Fig. 3). Two superlinacs at an energy 100 GeV and 1 km long each, fed by high power SHF sources installed about 10 m apart, "fire" at each other

Table 4

| Project/Laboratory | Particles | $\sqrt{S}$ (GeV) | | $L(\text{cm}^{-2}\text{s}^{-1})$ | Start |
|---|---|---|---|---|---|
| LEP (CERN) | $e^+e^-$ | Ist 100 | | $10^{32}$ | 1986 (?) |
| | | IIst 250 | | | |
| New Cornell Ring | $e^+e^-$ | | 100 | $3\cdot10^{31}$ | 1986 (?) |
| Stanford Single-pass Collider | $e^+e^-$ | | 100 | $1\cdot10^{30}$ | 1985 (?) |
| VLEPP (Novosibirsk) | $e^+e^-$ | Ist 200 | | $1\cdot10^{32}$ | 1989 (?) |
| | | IIst 600 | | $1\cdot10^{32}$ | |

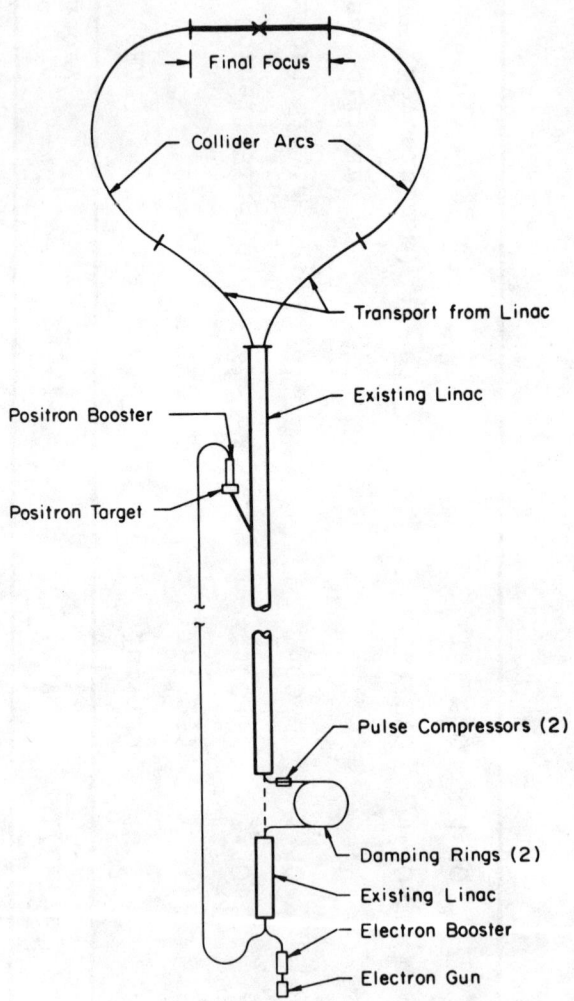

Fig. 2. General layout of the SLAC Linear Collider.

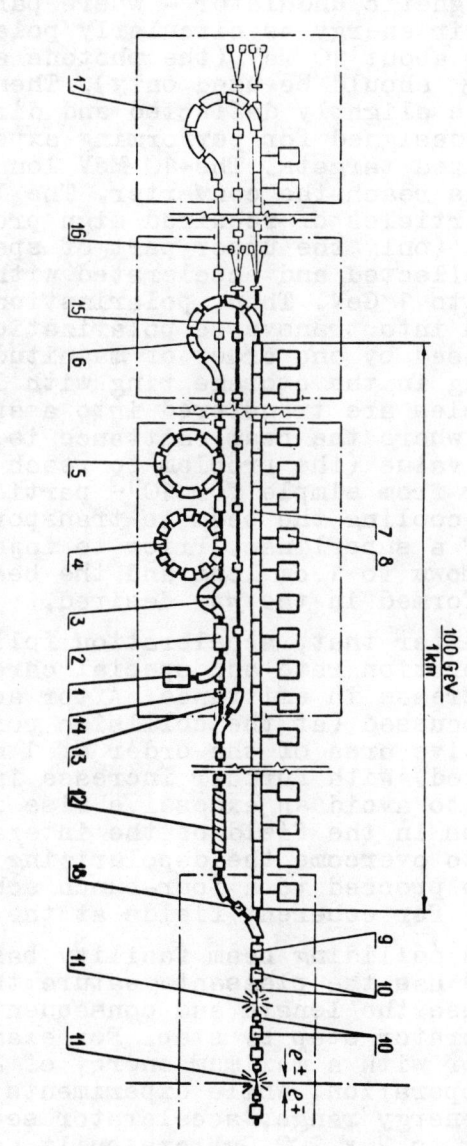

Fig. 3. The Novosibirsk linear $e^+e^-$ Collider (VLEPP).

1. INITIAL INJECTOR
2. INTERMEDIATE ACCELERATOR
3. DEBUNCHER
4. STORAGE RING
5. COOLER-INJECTOR
6. BUNCHER
7. ACCELERATING SECTIONS
8. SHF SOURCE
9. PULSE DEFLECTOR
10. FOCUSING LENSES
11. COLLISION POINTS
12. HELICAL ONDULATOR
13. THE BEAM OF $\gamma$-QUANTA
14. CONVERSION TARGET
15. RESIDUAL ELECTRON BEAM
16. ELECTRON (POSITRON) BEAM EXPERIMENTS WITH STATIONARY TARGET
17. THE SECOND STEP
18. SPECTROMETER

with single bunches 1 cm long with $10^{12}$ particles in each bunch of polarized electrons and positrons at a repetition rate of order 10 Hz. After collision the bunches are slightly deflected by a pulsed field into a small-angle analyzing system that makes it possible to measure the energy spectrum of the colliding particles. Then the bunch enters the conversion system - a long spiral magnetic undulator - where particles irradiate of 1% of their energy as circularly polarized photons with energy about 10 MeV (the photons emitted inside the angle $1/\gamma$ should be used only). Then remained polarized beam is slightly deflected and directed into special halls designed for performing experiments with stationary polarized targets. The ~10 MeV longitudinally polarized photons reach the converter. The longitudinally polarized particles of required sign produced in pairs on the target (only the upper part of spectrum should be taken) are collected and accelerated with high acceleration rate up to 1 GeV. Then, polarization of particles is transformed into transverse polarization and bunch length is increased by one order of magnitude. After radiation cooling in the storage ring with large acceptance the particles are transfered into a special storage ring-cooler, where the beam emittance is damped down to a very small value (the problem to reach a very small emittance is far from simple for $10^{12}$ particles in the beam). Upon total cooling the beam is transported to the injector end of a superlinac. Prior to injection the beam is shortened down to 1 cm long and the beam polarization is transformed in the way desired.

After that, acceleration follows with the highest acceleration rate and special care undertaken to avoid an increase in emittance. After acceleration the bunches are focussed (at the collision point) into ellipses with effective area of the order of 1 mkm$^2$, and the cycle is repeated. With further increase in collider energy, in order to avoid an excessive rise in synchrotron radiation loss in the field of the interacting bunch and especially to overcome the depolarizing effect, one apparently has to proceed to a four-bunch scheme with mutual compensation for coherent fields at the collision point / 9 /.

A colliding beam facility based on the scheme described has the pleasant feature that it is possible to increase the length and consequently the energy of the accelerator step by step. For example, initially an accelerator with a maximum energy of 2 x 100 GeV may be put into operation. While experiments are being conducted in this energy range, accelerator sections that raise the energy to 2 x 200 GeV are built, and so forth.

Finally, let us recall that VLEPP can be used as a

conventional double-energy accelerator with rather high average current of polarized $e^+e^-$. This is itself quite interesting.

A rough table of the parameters of an accelerator at energies of 2 x 100 and 2 x 300 GeV can be presented as follows:

Table 5

| | 2 x 100 GeV | 2 x 300 GeV |
|---|---|---|
| Energy | 2 x 100 GeV | 2 x 300 GeV |
| Length | 2 x 1 km | 2 x 3 km |
| Luminosity | $10^{32} cm^{-1} s^{-1}$ | $10^{32} cm^{-2} s^{-1}$ |
| Average beam power | 2 x 160 kW | 2 x 480 kW |
| Number of particles in a beam | $10^{12}$ | $10^{12}$ |
| Average mains power | 7 - 10 MW | 20 - 30 MW |
| Repetition frequency | 10 Hz | 10 Hz |

11.4. The first proton colliding beam facility (ISR) has been operating at CERN since 1971. Its maximum energy is 2 x 33 GeV, the maximum number of stored particles is up to $10^{14}$ in each beam, ultimate luminosity is $0.7 \cdot 10^{32} cm^{-2} s^{-1}$. During this period a number of important experiments have been conducted which provided valuable information.

Construction of the big superconducting storage rings is being carried out at Brookhaven proton-proton colliding beams at energy $\sqrt{s}$ = 800 GeV - ISABELLE - with very high design luminosity ($10^{33} cm^{-2} s^{-1}$). Implementation of proton-proton experiments on the Main Ring-Doubler facility is under consideration at Fermilab at an energy $\sqrt{s}$ = 1100 GeV (300 GeV on 1000 GeV). Proton-proton colliding beams are envisaged in the accelerating facility at Serpukhov (UNK) at an energy up to 2 x 3 TeV. So, we see, colliding beam energies will increase rather rapidly. But experimental feasibility of reactions, now of intense interest, with energy 0,1 megajoul in elementary interaction (beloved $10^{15}$ GeV in rest frame!) - is the question for not so near future (see Table 6).

11.5 The closest new installation with hadron colliding beams will be proton synchrotron SPS already in operation at CERN. This installation is being modified now for the mode of proton-antiproton colliding beams with energy up to $\sqrt{s}$ = 600 GeV /33 /. Next will be commissioning of the proton-antiproton installation at an energy up to $\sqrt{s}$ = 2000 GeV based on the superconducting proton

Table 6

| Project/Laboratory | Particles | $\sqrt{S}$ (GeV) | $L (cm^{-2} s^{-1})$ | Start |
|---|---|---|---|---|
| ISR (CERN) | pp | 62 | $0.7 \cdot 10^{32}$ | 1971 |
| ISABELLA (Brookhaven) | pp | 800 | $2 \cdot 10^{32} (1 \cdot 10^{33})$ | 1986 |
| Main Ring/Doubler (Fermilab) | pp | 1,100 | | |
| UNK (Serpukhov) | pp | 6,000 | | |
| ISR (CERN) | p$\bar{\text{p}}$ | 62 | | 1981 |
| SPS (CERN) | p$\bar{\text{p}}$ | 600 | $\geqslant 1 \cdot 10^{30}$ | 1981 |
| Tevatron, Phase I (Fermilab) | p$\bar{\text{p}}$ | 2,000 | $\geqslant 1 \cdot 10^{30}$ | 1984 |
| UNK (Serpukhov-Novosibirsk) | p$\bar{\text{p}}$ | 6,000 | $3 \cdot 10^{30}$ | 1990 |
| Pentavac (Fermilab) | p$\bar{\text{p}}$ | 10,000 | | |
| HERA (Hamburg) | e$^{\pm}$p | 300 (30e ⟷ 800p) | $4 \cdot 10^{31}$ | 1988 |
| CHEER (Fermilab) | e$^{-}$p | 200 (10e ⟷ 1,000p) | $5 \cdot 10^{31}$ | 1985 (?) |
| TRISTAN (KEK) | e$^{-}$p | 170 (25e ⟷ 300p) | $1 \cdot 10^{31}$ | 1988 |

synchrotron Doubler (Tevatron, Phase I) being built at Fermilab /34/. The p$\bar{\text{p}}$ project is designed for UNK (Novosibirsk-Serpukhov collaboration) at the energy up to $\sqrt{s}$ = 6 TeV /35,36/.

In the first years after announcing the first proton-antiproton colliding beam project (1966 VAPP-NAP, Novosibirsk /49,50/) the proton-antiproton experiments at maximum accessible energy were considered by many physicists as an exceedingly complicated addition to proton-proton experiments at the same energies. Even then, of course, it was evident that this addition is rather important. So, two classes of experiments are specific to proton-antiproton colliding beams: first, the study of hadron annihilation, second, the study of two-particle charge-exchange reactions, i.e., reactions with conservation of baryon charge of each colliding particle. The annihilation cross-section apparently decreases only inversely as the energy of the colliding beams and even at an energy 2 x 1000 GeV the cross-section will be of the order $10^{-30}$ cm$^2$. So, main problem will be the separation of annihilation processes from the vast majority of "the events of the total cross-section". At the same time, the cross-section of the process like

$$p\bar{p} \to \Lambda\bar{\Lambda}$$

decreases (in the energy region presently known) as $E^{-4}$ and only with a luminosity of the order $10^{32}$cm$^{-2}$s$^{-1}$ one can manage to get some data about these processes at energies above 100 GeV.

In recent years the attitude toward proton-antiproton colliding beams has changed greatly. The quark model is acquiring more and more dynamical content and more and more "public opinion" is inclined to consider hadrons as consisting of quarks interacting as point-like particles. Accordingly, processes with very large momentum transfer will occur through the interaction of quarks, the components of the colliding hadrons (Drell-Yan processes). Here protonproton collisions give quark-quark reactions, while proton-antiproton collisions give quark-antiquark reactions. In this sense one can say that in experiments in colliding proton-antiproton beams it is possible to obtain the same fundamental information as in colliding electron-positron beams of the same luminosity and with an energy of the order of one-sixth of the energy of the baryons. Similarly, proton-proton colliding beams are equivalent to electron-electron collisions. Of course, for strongly interacting particles such as protons and antiprotons we cannot say that they consist only of quarks of one "polarity". However, according to contemporary neutrino data the content of antiquarks in a proton is about 5% (this is also the estimate of the content of quarks in the antiproton). Therefore quark-antiquark interactions are dominant in proton-antiproton collisions, and quark-quark interactions provide only a small admixture. For proton-proton collisions the ratio will be the reverse. In addition, the average energy of quark-antiquark reactions in proton-proton collisions will be substantially lower than in proton-antiproton collisions.

Note, (see 10.) that it is feasible to obtain proton-proton polarized beams with full luminosity and also the proton-antiproton beam with luminosity one order of magnitude lower than that for unpolarized, including experiments with given helicities of initial particles.

Some interesting possibilities will open up when the cooling of high energy colliding $p\bar{p}$ beams using circulating electron beam /51-53/ will be developed. For high luminosity colliding beams the effective cooling achievement requires a solution of the very complicated technical problems and it is not so easy to predict correctly the prospects of this method.

But for luminosity about $10^{28}$ $cm^{-2}s^{-1}$ the circulating beam electron cooling looks like a not an extremely difficult problem. Using this cooling technique at energy more than 100 GeV one shall have very small equilibrium sizes of colliding beam. So, one shall have the possibility to measure precisely differential cross-section of $p\bar{p}$ elastic scattering at the angles of effective interference of strong and Coulomb interactions. Such measurement will give the information about behaviour of total cross-section at energies by the order of magnitude higher than the energy of $p\bar{p}$ collisions.

12. A few words about the "strategy" of advancing to the ultra-high energies. One can distinguish (quite schematically) four stages of exploring new regions at energies of hundreds GeV and higher /1/.

In the first stage only interactions of any point-like objects (nowadays - leptons, quarks) should be accessible which enable one to produce as large as possible momentum transfer both in scattering and in production of massive objects (space- and time-like momentum transfers). In the first stage it is not too important for which pairs it will be done. The question of primary importance is the question of having beams available for the first stage experiments. Colliding beams of particles and antiparticles seem to give more experimental information as the systems have less quantum numbers prohibitions for generation of new objects. From this point of view, the most advantageous variant for the experiments will be proton-antiproton colliding beams which will enable studying the fundamental quark-antiquark interactions at an energy of one-sixth of the energy of the proton-antiproton colliding pair.

Of course, when we are talking now about the study of fundamental interactions of different objects, it is just a way to classify the experiments over the initial states. Each certain class of experiments will also provide vast additional information.

At the second stage one can consider the experiments which cover the interactions of all fundamental particles i.e. the study of lepton-lepton, lepton-antilepton, quark-lepton, quark-antilepton, quark-quark and quark-antiquark interactions. In this case, the choice of concrete particles is still determined by which is most realistic to realize.

These problems will be solved soon, most probable, in the following colliding beam experiments:

a) lepton-lepton and lepton-antilepton - $e^- + e^-$ and $e^- + e^+$;

b) lepton-quark and antilepton-quark - $e^- + p$ and $e^+ + p$;

(the experiments of this kind are already planned at installations at superhigh energies which are being built and designed /54-57/;

c) quark- and antiquark interactions will primarily studied in pp and p$\bar{\text{p}}$ experiments. The feasibility of these experiments, the difficulties that will arise and the now visible ways to overcome them are sufficiently clear from the above discussions.

In the next stage it will apparently be important to obtain as complete as possible set pairs of fundamental particles in the initial state. And finally, for advance in understanding of fundamental interactions at ultimate high energies it will become necessary to study the collisions of all elementary particles and even nuclei.

12.1. In this connection, it is worth paying attention that many of those experiments which now seem exotic and unreal will become available in the not distant future.

So, quite soon after exploring proton-antiproton colliding beams deuteron-antideuteron experiments will become accessible (for the study of neutron-antineutron interactions): for the effectivness of stacking antideuterons is only four orders of magnitude lower than that for antiprotons, the luminosity of the order $10^{27}$ cm$^{-2}$s$^{-1}$

will be achieved immediately and one should not wait too long for progress in this field /3/.

With time, colliding beam experiments with unstable particles will become accessible. Good prospects for muons and pions acceleration are opened up with the use of intense beams of modern and future proton accelerators in proton klystron mode for superlinacs excitation (see s. 7) / 11 /.

Using the pions accelerated in this way with already up-to-date accelerators SPS and Main Ring it is possible to obtain pion-proton and pion-pion colliding beam luminosity of the order $10^{27}$ cm$^{-2}$s$^{-1}$ (with the use of a 100 KG pion magnetic track and the proton storage ring)/11/.

The muon colliding beam experiments will also be accessible /58,11/. For this purpose it is required to accelerate the cooled muon beams in the linear accelerator up to the required energy and make them collide in the sections with very strong focusing in a special ring with magnetic field as high as possible in order to increase the number of collisions during the life-time of the muons. Evaluations have shown that this way would enable one to achieve the satisfactory luminosity of the order $10^{31}$ cm$^{-2}$s$^{-1}$ at energies of hundreds GeV.

13. In conclusion, a rather trivial truth could be expressed: high energy physics nowadays not only advances rapidly but it is at the stage when qualitatively new accelerator and instrumentation possibilities are about to appear and their use will have crucial influence on elementary particle physics progress. The progress toward this goal is undoubtedly the main result of the efforts of our high energy physics community. But the use of current achievements becomes also more and more important for solution of applied problems put at present in front of us all.

14. I am grateful to V.N.Baier, V.E.Balakin, L.M.Barkov, A.I.Vainshtein, Ya.S.Derbenev, A.M.Kondratenko, G.N.Kulipanov, I.N.Meshkov, A.P.Onuchin, E.A.Perevedentsev, S.G.Popov, I.Ya.Protopopov, V.A.Sidorov, A.G.Khabakhpashev, S.I.Eidelman for helpful discussions, to A.A.Prokopenko for active assistance in preparing the manuscript and to D.Kline for care and attention.

# REFERENCES

1. Skrinsky A.N.: "Colliding beams - present and future"; IV All-Union Conference on Accelerators, Dubna, 1978; Preprint INP 79-12, Novosibirsk, 1979.

2. Budker G.I.: "Electron Cooling"; Atomnaya Energiya, 22, 346, 1967.

3. Budker G.I., Skrinsky A.N.: "Electron Cooling and New Possibilities in Elementary Particle Physics", UPhN, 124, v. 4, 1978.
Skrinsky A.N.: XVIII International Conference on Elementary Particle Physics, Tbilisi, 1976.

4. S.Van der Meer: CERN Intern. Report, CERN/ISR-PO/12-31, 1972.

5. S.Belyaev, G.Budker, S.Popov: "The possibility of using storage rings with internal thin target"; III Int. Conference on High Energy Physics and Nuclear Structure, 1970.

6. B.B.Voitsekhovsky et al.; Measurement of the "Radiative Tail" - Electron Spectrum in the reaction $ep \to e'p\gamma$", Pisma v ZhETF, 29, 105-109, 1979.

7. MacDaniel. "New Cornell $e^+e^-$ Project", this Conference, 1980.

8. E.Picasso; "LEP Project"; this conference, 1980.

9. Balakin V.E., Budker G.I., Skrinsky A.N.: "Feasibility of Creating a Superhigh Energy Colliding Electron-Positron Beam Facility"; VI All-Union Accelerator Conference, Dubna, 1978;
Balakin V.E., Skrinsky A.N.: "A Superhigh Enery Colliding, Preprint IYaF, 78-101, Electron-Positron Facility (VLEPP)", ICFA-II Proceedings, 1980.

10. Balakin V.E., Brezhnev O.N., Novokhatsky A.V., Semenov Yu.I.: "Accelerating Structure of a Colliding Linear Electron-Positron Beam (VLEPP)"; Preprint IYaF 79-83, Novosibirsk, 1979;
Proceedings of VI All-Union Accelerator Conference, Dubna, 1979.

11. Perevedentsev E.A. and Skrinsky A.N.: "On Possible Use of Intense Beams of big Proton Accelerators for Excitation of Linear Accelerator Structure", IV All-Union Accelerator Conference, Dubna, 1978; Preprint INP 79-80, Novosibirsk, 1979; ICFA-II Proceedings.

12. L.B.Leipuner et al.; "First experiences with a Fastbus system at Brookhaven"; this Conference, 1980.

13. B.Cortez et al.: "A Sensitive Search for the Time Evolution of an Electron Neutrino Beam", this Conference, 1980.

14. J.Blandino, U.Camerini, D.Cline et al.; "A Decay Mode Independent Search for Barion Decay Using a Volume Cherenkov Detector"; this Conference, 1980.

15. K.Deiters, A.Donat, K.Lanius et al.; "Test of a liquid argon multistrip ionization chamber"; this Conference, 1980.

16. A.E.Bondar et al., Proceedings of the III International Meeting on Proportional and Drift Chambers, Dubna, 1978, p. 184.

17. Yu. Pestov, G.Fedotovich: "Fast spark counter with lokalized discharge"; preprint INP 77-78, Novosibirsk, 1977.

18. A.A.Vorobyov et al., Nucl. Instr. Methods, 119 (1974) 509.
J.P.Burq et al.; "Measurements of the total $\pi$-He and $p$-He cross-sections", this Conference, 1980.

19. E.Albini et al.; "A live target for measuring the lifetime of charmed particles", this Conference, 1980.

20. D.Potter; "Two New Triggerable Track Chamber - Target"; this Conference, 1980.

21. M.Dykes et al.; "Test of high resolution small bubble chamber", this Conference, 1980.

22. The technique is being developed in Leningrad Nuclear Physics Institute.

23. R.S.Gilmore, D.W.G.S. Leith, S.H.Williams; "Development of high gain multigap avalanche detectors for Čerenkov ring imaging"; this Conference, 1980.

24. R.A.Astabatyan, M.P.Lorikyan, G.A.Manukyan, K.Zh.Markaryan; "On some possibilities of extraction of local ionization"; this Conference, 1980.

25. L.Barkov, M.Zolotorev: "Parity Violation in Atomic Transition"; Physics Letters, 85B, 308, 1979.

26. Yu.I.Bel'chenko et al.; "High Current Proton Beams at Novosibirsk", Proceedings of the X International Conference on High Energy Accelerators", Protvino, 1977, v. 1, p. 287.

27. Derbenev Ya.S. and Kondratenko A.M.: "Acceleration of Polarized Protons in Cyclic Accelerators"; Proceedings X International Accelerator Conference, Serpukhov, 1977; Sov. Dokl., v. 20, N 8.

28. Derbenev Ya.S., Kondratenko A.M., Skrinsky A.N.: "Spin Motion in Storage Ring with Arbitrary Field"; Preprint IYaF N2-70, Novosibirsk, 1970; Sov. Phys. Dokl. 192, 1255, 1970.

29. Z.Guiragossian, R.Rand: "A 300 GeV High Quality Electron and Pion Beam at the New Generation of Proton Accelerators"; San Francisko Particle Accelerators Conference, 1973.

30. Derbenev Ya.S., Kondratenko A.M., Saldin E.L.: "Polarization of the Electron Beam in a Storage Ring by Hard Circularly-Polarized Photons"; Nucl. Instr. Meth., 165, 15-19, 1979.

31. Derbenev Ya.S., Kondratenko A.M., Saldin E.L.: "Polarization of Electrons in a Storage Ring by Circularly Polarized Electromagnetic Waves"; NIM, 165, 201-208, 1979.

32. A.Kondratenko, E.Saldin: "Generation of Coherent Radiation by a Relativistic Electron Beam in an Undulator"; preprint INP 79-48, Novosibirsk, 1979.

33. "CERN SPS Used as $p\bar{p}$ Collider"; Proceedings of the Workshop on Producing High Energy Proton-Antiproton Collisions, Berkeley, 1978.
K.Rubbia. "$p\bar{p}$ at CERN"; this Conference, 1980.

34. "Tevatron Used as $p\bar{p}$ Collider"; Proceedings of the Workshop on Producing High Energy Proton-Antiproton Collisions, Berkeley, 1978.
"Tevatron: Phase-I" Fermilab, 1980.
D.Young. "$p\bar{p}$ at Fermilab"; this conference, 1980.

35. T.Vsevolojskaya et al.: "Antiproton Source for UNK"; preprint INP 80-182, Novosibirsk, 1980.

36. T.Vsevolojskaya et al.: "Proton-Antiproton Colliding Beams in UNK"; paper presented at VII All-Union Accelerator Conference, Dubna, 1980.

37. Perl M.L.: "Heavy Lepton Phenomenology"; SLAC-PUB--2219, 1978.

38. Bukin A.D. et al.: "$\phi$ - Meson: $\omega$ - $\phi$ Interference and Precision Mass Measurements for $\phi$ - Resonance"; Yadernaya Fisika, v. 27, 1978.

39. Zholentz A.A. et al.: "High Precision Measurement of the $\psi$ - and $\psi'$ - Meson Masses"; Preprint IYaF, 80-156, Novosibirsk, 1980.

40. Zholents A.A., Protopopov I.Ya., Skrinsky A.N.: "Energy Monochromatization of Particle Interaction in Storage Rings"; VI All-Union Accelerator Conference, Dubna, 1978. Pr. IYaF, 79-6, Novosibirsk.

41. Derbenev Ya.S., Kondratenko A.M., Skrinsky A.N.: "Radiative Polarization at Ultrahigh Energies", Particle Accelerators, v. 9, p. 247, 1979.

42. B.Richter et al.: "The SLAC Single-Pass Collider"; XI International Conference on high energy accelerators, CERN, 1980.
43. Skrinsky A.N.: "Colliding Beams in Novosibirsk"; International Seminar on Perspectives in High Energy Physics, Morges, Switzerland, 1971.
44. M.Tigner. Private communication.
45. U.Amaldi: "Collinear accelerators for high energy $e^+e^-$ collisions"; Phys. Lett., <u>61B</u>, 1976.
46. Gerke H., Steffen K.: "Note on a 45-100 GeV "Electron Swing" Colliding Beam Accelerator", Int. Rep., DESY PET-79/04, Hamburg, 1979.
47. Balakin V.E., Koop I.A., Novokhatsky A.V., Skrinsky, Smirnov V.P. "Beam Dynamics of a Colliding Linear Electron-Positron Beam (VLEPP)"; Preprint IYaF 79-79, Novosibirsk, 1979; Proceedings of VI All-Union Conf. on Accelerators, Dubna, 1978.
48. Balakin V.E., Mikhailichenko A.A.: "The Conversion System for Obtaining High Polarized Electrons and Positrons", Preprint IYaF 79-85, Novosibirsk, 1979.
49. G.Budker, A.Skrinsky: "Electron cooling and proton-antiproton colliding beams"; ORSAY Symposium on colliding beams, 1966.
50. VAPP-NAP Group Report; Proc VIII Intern. Conference on High Energy Accelerators, Geneva, CERN, 1971.
51. C.Rubbia: "Relativistic electron cooling to increase the luminosity of the $p\bar{p}$ collider", Workshop on Cooling of High Energy Beams, Madison, 1978.
52. A.Skrinsky. "On High Energy Electron Cooling"; Workshop on Cooling of High Energy Beams, Madison, 1978.
53. Derbenev Ya.S., Skrinsky A.N.: "On High Energy Electron Cooling", Preprint IYaF 79-87, Novosibirsk, 1979; Particle Accelerators, 1980.
54. Wiik. "HERA-Electron-Proton Colliding Beam Facility at DESY"; this Conference, 1980.
55. Y.Kimura: "Tristan-the Japanese Electron-Proton Colliding Beam Project"; this Conference, 1980.
56. "CHEER-Canadian Electron-Proton Project"; this Conference, 1980.
57. R.Wilson et al.: "Electron-proton colliding beams at Fermilab"; this Conference, 1980.
58. Budker G.I.: in Proceedings of the International Conference on High Energy Physics, Kiev, 1970.

59. "Design study of a facility for experiments with low energy antiprotons", CERN/PS/DL, 80-7, 1980.
60. A. Kusumegi, K. Kondo: "Possibilities of using heavy liquid materials as a radiator of total absorption calorimeters", KEK Preprint 80-5, 1980.

## DISCUSSION

Q1: Wallraff, RWTH Aachen: Would you please show again the numbers on the mass of the $\Psi$ and $\Psi'$ measured at Novosibirsk?

Q2: Schuler, Yale: Do you know of anyone who has tried the "Siberian Snake" experimentally?

A2: (no response found)

Chapter 11

Weak Interactions

# QFD AND UNIFICATION

H. Sugawara, KEK, Speaker

L. van Hove, CERN, Chairman

J. Kiskis, Los Alamos,

C. Zachos, Wisconsin-Madison,

Scientific Secretaries

QFD AND UNIFICATION

Hirotaka Sugawara
National Laboratory for High Energy Physics, Oho-machi
Tsukuba-gun, Ibaraki-ken, 305 Japan

INTRODUCTION

Attempts aimed at the unification of all the interactions are reviewed at the level of gravity being still excluded. While the degree of aesthetic anxiety is reduced compared with the electro-weak theory[1] by the inclusion of strong interaction, the physics output of grand unified theories are much harder to confront with the observation. This is caused by the fact that the scale parameter involved is $10^{-29}$ cm compared to $10^{-16}$ cm of electro-weak theory. This will be even more true in the unified theory of all interactions. Such a theory would be aesthetically complete to the human mind but the physics in general is concerned either with the very early universe before the Planck time or the transition amplitudes which are smaller by at least a factor of $10^{-8}$ than the proton decay amplitude. The physics motivation for unification of all the interactions is somewhat obscure by this fact save for its aesthetic completeness. One of the promising approaches along this line is, no doubt, the supergravity theory[2]. But in spite of all the efforts to find a link between the theory and physical reality it is still short of nothing more than 'a phenomenon in theoretical physics'*. On the other hand we now do believe that a grand unified theory is physically motivated: first by its 'prediction' of the baryon number asymmetry in the universe and second by its correct prediction of $\sin^2\theta_w$. Proton decay experiments would surely provide us a firmer basis for this belief.

In view of the fact that the physical motivation for the unification of gravity with the other interactions is so weak, we are still allowed to question if gravity should be treated on an equal basis with the other interactions. A theory of induced gravity is one such example[4]. The basis of this approach is the fact that the form of radiative corrections is strongly restricted by both general covariance and other symmetries, such as grand unified gauge invariance. This very fact has been used frequently in several spinor unified theories[5], since the work of Heisenberg[6], where all interactions are to be induced from a single spinor interaction. While this approach still attracts some workers it suffers from a serious drawback: non-renormalizability. Renormalizability may not be a principle of physics but it may well be a property which is

---

* At this Conference M. Gell-Mann has pointed out the possible physical relevance of N=8 supergravity theory developed by J. Ellis, M. Gaillard and B. Zumino.[3]

shared by all interactions.

Leaving aside gravity we have yet another interaction which we definitely wish either to induce from other interactions or 'unify' with them. This is the interaction of Higgs bosons with fermions or with themselves. Our understanding of this problem is still in its early days although we observe that a particular model is gaining popularity[7].

So far I have been discussing the materials which I do not deal with in the text. There are also some other topics I do not include in this review just for technical reasons (page limitation etc.) which otherwise deserve much attention. They are, for example, the properties of W's and Z's, $\nu$-masses and neutral currents ($\nu$-mass was included in my talk but will be omitted here).

I will discuss in the following sections the problem of hierarchy, some grand unified models, the family or generation problem, CP-violation, proton decay, baryon asymmetry in the universe and finally the monopole. I will try to clarify the relation between the assumptions and the results which can be derived from them. Sometimes the discussion becomes inevitably introductory. Sometimes I explain the material in detail (hierarchy for example) and sometimes I just quote the work of other people (proton decay kinematics for example). It is by no means my intention to discuss the materials in an equal proportion. I spare much effort in explaining the material which bothers me or interests me most since these are the subjects which I have spent most time in studying. I only hope that my interests are not too distinct from those of the readers.

## HIERARCHY OF SYMMETRY BREAKING

The unification of particle interactions can be achieved only when we realize that one interaction strength is the same as another in spite of their apparent mismatch. The electromagnetic interaction, for example, is governed by the fine structure constant $\alpha$ which takes the value 1/137. On the other hand the weak interaction is usually described by the Fermi constant $G_F$ with the value $G_F = 10^{-5} \times m_p^{-2}$, where $m_p$ stands for the proton mass. By requiring that we use the same interaction strength as in electromagnetism, i.e. $G_F/\sqrt{2} = m_W^{-2} \times e^2$, we are led to a new mass scale $m_W$:

$$m_W = \sqrt{(10^5 \times 4\pi\alpha)} m_p. \qquad (1)$$

This mass scale corresponds to the vector boson mass which mediates the weak interaction. In the standard model of Glashow, Salam and Weinberg the relation (1) is slightly modified to

$$m_W = \sqrt{(10^5 \times 4\pi\alpha)} m_p (2\sqrt[4]{2} \sin\theta_W)^{-1}, \qquad (1)'$$

where $\theta_w$ stands for the weak (Weinberg) angle.

We encounter new situation in trying to unify the strong interaction with electromagnetism. The Q.C.D. interaction strength $\alpha_s$ in the region of available accelerator energy is much larger than $\alpha=1/137$. While the interaction strength of electromagnetism has only a mild energy dependence, the Q.C.D. strength has appreciable energy dependence and we can estimate the energy where Q.C.D. coupling coincides with Q.E.D. strength. For this purpose let us write down equations which express the energy dependence of $\alpha$ and $\alpha_s$[8]:

$$1/\alpha(\mu)=(8/3)[1/\alpha_0 +(L/3\pi)N]-(L/3\pi)(11/2), \qquad (2a)$$

$$1/\alpha_s(\mu)=1/\alpha_0+(L/3\pi)N-(L/3\pi)(33/4). \qquad (2b)$$

These are obtained by solving the lowest order renormalization group equation in the standard $SU(3)_c \times SU(2) \times U(1)$ theory with N generations of fermions. Here $L=\ln(M^2/\mu^2)$. $\alpha(\mu)$ and $\alpha_s(\mu)$ coincide with each other (except for the 'leakage factor 3/8' to the neutral current) at $\mu=M$ and the common value is $\alpha_0$. If we use $\alpha_s(\mu=1GeV) \simeq 0.1$ and $\alpha(\mu=1GeV) \simeq 1/137$ as input we get $M \simeq 10^{15}$ GeV. This value of M roughly corresponds to the mass of the gauge bosons in a grand unified group which are not associated with the $SU(3)_c \times SU(2) \times U(1)$ decomposition.

Since we know that $M_w < 10^2$ GeV we are forced to have two vastly separated mass scales in a grand unified theory. Inverting the mass to the length we are going from $10^{-16}$ cm down to $10^{-29}$ cm.

The clue to understand this situation is provided by the statistical theory of atoms and molecules. Here the length parameter is of the order of $10^{-8}$ cm. Correlation lengths, however, can be large on a macroscopic scale depending on how close we are to the critical point. Figure 1 illustrates this situation. The phase I corresponds to the symmetric phase where the order parameter $\phi$ vanishes and phase II corresponds to that of broken symmetry where $\phi \neq 0$. T is the temperature and $T_c$ is the critical temperature. When the state A approaches the critical point from below $<\phi>$ becomes smaller and smaller and eventually vanishes at $T=T_c$ if the transition to the other phase is of the second order. $\phi$ will stay finite even at $T=T_c$ if the transition is of first order. Let us consider the case of two

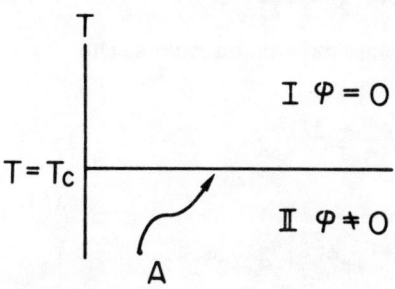

Fig. 1. Phase diagram with two phases

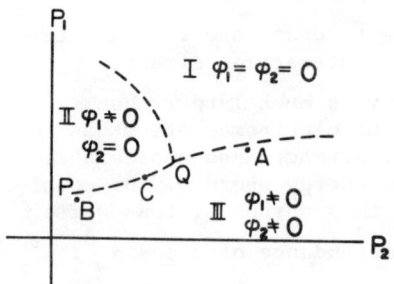

Fig. 2. Three phases
I. G
II. $SU_c(3) \times SU(2) \times U(1)$
III. $SU_c(3) \times U(1)$

phases as is illustrated in Fig. 2. Here $p_1$ and $p_2$ are appropriate parameters in the theory. They can be pressure versus temperature etc.. Three phases are characterized by two order parameters $\phi_1$ and $\phi_2$. The phase I is the symmetric one with vanishing order parameters, phase II the intermediate one and phase III the most ordered one with both $\phi_1$ and $\phi_2$ non-vanishing. In grand unified theory phase I corresponds to the state with unbroken unified group G, phase II corresponds to the state with $SU(3)_c \times SU(2) \times U(1)$ symmetry and phase III describes the real world where only $SU(3)_c \times U(1)_{E.M.}$ remains unbroken. Suppose we are at point A in this figure. Both $\phi_1$ and $\phi_2$ will be non-vanishing but they will both be small since A is close to the phase I where both $\phi_1$ and $\phi_2$ vanish. In this case the ratio $\phi_1/\phi_2$ can be anything. On the other hand if we are at point B, $\phi_2$ will be small but $\phi_1$ will in general be much larger and so the ratio $\phi_2/\phi_1$ can be small. Suppose we are at point C which is precisely on the critical line PQ. $\phi_2/\phi_1$ is identically zero if the transition is second order and it is non-zero but can be very small in case of first order transition. I will demonstrate that this situation is realized in the following simple model:

$$L = \tfrac{1}{2}\partial_\mu \phi_1 \partial^\mu \phi_1 + \tfrac{1}{2}\partial_\mu \phi_2 \partial^\mu \phi_2 - \tfrac{1}{2}m_1^2 \phi_1^2 - \tfrac{1}{2}m_2^2 \phi_2^2$$
$$-(\lambda_1/4)\phi_1^4 - (\lambda_2/4)\phi_2^4 - (\lambda_{12}/2)\phi_1^2 \phi_2^2 . \quad (3)$$

The effective potential in the one loop approximation takes the following form:

$$V(\phi_1,\phi_2) = \tfrac{1}{2}m_1^2 \phi_1^2 + \tfrac{1}{2}m_2^2 \phi_2^2 + \tfrac{1}{4}\lambda_1 \phi_1^4 + \tfrac{1}{4}\lambda_2 \phi_2^4$$
$$+ \frac{\lambda_{12}}{2}\phi_1^2 \phi_2^2 + \frac{1}{64\pi^2} \mathrm{Tr} W^2 \ln W , \quad (4)$$

where $W = \{\partial^2 V_{tree}/\partial \phi_i \partial \phi_j\} = \begin{pmatrix} m_1^2 + 3\lambda_1 \phi_1^2 + \lambda_{12}\phi_2^2 & 2\lambda_{12}\phi_1\phi_2 \\ 2\lambda_{12}\phi_1\phi_2 & m_2^2 + 3\lambda_2 \phi_2^2 + \lambda_{12}\phi_1^2 \end{pmatrix}$.
(5)

The renormalization has been performed in such a way that the

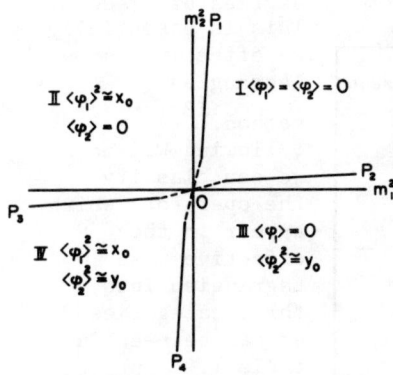

Fig. 3. Phase diagram for the system described by equation (4) with $\lambda_{12} > 0$

potential takes the above form. An arbitrary scale is introduced in this process and we set this to be 1. Detailed analysis of this potential is presented in appendix A. We have four phases in this case as is shown in Fig. 3. The line $P_2 P_3$ except near the origin is given by

$$m_2'^2 = m_2^2 + \lambda_{12} x_0 = (\frac{9\lambda_2}{64\pi^2})\exp\{-1 - (\frac{16\pi^2}{9\lambda_2})\}, \quad (6a)$$

and likewise the line $P_1 P_4$ by

$$m_1^2 + \lambda_{12} y_0 = (\frac{9\lambda_1}{64\pi^2})\exp\{-1 - (\frac{16\pi^2}{9\lambda_1})\}, \quad (6b)$$

where

$$x_0 = (-\lambda_2 m_1^2 + \lambda_{12} m_2^2)/(\lambda_1 \lambda_2 - \lambda_{12}^2), \quad (7a)$$

and

$$y_0 = (-\lambda_1 m_2^2 + \lambda_{12} m_1^2)/(\lambda_1 \lambda_2 - \lambda_{12}^2). \quad (7b)$$

Due to radiative corrections all the phase transitions are of the first order.* If we are at point A, for example, which is exactly on the critical line $OP_3$ we have:

$$<\phi_1>^2 \simeq x_0, \quad (8a)$$

and

$$<\phi_2>^2 = (\frac{1}{3\lambda_2})\exp\{-1 - (\frac{16\pi^2}{9\lambda_2})\}. \quad (8b)$$

If we take, for example, $x_0 \simeq O(1)$ and $\lambda_2$ to be much less than 1 we can easily get astronomical values for $<\phi_1>/<\phi_2>$.

The assumption that the system should be on the critical line is of course a dynamical one and therefore needs to be justified. We have no deeper understanding of this problem yet.

Accepting the idea that our system is right on the critical point we notice that we can either freeze (Kadanoff)[10] or more accurately integrate away (Wilson)[11] all the high frequency (high mass) components (compared to $<\phi_2>$) to obtain a system of fewer

---

* If we include the contributions from higher loops or from the renormalization effect, there is a chance that we end up with the second order transition. Gauge boson loop contribution may take it back to first order again.[9]

TABLE 1  Operators in $L_{eff}$ ($M=\langle\phi_i\rangle$)

| relevant operators | marginal operators | irrelevant operators |
|---|---|---|
| Absent | D = 4 or less<br><br>~ log M | D = 5, 6, 7, .........<br><br>~ $M^{4-D}$ |
| | Coupling constants<br>fermion masses<br>( except for $\nu$ )<br>$\theta_C$, KM matrix | D = 6  $p \to e^+ + \gamma$<br>$\nu_e + p \to \mu^+ + n$<br><br>D = 9  $n \leftrightarrow \bar{n}$ |

degrees of freedom. This is essentially an effective Lagrangian method.[12] Following Wilson we can classify the operators which appear in the effective Lagrangian into three categories as can be seen in table 1.* In this table irrelevant operators decrease as M tends to infinity with negative power of M. Marginal operators depend logarithmically on M. In later sections we discuss some of the marginal operators (fermion masses and CP-violating phase) and some typical irrelevnat operators (proton decay).

The analysis of hierarchy problem presented here is similar to but not quite identical to that of Gildener and Weinberg.[13] Their assumption is that there should be no explicit mass scale in the effective Lagrangian which one obtains after integrating out the fields corresponding to heavy particles ($m'^2_2=0$ in equation (6a)). In other words their assumption says that the system must be very close to the critical point but not necessarily be right on it. Quite independently of whichever dynamical assumption one may take many authors[14] investigated the possibility of an inherent bound on the ratio $\langle\phi_2\rangle/\langle\phi_1\rangle$. In our words the question is whether the strength of the first order transition remains as small as we want in the exact theory as has been the case in the one loop approximation. We can conclude that we have a physical problem to justify the dynamical assumption of the system being at the critical point and a mathematical problem of justifying the validity of the loop expansion.

## GRAND UNIFIED MODELS

There exist many proposed candidates for the grand unified gauge group. They include $SU(5)$[15], $SU(6)$[16], $SU(7)$[17], $SU(8)$[18], $SU(N)$[19] ($N \geq 9$); $SO(10)$[20]; $E(6)$[21], $E(7)$[22] etc., etc.. (It is somewhat puzzling why nobody ever came up with a unified model with a simplectic group.) Each model is interesting in its own way but here let me pick up and discuss one example from each category of simple Lie groups mentioned above. Namely I choose one from unitary groups, one from

---
* Acutually the meaning of 'relevant' or 'irrelevant' is reversed in our case from that of Wilson.

orthogonal groups and one from exceptional groups. They are SU(5), SO(10) and E(6), satisfying

$$SU(5) \subset SO(10) \subset E(6). \qquad (9)$$

All three groups provide us a consistent renormalizable gauge theory which can eventually be broken down to $SU(3)_c \times U(1)_{e.m.}$ due to the Higgs mechanism. The absence of anomalies can be shown in all cases, although the SU(5) case may look somewhat artificial[23]. There is even a discussion [24] to single out SU(5) or SO(10) as the leading candidates among others based on the assumption called 'naturality'[25] even though this assumption tends to give us such an unnatural relation as $m_e/m_\mu = m_d/m_s$.[24]

Once we start asking the origin of these groups* we immediately notice that the physical implication of each group differs from the rest in a fundamental manner. The unitary group is the largest invariance group of $\Sigma_i \phi_i^* \phi_i$ where $\phi_i$ is a complex number. $\phi_i$ can be interpreted as a quantum mechanical state corresponding to a particle. On the other hand the orthogonal group is the largest invariance group of $x_i x_i$ where $x_i$ is a real quantity. $x_i$ cannot be interpreted as a quantum mechanical particle state but rather as a component of space-time. E(6) provides us with the most interesting possibility. It is the invariance group of $Tr \bar{J}^{*t} J$, where $\bar{A}^t$ denotes the complex conjugation and transposition of matrix A and A* means the octonion conjugation with J given by [32]

$$J = \begin{pmatrix} \alpha & c & b^* \\ c^* & \beta & a \\ b & a^* & \gamma \end{pmatrix}, \qquad (10)$$

─────────
* We learned this lesson from the fate of SU(3) of Ikeda, Ogawa and Ohnuki[26], Gell-Mann[27] and Neeman[28]. The basic physical object which defines the group SU(3) was first introduced as a mere mathematical device to understand SU(3). They are 'quarks'[29] or 'aces'[30]. Quarks later turned out to be a physical entity but SU(3) remains only as an accidental but convenient mathematical device to understand low energy phenomena. By the same token, for example, the basic object on which the super transformation is defined, i.e. the superspace[31] should not be devoid of physical significance in order for super-symmetric theory to be a physical theory.

where $\alpha, \beta$ and $\gamma$ are ordinary (complex) numbers and a,b and c are octonions. We never encounter this kind of object in usual formulation of quantum theory nor in space-time theory. As was first noted by Jordan, von Neumann and Wigner[33] and was reintroduced by Gürsey[34] and his collaborators recently, the generalization of quantum theory to include non-associative algebra leads to the use of the octonions. A full development of field theory based on octonions still remains to be worked out. In brief the unitary group is a symmetry of fundamental particles, the orthogonal group requires a kind of internal space and the exceptional group leads to an extension of quantum theory.

Putting aside for the moment this fundamental problem of physical interpretation, let me summarize some of the properties of SU(5), SO(10) and E(6) theories based on the tables 2 to 10. We assume $SU(3)_c \times SU(2) \times U(1)$ symmetry.

(a) In table 2 we have a list of all the fermions of the first generation which belong to $\underline{27}$ of E(6). Small-cap letters correspond to the ordinary light fermions except for $(\nu^c)_L$ and capital letters correspond to heavy ($\sim 10^{14}$ GeV?) fermions. They do not have well defined B(baryon number) nor L(lepton number). We also note that $X_s$ is a quantum number[35] of SO(10) but not of SU(5).

(b) All the two body channels of light fermions $(+\nu_L^c)$ which couple to Higgs bosons $(F_L F_R)$ are listed in table 3.

(c) Inverting table 3 we get table 4 where the first entry shows $SU(3)_c \times SU(2) \times U(1)$ quantum numbers of all

TABLE 2

FERMIONS

| SO(10) | SU(5) | | $SU(3) \times SU(2) \times U(1)$ | | B | L | B−L | $X_S$ |
|---|---|---|---|---|---|---|---|---|
| 16 | $5^*$ | $d_L^c$ | $3^*$ | 1 | $\frac{2}{3}$ | $-\frac{1}{3}$ | 0 | $-\frac{1}{3}$ | $-\frac{3}{5}$ |
| | | $(\nu, e^-)_L$ | 1 | 2 | $-1$ | 0 | 1 | $-1$ | $-\frac{3}{5}$ |
| | 10 | $u_L^c$ | $3^*$ | 1 | $-\frac{4}{3}$ | $-\frac{1}{3}$ | 0 | $-\frac{1}{3}$ | $\frac{1}{5}$ |
| | | $(u, d)_L$ | 3 | 2 | $\frac{1}{3}$ | $\frac{1}{3}$ | 0 | $\frac{1}{3}$ | $\frac{1}{5}$ |
| | | $e_L^+$ | 1 | 1 | 2 | 0 | $-1$ | 1 | $\frac{1}{5}$ |
| | 1 | $\nu_L^c$ | 1 | 1 | 0 | 0 | $-1$ | 1 | 1 |
| 10 | $5^*$ | $D_L^c$ | $3^*$ | 1 | $\frac{2}{3}$ | | | $\frac{2}{3}$ | $\frac{2}{5}$ |
| | | $(N, E)_L$ | 1 | 2 | $-1$ | | | 0 | $\frac{2}{5}$ |
| | 5 | $D_L$ | 3 | 1 | $-\frac{2}{3}$ | | | $-\frac{2}{3}$ | $-\frac{2}{5}$ |
| | | $(N^c, E^c)_L$ | 1 | 2 | 1 | | | 0 | $-\frac{2}{5}$ |
| 1 | 1 | $L_L$ | 1 | 1 | 0 | 1 | 1 | 0 | 0 |

$E(6) = SO(10)(SU(5))$

$27 = 1 + 10\ (5, 5^*) + 16\ (5^*, 10, 1)$

## TABLE 3

### 2-BODY CHANNELS (HIGGS COUPLING)

| | Channels | SU(3)×SU(2) × U(1) | | | | B | L | B−L | $X_s$ |
|---|---|---|---|---|---|---|---|---|---|
| 1 | $\overline{(e_L^+)^c \cdot e_L^+}$ | 1 | 1 | 4 | | 0 | −2 | 2 | 2 |
| 2 | $e_L^+ \cdot (\nu, e)_L$ | 1 | 2 | 1 | | 0 | 0 | 0 | $-\frac{2}{5}$ |
| 3 | $e_L^+ \cdot d_L^c$ | $3^*$ | 1 | $\frac{8}{3}$ | | $-\frac{1}{3}$ | −1 | $\frac{2}{3}$ | $-\frac{2}{5}$ |
| 4 | $e_L^+ \cdot u_L^c$ | $3^*$ | 1 | $\frac{2}{3}$ | | $-\frac{1}{3}$ | −1 | $\frac{2}{3}$ | $\frac{2}{5}$ |
| 5 | $e_L^+ \cdot (u,d)_L$ | 3 | 2 | $\frac{7}{3}$ | | $\frac{1}{3}$ | −1 | $\frac{4}{3}$ | $\frac{2}{5}$ |
| 6 | $(\nu, e^-)_L \cdot (\nu, e^-)_L$ | 1 | 3 | −2 | 1 | 1 | −2 | 0 | 2 | −2 | $-\frac{6}{5}$ |
| 7 | $(\nu, e^-)_L \cdot d_L^c$ | $3^*$ | 2 | $-\frac{1}{3}$ | | $-\frac{1}{3}$ | 1 | $-\frac{4}{3}$ | $-\frac{6}{5}$ |
| 8 | $(\nu, e^-)_L \cdot u_L^c$ | $3^*$ | 2 | $-\frac{7}{3}$ | | $-\frac{1}{3}$ | 1 | $-\frac{4}{3}$ | $-\frac{2}{5}$ |
| 9 | $(\nu, e^-)_L \cdot (u,d)_L$ | 3 | 3 | $-\frac{2}{3}$ | 3 | 1 | $-\frac{2}{3}$ | $\frac{1}{3}$ | 1 | $-\frac{2}{3}$ | $-\frac{2}{5}$ |
| 10 | $d_L^c \cdot d_L^c$ | $6^*$ | 1 | $\frac{4}{3}$ | 3 | 1 | $\frac{4}{3}$ | $-\frac{2}{3}$ | 0 | $-\frac{2}{3}$ | $-\frac{6}{5}$ |
| 11 | $d_L^c \cdot u_L^c$ | $6^*$ | 1 | $-\frac{2}{3}$ | 3 | 1 | $-\frac{2}{3}$ | $-\frac{2}{3}$ | 0 | $-\frac{2}{3}$ | $-\frac{2}{5}$ |
| 12 | $d_L^c \cdot (u,d)_L$ | 8 | 2 | 1 | 1 | 2 | 1 | 0 | 0 | 0 | $-\frac{2}{5}$ |
| 13 | $u_L^c \cdot u_L^c$ | $6^*$ | 1 | $-\frac{8}{3}$ | 3 | 1 | $-\frac{8}{3}$ | $-\frac{2}{3}$ | 0 | $-\frac{2}{3}$ | $\frac{2}{5}$ |
| 14 | $u_L^c \cdot (u,d)$ | 8 | 2 | −1 | 1 | 2 | −1 | 0 | 0 | 0 | $\frac{2}{5}$ |
| 15 | $(u,d)_L \cdot (u,d)_L$ | 6 | 3 | $\frac{2}{3}$ | $3^*$ | 3 | $\frac{2}{3}$ | | | | |
| | | 6 | 1 | $\frac{2}{3}$ | $3^*$ | 1 | $\frac{2}{3}$ | $\frac{2}{3}$ | 0 | $\frac{2}{3}$ | $\frac{2}{5}$ |
| 16 | $\nu_L^c \cdot \nu_L^c$ | 1 | 1 | 0 | | 0 | −2 | 2 | 2 |
| 17 | $\nu_L^c \cdot d_L^c$ | $3^*$ | 1 | $\frac{2}{3}$ | | $-\frac{1}{3}$ | −1 | $\frac{2}{3}$ | $\frac{2}{5}$ |
| 18 | $\nu_L^c \cdot (\nu e^-)_L$ | 1 | 2 | −1 | | 0 | 0 | 0 | $\frac{2}{5}$ |
| 19 | $\nu_L^c \cdot u_L^c$ | $3^*$ | 1 | $-\frac{4}{3}$ | | $-\frac{1}{3}$ | −1 | $\frac{2}{3}$ | $\frac{6}{5}$ |
| 20 | $\nu_L^c \cdot (u,d)_L$ | 3 | 2 | $\frac{1}{3}$ | | $\frac{1}{3}$ | −1 | $\frac{4}{3}$ | $\frac{6}{5}$ |
| 21 | $\nu_L^c \cdot e_L^+$ | 1 | 1 | 2 | | 0 | −2 | 2 | $\frac{6}{5}$ |

the possible Higgs bosons which couple to light fermions. Only the underlined channels can have non-zero vaccum value since they are the only $SU(3)_c \times U(1)_{e.m.}$ invariants. Channel (1,1,0) is the only component which can contribute to the Majorana mass of right handed neutrino $(\bar{\nu}_L \nu_L^c)$. Channel (1,2,1) is the usual Weinberg-Salam doublet and (1,3,1) is the SU(2) triplet component which contributes to the I≠1/2 component of weak boson masses. Experimentally this is known to be small[36] and theoretically it is likely to be very small[37] ($\simeq <\phi(1,2,1)>^2 / <\phi(1,1,0)>$). Three channels inside the box contribute to the baryon number non-conserving processes, since they are the only ones which couple to two channels with different baryon numbers. Among the channels 11($d_L^c, u_L^c$), 15(($u,d)_L \cdot (u,d)_L$) and 13($u_L^c, u_L^c$) only 11 and SU(2) singlet component of channel 15 are allowed because of statistics. Only the channel (3,1, −2/3), therefore, can contribute to the Cabibbo allowed proton decay. All three can contribute to the Cabibbo suppressed proton decay. Needless to say all of them can contribute to the baryon asymmetry of the universe[38]. In table 4 the entry under SU(5) or SO(10) is restricted to the representations which can be made of two fermions (27*× 27 of E(6)).

  (d) Table 5 shows the usual decomposition of SU(5) into SU(3) × SU(2) × U(1).

  (e) In table 6 one finds all the Higgs bosons which are used to violate the unifying group down to SU(3) × SU(2) × U(1).

## TABLE 4
## HIGGS CHANNELS

| SU(3) x SU(2) x U(1) | | | B | L | B−L | $X_S$ | Fermion Coupling (channels in table 3) | | | SU(5) | SO(10) | E(6) |
|---|---|---|---|---|---|---|---|---|---|---|---|---|
| 1 | 1 | 0 | 0 | −2 | 2 | 2 | 16 | | | 1 | 126 | 351 |
| 1 | 1 | 2 | 0 | −2 | 2 | 6/5 | 6* | 21 | | 10 | 120, 126 | 351, 351' |
| 1 | 1 | 4 | 0 | −2 | 2 | 2 | 1 | | | 50* | 126 | 351 |
| 1 | 2 | 1 | 0 | 0 | 0 | −2/5 | 2* | 18 | 12  14 | 5, 45 | 10, 120, 126; 120, 126* | 27, 351', 351; 351', 351 |
| 1 | 3 | 2 | 0 | −2 | 2 | 6/5 | 6* | | | 15 | 126 | 351 |
| 3 | 1 | −2/3 | 1/3 | 1 | −2/3 | 2/5 | 4* | 9  17* | | 5, 45 | 10, 120, 126*; 120, 126, | 27, 351', 351; 351', 351 |
| | | | | | | | 11* | 15 | | 50 | 126 | 351 |
| 3 | 1 | 4/3 | 1/3 | 0 | −2/3 | 6/5 | 10 | | | 10 (40) | 120, 126 (144) | 351', 351 |
| 3 | 1 | −8/3 | 1/3 | 1 | −2/3 | 2/5 | 3* | | | 45* | 120, 126 | 351', 351 |
| | | | | | | | 13 | | | | | |
| 3 | 1 | 4/3 | 1/3 | 0 | −2/3 | 6/5 | 19* | | | 10 (40) | 120, 126 (144) | 351', 351 |
| 3 | 2 | −1/3 | 1/3 | −1 | 4/3 | 6/5 | 7 | 20 | | 10, 15 (40) | 120, 126; 126 | 351, 351'; 351 |
| 3 | 2 | 7/3 | 1/3 | −1 | 4/3 | 2/5 | 5 | 8* | | 45*, 50 | 120, 126; 126 | 351, 351'; 351 |
| 3 | 3 | −2/3 | 1/3 | 1 | −2/3 | 2/5 | 9* | | | 45 | 120, 126* | 351', 351 |
| | | | | | | | 15* | | | | | |
| 6 | 1 | −4/3 | 2/3 | 0 | −4/3 | 6/5 | 10* | | | 15 | 126* | 351 |
| 6 | 1 | 2/3 | 2/3 | 0 | −4/3 | 2/5 | 11* | 15 | | 45* | 120, 126 | 351, 351 |
| 6 | 1 | 8/3 | 2/3 | 0 | −4/3 | 2/5 | 13* | | | 50* | 126* | 351 |
| 6 | 3 | 2/3 | 2/3 | 0 | −4/3 | 2/5 | 15 | | | 50* | 126 | 351 |
| 8 | 2 | −1 | 0 | 0 | 0 | −2/5 | 12 | 14 | | 45, 50 | 120, 126; 126 | 351', 351; 351 |

TABLE 5
SU(3) x SU(2) x U(1) DECOMPOSITION OF SU(5)

$5 = (3, 1, -\frac{2}{3}) + (1, 2, 1)$

$10 = (3^*, 1, -\frac{4}{3}) + (3, 2, \frac{1}{3}) + (1, 1, 2)$

$45^* = (6, 1, \frac{2}{3}) + (8, 2, -1) + (1, 2, -1) + (3, 1, -\frac{8}{3}) + (3, 2, \frac{7}{3}) + (3^*, 3, \frac{2}{3}) + (3^*, 1, \frac{2}{3})$

$50^* = (6^*, 1, -\frac{8}{3}) + (8, 2, -1) + (6, 3, \frac{2}{3}) + (3^*, 1, \frac{2}{3}) + (3, 2, \frac{7}{3}) + (1, 1, 4)$

$15 = (6, 1, -\frac{4}{3}) + (3, 2, \frac{1}{3}) + (1, 3, 2)$

$40 = (8, 1, -2) + (6, 2, -\frac{1}{3}) + (3^*, 2, -\frac{1}{3}) + (3, 3, \frac{4}{3}) + (3, 1, \frac{4}{3}) + (1, 2, 3)$

$24 = (8, 1, 0) + (3^*, 2, \frac{5}{3}) + (3, 2, -\frac{5}{3}) + (1, 3, 0) + (1, 1, 0)$

$1 = (1, 1, 0)$

TABLE 6
HIGGS BOSONS WHICH DO NOT COUPLE TO LIGHT FERMIONS

Light − Heavy ( 16 × 10 of SO(10) )          Vacuum Value

| E(6) | SO(10) | SU(5) | $X_S$ | SU(3) x SU(2) x U(1) x (B−L) |
|---|---|---|---|---|
| 351, 351' | 144* | 5, 45 | $\frac{3}{5}$ | (1, 2, 1, 1) |
|  |  | 5* | $\frac{7}{5}$ | (1, 2, −1, 1) |
|  |  | 10*, 15, 40 | $-\frac{1}{5}$ | 10(1, 1, −2, −1), 15(1, 3, −2, −1) |
|  |  | 24 | −1 | (1, 3, 0, −1), (1, 1, 0, −1) |
| 27, 351' | 16 | 1 | 1 | (1, 1, 0, 1) |
|  |  | 5* | $-\frac{3}{5}$ | (1, 2, −1, −1) |
|  |  | 10 | $\frac{1}{5}$ | (1, 1, 2, 1) |
| 351, 351' | 16* | 1 | −1 | (1, 1, 0, −1) |
|  |  | 5 | $\frac{3}{5}$ | (1, 2, 1, 1) |
|  |  | 10* | $-\frac{1}{5}$ | (1, 1, −2, −1) |

Heavy − Heavy ( 10 × 10 of SO(10) )

| | | | | |
|---|---|---|---|---|
| 27, 351 | 1 | 1 | 0 | (1, 1, 0, 0) |
| 351 | 54 | 15 | $-\frac{4}{5}$ | (1, 3, 2, 0) |
|  |  | 15* | $\frac{4}{5}$ | (1, 3, −2, 0) |
|  |  | 24 | 0 | (1, 3, 0, 0), (1, 1, 0, 0) |
| 351' | 45 | 1 | 0 | (1, 1, 0, 0) |
|  |  | 10 | $-\frac{4}{5}$ | (1, 1, 2, 0) |
|  |  | 10* | $\frac{4}{5}$ | (1, 1, −2, 0) |
|  |  | 24 | 0 | (1, 3, 0, 0), (1, 1, 0, 0) |

A Higgs boson which couples to light-heavy channels of fermions with SO(10) (SU(5)) content 144*(24) is the usual one breaking down SU(5) to SU(3)×SU(2)×U(1). There exists 24 of SU(5) in the heavy-heavy Higgs bosons but one notices that all four heavy-heavy bosons with the vacuum expectation value conserve B-L. To understand the situation more clearly let us look at the table 7 where SU(4) deduction is performed. We notice that one component available in 54 of SO(10) leaves $SU(4)_c \times SU(2)_L \times SU(2)_R$ invariant and two components from 45 of SO(10) leave either $SU(4)_c \times SU(2)_L \times U(1)_{T_{3R}}$ or $SU(3) \times SU(2)_L \times SU(2)_R \times U_{B-L}$ invariant. This way of symmetry breaking has been extensively used by Marshak and Mohapatra[39] in their L-R symmetric theories and by Lazarides, Magg and Shafi[40] in connection

## TABLE 7

### SU(4) CONTENT OF SO(10) HIGGS BOSONS

$$SU(4)_c \times SU(2)_L \times SU(2)_R \qquad Y = B-L + 2I_{R3}$$

$10 = (6,1,1) + (1,2,2)$

$16 = (4,2,1) + (4^*,1,2)$

$10 \times 10 = 1 + 45 + 54$  (← Heavy-Heavy Higgs Bosons)

$45 = (15,1,1) + (1,3,1) + (1,1,3) + (6,2,2)$

$54 = (20,1,1) + (6,2,2) + (1,1,1) + (1,3,3)$

$1 = (1,1,1)$

under the condition that $SU(3)_c \times SU(2)_L \times U(1)$ is kept.

$54$ cannot violate $SU(4)_c \times SU(2)_L \times SU(2)_R$

$45$ cannot violate $SU(3)_c \times SU(2)_L \times U(1)$

$\left.\begin{array}{l}(1,1,3)\ SU(4)_c \times SU(2)_L \times U(1)_{T_3R} \\ (15,1,1)\ SU(3)_c \times SU(2)_L \times SU(2)_R \times U_{B-L}\end{array}\right\}$

B−L cannot be violated

## TABLE 8

### LIGHT FERMION 2-BODY CHANNEL

| Channels | $SU(3) \times SU(2) \times U(1)$ | | | | B | L | B−L | $X_{S'}$ |
|---|---|---|---|---|---|---|---|---|
| 1) $e_L^+ \cdot e_L^+$ | ( 1 | 1 | 0 ) | | 0 | 0 | 0 | 0 |
| 2) $(\nu,e^-)_L \cdot e_L^+$ | ( 1 | 2 | 3 ) | | 0 | −2 | 2 | $\frac{4}{5}$ |
| 3) $(d^c)_L \cdot e_L^+$ | ( 3 | 1 | $\frac{4}{3}$ ) | | $\frac{1}{3}$ | −1 | $\frac{4}{3}$ | $\frac{4}{5}$ |
| 4) $u_L^c \cdot e_L^+$ | ( 3 | 1 | $\frac{10}{3}$ ) | | $\frac{1}{3}$ | −1 | $\frac{4}{3}$ | 0 |
| 5) $(u,d)_L \cdot e_L^+$ | ( 3, 3* | 2 | $\frac{5}{3}$ ) | | −$\frac{1}{3}$ | −1 | $\frac{2}{3}$ | 0 |
| 6) $(\nu,e^-)_L \cdot (\nu,e^-)_L$ | ( 1 | 3 | 0 ) | | 0 | 0 | 0 | 0 |
| 7) $d_L^c \cdot (\nu,e^-)_L$ | ( 3 | 2 | −$\frac{5}{3}$ ) | | $\frac{1}{3}$ | 1 | −$\frac{2}{3}$ | 0 |
| 8) $u_L^c \cdot (\nu,e^-)_L$ | ( 3 | 2 | $\frac{1}{3}$ ) | | $\frac{1}{3}$ | 1 | −$\frac{2}{3}$ | −$\frac{4}{5}$ |
| 9) $(u,d)_L \cdot (\nu,e^-)_L$ | ( 3, 3* | 3 | −$\frac{4}{3}$ ) ( 3 | 1 | −$\frac{4}{3}$ ) | −$\frac{1}{3}$ | 1 | −$\frac{4}{3}$ | −$\frac{4}{5}$ |
| 10) $d_L^c \cdot d_L^c$ | ( 8 | 1 | 0 ) | | 0 | 0 | 0 | 0 |
| 11) $u_L^c \cdot d_L^c$ | ( 8 | 1 | 2 ) | | 0 | 0 | 0 | −$\frac{4}{5}$ |
| 12) $(u,d)_L \cdot d_L^c$ | ( 6, 2* | $\frac{1}{3}$ ) ( 3 | 2 | $\frac{1}{3}$ ) | −$\frac{2}{3}$ | 0 | −$\frac{2}{3}$ | $\frac{4}{5}$ |
| 13) $u_L^c \cdot u_L^c$ | ( 8 | 1 | 0 ) | | 0 | 0 | 0 | 0 |
| 14) $(u,d)_L \cdot u_L^c$ | ( 6, 2* | −$\frac{5}{3}$ ) ( 3 | 2 | −$\frac{5}{3}$ ) | −$\frac{2}{3}$ | 0 | −$\frac{2}{3}$ | 0 |
| 15) $(u,d)_L \cdot (u,d)_L$ | ( 8 | 3 | 0 ) ( 8 | 1 | 0 ) ( 1 | 3 | 0 ) ( 1 | 1 | 0 ) | 0 | 1 | 0 | 0 |

TABLE 9

GAUGE BOSONS

| Channels | B | L | B-L | $X_S$ | | Fermion Channels | | | | SU(5) | SO(10) | E(6) |
|---|---|---|---|---|---|---|---|---|---|---|---|---|
| U(1) 1 | 1 | 0 | 0 | 0 | 0 | 0 | 1 | 10 | 13 | 15 6 | 24 | 45 | 78 |
| 1 | 1 | 2 | 0 | 0 | 0 | $-\frac{4}{5}$ | 11 | | | | 10 | 45 | 78 |
| 1 | 2 | 3 | 0 | -2 | 2 | $\frac{4}{5}$ | 2 | | | | | | |
| SU(2) 1 | 3 | 0 | 0 | 0 | 0 | 0 | 6 | 15 | | | 24 | 45 | 78 |
| 3 | 1 | $\frac{4}{3}$ | $\frac{1}{3}$ | −1 | $\frac{4}{3}$ | $\frac{4}{5}$ | 3 | 9* | | | 10* | 45 | 78 |
| 3 | 1 | $\frac{10}{3}$ | $\frac{1}{3}$ | −1 | $\frac{4}{3}$ | 0 | 4 | | | | | | |
| X,Y  3 | 2 | $-\frac{5}{3}$ | $\frac{1}{3}$ | 1 | $-\frac{2}{3}$ | 0 | 5* | 7 | | | (24) | 45 | 78 |
| | | | $-\frac{2}{3}$ | 0 | $-\frac{2}{3}$ | 0 | 14 | | | | | | |
| X',Y'  3 | 2 | $\frac{1}{3}$ | $\frac{1}{3}$ | 1 | $-\frac{2}{3}$ | $-\frac{4}{5}$ | 8 | | | | (10) | 45 | 78 |
| | | | $-\frac{2}{3}$ | 0 | $-\frac{2}{3}$ | $-\frac{4}{5}$ | 12 | | | | | | |
| 3 | 3 | $\frac{4}{3}$ | $\frac{1}{3}$ | −1 | $\frac{4}{3}$ | $\frac{4}{5}$ | 9* | | | | | | |
| 6 | 2 | $-\frac{1}{3}$ | $\frac{2}{3}$ | 0 | $\frac{2}{3}$ | $\frac{4}{5}$ | 12* | | | | | | |
| 6 | 2 | $\frac{5}{3}$ | $\frac{2}{3}$ | 0 | $\frac{2}{3}$ | 0 | 14* | | | | | | |
| SU(3)$_C$ 8 | 1 | 0 | 0 | 0 | 0 | 0 | 10 | 13 | 15 | | 24 | 45 | 78 |
| 8 | 1 | 2 | 0 | 0 | 0 | $-\frac{4}{5}$ | 11 | | | | | | |
| 8 | 3 | 0 | 0 | 0 | 0 | 0 | 15 | | | | | | |

TABLE 10

GAUGE BOSONS

$$E(6): 78 = 1 \quad + \quad 45 \quad + \quad 16 \quad + \quad 16^* \leftarrow SO(10)$$

$$\parallel \qquad \parallel \qquad \parallel \qquad \parallel$$

$$1(0) \qquad 1(0) \qquad 1(1) \qquad 1(-1)$$
$$+ \qquad + \qquad +$$
$$24(0) \quad 5^*(-\tfrac{3}{5}) \quad 5(\tfrac{3}{5}) \leftarrow SU(5)(X_S)$$
$$+ \qquad + \qquad +$$
$$10(-\tfrac{4}{5}) \quad 10(\tfrac{1}{5}) \quad 10^*(-\tfrac{1}{5})$$
$$+$$
$$10^*(\tfrac{4}{5})$$

seems to be hard except perhaps through a detailed study of E(6) prediction of CP-violation.

with the monopole production in the early universe.
(f) Table 8 is vector boson version of table 3.
(g) Table 9 shows that there are only two vector bosons ((X,Y) of 24 in SU(5) and (X',Y') of 10 in SU(5)) which can contribute to the proton decay. They both belong to the 45 of SO(10)). One notices that B-L is conserved in this case as well as in case of Higgs bosons.[41]
(h) Table 10 provides all the possible gauge bosons available.
The crucial test to distinguish SU(5) from SO(10) will be in neutrino experiments. In SU(5) the neutrinos have two components and the only possible way for them to be massive is to be a Majorana particle. In SO(10) as well as in E(6) the neutrinos have four components and both Majorana and Dirac masses are possible. Distinction between SO(10) and E(6) in low energy phenomena

## GENERATION

This is a poorly understood subject in spite of large number of contributions. The understanding of the subject promises to be fruitful: quark and lepton mass spectra and the KM matrix[42] including the CP violating phase will be determined. For this purpose we have to know why and how many of the multiplets:

$$(\nu_e, e, d, u), (\nu_\mu, \mu, s, c), (\nu_\tau, \tau, b, t),\ldots \quad (11)$$

are repeated. Here let me mention three available approaches to this problem. First is the most obvious one and we assume that the multiplets in (11) constitute a gigantic multiplet of a large unification group. Some of the examples are SU(9) or SU(11)[43] out of unitary groups, O(14), SO(15) or SO(16)[44] out of orthogonal groups and E(8)[45] out of exceptional groups. There are some restrictions to these theories due to observed $SU(3)_c$ symmetry and V-A character of the weak current. The requirement of absolute $SU(3)_c$ symmetry tells us that the fermion multiplet which contains $f_L$ must also contain $f_L^c$ which belongs to the conjugate representation of $f_L$ under $SU(3)_c$. This is because chiral symmetry is either spontaneously or explicitly (by Higgs) broken and $f_L$ will look for a chiral partner to become massive. If, therefore, $f_L^c$ belongs to representation other than the conjugate representation of $f_L$, then $SU(3)_c$ will be broken which we do not want. On the other hand $f_L^c$ should not belong to the conjugate representation of $f_L$ under $SU(2)_W$ since we know that the W couples only to left handed fermions. We can have self-conjugate representations of $SU(2)_W$ in a fermion multiplet but they must correspond to heavy fermions. An example has already been provided by heavy fermions in E(6).[21] If we discard the possibility of self conjugate representations of $SU(2)_W$ there will be no need for heavy fermions. O(14), SO(15), O(16) and E(8) are not in this category since the spinor representations of these groups contain equal numbers of 16 and 16* of SO(10). But of course there is no apriori reason to restict ourselves to such theories.

The second approach is to impose a global symmetry (in most cases a product of U(1)) only to the D=4 part of the Lagrangian.[47] To avoid the appearance of massless Nambu-Goldstone bosons the symmetry is explicitly broken by the D=2 or D=3 parts of the Lagrangian. Symanzik's theorem[48] guarantees the symmetry of the D=4 part even after radiative corrections. Let us discuss the case of SO(10) in some detail.[49] Some necessary mathematical formulas are given in Appendix B. From equation (B10) of Appendix B we get a mass matrix in the case of 3 generations

$$(m_\pm)_{ij} = (g^{i,j} + ih^{i,j})(<\phi_0> \pm i<\phi_3>), \quad (i,j,=1,2,3) \quad (12)$$

where the + sign is for charge 2/3 quarks and for $\nu$'s and the − sign for charge −1/3 quarks and for leptons with charge (−1). If there is only one 10 of Higgs, we can rotate $<\phi_3>$ away and we obtain

$$m_\nu = m_u = m_d = m_e, \quad (13)$$

which is not desirable. At this point we speculate on a possible desirable mass matrix. For example, we can introduce $\phi^{(1)}$ and $\phi^{(2)}$ and assume that

$$h^{(2)ij} \equiv g^{(2)ij} \equiv 0 \text{ for } \phi^{(2)} \text{ except for } (i,j)=(2,3) \text{ or } (3,2),$$

and

$$h^{(1)ij} \equiv g^{(1)ij} \equiv 0 \text{ for } \phi^{(1)} \text{ except for } (i,j)=(1,2), (2,1) \text{ and } (3,3).$$

In this case we have

$$m_\pm = \begin{bmatrix} 0, & f^{(1)}(v_1 \pm iv_3), & 0 \\ f^{(1)}(v_1 \pm iv_3), & 0, & f^{(2)}(u_1 \pm iu_3) \\ 0, & f^{(2)}(u_1 \pm iu_3), & f^{(3)}(v_1 \pm iv_3) \end{bmatrix}, \quad (14)$$

where $f^{(1)} = h^{(1)1,2} + ig^{(1)1,2}$, $f^{(2)} = h^{(2)2,3} + ig^{(2)2,3}$, and $f^{(3)} = h^{(1)3,3} + ig^{(1)3,3}$ and $v_{0,3} = (\phi^{(1)}_{0,3})$ and $u_{0,3} = <\phi^{(2)}_{0,3}>$.

This is shown to result from following $U(1) \times U(1)$ invariant terms in the Lagrangian:

$$L_Y = f^{(1)}\bar{\psi}^{(1)}C(B\gamma_0\gamma_\mu)\psi^{(2)}\phi^{(1)}_\mu + f^{(2)}\bar{\psi}^{(2)}C(B\gamma_0\gamma_\mu)\psi^{(3)}\phi^{(2)}_\mu$$
$$+ f^{(3)}\bar{\psi}^{(3)}C(B\gamma_0\gamma_\mu)\psi^{(3)}\phi^{(1)}_\mu. \quad (15)$$

This is invariant under $\phi^i \to e^{i\theta_i}\phi^i$ together with $\psi_i \to e^{i\alpha_i}\psi_i$

where $\alpha_1 = \theta_2 - (3/2)\theta_1$, $\alpha_2 = -\theta_2 + (1/2)\theta_1$ and $\alpha_3 = (-1/2)\theta_1$.

The form (14) is due to speculations based on relations like

$$\sin\theta_c = \sqrt{(m_d/m_s)}, \quad (16)$$
$$\text{or } m_e/m_d = m_\mu/m_s = m_\tau/m_b. \quad (17)$$

(17) is not good even after renormalization effect.

There are many other versions of this kind including a scheme which gives rise to more satisfactory mass relations like[47]

$$3m_e/m_d = m_\mu/3m_s = m_\tau/m_b = 1. \tag{18}$$

A weak point of this approach is the obscure physical meaning of the symmetry $U(1) \times U(1) \times \ldots$ of only the D=4 part of Lagrangian.

A third approach is to assume horizontal descrete symmetry[50]. The advantage of this approach compared to the second one is that we can violate the symmetry spontaneously without worrying about a massless Nambu-Goldstone boson. But the symmetry so far suggested is as ad hoc as the second approach. What we hope for in this approach is to find a descrete transformation which is intimately connected to the continuous group of grand unification just as P, T or C is to the Lorentz group. The extension of SO(10) to O(10), for example, introduces 16* in addition to 16 in the spinor representation. What we want is 16 rather than 16* and we want three or more of them not just one. Obviously O(10) is not the solution of generation problem. So far we have been unable to find any plausible candidate which comes close to solving the problem.

## CP VIOLATION

There are problems of the $\theta$-vacuum, spontaneous versus explicit CP violation, hard versus soft CP violation, etc.etc.. Here I concentrate on a rather introductory discussion of the $\theta$-vacuum problem since other topics will be touched in later sections (especially in the section on baryon asymmetry in the universe) and also in other talks[51].

Let me start with a gauge invariant formulation[52] of Q.C.D. theory. In this formulation the gauge condition will not be introduced thus leaving the Gauss law as one of the first class constraints[53] which, after quantization, is interpreted as a condition on the physical states. The condition is equivalent to

$$\Omega \Psi = \Psi, \tag{19}$$

where $\Omega$ is an arbitrary gauge transformation which is homotopic to 1 and $\Psi$ is an arbitrary physical state. Homotopic groups of gauge transformations differ depending on which boundary condition we take for the gauge fields. Here we take:

$\lim_{|x| \to \infty} A_\mu(x)$ is independent of the direction (we are considering Euclidean gauge theory)*

In this case the classification group is ;

---

* Another example is the periodic boundary condition[54] which leads to the classification group $\pi$(projection space $\to$ SU(3)).

$$\pi_3(G) \equiv \pi(S_3 \to G) = \{Z\}, \tag{20}$$

where $G = SU(3)$ and $\{Z\}$ is a group of all integers. Although $\Omega_N(N \neq 0)$ does not leave the vacuum invariant it is still an invariance group of the system. We, therefore, must have $\Omega_N|0\rangle = e^{i\omega_N}|0\rangle$. Uniqueness of the vacuum together with the relation

$$\Omega_{N+N'} = \Omega_N \cdot \Omega_{N'},$$

thus gives:

$$\Omega_N|\theta\rangle = e^{i\theta N}|\theta\rangle, \tag{21}$$

where $|\theta\rangle$ stands for the vacuum state. In pure gauge theory (theory with only gluons without quarks) there is no operator which brings $|\theta\rangle$ to $|\theta'\rangle$. Thus each $|\theta\rangle$ is a ground state of inequivalent representation. On the other hand we can construct an operator explicitly which brings $|\theta\rangle$ to $|\theta'\rangle$ when we add quark fields to the theory. Let us first discuss when at least one quark has zero mass. As is well known,

$$J^5_\mu(x) = \sum_{i=1}^{n} \bar\psi_i \gamma_\mu \gamma_5 \psi_i, \tag{22}$$

is not conserved because of an anomaly[55] (n is the number of massless quark) but

$$\tilde J^5_\mu(x) = J^5_\mu(x) - n\left(\frac{4}{\pi^2}\right)\varepsilon_{\mu\nu\rho\kappa} \operatorname{Tr}(A_\nu \partial_\rho A_\kappa + \left(\frac{2}{3}\right) A_\nu A_\rho A_\kappa), \tag{23}$$

is conserved.[56] While $J^5(x)$ is gauge invariant $\tilde J^5(X)$ is not. Actually if we define,

$$\tilde Q_5 = \int d^3 x\, \tilde J^5_0(x), \tag{24}$$

we get,

$$\Omega_N \tilde Q_5 \Omega_N^{-1} = Q_5 + 2N. \tag{25}$$

(21) and (25) then yield

$$\exp\left\{-\left(\frac{i}{2}\right)\theta' \tilde Q_5\right\}|\theta\rangle = |\theta + \theta'\rangle. \tag{26}$$

All these manipulations do not stand of course without objection. If we continue this formal discussion we get $\tilde Q_5|\theta\rangle = 0$ from just the Lorentz invariance of $\tilde Q_5$ and of $|\theta\rangle$ [57] clearly in contradiction with equation (26). What is happening here is that we have a massless Goldstone particle and equation (24) is defined only when sandwitched between states which belong to different representations (different Hilbert spaces). Usually we think that this Goldstone boson either does not exist or that it is some kind of phantom[58].

Whatever it is it has nothing to do with the Goldstone boson which results from the violation of chiral invariance (related to $J_\mu^5$) which already picks up some mass from the anomaly. The vacuum depends on the chiral phase $\alpha$ in addition to $\theta$ and equation (26) can be written:

$$\exp\{(-\tfrac{i}{2})\theta'\tilde{Q}_5\}|\theta,\alpha\rangle = |\theta+\theta', \alpha-\tfrac{\theta'}{2}\rangle . \qquad (26)'$$

The phase $\alpha$ is determined as a function of $\theta$ by minimizing the effective potential as a function of an order parameter* corresponding to the spontaneous chiral breaking. This shows that we can choose the value of $\theta$ to be zero by redefining the vacuum through (26)' if we have at least one massless quark.

Let me turn to the case when there is no massless quark. Even for this case Peccei and Quinn[59] were able to construct a chiral current which plays the role of $J_\mu^5$. The Nambu-Goldstone boson which results from the spontaneous breaking of $J_\mu^{P \cdot Q}$ can pick up a mass only from the anomaly which makes it very light[60] in contradiction with experiment. Let us assume that there is no such current. At this stage we make use of the Euclidean classical solution -instanton- and write down the relevant vacuum to vacuum amplitude, approximating $A_\mu^i$ by $A_\mu^i$(instanton):[61]

$$\langle\theta|\theta\rangle = \int d\psi_i d\bar{\psi}_i \exp i [\int \{\bar{\psi}_i i\gamma^\mu D_\mu \psi_i - \bar{\psi}_i M_{ij} \psi_j\} + \theta\nu] , \qquad (27)$$

where $D_\mu$ is the covariant derivative with instanton configuration of topological quantum number $\nu$. Before discussing the calculability of $\theta$ let us rewrite (27) using the following change of variables:

$$\psi_i \to \psi_i' = \exp\{(\tfrac{i}{2})\theta'\gamma_5\}\psi_i , \qquad (28)$$

we get

$$\langle\theta|\theta\rangle = \int J d\psi_i' d\bar{\psi}_i' \exp i [\int \{\bar{\psi}_i' i\gamma^\mu D_\mu \psi_i' - \bar{\psi}_i' \{\exp(-\tfrac{i}{2}\theta'\gamma_5) M_{ij}$$

$$\times \exp(-\tfrac{i}{2}\theta'\gamma_5)\}\psi_j' + \theta\nu] . \qquad (29)$$

J can be calculated noting that the difference of the number of zero-modes of the left-handed and right-handed solutions is given by $\nu$:[62]

$$J = \exp(-i\theta' n_f \nu), \qquad (30)$$

with $n_f$ the number of quark flavors. Combining J with the last term in the action $\theta\nu$ we see that $(M_{ij}, \theta)$ is transformed to:

$$(\exp(-\tfrac{i}{2}\theta'\gamma_5) M_{ij} \exp(-\tfrac{i}{2}\theta'\gamma_5), \theta-\theta' n_f) . \qquad (31)$$

By choosing $\theta'=\theta/n_f$ we prove the equivalence of $(M_{ij},\theta)$ with $(M_{ij}(\theta), 0)$, where $M_{ij}(\theta)$ is given by the first entry of equation (31). Let us calculate the contribution of $\theta$ dependent fermion loop to the effective potential as shown in Figure 4.

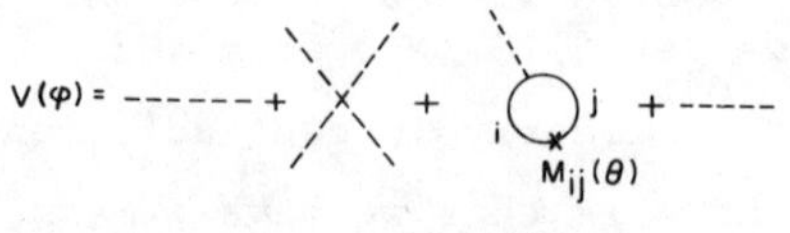

Fig. 4. Fermion loop contribution to the effective potential

---

* A non-local bilinear form in the quark fields is considered as a possibility.

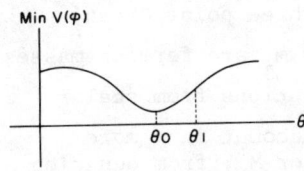

Fig. 5. θ dependence of Minimum of V(φ)

Here the Higgs bosons are explicitly introduced to avoid any ambiguity in defining the order parameter. The minimum of $V(\phi)$ will in general depend on $\theta$ as is illustrated in Figure 5. Here we must ask the crucial question:

Is $\theta_0$ the only allowed value of $\theta$ or is any other value $\theta_1$ also allowed? In the latter case <u>$\theta$ is a free parameter of the theory which can only be determined by experiment</u>. The possibility of first case is not generally accepted because of the absence of genuine physical Nambu-Goldstone boson.[58] It is extremely interesting to invent some mechanism which makes the vacuum $<\theta_1>$ unstable. There is a correction to $M_{ij}(\theta)$ from higher orders and $\theta$ will be changed if the correction violates CP. For example a contribution from higher orders of

$$L_Y = g_{ijk}\bar{\psi}_i\psi_j\phi_k + \text{h.c.}, \qquad (32)$$

with complex $g_{ijk}$ or contribution from soft CP violation have been calculated by a number of authors.[63]

## PROTON DECAY

In any grand unified model in which leptons and quarks belong to the same representation we usually end up with an unstable proton.[64] From table 9 we see that only (X,Y) and (X',Y') can mediate proton decay (Figure 6) if we neglect the contribution from Higgs bosons. There are number of papers on both kinematical[65] (branching ratios etc.) and dynamical[66] (decay rate) analyses. Let us start with the attempts at dynamical calculations. Figure 6 gives

$$|\text{Proton decay amplitude}|^2 \sim m^5/M_{X(X')}^4, \qquad (33)$$

where m is the proton mass and $M_{X(X')}$ is the mass of X(X') boson. The main issue involved in the calculation of the proton lifetime is the estimation of $M_X(M_{X'})$. The basic ingredients which make this estimation possible are the following set of renormalization group equations for the $SU_c(3) \times SU(2) \times U(1)$ gauge theory:

$$\mu(dg_i/d\mu) = \beta_i(g(\mu), m(\mu), \alpha(\mu), \mu, M_X, M_{X'}),$$

$$(\mu/m_j)(dm_j/d\mu) = \gamma_j(g(\mu), m(\mu), \alpha(\mu), \mu, M_X, M_{X'}),$$

$$\mu(d\alpha_k/d\mu) = \delta_k(g(\mu), m(\mu), \alpha(\mu), \mu, M_X, M_{X'}). \qquad (34)$$

where $g_1(\mu)$, $g_2(\mu)$, $g_3(\mu)$ are U(1), SU(2) and SU(3) coupling constants defined in terms of an appropriate three point Green's function at external momentum squared of $-\mu^2$. $m_j$ are fermion masses and $\alpha_k$ are gauge parameters of SU(k). Contributions from scalar bosons are neglected here but are taken into account in a more serious calculation. How can we determine $M_X$ or $M_{X'}$ from equation (34)? First of all we must fix the intial condition. We use

$$\alpha = e^2/4\pi = 1/137.036^*, \qquad (35)$$

where $e = g(0)g'(0)/\sqrt{(g(0)^2+g'(0)^2)}$,
with $g(\mu) = g_2(\mu)$ and $g'(\mu) = \sqrt{(3/5)}g_1(\mu)$. We can also use

$$\sin^2\theta_W = g'(\mu)^2/(g^2(\mu) + g'(\mu)^2) \simeq 0.23 \pm 0.02^{**},$$

$$\text{at } \mu \simeq m_W. \qquad (36)$$

For $g_3$ we have a value from the analysis of deep inelastic scattering:

$$\alpha_s^{-1}(\mu) = (g^2(\mu)/4\pi)^{-1} = (\frac{\beta_0}{4\pi})\ln(\mu^2/\Lambda^2) + (\frac{\beta_1}{4\pi})\ln(\ln\mu^2/\Lambda^2).^{***} \qquad (37)$$

For fermion masses we have a rather ambiguous definition

$$m_i(\mu \simeq 2m_i) = \text{current algebra mass}.$$

---

\* In the calculations of Marciano and of Goldman and Ross[66] the extrapolation to $e^2(4m_W^2)/4\pi$ is done first by taking light quarks and leptons into account in the calculation of $Z_3$:

$$\alpha^{-1}(4m_W^2) \simeq \alpha^{-1}(0) - 1/3\pi \, \Sigma e_i^2 \ln(m_W^2/m_i^2).$$

They obtain
$$\alpha^{-1}(4m_W^2) \simeq 128.8.$$

\*\* Analysis by Langacker et al. gives $0.232\pm0.012$. This value is quoted by Marciano and by Ellis et al.

\*\*\* The analysis of Bardeen et al.[67] uses the minimal subtraction scheme in a slightly modified form from the original 'tHooft[68] version. Goldman and Ross[66] note that $\Lambda^{sym}$, which defines (through (37)) $\alpha_s(\mu)$ as the three point Greens function at the symmetric point is related to $\Lambda^{\overline{m.s.}}$ of the modified minimal subtraction scheme by $\Lambda^{sym} = 3.6 \, \Lambda^{\overline{m.s.}}$. The estimation of $\Lambda^{\overline{m.s.}}$ varies appreciably depending on the process and the method of analysis.[69] (see table 11).

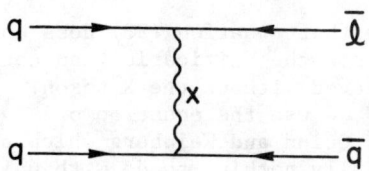

Fig. 6. Proton decay amplitude

Gauge parameters will be fixed by using the Feynman gauge in the low energy region*. With these initial values we can solve equation (34) as long as the β's are given. The solution will in general depend on the value of $M_X$ and $M_{X'}$. The key assumption (grand unification assumption) to be made is the following

$$g_1^2(\mu) = g_2^2(\mu) = g_3^2(\mu) = 1/\log(\mu/\mu_0) \quad \text{(when } \mu \to \infty\text{)} . \tag{38}$$

When $\mu$ is larger than any particle mass which goes in equation (34) we assume that equation (38) is a good approximation with an error of at most $(\frac{1}{\log\mu/m})^2$ where m is the largest of all the particle masses. One out of three equations in (38) is used to fix the value of $\mu_0$. We are, therefore, left with two equations to fix two parameters $M_X$ and $M_{X'}$. In case there is no X'boson (SU(5) model for example) or the X'boson does not contribute to equation (34) ($M_{X'} \gg M_X$) the extra equation can be used to determine the value of $\sin^2\theta_w$ in equation (36).

Lowest order calculation of β includes only one loop diagrams of light particles. $M_X$ does not enter the equation. In this case we can still estimate the value of $M_X$ in the following way.

* The effect of a gauge change is estimated to be small[70].

TABLE 11  Estimation of $\Lambda$ from various sources

| reference | group | $\Lambda$ (Mev) | $\alpha_s$ |
|---|---|---|---|
| Session C10 of this conference (P. Norton) | EMC ($\mu$A) | ~ 100 | |
| | SLAC (ep, ed) | ~ 300 | |
| Session T3 (L.E. François) | CFS (pp–$\mu\mu$) | 257 ~ 321 | 0.27 ± 0.01 |
| Session T5 (B. Wiik) | Mark J ($e^+e^- \to$ 3 jets) | 723~987 ($\theta^2=30^2$)<br>289~395 ($\theta^2=12^2$) | ← 0.23 ± 0.01 |
| | JADE ( " ) | 55~606 ($\theta^2=30^2$)<br>22~242 ($\theta^2=12^2$) | ← 0.17 ± 0.04 |
| | TASSO ( " ) | 32~723 ($\theta^2=30^2$)<br>13~289 ($\theta^2=12^2$) | ← 0.17 ± 0.02<br>± 0.03 |
| A. Gonzalez-Arroyo<br>C. Lopez and<br>F.J. Yndurain CERN<br>Preprint 2728 | | 170 ± 80 | |
| D.W. Duke &<br>R.G. Roberts<br>Rutherford<br>Preprint<br>RL-80-016 | | 790 ± 40 | |

Fig. 7. Contribution of heavy boson

We note that equation (38) does not hold since the unification can not be achieved without the X boson. Instead we use the equation of Georgi, Quinn and Weinberg which has essentially nothing to do with unification models but gives a rough extimate of $M_X$ as the mass where coupling constants coincide. Genuine calculation must take into account the effect of $M_X$ or $M_{X'}$, at least at the one-loop level which turns out to be comparable to 2-loop contributions of light particles. At this point there are at least three ways to proceed:

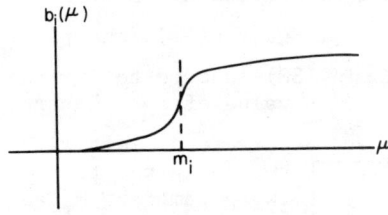

Fig. 8. $\mu$ dependence of $b_i(\mu)$

(a) We can solve equation (34) numerically.[71]

(b) Considering that $\beta$ has typically the form illustrated in figure 8, we can approximate it by a $\theta$ function.[72]

(c) We use the effective Lagrangean method combined with the minimal subtraction scheme.[73]

Let me only briefly discuss the third approach since the other approaches are rather straightforward. $L_{eff}$ is defined to be

$$\exp(iL_{eff}) = \int d\psi_h e^{iL}, \qquad (39)$$

where $\psi_h$ stands for all the heavy particles. We use dimensional regularization and then dimensionless bare coupling constants become dimensional when $D \neq 4$. If we denote those coupling constants in $L_{eff}$ by $g_b \mu^{(2-D/2)}$ and the corresponding ones in L by $g_a \mu^{(2-D/2)}$ we get (through the integration (39)) in general,

$$g_b = (Z)^{-1/2} g_a . \qquad (40)*$$

Z in our convention will have the following form:

$$Z = 1 - g_a^2 \{\lambda(\mu) + \lambda'(\mu)(1/(D-4) + \gamma_E/2 - (\log 4\pi)/2)\}, \qquad (41)$$

---

* We must include in this formula contributions from a loop diagram only a part of which corresponds to heavy particles. These contributions arise when we take the irrelevant operators (non-local) in $L_{eff}$ into account.

where $\gamma_E=0.577$. $\lambda$ and $\lambda'$ are model dependent quantities which depend on $\ln(m_h/\mu)$ with $m_h$ being a heavy particle mass**.
Let us now transform equation (40) into a relation between the renormalized coupling constants. For this purpose we use the following modified minimal subtraction scheme of Bardeen et al.[67]:

$$g_{a(b)}\mu^{2-D/2} = g_{a(b)}(\mu) - bg_{a(b)}^3(\mu) \times [1/(D-4) + \gamma_E/2 - \ln 4\pi/2], \quad (42)$$

where $g_{a(b)}$ is a renormalized constant. Substituting the equation (42) on both sides of (40) we obtain the extrapolation formula of Weinberg:[12]

$$g_b(\mu) = g_a(\mu) + (\lambda(\mu)/2) \times g_a^3(\mu). \quad (43)$$

The $L_{eff}$ can be defined in each step when we cross the threshold of a heavy particle. Each time the extrapolation formula (43) can be used. Between each threshold $g(\mu)$ satisfies the mass independent renormalization group equation. This method, therefore, does not seem to be much different from the $\theta$-function approximation in spite of its more solid formulation.

Taking into account other factors such as the quark wave function inside the proton[75] or the Q.C.D. enhancement factor[76] the final value of the proton

TABLE 12  Branching ratios in various models

| reference | branching ratio | X,Y dominant | (X,Y)(X,Y) comparable | (X,Y) dominant |
|---|---|---|---|---|
| M. Machacek | $p \to e^+ X_{ns}/p \to$ all | 83 | 54 | 70 |
| | $p \to e^+ X_s/p \to$ all | 0 | 0 | 0 |
| | $p \to \mu^+ X_{ns}/p \to$ all | 1 | 4 | 1 |
| | $p \to \mu^+ X_s/p \to$ all | 1 | 0 | <1 |
| | $p \to \bar{\nu}_e X_{ns}/p \to$ all | 13 | 41 | 27 |
| | $p \to \bar{\nu}_e X_s/p \to$ all | 0 | 0 | 0 |
| | $p \to \bar{\nu}_\mu X_{ns}/p \to$ all | <1 | <1 | <1 |
| | $p \to \bar{\nu}_\mu X_s/p \to$ all | <1 | 0 | <1 |
| A. De Rujula | $p \to e^+K/(p \to e^+K) + (p \to \mu^+K)$ | 12, 2 | 20, 4 | 100, 90 |
| H. Georgi | $p \to \mu^+\pi/(p \to e^+\pi) + (p \to \mu^+\pi)$ | 3.2, 1 | 2.5, 1.3 | 0.6, 5.2 |
| S. L. Glashow | $p \to \bar{\nu}\pi^+/(p \to \bar{\nu}\pi^+) + (p \to e^+\pi)$ | ~28, ~28 | ~50, ~50 | ~68, ~68 |
| J.F. Donoghue | $p \to e^+\pi^\circ/p \to$ 2 body | 9 | | |
| | $p \to e^+\rho^\circ/p \to$ 2 body | 21 | | |
| | $p \to e^+\eta^\circ/p \to$ 2 body | 3 | | |
| | $p \to e^+\omega^\circ/p \to$ 2 body | 56 | | |
| | $p \to \bar{\nu}\pi^+/p \to$ 2 body | 3 | | |
| | $p \to \bar{\nu}\rho^+/p \to$ 2 body | 8 | | |
| G. Kane G. Karl | $p \to e^+\omega/p \to$ 2 body | 21.4, 24.9, 25.9 | | |
| | $p \to e^+\rho^\circ/p \to$ 2 body | 2.4, 6.6, 10.5 | | |
| | $p \to e^+\pi^\circ/p \to$ 2 body | 35.7, 39.8, 38.4 | | |
| | $p \to e^+\eta/p \to$ 2 body | 6.9, 1.5, 0 | | |
| | $p \to \bar{\nu}\rho^+/p \to$ 2 body | 1.0, 2.6, 4.2 | | |
| | $p \to \mu^+K^\circ/p \to$ 2 body | 18.3, 8.4, 4.9 | | |
| | $p \to \bar{\nu}_\mu K^+/p \to$ 2 body | 0, 0.2, 0.6 | | |

** The gauge invariance greatly simplifies the computation.[74] The Feynman gauge calculation shows that contributions from irrelevant operators ($Z_1$ and $Z_2$) cancel exactly and we are left with only the contribution from $Z_3$ as illustrated in figure 7. $Z_3$ from this diagram has the form

$$\sqrt{Z_3^{AB}} = 1 + (g^2/(4\pi)^{D/2})((26-D)/12)\Gamma(2-(D/2))\text{Tr}[t_x^A t_x^B (M_x^2)^{D/2-2}]),$$

where A or B denotes the gauge boson type and $t_x^A$ or $t_x^B$ is the Clebsch-Gordan coefficient matrix for the AXX or BXX coupling. Expanding this into a series in powers of D-4 gives equation (41).

life time ranges from approximately $10^{30}$ years to $10^{32}$ years in the SU(5) model (only X boson). Let me now turn briefly to the discussion of kinematical analyses. There are several works[65] for this topic as are summarized in table 12. The work by M. Machacek[65] is based on non-relativistic SU(6) arguments and the calculations of G. Kane & G. Karl[65] also utilize SU(6), but it has three versions: static, with recoil and relativistic. The work by A. DeRújula, H. Georgi and S. L. Glashow[65] has specific SO(10) models: one with 10 dimensional Higgs only and one with 10 dimensional and 126 dimensional Higgs. J. F. Donoghue[65] makes use of the bag model wave functions to estimate the 2-body channel branching ratios. I leave all the detailed discussion to each reference.

## BARYON NUMBER ASYMMETRY IN THE UNIVERSE

Unified theory naturally leads to the baryon number non-conservation although not necessarily to the proton instability[77]. Yoshimura[78] was the first to discuss the baryon asymmetry of the universe in the context of unified theory. The fact that the following three conditions are to be satisfied for the non-vanishing baryon asymmetry to exist in the universe was known to Sakharov already in 1967[79].
(1) There exists an elementary process which violates baryon number conservation.
(2) Elementary process also violates CP conservation.
(3) The process occurs in the early universe in a non-equilibrium condition.

These conditions are more or less self-evident[*] unless, of course, the initial condition of our universe itself is asymmetric and the asymmetry does not get wiped out by the physical processes of later time. The usual method of studying the non-equilibrium processes is to make use of the Boltzmann equation or its modification in various situations[81]. In our case it reads:

$$Dn_i/Dt = 1/2 \sum_{j,k,l} [\omega_{ij,kl} n_k n_l \tilde{n}_i \tilde{n}_j - \omega_{kl,ij} n_i n_j \tilde{n}_k \tilde{n}_l], \qquad (44)$$

where $n_i$ stands for the density of particle i, $\tilde{n}_i = 1 \mp n_i$ with -sign

---

[*] The role of the unitarity condition[80] is to guarantee the existence of equilibrium solution to the Boltzmann equation when D/Dt of equation (44) is the ordinary time derivative. It also guarantees the H-theorem in this particular case.

for fermion and +sign for boson. $\omega_{A,B}$ is the transition rate of the $B \to A$ process. (44) can easily be generalized to other processes which involve more (or less) than 2 particles in the initial or final state. D/Dt is the Liouville's operator[81] which takes the form

$$D/Dt = P^\alpha(\partial/\partial X^\alpha) - \Gamma^\alpha_{\beta\gamma} P^\beta P^\gamma (\partial/\partial p^\alpha), \qquad (45)$$

where $\Gamma^\alpha_{\beta\gamma}$ is given in the isotropic and homogeneous universe of Robertson-Walker by

$$\Gamma^\alpha_{\beta\gamma} = (1/2) g^{\alpha\delta} [\partial g_{\delta\beta}/\partial x^\gamma + \partial g_{\delta\gamma}/\partial x^\beta - \partial g_{\beta\gamma}/\partial x^\delta], \qquad (46)$$

with $g_{tt} = -1$, $g_{it} = 0$ $(i=r,\theta,\phi)$, $g_{kj} = R^2(f)\tilde{g}_{kj}$,

$$\tilde{g}_{rr} = (1-kr^2)^{-1}, \quad \tilde{g}_{\theta\theta} = r^2 \text{ and } \tilde{g}_{\phi\phi} = r^2 \sin^2\theta. \qquad (47)$$

Let us first solve $Dn_i/Dt = 0$ for the particle i with mass $m_i$. It turns out that

$$n_i = f_i(R(t)|p|/T_0 R_0), \qquad (48)$$

is a solution[82] with an arbitrary function $f_i$ and arbitrary constants $T_0$ and $R_0$. With the equilibrium boundary condition at $t = t_0 (R(t_0) = R_0)$

$$f_i^{-1} = \exp\{\sqrt{(m_i^2 + |p|^2)}/KT_0\} \mp 1,$$

we get

$$n_i^{-1} = \exp[\sqrt{\{m_i^2 + (R^2(t)/R_0^2)|p|^2\}}/KT_0] \pm 1. \qquad (49)$$

The expression (49) is not a solution of equation (44) except for the case $m_i \equiv 0$. In this case both left and right hand sides of equation (44) vanish when equation (49) is substituted since this reduces to the equilibrium form:

$$n^{-1} = \exp(E/KT) \mp 1, \qquad (50)$$

with the temperature $T = (R_0/R(t))T_0$. This leads to the important consequence first noted by Toussaint et al.[83] that the processes which involve only massless particles do not lead to the baryon asymmetry of the universe. Based on this observation Weinberg[84] and Toussaint et al.[83] suggest that the asymmetry is caused by the decay of very heavy particles such as X bosons. From table 4 and table 9 we learn that Higgs bosons and gauge bosons which can contribute to baryon asymmetry are (3,1,-2/3,-2/3),(3,1,-8/3,-2/3),

(3,3,-2/3,-2/3) and (3,2,-5/3,-2/3),(3,2,1/3,-2/3) respectively. Heavy fermions in table 2 must also be taken into account if they exist. Let me denote one of the above-listed bosons and its anti-particle by x and $\bar{x}$ respectively. Partial decay rates of x into $q\bar{l}$ and $\bar{q}\bar{q}$ are $\gamma_q$ and $\gamma_{tot} - \gamma_q$ and of $\bar{x}$ into $\bar{q}l$ and $qq$ are $\gamma_{\bar{q}}$ and $\gamma_{tot} - \gamma_{\bar{q}}$ respectively. The equality of total decay rate ($\gamma_{tot}$) of x and $\bar{x}$ comes from CPT conservation. With these parameters in mind we make the following assumption to solve equation (44):

$$\int \omega n \, dp \cong \int \omega dp \int n dp, \qquad (51)$$

where momentum integration in equation (44) is appropriately split into two parts and the assumption is made that it factorizes. This will at least give a zeroth order approximation to equation (44). We also assume that $n_x = n_{\bar{x}} = n$ (after momentum integration) and take into account only the decay process in equation (44). Inverse process can be neglected if the temperature is much lower than $M_x$ due to the Boltzmann factor*. We also note that the expansion rate of the universe must be much larger than $\gamma_{tot}$ around this temperature for the consistency since otherwise the equilibrium will be reached at $1/\gamma_{tot}$ and there will be no asymmetry. Under these conditions equation (44) reduces after integration over the momentum to [85]:

$$dn/dt + \gamma_e n = -\gamma_{tot} n, \qquad (52)$$

where $n = n_x = n_{\bar{x}}$ and $\gamma_e$ is the expansion rate defined as

$$\gamma_e = 3\dot{R}/R.$$

We also get

$$dn_B/dt + \gamma_e n_B = (\gamma_q - \gamma_{\bar{q}})n, \qquad (53)$$

where $n_B$ stands for the baryon number density. Equations (52) and (53) can be solved with the final result

---

* Since $M_x$ depends on the temperature like $M_x \sim (T_0-T)^a$ the temperature must be lower than the critical value $T_0$ of the $G \rightarrow SU(3)_c \times SU(2) \times U(1)$ phase transition.

$$n_B/n_\gamma|_{present} = (n(T_0)/n_\gamma(T_0))(\gamma_q - \gamma_{\bar{q}})/\gamma_{tot}, \qquad (54)$$

where $n_\gamma(T)$ is the photon density which is used only as a normalization and the initial temperature $T_0$ must be taken to be around critical temperature of $G \to SU_c(3) \times SU(2) \times U(1)$.

The calculation of $\gamma_q$, $\gamma_{\bar{q}}$ and $\gamma_{tot}$ has been performed by several groups[86]. They all seem to agree that the dominant contribution comes from Higgs bosons rather than gauge bosons[87].

A more serious attempt to solve the Boltzmann equation (44) has been performed recently by Fry, Olive and Turner[88]. They make the following ansatz for the $n_i$ in equation (44):

$$n_i = n_i^0(t)/[\exp(|p|/T)+1], \qquad (55a)$$

for quarks and leptons and

$$n_x = x(t)f(P/T), \qquad (55b)$$

for heavy bosons. Then the equation (44) can be transformed into a set of differential equations and can easily be solved. The result of their calculation is shown in Figure 9.

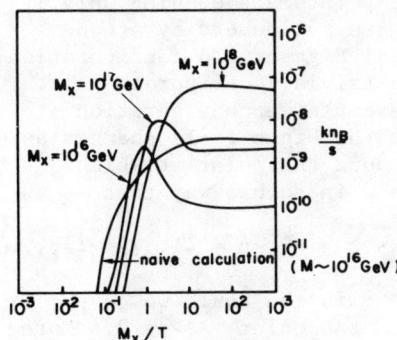

Fig. 9. Baryon number-entropy ratio (J.N.Fry, K.A.Olive and M.S.Turner, E.F.I.preprint)

$kn_B/s$ is the baryon number to entropy ratio where s is related to $n_\gamma$ by $s = (2\pi^4/45\zeta(3))n_\gamma$.[89] The decay parameters are assumed to be

$$(\gamma_q - \gamma_{\bar{q}})/\gamma_q = 10^{-5}, \qquad (56)$$

in this diagram. The decrease of $Kn_B/s$ in case of $M_x = 10^{16}$ GeV or $M_x = 10^{17}$ GeV reflects the fact that a smaller $M_x$ gives smaller Boltzmann factor which is responsible for pushing the process into equilibrium.

One of the most important remaining problems in the analysis of baryon asymmertry is its sign. The sign makes sense only when we can relate this to the CP-violating amplitude of other processes such as K decay[90]. If the CP violation is caused by a non-zero vacuum value of Higgs bosons the only possibility of relating CP violation in the low energy region to that in heavy particle decay is when the relevant Higgs bosons have large (~$10^{16}$GeV) vacuum expectation value. Otherwise the high temperature will, in general, wipe out the CP violation[91]. In SU(5) or SO(10) models the Higgs bosons with large vacuum value decouple from usual quarks. On the other hand in E(6) model with the assignment due to Barbieri and Nanopoulos[21] light fermions are mixed with heavy fermions which can couple to Higgs bosons with large vacuum value. There is, therefore, a chance that the sign of asymmetry can be related to low energy CP violating phenomena.

Hard CP violation does not disappear at high temperature but it has its own demerit (such as diverging $\theta$[63] as discussed in the section on CP violation).

## MONOPOLES[92]

At the very early stage of the universe when the grand unified group was still not violated the statistical average of all the Higgs fields $\langle\phi\rangle$ was zero. When the temperature decreased to $T \sim 10^{15}$GeV the phase transition to a state with lower symmetry took place. There were small regions or points, where $\langle\phi\rangle = 0$[93], scattered all over the universe at this stage. These were remnants from the previous stage of the universe (see Figure 10).

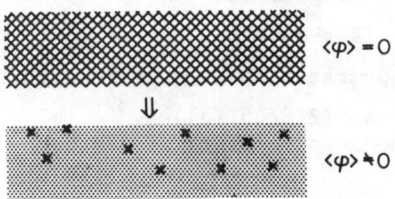

Fig. 10. Phase transition from symmetric universe to asymmetric one. x stands for monopole

These points are stable* only if they are surrounded by a non-trivial Higgs field configuration. A non-trivial configuration is, for example, a configuration where the Higgs field changes as we go around the point at which $\langle\phi\rangle = 0$ in such a way that

$$\arg \langle\phi\rangle = n\theta, \quad n = \pm 1, \pm 2, \pm 3, \ldots \quad (57)$$

where $\theta$ is the angle we cover around the point $\langle\phi\rangle = 0$. More rigorously[94], $\pi_1(H)_G$ must be a non-trivial group for a stable monopole to exist. H is the group which remains unbroken after the phase transition (it can be $SU(3)_c \times SU(2) \times U(1)$) and G is the grand unifying group. $\pi_1(H)_G$ is defined by considering only a trivial loop in G and asking its topological property in H. If G is simply connected like SU(5) and H is not simply connected like $SU_3 \times SU_2 \times U(1)$ (U(1) is not simply connected) then $\pi_1(H)_G$ is

---

* Due to a rather paradoxical nature of path integral, the rigorous proof of stability of monopole after quantum effect may not be so easy.

non-trivial. Suppose that we have this situation. The mass of the monopole has been calculated exactly[95] in the Prasad-Sommerfield limit[96] and it gives a lower bound for the general case:

$$M_m \geq (4\pi/g^2 \times g <\phi>), \qquad (58)$$

where g is the grand unified gauge coupling. This gives,

$$M_m \gtrsim 10^{16} \text{GeV} \times (1-T^2/T_0^2)^{1/2}, \qquad (59)$$

where $T_0$ is the transition temperature.

Let us now try to estimate the monopole density assuming that the monopoles and anti-monopoles should stop annihilating when the annihilation rate becomes smaller than the expansion rate H of the universe:

$$\Gamma = H(=10T^2/M_p), \qquad (60)$$

where $M_p$ is the Planck mass ( $\sim 10^{19}$ GeV) and $\Gamma$ can be written as

$$\Gamma = (1/\alpha_G^2)T^2 n_m(T), \qquad (61)$$

where $n_m(T)$ is the monopole density at temperature T. From (60) and (61) we get:

$$n_m/n_\gamma = 10^{-6}. \qquad (62)$$

This number is independent of time and corresponds to the cosmological upper bound of:

$$(m_p/M_m)(n_p/n_\gamma) = 10^{-16} \times 10^{-8} = 10^{-24}, \qquad (63)$$

where $m_p$ is the proton mass and $n_p/n_\gamma$ is the proton-$\gamma$ density ratio. Discrepancy between two numbers in equation (62) and equation (63) is enormous[97]. To invalidate the estimation of $n_m$ through equation (60) we must suppress the production rate of monopoles. There are already several proposals as a possible solution to this problem. Let me discuss some of them briefly in the following:

(1)[98] The universe may have gone through a rather strong first order phase transition when it passed from the grand unified phase to a phase of lower symmetry or from the latter to a phase of even lower symmetry. This resulted in a super cooling of the universe. In this case the bubble formation (monopole production) occurred at much later time. We do not know any reliable way to estimate the bubble density. Guth and Tye discuss this problem assuming the

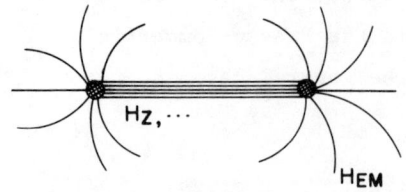

Fig. 11. Monopole-anti-monopole pair with magnetic flux of heavy bosons

existence of horizon which, however, is not without debate.*

(2) If the monopole carries 'magnetic field' corresponding to heavy gauge bosons in addition to the usual electromagnetic one, it will be attached to an anti-monopole by a string of the field due to Meisner effect(fig. 11)[99]. G. Lazarides et al.[100] argue that at least some of the monopoles are of this kind. The monopole-anti-monopole pair will eventually annihilate.

(3) Langacker & Pi[101] suggest a more drastic measure to reduce the monopole production. They suggest that the universe underwent a stage where electromagnetic charge is not conserved although in the end the symmetry is recovered again:

$$G \xrightarrow{T_1} H_1 \to \ldots \to \underset{T_n}{SU_3^c} \to \ldots \to \underset{T_{n+1}}{SU_3^c} \times U(1)_{e.m.}.$$

Between the temperatures $T_n$ and $T_{n+1}$ the only remaining symmetry is $SU_3^c$ which is simply connected. No monopole, therefore, is stable in this stage. Langacker and Pi construct an explicit model within the $SU_c(3) \times SU(2) \times U(1)$ framework. In their model there are more than one Higgs doublet: $\phi_1, \phi_2, \ldots$. The vacuum values at T=0 have the form,

$$<\phi_i> = \begin{pmatrix} v_i \\ 0 \end{pmatrix}, \qquad (64)$$

consistent with the charge conservation. At some finite temperature they get

$$<\phi_2> = \begin{pmatrix} v_2(T) \\ v'_2(T) \end{pmatrix}, \qquad (65)$$

where $<\phi>$ denotes the statistical average. The fact that $v'_2 \neq 0$ means the non-conservation of electric charge being statistically realized: States with different electric charges form a statistical mixed state.

Unfortunately there is no reliable way so far to estimate the monopole density. All the above suggestions still are of qualitative nature. In this circumstance I would like to urge

---

* Transition to a homogeneous and isotropic universe starting from a horizonless universe is under intensive study.[102]

TABLE 13   Monopole Search in Matter

[ taken from G. Barbiellini et al " Quarks and Monopoles at
LEP"; DESY Preprint, DESY 8142 ]

| Matter | Maximum Monopole Mass (GEV) | Upper limit of $\sigma$ (cm) | Method | Authors |
|---|---|---|---|---|
| Mag. outcrop Meteorite | 1<br>$10^2$ | $10^{-37}$<br>$10^{-32}$ | $\frac{dE}{dx}$ | E. Goto et al. |
| Manganese Module | 1<br>$10^2$ | $10^{-34}$<br>$10^{-34}$ | $\frac{dE}{dx}$ | R.L. Fleischer et al. |
| " | 1<br>$10^2$ | $10^{-42}$<br>$10^{-37}$ | $\frac{dE}{dx}$ | " |
| Mica Obsidian | $10^5$<br>$10^3$ | $10^{-28}$<br>$10^{-33}$ | $\frac{dE}{dx}$ | " |
| Sediment | 10<br>$10^3$ | $10^{-40}$<br>$10^{-33}$ |  | H.H. Kolm et al. |
| lunar Material | $<10^2$ | $10^{-38}$ | $\frac{dH}{dt}$ | P.H. Eberhard et al, L. Alvarez et al. |
| lunar Material | 1<br>$10^2$ | $10^{-37}$<br>$10^{-32}$ | $\frac{dH}{dt}$ | R.R. Ross et al. |

experimentalists to try to look for monopoles with the mass of $10^{16}$GeV seriously. Previous monopole searches in matter (table 13) were not aimed at such heavy monopole*. The velocity of monopoles will be too small to give a detectable ionization. The dE/dx method, therefore, may have to be abandoned in searching for a monopole of $10^{16}$GeV**. Here I would like to suggest an elementary method of monopole search making use of the fact that monopole is very heavy***. $10^{16}$GeV is approximately $10^{-8}$g and it is as heavy as a powdered grain of iron with diameter of $10^{-2}$mm. If, therefore, a momopole is attached to this grain of iron the specific mass will be increased by several factors. It will sink when put into a container filled with mercury as shown in Fig. 12****. The magnetic force which attracts the monopole on the surface of the iron is much stronger than the earth's gravity which pulls the monopole to the center of the earth. If the magnetic field of the iron is $\simeq 1$ tesla, We get

---

* $\frac{dH}{dt}$ method may be suitable for this purpose if the purpose is to find monopoles but not to collect them.
** Of course it is a different matter if one wants to look for it in cosmic ray.
*** The details will be published elsewhere (S. Pakvasa & H. Sugawara to be published.)
**** When we actually process tons of iron we may use a centrifuge.

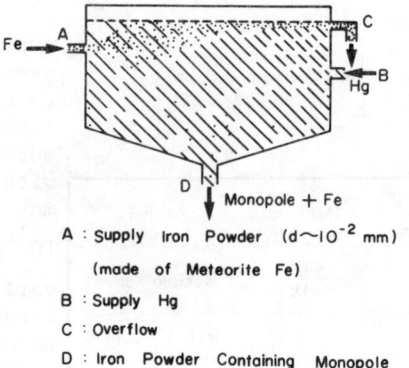

Fig. 12. Device to collect monopoles

A : Supply Iron Powder (d ~ $10^{-2}$ mm) (made of Meteorite Fe)
B : Supply Hg
C : Overflow
D : Iron Powder Containing Monopole

magnetic force/earth's gravity
$$= (1/\alpha)eH/M_m \cdot g \simeq 50, \quad (66)$$
where $M_m = 10^{16}$ GeV.

Figure 13 shows how a monopole is attached to the surface of ferromagnetic substance. We do not know at the moment how to calculate the probability of this trapping to happen.

I insist on the importance of looking for monopoles in matter rather than in cosmic rays because we have a chance, however remote, to witness the reactions in our own laboratory which otherwise took place only in a very early universe. If we find a sufficiently large number of monopoles and anti-monopoles and we put them in nuclear emulsion under the strong magnetic field, processes like,

$$M + \bar{M} \to X + 'x' \quad , Y + 'x', H_X + 'x',$$

will be observed. Since X is a color triplet it will be seen in the form of the $\bar{X}X$ or Xq. These reactions will give us genuine direct test of the grand unified theories.

Fig. 13. A monopole trapped on the surface of ferromagnetic material

## CONCLUSION

As we have seen above we go from the physics of $\simeq 10^{-13}$ cm to that of $\simeq 10^{-29}$ cm when we unify the strong, electromagnetic and weak interactions. This is more than going from the physics (atomic-physics) of $10^{-8}$ cm to physics of macroscopic size $1 \sim 100$ cm. I mentioned an analogy between the frequency and the mass in the section of hierarchy. It is important that all the frequencies from 0 to $\infty$ exist inherently in the case of a statistical theory. This is the reason why we can explain macroscopic scales in a theory with the parameter of $10^{-8}$ cm. We do not know whether we have a series of mass scales between 1GeV and $10^{16}$ GeV. We cannot prove nor disprove the existence of these scales although there are some conditions to be satisfied by these scales if they exist. For example, baryon asymmetry should not be wiped out due to the baryon number non-conserving processes with lighter (1GeV < M < $10^{16}$ GeV) particle exchange. Higher order processes should not lead to large B-L conserving proton decay. We must also guarantee that not too many of possible monopoles in the mass range of ≃ 100GeV to ≃ $10^{16}$ GeV are produced.

The unified theory is a theory of physics with the scale parameter of ≃ $10^{-30}$ cm. The existence of an intermediate scale is related to the dynamical properties of this theory, in particular its Higgs structure.

The unified theory itself has not been established yet although there are some candidates with attractive properties. In this connection I would like to emphasize the importance of studying the physical object through which the unification group is defined.

The problem of generations is in a sense distinct from that of unification. It exists at the level of SU(2) × U(1) although its solution may be closely related to the unification. The solution of this problem is most urgent from the experimental point of view: Large number of data are being accumulated which are related to quark masses, KM matrices etc.

CP violation deserves special attention due mostly to its obscure origin at the present level of our understanding.

Finally a word on the experimental checks of a unified theory. ν - mass experiment will give a partial test and so will the proton decay experiment. But as in the case of the SU(2) × U(1) model where the genuine test was(is) the discovery of W bosons and Z bosons, the real test of unified theory will be the discovery of X,Y and/or X',Y' bosons. With abundant monopoles we can create these bosons easily inside our laboratory by monopole anti-monopole annihilation process. The alternative is to construct bigger and bigger accelerators until we reach the energy of ≃ $10^{16}$ GeV. A formidable challenge to accelerator physicists. I hope God (or Buddha) is kind enough that we need not take the second alternative.

## ACKNOWLEDGMENT

I would like to thank S. Pakvasa and M. Suzuki for many useful discussions. I am also indebted to members of KEK theory group; to T. Kaneko for checking equations and for preparing some of the tables, to M. Kobayashi for many enlightening discussions and suggestions, to M. Yoshimura for numerous informative discussions and to R. J. Cashmore, M. Fukugita, T. Kinoshita and G. Rajasekaran for reading the manuscript.

## REFERENCES

1. S. Weinberg, Phys. Rev. Letters 13, 168 (1967); Abdus Salam, Nobel Symposium, Stockholm, 1968 p. 367; S. L. Glashow, J. Illiopulos, and L. Maiani, Phys. Rev. D2, 1985 (1970)
2. See for example, D. Z. Freedman, Proc. 19th Int. Conf. High Energy Phys., Tokyo 1978, Edited by S. Homma, M. Kawaguchi and H. Miyazawa, p. 535; Abdus Salam, Ibid p. 933
3. J. Ellis, M. K. Gaillard and B. Zumino, CERN Preprint, TH. 2842 (1980); E. Cremmer and B. Julia, Phys. Letters 80B, 48 (1978); Nucl. Phys. B159, 141 (1979)
4. A. D. Sakharov, Dokl, Akad, Nauk, CCCP 177, 70 (1967) (Sov. Phys. Dokl. 12, 1040 (1968); A. Zee, Phys. Rev. Letters 42, 417 (1979); L. Smolin, Nucl. Phys. B 160, 253 (1979); S. L. Adler, Phys. Rev. Letters, 44, 1567 (1980)
5. See for example, H. Terazawa, K. Akama and Y. Chikashige, Prog. Theor. Phys. 56, 1935 (1976), Phys. Rev. D15, 480 (1977)
6. W. Heisenberg, "Introduction to the Unified Theory of Elementary Particles." Interscience, London (1966)
7. L. Susskind, plenary talk on 'New Ideas', this conference.
8. H. Georgi, H. R. Quinn and S. Weinberg, Phys. Rev. Letters 33, 451 (1974)
9. S. Coleman and E. Weinberg, Phys. Rev. D7, 1888 (1973)
10. L. P. Kadanoff in "Phase Transitions and Critical Phenomena" (C. Domb and M. S. Green, ed) Academic Press, London (1976)
11. K. Wilson, Rev. Mod. Phys. 47, 773 (1975)
12. B. Ovrut and H. J. Schnitzer, a contributed paper to session B3 and B12, #0398; T. Hagiwara and N. Nakazawa, Ibid., #0024; Y. Kazama and Y. P. Yao, Univ. of Michigan Preprint UMHE 79-40 (1979), S. Weinberg, Phys. Letters, 91B, 51 (1980)
13. E. Gildener and S. Weinberg, Phys. Rev. D13, 3333 (1976); S. Weinberg, Phys. Letters 82B, 387 (1979); See also J. Ellis, M. K. Gaillard, A. Petermann and C. Sachrajda CERN Preprint, TH-2696 (1979)
14. R. N. Mohapatra and G. Senjanovic, Hadronic J. 1, 903 (1978); O. K. Kalashnikov and V. V. Klimov, Phys. Letters 80B, 75 (1978); K. T. Mohapatra and D. G. Unger, Phys. Letters, 78B 604 (1978); M. A. Namazie and W. A. Sayed, Preprint ICTP 178-7919; I. Bars and M. Serdaroglu, Yale preprint COO-3075-188; K. T. Mohantappa, M. A. Sher and D. G. Unger, Phys. Letters, 84B, 113 (1979); V. Elias and T. N. Sherry, Univ. of Toronto preprint; T. P. Cheng and L. F. Li, Preprint COO-3066-143 (1980); See also J. M. Frere, CERN preprint TH-2783 (1979); P. J. O'Donnell, Z. Physik C, P. and F. 5, 43 (1980)
15. H. L. Georgi and S. L. Glashow, Phys. Rev. Letters, 32, 438 (1974); H. J. Lipkin, contributed paper to sessions B3 and B4 #0354, H. Komatsu, contributed paper to sessions B3, and B4

#0012; M. Kaniel and C. E. Vayonakis, contributed paper to sessions B3 and B4, #0694

16. K. Inoue, A. Kakuto and Y. Nakano, Progr. Theor. Phys. 58, 630 (1977); M. Abud, F. Buccella, H. Ruegg and C. A. Savoy, Phys. Letters, 67B, 313 (1977); B. W. Lee and S. Weinberg, Phys. Rev. Letters, 38, 1237 (1977); M. Yoshimura, Progr. Theor. Phys., 58, 972 (1977)

17. Z. Ma, T. Tu, P. Xue and Z. Yue, contributed paper to session B2, #0786

18. T. L. Curtright and P. G. O. Freund, Enrico Fermi Institute preprint, EFI 79/25 (1979); N. G. Deshpande and P. D. Manheim, contributed paper to sessions B3 and B4, #0820, #0774

19. For example talk by J. Pati, session B; J. Chakrabarti, M. Popovic and R. N. Mohapatra, contributed paper to sessions B3 and B4, #0006

20. H. Georgi, Particles and Fields, ed. C. E. Carlson (AIP, New York, 1975); H. Fritzsch and P. Minkowski, Ann. Phys. 93, 193 (1975); M. S. Chanowitz, J. Ellis and M. Gaillard, Nucl. Phys. B128, 506 (1977); H. Georgi and D. V. Nanopoulos, Nucl. Phys. B155, 52 (1979); Q. Shafi, M. Sondermann and C. Wetterich, Phys. Letters 92B, 304 (1980); F. Buccella, H. Ruegg and C. A. Savoy, Universite de Geneve preprint (1980); M. Yasue, contributed paper to session B3, #0323 R. N. Mohapatra and B. Sakita, CCNY preprint, CCNY-HEP-7919; J. A. Harvey, P. Ramond and D. B. Reiss, Caltech preprint CALT-68-758; N. P. Chang and J. Perez-Mercader, CCNY preprint CCNY-HEP-78-15

21. F. Gürsey, P. Ramond and P. Sikivie, Phys. Letters 60B, 177 (1976); Y. Achiman and B. Stech, Phys. Letters 77B, 389 (1978); B. Stech, contributed paper to session B3, #0584 R. Slansky, Los Alamos Scientific Laboratory preprint LA-UR-80-591; G. L. Shaw and R. Slansky, Ibid. LA-UR-80-1001; R. Barbieri and D. V. Nanopoulos, CERN preprint, TH 2810

22. P. Ramond, Nucl. Phys. B110, 214 (1976)
23. J. Banks and H. Georgi, Phys. Rev. D14, 1159 (1976); S. Okubo, Phys. Rev. D16, 3528 (1977)
24. Third paper in reference 20
25. S. L. Glashow and S. Weinberg, Phys. Rev. D15, 3416 (1977)
26. Ikeda, S. Ogawa and Y. Ohnuki, Progr. Theor. Phys. 22, 715 (1959), 23, 1073 (1960)
27. M. Gell-Mann, Phys. Rev. 125, 1067 (1961); Caltech CTSL-20 (1961)
28. Y. Neeman, Nucl. Phys. 26, 222 (1961)
29. M. Gell-Mann, Phys. Letters 8, 214 (1964)
30. G. Zweig, CERN preprints TH401, 412 (1964) (unpublished)
31. Abudus Salam and J. Strathdee, Phys. Rev. D11, 1521 (1975)
32. A. A. Albert, Ann. Math. 35, 65 (1934)
33. P. Jordan, J. v. Neumann and E. Wigner, Ann. Math. 35, 29 (1934)

34. M. Gunaydin and F. Gürsey, Phys. Rev. D9, 3387 (1974)
35. F. Wilczek and A. Zee, Phys. Letters 88B, 311 (1979)
36. See rapporteur's talk by C. Baltay, Proc. Tokyo conference, Tokyo 1978, edited by S. Homma, M. Kawaguchi and H. Miyazawa, p. 882
37. M. Magg and Ch. Wetterich, contributed paper to sessions B3 and B4, #0071
38. See the section on the baryon asymmetry in the universe.
39. R. N. Mohapatra and R. E. Marshak, Virginia polytechnique preprint VPI-HEP-8014; R. E. Marshak, R. N. Mohapatra and Riazuddin Ibid., VPI-HEP-8017; R. N. Mohapatra and R. E. Marshak, contributed paper to B3 and B4, #015 and #0152 R. N. Mahapatra and R. E. Marshak Phys. Rev. Letters 44, 1316 (1980); R. E. Marshak and R. N. Mohapatra, Phys. Letters, 91B, 222 (1980); See also J. C. Pati and A. Salam, Phys. Rev. D10, 275 (1974); R. N. Mohapatra and J. C. Pati, Phys. Rev. D11, 566, 2559 (1975); G. Senjanovic and R. N. Mohapatra, Phys. Rev. D12, 1502 (1975); L. N. Chang and N. P. Chang, Phys. Letters, 92B, 103 (1980)
40. C. Lazarides, M. Magg and Q. Shafi, CERN preprint, TH-2856
41. F. Wilczek and A. Zee, Phys. Rev. Letters, 88B, 311 (1979); S. Weinberg, Phys. Rev. Letters, 43, 1566 (1979)
42. M. Kobayashi and T. Maskawa, Progr. Theor. Phys., 652, 49 (1973)
43. There are also SU(7) models: P. H. Frampton, Phys. Letters. 88B, 299 (1979), Z. Ma, T. Tu, P. Xue and Z. Yue, contributed paper to session B3, #0787; SU(9) models include P. H. Frampton and S. Nardi, Phys. Rev. Letters 43, 1460 (1979); P. H. Frampton, Phys. Letters 89B, 352 (1980); SU(11) model is suggested by H. Georgi, Nucl. Phys. B156 126 (1979); See also A. Davidson and D. C. Wali, contributed paper to session B3, #0315
44. H. Sato, contributed paper to sessions B2 and B3, #0464; M. Ida, Y. Kayama and T. Kitazoe, contributed paper to session B3, #0528; Z. Ma, T. Tu, P. Xue and X. Zhou, contributed paper to session B3, #0785; F. Wilczek and Z. Zee, Univ. of Penn. preprint
45. I. Bars and M. Gunaydin, Yale preprint (1980)
46. H. Georgi, reference 43.
47. H. Georgi and C. Jarlskog, Phys. Letters 86B, 297 (1979)
48. K. Symanzik, in Coral Gables Conf. II, ed. A. Perlmutter, G. J. Iverson and R. M. Williams (Gordon and Breech, New York 1970)
49. H. Georgi and D. V. Nanopoulos, Nucl. Phys. B159,16 (1979) See also P. Ramond, Caltech preprint CALT-68-770; J. A. Harvey, P. Ramond and D. B. Reiss, Caltech preprint, CALT-68-758
50. There are many works in the SU(2) × U(1) level; S. Pakvasa and H. Sugawara, Phys. Letters 73B, 61 (1978); E. Derman, Phys. Letters 78B 497 (1978); W. D. Di, Phys. Letters 85B,

463 (1979); G. Ségre, H. A. Weldon and J. Weyers, Phys. Letters 83B, 351 (1979); G. Ségre and H. A. Weldon, Phys. Rev. Letters 42, 1191 (1979); D. Wyler, Phys. Rev. D19, 330 (1979); H. Sato, Nucl. Phys. B148, 433 (1979); K. Uehara, Progr. Theor. Phys. 61, 1426 (1979); D. Grosser, Phys. Letters 83B, 355 (1979); D. Weyler, Phys. Rev. D19, 3369 (1979); S. Pakvasa and H. Sugawara, Phys. Letters 82B, 105 (1979); E. Derman and H. S. Tsao, Phys. Rev. D20, 1207 (1979); Some general results are obtained by R. Barbieri, R. Gatto and F. Strocchi, Phys. Letters 74B, 334 (1978) R. Gatto, G. Morchio and F. Strocchi, Phys. Letters 80B, 265 (1979), Phys. Letters 83B, 348 (1979); G. Sartori, Phys. Letters 82B, 255 (1979); G. Ségre and H. A. Weldon, UPR-0129T, UPR-0125T; G. Ecker and W. Konetschng; UW. Th. ph-79-17

51. For example S. Pakvasa, Rapporteur talk in session M4.
52. There is an attempt to prove the equivalence of gauge-invariant and gauge-fixing formulations in Abelian case, S. S. Chang. P. R. D17, 2611 (1978)
53. P. A. M. Dirac, Can. J. Math., 2, 129 (1950)
54. G. 'tHooft, Nucl. Phys., B153, 141 (1979)
55. S. L. Adler Phys. Rev. 177, 2426 (1969); J. S. Bell and R. Jackiw, Nuovo Cimento 60, 47 (1969); W. Bardeen, Phys. Rev. 184, 1848 (1969)
56. J. Kogut and L. Susskind, Phys. Rev. D11, 3594 (1975)
57. J. Goldstone, Abdus Salam and S. Weinberg, Phys. Rev. 127, 965 (1962)
58. Reference 56 and T. Kugo, Nucl. Phys. B155, 368 (1979)
59. R. Peccei and H. Quinn, Phys. Rev. Letters, 38, 1440 (1977)
60. S. Weinberg, Phys. Rev. Letters 40, 20 (1978); F. Wilczek, Ibid. 40, 279 (1978)
61. A. A. Belavin, A. M. Polyakov, A. S. Schwartz and Yu. S. Tyupkin, Phys. Letters., 59B, 85 (1975); G. 'tHooft, Phys. Rev. Letters 37, 8 (1976); R. Jackiw and C. Rebbi, Phys. Rev. Letters 37, 172 (1976); C. Callan, R. Dashen and D. Gross, Phys. Letters 63B, 334 (1976)
62. K. Fujikawa, Phys. Rev. Letters 42, 1195 (1979); M. Kobayashi (private communication)
63. M. A. Bég and H. S. Tsao, Phys. Rev. Letters 41, 278 (1978); H. Georgi, Had. Journal 1, 155 (1978); G. Ségre and H. Weldon, UPR-0112T (1978) ; S. Barr and P. Langacker, Phys. Rev. Letters 42, 1654 (1979); J. Ellis and M. K. Gaillard, Nucl. Phys. B150, 141 (1979); R. N. Mohapatra and G. Senjanović, Phys. Letters 79B, 283 (1978)
64. J. C. Pati and Abdus Salam, Phys. Rev. D10, 275 (1974); See also R. N. Mohapatra and J. C. Pati, Phys. Rev. D11, 566, 2558 (1975); G. Senjanović and R. N. Mohapatra, Phys. Rev.D12, 1502 (1975)
65. Reference 41; M. Machacek, Nucl. Phys. B159, 37 (1979); A. Hurlbert and F. Wilczek, Phys. Letters 92B, 95 (1980);

J. F. Donoghue, Phys. Letters 92B, 99 (1980); G. Kane and G. Karl, Univ. of Michigan preprint; R. N. Mohapatra, Phys. Rev. Letters 43, 893 (1979); R. E. Marshak and R. N. Mohapatra, Phys. Rev. Letters 893, 43 (1979); R. E. Marshak and R. N. Mohapatra, Virginia tech. preprint, VPI-HEP-8012; J. Ellis, M. K. Gaillard and D. V. Nanopoulos, CERN preprint TH 2749; H. A. Weldon and A. Zee, contributed paper to session B1, #0759; H. J. Lipkin, contributed paper to sessions B3 and B4, #517; A. De Rujula, H. Georgi and S. L. Glashow, Phys. Rev. Letters 45, 413 (1980)

66. A. J. Buras, J. Ellis, M. K. Gaillard and D. V. Nanopoulos, Nucl. Phys. B135, 66 (1978); W. J. Marciano, Phys. Rev. D20, 274 (1979); T. J. Goldman and D. A. Ross, Phys. Letters 84B, 208 (1979) and Caltech. preprint, CALT-68-759; D. A. Ross, Nucl. Phys. B157, 273 (1979); W. J. Marciano, Rockefeller Univ. preprint, COO-2232B-192, 195; A. M. Din, G. Girardi and P. Sorba, Phys. Letters 91B, 77 (1980); G. Lazarides, O. Shafi and C. Wetterich, contributed paper to sessions B3 and B4, #0363; Y. Tomozawa, contributed paper to session B4, #0337; G. Senjanovic, Invited talk at the Workshop on Weak Interactions, Virginia Polytechinic Institute; J. Ellis, M. K. Gaillard, D. V. Nanopoulos and S. Rudaz, CERN preprint, TH2833 (1980); L. Hall, Harvard preprint, HUTP-80/A204

67. W. A. Bardeen, A. J. Buras, D. W. Duke and T. Muta, Phys. Rev. D18, 3998 (1978)

68. G. 'tHooft, Nucl. Phys. B61, 455 (1973), B62, 444 (1973)

69. A. Gonzalez-Arroyo, C. Lopez and F. J. Yndurain, CERN preprint TH2728; D. W. Duke and R. G. Roberts, Rutherford Lab. preprint, RL-80-016

70. Second from last paper of reference 66.

71. K. Matsuki and N. Yamamoto (to be published)

72. T. J. Goldman and D. A. Ross, papers in reference 66

73. Papers in reference 12; L. Hall, last paper in reference 66.

74. T. Kaneko (private communication)

75. A. J. Buras, J. Ellis, M. K. Gaillard and D. V. Nanopoulos first paper in reference 66; A. M. Din, G. Girardi and P. Sorba, paper in reference 66.

76. First paper of reference 75.

77. M. Gell-Mann, P. Ramond and R. Slansky, Rev. Mod. Phys. 50, 721 (1978); M. Yoshimura, Prog. Theor. Phys. 58, 972 (1977); G. Sègre and H. A. Weldon, Phys. Rev. Letters 44, 1737 (1980)

78. M. Yoshimura, Phys. Rev. Letters, 41, 281 (1978); 42, 746 (E) (1979)

79. A. D. Sakharov, ZhETF Pis. Red. 6, 772 (1976) [JETP Letters, 6, 236 (1967)]; See also A. Yu. Ignatiev, N. Y. Krosnikov, V. A. Kuzmin and A. N. Tavkhelidze, Phys. Letters, 76B, 436 (1978)

80. A. Ahrony, in Modern Developments in Thermodynamics (Wiley,

81. J. Ehlers, in General Relativity and Cosmology, ed. by R. K. Sachs P.1-70 (Academic Press; New York, 1971)
82. J. Ehlers, P. Green and R. K. Sachs, Journ. Math. Phys. 9, 1344 (1968)
83. B. Toussaint, S. B. Treiman, F. Wilczek and A. Zee, Phys. Rev. D19, 1036 (1979)
84. S. Weinberg, Phys. Rev. Letters, 42, 850 (1979); See also S. Dimopoulos and L. Susskind, Phys. Rev. D18, 4500 (1978); J. Ellis, M. K. Gaillard and D. V. Nanopoulos, Phys. Letters 80B, 360 (1979), 82B 464(E) (1979)
85. For more detailed discussions, see, for example, E. W. Kolb and S. Wolfram, Caltech preprint, CALT-68-754
86. D. V. Nanopoulos and S. Weinberg, Phys. Rev. D20, 2484 (1979); T. Yanagida and M. Yoshimura, Nucl. Phys. B168, 534 (1980); A. Yildiz and P. Cox, Phys. Rev. D21, 906 (1980); S. Barr, G. Segrè and H. A. Weldon, Phys. Rev. D20, 2494 (1979)
87. Fermions have also been taken into account; T. Yanagida and M. Yoshimura, Phys. Rev. Letters 45, 71 (1980) and to be published.
88. J. N. Fry, K. A. Olive and M. S. Turner, Enrico Fermi Institute preprint No. 80-07
89. See for example S. Weinberg, Gravitation and Cosmology (Wiley, New York, 1972)
90. First paper in reference 86.
91. See, however, R. N. Mohapatra and G. Senjanovic, Phys. Rev. Letters 42, 1651 (1979)
92. G. 'tHooft, Nucl. Phys. B79, 276 (1974); A. M. Polyakov, JETP letters 20, 194 (1974)
93. J. Arafune, P. G. O. Freund and C. J. Goebel, J. Math. Phys. 16, 433 (1975)
94. P. Goddard, H. Nuyts and D. Olive, Nucl. Phys. B125, 1 (1977); P. Goddard and D. Olive, Rep. Progr. Phys. 41, 1357 (1978)
95. E. B. Bogomol'nyi, SJNP 24, 449 (1976); S. Coleman, S. Parke, A. Neveu and C. M. Sommerfield Phys. Rev. D15, 544 (1977)
96. M. Prasad and C. Sommerfield, Phys. Rev. Letters 35, 760 (1975)
97. J. P. Preskill, Phys. Rev. Letters 43, 1365 (1979); See also Ya. B. Zeldovich and M. Yu. Khelopov, Phys. Letters 79B, 239 (1978)
98. A. Guth and S. -H. H. Tye, Phys. Rev. Letters 44, 631 (1980); M. B. Einhorn, D. L. Stein and D. Toussaint, Univ. of Michigan preprint, UM HE 80-1
99. Y. Nambu, Nucl. Phys. B130, 505 (1977)
100. G. Lazarides and Q. Shafi, CERN preprint TH 2821; See also M. Daniel, G. Lazarides and Q. Shafi, Ibid. 2800; G. Lazarides, M. Magg and Q. Shafi, CERN preprint TH 2856
101. P. Langacker and S. Pi, Phys. Rev. Letters 45, 1 (1980)
102. C. W. Misner, Phys. Rev. Letters 22, 1071 (1969); L. Parker, Phys. Rev. 183, 1057 (1969); J. B. Hartle and B. L. Hu, Phys. Rev. D20, 1772 (1979)

## APPENDIX A: ANALYSIS OF EFFECTIVE POTENTIAL IN THE ONE LOOP APPROXIMATION

Before discussing the potential given by equation (4) of the text let us discuss the following potential where we have only one Higgs field:

$$V = \frac{1}{2} m_2^2 y + \frac{\lambda_2}{4} y^2 + \frac{1}{64\pi^2} (m_2^2 + 3\lambda_2 y)^2 \ln(m_2^2 + 3\lambda_2 y), \quad (A1)$$

where $y = \phi_2^2$. We know that $m_2^2 < 0$ corresponds to $<\phi_2> \neq 0$ and so we consider only the case $m_2^2 \geq 0$ in the following. We note that

$$V''(\phi) = \frac{\lambda_2}{2} + \frac{(3\lambda_2)^2}{64\pi^2} (2 \ln (m_2^2 + 3\lambda_2 y) + 3), \quad (A2)$$

is a monotocially increasing function of y as is illustrated in figure A1.
We have

$$y_0 = \frac{1}{3\lambda_2} [\exp(-\frac{3}{2} - \frac{16\pi^2}{9\lambda_2}) - m_2^2]. \quad (A3)$$

Solving $V'(0) = 0$ we obtain

$$m_2^2 = \exp\{-\frac{1}{2}(1 + \frac{32\pi^2}{3\lambda_2})\}, \text{ or } m_2^2 = 0, \quad (A4)$$

and $V'(y_0) = 0$ gives

$$m_2^2 = \frac{9\lambda_2}{32\pi^2} \exp(-\frac{3}{2} - \frac{16\pi^2}{9\lambda_2}). \quad (A5)$$

Taking these into account we have three distinct cases as illustrated in figure A2a to figure A4b. In figure A4a the graph ($\beta$) corresponds to the case:

$$m_2^2 > \exp(-\frac{3}{2} - \frac{16\pi^2}{9\lambda_2}). \quad (A6)$$

The critical point is given by graph ($\beta$) of figure A3b and the critical value for $m_2^2$ and the corresponding vacuum value y are obtained by solving

$$V'(y) = \frac{1}{2} m_2^2 + \frac{\lambda_2}{2} y + \frac{3\lambda_2}{64\pi^2} \zeta_2 (2\ln\zeta_2 + 1) = 0, \quad (A7a)$$

$$V(y) - V(0) = \frac{1}{2} m_2^2 y + \frac{\lambda_2}{4} y^2 + \frac{1}{64\pi^2} (\zeta_2^2 \ln\zeta_2 - m_2^4 \ln m_2^2) = 0, \quad (A7b)$$

with $\zeta_2 = m_2^2 + 3\lambda_2 y$.
As a unique solution which satisfies the constraint

$$\exp\{-\frac{1}{2}(1 + \frac{32\pi^2}{3\lambda_2})\} \leq m_2^2 \leq \frac{9\lambda_2}{32\pi^2} \exp(-\frac{3}{2} - \frac{16\pi^2}{9\lambda_2}), \quad (A8)$$

we get
$$m_2^2 \cong \frac{9\lambda_2}{64\pi^2} \exp\left(-1 - \frac{16\pi^2}{9\lambda_2}\right), \tag{A9}$$

with
$$y \cong \frac{1}{3\lambda_2} \exp\left(-1 - \frac{16\pi^2}{9\lambda_2}\right). \tag{A10}$$

The case of two Higgs fields (equation (4)) can be solved easily if the following two conditions are satisfied:

(1) $\lambda_{12} < \lambda_1$ or $\lambda_2$ and

(2) $\langle\phi_1\rangle$ is so large that its value can be approximated by its tree value: $\langle\phi_1\rangle^2 = x_0 = (\lambda_1\lambda_2 - \lambda_{12}^2)/(-\lambda_2 m_1^2 + \lambda_{12} m_2^2)$. (A11)

Under these conditions equation (4) becomes

$$V(y) = \{\tfrac{1}{2} m_1^2 x_0 + \tfrac{1}{4}\lambda_1 x_0^2 + \tfrac{1}{64\pi^2}\zeta_1^2 \ln\zeta_1\}$$
$$+ \tfrac{1}{2} m_2'^2 y + \tfrac{1}{4}\lambda_2 y^2 + \tfrac{1}{64\pi^2}\zeta_2^2 \ln\zeta_2, \tag{A12}$$

where $\zeta_1 = m_1^2 + 3\lambda_1 x_0$ and $\zeta_2 = m_2'^2 + 3\lambda_2 y$ with $m_2'^2 = m_2^2 + \lambda_{12} x_0$.

This potential is essentially the same as in equation (A1) if we replace $m_2^2$ by $m_2'^2$. We, therefore, get

$$m_2'^2 = \frac{9\lambda_2}{64\pi^2} \exp\left(-1 - \frac{16\pi^2}{9\lambda_2}\right), \tag{A13}$$

as the critical condition and

$$y = \langle\phi_2\rangle^2 = \frac{1}{3\lambda_2} \exp\left(-1 - \frac{16\pi^2}{9\lambda_2}\right), \quad \text{as the vacuum value on the}$$
critical line.

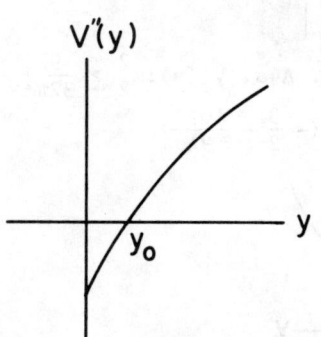

Fig. A1.  $V''(y)$ vs. $y$

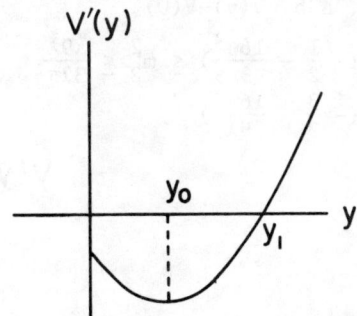

Fig. A2a.  $V'(y)$: $m_2^2 \leq \exp\left(-\tfrac{1}{2} - \tfrac{16\pi^2}{3\lambda_2}\right)$

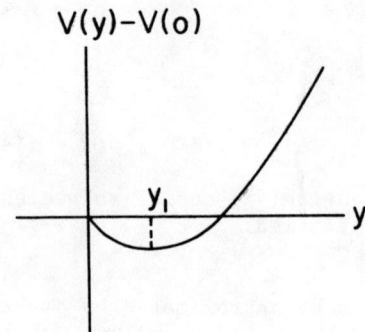

Fig. A2b. $V(y)-V(0): m_2^2 \le \exp(-\frac{1}{2} - \frac{16\pi^2}{3\lambda_2})$

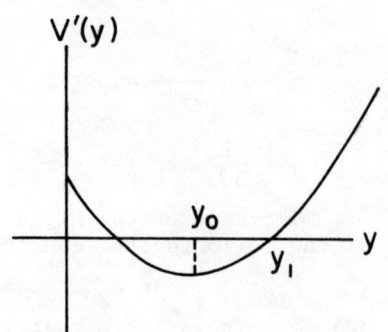

Fig. A3a. $V'(y): \exp(-\frac{1}{2} - \frac{16\pi^2}{3\lambda_2}) \le m_2^2 \le \frac{9\lambda_2}{32\pi^2} \exp(-\frac{3}{2} - \frac{16\pi^2}{9\lambda_2})$

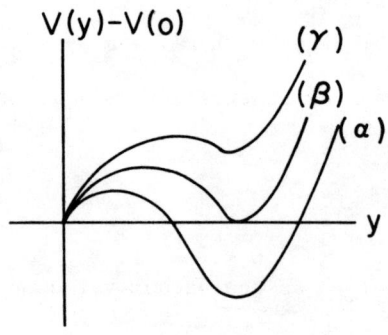

Fig. A3b. $V(y)-V(0)$
$\exp(-\frac{1}{2} - \frac{16\pi^2}{3\lambda_2}) \le m_2^2 \le \frac{9\lambda_2}{32\pi^2}$
$\exp(-\frac{3}{2} - \frac{16\pi^2}{9\lambda_2})$

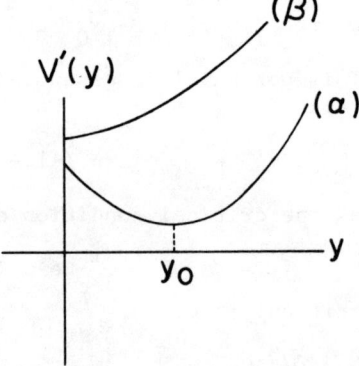

Fig. A4a. $V'(y): m_2^2 \ge \frac{9\lambda_2}{32\pi^2}$
$\exp(-\frac{3}{2} - \frac{16\pi^2}{9\lambda_2})$

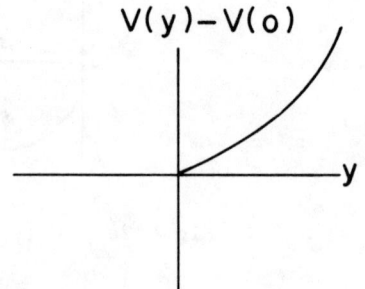

Fig. A4b. $V(y)-V(0): m_2^2 \ge \frac{9\lambda_2}{32\pi^2}$
$\exp(-\frac{3}{2} - \frac{16\pi^2}{9\lambda_2})$

## APPENDIX B: SPINOR REPRESENTATION OF SO(10)

We have 10 basic 32 by 32 $\gamma$ matrices:

$$\Gamma_\mu = \begin{pmatrix} 0 & \gamma_\mu^+ \\ \gamma_\mu & 0 \end{pmatrix} \quad \mu = 0,\ldots,9 \tag{B1}$$

where $\gamma_0 = 1$, $\gamma_j = -i\rho_3\eta_j$, $\gamma_{j+3} = -i\rho_1\sigma_j$ and $\gamma_{j+6} = -i\rho_2\tau_j$ (j=1,2,3).

$\sigma, \tau, \eta$ and $\rho$ are 2 by 2, 4 by 4, 8 by 8 and 16 by 16 Pauli matrices respectively. Generators are

$$\sigma_{\mu\nu} = \frac{1}{2i}[\Gamma_\mu, \Gamma_\nu] = -\sigma_{\nu\mu} . \tag{B2}$$

We immediately notice that this has the diagonal form

$$\sigma_{\mu\nu} = \begin{pmatrix} S_{\mu\nu} & , & 0 \\ 0 & , & BS_{\mu\nu}^T B^{-1} \end{pmatrix} , \tag{B3}$$

where $S_{\mu\nu} = \frac{1}{2i}(\gamma_\mu\gamma_\nu^+ - \gamma_\nu\gamma_\mu^+)$ and $B = (i\sigma_2)(i\tau_2)(i\eta_2)(i\rho_2)$.

S and $BS^T B^{-1}$ are 16 and 16* representations respectively. We also define $B_L = \begin{pmatrix} B_s & 0 \\ 0 & B_s \end{pmatrix}$ and $B_s = (i\sigma_2)(i\tau_2)(i\eta_2)$. Fermions are assigned to

$$\psi_{32} = \begin{pmatrix} \psi \\ B\psi^c \end{pmatrix} \text{ with } \psi = \begin{pmatrix} f \\ g \end{pmatrix} . \tag{B4}$$

Here $f^T = (\nu_L, u_L, e_L, d_L)$ for the first generation and $g = B_s f_L^c$ ($\psi^c = c\bar\psi^T$ as usual).

There are two possible ways to couple spinors to 10 dimensional Higgs:

$$\psi_{32}^T CB_L \Gamma_0 \Gamma_\mu \psi_{32} \phi_\mu = (\bar\psi CB\gamma_\mu\phi + \bar\psi\gamma_\mu B\psi^c)\phi_\mu , \tag{B5}$$

and

$$i\psi_{32}^T CB_L \Gamma_0 \Gamma_\mu \Gamma_\chi \psi_{32} \phi_\mu = i(\bar\psi CB\gamma_\mu\psi - \bar\psi\gamma_\mu B\psi^c)\phi_\mu , \tag{B6}$$

where $\Gamma_\chi = -i\prod_{\mu=0}^{9} \Gamma_\mu$ .

B5 is the scalar coupling and B6 is the pseudo-scalar coupling under the parity operation

$$P\psi_{32} = \tilde\gamma_0 \Gamma_0 \psi_{32} , \tag{B7}$$

where $\tilde{\gamma}_0$ stands for usual Dirac $\gamma$ in Lorentz space. Usual charge conjugation turns out to be

$$C\psi_{32} = B_L \psi_{32} . \tag{B8}$$

We note that under the CP transformation obtained by combining equation (B7) and equation (B8)

$$\psi_{32}^T CB_L \Gamma_0 \Gamma_\mu \psi_{32} \to \pm \psi_{32}^T CB_L \Gamma_0 \Gamma_\mu \psi_{32} \tag{B9}$$

where + is for $\mu = 1,3,4,6,8$ and − for $\mu = 0,2,5,7,9$

The most general $SU(3)_c \times U(1)_{e.m.}$ invariant mass can be derived from B5 and B6:

$$m_f = (g^{a,b} + ih^{a,b})[<\phi_0> (\overline{f_R^a} f_L^b + \overline{f_R^b} f_L^a) + i<\phi_3>(\overline{f_R^a}\eta_3 f_L^b$$
$$+ \overline{f_R^b}\eta_3 f_L^a)] + h.c , \tag{B10}$$

where g or h stands for coupling strength in equation (B5) or in equation (B6) respectively. a and b stand for generations. We see that equation (B10) gives symmetric (with respect to generations) mass matrix.

## DISCUSSION

Question: L. M. Jones
What is the status of theories where quarks and leptons are composites made out of subunits?

Answer:
Well, I know there are several proposals to describe quarks and leptons in terms of more elementary constituents, which is an economical way of understanding present physics, but at the moment there is nothing compelling about them, and quarks and leptons appear point like. What might be happening at distances of order of $10^{-33}$ cm is, of course, another matter.

# LOW ENERGY WEAK INTERACTIONS AND DECAYS

G.H. Trilling, UC-Berkeley, Speaker

V. Fitch, Princeton, Chairman

G. Thomson, Wisconsin,

L. Schachinger, Enrico Fermi Institute,

M. Kreisler, Massachusetts,

Scientific Secretaries

# LOW ENERGY WEAK INTERACTIONS AND DECAYS

G. H. Trilling
Department of Physics and Lawrence Berkeley Laboratory
University of California, Berkeley, California 94720

## I. INTRODUCTION

My task in this review is to discuss results presented to the Conference during Sessions B5 - 7 which cover various aspects of low energy weak interactions including recent work on neutrino oscillations. One topic whose subject matter might properly place it here, namely the weak decays of mesons containing b quarks, was not discussed in these sessions and will be reviewed in Professor Berkelman's paper. I shall try to summarize the results from essentially all of the material which was presented at Sessions B5 - 7.

## II. CP-INVARIANCE VIOLATION

A Yale-BNL group has been engaged in an ambitious program to measure with high precision CP-invariance violation parameters in K-decay in the hope of distinguishing between milliweak and superweak violation effects.[1] An intermediate step in their program has been an accurate determination of the muon polarization $P_n$ perpendicular to the decay plane in the decay $K_L^0 \to \pi^- + \mu^+ + \nu_\mu$. Their beautiful experimental work has already been published, and I will therefore only give the results presented at the Conference which differ just very slightly from those in the published article:

$$P_n = (1.6 \pm 5.3) \times 10^{-3}$$

$$\text{Im } \xi = 0.009 \pm 0.028$$

where $\xi$ is the usual ratio of form factors. These results are not yet precise enough to differentiate between milliweak and superweak, but the group is preparing further experiments at Brookhaven with substantially improved sensitivity.

## III. HIGH STATISTICS STUDY OF Λ BETA DECAY

A University of Massachusetts-BNL collaboration[2,3] working at the AGS has been doing a high statistics study of the decay mode,

$$\Lambda^0 \to p + e^- + \bar{\nu}_e .$$

The rate measurement, based on a sample of 10,000 beta decays, has already been published,[2] and interested readers can look up the experimental details. We give here the result,

$$\frac{\Gamma(\Lambda^0 \to pe\bar{\nu})}{\Gamma(\Lambda^0 \to p\pi^-)} = (1.318 \pm 0.024) \times 10^{-3}$$

from which, using the $\Lambda^0$ lifetime and branching ratio into $p\pi^-$, one finds,

$$\Gamma(\Lambda^0 \to pe\bar{\nu}) = (3.215 \pm 0.068) \times 10^6 \text{ s}^{-1} .$$

In a paper submitted to the Conference the group has analyzed in its 10,000-event sample the $e$-$\bar{\nu}$ angular correlation, $dN/d(\cos\theta_{e\nu})$. One can write down the weak hadronic current for the $\Lambda$ beta decay, dropping terms of order $m_e/M$, in the form

$$J_\mu^h = \bar{\psi}_p \{f_1(q^2)\gamma_\mu + \frac{i}{M}f_2(q^2)\sigma_{\mu\nu}q_\nu + g_1(q^2)\gamma_\mu\gamma_5 + \frac{i}{M}g_2(q^2)\sigma_{\mu\nu}q_\nu\gamma_5\}\psi_\Lambda$$

where $f_1$ is the vector coupling constant, $g_1$ is the axial-vector coupling constant, $f_2$ is the weak magnetism term, and $g_2$ is the second class current. In its analysis the Mass-BNL group focused on the precise determination of the ratio of axial vector to vector coupling constants $|g_1/f_1|$, assuming the theoretical value $g_2(0) = 0$. With appropriate radiative corrections, and use of the expected dipole $q^2$ dependence in the analysis, the group obtains from the angular correlation study,

$$|g_1(0)/f_1(0)| = 0.734 \pm 0.031 .$$

This result is relatively insensitive to the value of $g_2$, and is independent of any assumptions about the value of $f_2/f_1$.

From the absolute decay rate and the ratio $|g_1/f_1|$ one can calculate $|f_1(0)|$ and $|g_1(0)|$,

$$|f_1(0)| = 1.229 \pm 0.024$$
$$|g_1(0)| = 0.903 \pm 0.030 .$$

The result for $|f_1(0)|$ agrees well with the naive Cabibbo-model prediction of $\sqrt{3/2} = 1.22$.

It is worth noting that this experiment at its present state of analysis has statistics more than an order of magnitude larger than any previous experiment. Further information will be coming from analysis of the full data sample ($\sim 100,000$ events) and from polarization information available in an appropriate subclass of the event sample.

## IV. PARITY VIOLATION IN PROTON-NUCLEUS SCATTERING AT 6 GeV/c

Lockyer et al.,[4] a collaboration from six institutions, have presented final results from an experimental program carried on over several years to measure a parity-violating asymmetry in proton-nucleus scattering at 6 GeV/c. The experiment measures the asymmetry parameter,

$$A_L \equiv (\sigma_+ - \sigma_-)/(\sigma_+ + \sigma_-)$$

where $\sigma_+$ ($\sigma_-$) is the total cross section for positive (negative) helicity protons on a water target. A very rough estimate of the expected asymmetry is $A_L \sim \sqrt{\sigma_{weak}/\sigma_{strong}} \sim 10^{-6}$. Theoretical estimates by Henley and Krejs[5] suggest a much smaller value, $A_L \sim 10^{-7}$ with considerable quantitative uncertainty (because of unknown or inaccurately known parameters in the representation of both the strong and the weak amplitude in pp and pn scattering).

Needless to say, experiments to measure such extremely small asymmetries are exceedingly difficult. The experiment of Lockyer et al. used the ZGS polarized proton beam facility at Argonne; and, through a vertical magnetic deflection of the beam, produced the required longitudinal polarization. This longitudinal polarization was reversed each beam pulse. The transverse polarization prior to the rotation, plus any residual transverse polarization after rotation, can give rise to parity-allowed asymmetries which mask the real effect. The experimenters have made careful studies of such systematic effects by accentuating them to a known extent and thereby determining their impact on the measurements. They have also, through magnetic analysis, removed any contributions introduced by parity-violating hyperon decays.

The final result of the analysis is,

$$A_L = (3.38 \pm 0.65) \times 10^{-6} .$$

It is interesting to note that the raw asymmetry is $(5.90 \pm 0.58) \times 10^{-6}$, and that the removal of the various systematic effects leads to the final result which is nearly a factor of two smaller.

This result can be compared to lower energy measurements on p-p scattering, namely $A_L = (-3.2 \pm 1.1) \times 10^{-7}$ at 45 MeV (Ref. 6) and $(-1.7 \pm 0.8) \times 10^{-7}$ at 15 MeV.[7] These previous low energy measurements are in reasonable agreement with respect to both sign and magnitude with recent calculations by Desplanques et al.[8] Taken at face value, the new measurement suggests an increase of about 10 in the magnitude of $A_L$ and a change in its sign as the incident proton momentum is increased to 6 GeV/c and the target is changed from protons to a mixture of protons and neutrons. While the theoretical work of Henley and Krejs does not allow sharp quantitative predictions, it does not suggest any large increase in $|A_L|$ due either to the much higher energy or to the presence of target neutrons.

Since the ZGS has been turned off, no further work with very high energy polarized proton beams is possible in the near future, but an experiment at 1.5 GeV/c is underway at LAMPF and should shed further light on this problem.

## V. NEW RESULTS ON THE $\tau$

The SLAC-LBL Group presented a branching ratio determination for the decay mode

$$\tau \to \rho \nu \qquad (1)$$

based on statistics larger than those used for its previously published results[9] and also reported the first observation of the Cabibbo-suppressed decay mode,[10]

$$\tau \to K^*(890)\nu . \qquad (2)$$

These measurements were made with the Mark II detector at SPEAR. The details of event selection and analysis procedure for the mode (1) have been discussed in Ref. 9, and I shall just give the final result

$$B(\tau \to \rho\nu) = (21.6 \pm 1.8 \pm 3.2)\%$$

where the two quoted uncertainties are the statistical and systematic errors respectively.

The decay mode (2) was identified by observation of the sequence,

$$e^+ + e^- \to \tau^+ + \tau^- \qquad (3a)$$

$$\tau^\pm \to e^\pm, \mu^\pm + 2\nu \qquad (3b)$$

$$\tau^\mp \to K_S^0 + \pi^\mp + \nu \qquad (3c)$$

$$K_S^0 \to \pi^+ + \pi^- \ . \quad (3d)$$

The leptons in (3b) are identified in the liquid-argon calorimeters or muon identifiers, and the $K_S^0$ is reconstructed from its $\pi^+\pi^-$ decay. The $K_S^0\pi^\mp$ mass spectrum from events exhibiting the sequence (3), shown in Fig. 1, has a clear $K^*(890)$ peak. The corresponding branching ratio determined after appropriate background correction and Monte Carlo evaluation of efficiencies is

$$B(\tau \to K^*\nu) = (1.7 \pm 0.7)\% \ .$$

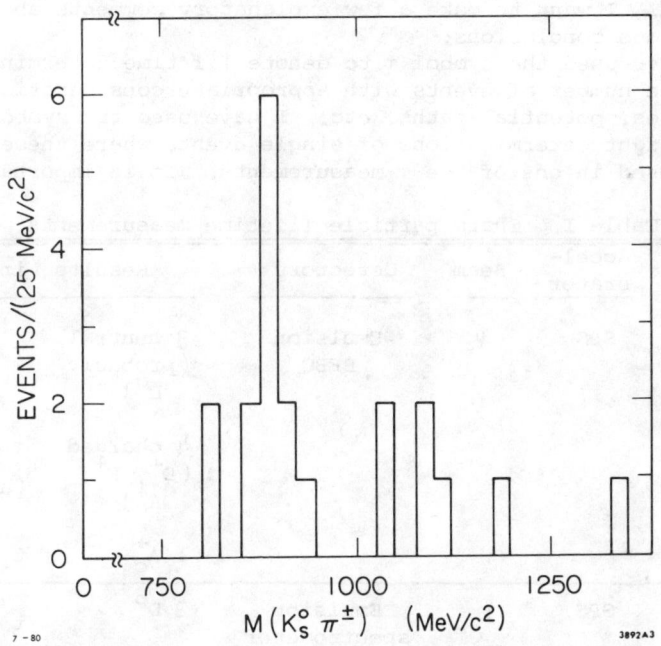

Fig. 1. $K_S\pi$ mass spectrum for $\tau \to K_S\pi\nu$ candidates.

The theoretical predictions based on the standard weak interaction model, as calculated by Tsai,[11] are

$$B(\tau \to \rho\nu) = (21.5 \pm 1.8)\% \ ,$$
$$B(\tau \to K^*\nu) = \tan^2\theta_c \, f_{ps} \, B(\tau \to \rho\nu) = (1.0 \pm 0.1)\% \ ,$$

where $\theta_c$ is the Cabibbo angle and $f_{ps} = 0.93$ is a phase space factor. Thus the experimental results continue to be in full agreement with the theoretical expectations based on the interpretation of the $\tau$ as a sequential lepton.

## VI. CHARM PARTICLE DECAYS

### A. Direct Lifetime Determinations

Over the last few years there have been several experiments designed to detect and measure the finite distances traveled by charm particles prior to their decay. The original intent of such experiments was to provide compelling evidence for the weak character of charm particle decay and rough estimates of lifetime.[12] The more recent experiments have been aimed at the identification of specific decay modes, and the quantitative determination of lifetimes of $D^+$, $D^0$, $F^+$, and $\Lambda_c^+$ particles.[13,14,15] Although the statistics are still almost as weak as the particle decays, the recent experiments have yielded some very interesting insights with respect to both lifetime and decay modes.

In Table I I have summarized the results from four emulsion-plus-downstream-detector experiments in which specific decay modes have been identified. I want to make a few explanatory comments about the Table and draw some conclusions:

(1) I have used the symbol $\tau$ to denote lifetime determinations based on a finite number of events with appropriate consideration of biases, efficiencies, potential paths, etc. I have used the symbol t to denote time-of-flight determinations of single events where these events have not been used in one of the $\tau$ measurements. It is important to note

Table I. Charm particle lifetime measurements.

| Group | Accelerator | Beam | Detectors | Results (in $10^{-13}$ s) | |
|---|---|---|---|---|---|
| WA-17[12] | SPS | $\nu$ | Emulsion BEBC | 3 neutral (probably $D^0$) | $\tau = 0.53^{+0.57}_{-0.25}$ |
| | | | | 4 charged ($D^+$, $F^+$, $\Lambda_c^+$) | $\tau = 2.5^{+2.2}_{-1.1}$ (assuming $D^+$) |
| | | | | 1 $\Lambda_c^+$ | t = 7.3 |
| WA-58[13] | SPS | $\gamma$ | Emulsion spectrometer | 2 $D^0$ | $t = \begin{array}{c}0.84\\0.45\end{array}$ |
| | | | | 1 $\Lambda_c^+$ | t = 0.57 |
| | | | | 1 $D^-$ | t = 1.0 |
| E-531[14] | FNAL | $\nu,\bar{\nu}$ | Emulsion spectrometer | 10 $D^0$ | $\tau = 1.01^{+0.43}_{-0.27}$ |
| | | | | 5 $D^\pm$ | $\tau = 10.3^{+10.5}_{-4.1}$ |
| | | | | 5 $\Lambda_c^+$ | $\tau = 1.36^{+0.84}_{-0.46}$ |
| | | | | 2 $F^\pm$ | $\tau = 2.2^{+2.8}_{-1.0}$ |
| Ammar et al.[15] | FNAL | $\nu,\bar{\nu}$ | Emulsion 15' chamber | 1 $F^+$ | t = 1.4 |

that the determination of a reliable value of τ goes well beyond the process of averaging a group of individual time-of-flight measurements. The efficiency for finding such events depends on their time-of-flight in a manner which is highly sensitive to the event search techniques used. Furthermore since such techniques obviously have to be different for neutral and charged particle decays, the ability to compare neutral and charged particle lifetimes depends on careful correction for detection biases. For these reasons, I have not attempted to incorporate data from single time-of-flight measurements into the lifetime determinations, with just one exception (see next paragraph).

(2) The three F events identified in emulsion are of great interest and I have therefore provided more detail on those events in Table II, including a best-fit lifetime for all three events. I shall discuss aspects of these events other than their lifetime in a later section.

Table II. F decays.

| Decay mode | P (GeV/c) | Mass (MeV/c$^2$) | Proper time ($10^{-13}$ s) | Group |
|---|---|---|---|---|
| $\pi^-\pi^-\pi^+\pi^0$ | 12.2 | 2026 ± 56 | 3.70 | E-531[14] |
| $K^+\pi^-\pi^+\bar{K}^0$ | 9.7 | 2089 ± 121 | 0.91 | E-531[14] |
| $\pi^+\pi^+\pi^-\pi^0$ | 2.37 | 2017 ± 25 | 1.4 | Ammar et al.[15] |

Overall $\tau = (2.0^{+1.8}_{-0.8}) \times 10^{-13}$ sec

(3) Although obviously the statistics are still very limited, it appears that the $D^{\pm}$ lifetime is substantially larger than the $D^0$ lifetime. This point is particularly clear in the E-531 data but is also suggested by the WA-17 results. The same conclusion has been drawn from other inputs which will be discussed in the next section. On the basis of rather less statistical strength, it also appears that the F and $\Lambda_c$ lifetimes may be intermediate between those of the $D^0$ and the $D^{\pm}$.

B. Semileptonic Branching Ratios

Pais and Treiman[16] pointed out several years ago that since Cabibbo-allowed semileptonic decay ($c \rightarrow s + e^+ + \nu_e$) satisfied the isospin rule $|\Delta\vec{I}| = 0$, the relation

$$\Gamma_{SL}(D^+) = \Gamma_{SL}(D^0) \tag{4}$$

for any semileptonic decay mode (or for the totality of such modes) had to hold, subject only to small phase space corrections arising from mass differences between isospin multiplet members. It follows therefore that

$$\frac{\tau(D^+)}{\tau(D^0)} = \frac{B_{SL}(D^+)}{B_{SL}(D^0)}, \tag{5}$$

where $B_{SL}$ represents the semileptonic branching ratio. The DELCO and the Mark II Groups at SPEAR have both made measurements of the total semileptonic branching ratios by means of very different analysis

procedures. The DELCO work has now been published[17] and I confine myself to quoting its main results:
DELCO:
$$\tau(D^+)/\tau(D^0) > 4.3 \quad (95\% \text{ C.L.})$$
$$B_e(D^+) = 22^{+4.4}_{-2.2}\%$$

The Mark II analysis is based on the inclusive study[18] of $D^+$ and $D^0$ decays tagged by the identification through a known exclusive channel of an accompanying $D^-$ or $\bar{D}^0$ decay, the $e^+e^-$ total energy being at the $\psi''(3770)$ where $D\bar{D}$ pair production is known to be the dominating process. The Mark II results are as follows:
$$\tau(D^+)/\tau(D^0) = 3.1^{+4.6}_{-1.4}$$
$$B_e(D^+) = 16.8 \pm 6.4\%.$$

Combining the DELCO and Mark II data to define a best estimate for $B_e(D^+)$, I obtain,
$$B_e(D^+) = 21^{+4}_{-2}\%.$$

I have not attempted to put together all the information on $\tau(D^+)/\tau(D^0)$ from both direct lifetime data and semileptonic branching ratio determinations, but it is clear that a numerical value in the range 4 - 20 would be consistent with all the experimental information.

## C. Comparison of Semileptonic Rate with Theoretical Expectations

If one combines the measured value of $\tau(D^+)$ from Table I with the above determination of $B_e(D^+)$ one obtains,
$$\Gamma_e(D) = \frac{B_e(D^+)}{\tau(D^+)} = (2 \pm 1) \times 10^{11} \text{ sec}^{-1}.$$

This result can be compared with the prediction given by Cabibbo and Maiani, and Cabibbo, Corbo and Maiani[19] based on analysis of the decay $c \to s + e^+ + \nu_e$,
$$\Gamma_{SL} = \frac{G^2}{192 \pi^3} M_c^5 g(\epsilon)[1 - \frac{2\alpha_s}{3\pi} f(\epsilon)], \quad (6)$$

where $g(\epsilon)$ is a phase space correction arising from the mass of the s quark, $\epsilon = M_s/M_c$, and the term in the brackets is a QCD correction. The charmed quark mass $M_c$ to which $\Gamma_{SL}$ is obviously very sensitive is determined to be 1.75 GeV/c² from a study of the D semileptonic decay electron spectrum, as measured by the DELCO group, and the corresponding prediction for $\Gamma_{SL}$ is $2 \times 10^{11}$ sec$^{-1}$ in excellent agreement with the experimental value.

## D. Further Study of Charm Meson Decays

The lifetime difference between $D^+$ and $D^0$ decays provides an important clue on the mechanism for the Cabibbo-allowed hadronic decay process. Perhaps the most natural mechanism is the one shown in the diagram of Fig. 2a in which the light quark bound to the decaying c quark acts in a spectator role. Similar diagrams can be constructed

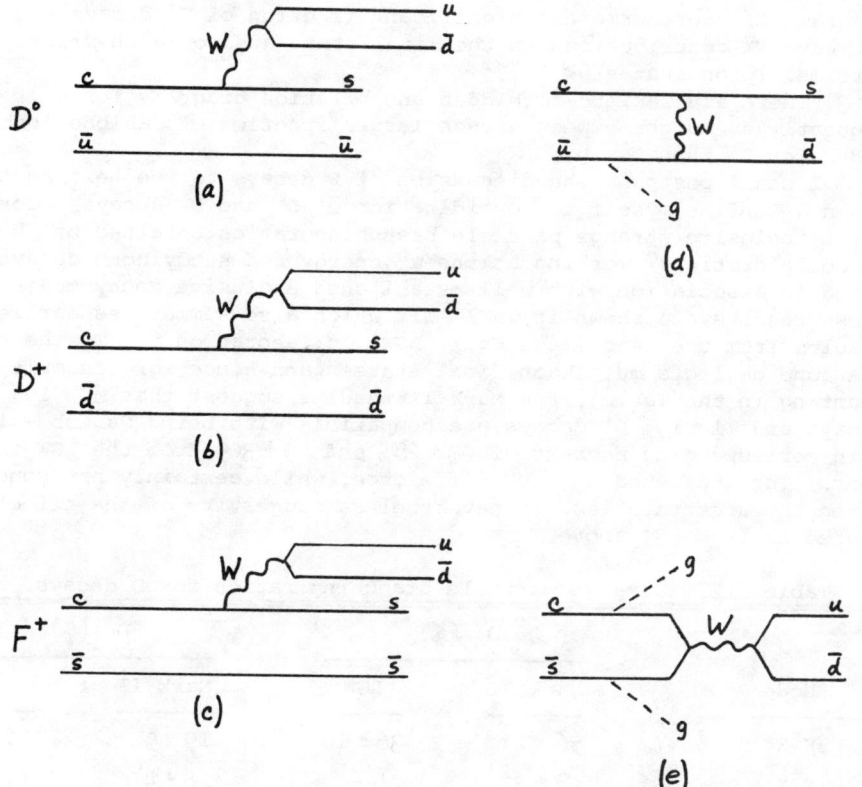

Fig. 2. Diagrams for D and F decay.

for $D^+$ and $F^+$ decays (Fig. 2b,c) and lead to the expectation of equal lifetimes for $D^0$, $D^+$ and $F^+$. The actual different lifetime values suggest a significant enhancement factor for the $D^0$ hadronic decays. Simple mechanisms for such enhancements (which would also help explain the $|\Delta \vec{I}| = 1/2$ enhancement in K decay) are the W annihilation mechanisms shown for $D^0$ decay in Fig. 2d and $F^+$ decay in Fig. 2e, but not possible for Cabibbo-allowed $D^+$ decay.[20] The presence of gluons is essential in these processes to remove helicity suppression factors which would otherwise strongly inhibit them. The enhancements provided by the annihilation diagrams plus perhaps the slight inhibitions induced by the Pauli principle in the $D^+$ decay of Fig. 2b, due to the two $\bar{d}$ in the final state,[21] can possibly account for a factor of perhaps 5 in $\tau(D^+)/\tau(D^0)$, although there might be theoretical difficulty if the experimental ratio turned out substantially larger.

The further experimental consequences of the dominance of diagrams such as those of Fig. 2d and 2e would be the following:

(1) D hadronic final states would be dominated by isospin 1/2. One would expect the rates of individual $D^+$ final state channels (which have to be $I = 3/2$) to be substantially less than the individual rates for the corresponding $D^0$ channels.

(2) The $F^+$ would have a shorter lifetime than the $D^+$ in view of diagram 2e. Furthermore a significant fraction of $F^+$ decays would not have $K\bar{K}$ contributions in the final state and could therefore go into multipion states ($\geq 3\pi$).

(3) There are Cabibbo-forbidden annihilation diagrams for $D^+$. Consequently one might expect a much larger fraction of Cabibbo-forbidden modes for $D^+$ than for $D^0$.

I shall postpone the discussion of F decays to the next section, and now confine myself to consideration of $D^0$ and $D^+$ decays. Consider first inclusive strange particle branching ratios obtained by the Mark II Collaboration[18] working at the $\psi''$ energy and studying D decays produced in association with well-established exclusive decay modes. These results are shown in Table III which also summarizes earlier results from the Lead Glass Wall (LGW) Collaboration.[22] To the extent that one neglects multikaon final states (and hence ignores multiple counting in the Table), the Mark II results suggest that $85 \pm 15\%$ of $D^0$ decays and $71 \pm 19\%$ $D^+$ decays are compatible with being Cabibbo-allowed, with corresponding numbers of $93 \pm 28\%$ and $49 \pm 30\%$ from the LGW experiment. The indicated $D^+ - D^0$ difference, while certainly not conclusive given the uncertainties, is nevertheless suggestive of the effect mentioned in item (3) above.

Table III. Strange particle branching ratios for D decays.

| Mode | $D^0$ (%) | | $D^+$ (%) | |
|---|---|---|---|---|
| | Mark II | LGW | Mark II | LGW |
| $D \to K^- X$ | $56 \pm 11$ | $36 \pm 10$ | $19 \pm 5$ | $10 \pm 7$ |
| $D \to K^+ X$ | $8 \pm 3$ | -- | $6 \pm 4$ | $6 \pm 6$ |
| $D \to \bar{K}^0 X$ | $29 \pm 11$ | $57 \pm 26$ | $52 \pm 18$ | $39 \pm 29$ |
| Total Cabibbo-favored | $85 \pm 15$ | $93 \pm 28$ | $71 \pm 19$ | $49 \pm 30$ |

To test item (1), I consider briefly exclusive final states of D decay and their isospin character. The most recent branching ratio information from the Mark II experiment[18] is summarized in Table IV along with published results from the LGW experiment.[23] The agreement between the two sets of data is not overwhelmingly good, partly because of differences in the total cross section measurements at 3770 MeV, and in my further considerations I have just used the Mark II results.

In Table V, I have listed $K^*\pi$ and $K\rho$ branching ratios obtained from Dalitz plot fits to the $K\pi\pi$ final states.[18] The errors quoted in Table V combine quadratically systematic and statistical uncertainties, but do not include the errors in the $D\bar{D}$ cross sections which are common to all the measurements (and hence do not affect comparisons of the branching ratios). The Dalitz plots and interesting projections are shown in Fig. 3.

In interpreting the branching ratios for the $K\pi$, $K^*\pi$, $K\rho$, etc. final states one should keep in mind that the Cabibbo-allowed hadronic charm decay is expected to satisfy a $|\Delta I| = 1$ selection rule (note

Table IV. σ·B and B for Cabibbo-favored D decays.

| Mode | σ·B (nb) | B (%) | LGW B (%) |
|---|---|---|---|
| $K^-\pi^+$ | 0.24 ± 0.02 | 3.0 ± 0.6 | 2.2 ± 0.6 |
| $\bar{K}^0\pi^0$ | 0.18 ± 0.08 | 2.2 ± 1.1 | -- |
| $\bar{K}^0\pi^+\pi^-$ | 0.30 ± 0.08 | 3.8 ± 1.2 | 4.0 ± 1.3 |
| $K^-\pi^0\pi^+$ | 0.68 ± 0.23 | 8.5 ± 3.2 | 12.0 ± 6.0 |
| $K^-\pi^+\pi^-\pi^+$ | 0.68 ± 0.11 | 8.5 ± 2.1 | 3.2 ± 1.1 |
| $\bar{K}^0\pi^+$ | 0.14 ± 0.03 | 2.3 ± 0.7 | 1.5 ± 0.6 |
| $K^-\pi^+\pi^+$ | 0.38 ± 0.05 | 6.3 ± 1.5 | 3.9 ± 1.0 |
| $\bar{K}^0\pi^0\pi^+$ | 0.78 ± 0.48 | 12.9 ± 8.4 | -- |
| $\bar{K}^0\pi^+\pi^-\pi^+$ | 0.51 ± 0.18 | 8.4 ± 3.5 | -- |
| $K^-\pi^+\pi^-\pi^+\pi^+$ | < 0.23 | < 4.1 | -- |

Table V. B for quasi-two-body $K\pi\pi$ states.

| Mode | B (%) |
|---|---|
| $K^-\rho^+$ | 7.2 ± 2.5 |
| $\bar{K}^{*0}\pi^0$ | $1.4^{+1.9}_{-1.4}$ |
| $K^{*-}\pi^+$ | 3.2 ± 1.0 |
| $\bar{K}^0\rho^0$ | $0.1^{+0.4}_{-0.1}$ |
| $\bar{K}^{*0}\pi^+$ | < 4 |

that $c \to s + u + \bar{d}$), from which one easily derives the triangular amplitude relations,

$$A(-+) + \sqrt{2}\, A(0\,0) = A(0+) \tag{7}$$

where $A(+-) \equiv$ decay amplitude for $D^0 \to K^-\pi^+$, $K^{*-}\pi^+$, $K^-\rho^+$
$A(0\,0) \equiv$ decay amplitude for $D^0 \to \bar{K}^0\pi^0$, $\bar{K}^{*0}\pi^0$, $\bar{K}^0\rho^0$
$A(0+) \equiv$ decay amplitude for $D^+ \to \bar{K}^0\pi^+$, $\bar{K}^{*0}\pi^+$, $\bar{K}^0\rho^+$ .

The dominance of $I = 1/2$ final states, appropriate to the annihilation diagram of 2d implies that $|A(0+)|^2$ is typically a few times smaller than $|A(-+)|^2 + |A(0\,0)|^2$, but since the enhancement factors involved seem to be at most of order 10, one can hardly argue that $A(0+)$ should be negligible in (7). It follows that relations like $|A(0\,0)|^2/|A(-+)|^2 = 1/2$ whose counterparts in strange particle decay are accurately obeyed should only show rough experimental agreement.

I have attempted to summarize the experimental situation in Table VI, using as my inputs the Mark II numbers of Tables IV and V. Since $\tau(D^0)/\tau(D^+) \sim 1/5 - 1/10$ both $K\pi$ and $K^*\pi$ final states satisfy the

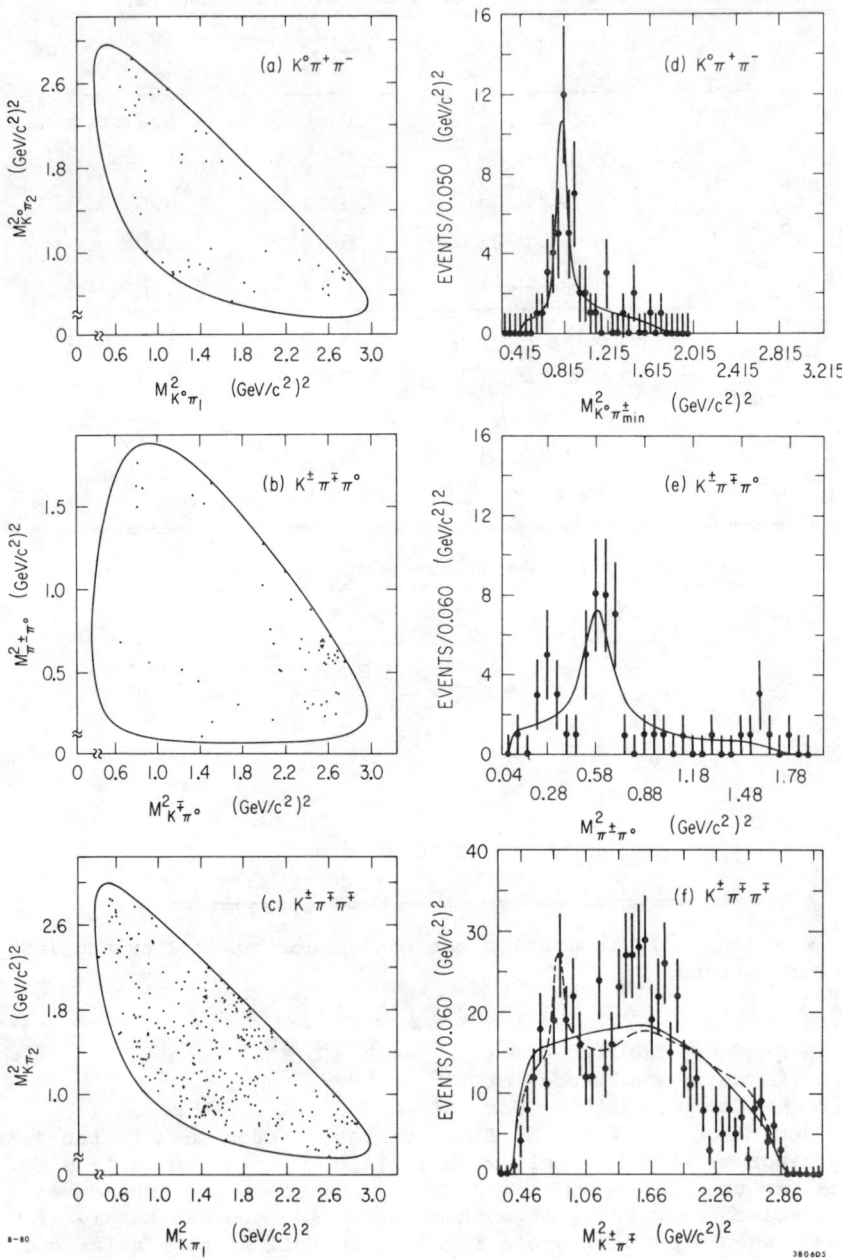

Fig. 3. Dalitz plots and projections for $D \to K\pi\pi$ decay.[18]

Table VI.  I = 1/2  final state tests.

| Decay mode | $\dfrac{\Gamma(0\,0)}{\Gamma(-\,+)}$ | $\dfrac{\Gamma(0\,+)}{\Gamma(-\,+)+\Gamma(0\,0)}$ |
|---|---|---|
| $K\pi$ | $0.7^{+0.5}_{-0.3}$ | $(0.4\pm 0.2)\,\dfrac{\tau(D^0)}{\tau(D^+)}$ |
| $K^*\pi$ | $0.4^{+0.7}_{-0.4}$ | $< 0.8\,\dfrac{\tau(D^0)}{\tau(D^+)}$ |
| $K\rho$ | $0.01^{+0.06}_{-0.01}$ | $\sim 1$ (from Eq. 7) |

expected I = 1/2 dominance, but the $K\rho$ state seems to pose a problem. Indeed the apparent absence of $\bar{K}^0\rho^0$ implies via (7) above that $\Gamma(\bar{K}^0\rho^+) \approx \Gamma(K^-\rho^+)$, hence that the $\bar{K}^0\rho^+$ final state should have a very large branching ratio. These indications are barely compatible with the rather poorly measured $\bar{K}^0\pi^+\pi^0$ branching ratio of $12.8\pm 8.4\%$. If confirmed, these $D \to K\rho$ results pose problems for both the annihilation diagram dominance and also for the so-called sextet dominance (based on analogy with octet dominance for $|\Delta\vec{I}| = 1/2$ in strange particle decay) according to which $\bar{K}^{*0}\pi^+$ and $\bar{K}^0\rho^+$ rates can be enhanced, but should be equal. Clearly much more extensive data on $K\pi\pi$ decay modes, hopefully to be obtained in the future from the Mark III detector, will be required to clear up this question.

### E.  F Decays

Cabibbo-allowed diagrams relevant to F decay have already been exhibited in Fig. 2c and 2e. The W-radiation diagram of Fig. 2c would lead dominantly to decay modes containing a $K\bar{K}$ component such as

$$F^+ \to K + \bar{K} + \pi\text{'s} \qquad (8a)$$
$$\to \eta + \pi\text{'s} \qquad (8b)$$
$$\to \eta' + \pi\text{'s} \qquad (8c)$$

and would have a rate comparable to that for $D^+$ decay. Dominance of the annihilation diagram of Fig. 2e would lead to multipion final states and lifetimes perhaps more comparable to those of $D^0$.

Most of our past information on F decay has come from the DASP experiment[24] at DORIS which reported a substantial inclusive $\eta$ cross section ($\sigma_\eta \approx 4$ nb) in the neighborhood of the 4.4 GeV $e^+e^-$ annihilation cross-section bump. The DASP group claimed a threshold-like behavior for $\sigma_\eta$ (see Fig. 4) in the region of 4.1 GeV which they ascribed to the onset of $F\bar{F}$ production accompanied by decay modes (8b). In addition, explicit observation of the exclusive mode

$$F^+ \to \eta + \pi^+ \qquad (9)$$

was reported, with an F mass determination $M_F = 2.03\pm 0.06$ GeV/$c^2$.

In my view, the most striking new results on F decay come from the three events observed in emulsion and described in Table II. The measurements on all these events appear to be very complete and represent the most compelling experimental evidence for the existence of F mesons.

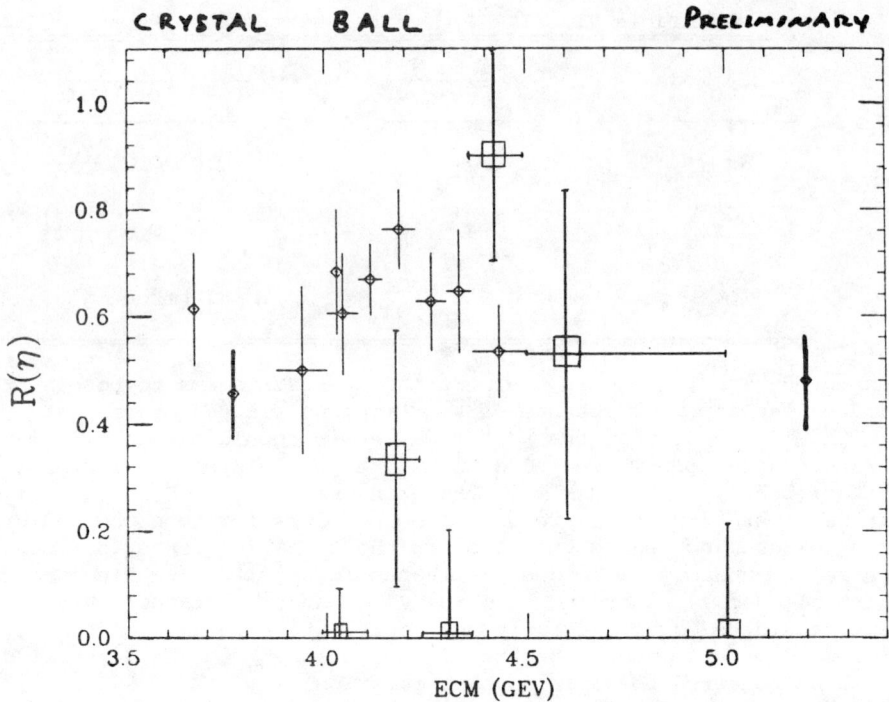

Fig. 4. Plot of ratio $R(\eta) \equiv \dfrac{\sigma(e^+e^- \to \eta X)}{\sigma(e^+e^- \to \mu^+\mu^-)}$ as a function of total energy. Diamonds are preliminary Crystal Ball data,[25] and squares are earlier DASP data.[24]

The lifetime, $\tau = (2.0^{+1.8}_{-0.8}) \times 10^{-13}$ sec, though in obvious need of statistical strengthening, appears intermediate between $D^0$ and $D^+$, and certainly compatible with the idea of W-annihilation dominance (where two gluons must here be emitted because of the W being a color singlet). Furthermore, out of three reconstructed decays, two are of the form $\pi^+\pi^-\pi^\pm\pi^0$ in which neither of the $\pi^+\pi^-\pi^0$ combinations fit the $\eta$ mass.

There are also new results from the Crystal Ball Collaboration on the measurement of inclusive $\eta$ production in $e^+e^-$ annihilation between 3.6 and 4.5 GeV.[25] The Crystal Ball is a $4\pi$ sodium iodide detector plus inner tracking chambers for the detection of charged particles, with very good photon and electron energy resolution. The reconstruction of $\eta$'s is based on detection of their $2\gamma$ decay mode. The major problem is the development of techniques to bring out the rather small $\eta$ signal in the presence of a prodigious $\pi^0$ background. The preliminary Crystal Ball results are compared to the earlier DASP data in Fig. 4 which shows the ratio

$$R(\eta) \equiv \frac{\sigma(e^+e^- \to \eta X)}{\sigma(e^+e^- \to \mu^+\mu^-)}$$

as a function of total c.m. energy. The data in the Figure lead to the following conclusions:

1) The Crystal Ball measurements of $R(\eta)$ show no evidence of threshold behavior near 4.1 GeV and no indication of peaking at 4.4 GeV. These preliminary data thus do not confirm the DASP indications of large $B(F \to \eta X)$, although they say nothing about the exclusive $F^{\pm} \to \eta \pi^{\pm}$ decay mode.

2) There are about 0.15 $\eta$/hadronic event for all c.m. energies between 3.68 and 4.5 GeV.

3) From data at the $\psi''$, one can set a limit $B(D \to \eta X) < 0.1$.

Overall these various results on F decay are all remarkably consistent in supporting the interpretation of the $\tau(D^0)/\tau(D^+)$ ratio through enhancement from W annihilation diagrams. A few words of caution are however in order. Firstly, as often mentioned, the statistics on which these various results hang are still very limited -- for example, the lifetime results themselves are still in the state where one additional event can make a significant difference in the overall lifetime. Secondly, the photoproduction experiment on the Omega Spectrometer[26] at the SPS has reported evidence for ($\eta$ + pions) bumps at masses near 2000 MeV, which they have rather naturally ascribed to F production and decay. My arguments above are not intended to suggest that such modes are not present, but only that they are perhaps not dominant in the same sense that modes involving K mesons overwhelmingly dominate $D^0$ decays. Incidentally I have not discussed here in any detail the F photoproduction results because they were not presented in those parallel sessions which it is here my task to summarize -- they will undoubtedly be described in Professor Wojcicki's summary paper.

## VII. NEUTRINO OSCILLATIONS

The possibility of neutrino oscillations was suggested quite some time ago,[27] but a number of recent indications, including some evidence for nonzero neutrino masses, and new results from experiments with reactor-produced $\bar{\nu}_e$ have lately focused renewed interest on this subject. I shall not discuss here the evidence on nonzero neutrino masses -- it is both experimental (from a new study of the endpoint region of the tritium beta spectrum[28]) and theoretical (from astrophysics considerations[29]), and suggests mass values in the range of 10-40 eV. I do want however to discuss in some detail the reactor experiments.

To understand the interpretation of these experiments in terms of neutrino oscillations, it is useful to put down some very simple phenomenology. We assume two sets of neutrino eigenstates, the weak charged-current eigenstates $\nu_\alpha$ ($\alpha = e, \mu, \tau, \ldots$) and the mass eigenstates $\nu_j$ ($j = 1, 2, 3, \ldots$) related to each other through a unitary transformation U,

$$\nu_\alpha = \sum_j U_{\alpha j} \nu_j . \qquad (10)$$

Initially a single weak eigenstate $\nu_\alpha$ is produced (for example, by beta decay); but, if U is not a unit matrix and the masses corresponding to the $\nu_j$ differ from each other, the various $\nu_j$ amplitudes will oscillate with different frequencies and change their phase relationships as time goes on. It is easy to show that after time t the probability that the initial neutrino $\nu_\alpha$ manifests itself as a new weak eigenstate $\nu_\beta$ is given by

$$P_{\alpha\beta} \equiv P(\nu_\alpha \to \nu_\beta) = \left|\sum_j U_{\alpha j} U^*_{\beta j} e^{-iE_j t}\right|^2$$

$$P_{\bar{\alpha}\bar{\beta}} \equiv P(\bar{\nu}_\alpha \to \bar{\nu}_\beta) = \left|\sum_j U^*_{\alpha j} U_{\beta j} e^{-iE_j t}\right|^2 \quad (11)$$

where $E_j$ is the total energy of state $\nu_j$.

As a simple example we consider the case in which only $\nu_e$ and $\nu_\mu$ mix. Formula (11) then reduces to the relations

$$P_{ee} = P_{\mu\mu} = P_{\bar{e}\bar{e}} = P_{\bar{\mu}\bar{\mu}} = 1 - \sin^2 2\theta \sin^2 \Delta_{12}/2 \quad (12a)$$

$$P_{e\mu} = P_{\mu e} = P_{\bar{e}\bar{\mu}} = P_{\bar{\mu}\bar{e}} = \sin^2 2\theta \sin^2 \Delta_{12}/2 \quad (12b)$$

where $\theta$ is the two-dimensional rotation angle (similar to the Cabibbo angle) which parametrizes the $2\times 2$ matrix U and

$$\frac{\Delta_{12}}{2} = \frac{m_1^2 - m_2^2}{4}\left(\frac{L}{E}\right) = 1.27 \,\delta m^2 \left(\frac{L}{E}\right). \quad (12c)$$

The right-hand expression in (12c) has the mass squared difference $\delta m^2 = |m_1^2 - m_2^2|$ in $eV^2$, the length L traveled between production and detection in meters, and the neutrino energy $E \gg m_1, m_2$ in MeV. For the purposes of discussing the reactor experiments, I shall assume the validity of (12). The consideration of complete three-dimensional mixing between $\nu_e$, $\nu_\mu$ and $\nu_\tau$ does not significantly affect the interpretation of these experiments, which only measure $P_{\bar{e}\bar{e}}$ in a limited range of L/E, but does have important impact on attempts to make consistent fits in terms of neutrino oscillations to a wider variety of pheonomena. I shall come back to this very briefly at the end of this section.

It is clear from (12) that a necessary condition for the observation of oscillations is that $\Delta_{12}$ not be too small; i.e., $\delta m^2(L/E) \gtrsim O(1)$. The $\delta m^2$ ranges probed by different kinds of experiments is illustrated[30] in Table VII. As shown in the Table, the reactor experiments for which L ~ meters, E ~ MeV probe values of $\delta m^2$ of the order of 1 $eV^2$ or higher.

We now move from these general remarks to a more specific discussion of reactor experiments, for which the experimentally determined values of $P_{\bar{e}\bar{e}}$ at distances L of the order of 6-11 meters from the core serve as the measure of possible oscillations. Although in principle any significant downward deviation of $P_{\bar{e}\bar{e}}$ from unity can be considered evidence for oscillations, it is clear that a really compelling demonstration of this phenomenon requires measurements at more than one value of L to establish the characteristic L/E dependence of (12).

The investigation of reactions induced by reactor-produced $\bar{\nu}_e$ has been largely pioneered by Reines and his collaborators ever since the first experiment which detected neutrino interactions.[31] In the search for neutrino oscillations, large reactors have the following nice features:

(i) Only $\bar{\nu}_e$ are initially produced;
(ii) The $\bar{\nu}_e$ have relatively low energies (a few MeV), and it is

Table VII. Sensitivity of various experiments to $\delta m^2$.

| Experiment type | $\delta m^2$ range (eV$^2$) |
|---|---|
| High energy accelerator | $> 10^2$ |
| Reactor, Meson Factory Low Energy accelerator | $> 1$ |
| Deep Mine (present) | $> 10^{-2}$ |
| Deep Mine (future) | $> 10^{-5}$ |
| Solar neutrinos | $> 10^{-10}$ |

possible therefore to set up conveniently experiments for which $L/E \sim$ 1 m/MeV to study the $\delta m^2 \approx 1$ eV$^2$ region;

(iii) The flux of $\bar{\nu}_e$ can be very large ($\sim 2 \times 10^{13}$ $\bar{\nu}_e$ cm$^{-2}$s$^{-1}$ from the Savannah River Reactor).

Ideally if the spectrum (both shape and magnitude) of $\bar{\nu}_e$ produced by the reactor were known accurately, measurements of the corresponding spectrum at a well-defined distance from the reactor core would permit a direct test for the existence of oscillations. Such measurements have indeed been made through detection of the inverse beta reaction,

$$\bar{\nu}_e + p \rightarrow n + e^+ \tag{13}$$

at 6 m by Nezrick and Reines,[32] at 11.2 m by Reines, Gurr and Sobel[33] and very recently at 8.7 m by the Caltech-Grenoble-Munich Collaboration.[34] Although comparison of experiments at various distances (for example, the 6 and 11.2 m experiments) can in principle provide information on oscillations independently of knowledge of the production spectrum, the three experiments listed are all somewhat different, and the systematic differences are perhaps too large to allow firm conclusions from such comparisons.

Alternatively, as mentioned above, one can search for oscillations by comparing any one of these measurements with the expected $\bar{\nu}_e$ spectrum produced by the reactor. The limitation here is that this spectrum is not well known, particularly in the upper end of the $\bar{\nu}_e$ energy range. The expected spectrum has been calculated independently by Avignone and Greenwood,[35] and by Davis et al.[36] The two calculated spectra differ by about 30% in the integrated rate predicted in the absence of oscillations for the reaction (13) and also differ slightly in shape, the Avignone spectrum predicting more rate particularly at the high $\bar{\nu}$ energies. Figure 5 summarizes both the Avignone and the Davis predictions for the inverse beta rate and shows the experimental results of Reines, Gurr and Sobel at 11.2 meters. From Fig. 5, one sees that the measurements, while in good agreement with the Davis spectrum over the lower part of the energy region, disagree with both spectra for positron energy $> 5$ MeV (or $\bar{\nu}_e$ energy $> 6.8$ MeV). Since the predicted spectrum is most unreliable at the high energy end, this theoretical uncertainty provides the natural interpretation of the discrepancy between expected and measured spectra in Fig. 5. However if one had strong confidence in one of the calculated spectra, the discrepancy could also be interpreted in terms of a neutrino oscilla-

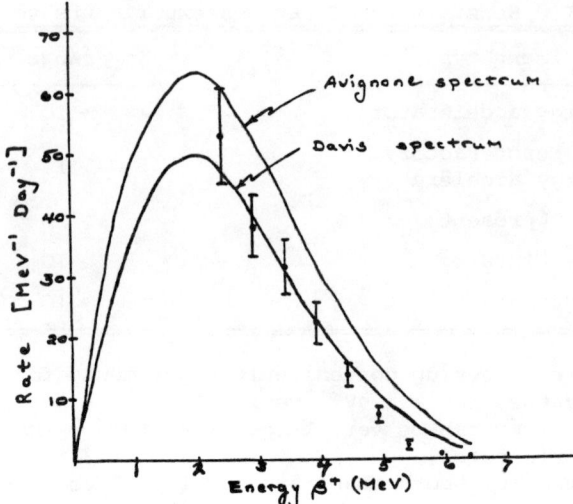

Fig. 5. Rate (at Savannah Reactor) for the process $\vec{\nu}_e + p \to n + e^+$ as a function of $e^+$ energy. The solid curves are from the calculations of Avignone and Greenwood[35] and Davis et al.[36] The data points are from the preliminary measurement at 11.2 m by Reines et al.[33]

tion between the production point and the detector at 11.2 m. As an example, Fig. 6 compares the ratio between the measured points and the Avignone spectrum with a plot of $P_{\bar{e}\bar{e}}$ calculated from (12) with $\delta m^2 = 1$ eV$^2$ and $\sin^2 2\theta = 1$. This plotted curve is not based on a fit, but simply illustrates the fact that if one literally believed the Avignone spectrum over the full $\bar{\nu}$ energy range, the neutrino oscillation hypothesis could provide a credible explanation for the discrepancy between experiment at 11.2 m and this spectrum seen in Fig. 5.

Unfortunately the calculated production spectrum is not sufficiently reliable to give much credence to the comparison in Fig. 6; and, therefore, Reines, Sobel and Pasierb (RSP) have searched for evidence of oscillations by an ingenious but very difficult experiment whose interpretation is rather insensitive to the spectrum.[37] Specifically they measure at 11.2 m the ratio,

$$r = \frac{S(\bar{\nu}_e + d \to n + n + e^+)}{S(\bar{\nu} + d \to n + p + \bar{\nu})} \equiv \frac{CCD}{NCD} \quad (14)$$

where $S(X)$ is the rate for the reaction X integrated over all $\bar{\nu}$ energies and CCD, NCD are abbreviations for "charged current on deuterium," "neutral current on deuterium" (the notation CCP will be used for reaction (13)). No subscript was put on the $\bar{\nu}$ in the denominator of (14) because that process occurs independently of the $\bar{\nu}$ flavor and its rate is therefore unaffected by the existence or non-existence of oscillations (as long as the total number of $\bar{\nu}$ is conserved).

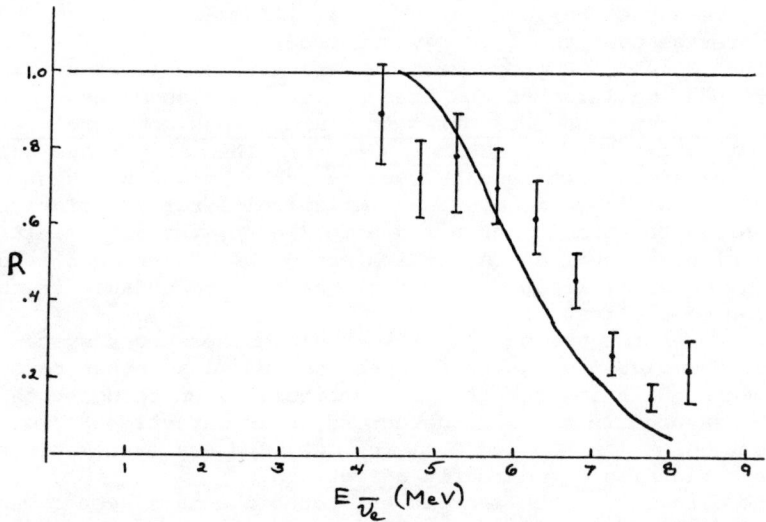

Fig. 6. Ratio of data in Fig. 5 to Avignone spectrum as a function of $\bar{\nu}_e$ energy. Curve corresponds to Eq. (12a) with $\sin^2 2\theta = 1$, $\delta m^2 = 1$ eV$^2$.

RSP measure the value of $r$, $r_{exp}$, by counting the relative numbers of events which give rise to two neutrons (CCD) and to a single neutron (NCD). They determine $r_{exp}$ from the relation

$$r_{exp} = \frac{\overline{\eta'}S_{2N}}{\overline{\eta^2}(S_{1N} - S_{1N}^{BKGND}) - 2(\overline{\eta'} - \overline{\eta'^2})S_{2N}} \qquad (15)$$

where $S_{2N}$, $S_{1N}$ are two-neutron and one-neutron event rates, $S_{1N}^{BKGND}$ is an experimentally measured reactor-associated one-neutron background and the $\eta$, $\eta'$ terms are various efficiencies which I shall not discuss in detail here. Roughly $\overline{\eta^2}$ (let us neglect the differences between $\eta$ and $\eta'$ which involve subtle details) is the average two-neutron efficiency, $\bar{\eta}$ is the average one-neutron efficiency, and the second term in the denominator of (15) is the rate of one-neutron events arising from the CCD process because only one of the two produced neutrons is detected.

The advantages and disadvantages of using this type of experiment to get at neutrino oscillations are as follows:

(1) The theoretical value of $r$, $r_{the}$ (calculated on the assumption of no oscillations, is independent of the absolute magnitude of the $\bar{\nu}$ flux and is insensitive to the shape of the spectrum. In particular, $r_{the}$ = 0.42 (Davis spectrum), 0.44 (Avignone spectrum). However there is no assurance that the Avignone and Davis calculations span the full range of possibility; and indeed if one takes the CCP measurements shown in Fig. 5 as the best measure of the spectrum $r_{the}$ = 0.36.

(2) The ratio

$$R \equiv r_{exp}/r_{the} \qquad (16)$$

directly measures the quantity $\langle P_{\bar{e}\bar{e}} \rangle$ at 11.2 m where $\langle P_{\bar{e}\bar{e}} \rangle$ is an appropriate average over the $\bar{\nu}$ energy spectrum.

(3) The CCD and NCD cross sections are very small, about two orders of magnitude smaller than the free proton cross sections. One consequence is that the uncertainties in R are almost completely dominated by the statistics of the CCD measurement. This is perhaps good in that statistical errors are usually more reliably known than systematic ones, but bad in that in this case the statistical error is relatively large.

(4) Unlike the measurements shown in Fig. 5, the determination of $S_{2N}$ gives a single global average which cannot be broken down into individual $\bar{\nu}_e$ energy bins. This obviously provides less redundancy in the interpretation of the results.

(5) Although the use of the ratio r diminishes those systematic uncertainties associated with the $\bar{\nu}_e$ spectrum and flux, other potential systematic uncertainties remain. In particular even though both CCD and NCD are measured in the same detectors their detection efficiencies are different (0.32 for NCD and 0.11 for CCD) and any systematic error in these efficiencies can directly affect $r_{exp}$.

The details of the experimental technique have been given in the paper of Pasierb et al.[37] It suffices to note here that the CCD and NCD signal rates are roughly 3 and 70 per day respectively, and the corresponding residual cosmic ray backgrounds (which must be removed by a reactor-on/reactor-off subtraction) are about 50 and 400 per day respectively. There is also a well understood and accurately known reactor-associated single neutron background of $10.2 \pm 0.7$ events per day which is subtracted from the single neutron rate.

The final result of RSP is as follows:

$$r_{exp} = 0.167 \pm 0.093$$

and $R \equiv (r_{exp})/(r_{the}) = 0.38 \pm 0.21$ (Avignone spectrum), $0.40 \pm 0.22$ (Davis spectrum). Thus $1 - R = 0.61 \pm 0.21$, and there seems to be a $3\sigma$ deviation from expectations in the absence of oscillations. This actually somewhat overstates the statistical significance of the effect, because of the coupling between the value of $r_{exp}$ and its error, arising from the presence of $S_{2N}$ in both numerator and denominator of (15). It turns out that an increase in the average CCD counting rate of close to a factor of 2 would take R from its measured value to the value of unity, and that this increase would actually represent a fluctuation of $2.3\sigma$. Furthermore as noted by RSP, the possible range of $r_{the}$ may not be bracketed by the Avignone and Davis spectra. If one uses the experimental spectrum of Fig. 5 as a measure of the shape of the production spectrum, the predicted $r_{the}$ becomes 0.36, the value of R goes to $0.46 \pm 0.26$, and the overall statistical significance of the deviation from unity goes from $2.3\sigma$ to $1.8\sigma$.

This factor of $1.8\sigma$ is based only on $r_{exp}$ and on the shape of the measured CCP spectrum at 11.2 m but not on the absolute magnitude of the $\bar{\nu}_e$ flux which it also provides. We now move to consider the further information which this flux measurement gives by showing in Table VIII a set of ratios of measured to predicted rates for the various experiments at 11.2 and 6 m supplied by RSP. The numbers given for the 11.2 m CCP data for both Avignone and Davis spectra provide average numerical representations of the behavior already exhibited in Fig. 5.

Table VIII. Ratios of measured to predicted values.

| Distance (meters) | Reaction | Neutrino threshold (MeV) | Spectra | | CCP measurement at 11.2 m |
|---|---|---|---|---|---|
| | | | Avignone | Davis | |
| 11.2 | NCD | 2.2 | 0.83 ± 0.13 | 1.1 ± 0.16 | 1.3 ± 0.22 |
| 11.2 | CCD | 4.0 | 0.32 ± 0.14 | 0.44 ± 0.19 | 0.61 ± 0.29 |
| 11.2 | CCP | 4.0 | 0.68 ± 0.12 | 0.88 ± 0.15 | ≡ 1.0 |
| 11.2 | CCP | 6.0 | 0.42 ± 0.09 | 0.58 ± 0.12 | ≡ 1.0 |
| 6 | CCP | 1.8 | 0.65 ± 0.09 | 0.84 ± 0.12 | --- |
| 6 | CCP | 6.0 | 0.81 ± 0.11 | 1.02 ± 0.15 | 1.19 ± 0.27 |

I want to particularly emphasize the contents of the last column in which the $\bar{\nu}_e$ spectrum used for determining the predicted rates is the spectrum <u>measured at 11.2 m</u> by Reines, Gurr and Sobel via the CCP reaction and exhibited by the experimental points of Fig. 5. In the absence of measurement errors, the CCD entry for the last column should be unity, independently of the existence or non-existence of oscillations, since both CCD and CCP at 11.2 m are measurements of the $\bar{\nu}_e$ spectrum at that point.[38] The actual entry is 0.61 ± 0.29 where again the dominating contribution to the uncertainty is the CCD statistical error. It therefore seems likely that the difference between 0.61 and unity is at least in part due to a downward statistical fluctuation in the measured CCD rate. This evidence for such a fluctuation reduces the strength of the case for oscillations, based on the fact that R is well below unity.

This can be put in a more precise way as follows. Instead of using the ratio $r \equiv CCD/NCD$, we can instead use another ratio $r' \equiv CCP/NCD$, and form the quantity $R' \equiv r'_{exp}/r'_{the}$, which, in the absence of measurement errors, should differ from R defined in (16) only to the extent that the average over the $\bar{\nu}$ spectrum may be very slightly different. The ratio R' has the same advantages as R with respect to insensitivity to assumptions about $\bar{\nu}$ spectrum. With regard to other uncertainties it has one disadvantage and one advantage with respect to R: the disadvantage is that the CCP and NCD data were not obtained at the same time, hence there is a systematic normalization uncertainty (which is included in all the quoted errors); the advantage is that the dominating error in the R measurement, namely the statistical uncertainties of the CCD rate has been completely removed. The results for R' derived from the data of Table VIII and from the counting rate numbers of RSP are as follows,

$R' = 0.77 \pm 0.20$ (Avignone or Davis spectrum)

$R' = 0.90 \pm 0.23$ (CCP spectrum)

It may seem surprising that the ratio of R' to R is larger than 1/0.61, the factor needed to raise the low CCD entry in the last column of Table VIII to unity; this is a consequence of the fact that $r_{exp}$ in (15) contains $S_{2N}$ in both numerator and denominator positively correlated

-- if $S_{2N}$ is increased by a factor $1/0.61 = 1.6$, $r_{exp}$ is increased by almost a factor of 2.

The various values of R and R' which I have quoted are succinctly summarized in Table IX. In the last column of the Table, I have quoted probability levels corresponding to the fluctuations given in the adjacent column, account being taken of the fact that only in one direction could these fluctuations have simulated the existence of neutrino oscillations.

Table IX. Summary of rate-ratio measurements.

| Ratio of rates used | Assumed $\bar{\nu}$ spectrum shape | $R = \dfrac{r_{exp}}{r_{the}}$ | Deviation from no oscillation expectation | Probability of one-sided fluctuation of equal or greater magnitude |
|---|---|---|---|---|
| $\dfrac{CCD}{NCD}$ | Avignone Davis | $0.39 \pm 0.21$ | $2.3\ \sigma$ | 0.01 |
| $\dfrac{CCD}{NCD}$ | CCP data | $0.46 \pm 0.26$ | $1.8\ \sigma$ | 0.04 |
| $\dfrac{CCP}{NCD}$ | Avignone Davis | $0.77 \pm 0.20$ | $1.2\ \sigma$ | 0.11 |
| $\dfrac{CCP}{NCD}$ | CCP data | $0.90 \pm 0.23$ | $0.4\ \sigma$ | 0.34 |

What do I conclude from Table IX? I want to emphasize those inputs which help decide the case for or against the existence of oscillations rather than those inputs which, if one assumes that oscillations exist, give the best measurement of $\langle P_{\bar{e}\bar{e}} \rangle$. In my view this dictates particular consideration of the second and fourth rows of the Table since only these are really independent of assumptions about the $\bar{\nu}_e$ spectrum. If we give full weight to the deuterium experiment, the 4% probability level in the second row gives the right measure (on the assumption that none of the systematic uncertainties have been grossly underestimated) of the chance that normal non-oscillating behavior gave rise to the observations. The fourth row does not give full weight to the deuterium experiment -- it ignores the CCD rate measurement which is the weakest statistical piece and replaces it by the CCP rate measurement which supposedly is sensitive to the same input. In that case, the neutrino oscillation indication disappears completely; and, in my view, this fact weakens whatever positive conclusion one may have drawn from the smallness of the 4% in the second row of the Table.

Sobel, in his presentation to the Conference, has quoted preliminary results from the CIT-Grenoble-Munich Collaboration at $L = 8.7\,m$. The ratios of measured to predicted rates for the CCP reaction integrated over $\bar{\nu}$ energy above 3 MeV are $0.81 \pm 0.18$ (Davis spectrum) and $0.63 \pm 0.14$ (Avignone spectrum). One can as yet say little about the presence or absence of oscillations from these numbers given the spectrum uncertainties.

I complete this discussion by taking note of the fact that there exist other experiments whose results have a bearing on the absence or presence of neutrino oscillations and on the relevant parameters (mass differences, mixing matrix). These have been very usefully summarized in several papers by Barger and collaborators[30] who have also discussed possible sets of parameters (using the full three-dimensional mixing matrix). I shall not discuss these here. In my opinion as a perennial skeptic, those effects which are consistent with no oscillations (for example, the absence of $\nu_\mu \to \nu_e$ transitions in various accelerator experiments) seem better established than most of those whose interpretation could require oscillations.

My final conclusions on the present state of the subject of neutrino oscillations are then as follows:

1. There is no compelling evidence at the present time for the existence of neutrino oscillations.

2. The recent reactor experiment of Reines, Sobel and Pasierb hints at a possible anomaly, but even if this anomaly is indeed present, further proof is needed to connect it to oscillations. Furthermore from the totality of existing reactor data, it appears likely that even if there are oscillations, the magnitude of $\langle P_{\bar{e}\bar{e}} \rangle$ for the relevant L/E values is much closer to unity than the $0.39 \pm 0.21$ result of that experiment.

3. I understand that the CIT-Grenoble-Munich group is planning to continue and extend its measurements and that Reines, Sobel and collaborators are preparing a new detector capable of measuring the CCP reaction at various distances from the reactor core. Hopefully these experiments will resolve the interesting issues raised by the present set of experimental results.

## DISCUSSION

**Isgur:** (Toronto) I would like to comment that it might be more useful to quote $\tau$ decay rates relative to $\tau \to e\nu\bar{\nu}$. Since the theoretical prediction of some of the hadronic modes like $\tau \to A_1 \nu$ is uncertain, the real test of a mode like $\tau \to \rho\nu$ is its ratio to $\tau \to e\nu\bar{\nu}$ and not its branching ratio.

**Trilling:** The relevant experimental ratios from the Mark II Collaboration results are:

$$\frac{B(\tau \to \rho\nu)}{B(\tau \to e\nu\bar{\nu})} = 1.24 \pm 0.13 \pm 0.23$$

$$\frac{B(\tau \to K^*\nu)}{B(\tau \to e\nu\bar{\nu})} = 0.091 \pm 0.039 \ .$$

**Rosner:** (Minnesota) Can you comment on any of the present limits on the proton lifetime?

**Trilling:** I have no comment on this.

**Kugler:** (Weizmann Institute) Could you comment on the direct neutrino mass measurement in the tritium experiment?

Trilling: I am not sufficiently familiar with the details of that experiment to comment intelligently on it. As far as I know it was not presented at the Conference.

Petroff: (Orsay) Apparently the situation on the F decay in $\eta + X$ seems unclear, but I would like to remind you that we presented results at this Conference on photoproduction of F mesons. We observed the F decay in three independent modes: $\eta\pi$, $\eta 3\pi$, $\eta' 3\pi$ at a mass which favored the DORIS result ($2.03$ GeV/$c^2$). This experiment has been done at the Omega spectrometer at CERN.

Trilling: I apologize for not mentioning them. Due to the organization of the parallel sessions, your results were not presented in the sessions which I covered, and I did not hear your presentation nor see your paper. I presume that this will be covered by one of the other speakers. [Note: The Omega work is mentioned in this written version of the talk, but was not mentioned in the oral version.]

## REFERENCES

1. M. P. Schmidt, Session B-5 of this Conference; M. P. Schmidt et al., Phys. Rev. Lett. **43**, 556 (1979).
2. J. Wise et al., Phys. Lett. **91B**, 165 (1980).
3. D. A. Jensen et al., Paper No. 0490A, Session B-5 of this Conference.
4. N. Lockyer et al., Paper No. 0370, Session B-5 of this Conference.
5. E. M. Henley and F. R. Krejs, Phys. Rev. **D11**, 605 (1975).
6. R. Balzer et al., Phys. Rev. Lett. **44**, 699 (1980).
7. D. E. Nagle et al., High Energy Physics with Polarized Beams and Polarized Targets - 1978, edited by G. H. Thomas, p. 224.
8. B. Desplanques, J. F. Donoghue, and B. R. Holstein, Ann. of Phys. **124**, 449 (1980).
9. G. S. Abrams et al., Phys. Rev. Lett. **43**, 1555 (1979).
10. J. Dorfan, SLAC-PUB-2566 (1980), Session B-7 of this Conference.
11. Y. S. Tsai, Phys. Rev. **D4**, 2821 (1971); Y. S. Tsai, SLAC-PUB-2450.
12. D. Allasia et al., CERN-EP/80-76, submitted to Nuclear Physics.
13. Photon-Emulsion Collaboration and Omega-Photon Collaboration, Paper No. 0755 presented by G. Diambrini-Polazzi at this Conference.
14. K. Niu, Session B-6 of this Conference.
15. R. Ammar et al., Paper 0265, Session B-6 of this Conference.
16. A. Pais and S. B. Treiman, Phys. Rev. **D15**, 2529 (1977).
17. W. Bacino et al., Phys. Rev. Lett. **45**, 329 (1980).
18. A. J. Lankford, Session B-7 of this Conference.
19. N. Cabibbo and L. Maiani, Phys. Lett. **79B**, 109 (1978); N. Cabibbo, G. Corbo, and L. Maiani, Nucl. Phys. **B155**, 93 (1979).
20. W. Bernreuther, O. Nachtmann, and B. Stech, Z. Physik C. Particles and Fields **4**, 257 (1980); H. Fritzsch and P. Minkowski, Phys. Lett. **90B**, 455 (1980); M. Bander, D. Silverman, and A. Soni, Phys. Rev. Lett. **44**, 7 (1980).
21. N. Cabibbo, Session B-8 of this Conference.
22. V. Vuillemin et al., Phys. Rev. Lett. **41**, 1149 (1978).
23. I. Peruzzi et al., Phys. Rev. Lett. **39**, 1301 (1977).
24. R. Brandelik et al., Phys. Lett. **80B**, 412 (1979).

25. F. Porter, Session B-7 of this Conference.
26. B. d'Almagne, Session A-11 of this Conference.
27. A recent review with complete references to the earlier work is given in S. M. Bilenky and B. Pontecorvo, Phys. Reports $\underline{41}$, 225 (1978).
28. V. A. Lubimov et al., Phys. Lett. $\underline{94B}$, 266 (1980).
29. E. Witten, Session B-4 of this Conference.
30. V. Barger et al., Paper 0122, submitted to Physics Letters; V. Barger et al., Paper 0504 submitted to this Conference.
31. Much of what I say concerning the reactor experiments is based on the talk by H. W. Sobel given at the Conference in Session B-5.
32. F. Nezrick and F. Reines, Phys. Rev. $\underline{142}$, 852 (1965).
33. The measurement of the $e^+$ spectrum was measured concurrently with the $\bar{\nu}_e$ - e scattering experiment reported in F. Reines, H. S. Gurr, and H. W. Sobel, Phys. Rev. Lett. $\underline{37}$, 315 (1976).
34. Preliminary results of the Caltech-Grenoble-Munich Collaboration reported by H. W. Sobel.
35. F. T. Avignone III and Z. D. Greenwood, University of South Carolina Preprint, February 1979.
36. B. R. Davis et al., Phys. Rev. $\underline{D19}$, 2259 (1979).
37. Initial results of this experiment have been published in E. Pasierb et al., Phys. Rev. Lett. $\underline{43}$, 96 (1979).
38. This point has strongly been made in the unpublished note, "Comments on the Evidence for Neutrino Instability," by R. P. Feynman and P. Vogel.

# LOW-ENERGY WEAK INTERACTIONS: THEORY

S. Pakvasa, Hawaii, Speaker

P. Rosen, Purdue, Chairman

D. Toussaint, UC-Santa Barbara,

P. Frampton, Harvard,

D. Unger, Michigan,

Scientific Secretaries

LOW-ENERGY WEAK INTERACTIONS: THEORY

Sandip Pakvasa
Department of Physics and Astronomy
University of Hawaii at Manoa, Honolulu, Hawaii 96822

ABSTRACT

The state of the art in weak interactions is reviewed. The present understanding of non-leptonic decays of strange, charmed, and b-flavored particles is summarized. Models of CP-nonconservation are compared and contrasted. Neutrino masses and mixing and consequences thereof, such as oscillations, decays, CP-nonconservation, etc., are discussed.

## 1. INTRODUCTION

My charge is to review theoretical and phenomenological developments in Low-Energy Weak Interactions. I shall interpret low energy loosely to reach ∼100 GeV when needed. The topics I will cover are weak decays of old flavors viz. strange and charm; parity violation in nuclei; weak decays of new flavors viz. b and t; CP violation; neutrino masses and mixings; and the consequent phenomena such as oscillations, decays, CP violation, and such.

Except when stated explicitly I shall keep within the SU(2) x U(1) minimal model[1] with the standard assignments.[2] For the most part we will be concerned with the charged-current weak interactions. The charged current in the standard model is

$$J_\mu = (\overline{u}\overline{c}\overline{t})_L \gamma_\mu U_{KM} \begin{pmatrix} d \\ s \\ b \end{pmatrix}_L + (\bar{\nu}_1 \bar{\nu}_2 \bar{\nu}_3)_L \gamma_\mu U^\dagger \begin{pmatrix} e \\ \mu \\ \tau \end{pmatrix}_L , \quad (1)$$

where $U_{KM}$ and $U$ are unitary matrices with one free phase, and we have allowed for non-zero masses for neutrinos.

## 2. STRANGE PARTICLE DECAYS

Consider the terms in the effective Hamiltonian which have $|\Delta S| = 1$, $\Delta C = 0$, $|\underline{\Delta T}| = 1/2$ and $3/2$. These are responsible for non-leptonic decays of hyperons, kaons, and also the $\Omega^-$. Let me recapitulate briefly some ancient history of the subject. In 1954 it was proposed[3] that the $\underline{\Delta T} = 1/2$ amplitude dominated over the $\Delta T = 3/2$ one. This accounted for the suppression of $K^+ \to 2\pi$ compared to $K_S \to 2\pi$. Later it was found to work in $\Lambda$, $\Sigma$ and $\Xi$ decays as well. In 1964 a sum rule for the hyperon decays was proposed by Sugawara and by Lee based on assuming dominance of $\underline{8}$ over $\underline{27}$ component plus other assumptions.[4] This sum rule was also found to work well. The next significant step in understanding the properties of hyperon decays was when soft pion limit and current algebra were applied to them.[5] The end result was that s-wave amplitudes were given by the baryonic matrix elements of the effective Hamiltonian and p-waves by the baryon pole terms[6] and so were related to the s-waves. The $\Lambda$ and $\Xi$ decay amplitudes have to be pure $\Delta T = 1/2$ in this limit. The $\Delta T = 1/2$ rule for $\Sigma$ decay was not explained, and quantitatively the relationship between s- and p-waves was off by a factor of 2;[7] i.e., if the s-wave amplitudes are fit to the data, the p-wave amplitudes are too small. It was noticed in 1966-67 that if in the evaluation of baryon matrix elements SU(6) wave functions were used, then $\Delta T = 1/2$ rule is obtained for $\Sigma$ decays also.[8] This was finally understood[9,10] in terms of the color symmetry properties of the currents and the baryon wave function. At the same time, this predicts a $D/F = -1$ for the baryon matrix element. This makes even the s-waves difficult to fit as the observed amplitude prefers a predominantly F-type coupling.

One way to try to correct both the problem of the magnitude and the correct D/F ratio is to add contributions that vanish in the soft pion limit. These are, e.g., K-pole (with derivative coupling) in the p-wave amplitudes and $K^*$-"pole" in the s-wave amplitudes. The overall fit is improved now and almost all the amplitudes (except perhaps $\Sigma^+ \to n\pi^+$) can be fit.

In an ambitious, first principles approach one would like to explain all features of non-leptonic decays with no free parameters. In particular one would like to understand i) the overall strength, ii) the pattern i.e. the relative strengths between multiplets, iii) s-wave to p-wave relative strengths, iv) $\Delta T = 3/2$ amplitudes which vanish in the soft pion limit and v) K-decays.

There has been some progress in this direction in the post-QCD era following suggestions by Wilson[11] that the $\Delta T = 1/2$ amplitudes may be more singular than $\Delta T = 3/2$ ones. It was found[12] that short distance correction due to hard gluons to the W-exchange diagram enhanced $\Delta T = 1/2$ amplitudes by 2.5 to 3 and suppressed $\Delta T = 3/2$ ones by 0.6. That both are in the right direction was encouraging but neither is near enough. Then it was pointed out by Shifman et al.[13] that there are additional diagrams ("Penguin") in which $s \to d$ by W loop and a gluon is emitted by the intermediate u (or c) quark. These are pure $\Delta T = 1/2$ and, if sufficiently enhanced, could account for the bulk of the $\Delta T = 1/2$ amplitudes. Unfortunately, the short distance enhancement factor is found to be very small, and one has to resort to assuming large matrix elements.[13] With the method of evaluating advocated[13] the end result is precisely the one mentioned above viz. that "Penguins" simulate the K-pole and $K^*$-"pole" terms needed to improve the current algebra soft pion results with the additional bonus that the two are related, hence reducing the number of free parameters by one. There are many evaluations[14] of the absolute strength i.e. the baryon matrix elements of $H_W$. These depend on the choice of one's favorite model, e.g., M.I.T. bag model, Harmonic Oscillator, relativistic quark model, etc.

However, as has been known to cognoscenti for a long time, the p-wave amplitudes (in particular $\Sigma^+ \to n\pi^+$) cannot be fit well.

One way to see this[15] is to note that $K \to \pi$ pole cannot contribute to $\Sigma^+ \to n\pi^+$. Hence the old problem of p-wave hyperon decays is still with us.[16] It seems a new idea is needed here. Perhaps the resemblance to the strong decays (SU(3) breaking part)[17] is a clue.

What about the $\Delta T = 3/2$ amplitudes? We know they should vanish in the soft pion limit. One possibility that suggests itself is to assume factorization, i.e., for example,

$$\langle B\pi | H_W^{p.v.} | A \rangle_{\Delta T=3/2} \sim \langle B | V_\mu | A \rangle \langle \pi | A_\mu | 0 \rangle G \sin\theta_C \cos\theta_C. \quad (2)$$

The fact that factorization diagrams contain comparable amounts of $\Delta T = 3/2$ and $1/2$ and that the $\Delta T = 1/2$ part is too small was first observed around 1960 by Oneda and collaborators.[18] In 1964 Feynman asked the question whether the $\Delta T = 3/2$ amplitudes are given correctly by factorization. He found that they were too large by about a factor of 2.5 in K decay and 4 in $\Lambda$ decay. Now the suppression factor of 0.6, due to short-distance QCD correction, brings it closer[13] to 1. Further, one predicts (to within SU(3) Clebsch'es) that

$$\begin{aligned} \text{i)} & \quad (p.v./p.c.)_{\Delta T=3/2} \sim \frac{m_A - m_B}{m_A + m_B} \sim 1/10, \\ \text{ii)} & \quad \text{sgn}[(p.v.)(p.c.)] > 0. \end{aligned} \quad (3)$$

Now, the first one is satisfied but the second prediction fails in $\Lambda$ decays.[20] Better data and more theoretical effort are called for.

Non-leptonic decays of $\Omega^-$ were first considered in the pioneering papers of Glashow and Socolow[21] and of Suzuki.[22] Relating $\Omega^- \to \Xi^{*-}$ to $\Xi^0 \to \Lambda$ by SU(6) and $\Xi^* \to \Xi\pi$, $\Xi^* \to \Lambda K^-$, $\Omega^- \to \Xi^0 K^-$ by SU(3) and using the baryon pole terms, branching ratios of $\Omega^-$ into $\Lambda K^-$, $\Xi\pi$, and total rates can be calculated.[23] In this limit all amplitudes conserve parity, and all $\alpha$'s are zero. However, if "Penguins" are important, then $K^* \to \pi$ can give a contribution to parity-violating amplitude for $\Omega^- \to \Xi\pi$ and $\alpha_{\Xi\pi} \neq 0$. Table I shows the comparison of predictions with recent data.[24]

Table I.  $\Omega^-$ Decays

|  | Observed | Calculated |
|---|---|---|
| $R(\Lambda K^- / \Xi^0 \pi^-)$ | 2.93±0.3 | $\sim 3$ |
| $R(\Xi^0 \pi^- / \Xi^- \pi^0)$ | 2.92±0.35 | 2.1 |
| $\tau_\Omega$ | $0.8 \times 10^{-10}$ sec | $(0.5 \text{ to } 1.2) 10^{-10}$ sec |
| $\alpha_\Lambda$ | 0.06±0.14 | 0 |
| $\alpha_{\Xi\pi}$ | ? | 0 (no Penguin) |

The only discrepancy is the possible deviation from $\Delta T = 1/2$ rule in the $\Xi\pi$ modes. It has been suggested[25] that contributions from the factorized diagrams to the $\Delta T = 3/2$ amplitudes are large.

Finally, I turn briefly to radiative decays of hyperons. In 1964 Hara[26] showed that in the standard model, CP invariance implied vanishing p.v. amplitudes (and hence $\alpha$'s) for $\Sigma^+ \to p\gamma$ and $\Xi^- \to \Sigma^-\gamma$. Branching ratio and $\alpha$ for $\Sigma^+ \to p\gamma$ were found to be[20] $2.4 \times 10^{-3}$ and $-1.03^{+0.5}_{-0.4}$, respectively. Simple calculations based on pole dominance[27] gave reasonably good values for branching ratios, but $\alpha$ remained close to 0. In the meantime, there have been many proposals to explain $\alpha(\Sigma^+ \to p\gamma) \sim -1$. A new measurement[28] yielded $\alpha(\Sigma^+ \to p\gamma) = -0.53^{+0.36}_{-0.38}$. Both values are less than 2 standard deviations from zero. One should wait and see. Other expectations of the simple model are that $\Xi^- \to \Sigma^-\gamma$ has a very small rate (B.R. $\lesssim 10^{-5}$), whereas $\Xi^0 \to \Sigma^0 + \gamma$, $\Lambda^0 + \gamma$ and $\Lambda \to n\gamma$ have sizable rates $\sim 10^{-3}$ to $10^{-4}$.

## 3. PARITY VIOLATION IN NUCLEI

The isospin properties of parity-violating $\Delta S = 0$ amplitudes can be observed in nuclear transitions and were suggested as a probe of the structure of the weak Hamiltonian a long time ago.[29] With only charged currents the $\Delta S = 0$ part has $\Delta I = 0$ and 2 components with coefficient $\cos^2\theta_C$, whereas $\Delta I = 1$ component has a coefficient $\sin^2\theta_C$. The $\Delta I = 0$ and 2 can give p.v. NN$\rho$ couplings, whereas p.v. NN$\pi$ can only come from $\Delta I = 1$. And so p.v. in $\pi$-exchange one can probe the $\Delta I = 1$ component which in the charged-current couplings is suppressed by $\sin\theta_C$ and by the smallness of the matrix element due to the presence of heavy quarks ($s\bar{s}$ and $c\bar{c}$) in the $\Delta I = 1$ terms. When the Z exchange term is included, it adds a $\theta_C$-independent contribution to the $\Delta I = 1$ p.v. part. The total $\Delta I = 1$ contribution after short-distance correction due to hard gluon exchange looks like[30]

$$H = H_W + H_Z \qquad (4a)$$

$$H_W = \frac{G}{2\sqrt{2}} \sin^2\theta_C [L_{84} O_{84} - L_{20} O_{20}] \qquad (4b)$$

$$H_Z = -\frac{G}{2\sqrt{2}} (1-2x_W^2) [L_{84} O_{84} + L_{20} O_{20} + \frac{2x^2}{3} L_{15} O_{15}] \qquad (4c)$$

where 
$$O_{\binom{84}{20}} = \frac{1}{2}[(\bar{u}Au - \bar{d}Ad)(\bar{s}Vs - \bar{c}Vc)$$
$$+ \frac{1}{2}(uVu - dVd)(sAs - cAc) \pm \text{Fierz}] \qquad (4d)$$

$$O_{15} = (\bar{u}Au - \bar{d}Ad)(\bar{u}Vu + \bar{d}Vd + \bar{c}Vc + \bar{s}Vs). \qquad (4e)$$

The coefficients $L_i$ are[30] $L_{84} \sim 1/3$, $L_{15} \sim 2$, and $L_{20} \sim 5$. Then with $x_W^2 \sim 1/4$, $\sin^2\theta_C \sim 1/20$,

$$H = -\frac{G \sin^2\theta_C}{2\sqrt{2}} O_{20}\left[1 + 10 + \frac{10}{3}\frac{O_{15}}{O_{20}} + \frac{1}{6}(-1 + 10)\frac{O_{84}}{O_{20}}\right]. \quad (5)$$

It is reasonable to expect $\langle O_{84}\rangle/\langle O_{20}\rangle \ll 1$ (from smallness of $\Delta T = 3/2/\Delta T = 1/2$). Then the expected enhancement over charged-current value is

$$\frac{\langle H_W + H_Z\rangle}{\langle H_W\rangle} = 11 + \frac{10}{3}\frac{\langle O_{15}\rangle}{\langle O_{20}\rangle}. \quad (6)$$

Now we expect $|\langle O_{15}\rangle/\langle O_{20}\rangle| $ to be $> 1$ as $\langle O_{20}\rangle$ contains heavy quarks. However, the relative sign is unknown, hence the precise amount of enhancement is hard to predict, e.g., if $\langle O_{15}\rangle/\langle O_{20}\rangle$ is positive, we may expect enhancement factor of $\sim 20$, if negative $\sim 5$. Table II shows a compilation of $\Delta I = 1$ p.v. transitions with the predicted values[31] from $H_W$ to be compared to data.[32]

Table II. Nuclear Parity-Violating $\Delta I = 1$ Transitions

| Transition | Measured Quantity | Experimental Value | Predicted Value (charged current) |
|---|---|---|---|
| $np \to d\gamma$ | Asymmetry | $(6\pm 21) \times 10^{-8}$ | $0.5 \times 10^{-8}$ |
| $^{18}F^*(0^-, I=1) \to {}^{18}F(0^+, 0)\gamma$ | Circ. Pol. | $(-7\pm 2) \times 10^{-4}$ $(1\pm 5) \times 10^{-3}$ | $3.6 \times 10^{-4}$ |
| $^{20}Ne(1^+, I=1) \to {}^{16}O(0^+, 0) + \alpha$ | Rate | $1.4 \times 10^{-4}$ eV | $4 \times 10^{-6}$ eV |

In the last entry, the rate for $\alpha$ decay of $^{20}$Ne depends on the square of the enhancement factor. It is still not possible to say how large the enhancement is, but there seems to be some enhancement definitely.

A more serious problem may be in the offing in nuclear parity-violating transitions. This is the value for circular polarization of photons in $np \to d\gamma$ which has been measured[33] to be $(-1.3\pm 0.45) \times 10^{-6}$, whereas the expected value (from $H_W$) is $2 \cdot 10^{-8}$. Since this depends solely[34] on $\Delta I = 0$ and 2, and there $\langle H_Z \rangle$ is comparable to $\langle H_W \rangle$, it is difficult to raise the predicted value above $\sim 5 \cdot 10^{-8}$. The simplest way to explain this result, consistent with all other data on parity violation in nuclear transitions, is to assume that $\Delta I = 2$ amplitude is enhanced considerably.[35] Obviously a new and independent measurement of circular polarization in $np \to d\gamma$ would be very welcome.

## 4. CHARM DECAYS

In the last few years we had come to expect naively that non-leptonic decays of charm will be correctly described by the simple heavy quark decay model (impulse or spectator model). This leads to many specific predictions[36] about rates and branching ratios of charmed particles, e.g., equality of lifetimes of charmed particles $\tau(D^o) = \tau(D^+) = \tau(F^+) = \tau(\Lambda_c^+) = \ldots$ or color suppression of modes such as $D^o \to \bar{K}^o \pi^o$ and $\Lambda_c^+ \to \bar{K}^o p$ leading to expectation that $\Gamma(D^o \to \bar{K}^o \pi^o)/\Gamma(D^o \to K^- \pi^+) \ll 1$ and $\Gamma(\Lambda_c^+ \to \bar{K}^o p)/\Gamma(\Lambda_c^+ \to \Lambda \pi^+) \ll 1$. Over the last year or so we have learned[37] that $\tau(D^+) \gg \tau(D^o) \sim \tau(\Lambda_c^+)$ and that $\bar{K}^o \pi^o$, $K^- \pi^+$ modes of $D^o$ and $\bar{K}^o p$, $\Lambda \pi^+$ modes of $\Lambda_c^+$ have comparable rates. Thus the simple naive picture of decays of charm is shattered. There has been considerable theoretical activity in trying to understand these new-found properties of charm decays.

Before discussing recent work[38] on charm decays, it should be mentioned that in a remarkable paper[39] written in 1972, Ogawa and collaborators predicted that $\tau(D^+) \gg \tau(D^o)$ just as $\tau(K^+) \gg \tau(K_s^o)$. They based their prediction on a quark number conservation rule which is qualitative in nature and asserts that diagrams such as W-exchange, annihilation, and penguin dominate over the quark decay diagram.[40] Most of the recent work is based on actually trying to estimate the contribution of each diagram in the presence of strong interactions. Since the calculations are perturbative and most likely the bulk of the strong interaction effect is a low-energy, non-perturbative phenomenon, it is fair to say that even now the understanding is qualitative rather than quantitative. What seems to be going on is as follows. The exchange diagram for $D^o$ decay was originally neglected[36] on the grounds of helicity suppression (analogous to suppression of $\pi \to e\nu$) for light quarks in the final state. Including a gluon[41] (either emitted or already present) removes the helicity constraint and makes this amplitude as large or larger than the c-quark decay amplitude. Similar arguments apply to the annihilation diagram for F and the exchange diagrams in $\Lambda_c^+$ decay.[42] If these diagrams dominate, then the relative rates for $D^o \to \bar{K}^o \pi^o$ and $K^- \pi^+$ are governed by $I = 1/2$ final state and $\bar{K}^o \pi^o / K^- \pi^+ \ll 1$ is no longer expected and similarly for $\Lambda_c^+$, $\bar{K}^o p / \Lambda \pi^+ \ll 1$ is not expected. The quark decay model should work and does work when it <u>has to</u> viz. when the decay cannot proceed any other way, i.e.

for the Cabibbo allowed decays of $D^+$. The lifetime of $D^+$ is close to the expected value. In fact there may be a slight suppression[43] of the $D^+$ decay rates due to the Pauli principle suppressing the phase space in $D^+ \equiv cd \to sudd$.

General expectations for charm decays are: a) that $\tau(D^+) > \tau(F^+) \gtrsim \tau(D^o)$, b) $F^+$ will decay into pions significantly, c) baryons such as (cus), (css), (ccu), (ccs), and (ccc) will be long-lived like $D^+$ and (cds), (ccd) will be short-lived like $\Lambda_c^+$. Whether penguin diagrams play an important role can be learned only by studying Cabibbo-suppressed decay modes in great detail. In conclusion, decays of charm turned out to be as complicated as strange particles. Perhaps $m_c$ is in an awkward range, i.e. large enough to make one think that perturbative QCD may be applicable but too small for that to be really true.

## 5. B DECAYS

Let me turn to the next heavier flavor. Assuming in the first approximation that quark decay dominates there are some simple predictions that can be made about decays of B. If $b_L$ is part of a $(t,b)_L$ doublet and $m_t > m_b$, b decays by mixing. If the mixing is constrained by requiring $K_L$-$K_S$ mass difference (real as well as imaginary parts) to be given correctly by the K-M matrix, then $\tau_b$ is constrained to lie between[44] $10^{-13}$ and $10^{-14}$ sec. Branching ratio for b into direct e's or $\mu$'s (taking phase space for $\bar{c}s$ modes to be roughly 1/2 that of the others) is simply $1/7.5 = 13\%$ each. Finally the number of kaons (or the number of strange quarks) is $3/7.5 = 0.4$ if $b \to u$ and $9/7.5 = 1.2$ if $b \to c$. The variation in number of s-quarks, as relative strengths are varied,[45] is given in Fig. 1. These results are compared in Table III to the experimental results from CESR presented here by Thorndyke[46] and by Berkelman.[47] The expected kaon number includes the ($\sim 15\%$) probability of kaon pair production. The numbers do not change much[48] on inclusion of annihilation graph for $B^+(b\bar{u}) \to c\bar{s}$ which seems at first sight to enhance $N_K$ for $b \to u$ coupling.

Table III. B-Decay Properties

| Quantity | Expected in Standard Model | Experimental Value |
|---|---|---|
| $\tau_{B^o} = \tau_{B^+}$ | $10^{-14} < \tau_B < 10^{-13}$ sec | $\tau_B < 3 \cdot 10^{-11}$ sec |
| B.R.$(B \to e/\mu + X)_{direct}$ | 0.26 | $0.235 \pm 0.075$ |
| $N_K$ per B-decay | $b \to c$   1.5<br>$b \to u$   0.7 | $2.5 \pm 0.5 \pm 0.5$ |

It is clear that experiments are in good general agreement with the expectations of standard model and indicate that $b \to u \ll b \to c$. Other means to determine $b \to u/b \to c$ more accurately are, e.g., $\Gamma(B \to \pi e \nu)$ compared to $\Gamma(B \to De\nu)$ (Fig. 2)[45] or the electron momentum spectrum in $e^+e^- \to B\bar{B} \to eX$ to pick out the increased range allowed by $b \to u$ (Fig. 3).[49]

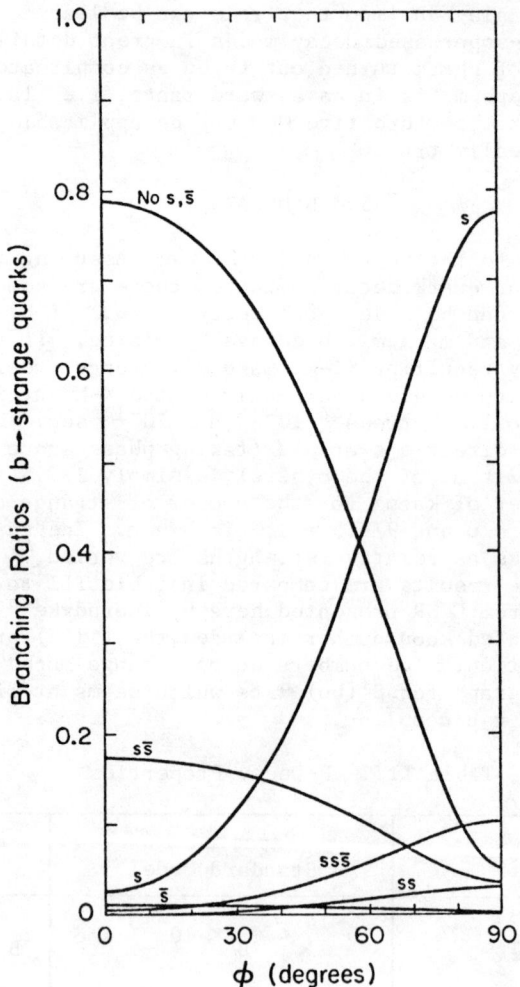

Fig. 1. Probability of strange quarks produced in b decay; $\tan\phi = |b \to c|/|b \to u|$.

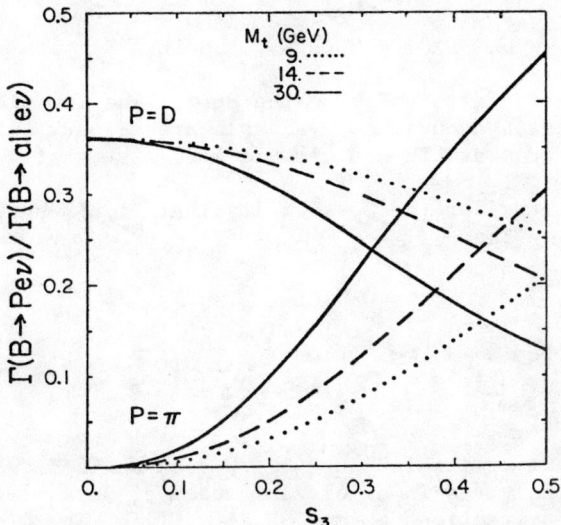

Fig. 2. Branching ratios of $B \to De\nu$ and $B \to \pi e\nu$ as functions of $s_3$ and $m_t$. $|b \to u|/|b \to c|$ goes from 0 to about 1.8.

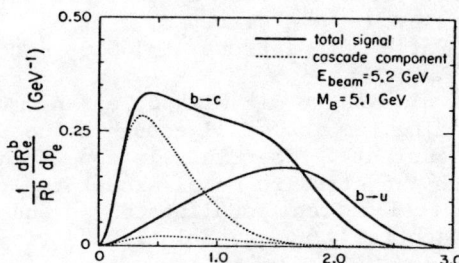

Fig. 3. Single electron spectra for $e^+e^- \to B\bar{B} \to eX$; $R^b \equiv \sigma(e^+e^- \to B\bar{B})/\sigma(e^+e^- \to \mu^+\mu^-)$ and $R^b_e \equiv \sigma(e^+e^- \to B\bar{B} \to eX)/\sigma(e^+e^- \to \mu^+\mu^-)$.

However, it is clear that when $b \to u$ is small compared to $b \to c$, all three methods viz. measuring $N_K$, $B \to Pe\nu$ or $dN/dp_e$ are rather insensitive, and it may be a long wait before we will know the precise value of $|b \to u|/|b \to c|$.

## 6. EXOTIC MODELS FOR B DECAYS

Confront hard facts and bite the dust. The few pieces of information about B decays now available are already sufficient to rule out several exotic and bizarre models suggested for decays of B.

(1) First consider the possibility that $(t,b)_L$ has no mixing with $(u,d,c,s)_L$. Then

$$U_{KM} = \begin{pmatrix} c & s & 0 \\ -s & c & 0 \\ 0 & 0 & 1 \end{pmatrix}, \quad (7)$$

and (1a) b is either stable[50] or (1b) b decays via flavor-changing Higgs coupling only (as in an elegant model of Derman[51]) or (1c) b decays by new interactions beyond $SU(2) \times U(1)$. Now (1a) implies $\tau_b \to \infty$ and hence is ruled out; (1b) implies b can only decay as $b \to de\mu$, $se\tau$, etc., implying a 100% branching ratio into leptons and hence is now ruled out. If b decays as in (1c) by a new interaction beyond $SU(2) \times U(1)$, obviously it cannot be ruled out without a specific proposal.

(2) Next consider the possibility that there is no t-quark and $b_L$ is a singlet under $SU(2)_L$. Then there are two possibilities.

(2a) To preserve natural flavor conservation forbid $b_L$ to mix with $d_L$ and $s_L$. In this case b has to decay by a new interaction beyond $SU(2) \times U(1)$. Again a general statement cannot be made. But several specific proposals made by Georgi et al.,[52] based on $E_6$ or $SU(7)$, predict $b \to \tau^+ e^- d$, $\nu_\tau \nu_e d$... or $b \to \tau^+ \bar{u}u$... and hence again a branching ratio into leptons of 100%. Hence these models are ruled out.

(2b) If $b_L$ is allowed to mix with $d_L$ and $s_L$, then one has to insure that $d,s$ mixing remains small enough to not upset the limits on $\Delta S = 1$ neutral currents. When this is done by hand, the results are very similar to the standard model except the prediction of flavor-changing neutral-current couplings $b \to d$ and $b \to s$. Suppressing $d \to s$ gives two essentially unique solutions[53] viz. either $b \to u$ is dominant with $(b \to d) \sim \frac{1}{2}(b \to u)$ or $b \to c$ is dominant with $(b \to s) \sim \frac{1}{2}(b \to c)$. Hence in addition to all the usual charged-current decay modes, we expect neutral-current modes with clear signatures such as $b \to se^+e^-$, $s\mu^+\mu^-$, and $s\bar{\nu}_i\nu_i$ (for the case $b \to c$). The branching ratios are fixed to be about 2%, 2%, and 12% each, respectively, independently of the mixing angles. <u>It is interesting that the existence of a t-quark can be established in decays of b by ruling out these decay modes at this level</u>.[53] Data from CESR and PETRA should accomplish this soon.

Now in unified models such as $E_6$,[54] which accommodate a singlet b, the b is accompanied by a second -1/3 charge singlet, say b'. Then $b_L$, $b'_L$, $d_L$, $s_L$ all mix and the unique solutions for u → b and c → b above will no longer obtain in general.[55] However, it may be that b → u and b' → c are essentially zero and their couplings are determined by b → c and b' → u alone--certainly the data are consistent with this.

## 7. NON-LEPTONIC DECAYS OF B

The general expectation about the non-leptonic decays of b-flavored hadrons are that quark decay should be dominant, and exchange, annihilation, and penguin diagrams should be unimportant or at least not dominant. Much of the phenomenology has been done on this basis. There is some support for this optimistic view, both from first-order QCD calculations,[41-43] as well as from attempts[56] at summing up many diagrams. On the other hand, perhaps we should be prepared for surprises--after all, we had thought charm decays would be simple also!

Perhaps the most exhaustive thing to do is consider all diagrams with arbitrary strengths[57] and wait for data to tell us which dominate. I have prepared a table (Table IV) in the most simple-minded way. For each mechanism I have listed what final state can be reached. This can then be used in two ways. One can look for final states which have to be there independently of detailed dynamics (Table V). Or one can look at detailed properties of final states to learn something about the mechanism: e.g., Table VI lists properties the final states should satisfy if the quark decay mechanism dominates. Similar tables can be constructed for other possibilities.[58]

## 8. TOP AND BEYOND

If a t-quark exists and has a mass beyond 20 GeV, what are its expected properties? Some are quite obvious and can be listed right away.

We expect[44] t → b to be the dominant coupling (t → s and t → d less than about 14%). We expect that by $m_t$ surely quark decay is dominant and should work, i.e., $\tau(T^0 = t\bar{u}) = \tau(T^+ = t\bar{d}) = \tau(T_s^+ = t\bar{s})$ and $\tau(T) < 10^{-17}$ sec for $m_t > 20$ GeV. We expect $B(T \to eX) \sim 11\%$. In $e^+e^- \to t\bar{t} \to X$, X contains: 1 lepton 40% of the time, up to 10 leptons (via t → b$\bar{b}$c) and up to 12 kaons.[59]

There are more subtle and interesting results in decays of topsilon (topponium?). When $m_t$ becomes large enough, the weak decay rates of $t\bar{t}$, by i) $t\bar{t} \to t\bar{b} + X$ ($W_1$) by W radiation, ii) $t\bar{t} \to b\bar{b}$ ($W_2$)

Table IV.  Final States in b Decay

(a)  b → c dominant

| Diagram Meson | An./Ex. | Peng. | Q Decay | |
|---|---|---|---|---|
| $B^0$ | $c\bar{u}$<br>$D\pi\ldots$<br>$I=1/2$ | $s\bar{u}$<br>$K\pi\ldots$<br>$I=1/2$ | $c\bar{u}d\bar{d}$,<br>$D\pi\ldots$,<br>$I=1/2$ and | $c\bar{c}s\bar{d}$<br>$K\pi\ldots$<br>$3/2$ |
| $B^-$ | --- | $s\bar{u}$<br>$K\pi\ldots$<br>$I=1/2$ | $c\bar{u}\bar{u}d$,<br>$D\pi\ldots$,<br>$I=3/2$, | $c\bar{u}\bar{c}s$<br>$K\pi\ldots$<br>$I=1/2$ |
| $B_s^0$ | $c\bar{c}$<br>$D\bar{D}\pi$ (no $\psi$)<br>$I=0$ | $s\bar{s}$<br>$K\bar{K}\ldots$<br>$I=0$ | $c\bar{u}d\bar{s}$,<br>$DK,F\pi\ldots$,<br>$I=1$ | $c\bar{c}s\bar{s}$<br>$F\bar{F}\ldots$<br>$I=0$ |
| $B_c^-$ | $\bar{u}d$, $\bar{c}s$<br>$\pi\pi\ldots$, $\bar{D}K\ldots$<br>$I=1$, $I=0$ | $\bar{c}s$<br>$\bar{D}K\ldots$<br>$I=0$ | $c\bar{c}\bar{u}d$,<br>$D\bar{D}\pi\ldots$<br>$I=1$ | $c\bar{c}\bar{c}s$<br>$\psi F\pi\ldots$<br>$I=0$ |

(b)  b → u dominant

| | | | | |
|---|---|---|---|---|
| $B^0$ | $u\bar{u}$<br>$\pi\pi\ldots$<br>$I=0,1$ | $d\bar{d}$<br>$\pi\pi\ldots$<br>$I=0,1$ | $u\bar{u}d\bar{d}$,<br>$\pi\pi\ldots$,<br>$I=0,1,2;$ | $u\bar{d}c\bar{s}$<br>$D\bar{K}\ldots$<br>$I=1$ |
| $B^-$ | $d\bar{u}$, $s\bar{c}$<br>$\pi\pi\ldots$, $\bar{D}K\ldots$<br>$I=1$, $I=0$ | $d\bar{u}$<br>$\pi\pi\ldots$<br>$I=1$ | $u\bar{u}\bar{u}d$,<br>$\pi\pi\ldots$<br>$I=0,1,2;$ | $u\bar{u}\bar{c}s$<br>$D\bar{K}\pi\ldots$<br>$I=0,1$ |
| $B_s^0$ | $u\bar{c}$<br>$D\pi\ldots$<br>$I=1/2$ | $d\bar{s}$<br>$\bar{K}\pi\ldots$<br>$I=1/2$ | $u\bar{s}u\bar{d}$,<br>$\bar{K}\pi\ldots$<br>$I=1/2,3/2;$ | $u\bar{s}c\bar{s}$<br>$\bar{D}\bar{K}K\ldots$<br>$I=1/2$ |
| $B_c^-$ | --- | $d\bar{c}$<br>$\bar{D}\pi\ldots$<br>$I=1/2$ | $u\bar{c}\bar{u}d$,<br>$\bar{D}\pi\ldots$,<br>$I=1/2,3/2;$ | $u\bar{c}\bar{c}s$<br>$\bar{D}\bar{D}K\pi\ldots$<br>$I=1/2$<br>$C=-2$<br>$S=-1$ |

Table V. Final States in B Decays Independent of Dynamics

|  | $b \to c$ | $b \to u$ |
|---|---|---|
| $B^o$ | $D\pi...$ or $K\pi...$ | $\pi\pi...$ |
| $B^-$ | $K\pi...$ | $\pi\pi...$ or $\bar{D}K...$ |
| $B_s^o$ | $D\bar{D}...$ or $K\bar{K}...$ | $D\pi...$ or $\bar{K}\pi...$ |
| $B_c^-$ | $\bar{D}K...$ | $\bar{D}\pi$ |

Table VI. Tests that Q Decay Dominates

|  | $b \to c$ | $b \to u$ |
|---|---|---|
| $B^o$ | I=3/2 and 1/2 comparable | c=-1, s=-1 comparable to s=c=0 |
| $B^-$ | c=0, s=1<br>c=1, s=0 comparable | --- |
| $B_s^o$ | c=1, s=-1<br>c=s=0 comparable | s=1, c=0 and<br>c=-1, s=0 comparable |
| $B_c^-$ | --- | c=1, s=0 and<br>c=-2, s=-1 comparable |

by W exchange, and iii) $t\bar{t} \to q_i \bar{q}_i$ or $\ell_i \bar{\ell}_i$ (Z) by Z exchange, become substantial[60,61] as seen in Table VII.

Table VII. Weak Branching Ratios of Topsilon

| $m_t$ | $B_1$ | $B_Z$ | $B_2$ |
|---|---|---|---|
| 30 GeV | 10% | 0 | 3% |
| 45 GeV | 14% | 60% | 0 |
| 60 GeV | 58% | 5% | 7% |

One use[61,62] of these high rates for weak decays is that naked t can be observed <u>below</u> the threshold for $T\bar{T}$ production.

These remarks apply even more strongly for a hypothetical fourth doublet $(t',b')$ with masses assumed to go as $m_{t'} > m_{b'} \sim 100$ GeV. In this case the <u>dominant</u> decay modes of the onium $b'\bar{b}'$ are <u>weak</u>. In particular[61] $\Gamma_1 \sim \Sigma_i |\varepsilon_i|^2 \times 5$ MeV, where $\varepsilon_i$ are the $|b' \to q_i|$ etc. mixings; e.g., if $\Sigma_i |\varepsilon_i|^2 \sim 0.01$, $\Gamma_1 \sim 50$ KeV, and $\Gamma_Z \sim 26$ KeV, whereas $\Gamma_{tot} \sim 95$ KeV. Decays of naked $b'$, of course, are striking: $b' \to q_i W$, with spectacular signatures as emphasized by Bjorken.[63]

## 9. CP VIOLATION

I shall ignore the problem of "strong" CP violation completely and stick to phenomenology of CP violation in weak interactions. Before turning to various proposals, I would like to tell you the contents of a famous paper[2] that everybody quotes but nobody (presumably) reads. In their 1973 paper, Kobayashi and Maskawa observed that the standard model with four quarks and one Higgs doublet has no possibility of CP violation. They suggested several ways to enlarge the number of degrees of freedom so that CP violation becomes possible, in particular one can

   i) enlarge the Higgs sector, or
   ii) include right-handed currents, or
   iii) add more fermions.

Any one of these would allow for CP violation and eventually all of these avenues were studied in detail. The alternative (iii) with six quarks has come to be called the Kobayashi-Maskawa (K-M for short) model.

Consider first the possibility of enlarging the Higgs sector. Then there are two distinct classes of such models. One, to be called Class A, is one in which flavor-changing couplings are allowed to occur. In this case only two Higgs doublets are sufficient.[64] The detailed predictions depend on the specific model, but the following features emerge.[65] If the additional neutral Higgses couple to $\bar{s}d$, they have to be very heavy--several hundred GeV, but if they do not (i.e. there is partial flavor conservation[66]), then the limits on $m_{H^0}$ are not very severe. In $K_L \to 2\pi$, $\varepsilon' \approx 0$; electric dipole moment of the neutron can be between $10^{-24}$ and $\underline{10^{-30}}$ ecm. It is possible to have large CP-violating effects in $D^0$-$\bar{D}^0$, $B^0$-$\bar{B}^0$ systems.

The other class of models (Class B) in which CP violation occurs in the Higgs sector, but no flavor-changing coupling to neutral Higgses are allowed to occur. In this case a minimum of three Higgs doublets is needed. A prototype model was proposed by Weinberg[67] and has been much studied.[68] The most exhaustive study is by Anselm[69] who considered a six-quark model with three Higgs doublets, where CP invariance is broken spontaneously and $U_{KM}$ is real. In this class of models CP violation is dominated by the charged Higgses $H^{\pm}$ and they cannot be too heavy, typically, $m_H \sim$ a few GeV. Hence it is of interest to look for $H^{\pm}$ in decays such as[70] $t \to bH^+$ or $b \to cH^-$ and indirectly in enhanced rates for $F \to \tau\nu_\tau$ or $B_c \to \tau\nu_\tau$. Anselm estimates $|\eta_{+-}/\eta_{oo}| - 1$ to be about 0.05, electric dipole moment of neutron to be about $2 \times 10^{-25}$ ecm, e.d.m. of $\Lambda$ to be $5 \times 10^{-23}$ ecm and $\varepsilon_D \gtrsim 3 \varepsilon_K$. The result for e.d.m. of neutron depends on neglecting the $s\bar{s}$ component in the neutron.

CP violation caused by the presence of right-handed currents has been pursued by Mohapatra[71] and others. A specific example is the $SU(2)_L \times SU(2)_R \times U(1)$ model discussed by Mohapatra and Pati.[72] In this model CP violation is due to phase clash between the parity-conserving and parity-violating parts of the effective weak Hamiltonian, i.e.

$$H_W = e^{i\phi} H_{p.c.} + H_{p.v.} .$$

Predictions of this particular model are $|\eta_{+-}/\eta_{oo}| = |\eta_{+-o}/\eta_{ooo}| = 1$, electric dipole moment of the neutron is $\sim 10^{-29}$ ecm and $|\eta_{+-o}/\eta_{+-} - 1| \sim 0(1)$. Here $\eta_{+-o}$, $\eta_{ooo}$ are the analogous of $\eta_{+-}$ and $\eta_{oo}$ for the $3\pi$ modes of $K_S$.

Finally, let us consider the popular K-M model, where the phase in K-M matrix is solely responsible for all observed CP violation. With K-M parameterization the mixing matrix in Eq. (1) is

$$U_{KM} = \begin{pmatrix} c_1 & s_1 c_3 & s_1 s_3 \\ -s_1 c_2 & c_1 c_2 c_3 + s_2 c_3 e^{i\delta} & c_1 c_2 s_3 - s_2 c_3 e^{i\delta} \\ -s_1 s_2 & c_1 s_2 c_3 - c_2 s_3 e^{i\delta} & c_1 s_2 s_3 + c_2 c_3 e^{i\delta} \end{pmatrix} . \quad (8)$$

There are other,[73] sometimes more convenient, ways to parameterize $U_{KM}$. If $m_{\nu_i}$ are small enough, then from $\beta$-decay, $\mu$-decay, and $K_{e3}$ (including radiative corrections), it has been shown that[74] $s_1 \sim 0.23$ and $s_3 < 0.5$. Then $\delta m_{K_L - K_S}$ (both real and imaginary parts) can be calculated[75] in terms of $s_2$, $s_3$, $\delta$, and $m_t$. Requiring that $\delta m_{K_L - K_S}$ has the observed values then fixes $s_2$ and $\delta$ for every choice[44] of $s_3$ and $m_t$. One can then calculate other CP-violating amplitudes such as D, B, T decays or electric dipole moment of the neutron.

In K-decays the naive expectation is that since $\varepsilon'$ depends on $\bar{s}d \to t\bar{t}$, $c\bar{c} \to u\bar{u}$, $d\bar{d}$ only, it should be very small compared to $\varepsilon$ by IOZ suppression.[76] However, if the penguin graphs are responsible for the dominant $\Delta T = 1/2$ $K_S \to 2\pi$ amplitudes, they will also contribute to $\varepsilon'$. The actual estimate[77] for $|\varepsilon'/\varepsilon|$ varies from $5 \times 10^{-2}$ to $2 \times 10^{-3}$. Another reason for interest in $\varepsilon'$ is that it can fix the quadrant $\delta$ it is in. Re$\varepsilon$ goes as Re$\varepsilon \sim A \sin\delta + B \sin 2\delta = A \sin\delta (1 + c\cos\delta)$, with Re$\varepsilon > 0$, $A > 0$, and $C > 0$, the possible quadrants for $\delta$ are i) $\sin\delta > 0$, $\cos\delta > 0$; ii) $\sin\delta > 0$, $\cos\delta < 0$; and iii) $\sin\delta < 0$, $\cos\delta < 0$. Re$\varepsilon'$ goes as Re$\varepsilon' \sim A \sin\delta$. Hence a measurement of sign of Re$\varepsilon'$/Re$\varepsilon$ can decide between $\sin\delta$ positive or negative. The amount of CP violation in $K \to 3\pi$ system is similar[78] to that in $K \to 2\pi$, e.g., $|\eta_{+-o}/\eta_{+-} - 1| \gtrsim 0.01$, if $\varepsilon'$ is given correctly by Penguinology. The electric dipole moment of the neutron is probably much smaller than originally thought.[79] Estimates range from[80] $10^{-30}$ to less than[81] $10^{-34}$ ecm. I would bet on $10^{-33}$ ecm. or less.

How much CP violation to expect in D decays? Unfortunately, not very much. And what little there is will be difficult to see because the GIM mechanism makes mixing in $D^o-\bar{D}^o$ system so small.[82] Consider $e^+e^- \to D^o\bar{D}^o$ and $D^o\bar{D}^o \to e^+e^-X$ ($N^{+-}$) or $e^+e^+Y$ ($N^{++}$) or $e^-e^-Z$ ($N^{--}$) (via $D^o \leftrightarrow \bar{D}^o$ mixing). Then $\Delta = \dfrac{N^{++} + N^{--}}{2N^{+-}} \gtrsim 10^{-5}$ to $10^{-6}$.

The asymmetry[82,83] due to CP violation is $a = \dfrac{N^{++} - N^{--}}{N^{++} + N^{--}} \gtrsim 10^{-4}$.

So you see why a is so hard to measure. Similarly one can look for $D\bar{D} \to K^+K^+X'$ ($M^{++}$) and $K^-K^-Y'$ ($M^{--}$) and here $\Delta_K \sim \tan^4\theta_C \sim 10^{-3}$.

The corresponding asymmetry $b = \dfrac{M^{++} - M^{--}}{M^{++} + M^{--}}$ then may be easier to measure[84] and also may be larger than a since it has on-shell contributions.

There is the general feeling and hope that somewhere CP-violating effects must be large. Perhaps B decays is the right place.[85] If $b \to u$ is large, the $B^o-\bar{B}^o$ system is a good place, $\Delta$ and a vary from 0.1 to 0.8 and $10^{-3}$ to $10^{-4}$ as $s_3$ goes from 0 to 0.5. If $b \to c$ is large, $B^o_s-\bar{B}^o_s$ is better suited, $\Delta$ going from 1 to 0.5 and a from $10^{-2}$ to $10^{-4}$. Again one can look for asymmetries between particle-antiparticle decay rates[86] such as $(\Gamma(B \to DX) - \Gamma(\bar{B} \to \bar{D}X))/\Gamma(B \to DX) + \Gamma(\bar{B} \to \bar{D}X)$ which depend mostly ($B^o$) or completely ($B^\pm$) on on-shell effects. Depending on the precise values of the K-M angles, these can become as large as 20% in some cases.[87]

A summary of predictions of these various models of CP violation is given in Table VIII. I would like to stress that it is quite possible that CP violation is occurring due to several sources at once, i.e. there are several Higgses violating CP, there is a phase in K-M matrix, etc. Thus, e.g., neutron may have an electric dipole moment at a level between $10^{-25}$ and $10^{-30}$ ecm., but everything else may be as in the K-M model. To my mind, this is a perfectly sensible scenario.

Table VIII. Comparison of Predictions of
Various Mechanisms of CP Violation.

| | Higgs A | Higgs B | RHC | KM |
|---|---|---|---|---|
| $\left|\dfrac{\eta_{+-} - \eta_{oo}}{\eta_{+-}}\right|$ | $\sim 0$ | 0.05 | 0 | 0.002 to 0.02 |
| $\left|\dfrac{\eta_{+-o}}{\eta_{+-}} - 1\right|$ | $\sim 0$ | $\sim 0$ | $O(1)$ | less than 0.01 |
| neutron e.d.m. | $10^{-24} - 10^{-30}$ | $2 \cdot 10^{-25}$ | $10^{-29}$ | $\sim 10^{-33}$ |
| $\Lambda$ e.d.m. | --- | $5 \cdot 10^{-23}$ | --- | --- |
| CPV in D | could be large | small | --- | small |
| CPV in B | could be large | small | --- | small to medium |

## 10. NEUTRINO MASSES, MIXINGS, ETC.

In the standard $SU(2) \times U(1)$ the neutrino mass is guaranteed to be zero by making two assumptions: i) non-existence of $\nu_R$ and ii) lepton conservation. So obviously $m_\nu$ can be generated by dropping one or both of these assumptions. (a) If a $\nu_R$ is allowed, there can be a Dirac mass term for $\nu$ just as for other fermions from the usual $I = 1/2$ Higgs. (b) If there is no $\nu_R$ but lepton number need not be conserved, an $I = 1$ Higgs ($H^{++}$, $H^+$, $H^o$) coupling to $(\nu, e)_L$ and $(\bar{\nu}^c \bar{e}^c)_R$ can give a Majorana mass term to $\nu_L$. (c) A $\nu_R$ can get Majorana mass from a $I = 0$ Higgs or as a bare mass term. (d) All (a), (b), (c) simultaneously present. The phenomenology, especially in oscillations, is almost identical in these cases. The case when all mass terms are present can be distinguished and is discussed by Barger[88] and Langacker[89] in the parallel sessions.

What can be said about the expected values of the neutrino masses? Very little, I am afraid. Like other fermion masses they are not predicted. One can speculate wildly or wisely. A wild speculation[90] would be to recall that

$$m_e/m_\mu \sim O(\alpha)$$

and ask whether

$$m_{\nu_e}/m_e \sim O(\alpha^2)$$

leading to

$$m_{\nu_e} \sim 25 \text{ eV?} \tag{9}$$

More conservatively one may anticipate a scaling such as

$$m_{\nu_e}/m_{\nu_\mu}/m_{\nu_\tau} \sim m_e/m_\mu/m_\tau \sim m_u/m_c/m_t . \tag{10}$$

This can be arranged in theories[91] which have fermion mass matrices in different charge sectors proportional. Note the prediction of t-quark mass between 20 and 30 GeV. In this case the mixing matrices are proportional.

$$U^\dagger = U_{KM} . \tag{11}$$

There are other proposals based on specific Grand Unified Models. In an SO(10) scheme[92] $\nu_R$ gets a very large Majorana mass M, and the Dirac masses are like other fermion masses. Then $m_\nu \sim m_q^2/M$, $m_{\nu_e}/m_{\nu_\mu}/m_{\nu_\tau} \sim m_e^2/m_\mu^2/m_\tau^2$ and if $M \sim 10^{15}$ GeV, the expected neutrino masses are $m_{\nu_e} \sim 10^{-12}$ eV to $m_{\nu_\tau} \sim 10^{-4}$ eV. In a modification of this scheme[93] $\nu_R$ gets a mass at the one-loop level, as a result

$$m_\nu \sim \frac{m_q m_W}{\alpha^2 M}$$

and $\nu$ masses and mixings satisfy the scaling laws (10) and (11). The actual values now range from 1 eV for $\nu_e$ to 3 KeV to $\nu_\tau$. In SU(5) there is no $\nu_R$ in the $10 \oplus \bar{5}$ used to classify the fermions. Then Majorana mass for $\nu_L$ can be generated by higher-order corrections which simulate an effective I = 1 Higgs. This can be arranged in a variety of ways.[94] In theories with left-right symmetry, it is possible[95] to relate the small mass of $\nu$ to the large mass of right-handed W's, e.g.,

$$m_{\nu_\ell} \sim m_\ell^2/gm_{W_R} . \tag{12}$$

As soon as $\nu$'s have unequal masses and mixings, a large number of flavor-changing processes become possible; e.g., decays of $\nu$'s, $\nu$ oscillations, $\mu \to e\gamma$, etc., CP violation in leptonic processes, etc.

These were first discussed in 1962-63 by the Nagoya group[96] and rediscovered by Pontecorvo[97] in 1967. Finally in the context of gauge theories, there was extensive discussion of the whole subject starting in 1974.[98]

To discuss these phenomena, recall the leptonic charged current which is

$$J_\mu^\dagger = (\bar{e}\bar{\mu}\bar{\tau})_L \gamma_\mu U \begin{pmatrix} \nu_1 \\ \nu_2 \\ \nu_3 \end{pmatrix} \tag{13}$$

or

$$\begin{pmatrix} \nu_e \\ \nu_\mu \\ \nu_\tau \end{pmatrix} = U \begin{pmatrix} \nu_1 \\ \nu_2 \\ \nu_3 \end{pmatrix}, \tag{14}$$

where $U$ is a unitary matrix and $\nu_i$ are the mass eigenstates. For three Dirac neutrinos $U$ has only one non-trivial phase just like $U_{KM}$ in the quark sector. Limits on $\nu$ masses have to be reinterpreted since weak decays do not produce mass eigenstates as, e.g., $m_{\nu_e}^2 \to \Sigma_i |U_{ei}|^2 m_i^2$. Attempts to measure $m_{\nu_e}$ (or others) should also take these superpositions into account. If one of the $m_\nu$'s (e.g., $\nu_\tau$) is in the range 2 $m_e$ to 200 MeV, it has many decay modes[99]

$$\nu_\tau \to e^+ e^- \nu, \mu^- e^+ \nu \tag{15}$$
$$\to e\pi$$

with a lifetime $\tau < 10^{-4}$ sec. Also in this range $m_\nu$ can be measured[100] by the ratio of rates $\Gamma(F \to \tau\nu_\tau)/\Gamma(F \to \mu\nu_\mu)$ which depends only on $m_{\nu_\tau}$ and goes from 13.6 to 5 as $m_{\nu_\tau}$ goes from 0 to 200 MeV. Other effects which can be used to measure masses in the MeV range are kinks in Kurie plots[96] or in two-body decay modes such as $K \to e\nu, \mu\nu$, etc.[101]

Rates for flavor-changing processes such as $\mu \to e\gamma$ have been calculated in the standard model.[102] One finds

$$\text{B.R.} (\mu \to e\gamma) = \frac{3\alpha}{32\pi} \left| \sum_i \frac{m_i^2}{m_W^2} U_{\mu i}^* U_{ie} \right|^2, \qquad (16)$$

$$\text{B.R.} (\mu \to e e \bar{e}) = \frac{\alpha}{4\pi} \text{B.R.} (\mu \to e\gamma), \qquad (17)$$

and similarly for $\tau \to \mu\gamma$, $e\gamma$, $ee\bar{e}$, etc. These branching ratios are extremely small, e.g., for $m_{\nu_i} \sim 100$ eV and maximal mixing,

$$\text{B.R.} (\mu \to e\gamma) \sim 10^{-40}. \qquad (18)$$

Radiative decays of neutrinos can be calculated in the same way. For $\nu_i \to \nu_j + \gamma$ one finds[103]

$$\Gamma(\nu_i \to \nu_j + \gamma) = \frac{G^2 m_i^5}{128\pi^3} \left(\frac{9}{16}\right) \frac{\alpha}{\pi} \left| \sum_\alpha \frac{m_\alpha^2}{m_W^2} U_{\alpha i} U_{j\alpha}^* \right|^2, \qquad (19)$$

where $\alpha$ runs over $e$, $\mu$, $\tau$..., and $m_i \gg m_j$. For $\alpha = \tau$ and maximal mixing, these decay rates are, e.g.,

$$\begin{array}{cc} m_i = 1 \text{ eV} & m_i = 1 \text{ MeV} \\ \Gamma(\nu_i \to \nu_j + \gamma) \sim 10^{-45} \text{ sec}^{-1}, & 10^{-15} \text{ sec}^{-1} \end{array} \qquad (20)$$

It has been suggested[104] that if $m_{\nu_i} \sim$ few eV and these rates are not as low, one could look for the decay photons in the UV range from the Galactic Halos. Neutrinos could decay into three neutrinos[90] $\nu_i \to \nu_j \nu_k \bar{\nu}_k$ either radiatively or via flavor-changing neutral current. In the latter case the rate is simply

$$\Gamma = \frac{\varepsilon^2 G^2}{192\pi^3} m_i^5 \qquad (21)$$

and for maximal strength ($\varepsilon \sim 1$), this rate is $10^{-35}$ sec$^{-1}$ for $m_i \sim 1$ eV and $10^{-5}$ sec for $m_i \sim 1$ MeV. Limits on $\varepsilon$ from study of $\bar{\nu}\nu\bar{\nu}\nu$ coupling[105] (in $K \to \mu\nu\nu\nu$, etc.) are not very good, e.g., $\varepsilon < 10^5$!

## 11. NEUTRINO OSCILLATIONS AND CP VIOLATION

Once neutrinos have unequal masses and non-zero mixings, neutrino oscillations can take place. In a weak decay process, one produces a weak eigenstate at $t = 0$, say $\psi(0) = \nu_e$. The mass eigenstates propagate differently and so at a later time

$$\psi(t) = \sum_i U_{ei} \nu_i e^{-iE_i t} . \quad (22)$$

The probability of finding $\nu_\alpha$ ($\alpha = e$, $\mu$ or $\tau$) at $t$ is

$$P(\nu_e \to \nu_\alpha, t) = \left| \sum_i U_{ei} U_{i\alpha}^* e^{-iE_i t} \right|^2 \quad (23)$$

$$= \sum_{i,j} \{ |U_{ei}|^2 |U_{i\alpha}|^2 + 2 \, \text{Re}(U_{ei} U_{i\alpha}^* U_{ej}^* U_{j\alpha}) \cos\delta E_{ij} t$$
$$+ 2 \, \text{Im}(U_{ei} U_{i\alpha}^* U_{ej}^* U_{j\alpha}) \sin\delta E_{ij} t \} . \quad (24)$$

Here it is assumed that the neutrino mass differences are very small compared to the Q value of the weak decay that produced $\nu_e$. Otherwise, $\nu_1$, $\nu_2$, $\nu_3$ are produced with different phase space, and this has to be taken into account. Also if the mass differences are so large that the wave-packets of $\nu_i$ do not overlap, then $\nu_i$ propagate incoherently, the oscillating terms average to zero, and $P(\nu_e \to \nu_\alpha, t)$ is given simply by

$$P(\nu_e \to \nu_\alpha, t) = \sum_i |U_{ei}|^2 |U_{i\alpha}|^2 . \quad (25)$$

The last term in Eq. (24) is a measure of CP violation in the leptonic sector[106] and has so far been ignored in the analysis of data. Since the phase in U may be large, this is not justified.

The tests of CP conservation are easier said than done. E.g., $P(\bar{\nu}_e \to \bar{\nu}_\alpha, t)$ is given by the expression in Eq. (24) with the last term changing sign. So the deviation of $P(\nu_\alpha \to \nu_\beta, t) - P(\bar{\nu}_\alpha \to \bar{\nu}_\beta, t)$ from 0 is a measure of CP violation. This should be checked carefully in $\nu_\mu \to \nu_\tau$, $\bar{\nu}_\mu \to \bar{\nu}_\tau$, $\nu_\mu \to \nu_e$, $\bar{\nu}_\mu \to \bar{\nu}_e$, etc. It is easy to see that if the phase in U is large, it is possible that even if $P(\nu_\alpha \to \nu_\beta, t) \approx 0$ for some $t$, $P(\bar{\nu}_\alpha \to \bar{\nu}_\beta, t)$ can be $O(1)$. One can also look at $P(\nu_\beta \to \nu_\alpha, t)$ which should be the same as $P(\bar{\nu}_\alpha \to \bar{\nu}_\beta, t)$

by CPT but not necessarily equal to $P(\nu_\alpha \to \nu_\beta, t)$. Another way to check CP violation is to measure a single probability over a long enough time (distance), say $P(\nu_\alpha \to \nu_\beta, t)$ to do a Fourier analysis. If $P(\nu_\alpha \to \nu_\beta, t)$ contains a sine and cosine of the same argument, i.e. it contains $\ldots + B \cos\alpha t + C \sin\alpha t + \ldots$, then $C \neq 0$ implies CP and T violation.

In the case of three neutrinos, there is the following result.[107] If $P(\nu_e \to \nu_e, t \to \infty)$ takes the minimum value $1/3$, then CP conservation (real U) forces $P(\nu_\mu \to \nu_\mu, t \to \infty) = P(\nu_\tau \to \nu_\tau, t \to \infty) = 1/2$. Whereas if CP is violated, all diagonal probabilities can reach $1/3$. So if

$$\sum_{\alpha=e,\mu,\tau} P(\nu_\alpha \to \nu_\alpha, t \to \infty) < 4/3, \qquad (26)$$

then either CP is not conserved or there are more than three neutrino mass eigenstates.

If there is CP violation in the lepton sector, there are other possible effects: $e$, $\mu$, $\tau\ldots$ will have non-zero electric dipole moments; in decays such as $\tau \to e\mu\bar{\mu}$, $\mu \to ee\bar{e}$, etc., there are T-violating correlations[108] in the final state, e.g., $\underline{\sigma}\cdot(\underline{p}_1 \times \underline{p}_2)$.

When there are both Majorana and Dirac masses present with the two resulting eigenvalues comparable, the phenomenology of oscillations[109] is as if there is doublet-singlet mixing. Since the singlet component does not interact, part of the beam is effectively lost into a "sterile" component. In this case there is also a small probability for "true" neutrino-antineutrino ($\nu_{e_L} \to \bar{\nu}_{e_R}$) oscillations as originally proposed by Pontecorvo[110] in 1958. However, they are suppressed by the smallness of V + A coupling and of $m_\nu^2/E_\nu^2$.

## 12. CONCLUSION

It seems to me that it is obvious what is to be done over the next few years.

Experimentally, we would like to do the following. Determine the $c \to d$ and $c \to s$ couplings by studying in detail $\nu$-production of charm and the semi-leptonic decays of charm. Similarly, learn $b \to c$ and $b \to u$ as precisely as possible. Establish the detailed patterns in non-leptonic decays of D, F, $\Lambda_c^+\ldots$ and then of $B^0$, $B^-\ldots$. Measure $\varepsilon'$ in K decays accurately and "find" CP violation elsehwere: D and B systems, for example. Measure neutrino masses, mixings, etc.... In short, fix the parameters of the quark and lepton mixing matrices, all masses and clarify the dynamics of non-leptonic decays.

Theoretically, it would be most satisfying to be able to predict all fermion masses and mixings. That is a tall order. In the meantime, for those theorists who can't sleep nights, I have a suggestion: think about p-wave hyperon decay!

I would like to thank Vernon Barger, Jacques Leveille, Peter Rosen, Hirotaka Sugawara, and Mahiko Suzuki for many stimulating discussions and David Unger for assistance. This work is supported in part by the U. S. Department of Energy under Contract DE-AC03-76ER00511.

* * * * *

DISCUSSION (M4 -- S. Pakvasa)

Q1: *M. Schmidt*, Yale University: First a comment: In the CP-violating models of the Higgs type B, I understand that one may expect a measurable T-violating polarization in $K_{\mu 3}$ (an out-of-plane component of $\mu$ polarization). This is due to Weinberg. My question is whether the production of a Majorana $\nu$ mass by a triplet of Higgs upsets the predicted relationship of the mass of the W and the Z

$$m_W = \cos\theta_W \, m_Z$$

which is preserved under the introduction of any number of Higgs doublets?

A1: *Pakvasa*: Cheng and Li have estimated how large a vev is allowed for a complex I = 1 Higgs by the experimental limit on deviation of $\rho$ from 1. They find it can be as large as 1/4 of the vev for I = 1/2. So the limit is not that tight. About T violation in $K_{\mu 3}$ I should point out that (once there is CP violation in the lepton sector) all models expect some T violation in $K_{\mu 3}$ but none (including the Weinberg model) predicts the amount.

Q2: *M. Chen*, MIT: You said there is still one topless B decay model which has not been ruled out. In that model B → hadrons + $\mu^+\mu^-$ at a % level. I would like to comment that the fact at PETRA we haven't observed any dimuon-hadronic events out of more than 200 expected $B\bar{B}$ events from each group should set a stringent limit on that model.

A2: *Pakvasa*: So you think you can rule it out already? That's great - I'd be delighted to see that.

Q3: *S. P. Rosen*, Purdue University: Does $E_6$ allow the b-quark to decay non-leptonically, or only viz semi-leptonic modes?

A3: *Pakvasa*: In the topless, minimal version of $E_6$ the quarks are assigned as (udb) and (csb') in the 27. If b is allowed to mix with d and s, it will have all the usual decay modes (including non-leptonic ones) and in addition the neutral current ones like $b \to s\mu^+\mu^-$, etc., just mentioned. One can forbid b from mixing with d and s and then it is possible to have schemes in which b has no non-leptonic decay modes. So the answer to your question, in short, is that it depends!

Q4: *R. Strub*, CRN Strasbourg: Are the predictions for the decay asymmetry parameter in $\Omega^- \to \Xi^0 \pi^-$ modes different in this simple pole model and in S.V.Z. model?

A4: *Pakvasa:* Yes, I think so because if you don't have the penguin contributing, then the $\alpha$ is zero for all decay modes ($\Lambda K^-$ as well as $\Xi\pi$).

Q5: *Y. Achiman*, Wuppertal: I would like to give more details concerning the topless $E_6$ model, which you presented in the talk and in the answer to Prof. Rosen's question. In addition to the $\begin{pmatrix} u \\ d \\ b \end{pmatrix}_L \begin{pmatrix} c \\ s \\ b' \end{pmatrix}_L$ ordering, one can exchange $b \leftrightarrow b'$ or mix them in an arbitrary way. One can suppress strangeness-changing neutral currents naturally and one has non-leptonic as well as semi-leptonic decays. By the way, the exchanged version involves $b \to s$ decay that may explain the excess of K's observed experimentally.

A5: *Pakvasa:* I agree with you and in fact I did flash this point in one of the transparencies.

## REFERENCES

The list of references given here may seem excessively long. Even then, it is probably not exhaustive, although I have tried to be complete and follow Dickens[111] as much as possible: "Many authors entertain, not only a foolish, but a really dishonest objection to acknowledge the sources from whence they derive much valuable information. We have no such feeling. We are merely endeavouring to discharge, in an upright manner, the responsible duties of our functions; and whatever ambition we may have felt...a regard for truth forbids us to do more than claim the merit of their judicious arrangement and impartial narration.... The labours of others have raised for us an immense reservoir of important facts. We merely lay them on, and communicate them, in a clear and gentle stream... to a world thirsting for...knowledge."

1. S. L. Glashow, Nucl. Phys. <u>22</u>, 579 (1961); A. Salam and J. C. Ward, Phys. Lett. <u>13</u>, 1684 (1964); S. Weinberg, Phys. Rev. Lett. <u>19</u>, 1264 (1967); A. Salam, in <u>Elementary Particle Theory</u> (Nobel

Symposium No. 8), ed. by N. Svartholm (Almquist and Wiksell, Stockholm, 1968), p. 367; S. L. Glashow, J. Iliopoulos, and L. Maiani, Phys. Rev. D $\underline{2}$, 1285 (1970).
2. M. Kobayashi and T. Maskawa, Progr. Theoret. Phys. (Kyoto) $\underline{49}$, 652 (1973).
3. M. Gell-Mann and A. Pais, Proceedings of the 1954 Glasgow Conference on Nuclear and Meson Physics (Pergammon Press, London, 1955), p. 349; M. Gell-Mann and A. H. Rosenfeld, Ann. Rev. Nucl. Science $\underline{7}$, 407 (1954).
4. H. Sugawara, Progr. Theoret. Phys. (Kyoto) $\underline{31}$, 212 (1964); B. W. Lee, Phys. Rev. Lett. $\underline{12}$, 83 (1964).
5. H. Sugawara, Phys. Rev. Lett. $\underline{15}$, 870, 997 (1965); M. Suzuki, ibid. $\underline{15}$, 986 (1965).
6. Y. Hara, Y. Nambu, and J. Schechter, Phys. Rev. Lett. $\underline{16}$, 380 (1966); L. S. Brown and C. M. Sommerfeld, ibid. $\underline{16}$, 751 (1966); C. Bouchiat and Ph. Meyer, Phys. Lett. $\underline{20}$, 329 (1966).
7. A. Kumar and J. C. Pati, Phys. Rev. Lett. $\underline{18}$, 1230 (1967); M. Hirano, K. Fujii, and O. Terazawa, Progr. Theoret. Phys. (Kyoto) $\underline{40}$, 114 (1968); C. Itzykson and M. Jacob, Nuov. Cim. $\underline{48A}$, 655 (1967).
8. A. P. Balachandran, M. Gundzik, and S. Pakvasa, Phys. Rev. $\underline{153}$, 1553 (1967); K. Miura and S. Minamikawa, Progr. Theoret. Phys. (Kyoto) $\underline{38}$, 954 (1967).
9. T. Goto, O. Hara, and S. Ishida, Progr. Theoret. Phys. (Kyoto) $\underline{43}$, 849 (1970); C. H. Llewellyn-Smith, Ann. Phys. (N.Y.) $\underline{53}$, 52 (1969); G. Feldman, T. Fulton, and P. T. Mathews, Nuov. Cim. $\underline{50A}$, 349 (1967); R. P. Feynman, M. Kisslinger, and F. Ravndal, Phys. Rev. D $\underline{3}$, 2706 (1971).
10. J. C. Pati and C. H. Woo, Phys. Rev. D $\underline{3}$, 2920 (1971).
11. K. Wilson, Phys. Rev. $\underline{179}$, 1499 (1969).
12. G. Altarelli and L. Maiani, Phys. Lett. $\underline{52B}$, 351 (1974); M. K. Gaillard and B. W. Lee, Phys. Rev. Lett. $\underline{33}$, 108 (1974).
13. A. I. Vainshtein, V. I. Zakharov, and M. A. Shifman, Pisma Zh. Eksp. Teor. Fiz. $\underline{22}$, 123 (1975) (English translation JETP Lett. $\underline{22}$, 55 (1975)); Nucl. Phys. $\underline{B120}$, 316 (1977); Zh. Eksp. Teor. Fiz. $\underline{72}$, 1275 (1977) (English translation Sov. Phys. JETP $\underline{45}$, 670 (1977). See C. T. Hill, these Proceedings.
14. C. Schmid, Phys. Lett. $\underline{66B}$, 353 (1977); A. Le Yaouane, Phys. Lett. $\underline{72B}$, 53 (1977); J. F. Donoghue et al., Phys. Rev. D $\underline{21}$, 186 (1980); Y. Abe et al., Conference Papers #300 and #320; D. Tadic and J. Trampetic, Conference Papers #521 and #522; Riazuddin and Fayyázudin, Phys. Rev. D $\underline{18}$, 1578 (1978).
15. Most recently this has been stressed by M. K. Gaillard and J. Finjord, LAPP-TH-11 (1979).
16. For a dissenting point of view see M. Scadron, Conference Paper #703 and references cited therein.
17. See e.g. P. N. Dobson, S. Pakvasa, and S. F. Tuan, Hadronic Journal $\underline{1}$, 476 (1978) and references cited therein.
18. S. Oneda and A. Wakasa, Nucl. Phys. $\underline{1}$, 445 (1956); S. Oneda, J. Pati, and B. Sakita, Phys. Rev. $\underline{119}$, 482 (1960).

19. R. P. Feynman, Symmetries in Elemetary Particle Physics, ed. by A. Zichichi (Academic Press, New York, 1965), p. 111.
20. Particle Data Group, Rev. Mod. Phys. 52, No. 2 (1980).
21. S. Glashow and R. H. Socolow, Phys. Lett. 10, 143 (1964).
22. M. Suzuki, Progr. Theoret. Phys. (Kyoto) 32, 138 (1964).
23. A. P. Balachandran et al., Phys. Rev. 153, 1553 (1967); L. R. Ram Mohan, Phys. Rev. D 1, 266 (1970).
24. M. Bourqin et al., CERN Report CERN-EP/79-107; G. Sauvage, in Proceedings of XIX International Conference on High Energy Physics (Tokyo, 1978), ed. by S. Homma et al., Physical Society of Japan, 1979, p. 427.
25. J. Finjord, Phys. Lett. 76B, 116 (1978); H. Galic, D. Tadic, and J. Trampetic, ibid. 89B, 249 (1980).
26. Y. Hara, Phys. Rev. Lett. 12, 378 (1964).
27. R. H. Graham and S. Pakvasa, Phys. Rev. 140, B1144 (1965); L. R. Ram Mohan, Phys. Rev. D 3, 785 (1971); G. R. Farrar, Phys. Rev. D 4, 212 (1971).
28. A. Manz et al., CERN Report CERN-EP/80-59. This paper contains a list of many theoretical proposals.
29. R. Dashen, M. Gell-Mann, S. Frautschi, and Y. Hara, Proceedings of XII International Conference on High Energy Physics (Dubna, 1964), ed. by Ya A. Smorodinsky et al., Atomizdat, Moscow, 1966, Vol. 2, p. 192.
30. G. Altarelli, R. K. Ellis, L. Maiani, and R. Petronzio, Nucl. Phys. B88, 215 (1975).
31. The more recent calculations are by: M. Gari and J. Schlitter, Phys. Lett. 59B, 118 (1975); K. R. Lassey and B.H.J. McKellar, Nucl. Phys. A260, 413 (1976); J. P. Leroy et al., ibid. A280, 377 (1977); M. Gari et al., Phys. Rev. D 11, 1485 (1975); M. A. Box et al., Nucl. Phys. A271, 412 (1976); E. M. Henley and L. Wolfenstein, Nucl. Phys. A300, 265 (1978); B. Desplanques and J. Missimer, Phys. Lett. 84B, 363 (1979); Nucl. Phys. A300, 286 (1978). A complete list of references can be traced from the following review articles: D. Tadic, Rep. Progr. Phys. 43, 67 (1980); B. Desplanques, J. F. Donoghue, and B. R. Holstein, Ann. Phys. (N.Y.) 124, 449 (1980); E. Fischbach, and D. Tadic, Phys. Rep. C6, 125 (1973); M. Box et al., J. Phys. G 1, 493 (1975). See also D. Tadic, these Proceedings.
32. C. A. Barnes et al., Phys. Rev. Lett. 40, 840 (1978); H. Wäffler, unpublished (1978); J. F. Cavaignac et al., Phys. Lett. 67B, 148 (1977); P. R. Maurenzig, Proceedings of Neutrino 79 (Bergen, June 1979), ed. A. Haatuft and C. Jarlskog, Vol. 2, p. 179.
33. V. M. Lobashov et al., Nucl. Phys. A197, 241 (1972).
34. G. S. Danilov, Phys. Lett. 18, 40 (1965).
35. K. R. Lassey and B.H.J. McKellar, Phys. Rev. C 12, 721 (1975).
36. J. Ellis, M. K. Gaillard, and D. V. Nanopoulos, Nucl. Phys. B100, 313 (1975); D. Fakirov and B. Stech, Nucl. Phys. B133, 315 (1978); N. Cabibbo and L. Maiani, Phys. Lett. 73B, 418 (1978).
37. See the review by G. Trilling, these Proceedings.
38. This discussion is brief as there are several excellent reviews at this Conference: by N. Cabibbo, by J. Leveille, and by L. L. Wang, these Proceedings.

39. T. Hayashi, M. Nakagawa, H. Nitto, and S. Ogawa, Progr. Theoret. Phys. (Kyoto) 49, 351 (1973); ibid. 52, 636 (1974).
40. E. Ma, S. Pakvasa, and W. A. Simmons, Z. Physik C5, 309 (1980).
41. M. Bander, D. Silverman, and A. Soni, Phys. Rev. Lett. 44, 7 (1980); R. Cahn, M. Suzuki, J. Leveille, V. Barger, Y. Kang (unpublished); H. Fritzsch and P. Minkowski, Phys. Lett. 90B, 455 (1980); W. Bernreuther, O. Nachtmann, and B. Stech, Z. Physik C4, 257 (1980).
42. V. Barger, J. Leveille, and P. Stevenson, Phys. Rev. Lett. 44, 226 (1980); S. P. Rosen, ibid. 44, 4 (1980).
43. K. Jagannathan and V. S. Mathur, Phys. Rev. D 21, 3165 (1980); Conference Papers #60 and #66; B. Guberina et al., Phys. Lett. 89B, 111 (1979).
44. V. Barger, W. F. Long, and S. Pakvasa, Phys. Rev. Lett. 42, 1585 (1979); J. of Phys. G5, L147 (1979); R. Shrock, L. L. Wang, and S. B. Treiman, Phys. Rev. Lett. 42, 1589 (1979); J. Hagelin, Phys. Rev. D 20, 2893 (1979). Such estimates are subject to the usual uncertainties in evaluation of hadronic matrix elements as emphasized by L. Wolfenstein, Nucl. Phys. B160, 50 (1979).
45. V. Barger, W. F. Long, and S. Pakvasa (unpublished).
46. E. Thorndyke, these Proceedings.
47. K. Berkelman, these Proceedings.
48. J. Leveille, private communication.
49. V. Barger, T. Gottschalk, and R.J.N. Phillips, Phys. Lett. 82B, 445 (1979); A. Ali, Z. Physik C1, 25 (1978).
50. R. N. Cahn, Phys. Rev. Lett. 40, 80 (1978); R. N. Mohapatra and G. Senjanovic, Phys. Lett. 73B, 176 (1978).
51. E. Derman, Phys. Rev. D 19, 133 (1979).
52. H. Georgi and M. Machacek, Phys. Rev. Lett. 43, 1635 (1979); H. Georgi and S. L. Glashow, Nucl. Phys. B167, 173 (1980).
53. V. Barger and S. Pakvasa, Phys. Lett. 81B, 195 (1979); see also G. L. Kane, Michigan preprint, June 1980.
54. F. Gürsey, P. Ramond, and P. Sikivie, Phys. Lett. 60B, 177 (1976); F. Gürsey and M. Serdaroglu, Lett. Nuovo Cimento 21, 28 (1978); Y. Achiman and B. Stech, Phys. Lett. 77B, 389 (1978).
55. This point was raised to me by Y. Achiman.
56. M. Suzuki, Conference Paper #183.
57. As proposed for charm decays by T. Rizzo and L. L. Wang, Conference Paper #658.
58. For exchange diagrams such tests have been discussed by E. Ma et al., Ref. 40; I. Bigi and M. Fukugita, CERN Reports; S. P. Rosen, Conference Paper #113; and J. Leveille, these Proceedings.
59. N. Cabibbo and L. Maiani, Phys. Lett. 87B, 366 (1979).
60. K. Fujikawa, Progr. Theoret. Phys. (Kyoto) 61, 1186 (1979).
61. S. Pakvasa, F. Halzen, M. Dechantsreiter, and D. Scott, Phys. Rev. D 20, 2862 (1979).
62. G. J. Tarnopolsky, Phys. Lett. 79B, 451 (1978).
63. J. Bjorken, SLAC-PUB-2281, 1979.

64. T. D. Lee, Phys. Rev. D $\underline{8}$, 1226 (1973); Phys. Rep. $\underline{9C}$, 143 (1979); P. Sikivie, Phys. Lett. $\underline{65B}$, 141 (1976).
65. A. B. Lahanas and N. J. Papadopoulos, Phys. Rev. D $\underline{19}$, 2158 (1979); J. Jacquot, Lett. Nuovo Cimento $\underline{26}$, 155 (1979).
66. S. Pakvasa and H. Sugawara, Phys. Lett. $\underline{73B}$, 61 (1978).
67. S. Weinberg, Phys. Rev. Lett. $\underline{37}$, 657 (1976).
68. N. Deshpande and E. Ma, Phys. Rev. D $\underline{16}$, 1583 (1977); A. Ali and Z. Aydin, Nucl. Phys. $\underline{B148}$, 165 (1979); A. A. Anselm and D. I. Dyakonov, Nucl. Phys. $\underline{B145}$, 271 (1978).
69. A. A. Anselm and N. G. Uraltsev, Yad. Fiz. $\underline{30}$, 465 (1979) (English translation Sov. J. Nucl. Phys. $\underline{30}$, 240 (1979)).
70. E. Golowich and T. C. Yang, Phys. Lett. $\underline{80B}$, 245 (1979).
71. R. N. Mohapatra, Phys. Rev. D $\underline{6}$, 2023 (1972); R. N. Mohapatra, J. Pati, and L. Wolfenstein, Phys. Rev. D $\underline{11}$, 3319 (1975).
72. R. Mohapatra and J. C. Pati, Phys. Rev. D $\underline{11}$, 566 (1975).
73. L. Maiani, Phys. Lett. $\underline{68B}$, 183 (1976).
74. M. Roos, Nucl. Phys. $\underline{B77}$, 420 (1974); R. Shrock and L. L. Wang, Phys. Rev. Lett. $\underline{41}$, 1692 (1978); A. Sirlin, S. S. Shei, and H. S. Tsao, Phys. Rev. D $\underline{19}$, 981 (1979).
75. A complete calculation of all the graphs in $\bar{s}d \leftrightarrow \bar{d}s$ for arbitrary masses of the internal quarks has now been done by several authors. Vysotsky (ITEP Preprint #113), V. Goffin (unpublished), E. Ma and A. Pramudita (unpublished), F. Gilman and M. Wise, Phys. Rev. D $\underline{20}$, 2392 (1980); T. Inami and C. S. Lim, Conference Paper #238. See M. B. Wise, these Proceedings.
76. S. Pakvasa and H. Sugawara, Phys. Rev. D $\underline{14}$, 305 (1976).
77. F. Gilman and M. Wise, Phys. Lett. $\underline{83B}$, 83 (1979); Y. Prokhorov, Yad. Fiz. $\underline{30}$, 111 (1979); B. Guberina and R. D. Peccei, Nucl. Phys. $\underline{B163}$, 289 (1980); J. S. Hagelin, Harvard Report (1980). See also M. B. Wise, these Proceedings.
78. L. F. Li and L. Wolfenstein, Phys. Rev. D $\underline{21}$, 178 (1980).
79. J. Ellis, M. Gaillard, and D. Nanopoulos, Nucl. Phys. $\underline{B131}$, 285 (1977).
80. D. V. Nanopoulos, A. Yildiz, and P. H. Cox, Harvard Preprint HUTP-79/A048; B. F. Morel, Harvard Preprint HUTP-79/A009.
81. E. P. Shabalin, ITEP Preprints #12 and #13 (1980).
82. J. Ellis et al., Ref. 79; A. Ali and Z. Aydin, Ref. 68.
83. E. Ma, W. Simmons, and S. F. Tuan, Phys. Rev. D $\underline{20}$, 2883 (1980); V. Barger et al. (unpublished).
84. D. Dorfan, private communication.
85. J. Ellis, M. Gaillard, D. Nanopoulos, and S. Rudaz, Nucl. Phys. $\underline{B131}$, 285 (1977); A. Ali and A. Aydin, Ref. 68; V. Barger, W. F. Long, and S. Pakvasa, Phys. Rev. D $\underline{21}$, 174 (1980); E. Ma, W. A. Simmons, and S. F. Tuan, Ref. 83; J. S. Hagelin, Phys. Rev. D $\underline{20}$, 2393 (1980).
86. M. Bander, D. Silverman, and A. Soni, Phys. Rev. Lett. $\underline{43}$, 242 (1979).
87. A. Carter and A. Sanda, Phys. Rev. Lett. $\underline{45}$, 952 (1980).
88. V. Barger, these Proceedings.
89. P. Langacker, these Proceedings.

90. K. Tennakone and S. Pakvasa, Phys. Rev. Lett. $\underline{13}$, 757 (1971).
91. S. Pakvasa and H. Sugawara, Phys. Lett. $\underline{82B}$, 105 (1979).
92. M. Gell-Mann, P. Ramond, and R. Slansky, in Supergravity, ed. by P. van Nieuvenhuizen and D. Z. Freedman, N. Holland (1979); P. Ramond, CALT-68/1709 (1979); R. Barbieri et al., Phys. Lett. $\underline{90B}$, 91 (1980). The properties of the mixing matrix U in this case has been studied by S. Hama, K. Milton, S. Nandi, and K. Tanaka, Conference Paper #783. Such a suggestion for generating neutrino mass but without grand unification was made independently by T. Yanagida, in Proceedings of the Workshop on the Unified Theory and Baryon Number in the Universe, ed. by O. Sawada and A. Sugamoto, KEK (1979).
93. E. Witten, Harvard Preprint HUTP-79/A076.
94. J. Ellis and M. K. Gaillard, CERN Preprint TH-2787 (1979); A. Zee, Penn. Preprint UPR-0150T (1980); L. Wolfenstein, CMU Preprint COO-0150T (1980).
95. R. N. Mohapatra and G. Senjanovic, Phys. Rev. Lett. $\underline{44}$, 912 (1980).
96. Z. Maki, M. Nakagawa, and S. Sakata, Progr. Theoret. Phys. (Kyoto) $\underline{28}$, 870 (1962); M. Nakagawa et al., ibid. $\underline{30}$, 258 (1963).
97. B. Pontecorvo, JETP $\underline{53}$, 1717 (1967) (English translation, Sov. Phys. JETP $\underline{26}$, 989 (1968)).
98. S. Eliezer and D. A. Ross, Phys. Rev. D $\underline{10}$, 3088 (1979); S. Eliezer and A. Swift, Nucl. Phys. $\underline{B105}$, 45 (1976); A. Szymancha, J. Phys. $\underline{G2}$, 193 (1977); H. Fritzsch and P. Minkowski, Phys. Lett. $\underline{62B}$, 721 (1976); T. P. Cheng and L. F. Li, Phys. Rev. D $\underline{17}$, 2375 (1978).
99. E. W. Kolb and T. Goldman, Phys. Rev. Lett. $\underline{43}$, 897 (1979).
100. M. P. Rekalo, Pis'ma Zh. Eksp. Teor. Fiz. $\underline{27}$, 588 (1978) (English translation JETP Lett. $\underline{27}$, 555 (1978)).
101. R. Shrock, Stony Brook Preprint (1980).
102. B. W. Lee, S. Pakvasa, R. Shrock, and H. Sugawara, Phys. Rev. Lett. $\underline{38}$, 937 (1977); H. Fritzsch, Phys. Lett. $\underline{67B}$, 451 (1977).
103. S. T. Petcov, Yad. Fiz. $\underline{25}$, 641 (1977) (English translation, Sov. J. Nucl. Phys. $\underline{25}$, 340 (1977); Erratum $\underline{25}$, 698 (1977)).
104. A. de Rújula and S. L. Glashow, Phys. Rev. Lett. $\underline{45}$, 942 (1980). See also S. Pakvasa and K. Tennakone, Phys. Rev. Lett. $\underline{28}$, 1415 (1972).
105. D. Yu. Bardin, S. M. Bilenky, and B. Pontecorvo, Phys. Lett. $\underline{32B}$, 121 (1970).
106. N. Cabibbo, Phys. Lett. $\underline{72B}$, 333 (1978).
107. L. Wolfenstein, Phys. Rev. D $\underline{18}$, 958 (1978); S. Nussinov, Phys. Lett. $\underline{63B}$, 201 (1976).
108. S. B. Treiman, F. Wilczek, and A. Zee, Phys. Rev. D $\underline{16}$, 152 (1977); J. F. Donoghue, Phys. Rev. D $\underline{18}$, 1632 (1978).
109. The presence of both Majorana and Dirac masses was discussed by S. M. Bilenky and B. M. Pontecorvo, Lett. Nuovo Cimento $\underline{17}$, 569 (1976). In the context of gauge models, there has been much discussion of this phenomenon recently: V. Barger,

P. Langacker, J. Leveille, and S. Pakvasa, Phys. Rev. Lett. $\underline{45}$, 692 (1980); S. M. Bilenky, J. Hosek, and S. Petcov, Dubna Preprint (1980); Wu Dan Di, Harvard Preprint HUTP-80/A032; T. Yanagida and M. Yoshimura, KEK Preprint TH-14 (1980); J. Schechter and J. P. Valle, Syracuse Preprint SU-4217-167 (1980); T. P. Cheng and L. F. Li, CMU Preprint COO-3066-152 (1980); M. Gell-Mann, R. Slansky, and G. Stephenson, unpublished (1980).

110. B. Pontecorvo, JETP $\underline{34}$, 247 (1958) (English translation, Sov. Phys. JETP $\underline{7}$, 172 (1958)); J. N. Bahcall and H. Primakoff, Phys. Rev. D $\underline{15}$, 3463 (1978).

111. C. Dickens, <u>The Posthumous Papers of the Pickwick Club</u> (Chapman and Hall, London, 1837), Oxford University Press, London (1948), p. 45.

Chapter 12

Hadron Dynamics, Experiment and Theory

# LIGHT QUARK-HADRON SPECTROSCOPY

L. Montanet, CERN, Speaker

Y. Goldschmidt-Clermont, CERN, Chairman

P. Estabrooks, Carleton,

I. Gaines, Fermilab,

Scientific Secretaries

LIGHT QUARK-HADRON SPECTROSCOPY

L. Montanet
CERN, Geneva, Switzerland

1. INTRODUCTION

The aim of hadron spectroscopy is not merely to catalogue the states in the quark model, but also to provide some information about basic questions such as the role of spin, orbital, and radial excitations of the quark system, the quark confinement mechanism, etc. Still, the main guiding line of this review will be to confront the experimental observations with the non-relativistic quark model, in the hope that some facts will force us to introduce new concepts, the MIT bag model, glueballs, etc.

Our plan is straightforward: we shall discuss, in turn, systems which may consist of one, two, three, four ... quarks.

2. SEARCHES FOR QUARKS

Quarks have been looked for using accelerators (and cosmic rays) or searched for in matter. The accelerator searches are limited by the mass of the free quarks, those in matter by their abundance. Therefore these two series of attempts are complementary.

Yock[1] has used a cosmic-ray telescope at the sea level. The telescope consists of six scintillator planes, two wide-gap spark chambers, and steel absorbers. The scintillators provide a time of flight over a 2 m path length, and pulse-height measurements before and after the absorber. The trigger accepts slow particles ($\beta < 0.5$) which have penetrated the absorber. Three possible examples of fractionally charged particles are given. The measured masses and charges are $m_1 > 5.3\ m_p$, $m_2 > 2.8\ m_p$, $m_3 > (9.3 \pm 3.0)\ m_p$, and $q_1 = (\pm 0.75 \pm 0.05)$, $q_2 = (\pm 0.70 \pm 0.05)$ and $q_3 = (\pm 0.89 \pm 0.06)$, respectively. Before considering these events as convincing evidence of fractionally charged particles, the multitrack efficiency of the spark chambers has to be demonstrated.

Bussière et al.[2] have searched for long-lived particles in 200-240 GeV/c proton-nucleon collisions. Pulse heights and times of flight of eight scintillation counters and of four Čerenkov counters (of which two were of the DISC type) were recorded. Advantage was taken of a superconducting RF separator to remove high-velocity particles, bringing the acceptance to more than $2 \times 10^7$ particles per burst. The time measurement accuracy of each scintillator was ±0.15 ns. The 200 MHz RF structure of the CERN Super Proton Synchrotron (SPS) proton beam was used to give a time-of-flight resolution of ±0.6 ns over the 210 m beam length. Several hundreds of antitritons and $^3\overline{\text{He}}$ have been recorded. No quark candidate was detected, the upper limit at 95% confidence level for the ratio $q/\pi$ for particles q with charge $\pm^2/_3$ and masses $1 < m < 5$ GeV being $10^{-10}$.

The CBFRB Collaboration[3] has been looking for quarks in $\nu, \bar{\nu}$ interactions on lead nuclei. Candidates from an electronic pulse-height analysis were examined more closely by looking at the tracks left in an avalanche chamber. No good candidate has yet been found. An upper limit for charged $\pm\frac{1}{3}$ particles to normal secondary hadrons of $10^{-4}$ is achieved.

Bartel et al.[4] have searched for free quarks in $e^+e^-$ annihilations at PETRA in the energy range 30 to 35 GeV, using the JADE detector. The particle identification is given by 48 measurements of dE/dx in the drift chamber gas. Exclusive and inclusive quark production were both studied. There is no evidence for tracks with abnormal ionization, an upper limit for $\sigma(e^+e^- \to q\bar{q})/\sigma(e^+e^- \to \mu^+\mu^-)$ being of the order of $10^{-2}$ and $10^{-1}$ for exclusive and inclusive production, respectively, with masses in the range $1 < m < 15$ GeV.

At this conference, the results of two experiments on the search for quarks in matter have been reported. Marinelli and Morpurgo[5] have studied the levitation of small steel balls (Ø of 0.2 and 0.3 mm) suspended magnetically between two capacitor plates. It is shown that, with a very good precision, the force acting on the balls takes the form of

$$F_x = \left(q + \alpha \frac{\partial E_x}{\partial x} + \beta \frac{\partial H_x}{\partial x}\right) E_x ,$$

where q is the electric charge to be measured, $\alpha$ is the polarizability of the balls $\alpha = 4\pi\epsilon_0 r^3$, $E_x$ is the applied electric field, and $H_x$ is the magnetic field in which the ball is levitating; $\alpha(\partial E_x/\partial x)$, known as the patch effect, is due to non-uniformities of the electric potential on the surface of the capacitor plates. This patch effect can induce large errors and must be kept under careful control during the experiment. To work at constant patch, the levitation chamber was built in such a way that several balls could be measured in succession without opening the chamber. Marinelli and Morpurgo find that the drift observed for several balls cannot be attributed to this patch effect, but to a new force, of magneto-electric nature. This is the term $\beta(\partial H_x/\partial x)$ which appears in the above formula, $\beta$ being a coefficient which is ball-dependent. To isolate this effect, Marinelli and Morpurgo measure the apparent charge under two different conditions, given by

$$\frac{\partial H_x^\pm}{\partial x} = \left(\frac{\partial H_x}{\partial x}\right)_0 \pm \Delta\left(\frac{\partial H_x}{\partial x}\right) .$$

Then the apparent charge can be written:

$$Q_R^\pm = q + \alpha \frac{\partial E_x}{\partial x} + \beta \frac{\partial H_x^\pm}{\partial x}$$

The results of the measurements for 23 balls with Ø = 0.2 mm and for 24 balls with Ø = 0.3 mm are shown in Figs. 1 and 2, respectively. Note that a patch effect is visible for the larger balls, at least for the run corresponding to the symbol ⊙, but all the measurements are compatible with Q = 0 ± 0.05e, i.e. no quark is found in 3.4 mg of steel.

Instead of using a ferromagnetic material, La Rue, Phillips and Fairbanks[6] use a superconducting material, niobium. In previous publications[7,8], evidence for the existence of fractional charge was presented. The new measurements, which consist in 21 independent measurements on 5 different balls, support the previous results. During these measurements it has been found that the charge of the same ball can jump by $\frac{1}{3}$ e units, indicating that the fractional charge can be transferred either to or from the surface of the ball when it is in contact with other substances. Altogether, 39 independent measurements have been made on 13 balls. The results are shown in Fig. 3. It seems that the magneto-electric force which explains the drifts observed by Marinelli and Morpurgo cannot account for the observations made by La Rue et al., since they find charges consistent with 0 or $\pm\frac{1}{3}$ and observe no continuous drift with time.

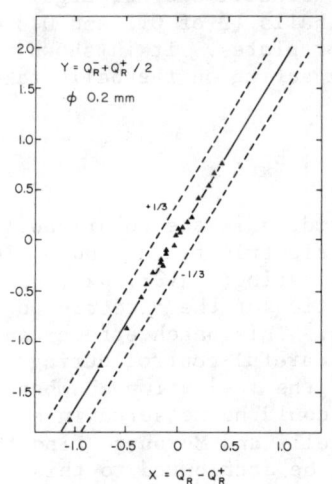

Fig. 1

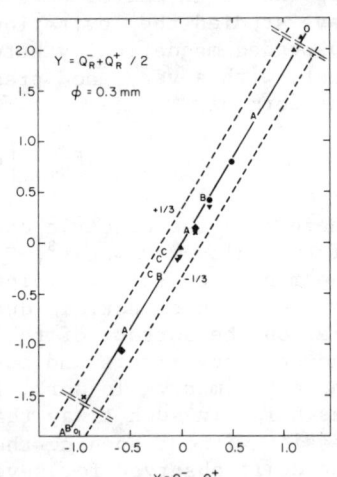

Fig. 2

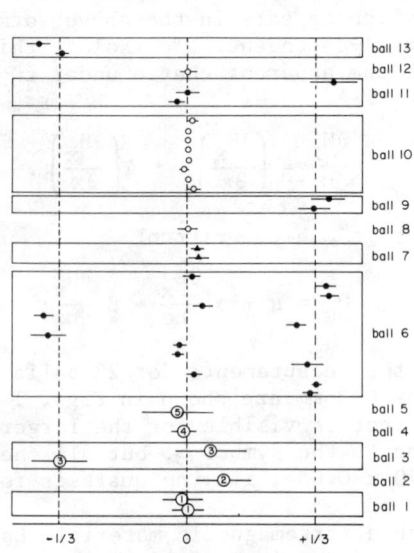

Fig. 3

Should we attribute the differences in these results to the differences in the nature of the tested material? Are there solid-state physics effects that are not yet completely understood? Or, as suggested by Lackner and Zweig[9], are there atomic chemistry effects? If electric charges of $\pm 1/3$ have effectively been observed, should they be attributed to quarks? It is now too early to answer these questions, but the measurements and investigations now in progress will hopefully shed more light on this important problem in the near future.

## 3. MAGNETIC MOMENTS OF THE QUARKS

Handler et al.[10] have reported new accurate measurements on the magnetic moment of the $\Xi^0$ and $\Xi^-$. They are given in Table I, together with recently published results on the $\Lambda^0$ [11] and $\Sigma^+$ [12], as well as with the known values of the magnetic moments of other stable baryons.

Table I  Magnetic moments of the baryons

(unit: $\mu_N = e\hbar/2m_p c$)

| Baryon | Experiment | Quark model |
|---|---|---|
| p | +2.7928456 | Input |
| n | -1.91304184 | Input |
| $\Lambda$ | -0.6138 ± 0.0047 | Input |
| $\Sigma^+$ | +2.33 ± 0.13 | 2.67 |
| $\Sigma^-$ | -1.48 ± 0.37 | -1.09 |
| $\Xi^0$ | -1.237 ± 0.016 | -1.44 |
| $\Xi^-$ | -0.75 ± 0.06 | -0.50 |

Using the static quark model relations,

$$p = 4/3\, u - 1/3\, d ,$$
$$n = 4/3\, d - 1/3\, u ,$$
$$\Lambda = s ,$$
$$\Sigma^+ = 4/3\, u - 1/3\, s ,$$
$$\Sigma^- = 4/3\, u - 1/3\, s ,$$
$$\Xi^0 = 4/3\, s - 1/3\, u ,$$
$$\Xi^- = 4/3\, s - 1/3\, d ,$$

where the particle symbols stand for their magnetic moments, one can predict the magnetic moments of $\Sigma^+$, $\Sigma^-$, $\Xi^0$, $\Xi^-$ (Table I).

We see that the static quark model gives a qualitative understanding but with the new accurate measurements of the $\Sigma^+$, $\Xi^0$, and $\Xi^-$, any attempt to give a good quantitative interpretation fails badly.

Non-static effects (orbital, relativistic, exchange couplings) can now be usefully studied[13-17]. Relativistic effects are suggested[13] as the dominant symmetry-breaking mechanism. It will be interesting to watch the results of new measurements undertaken on the $\bar{\Sigma}^-$, $\langle\Lambda|\Sigma^0\rangle$, and $\bar{\Omega}^-$.

Nevertheless, using as first approximation the quark magnetic moments deduced from the baryon magnetic moments,

$$u = +1.85, \quad d = -0.97, \quad s = -0.61,$$

one can predict the radiative decay of the vector mesons into pseudo-scalars $V \to P\gamma$, which are pure M1 transitions (Table II).

Table II  Radiative width of vector mesons (keV)

| Meson | Experiment | Ref. | Quark model |
|---|---|---|---|
| $\rho^-$ | 67 ± 7 | (19) | 76 |
| $\omega^0$ | 789 ± 92 | (20) | 798 |
| $K^{*-}(892)$ | 60 ± 15 | (18) | 83 |
| $K^{*0}(892)$ | 75 ± 35 | (21) | 137 |

In Table II these predictions are compared with recent measurements[18,19]. The new value observed for the radiative decay of the $\rho^-$ eliminates one of the most striking disagreements encountered previously by the quark model. Kamal and Kane[22] have suggested that a possible explanation of the earlier results, i.e. $\Gamma(\rho \to \pi\gamma) = 35 \pm 10$ keV obtained with 23 GeV/c pion beams, instead of the 200 GeV/c used in Ref. 18, may be found in the presence of isovector hadronic exchange (in addition to the Primakoff effect). They also note that the importance of this exchange could invalidate also the $K^{*0}(892)$ measurements.

## 4. MESON SPECTROSCOPY

If a simple hadronic oscillator potential is assumed for the $q\bar{q}$ model of the mesons, their spectroscopy can be reduced to a two-parameter classification of nonets,

$$M \sim (2n + L)M_0 = N \cdot M_0 ,$$

where n is the radial degree of excitation, L the internal orbital angular momentum.

For $N = 0$ we have the two well-known nonets of pseudoscalar and vector mesons ($J^{PC} = 0^{-+}, 1^{--}$).

For $N = 1$ we expect four nonets, with $J^{PC} = 0^{++}, 1^{+-}, 1^{++}, 2^{++}$. The $2^{++}$ nonet is well identified. This is not yet the case for the other $N = L = 1$ nonets.

## 4.1 The $0^{++}$ mesons

The $J^{PC} = 0^{++}$ mesons are exceptionally important for testing our ideas on hadron structure. In addition to the conventional $^3P_0$-wave $q\bar{q}$ nonet, a low-lying $q^2\bar{q}^2$ nonet has been proposed[23]. The $0^{++}$ states built from gluons have also been proposed[24,25]. Inspired by the MIT bag model, Aerts et al.[26] have calculated the $q\bar{q}$ spectrum and have found a $0^{++}$ nonet with masses $\sim$1285 MeV for $I = 0$ and $I = 1$, $\sim$1475 MeV for the $s\bar{s}$ isovector, and $\sim$1380 MeV for $I = \frac{1}{2}$.

The experimental situation has been clarified by a series of new results. The existence of an $I = 0$ $J^{PC} = 0^{++}$ resonance in the 1300-1400 MeV mass range, mainly coupled to $\pi\pi$, is now firmly established. Not only has it been shown[27] that the $\pi^+\pi^-$ S-wave has a resonant behaviour with an intensity peaking $\sim$ 1300 MeV, but Cason et al.[28] have produced $\pi^0\pi^0$ data, i.e. $\pi^+p \to \Delta^{++}\pi^0\pi^0$, which show unambiguously a large S-wave contribution in the broad $\varepsilon(700)$ and $\varepsilon(1300)$ mass regions, with substantial S- and D-wave interferences (Fig. 4).

Moreover, independent evidence for an $I = 0$ $\varepsilon(1300)$ comes from the study of the $K\bar{K}$ system. Görlich et al.[29] observe a $K^+K^-$ S-wave enhancement in the 1100-1500 MeV mass region which they tend to attribute to $I = 0$. Loverre et al.[30] show that the $K_S^0 K_S^0$ S-wave has production properties similar to those of the D-wave at small t and for $M \sim 1300$ MeV (Fig. 5), suggesting that it is produced by $\pi$ exchange in the reaction $\pi^-p \to K_S^0 K_S^0 n$ and therefore must be $I = 0$. Cohen et al.[31] have now carried out a complete amplitude analysis of the $K^+K^-$ system produced in $\pi^-p \to K^+K^-n$ and $\pi^+n \to K^+K^-p$. Having large statistics for both channels, they can separate the $I = 0$ from the $I = 1$ $K^+K^-$ contributions. The analysis leads to several ambiguous solutions, but only one exhibits t dependences consistent with the expected production mechanisms (Fig. 6). This solution presents a large $I = 0$ enhancement at 1300 MeV. Figure 7 shows the $\pi\pi \to K\bar{K}$ scattering amplitude. The speed variations indicate a resonance at $\sim$1425 MeV with $\Gamma \sim 160$ MeV. The Argand plot also shows that the resonance couples asymmetrically to the two channels, $\pi\pi$ being the dominant one, $0.28 < (x_\pi x_K)^{1/2} < 0.40$. The OZI rule predicts $(x_\pi x_K)^{1/2} \sim 0.40$, while a pure SU(3) singlet such as a glueball would have $(x_\pi x_K)^{1/2} \sim 0.50$.

A complete coupled-channel analysis of the $I = 0$ $J^{PC} = 0^{++}$ scattering amplitudes $\pi\pi \to \pi\pi$, $\pi\pi \to K\bar{K}$, and $K\bar{K} \to K\bar{K}$ is still necessary in order to fully understand the scalar mesons. It has been shown[32], in particular, how some experimental information on $K\bar{K} \to K\bar{K}$ could constrain the solutions. Nevertheless, the new data and analyses of Cohen et al.[31] provide convincing evidence for an $I = 0$ $J^{PC} = 0^{++}$ resonance with $M = 1425 \pm 15$ MeV, $\Gamma = 160 \pm 30$ MeV, the $\pi\pi$ coupling[27,33] being dominant. In addition, a large background below 1000 MeV and the $S^*(980)$ are both necessary for interpreting the data.

Aston et al.[34], using the LASS spectrometer at SLAC, have made a partial-wave analysis of the $K^-\pi^+$ system. The interaction analysed is

$$K^-p \to K^-\pi^+n \text{ at } 11 \text{ GeV/c}.$$

The $K\pi$ S-wave shows a clear elastic loop on the Argand plot (Fig. 8), which can be interpreted as a resonance with $M \sim 1500$ MeV, $\Gamma \sim 250$ MeV.

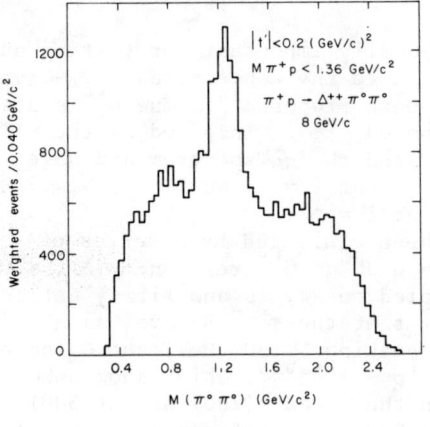

Fig. 4

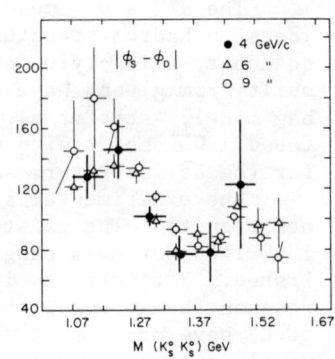

Fig. 5

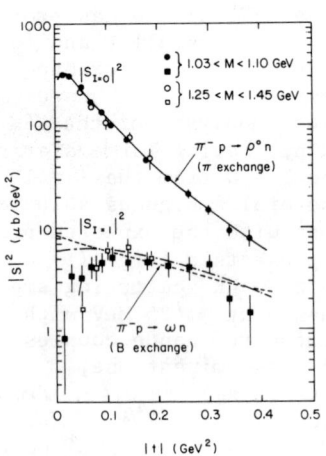

Fig. 6

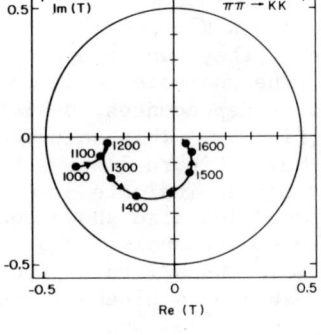
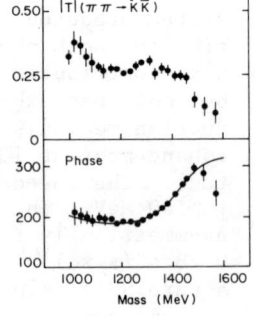

Fig. 7

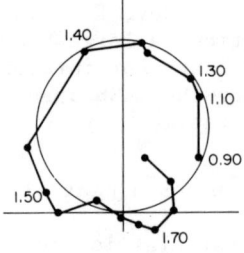

Fig. 8

In conclusion, it seems possible that a $0^{++}$ nonet emerges in the 1300-1600 MeV mass region. Of course it remains to identify the I = 1 and I = 0 ($s\bar{s}$) constituants. The simplest decay mode for the I = 1 scalar would be $\eta\pi$, a two-body system not yet well known. In this picture, the $\delta(980)$ and $S^*(980)$ could be the $q^2\bar{q}^2$ states $s\bar{s}$ ($u\bar{u} \pm d\bar{d}$) advocated by Jaffé[23]. The quark configuration of the $\delta(980)$ and $S^*(980)$ is, however, still open to debate[35,36].

## 4.2 The $1^+$ mesons

Our experimental knowledge of the $J^{PC} = 1^{+\pm}$ mesons has recently made important progress; but this is not to say that everything is clear for the two expected nonets.

It is now well established that two resonances are observed in the $K\pi\pi$ system with $J^P = 1^+$: the $Q_1$ with M $\sim$ 1270 MeV, $\Gamma \sim$ 80 MeV decaying mainly into $K\rho$; the $Q_2$, with M $\sim$ 1400 MeV, $\Gamma \sim$ 200 MeV, decaying mainly into $K^*\pi$. These results come mainly from the diffractively produced $K\pi\pi$ system[37,38]. In fact, the $Q_1$ was first observed in $\bar{p}p$ annihilations at rest[39], and called at that time the C(1240) meson. The $Q_1$ is also observed in non-diffractive reactions, such as

$$\pi^- p \rightarrow (K\pi\pi)^0 \Lambda^0 .$$

New results are available on this reaction at 3.9 GeV/c [40]. A $Q_1$ is clearly observed (Fig. 9) with M = 1294 ± 10, $\Gamma$ = 66 ± 15 MeV, and a partial-wave analysis shows that it is mainly a $J^P = 1^+$ $K\rho$ S-wave. The peak observed at 1435 MeV is attributed as a whole to the $K^*(1435)$. The same collaboration gives an upper limit for the ratio $K\omega/K\rho$ of 30%, to be contrasted with the observation of a $K\omega$ peak in the diffraction data which is claimed[38] to be mainly associated with the $Q_1$. This uncertainty on the $K\omega$ partial width of the $Q_1$ must be kept in mind when discussing the $1^\pm$ nonets' characteristics. For the first time, the ACCMOR Collaboration[38] sees the $K^*\pi$ D-wave contribution to the $Q_1$-$Q_2$ resonances, which fits nicely with the absence of $\rho\pi$ D-wave in the $A_1(1100-1300)$ meson and the D/S ratio of 30% observed for the $\omega\pi$ decay of the B(1235) meson.

Recently, the ACCMOR Collaboration[41] has published evidence showing that the diffractively produced $1^+$ $\rho\pi$ S-wave in the reaction $\pi^- p \rightarrow \pi^+\pi^-\pi^- p$ at 63 and 94 GeV/c incident momenta contains a resonant component with a mass of 1280 ± 30 MeV and a width of 300 ± 30 MeV. It is well known that the production of $A_1$ in diffractive reactions is confused by the presence of Deck processes. The mass and width quoted above have been reached with the use of a formulation proposed by Bowler[42]. New results on the $A_1$ have been presented at this conference[43], using large statistics on the neutral ($\pi^+\pi^-\pi^0$) spectrum produced in forward exchange at 8.45 GeV/c,

$$\pi^- p \rightarrow \pi^+\pi^-\pi^0 n .$$

This analysis is particularly interesting since it can separate four incoherent contributions having natural or unnatural parity exchange and nuclear helicity flip or non-flip. Both partial waves with I = 1 and I = 0, $J^P = 1^+$ $\rho\pi$ S-waves, have a significant contribution and both peak at $\sim$ 1130 MeV in the $3\pi$ mass spectrum (Fig. 10a and 10c) for the natural parity exchange, nuclear helicity flip sample. Figure 10 also shows the exotic I = 2 $J^P = 1^+$ $\rho\pi$ S-waves: the M = 1 (Fig. 10b) is barely detectable, but the M = 0 component (Fig. 10d) is sufficiently robust to serve as a phase reference. Figure 10e shows the $A_2$(1310) meson. The phases of the I = 0 and I = 1 $J^P = 1^+$ $\rho\pi$ partial waves, relative to the exotic I = 2 or to the $A_2$(1310), show a large forward motion of $\sim$ 150°, establishing the resonant character of these $J^P = 1^+$ waves. This is the first convincing evidence for the existence of the isoscalar H(1190) $J^P = 1^+$ $\rho\pi$ resonance.

To obtain the resonance parameters of the H and $A_1$, Dankowych et al.[43] have applied, as did the ACCMOR Collaboration, the formulation of Bowler which treats the Deck-like background in a coherent way. As in the purely diffractive case, a consequence of this formulation is to shift the mass from $\sim$ 1100 MeV, where the peaks are observed, to $\sim$ 1200 MeV. More precisely, Dankowych quotes the following masses and widths:

$$I = 1 \quad A_1 \quad M = 1240 \pm 80 \text{ MeV} \quad \Gamma = 280\text{-}480 \text{ MeV}$$
$$I = 0 \quad H \quad M = 1190 \pm 60 \text{ MeV} \quad \Gamma = 270\text{-}370 \text{ MeV}.$$

Other observations of the I = 1 $J^{PC} = 1^{++}$ ($A_1$) come from baryon-exchange reactions, which are free from Deck processes (but of course may have other kinds of background).

In particular, Gavillet et al.[44] observe a peak in the $\pi^+\pi^-\pi^-$ mass spectrum of the reaction

$$K^-p \rightarrow \Sigma^-\pi^+\pi^-\pi^- \quad \text{at 4.2 GeV/c}.$$

This collaboration has now achieved a new partial-wave analysis of the $3\pi$ system, and finds a phase variation of the $1^+$ $\rho\pi$ S-wave which supports the resonance hypothesis for the peak observed at 1050 MeV with a width of 200 MeV. Although this resonance has all the quantum numbers and decay properties observed in diffractive and charge-exchange processes, the mass (1050 compared to more than 1200 MeV) are not compatible. It is possible that the diffractive (and to some extent the charge-exchange) data, although strongly

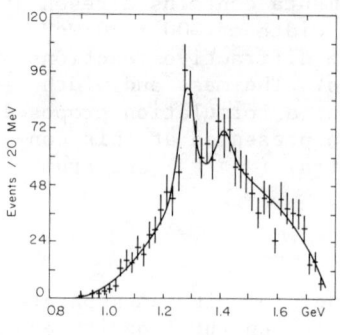

Fig. 9

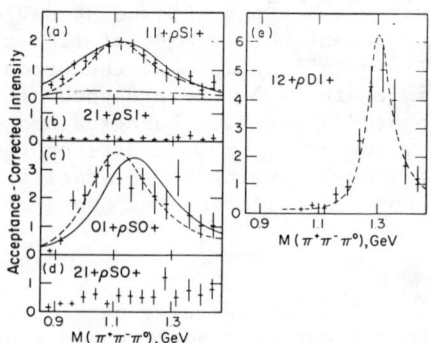

Fig. 10

indicating the presence of a resonant contribution to the $A_1$ enhancement, do not give a precise determination of the mass, since the presence of Deck contributions makes it necessary to solve this problem in a model-dependent way. It is also possible that baryon exchange mechanisms have their own complexities which would require much larger statistics in order to be fully appreciated. A more attractive hypothesis may be that both results are correct, and that the $A_1$ observed with M ∼ 1240 MeV is the $q\bar{q}$ partner of the $1^{++}$ nonet, whereas the $A_1$ observed with M ∼ 1050 MeV is of a more exotic nature, a $2q2\bar{q}$ object being more naturally expected in these reactions if the t-exchange is non-exotic. Note finally that Aerts et al.[26] predict a mass of 1285 MeV for the $J^{PC} = 1^{++}$ isovector $q\bar{q}$ (non-strange) meson.

To complete the $1^+$ meson review, we must now discuss the status of the E meson. In "classical" hadron interactions, ($\pi^-p$, $K^-p$, $\bar{p}p$ annihilations in flight), one observes two relatively narrow peaks in the $K\bar{K}\pi$ mass spectrum: the D(1285) and the E(1420). See, for instance, the new results of Dionisi et al.[45] and of Bromberg et al.[46] (Figs. 11 and 12).

The quantum numbers of the D are well established: $I^G J^P = 0^+ 1^+$. Those of the E have been uncertain for several years. A new analysis[45] now gives convincing evidence that, for the E also, $I^G J^P = 0^+ 1^+$.

The D has not enough phase space to decay into $K^*K$. Most authors attribute its $K\bar{K}\pi$ decay to D → δπ, δ → $K\bar{K}$. Since the δ(980) has a large ηπ decay mode, the D should also be observed in the ηππ system -- and it is. This is not the case for the E, the $K^*K$ decay mode being dominant (86% according to Ref. 45, 76% according to Ref. 46). Note that, in the $K_S^0 K^\pm \pi^\mp$ Dalitz plot of the E, the two $K^*$ interfere constructively (C = +1) and therefore create a large accumulation near the $K\bar{K}$ threshold. Therefore a δ cut is an efficient selection for enhancing the E (and the D) over the $K_S^0 K^\pm \pi^\mp$ background (Fig. 12).

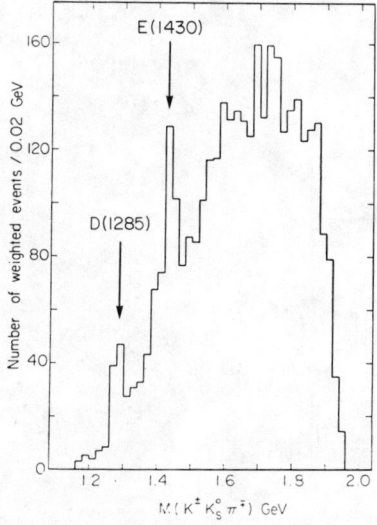

Fig. 11

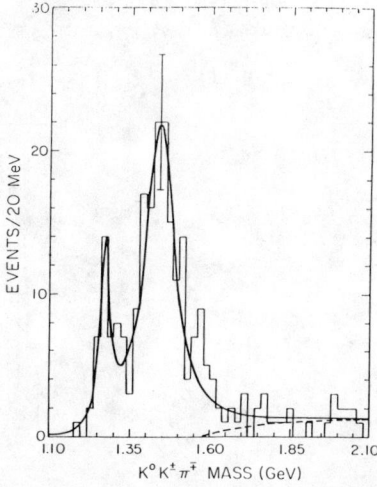

Fig. 12

With these observations it makes sense to attribute the D and the E to the $J^{PC} = 1^{++}$ nonet, the E being a "$\phi$-like" object made primarily of $s\bar{s}$ quarks, with the following new determinations of its mass and width:

Ref. 45:   M = 1426 ± 6 MeV    Γ = 40 ± 15 MeV ,

Ref. 46:   M = 1440 ± 5 MeV    Γ = 62 ± 14 MeV .

However, $\bar{p}p$ annihilation at rest[47] and the radiative decay modes of the $J/\psi$ [48,49] also show a $K\bar{K}\pi$ enhancement at 1425 MeV with a width which may be slightly larger than the one observed in classical hadronic interactions, i.e. Γ ∼ 80 MeV, but which is not accompanied by the D(1285) (Figs. 13 and 14). This is particularly remarkable for annihilations at rest, where the 1425 MeV $K\bar{K}\pi$ enhancement was indeed originally observed with no visible D(1285), whereas for $\bar{p}p$ annihilations in flight, i.e. at 700-750 MeV/c [50], both D(1285) and E(1425) are clearly observed. Moreover, this E seems to have a larger branching ratio into $\delta\pi$ than the one observed in $\pi^-p$ and $\bar{p}p$ annihilations in flight: the $\eta\pi^+\pi^-$ mode is observed (Fig. 15) and a branching ratio E → $\delta\pi$ as large as 50% is given[47]. A spin-parity determination of this E favours $0^{-+}$ over $1^{++}$.

It may not be a surprise that $\bar{p}p$ annihilations at rest show some similarities to the $J/\psi$ decays, since the dominant $\bar{p}p$ initial state is $^3S_1$, a state which has the quantum numbers of the $J/\psi$.

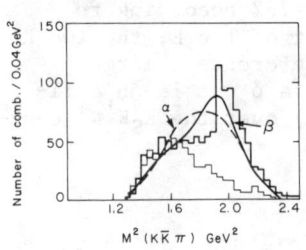

Fig. 13

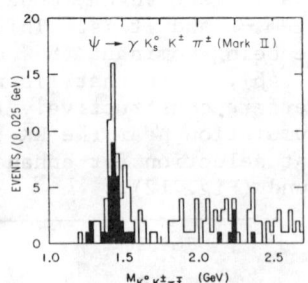

Fig. 14

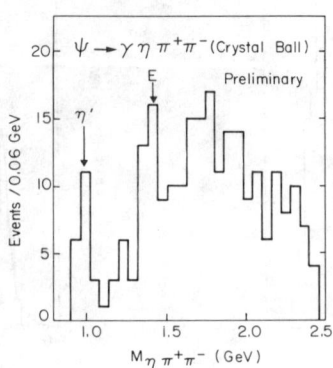

Fig. 15

An attractive interpretation of these observations is to assume that in $\bar{p}p$ annihilations at rest and in the radiative decay of $J/\psi$ one produces a resonance which is not the $I = 0$ $s\bar{s}$ $J^{PC} = 1^{++}$ of the quark model, but may be a $2q2\bar{q}$ or a glueball object.

## 4.3 Higher-L excitations

The partial-wave analysis of high statistics experiments performed at high energy have succeeded in establishing the existence of several higher-spin states schematically represented in Fig. 16.

Of the four $q\bar{q}$ nonets corresponding to $L = 2$ ($3^{--}$, $2^{--}$, $2^{-+}$, $1^{--}$), only one ($3^{--}$) is relatively well known. The ACCMOR Collaboration[51] has presented conclusive evidence for the existence of an $A_3(1670)$ meson, with $\Gamma \sim 210$ MeV, $J^{PC} = 2^{-+}$, a peak and corresponding phase variation being observed in three partial waves: $2^-S(f\pi)$, $2^-P(\rho\pi)$, and $2^-D(\epsilon\pi)$. The same collaboration has also shown that the $K\pi\pi$ diffractive system contains at least one resonant component, at $M \sim 1820$ MeV, $\Gamma \sim 200$ MeV, with $J^P = 2^-$, peaks and phase variations again being observed in three partial waves, i.e. $2^-S(K^{**}\pi)$, $2^-P(K^*\pi)$ and $2^-D(Kf)$. However, even with the large statistics available in this experiment, it is not yet possible to disentangle the two resonant states ($2^{--}$, $2^{-+}$) which are expected in this mass region.

The $F_1(1540)$ has been considered for some time as a possible $2^-$ candidate. It has now been shown[50] that this $K\bar{K}\pi$ enhancement is not a genuine resonance.

For the $L = 3$ nonets, we now have to add three recently observed resonances to the $I^G J^P = 0^+ 4^+$ $h(2040)$. The ACCMOR Collaboration[38] observes a peak and a phase variation for the $2^+P(f\pi)$ partial wave with $M \sim 1700$ MeV and $\Gamma \sim 200$ MeV, which may be interpreted as the $I = 1$ member of the $^3F_2$ $q\bar{q}$ nonet (Fig. 17). Cleland et al.[52] observe a peak and phase variation for the $I^G J^P = 1^- 4^+$ ($K_S^0 K^\pm$) partial wave with $M \sim 2000$, $\Gamma \sim 250$ MeV, and for the $I^G J^P = 1^- 6^+$ ($K_S^0 K^\pm$) partial wave with $M \sim 2515$, $\Gamma \sim 450$ MeV (Fig. 18). The same collaboration, analysing the $K_S^0 \pi^\pm$ system produced at small t in the reactions

$$K^\pm p \to (K_S^0 \pi^\pm) p$$

at 50 GeV/c, finds also a peak and a phase variation for the $J^P = 4^+$ partial wave with $M \sim 2060$ MeV, $\Gamma \sim 150$ MeV, in addition to the tail of the $K^*(1425)$ and to the $J^P = 3^- K^*(1780)$ resonance (Fig. 19).

Finally, the ACCMOR Collaboration[53] observes a bump in the $K^+K^-$ system produced at small t in the reaction

$$\pi^- p \to K^+ K^- n$$

at 60 GeV/c, the decay angular distribution indicating the dominance of a $J^P = 5^-$ wave at $M \sim 2300$, $\Gamma \sim 260$ MeV to which it seems reasonable to assign $I^G = 1^+$ (Fig. 20).

All these new resonances fall nicely on linear spin-mass squared meson trajectories (Fig. 16).

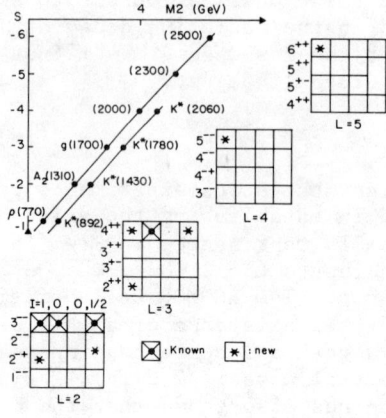

Fig. 16

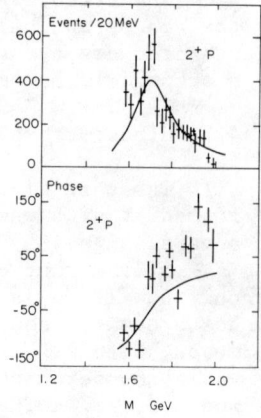

Fig. 17

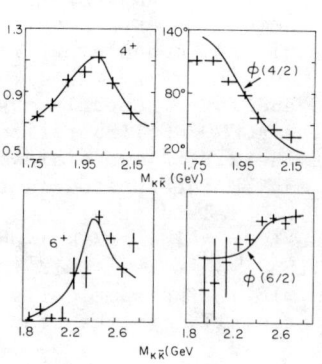

Fig. 18

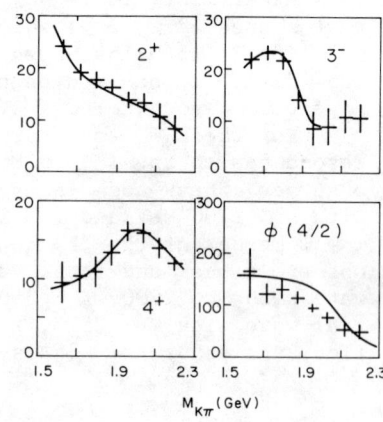

Fig. 19

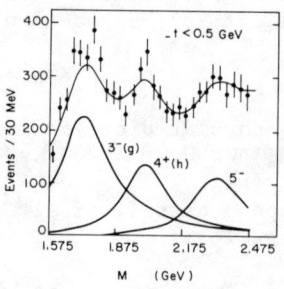

Fig. 20

## 4.4 Heavy vector mesons

Recent results from photoproduction and $e^+e^-$ experiments have very much increased our information on heavy vector mesons. Note that the quark model predicts the existence of two nonets of $J^{PC} = 1^{--}$ in the 1200-1800 MeV mass region, the $n = 0$, $L = 2$ $^3D_1$ and the $n = 1$, $L = 0$ $^3P_1$ nonet.

It has been known for some time that the annihilation

$$e^+e^- \rightarrow 2\pi^+ 2\pi^-$$

presents a maximum around 1600 MeV, with $\Gamma \sim 500$ MeV, called the $\rho'(1600)$. Bizot et al.[54] have combined their measurements, made between 1.4 and 2.2 GeV, with those of Novosibirsk[55] (Fig. 21), and get a mass which varies from 1580 to 1630 MeV with $\Gamma$ varying from 465 to 540 MeV, depending on the model used for the $4\pi^\pm$ dynamics. This is in good agreement with the results presented by d'Almagne[56] on the reaction

$$\gamma p \rightarrow 2\pi^+ 2\pi^- p \, .$$

The $\rho\epsilon$ decay mode is presumably dominant. However, d'Almagne claims that he needs a contribution from the $A_1\pi$ channel to interpret his data. The $\rho\epsilon$ decay mode produces a bump in the reaction

$$e^+e^- \rightarrow \pi^+\pi^-\pi^0\pi^0, \quad \text{with } \sigma_{+-00} \sim \tfrac{1}{2} \sigma_{++--} \, ,$$

but, in addition, one observes for this reaction a maximum around 1300 MeV which is attributed to another $\rho'$, the $\rho'(1250) \rightarrow \omega^0 \pi^0$, with

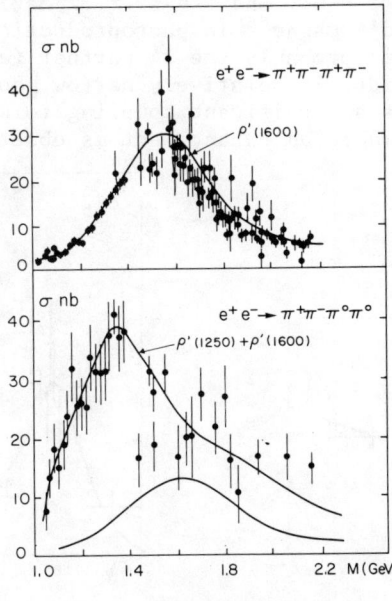

Fig. 21

$M \sim 1250$ MeV, $\Gamma \sim 250$ MeV (Fig. 21). This $\rho'(1250)$ is also clearly observed in the photoproduction data[56], the decay angular distributions supporting the hypothesis $J^P = 1^-$, and not $1^+$ as expected for the $B(1235) \to \omega^0\pi$ meson. However, the resonant character of this $\rho'(1250)$ is not yet firmly established.

The problem raised by the $\pi^+\pi^-$, $K_S^0 K_L^0$, $K^+K^-$ possible decay modes of these $\rho$'s is not easy to solve. In photoproduction experiments[56,57] a $\rho' \to \pi^+\pi^-$ is observed, but with a width which is much smaller than the one in the $4\pi$ decays: $\Gamma \sim 250$ MeV, suggesting that simple fits adding Breit-Wigners incoherently on a "smooth" background are misleading. In these conditions, it is difficult to quote a branching ratio for the $\rho'(1600)$. Photoproduction experiments tend to favour a relatively small branching ratio $2\pi^{\pm}/4\pi^{\pm} \sim 10\text{-}20\%$. which is in agreement with the inelasticity measured for the $1^-$ partial wave observed in the phase-shift analyses of the $\pi^- p \to \pi^+\pi^- n$ reaction.

Bizot et al.[54] note that one cannot interpret simultaneously the annihilations

$$e^+e^- \to K_S^0 K_L^0 \quad \text{and} \quad e^+e^- \to K^+K^-$$

without the introduction not only of a $\rho'(1600)$ but also of an isoscalar resonance which interferes constructively with the $\rho'$ in the $K^+K^-$ channel and destructively in the $K_S^0 K_L^0$ one. A fit for this isoscalar gives $M = 1674 \pm 6$ MeV, $\Gamma = 94 \pm 30$ MeV. These numbers are similar to those obtained for the enhancement observed in the annihilations

$$e^+e^- \to \omega^0\pi^+\pi^-$$

(Fig. 22), ($M = 1634 \pm 9$ MeV and $\Gamma = 89 \pm 32$ MeV)[54]. The same structure is observed by d'Almagne[56] in photoproduction.

This isoscalar is probably the $\phi'$ partner (and not the $\omega'$) of the $\rho'(1600)$: its width is relatively narrow [compared to the $\rho'(1600)$], and it has a significant coupling to $K\bar{K}$ and an even larger coupling to $K_S^0 K^{\pm}\pi^{\mp}$, where an enhancement is observed (Fig. 23), with

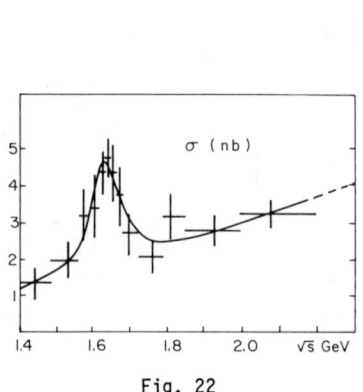

Fig. 22

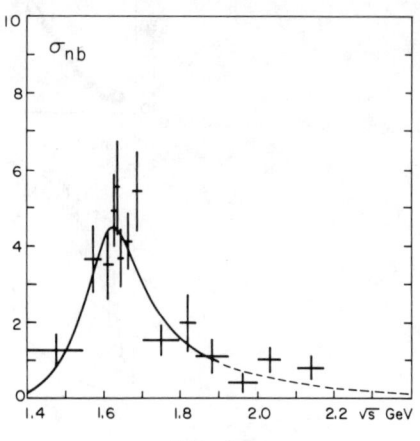

Fig. 23

$K^{*0}K^0 - K^{*\mp}K$ interference effects which require a large I = 0 contribution. All these observations are made with $e^+e^-$ annihilations[54] as well as with photoproduction[56]. In this last case, the 5π structure observed at 1600 MeV with Γ ∼ 100 MeV is superimposed on a large $\omega^0\pi^+\pi^-$ threshold enhancement which extends from 1500 to 2000 MeV: one may speculate that the ω' partner of the ρ'(1600) is hidden in this enhancement.

To complete this short review on heavy vector mesons, one may add the information which comes from two experiments, both of them studying the $K^+K^-$ final state. Amirzadeh et al.[58] have analysed the reaction $K^-p \to \Lambda^0 K^+K^-$ at 8 GeV/c, and see two peaks in the $K^+K^-$ mass spectrum above 1200 MeV, the f'(1515) and a second peak at 1851 MeV (Fig. 24). This is supported by the observations made by Frame et al.[59], who have analysed the reaction $K^+p \to K^+K^-K^+p$ at 13 GeV/c, and see also two enhancements, the f'(1515) and a peak at 1830 ± 9 MeV, Γ = 72 ± 3 MeV. Although we present these results in the context of heavy vector mesons, the spin parity of this new resonance has not been determined: it could be, indeed, the $J^P = 3^-$ state which is still missing in the $^3D_3$ nonet.

Finally, the partial-wave analysis performed by Aston et al.[34] on the $K^-\pi^+$ system produced in

$$K^-p \to K^-\pi^+n \quad \text{at 11 GeV/c}$$

shows two loops in the Argand plot of the P-wave, a large elastic loop corresponding to the $K^*(892)$ and an inelastic one, with maximum speed around 1650 MeV (Fig. 25).

*Conclusions:* The heavy vector mesons exist. With the ρ'(1600) one may associate a ϕ'(1670) and a $K^{*\prime}$(1650). In addition, hints of an ω' → ωππ and of another ρ', with M ∼ 1250 MeV and $\omega^0\pi^\pm$ as a dominant decay mode, come from $e^+e^-$ annihilations, photoproduction, and $\bar{p}p$ annihilations at rest[60].

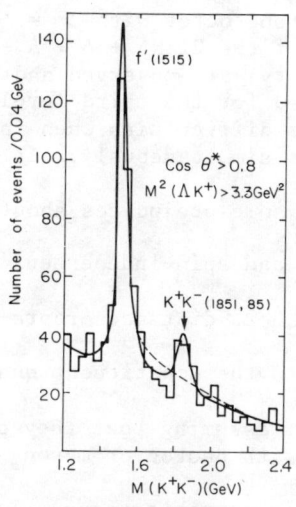

Fig. 24

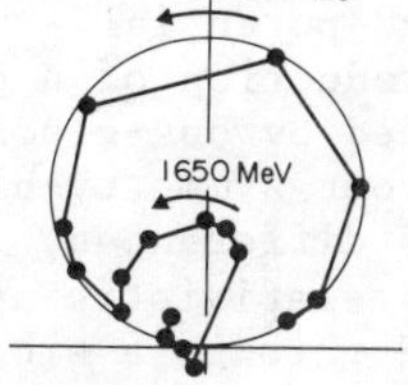

Fig. 25

## 5. BARYON SPECTROSCOPY

The experimental information on baryons being much more complete than for other hadrons, the baryons have been considered as the best field for testing our ideas on the quark model, although they are made of three quarks and therefore require, in principle, the solution of a three-body problem.

The usual non-relativistic quark model for baryons starts from the wave function

$$|3\rangle = \underbrace{|\text{Flavour}\rangle|\text{Spin}\rangle|\text{Space}\rangle}_{SU(6) \otimes O(3)}|\text{Colour}\rangle \otimes 1_c$$

and approximates the quark binding forces with a two-body harmonic interaction which leads to the famous harmonic oscillator shell model with radial and orbital excitation[61]. Using the notation $[SU(6), L^P]$, L being the orbital excitation and P the parity of the states, we get, for the first three levels:

$$N = 2 \quad [56, 2^+], [70, 2^+], [56, 0^+], [70, 0^+], [20, 1^+]$$

$$N = 1 \quad [70, 1^-]$$

$$N = 0 \quad [56, 0^+]$$

This model works remarkably well. Not only have we been able to identify all the members of the N = 0 level (one octet with $J^P = \frac{1}{2}^+$ and one decuplet with $J^P = \frac{3}{2}^+$), but 18 out of the 21 $N^* + \Delta + \Lambda + \Sigma$ predicted for the second level (negative parity) are observed and a number of good candidates have been identified for the third level. However, this simple model encounters serious difficulties when interpreting the mass spectra and decay properties simultaneously. It predicts too many states.

Starting from this framework and adding some prejudices about quark dynamics,
- confinement of the quarks in a flavour- and spin-independent long-range potential,
- introduction of a perturbative term at short distances represented by one-gluon exchange,
- flavour symmetry-breaking entirely due to the constituent quark mass differences,
- non-relativistic approach to the decays, assuming that they proceed through a single quark transition with photon or meson emission,

Isgur, Karl and Koniuk[62] have shown that it is possible to go a long way inside a detailed interpretation of the mass spectra and decay

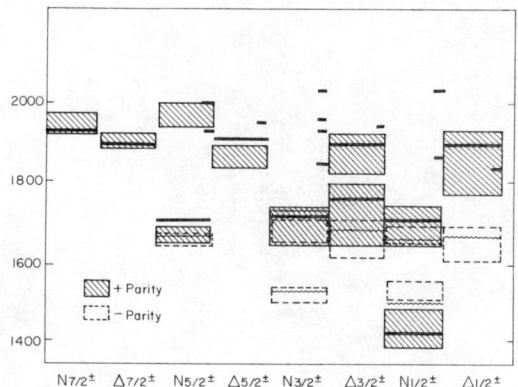

Fig. 26

modes of the baryons with a non-relativistic quark model. Figure 26 shows schematically some of the achievements of this model for the S = 0 baryons with positive and negative parity. The predictions are represented by thick lines of length proportional to the coupling to the initial state. The experimental results are represented by boxes giving the regions in which the masses of the resonances should most likely lie. Most predicted resonances with a large coupling fall inside or close to a box. Those which have a small coupling to the initial state are not, as expected, observed.

5.1 The S = 0 sector

For the S = 0 sector, most of our experimental information comes from $\pi^{\pm}p$ partial-wave analyses made by the CMU-LBL Collaboration[63], the Helsinki-Karlsruhe Collaboration[64], and the recent analysis of Chew[65]. Although these analyses use different approaches, their conclusions, as shown for example in Fig. 27 for the $\Delta$'s and Fig. 28 for $N^*$'s, are generally in good agreement. The CMU-LBL partial-wave analysis concentrates on the energy region above the $\Delta(1232)$, combines data from various experiments, and performs energy-independent fits, using a parametrization which incorporates t- and u-channel analyticity. Several minima are then combined into clusters characterized

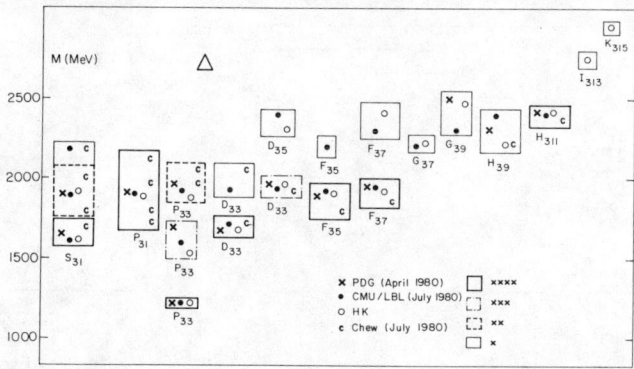

Fig. 27

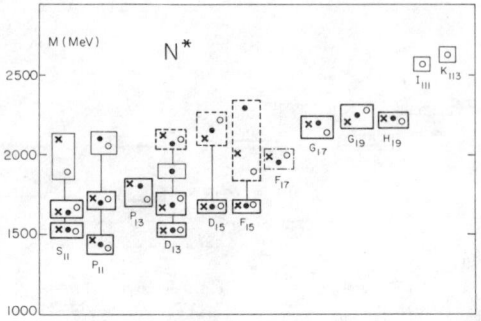

Fig. 28

by a correlated set of partial-wave amplitudes. The results on individual partial-wave amplitudes are then fitted with a coupled channel plus background parametrization. The Karlsruhe-Helsinki analysis extends to 10 GeV/c. Energy-independent fits are performed. Analyticity and dispersion relation constraints are introduced for fixed s, fixed t, and fixed $\phi$ amplitudes. Chew has analysed the $\pi^+p$ elastic scattering data between 0.6 and 2.3 GeV/c, using the method of Barrelet's zeros. Transversity amplitudes at fixed energy are parametrized in terms of polynomials in $w = e^{i\phi}$. The most recent results of this analysis suggest that four $S_{31}$ and four $P_{31}$ resonances coexist in the 1700-2200 MeV mass range. The exact status of the $S_{31}(1620)$ resonances is important since, in the harmonic oscillator quark model, it degenerates with the $D_{33}(1700)$. The CMU-LBL analysis suggests that these two states have a definite mass difference.

Hendry[66] has analysed the 2 to 4 GeV mass region. He finds numerous candidates with spins going from $G_{3,9}$ to $N_{3,21}$. These results are in fair agreement with K-H [64] up to $K_{3,15}$. These resonances do not lie on a linear Regge trajectory (Fig. 29) but if drawn, instead, on a spin-k (the c.m. momentum of the coupled pion-nucleon channel) plot, they do lie on a straight line (for $N^*$ as well as for $\Delta$), suggesting that centrifugal barrier effects are important (Fig. 30). It seems difficult[66] to reconcile the large widths observed for these high-spin states with estimates based on the quark model.

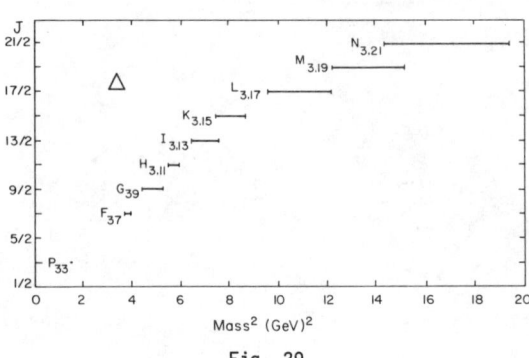

Fig. 29

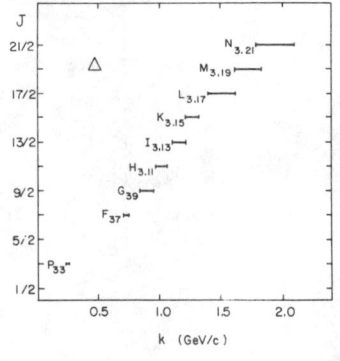

Fig. 30

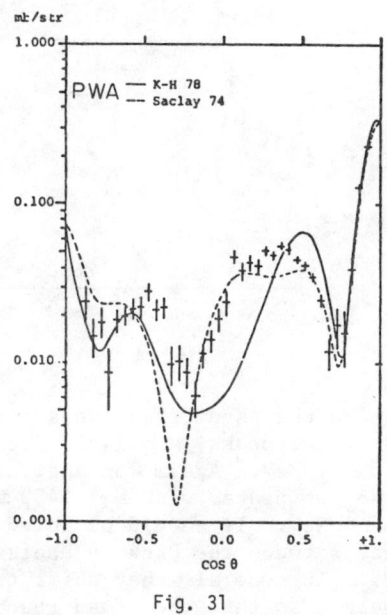
Fig. 31

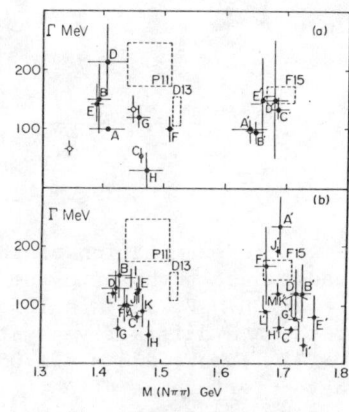
Fig. 32

The agreement between the partial-wave analysis and the remarkable success of the non-relativistic model of Isgur-Karl-Koniuk should not distract our attention from the weaknesses which need to be cured.

New measurements on $\pi^-p$ elastic scattering and charge-exchange measurements between 2 and 4 GeV/c made at KEK[67] have revealed the weaknesses of the partial-wave analysis predictions, which are not even qualitatively good for the new polarization data (Fig. 31).

Some measurements of the $\pi^{\pm}p$ elastic scattering at large angles in the 2-6 GeV/c momentum range observe narrow structures which seem qualitatively consistent with the Ericson fluctuation mechanism[68]. Other very accurate measurements of the same reactions in the 6-14 GeV/c momentum range do not show these narrow structures[69]. Narrow baryon resonances seem to be observed in the diffraction processes. Fukunaga et al.[70], who have compiled the $\pi N$ and $\pi\pi N$ data observed in diffraction dissociation (Fig. 32), show that these data do not match very well with the partial-wave resonances $P_{11}(1470)$ and $F_{15}(1690)$ with which they are usually associated. A detailed analysis of the diffractive dissociation processes

$$\pi^+ p \to \pi^+(\pi^+ n)$$

and $\quad \pi^+ p \to \pi^+(\pi^0 p)$

at 16 GeV/c [71], with a mass resolution of $\sim$ 10 MeV, shows that the $\pi N$ diffractive enhancement is best interpreted, at $t < 0.1$ GeV$^2$, in terms of two narrow peaks, with M = 1344 ± 5, Γ = 66 ± 14, and M = 1451 ± 7, Γ = 132 ± 22 MeV (Fig. 33), whereas, at $t > 0.2$ GeV$^2$, a two-peak structure shows up with M = 1450, Γ = 130 MeV and M = 1639 ± 7, Γ = 96 ± 20 MeV. (Fig. 34).

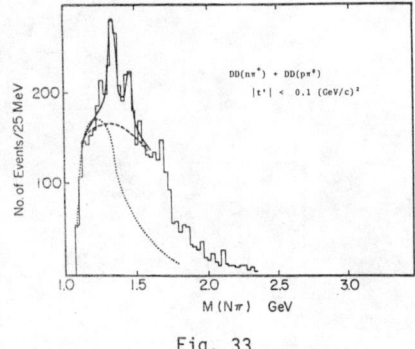

Fig. 33

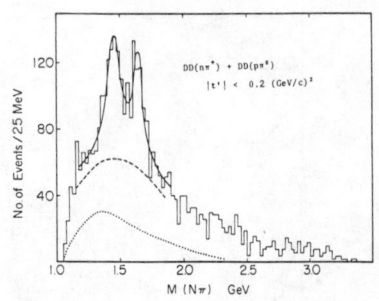

Fig. 34

A re-examination of the world data on the πN diffractive system leads these authors to the observation of two peaks, M = 1471 ± 3, Γ = 86 ± 8 MeV, and M = 1655 ± 5, Γ = 110 ± 11 MeV. A similar analysis of the ππN diffractive system gives also two peaks, with M = 1450 ± 3, Γ = 98 ± 7 MeV, and M = 1705 ± 2, Γ = 84 ± 4 MeV. It should be noted that interference effects may take place between the Deck mechanism and the direct production of resonances. In general, they shift the mass determinations and enlarge the widths, so that the broad resonances observed in formation cannot become narrower with Deck interference effects. Moreover, the three peaks (1.34, 1.46, and 1.70 GeV) do not move with $t'$, whereas Deck effects would. One is led to the conclusions that $N^*(1.34)$ with Γ = 66 ± 14 MeV has no analogue in formation experiments, and that $N^*(1.46)$ with Γ ∼ 90 MeV cannot be $P_{11}(1470)$, which has a width between 180-240 MeV, nor $D_{13}(1520)$, which has a mass between 1510 and 1530 MeV. If the difference in mass between $N^*(1.66) \to \pi N$ and $N^*(1.71) \to \pi\pi N$ is statistically significant, it becomes unrealistic to consider both of them as being due to a single resonance $F_{15}(1890)$. The $N^*(1.71) \to \pi\pi N$ could be associated with $D''_{13}(1700)$, 90% of its decay being ππN.

These results open up the interesting speculation that Pomeron exchange creates a new type of resonance[72]. It may also be appropriate to underline that such low-mass $N^*$ resonances would not fit easily into the simple quark model, but may find a less difficult interpretation within the framework of the MIT bag model[73].

Extensive measurements of $\pi^-p$ inelastic reactions have been made at RHEL[74], and the isobar model has been applied to the channel πN → ππN [75,76], bringing us new information on the $P_{11}$, $S_{31}$, $P_{31}$, $P_{33}$, and $D_{33}$ waves. One of the most interesting results of these analyses is the claim that there is a narrow $P_{31}$ state around 1525 MeV, Γ ∼ 40 MeV, again rather difficult to accommodate in the simple quark model. A more detailed discussion of these results may be found elsewhere[77,78].

One of the most significant threshold enhancements which may be interpreted as an inelastic $N^*(1810)$ resonance is the ωp threshold peak observed by Arenton et al.[79], using the polarized proton beam of the ZGS:

$$p{\uparrow}p \to p\omega p \ .$$

The t-distribution suggests ω-exchange, and the ωp angular distributions are qualitatively consistent with $J^P = 3/2^-$ for an $N^*$ which would have a mass of 1810 ± 25 and a width of 140 ± 40 MeV. As is often the case for threshold enhancements, the resonance nature of the peak is difficult to establish with confidence.

5.2  The S = −1 sector

We have already underlined the general good agreement between the experimental observations and the simple non-relativistic quark model[62] for the S = 0 sector. The same agreement applies for the S = 1 sector (Fig. 35). The model does not reproduce the mass splitting of the $\Lambda(1405)$ and $\Lambda(1520)$. However, it should be noted that the model ignores spin-orbit forces. Another apparent difficulty is that the model does not predict large enough widths. This, however, may be partially due to the experimental methods, the widths generally quoted being those observed in the partial-wave analyses of formation experiments, notably larger than those observed in production experiments. In addition, several production experiments report the observation of generally narrow $Y^*$'s which have no obvious counterparts in the $Y^*$ spectroscopy deduced from formation experiments.

The ACNO Collaboration has reported three such $Y^*$'s: i) A $\Sigma(1550)$ produced in a rather complex final state: $K^-p \to \Sigma^\pm(1550)K_S^0 K^\mp$ at 4.2 GeV/c incident momentum[80], the $\Sigma^+(1550)$ being produced mostly backward, the $\Sigma^-(1550)$ mostly forward. The mass and width are 1553 ± 7 and 80 ± 30 MeV, respectively. The main decay modes seem to be $\Lambda\pi^\pm$ and $\Sigma^0\pi^\pm$. ii) A $\Sigma(2230)$, observed[81] in the reactions $K^-p \to \Sigma^\pm\omega^0\pi^\mp$, $\Sigma^\pm(2230) \to \Sigma^\pm\omega^0$, the $\Sigma^+(2230)$ again being produced mostly backward, the $\Sigma^-(2230)$ mainly forward. The mass and width are 2230 ± 30 and 60 ± 20 MeV, respectively. iii) A $Y^*(2580)$, observed[82] in the reaction $K^-p \to Y^*(2580)\pi^0$, with a rather complex decay mode, involving three strange particles,

$$Y^*(2580) \to \Sigma^\pm K^\pm K_S^0 \ .$$

The mass and width of this peak are 2576 ± 10 and 37 ± 11 MeV (Fig. 36). No charge decay is observed, suggesting that this $Y^*(2580)$ is a $\Lambda^0$.

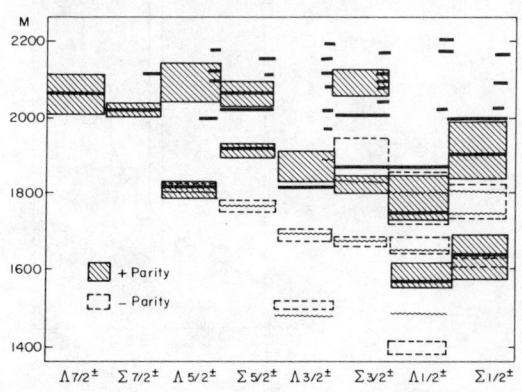

Fig. 35

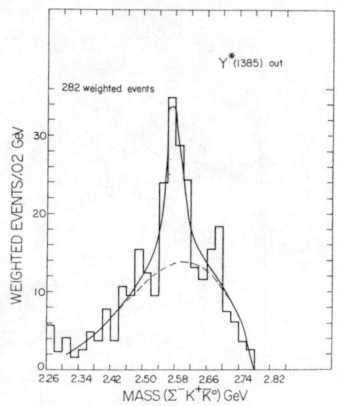

Fig. 36

The CCMS Collaboration[83] has presented evidence for a low-mass $Y^*$, the $\Sigma(1460)$, observed at the $K^-n$ threshold of the reaction

$$\pi^- p \to (K^- n) K^0_S \pi^+ \quad \text{at 3.9 GeV/c},$$

with also some evidence for a $(\bar{\Sigma}^0 \bar{\pi}^0)$ decay mode (Fig. 37). The mass and width of this peak are 1464 ± 8 and 44 ± 24 MeV, respectively. This resonance may have been observed previously by Yu-Li Pan[84] in the reactions

$$\pi^+ p \to (\Lambda\pi)^+ K^+ , \quad (\Sigma\pi)^+ K^+ \quad \text{at 1.7 GeV/c},$$

but in this case, it seems that the predominant decay mode was $\Lambda\pi^+$, which is not observed by CCMS.

None of these four $Y^*$ having an obvious candidate among the $Y^*$'s observed in formation experiments, and most of them being produced in complex reactions or having complex decay modes, they may be seriously considered as candidates for more complex objects than those predicted by the three-quark model.

The most remarkable $Y^*$ observed in production experiments may be the R(3170), reported by the BCGMP[85] Collaboration, the data coming from two experiments, both studying the reaction

$$K^- p \to \pi^- R^+ \quad \text{at 6.5 and 8.25 GeV/c}.$$

The $R^+(3170)$ is a relatively narrow peak observed in five- and six-body combinations containing several strange particles (Fig. 38). The $R^+(3170)$ is produced forward, implying I = 3/2 baryon exchange in the t-channel. I = 1 is consistent with the observation of a small $R^0$ signal. Its mass and width are 3170 ± 5 MeV and $\Gamma$ < 20 MeV, respectively. With these characteristics, the R(3170) could be a new type of hadron: H. Lipkin has suggested that it could be a bound $BB\bar{B}$ state.

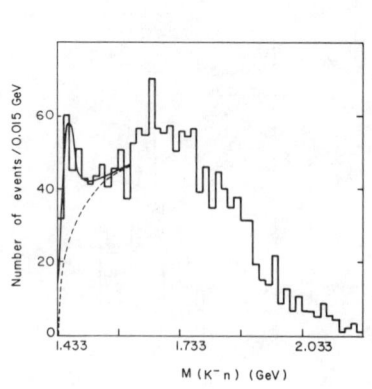

Fig. 37

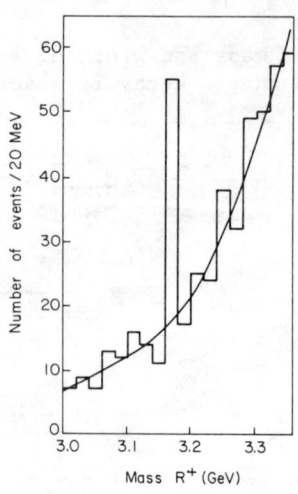

Fig. 38

## 5.3 The S = -2 sector

It is well known that many $\Xi^*$ are still missing in the SU(6) classification. Apart from the $\Xi^-(1321)$ and $\Xi^*(1530)$, which are the two members of the $[56, 0^+]$ N = 0 level, two $\Xi^*$ are reasonably well established, although we do not know the parity of the $\Xi^*(1820)$, which however has a spin $J \geq 3/2$, and do not know the spin and parity of the $\Xi^*(2030)$. Claims have been made for two other $\Xi^*$, the $\Xi^*(1630) \to \Xi\pi$ and the $\Xi^*(1940) \to \Xi\pi$, $\Xi(1530)\pi$, but, strangely enough, no new information has come from recent high-statistics experiments on these two evasive resonances.

Two other $\Xi^*$ have recently been observed. The ACNO Collaboration[86] has presented evidence for an enhancement at the $\Sigma\bar{K}$ threshold, which appears both in the neutral and negative charge states. Weak evidence for a corresponding $\Lambda\bar{K}$ enhancement is also presented, and a $(\Sigma\bar{K})$-$(\Lambda\bar{K})$ coupled channel analysis gives results compatible with the existence of a $\Xi^*(1680)$: M = (1684 ± 5) MeV, $\Gamma$ = (20 ± 4) MeV. Its closeness to the $\Sigma\bar{K}$ threshold favours qualitatively the $J^P = 1/2^-$ assignment. The other new $\Xi^*$ comes from the observation made by the BCGMP Collaboration[85] of $\Xi^*(2370)$, which has the remarkable feature of having a dominant three-body decay, $Y\bar{K}\pi$ (Fig. 39). However, 48% of its observed decay modes can be attributed to quasi-two-body decays, including 24% of $\Lambda K^*(892)$, 15% of $\Sigma^*(1385)$ $K^0$, and 9 ± 4% of $\Omega^-K^0$. This $\Xi^*(2370)$, with M = (2373 ± 7) MeV and $\Gamma$ = (85 ± 20) MeV, seems to be produced by I = 0 exchange ($\Lambda$ exchange), so that I = 1/2 is preferred.

As we see, the $\Xi^*$ spectroscopy is still at an embryonic stage. New techniques are certainly necessary to change this situation: this may be the case with the advent of good hyperon beams at Fermilab and CERN.

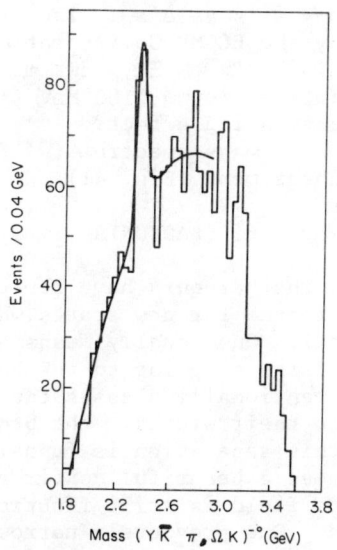

Fig. 39

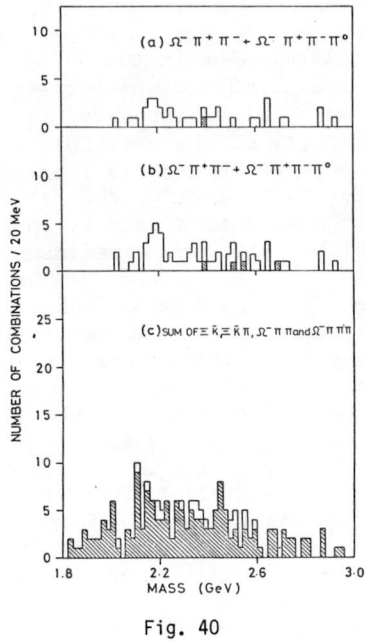

Fig. 40

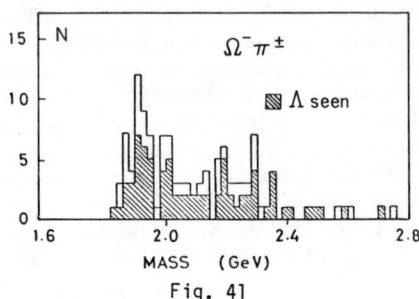

Fig. 41

## 5.4 The S = -3 sector

If the last two years have been very successful for the $\Omega^-$ (lifetime, decay modes, weak decay parameters, etc.), no $\Omega^+$ have yet been observed. In contrast, five $\Delta$'s are well known [in SU(3), we expect to observe as many $\Omega^*$'s as $\Delta$'s]. The most significant work on $\Omega^*$ has been presented by the BCGMP Collaboration[85] (Fig. 40). The $\Omega^*$ have been searched for in $\Omega^-\pi^+\pi^-$, $\Xi\bar{K}$, $\Xi\bar{K}\pi$ mass spectra. There may be a hint of an accumulation around 2180 MeV (11 ± 4 events) in the $\Omega^-\pi^+\pi^-$, but no really convincing effect.

It seems that the $\Omega^-\pi^\pm$ mass spectrum ($\Omega^*$ forbidden by I-spin) presents a threshold enhancement (Fig. 41).

## 6. BARYONIA

In the past years, the baryonia have represented one of the most favoured fields in the search for new states which would not fit with the simple ($q\bar{q}$, 3q) model. One usually means by "baryonium" a state which has a relatively large coupling to the baryon-antibaryon system. Experimentally, one conventionally classes the baryonia into two categories, according to their widths: the broad ones and the narrow ones. Theoretically, this separation is fundamental since the narrow baryonia could be taken as a beautiful confirmation of QCD[87].

The broad baryonium field is still flourishing. The narrow baryonium field is dead. One previously narrow baryonium [the S(1936)] has become broad. A recent theoretical and experimental review has been published elsewhere[88].

It is well known that, in total, elastic, and $\bar{p}p$ annihilation cross-sections, broad bumps have been observed in the 2100-2400 MeV mass region. Three broad enhancements have been identified with the following properties:

$T_1(2190)$   $M = (2185 \pm 5)$ MeV   $\Gamma = (130 \pm 30)$ MeV   $I = 1$

$U_1(2350)$   $M = (2350 \pm 5)$ MeV   $\Gamma = (185 \pm 20)$ MeV   $I = 1$

$U_0(2390)$   $M = (2385 \pm 10)$ MeV   $\Gamma = (80 \pm 30)$ MeV   $I = 0$.

Partial-wave analyses of the two-body annihilations $\bar{p}p \to \pi^+\pi^-$ and $\bar{p}p \to \pi^0\pi^0$ have shown that nearly all the partial waves, from $^3D_1$ to $^3G_5$, have a resonant behaviour in this mass range, with an indication that the total spin increases with the mass[89,90]. However, these resonances contribute to only a small fraction (10 to 20%) of the bumps observed in the total cross-section. In other words, these resonances could have relatively large elasticity, giving some support to the baryonium interpretation. In Fig. 42 we sketch what may happen in the T-U region.

The S(1936) seems to present a similar situation. New high-statistics measurements on the total, charge-exchange, and annihilation $\bar{p}p$, $\bar{p}n$ cross-sections[91,92] conclude that an enhancement is present in the total and annihilation cross-sections, with the following features (Fig. 43):

$\sigma_{tot} = 3.0 \pm 0.7$ mb   $M = (1939 \pm 2)$ MeV   $\Gamma = (22 \pm 6)$ MeV

$\sigma_{ann} = 2.5 \pm 0.8$ mb   $M = (1937 \pm 2)$ MeV   $\Gamma = (21 \pm 10)$ MeV

Most of the recent experiments have concentrated their efforts on a search for a resonance with $\Gamma < 10$ MeV but have failed to find it. This is, of course, not in contradiction with the results of Hamilton et al.[91]. Lowenstein et al.[93] measure the $\bar{p}p$ annihilation cross-section between 1925 and 1955 MeV with an r.m.s. resolution of ±2.5 MeV, and have set an upper limit of 9 mb MeV (2 standard deviations) for any resonance with $\Gamma \lesssim 4$ MeV (Fig. 44). Jastrzembski et al.[94] also measure the $\bar{p}p$ annihilation cross-section in the mass range 1932-1942 MeV and set an upper limit of 12 mb MeV for any resonance with $\Gamma \sim 4$ MeV (Fig. 45). Cresti et al.[95] have repeated the measurements of Chaloupka et al.[96], using a hydrogen bubble chamber as a spectrometer but with initial $\bar{p}$ momenta slightly different from those used in the original experiment. They do not see a resonance with $\Gamma \lesssim 10$ MeV in the total cross-section (Fig. 46); and the reinterpretation of the original data, leaving aside a few data points at the entrance of the bubble chamber, and without merging the data with others (to avoid normalization problems), gives the numbers: $M = (1935 \pm 3)$, $\Gamma = (16 \pm 11)$ MeV, $\sigma_{tot} = (5.2 \pm 1.5)$ mb, in agreement with Hamilton et al.[91]. Several experimental results indicate that the S(1936) may be, like the T and U bumps, of a complex nature: absence of a peak in charge-exchange cross-section[91], angular distribution of the $K_S^0 K_L^0$ annihilation channel[50], and the enhancement observed in $5\pi$ annihilations by Defoix et al.[97], who assign to this enhancement

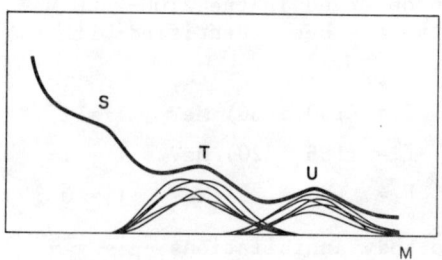

Fig. 42

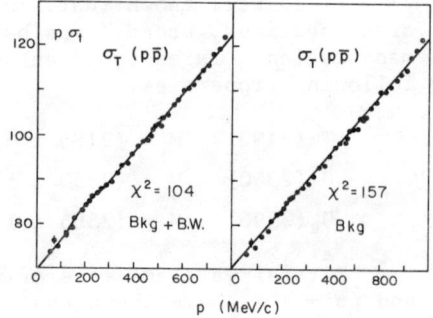

Fig. 43

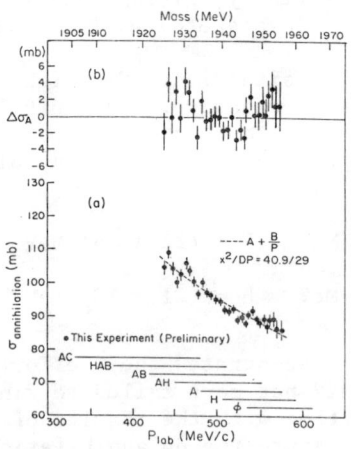

Fig. 44

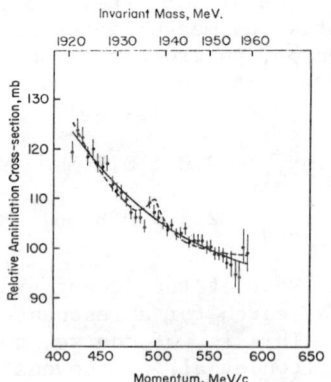

Fig. 45

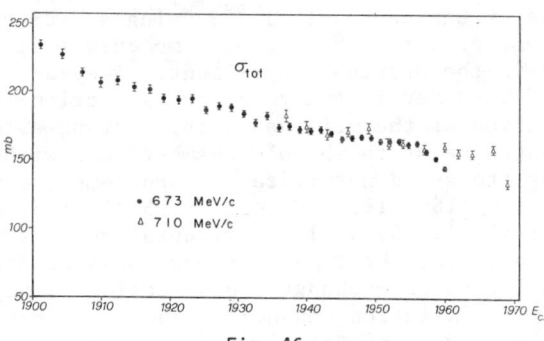

Fig. 46

a mass of $(1949 \pm 10)$ MeV, a width $\Gamma = (80 \pm 20)$ MeV, and $\sigma \sim 5$ mb $(I = 1)$. Note that Igi[98] favours a width of 20 to 30 MeV for the S(1936) in a model based on duality and t-channel vector meson exchange.

A recent result on the observation of the S(1936) in a production experiment[99] is not confirmed by its authors. Finally, the S(1936) may have been observed in photoproduction[100], but the mass and width quoted agree only marginally with the values observed in $\bar{p}p$ formation experiments (Fig. 47). Photoproduction may involve complex threshold behaviour which makes difficult the interpretation of the $\bar{p}p$ mass spectrum.

Broad resonances have also been observed in the $(\bar{\Lambda}p)$ and $(\Lambda\bar{p})$ system, with no presently known counterpart in purely mesonic channels, suggesting a baryonium nature. Cleland et al.[52] have made a partial-wave analysis of the $\Lambda p$ system produced diffractively in the reactions

$$K^+p \to (\bar{\Lambda}p)p \quad \text{and} \quad K^-p \to (\Lambda\bar{p})p \text{ at 50 GeV/c },$$

with $0.05 < -t < 1$ GeV$^2$, and produce evidence for three broad states with $J^P = 2^-$, $3^+$, and $4^-$ at $(2260 \pm 20)$, $(2320 \pm 30)$, and $(2490 \pm 20)$ MeV, respectively. The widths are of the order of 250 MeV (Fig. 48). On a Chew-Frautschi plot, these resonances fall significantly below the dominant Regge trajectories.

None of the previous claims for the observation of narrow baryonium resonances have survived a variety of high-statistics and precise experiments. The M(2020) and M(2200), originally observed in $\pi^-p$ reactions at 9 and 12 GeV/c, produced in association with $\Delta^0$ or $N^{*0}(1520)$, and having cross-sections of the order of 10 to 30 nb [101], have not been observed in a similar experiment performed at BNL[102] with 16 GeV/c incident $\pi^-$ (Fig. 49). An upper limit for the cross-sections is set at 3 nb and would correspond, to be compatible with

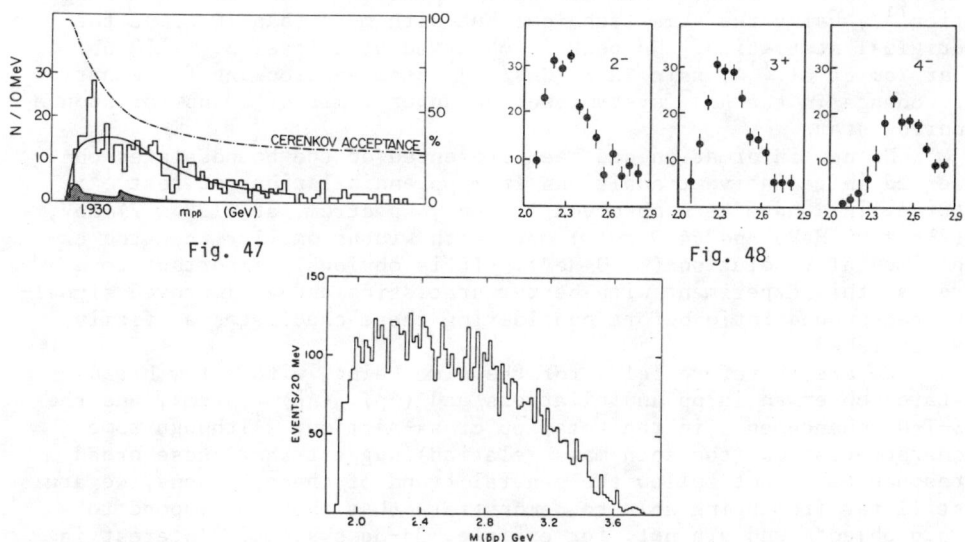

Fig. 47

Fig. 48

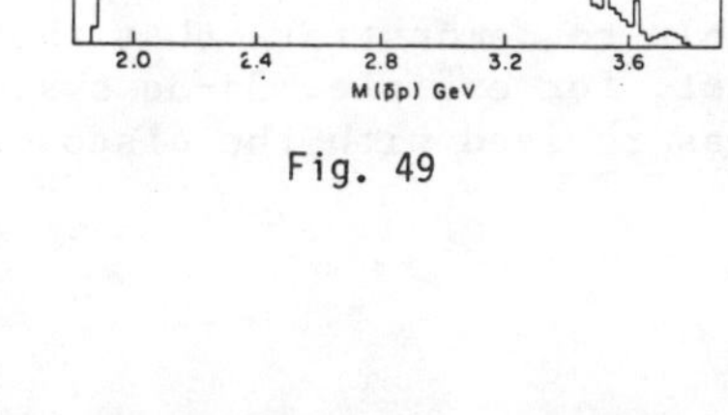

Fig. 49

Benkheiri et al., to a laboratory momentum dependence of the order of $p^{-10}$. Using a $\pi^+$ beam and looking for the M's in the reaction $\pi^+ p \to \Delta^{++}\bar{p}p$ at 10 GeV/c, with a forward $\Delta^{++}$, Bionta et al.[103] set an upper limit of 15 nb, although the production of M's was expected to be enhanced, in this reaction, with respect to $\pi^-$-induced reactions, since the t-channel exchange seems to be dominated by $I = \frac{1}{2}$ [101]. The M's have also been looked for in $\bar{p}$-induced reactions, which are *a priori* more favourable and can be performed at relatively low incident momenta. Chung[104], using a 5 GeV/c $\bar{p}$ beam, sets an upper limit of 100 nb, whereas Armstrong et al.[105], using a 12 GeV/c $\bar{p}$ beam at CERN, set an upper limit of 40 nb. Frame et al.[106], studying the reaction

$$K^+ p \to \bar{p}_f p K^+ p \text{ at 13 GeV/c },$$

give an upper limit of 30 nb. Finally, Kooijman et al.[107] have searched for baryonium states in the reaction

$$pp \to pp\bar{p}p \text{ at 12 GeV/c}$$

and set an upper limit of 10 nb for $\bar{p}p$ states with $M < 2200$ MeV, $\Gamma < 10$ MeV, and for $pp\bar{p}$ states with $M < 3400$ MeV.

This laconic enumeration of negative results does not of course, "explain" the original results of Benkheiri et al.[101]. More work is indeed continuing at the Omega spectrometer, at CERN, with 20 GeV/c incident $\pi^+$ and 12 GeV/c incident $\pi^-$, in an attempt to reproduce the original claims. Perrin[108] has given a recent review of the status of these experiments.

Another narrow baryonium had previously been reported, in the exotic channel $\Lambda p \pi^+$, with $M = 2460$ MeV[109]. This experiment, studying $K^+ p \to \bar{\Lambda} p \pi^+ n$ at 13 GeV, has now been repeated by the same collaboration[110], using the same technique but with more than 10 times the original statistics. No peak is observed at a level of $\sim 150$ nb. Bar Yam et al.[111], using a 16 GeV/c $\pi^-$ beam and looking for a narrow resonance in the $\Lambda p \pi^-$ system, set an upper limit of 20 nb for such a narrow state.

No new information has been presented on the bound states observed in radiative transitions from $\bar{p}p$ annihilations at rest[112]: three lines have been observed in the $\gamma$ spectrum, at $(183 \pm 7)$ MeV, $(216 \pm 9)$ MeV, and $(420 \pm 17)$ MeV, with widths smaller than the experimental resolution ($\sim 20$ MeV). It is obviously important to repeat this experiment with better statistics and an improved signal-to-background ratio before considering these candidates as firmly established.

We are therefore left, for the time being, with a few broad states observed in $\bar{p}p$ annihilations and $(\bar{\Lambda}p)$, $(\Lambda\bar{p})$ systems, and the S-T-U enhancements in the total $\bar{p}p$ cross-section. Although some characteristics (the spin-mass relation) suggest that these broad resonances do not follow the general trend of the $\bar{q}q$ mesons, we are still far from being able to demonstrate that they correspond to $2q2\bar{q}$ objects and are not, for example, $3q$-$3\bar{q}$ systems. Interest in this last hypothesis has revived with the discovery of dibaryon resonances.

## 7. FIVE-QUARK RESONANCES?

We summarize here the experimental observations which may be related to the existence of ($q^4\bar{q}$) systems. Most of these observations have already been mentioned in the baryon spectroscopy section. Strottman[113] has worked out a $q^4\bar{q}$ spectrum which predicts a $J^P = \frac{1}{2}^-$ I = 0 resonance with M = 1420 MeV, whereas the simple quark model of Isgur and Karl[62] predicts degenerated $J^P = \frac{1}{2}^-$ and $\frac{3}{2}^-$ at M = 1490 MeV. Experimentally, the $J^P = \frac{1}{2}^-$ object has a mass of 1405 MeV, the $J^P = \frac{3}{2}^-$ a mass of 1520 MeV. The $q^4\bar{q}$ model predicts also a $J^P = \frac{1}{2}^-$, I = 1 resonance at 1420 MeV, which may be compared to the $\Sigma(1460)$ observed by the CCMS Collaboration[83]. No such low-mass state is predicted in the simple quark model.

On the other hand, Högaasen and Sorba[114], starting from a $(3q)_8 - (q\bar{q})_8$ model and using as input the $\Lambda(2130)$ observed in proton-proton interactions at the CERN ISR[115] and the $\Sigma(2230) \to \Sigma\omega$ observed by the ACNO Collaboration[81], predict a $\Lambda(2550)$ which contains three strange quarks and could therefore by the $Y^*(2580)$ discussed above[82]. The same model predicts a $\Xi(2375)$ which should have a dominant three-body decay, as observed for the $\Xi(2370)$ [85].

Before considering these states as solid evidence for multiquark states, much more experimental and theoretical work must be done. It is interesting, however, to see how well some predictions agree with recent experimental results which do not easily find a natural interpretation in the classic hadron spectroscopy.

## 8. DIBARYONS

Yokosawa[116] has reviewed here the new experimental data related to the existence of dibaryon states. A striking energy dependence has been observed in the difference between the nucleon-nucleon total cross-sections for pure spin states:

$$\Delta\sigma_L = \sigma_{tot}(\rightleftarrows) - \sigma_{tot}(\rightrightarrows)$$

and

$$\Delta\sigma_T = \sigma_{tot}(\uparrow\downarrow) - \sigma_{tot}(\uparrow\uparrow) .$$

Several other observables have been measured, giving information on the spin-spin correlation parameters, polarization data in elastic scattering, etc. The interpretation of these data leads to the conclusions that three dibaryons are necessary in order to fit the data, and five more candidates may be present. The three best established dibaryons have the following properties:

$B_2^1(2140)$: I = 1, $J^P = 2^+$, 2140 < M < 2170 MeV, 50 < $\Gamma$ < 100 MeV ($^1D_2$)

$B_2^1(2220)$: I = 1, $J^P = 3^-$, 2200 < M < 2250 MeV, 100 < $\Gamma$ < 200 MeV ($^3F_3$)

$B_2^0(2220)$: I = 0, $J^P = 3^-$, 2200 < M < 2260 MeV, 100 < $\Gamma$ < 200 MeV ($^1F_3$).

Data on the photodissociation of the deuteron are accumulating rapidly, with polarized beams, targets, or recoils. The consistency

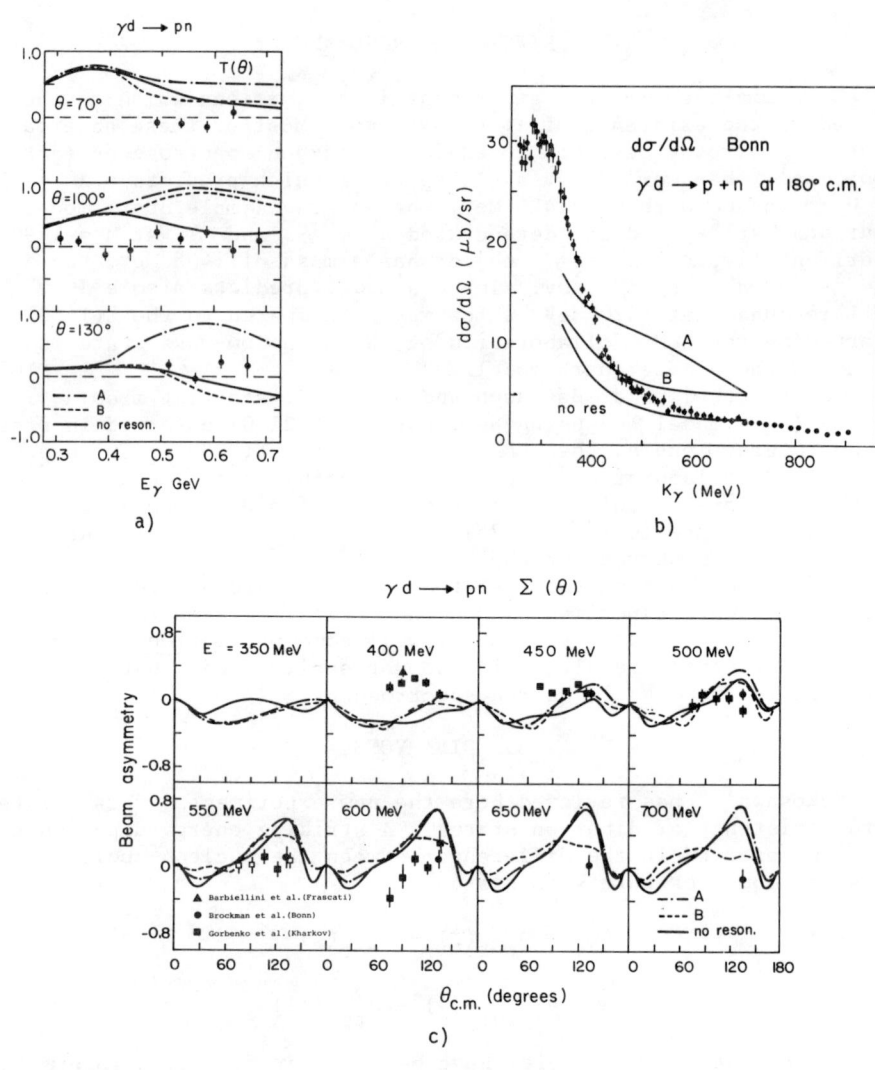

Fig. 50

between these data, coming from Tokyo, Kharkov, Stanford and Bonn, is good[117], but the "old" Tokyo analysis[118], which suggested the presence of two dibaryons, $B_2(2260)$ with $I(J^P) = 1(3^-)$ and $B_2(2380)$ with $0(1^+)$ (solution "A") or $B_2(2260)$ with $1(3^-)$ and $B_2(2380)$ with $0(3^+)$ (solution "B") is ruled out by the new extended high-precision data (see, for example, Fig. 50). A new analysis is therefore necessary to give a good interpretation, not only of $d\sigma/d\Omega$ and $P(\theta)$ data but also of $\Sigma(\theta)$ and $T(\theta)$ (new polarized beam and target data). A detailed review of the problems raised by the use of deuteron targets to study the dibaryon resonances induced by pion beams has been prepared for this conference[119]. It is found that the elastic π-deuteron scattering can

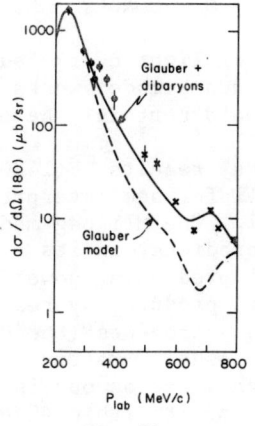

Fig. 51

be very well described by the Glauber model, introducing three different dibaryons with the following characteristics: $B_2(2130)$ with $J^P = 2^+$, $B_2(2260)$ with $J^P = 3^-$, and $B_2(2510)$ with $J^P = 0^+$ or $4^+$ (Fig. 51).

It is of fundamental interest to study the high-energy behaviour of the deuteron in order to learn about the transition of a weakly bound N-N system into a many-quark state. In this respect, Arnold et al.[120] have measured the reaction

$$e + D \to e' + X$$

near $M_X^2 = M_0^2$ and at $Q^2 = 8$ GeV$^2$. These data continue the approximate scaling observed at lower $Q^2$, and their confrontation with various models shows factor of 10 discrepancies, underlying the need for further theoretical developments.

Dibaryons with S = -1 have been looked for in reactions of the type

$$K^-d \to \pi^{\pm}(B_2)^{\mp} ,$$

$$B_2 \to \Lambda p, \Sigma n \ldots .$$

Braun et al.[121] and others have observed a $B_2 \to \Lambda p$ with M = 2129.0 ± 0.4 MeV, $\Gamma$ < 6 MeV. This peak is observed with larger statistics in a new experiment[122] which is designed to look not only for $B_2$'s with S = -1 but also for $B_2$ with S = -2 ($K^-d \to K^+B_2^-$). Ströbele et al.[123] have submitted results on the reaction

$$K^-d \to \pi^+ d_{\Sigma}^-$$

$d_{\Sigma}^-$ being a hyperdeuteron with a mass below the $\Sigma^-$ threshold. Although the signal is significant, the interpretation of events which do not fit the dominant channel $K^-d \to \pi^+\Sigma^0 n$ is a delicate matter. Confirmation of these results by experiments now in progress[122] would be most useful.

## 9. CONCLUSIONS

This bird's-eye view of light quark-hadron spectroscopy shows that the non-relativistic quark model works amazingly well. The work of Isgur-Karl-Koniuk[62] should certainly be extended to mesons, $\Xi^*$ and $\Omega^*$.

Remarkable experimental results [E(1420), Y*(2580), Σ(1460), R(3170), $\Xi^*$(2370)] may call for new interpretations which differ from the classical $q\bar{q}$, 3q model. The MIT bag model and its extensions is a good candidate, but it predicts, in its present formulation, too many states to have a real predicting power.

One is still unable to produce any real "exotic" (double charged, $I = 5/2$ ...) state, but new approaches (the P-matrix) may reveal exotics which do not appear in the usual S-matrix analysis.

The light quark-hadron spectroscopy is such a lively field that it is impossible to summarize it within 40 pages!

There are so many of my colleagues whom I would like to thank for help, advice, and information that I cannot list all their names here. Let me, however, mention P. Eastabrooke and I. Gaines, my scientific secretaries, who did their best during the Conference to ease my way.

\* \* \*

## DISCUSSION (M5 -- L. Montanet)

Q1: *Sorenson*, Argonne National Laboratory: You reported on various meson resonances with high mass and high spin. Is it possible to extract from the data any information about S-waves for masses between, say, 1.7 GeV and 2.0 GeV? If so, this could be relevant to various puzzles associated with the decay of (charmed) D-mesons.

A1: *Montanet:* I have perhaps been very brief on the S-wave. I think we still have a lot to understand about the S-wave; for instance, the apparently narrow S* and δ peaks near $K\bar{K}$ threshold may, in fact, be very wide objects. I think we come to the conclusion that we have at least isosinglet and $I = 1/2$ resonances around 1300 MeV -- that is rather well established now.

Q2: *Gago*, IST-Lisbon and WA56 Omega Collaboration: This is a comment about the so-called "death" of the narrow baryonium states. The $p\bar{p}$ peaks at 2020 and 2200 MeV that have been formed in $\pi^- p \to \Delta_f^0 (p\bar{p})$ some years ago at Omega are still the most convincing evidence for narrow $p\bar{p}$ peaks. It seems premature to affirm that they do not exist. In fact, the results that have been presented either do not look at the $\bar{p}p$ system backwards produced or they do not have a fast $\Delta^0/N^*$ associated.

We have made a high statistics experiment at Omega this year looking for baryonium exotic states in $\pi^+p$ at 20 GeV/c and we also repeated the old $\pi^+p$ experiment at 12 GeV/c. As the analysis is in progress, we expect results by the beginning of 1981.

A2: *Montanet:* It is true that, strictly speaking, nobody has reproduced an experiment at 9 or 12 GeV/c with $\pi^-$ and looked for the backward produced $\bar{p}p$ system. However, we have seen a result which I had no time to present in detail -- results obtained in 16 GeV/c $\pi^-p$, for instance, at Brookhaven, using exactly the same technique -- the same kind of trigger, the same experimental conditions, with good resolution and convincing statistics; they find an upper limit of 3 nb. Of course, we do not know the momentum dependence of these effects, so we cannot extract from this upper limit what one could have expected if one had done the experiment at 12 GeV/c.

They have an $N^*$ and a $\Delta$, as well as the Omega group. [Figure comparing $N^*$, $\Delta$ at 16 GeV/c (BNL) with 9 and 12 GeV].

Q3: *Lindenbaum*, BNL/CCNY: The question is whether there is any doubt in the experiment. There is always doubt in every experiment, but much less doubt in ours.

Q4: *Weiner*, Marburg: Recent theoretical arguments[*)] suggest that the hadronic mass spectrum cannot increase exponentially at large m, as was assumed so far. What is the present experimental situation? and what are the prospects of clarifying this problem?

A4: *Montanet:* I do not know. The diffractive peak observed in pp at 400 GeV at Fermilab (Francini et al.) may tell us something. The authors tried to extract the diffractive inelastic part, and they see that that can interpret this diffractive peak of the proton using the 12 $N^*$ known in this mass range up to 10 GeV.

Q5: *Roy*, TIFR: Narrow states had been reported in the radiative decay of $\bar{p}p$ at rest. Could you comment on the current status of these states?

A5: *Montanet:* There is no new information -- for the time being I think we just have to wait for the complete analysis of the new data. You refer to the peaks observed in the photon spectrum when you study $\bar{p}p$ annihilation at rest -- they could correspond to bound states of the $\bar{p}p$ system. The experiment has been repeated with better conditions, better statistics; but the group, for the time being, refuses to say anything on the new experiment.

---

*) G. Fowler and R. Weiner, Physics Letters, January 1980.

Q6: *Weiss,* SLAC: Is there any understanding of why some of the high-mass $Y^*$'s have large branching ratios into two- and three-body modes while charmed objects of similar masses do not?

A6: *Montanet:* There are, of course, models or interpretations. For instance, Högaasen and Sorba have proposed that, using the bag model, you could easily explain some of these observations. In fact, they have predicted some of the states which are now observed. Using this five-quark bag model they are able to understand why they have exotic decays.

Q7: *O'Donnell,* Toronto: If we take the width of the $\rho'$ to be $\sim$ 250 MeV as measured in $\gamma N \rightarrow \pi\pi N$, then the cross-section for $2\pi$ production in the colliding beam experiments is predicted to be much larger than that which is presently observed (although the statistics are, admittedly, poor).

A7: *Montanet:* There are two experiments which see the same effect. The experimental observation cannot be disregarded. The interpretation, however, is difficult. There may be not one $\rho'$ but several $\rho'$'s, and you may have to build up a rather complex amplitude with interference effects to understand these effects. In $e^+e^-$ it is completely clear that you really need to have isovector and isoscalar to interpret the set of data presented at this conference by Bizot.

Q8: *Lindenbaum,* BNL/CCNY: In response to the comment from the audience that the original CERN experiment that discovered the 2020 and 2200 MeV narrow baryonium resonances had not been repeated, Montanet correctly pointed out that the 16 GeV/c $\pi^-$ + p experiment was not in effect a good BNL repetition and contradicted it. In response to the further statement -- could there still be doubt in this repeated result? I would say in any experiment there is some doubt, but the doubt in our BNL experiment and its conclusion is quite small -- certainly there is much less doubt in our experiment than the original CERN experiments it contradicted.

* * *

## REFERENCES

1. P.C.M. Yock, paper 95, this conference.
2. A. Bussière et al., paper 14, this conference.
3. G. Valenti (CERN-Bologna-Frascati-Rome-Bari Collab.), Communication at this conference.
4. W. Bartel et al., paper 674, this conference.
5. M. Marinelli and G. Morpurgo, papers 155 and 156, this conference.
6. G.S. La Rue, J.D. Phillips and W.M. Fairbanks, paper 622, this conference.

7. G.S. La Rue, W.M. Fairbanks and A.F. Hebard, Phys. Rev. Lett. $\underline{38}$, 1011 (1977).
8. G.S. La Rue, W.M. Fairbanks and J.D. Phillips, Phys. Rev. Lett. $\underline{42}$, 142 and 1019 (1979).
9. K.S. Lackner and G. Zweig, paper 612, this conference.
10. R. Handler et al., paper 801, this conference.
11. L. Schachinger et al., Phys. Rev. Lett. $\underline{41}$, 1348 (1978).
12. R. Seales et al., Phys. Rev. D $\underline{20}$, 2154 (1979).
13. J. Franklin, Phys. Rev. D $\underline{20}$, 1742 (1979) and preprint from Temple University, Philadelphia, Penn.
14. I. Cohen and H.J. Lipkin, paper 353, this conference.
15. N. Isgur and G. Karl, paper 20, this conference.
16. H.J. Lipkin, paper 356, this conference.
17. J. Rosner, Communication at this conference.
18. P. Thompson et al. (Rochester-Minnesota-Fermilab Collab.), Communication at this conference.
19. D. Berg et al., Phys. Rev. Lett. $\underline{44}$, 706 (1980).
20. Particle Data Group, revised by T. Oshima, Rochester preprint, UR 733-1980.
21. W.C. Carithers et al., Phys. Rev. Lett. $\underline{35}$, 349 (1975).
22. A.N. Kamal and G.L. Kane, Phys. Rev. Lett. $\underline{43}$, 551 (1979).
23. R. Jaffé, Phys. Rev. D $\underline{15}$, 267 (1977).
24. R. Jaffé and K. Johnson, Phys. Lett. $\underline{60B}$, 201 (1975).
25. J.F. Donoghue, Communication at this conference.
26. A.T. Aerts, P.J. Mulders and J.J. de Swart, Phys. Rev. $\underline{21}$, 1370 (1980).
27. H. Becker et al., Nucl. Phys. $\underline{B151}$, 46 (1979).
28. N.M. Cason et al., paper 473, this conference.
29. L. Görlich et al., to be published in Nucl. Phys.
30. P.F. Loverre et al., paper 644, this conference.
31. D. Cohen et al., ANL-HEP-PR80-23, and A.B. Wicklund, paper 769, this conference.
32. M. Aguilar-Benitez et al., Nucl. Phys. $\underline{B140}$, 73 (1978).
33. A.D. Martin and M. Pennington, Ann. Phys. $\underline{114}$, 1 (1978).
34. D. Aston et al., paper 447, this conference.
35. A. Bramon and E. Masso, paper 712, this conference.
36. N.N. Achasov et al., paper 595, this conference.
37. G.W. Brandenburg et al., Phys. Rev. Lett. $\underline{36}$, 703 and 706 (1976).
38. R.J. Cashmore (Amsterdam-CERN-Cracow-Munich-Oxford-RHEL Collab.), talk at EMS Conf., Brookhaven, 1980.
39. R. Armenteros et al., Phys. Lett. $\underline{9}$, 207 (1964).
40. P.F. Loverre et al. (CERN-Collège de France-Madrid-Stockholm Collab.), unpublished data.
41. C. Daum et al., Phys. Lett. $\underline{89B}$, 281 (1980).
42. M. Bowler et al., Nucl. Phys. $\underline{B97}$, 227 (1975).
43. J.A. Dankowych et al., paper 474, this conference.
44. Ph. Gavillet et al., Phys. Lett. $\underline{69B}$, 119 (1977).
45. C. Dionisi et al., paper 625, this conference, and Nucl. Phys. $\underline{169}$, 1 (1980).
46. C. Bromberg et al., paper 761, this conference.
47. P. Baillon et al., Nuovo Cimento $\underline{50A}$, 393 (1967).
48. G. Feldman, XV Rencontre de Moriond, Les Arcs, 1980.
49. D. Ashman, XV Rencontre de Moriond, Les Arcs, 1980.

50. S. Ganguli et al., submitted to Nucl. Phys.
51. C. Daum et al., Phys. Lett. 89B, 285 (1980) and paper in preparation; see also ref. 38.
52. W. Cleland et al., University of Geneva preprint and talk by C. Nef at EMS Conf., Brookhaven, 1980.
53. B. Alper et al., to be published in Phys. Lett.
54. J.C. Bizot et al., paper 735, this conference.
55. L.M. Kurdaze et al., Int. Symposium on Lepton and Photon Interactions, Batavia, 1979.
56. B. d'Almagne, Communication at this conference (data from the WA4 Collaboration).
57. M.S. Atiya et al., Phys. Rev. Lett. 43, 1691 (1979).
58. J. Amirzadeh et al., paper 605, this conference.
59. D. Frame et al., paper 638, this conference.
60. For a review, see L. Montanet, in Laws of hadron matter, Erice, 1973 (Acad. Press, NY, 1975).
61. O. Greenberg, Phys. Rev. Lett. 13, 598 (1964) and Phys. Rev. 163, 1844 (1967).
    R. Dalitz, in High-Energy Physics, Ecole d'été, Les Houches, 1963.
62. R. Koniuk and N. Isgur, Phys. Rev. D 21, 1868 (1980).
    N. Isgur and G. Karl, Phys. Rev. D 18, 4187 (1978).
63. R.E. Cutkosky, Talk at the Int. Baryon Conf., Toronto, 1980.
64. R. Koch, Talk at the Int. Baryon Conf., Toronto, 1980.
65. D. Chew, Talk at the Int. Baryon Conf., Toronto, 1980.
66. A. Hendry, Talk at the Int. Baryon Conf., Toronto, 1980.
67. K. Miyake, Talk at the Int. Baryon Conf., Toronto, 1980.
68. K. Jenkins et al., Phys. Rev. D 21, 2445 (1980).
69. P. Baillon et al., to be published in Phys. Lett.
70. C. Fukunaga et al., Tokyo Univ. TMUP-HEP-8005/Exp.
71. T. Hirose et al., Nuovo Cimento 50A, 120 (1979) and Proc. INS Symposium on Particle Physics, Tokyo, 1979.
72. S. Ishida and M. Oda, Prog. Theor. Phys. 61, 1401 (1979).
73. P.J. Mulders et al., Phys. Rev. D 19, 2635 (1979).
74. D.H. Saxon et al., Nucl. Phys. B162, 522 (1980).
75. R.A. Arndt et al., Phys. Rev. D 20, 651 (1979).
76. K.W. Barnham et al., Imperial College report IC/HENP/78/3 (1978).
77. R.L. Kelly, Review talk at the Int. Baryon Conf., Toronto, 1980.
78. P. Litchfield, Review talk at the Int. Baryon Conf., Toronto, 1980.
79. M.W. Arenton et al., paper 748, this conference.
80. C. Dionisi et al., Phys. Lett. 78B, 154 (1978).
81. Ph. Gavillet et al. (Amsterdam-CERN-Nijmegen-Oxford Collab.), EPS Conf. on High-Energy Physics, Budapest, 1977.
82. M. Mazzucato et al. (Amsterdam-CERN-Nijmegen-Oxford Collab.), Int. Conf. High-Energy Physics, Tokyo, 1978.
83. P.F. Loverre et al. (CERN-Collège de France-Madrid-Stockholm Collab.), unpublished data.
84. Yu-Li Pan et al., Phys. Rev. D 2, 49 (1970).
85. J. Kinson (Birmingham-CERN-Glasgow-Michigan S.U.-Paris Collab.), Int. Baryon Conf., Toronto, 1980.
86. C. Dionisi et al. (Amsterdam-CERN-Nijmegen-Oxford Collab.), Phys. Lett. 80B, 145 (1978).

87. Chan Hong Mo, IV European Antiproton Symposium, Strasbourg, 1978.
88. L. Montanet, G.C. Rossi and G. Veneziano, Physics Reports C $\underline{56}$, (1979).
89. A.D. Martin and M.R. Pennington, to be published in Nucl. Phys.
90. B.R. Martin and D. Morgan, Rutherford Lab. report RL 80-009.
91. R.P. Hamilton et al., Phys. Rev. Lett. $\underline{44}$, 1182 and 1179 (1980).
92. M. Alston-Garnjost et al., Phys. Rev. Lett. $\underline{43}$, 1901 (1979).
93. D.I. Lowenstein et al., Paper contributed to this conference. See also G.A. Smith, Review talk at EMS Conf., Brookhaven, 1980.
94. E. Jastrzembski et al. (Temple UC Irvine-New Mexico Collab.), EMS Conf., Brookhaven, 1980.
95. M. Cresti (Frascati-Padova-Rome-Trieste Collab.), talk presented at the V European Antiproton Conf., Bressanone, 1980.
96. V. Chaloupka et al., Phys. Lett. $\underline{61B}$, 487 (1976).
97. Ch. Defoix et al., Nucl. Phys. $\underline{B162}$, 12 and 41 (1980).
98. K. Igi, paper 180, this conference.
99. C. Daum et al., Phys. Lett. $\underline{90B}$, 475 (1980).
100. D. Aston et al., Phys. Lett. $\underline{93B}$, 517 (1980).
101. P. Benkheiri et al., Phys. Lett. $\underline{68B}$, 483 (1977) and $\underline{81B}$, 380 (1979).
102. S.U. Chung et al., paper 696, this conference.
103. R.M. Bionta et al., paper 484, this conference.
104. S.U. Chung et al., paper 697, this conference.
105. T. Armstrong et al. (CERN-Liverpool Collab.), contribution to the Int. Baryon Conf., Toronto, 1980.
106. D. Frame et al., paper 514, this conference.
107. S. Kooijman et al., paper 533, this conference.
108. D. Perrin, Review talk given at the V European Antiproton Conf., Bressanone, 1980.
109. T. Armstrong et al., Phys. Lett. $\underline{77B}$, 447 (1978).
110. D. Frame et al., paper 515, this conference.
111. Z. Bar Yam et al., paper 698, this conference.
112. P. Pavlopoulos et al., Phys. Lett. $\underline{72B}$, 415 (1978).
113. D. Strottman, Phys. Rev. D $\underline{20}$, 748 (1979).
114. H. Högaasen and P. Sorba, paper 725, this conference.
115. W. Löckman et al., Saclay DPHDE 78-01 (1978) and UCLA 1109 (1978).
116. P. Auer et al., paper 581, this conference.
117. R. Kajikawa, Talk at the Int. Baryon Conf., Toronto, 1980.
118. H. Ikeda et al., Phys. Rev. Lett. $\underline{42}$, 1321 (1979).
119. K. Kanai et al., paper 534, this conference.
120. R.G. Arnold et al., paper 592, this conference.
121. O. Braun et al., Nucl. Phys. $\underline{B160}$, 467 (1979).
122. G. Auriemma et al. (Rome-Saclay-Vanderbilt Collab.), Talk at the Int. Baryon Conf., Toronto, 1980.
123. M. Ströbele et al., paper 582, this conference.

# HADRON DYNAMICS

V. Zakharov, ITEP, Speaker

H. Lipkin, Fermilab, Chairman

R. Giles, MIT,

T. DeGrand, UC-Santa Barbara,

W. Stirling, Washington,

Scientific Secretaries

# HADRON DYNAMICS

V. Zakharov

Institute of Theoretical and Experimental Physics,
Moscow 117259.

## Abstract.

Consequences of quark models and of QCD for the hadron dynamics, mostly hadron spectroscopy, are considered. Both ordinary and exotic states are discussed. The general emphasis is on nonperturbative QCD.

## INTRODUCTION

The overwhelming belief now is that hadrons are made of quarks and gluons. The consensus is not so unanimous, however, once we come to specify what the "quarks and gluons" mean. Indeed, there are various kinds of quarks (gluons) discussed around:

current quarks
constituent quarks
bagged quarks
Regge-pole quarks    and so on.

Let me first summarize the differences (suppressing all the references as too numerous to be listed here).

<u>Current</u> quarks enter, say, the definition of the electromagnetic current

$$j_\mu^{el} = \tfrac{2}{3}\bar{u}\gamma_\mu u - \tfrac{1}{3}\bar{d}\gamma_\mu d - \tfrac{1}{3}\bar{s}\gamma_\mu s \ldots$$

and are produced therefore in $e^+e^-$ annihilation (as well as in other short distance processes).

<u>Constituent</u> quarks interact among themselves via some kind of a confining potential and are mostly nonrelativistic. The constituent quarks are readily counted, e.g.,

$$K^+ \sim u\bar{s}, \quad p \sim uud \ldots$$

<u>Bagged</u> quarks are also very limited in number inside hadrons but they interact primarily with the bag. These quarks are ultrarelativistic (as far as u- and d-

quarks are concerned). New degrees of freedom are introduced through the bag.

   Regge-pole quarks enter the topological expansion, ladder-type graphs and are uncovered in high energy collisions. They do not interact if are fast enough and the number of quarks ascribed to a given hadron depends on the coordinate system.

   The multitude of the languages used by the particle physicists was emphatically exposed by Gell-Mann in 1972 [1] Since then the division has simplified, to my mind, and now looks as

   current quarks  = QCD quarks
   others          = QCD motivated quarks.

Indeed, only the current quarks have the privilege to enter the fundamental Lagrangian. Moreover, it is not easy to find a model which ignores QCD altogether so that almost everybody relies on QCD for inspiration. Actually, with QCD getting mature, there is a danger of a split within QCD itself. Thus, one can think of a quark on the lattice and in the continuum limit. But, hopefully we can disregard the problem at the moment.

   The special status acquired by the current quarks does not necessarily imply that everybody starts with that notion nowadays. The point is that there exist phenomenological models which are experienced in describing experimental data in different terms.

   Thus, it seems logical to organize the talk as follows:
   I.   Overview of the QCD motivated models.
   II.  Exotics (as applications of the models).
   III. QCD vs. models.
   IV.  Consequences of QCD (safe, first).
   V.   Consequences of QCD (more speculative).

## I. QUARK MODELS.

Quark models are older than QCD and have already provided us with important keys to the hadron dynamics. More or less complete review of the models would take hours. Instead, I will concentrate on recent developments and, moreover, confine myself to the papers presented or submitted to this Conference. Even with these limitations the material is still too extensive and I merely exemplify the general trend by a few papers whose claims look most impressive.

<u>Constituent quarks.</u> An outstanding effort to transform the nonrelativistic quark model for baryons into a real quantitative framework has been made by Isgur and Karl [2]. The early development was summarized by Karl at the Tokyo Conference. Since then the calculations of the properties of the excited baryons have been pushed further within one and the same framework and brought new confirmation to the model. The success is indeed impressive and it is no accident that the recent note in the "CERN Courier" has compared the development to a "quiet revolution".

The starting assumptions concern with
        quark potential and
        quark selection rules.
The confining potential is taken to be flavour and spin independent. Moreover, it is close to that of harmonic oscillator:

$$V(r) = \frac{1}{2} K r_{ij}^2 + (anharmonic\ piece) \qquad (1)$$

where indices i,j label quarks and $r_{ij}$ is the distance between the two quarks. The anharmonic piece is considered to be small and is treated as a perturbation. No specific assumption on its form is needed since its effect is reduced to a single matrix element.

Spin forces are inferred from the one-gluon exchange between static quarks:

$$H_{hyp} = \alpha_s \frac{\lambda_i^a}{2} \frac{\lambda_j^a}{2} \left\{ \frac{8\pi}{3} \delta^3(r_{ij}) \vec{s}_i \vec{s}_j + \right. \quad (2)$$
$$\left. + r_{ij}^{-3} [3 r_{ij}^{-2} (\vec{s}_i \vec{r}_{ij})(\vec{s}_j \vec{r}_{ij}) - (\vec{s}_i \vec{s}_j)] \right\}$$

where $\alpha_s$ is the strong interactions coupling constant and $\lambda^a$ are the SU(3) matrices in the color space. Note the absence of spin-orbit interaction.

The decay selection rule is also very simple. Imagine an excited baryon decaying into a baryon and a meson (see Fig.1). The wavy lines in the figure indicate the two quarks which are in an excited state. Let it be the third quark who is involved in the decay. The rule is that the final baryon cannot be a nucleon since there are no excited quarks inside the nucleon.

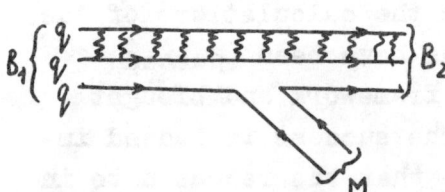

Fig.1 Quark selection rule for the decay $B_1 \to B_2 M$. Baryon $B_2$ is <u>not</u> a nucleon.

As a result, many higher states decouple from $\pi N$ or $KN$ channels. Thus, the standard SU(6) sample of excited baryons is drastically reduced as far as only the states revealed by the phase shift analysis are counted.

It is amusing to watch how the selection rule does remove from the theoretical list all the states which have not been observed so far. The others are seen experimentally and their masses fall close to the predicted values.

All the parameters are fixed by the fit to the lowest states. The outcome of the calculation is a successful description of dozens of masses and of hundreds of decay and photoproduction amplitudes.

<u>Regge-pole quarks</u>. Combining the quark-gluon picture with the reggeon calculus proves very successful [3]

and leads to deeper understanding of the hadron dynamics. In submitted to the Conference papers [4)] Kaidalov utilizes the ideas to calculate the masses of mesons made of different quarks $\bar{q}_i q_j$ ($i \neq j$) in terms of masses of "diagonal" states $\bar{q}_i q_i$.

Consider a process with quantum numbers exchange in the t-channel. In the quark language the quantum numbers of a hadron are carried by the valence quarks. Usually, all the valence quarks are fast but only slow quarks interact strongly. Therefore, we need a configuration with a large rapidity gap $\Delta y$ between the quarks. The probability to find such a configuration is given by

$$\omega(\Delta y, \Delta \vec{b}) \sim \exp[-(1-\alpha(0))\Delta y] \cdot \qquad (3)$$
$$\cdot \exp[-(\Delta \vec{b})^2/4\alpha' \Delta y]$$

where $\Delta \vec{b}$ is difference of the impact parameters of the quarks and $\alpha'$, $\alpha(0)$ are the parameters of the corresponding Regge trajectory. Eq.(3) describes the quark random walk in the rapidity and impact parameter spaces.

The central point of the Kaidalov's approach is the factorization of the S-matrix in the s-channel:

$$\operatorname{Im} T_{ab \to cd} \sim \omega^i_{ab}(\Delta y_1, \Delta \vec{b}) \cdot \omega^j_{cd}(\Delta y_2, \Delta \vec{b}) \qquad (4)$$

where a,b and c,d refer to the initial and final hadrons, respectively; the quark quantum numbers i,j are determined by the process considered; $\Delta y_1 + \Delta y_2 \sim \ln \frac{s}{m^2}$ where s is the total energy.

The physical picture underlying the factorization hypothesis is that hadrons are produced via a color tube and, as a result, all the memory of the initial state is washed away. It is only the probability to find the relevant quarks that counts.

The s-channel factorization is consistent with the

standard Regge-pole description only if the intercepts and slopes satisfy the following relations:

$$\alpha_{ii}(0) + \alpha_{jj}(0) = 2\alpha_{ij}(0)$$
$$[\alpha'_{ii}(0)]^{-1} + [\alpha'_{jj}(0)]^{-1} = 2[\alpha'_{ij}(0)]^{-1} \quad (5)$$

where $\alpha_{ij}, \alpha'_{ij}$ refer to the Regge trajectory incorporating the $\bar{q}_i q_j$ states.

Eqs.(5) allow to find the masses of mesons with open strangeness, charm, beauty.. Some of the predictions are confronted with the data in Table 1.

Table 1

| Meson | Quark content | $J^P$ | Mass (MeV) Theory | Mass (MeV) Experiment |
|---|---|---|---|---|
| K | $u(d)\bar{s}$ | $0^-$ | 474 | 493 |
| K* | $u(d)\bar{s}$ | $1^-$ | 897 | 892 |
| K** | $u(d)\bar{s}$ | $2^+$ | 1423 | 1434 ± 5 |
| D | $u(d)\bar{c}$ | $0^-$ | 1860 | 1868 ± 1 |
| D* | $u(d)\bar{c}$ | $1^-$ | 2010 | 2009 ± 1 |
| F | $s\bar{c}$ | $0^-$ | 2010 | 2030 ± 60 |
| F* | $s\bar{c}$ | $1^-$ | 2150 | 2140 ± 60 |
| B | $u(d)\bar{b}$ | $0^-$ | 5220 | 5170—5270 |

**Bagged quarks** at this Conference are discussed mostly in connection with exotics and we will come back to them in the next section.

Let me also note that there are several interesting papers dealing with quarks in nuclei [5]. The subject is reviewed by Miettenen.

In view of the apparent success of the QCD inspired models there can be no other conclusion to part I but

- Quark models describe a huge amount of data and are of great heuristic value.

It is a different and difficult question, however, whether the models can be taken as an ultimate answer to the hadron dynamics. A sceptic could record, for example, that both the nonrelativistic quark model and the bag model turn to be successful although, in a way, they are just opposite to each other. It is not clear, to my mind, what are the fundamental lessons of the successful model predictions.

## II. EXOTICS.

It is only natural for those who are confident of their calculations of the properties of the known hadrons to apply the art to the states not yet established experimentally. On general grounds alone it is difficult to understand why all the hadrons are to be organized into the $(qqq)$ and $(\bar{q}q)$ series, and existence of hadrons with richer constituent structure has been professed by many.

It is convenient, to say the least, to classify the new possibilities in terms of the bag model. Then one can imagine having in the same bag

$\qquad qq\bar{q}\bar{q},\ldots$ (exotic mesons)
$\qquad qqqq\bar{q},\ldots$ (exotic baryons)
$\qquad GG, GGG,\ldots$ (glueballs)

where G stands for a gluon.

And, indeed, the straightforward application of the bag model leads to the conclusion that all such states should exist and their masses are not particularly high[6]. Thus, the $(q^2\bar{q}^2)$ states fall between .65 GeV and 2.3 GeV[7].

Let me review first the status of this, <u>bag model exotics</u> (I widely use here the Hey's talk at Session A2)

$(q^2\bar{q}^2)$ mesons. Recently the theoretical understanding of these states has drastically changed, new insight coming from the paper by Jaffe and Low [9].

The starting observation of these authors is that the states considered have a component which is actually not confined. That corresponds to the grouping of the $q\bar{q}$

pairs into color singlets.

On the other hand, imposing the bag boundary condition one assumes that the quarks cannot be separated beyond some distance b, and this assumption is no longer true in the case considered.

Although outside the bag radius b the standard description certainly fails, it could be correct inside the "would be" bag. Thus, we are set up to detect the bag effects without the bag itself.

To this end supplement the bag-like picture inside the radius b by the no-interaction hypothesis outside this region. The resulting potential then looks like that in Fig.2b which is to be compared to the "bag approximation" of Fig.2a.

It is no surprise that measuring the phase shifts one can check the emerging picture for the potential. Jaffe and Low choose to work in terms of the so called P-matrix which is designed in such a way as to have poles exactly at the same positions as the standard S-matrix would have in the "bag approximation" of Fig. 2a.

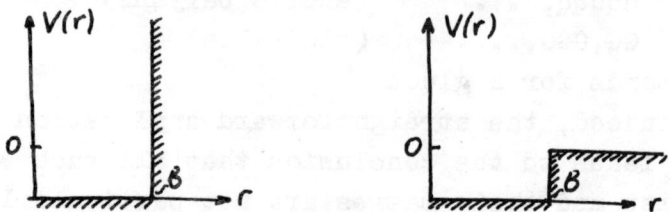

Fig.7. a) "Bag approximation" to the potential    b) Realistic potential.

In the simplest case of the square well potential the P-matrix is given by

$$P(\kappa) = \kappa \cot[b\kappa + \delta(\kappa)] \qquad (6)$$

Note that the P-matrix has poles even if there is no interaction at all, $\delta(\kappa) \equiv 0$, ($\delta(\kappa)$ is the phase shift for the realistic potential and is directly measurable).

On the other hand, if the realistic potential has a deeply bound state, the P-matrix has a pole at the same energy as the S-matrix does.

The results for the $\pi\pi$ scattering with total isospin I=2 are represented in Fig.3. The measured phase shift is rather small and corresponds to repulsion. On the other hand, the P-matrix has a pole at the mass predicted first [7] to be the mass of the corresponding exotic meson with I=2. This confirms the bag-like picture at relatively short distances.

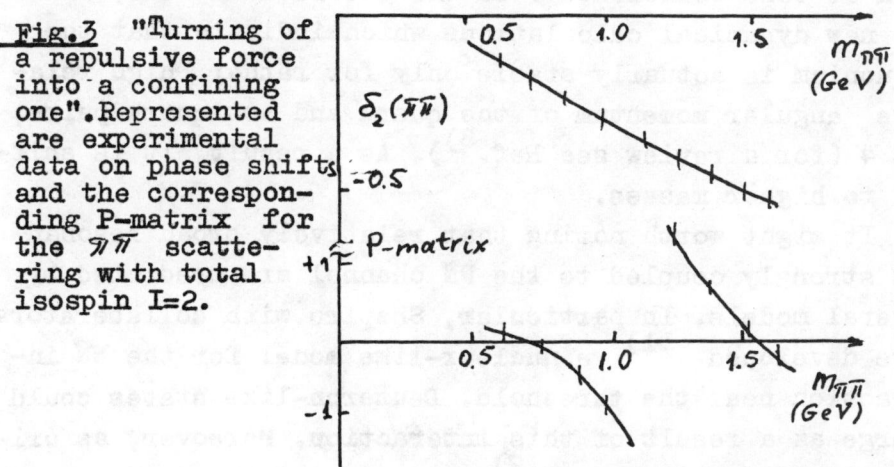

Fig.3 "Turning of a repulsive force into a confining one". Represented are experimental data on phase shifts and the corresponding P-matrix for the $\pi\pi$ scattering with total isospin I=2.

There are several examples of this kind. In all the cases the P-matrix poles follow the pattern predicted first for the real states. Note that the relative position of the poles depends on the relative depth of the potential in the inside region. According to the bag model the shifts in the potential depths are controlled by one-gluon exchange.

Thus, we are both relieved of the search for the real ($q^2\bar{q}^2$) states and get more confidence in the bag model (the former conclusion refers to the so called "round" states, with no orbital momentum involved).

$(qq)(\overline{qq})$, or diquonium states. There is an ingenious guess [10] that there exist exotic states made of $(qq)$ and $(\overline{qq})$ parts protected against collapse by a centrifugal barrier. These states are expected to exhibit some very specific features. In particular, the so called M-baryonium is expected to be very narrow and decay preferably into $N\overline{N}\pi$ final state.

The latest news doesn't look favorable, however. Several candidates for the diquonium seem to be rejected by the most recent experimental data (see the Montanet's talk at this Conference). On the theoretical side, there are new dynamical calculations which indicate that the diquonium is actually stable only for rather high relative angular momentum of the quark and antiquark pairs, $L \geqslant 4$ (for a review see Ref.[8]). As a result, it is shifted to higher masses.

It might worth noting that relatively broad resonances strongly coupled to the $N\overline{N}$ channel are predicted by several models. In particular, Shapiro with collaborators have developed [11] a nuclear-like model for the $N\overline{N}$ interaction near the threshold. Deuteron-like states could emerge as a result of this interaction. Moreover, as originally noted by Rosner [12] existence of the mesons coupled primarily to the nucleon-antinucleon pair is suggested by the dual models. There was no discussion of the models [11,12] at this Conference.

$(qqqq\overline{q})$ baryons. These were scarcely discussed at the Conference. All the previous considerations seem to be subject to a revision in view of the Jaffe and Low paper[9] but, to my knowledge, the P-matrix formalism has not been applied yet to the meson-baryon scattering.

Another remark concerns possible interplay of the models. If one believes that the success of the Isgur-Karl model for the excited baryons is not accidental, then there exists a reliable framework to see what baryons are left as candidates for five quark states. There seem to be very few of them, if any.

(GG), or gluonium. The mesons made of gluons, with no quarks, seem so specific to QCD that they attract more and more attention in recent times [13]. The calculation of the gluonium spectrum was tried within the bag model several years ago [6] and none has disproved the results within the model itself. There is a subtle point about these states, however. The bag model predicts vector mesons made of two gluons. The result is not easy to assess if one considers glueballs as resonances revealed in the gluon-gluon scattering or tries excitation of glueballs by the corresponding gluonic currents. In both cases, say, scalar and vector glueballs are not on the same footing since two massless gluons do not go into a vector state (the observation is due to Donoghue (see Session A2) and Novikov et.al.[14]). This remark might trigger some revision of the original analysis ( see also footnote 43 in the Bjorken's lectures [13]).

All the models aside, there are good perspectives to observe glueballs experimentally through study of radiative decays of heavy-quark-built states like $J/\Psi$ or $\Upsilon$:

$$J/\Psi \rightarrow X + \gamma$$

which proceed via a two-gluon stage, $(c\bar{c}) \rightarrow 2G + \gamma$ [*].

---

[*] There is an apparent analogy to the case of the $e^+e^-$ annihilation which serves as a source of quarks. The gluonic source is not so perfect, however : for a fixed gluons mass their relative energy is not uniquely determeined. Moreover, the gluons can be in different states (see next page). For some sets of quantum numbers the mixing of gluon and quark currents at short distances is strong. In particular, the latter remark refers to tensor currents. Thus, what seems to be a gluon source can actually be the quark one as well. I do not think, however, that all these "QCD refinements" modify the naive picture in its salient features.

Moreover, the recently discovered decay

$$J/\psi \to E(1420) + \gamma$$

(see Sharre's talk at Session C5) turns to be quite abundant and might eventually lead to identifiaction of the E-meson with a glueball. I would not like to go into detailed discussion of the dynamics of the E meson since we have had interesting talks by Meshkov and Carlson on the subject and confine myself to a single remark which seems to be of quite a general nature although is not widely recognized.

The remark concerns spin assignment of the E-meson. It is more or less established experimentally that the observed total radiative decay rate is dual to the corresponding quark transition:

$$\Gamma(J/\psi \to X\gamma) \simeq \Gamma[(c\bar{c})_{J/\psi} \to 2G + \gamma] \qquad (7)$$

where the elementary transition is calculated within the lowest order perturbation theory.

But in perturbation theory one can read off the total angular momentum of the gluons as well [15]. There are four different states:

$$(c\bar{c})_{J=1} \to (0^{\pm}, 2^{\pm} \text{ gluons}) + \gamma$$

and spin 2 dominates strongly over the spin 0 state.

The final state interaction could affect drastically the mass spectrum of the gluons but, by virtue of the duality argument, we still expect abundant production of tensor mesons ( the $J/\psi \to f + \gamma$ decay could be due to the strong mixing of the gluon and quark currents, see the footnote on the previous page). Spin 1 assignment for the E meson would be very embarrasing while a pseudoscalar meson might be not ruled out but certainly not welcome. The reason is that the $\eta$ and $\eta'$ are already dual to the corresponding continuum. What we really need, is a tensor meson.

Rather ironically, the E meson produced in hadron--hadron collisions most probably has spin 1. It would be much better, therefore, if the meson observed in the radiative decays of the J/ψ is not the same one. There are some experimental indications to this effect (see Session C5) but we should wait for a definite result.

So far we discussed "old exotics" which have been under consideration for a few years. Several papers submitted to the Conference advocate existence of "new exotics".

Heavy-quark-bag exotics. Heavy quarks in the bag are considered by Hasenfratz et.al.[16] (see also Kuti's talk).

The basic idea is that heavy quarks move slowly inside the bag and one can apply the Born-Oppenheimer approximation for this reason. First, the gluon field configuration is found for fixed heavy quark positions and then the quark Schrodinger equation is solved for the effective quark potential.

All the parameters of the model are fixed by a fit to the J/ψ series and some new states are predicted. Notably, the bag constant B differs substantially from the standard fit to the light mesons and we will come back to the discussion of this point in part V.

One of the interesting conclusions is that there exist exotic mesons with quantum numbers $J^{PC} = 1^{-+}$. These quantum numbers cannot be realized in the $Q\bar{Q}$ system and their discovery would prove excitation of the gluon degree of freedom. What is remarkable, is that for b quarks the exotic meson is predicted below the threshold of open beauty production:

$$M(b\bar{b}G, J^{PC} = 1^{-+}) < 2M_B \qquad (8)$$

I do not think that the model can be taken too literally (for a relevant discussion see also part III ) but the overall picture for the levels could well be true.

<u>Regge-pole exotics.</u> In a submitted paper [17] Grigoryan and Kaidalov argue in favor of exotic baryons with isotopic spin I = 5/2,7/2... basing on the reggeon sum rules. The masses are not too high and discovery of such states, if they exist, is within the grasp of present experiment.

Now we come to the <u>conclusion</u> to part II:

- no ($q^2\bar{q}^2$) "round" exotics but the "would be" bag is confirmed via the P-matrix formalism;
- diquonium seems to be pushed to higher orbital momentum, higher mass;
- glueball business is active but speculative at the moment;
- new exotics are in sight ( heavy $\bar{Q}QG$ states ).

## III. QCD vs. MODELS.

An important question is whether the quark models can be incorporated into quantum chromodynamics. In this Section I will discuss several statements which can be proven within QCD but seem completely foreign to the models. Thus, the relation between the models and QCD is not simple. Here are several "flat no" of QCD to the models:

1. The pion is not a simple $\bar{q}q$ state:

$$m_\pi^2 \simeq 0 \Longleftrightarrow \langle \bar{q}q \rangle \neq 0$$

2. Gluons can never be neglected even in the classical quark states, e.g.:

$$\langle N | - \frac{9\alpha_s}{8\pi} G^a_{\mu\nu} G^a_{\mu\nu} | N \rangle \simeq m_N \bar{N}N \qquad (9)$$

( $G^a_{\mu\nu}$ is the gluon field strength tensor and N is nucleon).

3. Even for infinitely heavy quarks, $m_Q \to \infty$, there is no potential beyond the Coulomb-like:

$$V(r) = -\frac{4}{3}\frac{\alpha_s}{r} + (NO\ potential) \qquad (10)$$

Let me now <u>comment</u> on the statements made.

1. The problem of the pion is a widely recognized one. On one hand, the pion is a collective excitation due to the spontaneous chiral symmetry breaking. On the other hand, quark models consider it as a standard $\bar{q}q$ state.

The attempts to modify the bag model in a way that the problem gets resolved or nearly so were reported to the Conference by Donoghue[13] and Haymaker[18]. According to Ref.[18] the pion is bound mostly by the QCD invariant interaction and is a solution to the corresponding Bethe-Salpiter equation. The bag boundary condition which violates the chiral symmetry by "brute force" removes only the long-range tail of the wave function which might be not so important.

2. Eq.(9) is derived[19] through a simple chain of arguments. First, the trace of the energy-momentum tensor is given by

$$\Theta_{\mu\mu} = \frac{\beta(\alpha_s)}{4\alpha_s}(G_{\mu\nu}^a)^2 + \sum_q m_q \bar{q}q \qquad (11)$$

where $\beta(\alpha_s)$ is the Gell-Mann-Low $\beta$ function and the first term in the r.h.s. is the so called triangle anomaly[20]. The product $\beta/\alpha_s \cdot G^2$ is renorminvariant and we choose such a normalization point for the operator $G^2$ that $\alpha_s$ is small and $\beta(\alpha_s)$ can be expanded in $\alpha_s$.

Averaging operator eq.(11) over the nucleon state brings further simplifications. Heavy quarks are integrated out and, as a result, one can forget both the $m_Q \bar{Q}Q$ term and the heavy quark contribution to the $\beta(\alpha_s)$. As for the light quark masses, $m_{u,d,s}$, they can be safely neglected because of the approximate chiral symmetry.

On the other hand, on purely phenomenological grounds we know that

$$\langle N| \theta_{\mu\mu} |N\rangle \sim m_N \bar{N}N \tag{12}$$

and combining all the factors gives eq.(9).

Thus, the total energy of nucleon can be rewritten in terms of the gluon field alone ,with no large coefficients involved. I would be very much surprised if any constituent model could reproduce this QCD result although I would not insist that the failure to evaluate some matrix element represents a fundamental setback for a model.

3. A formal proof of the "no potential" theorem is given by Voloshin [21] and is based on consideration of heavy quark-antiquark Green function G( x,y, $\varepsilon$ ). The common wisdom tells us that

$$G(x,y,\varepsilon) = {\sum_n}' \frac{\Psi_n(x)\,\Psi_n(y)}{\varepsilon_n - \varepsilon} \tag{13}$$

where $\varepsilon_n$ are the eigenvalues and $\Psi_n(x)$ are the corresponding eigenfunctions of the quark-antiquark Hamiltonian. Moreover, one usually assumes that $\{\Psi_n\}$ is a complete set of orthonormalized functions. Then it is readily seen that

$$\frac{\partial}{\partial \varepsilon} G(x,y,\varepsilon) = \int G(x,z,\varepsilon) G(y,z,\varepsilon)\,dz \tag{14}$$

and this condition can be considered as the most general one for validity of a two-body , or potential picture .

On the other hand, the Green function at short distances can be evaluated explicitly in QCD. To zero approximation it coincides with the free particles Green function . The correction comes from interaction with the vacuum fields. Taking the quark mass to be large simplifies the calculation of the correction since one can enjoy both nonrelativistic picture and asymptotic freedom in this case. Indeed, the momentum p of the quarks can be large in the hadronic mass scale, $|p| \gg \mu$, but small compared to the quark mass, $|p| \ll m_Q$. The calculation shows that eq.(14) does <u>not</u> hold for the physical vacuum state with $\langle \alpha_s G^2 \rangle \neq 0$ .

Although the result might look unexpected, it is very natural from the point of view described in Ref.22: quarks, injected into the physical vacuum at short distances, find the long-wave gluon fields present in the vacuum medium and interact primarily with these fields, not between themselves. A bit more formally [21], the potential picture assumes that

$$\omega_q \ll \omega_{vacuum}$$

where $\omega_q$ is the quark frequency and $\omega_{vacuum}$ is that of the vacuum field. But for heavy quarks bound by the Coulomb-like potential

$$\omega_q \sim \alpha_s m_Q \gg \omega_{vacuum} \qquad (15)$$

and just the opposite relation holds for $m_Q \to \infty$.

The argument does not imply, of course, that one cannot introduce and calculate static energy of two quarks since then $\omega_q \to 0$. Moreover, if the distance between the quarks becomes large, much larger than characteristic scale of the vacuum fluctuations, the quarks organize vacuum field "granules" between them and one can think again in terms of potential (which is linear, by all the bets).

But as far as short distances are concerned there is no potential beyond the Coulomb-like. This no--potential region extends to $r \sim (200 \text{ MeV})^{-1}$

I would not like to imply that all these conflicts between QCD and the models rule out the models. The models can be still safe in phenomenological applications. But as far as we have no idea on the relation between the fundamental fields and those introduced by models we cannot claim successes of the models to be successes of QCD.

Thus, the conclusion to part III is:
- identification of the constituent, bagged... quarks and gluons with the fundamental fields is dangerous and can be proven to be wrong in some cases.

## IV. QCD CERTAINTIES.

Once we abandon the idea to reduce QCD in some approximation to a well developed model we must start with QCD itself from the very beginning. In this part several implications of QCD are discussed which I would risk to call "QCD certainties" since I do not see how they could be changed by further developments. Without pretending to be complete, here is the list:

1. Pion is a Nambu-Goldstone boson.
2. $SU(3)_{flavor}$ is due to $m_{u,d,s} \simeq 0$ but $m_{u,d} \ll m_s$.
3. The $\Delta I = 1$ effects (like the $\omega - \rho$ mixing) involve both $m_d - m_u \neq 0$ and $\langle \bar{q}q \rangle \neq 0$.
4. Nearly massless $\eta'$ is removed from the physical spectrum by the gluon anomaly in the divergence of the axial-vector current:
$$\partial_\mu (\bar{u}\gamma_\mu \gamma_5 u + \bar{d}\gamma_\mu \gamma_5 d + \bar{s}\gamma_\mu \gamma_5 s) = \frac{3\alpha_s}{4\pi} G^a_{\mu\nu} \tilde{G}^a_{\mu\nu} \quad (16)$$
5. Asymptotic freedom is taken over by nonperturbative effects at low $\alpha_s$:
$$\alpha_s^{crit} \sim 0.2 \div 0.3 \quad (17).$$

### Comments:

1. Although the quark vacuum condensate, $\langle \bar{q}q \rangle \neq 0$, is known for a long time its connection with confinement has been established only recently[23] and I will outline the argument due to 't Hooft [23].

Consider the triangle graph associated with the Adler anomaly [24]. If $q_\mu$ is the momentum carried by the axial vector current and the photons are real ($k_1^2 = k_2^2 = 0$) then the anomaly implies the singularity in the amplitude of the kind

$$M_\mu \sim \frac{q_\mu}{q^2} \varepsilon_{\alpha\beta\gamma\delta} F^{(1)}_{\alpha\beta} F^{(2)}_{\gamma\delta} \quad (18)$$

where $F^{1,2}$ are the photons wave functions. Indeed, multiplying $M_\mu$ by $q_\mu$ we should get $F\tilde{F}$ -this is just what the "anomaly" means. On the other hand, the matrix element $M_\mu$ for the transition of the axial current into the photons is gauge invariant and there is no way out but the singularity in the amplitude.

The presence of singularity can be checked in the perturbation theory explicitly and the singularity is indeed known to happen [25]. The meaning of the singularity in the perturbation theory is simple: we have massless quarks and the quark-antiquark pair propagates freely over a large distance.

Now we assume that non-abelian gauge theory ensures confinement. Then confined quarks cannot escape too far and therefore cannot be responsible for a singularity in the amplitude.

But the singularity is still there since it is required by the equations of motion (triangle anomaly). Therefore we must look for a substitution of massless quarks by massless physical particles. The alternatives are:
massless pion or
massless baryons.

Thus, if one is able to rule out the second alternative then the presence of a massless pion is proven.

It is rather evident that massless baryons are indeed a clumsy possibility. The point is that the anomaly is generated by quarks which form a flavour triplet $\{3\}$. But physical bound states can come only in octets, $\{8\}$, decuplets, $\{10\}$, and so on. It is not easy to imitate the algebra of the triplets by these, more complicated representations. As 't Hooft does show, it is impossible to satisfy the algebra in a natural way. Naturalness in this case includes some simple ideas on how the spectrum evolves when the quark masses are changed.

Once massless baryons are proven not to satisfy the triangle anomaly algebra, we are left with a massless pion and the spontaneous breaking of the chiral symmetry is seen to be a consequence of the confinement. What is a bit disturbing about this argument is that, as noted by 't Hooft, for two massless quarks the algebra is trivially satisfied since both quarks and nucleons then belong to the same representation of the flavour group. On the other hand, it is difficult to imagine that the s-quark is so crucial to the fate of the pion.

According to Coleman and Witten[23] the massless baryons are ruled out, independent of the flavour group, if the number of colours, $N_c$, is large. The point is that baryons contain $N_c$ quarks and graphs with many loops decouple in the large $N_c$ limit. Saturation of the triangle anomaly condition with massless baryons would correspond to a highly exotic state being dual to a simple quark graph. Such a possibility is against all the experience with the hadron dynamics that we have and, formally, is ruled out by the $N_c$ counting. Note that a massless pion is a nonexotic solution.

2. It is widely accepted now that the current u-, d- and s- quarks have small and very different masses. Moreover the estimates[27]

$$m_u \sim 7\, MeV, \quad m_d \sim 4\, MeV, \quad m_s \sim 150\, MeV \tag{19}$$

have also become common (the estimates seemingly refer to a low normalization point, formally, $\alpha_s(\mu) \sim 1$ [22]).

The only recent development which I would like to mention here is that the charmonium physics also provides with an opportunity to measure a ratio of the quark masses[28]:

$$\frac{\Gamma(\psi' \to J/\psi\, \pi^0)}{\Gamma(\psi' \to J/\psi\, \eta)} = \frac{27}{16}\left(\frac{m_d - m_u}{m_s}\right)^2 \left(\frac{p_\pi}{p_\eta}\right)^3 \tag{20}$$

where $p_\pi$, $p_\eta$ are the $\pi$ and $\eta$ 3-momenta, respectively. The standard quark masses are consistent with the data.

3. Among other things, the smallness of the quark masses implies that spontaneous chiral symmetry breaking has deeper impact on the observed mass spectrum than is commonly recognized. In particular, it follows from dimensional consideration alone that

$$m_{\rho\omega}^2 \sim (m_d - m_u) <\bar{q}q> m_\rho^{-2} \qquad (21)$$

where $m_{\rho\omega}^2$ is the $\rho$-$\omega$-mixing mass.

Indeed, it is well known that virtual electromagnetic interactions do not explain the observed pattern of the $\Delta I=1$ effects. Therefore one must invoke the bare $(m_d-m_u)$ mass difference. Smallness of the quark masses makes $m_d^2 - m_u^2$ negligible. The only way to get linear in $(m_d-m_u)$ effect is to multiply it by $<\bar{q}q>$.

An explicit way to see how the product $m_q <\bar{q}q>$ emerges is to study the QCD sum rules [22]. As for constituent or bag models, I believe, calculation of the $\Delta I=1$ effects does represent a serious problem.

4. Gluon anomaly (16) leads to a number of consequences. First, the anomaly is a necessary condition to shift the ninth pseudoscalar meson to a higher mass. It is much more difficult to follow the details of dynamics of this mass generation. As first shown by 't Hooft, instantons do induce effective interaction which explicitly violates the U(1) symmetry and could, in principle, be responsible for the $\eta'$ mass. It is not clear at the moment whether this particular inteaction does play the leading role. I feel that more complicated vacuum fluctuations are involved but this conclusion can be challenged and is not discussed here.

The anomaly also allows to calculate some gluonic matrix elements in a model independent way:

$$\langle 0| \tfrac{3 d_s}{4\pi} G^a_{\mu\nu} \tilde{G}^a_{\mu\nu} |\pi^0\rangle = \tfrac{3}{\sqrt{2}} \tfrac{m_d - m_u}{m_d + m_u} m_\pi^2 f_\pi \quad (22a)$$

$$\langle 0| \tfrac{3 d_s}{4\pi} G^a_{\mu\nu} \tilde{G}^a_{\mu\nu} |\eta\rangle = \sqrt{\tfrac{3}{2}} f_\pi m_\eta^2 \quad (22b)$$

The derivation [30)31)] relies only on the anomaly and such well established flavor symmetries as SU(2)xSU(2) and SU(3). (Eq.(22b) actually assumes the $\eta$ to be a component of an SU(3) octet; more involved Ward identities can be found in Ref.[32)]).

In the case of $\eta'$ the anomaly is not powerful enough to fix the gluonic matrix element but combined with the large $N_c$ limit it leads to new and beautiful results [34)].

First, the $\eta'$ mass vanishes in this limit. To demonstrate this consider the correlator

$$K(q) = i\int e^{iqx} \langle 0|T\{\tfrac{3 d_s}{4\pi} G\tilde{G}(x), \sum \bar{q} \gamma_5 q(0)\}|0\rangle dx \quad (23)$$

The standard counting gives (23)
$$K(q) \sim (N_c)^0$$
while the exact low energy theorem reads [35)]:

$$K(0) = -\sum_{q=u,d,s} 2\langle \bar{q} q\rangle + O(m_q) \text{ or } K(0) \sim N_c \quad (24)$$

Thus, the correlator is anomalously large at low q and indicates in this way existence of a particle with vanishing mass:

$$m_{\eta'}^2 \sim (N_c)^{-1} \quad (24)$$

For comparison, consider Ward identities in the scalar case [14)]:

$$i\int dx \langle 0|T\{\tfrac{3 d_s}{4\pi} G^2(x), \tfrac{3 d_s}{4\pi} G^2(0)\}|0\rangle = \tfrac{18}{8} \langle \tfrac{d_s}{\pi} G^2\rangle \sim (N_c)^0$$

$$i\int dx \langle 0|T\{\bar{q} q(x), \tfrac{3 d_s}{4\pi} G^2(0)\}|0\rangle = \tfrac{18}{8} \langle \bar{q} q\rangle \sim (N_c)^0 \quad (25)$$

These imply two scalar states, $\sigma_q$ and $\sigma_G$ ("quarkonium" and "gluonium", respectively) with residues given by the ordinary counting, say,

$$m_{\sigma_q} \sim (N_c)^0 ; \quad \langle 0| \tfrac{3 d_s}{4\pi} G\tilde{G} |\sigma_q\rangle \sim N_c^{-1/2} ; \quad \langle 0|\bar{q}q|\sigma_q\rangle \sim (N_c)^{1/2} \quad (26)$$

An explicit realization of eq.(24) is provided by the mass relation advocated in ref.[14]:

$$f_{\eta'}^2 \, m_{\eta'}^2 \simeq \tfrac{9}{b} \langle \tfrac{d_s}{\pi} G^2 \rangle$$

where b is the coefficient in the Gell-Mann-Low beta function, $b = 11/3 \cdot N_c - 2/3 \, n_f$, and $f_{\eta'}$ is the coupling of $\eta'$ to the singlet axial vector current,

$$\langle 0| \bar{u}\gamma_\mu\gamma_5 u + \bar{d}\gamma_\mu\gamma_5 d + \bar{s}\gamma_\mu\gamma_5 s |\eta'\rangle = i f_{\eta'} P_\mu$$

Note that the large $N_c$ limit fixes $f_{\eta'}$ to be $f_{\eta'} = \sqrt{3} f_\pi$.

Smallness of the $\eta'$ mass allows to develop the formalism of effective Lagrangians [34,32] and the physics of $\eta'$ becomes at large $N_c$ as simple as that of pions. Of course, there remains a question on how important are the $1/N_c$ corrections in the real case of $N_c=3$. This question cannot be answered with confidence yet (see, however, Ref.[14]).

5. Finally, we come to a discussion of the mechanism of asymptotic freedom violation. The alternatives are: either the breaking of expansion in $\alpha_s$ is signalled by the perturbation theory itself, or it occurs at such low $\alpha_s$ that, on its own ground, the perturbation theory seems to be safe yet.

The QCD sum rules provide an ample evidence of an early breaking of asymptotic freedom. Namely, the nonperturbative terms blow up at such low $\alpha_s$ that the $\alpha_s$ corrections themselves are very moderate.

It is encouraging that a very different approach leads to the same conclusion. Fig.4, borrowed from Ref.[36], illustrates the matching of the weak and strong coupling expansions in pure gluodynamics.

Two points are worth emphasizing. First, the matching is quite smooth at critical $\alpha_s$, $\alpha_s \sim .2$. Moreover, as reported by Pearson to this Conference, inclusion of the next order term in the 1/g expansion makes the matching even smoother. Second, the weak coupling expansion breaks down at low $\alpha_s$ and, by itself, does not indicate that something is going wrong.

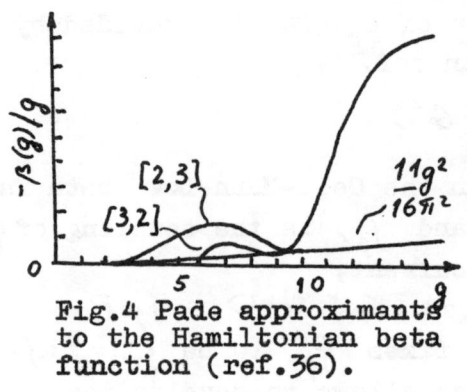

Fig.4 Pade approximants to the Hamiltonian beta function (ref.36).

Thus, I would conclude that any use of $\alpha_s$ larger than .2-.3 implies that we are dealing with a model calculation effectively incorporating the nonperturbative terms.

The overall conclusion to part IV:
- well established phenomenological implications of QCD are nontrivial, sometimes conceptionally new but apply to quite a limited number of cases.

### V. MORE ON QCD.

At present, if we want to extract more consequences from QCD we should be ready to make some numerical approximations. In this part I will discuss a few attempts of this kind.

#### 5.1 QCD Sum Rules

The basic idea underlying the sum rules [22] is that asymptotic freedom is violated through interaction of quarks and gluons with the vacuum fields. Consider a quark-antiquark pair injected into the vacuum by a virtual photon. First the quarks propagate as free particles. Then, with the distance between the quarks growing, they acquire a color dipole moment: $\vec{d}^a \sim Q^a \vec{r}$
This dipole moment interacts with the fluctuating gluon field around. If one succeeds to evaluate this effect it should reduce to nothing else but

$$\langle \alpha_s G^a_{\mu\nu} G^a_{\mu\nu} \rangle$$

since it is the simplest gluonic vacuum expectation value consistent with the Lorentz and color invariance.

Similarily, the quark exchange interaction leads to the effects proportional to the quark condensate, $\langle \bar{q}q \rangle$.

For example, sum rules for $e^+e^-$ annihilation into hadrons with total isospin I=1 look as :

$$\frac{2}{3M^2}\int R^{I=1}(s) \exp(-s/M^2)\, ds = \qquad (28)$$

$$= 1 + \frac{\alpha_s}{\pi} + \frac{1}{3M^4}\langle \frac{\alpha_s}{\pi} G^2 \rangle - \frac{64\pi^3}{9M^6}\alpha_s (\langle \bar{q}q \rangle)^2 + O(M^{-8})$$

where $\quad R^{I=1} = \sigma(e^+e^- \to X, I=1)/\sigma(e^+e^- \to \mu^+\mu^-)$
and $M^2$ is variable.

The unit in the r.h.s. corresponds to asymptotic freedom while the other terms are corrections due to various type of interactions: perturbative, scattering off nonperturbative gluon and quark fields, respectively.

In the language of quantum mechanics the sum rules correspond to the Born series for the imaginary time evolution operator [21,37]. Thus for the harmonic oscillator:

$$\left(\frac{m}{2\pi\tau}\right)^{-3/2} \sum_n |\psi_n(0)|^2 e^{-\varepsilon_n \tau} = 1 - \frac{(\omega\tau)^2}{4} + \frac{19}{480}(\omega\tau)^4 + \ldots \quad (29)$$

There is no point, in general, to check the quantum mechanical sum rules since they are certainly true. One numerical lesson is remarkable, however. There exists a range of the time variable $\tau$, $\tau \sim \omega^{-1}$, where, on one hand, the evolution still follows closely that of free particles and, on the other hand, the sum over the eigenstates is actually dominated by the lowest level. In other words, matching of the resonance and asymptotic regions is very smooth, see Fig.5. It is even smoother for steeper potentials.

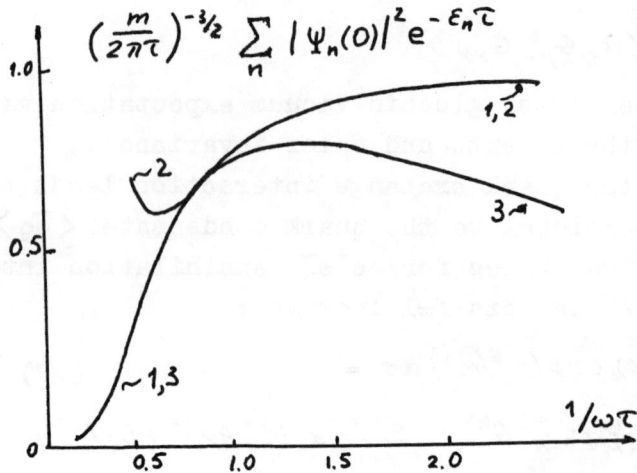

Fig.5. Sum rules for the harmonic oscillator [37]. Curve 1 is the exact sum over the levels. Curve 2 is the asymptotic approximation to it with the first two power corrections left. Curve 3 is the contribution of the first two levels. Around $\omega \sim \tau^{-1}$ where the power corrections become noticable all the curves are close.

The QCD sum rules (28) cannot fail as far as quantum chromodynamics is correct. As for the smooth matching of the resonance and asymptotic expansions it could be true or, generally, could be not true.

But at least for some resonance quantum numbers the matching is extremely smooth in QCD. As a result the resonance masses and residues are related to the properties of the QCD vacuum state. For light hadrons the major role is played by the quark condensate while for heavy hadrons the gluon condensate is most important. Say, the $\rho$ meson mass is related to $<\bar{q}q>$. Approximately,

$$m_\rho^2 \simeq 9|<\alpha_s^{4/3}\bar{q}q>|^{2/3}$$

Because of the smooth matching, the analysis of the sum rules simplifies greatly. All the QCD certainties discussed in the previous part are checked and rechecked; the salient features of the resonance spectra in the vector and axial vector channels are seen to be related to the interaction of quarks and gluons with the vacuum fields.

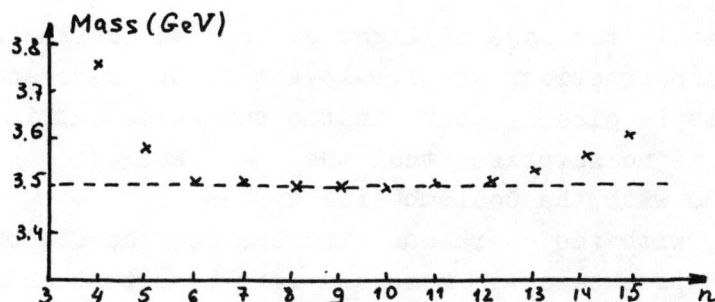

Fig.6 Sum rule determination of the mass of the $^3P_1$ state of charmonium [38]. The dashed straight line corresponds to single resonance saturation of the sum rules. Crosses are theoretical points (asymptotic freedom + first power correction due to the gluon condensate). For high values of the moment n further power corrections become important.

In the paper submitted to the Conference Reinders et.al.[38] calculated the mass splitting within the family of the $\bar{c}c$ bound states. Fig.6 exemplifies the results obtained. The crosses are the theoretical predictions for the moments of the cross section associated with the charmed quark current $\bar{c}\gamma_\mu\gamma_5 c$. The moments are defined as the integrals
$$\int \sigma(s)(s+Q^2)^{-n}ds$$
and the ratio $\xi = Q^2/4m_c^2$ is taken to be 2.5 in this particular case. The theoretical numbers depend on the value of $\langle \alpha_s G^2 \rangle$ which is fixed by the sum rules for the $J/\psi$.

The straight line corresponds to the same moments with a single resonance left in the cross section. The mass of the resonance is read off as 3.5 GeV, in a remarkable agreement with the experiment. No assumption on the potential is involved (moreover, there is no potential at all, see part III) and there is no fit parameter.

The picture gives feeling on how smooth is the matching of the resonance and asymptotic regions: for low n more than one resonance contribute, for high n higher power corrections become important, but at intermediate n the straight line is very close to the asymptotic expansion .

Note that in the case of light or charmed quarks the perturbative corrections are irrelevant to the resonance physics. This is clearly seen in the sum rules and, indeed, there is no surprise that the $\rho$ and $J/\psi$ have nothing to do with the Coulomb-like system.

However, with the quark mass increasing, the Coulomb-like interaction becomes more and more important and eventually dominates the picture. Already in the case of the $\Upsilon$ family one has to trace modification of the power corrections by the Coulomb-like interaction. As a result, derivation of sum rules sensitive to a single resonance becomes much more involved.

The challenge was met by Voloshin [21] who considered the $\Upsilon$ sum rules in great detail. In particular, the sum rules demonstrate that one cannot rely on sayptotic expansion for $\alpha_s$ if $\alpha_s \gtrsim .3$. Another interesting observation is that at $m_Q \to \infty$ the sum rule technique is taken over by exact theory of a single level since the quarks are kept at short distances by the Coulomb-like force. The corrections due to the gluon condensate are calculable. For example, the energy of the lowest level is given by:

$$\frac{\delta \varepsilon_0}{|\varepsilon_0|} = 0.906 \left\langle \frac{\alpha_s}{\pi} G^2 \right\rangle \frac{m_Q^2}{K_1^6} \tag{30}$$

where $K_1 = \frac{2}{3} m_Q \alpha_s(K_1)$.

## 5.2 Looking for Glue in Hadrons.

As is emphasized above, QCD fixes some gluonic matrix elements (see eqs.(9),(22)). The results are specific for QCD which accounts for transitions between quarks and gluons and makes any distiction between "quarkonium " and "gluonium" somewhat conditional ( see also sect. 5.4). It would be , therefore, instructive to test these results.

First, once the matrix elements
$$\langle 0 | \alpha_s G\tilde{G} | \pi, \eta, \eta' \rangle$$
are known one may hope for a theory of the corresponding radiative decays of heavy mesons, like $J/\psi \to \pi\gamma$.

Intuitively, one expects that the ratios of these decay rates are given by the ratios of the corresponding matrix elements, $\langle 0 | G\tilde{G} | meson \rangle$, squared. More sophisticated studies confirm the guess[39]. Since the matrix element $\langle 0 | G\tilde{G} | \eta \rangle$ is fixed by the SU(3) symmetry alone, measuring the ratio $\Gamma(J/\psi \to \eta\gamma)/\Gamma(J/\psi \to \eta'\gamma)$ checks various theories of the $\eta'$ ( see sect. 4.4 ). In particular, large $N_c$ justify the naive quark model while the QCD sum rules[30] tend to a lower value by about factor of 2 in the amplitude. The data reported to this Conference by Sharre (Session C5) are in favour of the sum rules.

Moreover, the sum rules turn to be powerful enough to determine not only the ratios but the radiative decay widths themselves (Novikov et.al.[39]) :

$$\Gamma(J/\psi \to \eta\gamma) = \frac{2^6 \pi \alpha^3}{5^2 3^{11}} \left( \frac{m_\eta m_\psi}{m_c^2} \right)^2 \frac{f_\pi^2}{\Gamma(J/\psi \to e^+e^-)} \tag{31}$$

This prediction, with no free parameters involved, agrees with the data reasonably well ($m_c = 1.25$ GeV).

As mentioned in part III, the matrix elements of the operator $G^2$ for zero momentum transfer can be rewritten in terms of the total energy. For pions, because of their Goldstone nature, one can go further and include first term of the expansion in the momentum transfer squared[40]:

$$\langle 0 | -\frac{\alpha_s}{8\pi} G^a_{\mu\nu} G^a_{\mu\nu} | 2\pi \rangle = m^2_{\pi\pi} \tag{32}$$

where $m_{\pi\pi}$ is the two pion invariant mass.

The operator $G^2$ entering eq.(32) is a local one. In nature we do not have local sources of gluons. But imagine that the gluons are emitted at relatively short distances so that the source is not local but rather compact in the characteristic hadronic scale. Does evolution of the gluons obey eq.(32) or this equation should be understood in a subtle way?

Actually, rather compact source of gluons does exist, that is, heavy quarkonium. As noted first by Gottfried (Ref.[41]), say, the $\Psi' \to \eta \Psi$ transition can be considered in the dipole approximation[41,21,42] with respect to the gluon field. Therefore, there arises a possibility to check eq.(31) [40]:

$$\frac{\Gamma(\Psi' \to J/\psi + \eta)}{\Gamma(\Psi' \to J/\psi + \pi\pi)} \simeq \frac{16\pi^2 f_\pi^2 P_\eta^3 m_\eta^4}{M_\Psi (M_{\Psi'} - M_\Psi)^7} \left(\frac{9}{b}\right)^2 19.6 \simeq 0.10 \div 0.14 \quad (33)$$

where b is the coefficient in the Gell-Mann-Low $\beta$-function, b = 9. The prediction is consistent with the data. Since we started with the matrix element (32) we now know experimentally that the gluon transition into ordinary hadrons is large in some cases.

A crucial question is whether this large value of the gluonic matrix element a rule or exception. The answer is not known yet since we have little experience with the gluonic currents. The point is that all the matrix elements considered so far are somewhat special: they are associated with the anomalies (otherwise we would not be able to find them). The only comment which I can add is that from the same decay $\Psi' \to \Psi \pi\pi$ one can extract [43] the matrix element

$$\langle 0 | \vec{E}_a^2 + \vec{B}_a^2 | (\pi\pi) \rangle$$

where $\vec{E}_a$ and $\vec{B}_a$ are the color electric and magnetic fields. This matrix element is substantially smaller than the matrix element of $\vec{B}_a^2 - \vec{E}_a^2$ discussed above. Note that for the vacuum-to-vacuum transition the matrix element of $\vec{B}_a^2 + \vec{E}_a^2$ vanishes by virtue of the Lorentz invariance while that of $\vec{B}_a^2 - \vec{E}_a^2$ does not. We see that the general structure of

the fields is not greatly disturbed inside the pions.
This remark brings us to the discussion of the next
problem.

### 5.3 Bag Made of Vacuum.

The so called QCD bag [44,14] was discussed first
at the Tokyo Conference. The idea is simple : the color
field of quarks inside hadrons suppresses (nonperturbative) fluctuations which are usually present in the vacuum. This costs energy since the nonperturbative vacuum
m fluctuations lower the energy. Thus, there arises the
energy difference between the vacuum and the inside region of hadrons which is described phenomenologically
in terms of the abg constant B. The idea is elaborated
in great detail by Callan et.al.[44] on the example of the
instanton gas as a model for the physical vacuum state,
with a far reaching conclusion on the phase transition.

I feel that in its gross features the picture is
true beyond any doubt. There is a lot of questions left,
however . Thus, we know that the quark condensate is crucial for the light hadrons and, therefore, the empirical
value of the constant B depends on $\langle \bar{q}q \rangle$ . Theoretically
this dependence has never been traced and all the calculations refer to the pure gluodynamics.

In particular , it is an open problem whether the suppression of the vacuum fluctuations inside hadrons is
strong or relatively weak. The phase transition advocated in Ref.[44] implies complete suppression of the vacuum
fluctuations by the quark field. However, even if we take
the theoretical derivation as perfect, there are some assumptions, such as static quarks, which can be wrong for the
real hadrons. Thus, it is better turn to the phenomenology.
The central statement of this section (sects.5.3,5.4 are
based on Ref.[14]) is that ordinary hadrons are relatively shallow structure on the vacuum energy sea.
The vacuum energy is evaluated using eq.(11) which has

been already used above for similar purposes [22]:

$$|\varepsilon_{vacuum}^{non\ pert.}| \simeq \langle 0| \frac{9\alpha_s}{32\pi} G_{\mu\nu}^a G_{\mu\nu}^a |0\rangle \simeq (0.25\ GeV)^4 \quad (34)$$

On the other hand, the bag constant is fitted as

$$B \simeq (0.1 \div 0.12\ GeV)^4$$

or  (35)

$$B \simeq (\tfrac{1}{10} \div \tfrac{1}{20}) |\varepsilon_{vac.}^{non\ pert.}|$$

There are some intriguing qualitative consequences of eq.(35):
- QCD origin of SU(6): both $\pi$ and $\rho$ disturb the vacuum only slightly ,
- QCD exotics may exist with much more substantial destruction of the vacuum inside hadrons.

Let me now <u>comment</u> on these.

As is mentioned in part III, there is a long standing problem of the pion. It is aggravated by the fact that SU(6) unifying pion and $\rho$ meson works quite satisfactory. The reason might be that in both cases the vacuum is disturbed only slightly. Indeed, because of the chiral symmetry it costs no energy to produce a pion at rest. Now, we come to the conclusion that the $\rho$ meson is also a minor perturbation on the vacuum. Thus, SU(6) could be a reflection of $B \ll |\varepsilon_{vac}|$ much in the same way as $SU(3)_{flavor}$ is a reflection of smallness, not likeness of the quark masses.

Suppression of the vacuum fluctuations inside hadrons is not necessarily universal. For example, the color electric field suppresses instantons but the magnetic field stimulates them. For light, or ultrarelativistic quarks the balance can be delicate while for heavy quarks the effect of the electric field prevails strongly.

As a final remark, let me note that actually variation of the bag constant was introduced first phenomenologically in Ref.[45]. Moreover, for heavy quarks B turns to be larger than for the light ones (see Ref.[16] and sect. II). Thus, one might say that there exist strong phenomenological indications that the suppression of the vacuum fluctuations inside hadrons is not universal. Unfortunately I do not know how firm the conclusion is since the models still differ from the fundamental QCD in such important for any numerical analysis issues as the value of the coupling constant $\alpha_s$ or treatment of the quark vacuum condensate.

### 5.4 Are All Hadrons Alike?

The QCD sum rules indicate that the QCD exotics are not a mere possibility but an expectation too.

The sum rules can be derived for various two-point functions associated with different currents. It is instructive to compare the sum rules for the currents having the quantum numbers of $\rho$ meson and scalar gluonium:

$$j_\mu^\rho = \tfrac{1}{2}(\bar{u}\gamma_\mu u - \bar{d}\gamma_\mu d) \quad \text{and} \quad j_s^G = \alpha_s G_{\mu\nu}^a G_{\mu\nu}^a$$

respectively.

One calculates the asymptotic contribution due to the simplest loop graph and power corrections arising from interaction with the vacuum fields. In both cases the correction is expressed in terms of $\langle \alpha_s G^2 \rangle$. Nevertheless, for a well defined reason, the mass scale where the power correction violates strongly the asymptotic behaviour turns to be quite different:

$$\begin{aligned} M_{crit}^2 &\simeq 0.6 \text{ GeV}^2 \quad \text{for } j_\mu^\rho \text{ and} \\ M_{crit}^2 &\simeq 6 \text{ GeV}^2 \quad \text{for } j_s^G. \end{aligned} \qquad (36)$$

( The sum rules for the quark vector current are quoted in eq.(28) while the sum rules for the gluon current can be found in the materials of Session D9 or in the corresponding original paper [14]).

Eq.(36) means that the annihilation cross section in the gluonic channels becomes parton-like at higher energy than in the familiar case of $e^+e^-$ annihilation into hadrons. The value of $M^2_{crit}$ does vary with the quantum numbers, however, and ,say, for the tensor gluon channel falls in between the extremes ( 36 ).

For the quark vector current the value of $M^2_{crit}$ coincides with the mass of the lowest resonance, that is $m^2_\rho$. One might expect similar things happen in other channels as well. Although the guess looks plausible in general, there are some exceptions to the rule. Thus, the pion is almost massless but implies the $M^2_{crit}$ to be around 1.5 GeV$^2$ in the pseudoscalar quark channel. A more detailed analysis of the sum rules indicates the reasons to expect the scalar gluonium to be relatively light, $M^2_{0^+, GG} \simeq$ (1-2) GeV$^2$. As for the pseudoscalar gluonium it is indeed expected to be quite heavy:

$$M^2_{pseudoscalar\ glueball} \simeq 6\ GeV^2$$

Had we decided to mimic these predictions within the bag model we would have varied the energy density B strongly from channel to channel. Therefore, in terms of the previous section glueballs could be QCD exotics.

While awaiting experimental data to confirm or discourage the theory developed we may notice that even now eq.(36) does resolve[46] what I would call the $\eta'$ problem. The point is that it appears impossible to avoid quite a strong $\eta' \to$ gluons transition, on one hand, while, on the other hand, $\eta'$ behaves in many respect as a true quark resonance.

Indeed, start with the idea that $\eta'$ is a pure quark state. Then $\langle 0| \sum_{u,d,s} \bar{q} \gamma_\mu \gamma_5 q |\eta'\rangle = \sqrt{3} f_\pi p_\mu$

The QCD triangle anomaly turns this number into a prediction for the gluonic matrix element , $<0|\frac{3\alpha_s}{4\pi} G\tilde{G}|\eta'> = \sqrt{3} f_\pi m_\eta^2$. Thus, the very notion of a "pure" quark made state cannot be pushed too far in QCD. The crucial point still is whether this coupling to the gluons is strong or relatively weak. To answer the question calculate the corresponding decay rate $\Gamma(J/\psi \to \eta' \gamma)$ and normalize it to the gluon continuum in the same decay, $(c\bar{c})_{J/\psi} \to 2G + \gamma$. Then one can see that naive quark model would correspond to $\eta'$ meson dual to more than $2 GeV^2$ of the gluon continuum. Therefore $\eta'$ would look more powerful in the gluon channel than the $\rho$ in the quark channel ($\rho$ is dual to about $1.5 GeV^2$ of the quark continuum).

All the puzzles are resolved if one allows for a possibility of different hadronic scales, depending on the quantum numbers. This possibility is suggested by the sum rules. Note also that the sum rules tend to a lower value of the matrix element $<0|G\tilde{G}|\eta'>$, somewhat diminishing the duality interval for the $\eta'$ in the gluon channel.

It is worth emphasizing that the difference in the mass scales discussed is not accounted for by the $N_c$ counting. It is due to the difference in interaction of quark and gluon currents with the vacuum fields which can be readily classified too.

### 5.6 QCD Hints on "Ordinary Exotics".

As final topic, let me consider what QCD has to say on "ordinary" exotics discussed in part II. The general approach, suggested by QCD, is to study two-point functions associated with the corresponding exotic currents.

$(q^2\bar{q}^2)$ mesons. As noted by Witten[26] correlators for such currents factorize in the limit of large $N_c$ and reduce in this sense to the correlators of their colorless components ($\bar{q}q$). The factorization indicates that there might be no speicifc four-quark states. I think that this remark is in line with the development started by Ref.9.

(GG), or gluonium. Actually, we have discussed these states in previous sections. In brief, there might be no universality of the mass scales and some states can be shifted to a higher mass. Moreover, it is difficult to recoincile the OZI rule for gluons with the value of the matrix element for the two gluon transition into two pions (see eq.(31)) which is not suppressed by any standard. Therefore, we must be cautious about naive estimates of the glueball widths.

($Q\bar{Q}G$), or heavy quark exotics. In this case it is difficult to escape conclusion that the exotic states do exist. Indeed, the general structure of the sum rules looks similar, say, for the qqq and $Q\bar{Q}G$ currents. Thus, since nucleon exists I do not see any reason why the exotics should evade. Thus, my feeling is that conclusions of Ref.16) are in general supported by QCD.

In conclusion to part V it seems fair to say that
- with time going on, more and more consequences of QCD are being accumulated, and none has failed so far.

Conclusions to the talk:
- quark models are successful;
- QCD is illuminating, sometimes confirmed;
- some exotic states or phenomena should exist; nothing boring in sight.

This is in positive. On the other hand:
- the models are difficult to interpret on a fundamental level;
- QCD lacks systematic way to calculate observables.

## DISCUSSION

Q1: Szymacha, Warsaw: If $\bar{q}q$ changes the properties of vacuum only weakly, does that mean that the short GN pion may enter inside the hadron?

A1: Zakharov: I mentioned that the pion, as a GN-boson, does not disturb vacuum in any way. The bag-like picture is apparently too crude for the pion. As for the coupling of the pion to other hadrons, its specific nature is manifested in current algebra relations between the pion-hadron vertices. Any particle is coupled to the pion by virtue of the chiral symmetry; effective Lagrangians demonstrate this most explicitly. Whether it is appropriate to interpret these facts as evidence for free entry of the pion inside hadrons, I do not know.

Q2: Pasupathy, Indian Institute of Science: Have you compared the values of QCD parameters obtained from charmonium and upsilon states? Parameters like $\alpha_s$, $<0|G_{\mu\nu}G^{\mu\nu}|0>$ ?

A2: Zakharov: The results of the charmonium sum rules are fully confirmed by analysis of the $\Upsilon$ family within the same framework. The $\Upsilon$ sum rules were considered in detail by M. Voloshin (Yad. Fiz., $\underline{29}$, 1368, (1978)), preprint ITEP-21, (1980).

Q3: Blondel, LBL: You spoke about $\bar{q}qG$ bound states, but I never heard about any, say $e^+e^-\gamma$ or such bound state – could you clarify in my mind what would make the difference?

A3: Zakharov: Photons interact weakly with electrons. This makes the difference. Technically, if you consider the sum rules, say, for the current $\bar{q}\sigma^{\mu\nu}\lambda^a qG_{\mu\nu}^a$, then there arise power terms due to the nontrivial structure of the QCD vacuum. These terms are absent in QED. Nobody would put electrons in the bag, either.

Q4: Dasgupta, Kalyani, India: In actual applications of the QCD sum rule, you always restrict your calculation of the sum-rule coefficients to the lowest order in perturbation theory. Perhaps such calculations are successful because the multiquark exotics do not make any important contribution. Is this right?

A4: Zakharov: Expansion in $\alpha_s$ and $M^{-2}$ is justified for short distances. The smooth matching with the resonance region might indicate that $\bar{q}qG$ mixture in, say, $\rho$ is small. I think I should agree with you. A direct way to study exotics would be to introduce $\bar{q}qG$ currents and derive the corresponding sum rules.

Q5: Rosen, Purdue: In the light of your remarks on SU(6), would you comment on the following. To explain the difference in $D^0$ and $D^+$ lifetimes, Fritzsch and Minkowski proposed that the D contains a large component of ($c\bar{q}$ Gluon) in its wave function in addition to the simplest $^1S_0$ state ($c\bar{q}$). Baryon magnetic moments give similar hints of the presence of other than the simplest SU(6)-type wave functions.

A5: Zakharov: There are several examples of gluonic matric elements over ordinary hadrons which are guaranteed by QCD to be large by any standard. Recent remarks on a possibility of a relatively large admixture of gluons inside hadrons which you mention are made within the context of composite models, not fundamental QCD. So, there is no simple translation from one language into the other. Still, model observations and QCD relations could go in the same direction.

Q6: Gasiorowicz, Minnesota: Could you comment on your statement that the QCD sum rules also indicate a rapid transition from the A.F. domain to the strong coupling region?

A6: Zakharov: Sum rules start from short distances and keep both perturbative and nonperturbative corrections so that one can watch which one blows up first. The statement is that nonperturbative contributions become important when $\alpha_s$ is small. The simplest way to demonstrate this is to consider sum rules for gluonic currents, say, for $\Pi(q)$ where $\Pi(q) = <0|i\int d^4x\, T\{G^2(o)\}|0>$. Then one can see (V. Novikov et al., Nucl. Phys. B165, 67 (1980)) that power corrections become important at $Q^2 \sim 6$ GeV$^2$. There is no doubt that $\alpha_s$ is still small at such distances. Actually, the conclusion on the leading role of power corrections was made earlier, from the analysis of the $\rho$ meson sum rules. M.A. Shifman et al., Nucl. Phys. B147, 385, 448), but in this case you need to be a bit more detailed and, say, keep track of the signs of different corrections. Depending on the channel considered, asymptotic freedom is taken over by the power corrections at $\alpha_s = 0.1 - 0.3$, in amusing agreement with the results of the lattice type calculations. The transition from A.F. to confinement is rapid in the sum rules, since it is due to power-like terms.

Q7: Lindenbaum, BNL/CCNY: I would like to make a comment. Over two years ago the BNL/CCNY collaboration published the Physical Review letters on the reaction 23 GeV/c $\pi^- + p \to \phi\phi n$. We showed that the OZI rule was badly violated in this forbidden hairpin quark line diagram reaction. The necessity of relatively weakly coupled 2 or 3 gluon exchange is the standard explanation of why the OZI rule was expected to work in this reaction. However if one imagines a series of glueball resonances as the exchanged gluon system between the two disconnected parts of the diagram the coupling could well become effectively strong

and the OZI suppression would be expected to fail. The ϕϕ
hairpin is different from a single ϕ hairpin in that the ϕϕ
can have a range of variable effective mass from the 2ϕ
threshold to the kinematic limit. Thus a range of glueball
state energies extending above the ground state could be
involved in this I=0, flavorless, colorless system. At the
same time, the OZI suppression would still be expected to
suppress the ordinary hadronic background associated with
connected diagrams and thus enhance glueball states prominence
in this channel. In the last experiment we only obtained
∼ 180 ϕϕ events. We are planning to redo the experiment to
obtain ∼ 4,000 ϕϕn and study related $\pi^-$ and $K^-$ induced single
and double ϕ reactions. The ∼ 4,000 ϕϕn events will allow
a moment analysis and critical search for resonant states
including glueballs. Since the first experiment, I have
believed this is a particularly promising channel in which
to look for glueballs.

### Acknoledgements.

I am grateful to T.DeGrand, J.Donoghue, R.Jaffe, A.Kaidalov, I.Kobzarev, V.Novikov, L.Okun, M.Shifman, A.Vainshtein and M.Voloshin for extremely useful discussions.

### References.

1. M.Gell-Mann, Proc. of XVI Int.Conf. on High Energy Physics, edited by J.D.Jackson, Batavia,(1972), vol.4, p.357.
2. N.Isgur and G.Karl, Phys.Lett., $\underline{72B}$, 109,(1977),$\underline{74B}$, 353 ,(1978); Phys.Rev.,$\underline{D21}$, 3175,(1980) ;G.Karl, Proc. of the Tokyo Conf. on High Energy Physics,(1980),p.135 N.Isgur and R.Koniak, Phys.Rev.Lett., $\underline{44}$, 485,(1980) and paper No.19 submitted to this Conference; N.Isgur, Talk at A2 Session.
3. G.Veneziano, Proc. of XIX Int. Conf. on High Energy Physics, Tokyo, (1978),p.725.
4. A.B.Kaidalov, preprint ITEP-78, (1980), Moscow.
5. L.Frankfurt and M.Strikman, Leningrad preprint,LINP, (1980); Yu.P.Nikitin, P.I.Porfiriev, and S.A.Voloshin, paper No.561 (1980).
6. R.Jaffe and K.Johnson, Phys.Lett., $\underline{60B}$, 201, (1976).
7. R.L.Jaffe, Phys.Rev., $\underline{D15}$, 267, (1977).
8. A.J.C.Hey, preprint SHEP 79/80, Southhampton University,(1980), and Talk at A2 Session.
9. R.L.Jaffe and F.E.Low, Phys.Rev., $\underline{D19}$,2105,(1979).
10. Chan Hong-Mo and Hagasen, Phys.Lett., $\underline{72B}$, 121, (1977), Nucl.Phys., $\underline{B136}$ , 401, (1978).
11. I.S.Shapiro, Phys.Repts., $\underline{C35}$, 129, (1978).
12. J.L.Rosner, Phys.Rev.Lett., $\underline{21}$, 950, (1968).
13. H.Fritzsch and P.Minkowski, Nuovo Cim., A30, 393,(1975) D.Robson, Nucl.Phys., $\underline{B130}$, 328, (1977), J.D.Bjorken, SLAC-PUB-2372 (1979); J.F.Donoghue, Talk at A2 Session; S.Meshkov, Talk at C7 Session; C.E.Carlson , Talk at D8 Session.
14. V.A.Novikov et.al., preprints ITEP-87,88 ,(1980), Moscow.

15. A.Billoire et.al., Phys.Lett., B80, 381, (1979).
16. P.Hasenfratz et.al., preprint CERN TH -2837,(1980).
17. A.A.Grigorian and A.B.Kaidalov, preprint ITEP-100, (1979), Moscow.
18. T.Goldman and R.W.Haymaker, Conference paper No.382, R.W.Haymaker, Talk at D8 Session.
19. M.Shifman, A.Vainshtein, and V.Zakharov, Phys.Lett., 73B, 43, (1978).
20. R.Crewter, Phys.Rev.Lett., 28, 1421, (1976); M.Chanowitz and J.Ellis, Phys.Lett.,B40, 397, (1972), J.Collins, L.Duncan, and S.Joglekar, Phys.Rev., D16, 438, (1977).
21. M.B.Voloshin, Nucl.Phys., B154, 365,(1979), preprint ITEP- 21 (1980),Moscow.
22. M.Shifman, A.Vainshtein,and V.Zakharov, Nucl.Phys., B147, 385,447,519 (1979).
23. A.Casher, preprint TAUP 734/79 (1979), G.'tHooft, Lectures given at Cargese Summer Institute, (1979), Lecture III; S.Coleman and E.Witten, SLAC PUB 2493 (1980); E.Witten, Talk in session D9.
24. S.L.Adler, Phys.Rev., 177, 2622, (1969); J.S.Bell and R.Jackiw, Nuovo Cim.,60, 47, (1969).
25. A.D.Dolgov and V.I.Zakharov, Nucl.Phys.,B27, 525, (1971).
26. E.Witten, Nucl.Phys., B160, 57, (1979).
27. H.Leutwyller, Phys.Lett., 48B, 45,(1974); Nucl.Phys., B76, 413, (1974).
28. M.B.Voloshin in Proc. of Workshop on $e^+e^-$ Rings, Novosibirsk preprint INP -4/80. B.L.Ioffe, Yad.Fiz., 29, 1611, (1979), B.L.Ioffe and M.A.Shifman, preprint ITEP-53 (1980).
29. G. 't Hooft, Phys.Rev., D14, 3432,(1976).
30. V.A.Novikov et.al., Phys.Lett., 86B, 347,(1979).
31. D.J.Gross, S.B.Treiman ,and F.Wilczek, Phys.Rev., D19, 2188,(1979).
32. H.Goldberg, Phys.Rev.Lett., 44 ,363, (1980), R.Arnowitt and P.Nath, Northeasten University preprints Nos.2417,2445 (1979).
33. G.'t Hooft, Nucl.Phys., B72, 461, (1974).
34. E.Witten, Nucl.Phys., B160 , 57, (1979), G. Veneziano, Nucl.Phys., B159, 213, (1979). P. Di Vecchia, Phys.Lett., 85B , 357, (1979).

35. R.Crewther, Phys.Lett., B70, 349, (1977).
36. J.B.Kogut, R.B.Pearson, and J.Shigemitsu, Phys.Rev.Lett., 43, 484, (1979);
    R.B.Pearson, Talk at this Conference.
37. A.Vainshtein et.al., preprint ITEP-82 (1980), Moscow.
38. L.J.Reinders, H.R.Rubinstein, and Yazski, preprint RL-80-031, Conference paper No.826.
39. H.Goldberg, Phys.Rev.Lett., 35, 605, (1975), 44, 363, (1980),
    V.A.Novikov et.al., Nucl-Phys., B165, 55, (1980).
40. M.B.Voloshin and V.I.Zakharov, preprint DESY-80/28, (1980).
41. K.Gottfried, Phys.Rev.Lett., 40, 538, (1978).
42. M.Peskin, Nucl.Phys., B156, 365, (1979).
    Tung-Mow Yan, preprint CLNS, Cornell, (1980).
43. V.A.Novikov and M.A.Shifman, preprint ITEP-93, (1980).
44. C.Callan, R.Dashen, and D.Gross, Phys.Lett., B78, 1307, (1978), D.J.Gross, Proc. of Tokyo Conference on High Energy Physics, Tokyo, (1978), p.486.;
    E.V.Shuryak, Phys. Lett., B79, 135, (1978).
45. I.Yu.Kobtarev, B.V.Martemyanov, and M.G.Schepkin, Yad.Fiz., 29, 1620, (1979).
46. W. Bardeen and V.Zakharov, Phys.Lett., B91, 111, (1980).

CHARACTERISTICS OF INCLUSIVE AND EXCLUSIVE
FINAL STATES AT LOW $p_T$, ALL BEAMS, EXPERIMENT

H. Miettinen, Helsinki, Speaker

H. Harari, Weizmann Institute, Chairman

M. Corcoran, Rice,

R. Wilkes, Washington,

G. Fanourakis, Wisconsin-Madison,

Scientific Secretaries

Written Version Not Available

# HADRON STRUCTURE FROM LEPTON BEAMS

F. Sciulli, Caltech, Speaker

R. Hofstadter, Stanford, Chairman

L. Stutte, Fermilab,

S. Kahn, Brookhaven,

Scientific Secretaries

# HADRON STRUCTURE FROM LEPTON BEAMS

Frank Sciulli
California Institute of Technology,
Pasadena, California

We are to summarize our knowledge of hadron structure using electron, muon, and neutrino probes. In our present thinking, the electromagnetic and weak interactions of these particles provide information about the constituents of nucleons in the processes:

$$e(\mu) + N \rightarrow e(\mu) + X \qquad (1)$$

$$\nu_\mu + N \rightarrow \mu + X \qquad (2)$$

A few words about the interactions utilized by these probes are in order. The electromagnetic mechanism underlying reaction (1) is well-understood; the limits on two-photon exchange that might complicate the analysis are such that their contributions in the present energy regime are extremely small.[1] The weak neutral current process, or Z-exchange, is also expected to be small. We will return briefly to the existing information on the neutral current interaction. The charged current process of reaction (2) is described on an elementary level by the famous V-A theory. This states that the interaction of neutrinos or antineutrinos with elementary spin one-half fermions is given by

$$H = \frac{G_F}{\sqrt{2}} \left[ \bar{\mu} \, \gamma_\alpha (1+\gamma_5) \, \nu_\mu \right] \left[ \bar{T}' \, \gamma_\alpha (1+\gamma_5) \, T \right] \qquad (3)$$

where T(T') is the target (recoil) quark or lepton. Existing information on pion decay, muon decay, and nuclear beta decay provides limits at approximately the one percent level on deviations from this simple hamiltonian at low energies.

## I. NEUTRAL CURRENTS

The neutral current process makes only small contributions to reaction (1) except at higher energies, but it is the entire mechanism for reactions of the type

$$\nu_\mu + N \rightarrow \nu_\mu + X \qquad (4)$$

These are, according to our best information, most simply described by the hamiltonian

$$H = \frac{G_F}{\sqrt{2}} \left[ \bar{\nu}_\mu \, \gamma_\alpha (1+\gamma_5) \, \nu_\mu \right] \left[ \bar{T} \gamma_\alpha (g_V^T + g_A^T \gamma_5) T \right] \qquad (5)$$

for scattering of neutrinos or antineutrinos from elementary fermions. Here $g_V^T$ and $g_A^T$ are the vector and axial-vector couplings of the neutral current to the target, T. In the SU2 x U1 picture, they are given by

$$g_V^T = I_3^T - 2Q_E^T \sin^2\theta_w \qquad (6)$$
$$g_A^T = I_3^T$$

where $I_3^T$ is the 3rd component of weak isospin and $Q_E^T$ is the electric charge of the target. $\theta_w$ is the supposedly universal and energy independent angle of the theory. In the Weinberg-Salaam-Ward model the coefficient, $\rho$, is given by

$$\rho = \frac{M_W^2}{M_Z^2 \cos^2\theta_w} = 1 \qquad (7)$$

where $M_W(M_Z)$ is the charged (neutral) boson mass.

TABLE I

| Reaction | Quantity Measured | Data ± 1σ | WSW prediction with $\rho = 1$ $\sin^2\theta_w = .232$ |
|---|---|---|---|
| $\nu N \to \nu X$ | $R_\nu$ | 0.307 ± 0.008 | 0.305 |
|  |  | 0.30 ± 0.04 | 0.325 |
|  |  | 0.28 ± 0.03 | 0.304 |
| $\bar\nu N \to \bar\nu X$ | $R_{\bar\nu}$ | 0.373 ± 0.025 | 0.386 |
|  |  | 0.33 ± 0.09 | 0.365 |
|  |  | 0.35 ± 0.11 | 0.399 |
| $\nu N \to \nu X$ | $g_L^2$ | 0.32 ± 0.03 | 0.298 |
| $\bar\nu N \to \bar\nu X$ | $g_R^2$ | 0.04 ± 0.03 | 0.030 |
| $\nu p \to \nu X$ | $R_\nu^p$ | 0.52 ± 0.06 | 0.448 |
|  |  | 0.48 ± 0.17 | 0.414 |
| $\bar\nu p \to \bar\nu X$ | $R_{\bar\nu}^p$ | 0.42 ± 0.13 | 0.383 |
| $\nu_p^n \to \nu X$ | $R_\nu^{n/p}$ | 1.22 ± 0.35 | 1.12 |
| $\bar\nu_p^n \to \bar\nu X$ | $R_{\bar\nu}^{n/p}$ | 0.64 ± 0.18 | 0.935 |
| $\nu N \to \nu\pi^\pm X$ | $R_\nu^{+/-}$ | 0.77 ± 0.14 | 0.84 |
| $\bar\nu N \to \bar\nu\pi^\pm X$ | $R_{\bar\nu}^{+/-}$ | 1.65 ± 0.33 | 1.16 |
|  |  | 1.27 +0.36 −0.27 | 1.01 |
| $\nu_\mu e \to \nu_\mu e$ | $\sigma/E$ | 2.4 +1.2 −0.9 (a) | 1.52 |
|  |  | 1.8 ± 0.8 (a) | 1.52 |
|  |  | 1.1 ± 0.6 (a) | 1.52 |
| $\bar\nu_\mu e \to \bar\nu_\mu e$ | $\sigma/E$ | 2.2 ± 1.0 (a) | 1.32 |
|  |  | 1.0 +1.3 −0.6 (a) | 1.32 |
| $\bar\nu_e e$ (low E) | $\sigma$ | 7.6 ± 2.2 (b) | 6.37 |
| $\bar\nu_e e$ (high E) | $\sigma$ | 1.86 ± 0.48 (b) | 1.21 |

(a) units of $10^{-42} E_\nu$ (cm$^2$/GeV)
(b) units of $10^{-46}$ cm$^2$

The information on the neutral current interaction as of about one year ago was nicely summarized in a U. of P. preprint.[2] Their conclusion was that all existing data could be described by the expression (5) with $\rho = 1$ and

$$\sin^2\theta_w = .232 \pm .009. \tag{8}$$

This is illustrated by Table I, taken from that paper. Additional data, not shown in the table, come from experiments on the $Q^2$ dependence of elastic neutrino and antineutrino scattering and from the y-dependence of the neutral current-electromagnetic interference term of deep inelastic electron deuteron scattering. The latter data alone give $\sin^2\theta_w = .224 \pm .020$.[3] All of this data was used in the fit. The data shown consist of inclusive and several different exclusive neutrino processes. The predictions, using the WSW model and the best fit value (8) gives good quantitative agreement with all the various measurements. An overall fit to the data, leaving both $\rho$ and $\theta_w$ as free parameters yields

$$\sin^2\theta_w = .235 \pm .016 \tag{9a}$$

$$\rho = 1.004 \pm .019 \tag{9b}$$

This is in good agreement with the Weinberg-Salam-Ward model.

There have been three new relevant contributions to this conference which continue to corroborate this simple picture. A new measurement of $R_\nu^{n/p}$ of Table I from a 15 ft. bubble chamber exposure with deuterium gives[4]

$$R_\nu^{n/p} \equiv \frac{(\nu + n \to \nu + X)}{(\nu + p \to \nu + X)} = 1.02 \pm .19 \tag{10}$$

which halves the error on the new world average

$$R_\nu^{n/p} = 1.07 \pm .17 \text{ (world average)} \tag{11}$$

in good agreement with the value $R_\nu^{n/p} = 1.12$ predicted by the WSW model with $\sin^2\theta_w = .232$.

The CHARM collaboration[5] has presented y-distributions from neutral and charged current data taken in the CERN-SPS narrow band neutrino beam. Assuming approximate scaling, and scattering from fermion constituents only, these distributions for neutrino and antineutrino beams have the following general forms

$$\frac{dN}{dy} = C\left[(1-\alpha) + \alpha(1-y)^2\right] \quad (12)$$

$$\frac{d\bar{N}}{dy} = C\left[\alpha + (1-\alpha)(1-y)^2\right] \quad (13)$$

where $\alpha$ has contributions from antiquarks with V-A scattering and from quarks with V+A scattering. Differences in the charged current and neutral current data, assuming that the scattered constituents are identical, can be attributed to the small right-handed (V + A) part of the neutral current. The y-distributions are shown in Fig. 1. They find

$$\alpha_{cc} = 0.17 \pm .03 \quad (14)$$

$$\alpha_{nc} = 0.23 \pm .04 \quad (15)$$

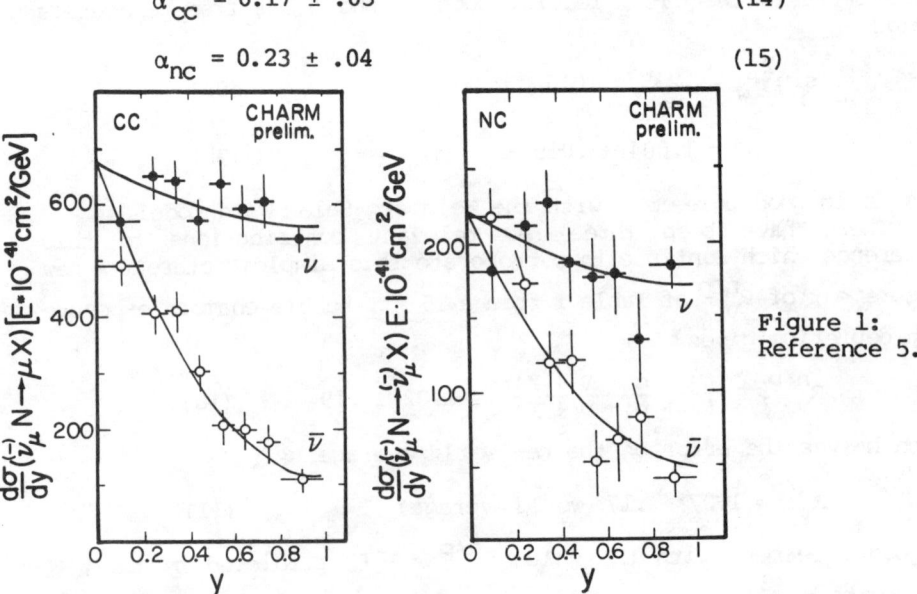

Figure 1: Reference 5.

A fit to the best weak angle for this data yields a best fit $\sin^2\theta_W = .218 \pm 0.16 \pm .02$, where the last error is the model-dependent error involving uncertainties in nucleon structure. A fit using the Paschos-Wolfenstein method, about which we will have more to say, has less systematic error and yields $\sin^2\theta_W = .230 \pm .023$.

Another new result comes from the newly commissioned CFRR[6] experiment running in the FNAL narrow band beam. Because the deep-inelastic data on neutral currents lends itself to very high statistical accuracy, such experiments are tending to be limited by systematic errors, both from the experiments themselves and from the models assumed in the analysis. Much of the latter uncertainty is removed, at the expense of statistical precision, by using the Paschos-Wolfenstein technique. This approach utilizes the ratio of neutral current to charged current total cross-sections. A slightly modified version of the technique would measure the ratios

$$R^{\pm} = \frac{\frac{d\sigma}{dy}^{nc}_{\nu} \pm \frac{d\sigma}{dy}^{nc}_{\bar{\nu}}}{\frac{d\sigma}{dy}^{cc}_{\nu} \pm \frac{d\sigma}{dy}^{cc}_{\bar{\nu}}} \qquad (16)$$

which, in SU2 x U1, should be independent of y and equal to

$$R^{+} = \rho \left(\frac{1}{2} - \sin^2\theta_w + \frac{10}{9}\sin^4\theta_w\right) \qquad (17)$$

$$R^{-} = \rho \left(\frac{1}{2} - \sin^2\theta_w\right) \qquad (18)$$

for an isoscalar target. The CFRR group has used data only over a restricted y-region, where there is little or no confusion between neutral and charged current events. This selection of data, combined with a more favorable geometry in the Fermilab Neutrino Area, has allowed background corrections which are about half that of the CHARM experiment and one-fourth that of the older CDHS experiment. From a preliminary analysis of about 30% of the data, the CFRR group finds

$$\sin^2\theta_w = .239 \pm .023 \qquad (19)$$

for $\rho = 1.0$. Fig. 2 shows the result of a simultaneous fit to $\rho$ and $\sin^2\theta_w$.

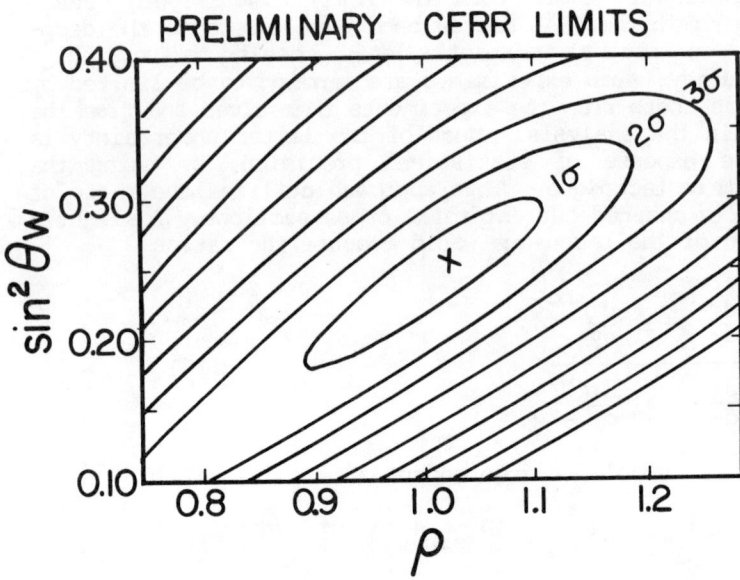

Figure 2: Reference 6.

Quite clearly, there is now essentially universal agreement between data and the simplest WSW model. Even the older discrepancies with low energy atomic physics experiments seem to be open to experimental question with the advent of newer results.[7] In addition, evidence is accumulating that the value of $\sin^2\theta_w$ is significantly different from the SU5 grand unification prediction of 0.20.

One very important point, for which there exists some new qualitative information, is the equivalence of the x-distributions in neutral and charged current interactions. While these may not be identical, they should be very similar for scattering from isoscalar targets. The first measurement of x-distributions in neutral current interactions has recently been published,[8] and indeed is found to be qualitatively similar to that found in charged current scattering (see Fig. 3). Preliminary results on an other such measurement have been presented previously.[9] Experiments at the SPS(CHARM) and at FNAL (Lab C) are expected to make this comparison more precise in the future.

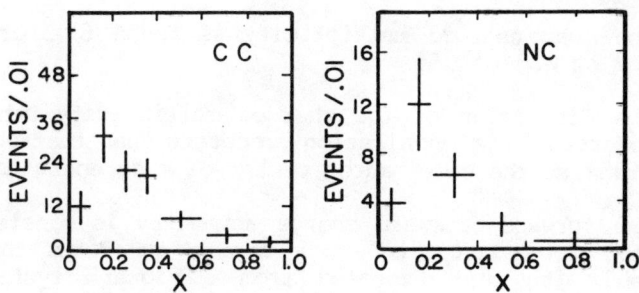

Figure 3:   Reference 8

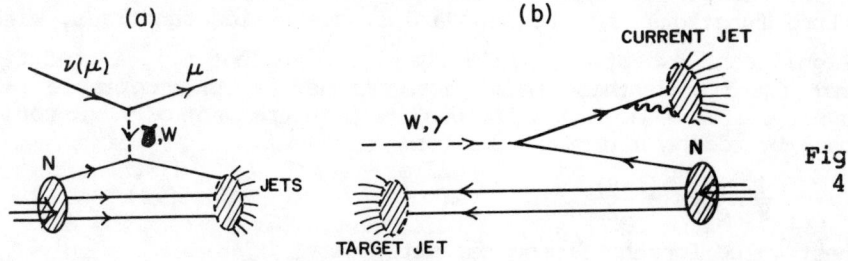

Fig 4

## II.   FINAL STATE JETS

In the standard picture for inelastic scattering, the reaction proceeds as shown in Fig. 4. The bound quarks in Fig. 4b incident from the left encounter the photon or weak boson propogator. A single quark is scattered, absorbing the virtual boson and reemerges to the right in this frame. Because the struck quark has had its color accelerated, it is more likely to emit a gluon brehmstrahlung than the spectator quarks which continue traveling to the left in the figure. One might anticipate, therefore, that the hadronic jet associated with the struck quark ("current" or forward jet) will have some characteristics that are different from those of the spectator ("target" or backward jet).

Some specific characteristics of the hadrons associated with forward jets found in data presented at this conference are the following:

(1) The mean charged multiplicity is found to increase with the logarithm of W. [10,11,12]

(2) The dispersion of the charged multiplicity behaves more like that observed in $\bar{p}p$ annhilation processes than that observed in pp interactions at the same hadron center-of-mass energies. [10,11,12]

(3) The forward-backward charge asymmetry is consistent with that expected for scattering of a single quark in the forward direction and with that expected from a diquark system in the backward direction. [11]

(4) The distribution in fractional energy of single pions ($z = E_\pi/E_{tot}$) in the process

$$\nu_\mu + N \to \mu^- + \pi^\pm + X \qquad (20)$$

defined as

$$D(z) = \frac{1}{N_{ev}} \frac{dN^\pm(z)}{dz} \qquad (21)$$

typically behaves as expected assuming the standard deep-inelastic structure functions and the standard fragmentation functions, with one significant exception. At large z for $\pi^+$ (not $\pi^-$), a good fit was not found with these usual assumptions. [13] An acceptable fit could not be found with an additional term independent of y but could be found by adding a term of the form

$$\frac{4}{9} K_{eff}^2 \frac{(1-y)}{Q^2} \qquad (22)$$

The best value for the fitted parameter is

$$K_{eff}^2 = (1.73 \pm .21) \text{ GeV}^2 \qquad (23)$$

This y-dependence is expected from a scale-breaking interaction at the hadron vertex induced by a longitudinal virtual propagator. [14] That is, it requires a finite $\sigma_L/\sigma_T$ term. Caution should be exercized in interpreting $K_{eff}$, however, since mass terms in addition to transverse momentum terms are expected to contribute.

(5) The mean-square transverse momentum of final state hadrons in an individual event is found to most simply be parametrized in terms of the Bjorken scaling variable, x, and the invariant mass of the hadronic system, W.[15,16] That is,

$$\langle p_T^2 \rangle = f_1(x)\, f_2(W^2) \qquad (24)$$

whereas the $\langle p_T^2 \rangle$ is not found to be separable in terms of x and the invariant 4-momentum, $Q^2$. Additionally, there appears to be little azimuthal effect to the $p_T$-broadening, as one expects at these energies. This serves as a check on any experimentally induced effects.

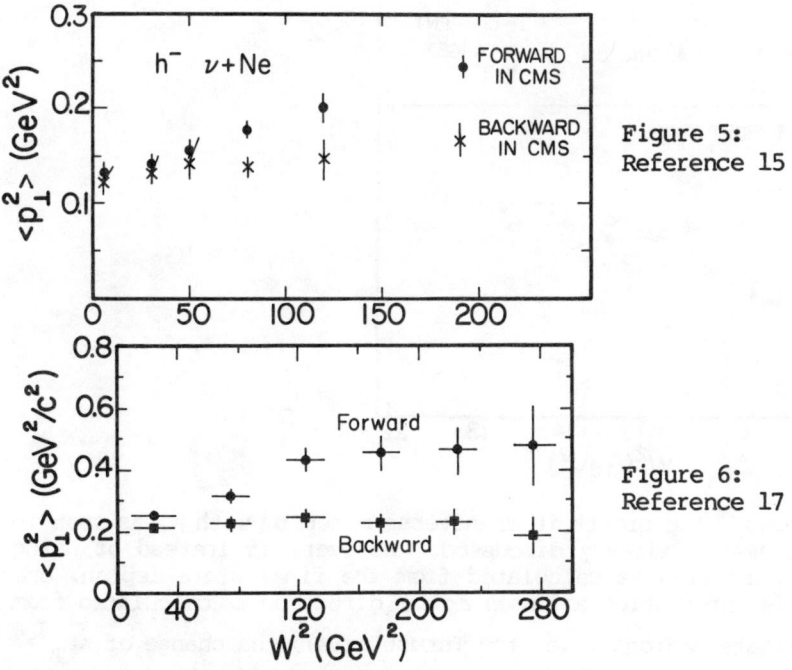

Figure 5: Reference 15

Figure 6: Reference 17

The picture of the scattering process shown in Fig. 4b has had several corroborations. Fig. 5 and 6 show neutrino data from two different bubble chamber collaborations[15,17] that illustrate the point. There have been somewhat different z-cuts made on the data. Both experiments show no significant growth in mean $P_T$ with W for the target jet, but do show a clear increase in mean $P_T$ with W for the current jet. This is interpreted as demonstrating the presence of gluon brehmstrahlung in the struck quark system, whereas the target quarks show no such effect. Here, $P_T$ is measured relative to the struck quark direction as determined by the parameters of the outgoing lepton and the kinematics of the inelastic process. The importance of this information is illustrated in Fig. 7.

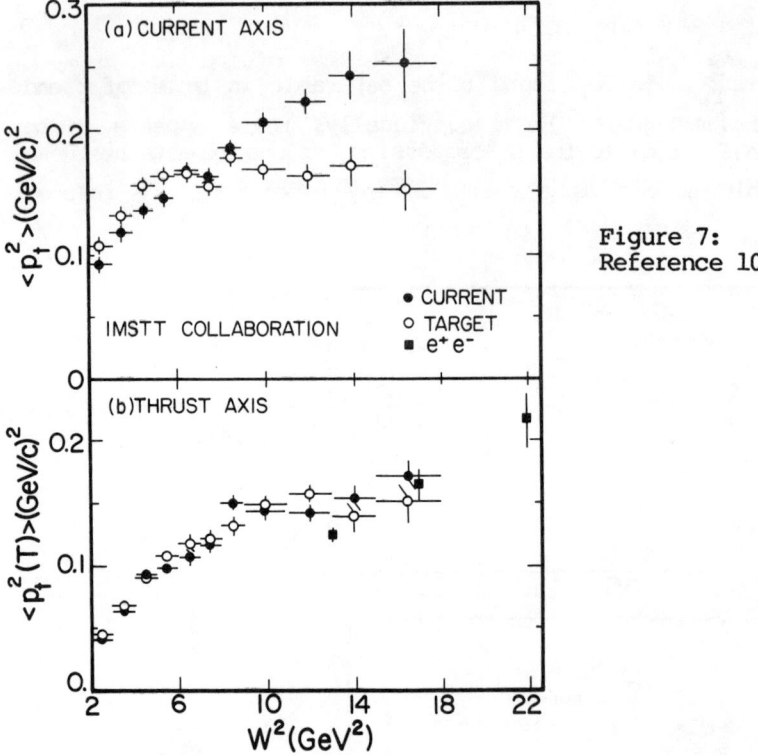

Figure 7: Reference 10

Fig. 7a shows[10] the growth in transverse momentum with W, as seen in other experiments already discussed. However, if instead of using the quark direction as calculated from the final-state lepton, one ignores this information and uses a mean direction as calculated from the final-state hadrons (i.e. the Thrust axis), the change of $<P_T^2>$ with W is as shown in Fig. 7b. The growth of transverse momentum is competely washed out by the intrinsic uncertainty of the struck quark direction. The rise in $<P_T^2>$ can be seen at these low energies in neutrino and muon scattering only because the direction of the struck quark is known independently of the jet itself. There are encouraging indications[17] that even the multi-jet structure of gluon brehmstrahlung as observed at PETRA may be seen at these lower energies in the deep-inelastic process.

It should be noted that the increase in mean $P_T$ with $W^2$ is similar to that seen in hadron-hadron collisions. Fig. 8 shows the distribution in $p_T$ as seen in one experiment.[17] Note that the increase has its origin as a rise in the tail at large $P_T$. One interesting question is whether sensitivity to perturbative QCD effects would be improved by using higher moments of $P_T$. This would emphasize the behaviour of this distribution at the larger values.

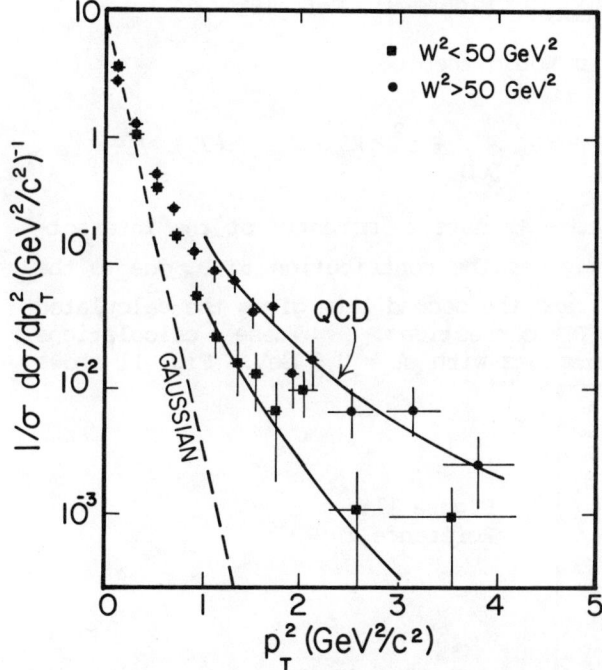

Figure 8: Reference 17

We conclude that the behaviour of hadrons from final state jets in deep-inelastic scattering have a "QCD-like" behaviour: their transverse momentum has non-scaling behaviour that can be related to the quark-gluon coupling constant measured with deep-inelastic structure functions through first-order QCD calculations and empirically-determined fragmentation functions.

(6) A measurement of the intrinsic transverse momentum of the quark inside the nucleon has been obtained from the hadronic final states in muon-inelastic scattering. The muon data[16] show a similar rise of $<P_T^2>$ to that seen in the neutrino data (see Fig. 9); that is, a rise with $W^2$ at fixed z. An observable z-dependence beyond that predicted by the fragmentation function of the quark and the perturbative QCD effects would be a demonstration of intrinsic transverse momentum, as illustrated in Fig. 10.

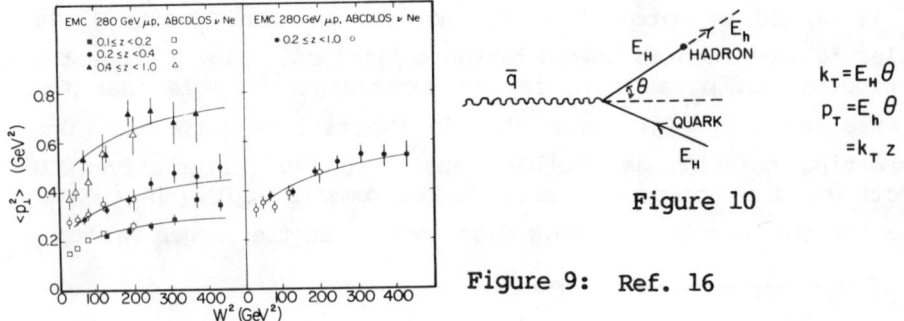

Figure 9: Ref. 16

Figure 10

The mean transverse momentum would then be

$$\langle P_T^2 \rangle = \langle P_T^2 \rangle_{FRAG} + \langle P_T^2 \rangle_{QCD} + z^2 \langle k_T^2 \rangle \qquad (25)$$

where $\langle k_T^2 \rangle$ is the mean-square transverse momentum of the interacting quark. The first term gives the contribution to $P_T$ due to the fragmentation of the quark, and the second term gives the calculated W-dependence due to QCD corrections. These calculations incorporated first-order diagrams with $\Lambda = 0.5$ GeV. Fig. 11 shows

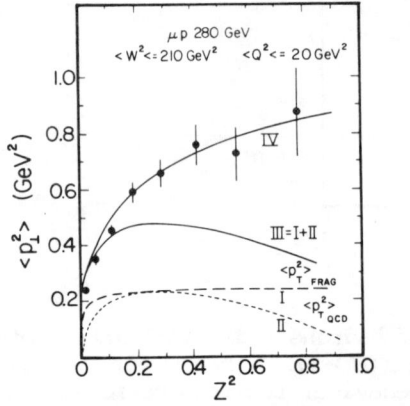

Figure 11: Reference 19

the remaining contribution depending linearly on $z^2$. The coefficient obtained in the fit is

$$\langle k_T^2 \rangle = (.63 \pm .10) \text{ GeV}^2 \qquad (26)$$

An earlier analysis using data with z<0.4 found no evidence for this effect.[18] (See discussion.)

(7) An independent attack on this same question has been made using neutrino data[19] with opposite-sign dimuon final states. These events have been shown to be consistent with having their origin in the production and semi-leptonic decay of charmed particles. The mean transverse momentum of the outgoing second muon perpendicular

to the production scattering plane should obey the following equation:

$$\langle P^2_{T,OUT} \rangle = \frac{1}{2}(z^2 \langle k^2_T \rangle + C W^2 + \text{const}) \quad (27)$$

The last term is the contribution from the charm decay which should be independent of W and z, the second term is the calculated QCD contribution, and the first term measures the intrinsic $k_T$ of the interacting quark. Fig. 12 shows the mean transverse momentum

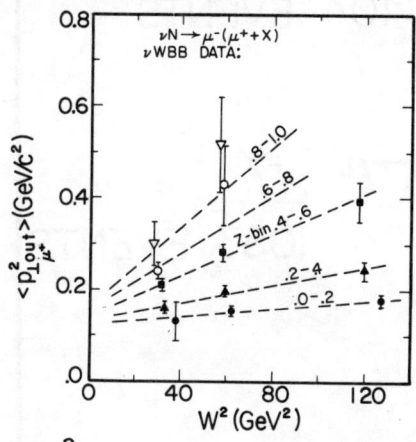

Figure 12: Reference 19

versus $W^2$ at fixed z. The intrinsic $k_T$ obtained is

$$\langle k^2_T \rangle = (0.6 \pm 0.2 \pm 0.2) \text{ GeV}^2 \quad (28)$$

in agreement with that obtained from pion final states. (See II-6)

An investigation of multimuon events by neutrinos and anti-neutrinos clearly show a difference in the mechanisms for the production of charm by the two different initiating particles. We expect opposite-sign dimuons to be produced as follows:

$$\nu_\mu + \begin{Bmatrix} s \\ d \end{Bmatrix} \rightarrow \mu^- + c \underset{\hookrightarrow \mu^+}{} \quad (29)$$

$$\bar{\nu}_\mu + \begin{Bmatrix} \bar{s} \\ \bar{d} \end{Bmatrix} \rightarrow \mu^+ + \bar{c} \underset{\hookrightarrow \mu^-}{} \quad (30)$$

The strange sea ($s \approx \bar{s}$) should be dominant at small x-values, while the contribution of d(not $\bar{d}$) should show the usual valence behaviour by extending to larger x values. Therefore, the shapes of the x-distributions of opposite sign dimuons should have qualitatively different behaviour when comparing neutrino data with antineutrino data. Fig. 13 illustrates this behaviour dramatically.[19] To go

Figure 13: Reference 19

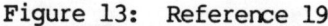

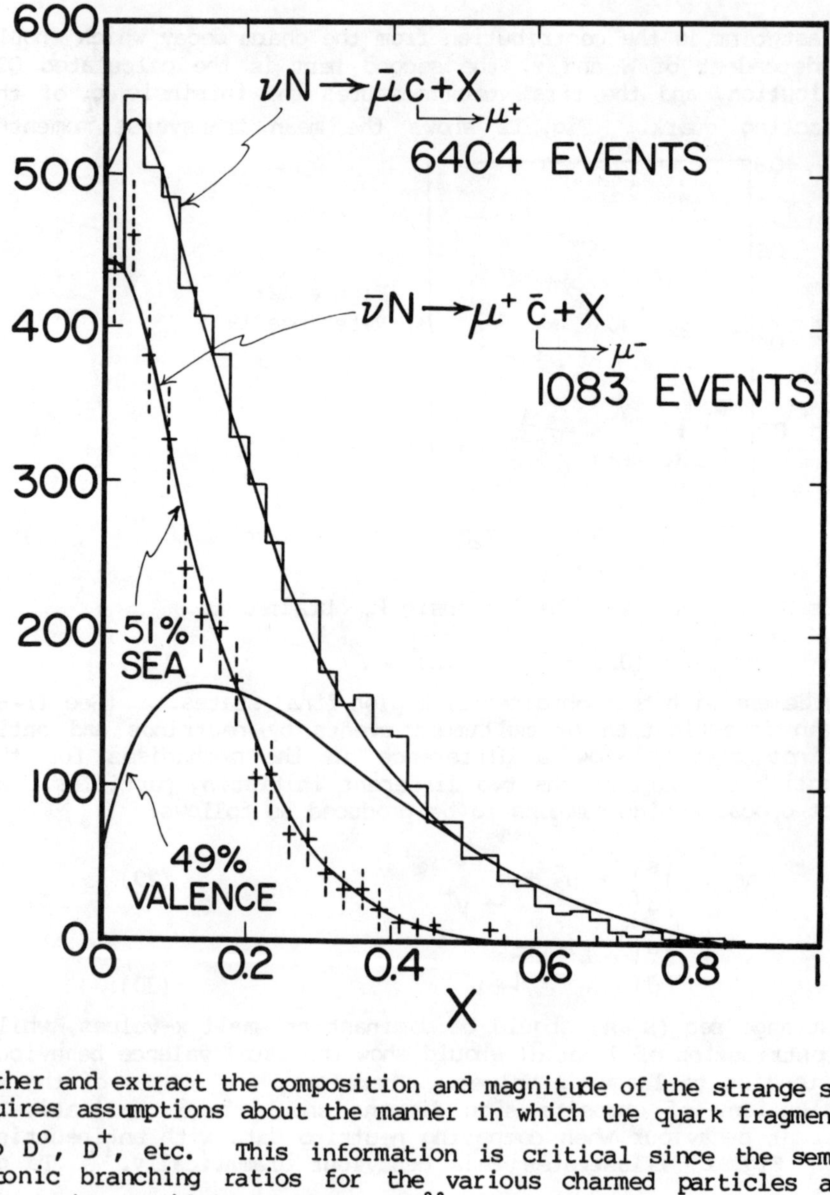

further and extract the composition and magnitude of the strange sea requires assumptions about the manner in which the quark fragments into $D^o$, $D^+$, etc. This information is critical since the semi-leptonic branching ratios for the various charmed particles are known to be considerably different.[20]

An investigation of same-sign dimuon events[21,22] gives strong evidence for their existence; their interpretation is consistent with resulting from associated-production and decay of charm, based on their observed kinematic characteristics. The measured cross-sections are much higher than first-order QCD calculations would predict. This topic will be discussed in more detail by S. Wocjicki.

### III. CHARACTERISTICS OF THE CONSTITUENT QUARKS

In this section we deal with the primary topic of the session: the evidence dealing with the properties of the hadron constituents from reactions initiated by neutrino, muon, and electron beams.

(1) Spin Constitution of Nucleon

A measurement[23] has been made of the deep-inelastic process

$$e^- + p \to e^- + X \qquad (31)$$

with both the electron beam and the target protons polarized. The asymmetry in the experiment is (with small corrections) given by

$$A \approx \frac{\sigma_{1/2} - \sigma_{3/2}}{\sigma_{1/2} + \sigma_{3/2}} \qquad (32)$$

where $\sigma_j$ is the scattering cross-section for the virtual photon and the proton with total angular momentum, j. The experiment was performed at SLAC, so that the data typically have $Q^2 < 10$ GeV$^2$. Fig. 14 shows the x-dependence of the asymmetry. We see essentially zero asymmetry at small-x, rising to nearly unit asymmetry at large x.

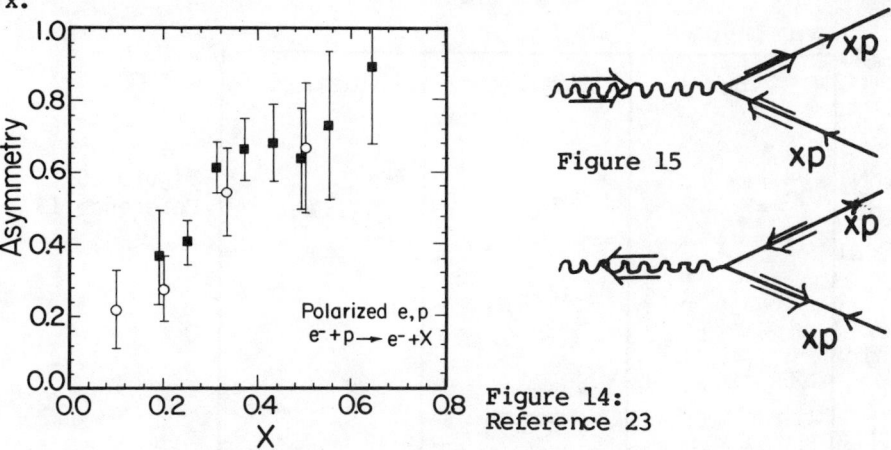

Figure 15

Figure 14: Reference 23

We can picture the scattering on a constituent level as shown in Fig. 15. A hard scattering (i.e. large y) should scatter only with

$j = 1/2$ independent of the helicity of the photon. The interpretation follows that the asymmetry reflects the fraction of the time that the struck quark (vs x) has the proton spin. At small x, where the sea dominates, the quarks have no correlation with the proton spin. At larger x, on the other hand, the struck valence quark has the proton spin a large fraction of the time. As we will see later, these large-x quarks in the proton are mainly u-quarks.

(2) Charge of Struck Quark

New data[15] on single hadrons in the final state from the reaction

$$\mu^- + p \rightarrow \mu^- + h^{\pm} + X \qquad (33)$$

corroborate the now generally accepted picture that these hadrons originate from the fragmentation of scattered quarks. The hadron with the highest energy (leading hadron) in the shower will contain the struck quark most often. Hence, the charge distribution of leading particles tells us about the charge of the struck quark. At small x, where the scattering occurs from the sea of quark-antiquark pairs, the leading pions are expected to be charge symmetric. At large x, we expect the positive pions to dominate for several reasons: (i) there are twice as many u-quarks as d-quarks forming the valence quark distribution in the proton; (ii) the u-quark charge is twice that of the d-quark charge, and hence the scattering from u-quarks should be four times as effective; and (iii) the u-quark valence distribution dominates over the d-quark distribution as x becomes larger. (See below)

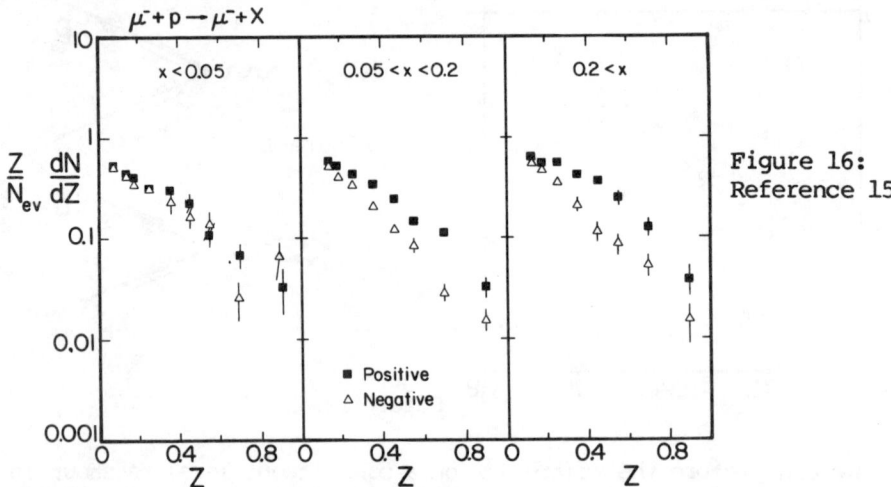

Figure 16: Reference 15

Fig. 16 shows the z-distributions (see eqn. 21) of positive and negative final state hadrons in three different bins of x. The leading particles show no asymmetry at small x, but do show a clear asymmetry at large x in agreement with this general picture.

(3) The d/u Constituency of the Proton

A comparison of deep-inelastic electron and neutrino scattering data on proton targets permits a measurement of the d/u ratio at larger x-values. Here d(x) and u(x) refer to the fractional momentum (x) distributions of d and u quarks, respectively, in the proton (see Section IV-1). For large x (x>.2) where $\bar{q}(x)$ is small, neutrino-proton scattering measures $2d(x)$ and electron-proton measures $(4u(x)+d(x))/9$, where the normalization is such that $\int (d(x)/x)dx = \int (u(x)/x)dx = 1$. Hence, in this region, we expect

$$\frac{F_2^{\nu p}(x)}{F_2^{ep}(x)} = \frac{18 \frac{d(x)}{u(x)}}{4 + \frac{d(x)}{u(x)}} \quad (34)$$

Note that for d and u carrying equal momentum, as one would have for an isoscalar target, we obtain the usual 5/18 ratio. New νp data[24] from a BEBC group has been compared with older SLAC ep data to obtain this ratio as shown in Fig. 17. We see clearly that u-quarks

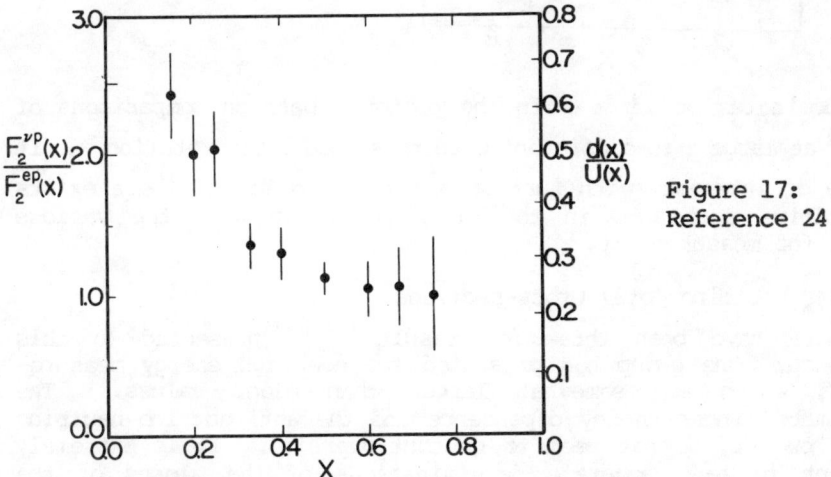

Figure 17: Reference 24

dominate at larger x-values in the proton. More quantitatively, the authors conclude that for x>.5, the u-quarks are four times more abundant than d-quarks.

A similar conclusion would be drawn from data[12] taken with antineutrinos incident on deuterium. This corraborates the old quark-parton model prediction that the ratio of integrated antineutrino scattering cross-sections on neutron and proton targets should be approximately 1/2 at high energy. They find

$$\frac{\sigma\,(\bar{\nu}n \to \mu^+ x)}{\sigma\,(\bar{\nu}p \to \mu^+ x)} = .47 \pm .06 \qquad (35)$$

consistent with the simple prediction, and with older, less precise, values. Since antineutrinos scatter only from u-quarks, the scattering from the proton target measures u(x). The u(x) distribution from neutrons is, by charge symmetry, equal to the d-distribution in protons, or d(x) in our notation.

The data on the unit-normalization x-distributions of events seen in Fig. 18 shows the qualitative behaviour expected if the u-

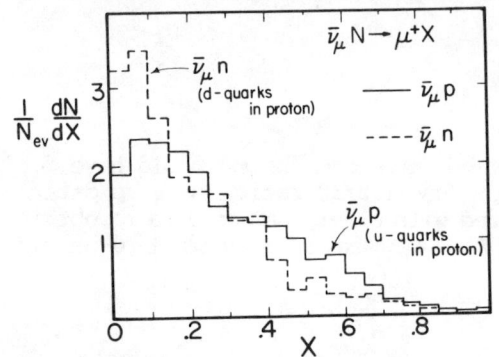

Figure 18: Reference 12

quark dominates at large x in the proton. Data on comparisons of $F_2^{en}/F_2^{ep}$ at large x is consistent with this same interpretation[25]. It will be of interest, with further analysis, to see if there exists quantitative agreement in the d(x)/u(x) ratio with the various methods for measuring it.

(4) Neutrino Total Cross-sections

There have been three new results[5, 21, 26] presented to this conference. One group has presented two new high energy measurements[21], which are somewhat larger than older values. The approximate linear energy dependence and the antineutrino-neutrino ratio, however, do not seem to be controversial. It is extremely important to have accurate determinations of the slopes of the neutrino and antineutrino cross-sections. They are linearly related to (i) the momentum fraction carried by struck quarks in the nucleon; (ii) the experimental values of the Gross-Llewelyn Smith sum rule and other sum rules; and, (iii) the normalization of all neutrino structure functions. Specific comparisons with electron and muon structure functions must be done with care if there is uncertainty about the neutrino cross section slopes.

A result from the SKAT bubble chamber[26] operating in the wide band beam at Serpukhov gives low energy values (E<30 GeV)

$$(\sigma/E)^\nu = (.70 \pm .06) \times 10^{-38} \text{ cm}^2/\text{GeV} \qquad (36)$$

$$(\sigma/E)^{\bar\nu} = (.23 \pm .05) \times 10^{-38} \text{ cm}^2/\text{GeV} \qquad (37)$$

These are consistent with the older low energy values from the Cern PS.

Table II: High Energy Cross Section Slopes (units $10^{-38}$ cm/GeV)

| | Expt | Energy | Neutrino | Antineutrino |
|---|---|---|---|---|
| (a) | CITFR[27] | 45-205 GeV | .509±.03 | .290±.015 |
| (b) | BEBC[28] | 20-200 | .630±.05 | .290±.03 |
| (b) | CDHS[29] | 30-190 | .620±.05 | .300±.02 |
| (b) | CHARM[5] | 20-200 | .594±.027 | .292±.015 |
| (c) | CFRR-I[30] | 50-260 | .700±.015±.035 | |
| (c) | CFRR-II[21] | 40-225 | .733±.005±.073 | .371±.004±.037 |

The high energy data are summarized in Table II. Results labelled with common letters have common flux-monitoring instrumentation. Entry (a), the first experiment designed to measure cross-sections in a narrow band beam, has considerable overlap in group membership with the group performing the recent measurements (c); however, the flux-monitoring system was completely re-designed and constructed for these new runs. The redundancy in the newer measurements is much improved. The last entry (CFRR-II) is still preliminary, in that there is considerably more information and redundancy available in this data, taken just six months ago. Hence the quoted systematic error of 10% should be substantially reduced in the near future.

IV. STRUCTURE FUNCTIONS, SCALING, AND QCD

(1) Introduction

Of great interest are the inelastic structure functions as measured in the reactions

$$e^\pm(\mu^\pm) + N \to e^\pm(\mu^\pm) + X \qquad (38)$$

$$\nu_\mu(\bar\nu_\mu) + N \to \mu^-(\mu^+) + X \qquad (39)$$

As example, consider an electron (muon) beam incident on an isoscalar target. The general form for the electromagnetic reaction (38) is

$$\frac{d^2\sigma^{eN}}{dx\,dy} = \frac{8\pi\alpha^2}{Q^4}\{ME\ 2xF_1^{eN}(x,Q^2)\left[1+(1-y)^2\right]+2F_L^{eN}(x,Q^2)[1-y]\} \qquad (40)$$

where the spin 1/2 part in simple free-quark models is given by

$$2xF_1^{eN}(x,Q^2) \approx q^{eN}(x) + \bar q^{eN}(x) \qquad (41)$$

with $q^{eN}(x)$ and $\bar{q}^{eN}(x)$ representing the distribution in fractional momentum carried by quarks and antiquarks, respectively, multiplied by the fractional momentum and the square of the quark charge. For example, ignoring the contribution of the strange or charmed constituents of the nucleon: $q^{eN} = (4u+d)/9$. The structure function $F_L(x,Q^2)$ would be zero in such models. It would naturally arise in scaling models through interaction of spin 0 nucleon constituents. In general, $F_L$ can assume finite values and $F_1(x,Q^2)$ can have an explicit dependence on $Q^2$ because of known limitations in the scaling assumptions: e.g. finite target mass effects or finite transverse momenta for the bound state quarks. These effects are expected to fall like a power of $1/Q^2$, and with an uncertain x-dependence. Although the qualitative behaviour in this low-$Q^2$ (long range) regime is anticipated, there are no unambiguous predictions for the specific functional form for the structure functions.

At higher $Q^2$, or short range, perturbative QCD makes very specific predictions: the structure functions should fall logarithmically with $Q^2$. $F_L$ or

$$R \equiv \frac{F_L(x,Q^2)}{2xF_1(x,Q^2)} \quad (42)$$

should have predictable values that are nonzero. These predictions for $F_1$ and R make assumptions about the form of the distribution for gluons inside the nucleon.

The form of equation 40 usually fitted to data is

$$\frac{d^2\sigma^{eN}}{dx\,dy} = \frac{8\pi\alpha^2}{Q^4} ME \{F_2^{eN}(x,Q^2)\left[1+(1-y)^2\right] - F_L^{eN}(x,Q^2)y^2\} \quad (43)$$

where

$$F_2^{eN}(x,Q^2) \equiv 2xF_1^{eN}(x,Q^2) + F_L^{eN}(x,Q^2) \quad (44)$$

This form allows direct determination of $F_2(x,Q^2)$ from data with little dependence on the assumed value of R except at large y-values.

The analogous equation for fitting neutrino and antineutrino data is

$$\frac{d^2(\sigma^{\nu N} + \sigma^{\bar{\nu} N})}{dx\,dy} = \frac{G^2}{\pi} ME \{F_2^{\nu N}(x,Q^2)\left[1+(1-y)^2\right] - F_L^{\nu N}(x,Q^2)y^2\} \quad (45)$$

where the left-hand side refers to the sum of the cross-sections. The values of $F_2^{\nu N}$ and $F_2^{eN}$ should be directly related by the mean-square charges of the quarks involved in the scattering. The other linear combination of cross-sections measures a completely different structure function, $F_3$.

$$\frac{d^2(\sigma^{\nu N} - \sigma^{\bar{\nu} N})}{dx\,dy} = \frac{G^2}{\pi} ME\; xF_3^{\nu N}(x,Q^2)\left[1-(1-y)^2\right] \quad (46)$$

In free quark models, $xF_3 = q(x) - \bar{q}(x)$ is the difference in quark-antiquark composition of the nucleon. This structure function has the experimental advantage that its determination is essentially independent of the assumed value of R. It has the pleasant theoretical property that its prediction from asymptotic QCD does not require knowledge of the gluon distribution.

(2) Measurements of $R = F_L/2xF_1$

As mentioned in the last section, there are distinct expectations for R, depending on the $Q^2$ regime. In perturbative QCD the value of $F_L$ is expected to be large at small x and small at large x, with an approximately logarithmic dependence on $Q^2$. However, at lower $Q^2$, we anticipate that there will be terms that behave like $f(x)/Q^2$, due to the finite $k_T$ of the struck quarks, for example, which depends on the wave function of the bound quarks. Such dependence is often denoted as "higher twist" contributions by aficionados of QCD. Some specific diagrams[31] representing non-perturbative mechanisms can give rather perverse behaviour to f(x) at some x-values: e.g. $f(x) = 1/(1-x)$.

Our present knowledge of this structure function is poor. It is generally acknowledged by all concerned to be a difficult measurement. Table III summarizes the available measurements specifically quoted for R.

Table III: Measurements of Averaged Values of R

| GROUP-TECHNIQUE | | x-range | $Q^2$-range | R |
|---|---|---|---|---|
| SLAC-MIT[32] | eN | .1 - .9 | 2-20 GeV | 0.20±0.10 |
| Hrv,etc[33] | νN | 0. - .1 | 1-12.5 | 0.44±0.25±0.19 |
| BEBC[34] | νN | 0. - 1. | .1-50 | 0.15±0.10±0.04 |
| CDHS[29] | νN | 0. - 1. | 2-200 | -.03±.05 |
| CDHS[35,36] | νN | 0. - 1. | 2-200 | 0.03±.05±.1 |
| HPWF[37] | νN | 0. - 1. | 2-200 | 0.18±.06±.04 |
| CDHS[38] | νN | 0. - 1. | 2-200 | 0.10±.025±.07 |

A few remarks about this list are in order. A superficial scan shows a wide variation in the average value of R. However, the quoted statistical (first) and systematic (second) errors are also quite large. There does seem to be a tendency toward positive values. Also the largest value in the list is measured at small x-values[33] and small $Q^2$ values. Most of the non-zero effect seen by these authors was for x<.02. The newest result[38] represents a convergence by this group to finite positive values relative to their previous results[29,35,36]. Fig. 19 shows the x- and $Q^2$-dependence of R presently found in their data. No dramatic $Q^2$ or x dependence can be attributed as yet. The dependence on x and $Q^2$ in the earlier eN data had a similar behaviour, although their average value was higher. We can as yet draw no conclusion about R as it relates to perturbative

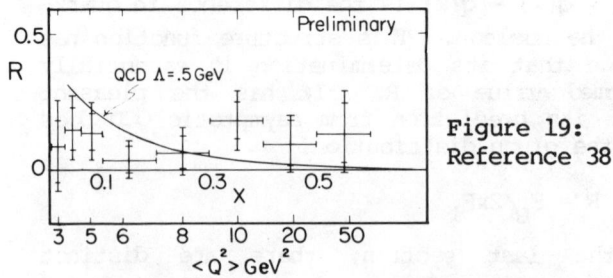

Figure 19:
Reference 38

QCD. While the $Q^2$-dependence does seem to be slow, as logarithmic behaviour would produce, the x-dependence does not exhibit the fast-falling expectations of perturbative QCD. There may be several mechanisms operating in this $Q^2$-range, as has been suggested.[31]

(3) Scaling - does it or doesn't it?

Do the structure functions ($F_1$, $F_2$, $F_3$) depend explicity on $Q^2$? In fact, there is absolutely no controversy that deep-inelastic scattering data do not scale. The very earliest SLAC-MIT data showed non-scaling behaviour. At that time, the approximate nature of scaling was enough to completely overhaul the picture that we had of nucleon structure. It was even projected that scaling might become exact if we went to $Q^2$-values that were large enough. There were many mechanisms that could be invented to explain deviations from exact scaling that would behave approximately like 1 GeV$^2$/$Q^2$. These include finite target mass effects, resonance production, and finite $k_T$ effects, where $k_T \approx 800$ MeV/c (see Section II). All theories will have such effects at low $Q^2$: they were anticipated, and probably observed, long before the advent of QCD.

Nowadays, we no longer expect exact scaling at high $Q^2$. However, the interpretation problem at low $Q^2$ has changed little. While some forms that specific higher twist terms may take can be anticipated[39] (and there is hope that specific predictions in this regime may be possible), we cannot now unambiguously make predictions from theory about the $1/Q^2$ behaviour of the structure functions. From the point of view of a skeptical experimentalist, so long as one can explain data in terms of corrections to the structure functions that go like $1/Q^2$, or as was done ten years ago, explain the data in terms of a scaling variable that differs from the asymptotic variable, $x = Q^2/2M\nu$, by corrections to the <u>variable</u>[40] of order $1/Q^2$, then we cannot categorically state that we are observing scaling violations that are anything other than that due to low-$Q^2$ mechanisms. It should be noted that the derivation of the variable, x, as representing the fractional quark momentum is only unambiguously correct in the limit, $Q^2 >> M^2$. There is no proof that parametrization in terms of x is in any sense more correct at low $Q^2$ than another variable that approaches x at large $Q^2$. Some mechanisms, it has been argued[40,41], are more naturally described in terms of different variables than by additional functions of the x-variable. No unique variable has been proposed to naturally describe all anticipated low-$Q^2$ scale-breaking.

Equally non-controversial with the non-scaling behaviour of the data are the high-$Q^2$ predictions of first-order perturbative QCD. The question is, "At what $Q^2$ can we reasonable expect the perturbative QCD (i.e. logarithmic) behaviour to dominate?" During the past several years, there has been considerable effort devoted to interpreting all data solely in terms of perturbative QCD. Some qualitative features of the data and of the predictions gave encouragement to this approach. Specifically, asymptotic QCD states that the x-dependent structure functions should shrink as $Q^2$ is increased. It seemed that the small x data did show an increase with $Q^2$ up to 5-10 GeV$^2$.[42] Large x data, as early as the SLAC-MIT experiments prior to 1970, fell with $Q^2$, but this could largely be removed by a redefinition of the scaling variable.[40]

The newest data at small x(e.g. x = .08) now extend to higher $Q^2$. They do not clearly show a continuation of the rise with $Q^2$ that was so evident for $Q^2$<5 GeV$^2$. (See Sections IV-5,6.) In addition, there is some reason to question the interpretation of scale-breaking in this region for other reasons. The effects of associated charm production may well be a significant fraction of any scaling violations that appear.[43] Taken together, these factors in no sense disprove QCD, but we cannot invoke the low x, higher $Q^2$ data, to support it.

Another way of arriving at this conclusion is shown in Fig. 20. This is preliminary data from the CFRR neutrino experiment.[21]

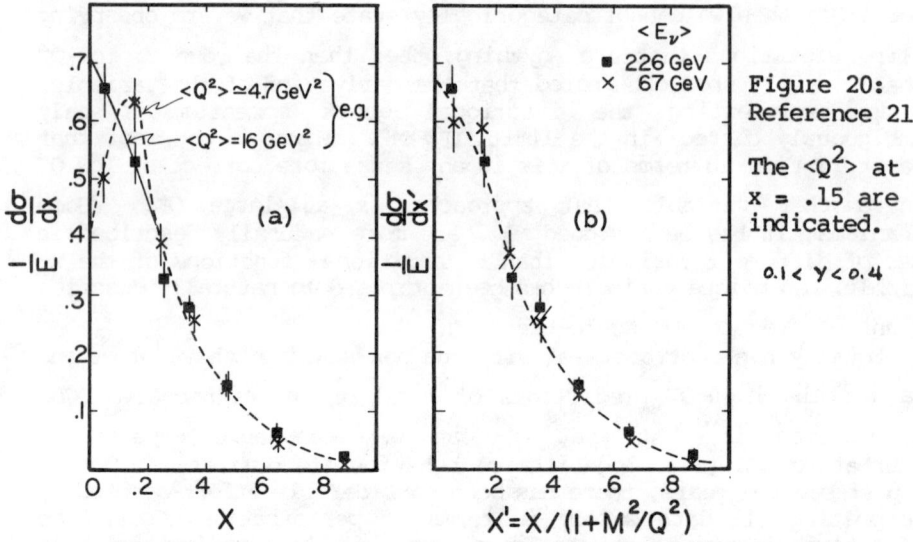

Figure 20: Reference 21

The $<Q^2>$ at $x = .15$ are indicated.

$0.1 < y < 0.4$

Fig. 20a shows the differential x-cross section, integrated over y as stated, for two regions that differ in energy by about a factor of 3. The non-scaling at low-x is quite visible. Fig. 20b shows the same data in which the differential cross-section is plotted against an empirical variable $x' = x/(1+M^2/Q^2)$, where M is the nucleon mass. The fact that the non-scaling disappears does not necessarily demonstrate anything fundamental about this particular variable, and certainly does not demonstrate that the low x non-scaling is not QCD. But the exercise does demonstrate that the small x scale-breaking can have explanations through mechanisms that have nothing to do with perturbative QCD.

We conclude that the present interpretation of scale-breaking according to the predictions of perturbative QCD rests on data at larger x-values. We will return to the high-statistics data at higher x in sections III-7,8.

(4) Structure Functions with Proton Targets

The use of hydrogen as an experimental target has some important advantages over using complex nuclei. The target is conceptually simpler, e.g. no complications of nuclear Fermi motion. In addition, there exists the original ep data from SLAC-MIT to compare directly. In the long term, there may be ep data from colliding beams that will even further extend the range of $Q^2$. There are also several compromises that are associated with hydrogen runs: the major one is the lower event rate, another is poorer resolution on total hadron energy, which is used in calculating the kinematic variables. Hence, data with hydrogen will often not extend as high in $Q^2$ as that from complex nuclei, but it nicely complements such data, and may turn out to be the most important in the long term.

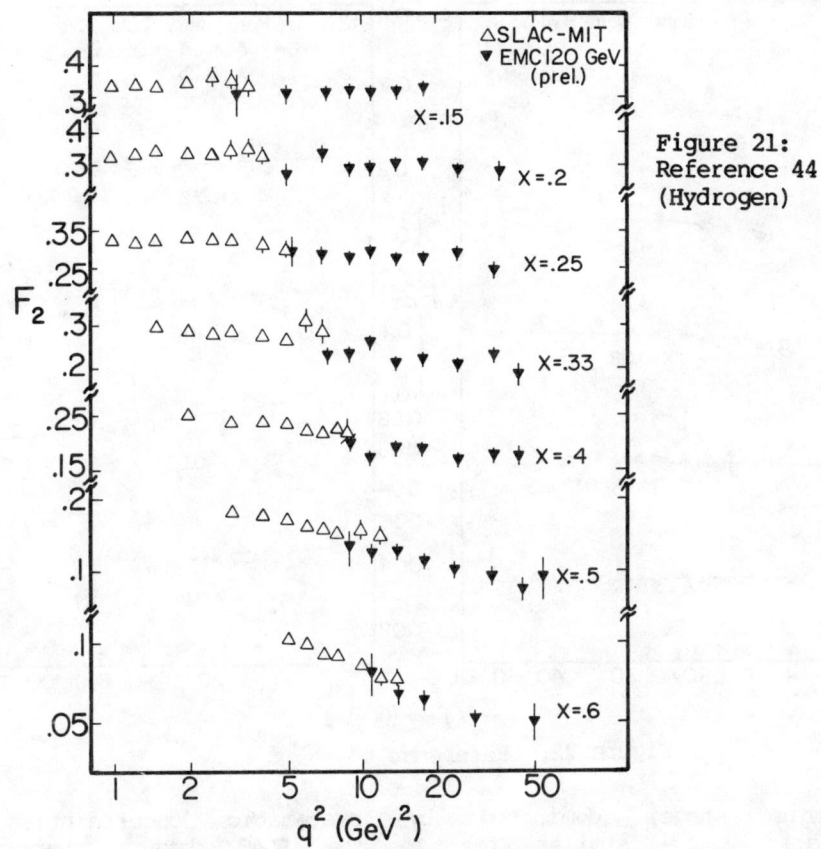

Figure 21: Reference 44 (Hydrogen)

New data has been presented by the EMC group for $\mu^-$ deep-inelastic scattering from hydrogen.[44] Fig. 21 shows their data at the lower energy setting compared with the SLAC-MIT data[45,32] at still lower energy. In regions of overlap, they tend to agree within quoted statistical errors, although they could be systematically lower by 5-10%. Fig. 22 shows the 120 GeV and 280 GeV muon data combined. The data do not clearly demonstrate scaling violations except in the small x region. The result of a $\chi^2$ test assuming there is no $Q^2$ dependence in the data of Fig. 22 with x>.2 gives $\chi^2$ = 80 for 70 degrees of freedom. A first-order QCD fit[43] to this same data gives $\Lambda$<100 MeV, considerably smaller than values that we have seen previously. While one should keep in mind that these data are preliminary, as are the analyses which we will mention, the experimenters have attempted to estimate the effects of statistical and systematic uncertainties. They ascribe an error on $\Lambda$ of 100 MeV

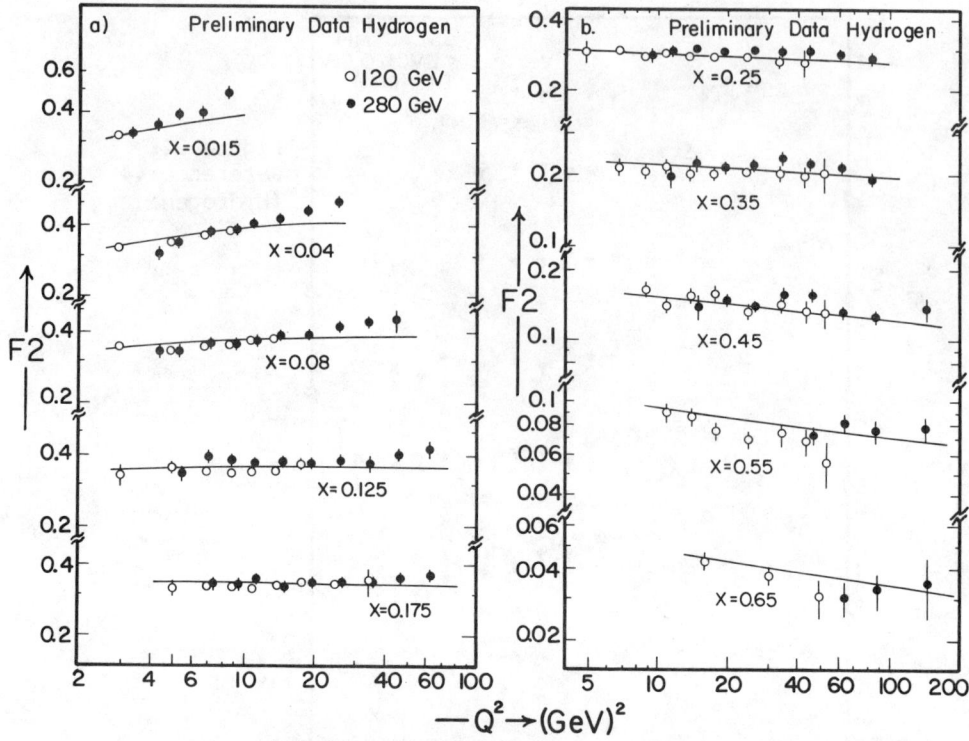

Figure 22: Reference 44

at this stage, dominated by systematic uncertainties. Interestingly, a similar fit to the SLAC data[45,32] gives $\Lambda$ = 350 MeV. We are dealing with values of $\Lambda$ here that are small enough so that it is unlikely that inclusion of next to leading order in the perturbation expansion will change the qualitative conclusions. The likelyhood is that the mechanisms below and above $Q^2$ of about 10 GeV are not <u>both</u> dominated by perturbative QCD.

As stated above, there is no unambiguous way to parametrize the higher-twist effects. A form that is popular, but neither necessary nor unique is

$$F_2(x,Q_o^2) = Ax^\alpha (1-x)^\beta + \frac{a}{Q^2} x^\alpha (1-x)^{\beta-1} \qquad (47)$$

where A, a, $\alpha$, $\beta$ are to be determined from the data along with the value of $\Lambda$ that occurs in the quark-gluon coupling constant. We shall see that this simple form is not adequate to bring a consistent interpretation to the μp and the ep data. There is a very strong correlation found between the magnitude of the higher-twist term, <u>a</u>, and the QCD parameter, $\Lambda$. Fig. 23 shows the pairs[43] of <u>a</u>, $\Lambda$ values for which fits where obtained for the two kinds of data. The inclusion of this single term does not make for consistent fits with

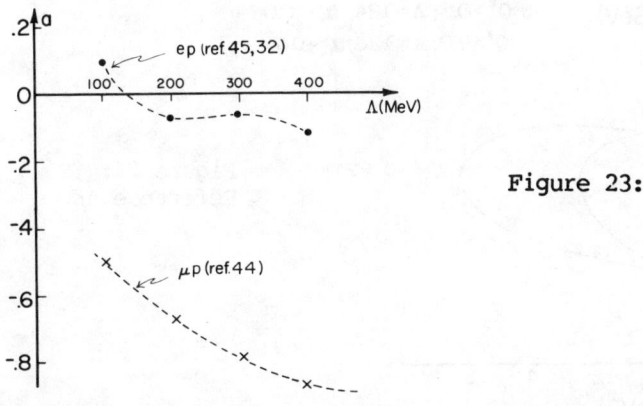

Figure 23:

a single set of parameters. Deep-inelastic electromagnetic scattering on hydrogen, it is concluded, cannot now globally be understood in terms of this simple parametrization. The data at higher $Q^2$ is not yet accurate enough by itself to see clear logarithmic dependence with $Q^2$. It is accurate enough, it would appear, to contradict the large values of $\Lambda$ that have been quoted in recent years.

(5) New Neutrino Data with Complex Targets

New data has been presented at this conference from both bubble chamber[46] and counter experiments[38,47] on the charged current structure functions. The bubble chamber results[46] contain approximately 7500 events, over half of which are from antineutrino interactions. The data were taken in the SPS wide band beam. An analysis was performed using the same higher-twist term assumed in equation 47, and evolving the perturbative QCD dependence of the structure functions using several techniques that incorporate second-order terms. Although there are some differences in the values of the parameters obtained, the authors conclude that in both 2nd-order schemes used, "A clear QCD dependence of the structure functions is observed...Higher twist terms cannot be excluded, but such terms alone cannot account for the $Q^2$ dependence of the structure functions." This is illustrated in Fig. 24, which shows the acceptable contours of a (higher twist coefficient) and $\Lambda_{ms}$, the QCD parameter of the fit. As can be seen the data seem to prefer a finite value of $\Lambda_{ms}$, independent of a. The best values of $\Lambda_{ms}$ are the same whether the data for $F_2$ or $F_3$ are used, and appear independent of whether a cut on $Q^2 < 5$ GeV$^2$ is applied. This may contradict the conclusions obtained from the muon and electron hydrogen data above, where the higher twist term showed strong correlation with the QCD parameter, and where the low and high $Q^2$ data did not indicate QCD as the dominant mechanism operating in both energy regions. The only identifiable difference in analysis is the

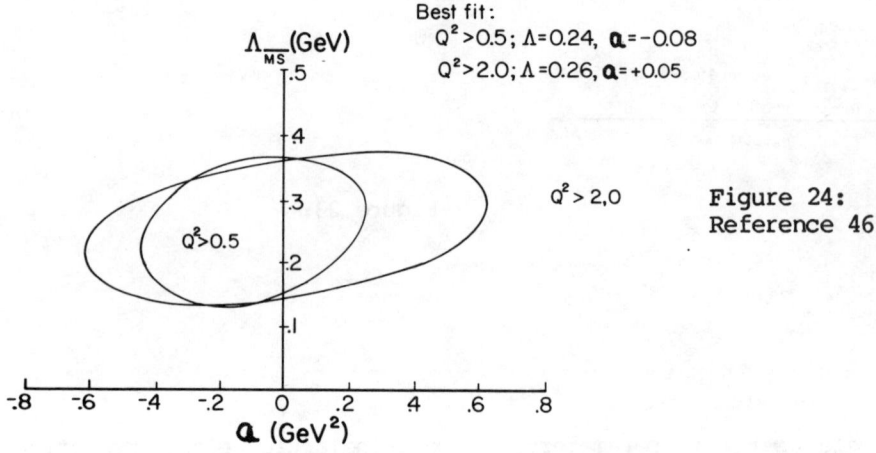

Figure 24: Reference 46

inclusion of next-order QCD in the neutrino analysis, but this is not expected to be so important.

It should be noted that the data used in this fit show somewhat different trends than other recent data. Fig. 25 shows the $xF_3$ data in two x-bins where the data of references 46 and 38 overlap. (We will discuss the data of the latter experiment presently.) The bubble chamber data have considerably fewer events, as expected. However, there is also a definite difference in the trend of the data, with the counter data showing substantially less $Q^2$-dependence. (It should be noted that both experiments have normalized their data to the same total cross-sections, so that the slopes are more relevant than the absolute values.) Besides the differences in experimental detectors, one important difference between the experiments is the use of a wide-band beam in one case,[46] and the use of a narrow-band neutrino beam in the other.[38]

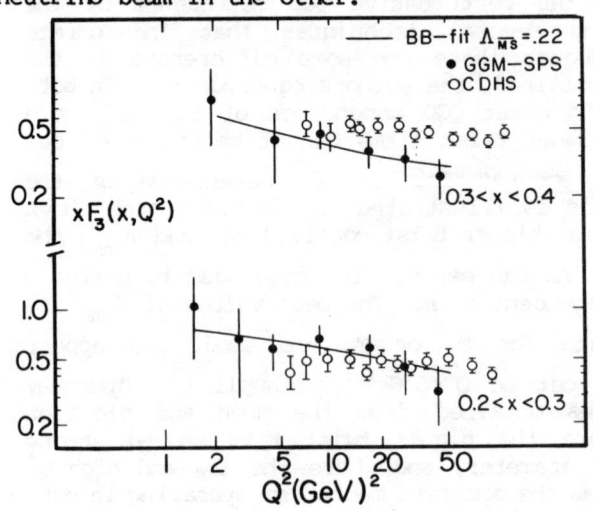

Figure 25
References 38, 46

New results from the HPWF counter experiment performed in the FNAL wide-band beam were reported.[47] The data are of similar statistical precision to the narrow-band CERN data reported previously,[29] and corroborate the gentle $Q^2$ dependence seen in that data.

New results presented at this conference from the CDHS experiment[38] at the SPS are based on a data sample approximately four times as large as formerly reported. Within statistical errors, the old results are in agreement. Fig. 26a,b show the results for $xF_3$ and $F_2$, respectively. The values of $F_2$ were obtained assuming that $R = 0.1$ and constant. As stated previously, the rise with $Q^2$ at low x is not nearly so apparent for the new data with $Q^2 > 10$ GeV$^2$. The fall with $Q^2$ at larger x-values is now statistically certain.

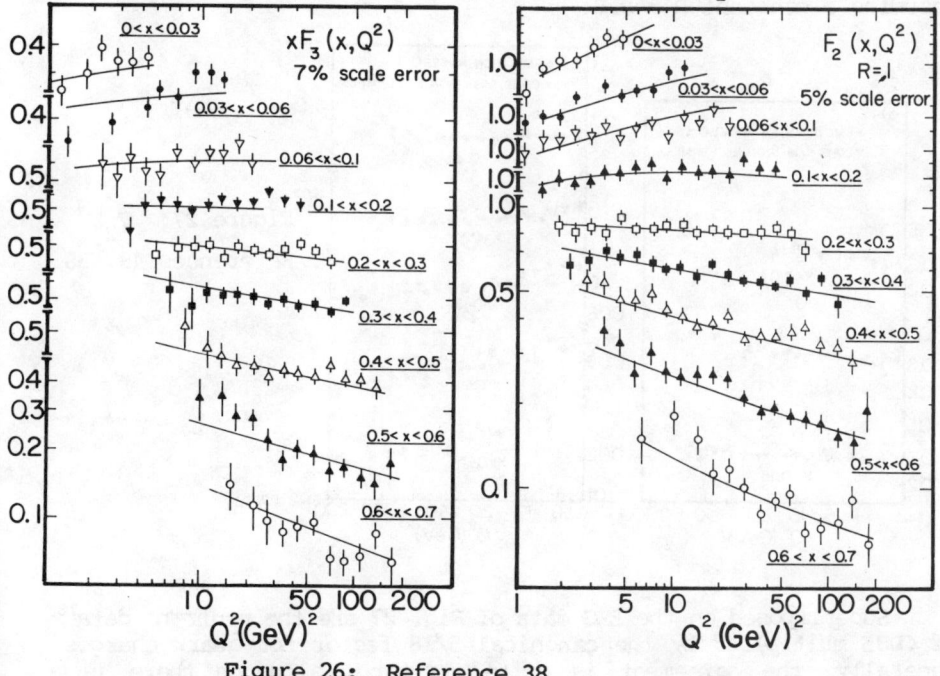

Figure 26: Reference 38

A fit to the $xF_3$ data of Fig. 26a, using standard forms for the x-dependence at fixed $Q^2$, and evolving with the Altarelli-Parisi equations gives

$$\Lambda = .3 {+ .17 \atop - .13} \text{ GeV} \qquad Q^2 > 10 \text{ GeV}^2 \qquad (48)$$

This value is somewhat lower, but consistent with the value obtained from fitting their previous data[48] on $xF_3$: $\Lambda = .55 \pm .15 \pm .10$ GeV.

As we shall see, this is typical of a general trend in the newer, high statistics data toward smaller values of $\Lambda$. We will defer discussion of fits to $F_2$ until we discuss the muon deep-inelastic data.

(6) New Muon Data with Complex Targets

Several groups have presented new $F_2$ structure functions from muon inelastic scattering experiment using heavy nuclear targets. Data[49] taken by the European Muon Collaboration are shown in Fig. 27, after corrections for resolution smearing and acceptance, but without corrections for Fermi motion. These data were taken with beam energies of 120, 250, 280 GeV; the values of $F_2$ were obtained assuming a constant R = 0.2.

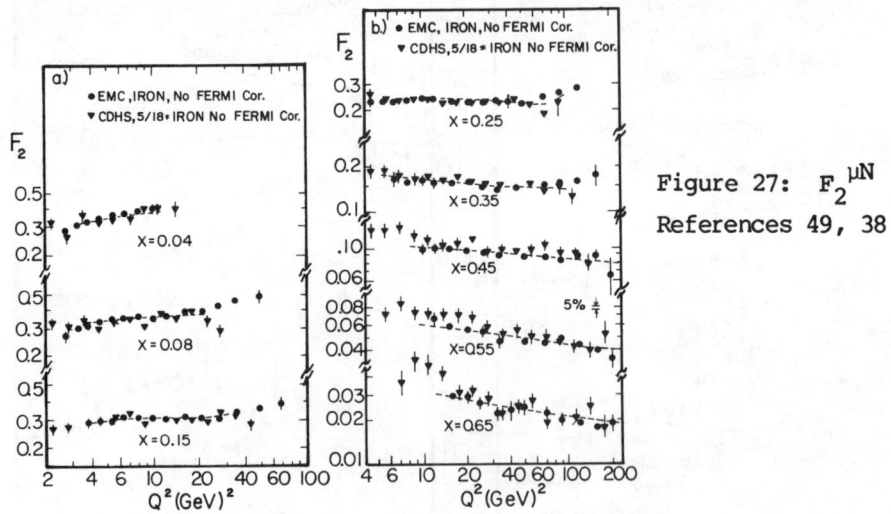

Figure 27: $F_2^{\mu N}$
References 49, 38

Superimposed on the EMC data of Fig. 27 are the neutrino data[38] of CDHS multiplied by the canonical 5/18 factor for quark charges. Generally, the agreement is quite striking, although there is a tendency for the muon data to be lower than the neutrino data at lower $Q^2$. At the lowest part of the $Q^2$ range, the neutrino data show the most dramatic rise but the muon data do not extend this far. (See, e.g., x = 0.65, $Q^2$<15 GeV$^2$ and x = 0.45, $Q^2$<10 GeV$^2$.) It should be noted that at such low $Q^2$ and large x values, the hadron energy is low (typically $\sim$ 10 GeV); in this region, resolution with calorimeters is not very precise, and experiments without direct measurements of hadron energy rely heavily on accurate knowledge of the beam energy.

The dashed lines in Fig. 27 have been sketched primarily to guide the eye through the muon data...they have the slopes of the best QCD curves fitted to the muon data after Fermi corrections (see Fig. 28). There are small, but finite, differences in slope between the two sets of data, with the muon data preferring smaller slopes.

The data after correction for Fermi motion are shown in Fig. 28 with the QCD fitted curves, which will be discussed in the next section. These corrections in general tend to enhance the observed slopes[50], and this is indeed the case here. Hence, the logarithmic slopes after correction are slightly larger than they would be if no Fermi corrections were applied.

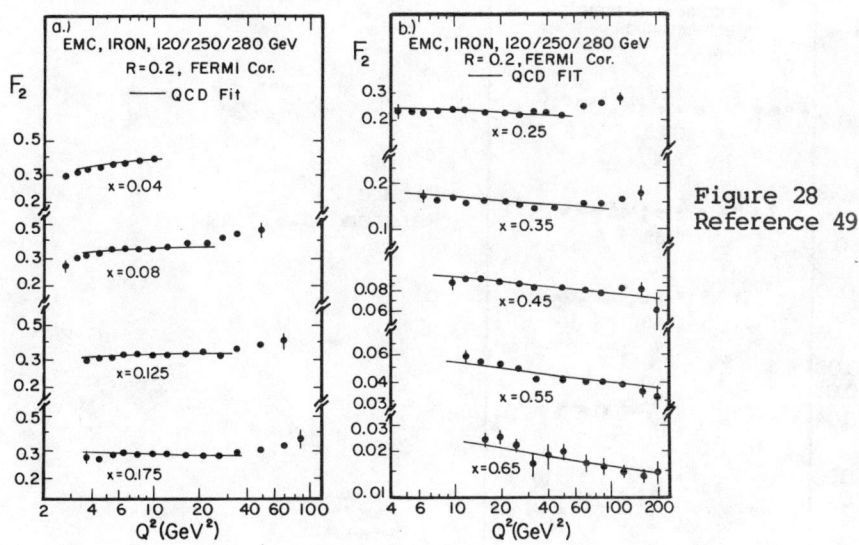

Figure 28
Reference 49

We conclude from this EMC iron data that the "raw" $F_2$ values from muon scattering data show slightly smaller slopes over the high $Q^2$ range than that seen at present in neutrino scattering. When comparing results of fit, it should be remembered that the fits to EMC data have been done after Fermi correcitons. (The neutrino data do not have Fermi corrections applied.) If uncorrected muon data were fit, $\Lambda$ would be smaller, not larger.

New data from the BCDMS group[51] covering the higher x-range is shown in Fig. 29. Superimposed are the EMC data,[49] which extend to lower $Q^2$. One relevant difference to note when making this comparison is that the BCDMS results assume R = 0 while those from EMC data assume R = .2. The effect of changing R in the BCDMS data, for example, to R = .2 would increase $F_2$ by at most ten percent at the highest $Q^2$ relative to the lowest $Q^2$, which would make the slope for their data in Fig. 29 slightly smaller. Both sets of data are uncorrected for Fermi motion. In any case, the data agree well. Both show slopes that are smaller than the CDHS neutrino data. This tendency to observe a very flat $Q^2$ dependence is shared by other muon data that has been presented at this conference.[52]

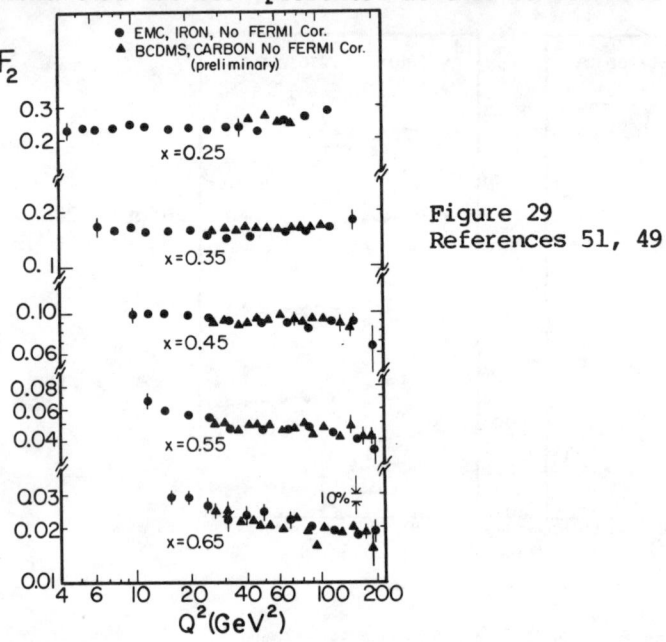

Figure 29
References 51, 49

(7) Fits to $F_2$ from Heavy Target Experiments

It <u>is</u> not unexpected that the fits to $F_2$ from the most recent EMC muon data would give a smaller value of the first-order parameter, $\Lambda$. It is somewhat surprising that this value is as small as 0.1 ± 0.1 GeV. This can be compared to the $\Lambda$ obtained from fits to the CDHS neutrino data (both $F_2$ and $xF_3$) which is 0.5 ± .05 ± 0.1 GeV. The $Q^2$ range is similar, though the neutrino data extend to lower $Q^2$ values at large x. In this large x region, the neutrino data do show larger slopes, as noted before. It is unlikely that the systematic differences in $\Lambda$ are solely due to different assumptions in extracting $F_2$ from data: changing the R-assumption has a small effect, and the Fermi correction made by EMC makes Λ larger. In extracting Λ, however, one must make an assumption about the form of the x-dependent gluon distribution. The EMC group uses a gluon distribution (whose parameters are obtained from the fit) that seems to be somewhat narrower than that used by the CDHS group. There may well be discrepancies in Λ arising from the assumptions made in the forms of these distributions.

Another possibility cannot be ignored. Since we do not know a priori the exact form for higher-twist scale-breaking, there is no way to rule it out. It may be especially a problem at large x-values. Suspicion must rear its ugly head if, as the large $Q^2$ range extends further, the value of Λ gets smaller and smaller. We are not yet at the point where this suspicion is a certainty. There are still unresolved differences among the experiments and among the analyses already done, and there will be additional data in this energy region soon. There can be little argument that, if there are differences in Λ that arise from identical processes and identical analyses done in different $Q^2$ ranges, the higher $Q^2$ data is more credible from a theoretical standpoint.

We should keep in mind that many of the results we have seen are preliminary. We may take heart that, at presently available energies, there is sensitivity at the level of Λ ≈ 0.1 GeV. One question that our theoretical colleagues should consider is "How small would Λ have to get before one re-evaluates the present picture of QCD?" Recall that the earliest values of Λ, approximately 0.7 GeV, were considered "natural" in scale. As experiments improve their sensitivity, if the accepted experimental values get smaller, at what point are they "unnatural"?

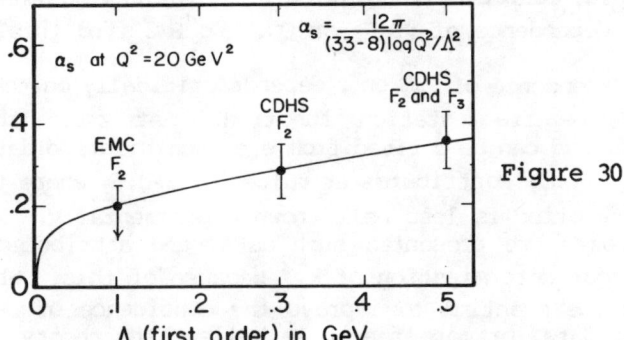

Figure 30

The story is not ended. Indeed, if there are arguments over Λ to decide whether the value is closer to 0.1 or to 0.5, the argument over the value of the quark-gluon coupling constant is not nearly so bizarre. Fig. 30 shows the range under discussion. The values of $\alpha_s$ at $Q^2$ = 20 GeV$^2$ change approximately by a factor of 2 over this range. Again we must reiterate: there are scaling violations. It begins to look unlikely that perturbative QCD will simply explain the data both for $1<Q^2<10$ GeV$^2$ and $10<Q^2<100$ GeV$^2$. It also looks unlikely that simple "guesses" for the higher twist term will simplify the problem. We hope for a firm theoretical framework to calculate the low-$Q^2$ behaviour. Even if this does not come to pass, we can be optimistic. We look forward to a consistent interpretation in the present high $Q^2$ range, and then corraboration at even higher $Q^2$. We do see scale-breaking. Is it perturbative QCD? We look to the present and the future experiments to tell us.

## Discussion

Q1: Wilson, Harvard: I wish to ask about the measurements of $P_T^2$ by the European Muon Collaboration. Firstly, in the Chicago, Harvard, Illinois, Oxford collaboration at Fermilab we believed we had a small dependence of $<P_T^2>$ on $Q^2$. Do EMC find this?

Secondly, the dependence of $P_T^2$ on z depends critically on the z dependence of the fragmentation function. At z<0.5 this increases with z and can be derived from $e^+e^-$ annihilation data. The term $z^2 <k_T^2>$ only contributes at values of z>0.5, where the fragmentation function is less well known experimentally. Two years ago at Tokyo, we presented such data, and attributed a large error to our determination of $k_T^2$ because of this. What has changed in the meantime to improve the confidence of EMC? Is there better data, or more reason to believe the theory?

A1: Brasse, DESY/CERN: The fragmentation has been calculated according to Field-Feyman with a cascade model taking into account the appropriate transverse momentum. In this way we get little z dependence for the fragmentation. As far as the $Q^2$ dependence of $<P_T^2>$ is concerned, for fixed range of W we do not see much $Q^2$ dependence.

Q2: Chaline, France: Recent results by rigorous non-perturbative methods show that radiative corrections in muon scattering at the current energies are much larger than thought, in the small x region. Although the one photon bremsstrahlung cross-section is know to be quite large, it turns out that two hard photon emission, one of which is collinear, can be an order of magnitude larger than the one photon emission gives. Since the new pure QED effects are not yet taken into account in the published experiments, one has to be careful concerning the scaling violations, at least in the small x and large $Q^2$ region.

Q3: Braccini, Pisa: Since all Drell-Yan experiments seem to agree in finding a factor ∽ 2 between the cross-sections they measure and the cross-sections which are derived from calculations (where products of structure are involved), I would like you to comment on the possibility that such discrepancy could be explained by the present uncertainties concerning the absolute normalization of the structure functions as derived by the different deep-inelastic experiments.

A3: Sciulli: I would say that the normalization of the neutrino structure functions is uncertain at the 20% level.

Q4: Bodek, Rochester: I have a question to the EMC Collaboration. What were the magnitude of the Fermi motion corrections? If these corrections are the cause of the difference between the CDHS and EMC data, then they are much larger than is ordinarily expected in the x range of those experiments.

A4: Wahlen, Wuppertal: Differences in normalization as shown by the speaker (at large ) are due to the fact that only in the EMC data a correction for Fermi motion is applied. If this is taken out, the data sets of CDHS, EMC, BCDMS agree within ± 5% in normalization. This does not affect the differences in slopes.

Q5: LoSecco, Michigan: To what do you attribute the new higher total neutrino cross-section?

A5: Sciulli: The first experiment was done five years ago with a rather rudimentary beam and a rather rudimentary monitoring system. We've learned a lot since then. If you ask me which one I believe, I will certainly believe the more recent results, especially after the analysis is complete. At the moment, I think the cross-sections will move up.

Q6: Jaffe, MIT: What, if anything, is known experimentally concerning the Adler Sum Rule?

A6: Sciulli: Not much.

Jaffe: Can you tell us when we can expect some information on this important sum rule?

Sciulli: Nezrick and HPWF have data but decline to comment.

Q7: Wosiek, Jagellonian: Should we be looking for scaling violations in the parameter W instead of $Q^2$?

A7: Sciulli: Theory tells us to look in $Q^2$ for deep-inelastic structure functions. If there is another way to look at this data, I'm sure people will be happy to plot it that way.

A7: Kleinknecht, Dortmund: Comment to the question on whether data have been analyzed in terms of $W^2$ by the CDHS group. The broadening of the charm jet in neutrino reactions has been analyzed in terms of $W^2$ (and $z^2$). The comparison with first order QCD gives a value $\alpha_s = 0.35$ at $<Q^2> = 25$ GeV$^2$/c$^2$.

A7: Brodsky, SLAC: The actual evolution in QCD is limited by $W^2$ rather than $Q^2$ so that it may make sense to analyze scale violations using $W^2$. This is discussed in papers by Lepage and myself, and Amati, et al.

Q8: Brodsky, SLAC: Could you comment on the fact that the power law scaling violation seen in $\nu p \rightarrow \mu^- \pi^+ X$ Gargamelle data at large z may account for the observed moment scaling violation of the fragmentation functions.

A8: Sciulli: Ed Berger has done this analysis. (See Ref. 14).

## References

[1] L.S. Rochester, et al, PRL **36**, 1284 (1976).
[2] P. Langacker, et al, U. of Pennsylvania preprint UPR-158T. Submitted to Review of Modern Physics.
[3] C.Y. Prescott, "Further Measurements of Parity Non-Conservation in Inelastic Electron Scattering", Proceedings of the Int'l Symposium on Lepton-Photon Interactions at High Energies, 1979, Batavia, Illinois, P. 271.
[4] S. Sommars, et al, "Neutral Current Interactions in $\nu$n and $\nu$p Collisions", Stony Brook, IIT, U of Maryland, Tohoku U., Tufts; contribution to this conference.
[5] M. Jonker, et al, "Experimental Study of Neutral and Charged Current Neutrino Cross-sections...CHARM collaboration", contribution to this conference.
[6] T. Kondo, et al, "Recent Neutrino Results from Caltech-Fermilab-Rochester-Rockefeller Collaboration", contribution to this conference (see reference 21).
[7] K. Winter, "Review of Experimental Measurements of Weak Neutral-Current Interactions", Proceedings of the Int'l Symposium on Lepton and Photon Interactions at High Energies, 1979, Batavia, Illinois, p. 258.
L.M. Barkov and M.S. Zolotoryov, PL **69B**, 323(1977).
P. Conti, et al, PRL **42**, 343(1979).
[8] C. Baltay, et al, PRL **44**, 916(1980).
[9] K.H. Mess, et al, Proceedings of the Int'l Conference on Neutrinos, 1979, Bergen, P. 371.
[10] T. Kitagaki, et al, "Current and Target Jets Produced in High Energy Neutrino-Deuterium Interactions", "Characteristics of Charged Particle Multiplicity Distributions in High Energy $\nu$n and $\bar{\nu}$p Interactions"; Tohoku U., IIT, U. of Maryland, Stony Brook, Tufts collaboration; contributions to this conference.
[11] P. Allen, et al, "Multiplicity Distributions in Neutrino-Hydrogen Interactions"; Aachen, Bonn, CERN, Munich, Oxford collaboration, contribution to this conference.
[12] D. Alassia, et al, "High Energy Antineutrino Interactions in Deuterium", Amsterdam, Bologna, Padova, Pisa, Saclay, Torino collaboration; contribution to this conference.
[13] C. Matteuzzi, et al, "Contributions from Higher Twist Effects to the Quark Fragmentation Functions in Neutrino Data", CERN, Milano, collaboration; contribution to this conference.
[14] E.L. Berger, PL **89B**, 241 (1980).
[15] L. Pape, et al, "Scaled Energy Distributions of Single Hadrons Observed in Muon-Proton Scattering", "Transverse Momentum of Hadrons Produced in $\nu$ and $\bar{\nu}$ Interactions on an Isoscalar Target in BEBC", Aachen, Bonn, CERN, Demokritos Athens, I.C. London, Oxford, Saclay; contribution to this conference.
[16] J.J. Aubert, et al, "Transverse Momentum of Produced Hadrons in Deep Inelastic Muon Scattering", EMC Collaboration: CERN, DESY, Freiburg, Kiel, Lancaster, LAPP, Liverpool, Oxford, Rutherford, Sheffield, Turin, Wuppertal; contribution to this conference.

[17] V.J. Stenger, et al, "Evidence for Gluon Radiation in High Energy Neutrino Interactions"; LBL, FNAL, Hawaii, U. of Washington, U. of Wisconsin; contribution to this conference.

[18] R. Wilson, et al, "Muon Scattering at Fermilab", Chicago-Harvard-Illinois-Oxford Collaboration, Proceedings of the 19th International Conference on High Energy Physics, Tokyo, 1978, P. 306.

[19] J. Knobloch, et al, "The Structure and Amount of the Sea and Charm Jet Broadening"; CDHS collaboration: CERN, Dortmund, Heidelberg, Saclay; contribution to this conference.

[20] See, e.g. R. Brock, PRL 44, 1027(1980).

[21] M. Shaevitz, et al, "Recent Results from the CFNRR Neutrino Experiments"; Caltech, Fermilab, Rochester, Rochefeller, Northwestern; paper presented to this conference.

[22] D. Bucholz, et al, presentation to Neutrino '80, Erice.

[23] G. Baum, et al, "Measurement of Asymmetry in Spin-Dependent e-p Resonance Region Scattering", Bielefeld, SLAC, Tsukuba, Yale; contribution to this conference.

[24] G. Myatt, et al, "Determination of the Quark Ratio $d(x)/u(x)$ in the Proton Using ep and $\nu p$ Data; Aachen, Bonn, CERN, Munich, and Oxford; contribution to this conference.

[25] A. Bodek, MIT thesis "Inelastic Electron-Deuteron Scattering and the Structure of the Neutron" (1972), SLAC-MIT group. I am indebted to J. Franklin for reminding me of this fact.

[26] Institute for High Energy Physics, Serphukov, USSR, "Investigation of Neutrino and Antineutrino CC-Interactions"; contribution to this conference.

[27] B.C. Barish, et al, PRL 39, 1595(1977).

[28] P.C. Bosetti, et al, PL 70B, 273(1977).

[29] J.G.H. deGroot, et al, Z. Physik C 1,143(1979).

[30] B. Barish, et al, Caltech, Fermilab, Rochester, Rockefeller collaboration, preprint CALT 68-734(1979).
D. Theriot, et al; Proceedings of the 1979 Int'l Symposium on Lepton and Photon Interactions at High Energies, Batavia, Ill; p. 337.

[31] L.F. Abbott, et al, SLAC-PUB-2327(1979).
L.F. Abbott, et al, SLAC-PUB-2400(1979).

[32] M. Mestayer, PhD Thesis, SLAC Report No. 214.

[33] B.A. Gordon, et al, PRL 41, 615(1978).

[34] P.C. Bosetti, et al, Nucl. Phys. B142, 1(1979).

[35] A. Savoy-Navarro, et al, Proceedings of Neutrino '79, Bergen, P. 253.

[36] J. Wotschack, et al, EPS Int. Conf. on High Energy Physics, Geneva, 1979.

[37] A. Benvenuti, et al, PRL 42, 1317(1979).

[38] J.G.H. deGroot, et al, "New Results on Neutrino Induced Charged Current Interactions", CDHS collaboration, contribution to this conference.

[39] See, e.g. R.M. Barnett, SLAC-PUB-2396, and references quoted there.

[40] E. Bloom and F. Gilman, PRL 25, 1140(1970).

[41] A. deRujula, et al, PL 64B, 428(1977); Ann Phys 103, 315(1977).
[42] H.L. Anderson, et al, PRL 38, 1450(1977).
[43] J.J. Aubert, et al, "Scaling Violation in $\mu$N and a Possible Interpretation", The European Muon Collaboration (EMC) contribution to this conference.
[44] P.R. Norton, et al, "Measurement of the Proton Structure Function $F_2$ in Muon-Hydrogen Interactions at 280 and 120 GeV", The European Muon Collaboration (EMC), contribution to this conference.
[45] A. Bodek, et al, PR D20(1979), 1471.
[46] H. Weerts, et al, "Measurement and Analysis of $F_2^\nu(x,Q^2)$ and $xF_3^\nu(x,Q^2)$ in the $Q^2$-region 0.5-40 GeV$^2$, GGM-SPS collaboration, contribution to this conference.
[47] S.M. Heagy, et al, "Nucleon Structure Functions from Measurements of Inelastic Neutrino and Anti-neutrino Interactions", HPWF collaboration, contribution to this conference.
[48] J.G.H. deGroot et al, PL 82B, 456 (1979).
[49] P.R. Norton, et al, "Measurement of the Nucleon Structure Function $F_2$ by Muon Iron Interactions at 280, 250, and 120 GeV, The European Muon Collaboration (EMC), contribution to this conference.
[50] A. Bodek and J. Ritchie, "Fermi Motion Effects in Deep Inelastic Lepton Scattering from Nuclear Targets", U. of Rochester preprint COO-3065-280(1980).
[51] M. Klein, et al, "Deep-Inelastic Muon-Nucleon Scattering at High $Q^2$"; Bologna, CERN, DUBNA, Munich, Saclay collaboration; contribution to this conference.
[52] R.C. Ball, et al, "Further Measurement of Nucleon Structure Functions in High Energy Muon-Iron Interactions", M.S.U.-FNAL, contribution to this conference.

# EXPERIMENTAL RESULTS IN DEEP INELASTIC HADRON INTERACTIONS

J. Lefrancois, France, Speaker

Y. Yamaguchi, Tokyo, Chairman

C. Kourkoumelis, Athens,

D. McCal, Wisconsin-Madison,

T. Weiler, Northeastern,

Scientific Secretaries

EXPERIMENTAL RESULTS IN DEEP INELASTIC HADRON INTERACTIONS
J. Lefrançois
Laboratoire de l'Accélérateur Linéaire, 91405 ORSAY France

The paper presents a summary of experimental results presented at the conference, or obtained in the previous year, in the subject of hadronic lepton pair production, direct photon production, high $p_T$ inclusive hadron production, and studies of jets produced in hadron collisions.

The subject of deep inelastic hadron interaction has evolved in the last few years. While in the preceeding years one was looking for qualitative comparison of experiment with models, we now have detailed quantitative comparison of experiments with QCD theory. I will reflect this evolution by giving more emphasis to Drell Yan physics (and to a lesser degree to direct photon physics) which have reached this quantitative comparison stage than to high $p_T$ hadron production (single particles and jets).

I. HADRONIC LEPTON PAIR PRODUCTION

The process of inclusive lepton pair production (the so called Drell-Yan process[1]) is described schematically in fig. 1. The lepton pair is created by annihilation of a quark antiquark pair in the incident and target particles.

If $X_1$ ($X_2$) are the fraction of the incident (target) particles momentum carried by the quarks one has two kinematical relations, when the masses of the quarks and their transverse momentum are neglected.

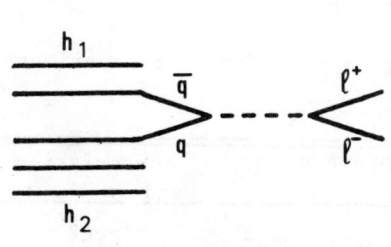

FIG. 1

$$M^2_{\mu\mu} = X_1 X_2 s$$

$$X_{\mu\mu} = \frac{2 P^*_{L\mu\mu}}{\sqrt{s}} = X_1 - X_2$$

s is the total center of mass energy
$X_1$ and $X_2$ can thus be calculated using these two equations
The annihilation cross-section is then given simply by

$$\frac{d^2\sigma}{dX_1X_2} = \frac{4\pi\alpha^2}{3sX_1X_2} \times \frac{1}{3} \times \sum_i \frac{Q_i^2}{X_1X_2} [f_i^{h_1}(X_1) f_i^{h_2}(X_2) + f_{\bar{i}}^{h_1}(X_1) f_{\bar{i}}^{h_2}(X_2)]$$

$\frac{4\pi \alpha^2}{3sX_1X_2} = \sigma_0$ is the point like electromagnetice annihilation cross section at an equivalent energy of $M^2\mu\mu$

$\frac{1}{3}$ is the colour factor

$Q_i$ the quark charge (1/3, 2/3)

$\frac{f_i(X_1)}{X_1}$ the probability for a quark to carry a fraction $X_1$ of the incident momentum in the center of mass system ($\sqrt{s}/2$).

The importance and the usefulness of Drell-Yan experiment comes from the fact than the $f(X)$ are identical to the one measured in deep inelastic scattering (D.I.S.) of electron muon or neutrino. The space like $Q^2$ variable of D.I.S. is identified to the time like $M^2\mu\mu$. The procedure to extract structure function in the Drell-Yan experiments are the following : if the apparatus acceptance is approximatively constant over a large range of $X_{\mu\mu}$ and $M_{\mu\mu}$ one obtains a two dimensional $X_1X_2$ away and it is possible to extract at the same time the structure function of the incident and target particles, $f^{h1}$, $f^{h2}$. When the experiment is more limited in $X_{\mu\mu}$ acceptance one uses D.I.S. results for one structure fonction and the data is used to extract the other structure function.

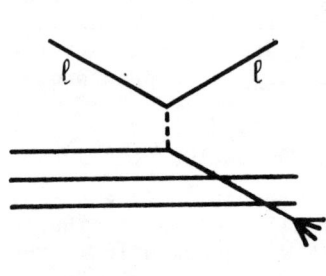

FIG. 2

The experimental result used in this report come from the experiments listed in table I.

TABLE I

| Incident particle | Energy GeV | Accelerator | Apparatus | Abbreviation used in the text |
|---|---|---|---|---|
| p | 200-400 | Fermilab | Two arms spectrometer | CFS[2] |
| p | 400 | " | Iron Toroids | MWNT[3] |
| p | $\sqrt{s}$ = 62 | ISR | Iron Toroids | CHAMNP[4] |
| p | $\sqrt{s}$ = 30,53,62 | ISR | Liquid Argon Cal. | A$^2$BCSY[5] |
| $\pi^\pm(K^\pm\bar{p}p)$ | 40 GeV | SPS | $\Omega$ Spectrometer | BCX[6] |
| $\pi^\pm K^\pm$ | 150 GeV | SPS | Dump+Spectrometer | NA3[7] |
| $\pi^\pm K^\pm \bar{p}p$ | 200 GeV | | | |
| $\pi^-$ | 280 GeV | | | |
| $\pi^\pm$ | 225 GeV | Fermilab | Dump+Spectrometer | CIP[8] |

The Drell-Yan model, and QCD corrections to it, makes some specific predictions which I will review comparing them to the experimental data.

<u>Scaling</u> : if there is no scaling violation in the structure functions one expects $\frac{M^3 d\sigma_{DY}}{dM}$ or $\frac{M^3 d\sigma_{DY}}{dM dX_{||}}$ to be independant of s and to be only a function of the scaling variable $M/\sqrt{s} = \sqrt{\tau}$.

When the gluon correction responsible for scaling violation in D.I.S. are computed in the DY process by QCD, it is found that, in the leading log approximation, the only effect is that the structure functions become $Q^2$ dependant but remain identical to the structure function obtained in D.I.S. : $F_{DY}(X, Q^2) \equiv F_{D.I.S.}(X, Q^2)$. If the $Q^2$ variation are unavailable from D.I.S. (i.e for the $\pi$ structure function) they can be calculated from Altarelli Parisi evolution equation using "reasonable" values of $\Lambda (\simeq 0.5$ GeV).

The predicted effects are then rather small : for example if at $\sqrt{\tau}$ = 0.5 one compares data at 40 GeV incident $\pi$ energy ($\Omega$ data at $M_{\mu\mu}$ = 4.5 GeV) with data at 280 GeV incident $\pi$ energy (NA3 data at $M_{\mu\mu}$ = 12 GeV) the predicted effect is only 25 %.

Comparison between NA3 data at three energies and the $\Omega$ data is shown in fig. 3. In fig. 4 is shown a comparison of the ISR data of CHMNP and A$^2$BCSY and the previous lower energy data of CFS.

In both case it is apparent that scaling is obeyed at a $\simeq$ 20 % level over a large range of incident energy. But statistical and systematic errors prevent a measurement of scaling violation.

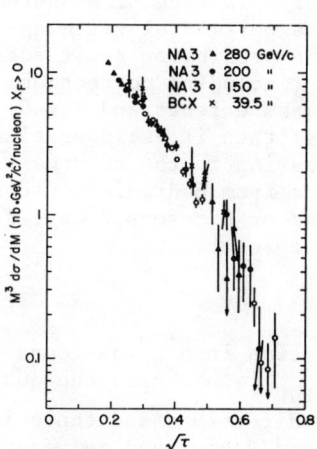

FIG. 3

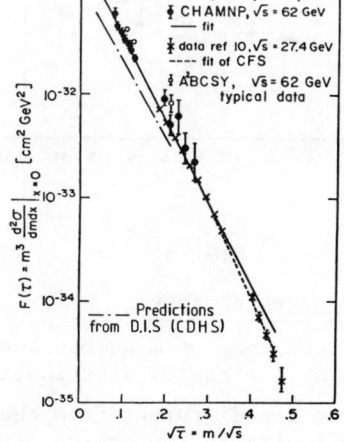

FIG. 4

## A DEPENDANCE

The Drell-Yan mechanism being a quark quark interaction one expects a linear A dependance. In the year 78-79 the situation was the following : CFS data at 400 GeV with incident proton on Baryllium, Copper and Platinum target measured a $A^\alpha$ dependance with $\alpha = 1.007 \pm .018 \pm .028$. CIP data at 225 GeV with incident pions on Carbon, Copper and Tungsten measured $\alpha = 1.12 \pm .05$. The NA3 data at 200 GeV with incident pions on hydrogen and platinum measured $\alpha = 1.03 \pm .03$. At this conference NA3 presented 150 GeV data with incident pions on hydrogen and platinum ; both the systematic and statistical error are reduced compared to their 200 GeV data. They find $\alpha = .994 \pm .015$.

Except for CIP which has a 2.5 $\sigma$ effect all experiments are compatible with $\alpha = 1$ and most experiments on nuclear target are analysed assuming $\alpha = 1$.

## $\pi^+/\pi^-$ RATIO

Because of the charges of the valence antiquark in the $\pi^-$ and $\pi^+$ mesons (- 2/3, + 1/3 respectively). Drell-Yan cross sections with incident $\pi^-$ and $\pi^+$ mesons should be in the ratio one to four, this is only strictly true on a I = 0 target and neglecting sea quarks contributions i.e. at large $X_1$, $X_2$. After the pioneering measurement of CIP more results were obtained by NA3 and more recently by the BCX experiment as shown in fig. 5.
I would like to stress the importance of this type of measurement : if the experimental ratio agrees with the prediction as it does on fig. 5 (including corrections for see quarks effects and I ≠ 0 effects) then it excludes sizeable contribution to the $\mu\mu$ cross sections from hadronic sources as B mesons or D mesons decay, J = 1 resonances ect...

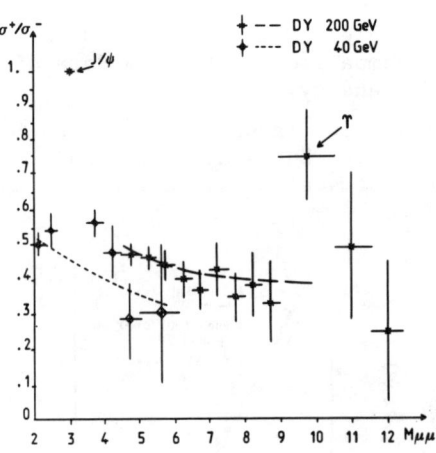

FIG. 5

## ANGULAR DISTRIBUTIONS

In the case of massless quark annihilation in $\mu^+\mu^-$ pair one expects a $(1 + \cos^2\theta)$ distribution. When $P_{T\mu\mu}$ is not zero the quark direction are different from the particles directions and there is some ambiguity in the choice of the reference frame used to measured $\theta$. One can use the incident particle direction (t channel frame or Gottfried Jackson) the nucleon direction (u channel frame) or the external bissectrix of these two directions (Collins Soper frame).

In all cases the general form of the distribution is the following

$$W(\theta,\phi) = W_T (1 + \cos^2\theta) + W_L \sin^2\theta$$

$$+ W_\Delta \sin\theta \cos\theta \cos\phi$$

$$+ W_{\Delta\Delta} \sin^2\theta \cos 2\phi$$

and integrating over the $\phi$ variable $W(\theta) = 1 + \lambda \cos^2\theta$.

Predictions are made with a QCD model of gluon emissions by Collins[9] and Kajantie[10], they predict small values of $W_L$ and $W_{\Delta\Delta}$ proportional to $P^2_{T\mu\mu}$. In the specific case of $\mu$ pair production with an incident pion and at large $X_1$. Berger and Brodsky[11,12] make the following predictions by computing the higher twist contributions.

$$\sigma(\theta,\phi) \simeq (1 - X_1)^2 (1 + \cos^2\theta) + 4/9 \frac{P_T^2}{M^2_{\mu\mu}} \sin^2\theta$$

$$+ (1 - X_1) \frac{4}{3} \frac{P_T}{M} \sin\theta \cos\theta \cos\phi$$

the $\sin^2\theta\, P_T^2/M^2$ term being due to the so called higher twist contributions.

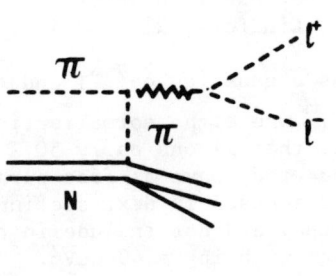

FIG. 6

In a contribution to this conference Sarma[13] calculates that a O.P.E. diagram gives rises to a $\sin^2\theta$ terms (fig. 6), the O.P.E term is only important at large $X_1$ and there is obviously a connection between this computation and the one from Berger and Brodsky.

In a proton proton ISR experiment the average distribution found (A²BCSY) is $1 + (1.15 \pm .34) \cos^2\theta$ (the data is averaged over the full $X_{\shortparallel}$ and M acceptance region).

In the case of pion induced reactions data was presented last year by the C.I.P. group ; the data is consistent with a $1 + \cos^2\theta$ distribution except in the large X region : at $X_1 = .9$ they find $1 - (.2 \pm .3) \cos^2\theta$ in agreement with Berger and Brodsky predictions.

## $P_T$ DISTRIBUTION

The fact that the average $P_T$ of the muon pair was larger than the simple quark model prediction ($\simeq$ 300 MeV) was one of the first effect of gluon emission seen in an experiment. Unluckily there is no simple parametrisation of the data. The critical calculation involves a convolution of the "primordial" $P_T$ distribution with distribution calculated from gluon emission (or gluon absorption) diagrams (fig.7).

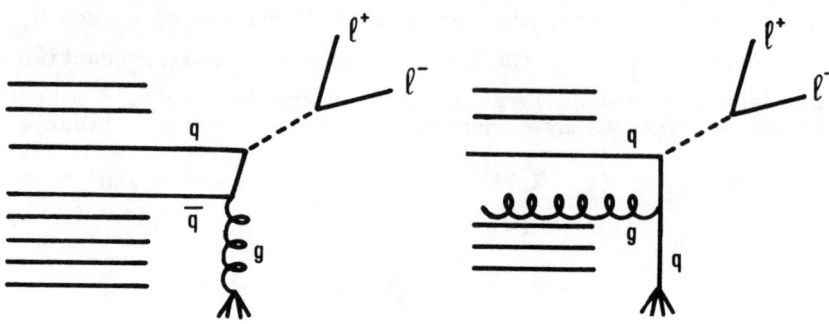

FIG. 7

In their final analysis CFS made a global fit of their data at three energies 200, 300, 400 GeV in bins of mass, rapidities, and $P_T$ using for the $P_T$ distribution the procedure described above (Altarelli Parisi, Petronzio[14], Kajantie, Lindfors, Raitio[10]).

They parametize the primordial $P_T$ as a gaussian $e^{-a\,K_T^2}$ and the gluon structure function $g(x)$ as $B(1-X)^m$ where the normalisation B is imposed from the measurement in D.I.S. that gluons carry 50 % of the nucleon momentum : $\int g(x)dx = .5$. Thus they have as free parameters the sea quarks structure function (discussed in next section) m, a, and the strong coupling constant $\alpha_s$. They did not include in their analysis any scaling violation the average $Q^2$ being = 40 GeV$^2$.

The results are presented in table II, the parameters $\alpha_s$, m and $\langle K_T^2 \rangle$ have quite reasonable value. The authors stressed that the errors given are statistical and do not include systematics or model uncertainties. Nevertheless I believe this to be a remarquable success of QCD.

TABLE II

| | | |
|---|---|---|
| $\alpha_s$ = 0.27 ± 0.1 | m = 4.1 ± .2 | $\langle K_T^2 \rangle$ = .88 ± .02 GeV$^2$ |

The separation of the $P_T$ in a primordial part and a QCD part can be checked more simply by looking at the $\sqrt{S}$ dependance ; one expects

$$<P_T> = a + b\sqrt{s}$$

or $<P_T^2> = a' + b'S$

where a or a' are the primordial part and $b\sqrt{s}$ or b'S is the QCD part.

Using their data and ISR data CFS finds for proton induced lepton pairs

$$<P_T> = .37 + .028\sqrt{s} \quad \text{at} \quad \sqrt{\tau} = .21$$

In the case of $\pi$ induced reaction the NA3 group analysing their data find

$$<P_T> = .49 \pm .08 + (.034 \pm .004)\sqrt{s} \quad \text{at} \quad \sqrt{\tau} = .27$$

these data are shown on fig. 8. The BCX data point falls nicely on the same line.

The $<P_T>$ of pion induced dimuons is larger than the $<P_T>$ of proton induced dimuons but it is not clear if this is linked to the QCD part, the intrincic part, or two both causes. We have to wait for a detailed analysis of the $\pi$ data similar to the one performed by the CFS group for the proton data.

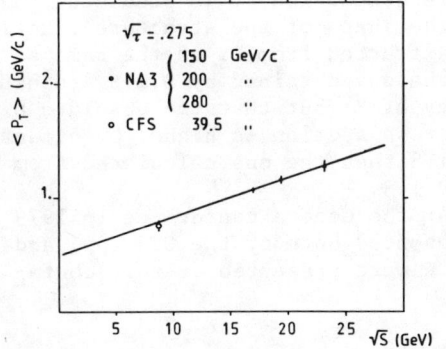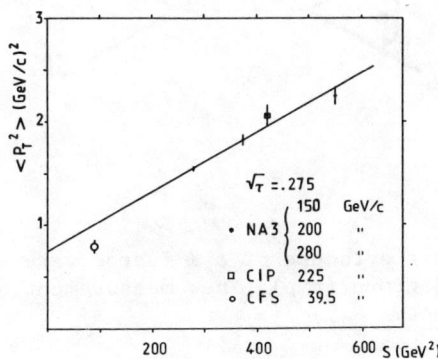

FIG. 8

## SHAPE OF F(X) AND ABSOLUTE CROSS SECTION PREDICTIONS

As explained in the introduction an absolute prediction of the lepton pair production by proton proton or antiproton proton collision can be calculated if the valence and sea quarks structure functions are known from D.I.S. measurements.

It was also proved that in the leading log approximation the QCD effects which are responsible for scaling violation in D.I.S. causes the same scaling violations in DY structure functions. In 1979 a full first order calculation of the QCD calculations was performed[15] (including non leading log terms n.l.l.). It was found that the n.l.l terms give a large correction to $\sigma_{\circ_{DY}}$ calculated with D.I.S. structure function, this correction is nearly independant of the rapidity of the μ pair and of the mass (in the range $M/\sqrt{s}$ = 2., .5 covered by experiments). This correction is about the same in the case of incident proton or incident π. In summary we can write

$$\sigma_{DY, QCD, \text{first order}} = K\, \sigma_{\circ_{DY}}$$

$$K \simeq \text{constant} = 1.8$$

the biggest contribution to this correction is from a vertex correction (fig. 9) which has different value in D.I.S. and DY contributing .6 to the correction.

Parisi conjectures[16] that all higher order terms of this type can be exponentiated $(1 + .6 \to e^{.6} = 1.82)$. If the conjecture is correct this would indicate the size of the higher order corrections.

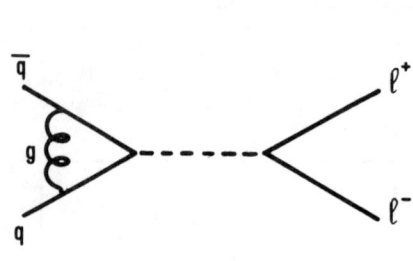

FIG. 9

In conclusion it is predicted that the shape of the structure function extracted from DY is the same as the one obtained by D.I.S. (constancy of K) but that the absolute DY cross section is higher by a factor 1.8 than the one calculated from D.I.S.

At the Geneva conference in 1979 first evidence of a K factor were presented both by the CFS (pp) and NA3 group (π p) ; new measurement on K were presented at this conference.

## ANTIPROTON DATA

The NA3 group has presented results obtained with 150 GeV incident $\bar{p}$ ($6 \times 10^5$ $\bar{p}$/burst identified by Cedars in a beam of $5 \times 10^7$ π/burst) they obtained 275 lepton pairs with masses greater than 4.1 GeV/c² they collected also 35 DY events obtained in similar experimental conditions with the same integrated beam intensity, but with incident protons.

By substracting the cross section ($\sigma_{DY\ \bar{p}N} - \sigma_{DY\ pN}$) the sea quarks or incident gluons contributions are cancelled and one is left with only a product of valence structure functions. Furthermore the $X_1$, $X_2$ values are in a range where there are good measurements of the valence quark structure function $F_3(X)$ obtained in D.I.S. $\nu$ interaction by the CDHS experiment[17].

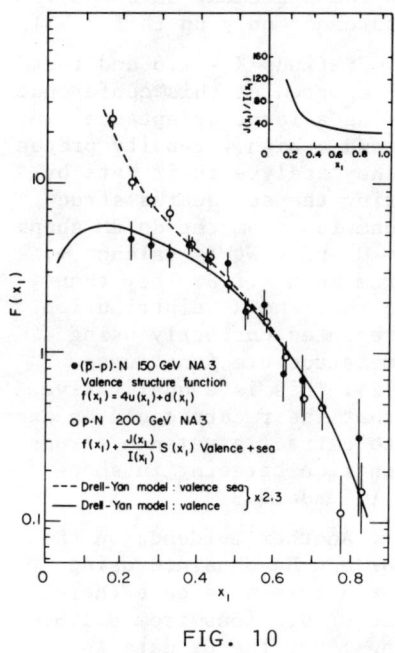

FIG. 10

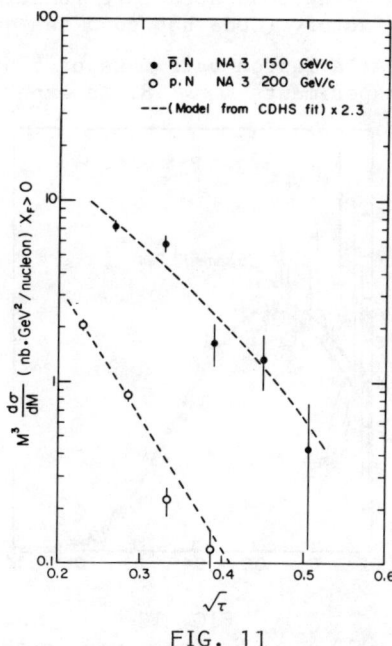

FIG. 11

First it should be noted that the shape of valence structure functions from DY agrees in shape with the one from D.I.S. (fig. 10).

If the u quark structure function at an average $Q^2$ of 20 GeV$^2$ is parametrized as $x^\alpha (1-x)^\beta$ values of $\alpha$ and $\beta$ are given in table III showing in a quantitative way this agreement in shape.

TABLE III

| EXP$^t$ | $\alpha$ | $\beta$ |
|---|---|---|
| CDHS | .51 | 2.8 |
| NA3 $\bar{p}_N - p_N$ | .8±.3 | 3.3±.5 |
|  | .5 fixed | 3.0±.3 |

PROTON DATA

Similarly NA3 analysed also their 200 GeV pN data ; the structure function extracted contains a valence and a sea part ; again the agreement in shape between the data and prediction from CDHS valence and sea structure functions is remarquable (fig. 10). Finally the shape of $d\sigma/dM$ is also reproduced quite well (fig. 11). But in all cases the DY cross section are larger by a factor K = 2.3 ± .4.

On figure 4 is represented data at $X_{||} = 0$ (i.e $X_1 = X_2 = \sqrt{\tau}$) from both ISR experiments (CHAMND, $A^2$BCSY) and CFS. Together with predictions using CDHS structure functions, again the agreement in slope is satisfactory (CDHS has good sea quarks measurement only up to $X \simeq .3$).

At a lepton pair mass of 5 or 6 GeV/c CFS finds $K \simeq 1.6$ and the ISR experiments $K \simeq 1.8$. An experiment was reported at this conference MNWT using a large acceptance toroid and a high intensity proton beam, they analyse their data by extracting the sea quarks structure function from the $d\sigma/dM$ shape at $X_{||} = 0$ and $\nu W_2$ ($\simeq$ valence + sea) from muon D.I.S. They then checked that the $X_F$ distribution is represented correctly using the same structure functions (fig. 12). This is a qualitative check that their data could be analysed to extract a valence structure function agreeing in shape with D.I.S. data.

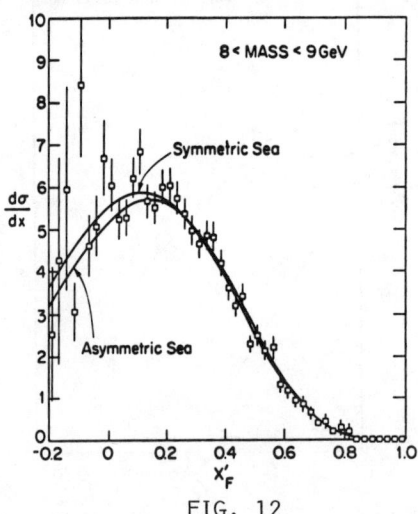

FIG. 12

π DATA : Another evidence on the K factor can be obtained using incident π ; in this case there exist no predictions from D.I.S. on $F_\pi(X_1)$, so the procedure is the following : use the DY data to extract the shape of $F_\pi(X_1)$ and $F_N(X_2)$ ; check the shape of $F_N(X_2)$ against nucleon structure function from D.I.S. (this is a verification of the constancy of the K factor). Normalize $F_\pi(X_1)$ using the sum rule :

$$\frac{F_\pi(X_1) dX_1}{X_1} = 1,$$ the number of $\bar{u}$ valence quark in the pion. Because of this normalization condition one can then obtain an absolute DY prediction, using the normalized $F_\pi(X_1)$, and $F_N(X_2)$ from D.I.S. And the K factor is computed as

$$K = \frac{\sigma_{DY} \text{ Experimental}}{\sigma_{DY} \text{ Predicted}}$$

$F_N(X_2)$ extracted from the π N data of NA3 is shown in fig.14 together with the quark structure function from CDHS ; the agreement in shape over the $X_2$ range .1 to .5 is remarquable.

The importance of the π data is that, using the substracted cross section $\sigma_{DY\pi^-} - \sigma_{DY\pi^+}$, contributions from hadronic background are cancelled. Results from NA3 and BCX are shown in Table IV, the BCX results are shown in fig. 13.

TABLE IV

| EXP$\underline{t}$ | Target | Particle | Energy | K |
|---|---|---|---|---|
| NA3 | $P_t$ | $\pi^-$ | 200 GeV | 2.05 ± .4 |
|  |  | $\pi^+$ | " |  |
| NA3 | $P_t$ | $\pi^-$ | 150 GeV | 2.4 ± .4 |
|  | $H_2$ | $\pi^-$ | 200 GeV | 2.4 ± .4 |
|  |  |  | 150 GeV | 2.4 ± .4 |
| BCX | W | $\pi^-$ | 39.5 | 2.45 |
| " | " | $\pi$ | " | 2.52 |
| " | " | $\pi^- - \pi^+$ | " | 2.22 ± .24 (statistical) |

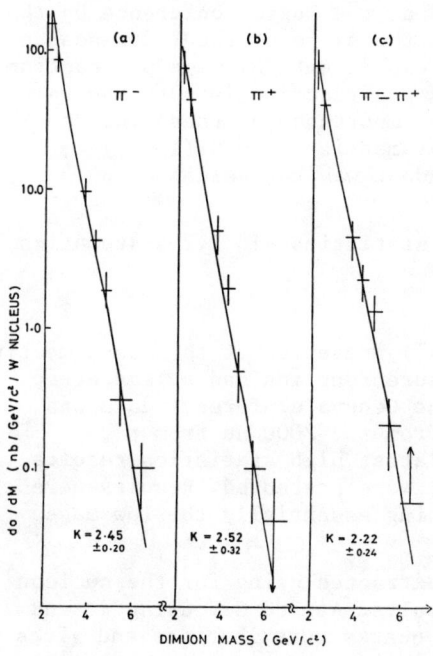

FIG. 13

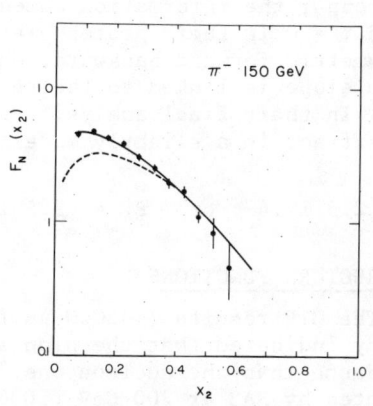

FIG. 14

Nucleon structure function determined from NA3 π N data. The dashed line is the valence nucleon structure function from D.I.S. (CDHS). The continuous line includes both the valence quark and sea quark contributions.

The CIP experiment, when analysed with a linear A dependance, gives a similar value of K.

In conclusion the K factor is constant within experimental errors (fig. 4, 10, 11, 14). It is not a nuclear effect since it is also seen in the π p data (NA3) and in both ISR pp experiments. It is not a "hadronic background effect" since it is seen in the π⁻ - π⁺ data of NA3 and BCX. It is clearly seen in $\bar{p}$ N data which is dominated by nucleon valence quark annihilation.

In conclusion I would say that lepton pair production has become an important testground of QCD, large effects are predicted and quantitatively verified (K factor, large $P_T$ effects). But it is very important that theorist provides us with a reliable estimate of higher order corrections to K. Also other detailed analysis of the $P_T$ behaviour (as done by the CFS group) should be performed (it should be available soon from NA3 and future high intensity pions experiments) and finally other accurate analysis of the angular effects are needed to check the higher twist contribution.

## NEW STRUCTURE FUNCTIONS

Having tested the DY mechanism one can now see how it has been used to measure structure functions unavailable from D.I.S. measurement.

## SU2 ASYMMETRY OF THE NUCLEON SEA

This result was already presented at the Tokyo conference by the CFS group ; the information comes from the slope of the Y dependance around Y = 0 in their proton-platinum experiment. Since a pp reaction is symmetric forward backward, while a p n reaction is not, the value of the slope is linked to the relative importance of these two reactions. In their final analysis they parametrize $S_{\bar{u}} = S_{\bar{d}}(1 - x)^\beta$ ; they extract in a slightly model dependant way two values of β

β = 2.5 ± .4   or   β = 3.5 ± .25 statistics ± 1.2 systematics

## π± STRUCTURE FUNCTIONS

The CIP results (≃ 2000 μμ from π⁻) presented at the Tokyo meeting already indicated that the pion structure function had a less steep dependance than the nucleon one. At the Geneva conference data was presented by NA3 at 200 GeV (5000 μμ from π⁻, 2000 μμ from π⁺, M > 4.1 GeV). At this conference the latest high statistics results from NA3 at 150 GeV (20000 μμ from π⁻) were presented. Results were also given from the BCX experiments using essentially the low mass data (2 GeV < Mμμ < 2.7 GeV).

The pion structure function are extracted using for the nucleon structure function the CDHS D.I.S. results. At 200 GeV using π⁻ and π⁺ data NA3 separates valence and sea quarks contributions and gives an estimate of the sea quarks of the pions.

Data is presented in Table V

### TABLE V

Valence = $x^\alpha (1 - x)^\beta$    Sea = $B(1 - x)^n$

| Experiment | $<Q^2>$ | α | β | B | n |
|---|---|---|---|---|---|
| NA3 200 GeV | ≈ 20 GeV² | .45±.1 | 1.04±1 | .25±.15 | 5.4±2.0 |
| NA3 150 GeV | ≈ 20 GeV² | .40±.1 | .90±.1 | " | " |
| BCX 39.5 GeV | ≈ 5 GeV² | .55±.1 | .9±.1 | " Fixed | " |

$F_\pi(X_1)$ together with the NA3 fit from the 150 GeV NA3 data is shown in fig. 15. Berger and Brodsky[11] have predicted at high X value the form

$$F_\pi(X) = \sqrt{X}\left[(1-X)^2 + 2/9\,\frac{K_T}{M^2_{\mu\mu}}\right]$$

were the second term represents higher twist contributions. NA3 has tried to represent their 150 GeV data with such a form (fig. 16). The continuous line represents the theoritical shape convoluted with experimental resolution, and normalized at X ≳ .9.

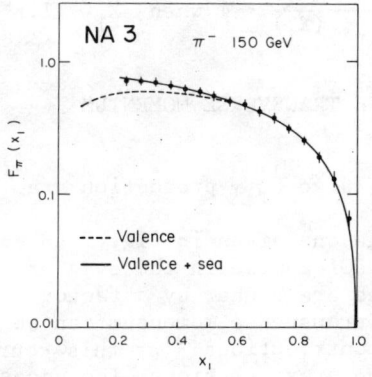

FIG. 15

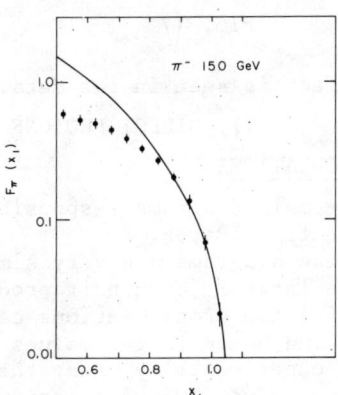

FIG. 16

It is apparent that the $(1 - X)^2$ dependance is too steep to represent the data over a large X interval (the interval of applicability, if one insists on this $(1 - X)^2 + 2/9\, K_T^2/M^2$ shape, is $X \gtrsim .7$).

## $\overline{K}$ STRUCTURE FUNCTION

NA3 has presented data at 150 GeV/c consisting in 700 dimuons events of mass $M_{\mu\mu} > 4.1$ GeV/c from $K^-$ platinum interaction collected simultanously with the 20 000 dimuons events from $\pi^-$ platinum interaction.

It can be shown that to a good approximation one can write

$$\frac{F_{\bar{u}\,K^-}(X_1)}{F_{\bar{u}\,\pi^-}(X_1)} = \frac{L_{\pi^-}}{L_{K^-}} \frac{(dN/dX_1)_{K^-}}{(dN/dX_1)_{\pi^-}}$$

where the $dN/dX_1$ are the experimental distribution of the dimuons events : $L_{\pi^-}$, $L_{K^-}$ are the experimental luminosity (beam × target) ; and the $F\bar{u}(X_1)$ the $\bar{u}$ antiquark structure functions. The ratio of the structure function is independant of detailed acceptance calculations or of the exact form of the nucleon structure function chosen. The data is presented in fig. 17 together with theoretical prediction[18]. On general ground one would expect that, since the mass of the $\bar{u}$ quark is smaller than the mass of the strange quark, it would carry a smaller share of the $K^-$ momentum and hence

$$\frac{F_{\bar{u}\,K^-}(X_1)}{F_{\bar{u}\,\pi^-}(X_1)} < 1 \text{ when } X_1 \to 1.$$

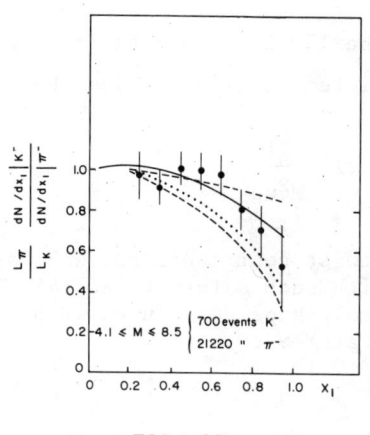

FIG. 17

This effect is seen in the data.

## II. DIRECT PHOTONS AT LARGE TRANSVERSE MOMENTUM

### QCD CALCULATIONS

The main diagrams responsible for large $P_T$ $\gamma$ production are given in fig. 18a, b.

These diagrams are very similar the one given in fig. 7 to calculate the large $P_T$ muon pair production. Of course in the case of $\gamma$ production the cross sections calculated are higher by a factor $\simeq 1/\alpha_{em}$ and hence larger values of the transverse momentum can be reached experimentally. Many theorist contributions[19] at this conference have given QCD calculations of the cross sections. The necessary ingredients are $\alpha_s$ the strong coupling constant, the quark and the gluon structure functions, and an intrinsic $p_T$ of the constituents. Some evaluation of other diagrams (quark scattering followed by Bermsstrallung have also been done).

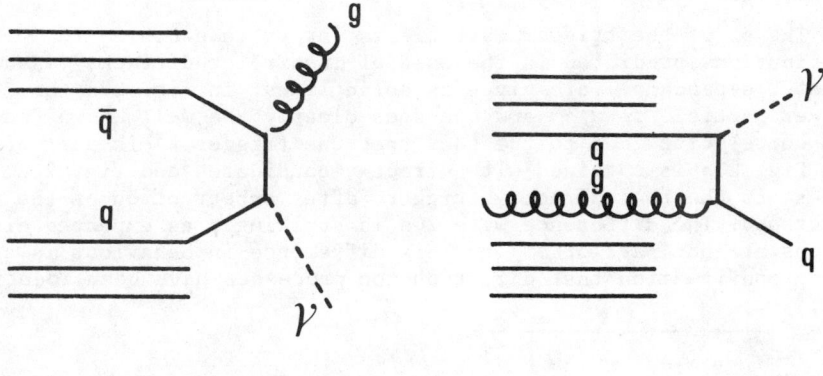

FIG. 18a        FIG. 18b

EXPERIMENTAL RESULTS

The experiments are very difficult because of the large background of photons which are produced indirectly by meson decay : $\pi^° \to \gamma\gamma$, $\eta^° \to \gamma\gamma$, $\omega^° \to \pi^°\gamma$, ect... Early results gave conflicting numbers because of large systematic errors. In 1979 at the Geneva and Fermilab conferences positive evidence were presented from two group $A^2BC$[20] at the ISR and F.-J.H.U[21] at FNAL.

In the F.-J.H.U data the photons are measured by two lead glass block arrays. They misidentify the $\pi^°$ either because one photon is lost or because the two decay photons are too close to be separated, they also have some background from neutron or $K^°$, finally they have to substract by Monte Carlo a background which is about ten times bigger than the direct photon signal. They measure a $\gamma/\pi^°$ ratio varying between .05 and .1 with a quoted systematic error of $\simeq$ .025. Their data are at $p_T$ values between 2 GeV/c and 4 GeV/c the experiment extends over backward angles (90° to 160°). Their data together with a theoritical prediction are shown on fig. 21.

In 1980 $A^2BC$ have published their final result. In their case the photons are detected by lead-liquid argon detectors ; the system has a fine granularity and can measure two showers separated by 5 cm. As a further rejection against merged photons from $\pi^°$, they require that the radius of the electromagnetic showers in the detector, are smaller than 1.35 cm. With this good rejection against background they see a clear effect in their raw data, above a $p_T$ of 5 GeV/c, with a background to signal ratio of one to one or less. The corrected $\gamma/\pi^°$ ratio are given in fig. 19.

The same group has also obtained interesting correlation results : they measure the probability of finding a charged track on the same side as the trigger particle, with a difference of azimuthal angle $\Delta\phi$ such that 20° < $\Delta\phi$ < 70°.

In fig. 20 this probability is plotted as a function of the pseudorapidity difference between the trigger particle and the observed charged particle.

The $p_T$ of the trigger particle is larger than 6.5 GeV/c. Distributions predicted in the case of uncorrelated tracks (assuming a flat Y dependance) are given as solid lines. In fig. 20a the trigger particle is a $\pi°$ and one sees clearly the well known "same side correlation" due to the fact that the trigger $\pi°$ is part of a jet. Fig. 20b is obtained with direct $\gamma$ candidates and fig. 20c represents the real direct $\gamma$ triggers after substraction of the $\pi°$ background. The difference with 20a is striking : as expected direct photons are not part of a jet. This difference in behaviour is a strong confirmation that direct photon processes have been identified.

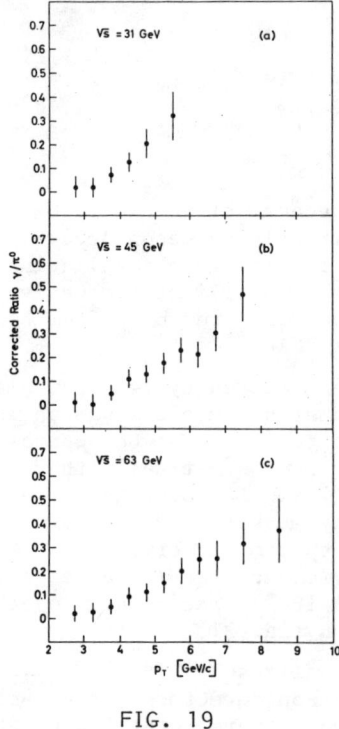

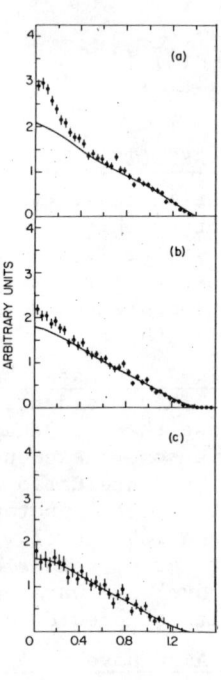

FIG. 19                    FIG. 20

Another ISR group CCOR[22] has presented results at this conference. Their apparatus consists of a thin superconducting solenoid filled with drift chambers. Outside of the solenoid they have two lead glass arrays. Because of the limited granularity of their neutral detector they use a statistical method to measure the direct photon cross section : they measure the neutral cross section with the lead glass array ; they also identify neutral events which have converted in the solenoid (one radiation length of aluminium) the probability of conversion is 53 % for a photon and 77 % for a π° or η°, since in that case either one of the two photons can convert. In the $p_T$ range 6 GeV/c to 10 GeV/c they find a ratio of single photons to all neutrals of .074 ± 0.12 with a systematic error of ± .053. When corrected by the π°/neutral ratio this would correspond to a cross section twice as small as the $A^2BC$ result, nevertheless, because of the large systematic error, no real disagreement is claimed. At $p_T$ larger than 10 GeV/c, CCOR finds γ/all = .26 ± .04 ± .05 in agreement with $A^2BC$.

CCOR has also done correlation measurement which show that the direct photon signal dissapears in events with an extra particle on the trigger side, in agreement with the $A^2BC$ result mentioned above.

They have also correlation measurements on the away side jet. In the case of proton proton production of direct photons, theory predicts a dominance of the Compton diagram (fig. 18b) the scattered quark is then more often a u quark, (because of the quark charge and abondance, the u/d ratio is 8/1) and one expects a strong charge asymmetry in the leading particle of the away side jet. Selecting in their data a trigger neutral with a $p_T$ of 7 GeV/c to 13 GeV/c not accompanied by a same side particle (i.e. direct photon candidates), they find that high $p_T$ away side charged particles ($X_E$ > .3) have a charge asymmetry R+/- = 3.7 ± 1.2 in agreement with the theoretical expectation.

In fig. 21 are presented the direct photon cross section of the F.-J.H.U experiment at Fermilab and the $A^2BC$ and CCOR experiment at

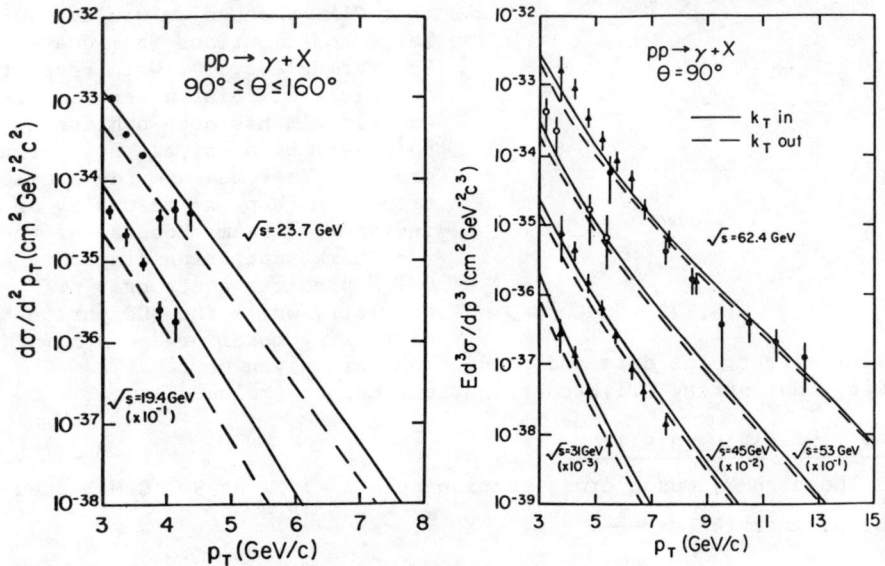

FIG. 21

the ISR together with one of the theoretical prediction (Cormell and Owens)[19] submitted at this conference ; the average primordial parton transverse momentum used is $< K_T^2 > = 1.0$ GeV$^2$.

The agreement is quite good especially when one remembers that the only ad-hoc parameter in the theory is the primordial $K_T$.

Direct photon measurement is certainly a very promising field. We already have some remarquable results but if one succeeds in reducing the systematic errors in the experiments and to understand better how to introduce the intrinsic $p_T$ in the theory, direct photon physics could then have as much success in measuring gluon structure function as Drell-Yan had in measuring quark structure function.

## III. HIGH $P_T$ HADRON PHYSICS

### INCLUSIVE

In the simplest QCD model the inclusive cross section of high $P_T$ particle production are believed to be explained by constituent scattering, quark-quark, quark-gluon, or gluon-gluon, followed by constituent fragmentation. This already complicated situation is made worse by the effect of scaling violations and primordial transverse momentum smearing. Finally there is the possibility that CIM mechanism would give large contributions ($\pi$ - quark scattering ect...). With respect to this last point a very important result has been published this year by a Chicago Princeton group[23]. They measure the production of high $p_T$ $\pi^\pm$ mesons by an incident $\pi^-$ beam. Because of the $\pi$ - quark scattering diagram C.I.M predicts a strong $\pi^-/\pi^+$ asymmetry while the QCD parton scattering model predicts a much smaller effect. The data and predictions are given on fig. 22 showing that strong C.I.M contributions can be excluded.

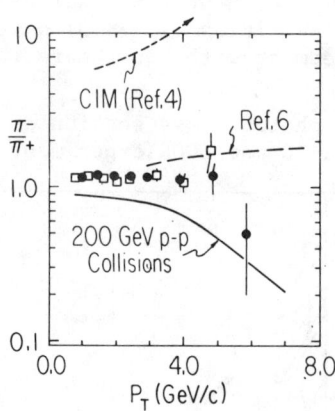

FIG. 22

### $P_T$, $X_T$, $X_P$ PARAMETRISATION

The high $P_T$ meson cross section $pp \rightarrow \pi^\pm + X$ (at 90° C.M.) used to be parametrise as $\dfrac{1}{P_T^n} \times (1 - X_T)^m$.

At $P_T$ values around 4 GeV/c to 8 GeV/c the fits gave $n \simeq 8$. In the last two years high $P_T$ $\pi°$ experiments done at the ISR by the CERN Saclay[24], the CCOR[25] and the $A^2BC$[26] collaborations reached much higher values of $P_T$ and saw a deviation from the usual parametrisation with n values of $\simeq$ 5 or 6 at large $P_T$ (8 to 14 GeV/c). The reference given are the latest publication of each group. The $A^2BC$ collaboration was the only one to positively identify $\pi°$ up to a $P_T$ value of 10 GeV/c and hence the presence of direct photon signal complicates the interpretation for the other experiments or for the $A^2BC$ data above $P_T > 10$ GeV/c.

When experiments were not done at 90° a convenient parametrisation was to replace $X_T = P_T/\sqrt{s}$ by $X_R = \sqrt{X_T^2 + X_{\shortparallel}^2}$ in the form $1/P_T^n \times (1 - X_T)^m$. No theoritical basis existed for this form but it represented the Fermilab and some of the ISR data quite well.

Data was presented at this conference by a Pisa Stony Brook[27] group. The experiment was done at the ISR at $\sqrt{s}$ = 24 GeV to $\sqrt{s}$ = 53GeV and at center of mass angles of 5°, 15°, 20°, and 90°. At the lower energy their data agrees quite well with the results of Carry et al[28] and with radial scaling but a $\sqrt{s}$ = 53 radial scaling predictions are wrong by a large factor. On the other hand the shape of the angular distribution are in good agreement with a Field Feynman prediction.

## JETS CROSS SECTION

There was some hope that the process hadron + hadron → large $P_T$ jets would be easier to compare to theory since one does not need in this case the fragmentation function of the constituents. Early study at FNAL[29] showed that the interpretation of the data depended critically on difficult corrections for the jet acceptance.

Preliminary results of the NA5 experiment (B,K,L, MPI, N collaboration[30]) were presented at this conference.

In this case the calorimeter used to measure the jet energy had a large angular acceptance : it covered Θ between 45° and 135° and had a 2π azimuthal acceptance.

It was divided in cells ΔΘ = 10°, Δϕ = 15°. A streamer chamber was used to measure the multiplicity. A trigger was performed on the total transverse energy $E_T = \Sigma |E_{Ti}|$ were $E_{Ti}$ is the energy measured in a cell i weighted by $\sin \Theta_i$. NA5 used both $\pi^-$ and proton beams of 300 GeV on a hydrogen target. The salient points are the following :

1) at large $E_T$(=10 GeV) the multiplicity is very high $<n_{ch}>$ = 26.5 of which $\simeq$ 12 are in the calorimeter.

2) A planarity analysis is used to try to isolate back to back jets by the azimuthal energy distribution in their calorimeter. They have no convincing evidence of jets in this way.

3) When they use, off line, only part of their calorimeter to simulate the acceptance of the previous FNAL experiment, they find

reasonable agreement within experimental errors.

4) The cross section increases rapidly with the azimuthal aperture of the calorimeter.

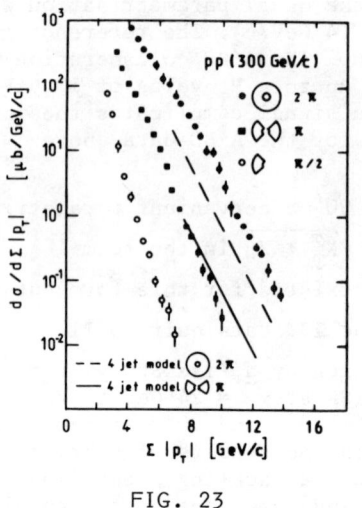

FIG. 23

Their Monte Carlo simulation of pp collisions giving 2 high $P_T$ jets and a forward and backward jet doesn't reproduce this large increase. It account for the two arm trigger but only for one fifth of the full calorimeter trigger Fig. 23.
Either a fraction of the event are simply phase space at high multiplicity (an isotropic explosion !) or, as argued by Selove in a paper to this conference[31], the jets are there in each event but distorted and shifted to higher energy by contamination of many low $P_T$ particles. More analysis is needed but the inescapable conclusion is that jet physics at $E_T$ of 10 or 12 GeV will be difficult. May be ISR or Collider energy will be needed.

<u>JET FRAGMENTATION</u>

Results were presented by the CCOR[32] group at this conference on the study of jet opposite to a high $P_T$ single particle trigger. Their detector was briefly described in the direct photon section. They trigger on large $P_T$ $\pi^\circ$ (more generally neutrals).

They have quite large statistics : 23000 events with $P_T$ from 7 to 9 GeV/c, $\simeq$ 2000 events with $P_T$ from 9 to 11 GeV/c. Charge particles are measured in the away side jet.

They try to divide $P_{out}$ the momentum perpendicular to the trigger plane in two components : $J_{TY}$ is the transverse momentum given, in the fragmentation process, to the trigger particle or to particles of the away side jet and $K_{TY}$ the transverse momentum of the hadrons that enter in the scattering process.

Following the suggestion of Levin et al[33] and Field Feynman and Fox[34] they use the following approximate relation

$$<|P_{out}|>^2 \ = \ 2 \ <|K_{TY}|>^2 \ X_E^2 \ + \ <|J_{TY}|>^2 \ (1 + X_E^2)$$

where $X_E = - \dfrac{\vec{P}_{trach} \cdot \vec{P}_{Trig}}{|P_{Trig}|^2}$

From a fit using only tracks with $P_{track} > 1.4$ GeV/c, they obtained, by extrapolating at $X_E = 0$ a value for $<|J_{TY}|>$, and then a value of $<|K_{TY}|>$. $J_{TY}$ is found to be constant as a function of $\sqrt{s}$ or $P_T$ of the trigger (Fig. 24) as expected by analogy to $e^+e^-$ results on the fragmentation $P_T$. On the other hand $<|K_{TY}|>$ has a rather strong variation with $P_T$ and $\sqrt{s}$ Fig. 25 as expected if $K_{TY}$ has two sources : primordial $K_T$ and gluon emission.

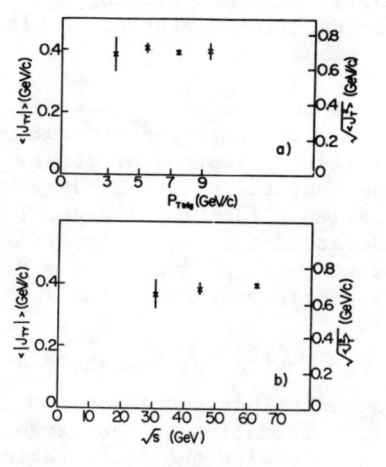

FIG. 24

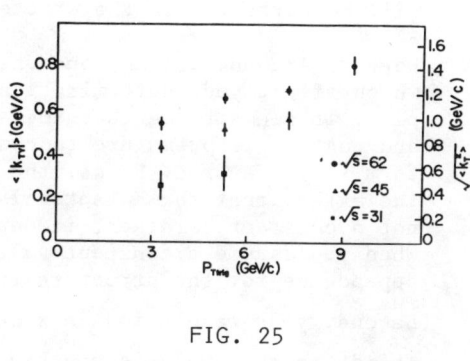

FIG. 25

These qualitative features are quite sugestive but we are still probably a long way from a rigourous and ambiguous free analysis which would allow quantitative comparison of results of this type with $J_T$ and $K_T$ results from $e^+e^-$ experiment and D.I.S. experiments.

## IV. CONCLUSIONS

The subject of deep inelastic hadron interaction as evolved considerably in the past two years this is especially true in the case of Drell-Yan which has seen a progress both in the quality of the data and in the predictive power of QCD theory.

A break trough has been achieved in the case a large $P_T$ direct photon production, it is hoped that in the coming years experiment and theory (higher order corrections) will reach the quality of Drell-Yan Physics.

In the case of high $P_T$ hadron or jets production a lot of excellent data as been accumulated over the past years. Because of the complexity of the dynamics involved the progress of our understanding requires confrontation between these results and input from $e^+e^-$, D.I.S, Drell-Yan, and direct photon experiments.

## DISCUSSION

Q1 : Resvanis, Athens : I would like to make an historical correction. The Athens-Brookhaven-CERN direct photon result was not published in 1980 as you say, but in 1979. In fact, it is the first published paper giving a definitive direct photon signal.

A1 : J. Lefrançois : I am sorry for the misinterpretation. I agree that your first results came in at Geneva 79. What I meant is that you also had new data in 80 while there was nothing new from JHU-FNAL since last year at the Fermilab conference. This will be corrected in the written version.

Q2 : Berger, Argonne : I am concerned about the systematic errors which affect your determination of the structure function of the pion. It seems to me that there are several sources of error, and that it is premature to conclude that the Berger-Brodsky form is disfavored. First, the so-called K factor is large. To the extent that the K factor is understood theoretically, it is not a constant ; rather, it depends strongly on x and tau = $M^2/s$. When you assume a constant value, you bias your extracted x dependence for the structure function.

Second, your resolution in x seems poor, with one bin above x = 1.

Third, in the analysis you have assumed values and forms for the nucleon structure function. However, information on the large x region for the pion is correlated tightly with the low x region of the nucleon structure function, since $x_F = x_\pi - x_N$ and $M^2 = s x_\pi x_N$. Thus, uncertainty about low $x_N$ affects your structure function at large $x_\pi$.

Fourth, the manner the analysis was done ties together the x and $M^2$ dependences. Thus, you would seem to be biased against seeing scaling violations.

I would like to know in detail your estimate of the size and $x_\pi$ dependence of these and other systematic errors.

A2 : I would like first to make clear that what I said in the talk was not that we exclude scaling violation terms like $K_T^2/Q^2$ but that a $\sqrt{X_1} \left[ (1 - X_1)^2 + \frac{2}{q} \frac{\langle K_T^2 \rangle}{Q^2} \right]$ form varies too rapidly with X, to agree with the data between $X_1$ = .5 and $X_1$ = .75. (In this range $\langle K_T^2 \rangle/Q^2$ is not dominant) it can be made to agree with the data only at $X_1$ > .7.
Now on the more technical questions.

First the K factor is not expected to depend strongly on X an τ as you state, it is expected (see ref. 15) to remain constant at a 10 %, 20 % level in our range on τ and X. I gave some mumbers (Part I paragraph absolute cross section, F(x) shape) to check this but the ultimate verification would need 10 000 DY events in pp collisions... far in the future.

Second, the resolution in X as $X \to 1$ is about 6 % it was folded in the theoritical curve in the comparison of fig. 16.

Third, we used <u>measured</u> forms of the nucleon structure function. Finally I insist that these points are anyhow not very relevant to the main effect : we find experimentally $\frac{F(.5)}{F(.75)} = 1.58$ the quadratic + $K_T^2/Q^2$ form predicts for this ratio $\simeq 2.72$ the linear fit predicts : $\simeq 1.63$.

Q3 : Besch, CERN : The differences between the results on single photons between R806 and R108 are partially due to the fact that R806 gives $\gamma/\pi$ while R108 gives $\gamma/\pi + z + \ldots$ This decreases the difference by a factor of $\sim 1.5$.

A3 : J. Lefrançois : I may be wrong but I believe that after you correct for that there is still a factor of 2 difference at low $P_T$.

Besch : Yes, I agree, in the very low $P_T$ data this is true.

J. Lefrançois : Around 6 or 7 GeV.

Q4 : Rosenberg, Iowa State : In fitting to the Berger-Brodsky prediction did you use $<K_T^2>$ or $K_T^2$ for each event ? (If mean $<K_T^2>$ was this taken as a constant or was input used ?)

What angular distribution was assumed when doing the Berger-Brodsky fit, since the high-x acceptance will change the shape of the pion structure function and thus the quality of the fit ?

A4 : J. Lefrançois : We used a $<K_T^2>$. We assume the $1+\cos^2\theta$ in the Collins-Soper frame. But again I would like to stress that in the region concerned by our statement ($.5 < X_1 < .75$) the ($1+\cos^2\theta$) term is dominant in the Berger-Brodsky form.

Q5 : Salvini, Rome : Measurement of direct production of single (jet independent) photons is very important. It will perhaps be even more important in future collisions at higher momentum and energy. It will be convenient to develop new detectors in order to identify single high energy photons without ambiguity. My questions are : Can we feel certain that isolated photons of high energy have been observed ? And how good do you consider the agreement with present theories ?

A5 : J. Lefrançois : I believe the agreement between theory and data is satisfactory. As to whether isolated photons have been seen is for people to judge. From my viewpoint, the fact that the $A^2BC$ data needs very little correction because it stands so far above background is quite convincing. The fact that you have correlation with other particles when you have a $\pi^\circ$ trigger and not when you have a single $\gamma$ trigger, is also very convincing.

Q6 : Garbincius, Fermilab : Two other phenomenological analyses give good agreement with the Drell-Yan measured pion structure functions. These are the recombination model applied to low $P_T$ meson fragmentation and a hard scattering model followed by fragmentation. The results for kaon structure functions don't agree quite as well, however.

A6 : J. Lefrançois : I agree with that. I did not have time to talk about all this.

Q7 : Yamadagni, Stockholm : In the NA5 experiment the $E_T > 10$ GeV is shared by $\sim$ 10-15 particles and not by 25 particles. In this case the average $P_T$ is 0.6-1.0 GeV. Is there any understanding why this high $P_T$ process has large cross section and no jets ?

A7 : J. Lefrançois : The only thing I can say is that it is obvious that the trigger is biasing in favor of having this high multiplicity. And as far as I know, there is no theoretical understanding of this cross section.

Q8 : Fields, Argonne : A comment on the decrease of n from 8 to $\sim$ 6 for $\pi°$ inclusive production at large $P_T$ : I believe that this decrease is based upon data points which are likely to include a substantial contribution from single photon production as well as from $\pi°$ production. So presumably the behavior of n should be redetermined at large $P_T$ by subtracting the measured contribution of single photons from the published cross sections. Is that correct ?

A8 : J. Lefrançois : This is not quite right, may be I should have gone into more detail. The final analysis of the $A^2BC$ group is done on recognized $\pi°$ and it is true that they have slightly larger statistical error, but they also confirm the fact that n decreases at large $P_T$. Correct ?

A8 : Resvanis, Athens : I would like to continue on your answer to Mr. Field's question. We identify and separate $\pi°$'s and $\gamma$'s up to 9.5 GeV/c in $P_T$. For higher values of $P_T$ we do not separate $\pi°$'s and $\gamma$'s.

J. Lefrançois : Added in the written version :
The statement on the variation of n in the written version of the talk takes these remarks into account.

Q9 : Wosiek, Jagellonian : Have people looked at the scaling violations in terms of the W-variable rather than $Q^2$ ?

In leading log approximation any large dimensional variable can be used. The problem is to choose one which minimizes higher order corrections. It follows from diagrammatic calculations that using variables controlling the phase space of undetected particles accounts for a part of the corrections.

## REFERENCES

1. S.D. Drell and T.M. Yan, Phys. Rev. Lett. $\underline{25}$ (1970) 316.
2. A.S. Ito et al., Fermilab Pub 80/19 EXP submitted to Phys. Rev. D.
3. Gustafson et al., paper 0823 A submitted to this conference.
4. D. Antreasyan et al., Phys. Rev. Lett. $\underline{45}$ (1980) 863.
5. Kourkoumelis et al., Phys. Lett. $\underline{91B}$ (1980) 475.
6. M.J. Corden et al., paper submitted to this conference and Cern preprint EP 80-152.
7. J. Badier et al., papers submitted to this conference, EP preprint 80-150, 80-148, 80-147.
   J. Badier et al., Phys. Lett. $\underline{89B}$ (1979) 145.
   J. Badier et al., Phys. Lett. $\underline{93B}$ (1980) 354.
8. K.J. Anderson et al., Phys. Rev. Lett. $\underline{42}$ (1979) 944.
   G.E. Hogan et al., Phys. Rev. Lett. $\underline{42}$ (1979) 948.
   C.B. Newman et al., Phys. Rev. Lett. $\underline{42}$ (1979) 951.
   K.J. Anderson et al., Phys. Lett. $\underline{43}$ (1979) 1219.
9. Collins, Phys. Rev. Lett. $\underline{42}$ (1979) 291.
10. K. Kajantie, J. Lindfors and R. Raitio, Phys. Lett. $\underline{74B}$ (1978) 384, and Nucl. Phys. $\underline{B144}$ (1978) 422.
11. E.L. Berger and S.J. Brodsky, Phys. Rev. Lett. $\underline{42}$ (1979) 940.
12. E.L. Berger, Invited talk at the 1979 EPS High Energy Physics Conference, Geneva, July 1979, and SLAC-PUB-2362.
13. Sarma, paper 0527 submitted to this conference.
14. G. Altarelli, G. Parisi, R. Petronzio, Phys. Lett. $\underline{76B}$ (1978) 351, Phys. Lett. $\underline{76B}$ (1978) 356.
15. J. Kubar-André and F.E. Paige, Phys. Rev. $\underline{D19}$ (1979) 221.
    G. Altarelli, R.K. Ellis and G. Martinelli, Nucl. Phys. $\underline{B157}$ (1979) 461.
    J. Abad and B. Humpert, Phys. Lett. 80B (1979) 286.
    B. Humpert and W.L. Van Neerven, Phys. Lett. $\underline{84B}$ (1979) 327 ; $\underline{85B}$ (1979) 293 ; $\underline{89B}$ (1979) 69.
    K. Harada, T. Kaneko and N. Sakai, Nucl. Phys. $\underline{B155}$ (1979) 169.
16. G. Parisi, Phys. Lett. $\underline{90B}$ (1980) 295.
17. A. Para, Proc. of the Lepton Photon Symposium Fermilab 1979 p. 343.
18. P. V. Chliapnikov et al., Nucl. Phys. $\underline{148B}$ (1979) 400.
    A. El Hassouni and O. Napoly, Dual model for parton densities ; DPhT. 11/80 (CEN-Saclay Preprint). The ratio $\bar{u}_K/\bar{u}_\pi$ is computed using $\alpha^\phi$ = 0.17 instead of 0. (O. Napoly private communication).
    F. Martin, (report on the work with A. De Rujula and P. Sorba) ; Paper presented at the XV Rencontre de Moriond, les Arcs, 1980 ; Cern Preprint TH 2845.
19. R. Baier, J. Engels and B. Peterson, paper 302, University of Bielefeld Germany BI-TP 80/07.
    A.P. Contogouris, S. Papadopoulos and C. Papavassiliou, paper 400 Mc Gill University Montreal Canada.
    L. Cormell and J.F. Owens, paper 245, University of Pensylvania and Florida state university Tallahassee, FSU-HEP-800307.
    F. Halzen, M. Dechantsreiter and D.M. Scott, paper 135, University of Madison Wisconsin preprint COO 881-140.
    K. Kato and H. Yamamoto, paper 469, University of Tokio, preprint UT 335.

20. Faljan, Proc. of the Geneva Conference, P. 742, Phys. Lett. 91B (1980) 296.
    C. Kourkoumelis et al., Cern Preprint EP 80-136.
21. B. Cox et al., Proc. of the Lepton Photon Symposium p. 602.
22. A.L.S. Angelis et al., paper submitted at this conference and Cern Preprint.
23. H.J. Frisch et al., Phys. Rev. Lett. 44 (1980) 511.
24. A.G. Clark et al., Phys. Lett. 74B (1978) 267.
25. A.L.S. Angelis et al., Phys. Lett. 79B (1978) 505.
26. C. Kourkoumelis et al., Cern Preprint EP/80-07, submitted to Zeitschrift für Physik C.
27. D. Lloyd Owen et al., paper submitted to this conference n° 589.
28. D. C. Carey et al., Phys. Rev. D14 (1976) 1196.
29. Bromberg et al., Nucl. Phys. B134 (1978) 189.
    Selove XIV Rencontre de Moriond, Quarks gluons and jets, p. 401.
30. C. Favuzzi et al., paper submitted to this conference.
31. W. Selove, paper submitted to this conference n° 663, and university of Pensylvania Preprint UPR-77E.
32. A.L.S. Angelis et al., paper submitted to this conference.
33. E.M. Levin et al., Soviet Journal JETP 69 (1975) 1537.
34. R.P. Feynman, R.D. Field and G.C. Fox, Nucl. Phys. B128 (1977) 1.

# PERTURBATIVE QCD

C.H. Llewellyn Smith, Oxford, Speaker

S. Gasiorowicz, Minnesota, Chairman

T. Gottschalk, Argonne,

M. Chase, Oxford,

W. Celmaster, Argonne,

Scientific Secretaries

PERTURBATIVE QCD

C.H. Llewellyn Smith
Dept. of Theoretical Physics, 1 Keble Rd, Oxford OX1 3NP, England.

ABSTRACT

Recent developments in perturbative QCD are reviewed.

## 1. INTRODUCTION

In the period preceding the 1978 Tokyo conference it was shown that QCD provides a justification for the parton model as a zeroth order description of most inclusive or semi-inclusive hard scattering processes [1] e.g. $pp \to \mu^+\mu^- x$, $pp \to$ large $p_T$ jets, $e^+e^- \to \pi x$ as well as deep inelastic lepton scattering and $\sigma_{e^+e^-}^{tot}$. Furthermore, QCD predicts calculable corrections to the "naive" parton model and also the existence of quite new phenomena e.g. three jet events in $e^+e^-$ annihilation and "hard" photon processes with specific characteristics.

In the last two years these new phenomena have been seen and many other tests of QCD have been made [2]. On the theoretical side, a tremendous amount of work has been done. I shall focus on three major topics in this talk:

1) the "classical" applications of QCD perturbation theory. Higher order corrections have now been calculated for many processes. I shall review these calculations and address the question of whether QCD perturbation theory is likely to converge at available energies. Next I will make some brief remarks on higher twist effects. Logically, I should then turn to phenomenology and discuss whether the predictions fit the data but, luckily, my instructions from the organisers excuse me from this difficult task.

2) extensions of the classical applications to "infrared sensitive quantities" and attempts to predict the complete structure of events. Examples of infrared sensitive quantities are the parton multiplicity, which counts the number of soft gluons, and the Drell-Yan cross-section for $Q_T^2 \ll Q^2$, a restriction which inhibits hard gluon radiation with the result that the cross-section suffers radiation damping.

3) applications of QCD perturbation theory to exclusive processes (e.g. form factors and elastic scattering of hadrons at large momentum transfer) and attempts to justify these applications.

The fact that $\alpha_s(E) \equiv \frac{g^2(E)}{4\pi}$ decreases like $(\ell n E)^{-1}$ in QCD is not sufficient to justify the use of perturbation theory at large energies since generally terms like $(\alpha_s(E) \ell n E/m)^n$ are found. What is needed is that

<u>either</u> no $\ell n E/m$ terms occur, as happens (e.g.) in $\sigma_{e^+e^-}^{tot}$
<u>or</u> that the $\ell n E/m$'s can be summed up and the m dependence factored

out, as in the case of deep inelastic structure functions whose $Q^2$ dependence can be predicted using perturbation theory. In either case (assuming the confining forces are merciful) we are left with a perturbation expansion in $\alpha_s$ with ("higher twist") corrections of order $m^2/Q^2$. If only one large variable is involved, the perturbation theory should be well behaved at large energies where $\alpha_s$ is sufficiently small. However, if several large variables are involved whose ratio becomes large, further "large logarithms" will be encountered, e.g. in deep inelastic scattering for $Q^2 \gg W^2$ or $x \approx 1$ powers of $\ln(1-x)$ occur in the expansion in $\alpha_s$ and in Drell-Yan for $Q^2 \gg Q_T^2$ there are powers of $\ln Q^2/Q_T^2$. When these logs are sufficiently large, perturbation theory will be useless unless the large terms can be summed up. This will be discussed in part III. First, however, I shall discuss the classic predictions for processes involving one large variable.

## 2. CLASSIC APPLICATIONS

a) <u>Convergence of the Perturbation Theory</u>. The processes which have now been calculated beyond leading order are catalogued in the accompanying table. In some cases (e.g. $\sigma_{e^+e^-}^{tot}$) the

| Process | Convergence of Pert. Series: | | Refs. |
|---|---|---|---|
| | 1st thoughts: | 2nd thoughts: | |
| $e^+e^- \to x$ | ✓ | ✓ | 3 |
| Nucleon structure functions | ? | ✓ | 4 |
| Photon structure functions | ? | ✓ | 5 |
| Para $Q\bar{Q}$ decay | ? | ✓ | 6 |
| P wave $Q\bar{Q}$ decay | ✓ | ✓ | 7 |
| Higgs decay | ✓ | ✓ | 8 |
| Fragmentation fnc. evolution | ✓ | ✓ | 9 |
| Drell Yan | ✓ | ✓ if... | 10 |
| $eN \to e\pi x$ | ✓ | ✓ | 11 |
| Large $p_T$ | ? | ? | 12 |
| $\nu q \to \mu Q$ | ✓ | ✓ | 13 |
| $e^+e^- \to 3$ jets | ? | ✓ if... | 14 |
| $qq \to$ large $p_T \gamma$ | ✓ | ✓ | 15 |
| $\gamma\gamma \to$ large $p_T$ jets | ✓ | ✓ | 16 |

coefficients in the perturbation expansion are small and there is general agreement that the expansion is under control. In many other cases, however, doubts have been cast on the convergence of the expansion somewhere in the literature. Before considering individual cases, it is necessary to discuss the question of convention dependence[17].

If an observable quantity O(M) which involves a large mass or energy M is given by

$$O(M) = A(\alpha_s(\chi M))^P [1 + R_1(\chi) \frac{\alpha_s(\chi M)}{\pi} + R_2(\chi) \frac{(\alpha_s(\chi M))^2}{\pi^2} + \ldots ,$$

the question we want to ask is: are the coefficients $R_i$ big? Unfortunately this question is doubly ambiguous because

1) the $R_i$ depend on how $\alpha_s$ is defined. In QED it is convenient to define $\alpha$ in the standard way as the coupling of a real photon to a real electron because low energy theorems allow certain cross-sections to be calculated exactly in terms of this quantity. Nevertheless, we could have defined a "coupling constant" $\alpha'$ in terms of the electron photon vertex evaluated at some off-shell point or, e.g., so that

$$(g-2)_\mu \equiv \frac{\alpha'}{\pi}.$$

Predictions in terms of the standard $\alpha$ and some other $\alpha'$ can easily be related using

$$\frac{\alpha'}{\pi} = \frac{\alpha}{\pi}(1 + \kappa \frac{\alpha}{\pi} + \ldots$$

(where, e.g., $\kappa = 1.5316$ if $\alpha'$ is defined in terms of $(g-2)_\mu$). Up to a given order these predictions will differ slightly - but the difference is of the next order and will be small numerically because $\alpha$ is small. In QCD, however, there is no natural definition of $\alpha_s$ and the difference between using different definitions is significant at energies now availabe because $\alpha_s$ is not particularly small.

2) The prediction for $O(M)$ does not depend on the energy $\chi M$ at which $\alpha_s$ is evaluated if the series is summed to all orders but the truncated series is extremely sensitive to $\chi$ e.g.

$$R_1(\chi) = \text{Const} + \frac{p\beta_0}{2} \ln \chi$$

so that for $p=2$ and four flavours

$$R_1(1/2) = R_1(1) - 5.8$$

Therefore if someone using $\chi=1$ finds that R is of order 6 and feels it is embarrassingly large - they can easily reduce it by cooking up arguments that $\chi=1/2$ would be a better choice!

Faced with this situation we must define $\alpha_S$ and choose $\chi$ in a way which seems likely a priori to minimise radiative corrections to all orders. If with such a choice $R_1$ is large, it seems unlikely that perturbation theory will be useful until very high energies (redefining $\alpha_s$ to reduce $R_1$ will simply cast the large coefficients into higher orders - which are likely to be large already). If $R_1$ is small, however, there is hope.

The choice of criteria likely to reduce radiation corrections is obviously highly subjective and non-unique. However, if plausible criteria lead to different conclusions this signals that perturbation theory is unlikely to be trustworthy. Two possibilities are:

1) "Momentum Subtraction" in which $\alpha_s(M)$ is defined in terms of a vertex function evaluated with space-like external legs having $p_i^2 = -M^2$. A large class of higher order corrections are then absorbed in $\alpha_s$ for external momenta Q of order M. However, there are still ambiguities: $\alpha_s$ depends on the vertex function used to define it (e.g. on whether the $q\bar{q}g$ vertex or the 3g vertex is used) and on the gauge[18]. Nevertheless, whichever vertex is used this should be a reasonably convergent scheme in a "reasonable" gauge (e.g. Landau gauge or Feynman gauge).

2) We can eliminate $\alpha_s$ and relate data to data, which is ultimately what is done in any case (the relation is, of course, scheme independent but it obviously depends on the energies at which the various pieces of data are measured i.e. we must choose a ratio of scales $\chi$ on which the convergence of perturbation theory will depend). For example, we could define

$$R_{e^+e^-} \equiv \Sigma_i Q_i^2 (1 + \alpha_s^{e^+e^-}/\pi) + O(m^2/Q^2).$$

Expansions in terms of $\alpha_s^{e^+e^-}$ will be independent of the scheme and gauge used to perform intermediate calculations and will relate all other observables to $R_{e^+e^-}$.

In the same spirit, the scale $\chi M$ should be chosen to be typical of the vertices etc. involved in the unknown higher order corrections which we want to minimize, in so far as there is a "typical" scale. An alternative (which leads to similar conclusions) is to minimize $R_2$ with respect to $\chi$, which is possible although $R_2$ is not known because its $\chi$ dependence is determined by $R_1$ and the first two coefficients in the $\beta$ function (this is a variant of a recent ingenious proposal by Stevenson for finding the optimal perturbation expansion [17]).

We can now consider some of the processes listed in table 1 (discussion of the others can be found in the references cited):

<u>Decays of Heavy Quark Bound States</u>. The ratios of hadronic to leptonic/photon widths can be calculated perturbatively for large $M_Q$. For para-quarkonium, for example,

$$\frac{\Gamma(Q\bar{Q}\to\text{hadrons})}{\Gamma(Q\bar{Q}\to\gamma\gamma)} = \frac{2}{9e_Q^4} \frac{\alpha_s(\chi M)^2}{\alpha^2}[1+R(\chi)\frac{\alpha_s(\chi M)}{\pi} + ..]$$

Barbieri et al.[6] calculated the correction R in the $\overline{MS}$ scheme (in which $\alpha_s$ is defined by a "modified minimal subtraction" prescription for the counter terms in the Lagrangian) and found

$$R^{\overline{MS}}(2) = 14.$$

However, it is clear that the scale involved in corrections to two gluon decays such as

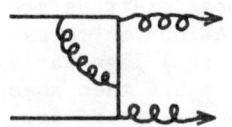
or multigluon decays is actually $M_Q$ or less [19]. Changing the scale reduces the correction factor to

$$R^{\overline{MS}}(1) = 8.2$$

There is no reason to expect radiative corrections to be particularly small in the $\overline{MS}$ scheme. In the more physical momentum subtraction scheme

$$R^{Mom}(1) = 1.8$$

while with $\alpha_s$ defined through the corrections to $R_{e^+e^-}$ (as discussed above)

$$R^{e^+e^-}(1) = 5.0.$$

Alternatively (following essentially the original discussion of Barbieri et al.) $\alpha_s$ could be defined so that the higher order corrections to the $n^{th}$ moment of $F_3$ vanish. Choosing n=6, for example,

$$R^6(1) = 2.4$$

We see that in "physical" schemes with a reasonable scale, the corrections are not outrageously large. Proceeding to minimize the next order corrections with respect to $\chi$, we find, e.g., in the $\alpha_s^{e^+e^-}$ scheme, that $\chi \approx 0.5$ - a reasonable result, since higher orders involving four gluon decays and corrections to three gluon decays would certainly seem to involve a scale less than $M_Q$.

I conclude that radiative corrections are under control in this case and perturbation theory should probably be useful for $b\bar{b}$ states. This conclusion is strengthened by the results of recent calculations for other states by Barbieri et al. [7]. Defining

$$B(J^{PC}) = \frac{\Gamma(J^{PC} \to \gamma\gamma)}{\Gamma(J^{PC} \to hadrons)}$$

they find

$$\frac{B(0^{-+})}{B(0^{++})} = 1 + 0.9 \frac{\alpha_s}{\pi}$$

$$\frac{B(2^{++})}{B(0^{++})} = 1 + 6.5 \frac{\alpha_s}{\pi}$$

$$\frac{\Gamma(0^{++} \to \gamma\gamma)}{\Gamma(2^{++} \to \gamma\gamma)} = \frac{15}{4} (1 + 5.5 \frac{\alpha_s}{\pi})$$

$$\frac{\Gamma(1^{--} \to e^+e^-)}{\Gamma(0^{-+} \to \gamma\gamma)} = \frac{3}{4} (1 - 1.96 \frac{\alpha_s}{\pi})$$

(the numbers here are for $c\bar{c}$ states - they change slightly for heavier quarks). Again none of the corrections involve very large coefficients and perturbation theory may be useful for $b\bar{b}$ states.

The Photon Structure Function. First measurements of the photon structure function were reported at this conference [20]. It is an object of great theoretical interest because it contains a component from diagrams of the type

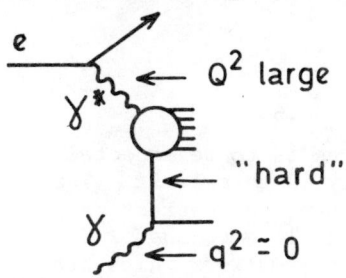

which is exactly calculable [21]. Here the quark coupled to the real "target" photon is very virtual — or "hard". The calculable contribution due to these diagrams increases with x and grows like $\ln Q^2$.

The remaining contribution from

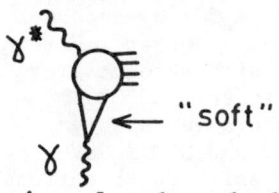

is hadron like (it might be given by the vector dominance model); it will fall with x and shrink to x=0 as $Q^2 \to \infty$.

Predictions for the calculable component are shown below. At first sight, the effect of higher order corrections might seem to be large. However, this conclusion is premature for several reasons:
1) for $0.2 \lesssim x \lesssim 0.8$ and $Q^2=20$ GeV$^2$ the corrections are actually quite small. As discussed below, large corrections are to be expected for $x \to 0$ and $x \to 1$ and we should not expect the lowest order calculation to work very well for $Q^2=3$ GeV$^2$.
2) the higher orders were included using $\alpha_s^{\overline{MS}}$. With a more physical definition they are reduced [23]. Defining

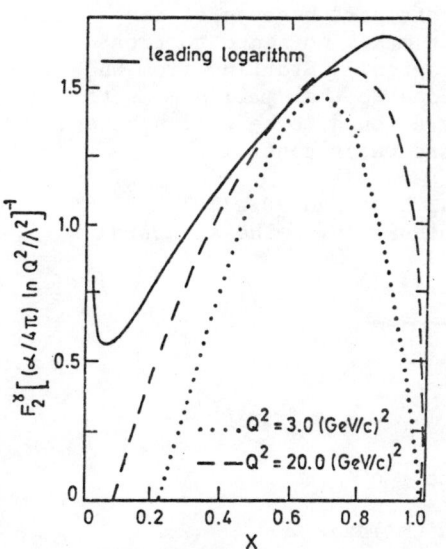

Fig.1. Predictions for the photon structure function [22].

$$\int F_2^\gamma(x,Q^2) x^{n-2} dx = \frac{A_n}{\alpha_s(Q^2)} \left(1 + R_n \frac{\alpha_s}{\pi} + \ldots\right)$$

it turns out that

| Scheme: | $\overline{MS}$ | Mom | $\alpha_s^{e^+e^-}$ |
|---|---|---|---|
| $R_4$ | -4.3 | -1.0 | -2.8 |
| $R_8$ | -5.1 | -1.85 | -3.6 |

3) The corrections become big as $n \to \infty$, $x \to 1$ - as is to be expected since $\ln(1-x)$ terms spoil perturbation theory in this limit. These corrections have a striking effect on $F_2^\gamma$ since, in contrast to hadron structure functions, it is large for x close to one. They arise partly from correcting kinematical approximations which are made in calculating the leading order, to which $F_2^\gamma$ is particularly sensitive because the real photon-quark vertex is not damped by $\alpha_s$. A large part of the corrections can be accounted for by an exact treatment of the kinematics at the lower vertex. Other important $\ln(1-x)$ terms can also be summed up in a way which accounts for them to all orders (see the references cited in section 3). Having done this, it should be possible to make a reasonably reliable absolute calculation of $F_2^\gamma$ for $x > 0.2$ and $Q^2 > 5$ GeV$^2$.
4) For $x \to 0$, the separation of the "hard" and "soft" contributions to the photon structure function is not unique beyond leading order (the 2nd order correction to the n=2 moment of the hard piece scales, as does the n=2 moment of the soft part because of the energy-momentum sum rule). In fact, it makes no sense to treat the former without the latter in this region. Starting from an input distribution at $Q_0^2$, the corrections to the small n moments (which control the behaviour as $x \to 0$), are found to be well behaved.

It seems that corrections to $F_2^\gamma$ are under control.

Production of Massive Muon Pairs.

Corrections to the Yamaguchi, Okun, Lederman, Drell-Yan [24] process have been considered by many authors [10]. The standard picture is

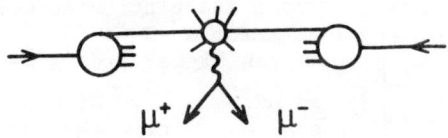

and the cross-section is proportional to

$$\int\int q(x,M^2) \hat{\sigma}(x,x',M^2,S) \bar{q}(x',M^2) dx dx'.$$

The separation between the quark (anti-quark) distributions $q(\bar{q})$ and the "hard scattering cross-section" is not unique beyond leading order. To specify q and $\bar{q}$ uniquely it is simplest to define [25]

$$F_2^{eN} = x \sum_i Q_i^2 (q_i + \bar{q}_i)$$

so that the muon pair cross-section will be predicted in terms of nucleon structure functions which are well measured. The $Q^2/s$ moments of the cross-section are proportional to the moments of $\hat{\sigma}$:

$$\hat{\sigma}_n = 1 + Z_n \frac{\alpha_s}{\pi} + \ldots$$

The first term is due to simple $q\bar{q}$ annihilation. The corrections (due to radiative corrections and gluon emission) are large [27] (e.g. $Z_n \approx 15$) and grow with n. However, a large part is due to a piece [26]

$$P_n \equiv \frac{2}{3} (\pi^2 + (\ell n n)^2)$$

in $Z_n$, whose origin is well understood. The mechanism which gives rise to this term will also produce large corrections in higher orders but it is thought that they can be summed to all orders, giving [28]

$$\hat{\sigma}_n = \exp(P_n \alpha_s(Q^2)/\pi)(1 + 0(\frac{\alpha_s}{\pi}) + \ldots)$$

where the residual perturbation theory is under control (the order $\alpha_s/\pi$ term, which is known, giving a correction of order 20%).

In deep inelastic scattering, the vertex correction/form factor diagram

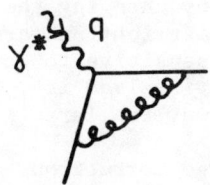

gives a contribution proportional to $-\ell n^2(-q^2)$ (where $-q^2=Q^2>0$), which cancels a $\ell n^2|q^2|$ term from real gluon emission. In Drell Yan, the vertex correction is proportional to $-\text{Re }\ell n^2(-q^2) = -\ell n^2|q^2| + \pi^2$ (since $q^2>0$ in this case). Again the $\ell n^2$ term cancels a contribution from real emission but the $\pi^2$ piece remains and gives the first part of $P_n$. When we compare deep inelastic scattering with Drell-Yan (as we do if $q/\bar{q}$ are defined through $F_2^{eN}$) we find, working all orders, a factor

$$\frac{\text{Drell-Yan}}{\text{Deep Inelastic}} \sim \frac{|F(q^2>0)|^2}{|F(q^2<0)|^2}$$

If, as is generally believed, the form factor has an exponential form similar to that found in the Abelian case, this ratio is equal to $\exp(2\pi\alpha_s/3)$. [28]

The $(\ell n n)^2$ term in $P_n$ reflects $\ell n(1-x)$ terms and appears because the kinematics are different in deep inelastic and Drell Yan. Similar large terms will appear in higher orders. They are associated with infra-red divergences in real gluon emission and have an exponential form when summed to all orders. [28]

Thus it is probably possible to identify and sum the large corrections to the Drell-Yan cross-section to all orders. They increase the lowest order predictions by about a factor of two but the residual corrections are expected to be small.

Large $P_T$ hadron reactions. The one particle inclusive cross-section is given by a convolution of constituent distributions and a fragmentation function with a hard scattering cross-section, the picture being

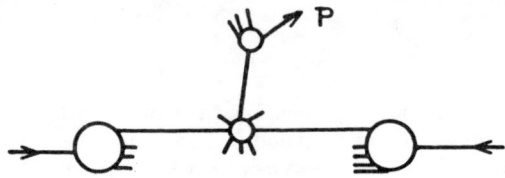

Up to now only the corrections to $qq \to qX$ have been calculated[12]. They turn out to be very large; if we parameterize them as

$$1 + \kappa \frac{\alpha_s}{\pi}$$

then in the $\overline{MS}$ scheme, with the "best" factorization scheme of reference 12, $\kappa$ is about 25 ($\kappa$ depends on all the variables involved in principle, but effectively the dependence is weak). In contrast to the cases considered above, there is no obvious reason for this big correction. It can be reduced somewhat by using a "physical" scheme but the relative reduction is small since $\kappa$ is so big in the first place. A further reduction can be made by changing the choice of "factorization scale" at which the quark distributions are evaluated[29]. However, the results are extremely sensitive to this choice and a good understanding of the physics involved is needed before this mechanism can be used to argue away the large corrections[30].

The phenomenological significance of these large corrections is not entirely clear since corrections to gluon-quark scattering etc. have not yet been calculated. However, it certainly appears that QCD perturbation theory may be useless qualitatively for large $p_T$ physics until extremely high energies, unless the origin of the large corrections can be understood and a way found to sum them to all orders.

3 jet production in $e^+e^-$ annihilation. In $e^+e^- \to ax$, a convenient measure of the jet structure is provided by the eigenvalues $\lambda_i$ of the quantity

$$\theta_{ij} = \sum_a P_i^a P_j^a / |\vec{P}^a| / \sum_a |\vec{P}^a|$$

where the sum a runs over all particles in a given event. In particular, the quantity

$$C = 3(\lambda_1\lambda_2 + \lambda_1\lambda_3 + \lambda_2\lambda_3)$$

provides a measure of the three jet contribution; it is zero for two jets and can be calculated perturbatively (for $C \neq 0$). Defining

$$\frac{1}{\sigma} \int_{\frac{1}{2}}^{1} \frac{d\sigma}{dC} dC = A \frac{\alpha_s}{\pi} [1 + R\frac{\alpha_s}{\pi} + \ldots$$

it is found [14] that, using $\alpha_s^{\overline{MS}}(Q^2)$, the radiative corrections to the lowest order three jet contribution and the four jet contribution give $R \simeq 18$. Consideration of the diagrams involved, such as

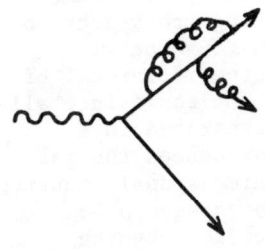

indicates that a better scale to use might be $Q/2$, in which case R is reduced to 15. A further reduction is achieved by using a physical scheme e.g.

$$R^{Mom}(Q/2) = 12, \quad R^{e^+e^-}(Q/2) = 13.$$

R is still large but, as in the Drell-Yan case, a large part is due to $\pi^2$ factors which can probably be summed to all orders [14]. If so, the result is

$$\frac{1}{\sigma} \int_{\frac{1}{2}}^{1} \frac{d\sigma}{dC} dC = A \frac{\alpha_s}{\pi} e^{17\pi\alpha_s/12}(1 + \tilde{R}\frac{\alpha_s}{\pi} + \ldots$$

where $\tilde{R}^{\overline{MS}}(Q/2) \simeq 4.5$, $\tilde{R}^{Mom}(Q/2) \simeq -1.5$, so the residual series is under control.

It therefore seems likely (although it has not really been proved) that, as in the case of Drell-Yan, the correction factor is large but the large part can be calculated to all orders leaving a small residual correction (the effect which this has on the value of $\alpha_s$ extracted from measurements of the three jet cross section is not clear; the measurements include the small C region - purposely excluded from the quantity considered above - where in perturbation theory there is a positive contribution which diverges as $C \to 0$, the divergent part being cancelled by a negative contribution proportional to $\delta(C)$ when there is any smearing, leaving a large negative net contribution).

Conclusions on the convergence of perturbation theory. It seems that with a sensible choice of scheme and of the scale of $\alpha_s$, perturbation theory is under control in most cases assuming that the large corrections associated with time-like form factors and infrared divergences as $x \to 1$ can be summed to all orders. However, it is all too easy to invent special a posteriori pleading to reduce the corrections in any given case so this conclusion must be treated with caution. Further work is needed to prove that the large corrections can indeed be summed to all orders and on large $p_T$

hadron reactions, where the corrections seem to make QCD perturbation theory useless quantitatively at present.

b) Higher Twist. A typical diagram which gives a "higher twist" contribution of order $(mass)^2/Q^2$ to nucleon structure functions is

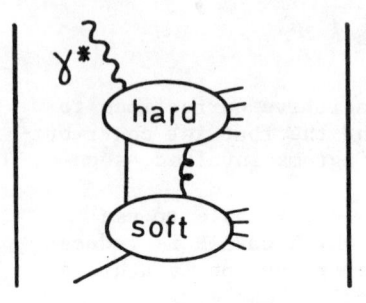

where the exposed quark and gluon separating the hard vertex (which only involves large momentum transfers) from the soft vertex are (almost) colinear and on mass-shell. Relative to the result which would have been obtained with real quark and gluon beams, the hard scattering cross-section involves a gluon flux factor proportional to the reciprocal of the area of the nucleon (since all the flux is concentrated in a region of this size close to the quark); this introduces the $(mass)^2$ factor into the final result (incidentally the dimensional counting rules for incident constituents in exclusive processes can be understood in this way; for constiuents of specific outgoing hadrons the dimensional factors arise because the second and subsequent constituents must scatter into a tube of cross-section $<p_T^2>$ around the first in momentum space). A contribution of the same order is given by the interference term in

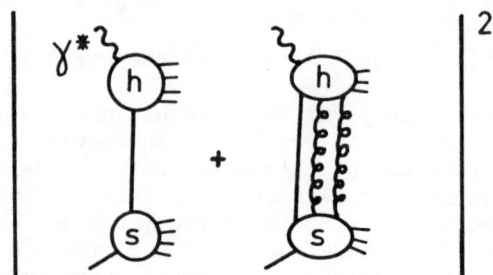

and by retaining the quark masses.

What can be said about higher twist?
1) It is not universal e.g. in Drell Yan both

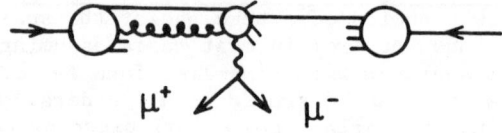

and

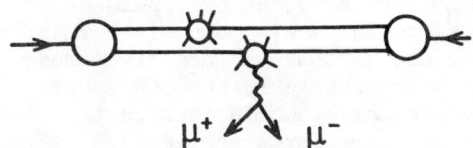

make contributions of order $(mass)^2/Q^2$. But the latter has no analogue in deep inelastic scattering – which shows that there is no hope of making a general universal parameterization of higher twist effects.

2) As indicated above, higher twist effects involve interference. It follows that a phenomenological representation of higher twist effects by "$k_T$ smearing" in processes like large $p_T$ inclusive hadron scattering can never be exact [31] (although it may be useful); nor should we expect it to be universal, as already indicated. A further complication is the discovery by Taylor, Frenkel et al.[32] that the Block Nordsieck theorem fails for certain higher twist effects in QCD.

3) Higher twist effects are expected to be relatively more important for x close to 1 [33]. This can be understood intuitively by noting that the promotion of a constituent to $x \simeq 1$ requires the other constituents to be stopped; this is a relatively easy task for higher twist which already involves multibody effects. In the case of nucleon structure functions it is generally expected that

$$\frac{F_2^{\text{non-leading}}}{F_2^{\text{leading}}} \sim \frac{(mass)^2}{Q^2(1-x)}.$$

For the pion, however, Ezawa and Jackson and Farrar [34] pointed out that $(1-x)$ will be replaced by $(1-x)^2$. This can be understood very roughly as follows. The point x=1 corresponds to elastic scattering so $\sigma_T$ must vanish for a spin zero particle such as the pion for $x \to 1$; this gives rise to a helicity selection rule at the constituent level which causes $\sigma_T$ to vanish one power of $(1-x)$ faster than would have been guessed by dimensional counting as $x \to 1$ (this has important implications for the theory of elastic form factors, as we shall see in section 4). Brodsky and Berger [35] have pointed out that this suppression of the leading twist should be easy to observe in $\pi N \to \mu^+ \mu^- X$ for $x \to 1$ where the non-leading twist will give rise to a pronounced longitudinal term and the differential cross section is expected to behave as

$$\frac{(1-x)^2 \cos^2\theta + C \sin^2\theta}{Q^2}$$

Brodsky reported the results of a calculation of the value of C in a parallel session [36]. This effect should also show up in inclusive pion lepto-production [37] and experimental evidence for it was presented to this conference [38,39].

4) Voloshin and Zakharov have recently pointed out [40] that higher twist effects could have a much greater influence on the properties of final states than on inclusive cross-sections. They argue that the gluon condensate in the QCD vacuum will produce substantial transverse momentum fluctuation in quark jets (possibly $\Delta <p_T^2> \sim O(1\ GeV^2)$ and more for gluon jets). Symbolically:

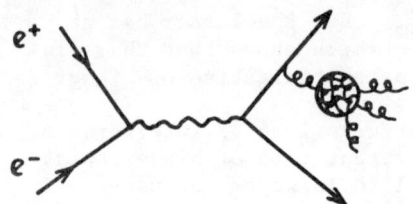

i.e. effectively the gluon field has a complicated spectral function at low $k^2$ which we can attribute to the existence of glueballs/expoxyons (this is similar in spirit to the proposal of Parisi and Petronzio [41], discussed further in section 4, to attribute a phenomenological mass $M_g$ to the gluon; however that would give corrections to $R_{e^+e^-}$ of order $M_g^2/Q^2$ whereas QCD vacuum effects are of higher order).
5) In the case of inclusive leptoproduction, the operator product expansion relates leading and non-leading twist contributions to local operator matrix elements. The leading contributions have been calculated in the bag model [42]. Calculations of the non-leading contributions in the bag and other models would be very interesting.

Not much else is known about higher twist. It needs to be stressed that discussions of scaling violations sometimes seem to suggest a false dichotomy between logarithmic QCD scaling violations and higher twist - both must occur in QCD. Clearly we should not be content until both are well understood, a task which will require much more work.

### 3. INFRARED SENSITIVE PROCESSES AND EVENT STRUCTURES.

Infrared sensitive processes are of two types
1. Processes in which hard gluon radiation is restricted in some way. Coloured particles want to radiate where they are accelerated; the probability that they do not radiate into certain regions of phase space is small i.e. if radiation is restricted there will be radiation damping. The investigation of these processes was pioneered by Dokshitser, Dyakanov and Troyan [43] who identified the essential physics (although the formulae they proposed turned out to be wrong in detail).
2. Processes which "count soft gluons" i.e. are sensitive to the small x region of the Bremsstrahlung $\frac{dx}{x}$ gluon spectrum.

In perturbation theory, the presence of radiation damping is signalled by large logarithms e.g. $\ln(1-x)$ as $x \to 1$ in lepto-production, $\ln Q^2/Q_T^2$ for $Q_T^2 \ll Q^2$ in Drell-Yan or $\ln \theta^2$ as $\theta \to 0$ in $e^+e^-$ acolinearity distributions (defined below). For example, taking the case of Drell Yan as a paradigm the perturbation series has (roughly) the form (omitting all constants)

$$\frac{Q_T^2}{\sigma}\frac{d\sigma}{dQ_T^2} \sim \alpha_s(Q_T^2)[1+\chi+\alpha_s(\chi^3+\chi^2+1) + \alpha_s^2(\chi^5+\chi^4+\ldots)]$$

where $\chi = \ln Q^2/Q_T^2$. The terms with the most powers of $\chi$ in each order of $\alpha_s$ (the "leading double logs") can be summed and give rise to a damping factor similar to the Sudakov form factor. However, in contrast to the classic applications where the renormalization group equations guarantee that the sum of terms which are subleading order by order is subleading so the sum of the leading logs dominates, the sum of the leading double logs is not necessarily dominant. The condition that individual neglected terms be unimportant relative to those which are summed is

$$(\alpha_s(Q_T^2))^n (\ln Q^2/Q_T^2)^{2n-1} \ll 1,$$

giving a minimum $Q_T^2$ which increases with $Q^2$. However, this condition is not sufficient to guarantee that the leading double log sum is a good approximation – especially since the sum of the leading terms is damped. Nor is it necessary – the sum of the subleading terms might be damped equally or even more strongly. Summing leading double logs does no harm in the "classic" region where they are small but detailed information about subleading terms is needed in order to know how far this procedure extends the utility of perturbation theory towards the boundary of phase space.

Turning to specific cases:

<u>Lepto-production as $x \to 1$</u>. The leading double logs can be summed by using the correct kinematics in the diagrams which dominate in the classic $x \neq 1$ region and choosing the argument of $\alpha_s$ carefully in the diagrams to be evaluted.[44] This procedure clearly sums certain subleading logs also. Ciafaloni has gone some way [45] towards proving that it sums all the important subleading terms, in which case it will work for $\ln((1-x)Q^2/\Lambda^2) \ll 1$ or $x < 1 - Q_0^2/Q^2$ with $Q_0$ of order a few GeV. For $x > 1 - Q_0^2/Q^2$ a different, form factor like behaviour, is expected to take over [44].

<u>Drell-Yan for $Q_T^2 \ll Q^2$</u>. We can distinguish three regions:

1. $Q_T^2 \sim O(Q^2)$. Perturbation theory can be used here (but there are few events).
2. $Q_0^2 < Q_T^2 \ll Q^2$; $\ln Q^2/Q_T^2$ is large here but the leading double logs can be summed [46]. This may make it possible to extend perturbation theory into this region.
3. $Q_T^2 \lesssim Q_0^2$. At first sight, perturbation theory can say nothing about this region.

Consider part of a typical diagram

If $Q_T^2 \ll Q^2$, the main contribution is from gluons with $(\vec{k}_i^T)^2 \ll Q^2$ (it is very unlikely for several emissions with $k^T \sim O(Q)$ to leave $Q_T \ll Q$). In calculating the leading double logs, the upper limit $k_i^T \lesssim Q_T$ can be imposed. This is quite in order as long as $Q_T$ is large but it is obviously silly as $Q_T \to 0$ (where the leading double log sum is not expected to be a good approximation anyway). To approach small $Q_T$, it is clearly necessary to take transverse momentum conservation into account exactly. Parisi and Petronzio have shown [46] how to do so by working in impact parameter space. They also argue that the $(\vec{k}_i^T)^2$ all grow with $Q^2$ (although $(\vec{k}_i^T)^2 \ll Q^2$) as does the multiplicity of emitted gluons. The result is that for sufficiently large $Q^2$, the multiple gluon emission can be treated perturbatively (since $\alpha_s(\langle \vec{k}_T^2 \rangle) \ll 1$) even for $Q_T < Q_0$. Furthermore, the effect of "non-perturbative smearing" (due to emission of gluons with $\vec{k}_T^2 \lesssim Q_0^2$ far from the virtual photon vertex etc.) is wiped out by the growing multiple emission with large $k_T$. Parisi and Petronzio find that the influence of the non-perturbative smearing function (which must be fitted to the data) is large at ISR energies but has become small ($O(10\%)$) by $\sqrt{s} = 800$ GeV (Isabelle energy).

The range of validity of the Parisi-Petronzio formalism, which sums the leading double logs (plus some non-leading terms), is unclear. However, in a so far unwritten contribution to this conference Collins and Soper claim [47] that they can also sum non-leading terms, leaving a residual perturbation series free from large logs (their result agrees with the Parisi-Petronzio form asymptotically). Assuming that this important result survives critical scrutiny, it will make it possible to calculate the Drell-Yan cross-section systematically for all $Q_T$. The phenomenological implications of the Collins-Soper formula have not yet been explored.

<u>Acolinearity Distributions in $e^+e^-$ annihilation</u>. Focussing on two particles a and b in an $e^+e^-$ annihilation event, if $\theta$ is the angle between their momenta, defined by

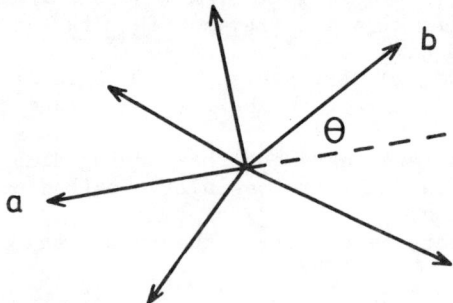

we can construct the quantity [48]

$$\frac{dW}{d\cos\theta} = \sum_{a,b} \int\int \frac{x_a x_b}{\sigma} \frac{d\sigma}{dx_a dx_b d(\cos\theta)} dx_a dx_b,$$

where $x_a = \frac{E_a}{E_a^{max}}$. This "energy-energy" correlation function is free from mass singularities as $m_{q,g} \to 0$ and can be reliably calculated perturbatively for $\theta$ away from $0$ and $\pi$. It provides a good way to test QCD [48,49]. However, perturbation theory reveals $\ln\theta$ terms which give rise to damping factors for $\theta \to 0$ when they are summed. The physics is very similar to Drell-Yan for $Q_T \to 0$. The $x_a x_b$ factor singles out particles which carry finite fractions of the available energy; asking that they emerge back to back ($\theta \to 0$) puts a restriction on transverse radiation. The leading log sum has been discussed by many authors[50]. As in the case of Drell Yan for $Q_T \ll Q$, it extends the region where perturbation theory is useful to smaller $\theta$ but it will certainly not work for $\theta \to 0$. Very recently, however, Baier and Fey[51] have extended the Parisi-Petronzio analysis [46] to acolinearity distributions, with the results shown in the accompanying figure. They fit an "intrinsic smearing factor" to

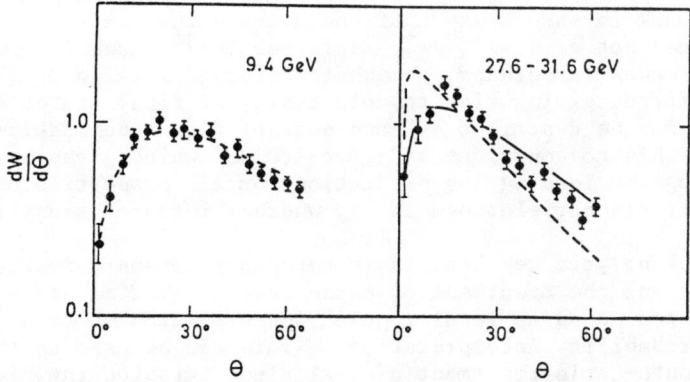

Fig. 2. Acolinearity distributions obtained [51] using a smearing function fitted to the data [52] at $\sqrt{s}$ = 9.4 (----) and then calculating for $\sqrt{s} \simeq 30$ with (——) and without (---) adding gluon radiation.

the data at $\sqrt{s}$ = 9.4 GeV to which they add radiation effects (which eventually wipe out the influence of the ad hoc smearing function). Gluon radiation is needed to fit the data at higher energies but it is too early to conclude that the damping factor or "quark form factor" has been seen, especially since the range of validity of the approximation is unclear. It will be very interesting to see how the improved theory of Collins and Soper fits the improved data now becoming available (in passing I would plead that $\frac{dW}{d\cos\theta}$ or $\frac{dW}{d\sin^2\theta}$ be plotted in future - not $\frac{dW}{d\theta}$; the latter quantity contains a kinematical zero as $\theta \to 0$, which obscures comparison of theory and experiment).

To conclude, radiation damped processes are extremely interesting but there is clearly much more work to be done on them, both theoretically and phenomenologically.

I shall now turn to predictions for event structures, multiplicities etc. which may be sensitive to soft gluon radiation. Presumably we can use perturbation theory to investigate parton distributions provided the invariant mass of every parton is greater than some value $Q_0$ (of a few GeV or more), which provides an infrared cut-off. Many interesting analytic results can then be obtained. For example
1. the parton multiplicity grows asymptotically [53] like

$$\left(\frac{\alpha_s(Q^2)}{\alpha_s(Q_0^2)}\right)^A \exp(B[\alpha_s^{-1/2}(Q^2) - \alpha_s(Q_0^2)^{-1/2}])$$

i.e. it grows faster than any power of $\ln Q^2$ but slower than any power of $Q^2$.
2. Every quark produced in, say, $e^+e^-$ annihilation can be paired with an antiquark in such a way that the average invariant mass of the pair does not grow as $Q^2 \to \infty$. This result [54] (dubbed "preconfinement") makes it, perhaps, somewhat easier to imagine non-perturbative forces giving rise to colour singlet final states (a process which may be described by some sort of flux tube model)[55].
3. It is possible to construct a "jet calculus" which gives rules for calculating the leading log prediction for all properties of jets (multiparticle correlations etc.), whether infrared sensitive or not.[56]

To obtain analytic results, it is necessary to approximate the matrix element and the treatment of phase space. The matrix element is evaluated using a "leading pole" approximation which allows a classical probability interpretation. This can be used as the basis for a Monte-Carlo treatment of real gluon emission in which phase space is treated (more or less) exactly [57]. Virtual processes are treated by enforcing the probability interpretation i.e. the (Sudakov-like) form factors are adjusted so that total probabilities add up to one. Monte Carlo studies show that the errors introduced by an approximate treatment of phase space can be big up to very large energies [57]. In particular, the evolution of jets is much slower than expected according to the analytic formulae (which are of course valid asymptotically). For example,
1. The multiplicity does not begin to rise faster than logarithmically until above PETRA energies, as shown in the accompanying figure (which shows the hadron multiplicity which is assumed to be equal to a constant multiple of the parton multiplicity).
2. The expectation that gluon jets are broader than quark jets is only borne out at large x and very large $Q^2$.[58]

The Monte Carlo technique has many applications. For example, it can be used to study the space-time evolution of jets [59]. It can also be used to study acolinearity distributions for

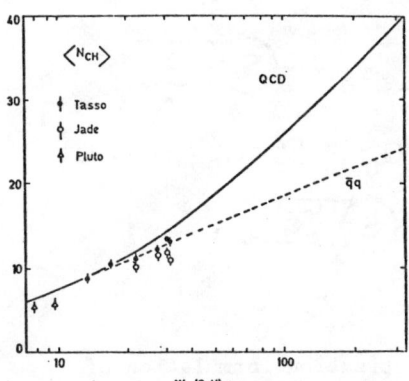

Fig. 3. The multiplicity predicted by a QCD Monte Carlo calculation [58](———) and the Feynman-Field two jet model (-----).

small θ with results which fit the data [60]. The approximation here is essentially that of Parisi, Petronzio, Baier and Fey - the leading pole approximation is used for the matrix element but phase space is treated exactly. As in the analytic treatment, this is not necessarily a good approximation for small θ - subleading logs and constant terms from the leading (ladder) diagrams are summed but leading logs and constants from other diagrams, which might be equally important or perhaps make cancelling contributions, are ignored.

This criticism can be levelled at all the Monte-Carlo calculations. The limitations of the method are not yet clear, nor have its proponents subjected each others work to critical scrutiny yet. However, I feel that it should be an improvement to treat phase space exactly and that even with a rough leading pole treatment of the matrix element fairly reliable predictions should be obtained in most cases (except perhaps in radiation damped processes). In any case, it seems that the beautiful analytic results available up to now are unreliable except at extreme energies in many cases, although further work may yet partially save us from thraldom to the computer. Even if the parton stage could be treated reliably analytically, it would seem that "hadronization" must be treated by Monte Carlo and it is clear that further thought is needed about how this should best be done [61].

It would be interesting to treat processes such as deep inelastic scattering and heavy muon pair production by Monte Carlo methods. There are many questions which could be studied numerically which have not yet been studied analytically (e.g. the transition between the region when $Q^2 \sim M_Q^2$ and $Q^2 >> M_Q^2$ in the electroproduction of heavy $Q$-$\bar{Q}$ pairs).

## 4. EXCLUSIVE PROCESSES AT LARGE MOMENTUM TRANSFER

The idea underlying all the work [62] on this subject is that the "hard" tail of the hadronic wave-function, for constituents which are very virtual or have large transverse momenta, is calculable from the "soft" wave-function (just as the quark distribution $q(x,Q^2)$ is calculable from $q(x,Q_0^2)$ for $Q^2 > Q_0^2$). For example, the hard $q\bar{q}$ wave function of the pion π─(H)─ is calculable from the soft wave function $\pi \to \boxed{S}$ perturbatively thus

$$\pi \to \boxed{H} \to q \quad = \quad \boxed{S} \to y,\bar{y} \to x \quad + \quad \boxed{S} \to \cdots$$

$$x, q_T^2 > Q_0^2 \qquad k_T^2 < Q_0^2 \qquad k_T^2 < Q_0^2 \quad k_T'^2 > Q_0^2$$

$$+ \quad \boxed{S} \to \bowtie \quad + \ldots$$

$$k_T^2 < Q_0^2$$

where I have followed[62] the light cone quantization formulation of Brodsky and Lepage so that the constituents are on mass shell and are characterised by $\vec{q}_T$ and $x = \dfrac{q_0 + q_3}{p_0 + p_3}$ where $p = (p_0, \vec{0}, p_3)$.

If an amplitude $\phi$ is defined in terms of the wave function $\psi$ by

$$\phi(x,Q) \sim \int^{Q^2} d\vec{q}_T^2 \, \psi(x, q_T^2)$$

it satisfies an evolution equation reminiscent of the Lipatov Altarelli Parisi [63] equation,

$$\frac{\partial \phi(x,Q)}{\partial \ln Q^2} = \int_0^1 V(x,y,\alpha_s(Q^2)) \phi(y,Q) \, dy$$

where the kernel V describes all possible ways of connecting a soft $q\bar{q}$ pair to a hard $q\bar{q}$ pair without going through a soft pair (which is already contained in the soft wave function). V contains large momenta and is free from mass singularities and may therefore be calculated perturbatively. Evaluating V to leading order, this equation is solved by

$$\phi(x,Q) = \sum_n a_n \, C_n^{3/2}(1-2x) \, (\alpha_s(Q^2))^{\gamma_n} (1 + O(\alpha_s))$$

where the C's are Gegenbauer polynomials. The $a_n$ are determined by the "starting" amplitude $\phi(x, Q_0)$, except that $a_0$ is proportional to the wave function at the origin and is therefore known in terms of $f_\pi$:

$$a_0 = \sqrt{3} \, f_\pi.$$

(other constraints on $\phi$ are discussed in a recent paper by Brodsky, Huang and Lepage)[62].

In a "physical" gauge, the lowest Fock state ($q\bar{q}$ for mesons, qqq for baryons) dominates the asymptotic behaviour of exclusive processes by a power [62] as a result of the usual dimensional counting arguments [64].

I will now discuss a variety of applications:
1) Form factors. First consider the large $Q^2$ behaviour of the pion form factor $F_\pi(Q^2)$. The virtual photon imparts a large transverse momentum to one constituent which must then recombine with the other to form another pion so the amplitude involves the large $q_T$ tail of $\psi$. In fact $F_\pi$ is given symbolically by

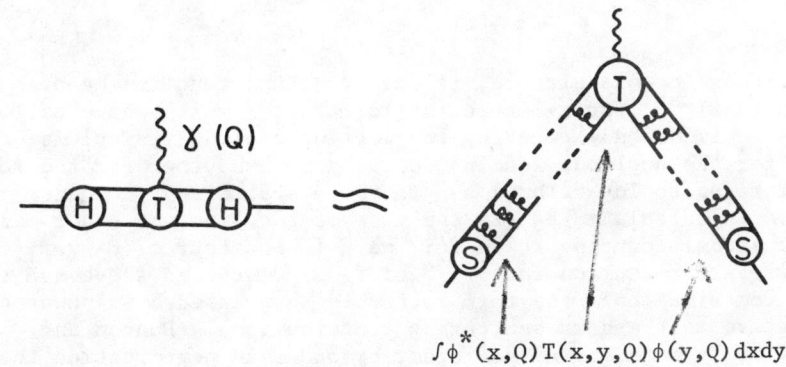

$$\int \phi^*(x,Q) T(x,y,Q) \phi(y,Q) \, dx \, dy$$

where

and the ladder diagrams give the leading contribution to $\phi$. Using the expression for $\phi$ above the result [62,65] is

$$F_\pi(Q^2) \xrightarrow{\text{large } Q^2} \frac{16\pi\alpha_s(Q^2)}{3Q^2} \left| \sum_n a_n (\alpha_s(Q^2))^{\gamma_n} \right|^2 (1+O(\alpha_s))$$

$$\rightarrow \frac{16\pi\alpha_s f_\pi^2}{Q^2}$$

where the last line follows from the fact that $\gamma_n > 0$ for $n \geq 1$ but $\gamma_0 = 0$.

The form factors of other mesons can be calculated in a similar way with many testable consequences [66] e.g. helicity selection rules in the case of vector mesons.

The calculation of baryon form factors requires an extra assumption. Drell and Yan and West (DYW) [67] long ago suggested that the nucleon form factor is controlled by diagrams in which a single constituent with $x \approx 1$ absorbs the virtual photon, the other "wee" partons in the incoming nucleon being able to realign themselves as "wees" in the outgoing nucleon without additional hard scattering:

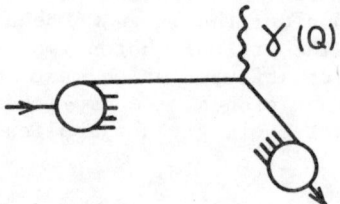

According to this picture, if the structure function behaves as $F_2 \sim (1-x)^{2a-1}$ for $x \to 1$ then the form factor will behave as $(Q^2)^{-a}$. Simple dimensional counting (neglecting scaling violations) gives $a=2$ for the nucleon, leading to the same behaviour for the form factor (up to logarithms) as "the hard scattering mechanism" used above to calculated $F_\pi$ (in the case of the pion and other mesons dimensional counting rules for the $x \to 1$ behaviour of $F_2$ fail by the Ezawa-Farrar-Jackson factor [34] of $(1-x)$ discussed above and the DYW contribution to the form factor is suppressed by a power of $Q^2$ relative to the hard scattering contribution). Duncan and Mueller [62] have shown that order by order in perturbation theory there is an important DYW like contribution; technically this makes it impossible to factorize the form factor into a nucleon amplitude and a hard scattering part which satisfy renormalization group/evolution equations. However, the nucleon amplitude calculated to all orders is drastically damped by a Sudakov factor for $x \to 1$. Brodsky and Lepage argue [62] that the DYW mechanism will therefore be unimportant and hard scattering will prevail. If so, factorization and renormalization group equations which are not true order by order would be true for the complete theory - an interesting possibility which will be hard to prove. Assuming it is true, Brodsky and Lepage have calculated $G_M$ [62]. The results, shown in the accompanying figure, are insensitive to $\phi$. A reasonable fit requires a rather small value of $\Lambda$ - but this is not ruled out by other experiments,

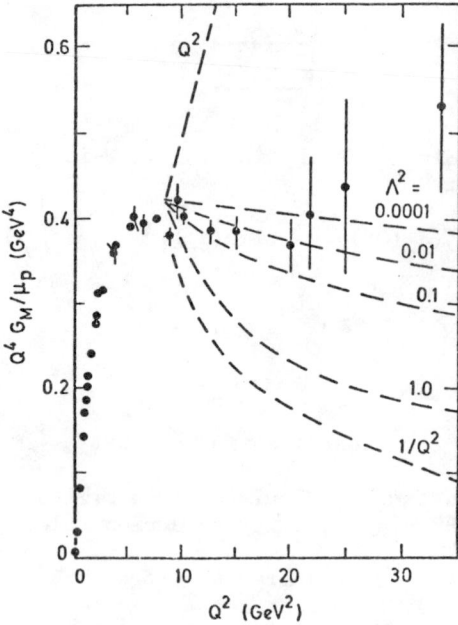

Fig.4. Predicted behaviour of the nucleon form factor $G_M$.

nor would the meaning of a disagreement be clear until higher order contributions have been calculated.
2) Exclusive decays of heavy $Q\bar{Q}$ bound states [68].

For large $M_Q$ the $Q\bar{Q}$ system becomes small and is coupled to hadron amplitudes $\phi$ by a "hard" kernel involving only short distances/large momenta which can be calculated perturbatively e.g.

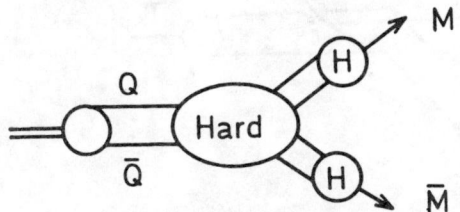

The $Q\bar{Q}$ wave function drops out of branching ratio so that, e.g., $\frac{\Gamma(^3P_0\to\pi\pi)}{\Gamma(^3P_0\to\text{all})}$ is exactly calculable (up to "higher twist" corrections of $O(\frac{M^2}{M_Q^2})$); for very large $M_Q$ it is controlled by $a_0$ and is given by $(\alpha_s(M_Q))^4 f_\pi^4/M_Q^4$ times a calculable constant.

3) Meson production in photon-photon collisions.
The amplitude is controlled by

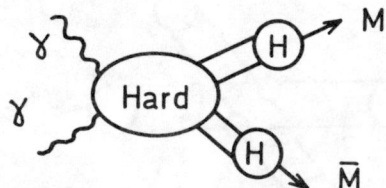

for large t and s. The differential cross-section is given by

$$\frac{d\sigma}{dt} = \frac{\alpha_s^2(s)}{s^4} f(\theta_{cm})$$

$f(\theta_{cm})$ is controlled by the meson amplitude $\phi_M$. In fact a measurement of f maps out $\phi_M$ in much the same way that a measurement of $F_2(x,Q^2)$ maps out $q(x,Q^2)$, so that $Q\bar{Q}\to\pi\bar{\pi}$ and the form factor $F_M$ could be predicted completely if f were measured.

4) Large angle exclusive hadron scattering.
The hard scattering mechanism involves diagrams such as

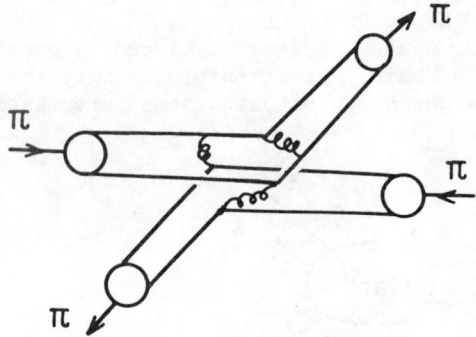

in which all the constituents are connected through a "hard" kernel and interact in a small volume of space-time. It gives

$$\frac{d\sigma}{dP_T^2} = \frac{1}{(P_T^2)^{n-2}} \alpha_s^{n-2} f(\theta_{cm}) \Sigma b_i(\alpha_s)^{\delta_i}(1+O(\alpha_s))$$

where n is the total number of constituents involved (e.g. n=8 for $\pi\pi \to \pi\pi$ and 12 for $pp \to pp$). Order by order in perturbation theory the Landshoff "pinch" mechanism [70]

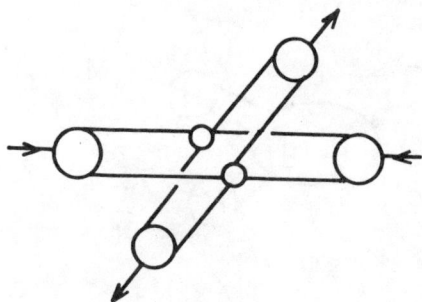

is actually more important than the hard scattering mechanism by a power of $P_T^2$. However, since it involves the scattering of almost real coloured constituents at different places, Sudakov damping is expected to occur [71] and explicit calculation suggests that it does [72]. If we are prepared to make the extra assumption, which was needed to calculate nucleon form factors, that the damping factor/"quark form factor" $F_q(t)$ calculated to all orders can be imposed on the perturbative constituent scattering amplitude, the result is

$$\frac{d\sigma}{dt} \sim \frac{1}{t^8} (\alpha_s(t))^6 |F_q(t)|^2$$

If this is correct, the damping factor will suppress the pinch contribution in the limit s→∞ with t/s fixed, and the hard scattering mechanism will dominate, while the pinch will prevail in the limit s→∞ with t fixed. However, in this case the hard scattering contribution will be difficult - perhaps impossible - to evaluate [73] because the relevant hard scattering diagrams can also give rise to a pinch contribution (which is undamped if the hard scattering kernel is evaluated perturbatively) and it will be a delicate matter to separate this contribution.

To first approximation the data exhibit the power behaviour predicted by the pinch and by hard scattering respectively for t large but much less than s and t of order s [74,75]. However, they show no trace of the rapidly varying factors $\alpha_s$ and $F_q$. One possible explanation [41] is that for small momentum transfers (and the momentum transfer between constituents, which is relevant here, is relatively small) $\alpha_s$ and damping factors like $F_q$ should be "frozen". The rationale behind this proposal, which would also help in fitting $G_M$ and understanding the value of $\alpha_s$ extracted from $\psi$ decay, is that at modest energies the effects of physical thresholds in gluon channels should be taken into account by giving the gluon a "phenomenological" mass $M_g$ of order 800 MeV which would cause $\alpha_s(t)$ to freeze for $|t|<2M_g$.

An additional curious feature is that fits inspired by the hard scattering mechanism require large dimensionless constants [75] e.g. in pp→pp the amplitude is approximately 5000 $4\pi\alpha_s G_M^2(t)$! Brodsky and Lepage attribute this to the fact that there are of order $10^6$ hard scattering Born terms in pp→pp, arguing that the value of the amplitude might be proportional to the square root of the number of diagrams. However, according to this argument the next order corrections to pp→pp [and also to $G_M$] are likely to be huge as the number of diagrams is bigger by a factor well in excess of $10^6$ [of order 100].

To conclude on exclusive processes:- the theory is well founded for meson form factors, $Q\bar{Q}$ decays and $\gamma\gamma\to MM$ but no amplitude has yet been calculated beyond leading order, so the reliability of the predictions is unknown, and not much relevant data is available. For baryon form factors and pp→pp etc. an extra assumption is needed. Whether or not the considerable body of data which is available support the theory is unclear.

## 5. FINAL CONCLUSIONS

1) Higher order corrections are probably under control in most of the classic applications but more work remains to be done on this question e.g. on checking that the $\pi^2$ factors in Drell Yan and $e^+e^-\to$jets really do exponentiate and calculating corrections to further processes. However, it seems that large corrections may spoil the application of QCD perturbation theory to large $P_T$ hadronic reactions.

2) The time has come to treat higher twist contributions as physics - not background - and attempt to find processes in which

they can be isolated experimentally and calculated theoretically.
3) Processes which exhibit radiation damping are a promising subject for future work; further investigation of the validity of the approximations involved and the corrections to them is clearly needed.
4) Soft gluon counting and the study of event structures may turn out to be best done numerically using Monte Carlo techniques.
5) The application of QCD perturbation theory to exclusive processes is an exciting development but further work is needed e.g. on higher order corrections and on the validity of the calculations for baryon form factors and $pp \to pp$ etc.

## ACKNOWLEDGEMENTS

Among the many colleagues with whom I have enjoyed fruitful discussion I am particularly grateful to Ed Berger, Stan Brodsky, Bill Celmaster, Marcello Ciafaloni, John Collins, Steve Ellis, Nathan Isgur, Roberto Odorico, Roberto Petronzio, Graham Ross, James Stirling and Steve Wolfram.

## DISCUSSION

Odorico, Bologna: This is a comment about differences between QCD results obtained by Monte Carlo methods and analytic methods. I would like to stress that there is no difference at all (analytic results are exactly recovered), if in the Monte Carlo, phase space constraints are not enforced exactly but with the same approximations which are used in analytic derivations. This has been extensively discussed in a paper of mine presented to the conference. Of course, I am talking of exact leading log Monte Carlos and not of model dependent ones.
Llewellyn Smith: Thank you for that comment.

Newman, DESY: Regarding QCD parton Monte Carlos: There have been several pieces of evidence from DESY that the evolution of gluon jets, and the corresponding jet broadening, is a slower function of $Q^2$ than expected naively. Examples of this are seen in the jet analysis of the $\Upsilon \to ggg$ decay by Pluto, and the comparison of jets by the MARK J at $\sqrt{s} = 12$ GeV and $\sim 35$ GeV which I presented in my talk.
The $e^+e^-$ data from PETRA and the values of $\alpha_s$ can be used to extract the QCD scale $\Lambda$ in a meaningful way assuming the complete 0 $(\alpha_s^2)$ theoretical calculation of the 2 and 3 jet cross sections is done. One should therefore be wary of switching to the use of QCD parton Monte Carlos in which one has only an approximate treatment of gluon emission, and in which a phenomenological treatment of hadronization is still a basic problem.
Llewellyn Smith: I was really interested in the theoretical part -- the parton part. I agree that for the hadronization part there is an awful lot of thought needed on how to do this. Present methods are certainly ad hoc in my mind.

Celmaster, Argonne: I want to make a remark about the corrections to high-$p_T$ scattering. It is not necessary to play games with $\alpha_s$ in order to dramatically reduce the size of the corrections. It suffices to do the calculation with distribution and fragmentation functions which are evaluated at, say, $p_T^2/2$ rather than $p_T^2$. The reason that such a change of scale changes the size of corrections is similar to the reason that changing the argument of $\alpha_s$ changes the size of corrections. This is "factorization prescription dependence" rather than "renormalization prescription dependence".
Llewellyn Smith: That is interesting. I was aware that it would have an effect, but I am surprised that it is so big.

Contogouris, McGill: I would like to comment on the large correction you mentioned in large-$p_T$ hadron production. Such a correction would be welcome to many theorists, in particular to account for the Fermilab data on $p+p \to \pi+X$. At intermediate $p_T$, to account for these data without QCD corrections one needs a huge primordial transverse momentum ($\sim$ 1 GeV); there are several objections to this. At very large $p_T$ errors in the data allow some increase of the inclusive cross-section. So if there is indeed an overall increase by a factor of 2 or 3, this is certainly not disturbing.
Llewellyn Smith: I think the point is that if the first order corrections are of order 1, then the next order is likely to be of order 1. It just means that you wouldn't expect the lowest order to work and in that sense it's not disturbing. However, it is disturbing because it means that you can't calculate anything. You don't know whether 1+1+... sums to infinity or $e^1$. But, if Celmaster is right, you don't have to worry anyway.

Lefrancois, Orsay: You said that higher order QCD corrections are probably under control while for high $p_T$ hadron physics they are not. What about high $p_T$ direct photons?
Llewellyn Smith: Aurenche and Lindfors and Contogouris et al. have looked at that and find the corrections are small. You have to be careful to distinguish the Bremsstrahlung from the direct terms but, with a careful definition, the pieces that have been looked at are under control.
Contogouris, McGill: The correction to $p+p \to \gamma+X$ due to photon Bremsstrahlung (QCD subprocess $q+q \to q+q+\gamma$) is significant. At $p_T$ = 8 GeV and at $\sqrt{s}$ = 53 Bremsstrahlung makes a 25-60% correction (depending on the form of the gluon distribution), whereas at $\sqrt{s}$ = 31 it makes a 35-100% correction. For details see my paper at the Parallel Session A8. (Direct Leptons and Photons in Hadronic interaction, TH).
Brodsky, SLAC: It might be emphasized that, even if one has the radiative corrections to leading twist under control in hadron production at high $p_T$, the higher twist terms must still be understood. This is especially obvious because of the trigger bias effect which strongly enhances contributions without fragmentation. Whenever "$k_T$-smearing" is invoked, one knows that higher twist terms must be considered.
Llewellyn Smith: I agree completely.

## FOOTNOTES AND REFERENCES

1. See the papers by G. Veneziano, H.D. Politzer and R.D. Field in Proc. 19th International Conference on High Energy Physics (Physical Society of Japan) and references therein. For the status of perturbative QCD as of one year ago see J. Ellis in Proc. 1979 International Symposium on Lepton and Photon Interactions at High Energies (Fermi National Accelerator Lab.)
2. See the review talks by J. Lefrancois, F. Sciulli and B.H.Wiik in these proceedings and references therein.
3. K.G. Chetyrkin, A.L. Kataev, F.V. Tkachov, Phys. Lett. $\underline{85B}$, 277 (1979).
   M. Dine and J. Sapirstein, Phys. Rev. Lett. $\underline{43}$, 688 (1979).
   W. Celmaster and R.J. Gonsalves, Phys. Rev. Lett. $\underline{44}$, 560 (1979) and Phys. Rev. $\underline{D21}$, 3112 (1980).
4. For a comprehensive review and references see A.J. Buras, Rev. Mod. Phys. $\underline{52}$, 199 (1980); see also A.J. Buras, parallel session C11.
   M. Moshe (Phys. Rev. Lett. $\underline{43}$, 1851 (1979)) has implied that perturbation theory may be badly behaved in this case but, as discussed by L.F. Abbott (Phys. Rev. Lett. $\underline{44}$, 1569 (1980)), the corrections only appear large if $\alpha_s$ is expanded in inverse powers of $\ln Q^2/\Lambda^2$ with an injudicious definition of $\Lambda$.
5. W.A. Bardeen and A.J. Buras, Phys. Rev. $\underline{D20}$, 166 (1979).
   D.W. Duke and J.F. Owens, Tallahasee preprint, contributed paper 244.
   For further discussion of the photon structure function see T.F. Walsh, DESY 80/52 (1980) and the contribution of W. Frazer to parallel session C1 (these proceedings) and references therein.
6. R. Barbieri et al., Nucl. Phys. $\underline{B154}$, 535 (1979).
   W. Celmaster and D. Sivers, Argonne preprint ANL-HEP-PR-80-29 (1980), contributed paper 720, presented by W. Celmaster in parallel session C7.
   P. Pascual and R. Tarrach, Barcelona preprint UBFT-11-80 (1980), contributed paper 407.
   A.J. Buras, Fermilab-Pub-80/43-Thy (1980).
7. R. Barbieri et al., University of Geneva preprint UGVA-DPT 1980/06-247.
8. T. Inami and T. Kubota, University of Tokyo preprint, Contributed paper 249.
   N. Sakai, Fermilab-Pub-80/51-Thy (1980).
   The leading order contribution is also discussed in E. Braaten and J.P. Leveille Wisconsin preprint COO-881-127 (1980), contributed paper 117.
9. G. Curci et al., CERN TH.2815 (1980)
   I. Antoniadis and C. Kounnas, Ecole Polytechnique preprint (1980).
10. G. Altarelli et al., Nucl. Phys. $\underline{B143}$, 521 (1978), $\underline{B146}$, 544(E)

10. (1978) and B157, 461 (1979).
    J. Kubar-Andre and F.E. Paige, Phys. Rev. D19, 221 (1979).
    K. Harada et al., Nucl. Phys. B155, 169 (1979).
    A.P. Contogouris and J. Kripfganz, Phys. Lett. 84B, 473 (1979) and Phys. Rev. D19, 2207 (1979).
    J. Abad and B. Humpert, Phys. Lett. 80B, 286 (1979) and 83B 371 (1979).
    A.N. Schellenks and W.L. van Neerven, Phys. Rev. D21, 2619 (1980) and Nijmegen preprint THEF-NYM-80.7 (1980), contributed paper 719.
    B. Humpert and W.L. van Neerven, Phys. Lett. 85B, 293 (1979).
    K. Harada and T. Muta, Phys. Rev. D22 (in press), contributed paper 5.
    For discussion of higher order corrections to transverse momentum distributions see M. Chaichian et al., Helsinki preprint HU-TFT-80-9 (1980), contributed paper 150 and P. Aurenche and J. Lindfors CERN TH 2877 (1980).
    Summation of large corrections to all orders is discussed in G. Parisi Phys. Lett. 90B, 295 (1980) and G. Curci and M. Greco, CERN TH 2786 (1979).
11. G. Altarelli et al., Nucl. Phys. B160, 301 (1979).
    N. Sakai, Phys. Lett. 85, 67 (1979).
12. R.K. Ellis et al., LBL preprint LBL-10304 (1979), presented by I. Hinchliffe in parallel session A6.
    See also M.A. Furman, Columbia preprint CU-TP-182 (1980).
13. T. Gottschalk, Argonne preprint ANL-HEP-PR-80-35 (1980), contributed paper 675.
14. R.K. Ellis, D.A. Ross and A.E. Terrano, Caltech preprint CALT68-785 (1980), presented by A.E. Terrano in parallel session C4.
    Partial results have been obtained by A. Ali et al., Nucl. Phys. B167, 454 (1980) and parallel session C4 and T. Chandramohan and L. Clavelli, Bonn preprint BONN-HE-80-8 (1980).
15. P. Aurenche and J. Lindfors, Nucl. Phys. B168, 296 (1980).
    A.P. Contogouris et al., McGill preprint, contributed paper 402.
16. F.A. Berends et al., DESY 80/08 (1980); this calculation is incomplete since the contribution of hard acollinear gluon bremsstrahlung is omitted.
17. W. Celmaster and R.J. Gonsalves, Phys. Rev. Lett. 42, 1435 (1979) and Phys. Rev. D20, 1420 (1980).
    P. Pascual and R. Tarrach, Barcelona preprint UBFT-1-80 (1980), contributed paper 406.
    P.M. Stevenson, Wisconsin preprint DOE-ER/00881-155 (1980), contributed paper 552A, and Wisconsin preprint DOE-ER/0881-153 (1980).
    W. Celmaster and D. Sivers, Argonne preprint ANL-HEP-PR-80-30 (1980).
18. Explicit formulae are given by Celmaster and Gonsalves (ref.17)
19. This was pointed out by G. Parisi and R. Petronzio, CERN TH 2804 (1980) and by Celmaster and Sivers (ref. 6)
20. See the contribution of W. Wagner to parallel session C1 and the rapporteur's talk by B.H. Wiik (these proceedings).

21. E. Witten, Nucl. Phys. B120, 189 (1977). For further references see T.F. Walsh and W. Fraser (ref. 5).
22. D.W. Duke and J.F. Owens (ref. 5).
23. W. Celmaster and D. Sivers (ref. 17).
24. Y. Yamagouchi, Nuovo Cimento 43, 193 (1966).
    L.B. Okun, Sov. Journal of Nucl. Phys. 3, 426 (1966).
    L.M. Lederman, Physics Reports 26, 149 (1976) and references therein.
    S.D. Drell and T.M. Yan, Phys. Rev. Lett. 25, 316 (1970).
25. G. Altarelli et al. (ref. 10), see the second paper cited for a particularly lucid discussion.
26. In some of the papers cited in ref.10, the quarks were taken off-shell to regulate infrared and mass singularities. This procedure gives a different coefficient of $\pi^2$ but it is now generally agreed to be wrong (J. Ambjorn and B. Sakai, NORDITA 80/14 (1980) - contributed paper 431, G.T. Bodwin et al., Urbana preprint ILL-(TH)-80-2 (1980), B. Humpert and W.L. van Neerven (CERN TH-2785 and 2805 (1980), contributed papers 142 and 145 ). The factorization theorem, which justifies the use of QCD perturbation theory, does not require confinement and it is obvious that the on-shell result is the correct one for unconfined quarks. The off-shell result does not satisfy all physical requirements, such as sum rules (see Ambjorn and Sakai, loc. cit.).
27. The $Z_n$ are scheme independent. The higher order corrections to the predicted $Q^2$ dependence of $q/\bar{q}$ are, of course, scheme dependent; Harada and Muta (ref. 10) have discussed the implications for Drell-Yan.
28. G. Parisi and G. Curci and M. Greco (ref. 10).
29. In ref. 12 the factorization scale ($M^2$) and the scale ($Q^2$) at which the running coupling constant is evaluated in the hard scattering cross-section were chosen to be different. This is allowed in principle but it makes the quark distribution depend on both these variables and this does not seem to have been taken into account. The "best" case quoted corresponded to $Q^2=\hat{t}$, $M^2= \hat{u}\hat{t}/\hat{s}$, where ^ denotes a parton variable ("best" means that $\kappa$ is much less than if $M^2$ is taken to be $\hat{s}$; a value of $M^2$ smaller than $\hat{u}\hat{t}/\hat{s}$ descreases $\kappa$). With $M^2=Q^2=\hat{t}$, $\kappa$ is increased by about 50% while with $Q^2=M^2= \hat{u}\hat{t}/\hat{s}$ it is decreased.
30. $\kappa$ is sensitive to the definition of the gluon distribution; it is conceivable that a case could be made for a choice which would reduce $\kappa$ substantially.
31. W.E. Caswell et al., Phys. Rev. D18, 2415 (1978).
    H.D. Politzer, Caltech preprint CALT-68-765 (1980) and addendum thereto.
32. R. Doria et al., Nucl. Phys. B168, 93 (1980).
    A. Andrasi et al., Oxford preprint 37/80 (1980).
    J. Frenkel, parallel session D8 (these proceedings).
33. A. De Rujula et al., Ann. of Physics 103, 315 (1977).
    For calculations of the anomalous dimensions associated with

higher twist operators see J. Gottlieb, Nucl. Phys. B139, 125 (1978). M.E. Peskin, Phys. Lett. 88B, 128 (1979). M. Okawa, University of Tokyo preprint OUT-337 (1980), contributed paper 347.

34. Z.F. Ezawa, Nuova Cimento 23A, 271 (1974).
    G.R. Farrar and D.R. Jackson, Phys. Rev. Lett. 35, 1416 (1975).
    See also A.I. Vainshtain and V.I. Zakharov, Phys. Lett. 72B, 368 (1978).
35. S.J. Brodsky and E.L. Berger, Phys. Rev. Lett. 42, 940 (1979).
36. S.J. Brodsky and G.P. Lepage, parallel session C11.
37. E.L. Berger, Phys. Lett. 89B, 241 (1980).
38. C. Matteuzzi, parallel session C10, see also the rapporteurs talk by F. Sciulli.
39. For further discussion of higher twist effects at large x see L.F. Abbott et al., Phys. Lett. 88B, 157 (1979) and E.L. Berger, Z. Physik C4, 289 (1980). The combination of the suppression of the leading twist contribution to the pion structure function for $x \to 1$ and trigger bias will make $PP \to \pi x$ at large $p_T$ particularly sensitive to higher twist: see T. Gottschalk, parallel session A6.
40. M. Voloshin and V. Zakharov, DESY 80/41 (1980).
    See also F.R. Ore and G. Sterman, Stony Brook preprint ITP-SB-80-27 (1980).
41. G. Parisi and R. Petronzio, CERN TH 2804 (1980).
42. R.L. Jaffe and G.G. Ross, Phys. Lett. and contribution to parallel session C11. For an interesting attempt to construct pion structure functions see A. DeRujula and F. Martin, MIT preprint CTP 851 (1980), presented in parallel session C11 by F. Martin.
43. Yu.L. Dokshitzer, D.I. Dyakanov and S.I. Troyan, Physics Reports 58, 269 (1980) and references therein.
44. S.J. Brodsky, SLAC-PUB-2447 (1979).
    D. Amati et al., CERN TH 2831 (1980), contributed paper 63.
45. M. Ciafaloni, PISA preprint SNS 1/80, contributed paper 149, presented in parallel session D8.
46. G. Parisi and R. Petronzio, Nucl. Phys. B154, 427 (1979).
    C.Y. Lo and J.D. Sullivan, Phys. Lett. 86B, 327 (1979).
    G.C. Fox and S. Wolfram, Caltech preprint CALT-68-723 (1979).
    S.D. Ellis and W.J. Stirling, Seattle preprint RLO-1388-821 (1980), contributed paper 410 presented by W. Stirling in parallel session C4.
    J.B. McKitterick, Urbana preprint ILL-(TH)-80-22 (1980).
    D.E. Soper, Oregon preprint OITS-134 (1980).
    H.F. Jones and N.S. Craigie, Imperial College preprint ICTP/79-80/21 (1980).
    H.F. Jones and J. Wyndham, Imperial College preprint ICTP/79-80/41 (1980).
47. J.C. Collins and D.E. Soper, contributed abstract 695 presented in parallel sessions A8 and D8.
48. Yu.L. Dokshitzer et al., (ref. 43).
    C.L. Basham et al., Phys. Rev. D19, 2018 (1979).

49. B.H. Wiik, these proceedings. See also F. Halzen and D.M. Scott, Wisconsin preprints COO-881-130 (1980), COO-88-141 (1980) and DOE-ER/0088/-154 (1980).
50. C.L. Basham et al., Phys. Lett. $\underline{85B}$, 297 (1979).
    G.C. Fox and S. Wolfram, ref. 46.
    G. Shirholz, DESY 79/71 (1979).
    P. Binetruy, CERN TH 2807 (1980).
    S.D. Ellis and W.J. Stirling, ref. 46.
    The leading log sum has been compared to the data by
    W. Marquart and F. Steiner, Phys. Lett. $\underline{93B}$, 480 (1980), and
    F. Halzen and D.M. Scott (ref. 49).
51. R. Baier and K. Fey, Bielefeld preprint BI-TP 80/10 (1980).
52. Ch. Berger et al., Phys. Lett. $\underline{90B}$, 312 (1980).
53. W. Furmanski et al., Nucl Phys. $\underline{B155}$, 253 (1979).
    A. Bassetto et al., Nucl. Phys. $\underline{B163}$, 477 (1980).
54. D. Amati and G. Veneziano, Phys. Lett. $\underline{83B}$, 87 (1979).
    A. Bassetto et al., Phys. Lett. $\underline{83B}$, 87 (1979) and ref. 53.
    D. Amati et al., ref. 44.
55. A. Casher et al., Phys. Rev. $\underline{D20}$, 179 (1980).
    See also C.B. Chiu, contributed paper 458.
56. K. Konishi et al., Phys. Lett. $\underline{78B}$, 243 (1979), $\underline{80B}$, 259 (1979) and Nucl. Phys. $\underline{B157}$, 45 (1979).
    A. Bassetto et al., Phys. Lett. $\underline{86B}$, 366 (1979).
    P. Cvitanovic et al., Phys. Lett. $\underline{85B}$, 413 (1979) and Nordita preprint 80/27 (1980).
57. G.C. Fox and F. Wolfram, Nucl. Phys. $\underline{B168}$, 285 (1980)
    S. Wolfram, Caltech preprint 68-778 (1980).
    R. Odorico, contributed paper 283, presented in parallel session C4.
    P. Mazzanti and R. Odorico, Bologna preprints IFUB 12/80 and 13/80 (1980).
    K. Kajantie and E. Pietarinan, Phys. Lett. $\underline{93B}$, 269 (1980).
    C.H. Lai et al., Niels Bohr Inst. preprint $\overline{NB1}$-HE-80-8 (1980), contributed paper 248.
58. P. Mazzanti and R. Odorico, ref. 57.
59. G.C. Fox and F. Wolfram (ref. 57). For analytic discussions of jet evolution see, for example, C.B. Chiu (ref. 55), S. Wolfram (ref. 57) and CALT-68-740 (1979) and L. Caneschi and A. Schwimmer, Phys. Lett. $\underline{86B}$, 179 (1979).
60. K. Kajantie and E. Pietarinan, ref. 57.
61. For some discussion see S. Wolfram, refs. 57 and 59 and R. Hwa, Phys. Rev. Lett. $\underline{44}$, 439 (1980) and V. Chang and R. Hwa, Oregon preprint OITS-138 (1980), presented in parallel session D10.
62. A.V. Efremov and A.V. Radyushkin, contributed paper 794, Riv. Nuovo Cimento $\underline{3}$, No.2 (1980), and Phys. Lett. $\underline{94B}$, 245 (1980) and references therein.
    G.R. Farrar and D.R. Jackson, Phys. Rev. Lett. $\underline{43}$, 246 (1976) and D.R. Jackson, Ph.D. Thesis, Caltech (1977).

G.P. Lepage and S.J. Brodsky in "Quantum Chromodynamics"
ed. W. Fraser and F. Henyey, AIP (1979), Phys. Lett. 87B,
359 (1979), Phys. Rev. Lett. 43, 545 (1979) and (E)43, 1625
(1979), SLAC-PUB-2478 (1980) and contribution to parallel
session C1.   See also S.J. Brodsky, ref. 44.
S.J. Brodsky et al., Phys. Lett. 91B, 239 (1980).
G. Parisi, Phys. Lett. 84B, 225, 1979.
A. Duncan and A.H. Mueller, Phys. Rev. D21, 1636 (1980),
Phys. Lett. 90B, 159 (1980) and 93B, 119 (1980).
M Chase, Oxford preprint 28/80 (1980), contributed paper 57.
S.J. Brodsky, T. Huang and G.P. Lepage, contributed paper
641 presented by T. Huang in parallel session D8.

63. L.N. Lipatov, Sov. J. Nucl. Phys. 20, 94 (1975).
    G. Altarelli and G. Parisi, Nucl. Phys. B126, 298 (1977).
64. V.A. Matveev et al., Lett. Nuovo Cimento 7, 719 (1973).
    S.J. Brodsky and G.R. Farrar, Phys. Rev. Lett. 31, 1153 (1973)
    and Phys. Rev. D11, 1309 (1975).
65. The asymptotic behaviour was first obtained by Farrar and
    Jackson (ref. 62) and A.V. Efremov and A.V. Radyushkin,
    Dubna preprint 11983 (1978).
66. See S.J. Brodsky and G.P. Lepage, ref. 62.
67. S.D. Drell and T.M. Yan, Phys. Rev. Lett. 24, 181 (1970).
    G. West, Phys. Rev. Lett. 24, 1206 (1970).
68. A. Duncan and A.H. Mueller, Phys. Lett. 93B, 119 (1980).
69. S.J. Brodsky and G.P. Lepage, parallel session C1.
70. P.V. Landshoff, Phys. Rev. D10, 1024 (1974).
71. G. Parisi, S.J. Brodsky and G.P. Lepage, ref. 62.
72. P.V. Landshoff and D.J. Pritchard, Cambridge preprint DAMTP
    80/4 (1980).
73. A. Duncan and A.H. Mueller, Phys. Lett. 90B, 159 (1980).
74. A. Donnachie and P.V. Landshoff, Z. Physik C2, 55 and 372 (1972)
75. S.J. Brodsky and G.P. Lepage, SLAC-PUB-2478 (1978).

Chapter 13

$e^+e^-$ Physics and New Particle Production

# NEW $e^+e^-$ PHYSICS

B.H. Wiik, DESY, Speaker

B. McDaniel, Cornell, Chairman

G. Zobernig, DESY,

J. Cleymans, Bielefeld,

D. Haidt, DESY,

Scientific Secretaries

# NEW $e^+e^-$ PHYSICS

B.H.Wiik

Deutsches Elektronen-Synchrotron DESY, Hamburg, Germany.

## I. INTRODUCTION

In this talk I will review data obtained in $e^+e^-$ collisions at PETRA during the past year using four large detectors, MARK J, JADE, PLUTO and TASSO. Each experiment has collected a total luminosity of about 5500 nb$^{-1}$ at c.m. energies between 12 GeV and 36.6 GeV with the bulk of the luminosity ($\sim$ 4800 nb$^{-1}$) at energies above 30 GeV. This corresponds to some 2500 multihadron events per experiment. PLUTO was replaced by a new experiment CELLO early this year.

Several new results have been obtained during the past year. The data[1] on various QED reactions agree with the theory up to $q^2$ = 1000 GeV$^2$ and s = 1200 GeV$^2$. Lepton universality is valid down to distances of 2 x 10$^{-16}$ cm. The cross sections for these reactions have been determined with a precision similar to the size of the effects expected from the electroweak interference terms. The experiments further show[2] that there is no new charged lepton with a mass less than 17 GeV.

Hadrons produced at high energies in $e^+e^-$ annihilation appear in two nearly collinear jets. The charged multiplicity and the charged particle composition of the jets have been determined[3] for momenta up to 5-6 GeV/c and long range charge[4] correlations among the fast particles in the two jets have been observed for the first time. The data[2] also show that the threshold for $t\bar{t}$ production where t is a quark with charge 2/3e must be above 36.0 GeV.

The outstanding experimental result has been the observation[5-9] of three jet events. Such events are evidence for hard gluon bremsstrahlung which is expected to occur in any field theory of strong interactions. Using different methods each group[4,10-12] has observed on the order of 200 clear three jet events. An analysis of these events shows that the gluon is most likely a vector particle and there are first indications that quarks and gluons fragment differently. The strength of the coupling between quarks and gluons has been determined.

A wealth of new information on photon-photon collisions have been reported[13] to the Conference. From the collisions of two real photons there are, besides data on QED and resonance production, also results[14] on $\gamma\gamma \to \rho^0\rho^0$ and on $\gamma\gamma \to$ hadrons. In the multihadron data there are evidence for two jet production as expected if the photon has a pointlike component. Data on deep inelastic electron-photon scattering have also been reported for the first time.

In this talk I'll describe these results in more detail. A more complete discussion can be found in the minirapporteur talks on PETRA physics referenced above.

## II. ELECTROWEAK REACTIONS

The Feynman graphs for Bhabha scattering, lepton pair production and two photon annihilation are shown in Fig. 1. Effects caused by

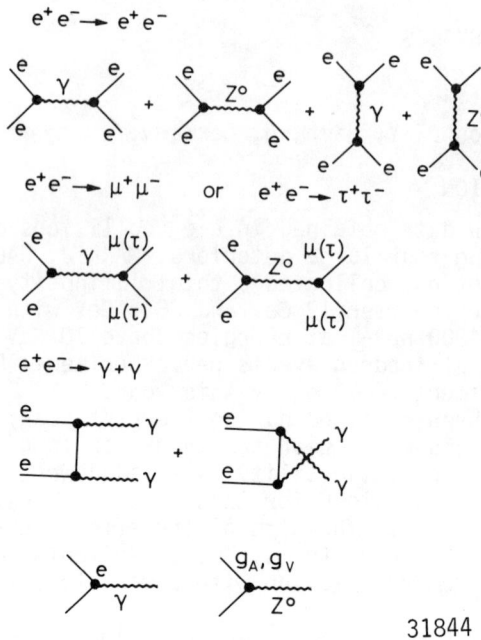

Fig. 1 - Feynman graphs for
a) $e^+e^- \to e^+e^-$
b) $e^+e^- \to \mu^+\mu^- (\tau^+\tau^-)$
c) $e^+e^- \to \gamma\gamma$

the interference of the weak and the electromagnetic currents start to become visible at the highest energies now available at PETRA and we will return to these effects after a brief discussion on QED limits.

## II.1 TEST OF QED

The QED predictions are based on the validity of the Maxwell equations and on the assumption that leptons are pointlike objects without excited states. The reactions above make it possible to test these assumptions at very small distances in a clean environment with only small corrections due to strong interactions.

The standard procedure used to compare data with the QED predictions can be summarized as follows:
1) Weak effects are neglected.
2) The measured cross section $d\sigma_0/d\Omega$ is corrected[15] for radiative effects $\delta_R$ and effects due to the hadronic vacuum polarization $\delta_H$.

$$\frac{d\sigma_c}{d\Omega} = \frac{d\sigma_0}{d\Omega}(1 + \delta_R + \delta_H). \qquad (1)$$

3) The corrected cross section is compared to the QED predicted cross section and deviations are parametrized[16] in terms of form factors. The formfactors used for Bhabha scattering and lepton pair production can be written as:

$$F_s(q^2) = 1 \mp \frac{q^2}{q^2 - \Lambda_{s\pm}^2} \qquad F_t(s) = 1 \mp \frac{s}{s - \Lambda_{t\pm}^2} \qquad (2)$$

where $F_s$ and $F_t$ are respectively the formfactors for spacelike and timelike momentum transfer squared.

The reaction $e^+e^- \to \gamma\gamma$ is modified by a form factor[17] of the type $F(q^2) \sim 1 \pm q^4/\Lambda_\pm^4$. Exchange of a heavy electronlike lepton would modify[18] the cross section as

$$F(s) \sim 1 + (s^2/2\Lambda^4)\sin^2\theta \qquad (3)$$

where $\Lambda$ is the mass of the heavy lepton.

All groups working at PETRA have data[2,19] on these reactions and some typical results are shown in Figs. 2-5.

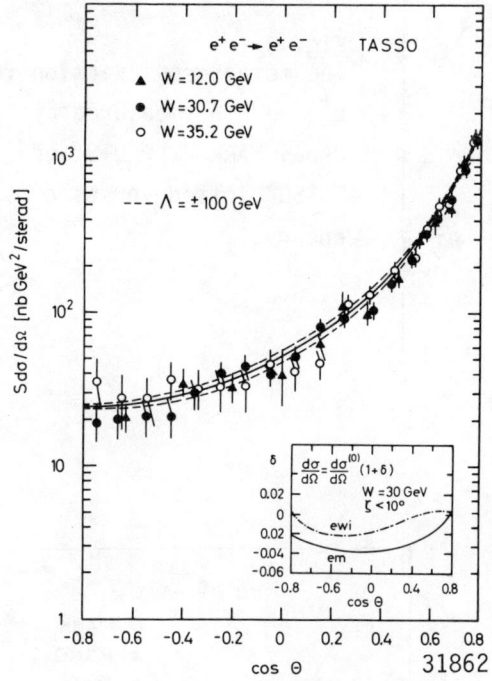

Fig. 2 - The cross section $s \cdot d\sigma/d\Omega$ for $e^+e^- \to e^+e^-$ measured by TASSO between 12.0 GeV and 35.2 GeV in c.m.

The cross section $s\, d\sigma/d\Omega(e^+e^- \to e^+e^-)$ measured by the TASSO Collaboration for c.m. energies between 12 and 35.2 GeV is plotted in Fig.2 versus scattering angle $\theta$. The data scatter around the QED prediction shown as the solid curve, the dotted curve indicates the limits corresponding to a cut off parameter $\Lambda = 100$ GeV.

The total cross section for $e^+e^- \to \mu^+\mu^-$ measured by JADE, MARK J, PLUTO and TASSO is plotted in Fig. 3 versus c.m. energy. The data agree well with the QED prediction shown as the solid curve and they are in general within the dotted curves corresponding to a cutoff parameter $\Lambda$ of 100 GeV.

The total cross section for $e^+e^- \to \tau^+\tau^-$ measured by MARK J, PLUTO and TASSO is plotted in Fig. 4 versus c.m. energy. This reaction has a very distinct signature at high energies and is easily separated from multihadron reactions. Again the data are in good agreement with the QED prediction shown as the solid curve.

The angular distribution for $s\, d\sigma/d\Omega$ ($e^+e^- \to \gamma\gamma$) measured by the JADE and PLUTO Collaborations is plotted in Fig.5 for c.m. energies between 12 and 31.6 GeV. The data are in good agreement with the QED prediction, the deviation corresponding to a cutoff parameter $\Lambda = 40$ GeV is shown as the dotted curve.

The data are therefore in agreement with QED and the limits on $\Lambda$ are summarized in Table 1.

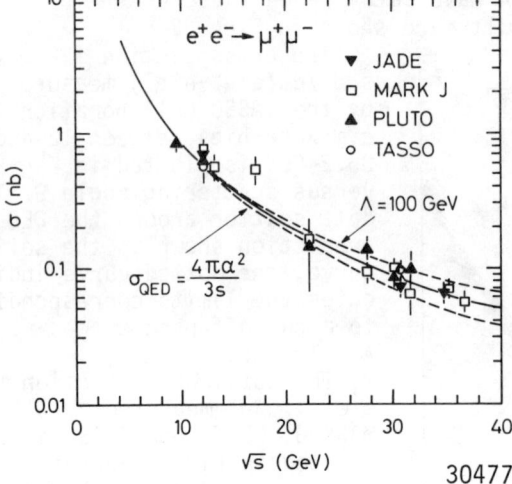

Fig. 3
The total cross section for $e^+e^- \to \mu^+\mu^-$ measured by JADE, MARK J, PLUTO and TASSO plotted versus c.m. energy.

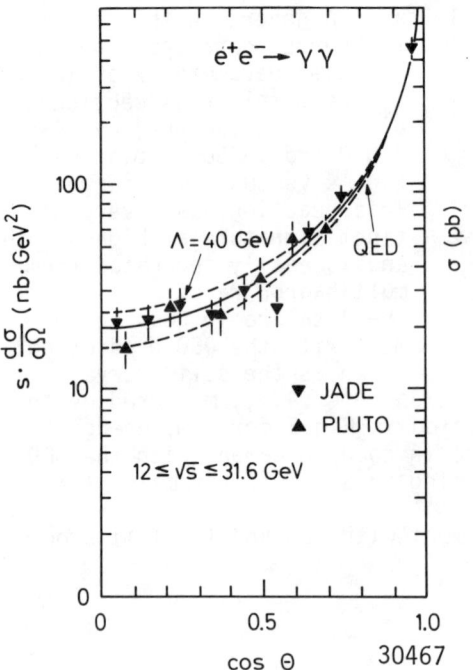

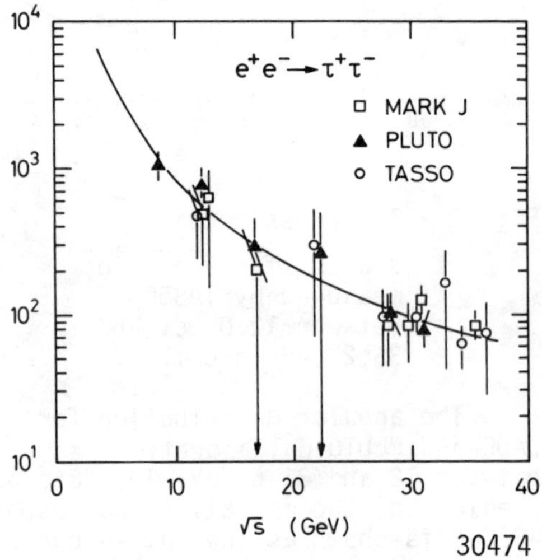

Fig. 4
The total cross section for $e^+e^- \to \tau^+\tau^-$ measured by MARK J, PLUTO and TASSO plotted versus c.m. energy.

Fig. 5
The cross section $s\frac{d\sigma}{d\Omega}$ for $e^+e^- \to \gamma\gamma$ measured by JADE and PLUTO plotted versus $\cos\theta$.

Table 1 - Limits on $\Lambda_{\pm}$

| | JADE | MARK J | PLUTO | TASSO |
|---|---|---|---|---|
| $e^+e^- \to e^+e^-$ | | | | |
| $\Lambda_+$ | 112 | 91 | 80 | 150 |
| $\Lambda_-$ | 106 | 142 | 234 | 136 |
| $e^+e^- \to \mu^+\mu^-$ | | | | |
| $\Lambda_+$ | 137 | 123 | 116 | 80 |
| $\Lambda_-$ | 96 | 142 | 101 | 118 |
| $e^+e^- \to \tau^+\tau^-$ | | | | |
| $\Lambda_+$ | - | 76 | 74 | 115 |
| $\Lambda_-$ | - | 154 | 65 | 76 |
| $e^+e^- \to \gamma\gamma$  $F(q^2) = 1 \pm q^4/\Lambda_{\pm}^4$ | | | | |
| $\Lambda_+$ | - | 44 | 46 | - |
| $\Lambda_-$ | - | 34 | 36 | - |
| heavy electron $\Lambda^*$ | | | | |
| $\Lambda_+$ | 47 | 55 | 46 | 34 |
| $\Lambda_-$ | 44 | 38 | - | 42 |

From this table we conclude the leptons are indeed pointlike down to distances of about $2 \times 10^{-16}$ cm. Furthermore there is no evidence for a charged electronlike lepton up to a mass of 40 - 50 GeV.

## II.2 ELECTROWEAK EFFECTS

The interference between the electromagnetic and the neutral weak current[20] will change the normalized QED cross section for muon and tau pair production by $\Delta R$ and lead to a forward backward asymmetry A in the angular distribution of the leptons in the final state. At present PETRA and PEP energies these effects can be written[21] as:

$$\Delta R = \frac{G_F}{\sqrt{2} \cdot \pi \cdot \alpha} \cdot \frac{s \cdot m_Z^2}{s - m_Z^2} g_V^2 + \left(\frac{G_F}{2\sqrt{2}\, \pi\alpha} \frac{s\, m_Z^2\, (g_V^2+g_A^2)}{s-m_Z^2}\right)^2 \quad (4)$$

$$A_{\mu\mu} = \frac{F - B}{F + B} = \frac{3}{4} \frac{G_F}{\sqrt{2} \cdot \pi \cdot \alpha} \frac{s \cdot m_Z^2}{s-m_Z^2} g_A^2 \simeq 2.7 \cdot 10^{-4} \frac{m_Z^2 \cdot s}{s - m_Z^2} g_A^2 \quad (5)$$

Here F and B denotes the number of negative muons (taus) in the forward, respectively in the backward hemisphere. In the standard model[20] with $\sin^2\theta_W = 0.23$, $m_Z = 89$ GeV, $g_V = 1/2(1-4\sin^2\theta_W) = 0.04$ and $g_A = -1/2$ where $g_V$ and $g_A$ denote the vector and the axial vector coupling of the neutral current to a pair of charged leptons. At $s = 1000$ GeV$^2$ this leads to $\Delta R = 0.002$ and $A_{\mu\mu} = -0.076$.

The change in ΔR cannot be observed at present energies whereas a measurement of the asymmetry is within reach.

The asymmetry data obtained by the various PETRA groups are listed in Table 2.

Table 2 - Forward-Backward asymmetry in $e^+e^- \to \mu^+\mu^-$

| Group | JADE | MARK J | PLUTO | TASSO |
|---|---|---|---|---|
| $A_{\mu\mu}$ (%) | -8 ± 9 | 0 ± 9 | 7 ± 10 | 1 ± 12 |

The systematic uncertainties are quite small and the data from the various groups can therefore be combined. The combined angular distribution is plotted in Fig. 6 and it yields $<A_{\mu\mu}> = -(0.9 \pm 4.9)\%$ to be compared to the predicted value of -6% including acceptance corrections.

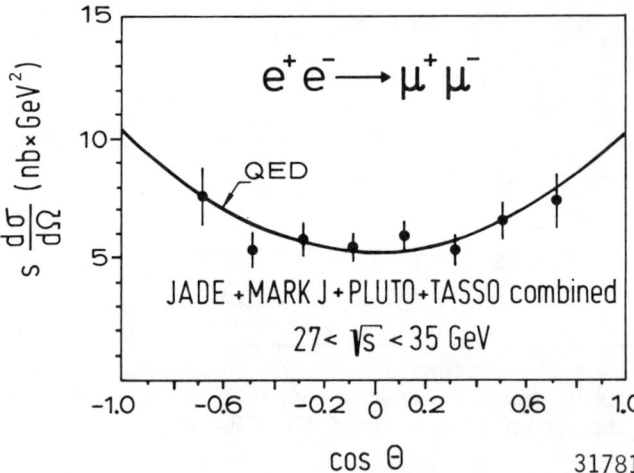

Fig. 6

The combined angular distribution of $S\, d\sigma/d\Omega$ for $e^+e^- \to \mu^+\mu^-$ at c.m. energies between 27 and 35 GeV.

The 95% upper confidence limit is $|A_{\mu\mu}| < 9\%$ i.e. $g_A^e \cdot g_A^\mu < 0.375$ compared to the theoretical value of 0.25 in the standard model.

The relative deviation of the Bhabha cross section due to weak effects, is plotted[22] in Fig. 7 versus scattering angle for various values of $m_Z$. Also indicated are the deviations expected for a cut off parameter $\Lambda$ of 250 GeV. It is clear that weak effects in $e^+e^- \to e^+e^-$ cannot be parametrized in terms of $\Lambda$ and they should be included in the theoretical cross section before extracting a value for $\Lambda$.

The deviation of Bhabha scattering from the lowest order QED prediction as measured by MARK J from c.m. energies between 29.9 and 35.8 GeV is plotted in Fig. 8 and compared to various predictions of the standard model with $\sin^2\theta_W$ = 0.25, 0.01 and 0.55 respectively. Note that they can only determine the cross section between 0° and 90° since they do not determine the charge. The data favours $\sin^2\theta_W = 0.25$ however, better data at higher energies are needed to set stringent limits on $\sin^2\theta_W$ from this reaction.

Using the standard model the data on $e^+e^- \to e^+e^-$, $e^+e^- \to \mu^+\mu^-$ can be used to extract values on $\sin^2\theta_W$. The results are listed in

Table 3 - Results on $\sin^2\theta_W$

| Group | limits on $\sin^2\theta_W$ | | $\sin^2\theta_W$ |
|---|---|---|---|
| | lower | upper | |
| MARK J | 0.07 | 0.42 | 0.24 ± 0.11 |
| JADE | - | 0.55 | 0.25 ± 0.18 |
| PLUTO | - | 0.57 | 0.23 ± 0.17 |
| TASSO | - | 0.52 | - |

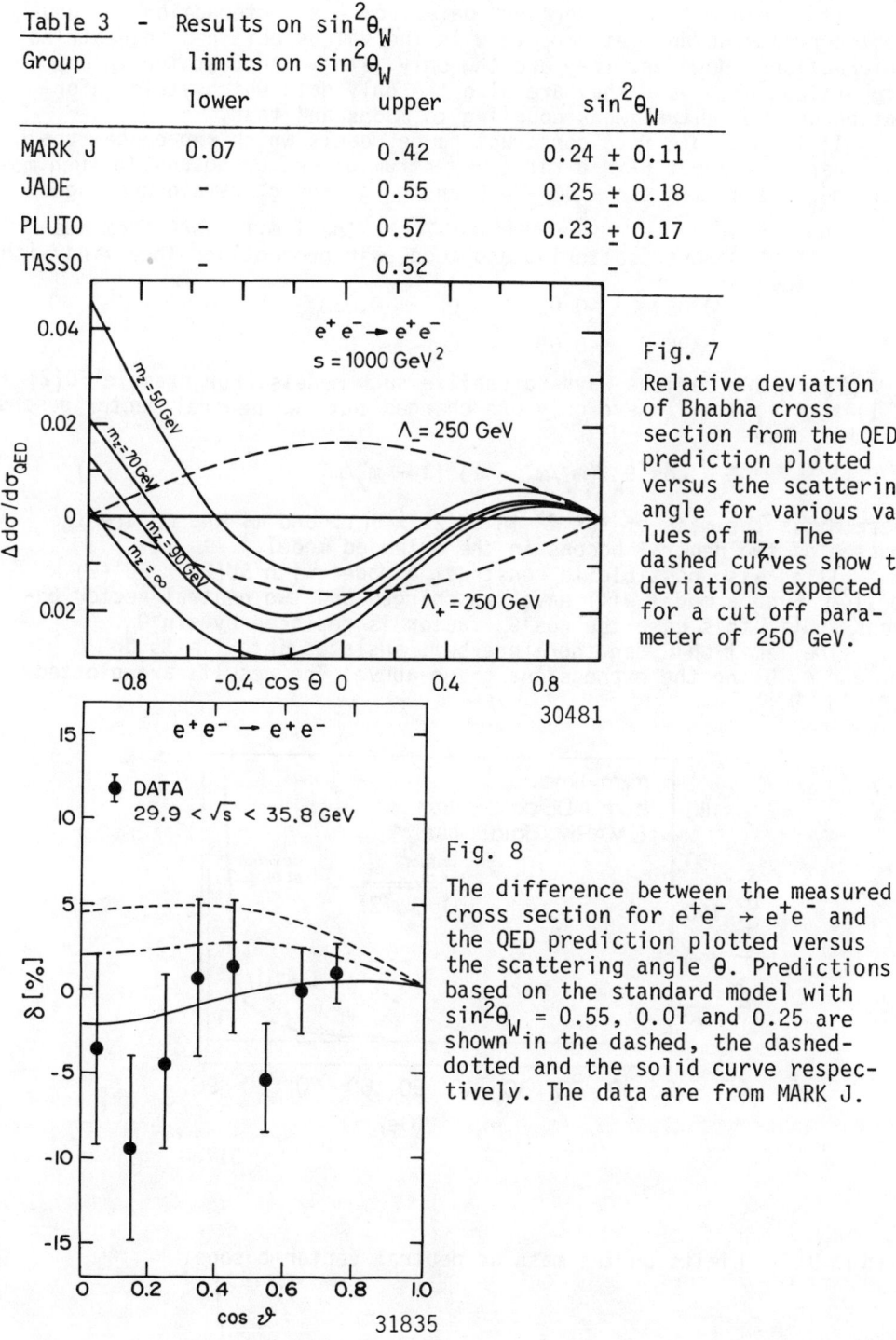

Fig. 7 Relative deviation of Bhabha cross section from the QED prediction plotted versus the scattering angle for various values of $m_Z$. The dashed curves show the deviations expected for a cut off parameter of 250 GeV.

Fig. 8 The difference between the measured cross section for $e^+e^- \to e^+e^-$ and the QED prediction plotted versus the scattering angle $\theta$. Predictions based on the standard model with $\sin^2\theta_W = 0.55$, 0.01 and 0.25 are shown in the dashed, the dashed-dotted and the solid curve respectively. The data are from MARK J.

It is clear that the present data from $e^+e^-$ interactions on neutral currents do not yet compete with the values obtained in neutrino interactions. However, they are the only data which test the theory at high values of $Q^2$ and they are also the only data which yield information on the neutral weak coupling to muons and taus.

It is possible[23] to construct gauge models which reproduce the low energy data but have a richer spectrum of vector bosons. In such models $g_V^2 = 1/4 \ (1-4 \sin^2\theta_W)^2 + 4 \ C$ and $g_V g_A$ and $g_A^2$ remain unchanged.

JADE and MARK J have determined[2,24,25] the limits on C from measurements of Bhabha scattering and muon pair production. They find with 95% confidence:

$$\text{JADE} \qquad -0.059 < C < 0.033$$
$$\text{MARK J} \qquad -0.097 < C < 0.027 \ .$$

There are various ways to realize such models. For example $SU(2) \times U(1) \times U'(1)$[26] will have only one charged but two neutral vector bosons. In this case

$$C = \cos^4\theta_W \ (m_Z^2/m_1^2 - 1) \ (1 - m_Z^2/m_2^2) \qquad (6)$$

here $m_Z$ is the mass of the $Z^0$ in $SU(2) \times U(1)$ and $m_1$ and $m_2$ are the masses of two neutral bosons in the extended model.

It is also possible to construct a model with $SU(2) \times SU'(2) \times U(1)$[27]. Such a model will have two charged and two neutral vector bosons, and in this case the $\cos^4\theta_W$ factor is replaced by $\sin^4\theta_W$.

The limit on C can therefore be translated into limits on $m_1$ and $m_2$ using the expressions given above. The results are plotted in Fig. 9.

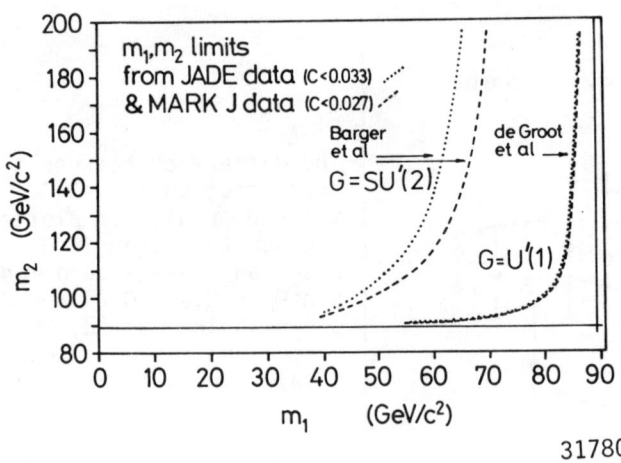

Fig. 9 - Limits on the mass of neutral vector bosons.

## II.3 HEAVY LEPTONS

It seems reasonable to expect that the charged lepton in a new generation of elementary fermions is lighter than the quarks. Leptons are pairproduced with a known cross section and decay either leptonically $L \to \ell \bar{\nu}_\ell \nu_L$ or semileptonically $L \to \nu_L$ hadrons. All the groups working at PETRA have searched[2,28] for new leptons. No evidence was found, and the limits set on the mass of a new lepton are summarized in Table 4.

Table 4 - Mass limits on new leptons

| Group | PLUTO | MARK J | TASSO | JADE |
|---|---|---|---|---|
|  | 14.5 | 16.0 | 15.5 | 17 |

PLUTO and MARK J have searched by selecting events in which a single high energy muon was recoiling against many hadrons. PLUTO demanded that the visible energy of the event should be greater than 3 GeV and the missing momentum greater than 2.5 GeV/c. Furthermore the thrust should be less than 0.95. MARK J required that the visible energy should be greater than 10% but less than 50% of the c.m. energy. The acoplanarity should be greater than $30^0$. The acoplanarity is defined as the absolute value of $(180^0 - \delta\phi)$, where $\delta\phi$ is the angle between the muon momentum vector and the total energy flow vector (see below) of the hadrons $\vec{E}_H$ projected on a plane perpendicular to the beam line. The energy deposited in the outer calorimeter should be 0.1 $E_{vis}$. The event should contain more than two charged tracks and the polar angle between the beam line and the energy flow vector should be between $30^0$ and $150^0$.

TASSO selected events in which a single isolated charged particle with momentum greater than 1.5 GeV/c was separated by at least $90^0$ from any other charged track. The event should contain at least 5 charged tracks and the charged energy should be greater than 8.0 GeV (9.3 GeV) at 30 GeV (35 GeV) in c.m. A similar search was also made requiring the track to be a lepton.

JADE considered events with a visible energy between 11 GeV and 32 GeV produced at 35 GeV in c.m. They searched for non coplanar events as follows: They defined two planes, the first plane was defined by the thrust of one of the "jets" and the $e^+$ direction, the second plane by the momentum of the remaining particles. The opening angle between the two planes should be greater than $45^0$, and the angle between the thrust axis and the $e^+$ direction should be at least $45^0$.

In the present phenemenology[29] of supersymmetric[30] theories all particles will have partners which differ in spin by half a unit. Thus there will be scalar electrons (muons) which can be pairproduced in $e^+e^-$ annihilation and which may decay into electrons (muons) and undetectable particles (photinos, goldstinos) leading to acoplanar two prong electron (muon) events. No evidence[2,28] was found resulting in the following limits on the mass of a scalar lepton: PLUTO $m_s > 13$ GeV, JADE $m_s > 16$ GeV, MARK J $m_s > 16$ GeV.

## III. HADRON PRODUCTION IN $e^+e^-$ ANNIHILATION

It has been conjectured[31-34] in the naive parton model that hadron production in $e^+e^-$ annihilation proceeds by quark-antiquark pair production as shown in Fig.10a, where the electromagnetic current couples proportional to the charge of a pointlike quark. The neutral weak current is expected to contribute on the order of 1% to the total cross section at $s = 1000$ GeV$^2$ energies and is neglected. The total cross section for hadron production in this approximation should therefore be proportional to the cross section for muon pair production with the constant of proportionality

$$R = \frac{\sigma(e^+e^- \to \text{hadrons})}{\sigma(e^+e^- \to \mu^+\mu^-)} = 3 \sum_i \left(\frac{e_i}{e}\right)^2 \qquad (7)$$

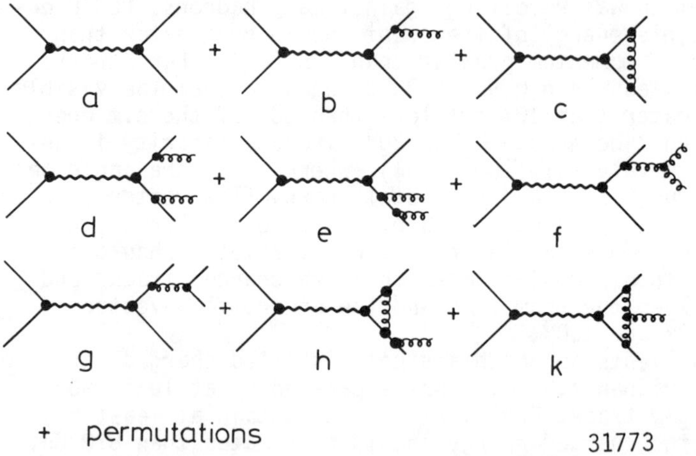

Fig. 10 Some of the diagrams for hadron production in $e^+e^-$ annihilation up to second order in $\alpha_s$.

+ permutations

Here $e_i$ is the charge of the ith flavour and the sum is over all flavours with masses $< E_{beam}$. The hadrons should then appear in two nearly collinear jets of hadrons with small and maybe constant momenta transverse and large and growing momenta parallel to the jet axis. The single particle distribution should scale i.e.

$$s \cdot \frac{d\sigma}{dx}$$

with $x = E_h/E_{beam}$ should become independent of energy at large energies. The charged particle multiplicity would be expected to increase logarithmically with $s = (2E)^2$. The data[35] from SPEAR and DORIS at lower energies support the gross features of this picture.

This naive parton picture will be modified in any field theory[36] of strong interactions. In a field theory $e^+e^-$ annihilation proceeds to lowest order by the Feynman graphs shown in Fig. 10 b, c. The produced quark radiate field quanta (gluons) and the gluons are expected to materialize as hadron jets in the final state.

This has well defined experimental implications[36-38]: The mean transverse momentum of the hadrons with respect to the jet axis will increase with energy. If the quark-gluon coupling constant is small only one of the two original jets will broaden. A primordial $q\bar{q}g$

state is necessarily planar and the final hadron configuration should retain the planarity. In a small fraction of the events the gluon may be radiated at an angle which is large compared to the angular spread of the hadron jet. Such events will be very striking with three visible jets of hadrons defining a plane. Higher order multiple gluon emission diagrams (Fig. 10d-f) are expected to become more visible at high energies since the angular spread of the hadrons resulting from the non-perturbative fragmentation of a single quark or gluon decreases rapidly with energy, enabling one to pick out at higher energies jets from gluons radiated at smaller angles relative to the primordial q and q̄ directions. Such multijet events are of course not planar in general and will lead to an increase of the momentum transverse to the event plane. A field theory of the strong interactions will also modify the value for R given above, the multiplicity will grow faster than ln s and the single particle distribution will no longer scale[36].

At present quantum chromodynamics (QCD)[39] is the leading candidate for a theory of strong interactions. The coupling strength in this theory depends on a characteristic strong interaction mass $\Lambda$ and a typical momentum transfer q in the process. The functional form is given by:

$$\alpha_s(q^2) = g^2/4\pi = \frac{12\pi}{(33-2N_f) \ln q^2/\Lambda^2} \qquad (8)$$

where $N_f$ is the number of flavours with mass below threshold.

Although the exact value of $\Lambda$ is still a subject of some controversy[40] it is presumably rather small, on the order of one to a few hundred MeV.

Here I will first discuss the gross properties of the final state, then summarize the evidence for gluon bremsstrahlung and finally discuss the properties of the gluon in some detail.

## IV GENERAL PROPERTIES OF THE FINAL STATE IN $e^+e^- \to$ HADRONS

The basic diagrams (Fig. 10) governing $e^+e^- \to q\bar{q}(g) \to$ hadrons are very simple. The properties of the hadrons in the final state can therefore be directly related to the properties of quarks and gluons and their fragmentation into hadrons.

### IV.1 THRUST AND SPHERICITY DISTRIBUTIONS

The energy dependences of the average sphericity $<S>$[33] and $(1 - <T>)$, where $<T>$ is the average thrust[41], are plotted versus c.m. energy together with data obtained at lower energies in Figs. 11 and 12.

Both quantities decrease with increasing energies as expected if the jets become more collimated with increasing energies. The jet cone half opening angle, as indicated from the sphericity distribution, shrinks from about 31° at 4 GeV in c.m. to 17° near 36 GeV. However, this decrease is slower than that expected in a pure $e^+e^- \to q\bar{q}$ model. The observed decrease is in agreement with computations including gluon bremsstrahlung. Note that the distributions are smooth indicating the absence of thresholds in the

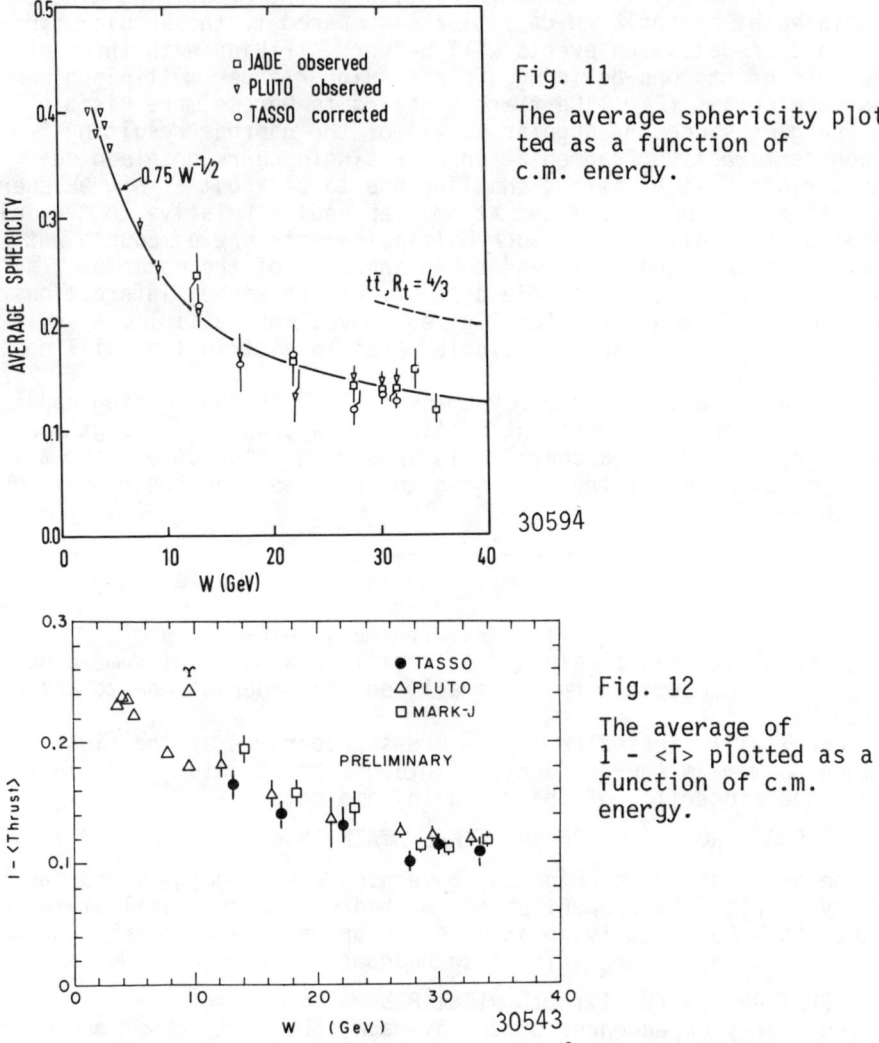

Fig. 11
The average sphericity plotted as a function of c.m. energy.

Fig. 12
The average of 1 - <T> plotted as a function of c.m. energy.

IV.2 THE TOTAL CROSS SECTION

The PETRA groups report data[2] on R for center of mass energies between 12 GeV and 36.5 GeV. The data, corrected for radiative effects including the vacuum polarization and with the contribution from τ pair production removed, are plotted versus c.m. energy in Fig. 13a together with data obtained at lower energies[42].

In the parton model with u, d, s, c and b quarks $R = 3\Sigma_i e_i^2 = 11/3$. This expression is modified[43] in first order QCD to

$$R = 3 \sum_i e_i^2 (1 + \alpha_s/\pi). \qquad (9)$$

At $s = 1000$ GeV$^2$, $\alpha_s/\pi$ is of the order of 5% yielding $R \simeq 3.9$. Higher order terms[43] depend on the renormalization scheme used but they are smaller than the first order term. The QCD prediction, plotted in Fig. 13a, is in agreement with the data.

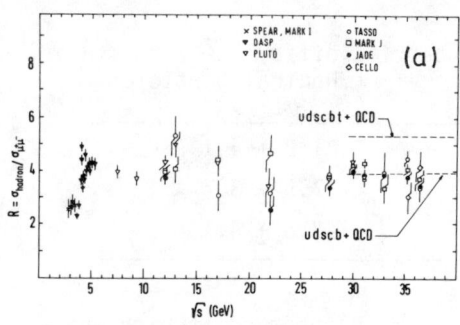

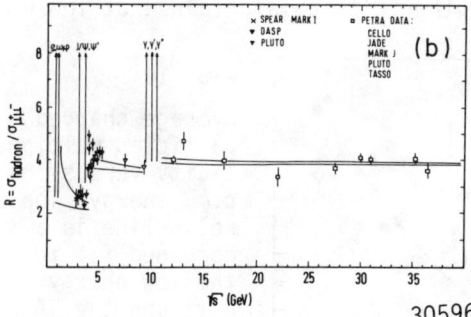

Fig. 13 -
a) The ratio R of the total hadronic cross section normalized to the muon pair cross section is plotted as a function of c.m. energy.
b) The value of R obtained by averaging the data of all the PETRA groups. The solid line represents two QCD predictions with $\Lambda = 1.0$ GeV (upper curve) and $\Lambda = 0.1$ GeV (lower curve).

Are the QCD corrections needed to fit the data ? Clearly not, since the systematic errors are believed to be of the order of 10%. However, note from Fig. 13a that the data which were collected using different trigger conditions and analysed using different cuts are in agreement within the statistical errors. This indicates that the systematic errors may be smaller than 10% and indeed there are good reasons to expect that R can eventually be measured rather well in the PETRA energy range.

It is therefore tempting to add the data from the various groups and the resulting cross section is plotted in Fig.13b. The solid lines are QCD predictions corresponding to $\Lambda = 1.0$ GeV and 0.1 GeV respectively. Averaging all the data above 20 GeV in c.m. yields $R = 3.97 \pm 0.06$. An error of 0.16 was computed from the fluctuations of the individual measurements. First order QCD at $s = 1000$ GeV$^2$ predicts $R = 3.87$ for $\Lambda = 100$MeV or $R = 3.94$ for $\Lambda = 500$ MeV.

## IV.3 THE NEUTRAL ENERGY FRACTION

The JADE Collaboration has determined[3] the fraction of the total energy converted into photons by a direct measurement of the photon energy deposited in lead glass counters surrounding the detector. They have also determined the total neutral energy fraction by measuring the energy carried away by charged particles and subtracting this from the known c.m. energy. The results, listed in Table 5 show that the neutral energy fraction, which includes $K^0_S$

and Λ's, increase with energy. Also the energy fraction carried off by photons seems to increase. However, in this case the errors are rather larger.

Table 5 - Energy fraction carried off by photons and by neutral particles

| $\sqrt{s}$ (GeV) | Energy fraction carried off by | |
|---|---|---|
| | Photons % | Neutral particles % |
| 12 | 21.3 ± 7.0 | 31.2 ± 4.1 |
| 30.4 | 26.1 ± 5.9 | 37.5 ± 3.7 |
| 34.9 | 30.7 ± 6.0 | 43.8 ± 4.1 |

## IV.4 CHARGED MULTIPLICITIES

The average charged multiplicity $<n_{ch}>$ observed at high energies[44-46] is plotted in Fig. 14 together with data obtained at lower

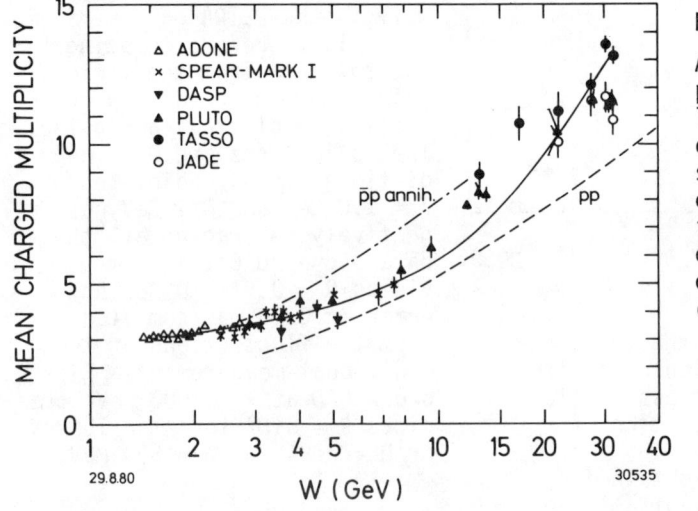

Fig. 14

Average charged particles multiplicity versus the c.m. energy. The solid line is a combined fit to the low energy data and the TASSO data at high energies.

energies[47]. The high energy data points from the various groups are in reasonable agreement and are well above the multiplicities obtained by extrapolating the lower energy data according to a + b ln s as predicted by the naive quark-parton model. For comparison, the multiplicities observed in pp[48] and p̄p[49] are also shown.

The average multiplicity in QCD may[50-51] increase as $n = n_0 + a \exp(b \sqrt{\ln s/\Lambda^2})$, and the data can indeed be fitted over the whole energy range using this form. The values of the parameters obtained by fitting the TASSO and the PLUTO data are listed in Table 6. The fit considers only the statistical errors and the results were obtained assuming Λ = 0.5 GeV/c.

Table 6. - Fits to the charged particle multiplicity

| Group | $n_o$ | a | b |
|---|---|---|---|
| TASSO | 2.92 ± 0.04 | 2.85 ± 0.07 | 0.0029 ± 0.0005 |
| PLUTO | 2.38 ± 0.09 | 1.92 ± 0.07 | 0.04 ± 0.01 |

The asymptotic value[51] of a = 2.4 in QCD. However, note that the fragmentation of the gluon in three jet events is expected to increase the average multiplicity by less than one unit at the highest energy.

## IV.5 INCLUSIVE PARTICLE SPECTRA

The scaled cross sections $s\,d\sigma/dx$ for inclusive charged particle production as determined by DASP[47], SLAC-LBL[52] and TASSO[3,53] for c.m. energies between 5 GeV and 36.6 GeV are plotted in Fig. 15 versus x.

The cross section for x > 0.2 scales to within 30% between 5GeV and 36.6 GeV. For x < 0.2 the cross section increases dramatically with energy and shows that the observed increase in multiplicity is due to slow particles. Gluon emission will lead[36] to a depletion of particles at large x and a corresponding increase in the yield at small x, since the energy is now shared between the quark and the gluon. In QCD these are rather small effects except at very large or very small x. In general the effects are only of the order of 10-20% at PETRA energies since $q^2$ is very large compared to $\Lambda^2$.

During the past year the PETRA experiments have succeeded in identifying hadrons over a considerable range in momentum. The available data[3] are summarized in Table 7.

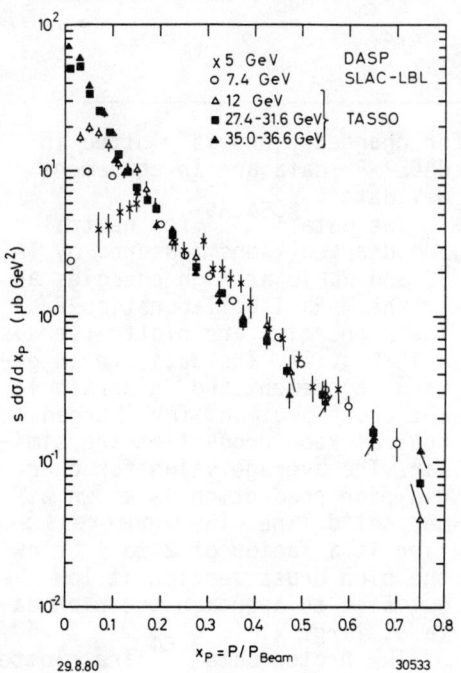

Fig. 15 - The scaled cross section $s\,d\sigma/dx$ ($x = p/p_{beam}$) for inclusive charged particle production.

Table 7 - Experiments measuring particle separated cross section

| Type of particle | Experiment | Technique | Momentum range(GeV/c) | Remark |
|---|---|---|---|---|
| $\pi^{\pm}$ | JADE | dE/dx [3] | < 0.7, 2-7 | preliminary |
| | TASSO | TOF [54] | < 1.1 | |
| | | Cerenkov [3] | < 5.0 | preliminary |
| $K^{\pm}$ | JADE | dE/dx [3] | < 0.7 | preliminary |
| | TASSO | TOF [54] | < 1.1 | |
| | | Cerenkov [3] | < 5.0 | preliminary |
| $K^0, \overline{K^0}$ | PLUTO | $K_S^0 \to \pi^+\pi^-$ [3] | all p | preliminary |
| | TASSO | $K_S^0 \to \pi^+\pi^-$ [3] | all p | |
| p, $\bar{p}$ | JADE | dE/dx [3] | < 0.9 | preliminary |
| | TASSO | TOF [54] | < 2.2 | |
| | | Cerenkov [3] | < 4.0 | |

The scaled cross section $s/\beta \, d\sigma/dx$ for charged pions is plotted in Fig. 16 versus x. The TASSO and the JADE[3,54] data are in agreement and seem to fall below the DASP 5.2 GeV data[47].

The data[3,54,55] for neutral and charged kaons measured by TASSO and PLUTO at high energies and by the MARK I Collaboration[56] at lower energies are plotted versus x in Fig. 17. The data are in general agreement, and in particular the cross sections for charged and neutral kaon production are similar. The average value for charged pion production is shown as the solid line. The kaon cross section is a factor of 2 to 4 below the pion cross section at low x but seem to approach the pion data at large x.

The proton data[3,54] are plotted in Fig. 18 and compared to the charged pion data represented by the solid lines. Within the rather large error bars the cross section for kaon and proton production are similar. The large p, $\bar{p}$ cross section seems surprising.

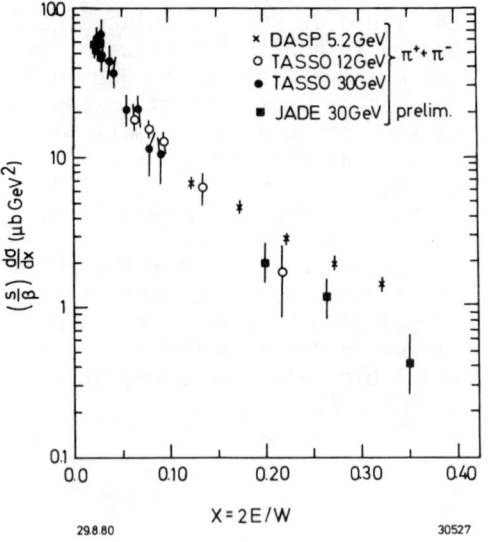

Fig. 16 - The scaled cross section $s/\beta \, d\sigma/dx$ for charged pions.

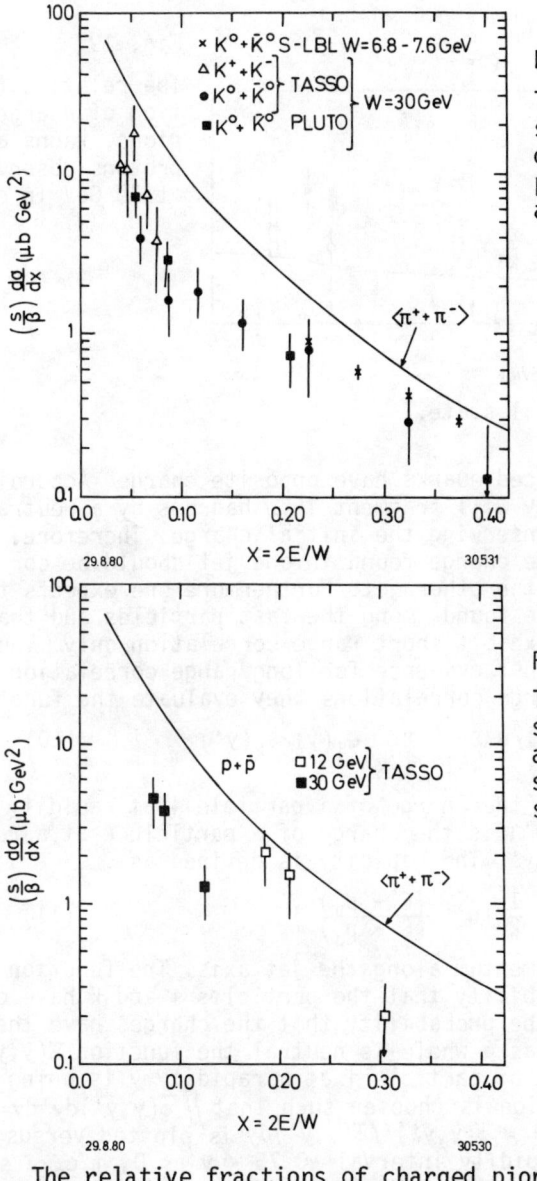

Fig. 17

The scaled cross section s/β dσ/dx for neutral and charged kaons. The average pion cross section is shown as the solid curve.

Fig. 18

The scaled cross section sβ dσ/dx for protons. The average charged pion cross section is shown as the solid curve.

The relative fractions of charged pions, kaons and protons observed at 30 GeV in c.m. are plotted in Fig. 19 as a function of particle momentum. At low momenta nearly all the particles are pions, however, the kaon and the proton yield rises rapidly with momentum such that at a momentum of 3.0 GeV/c the ratio of

$\pi^\pm$ to $K^\pm$ to $p^\pm$    is roughly    55 to 35 to 10.

An average event at a center of mass energy of 30 GeV consists of roughly 10 $\pi^\pm$, 1.4 $K^0\bar{K}^0$, 1.4 $K^+K^-$ and 0.4 $p\bar{p}$ i.e. about one out of 5 events

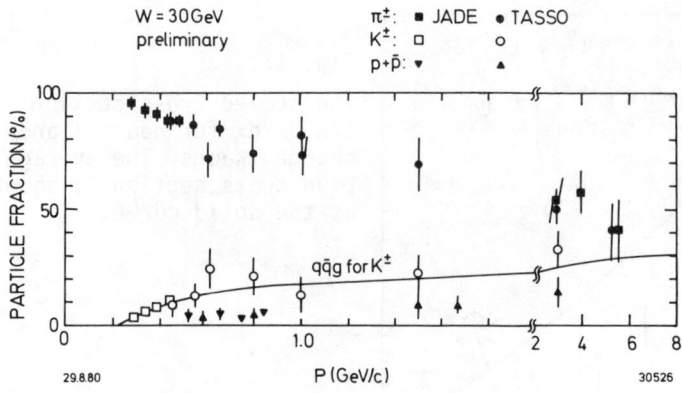

Fig. 19

The relative fraction of charged pions, kaons and protons observed at 30 GeV in c.m.

has an p, p̄ pair in the final state.

## IV.6 CHARGE CORRELATIONS

The back to back produced quarks have opposite charge. According to the standard picture they will fragment into hadrons by a neutral quark-gluon cascade thus conserving the initial charge. Therefore, apart from fluctuations, the charge found in one jet should be correlated with the charge of the other jet. Furthermore one expects this long range correlation to be found among the fast particles and that the slow particles should exhibit short range correlation only. The TASSO group reports[4] the first evidence for long range correlation.

To investigate the charge correlations they evaluate the function

$$\bar{\phi}(y,y') = -\frac{1}{\Delta y \Delta y'} \langle 1/n \sum_{k=1}^{n} \sum_{i \neq k} e_i(y) e_k(y') \rangle \quad (10)$$

In this expression $e_i(y)$ is the charge of a particle i at rapidity y in the interval $\Delta y$ and $e_k(y')$ is the charge of a particle k at a rapidity y' in the interval $\Delta y'$. The rapidity is defined as

$$y = \frac{1}{2} \ln \left( \frac{E + p_{\parallel}}{E - p_{\parallel}} \right) \quad (11)$$

where $p_{\parallel}$ is the particle momentum along the jet axis. The function $\bar{\phi}(y,y')$ is simply the probability that the particles i and k have opposite sign charges minus the probability that the charges have the same sign. Since the event as a whole is neutral the function $\bar{\phi}(y,y')$ simply shows how the charge of particle i at a rapidity y is being compensated. The normalization is choosen such that $\iint \bar{\phi}(y,y')dy'dy=1$. In Fig.20 the ratio $\phi(y,y') = \bar{\phi}(y,y')/\int \bar{\phi}(y,y')dy$ is plotted versus y with the test particle in the rapidity interval $-0.75 < y < 0$, i.e. a slow particle. This distribution peaks at small negative values of y' and shows that the charge of a slow particle is indeed compensated locally as expected if only short range correlations are present. The observed peak has an rms width of 1.3.

In Fig. 20 the same quantity is plotted as a function of y for the test particle at $-5 < y' < -2.5$. Although the bulk of the charge is compensated locally there is now a significant signal at the opposite end of the rapidity plot. Integrating the distribution for y > 2.5

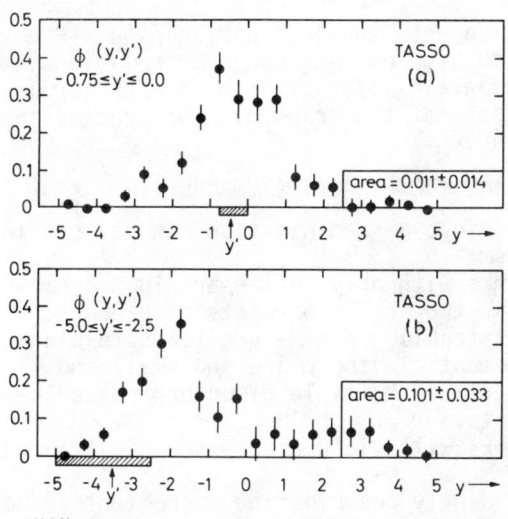

Fig. 20

The charge compensation function $\phi(y, y')$ with the test particle

a) at $-0.75 \leq y' \leq 0$ and

b) at $-5.0 \leq y' \leq -2.5$.

yields $0.101 \pm 0.033$ compared to the $0.011 \pm 0.014$ found for the test particle at $-0.75 \leq y' \leq 0$. There is therefore a clear signature for a long range correlation extending over some 7 units in rapidity.

It is interesting to compare the charge correlation to the particle density distribution defined by:

$$\bar{\rho}(y,y') = \frac{1}{\Delta y \Delta y'} < \frac{1}{n(n-1)} \sum_{k=1}^{n} \sum_{i \neq k} |e_i(y)||e_k(y')|> \quad (12)$$

The quantity $\rho(y,y') = \bar{\rho}(y,y') / \int dy \int dy' \bar{\rho}(y,y')$ with $\int dy \int dy' \bar{\rho}(y,y') = 1$ is the probability to find a charged particle with rapidity $y$ if there is another charged particle with rapidity $y'$. This particle density function is plotted in Fig. 21a and b for the test particle at $-0.75 \leq y' \leq 0$ and at $-5 \leq y' \leq -2.5$. Comparing Figs. 20 and 21 shows that the particle density function is wider than the charge correlation function - i.e. unlike sign particles are on the average closer in rapidity than like sign particles.

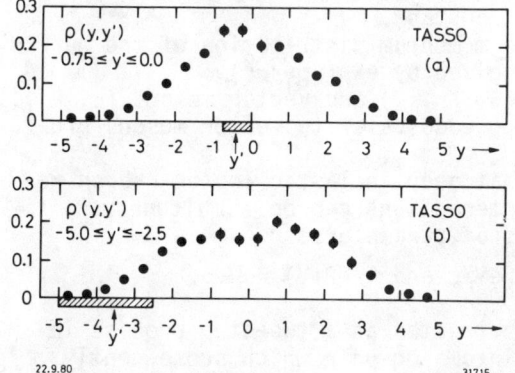

Fig. 21

The particle density function with the test particle

a) at $-0.75 \leq y' \leq 0$ and

b) at $-5.0 \leq y' \leq -2.5$.

The particle density function with the test particles at $-5 < y' \leq -2.5$ is rather smooth with no sign of long range correlations.
The correlation functions contain information on the primordial quarks and are more sensitive tests of the fragmentation process than the single particle distributions.

## V. EVENT TOPOLOGY AND THE FINAL STATE ANALYSIS

The topology of the hadrons in $e^+e^-$ annihilation can be used to identify the production mechanism
a) Pair production of light quarks with only collinear gluon bremsstrahlung will manifest itself as two collinear jets of hadrons.
b) Single wide angle gluon bremsstrahlung $e^+e^- \rightarrow q\bar{q}g$ leads to planar events with large and growing momenta in the plane and small and limited momenta transverse to the plane. Multiple gluon bremsstrahlung will lead to more isotropic events.
c) Pair production of heavy quarks will yield nearly spherical events close to threshold.

In the next paragraphs we briefly describe the ingredients used in the Monte Carlo simulations and the methods used to determine the topology of the hadrons in the final state.

### V.1 INPUTS TO THE MONTE CARLO SIMULATION

All groups have made extensive Monte Carlo computations to confront the various production mechanismns with the data. The inputs to these calculation are summarized below:
a) Quark pairs are pair-produced proportional to $e_i^2$. Light quark pairs are created from the vacuum in the ratio $u\bar{u} : d\bar{d} : s\bar{s} = 2 : 2 : 1$.
b) The basic gluon bremsstrahlung process (Fig. 10) is treated to first order in the strong coupling constant $\alpha_s$ by Hoyer et al.[58] whereas the computation by Ali et al.[59] includes all second order diagrams except those with internal gluon lines (Fig. 10 h, k).

The formalism of Field and Feynman[60] or the one set up by the Lund group[61] is then used to compute the fragmentation of the constituents.
c) The fragmentation of the quarks is described by 3 parameters in the Field-Feynman model:
i) $a_F$: The quark fragments $q \rightarrow q' + k$ according to a distribution function $f^h(z) = 1 - a_F + 3a_F(1-z)^2$ with $z = (p_{\parallel} + E)_h/(p_{\parallel} + E)_q$. $a_F$ is the same for u, d and s quarks and is determined experimentally. For the heavy quarks c and b: $f^h(z)$ = constant.
ii) $\sigma_q$: The primordial transverse momentum distribution of the quarks with respect to the jet axis is given by $\exp(-p_q^2/\sigma_q^2)$.
iii) P/(P+V): Only pseudoscalar ($\pi$, K ...) and vector mesons ($\rho$, K ...) are produced; P/V is the ratio of pseudoscalar to vector mesons produced in the primordial cascade.

Field and Feynman found[60] that deep inelastic lepton-hadron reactions and also hadron-hadron interactions can be simultaneously described by the following values of parameters:

$$a_F = 0.77, \quad \sigma_q = 0.30 \text{ GeV/c and } P/(P+V) = 0.5.$$

d) The fragmentation of gluons is treated as a two-step process in which the gluon first fragments into a $q\bar{q}$ pair which subsequently fragments into hadrons as outlined above. In the Hoyer et al. program[58] the gluon imparts all its energy to one of the quarks - i.e. in this model quark and gluon fragmentation are identical, Ali et al.[59] describe $\to q\bar{q}$ by the splitting function[62] $f(z) = z^2 + (1-z)^2$ where $z = E_g/E_q$.

## V.2 EVENT TOPOLOGY

The production mechanism can be delineated from the event shape. There are by now several methods used to determine the shape and the topology of an event. These methods are briefly discussed below.

The shape of an event is conveniently evaluated by constructing the second rank tensor[33,35]

$$M_{\alpha\beta} = \sum_{j=1} p_{j\alpha} \cdot p_{j\beta} \quad (\alpha, \beta = x, y, z) \tag{13}$$

where $p_{j\alpha}$ and $p_{j\beta}$ are momentum components along the $\alpha$ and $\beta$ axes for the jth particle in the event. The sum is over all charged particles in the event. Let $\vec{n}_1, \vec{n}_2, \vec{n}_3$ be the unit eigenvectors of this tensor associated with the normalized eigenvalues $Q_i$, where $Q_i = \Sigma(\vec{p}_j \cdot \vec{n}_i)^2/\Sigma p_j^2$. These eigenvalues are ordered such that $Q_1 < Q_2 < Q_3$ and are normalized with $Q_1 + Q_2 + Q_3 = 1$. The principal jet axis is then the $\vec{n}_3$ direction. The event plane is spanned by $\vec{n}_2$ and $\vec{n}_3$; and $\vec{n}_1$ defines the direction in which the sum of the square of the momentum component is minimized. Every event can be represented in a two dimensional plot of aplanarity $A = (3/2) Q_1$ (i.e. normalized momentum squared out of the event plane) versus sphericity $S = (3/2)(Q_1 + Q_2)$. In such a plot two jet events will cluster at small values of A and S, planar events have small values of A whereas both A and S will be large for spherical events. This is borne out by the Monte Carlo results shown in Fig. 22. This method has been used by TASSO[6] and JADE[9].

MARK J[7,63] uses a linear method based on energy flow where the coordinate system is defined as follows: the $\vec{e}_1$ axis coincides with the thrust axis which is defined as the direction of maximum energy flow. They next investigate the energy flow in a plane perpendicular to the thrust axis. The direction of maximum energy flow in that plane defines a direction $\vec{e}_2$ with a normalized energy flow

$$\text{major} = \sum_i |\vec{p}^i \cdot \vec{e}_2| / E_{vis} \tag{14}$$

where $E_{vis} = \Sigma |\vec{p}^i|$ the third axis $\vec{e}_3$ is orthogonal to both the thrust and the major axis $\vec{e}_2$, and it is very close to the minimum of the momentum projection along any axis i.e.

$$\text{minor} = \sum_i |\vec{p}^i \cdot \vec{e}_3| / E_{vis}. \tag{15}$$

The PLUTO group[64] has developed a two step cluster method to determine the event topology. The first stage associates all particles into preclusters irrespective of their momenta. Particles belong to the same precluster if the angles between any two tracks are less than a limiting angle $\alpha$. The momentum of a precluster is the sum of the momenta of all the particles assigned to that precluster. The preclusters are then combined to clusters if the angle between the momentum vectors is less than a given value $\beta$. The number of clusters

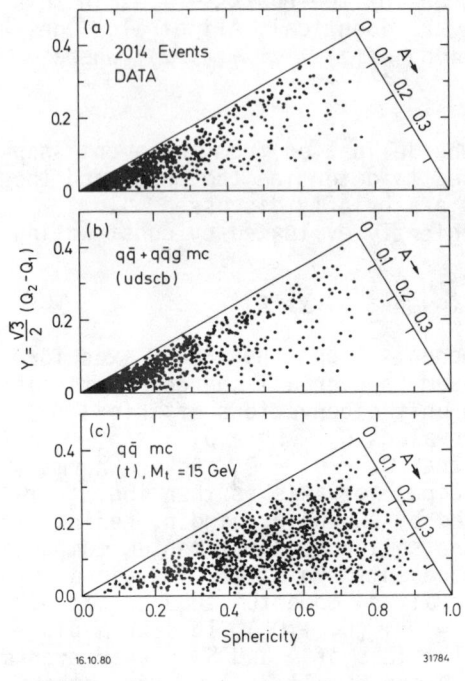

Fig. 22

a) The event distribution in aplanarity and sphericity observed by the TASSO Collaboration between 27.4 GeV and 36.6 GeV in c.m.

Monte Carlo created events in aplanarity and sphericity for

b) at 30 GeV in c.m.
$e^+e^- \to q\bar{q}g$ with
$q = u, d, s, c, b$.

c) $e^+e^- \to t\bar{t}$ with

$m_t$ = 15 GeV and at c.m. energy of 35 GeV.

n is defined as the minimum number of clusters which fulfil the inequalities

$$\sum_{i=1}^{n} E_{ci} > E_{vis} (1 - \varepsilon) \qquad (16)$$

where $E_{ci}$ is the cluster energy and $\varepsilon$ a small number. If the energy of a cluster, defined as the sum of the energies of all particles assigned to the cluster, exceeds a threshold energy $E_{th}$ then the cluster is called a jet. Typical values for the various parameters are $\alpha = 30°$, $\beta = 45°$, $\varepsilon = 0.1$ and $E_{th}$ = 2.0 GeV.

### V.3 EVIDENCE AGAINST NEW QUARKS

The distribution of events in the A, S plane observed[2] by TASSO at c.m. energies between 27.4 and 36.6 GeV is shown in Fig. 22a. The data cluster at small values of S and A with a long tail of planar events as expected for light quark production including gluon bremsstrahlung as shown in Fig. 22(b).

In the data at 35.0 ≤ W ≤ 36.6 GeV there are 2 events with A > 0.15 whereas we expect for a heavy quark a total of 57 events for charge 2/3e and 14 events for a charge 1/3e. Combining these data with similar data[2] from JADE and MARK J excludes a charge 2/3e heavy quark with a mass between the b quark and 18 GeV by some 12 standard deviations. The existence of a charge 1/3e quark is also rather unlikely.

Scanning the cross section has also failed to find any evidence for narrow states. The limit is $\Gamma_{ee} \cdot B_h$ < 0.4 keV for c.m. energies between 35.0 and 35.6 GeV. For a charge 2/3e quark or a 1/3e quark we expect to find respectively $\Gamma_{ee}$ = 5 keV or 1.3 keV.

## VI. GLUONS

At present QCD is the only theory of strong interactions available, and it is obviously crucial to carry out clean experiments which either support or refute this theory. The first step is to demonstrate that field quanta, gluons, indeed do exist. However, this is not sufficient since presumably any field theory of strong interactions contains gluons. To "prove" QCD one must demonstrate that the gluon is a flavour neutral, coloured vector particle with gauge couplings.

### VI.1 THE EVIDENCE FOR GLUONS

Gluon bremsstrahlung $e^+e^- \to q\bar{q}g$ (see Fig. 10) has well defined experimental signatures.

A) The average transverse momentum of the hadrons with respect to the jet axis will grow with energy. Normalized transverse momentum distributions, measured by TASSO and evaluated with respect to the sphericity axis are plotted in Fig. 23 versus $p_T^2$ for different c.m. energies. The observed $p_T^2$ distribution clearly broadens with energy. In QCD the growth is explained as hard non-collinear gluon emission. Fits based on this mechanism are shown in Fig. 23. However, it is also possible to fit the data up to moderate values of $p_T^2$ by increasing $\sigma_q$ as a function of c.m. energy.

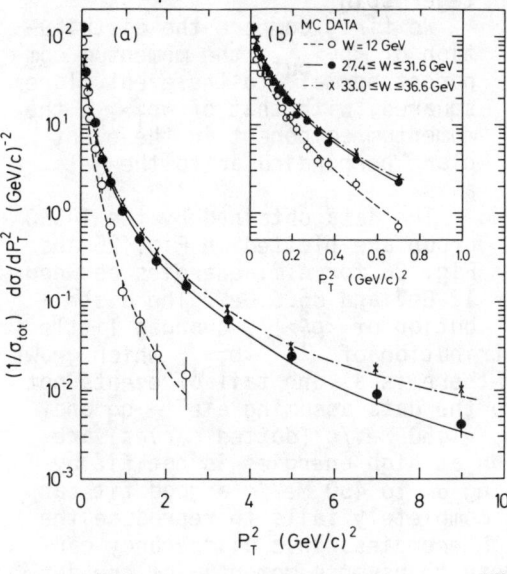

Fig. 23 - $1/\sigma \, d\sigma/dp_T^2$ at 12 GeV, 27.4-31.6 GeV and 33.0-36.6 GeV as a function of $p_T^2$. The curves are QCD fits to the data with $\sigma_q$ = 320 MeV/c.

B) If hard non-collinear gluon emission is a rare process as expected in QCD, then there should usually be only one wide-angle gluon per event: In fact the probability of emitting two gluons in one event compared to single gluon emission is proportional to $\alpha_s$. Hence only one of the jets should broaden.

To test this prediction the jets in an event are divided into a narrow and a wide jet. The data obtained by PLUTO[8] are shown in Fig. 24. Plotted are $<p_T^2>$ versus $z = p/p_{beam}$ at low and high energies for the wide and the narrow jet separately. A large asymmetry between the two jets is observed at high energies. Unlike the asymmetry observed at low energy the PLUTO group find that this cannot be explained by fluctuations in the two jet events.

c) Planarity. Regardless of the value of $\sigma_q$ (or the mean $p_T$), hadrons resulting from the fragmentation of a quark must on the average be uniformly distributed

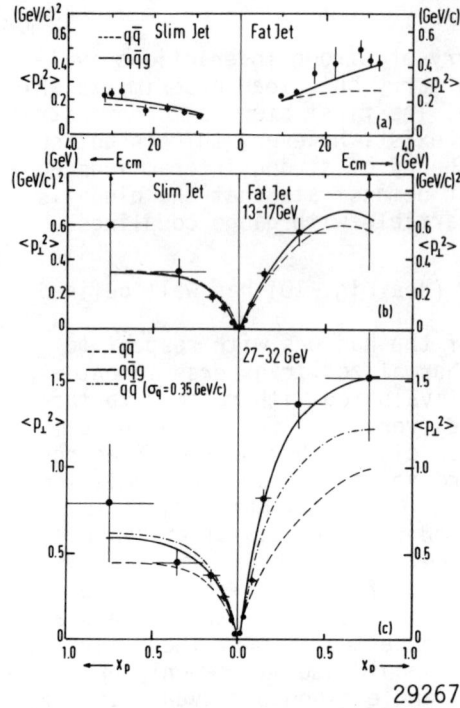

Fig. 24 - Data obtained by PLUTO on $\langle p_T^2 \rangle$ as a function of $z = p/p_{beam}$ for wide and narrow jets. The solid and the dashed curves are the $q\bar{q}g$ and $q\bar{q}$ predictions, respectively.

in azimuthal angle around the quark axis. Therefore, apart from statistical fluctuations, the two jet process $e^+e^- \to q\bar{q}$ will not lead to planar events whereas the radiation of a hard gluon, $e^+e^- \to q\bar{q}g$ will result in an approximately planar configuration of hadrons with large transverse momentum in the plane and small transverse momentum with respect to the plane. Thus the observation of such planar events, at a rate significantly above the rate expected from statistical fluctuations of the $q\bar{q}$ jets, shows in a model independent way that there must be a third confined particle in the final state. The third particle is not a quark since it has baryon number zero and cannot have 1/2 integer spin.

We first compare the distribution of $\langle p_T^2 \rangle_{out}$, the momentum component normal to the event plane squared, with that of $\langle p_T^2 \rangle_{in}$, the momentum component in the event plane perpendicular to the jet axis.

The data obtained by the TASSO group are plotted in Fig. 25 and Fig. 26 for c.m. energies between 12 GeV and 36.6 GeV. The distribution of $\langle p_T^2 \rangle_{out}$ changes little with energy in contrast to the distribution of $\langle p_T^2 \rangle_{in}$ which grows rapidly with energy, in particular there is a long tail of events not observed at lower energies. Fits to the data assuming $e^+e^- \to q\bar{q}$ and $\sigma_q$ = 300 MeV/c (solid curves) or $\sigma_q$ = 450 MeV/c (dotted curves) are also shown. The $\langle p_T^2 \rangle_{out}$ distribution at high energies is not fit by $\sigma_q$ = 300 MeV/c, however by increasing $\sigma_q$ to 450 MeV/c a good fit can be obtained. The $q\bar{q}$ model however, completely fails to reproduce the long tail observed in $\langle p_T^2 \rangle_{in}$ at high energies. This discrepancy cannot be removed by increasing the mean transverse momentum of the jet. Fig. 25 shows a fit assuming $\sigma_q$ = 450 MeV/c (which gives a good fit to $1/\sigma \, d\sigma/dp_T^2$ and to $\langle p_T^2 \rangle_{out}$). The agreement is poor. We therefore conclude that the data include a number of planar events not reproduced by the $q\bar{q}$ model independent of the average $p_T$ assumed.

Gluon bremsstrahlung offers a natural mechanism to explain the observed planarity of the events. Fig. 26 shows a second order QCD fit to the data using the Monte Carlo method outlined above. The fit assumed a constant value of $\sigma_q$ = 320 MeV and $\alpha_s$ = 0.17 (see below). The

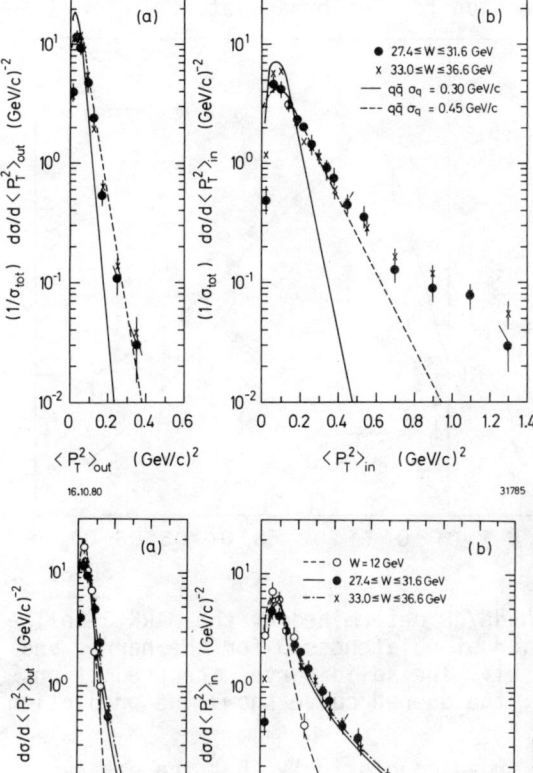

Fig. 25
Distributions of mean transverse momentum squared per event for charged particles, normal to ($<p_T^2>_{out}$) and in ($<p_T^2>_{in}$) the event plane measured by the TASSO Collaboration at low and high energies. The curves are the predictions for a $q\bar{q}$ final state with $\sigma_q$ = 300 MeV/c (solid lines) and $\sigma_q$ = 450 MeV/c (dashed lines).

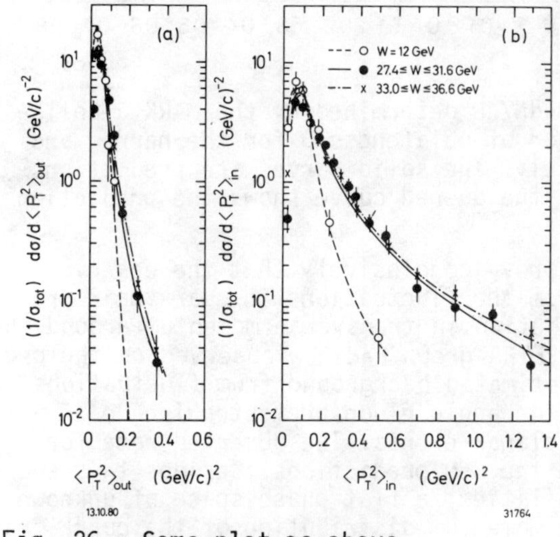

Fig. 26 - Same plot as above. The curves are the second order QCD predictions with $\alpha_s$ = 0.17 and $\sigma_q$ = 320 MeV/c.

long tail in $<p_T^2>_{in}$ is reproduced in this model. Note that the growth in $<p_T^2>_{out}$ is explained by the occurence at small fraction of 4 (or more) jet events which, in general, are not planar.

The data from PLUTO[8] and JADE[9] analyzed in a similar manner are in full agreement with the findings of the TASSO group.

The planarity of the events is also observed[7,10,63] by the MARK J group using a different technique. They divided each event into two hemispheres using the plane defined by the major and the minor axis (see above) and analyzed the energy distribution in each hemisphere as if it were a single jet. The jet with the smallest transverse momentum with respect to the thrust axis is defined as the narrow jet. the other as the broad jet. The oblateness defined as O = Major - Minor is a measure of the planarity of the event and is zero for phase space and two jet events and finite for three jet final states. The normalized event distribution measured for c.m. energies between 27 and 37 GeV is plotted versus oblateness in Fig. 27 for the narrow and the wide jet separately and compared to the predictions for $e^+e^- \to q\bar{q}$ (dashed curve) and $e^+e^- \to q\bar{q}g$ (solid line). A good fit is obtained with the $q\bar{q}g$ final state whereas the $q\bar{q}$ final state do not fit the oblateness distribution for the broad jet.

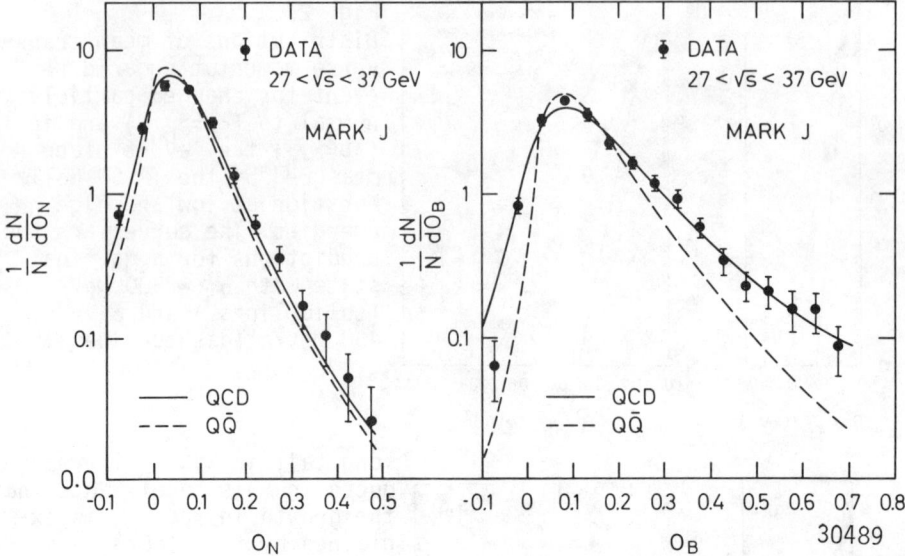

Fig. 27 - The distribution 1/N dN/dO determined by the MARK J Collaboration as a function of oblateness O for the narrow and the wide jet separately. The solid curves are predictions based on $e^+e^- \to q\bar{q}g$, the dashed curve shows the prediction for $e^+e^- \to q\bar{q}$.

The data discussed above prove conclusively that the observed planar events cannot result from the fluctuations in quark pair production with a Gaussian distribution in transverse momentum around the jet axis of the hadrons. Each PETRA group had now observed on the order of 200 planar events with an estimated background from fluctuations of two jet events of about 20%. Wide angle gluon bremsstrahlung $e^+e^- \to q\bar{q}g$ would naturally result in planar events. The observed rate for such events is consistent with the QCD predictions. Besides this source there are two ad hoc possibilities; a flat phase space of unknown origin, or that the transverse momentum distribution of the quark fragmentation has a long non-Gaussian tail. The first possibility can be excluded by observing events with 3 axes, the second by excluding the possibility that the 3 axes are defined by 2 multiparticle jets and a single high momentum particle at a large angle with respect to the jet axes.

D) Properties of planar events. The TASSO Collaboration use a generalization[65] of sphericity to define three-jet events. In this method the tracks are projected on to the event plane defined by $\vec{n}_2$ and $\vec{n}_3$ (see above). The projections are divided into three groups and the sphericity for each group $S_1$, $S_2$ and $S_3$ determined. The three axes and the particle assignment to the three groups are defined by minimizing the sum of $S_1$, $S_2$ and $S_3$. This defines the direction of the three jets and assigns the particles to these jet directions.

In Fig.28 the TASSO events are plotted versus tri-jettiness $J_3$ defined as

$$J_3 = \langle p_T^2 \rangle_{in} / (\tfrac{1}{2}(300 \text{ MeV}/c^2)^2).$$

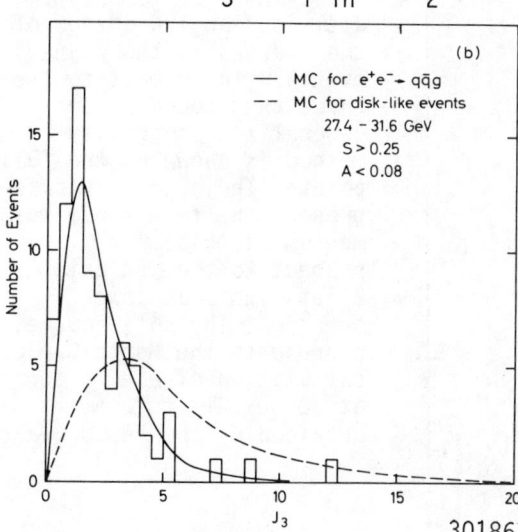

Here $\langle p_T^2 \rangle_{in}$ is evaluated for all charged tracks in an event with respect to their assigned axis. Thus for three jet events with a mean transverse momentum of 300 MeV with respect to the jet axis we expect to find the events clustered around $J_3=1$, compared with a wide distribution in $J_3$ in case of a flat phase space distribution. The data agree with the expectations for $e^+e^- \to q\bar{q}g$, shown as the solid line. The fit result in $\chi^2$/degree of freedom of 2.3/5. The data disagree strongly with a phase space calculation shown as the dashed line. This fit has a $\chi^2$/degree of freedom of 233/5. Thus the data are not consistent with a phase space distribution.

Fig. 28 - Planar events (S > 0.25, A < 0.08 measured by the TASSO Collaboration and plotted versus the tri-jettiness $J_3$. The Monte Carlo predictions for $e^+e^- \to q\bar{q}g$ (solid) and for $e^+e^- \to$ hadrons (dash).

The TASSO group has also evaluated the transverse momentum of charged particles from three jet events with respect to the jet axes to which they were assigned. This distribution $1/N \, dN/dp_T^2$ is plotted as the solid points in Fig. 29 versus $p_T^2$. It is compared to the $p_T^2$ distribution found with respect to the jet axis in two jet events shown as the open points. The agreement is very good and demonstrates that $\langle \sigma_q \rangle$ can be taken to be constant independent of energy, when the events are analyzed as three jet events.

The MARK J group observes a three jet structure in their energy flow analysis. To enhance effects resulting from gluon emission they select events with low thrust T < 0.8 and large oblateness O > 0.1. Fig. 30 shows the energy distribution of these events in the plane defined by $\vec{e}_1$ and $\vec{e}_2$. Plotted is the energy deposited in 5° bins as a function of angle. A clear three peak structure is seen. Plotted are the predictions from QCD, phase space + $q\bar{q}$ with $\sigma_q$ = 300 MeV/c and $q\bar{q}$ only with $\sigma_q$ = 500 MeV. Normalized to the total event sample, only the gluon bremsstrahlung hypothesis fits the data.

The JADE group[86] used an independent method suggested by Ellis and Karliner[67] to demonstrate the existence of three jet events. From the data taken at c.m. energies around 30 GeV they selected planar events which satisfied the condition $Q_2 - Q_1 > 0.1$ and determined the thrust axis for each event. The event was then divided into two jets by a plane normal to the thrust axis and the $p_T$ for each jet computed separately; the jet with the smallest $p_T$ was called the slim jet, the

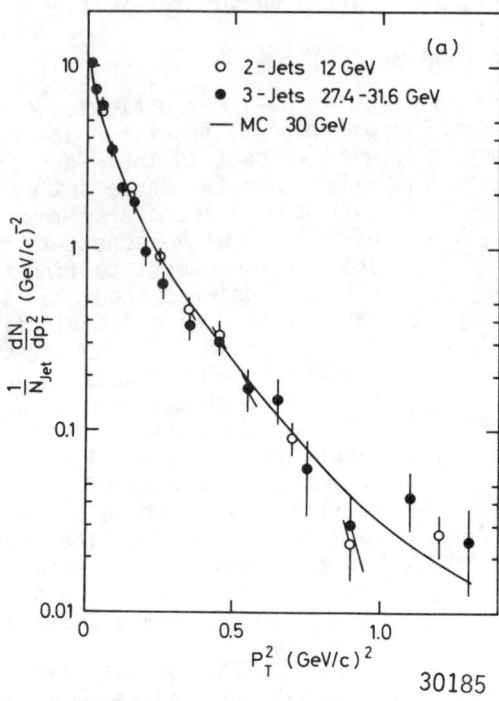

Fig. 29
The transverse momentum distribution $1/N\, dN/dp_T^2$ of the hadrons in the planar events with respect to the three axes found by the generalized sphericity method is shown as the full points. The open points represent the transverse momentum distribution with respect to the jet axis in 2 jet events at lower energies. The solid curve represents the Monte Carlo calculation of $e^+e^- \to q\bar{q}g$ at 30 GeV. The data were obtained by the TASSO group.

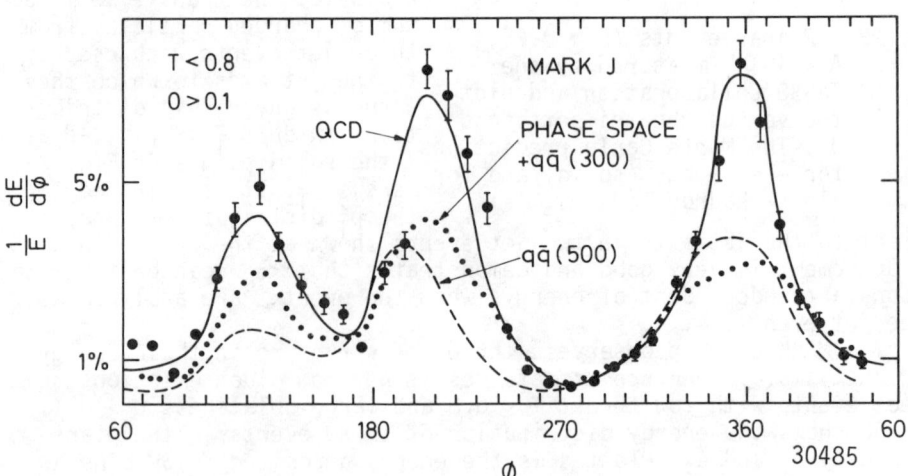

Fig. 30 - A plot of the energy distribution in the plane defined by the thrust and the major axes for all events with thrust < 0.8 and oblateness > 0.1. The measurements were done by the MARK J group. The QCD fit is shown by the solid line, a mixed phase space and $q\bar{q}$ model is shown as the dotted line, a pure $q\bar{q}$ model with $\sigma_q$ = 500 MeV is given by the dashed line.

other the broad jet. The particles in the broad jet are then transformed into its own rest system. If the broad jet consists of two jets they will now appear as two back to back jets along the new thrust axis T*. The distribution of T* in this system is plotted in Fig.31 together with the thrust distribution of two jet events measured at 12 GeV. The two distributions are in excellent agreement. Also other quantities like the invariant mass, mean $p_T$ and charged multiplicity evaluated for the broad jet in its own rest system are in agreement with the same quantities evaluated for a two jet event at 12 GeV. The data therefore exhibit a three jet structure as expected for gluon bremsstrahlung.

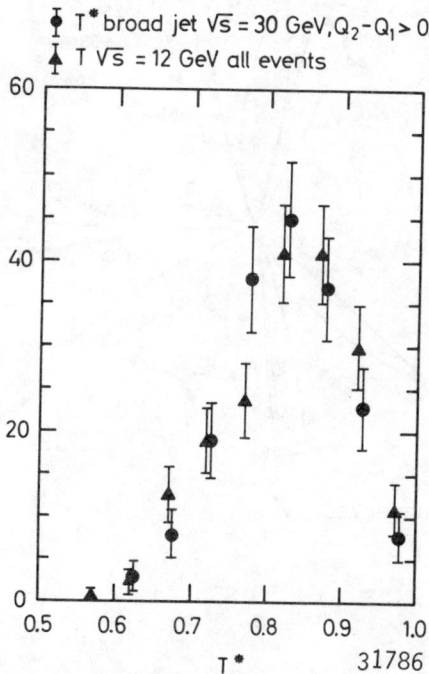

Fig. 31 - Distribution of thrust for the broad jet of planar events at 30 GeV compared with the thrust distribution at 12 GeV. The data were obtained by the JADE Collaboration.

The PLUTO group has analyzed[11,64] their data using the cluster method discribed above. The distribution of the observed number of jets per event are listed in Table 8 and compared to the predictions based on $q\bar{q}$, $q\bar{q} + q\bar{q}g$ ($\alpha_s = 0.15$) and phase space. The models are all normalized to the number of observed events. The data clearly favour a clustering of the particles around 3 axes.

The remaining question is then to decide whether the third jet is defined by a single particle or a group of particles. This can be done by examining the events. Figs. 32 a and b show typical candidates for three jet events observed by JADE and TASSO. Note that several tracks cluster around each axis.

In Fig. 33 the multiplicity distribution of the three jets determined by the TASSO group is shown. The jets are ordered according to $E_1 > E_2 > E_3$. The energies of the jets were computed from the observed opening angles between the jets neglecting parton masses. Furthermore only events for which the acceptance of the drift chamber is nearly complete were considered. It is clear that each jet in general consists of several charged particles and that the multiplicity distribution is reproduced by the QCD calculation.

In conclusion: all the properties observed in $e^+e^- \to$ hadrons can be naturally explained by gluon bremsstrahlung. No other alternative mechanism is known which explains all the data. However, to prove that the gluon observed is the QCD gluon we have to show that the gluon is a coloured, flavour neutral vector particle with gauge couplings. There are hints[40] from upsilon decays and from charm and beauty

Table 8 - Number of clusters

| $n_j$ | 1 | 2 | 3 | 4 | 5 | 6 | 7 |
|---|---|---|---|---|---|---|---|
| Data | 2 | 551 | 249 | 53 | 3 | 1 | |
| $q\bar{q}$ | 3 | 680 | 152 | 23 | 1 | | |
| $q\bar{q}+q\bar{q}g$ | 3 | 567 | 247 | 46 | 2 | | |
| phase space | 1 | 30 | 154 | 306 | 268 | 86 | 14 |

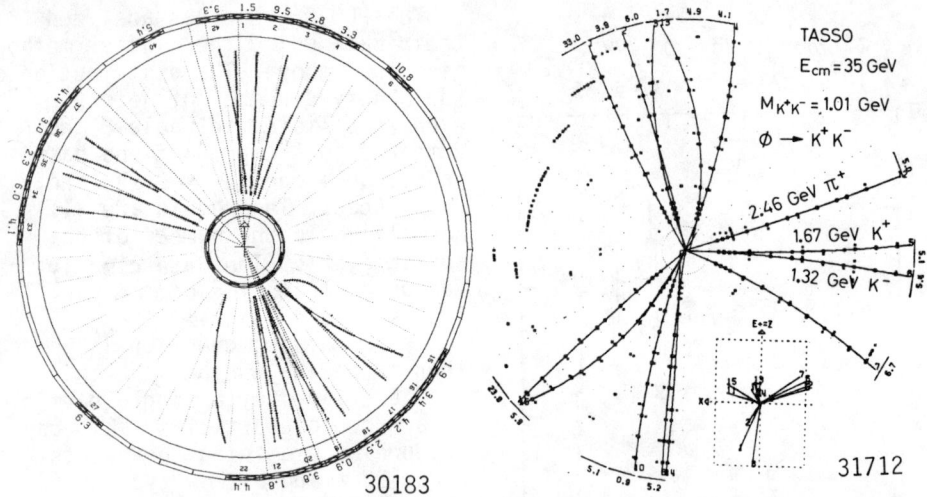

Fig. 32 - Example of three jet events observed by
a) JADE   and   b) TASSO

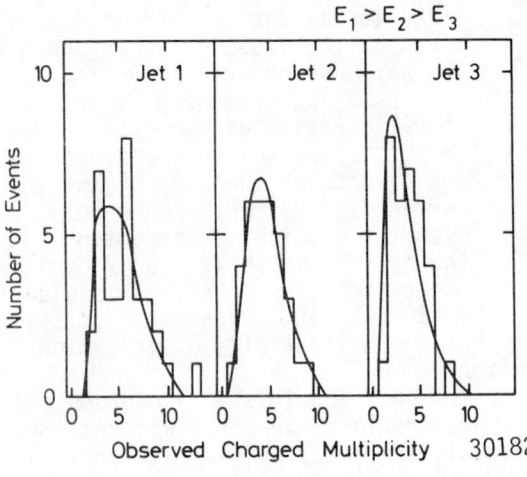

Fig. 33
The charged particle multiplicity distribution observed by TASSO in each of the three jets in a planar event with the jets ordered according to energy.

spectroscopy that the gluon might indeed be coloured and flavour neutral. Determinations of the gluon spin from three jet events will be discussed next.

## VI.2 THE SPIN OF THE GLUON

It is obviously crucial to determine the spin of the gluon from a sample of clean three jet events resulting from gluon bremsstrahlung. $e^+e^- \to q\bar{q}g$. This process can conveniently be described in a Dalitz plot using the variables $x_i = E_i/E_b$ where the energy carried off by the quark or the gluon $E_i$ is measured in units of the beam energy $E_b$. The variables are ordered such that $x_3 < x_2 < x_1$. The thrust of the $q\bar{q}g$ event is then given by $x_1$, and $x_1 + x_2 + x_3 = 2$.
The distribution of the events as function of $x_i$, averaged over production angles relative to the incident $e^+e^-$ directions, can be written as

$$\frac{1}{\sigma_0}\left(\frac{d\sigma}{dx_1 dx_2}\right)_V = \frac{2\alpha_s}{3\pi}\left(\frac{x_1^2 + x_2^2}{(1-x_1)(1-x_2)} + \text{cyclic permutations of 1,2,3}\right) \quad (17)$$

for the vector case and as

$$\frac{1}{\sigma_0}\left(\frac{d\sigma}{dx_1 dx_2}\right)_S = \frac{\tilde{\alpha}_s}{3\pi}\left(\frac{x_3^2}{(1-x_1)(1-x_2)} + \text{cyclic permutations of 1,2,3}\right) \quad (18)$$

for the scalar case.

The TASSO group[68] has determined the spin using the variable $\cos\tilde{\theta} = (x_2-x_3)/x_1$ suggested by Ellis and Karliner[67]. $\tilde{\theta}$ is the angle between the parton 1 and the axis of the parton 2 and 3 system boosted to its own rest frame. To ensure that the spin analysis is not affected by higher order terms one should avoid $x_1$ close to 1. Futhermore, for $x_1$ close to 1 the distributions are varying rapidly so that smearing effects caused by the non-perturbative fragmentation of gluons and quarks are important. For these reasons only events with $1 - x_1 > 0.1$ are used in the analysis.

A total of 248 events remained after this cut, with an estimated two jet event background of 17% and 18% for scalar and vector gluons respectively.

The distribution of the events as a function of $\cos\tilde{\theta}$ is plotted in Fig. 34 and compared with the distributions predicted for vector (solid) and scalar (dotted) gluons. The prediction was made using the model of Hoyer at al.[58]. Note that the distributions are normalized to the number of events in the plot i.e. the scalar and vector cases are discriminated using the shape only.

The data clearly favour the vector case. A fit to the data gives for three degrees of freedom $\chi^2 = 1$ for the vector gluon and $\chi^2 = 14.9$ for the scalar gluon - i.e. scalar gluons are disfavoured by 3.1 standard deviations.

One way to avoid binning effects is to evaluate the mean value of the $\tilde{\theta}$. The experimental value of $<\cos\tilde{\theta}>_{exp} = 0.349 \pm 0.013$ can be compared to the values $<\cos\tilde{\theta}>_V = 0.341 \pm 0.004$ and $<\cos\tilde{\theta}>_S = 0.298 \pm 0.003$ for vector and scalar gluons respectively. The experimental value differs from the vector gluon prediction by 0.6 standard deviations and by 3.8 standard deviations for the scalar case.

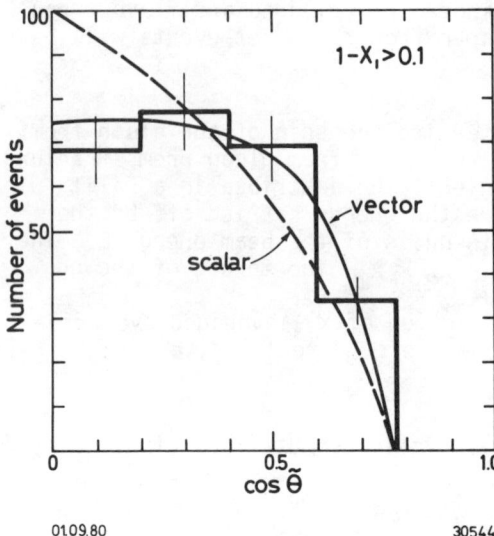

Fig. 34
Observed distribution of the data observed by TASSO in the region $1 - x_1 > 0.10$, as a function of the cosine of the Ellis-Karliner angle $\tilde{\theta}$ defined in Fig. 31b. The solid line shows the QCD prediction and the dotted line the prediction for scalar gluons, both normalized to the number of observed events.

This conclusion is remarkably insensitive both to the exact value of $\alpha_s$ and the details of the fragmentation. Varying the coupling constants by ± 20% changes the computed value of $\langle\cos\tilde{\theta}\rangle$ by about ±1%. Evaluating $\langle\cos\tilde{\theta}\rangle$ in the elementary model without fragmentation leave the scalar prediction unchanged and increases the predicted value for a vector gluon by about 2%.

Another analysis[11,64] of the gluon spin has been made by the PLUTO Collaboration. They investigate the $x_1$ distribution - i.e. the distribution of the most energetic jet in three jet events. This distribution is plotted in Fig. 35 with the $q\bar{q}$ contribution subtracted. The prediction for the vector and the scalar case normalized to the number of events is shown as the solid and the dotted curve respectively. The vector curve fits the data nicely whereas the scalar curve is clearly disfavoured. However, note that PLUTO consider events with $(1-x_1) > 0.05$. Removing the last bin would reduce the significance of this fit.

Both the TASSO and the PLUTO conclusions are based on a first order calculation in $\alpha_s$

## VI.3 DETERMINATION OF THE STRONG COUPLING CONSTANT $\alpha_s$

The strong coupling constant $\alpha_s(s)$ is directly related to the number of three jet events. After choosing a minimum angle between any pair of partons (q, $\bar{q}$ or g) the QCD cross section (Eq.17) can be integrated and normalized to the total $e^+e^-$ annihilation cross section. This ratio depends only on $\alpha_s$ and can be compared directly to the experimental ratio of three jet events to the total number of hadronic events. In practice the analysis must consider several effects:
1) The overlap between jets due to the hadronization process and to fluctuations which might cause a two-jet event to be classified as a three jet event. These effects are not crucial as long as the minimum angle between any two partons is large compared to the opening

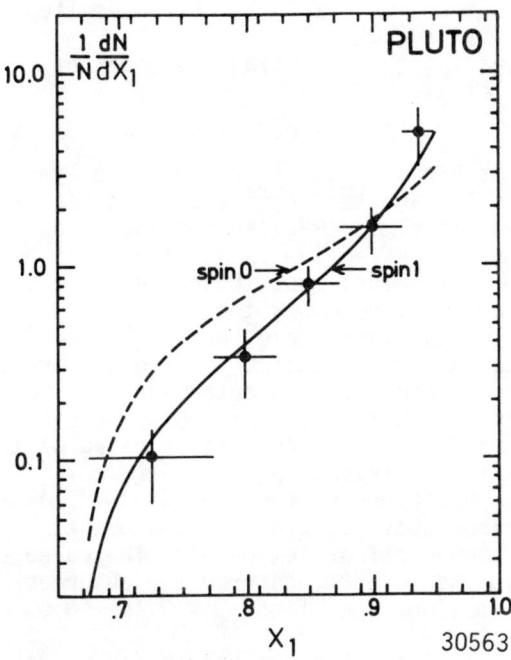

Fig. 35 - The thrust distribution observed by PLUTO in three jet events. The solid line represents the spin 1 case and the spin 0 case is shown by the dotted line. The $q\bar{q}$ contribution has been subtracted.

angle of the jet.
2) The omission of neutrals in some experiments.
3) Apparent multijet contributions resulting from b-decays.
4) Corrections from higher order processes in $\alpha_s$.
5) QED corrections[68] in particular hard photon emission in the initial state.

These and other effects have been taken into account using the elaborate Monte Carlo routines discussed above.

A first attempt to determine $\alpha_s$ at PETRA energies was made by the MARK J group[69] using data taken around 30GeV. They found, using the Ali et al. program, $\alpha_s = 0.23 \pm 0.02 \pm 0.04$ where $\pm 0.02$ is the statistical error and $\pm 0.04$ the systematic uncertainty. A recent analysis[10] based on further data and which also includes the hard photon correction omitted in the first analysis yields as a preliminary value $\alpha_s = 0.19 \pm 0.02 \pm 0.04$.

The TASSO group determined[70] $\alpha_s$ from the event distribution in the A, S plane. They found that $\alpha_s$ can be determined almost independently of the fragmentation parameters using events with $S > 0.25$. In this kinematical region three jet events dominate and non-perturbative effects are not important. The events with $S \geq 0.25$ were fitted for all allowed values of $a_F$ and $P/(P+V)$ with $\alpha_s$ and $\sigma_q$ as free parameters. This fit gave $\alpha_s$ values between 0.14 and 0.17 with a mean value of 0.16. Therefore $\alpha_s = 0.16 \pm 0.04$ is independent of the fragmentation parameters.

The fragmentation parameters were then determined in a further analysis using the events with low sphericity $S < 0.25$. This region is dominated by two jet events and is insensitive to $\alpha_s$.

Simultaneous fits were made to: i) The x distribution ($x = p_h/E_b$). ii) The $<p_{T\ out}^2>$ distribution. iii) The charged multiplicity distribution.

The fits yielded $a_F = 0.57 \pm 0.20$, $\sigma_q = 0.32 \pm 0.04$ GeV/c and $P/(P+V) = 0.56 \pm 0.15$ in agreement with the values found[60] earlier in lepton-hadron and hadron-hadron interactions. The quality of the fits is shown in Fig. 36. Using these parameter values as an input

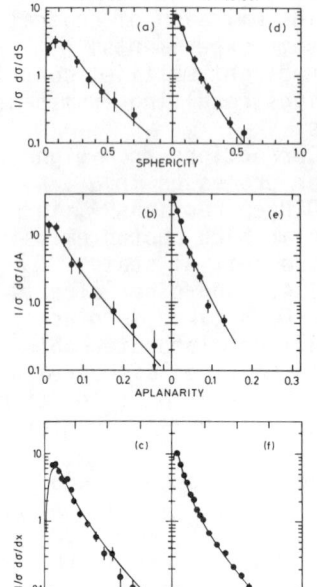

Fig. 36 - Comparison of the data with the QCD model (curves) at 12 GeV and 30 GeV in c.m.

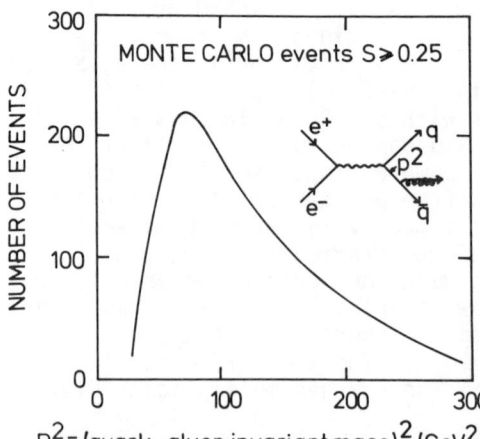

Fig. 37 - Distribution of the square of the quark-gluon mass as computed for events with large sphericity.

to fit the events with S > 0.25 resulted in $\alpha_s = 0.17 \pm 0.02 \pm 0.03$.
Repeating the fit using the model of Hoyer et al.[58], gave

$$\alpha_s = 0.19 \pm 0.02.$$

This is expected since this model only considers two and three jet events whereas the second order model[59] also considers four jet events. The effective value of $\alpha_s$ in the first order model must thus be about 10% larger. However, remember that diagrams with internal gluon lines have been neglected[71].

The JADE group basically used the procedure outlined above to determine $\alpha_s$. They find[12] $\alpha_s = 0.18 \pm 0.03 \pm 0.03$ consistent with an earlier[9] determination based on the planarity distribution.

The PLUTO group used the cluster method[11,64] described above to classify the events as two, three or four jet events. The value of $\alpha_s$ was then determined from the observed jet distribution. They find $\alpha_s = 0.15 \pm 0.03 \pm 0.02$.

It is clear that the values for $\alpha_s$ determined from the three jet events by various methods are in good agreement with a mean value $\alpha_s = 0.17$.

The value of $\alpha_s$ is related to $\Lambda$, the strong interaction mass scale (see Eq.8). It is still an open question what to use for $q^2$ in Eq.8. Maybe the best choice is $q^2 = p^2$ where $p^2$ is the quark-gluon effective mass squared. The $p^2$ distributions for the TASSO events with S > 0.25 are plotted in Fig. 37. Inserting the mean value $<p^2> = 140$ GeV$^2$ into the expression for $\alpha_s(q^2)$ yields

$$\Lambda = (95^{+65}_{-35}) \text{MeV for } \alpha_s = 0.17 \pm 0.02.$$

Including the systematic error yields $\Lambda < 290$ MeV.

From deep inelastic lepton hadron interactions values[40] for $\Lambda$ between 100 and 500 MeV are found.

## VI.4 SOFT GLUON EMISSION

So far we have primarily considered effects due to the emission of a single hard gluon at large angles. However, in the quark-gluon cascade leading to the final hadron jet most gluons are soft and emitted at small angles. The coupling constant $\alpha_s$ will thus be of order unity and many diagrams must be summed. It has been proposed[72] to relate the parton angular distribution within the cascade to two particle differential cross section:

$$1/\sigma \frac{d\sigma}{d\Omega} = \sum_{a,b} \int dz_a \int dz_b \cdot z_a \cdot z_b \frac{1}{\sigma} \frac{d^3\sigma}{dz_a \cdot dz_b \cdot d\Omega} \qquad (19)$$

where a and b are any two particles emitted in the event with normalized momenta $z_a$, $z_b$ ($z = p/p_{beam}$) and an opening angle of $\pi-\theta$.

The PLUTO Collaboration has determined[73] the two particle differential cross section for c.m. energies between 9.4 GeV and 31.6 GeV. The data at 9.4 GeV and 31.6 GeV are plotted in Fig. 38 versus $\theta$ for small angles - i.e. the particles belong to opposite jets. The dashed

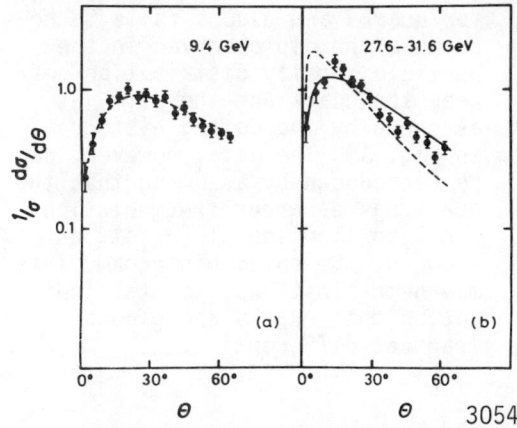

Fig. 38
$1/\sigma d\sigma/d\Omega$ measured by PLUTO plotted versus $\theta$. The dashed and solid curves represent theoretical fits to the data.

curve represents a theoretical fit[74] to the data without hadronization. The fit is based on the QCD leading logarithmic approximation which is used to evaluate the fragmentation function for a parton a to produce a parton b. The parameters were determined from the 9.4 GeV data and used to predict the cross section at 30 GeV. The trend of the data is reproduced by this fit, however, the fit was much improved by including the hadronization of the partons shown as the solid line.

## VI.5 DO QUARKS AND GLUONS FRAGMENT DIFFERENTLY ?

A gluon may fluctuate into pairs of quarks and gluons. Furthermore the ggg coupling is 9/4 times stronger than the $q\bar{q}g$ coupling such that gluon emission is expected to be more frequent for gluons than for quarks. This leads us to expect that a gluon and a quark will fragment into hadrons differently - the hadron spectrum from a gluon fragmentation will be softer with a correspondingly higher multiplicity.

Anderson, Gustafson and collaborators have suggested[75] studying the yields of low-energy particles emitted at large angles with respect to the jet axis. The JADE group has carried out this analysis using charged and neutral particles. Planar events with $Q_2 - Q_1 > 0.10$ were divided into a slim jet and a broad jet by the plane normal to the thrust axis. The broad jet is then boosted into its own rest system and the particles assigned to the two subjets. The softest jet is called the gluon jet. Monte Carlo calculations imply that this is true about 70% of the time and it simply reflects the softness of a bremsstrahlung spectrum. All the particles are then projected on to the plane defined by T, thrust axis of the event and T* the thrust axis of the boosted two jet system. They then plot the particle densities between the gluon jet and the slim jet and the quark jet and the slim jet in terms of normalized angles $\theta_i/\theta_{max}$, where $\theta_{max}$ is the opening angle between the gluon jet and the slim jet or the quark jet and the slim jet respectively. The data plotted in Fig. 39 show that the density of tracks is larger by a factor of 2 between the slim jet and the quark jet. The result of a Monte Carlo computation based on similar fragmentation functions for quarks and gluons fails to reproduce the dip observed in the particle density distribution between the quark and the slim jet, as shown by the dotted histogram in Fig. 39. The data, however, can be reproduced by assuming that the quark has a harder fragmentation function than the gluon jet, as shown by the solid histogram. This may be a first experimental indication that quarks and gluons fragment differently.

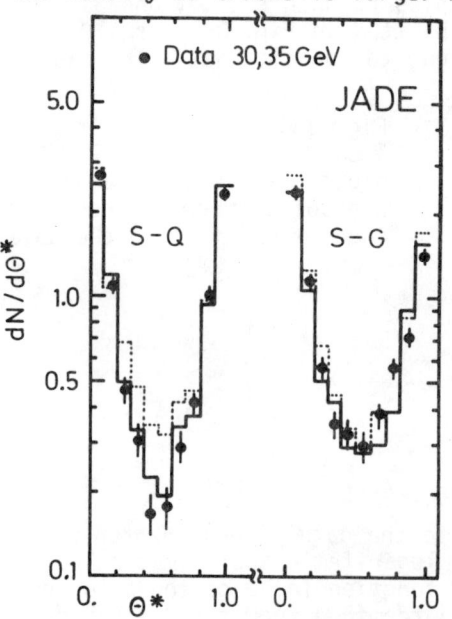

Fig. 39 - Angular distribution of charged particles between the slim jet and the gluon jet and between the slim jet and the quark jet as a function of the normalized angles $\theta/\theta_{max}$.

## VII. TWO PHOTON INTERACTIONS

Electron-positron collisions are also a source of photon-photon collisions[76] as shown in Fig. 40, where the mass and the energy of the

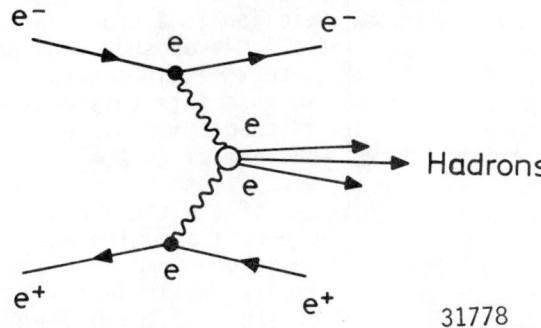

Fig. 40
Hadron production in
$e^+e^- \rightarrow e^+e^- X$

spacelike photon is determined from a measurement of the scattered lepton. These processes offer a unique opportunity to vary the mass of the target and the projectile over a wide range from collisions of two nearly real photons via deep inelastic electron scattering on a photon target to collisions of two heavy photons.

Experimentally, two photon events are separated from annihilation events by a cut on the observed energy. The c.m. energy of the $\gamma\gamma$ system is in general much lower than the available energy, reflecting the product of two bremsstrahlungsspectra. The background from beam-gas events is rather low as determined from the number of events which satisfy the selection criteria but originates outside of the inter-action point.

### VII.1 RESONANCE PRODUCTION

All hadrons with even charge conjugation and spin different from one can be produced[77] in real $\gamma\gamma$ collisions.

The MARK II Collaboration at SPEAR has published[78] data on $e^+e^- \rightarrow e^+e^-\eta'$. They found a clean $\eta'$ signal in the channel $\eta' \rightarrow \rho\gamma$ which gave $\Gamma^{\eta'}_{\gamma\gamma} = (5.8 \pm 1.1)$ keV with a systematic uncertainty of 20%. With the measured branching ratio of BR($\eta' \rightarrow \gamma\gamma$) = $(2.0 \pm 0.3)$% this yields a total width $\Gamma^{tot} = (293 \pm 76 \pm 59)$ keV in good agreement with the value of $(280 \pm^{n}_{-}100)$ keV determined[79] by D.M.Binnie et al. from the reaction $\pi^-p \rightarrow \eta'n$ near threshold.

Events of the type $e^+e^- \rightarrow e^+e^-t^+t^-$ have been selected by PLUTO and TASSO at PETRA and by MARK II and the San Diego group at SPEAR to search for $\gamma\gamma$ production of resonances decaying into pairs of charged hadrons. The effective mass distribution of untagged two prong events determined by the PLUTO group[80] is plotted in Fig. 41 assuming the particles to be pions. The data show a broad maximum near 1.2GeV decreasing steeply towards higher masses. The bulk of the two prong events results from QED reactions $e^+e^- \rightarrow e^+e^- (e^+e^- + \mu^+\mu^-)$ with an amplitude proportional to $e^4$. (In addition there is a small contribution from $\pi^+\pi^-$ Born events). This contribution has been evaluated[81] and is shown as the solid line in Fig. 41. It describes all the data except for an excess near 1.25 GeV. The difference between the observed distribution and the QED prediction is a peak near 1.25 GeV as shown in the insert. It is natural to associate this peak with $f^0$ production since the $f^0$ has a mass of 1.27 GeV and decays into a $\pi^+\pi^-$ pair 83% of the time. A fit to the data assuming the $f^0$ to be produced with a helicity amplitude $\Lambda = 2$ gave $\Gamma^{f^0}_{\gamma\gamma} = (2.3 \pm 0.5 \pm 0.35)$ keV where the first error is statistical and the second systematic.

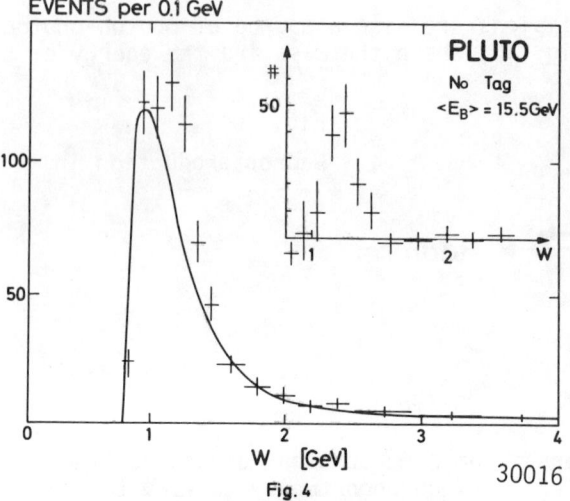

Fig. 41 - Untagged two prong events from PLUTO plotted versus the pair mass. The solid line shows the QED contribution. The difference between the measured two prong yield and the QED contribution is shown

The two prong mass distribution obtained[13,82] by TASSO with the QED contribution subtracted is plotted in Fig. 42. The data show a clear peak at 1.25 GeV which yields a preliminary value of $\Gamma^{f^0}_{\gamma\gamma} = (4.1 \pm 0.4 \pm 0.6)$ keV for $\Lambda = 2$, barely consistent with the PLUTO value.

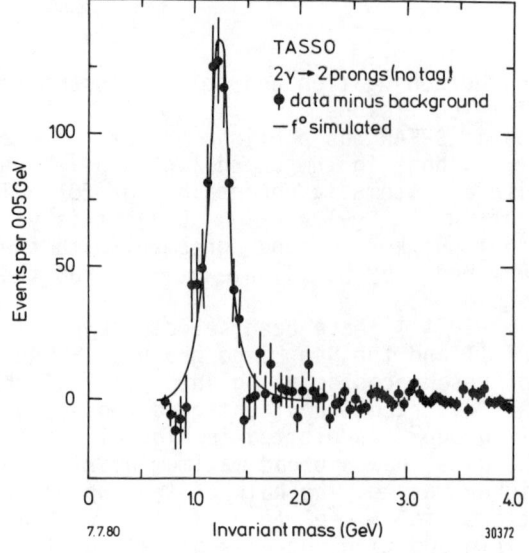

Fig. 42

Untagged two prong events from TASSO plotted versus the pair mass with the QED background subtracted. The data are preliminary.

Preliminary data[83] from MARK II are shown in Fig. 43 where the number of untagged two prong events is plotted versus the the pair mass. The QED contribution has been subtracted. Again a clear enhancement is seen at a mass around 1.2 GeV. However, the enhancement is not well fit by a single Breit-Wigner resonance and hence no value is quoted for the radiative width.

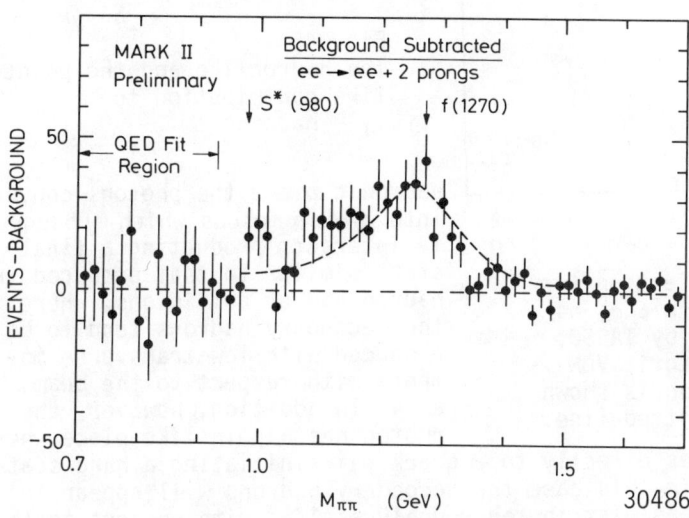

Fig. 43 Untagged two prong events from MARK II Collaboration at SPEAR plotted versus the pair mass with the QED background subtracted. The data are preliminary.

The San Diego group[84] using tagged events found 21 events above the QED background centered around 1.25 GeV. Assigning these events to $e^+e^- \to e^+e^- f^0$ yields $\Gamma^{f^0}_{\gamma\gamma} = 9.5 \pm 3.9 \pm 2.4$ keV assuming $\Lambda = 2$.

VII.2 OBSERVATION OF $e^+e^- \to e^+e^- \rho^0\rho^0$

The TASSO group reports[14,85] results on $\gamma\gamma \to \rho^0\rho^0$. The $\rho^0\rho^0$ cross section was extracted from the data by selecting neutral, four prong events with the sum of the transverse momentum with respect to the beam axis less than 0.15 GeV. The invariant $\gamma\gamma$ mass was required to be between 1.5 and 2.3 GeV. This results in 89 events, with a negligible background from beam gas events. They estimate one event from one photon annihilation and 15 events from events containing additional unobserved particles.

The cross section for $\gamma\gamma \to \rho^0\rho^0$ is plotted in Fig. 44 versus the c.m. energy of the $\gamma\gamma$-system. The cross section peaks strongly near threshold and drops rapidly with energy in disagreement with a simple V.D.M. asymptotic prediction. However close to threshold nonasymptotic effects are expected to be important

VII.3 HADRON PRODUCTION WITH REAL PHOTONS

The amplitude for $\gamma\gamma \to$ hadrons will presumably contain both the hadronlike[86] piece and the pointlike[87] piece shown in Fig. 45. In the

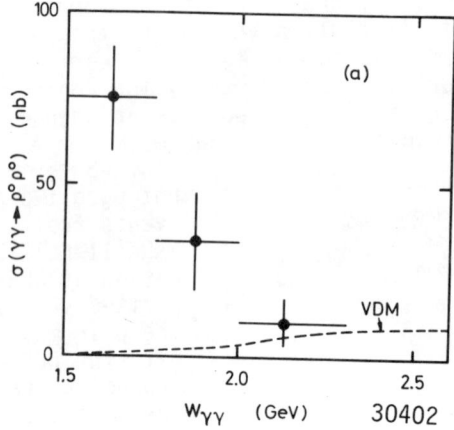
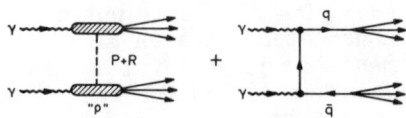

Fig. 45
The hadronlike and the pointlike contribution to $\gamma\gamma \to$ hadrons

Fig. 44 - Cross section for $e^+e^- \to e^+e^-\rho^0\rho^0$ as measured by TASSO. An asymptotic VDM prediction is shown as the dotted line.

hadronic piece the photons convert into vector mesons which subsequently interact producing a final state similar to that produced in hadron hadron collisions, where the secondary hadrons tend to be produced with low transverse momenta with respect to the beam axis. In addition, however, the photon has a pointlike piece where the photon couples directly to a quark pair initiating a hard scattering process. In this case the secondary hadrons will appear in two jets of hadrons distributed roughly as $1/p_T^4$ with respect to the beam axis.

The total cross section for $\gamma\gamma \to$ hadrons can be estimated from the imaginary part of the elastic scattering amplitude to

$$\sigma(\gamma\gamma \to \text{hadrons}) = 240 \text{ nb} + \frac{270 \text{ nb} \cdot \text{GeV}}{W} + \frac{C \text{ nb GeV}^2}{W^2} \quad (20)$$

The first term results from Pommeron exchange and was estimated from factorization $\sigma_{\gamma\gamma} \cdot \sigma_{pp} = (\sigma_{\gamma p})^2$. The second term involves f and $A_2$ exchange and leads to a cross section which decreases as $1/W$ where W is the energy of the hadronic system. The pointlike contribution is expected to decrease roughly as $1/W^2$.

The total hadronic cross section has been measured[13] both by the PLUTO and the TASSO Collaboration. Both groups collected data by detecting only one of the electrons leaving the other untagged. The total energy W of the produced hadron system was estimated from energy $W_{vis}$ observed in the detector. Only charged particles were observed in TASSO whereas PLUTO also measured photons. The observed cross sections were extrapolated to $Q^2 = 0$ using the $\rho$-propagator:

$$\sigma_{\gamma\gamma}(W) = \sigma_{\gamma\gamma}(W,Q^2) \cdot \left[\frac{m_\rho^2}{m_\rho^2 + Q^2}\right]^{-2} \quad (21)$$

Note that this simple Ansatz violates scaling and is not valid in electroproduction at large values of $Q^2$. The $Q^2$ dependence of the total cross section is shown in Fig. 46 together with the $\rho$-propagator fit.

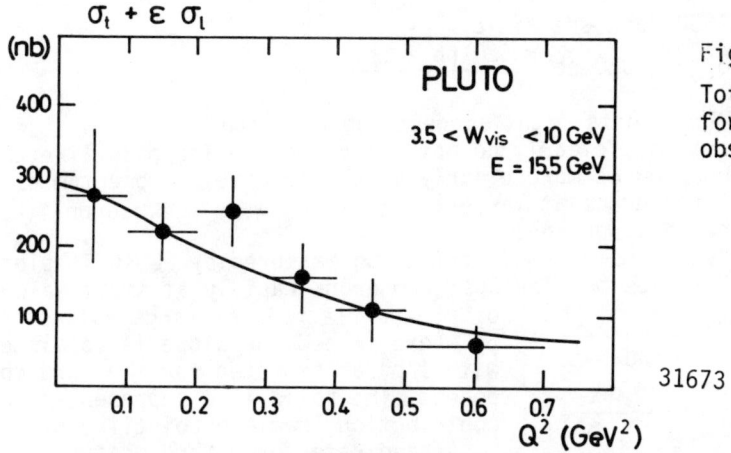

Fig. 46 Total cross section for $\gamma^*\gamma \to$ hadrons observed by PLUTO.

The cross sections extrapolated to $Q^2 = 0$ are plotted versus W in Fig. 47. The VDM contribution is also shown as the dotted line. Note that the TASSO data are preliminary.

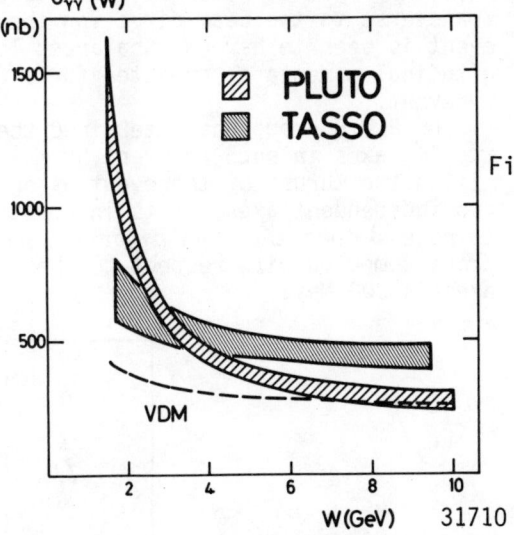

In addition to the statistical errors indicated by the bands, there are systematic errors of 15% and 25% for the PLUTO and the TASSO data respectively.

Fig. 47 - The total cross section for $\gamma\gamma \to$ hadrons plotted versus the mass of the produced hadron system. Note that the TASSO data are preliminary and that only statistical errors are shown.

The data from the two groups are marginally consistent within the errors. However, the PLUTO cross section clearly decreases steeper with energy than the TASSO cross section does:

A best fit to the PLUTO data gives:

$$\sigma_{\gamma\gamma} = A \left( 240 \text{ nb} + \frac{270 \text{ nb} \cdot \text{GeV}}{W} \right) + \frac{B \text{ nb GeV}^2}{W^2} \qquad (22)$$

with $A = 0.97 \pm 0.16$ and $B = 2250 \pm 500$

whereas the TASSO data are fit by

$$\sigma_{\gamma\gamma} = 380 \text{ nb} + \frac{520 \text{ nb} \cdot \text{GeV}}{W} \tag{23}$$

Although, the PLUTO data might suggest the presence of a pointlike term they clearly do not yet prove it. The pointlike contribution might show up more clearly in the transverse momentum distribution of the hadrons at large values of $p_T$ were the hadron-like contribution has disappeared.

The transverse momentum distribution measured by PLUTO is plotted in Fig. 48 versus $p_T^2$. The spectrum drops rapidly at small values of $p_T^2$ and flattens at large values of $p_T^2$ where indeed the slope is consistent with $1/p_T^4$ as expected for the hard component. The solid line represents the contribution from the pointlike diagram.

A candidate for a hard scattering event obtained by PLUTO is shown in Fig. 49. When viewed along the beam direction the event appears as two collinear jet of hadrons. When viewed transverse to the beam direction the event is seen to have a unbalanced longitudinal momenta as expected for a $\gamma\gamma$ event.

The PLUTO group has determined the two jet axes in such an event by maximizing the thrust of the event using two independent axes. It is interesting to note [88] that the mean $p_T$ of the hadrons computed with respect to these axes is 300 MeV.

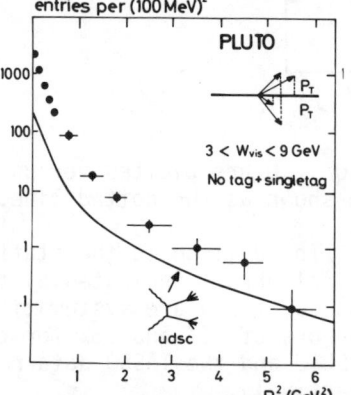

Fig. 48 - Number of tracks plotted versus the square of the transverse momentum with respect to the beam axis. The solid curve is a prediction based on $\gamma\gamma \to q\bar{q} \to$ hadrons with q = u,d, s, c quarks.

Fig. 49 - Candidate for $\gamma\gamma \to q\bar{q} \to$ hadrons observed by PLUTO.

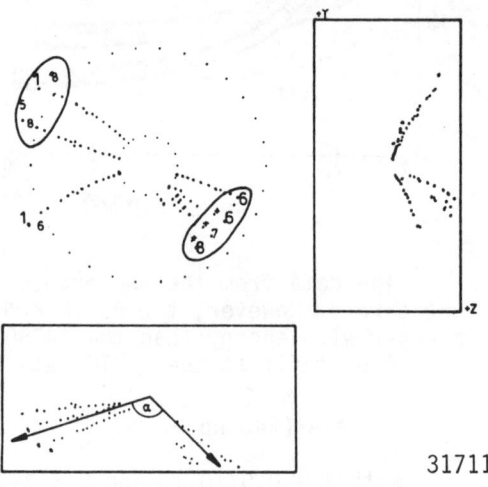

The properties of these events are strongly suggestive of a pointlike production mechanism. However, it should be remembered that ordinary one photon annihilation events with radiation in the initial state could lead to events with the same topology. This mechanism, however, can be excluded if the electron is tagged on the same side as the direction of the longitudinal momentum vector of the jets. The PLUTO group finds the expected number of 2-jet events with the tagged electron on the same side as the momentum vector, appearantly excluding this background source.

## VII.4 ELECTRON SCATTERING ON A PHOTON TARGET

The PLUTO Collaboration reports[13] the first data on deep inelastic electron-photon scattering. These data were collected by tagging one electron at scattering angles between 70 mrad and 250 mrad corresponding to values of $Q^2$ between 1 $(GeV/c)^2$ and 15 $(GeV/c)^2$ with a mean value of 5 $(GeV/c)^2$. The second electron was not detected yielding a nearly real target photon. A total of 120 multihadron events with this electron topology were observed.

Deep inelastic electron-photon scattering[89] as shown in Fig. 50 can be parametrized in terms of three structure functions. $F_L(x,Q^2)$ and $F_2(x,Q^2)$ corresponds to the longitudinal and to the transverse polarisation vector of the virtual photon respectively and $F_3(x,Q^2)$ to the transverse polarisation vector of the target photon in the scattering plane. $x = Q^2/(Q^2+W^2)$ and $W$ is the mass of the hadronic system. $F_3(x,Q^2)$ will average to zero since the scattering plane was not determined. Furthermore $F_L(x,Q^2)$ is expected to be smaller than $F_2(x,Q^2)$ and the PLUTO group therefore analyze their data in terms of $F_2(x,Q^2)$ only.

Both the hadronlike part and the pointlike part of the photon contributes to $F_2(x,Q^2)$ as indicated in Fig. 50b,c. In the hadronlike part the photon transforms into a vector meson and the virtual photon interacts with the quarks in the vector meson. This contribution cannot be calculated from first principles but it will have an x dependence similar to that observed in the structure function of the pion i.e. $F_2(x,Q^2)_{VDM} \sim (1-x)^{c_1+c_2 \ln(\ln Q^2)}$ and, just like lepton-hadron interactions, only its evolution with $Q^2$ is predicted in QCD. The pointlike piece (50c) can be calculated to all orders in a perturbation theory. The lowest order calculation gives:

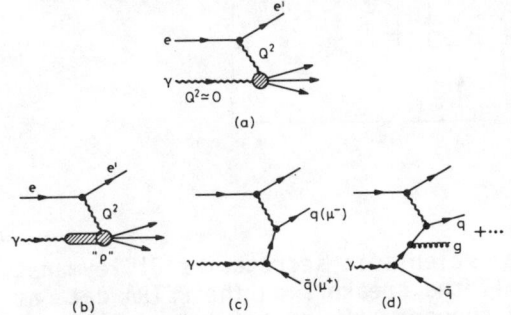

Fig. 50 - Diagrams contributing to $e\gamma \to e'$hadrons.

$$F_2(x,Q^2)_{point} = \frac{\alpha}{\pi} \sum_i e_i^4 \, x(x^2 + (1-x)^2) \cdot \ln Q^2/\Lambda^2. \qquad (24)$$

The pointlike contribution leads to a structure function which peaks at large values of x, whereas the VDM piece leads to a structure function which is large at small x and disappears at large x. The lowest order pointlike contribution is indeed proportional to the QED process $e^+e^- \to e^+e^- \mu^+\mu^-$.

Higher order QCD corrections (Fig. 50d), including the emission of soft gluons, will soften the Born spectrum by depleting the density of fast quarks and enchancing the density of slow quarks.

The values for $(1/\alpha)F_2(x)$, extracted from the hadron data, are plotted in Fig. 51a versus x and compared to the formfactor observed in the QED process $e^+e^- \to e^+e^- (e^+e^- + \mu^+\mu^-)$ (Fig. 51b). Both structure functions peak at large x demonstrating the existence of the pointlike piece. The Born prediction, shown as the solid curve, follows the general trend of the data quite well.

The evolution of the form factor with increasing $Q^2$ will make it possible to determine $\Lambda$ with good precision in a high statistics experiment. The simplicity of the target, with no finite mass effects and no higher twist effects caused by a premordial $p_T$ distribution, makes it possible to determine $\Lambda$ without the systematic uncertainties which have beset the determination in deep inelastic lepton-hadron processes. In particular it may be possible to use data in a $Q^2$ range such that the variation in $\ln Q^2/\Lambda^2$ are still quite large.

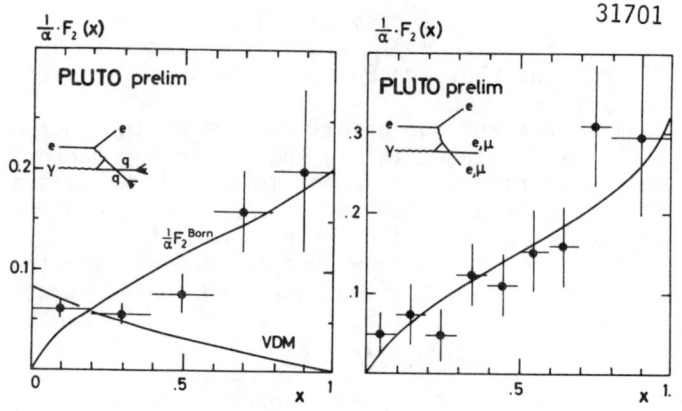

Fig. 51

The structure functions of the photon (a) and the lepton (b).

I would like to thank my scientific secretaries J.Cleymans, D.Haidt and G.Zobernig and all the speakers on the PETRA data at Wisconsin for their help and support. I'm grateful to J.Ellis, J. Freeman, T.Meyer, H.L.Lynch, T.F.Walsh, G.Wolf, S.L.Wu for support and discussions.

## DISCUSSIONS

Q. E.L.Berger (Argonne)

Do you have any information on the energy dependence of the very interesting back-to-back charge correlation which you showed ? Is it growing or falling with $Q^2$? It could be an inverse power $1/Q^2$ higher twist effect.

A. At present only the high energy data have sufficient statistics to establish the charge correlation effect.

Q. B.Barnett (John Hopkins Univ.)

In your charge correlation data, have you taken into account the fact that once you have chosen a positive particle there is an extra negative particle somewhere due to charge conservation. This effect will give an apparent correlation about the size you see, and, therefore your claim of seeing a positive effect is probably premature.

A. May be I didn't make myself clear. What we do is to pick a charge at a certain y. The remaining event is then of course oppositly charged and the question which we try to answer is how this compensating charge is distributed as a function of rapidity. If you pick the test particle at $y \sim 0$, then also the compensating charge is centered at $y = 0$. If the test particle has a large value of y then a significant part of the charge compensation is done by particles far removed in rapidity - i.e. they belong to the opposite jet.

Q. D.S.Narayan (Tata Institute, India)

1) Comment on oblateness events: At 12 GeV, Monte Carlo produces planar events in agreement with expectations. These planar events in Monte-Carlo are planar due to fluctuations. The fluctuations depend on the input parameters in the Monte Carlo. At 30 GeV, the Monte Carlo does not give adequate number of events. Does this show the existence of the gluon ? Or may be the inputs on the Monte Carlo are unrealistic.

2) Comment on charge correlation. Charge correlation (+, -) of the fastest particles on the opposite sides is partly kinematical since that charge = 0. An more appropriate thing to do is to look to the charge correlations of the fastest and next to fastest particles on the same side. In deep inelastic scattering, the uncorrelated Monte Carlo explains the data well but the parametrization of Feynman and Field give the wrong trend.

A. 1) Indeed the data at low energeis can be well fit in terms of $e^+e^- \rightarrow q\bar{q}$ and this can be used to determine the parameters for the quark fragmentation. This is not possible at high energies. The $q\bar{q}$ model cannot reproduce the observed final state for any values of the parameters in the Monte Carlo program. However, the data are very naturally explained as gluon bremsstrahlung with only a single new parameter $\alpha_s$ the strength of the coupling between a quark and a gluon.

2) See the answer given to Prof. Barnett.

Q. N.K.Yamdagni (Univ. of Stockholm, Sweden)

In view of fact the fraction of energy going to charged particles is varying significantly, perhaps the average charged multiplicity should be given as a function of the fraction of energy going in the charged mode.

Q. V.Lüth (SLAC)

How do you separate the untagged 2 jet events from $2\gamma$ interaction from background due to initial state radiation. How big is this background ? How do you measure the contribution ?

A. This is indeed one of the problems since radiation in the initial state will lead to the same topology as $e^+e^- \to e^+e^- q\bar{q}$ events. However this contribution can be computed and subtracted and the PLUTO group claim it is a small contribution. A collinear
two jet events where an electron is tagged on the same side as the direction of the momentum vector of the two jets cannot result from radiation in the initial state. PLUTO claim to find the correct number of such events.

Q. A.Roussarie (SLAC)

About the $\gamma\gamma$ production of $\pi\pi$ resonances. We are not quoting a limit on the partial width $\Gamma f^0 \to \gamma\gamma$ because as you have shown the $\pi\pi$ mass spectrum is not compatible with only f production. We think that we subtract the continuum in a better way than in PETRA because our spectrum begins at 0.6 GeV of mass and the PETRA spectrum due to the trigger begins only at 1 GeV. They may have acceptance problems.

A. Being able to measure down to 0.6 GeV is clearly an advantage. However, the DESY group do have a few points below the f mass and these observed cross sections can be explained by QED production only.


REFERENCES:
1) A.Böhm, Minirapporteur talk in section C8, XXth International Conference on High Energy Physics, July 17-23, 1980, University of Wisconsin, Madison, Wisconsin.
2) D.Cords, Minirapporteur talk in section C2, XXth International Conference on High Energy Physics, July 17-23, 1980, University of Wisconsin, Madison, Wisconsin.
3) D.Pandoulas, Minirapporteur talk in section C2, XXth International Conference on High Energy Physics, July 17-23, 1980, University of Wisconsin, Madison, Wisconsin.

4) S.L.Wu, Minirapporteur talk in section C2, XXth International Conference on High Energy Physics, July 17-23, 1980, University of Wisconsin, Madison, Wisconsin.
5) B.H.Wiik, Proceedings of the International Neutrino Conference, Bergen, Norway, June 1979, p. 113
   P.Söding, Proceedings EPS International Conference on High Energy Physics, Geneva, Switherland, July 1979, p. 271
6) TASSO-Collaboration, R.Brandelik et al., Phys.Lett. 83B,261 (1979)
7) MARK J Collaboration, D.P.Barber et al.,Phys.Rev.Lett.43,830(1979)
8) PLUTO Collaboration,Ch.Berger et al.,Phys.Lett.86B,418(1979)
9) JADE Collaboration, W.Bartel et al., Phys.Lett.91B,142 (1980)
10) MARK J Collaboration, H.Newmann, Minirapporteur talk in section C3, XXth International Conference on High Energy Physics, July 17-23, 1980, University of Wisconsin, Madison, Wisconsin
11) PLUTO Collaboration, V.Hepp, Minirapporteur talk in section C3, given at the XXth International Conference on High Energy Physics, July17-23, 1980, University of Wisconsin, Madison, Wisconsin.
12) JADE Collaboration, S.Yamada, Minirapporteur talk in section C3, given at the XXth International Conference on High Energy Physics, July 17-23, 1980, University of Wisconsin, Madison, Wisconsin.
13) W.Wagner, Minirapporteur talk in section C1, XXth International Conference on High Energy Physics, July 17-23, 1980, University of Wisconsin, Madison, Wisconsin.
14) E.Hilger, Minirapporteur talk in section C1, XXth International Conference on High Energy Physics, July 17-23, 1980 University of Wisconsin, Madison, Wisconsin.
15) F.A.Berends, K.J.F.Gaemers and R.Gastmans, Nucl.Phys.B57,381(1973), Nucl.Phys.B63, 381 (1973) and Nucl.Phys. B68, 541 (1979)
    F.A.Berends and G.J.Komen, Phys.Lett. 63B, 432 (1976) and private communication F.A.Berends and R.Kleiss.
16) H.Salecker, Zeitschr. für Naturforschung 8a, 16 (1953) and 10a, 349 (1955)
    S.D.Drell, Ann.Phys. 4, 75 (1958)
    T.D.Lee and G.G.Wick, Phys.Rev. D2, 1033 (1970)
17) J.A.McClure and S.D.Drell, Nuovo Cim. 37, 1638 (1965)
    N.M.Kroll, Nuovo Cim. 45A, 65 (1966)
18) A.Litke, Harvard University, Thesis 1970
19) JADE Collaboration, W.Bartel et al., Phys.Lett. 92B, 206 (1980)
    MARK J Collaboration, D.P.Barber et al., Phys.Rev.Lett.42,1110(1979) and 43, 1915 (1979) and Phys.Lett.95B, 149 (1980)
    PLUTO Collaboration, Ch.Berger et al.,Z.Physik C4, 269 (1980) and Phys.Lett. 94B, 87 (1980)
    TASSO Collaboration, R.Brandelik et al., Phys.Lett.92B, 199 (1980) and 94B, 259 (1980)
20) S.L.Glashow, Nucl.Phys.22, 579 (1961)
    S.Weinberg, Phys.Rev.Lett. 19, 1264 (1967)
    A.Salam, Proc. 8th Nobel Symposium, N.Svartholm editor, Wiley NY 1968
21) J.Ellis and M.K.Gaillard, CERN 76-18
22) N.Wright and J.J.Sakurai, Phys.Rev. D22, 220 (1980)
23) H.Georgi and S.Weinberg, Phys.Rev.D17, 275 (1978)
    J.D.Bjorken, Phys.Rev. D19, 335 (1979)

24) JADE Collaboration, R.Marshall, Private communication.
25) MARK J Collaboration, D.P.Barber et al., RWTH Aachen,
    PITHA Preprint 80/08 1980
26) E.H.de Groot, G.J.Gounaris and D.Schildknecht, Phys.Lett.85B,399(1979)
    Phys.Lett.90B,427 (1980) and Z.für Physik C5, 127 (1980)
27) V.Barger, W.Y.Kenny and E.Ma,Wisconsin-Hawaii Reports
    UW-COO-881-126(1980); US-C00881-133 (1980) and UW-COO-881-138 (1980)
28) MARK J Collaboration, MIT, Technical Report No. 113, 1980
    TASSO Collaboration, R.Brandelik et al., in preparation.
29) P.Fayet and S.Ferrara, Phys.Reports 32C, 249 (1977)
    P.Fayet, Phys.Lett. 69B, 489 (1977)
    G.R.Farrar - Proc.Int.School of Subnuclear Physics, Erice,Italy,
    Aug. 1978
    G.R.Farrar and D.Fayet, Phys.Lett. 76B, 578 (1970) and 79B,442 (1970)
    G.R.Farrar and D.Fayet, Searching for the spin 0 leptons of super-
    symmetry Rutgers / Caltech preprint 1979
    G.Barbellini et al., ECFA/LEP Study Group, DESY Preprint 79/67
30) Yu. A.Gal'fand and E.P.Likhtman, JETP Letters 13, 323 (1971)
    D.V.Volkov and V.P.Akulov, Phys.Lett. 46B, 109 (1973)
    J.Wess and B.Zumino Nucl.Phys.B70,39(1974)
31) S.D.Drell, D.J.Levy, and T.M.Yan, Phys.Rev.187, 2159 (1969) and
    Phys.Rev.D1, 1617 (1970)
32) N.Cabibbo, G.Parisi, and M.Testa, Lett.Nuovo Cimento 4,35 (1970)
33) J.D.Bjorken and S.J.Brodsky, Phys.Rev. D1, 1416 (1970)
34) R.P.Feynman, Photon-Hadron Interactions (Benjamin, Reading Mass.,
    p.166 (1972)
35) R.F.Schwitters et al., Phys.Rev.Lett. 35, 1230 (1975)
    G.G.Hanson et al., Phys.Rev.Lett. 35, 1609 (1975)
    G.G.Hanson, Proceedings of 13th Rencontre de Moriond, edited
    by J.Tran Thanh Van, Vol II, p.15 and SLAC-PUB 2118 (1978)
    PLUTO Collaboration, Ch.Berger et al., Phys.Lett.B78, 176 (1978)
36) J.Kogut and L.Susskind, Phys.Rev.D9, 697 , 3391 (1974)
    A.M.Polyakov, Proceedings of the 1975 International Symposium
    on Lepton and Photon Interactions at High Energies,
    Stanford, Aug. 21-27, 1975
37) The first quantitative discussion on the experimental implications
    of gluon bremsstrahlung in $e^+e^-$ annihilation was given by:
    J.Ellis, M.K.Gaillard and G.G.Ross, Nucl.Phys. B111,253 (1976)
    - erratum B 130, 516 (1977)
38) T.A.DeGrand, Yee Jack Ng, and S.-H.H.Tye, Phys.Rev.D16, 3251 (1977)
    A.de Rujula, J.Ellis, E.G.Floratos and M.K.Gaillard,
    Nucl.Phys.B 138, 387 (1978)
    G.Kramer and G.Schierholz, Phys.Lett. 82B, 102 (1979)
    G.Kramer, G.Schierholz and J.Willrodt, Phys.Lett.79B,249 (1978)
    P.Hoyer, P.Osland, H.G.Sander, T.F.Walsh and P.M.Zerwas,
    Nucl.Phys. B 161, 349 (1979)
    G.Kramer, G.Schierholz and J.Willrodt, Z.für Physik C4,149 (1980)
39) H.Fritzsch, M.Gell-Mann and H.Leutwyler, Phys..Report 47B,365(1973)
    D.J.Gross and F.Wilczek,Phys.Rev.Lett. 30, 1343 (1973)
    H.D. Politzer, Phys.Rev.Lett. 30, 1346 (1973)
    S.Weinberg, Phys.Rev.Lett. 31, 31 (1973)

H.D.Politzer, Phys.Reports 19, 129 (1979)
W.Marciano and H.Pagels, Phys. Reports 36, 137 (1978)
40) See reports given at the XX International Conference on High Energy Physics, Madison, Wisconsin, July 17-23, 1980
41) S.Brandt et al., Phys.Lett. 12, 57 (1969)
E.Farhi, Phys.Rev.Lett. 39, 1587 (1977)
S.Brandt and H.D.Dahmen, Z.Physik C1, 61 (1979)
42) A.Quenzer, thesis, Orsay Report LAL 1299 (1977)
A.Cordier et al., Phys.Lett. 81B, 389 (1979)
V.A.Sidorov, Proceedings of the XVIIIth International Conference on High Energy Physics, Tbilisi, USSR, B 13 (1976)
R.F.Schwitters, Proceddings of the XVIIIth International Conference on High Energy Physics, Tbilisi, USSR, B 34 (1976)
J.Perez-Y-Jorba, Proceedings of the XIXth International Conference on High Energy Physics, Tokyo, p.269 (1978)
PLUTO Collaboration, J.Burmester et al., Phys.Lett. 66B, 395(1977)
DASP Collaboration, R.Brandelik et al., Phys.Lett. 76B, 361 (1978)
43) T.Appelquist and H.Georgi, Phys.Rev. D8, 4000 (1973)
A.Zee, Phys.Rev. D8, 4038 (1973)
G.'t Hooft, Nucl.Phys. B 62, 444 (1973)
M.Dine and J.Sapirstein, Phys.Rev.Lett. 43, 668 (1979)
W.Celmaster and R.J.Gonsalves, UCSD Preprint UCSD-10P10-206,207(1979)
K.G.Chetyrkin, A.L.Kataev and F.V.Tkachov,Phys.Lett.85B,277(1979)
USSR Academy of Sciences, Institute of Nuclear Research Preprint D-0178 (1980)
44) TASSO Collaboration, R.Brandelik et al.,Phys.Lett.89B,418 (1980)
45) JADE Collaboration, W.Bartel et al.,Phys.Lett. 88B, 171 (1979)
46) PLUTO Collaboration, Ch.Berger et al., DESY Report 80/69 (1980)
47) C.Bacci et al., Phys.Lett. 86B, 234 (1979)
SLAC-LBL Collaboration, G.G.Hanson, 13th Rencontre de Moriond (1978) ed.by J.Tran Thanh Van, Vol. III - 1978
PLUTO Collaboration, Ch.Berger et al., Phys.Lett. 81B, 410 (1979) and V.Blobel private communication
DASP Collaboration, R.Brandelik et al., Nucl.Phys.B 148, 189 (1979)
48) W.Thomé et al., Nucl.Phys. B 129, 365 (1977)
See also review by E.Albini, P.Capiluppi, G.Giacomelli, and A.M.Rossi, Nuovo Cimento 32A, 101 (1976)
W.Thomé, Aachen preprint PITHA 80/4 (1980)
49) R.Stenbacka et al., Nuovo Cimento 51A, 63 (1979)
50) W.Furmanski, R.Petronzio and S.Pokorski, Nucl.Phys.B155,253 (1979)
A.Bassetto, M.Ciafaloni and G.Marchesini,Phys.Lett.83B,207 (1978)
51) K.Konishi, Rutherford Preprint RL 79-035 T 241 (1979)
52) G.J.Feldman and M.L.Perl, Phys.Reports 33, 285 (1977)
53) TASSO Collaboration, R.Brandelik et al., Phys.Lett.89B,418 (1980)
54) TASSO Collaboration, R.Brandelik et al., Phys.Lett.94B, 444 (1980)
55) TASSO Collaboration, R.Brandelik et al., Phys.Lett. 94B, 91 (1980)
56) V.Lüth et al., Phys.Lett. 70B, 120 (1977)
57) T.F.Walsh and P.Zerwas, Nucl.Phys. B 77, 494 (1974)
58) P.Hoyer, P.Osland, H.G.Sander, T.F.Walsh and P.M.Zerwas, Nucl.Phys. 161, 349 (1979)
59) A.Ali, E.Pietarinen, G.Kramer and J.Willrodt, Phys.Lett.93B,155(1980)

60) R.D.Field and R.P.Feynman, Nucl.Phys. B 136, 1 (1978)
61) The Lund Monte Carlo, T.Sjöstrand, B. Söderberg
    Lund Report LU TP 78-18 (1978)
    T.Sjöstrand, Lund Report LU TP 79-8 (1979)
62) G.Altarelli and G.Parisi, Nucl.Phys. B 126, 298 (1977)
    JADE-Collaboration, W.Bartel et al., Phys.Lett.91B,142 (1980)
63) Physics with High Energy, Electron-Positron Colliding Beams
    with MARK J Detector, Physics Report 63, 340 (1980)
64) PLUTO Collaboration, Ch.Berger et al., DESY Report 80/93
    H.J.Daum, H.Meyer and J.Bürger, DESY Report 80/101
65) S.L.Wu and G.Zobernig, Z.für Physik C2, 10) (1979)
66) JADE Collaboration, W.Bartel and A.Petersen, Talks given at the
    XVth Rencontre de Moriond, Les Arcs, March 9-21, 1980
67) J.Ellis and I.Karliner, Nucl.Phys. B 148, 141 (1979)
68) TASSO Collaboration, R.Brandelik et al., DESY Report 80/80 (1980)
    F.A.Berends and R.Kleiss - to be published.
69) MARK J Collaboration, D.P.Barber et al., Phys.Lett. 89B,139 (1979)
70) TASSO Collaboration, R.Brandelik et al., Phys.Lett. 94B, 437 (1980)
71) R.K.Ellis, D.A.Ross and A.E.Terrano, Caltech.Report 68-785 (1980)
    Z.Kunszt, DESY Report 80/79 (1980)
    K.Fabricius, I.Schmitt, G.Schierholz and G.Kramer,
    DESY Report 80/79 (1980)
72) Yu.L.Dokshitzer, D.I.D'yakonov and S.I.Troyan,
    Phys.Lett. 78B, 290 (1978)
73) PLUTO Collaboration, Ch.Berger et al., Phys.Lett. 90B, 312 (1980)
74) R.Baier and K.Fey, Univ. of Bielefeld preprint BI-TP 80/10 (1980)
75) B.Anderson and G.Gustavson, Lund Preprint LU TP 79-2 (1979)
    B.Anderson and G.Gustafson, Z.für Physik C3, 223 (1980)
    B.Anderson, G.Gustavfon and T.Sjöstrand, Lund Report LU TP80-1(1980)
    B.Anderson, G.Gustafson and C.Peterson, Nucl.Phys.B135, 273 (1978)
76) E.J.Williams, Kgl.Danske Videnskab.Selskab, Mat.Fys.Medd.
    13 No. 4 (1934)
    L.Landau and E.Lifshitz, Physik Z, Sovjetunion 6, 244 (1934)
    A.Jaccarini, N.Arteaga-Romero, J.Parisi and P.Kessler,
    Compt.Rend. 269B, 153, 1129 (1969)
    Nuovo Cimento 4, 933 (1970)
    S.J.Brodsky, T.Kinoshita and H.Terazawa, Phys.Rev. D4,1532 (1971)
    Phys.Rev.Lett.25, 972 (1970)
77) F.F.Low, Phys.Rev. 120, 582 (1960)
    F.Calogero and C.Zemach, Phys.Rev.120, 1860 (1960)
78) G.S.Abrams et al., Phys.Rev.Lett. 43, 477 (1979)
79) D.M.Binnie et al., Imperial College London, NO IC/HENP/79/2 (1979)
80) PLUTO Collaboration, Ch.Berger et al.,Phys.Lett. 94B, 254 (1980)
81) J.A.M.Vermaseren, private communication
    R.Bhattacharya, J.Smith and G.Grammer,Phys.Rev.D15, 3267 (1977)
82) E.Hilger, invited paper given at the International workshop on
    $\gamma\gamma$ collisions, Amiens, April 8-12, 1975 France
    DESY Report 80/75, Bonn-HE-80/5
83) A.Roussarie, Minirapporteur talk in section C1, XXth International
    Conference on High Energy Physics, July 17-23, 1980 University
    of Wisconsin, Madison, Wisconsin.

84) C.J.Biddick et al., Paper submitted to the Wisconsin Conference
85) TASSO Collaboration, R.Brandelik et al., DESY Report 80/77
86) J.J.Sakurai, Ann.Phys. 11, 1 (1960)
87) S.M.Berman, J.D.Bjorken and J.B.Kogut
    Phys.Rev. D4, 3388 (1971)
    S.J.Brodsky, F.E.Close and J.F.Gunion, Phys.Rev.D5,1384(1972)
88) H.Spitzer, talk given at the XV Recontre de Moriond, Les Arcs,
    France, March 15-21, 1980
89) T.F.Walsh, Phys.Lett. 36B, 121 (1971)
    S.B.Brodsky, T.Kinoshita, and H.Terazawa,Phys.Rev.Lett.27,280(1971)
    T.F.Walsh and P.Zerwas, Phys.Lett. 44B, 198 (1973)
    E.Witten, Nucl.Phys. B 120, 189 (1977)
    C.H.Llewellyn-Smith,Phys.Lett. 79B, 83 (1979)

# NEW FLAVOR PRODUCTION IN $\gamma$, $\mu$, $\nu$, AND HADRON BEAMS

S. Wojcicki, Stanford, Speaker

G. Salvini, CERN, Chairman

A.A. Carter, CERN,

F. Messing, Carnegie-Mellon,

T.Y. Ling, Ohio,

Scientific Secretaries

# NEW FLAVOR PRODUCTION IN $\gamma$, $\mu$, $\nu$, AND HADRON BEAMS*

Stanley Wojcicki
Physics Department and
Stanford Linear Accelerator Center
Stanford University, Stanford, California 94305

## INTRODUCTION

During the last few years, the main emphasis in the study of heavy particle production (i.e. mainly charm) by other means than $e^+e^-$ annihilation has been on the production mechanisms. Because of the relative cleanliness of the charm signal in the $e^+e^-$ process, most of the data on the properties of the charm particles has originated from that source. There are already indications, however, that this situation is changing. Improved detection techniques coupled with much higher intrinsic production rates suggest that in the future the study of the properties of charm particles will cease to be an exclusive domain of $e^+e^-$ machines.

This review, however, will concentrate mainly on the production data in the $\gamma$, $\mu$, $\nu$ and hadron beams. This is partly because the decay properties have been covered in the review talk of George Trilling and partly because up to now most experiments did emphasize mainly the production aspects. In addition there has been recently a considerable interest in trying to explain most of these data phenomenologically by use of first order QCD diagrams, i.e. photon gluon fusion diagram (Fig. 1a) in the case of photo and muon-production of charmed hadrons and gluon-gluon or quark-quark fusion (Fig. 1b,c) and charmed sea excitation (Fig. 1d,e) for hadronic production of charmed particles. These mechanisms relate the quark structure functions as measured in the massive di-lepton pair production experiments and the deep inelastic scattering experiments ($\mu$, e, and $\nu$) to the production distribution of the charmed hadrons. In addition the gluon diagrams, if dominant, allow one to measure the gluon distributions of the $\pi$, K, and the nucleon.

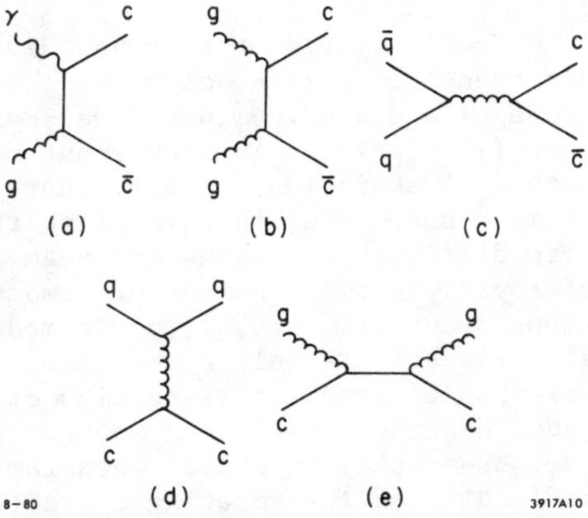

Fig. 1. Typical 1st order QCD diagrams for open charm production by photons and hadrons: a) $\gamma g \to c\bar{c}$, b) $gg \to c\bar{c}$, c) $q\bar{q} \to c\bar{c}$, d) $qc \to qc$, e) $gc \to gc$.

One can contrast this situation with the production of charmed particles in the neutrino interactions either via interaction of the W boson with a strange

---
*Work supported by the Department of Energy, contract DE-AC03-76SF00515, and the National Science Foundation.

0094-243X/81/681431-39$1.50 Copyright 1981 American Institute of Physics

quark from the sea (Fig. 2a), or alternatively via Cabibbo suppressed d → c quark transition (for antineutrinos only) (Fig. 2b). The interest here is mainly the x distribution of the strange sea, which can be extracted from the di-lepton production in the neutrino interactions. That subject (i.e. $\mu^+\mu^-$ and $\mu^\pm e^\mp$ production) has been covered adequately in Frank Sciulli's talk and will not be discussed further here. On the other hand QCD diagrams similar to the ones discussed above (e.g. Fig. 2c) are relevant to the question of <u>associated</u> production of the charmed particles in neutrino interactions, and the data relevant to that question will be summarized briefly.

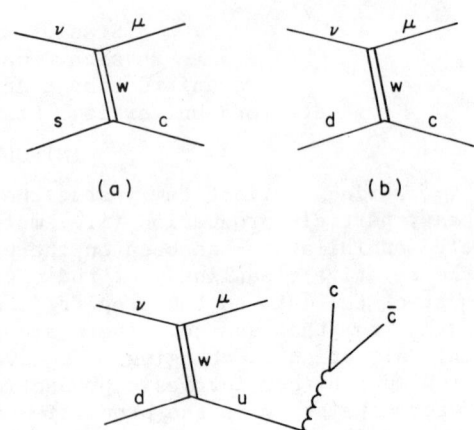

Fig. 2. Diagrams for charm production by neutrinos: a) off strange quark sea, b) off down quark, c) associated production.

## EXPERIMENTAL COMMENTS

The heavy flavor searches divide themselves naturally into 3 different categories, each one characterized by its own peculiar advantages and shortcomings. We shall summarize them here very briefly:

(1) Peaks in the invariant mass spectra. This is the classical method of searching for very short lived particles and has been extremely successful in unraveling the old spectroscopy. It becomes more difficult as masses and beam energies increase, mainly because of rapidly growing number of combinatorials. Furthermore, these kinds of experiments, if performed with electronic techniques, generally investigate only a very limited region of phase space, so extraction of total cross section or angular distribution becomes very model dependent. Finally another potential danger with this technique, especially important when the statistics are limited, is the difficulty of interpreting correctly the statistical significance of a peak in the presence of a large number of cuts. The cuts will be naturally chosen so as to maximize the peaks and thus raise the danger of overemphazing statistical fluctuations. On the other hand, the mass peak method provides the cleanest way to identify production of specific states (e.g. $\Lambda_c$, $F^+$, $D^o$, etc.).

(2) Semileptonic decay modes (i.e. detection of prompt ν, e, or μ and of muon polarization). Here most of the information on the parent particle is lost so identification of specific states is impossible. In addition, because of widely varying semileptonic branching ratios for different charm particles (see below) extraction of total cross section becomes difficult unless contribution of specific

states is known from other sources. Furthermore, the initial production features are somewhat degraded since one observes second generation particles. On the other hand, the important plus here is the possibility of obtaining rather good statistics with a good signal to noise ratio.

(3) Search for short tracks (emulsions, high resolution streamer chambers and bubble chambers, solid state detectors). Most of these detectors are at present undergoing vigorous development efforts and they will probably play a much more important role in the future. Except for the neutrino emulsion experiments, most of these techniques have so far only demonstrated feasibility of doing heavy flavor experiments but as yet their impact on the field has not been very great. Their obvious advantages are relatively bias-free identification of heavy particles, possibility to study in detail the systematics of these particles, and simultaneous exploration of the full $4\pi$ solid angle. One important shortcoming so far has been the relatively low event rate and a great deal of scanning effort necessary to extract the interesting events.

In practice, of course, these techniques are not orthogonal, and very frequently a given experiment will simultaneously rely on use of more than just one of these techniques.

Several additional experimental comments may be in order here.

(a) The relative ratio of charm to non charm hadron production is strongly dependent on the nature of the beam. The rough orders of magnitude for different beams are;

$$\begin{aligned}
\text{hadronic beams} &\sim 10^{-3} \\
\text{photon } (\mu) \text{ beams} &\sim 10^{-2} \\
\text{neutrino beams} &\sim 10^{-1} \\
e^+e^- \text{ annihilation} &\sim 1
\end{aligned}$$

(b) The evidence presented at this conference provides strong evidence that the lifetimes of different charm particles differ widely. The most systematic study of this question was presented in a report by Niu, who quoted[1])

$$\left. \begin{aligned}
D^+ &\to 10.3 \, ^{+10.5}_{-4.1} \\
D^0 &\to 1.01 \, ^{+.43}_{-.27} \\
F^+ &\to 2.2 \, ^{+2.8}_{-1.0} \\
\Lambda_c &\to 1.36 \, ^{+0.84}_{-0.46}
\end{aligned} \right\} \times 10^{-13} \text{ sec}$$

The significance of this result in the content of the present discussion is that the semileptonic branching ratios will be approximately proportional to the lifetime (that statement is rigorously true for $D^+$ and $D^0$). Thus all the cross section estimates extracted from the semileptonic experiments might be significantly in error if the production process is dominated by one single state.

(c) A dependence of the cross section is a relevant question here. Since most fixed target experiments use generally heavy nuclei as target material (e.g. iron, A = 56) and ISR experiments study p-p interactions, knowledge of A dependence is quite crucial to the comparison of different experiments. It is conventional now to assume linear A dependence for heavy flavors, in analogy with the $J/\psi$ production[2]. However, it should be stressed here that at present there are no experiments that bear on this question for unbound charm production, and that the A dependence could vary with x.

## CHARM PRODUCTION BY HADRONS

The field of hadronic production of unbound charm states (i.e. D, $\Lambda_c$, etc.) is still in its early infancy. Because of rather unfavorable signal to noise ratio only very sparce data on production rates are available and the information on x and $p_T$ distributions is even more scanty. Thus only very rough comparisons with phenomenological predictions can be made; this section, accordingly, shall emphasize mainly the experimental data and the outstanding experimental problems. More specifically, we shall address 3 separate topics here, i.e.

1) Central Production, Near x = 0
2) The Question of Forward Production
3) Anomalies and Disagreements Between Different Experiments.

1) <u>Central Production</u>. I shall try to summarize here the contributions of all those experiments that either concentrated on x = 0 region or had such acceptance that they were sensitive to the production in that region. No firm quantitative predictions and comparisons with the theory can be made here with any strong degree of assurance. This is at least partly due to potential contribution of several different diagrams (quark fusion, gluon fusion, flavor excitation by quark or gluon scattering), our ignorance about their relative importance,[3] and dependence of the calculations on the mass of the charmed quark. On the other hand we can make some reasonably intelligent guesses as to what the hadronic production of charm should look like if the diagrams discussed above were indeed the dominant ones. Specifically we would expect:

a) The cross section in the Fermilab and SPS region ($\sqrt{s} \approx 30$) to be about 5-20 μb for total charm production.
b) The increase between that energy domain and the ISR energy range ($\sqrt{s} \approx 60$) should be about a factor of 2-3.
c) The x distribution for production by nucleons should go roughly as $(1-x)^n$ with n being somewhere between 3 and 5, since that is the approximate dependence of the quark and gluon distributions in the nucleon. Mesonic production distribution might be expected to be slightly flatter.

The data available up to now are summarized in Table I. Several observations need and can be made regarding these data.

a) The comparison between different numbers should probably not be taken more seriously than up to a factor of 2. This is because of unknowns in A dependence, branching ratios, final states produced, and

TABLE I

Summary of charm cross section results

| Reaction | Group | Reference | Technique | Signature | $\sqrt{s}$ | $\sigma(\mu b)$ |
|---|---|---|---|---|---|---|
| $\pi^- p$ | BGRST | 4 | TST in BEBC | Single e | 11.5 | 19±11 |
| $\pi^-$Be | PSTB | 5 | $D^*$ only | $(K\pi)\pi$ | 19.4 | 10 ± 4 |
| $\bar{p}p$ | BHLMS | 6 | TST in BEBC | Single e | 11.5 | <24 |
| $\pi^- p$ | BCOPRRT | 7 | LEBC | Short tracks | 25.6 | 35-40 |
| pEm | Tata | 8 | emulsion | Short tracks | 27.4 | 160±40 |
| pp | CERN-Saclay-Zurich | 9 | spectrometer | $\mu^\pm e^\mp, e^\pm e^\mp$ | 53,63 | 22± 5 |
| pNe | Yale-Fermilab | 10 | streamer ch. | Short tracks | 25.8 | 20-50 |
| pEm | Nagoya-Aichi-Yokohama | 11 | emulsion | Short tracks | 27.4 | 30±20 |
| pFe | CIT-Stanford | 12 | total absorption | Single $\mu$ | 27.4 | 13-60 |
| pFe | CIT-Stanford | 13 | total absorption | $2\mu$ + ME | 27.4 | 7-20 |
| pFe | CFRS | 14 | total absorption | Single $\mu$ | 25.8 | 22 ± 9 |
| pFe | Serpukhov | 15 | beam dump | $\nu$ | 11.5 | 4 ± 3 |
| pW | Michigan | 16 | beam dump (test) | $\nu$ | 27.4 | 30-75 |
| pCu | Gargamelle | 17 | beam dump | $\nu$ | 27.4 | $80^{+40}_{-25}$ |
| pCu | BEBC | 18 | beam dump | $\nu$ | 27.4 | 11-22 |
| pCu | CDHS | 18 | beam dump | $\nu$ | 27.4 | 7-14 |
| pCu | CHARM | 19 | beam dump | $\nu$ | 27.4 | 12 ± 4 |
| pp | ACCDHW | 20 | SFM, $e^-$ trigger | $\Lambda_c \to K^-\pi^+ p$ | 63 | 140±60* |
| pp | ACCDHW | 20 | SFM, $e^-$ trigger | $D^o \to K^-\pi^+$ | 63 | 700±300* |

*$\frac{d\sigma}{dx_F}$ at $x_F = 0$

production mechanisms. No great effort has been made to insure that the assumptions used in extracting the final numbers for all the experiments have been entirely self consistent.

b) The data are dominated by the experiments near $\sqrt{s} \approx 27$. In that region, the total cross sections are consistent (up to a factor of 2) with $\sigma_{tot} \approx 20$ μb. The only point that appears to be slightly high is the preliminary result quoted by the Tata group[8] at this conference of $160 \pm 40$ μb.

c) To the extent that the data on this question are available the experiments are consistent with central production, i.e. x dependence of the form $(1-x)^n$ with $3 < n < 5$.

d) The situation in the ISR region near $x = 0$ is not clear as there appears some discrepancy between the 3 different ISR measurements quoted. The question as to whether the cross section at $x = 0$ rises dramatically between $\sqrt{s} = 27$ and 60 does not appear to be settled by these data.

2) <u>Forward Production</u>. There have now been several experimental programs that bear on this question.

a) at $\sqrt{s} = 53$ and 63 there are 3 experiments (by <u>S</u>plit <u>F</u>ield <u>M</u>agnet group,[21] <u>L</u>amp-<u>S</u>hade <u>M</u>agnet group,[22] and UCLA-Saclay group[23]) that study $\Lambda_c$ and D production at the ISR.

b) at $\sqrt{s} = 27$ there are the 3 beam dump experiments (CDHS,[24] BEBC,[25] and CHARM[19] collaborations) that study prompt $\nu$ interactions (presumably coming from the decay of short-lived particles). In addition, the Cal Tech-Stanford collaboration has studied the production of prompt forward muons[26] (presumably decay products of short-lived particles produced by the primary protons).

c) at $\sqrt{s} = 20$, a Fermilab experiment has studied forward diffractive production of D's in $\pi^- p$ interactions.[27]

d) at $\sqrt{s} = 7.4$ there have been 3 beam dump experiments performed at the Brookhaven AGS.[28]

Very briefly, the results of these experiments can be summarized as follows. Starting with the lowest energies, there appears to be no evidence for any prompt neutrino production in the BNL beam dump experiments. There is some discrepancy between the calculated $\nu_\mu$ fluxes (coming from $\pi$ and K decays from the original hadronic cascade) and the observed $\nu_\mu$ numbers,[29] but the majority belief is that the calculations probably are not reliable enough to make the discrepancy significant.[30] The cross-section limits for charmed particle production as obtained from these experiments[30] are still considerably above the interesting limits (12-20μb for $\sigma_{D\bar{D}} B_{D \to \nu_e}$).

For completeness one should mention here an older beam dump experiment performed at Serpukhov,[15] i.e. intermediate energy ($E_p = 70$GeV, $\sqrt{s} = 11.5$). They report evidence for prompt $\nu_e$ with a cross section, $\sigma_{D\bar{D}} B_{D \to \nu_e} = .5 \pm .4$μb.

A finite signal for diffractive $D\bar{D}$ production was obtained at Fermilab by the HFIOI collaboration[27] in $\pi^- p$ interactions at $\sqrt{s} = 20$. The evidence for production of roughly equal amounts of $D^0$ and $\bar{D}^0$ is displayed in Fig. 3b,c, where a narrow peak at the mass of the D is seen in both the $K^-\pi^+\pi^+$ and the $K^+\pi^-\pi^-$ mass spectra. Furthermore, the x distribution (Fig. 3a) supports the hypothesis that the D's are produced diffractively, although it should be pointed out that the trigger itself requires a slow proton thus favoring a forward mechanism. A model dependent cross section estimate yields $\sigma_{D\bar{D}} = (6-10)\pm 4\mu b$.

Turning now to $\sqrt{s} = 27$, there is an agreement (within a factor of 2) between the CERN beam dump experiments and the Stanford-Cal Tech experiment at Fermilab on the overall size of the prompt lepton signal (the discrepancies on the details will be discussed below). As an example, Fig. 4 compares the momentum distribution of the prompt lepton from the Fermilab and BEBC experiments. The techniques are totally different here; the comparison is relatively model independent. Some model dependence arises from the fact that the acceptance in the $p_T$-x space is quite different for these 2 experiments. Both experiments are sensitive to a large fraction of the forward x region; however the Stanford-CIT experiment has basically a flat 100% acceptance for $p_\mu > 60$ GeV; the relative detection efficiency for the beam dump experiments goes roughly as $p_\nu^3$. Furthermore the Stanford-CIT experiment accepts essentially all $p_T$; the neutrino experiments look only at very low $p_T$ ($\Delta\theta \leq 1.8$mr).

The gross features of the CERN experiments can be adequately explained by a central production.

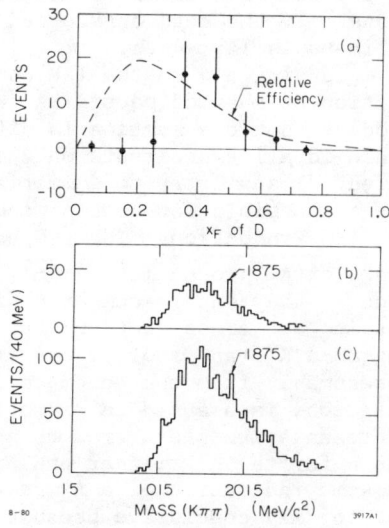

Fig. 3. a) $x_F$ distribution of the events in the D peak, b) $K^-\pi^+\pi^+$ mass spectrum, c) $K^+\pi^-\pi^-$ mass spectrum.

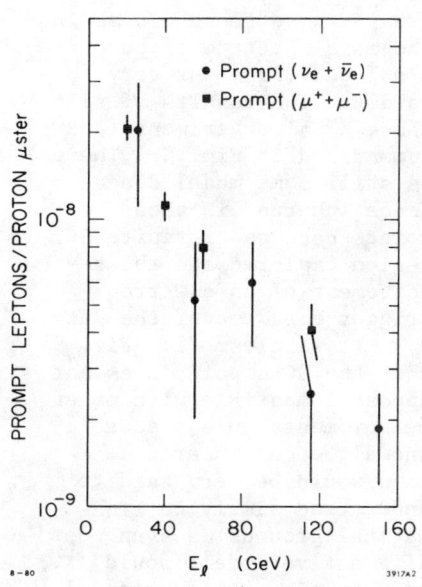

Fig. 4. Comparison of the prompt μ spectrum (CIT-Stanford) with prompt ν spectrum (BEBC).

mechanism that is consistent with that deduced by Stanford-CIT collaboration from their earlier experiments[12,13] that emphasized the central region. In addition, a diffractive mechanism is not a good fit to the Stanford-CIT $p_\mu$ distribution, predicting too many high energy muons. If the data are forced to a diffractive fit, the cross section estimate is $14 \pm 4$ μb.

Turning now to the ISR energies, we note the large forward production of charmed particles, especially $\Lambda_c$. The extraction of total production cross section is difficult and highly model dependent because in all experiments only a limited knematic region is investigated. I shall try to summarize the relevant facts in as coherent a way as possible for both $\Lambda_c$ and D production.

$\underline{\Lambda_c \text{ Production}}$. The LSM and SFM groups have presented 2 measurements of $\Lambda_c$ production.[21,22] Both experiments identify the $\Lambda_c \rightarrow K^-p\pi^+$ mode so they can be compared directly. Furthermore each experiment obtains one cross section measurement at a rather forward x by triggering on a $K^-$, and another one at a lower x by using an electron trigger (presumably from the accompanying $\bar{\Lambda}_c$ or $\bar{D}$). A universal 10% BR into electrons is assumed in extracting the cross section. Some of the representative plots from these experiments are displayed in Figs. 5 and 6 for the $K^-$ trigger and Figs. 7 and 8 for the e trigger. There is some indication of a $\bar{\Lambda}_c$ peak in the LSM data but the evidence is not totally conclusive because of low statistics and a slight downward displacement of the position of that mass peak.

One should also mention here an older published result by the UCLA-Saclay group[23] who found a peak in the mass spectrum of both $K^-p\pi^+$ and $(\Lambda 3\pi)^+$ at very forward x. The results of all those $K^-p\pi^+$ experiments are summarized in Fig. 9. There is still some model dependence inherent in those points because of finite $\Delta x$ region explored and the requirement of an electron trigger for some of the data.

It is clear from Fig. 9 that the UCLA point does not appear compatible with other measurements unless some anomaly occurs near x=1. This would be very hard to understand simply on kinematical grounds as even most diffractive models would tend to give suppression of

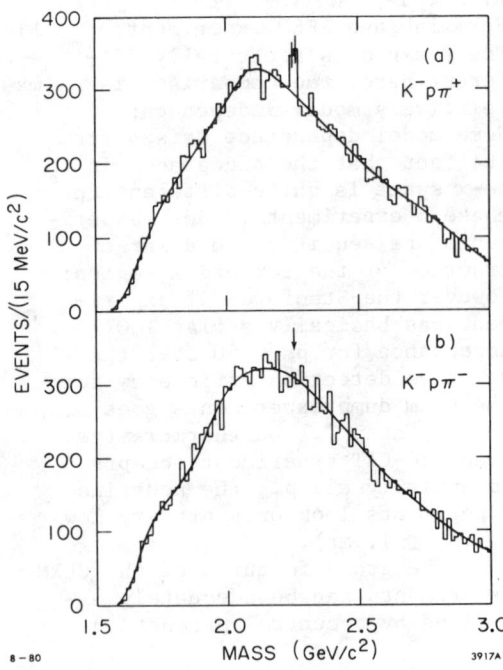

Fig. 5. $K^-p\pi^+$ and b) $K^-p\pi^-$ mass spectrum from the LSM experiment ($K^-$ trigger).

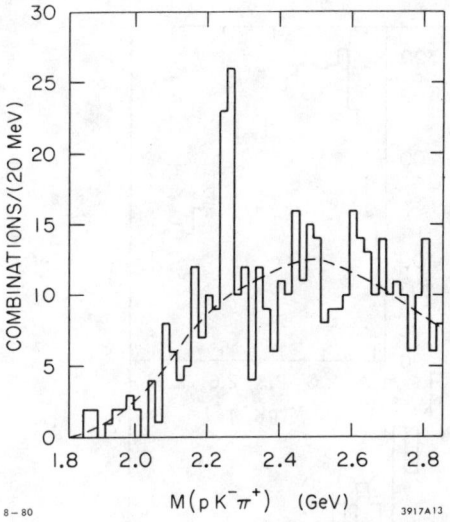

Fig. 6. $K^-p\pi^+$ spectrum from the SFM experiment ($K^-$ trigger).

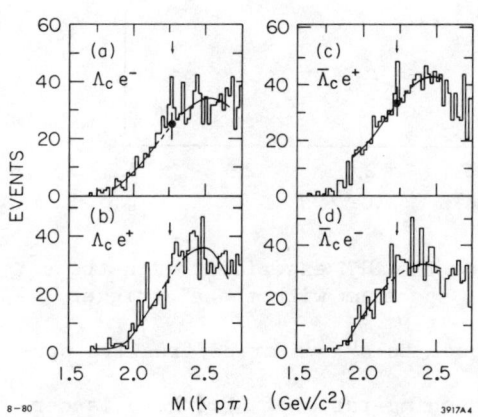

Fig. 7. $K^-p\pi^+$ mass spectrum from the LSM experiment with the $e^-$ (a) and $e^+$ (b) trigger and the $K^+\bar{p}\pi^-$ spectrum with $e^+$ (c) and $e^-$ (d) triggers.

$\Lambda_c$ production in that region due to the large mass of $\Lambda_c$ and accompanying $\bar{D}$. Ignoring the UCLA point we can extract a rough estimate of the total cross section by assuming a typical diffractive picture of flat x dependence up to $x \approx 0.7$. $\sigma_{tot}$ will then be given by $2 \cdot 0.7 \cdot B\frac{d\sigma}{dx}/B$. Taking 5μb for $B\frac{d\sigma}{dx}$ and 2.2±1.0% for $K^-p\pi^+$ branching ratio[31] we obtain 320μb for $\sigma_T$. This number should be compared with $\sigma_T B(\Lambda_c^+ \to \Lambda 3\pi) = (1.0\pm0.3)$μb extracted with the help of a diffractive production model by D. DiBitonto[32] from a different subset of the LSM data. No good data exist allowing one to relate branching ratios for these two decay modes but it is unlikely that $B(\Lambda_c \to 3\pi)$ is less than 1%. Thus we are faced with a discrepancy of at least a factor of 3.

We must remember that we have to add the D and F production to the above numbers to obtain total charm cross section. The indications from SFM are that the $\bar{D}$ cross section[20] is also around several hundred μb implying a total charm cross section in the vicinity of 1mb, i.e. 1½ orders of magnitude above the cross sections observed at $\sqrt{s} = 27$. Is that reasonable and can it be easily understood? Let us examine some of the possible mechanisms for this difference:

1. Standard QCD diagrams would predict only a factor of 2 or so between $\sqrt{s} = 30$ and 60. A possibility is contribution from a different process (e.g. hidden intrinsic charm in the nucleon[33]) with a sharp threshold near $\sqrt{s} = 30$. The assumption about a threshold appears slightly artificial.

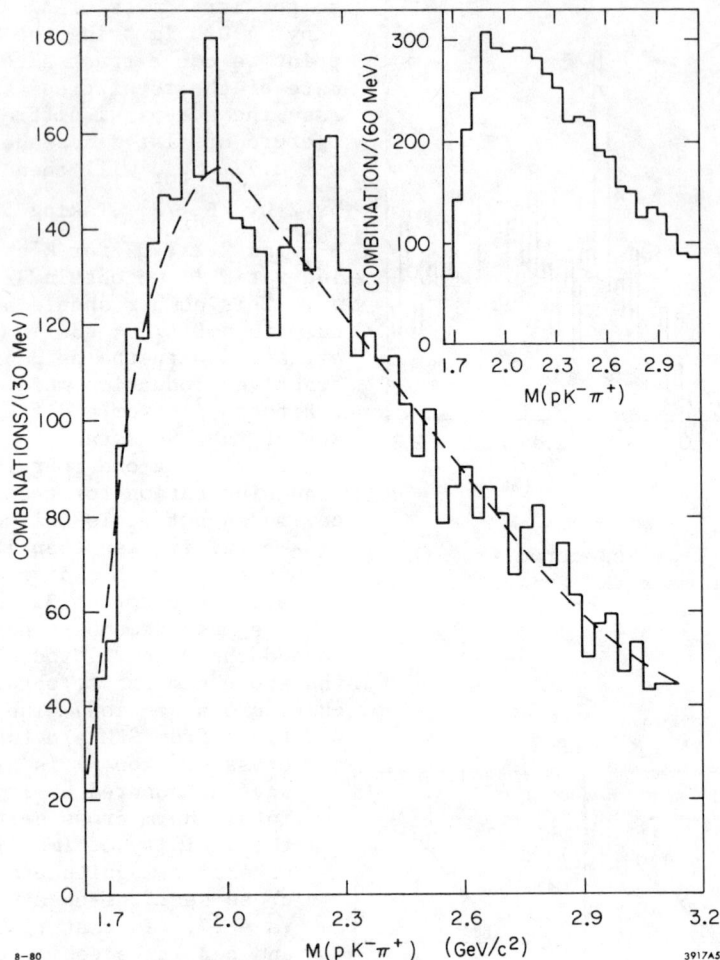

Fig. 8. $K^-p\pi^+$ spectrum from the SFM experiment with the $e^-$ trigger. Insert shows this spectrum with the $e^+$ trigger.

2. A dependence could be closer to $A^{0.75}$ for diffractive production.

3. Very low semileptonic branching ratio of $\Lambda_c$ (there is certainly some evidence for that - see above[1])) would decrease the sensitivity of the beam dump and CIT-Stanford experiments.

4. The error on $\Lambda_c \to K^-p\pi^+$ is still rather large. Thus a branching ratio larger by ~ 50% certainly cannot be excluded.

Each of these effects could certainly contribute a factor of 2-3 making the cross section difference much more reasonable. One outside possibility that has to be considered is whether the effect that is seen at the ISR is really a $\Lambda_c$ as opposed to a non-charmed resonant

state. It must be remembered
that $K^-p\pi^+$ is not an exotic
state, and thus does not constitute a <u>prima facie</u> evidence
for charm production. The
association with charm is
based on
   a)  narrow width of the state
   b)  absence of negative state
   c)  mass comparable to $\Lambda_c$ in $e^+e^-$ annihilation
   d)  association with electrons.
The first two pieces of evidence
are really not very strong.
Narrow non-charm states have
been seen, and positive states
are expected to dominate in pp
collisions in the diffractive
region. The latter is empirically observed for $\Sigma(1385)$.[23]
The mass question has been a
source of controversy for some
time and it might be worthwhile to consider the new
contributions on this subject.
The masses of $\Lambda_c$ have varied
from 2255 to 2290 MeV but the recent data appear to favor a value of
2285 MeV. To summarize the recent measurements we have

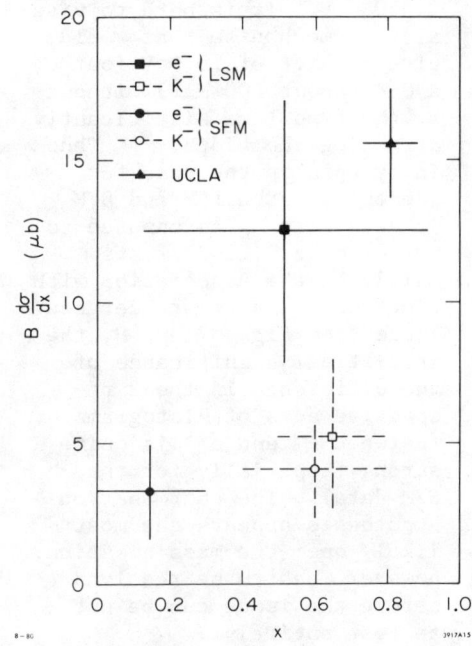

Fig. 9. Summary of the ISR $K^-p\pi^+$ data.

$$M_{\Lambda_c} = 2.285 \pm 0.006 \text{ GeV} \qquad e^+e^- \text{ SPEAR}^{31)}$$
$$M_{\Lambda_c} = 2.284 \pm 0.005 \text{ GeV} \qquad \gamma p \quad \text{(CIF)}^{34)}$$

In addition there have been 3 contributions to this conference from
the neutrino experiments that favor this value, i.e.

   a) a mass peak observed at $2.275 \pm 0.010$ in $\nu D$ interactions in
$\Lambda\pi^+$ and $K_S^0 p$ (see Fig. 10)[35]
   b) 2 completely fitted events from BEBC[36] giving $K^-p\pi^+$ masses
of $2.285 \pm 0.005$ and $2.280 \pm 0.003$ GeV.
   c) a fully reconstructed BEBC TST event[37]

$$\nu + p \to \mu^- \pi^+ \Sigma_c^+$$
$$\hookrightarrow \Lambda_c^+ + \pi^0$$
$$\hookrightarrow K^-p\pi^+$$

giving $M_{\Lambda_c} = 2290 \pm 0.003$.

   In contrast both LSM and SFM give consistently values around

2.260 GeV. It is hard to visualize a mechanism that would give a shift of 25 MeV (out of a Q of about 700 MeV) without at the same time significantly affecting the width.[38] Thus in my opinion the case for identifying the LSM and SFM effects with $\Lambda_c$ as opposed to for example $\Sigma(2250)$ rests mainly on the association with electrons. As the reader can judge from Figs. 7 and 8, the statistical significance of the difference in the 2 respective sets of histograms (between $e^-$ and $e^+$) is quite strong (especially for the SFM data). The charm baryon hypothesis appears the most likely one; the mass question, however, has to be resolved before the issue can be put to rest entirely.

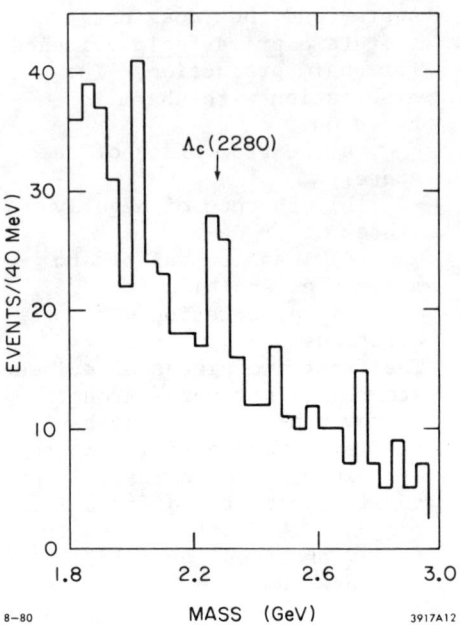

Fig. 10. $K^0 p$ and $\Lambda \pi^+$ mass spectrum from the $\nu D$ exposure.

D Production. The SFM group has previously published[21] evidence for $D^+$ mesons, as observed in the decay chain $D^+ \to K^{*0} \pi^+ \to K^- \pi^+ \pi^+$. The worrisome features of this result were the large cross section (150-2000 $\mu$b depending on the model used), strong association with the $K^*$ in contrast to the SPEAR results, and a slight mass shift: 1.91 GeV observed vs. 1.868 GeV accepted value.

The same group presented at this conference[20] a preliminary $4\sigma$ evidence for $D^0 \to K^- \pi^+$ at $\sqrt{s}$ = 63 GeV (Fig. 11) observed by using an $e^-$ trigger. The LSM group presented 95% confidence upper limits for D production from their data with $K^-$ trigger. Their 2 most stringent limits, the 2 positive D signals, and the result from an older lepton pair ($e\mu$ and ee) experiment[9] are summarized in Fig. 12. The comparison of the data is

Fig. 11. $K^- \pi^+$ mass spectrum from the SFM experiment obtained with the $e^-$ trigger. The smooth curve shows the shape of this spectrum taken with the $e^+$ trigger.

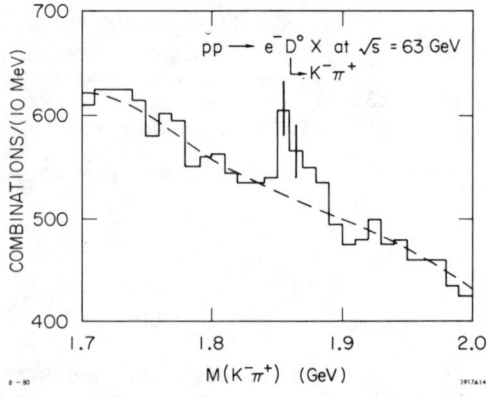

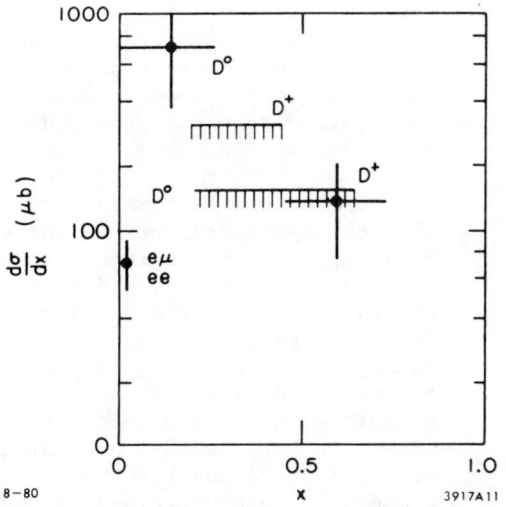

Fig. 12. Summary of the ISR $D^0$ and $D^+$ data. The shaded bars correspond to the upper limits from the LSM experiment.

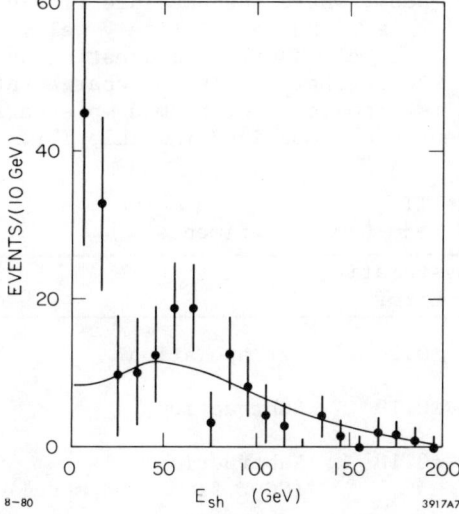

Fig. 13. Visible energy distribution of no-μ events from CHARM experiment (muon NC events have been subtracted). The curve shows expected contribution to electron neutrino interactions from $D\bar{D}$ decay.

made difficult by the fact that the semileptonic branching ratios of different charm particles are now known to be quite different.[1,39] Chilingarov et al., however, assumed 10% BR for both $D^0$ and $D^\pm$. A dominant production of $D^0(\bar{D}^0)$ near x=0 would be one easy way to resolve the apparent controversy. One should probably end this discussion by noting that the high e/π ratio at low x and $p_T$ observed at the ISR[40] is consistent with the charm production cross section of the order of several hundred microbarns.

3. <u>Anomalies and Discrepancies</u>. I would like to conclude this chapter by discussing 3 experimental results that are either anomalous in themselves or for which different experiments do not give a consistent answer.

a) The CHARM collaboration[19] in their beam dump experiment sees a 2.5σ excess of no-μ events for shower energies 2 < $E_{sh}$ < 20 GeV, above what one would expect from $D\bar{D}$ production normalized to $E_{sh}$ > 20 GeV (Fig. 13). Specifically the excess is 54±19 (statistical) ± 9 (systematic) events. Due to instrumental reasons, the other experiments cannot investigate identical region, the closest comparison being with the CDHS experiment who apply a lower cut on shower energy of $E_{SH}$ > 5 GeV. Within 1σ their data are consistent (Fig. 14) with the predictions based on $D\bar{D}$ production.

b) There are some indications that the $\nu_e/\nu_\mu$ ratio may not be equal to unity. The results of the 3 beam dump experiments are summarized in Table II. That ratio can be obtained either

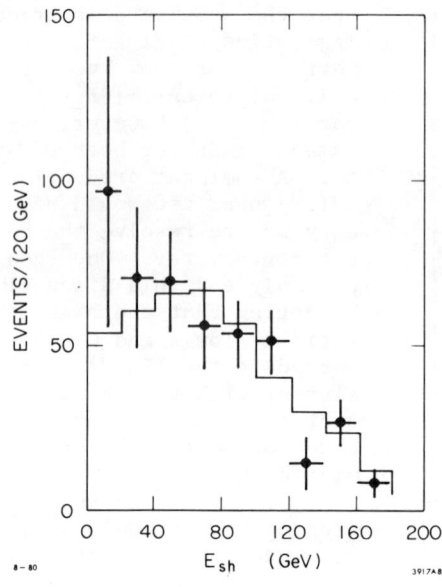

Fig. 14. No-μ events from CDHS experiment (muon NC events have been subtracted). Solid line indicates prediction from central $D\bar{D}$ production and decay.

by using prompt $\nu_\mu$ rate from extrapolation technique (comparison of rates at 2 different densities) or from subtraction method, where one uses a shower cascade calculation to obtain the contribution to $\nu_\mu$ flux from π and K decay.

One should emphasize here that the systematic errors for the 3 experiments are quite similar and are probably in the direction of overestimating the $\nu_\mu$ flux (beam scraping or hadronic cascade leakage from the front part of the target would certainly have this effect). The second relevant observation here is that according to the BEBC[25] and BEBC-TST groups,[41] this deficiency of $\nu_e$'s (if real) cannot be explained on the basis of $\nu_e \to \nu_\tau$ oscillations. The conclusion is based on the observation of the expected (within statistical errors) number of $\nu_e \to e$ events in the narrow band beam, where the absolute flux of $\nu_e$'s is known relatively well.

c) There is a question as to whether the lepton charge ratio is different from unity. The results from both the beam dump experiments and the CIT-Stanford experiment are summarized in Table III.

Table II
$\nu_e/\nu_\mu$ Ratios from CERN Beam-Dump Experiments

| Group | $\nu_e/\nu_\mu$ Ratio | Statistical Error | Systematic Error | Method |
|---|---|---|---|---|
| CDHS | 0.77 | ±0.18 | ±0.24 | Extrapolation |
| CDHS | 0.58 | ±0.07 | ±0.19 | Subtraction |
| CHARM | 0.48 | ±0.12 | ±0.10 | Subtraction ($CC\nu_e$ from prod model) |
| CHARM | 0.49 | ±0.21 | | Extrapolation ($CC\nu_e$ from prod model) |
| CHARM | 0.44 | ±0.11 | ±0.03 | Subtraction ($CC_e$ directly identified) |
| BEBC | 0.59 | +0.35 / −0.21 | | Subtraction |

Table III
Lepton charge ratio from different experiments

| Group | Ratio | Value | Statistical Error | Systematic Error | Method |
|---|---|---|---|---|---|
| CDHS | $\bar{\nu}_\mu/\nu_\mu$ | 0.12 | ±0.20 | ±0.12 | Extrapolation |
| CDHS | $\bar{\nu}_\mu/\nu_\mu$ | 0.56 | ±0.09 | ±0.13 | Subtraction |
| CHARM | $\bar{\nu}_\mu/\nu_\mu$ | 1.3 | ±0.5 | +0.4 / −0.2 | Subtraction |
| CHARM | $\bar{\nu}_\mu/\nu_\mu$ | 1.8 | | ±1.1 | Extrapolation |
| BEBC | $\bar{\nu}_\mu/\nu_\mu$ | 0.75 | | ±0.32 | Subtraction |
| BEBC | $\bar{\nu}_e/\nu_e$ | 0.76 | | ±0.35 | Subtraction |
| CIT-Stanford | $\mu^-/\mu^+$ | 1.3 | | ±0.4 | Extrapolation |

Clearly the largest, and the only really significant departure from unity, occurs for the CDHS extrapolation result. The dependence of the $\nu_\mu$ and $\bar{\nu}_\mu$ fluxes on energy is displayed in Fig. 15. Note that this Figure displays observed events, i.e. the ratio of $\bar{\nu}_\mu/\nu_\mu$ cross sections (0.48) has not been taken out. The discrepancy between the various beam dump experiments is due mainly to the low density $\mu^+$ point, as can be seen from Fig. 16.

It should be noted that there is nothing fundamental about the charge ratio deviating from unity. A variety of mechanisms, like $\Lambda_c$ production or unequal $D^+, D^-$ cross section could alter this ratio either by virtue of different semileptonic branching ratios or different x dependence.

In summary, several potentially interesting effects are suggested by the data. More detailed experiments are needed, however, to explore and answer these questions.

## CHARM PRODUCTION BY PHOTONS AND MUONS

We discuss these two topics together since the charm productions by muons occurs via virtual photon mechanism. Thus the physics explored by experiments with these 2 beams is quite similar. Schematically the outline of this chapter can be indicated as follows:

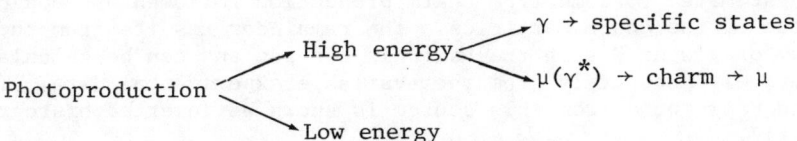

We shall review first the high energy experiments, discussing both the muon experiments that study charm production via their muonic decay modes and the photon experiments, in which specific states are studied via kinematical reconstruction. It is interesting to compare

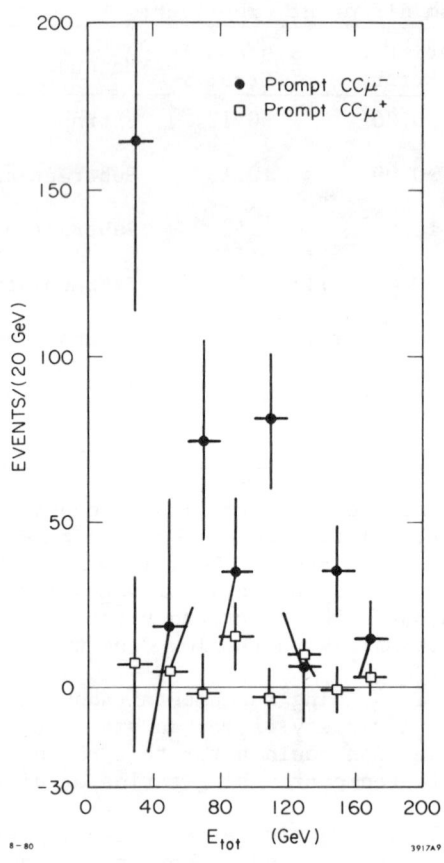

Fig. 15. Prompt CC $\mu^+$ and $\mu^-$ spectra from CDHS experiment using the extrapolation method.

the data with the predictions of the photon gluon fusion ($\gamma$gF) model.[42] This model, based on the lowest order QCD diagram (Fig. 1a) is the QCD analogue of the familiar Bethe-Heitler process, the only difference being the replacement of the virtual photon by a gluon. Thus with one vertex presumably determined by QED, the data can be used to extract the gluon distribution in the nucleon. A useful point of view is to look at this model as predicting a strong correlation between the c and $\bar{c}$ quarks in the sea, since they originate from the gluon dissociation into a $c\bar{c}$ pair.

The muon and photon experiments are complementary, in so far that the former are capable of providing very good statistical information at the expense of some of the detail; the latter, on the other hand can study the details of specific charm final states.

The muon data originate from 2 experiments, the Berkeley-Fermilab-Princeton collaboration[43] (BFP) and the European Muon Collaboration[44] (EMC). Because of the design meant to specifically emphasize multimuon final states, the BFP experiment has much better statistics ($\sim$20072 2$\mu$ events to be compared with 497 from EMC) and an experimental advantage of absence of any desensitized region in the detector. Both detectors emphasize the forward (i.e. diffractive) production region. The results from BFP[45] can be briefly summarized as follows:

a) (81 ± 10)% of the single extra muon (i.e. 2$\mu$) final states are estimated to come from charm production followed by muonic decay of one of the charm particles. The remainder results from the muonic decay of $\pi$'s or K's in the hadronic cascade and can be calculated relatively accurately from the available experimental data. The estimated background from this source is shown as inverted histograms in Fig. 17.

b) The data generally show good agreement with the $\gamma$gF model as seen from Fig. 17, where the predictions of the model (curves) are compared with the experimental distributions from which the $\pi$, K decay background has been subtracted.

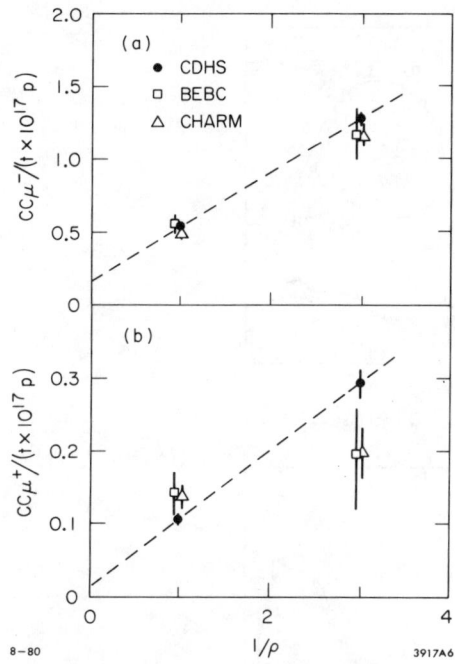

Fig. 16. Comparison of CCµ rates as a function of density for the 3 CERN beam dump experiments: a) $\mu^-$, b) $\mu^+$.

c) The observed diffractive charm production can account for about 1/3 of the total inclusive scale non-invariance in the kinematic region defined by $2 < Q^2 < 13$ GeV$^2$ and $50 < \nu < 200$ GeV.[46]

d) The data appear to require variation of the cross section with the photon energy ($\nu$), as demanded by the $\gamma gF$ model. Energy independent cross section does not reproduce the data (dashed line in Fig. 17a).

e) Photon charm cross section values have been extracted at two energy intervals, i.e. $\sigma^\gamma = 750^{+180}_{-130}$ nb at $\bar{E}_\gamma = 178$ GeV and $\sigma^\gamma = 560^{+200}_{-130}$ nb at $\bar{E}_\gamma = 100$ GeV. The rise with energy is statistically significant, $\Delta\sigma = 190^{+34}_{-52}$ nb because of common systematic errors.

f) It might be interesting to compare these numbers with the total photon hadronic cross section rise of about 4 µb between 40 and 150 GeV.[47] Of course, other processes are known to contribute also to this rise, one of which is presumably non-diffractive charm production to which the BFP experiment is insensitive.

Similar conclusions have been reached by the EMC collaboration from the analysis of their 2µ events.[48] They have compared their data both to the struck quark model (charmed quark density in the sea taken from the parametrization of Buras and Gaemers[49]) and to the $\gamma gF$ model. The first model predicted cross sections about a factor of 5 higher than observed, the latter gave excellent agreement.

The EMC collaboration has also analyzed their 3µ events with the goal of extracting $c\bar{c}$ production and their subsequent double muonic decay.[50] Experimentally, this is a more difficult problem because of the need to eliminate both the electromagnetic trident contribution as well as contributions due to vector meson ($\rho$, $\phi$, $\psi$ etc.) 2µ decay. These backgrounds can be eliminated to a large extent by two cuts, i.e.

$$1.0 < M_{\mu\mu} < \text{GeV}^2$$

and

$$z < 0.6 \text{ , with } z \equiv E_{\mu\mu}/\nu \text{ .}$$

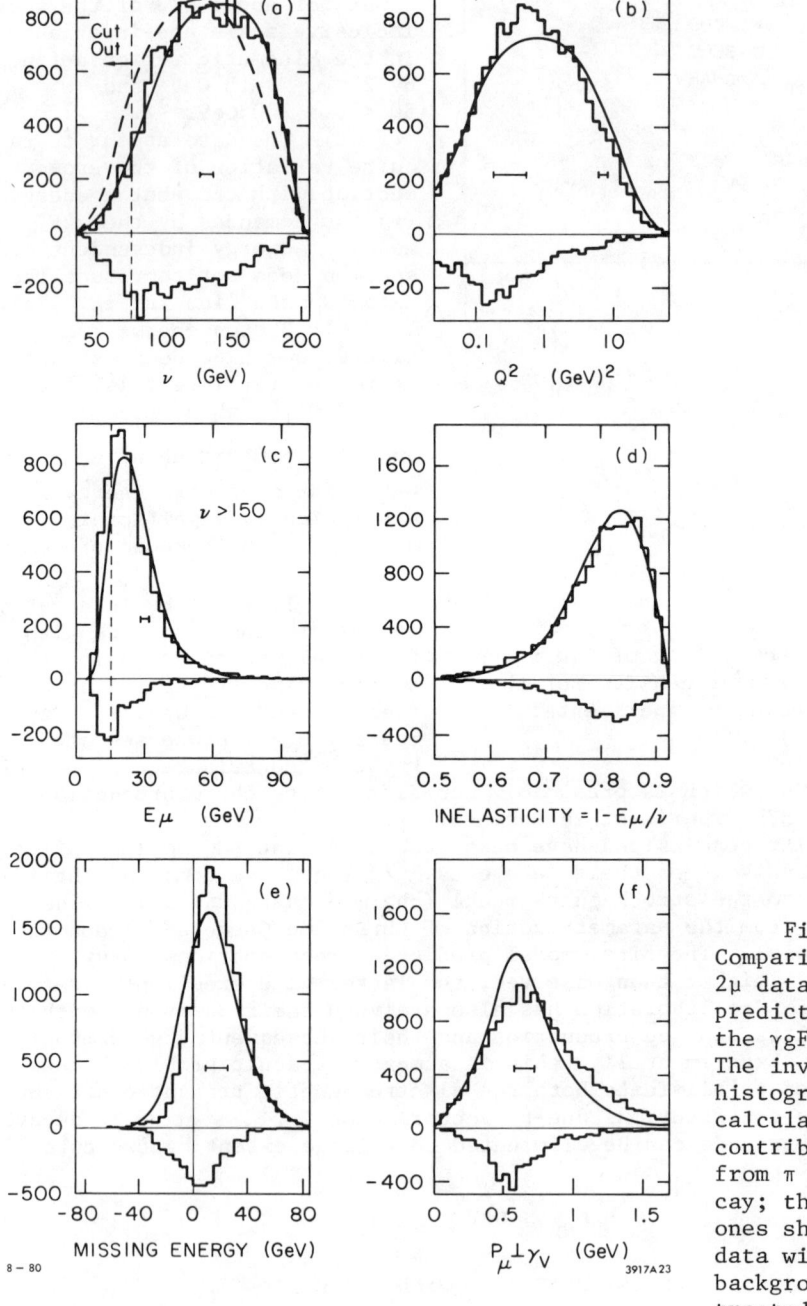

Fig. 17. Comparison of BFP 2μ data with the predictions of the γgF model. The inverted histograms show calculated contributions from π and K decay; the upright ones show the data with that background subtracted.

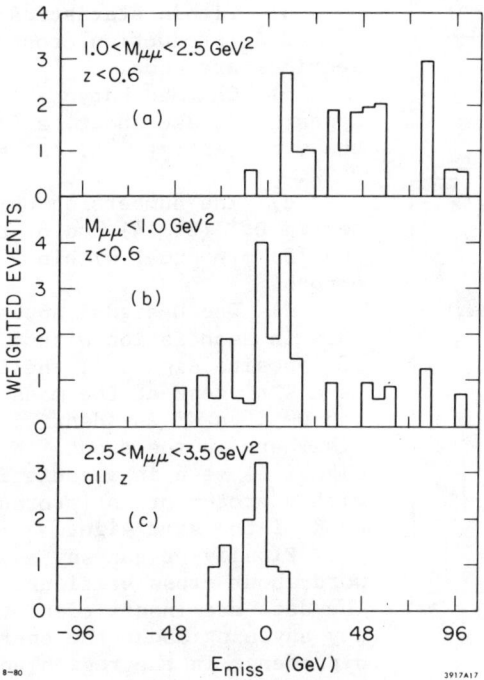

Fig. 18. Missing energy distribution for the 3 different $M_{\mu\mu}$ ranges from the EMC experiment.

The success of these cuts is demonstrated in Fig. 18 where one displays the missing energy distribution. The low mass events, dominated mainly by low mass vector meson and trident contributions, and high mass events, principally $\psi/J$, show a missing mass distribution reasonably consistent with zero. The events surviving the cuts, however, show a definite positive value of missing energy, $\overline{E_{miss}}$ = 44 GeV indicative of 2 neutrinos accompanying the 2 decay muons. 4 different kinematical quantities from the accepted events are displayed in Fig. 19. Again the fit to γgF model is very good; the dashed curves show the estimate of the background due to double π and K decay. The γgF fit used Λ = 0.5 GeV, $m_c$ = 1.5 GeV, and the conventional gluon distribution $\eta G(\eta) = 3(1-\eta)^5$.

These data can be compared with the results obtained by the Columbia-Illinois-Fermilab collaboration studying charm production by a broadband photon beam. The energy distribution of the beam, acceptance of the apparatus, and charm production cross section are such that most of the data come from events with $E_\gamma$ > 80 GeV. This is very similar to the ν > 75 GeV cut imposed by the BFP group. The apparatus is also sensitive mainly to the forward production of the charm particles.

Clear signals for $\Lambda_c$, $\overline{\Lambda}_c$ and $D^* \to D\pi$ are seen. The charmed baryons are identified by their $p(\overline{p})K_S^0$ decay made (Fig. 20). No significant peak is seen in any other final state with the same quantum numbers. The $D^* \to D\pi$ decay chain is identified by looking at the invariant mass difference between a K(nπ) and K(n-1)π system where n = 2 or 3. The events with the mass difference in the vicinity of 145 MeV are then candidates for this decay chain. The mass plot of the K(n-1)π system (i.e. $K^\mp\pi^\pm$ and $K_S^0\pi^+\pi^-$) for those events shows a clear peak (Fig. 21) at the masses of the $D^0$ and $D^\pm$. In addition, a 2σ signal is seen (not shown) for the inclusive $D^0$ production by looking at $K^\mp\pi^\pm$ mass distribution.

The details of the production process again appear to be consistent with the diffractive production of a charm-anticharm pair and can be understood within the framework of the γgF model. The specific observations that allow one to draw these conclusions are the

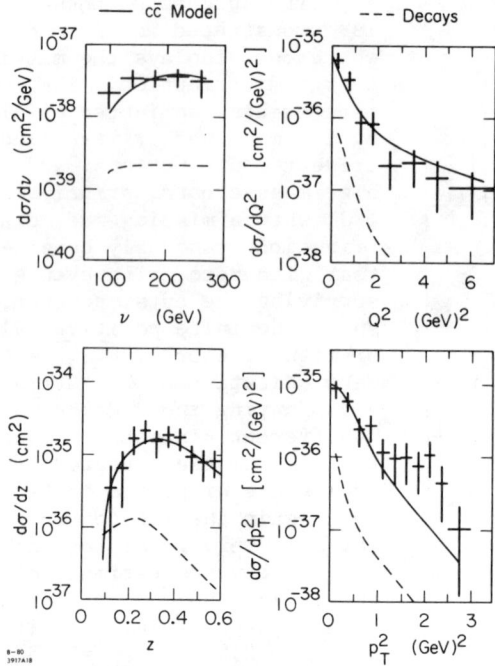

Fig. 19. Comparison of the EMC 3μ data with the predictions of the γgF model (solid curves). Calculation of background from π,K decays is indicated via dashed curves.

following:

a) within statistics $\Lambda_c$ and $\bar{\Lambda}_c$ production cross sections are equal.

b) Charmed baryon appears to take about half of the γ ray energy: $\bar{E}_{\Lambda_c}/\bar{E}_\gamma$ = 0.52.

c) The numbers of observed $D^{*+}$ (61±14) and of $D^{*-}$ (65±15) are equal within errors.

d) The D signal appears only in association with a K of opposite sign i.e. there is a $K^-\pi^+$ peak at the mass of the $D^0$ if a $K^+$ is identified elsewhere in the event. No signal is seen in association with a proton or antiproton, or K of the same sign.

Finally we can say a word about cross sections. The data are insufficient to say anything about the energy dependence in the region under study. For the purpose of extracting numbers, cross section was assumed to be flat over the whole energy range covered by the experiment. The deduced cross sections for the specific channels are:

$\sigma_{\Lambda_c} \approx 200$ nb (assuming BR for $\Lambda_c \to K^0 p$ = 1.5%)

$\sigma_{D^*} = 160 \pm 70$ nb

$\sigma_{D^0} = 390 \pm 190$ nb

If we make a reasonable assumption that the relative production rates $D^0 : D^+ : \Lambda_c : F = 2 : 1 : 1 : 1$ we obtain $\sigma^{tot}_{charm} \approx 1000$ nb at $\bar{E}_\gamma$ = 165 GeV. This number can be compared with BFP value of $750^{+180}_{-130}$ nb at $\nu$ = 178 GeV. We should stress, however, that both experiments are mainly sensitive to the forward production region.

The production mechanisms appear to be quite different at lower energies. The WA4 experiment at CERN has studied[52] the photoproduction of charm particles using a tagged photon beam with $E_\gamma <$ 70 GeV and the Ω apparatus that has considerably larger acceptance at wide angles than the CIF spectrometer. The most relevant features of their observations can be summarized as follows:

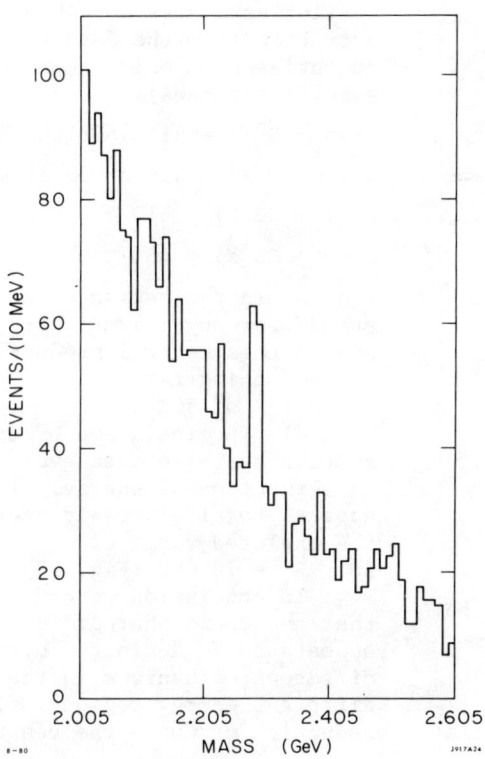

Fig. 20. $pK_S^0$ and $\bar{p}K_S^0$ mass spectrum from the CIF experiment.

a) A statistically significant $\bar{D}^0$ peak is observed in $K^+\pi^-$ and $K^+\pi^-\pi^0$ spectra but no comparable peak is seen for the charge conjugate states (Fig. 22).

b) This enhancement becomes especially pronounced when one looks only at the events with an associated proton (Fig. 23a). Similarly the $K_S^0\pi^+\pi^-$ (Fig. 23b) and $K_S^0\pi^-\pi^0$ combinations peak at the $D^0$ mass if one demands a similar association with the proton. Thus the natural explanation is the charm production via an associated production mechanism $\gamma p \to \Lambda_c \bar{D}$, $\Lambda_c$ subsequently decaying to a proton. No statistically significant enhancement at the $\Lambda_c$ mass is seen however in any $(\Lambda\pi\text{'s})$, $(Kp\pi)^+$, or $K^0 p$ combination.

c) There is evidence for F's in $\eta + \pi$'s channels (Fig. 24). The $\eta$'s are identified by their $2\gamma$ decay mode. The majority of the F signal in the $\eta5\pi$ system appears to come from the $\eta'3\pi$. The parameters of the F observations are summarized in Table IV. The best estimate for the F mass is $M_F = 2.020 \pm 0.010$ GeV. One should add here that of the 3 identified F decays in emulsions,[53] none is associated with an $\eta$. Thus $\eta + \pi$'s decay modes probably do not constitute more than 50% of the total decays implying a reasonably large F photoproduction cross section of about 200 nb.

Table IV
Summary of the $F \to \eta + \pi$'s observations

| Mode | Width (MeV) Expected | Observed | Observed Mass (GeV) | Efficiency | $B\sigma$ (nb) |
|---|---|---|---|---|---|
| $\eta\pi$ | 75 | 108±31 | 2.047±.025 | .07 | 12 ± 3 |
| $\eta3\pi$ | 50 | 38±24 | 2.021±.015 | .10 | 60 ± 15 |
| $\eta'3\pi$ | 40 | 48±34 | 2.008±.020 | .05 | 20 ± 8 |

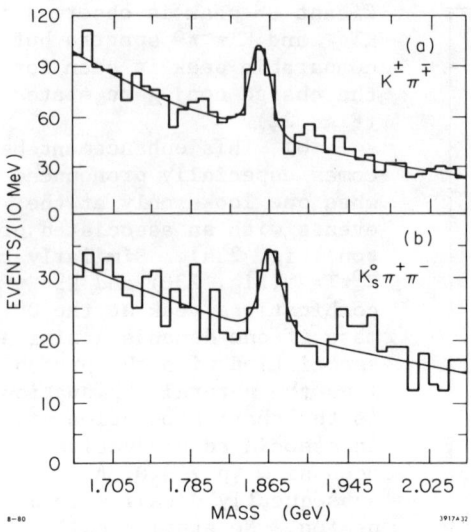

Fig. 21. $K^{\mp}\pi^{\pm}$ and $K_S^0\pi^+\pi^-$ mass spectra from the CIF experiment for events with $M_{K3\pi} - M_{K2\pi}$ (or $M_{K2\pi} - M_{K\pi}$) around 145 MeV.

d) The following cross section estimates have been extracted from the data (I summarize here only the most significant ones).

$\sigma(\gamma p \to \bar{D}^0 X) = 515 \pm 160 \pm 100$ nb

$\sigma(\gamma p \to D^0 X) < 450$ nb  $3\sigma$ level

$\sigma(\gamma p \to C\bar{D}^0 X) = 510 \pm 220$ nb

$\sigma(\gamma p \to CD^- X) = 450 \pm 310$ nb

(in the last 2 estimates C stands for any charmed baryon, one assumes central production, and branching ratio $C \to p + X$ of 50%).

e) The inclusive $\bar{D}^0$ cross section has also been evaluated as a function of energy. It appears to rise steeply over the explored range of $20 < E_\gamma < 70$ GeV (Fig. 25).

In conclusion we can say that the charm photoproduction appears to be dominated by different mechanisms in the different energy regions. In the lower energy range the associated production of $C\bar{D}$ in the central

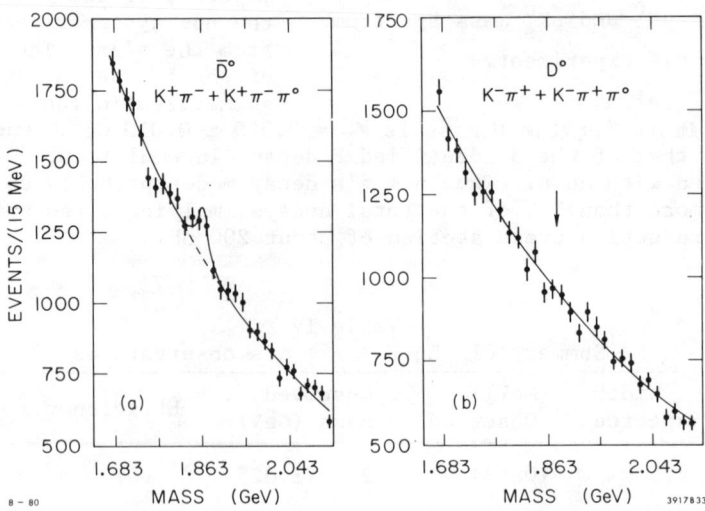

Fig. 22. $K^+\pi^-$ and $K^+\pi^-\pi^0$ (a) and $K^-\pi^+$ and $K^-\pi^+\pi^0$ (b) mass spectra from the WA4 experiment.

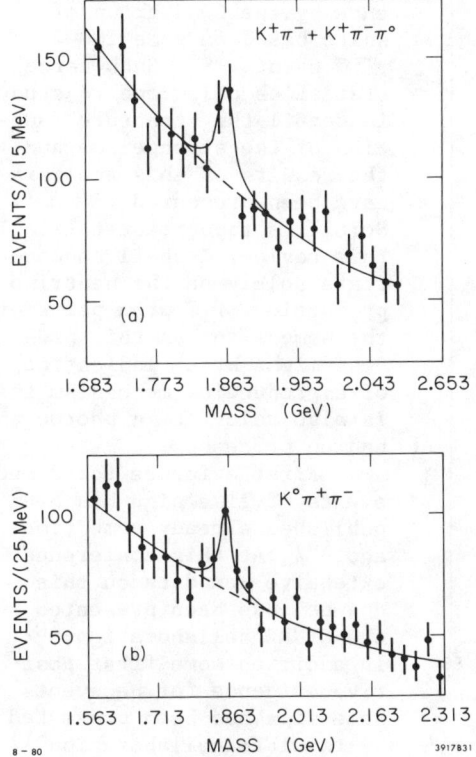

Fig. 23. $K^+\pi^-$ and $K^+\pi^-\pi^0$ (a) and $K^0\pi^+\pi^-$ (b) mass spectra from the WA4 experiment for the events in association with a proton.

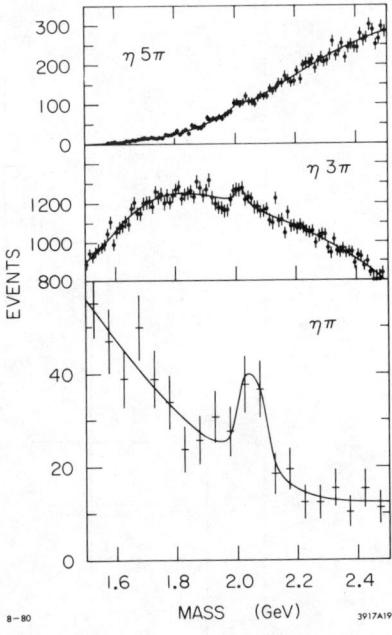

Fig. 24. Mass spectra of the $\eta + n\pi$ systems from the WA4 experiment.

region appears predominant; at higher energies the diffractive mechanism appears to take over. The amount of associated production at higher energies is uncertain because of the poor acceptance of the CIF spectrometer for this process. One should add here that the preliminary results from the FRAMM collaboration[54] working at medium energies support the diffractive production mechanism although their trigger and event selection criteria strongly bias them in that direction. On the other hand the only fully reconstructed emulsion event of charm photoproduction[55] is an example of associated production with $E_\gamma$ = 25 GeV. Finally, the Vector Meson Dominance hypothesis makes predictions about the ratio of elastic $\psi$ to open charm photoproduction.[56] The results of the BFP group would imply that non diffractive charm production must be at least comparable in magnitude to the diffractive production.

## NEW FLAVOR PRODUCTION BY NEUTRINOS

It has been only 6 years ago since the HPWF group reported at the London conference observation of 2 $\mu^+\mu^-$ events from neutrino interactions.[57] This first indication of charm production was the beginning of an intensive effort in this field which has led to the

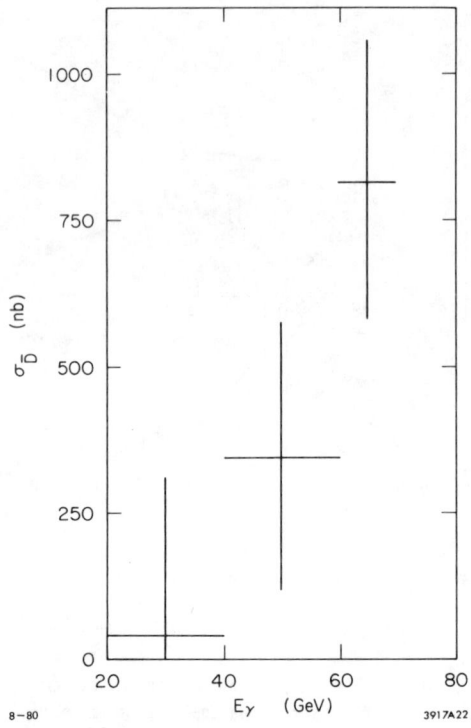

Fig. 25. $\bar{D}^0$ cross section as function of energy from WA4 experiment.

presentation at this conference by the CDHS group of results based on some 10000 $\mu^+\mu^-$ events.[58] Such large statistics allow one to study in detail the structure function of the strange sea and the results of this analysis have been presented in Sciulli's rapporteur talk. In this review, I shall concentrate solely on the neutrino production of lepton pairs of the same sign, as this process might be an indication of a production mechanism that is also relevant in photon and hadron processes.

First evidence for dimuon events of like sign has been published already some time ago.[59] At this conference extensive new data on this channel has been presented by the CFNRR collaboration;[60] in addition some first positive evidence for μe events of some sign has been presented by the IFIM collaboration[61] studying $\bar{\nu}$ interactions in the 15´ BC at Fermilab. The data on dimuon events as a function of neutrino energy is displayed in Fig. 26. We see that the ratio of $(\mu^\pm\mu^\pm)/\mu^+\mu^-$ is reasonably constant at about 0.1; in addition the rate of dimuon events is about an order of magnitude higher than the prediction based on associated charm production using a first order QCD diagram.[62]

Two contributions on same sign μe events have been received at this conference. The BFHWW collaboration[63] quotes an upper limit on $\mu^-e^-/\mu^- < 5 \times 10^{-4}$ from ν interactions, based on observation of 4 $\mu^-e^-$ events with a calculated background of 4 events. The cut on electron momentum is $P_e > 4$ GeV. 4 $\mu^+e^+$ events have been observed by the IFIM collaboration in $\bar{\nu}$ interactions ($P_e > 0.4$ GeV and expected background of 0.8 events) giving a ratio of $\mu^+e^+/\mu^+ = (6.4^{+3.6}_{-2.4})\times 10^{-4}$. A very interesting feature of these events is that 3 of them are associated with a $V^0$ (2Λ's and 1 $K^0_s$). The IFIM point is also displayed in Fig. 26 but it should be mentioned that it is not directly comparable to the μμ points since the electron momentum cut is considerably lower than the typical muon cut (generally $p_\mu \geq 9$ GeV).

The leptons of the same sign could be an indication of associated charm production with a rate considerably higher than expected on the naive grounds, a first evidence of new flavor production in neutrino interactions, or presence of as yet unexpected new

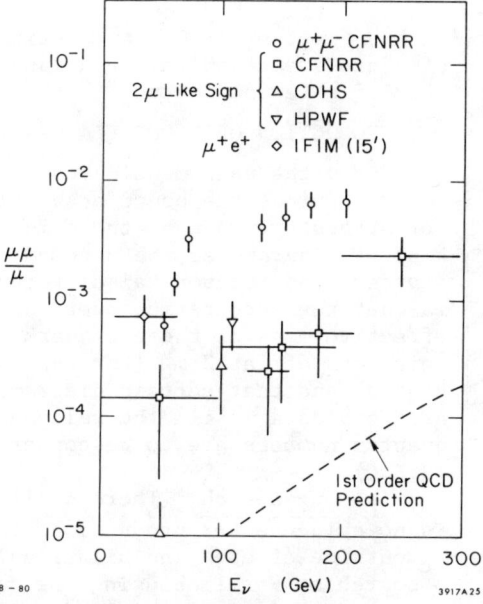

Fig. 26. Summary of like dilepton data from the neutrino interactions. CFNRR points for $\mu^+\mu^-$ are also shown for comparison. The curve is 1st order QCD prediction (ref. 62).

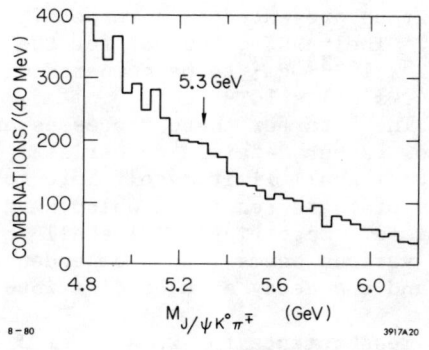

Fig. 27. Mass spectrum of the $J/\psi\ K^0\pi^\mp$ system from the WA11 experiment.

phenomenon. More detail and better statistics will be necessary to resolve this question.

## STATUS OF HEAVIER FLAVORS

A year ago there was presented a preliminary evidence[64] for the production of a bottom meson decaying via $B \to \psi K\pi$. The WA11 group has now increased their statistics fourfold to about 40000 $J/\psi$ events and find that the peak has disappeared[65] (Fig. 27). These data and $\psi K^0$ mass plot (not shown) give upper limits on these 2 decay modes i.e.

$\sigma_{B\bar{B}} \cdot BR(B \to J/\psi\ K^0\pi^\pm) < 0.51$ nb/nucleon

$\sigma_{B\bar{B}} \cdot BR(B \to J/\psi\ K^0) < 0.08$ nb/nucleon

This limit, obtained from 185 GeV $\pi^-$ interactions, can be compared with upper limits obtained using different techniques from 2 other experiments. The Princeton-Chicago group,[67] have obtained $\sigma_{B\bar{B}} \cdot BR(B \to J/\psi + X) \leq 0.24$ nb/nucleon for 225 GeV $\pi^-$ interactions. This limit is very sensitive to the $B \to \mu$ branching ratio (assumed to be 18%) as the experiment involves search for $J/\psi$ in association with a high $p_T$ muon. A reasonable assumption of 3% BR for the $J/\psi + X$ inclusive decay mode would translate the result into an upper limit of $\sigma_{B\bar{B}} < 8$ nb. Finally, the Cal Tech-Stanford experiment[68] has set a limit of $\sigma_{B\bar{B}} < 50$ nb for 400 GeV proton interactions by looking for a variety of multimuon final states. These numbers should be compared

There have been estimates[66] that the first decay mode should have a branching ratio of about 1%, which would translate into a total production upper limit of $\sigma_{B\bar{B}} < 51$ nb.

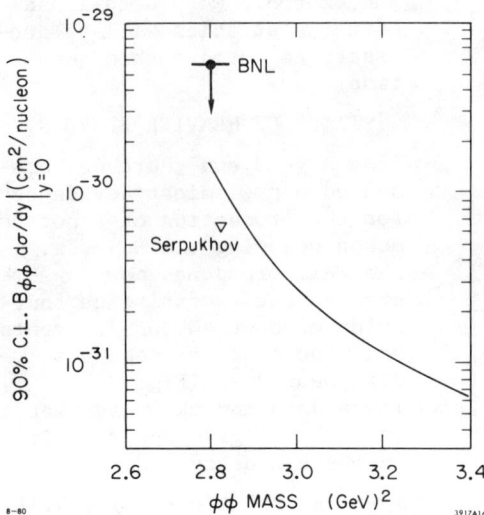

Fig. 28. Upper limits (as a function of $\eta_c$ mass) for $B(\eta_c \to \phi\phi) \left.\frac{d\sigma}{dy}\right|_{y=0}$ obtained by the Fermilab-Stonybrook collaboration.

with a first order QCD prediction of a cross section in the neighborhood of a few nanobarns.

PRODUCTION OF BOUND FLAVORS

From the phenomenological point of view, the bound heavy flavor states are made by the same kind of diagrams as the unbound states. The fundamental difference is that the integration over the effective mass of the $c,\bar{c}$ quark pair cuts off at $2 m_D$ (for charm states) and that certain diagrams are forbidden if all the relevant quantum numbers are to be conserved $(J,P,C)$.

a) $\underline{\eta_c \text{ search}}$. There still is no evidence for production of $\eta_c$ outside of $e^+e^-$ annihilations. A search at Fermilab using the decay mode $\eta_c \to \phi\phi$ has yielded negative results.[69] The limits as a function of mass are displayed in Fig. 28 together with earlier results from Brookhaven ($\pi^-p \to \phi\phi n$) and Serpukhov ($\pi^-p \to \gamma\gamma n$) experiments, scaled to the Fermilab energy region. Lipkin[71] has suggested that $\sigma \cdot BR(J/\psi \to e^+e^-) \approx \sigma \cdot BR(\eta \to \phi\phi)$. That would make $10^{-32}$ cm$^2$ an interesting goal to strive for. In addition, a low energy BNL experiment[71] (13 GeV) reported at this conference a limit of 260 pb for $\sigma \cdot BR$ for the process $\pi^-p \to \eta_c n$, $\eta_c \to \gamma\gamma$.

b) $\underline{T \text{ muonproduction}}$. The BFP group presented[72] an upper limit for T production via 208 GeV/c muons. Their 90% CL number for the process $\sigma(\mu N \to \mu T N)B(T \to \mu^+\mu^-)$ is $2.2 \times 10^{-38}$ cm$^2$ to be compared with the prediction of the $\gamma g F$ model of $(4.0 \pm 1.2) \times 10^{-39}$ cm$^2$.

c) $\underline{J/\psi \text{ and } T \text{ hadroproduction}}$. The data for these processes are now becoming quite extensive and allow rather detailed comparisons with various phenomenological models. I shall limit myself here to describing some very general features of these reactions which have a bearing on various production mechanisms. Specifically, I shall summarize the cross section data for various beams, the x dependence, information on intermediate states, and the decay angular distributions.

Additional total cross sections measurements for $x_F > 0$ for $J/\psi$ and T production by pions have been reported by the NA3 collaboration.[73] Together with the older measurements they are displayed as $M^2\sigma$ in Fig. 29. For comparison, I have also included lines indicating the approximate dependence of the same variable in the proton induced reactions.

The dependence of the cross section on the nature of the incident beam is interesting because it sheds light on the relative importance of the quark-antiquark vs. gluon-gluon fusion mechanisms. Naively,

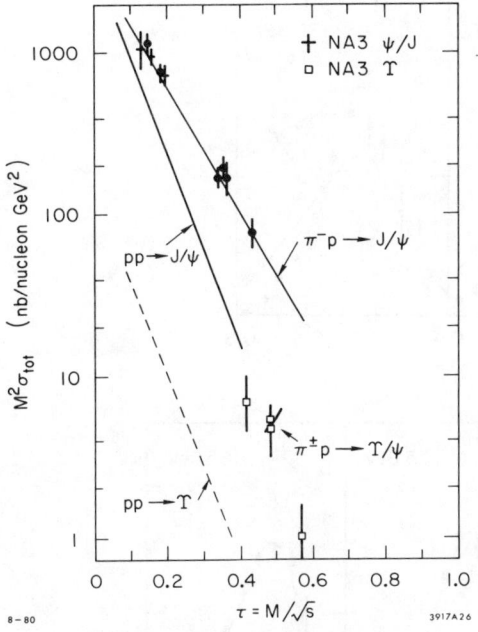

Fig. 29. Cross section for $J/\psi$ and $T$ production in $\pi$ nucleon interactions plotted in terms of scaling variables. For comparison, lines corresponding to production by protons are also shown.

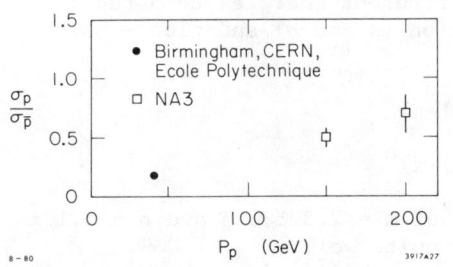

Fig. 30. Ratio of $\sigma(pN \to J/\psi + X)/\sigma(\bar{p}N \to J/\psi + X)$ as a function of incident energy.

because of the relationship $M^2 = sx_1x_2$, we would expect the gluon production mechanism to become more important at higher energy, since the gluon spectrum is rather soft. Furthermore, at lower energies,[74] where $q\bar{q}$ fusion might dominate, the cross sections for beams of particles containing appropriate valence antiquarks would be expected to be higher than for the particles without such valence quarks. These features are indeed demonstrated by the data shown in Fig. 30 where we plot the ratios for $J/\psi$ production in the proton and antiproton beams. For gluon fusion dominance the $\sigma_p/\sigma_{\bar{p}}$ ratio should approach unity. These qualitative features are also demonstrated in $J/\psi$ production by 40 GeV K beams where we have $\sigma_{K^+}/\sigma_{K^-} = 0.29 \pm 0.07$ and in the relative production of $T$ by $K^+$ and $\pi^+$ beams[75] at 200 and 280 GeV where $\sigma_{K^+}/\sigma_{\pi^+} = 0.10$. Note that the valence antiquark in $K^+$, i.e. $\bar{s}$, cannot annihilate with any valence quark in the nucleon to give a $J/\psi$ or an $T$. As might be expected, the $\pi^+$ - $\pi^-$ ratio for $J/\psi$ production is consistent with unity.

The importance of the gluon mechanism in 150 GeV $\pi^- p$ interactions is demonstrated by the analysis[76] of the $x_F$ distribution of the $J/\psi$. This distribution should be determined entirely by the pion and nucleon structure functions if $q\bar{q}$ annihilation is dominant. The data are compared to the theoretical expectations in Fig. 31a,b where NA3 structure functions have been used in calculating the expected curves. The agreement is quite poor and should be contrasted with the situation in Fig. 31c,d,e where the data were fitted to the gluon-gluon fusion mechanism assuming for the gluons the functional form

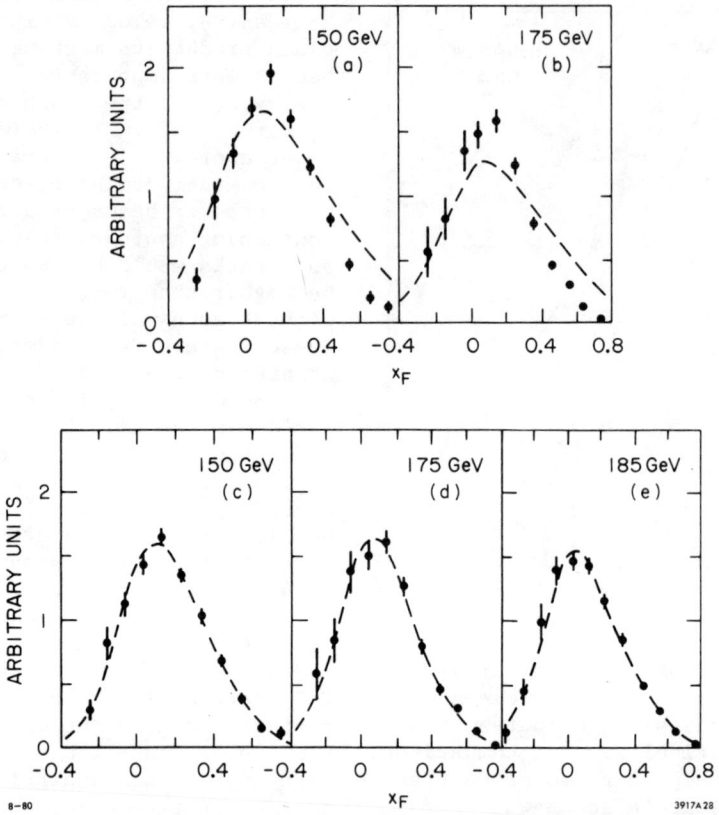

Fig. 31. $x_F$ distribution for J/$\psi$ production (points) from the WA11 experiment at different energies compared to quark-quark fusion prediction (a and b) and gluon-gluon fusion fits (c,d,e).

$$ng_\pi(n) \sim (1-x)^m$$

$$ng_N(n) \sim (1-x)^n$$

The fits gave very reasonable values of m = 2.3 ± 0.3 and n = 5.1 ± 0.6 and appear to reproduce the data quite well.

The same collaboration has also searched[77] for γ rays associated with the J/$\psi$ production. 2 experimental techniques were used to look for the photons: a Pb/scintillator sandwich calorimeter with a mean energy resolution of 50%/$\sqrt{E}$ FWHM and a γ → $e^+e^-$ conversion (22 MeV FWHM) either in the Be target or in downstream scintillators and chambers. Both methods give evidence for an intermediate χ state: the calorimeter shows 1 broad unresolved peak (Fig. 32a) in the J/$\psi$ γ mass spectrum between 3.5 - 3.6 GeV which corresponds to 36 ± 5% of the $\psi$'s resulting from the χ decay. The conversion technique gives

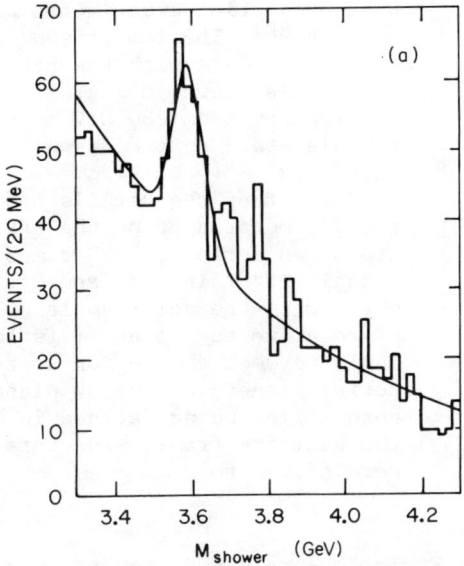

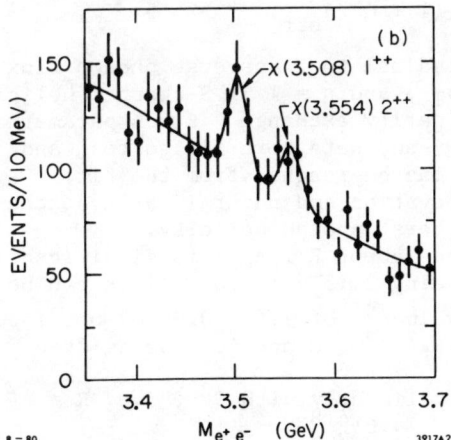

Fig. 32. Mass spectrum of the $J/\psi + \gamma$ system from the WA11 experiment using a calorimeter (a) and spectrometer (b).

two distinct peaks in this region (Fig. 32b): the first one identified with the $1^{++}$ $\chi$ (3.508) and corresponding to $19 \pm 4\%$ of all $J/\psi$ events, and the second with $2^{++}$ $\chi$ (3.554) and accounting for $12 \pm 4\%$ of all events. Using the known branching ratios for $\chi \to J/\psi + \gamma$ decay, these numbers correspond to a production cross section ratio of

$$\sigma_{2^{++}} / \sigma_{1^{++}} = 1.4 \pm 0.9.$$

The results of the calorimeter and conversion techniques are consistent with each other because of much different resolving power of the 2 methods. The total fraction of $J/\psi$ proceeding via $\chi$ intermediate state ($\sim 35\%$) is consistent with other measurements both at Fermilab and at the ISR.

The decay angular distribution of the $J/\psi$ has been studied[73] by the NA3 collaboration for $\pi^-$ production at 150 GeV. The most general distribution has to be of the form

$$\frac{dN}{\cos\Theta} = 1 + \lambda \cos^2\Theta.$$

For direct light quark annihilation $\lambda$ has to be near unity. The experimentally observed value $\bar{\lambda} = 0.05 \pm 0.07$ is consistent with previous measurements and argues that either quark annihilation proceeds via an intermediate state or is not a very important process at this energy.

In conclusion, the overall picture is consistent with the dominant mechanism being gluon-gluon fusion, the growth in importance of that mechanism with increasing energy, and an appreciable fraction of the $J/\psi$ produced via an intermediate $\chi$ state.

d) <u>$J/\psi$ muoproduction</u>. Partial results on this process have been published previously by both the BFP[78] and EMC[79] groups. In general the total cross section for this process, when extrapolated to $Q^2 = 0$ appears to agree quite well with the lower energy

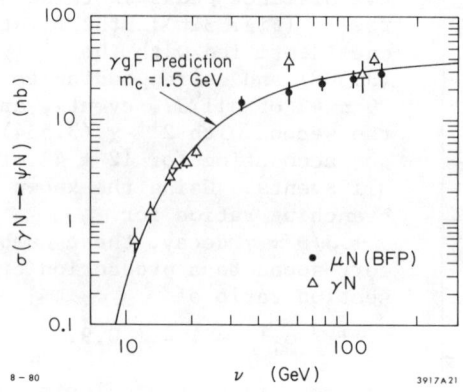

Fig. 33. Comparison of BFP J/ψ muoproduction points (extrapolated to $Q^2 = 0$) with the photon points and predictions of the γgF model (ref. 80).

photoproduction experiments and the γgF.[80] The comparison of the photon data with the BFP results is shown in Fig. 33; the EMC results from 280 GeV muon run also fall on the drawn curve.

To study the details of the J/ψ muoproduction the BFP group has performed a 3 dimensional fit[45] in $\Theta,\phi$, and $Q^2$, where $\Theta$ is the polar angle of $\mu^+$ relative to J/ψ and $\phi$ is the angle between the lepton scattering plane and ψ decay plane, both angles being defined in the helicity frame. The data were fitted to

$$W(\Theta,\phi) = \frac{1 + \cos^2\Theta + 2\varepsilon R \sin^2\Theta - \varepsilon\eta \sin^2\Theta \cos 2\phi}{(1 + \varepsilon R)(1 + Q^2/\Lambda_{eff})^2}$$

Where ε gives the ratio of longitudinal to transverse photon flux ($\bar{\varepsilon} = 0.8$ for BFP data), $R = \sigma_L/\sigma_T$, and $\eta = 1$ if S-channel helicity is conserved and we have natural parity exchange. R was parametrized as either constant or linear in $Q^2$ and data were fitted to η and $\Lambda_{eff}$. The following conclusions can be reached from the fit:

1 - s channel helicity conservation and natural parity exchange appear to be valid, i.e. ψ "remembers" photon helicity.

2 - independent of assumptions about R, $\Lambda_{eff}$ is significantly smaller than $m_\psi$, typical value being $2.15^{+0.19}_{-0.13}$ GeV. This can be compared with the published EMC value[79] of $2.4 \pm 0.3$ GeV obtained by fitting their data to $C(1 + Q^2/\Lambda_{eff}^2)^{-2}$, C and $\Lambda_{eff}$ being free parameters.

3 - when R is allowed to vary linearly with $Q^2$, i.e. $R = \xi^2 Q^2 m_\psi^2$, the best fit yields $\xi^2 = 4.6^{+4.8}_{-3.8}$.

The European Muon Collaboration has presented results[81] on inelastic J/ψ production with a total cross section approximately equal to that of the elastic J/ψ production. The inelastic events are defined as ones having more than 5 GeV deposited in the target. Those events tend to peak at high ν (Fig. 34) and have much broader $p_T^2$ distribution than the elastic events.

An effort has been made to estimate the contribution to these events from the production of higher lying bound charmonium states (ψ' and χ). The different methods indicate that only about half of the inelastic events come from that source. Their z distribution would tend to peak near high values in agreement with the data and the ν dependence can be calculated using the γgF model (Fig. 34b). It appears that the large fraction of events with ν > 100 GeV must

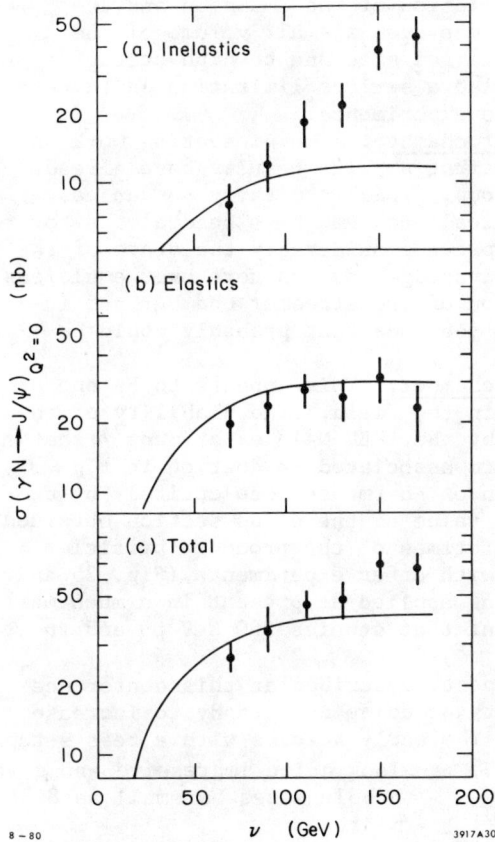

Fig. 34. $\nu$ distribution of elastic and inelastic $J/\psi$'s from the EMC collaboration. The curves are predictions of the $\gamma gF$ model (higher lying charmonium states are used as source of inelastic events for the purpose of the calculation).

come from another mechanism. Hard gluon emission by higher mass $c\bar{c}$ pair is postulated as one possible process that could account for these data.

PROSPECTS FOR THE FUTURE

There are several second generation experiments either in the planning stage or already taking data that will elucidate some of the questions posed above. However I feel that the main impact in the future on "naked" charm experiments will come from the technological development that is at present going on in the field of good spatial resolution detectors. The lifetimes of charm ground states are now established to lie between $10^{-13}$ and $10^{-12}$ secs; the estimates for the bottom states lie between $10^{-14}$ and $10^{-13}$ secs. Thus capability of "seeing" tracks in the range of 100-1000 microns would allow one to identify unambiguously presence of short lived particles. I would like to end this review with a few words about the present status of some of these detectors.

a) <u>emulsions</u>. This is the classical detector for looking at events with the ultimate spatial resolution. The price one pays, however, is quite severe - many hours of painful scanning. Much progress has been done towards reducing this time by placing sophisticated detector equipment downstream of the emulsion target which then allows us to reduce considerably the volume that needs to be scanned. This plan of attack, however, is clearly limited in its potential scope either to beams where the heavy flavor production constitutes a high fraction of the total cross section (e.g. neutrinos where the technique has proven to be very successful) or to experiments where the downstream detector can preferentially pick out charm events either at the trigger stage (very hard) or in off-line analysis. Otherwise the scanning effort again becomes quite prohibitive. Clearly at the root of all of these difficulties lies the intrinsic very poor time resolution of the emulsion and the great deal of time necessary to scan even a small volume of the emulsion. The relatively small target size due to high cost of emulsion and its processing is also a serious limitation on the use of this technique in the neutrino experiments.

b) <u>high resolution streamer chamber</u>. The pioneering work on this kind of a detector and the first physics results have already been published[10] by the Yale group. This is clearly not an easy technique but many complex technical problems have been already overcome and one can see a way to improve considerably the state of the art here.[82] Clearly, the big advantage one has here over emulsions is the much better time resolution of the streamer chamber and intrinsically much easier scanning job, one that probably could be adopted to full automation.

c) <u>high resolution bubble chamber</u>. This appears to be one of the most promising developments in the field. The viability of the technique has been demonstrated by the LEBC NA13 experiment[7] that observed several examples of charm associated production in $\pi^-$p interactions. The identification of charm was done entirely by detecting short lived decays. The value of the cross section obtained ($\sim 40$ µb) is dependent on the lifetimes of the produced particles but appears to agree quite well with other experiments (Fig. 35 and Table I). This technique is being applied at present in a much more fully instrumented NA16 experiment that studies 360 GeV pp and $\pi$p interactions.

Another very promising prospect, described at this conference by Montanet,[83] is the possibility of using holography to increase the depth of the field of view. The early results with a test setup look quite impressive and give bubble sizes as small as 8 microns.

d) <u>solid state detectors</u>. Development of high resolution solid state detectors would allow one to dispense with the scanning phase of the experiment, which appears crucial to the other 3 techniques discussed above. A silicon active target, composed of 40 300 µm wafers, has been used by the FRAMM collaboration[54] to study the diffractive charm photoproduction at the SPS. The technique relies on a sudden increase in pulseheight (Fig. 36) as one goes from one wafer to the next, corresponding to a multibody decay of a D meson. The potential candidate events are then fully analyzed by using the

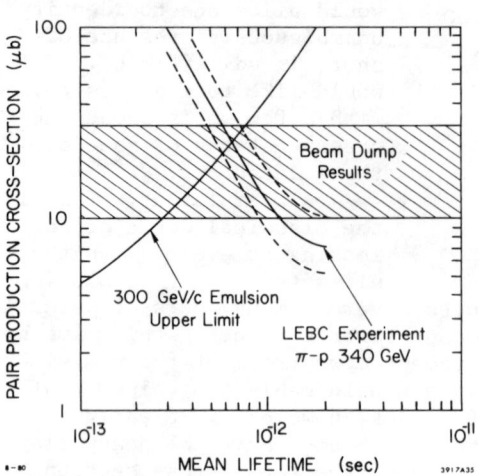

Fig. 35. Results of charm production cross section from the LEBC experiment and comparison with other results.

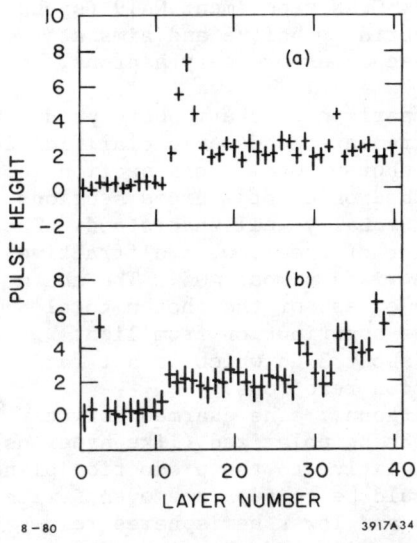

Fig. 36. 2 examples of pulse-height distribution from the active target in the FRAMM experiment: a) inelastic event, b) candidate for $D^o\bar{D}^o$, with the production occurring near wafer 3 and the 2 decays near wafer 11 and 34.

information from the downstream spectrometer, Čerenkov counters, and photon detectors. The preliminary analysis shows already evidence of a diffractive mass peak around 4 GeV which appears to decay into D's and/or D*'s.

The technique as used in this experiment is clearly specialized to diffractive photoproduction. On the other hand work is in progress on expanding the method by also reading out the transverse dimension which would increase the versatility of the detector. It should finally be noted that the 3 visual methods discussed above rely essentially on detecting the transverse displacement of the decay track from the production vertex and thus their efficiency is relatively beam energy independent. The FRAMM detector, however, actually "measures" the length of the decay track and thus its efficiency increases with the beam energy.

## ACKNOWLEDGEMENTS

I would like to thank my scientific secretaries: G. Fanourakis, T. Y. Ling and F. Messing for their help during the conference, and all the speakers in the relevant sessions for taking the time to explain their results to me.

## DISCUSSION

Q1: Conversi, Rome: I wish to make two short remarks. One refers to the mass of the $\Lambda_c^+$ baryon. The first example of neutrino induced production and decay of a $\Lambda_c^+$ was obtained in CERN experiment WA17 and published last year in Physical Review Letters together with the first estimate of the charmed particle lifetime. The mass of the $\Lambda_c^+$ reported there (and also at the Bergen and Geneva EPS Conferences) has been readjusted through a further analysis and it will appear in the final paper of that experiment, now in press in "Nuclear Physics." Also, this new mass value (only slightly smaller than that previously reported) agrees, within the error, with the average value of $M(\Lambda_c^+)$ reported in your talk.

As a second comment, I feel it is worth mentioning here that a search for associated production of beauty particles is being carried out now by an enlarged collaboration, after an exposure made at CERN just before the SPS shutdown. This is CERN experiment NA19 (spokesman P. Musset), which uses again a hybrid technique and aims at observing the decay sequence: beauty-charmed-ordinary hadrons, in the emulsion.

Q2: Prentice, Toronto: The comparison of charm photo production with photon total cross section measurements needs some clarification. The rise with energy of the hadronic photon total cross section is related to the rise of all the other hadronic total cross sections in the same energy range and is about equally well understood. The rise can be estimated from the behavior of the $\rho$, $\omega$, $\phi$ diffractive photoproduction. The $\phi$ cross section rises almost 40%. The charm cross section can be estimated by the excess of the photon total cross section over the estimate of the contribution from light quark vector mesons. This excess is about 2 μb which is not far from what is seen in the photoproduction reactions.

Q3: Lipkin, Weizmann/Fermilab/Argonne: The charmed baryons produced in hadronic interactions might be polarized (like hyperons) and give an asymmetry in the decays relative to the production plane. A simple check with low statistics would be to separate events with the decay baryon emitted in the upper and lower hemispheres relative to the production plane, i.e., according to the sign of $\vec{p}_i \times \vec{p}_{\Lambda_c} \cdot \vec{p}_d$, where $\vec{p}_i$, $\vec{p}_{\Lambda_c}$ and $\vec{p}_d$ denote the momenta of the incident beam, the charmed baryon and the decay baryon respectively. The difference between the two distributions would have automatic background subtraction. A signal would indicate a parity violation. If this effect exists, it would be both interesting and useful in analyses.

Q4: Jones, Michigan: The dramatic rise in $\sigma_c$ reported from the ISR (if confirmed) plus the A dependence of charm production suggests an engineering remark which may not have occurred to everyone; as we go to much higher energies (Tevatron and Pentavac) it is probable that the traditional means of generating μ and ν beams from π and K decay will give way to beam dump sources.

Q5: Devlin, Rutgers/Fermilab: I would like to add to your list of detection techniques a new device developed by Douglas Potter of Rutgers. It is a triggerable detector/target which has been operated in two modes. First, as a scintillation camera and, second, with a micro-channel plate. The parameters are available in preprint which was submitted to this conference.

Q6: Isgur, University of Toronto: Is the ISR $\Lambda_c^+$ signal consistent with the observed 50 - 100 MeV width of the $\Sigma(2250)$?

A6: Wojcicki: Maybe the people from ISR would like to comment. The width is very narrow and consistent with resolution. However the peak is removed from the value of 2.285 by the amount comparable to the resolution, or maybe more than the resolution. I'm told by the SFM people that such a shift is not inconsistent with their present understanding of systematics. On the other hand Lanmshade magnet people believe that their peak could not be displaced by more than 10 MeV from the true value. To make an intelligent experimental comment about this question of width and central value, I think one really has to look at things like the $K^0$ or $\Lambda$ peak in the same apparatus, in the same region, and try to extrapolate from that. But that has not really been done for a variety of reasons, and therefore the answer to these questions is still up in the air. It seems to me that some statistical fluctuation may be going on here; it is certainly not enough to generate a whole effect, but conceivably enough to confuse the questions of central values and the widths.

## REFERENCES

1. Alchi, Fermilab, Kobe, Seoul, McGill, Nagoya, Ohio State, Okayama, Osaka, Ottawa, Tokyo, Toronto, Yokohama Collaboration; paper presented by K. Niu in session B6 at this conference.
2. NA3 collaboration (Saclay, CERN, College de France, Ecole Polytechnique, Orsay) quote $0.935 \pm 0.025$ for the exponent of A from the study of $J/\psi$ production at 150 GeV.
3. See for example B. L. Cambridge, Nucl. Phys. B151, 429 (1979).
4. Bologna, Glasgow, Rutherford, Saclay, Torino Collaboration, paper #39 submitted to this conference.
5. Princeton, Saclay, Torino, BNL Collaboration, paper #477 submitted to this conference.
6. Brussels, Helsinki, Liverpool, Mons, Stockholm Collaboration, paper #516 submitted to this conference.
7. NA13 collaboration (Brussels, CERN, Oxford, Padova, Rome, Rutherford, Trieste), paper presented by C. Fisher at this conference; see also CERN/EP/80-49.
8. Tata Institute, paper presented by T. K. Malhotra in session B6 at this conference.
9. A. Chilingarov et al., Phys. Lett. 83B, 136 (1979).
10. J. Sandweiss et al., Phys. Rev. Lett. 44, 1104 (1980).
11. H. Fuchi et al., Phys. Lett. 85B, 135 (1979).
12. K. W. Brown et al., Phys. Rev. Lett. 43, 410 (1979).
13. A. Diamant-Berger et al., Phys. Rev. Lett. 43, 1774 (1979).
14. J. Ritchie et al., Phys. Rev. Lett. 44, 230 (1980).
15. A. E. Asratyan et al., Phys. Lett. 79B, 497 (1978).
16. B. P. Roe, University of Michigan report UMHE 79-2.
17. P. Alibran et al., Phys. Lett. 74B, 134 (1978).
18. H. Wachsmuth, Proc. of the Lepton-Photon Symposium, Fermilab (1979).
19. CHARM collaboration, paper presented in session A10 at this conference by F. Niebergall.
20. Annecy, CERN, College de France, Dortmund, Heidelberg, Warsaw Collaboration, paper presented in session A9 at this conference by G. Sajot; see also D. Drijard et al., Phys. Lett. 85B, 452 (1979).
21. Reference 20 and D. Drijard et al., Phys. Lett. 81B, 250 (1979).
22. K. L. Giboni et al., Phys. Lett. 85B, 437 (1979): see also papers #346 & 512 contributed by this group to this conference, summarized by F. Muller in session A9.
23. W. Lockman et al., Phys. Lett. 85B, 443 (1979).
24. CDHS collaboration, paper presented by K. Kleinknecht in session A10 of this conference.
25. BEBC collaboration, paper presented in session A10 of this conference.
26. Paper presented by K. W. B. Merritt in session A10 of this conference; see also K. W. B. Merritt, Ph.D. thesis, California Institute of Technology.
27. Illinois, Fermilab, Harvard, Oxford, Tufts Collaboration; paper #618 presented by L. J. Koester in session A9 of this conference.

28. These 3 experiments have been summarized in a paper presented in session A10 of this conference.
29. A. Soukas et al., Phys. Rev. Lett. <u>44</u>, 564 (1980).
30. P. Coteus et al., Phys. Rev. Lett. <u>42</u>, 1438 (1979).
31. G. S. Abrams et al., Phys. Rev. Lett. <u>44</u>, 10 (1980).
32. Daryl DiBitonto, Ph.D. thesis, Harvard University, October 1979.
33. S. J. Brodsky, P. Hoyer, C. Peterson and N. Sakai, NORDITA preprint, submitted to Phys. Lett.
34. Columbia, Illinois, Fermilab collaboration, paper presented by I. Gaines in session A11 at this conference.
35. IIT, Maryland, Stony Brook, Tohoku, Tufts Collaboration, paper #701 presented by T. Kitagaki in session A9 at this conference.
36. Padua, Bonn, CERN, Munich, Oxford Collaboration, paper presented by P. Bossetti in session A9 at this conference.
37. Bari, Birmingham, Brussels, CERN, Ecole Polytechnique, Rutherford, Saclay, London Collaboration, paper #819 presented to this conference; also available as Rutherford report RL-80-019.
38. The LSM collaboration feels that their mass calibration is correct to 10 MeV (private communication from F. Muller). The SFM group is much less certain of their absolute mass calibration because of the great complexity of the MWPC system and the difficulty of making sure that all components of the system are correctly aligned. Thus they do not exclude possibility of a 25 MeV shift (D. Drijard-private communication).
39. W. Bacino et al., Phys. Rev. Lett. <u>45</u>, 228 (1980).
40. M. Barone et al., Nucl. Phys. <u>B132</u>, 29 (1978); L. Baum et al., Phys. Lett <u>60B</u>, 485 (1976).
41. This group finds the ratio of the observed number of $\nu_e$ interactions to the expected number to be $1.2 \pm 0.2$ (BEBC TST group-private communication).
42. J. P. Leveille and T. Weiler, Nucl. Phys. <u>B147</u>, 147 (1979); M. Glück and E. Reya, Phys. Lett. <u>79B</u>, 453 (1978) and <u>83B</u>, 98 (1979).
43. Berkeley, Princeton, Fermilab Collaboration, paper presented by A. Clark in session A9 at this conference.
44. European Muon Collaboration (CERN, DESY, Freiburg, Kiel, Lancaster, LAPP, Liverpool, Oxford, Rutherford, Sheffield, Turin, Wuppertal), paper presented by R. Mount in session A9 at this conference.
45. A. R. Clark et al., paper #555 submitted to this conference; also available as LBL-10747; submitted to Phys. Rev. Lett.
46. A. R. Clark et al., paper #553 submitted to this conference; also available as LBL-10879; submitted to Phys. Rev. Lett.
47. D. O. Caldwell et al., Phys. Rev. Lett. <u>42</u>, 553 (1979).
48. J. J. Aubert et al., paper #740 submitted to this conference; also available as CERN-EP/80-61; submitted to Phys. Lett.
49. A. J. Buras and K. J. F. Gaemers, Nucl. Phys. <u>B132</u>, 249 (1978).
50. J. J. Aubert et al., paper #742 submitted to this conference; also available as CERN-EP/80-62; submitted to Phys. Lett.
51. Reference 34; for earlier published work see M. S. Atiya, Phys. Rev. Lett. <u>43</u>, 414 (1979).

52. Bonn, CERN, Ecole Polytechnique, Glasgow, Lancaster, Manchester, Orsay LAL, Paris VI, Paris VII, Rutherford, Sheffield Collaboration, paper presented by B. d'Almagne in session A11 at this conference.
53. Reference 1 and paper #265 presented by R. Ammer in session B6 of this conference (Kansas, Fermilab, Serpukhov, ITEP-Moscow, Krakow, Dubna, Gos Fotochem-Moscow, Washington Collaboration).
54. NA-1 (FRAMM) experiment - Frascati, Milano, Pisa, Roma, Torino, Trieste Collaboration; paper presented by E. Bertolucci in session A11 at this conference.
55. Paper #755 submitted to this conference by Photon-emulsion Collaboration (Bologna-CERN-Florence-Genova-Madrid-Moscow-Paris-Santander-Valencia) and Omega-photon Collaboration (Bonn-CERN-Glasgow-Lancaster-Manchester-Paris-Rutherford-Sheffield).
56. D. Sivers, J. Townsend, and G. West, Phys. Rev. $\underline{D13}$, 1234 (1976).
57. C. Rubbia (HPWF collaboration), XVII International Conference on High Energy Physics, London 1974, p. IV-118.
58. CDHS Collaboration (CERN, Dortmund, Heidelberg, Saclay); paper presented in session C10 at this conference.
59. A. Benvenuti et al., Phys. Rev. Lett. $\underline{41}$, 725 (1839).
60. CFNRR Collaboration (Cal Tech, Fermilab, Northwestern, Rochester, Rockefeller) paper presented by M. Shaevitz in session C9 at this conference.
61. Personal communication at this conference to the author from the ITEP, Fermilab, IHEP, Michigan Collaboration.
62. H. Goldberg, Phys. Rev. Lett. $\underline{39}$, 1598 (1977); also M. Shaevitz, private communication.
63. Paper #649 presented to this conference by LBL, Fermilab, Hawaii, Washington, Wisconsin Collaboration.
64. R. Barate et al., paper #184 presented at the 1979 International Conference on Photon and Lepton Interactions, Batavia, Ill.
65. Paper #975 presented by the WA11 Collaboration (Saclay, Imperial College, Southampton, Indiana) to this conference; also talk by J. G. McEwen in session A9 at this conference.
66. H. Fritzsch, Phys. Lett. 86B, 343 (1979).
67. R. N. Coleman et al., Phys. Rev. Lett. $\underline{44}$, 1313 (1980).
68. A. Diamant-Berger et al., Phys. Rev. Lett. $\underline{44}$, 507 (1980).
69. Paper #489 presented to this conference by the Fermilab-Stony Brook Collaboration.
70. H. J. Lipkin, in Prospects for Strong Interaction Physics at Isabelle, BNL 50701 (1977).
71. Paper #535 presented to this conference by the BNL-Illnois-Princeton Collaboration.
72. Paper #554 presented to this conference by the Berkeley-Fermilab-Princeton Collaboration; LBL-11009, submitted for publication.
73. NA3 experiment (Saclay, CERN, College de France, Ecole Polytechnique, Orsay Collaboration); paper presented by P. Delpierre in session A9 at this conference.
74. Paper #546 submitted to this conference by Birmingham, CERN, Ecole Polytechnique Collaboration.
75. Paper presented by the NA3 collaboration at the 1979 European Conference on High Energy Physics in Geneva, Switzerland.

76. Paper #342 submitted by the WA11 Collaboration to this conference; also talk by J. G. McEwen in session A9 at this conference.
77. Papers #807 (calorimeter method) and #982 (spectrometer method) submitted by the WA11 Collaboration to this conference.
78. A. R. Clark et al., Phys. Rev. Lett. $\underline{43}$, 187 (1979).
79. J. J. Aubert et al., Phys. Lett. $\underline{89B}$, 267 (1980).
80. See for example T. Weiler, Phys. Rev. Lett. $\underline{44}$, 304 (1980).
81. Paper #743 presented to this conference by the European Muon Collaboration; CERN-EP/80-84, submitted to Phys. Lett.
82. J. Sandweiss, paper presented at this conference in session D4.
83. L. Montanet, paper presented at this conference in session D4.

PHENOMENOLOGY OF NEW PARTICLE PRODUCTION

R.J.N. Phillips, Rutherford, Speaker

G. Kane, Michigan, Chairman

U. Sukhatme, Iowa,

F. Harris, Hawaii-Manoa,

J. LoSecco, Harvard,

Scientific Secretaries

# PHENOMENOLOGY OF NEW PARTICLE PRODUCTION

R.J.N. Phillips
Rutherford Laboratory, Chilton, Didcot, Oxon, England

## ABSTRACT

Charm and beauty production on hadron targets are compared to parton models with QCD interactions.

## INTRODUCTION

To make a manageable story this talk concentrates on the successes and failures of simple parton models for c and b production, via low-order QCD interactions. Theoretical papers have been appearing for years[1-24]; what's new in 1980 is that much better data have come along and much more significant comparisons can be made.

The special magic of charm and beauty is that they establish large mass scales, which we presume allow us to play with perturbative QCD (see Fig.1).

Fig. 1  Charm allows us to play with perturbative QCD

## $c\bar{c}$ PHOTOPRODUCTION

The lowest-order QCD process for both real[3-9] and virtual[8,15] photoproduction is gamma-gluon fusion: see Fig.2. The main ingredients for a calculation of $\gamma N \to c\bar{c}$ cross sections are the charmed quark mass (usually taken to be $m_c = 1.5$ GeV), the gluon distribution $G(\eta, \mu^2)$ for momentum fraction $\eta$ defined at mass scale $\mu$, and the QCD coupling constant $\alpha_s$

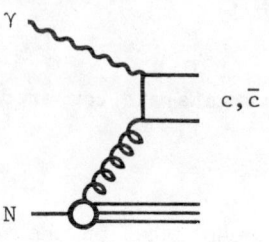

Fig. 2  $\gamma g$ fusion

$$\alpha_s(\mu^2) = 12\pi / [(33 - 2f) \ln(\mu^2/\Lambda^2)] \tag{1}$$

where $f = 4$ is the number of active flavours and the parameter $\Lambda$ is usually set around $\Lambda = 0.5$ GeV (an assumption, not too critical).

The mass scale $\mu$ is often taken to be the $c\bar{c}$ invariant mass $m \equiv m(c\bar{c})$, but the alternative $Q^2$-dependent form

$$\mu^2 = m^2 + Q^2 \tag{2}$$

is suggested by the following argument[17]. If a virtual photon of invariant mass square $-Q^2$ materializes as a pair of charmed quarks of mass square $m^2$, it is off-shell by an amount $\Delta E = \sqrt{Q^2 + m^2}$ in the Breit frame. This large $\Delta E$ defines a short time and distance scale $\Delta t = 1/\Delta E$ within which things have to happen, corresponding to the large mass and momentum scale of Eq.(2). Furthermore Eq.(2) correctly gives $\mu^2 \sim Q^2$ for very large $Q^2$. In principle we can find the best mass scale by calculating higher order contributions (see the talk by Llewellyn Smith) but this has not yet been done.

**Bound $c\bar{c}$ production.** A semilocal duality argument[1] suggests that production of $c\bar{c}$ pairs with invariant mass below the open charm threshold (i.e. with $2m_c < m < 2m_D$) corresponds to charmonium production of which a fixed fraction F is $\psi$. This gives

$$\sigma(\psi) = F \int_{2m_c}^{2m_D} \sigma(m) \, dm \tag{3}$$

where F is an empirical constant to be found from experiment. However, the $c\bar{c}$ invariant mass is not sacred[10] and can very well be changed by the final hadronization process (we routinely allow massless quarks and gluons to materialize as massive hadrons). So the narrow duality interpretation seems out of place. Instead it is plausible to assume that the production of $\psi$ is proportional to the density of $c\bar{c}$ states produced near the $\psi$ mass, which gives effectively the same formula without dogmatic constraints (ref.22 gives a particular realization). In particular, the remaining fraction 1 - F can appear mostly as unbound charm states; rather small fractions may be expected for $\psi'$ (almost unbound) and $\chi$, $\eta_c$ (unfavoured quantum numbers).

Another approach [10,18] is to assume charmonia are produced only by direct hard scattering processes such as $\gamma g \to \psi g$. Today however we shall use Eq.(3), and compare various $\gamma N \to \psi X$ data, looking at both $\nu$ and $Q^2$ dependence.

"Elastic" $\gamma N \to \psi X$. The muon experiments[25,26] distinguish "elastic" events with little accompanying hadronic energy, typically $E_{HAD} < 5$ GeV; low energy real photon experiments[27] have similar cuts in acceptance. These data are well fitted by $\gamma g$ fusion plus Eq.(3); presumably this mechanism leads mostly to N and low N* final states (a coherent nuclear peak is seen).

Figure 3 shows the cross section versus energy $\nu$ for real photons ($Q^2 = 0$); the two high-energy FNAL photoproduction points include inelastic events. The solid line is $\gamma g$ fusion with

$$G(\eta) = 3(1 - \eta)^5/\eta \qquad (4)$$

normalized to carry half the nucleon momentum as required by the momentum sum rule. With $F \simeq 1/6$ we fit the $\psi$ data[20].

Kinematics relates the gluon momentum fraction to the $c\bar{c}$ invariant mass m by $m^2 = 2m_N\nu\eta - Q^2$. Hence by Eq.(3) production at given $\nu$, $Q^2$ is controlled by $G(\eta)$ in a narrow band of $\eta$: thus the $\nu$-dependence of $\sigma(\psi)$ directly determines[19,20] the $\eta$-dependence of $G(\eta, m_\psi^2)$. This is shown on the axes in Fig.3, with G unnormalized. The dashed line is the broad gluon distribution[29]

$$G(\eta) = (1 + 9\eta)(1 - \eta)^4/\eta \qquad (5)$$

believed to be approximately correct for mass scale $\mu^2 \simeq 4$ GeV$^2$; QCD evolution brings this to a shape somewhat similar to Eq.(4) at the scale $\mu^2 \simeq 10$ GeV$^2$ of present interest[23,29].

The shape of $\eta G$ falls between the CDHS[58] result $(1-\eta)^{5.3}$ at $Q^2 \simeq 20$ and the CFS result $(1-\eta)^{4.1}$ at $Q^2 \simeq 36$ (see Lefrancois talk).

Figure 4 shows the $Q^2$ dependence of $Q^2 d\sigma/dQ^2$ for muoproduction[25,26], averaged over $\nu$ and normalized to 1 at $Q^2 = 0$. The dotted line is the VMD prediction $(1 + Q^2/m_\psi^2)^{-2}$; the dashed line is $\gamma g$ fusion with Eq.(4) and $\alpha_s(m_\psi^2)$; the solid line has the extra $Q^2$-dependence from taking $\alpha_s(m_\psi^2 + Q^2)$. In the narrow-mass-window approximation[20], the formulas for $\psi$ production reduce to

$$Q^2 \frac{d\sigma}{dQ^2} \simeq (constant)\alpha_s \frac{\eta G(\eta)}{(m_\psi^2 + Q^2)^2} \left[\frac{\nu}{4E^2} + \frac{E - \nu}{2\nu E}\right] \qquad (6)$$

with $\eta = (m_\psi^2 + Q^2)/(2m_N\nu)$. Beside the standard VMD factor, this formula has $Q^2$ dependence from $\eta G(\eta)$ that looks after the threshold behaviour (ignored by VMD), and possible further $Q^2$-dependence from $\alpha_s$. There is excellent agreement with data.

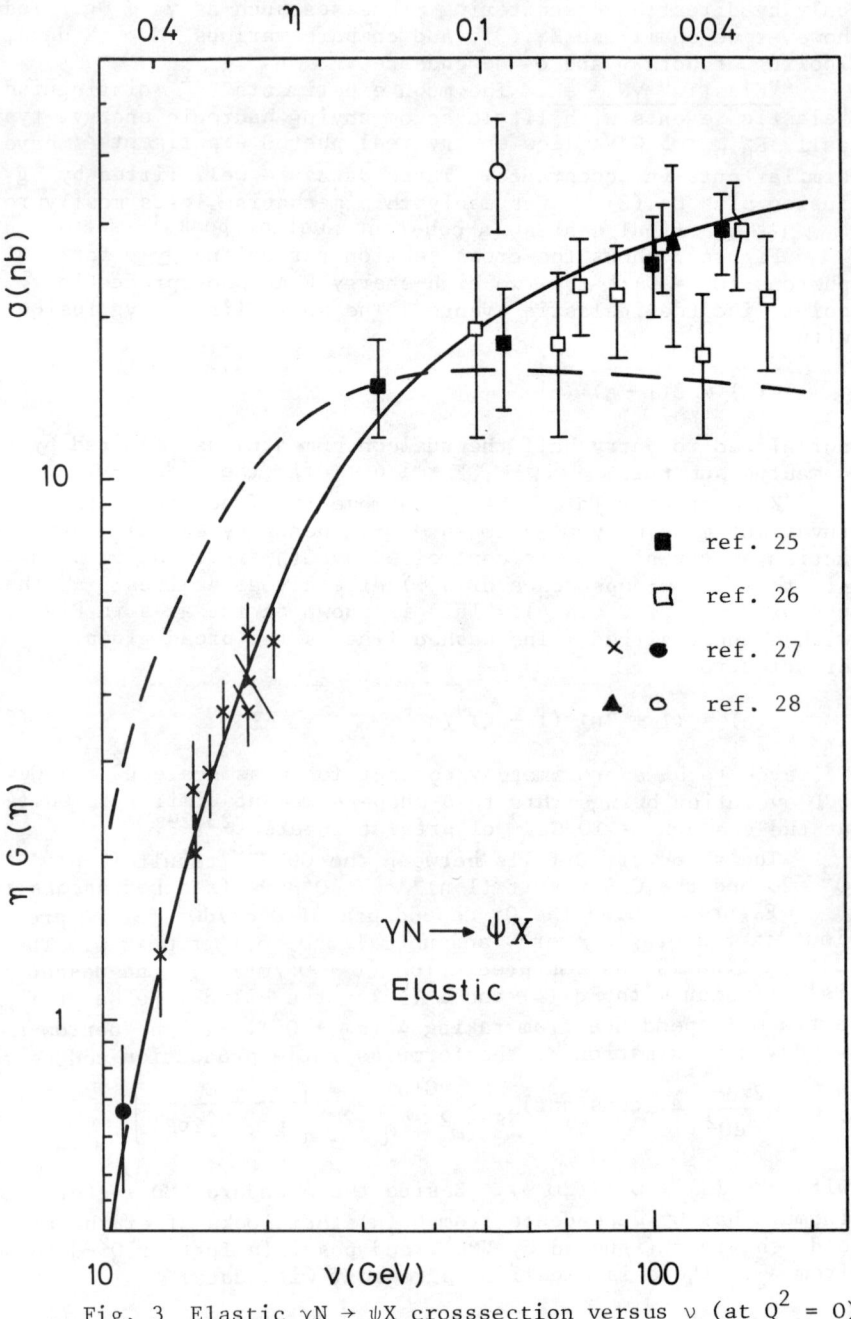

Fig. 3 Elastic $\gamma N \to \psi X$ crosssection versus $\nu$ (at $Q^2 = 0$)

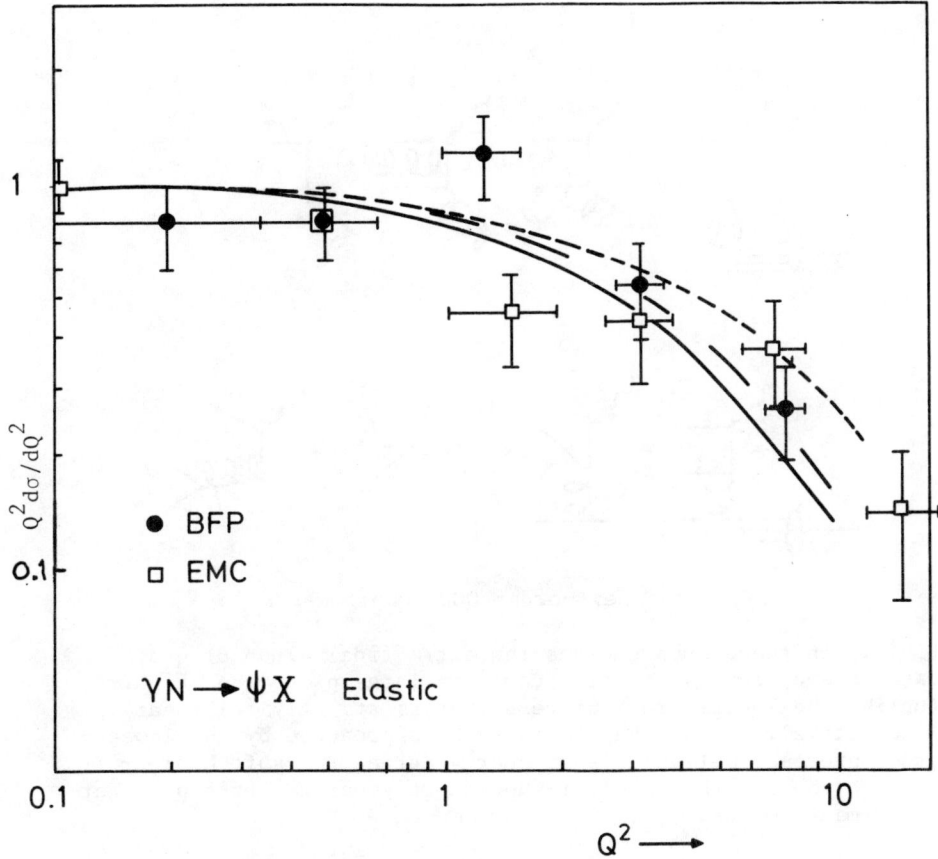

Fig. 4  $Q^2 d\sigma/dQ^2(\mu N \to \mu\psi X)$ normalized to 1 at $Q^2 = 0$

In principle the $Q^2$-dependence could provide an independent determination of $G(\eta)$ but the data shown are not adequate for this. However both the BFP and EMC groups find significant $Q^2$ dependence beyond the VMD factor.

"Inelastic" $\gamma N \to \psi X$.  EMC find inelastic ($E_{HAD} > 5$) production rates comparable to the elastic rates[26]. Some of these events presumably come from $\psi'$ production with $\psi\pi\pi$ decay, but this can only explain a small fraction. The remainder may come from the next-order QCD subprocesses $\gamma g \to c\bar{c}g$ and $\gamma q \to c\bar{c}q$: see

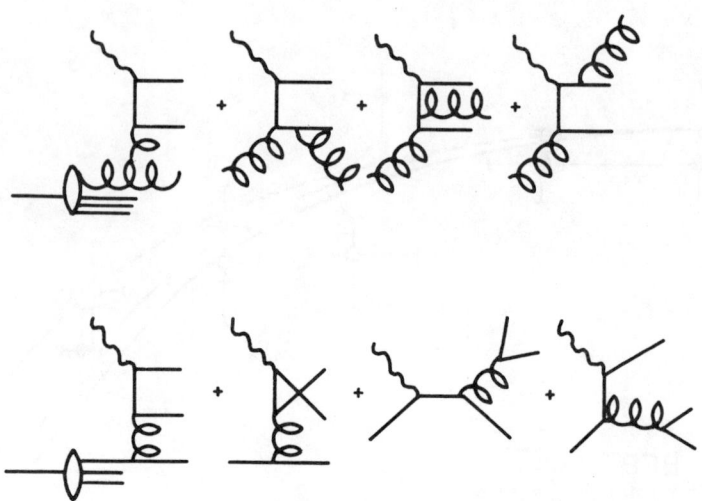

Fig. 5  Next-order QCD subprocesses

Fig.5. In these subprocesses the extra final gluon or quark carries away energy and also can give large $p_T$ to the $c\bar{c}$ pair (unlike the lowest order process that is strictly collinear). Alternatively we can imagine that $c\bar{c}$ is produced by the lowest-order process, but sheds some energy during the soft hadronization phase; this however should not generate large $p_T$. Let experiment decide.

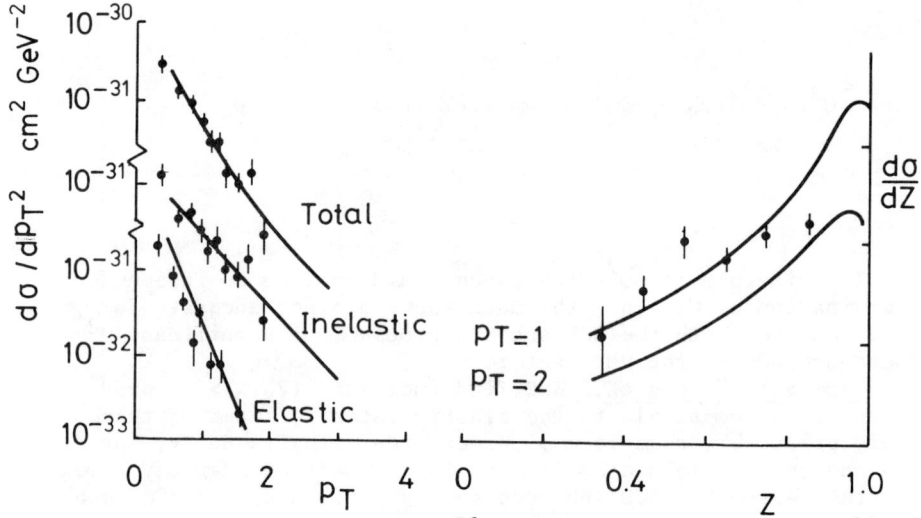

Fig. 6  Inelastic $\gamma N \to \psi X$ data[26] compared to next-order QCD[24]

Figure 6 shows the $p_T$ and $z$ dependence of the EMC data; note there is quite a lot of $p_T$. The calculation of the next-order diagrams has a singularity near $p_T = 0$ and needs careful handling there. However, ref.24 gives results away from $p_T = 0$ that show a striking similarity to the data (the normalization shown is arbitrary but common to elastic and inelastic events, distinguished by $E_{HAD}$). It seems that inelastic and high-$p_T$ elastic $\psi$ production can probably be explained this way.

<u>Open charm production</u>. The lowest order process is still $\gamma g \to c\bar{c}$ but we now have to describe the final hadronization to $D\bar{D}X$, $\Lambda_c \bar{D}X$, etc. At least two distinct types of hadronization pattern are imaginable, as shown in Fig.7:

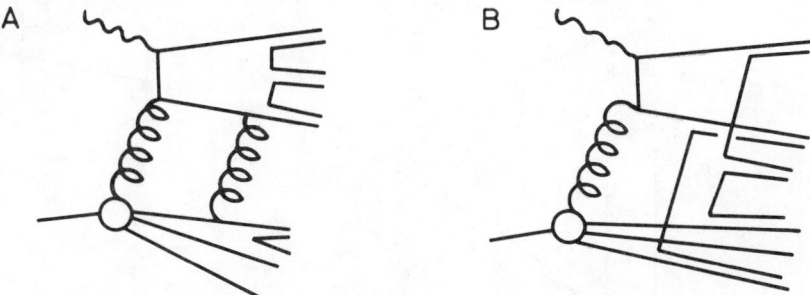

Fig. 7 Possible hadronization patterns

A) Patterns that are flavour-disconnected from the target. In such cases we expect the $c\bar{c}$ pairs to hadronize rather like those produced in $e^+e^-$ collisions; i.e. we should borrow fragmentation functions from SPEAR and work in the $c\bar{c}$ rest frame.

B) Patterns that are flavour-connected to the target. Such cases are more similar to neutrino production of charm, where a single produced c or $\bar{c}$ hadronizes on its own; i.e. we should borrow fragmentation functions from neutrino analyses and work in the c.m. or lab frame.

Which to choose? Data help us to decide. At large $\nu$ pattern A seems to dominate. The multimuon events of BFP and EMC collaborations apparently require $c\bar{c}$ to hadronize frequently into fast $D\bar{D}$ pairs carrying nearly all the energy[25,26]. Also the E87A experiment[30] at the Fermilab wideband photon beam finds fast $D\bar{D}$ and $\Lambda_c\bar{\Lambda}_c$ pair production (the latter at the 25% level similar to SPEAR) with $\langle E(\Lambda_c)\rangle \simeq \frac{1}{2}\langle E(\gamma)\rangle$. Pattern B seems to be important at small $\nu$ however, not surprisingly since it offers a lower threshold. The CERN WA4 experiment[31] at $\nu = 40-70$ GeV finds $\bar{D}$ mesons and evidence of $\Lambda_c\bar{D}$ production but no sign of $D\bar{D}$ pairs. Also the WA58 emulsion experiment has a $\Lambda_c^+\bar{D}^0$ event in this range[32].

$\gamma N \to c\bar{c}X$ data. We shall look at $\nu$ and $Q^2$ dependence - and also at normalization, since there is now no empirical factor F. The BFP group[25] have modelled their dimuon acceptance on the $\gamma g$ fusion model with D production and semileptonic decay, taking relative multiplicities and the fragmentation function from SPEAR data. They then convert their results directly into $c\bar{c}$ production cross sections and structure functions.

Figure 8 shows the BFP results for $\sigma(\gamma N \to c\bar{c}X)$ at $Q^2 \simeq 1$ versus energy $\nu$. The curve is a $\gamma g$ fusion calculation using Eq.(4) and including a contribution from the range $2m_c < m < 2m_D$; the absolute normalization has been adjusted by less than 20% - well within the theoretical uncertainties - to make a prettier picture.

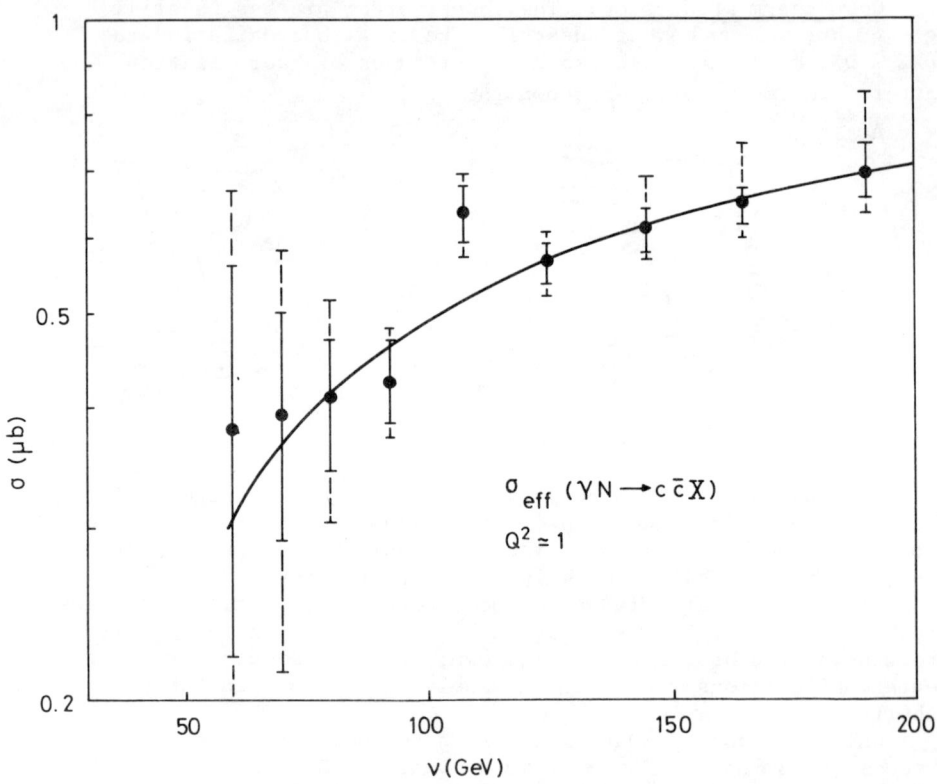

Fig. 8  $\sigma(\gamma N \to c\bar{c}X)$ at $Q^2 = 1$ from ref. 25, compared to $\gamma g$ fusion.

Figure 9 shows the same cross section at $Q^2 = 0$, where the real photoproduction data can be included[30,31]. The WA4 results have been split in three bins, pro rata to their $\bar{D}^0$ results that show a threshold behaviour[31]; apart from their highest point there is a good agreement again with the $\gamma g$ fusion calculation (unadjusted).

Figure 10 gives $F_2(\gamma N \to c\bar{c}X)$ versus $Q^2$ at $\nu = 100$ and $\nu = 178$ GeV from BFP[25]. The dotted line is the prediction of the Buras-Gaemers model[33] (using the empirical slow-rescaling variable $x' = (Q^2 + 4m_c^2)/(2m_N \nu)$); here it was assumed that the $c\bar{c}$ sea in the nucleon is negligible for $Q^2 < 1.8$ but evolves thereafter according to the QCD relations with $m_c = 0$, which evidently gives too much charm. The dashed line is a somewhat old-fashioned $\gamma g$ fusion

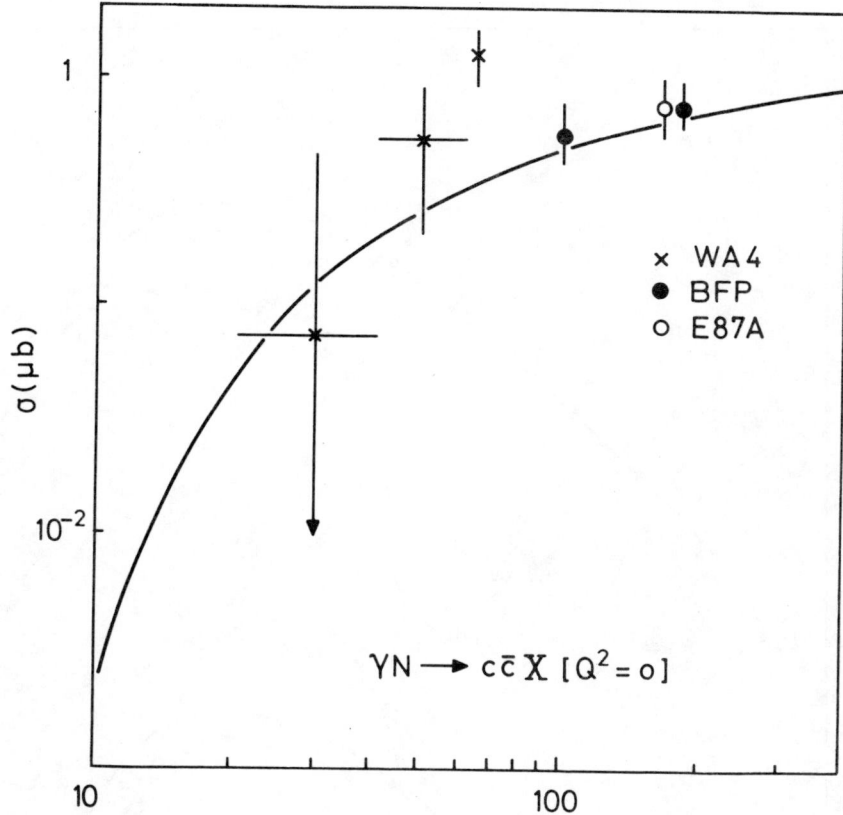

Fig.9 $\sigma(\gamma N \to c\bar{c}X)$ at $Q^2 = 0$ from refs.25,30,31 compared to $\gamma g$ fusion prediction, in which contributions from $m < 2m_D$ are excluded and $\alpha$ is defined at $\mu^2 = m^2$; as noticed by the experimenters[25], this is a bit unsatisfactory at large $Q^2$. The solid line is a more modern $\gamma g$ fusion prediction in which 5/6 of the cross section from $m < 2m_D$ is included and $\alpha$ is defined at $\mu^2 = m^2 + Q^2$; this does much better. The $\gamma g$ fusion curves are independently normalized to the data and both assume Eq.(4); ideally the $Q^2$-dependence of $G(\eta)$ should be included too, and this would add a little more curvature still. Incidentally, the linear rise of $F_2$ at small $Q^2$ is simply kinematics and there are no prizes for fitting this part; a small departure from linearity at very small $Q^2$ may be an effect of nuclear shadowing[34].

The EMC results[26] for dimuon and trimuon production by muons are also well explained by the $\gamma g$ fusion model, when conventional backgrounds have been subtracted. This group do not extract $F_2$ but instead put the model into a Monte Carlo calculation of their acceptance, and compare directly to cut cross sections: Fig. 11 shows some of their dimuon data with $\gamma g$ fusion comparison.

The overall agreement is very encouraging, but remember
(i) Higher orders are not yet included for open charm; they

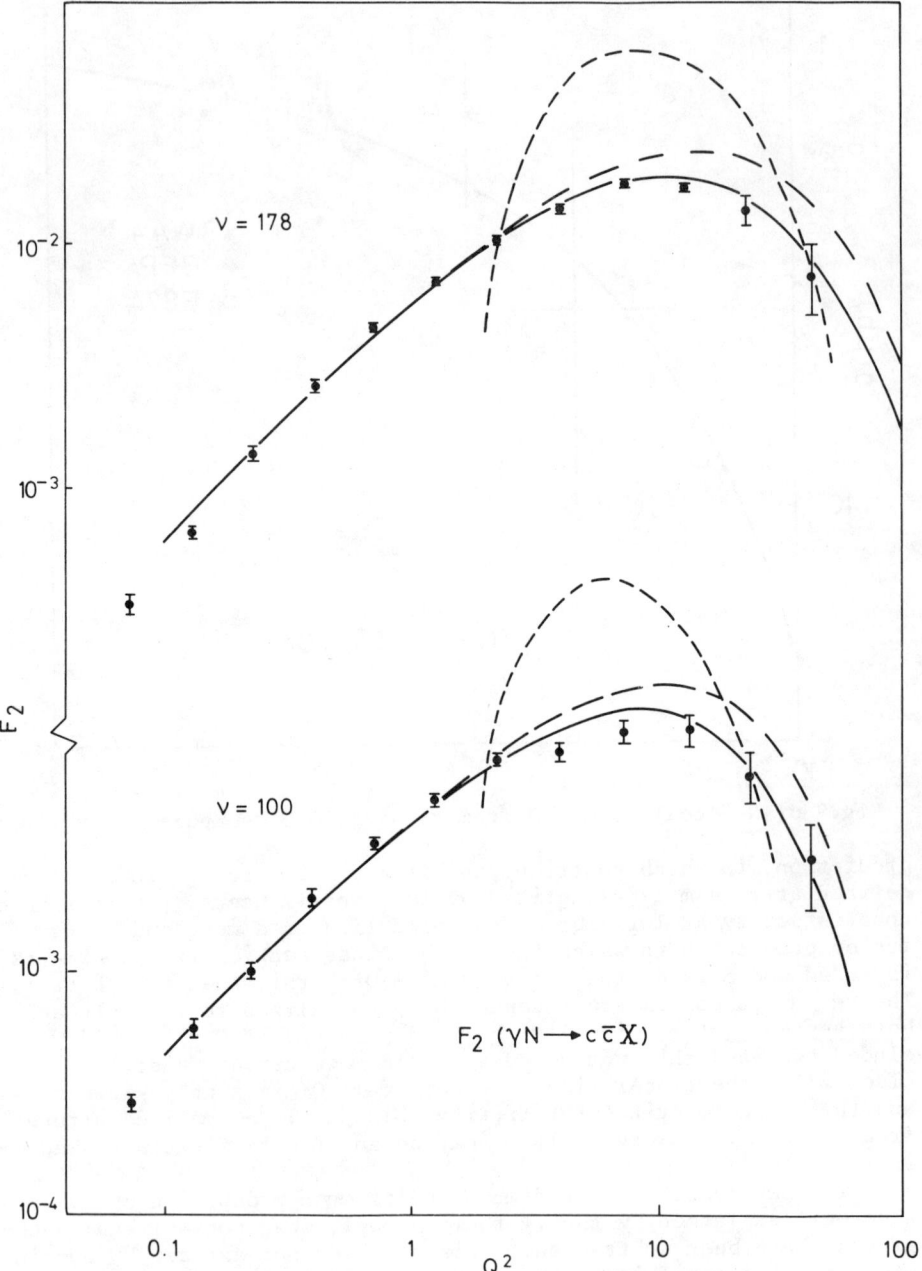

Fig. 10  $F_2(\gamma N \to c\bar{c}X)$ from ref. 25 compared to models.

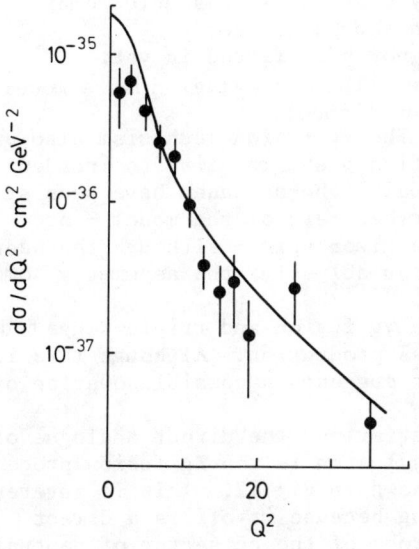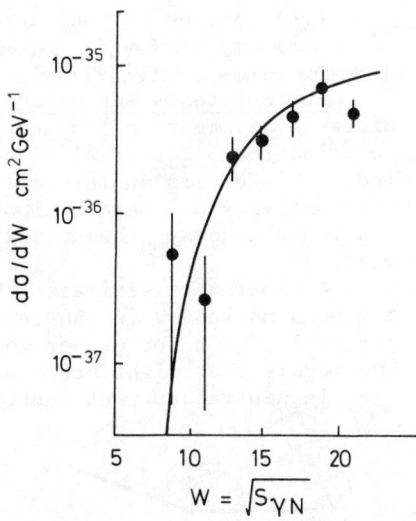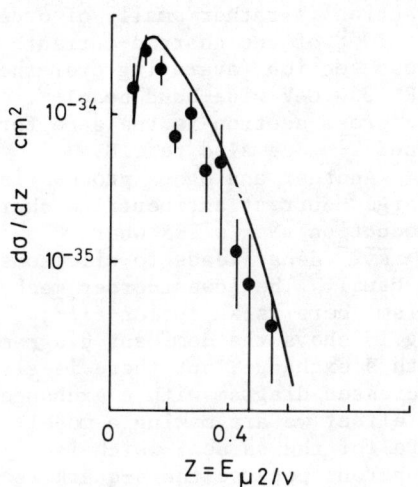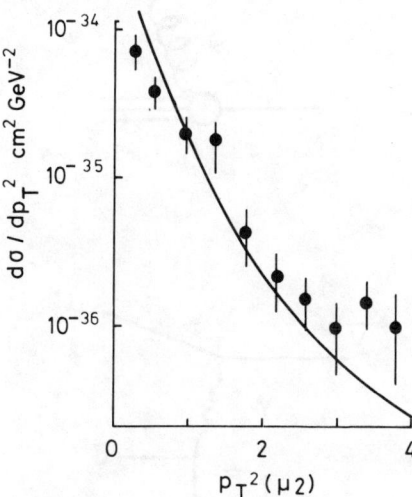

Fig. 11 Dimuon cross sections from ref. 26 compared to γg fusion

account for perhaps half of $\psi$. Many distributions should remain rather similar but the normalization should change.

(ii) B-type hadronization has not been fitted in yet.

(iii) $\gamma g$ fusion is not a complete theory: it is just a model for the cross section (not for the amplitude).

Further tests and extensions. The $\gamma g$ fusion mechanism also predicts an asymmetry of the $c\bar{c}$ production plane relative to incident $\gamma$ polarization. Both real[21] and virtual[35] photon cases have been studied. This offers an interesting further test of the model - or alternatively a determination of the gluon spin - although the hadronization and decay processes will presumably blur the asymmetry somewhat.

A remarkable similarity between $\gamma g$ fusion and triple-Regge models has been noticed[36] in the case of $s\bar{s}$ production. Although this is a marginal region for either model, it suggests a possible overlap of the models that might prove useful.

In neutral current neutrino scattering, the direct analogue of $\gamma g$ fusion is the $Zg$ fusion process shown in Fig.12. This is interesting because it offers a direct probe of the $c\bar{c}$ sector of neutral currents, difficult to reach otherwise. Elastic $\psi$ production can be calculated with the same input that successfully describes muoproduction. Unfortunately the predicted cross section times branching fraction is rather small, of order $2 \times 10^{-6}$ of the charged-current cross section (averaging over the CERN 350 GeV wide-band beam)[20]. The cross section is the same for $\nu$ and $\bar{\nu}$. See also ref.18.

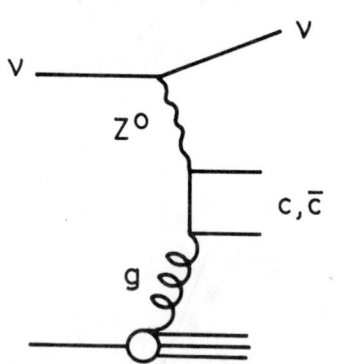

Fig. 12  Zg fusion

Another analogous process is charged-current antineutrino charm production $\bar{\nu}N \rightarrow \mu^+\bar{c}sX$ where $\bar{c} \rightarrow \bar{s}\bar{\nu}\ell^-$ decay leads to dileptons as usual. The lowest-order mechanism here is Wg fusion[12,15]; Fig.13 shows the dominant diagram with $\bar{s}$ exchange, but there is also a crossed diagram with c exchange. In effect we are making a model here for the $s\bar{s}$ sea, which is dangerous because the s-quark is relatively light and nonperturbative contributions may be expected. Nevertheless it is known to give cross sections in the right ballpark, and also has reasonable threshold behaviour automatically (no need for slow rescaling

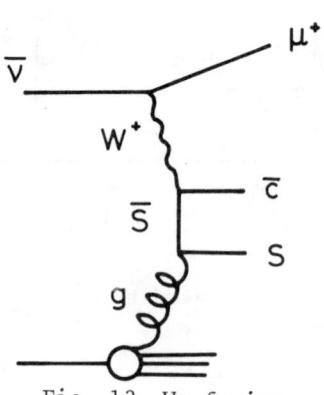

Fig. 13  Wg fusion

variables or other cookery). Since we now have better data, it is interesting to make one more comparison. Figure 14 shows $d\sigma/dxdy$ in units $G^2ME/\pi$ which is the same as $2x\bar{s}(x)$ in the parton model of charm production from the strange sea; the points are CDHS data[37] from dimuon events in the band $\nu = 40-80$; the curves are Wg fusion calculations using Eq.(4), with $m_s = 0.3$ and $0.5$ GeV, calculated at $\nu = 63$ and $E = 100$ GeV. The model contains a small contribution from the crossed diagram; it also has an energy-dependence different from that of the naive parton model that was partly used in binning the data. Nevertheless the agreement is remarkable and deserves further study. The analogue of $\gamma \to \psi$ production here is $W^+ \to F^{*+}$ production, that would also be interesting to study when there are data. For $O(\alpha_s)$ corrections in general see ref.38.

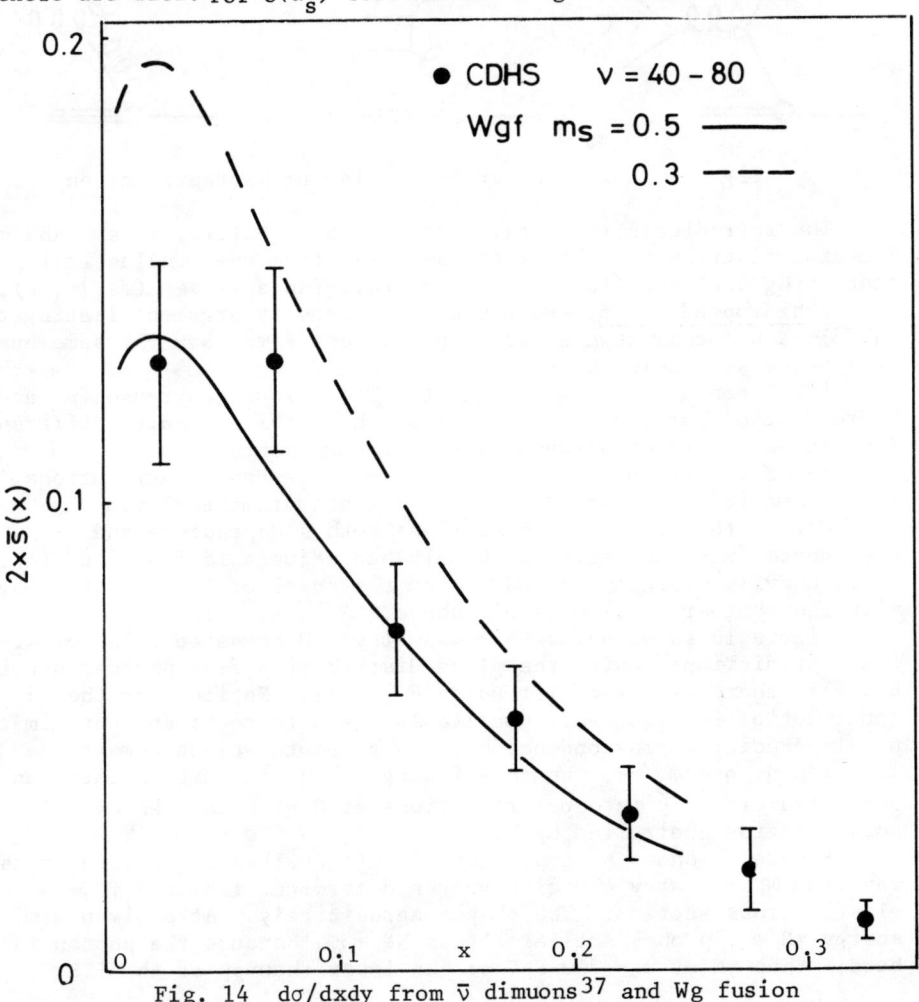

Fig. 14  $d\sigma/dxdy$ from $\bar{\nu}$ dimuons[37] and Wg fusion

## $c\bar{c}$ AND $b\bar{b}$ HADROPRODUCTION

The lowest-order QCD subprocesses for charm and beauty hadroproduction are shown in Fig.15 and variously calculated in refs.1-5, 9-11, 13, 14, 16, 18, 20, 22, 23. They include light $q\bar{q}$ fusion similar to Drell-Yan, and gg fusion similar to $\gamma$g fusion except for the addition of a three-gluon vertex diagram. The $q\bar{q}$ cross section is added incoherently to gg since it gives different kinds of final state.

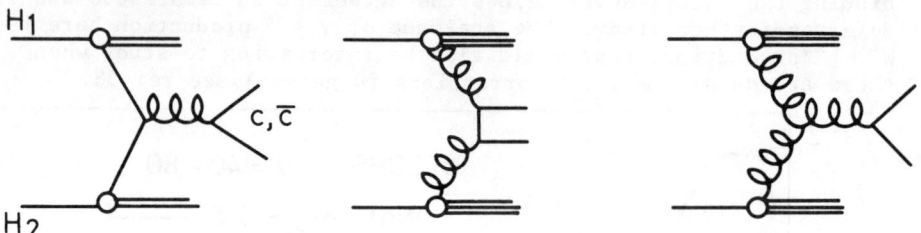

Fig. 15 Lowest-order heavy flavour hadroproduction

The ingredients for a calculation are as before, plus q and $\bar{q}$ momentum distributions, that can be taken from deep inelastic lepton scattering analyses (for N) or from Drell-Yan analyses (for $\pi$, K).

$\psi$ hadroproduction. We use a local-density argument leading to Eq.(3) as before. However we do not expect F to have the same numerical value as before, because[20]

i) F can be process-dependent. The colour-rearrangement and hadronization patterns in $\gamma$N → $c\bar{c}$X and hN → $c\bar{c}$X are quite different. They can also differ between $q\bar{q}$ and gg components.

ii) F can include $\psi'$ → $\psi\pi\pi$ and $\chi$ → $\psi\gamma$ cascade contributions; the kinematics here differ a little but not dramatically.

Notice that the model prescribes both s-dependence and y-dependence ($x_F$-dependence) with only one adjustable function $G(\eta)$, so it is heavily overconstrained. Also the shape of $G(\eta)$ ought to agree with the photoproduction result above.

Figure 16 shows $d\sigma/dy(NN \to \psi X)$ at y = 0 compared to a lowest-order prediction[20] using the gluon distribution from photoproduction Eq.(4); there is a good fit using $F \simeq 1/12$. Notice that the $q\bar{q}$ contribution is relatively small. At y = 0 there is an approximate point-to-point correspondence between s and the gluon momentum fraction $\eta$, $\eta \simeq m_\psi/\sqrt{s}$; hence allowing for $q\bar{q}$ the data points can be converted directly into determinations of $G(\eta)$ within an overall normalization controlled by F.

Figure 17 shows a comparison of $\eta G(\eta)$ values determined in this way from NN → $\psi$ at y = 0 with values determined from the $\gamma$N → $\psi$ elastic cross section. The shapes agree nicely. At a given c.m. energy $\gamma$N → $\psi$ probes smaller $\eta$ than NN → $\psi$, because the photon is hard; thus $\gamma$N at $\nu$ = 200 GeV probes lower than pp at the ISR.

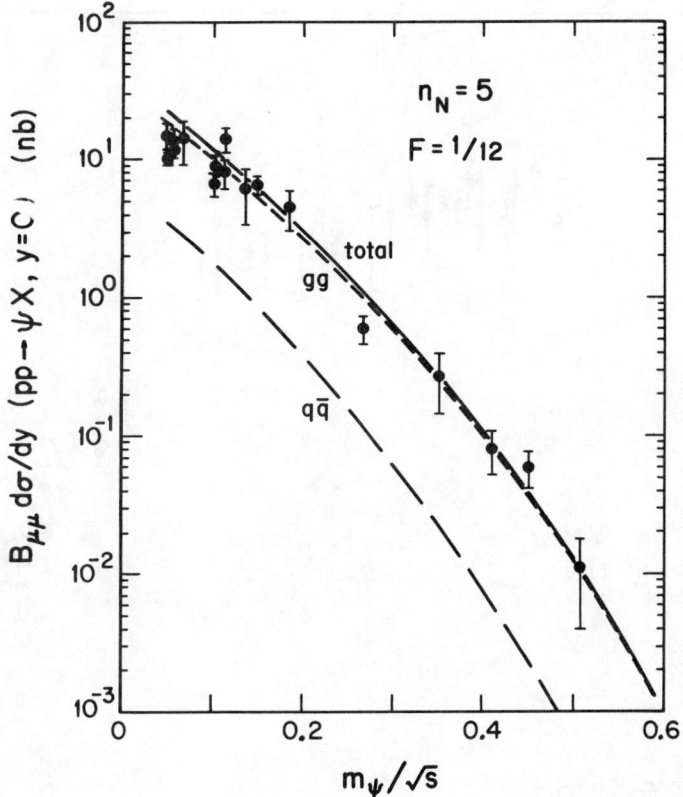

Fig. 16   s-dependence of $B_{\mu\mu} d\sigma/dy(NN \to \psi X)$ at $y = 0$ compared[20] to the fusion model

Figure 18 shows that the $x_F$-dependence in NN scattering at 225 and 400 GeV/c and $\bar{N}N$ scattering at 225 GeV/c is consistent with the same model using $F = 1/9, 1/12, 1/12$ respectively[20]. Fluctuations in F may be attributed to normalization problems in the data (notice for example the data fluctuations in Fig.16). When we take account of the intrinsic $p_T$ carried by the initial quarks and gluons, the simple $q\bar{q}$ plus gg fusion mechanisms above only make sense in the range[39]

$$s \langle p_T^2 \rangle x_F^2 < m_\psi^4 \tag{7}$$

(A similar limit applies to Drell-Yan). Conservatively assuming $\langle p_T^2 \rangle \lesssim 1$ GeV$^2$, this requires e.g. $|x_F| < 0.5$ for 200 GeV beams on a fixed target. At larger $x_F$, processes where both partons come from the same hadron become important.

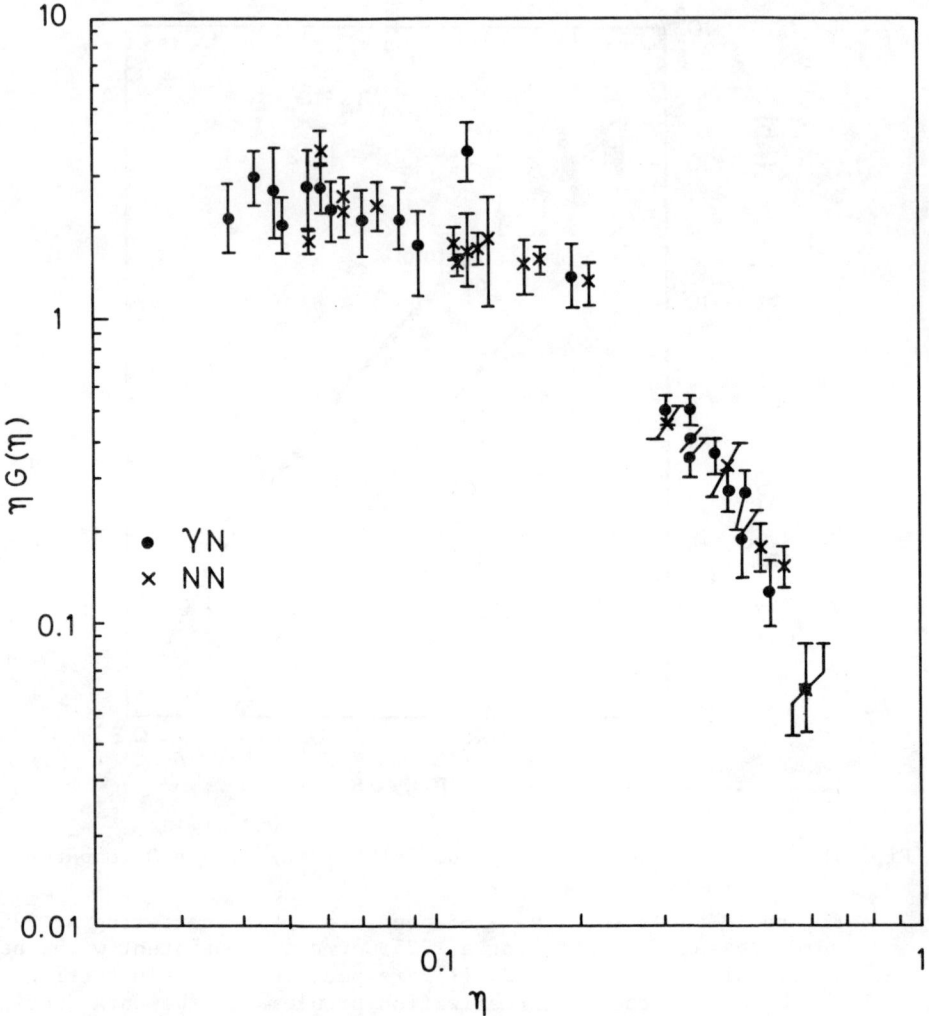

Fig. 17  Comparison of $\eta G(\eta)$ extracted from $\gamma N$ and $NN$ data

For $\pi N \to \psi X$ and $KN \to \psi X$ the new ingredients are the gluon distributions within $\pi$ and $K$, that we can hope to extract. There are not enough data to exploit the s-dependence, but we can extract $G(\eta)$ from the $x_F$-dependence instead. $K^+$ is better than $K^-$ because the $q\bar{q}$ term is minimized. Figure 19 shows a log-log plot of $\eta G(\eta)$ versus $1 - \eta$, where $d\sigma/dx_F$ data have been translated directly into $G$ for $\pi$ and $K^+$ separately, taking the $q$, $\bar{q}$ and $G_N$ distributions as known. On this plot simple power parameterizations

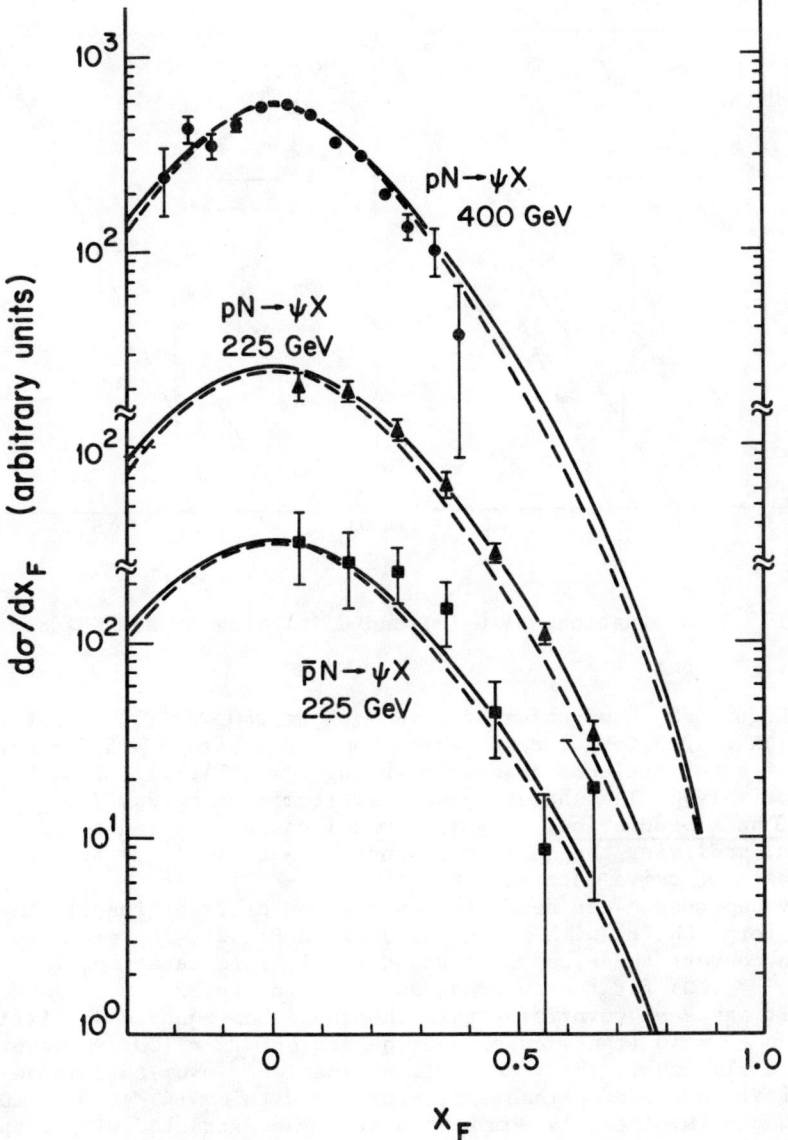

Fig. 18  $x_F$-dependence of NN → ψX, $\bar{N}$N → ψX in the fusion model[20].

$$G \sim (1-\eta)^n/\eta \qquad (8)$$

appear as straight lines; the value n = 3 gives approximate fits to both π and $K^+$ cases[20].

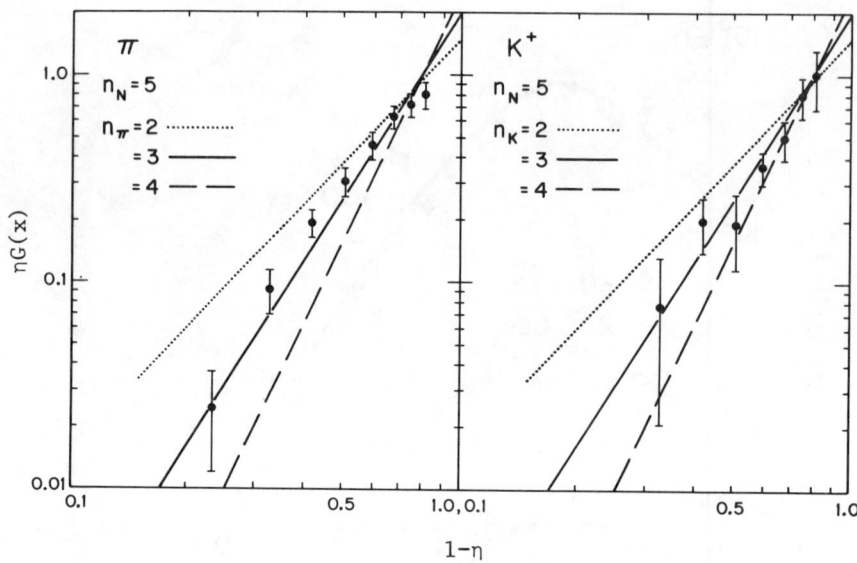

Fig. 19  Determinations of $G_\pi(\eta)$ and $G_K(\eta)$ from $\pi N$ and $KN$ data[20]

Figure 20 shows $d\sigma/dx_F$ for $\pi N \to \psi X$ at 240, 225, 39.5 GeV/c and $K^+ N \to \psi X$ at 225 GeV/c, compared to the model with n = 3 for mesons and n = 5 for nucleons as before (using F = 1/14, 1/12, 1/6, 1/12 respectively). The shapes agree well, remembering Eq.(7).

$\Upsilon$ hadroproduction. The same model can be applied to $b\bar{b}$ production, modifying the quark mass and the mass scale $\mu^2$ accordingly. The most extensive data are $NN \to (\Upsilon + \Upsilon' + \Upsilon'')X$ near y = 0. The energy dependence can be fitted by a gluon distribution of simple power form Eq.(8) with n = 6, as seen in Fig.21. Alternatively we can convert the cross sections directly into determinations of $G(\eta, \mu^2 = 100)$ for the nucleon, as shown in Fig.22.

It may seem surprising that the shape has changed so little between $\mu^2 = 10$ (the $\psi$ case) and the present $\mu^2 = 100$. However Fig. 22 also shows the $\mu^2$-evolution expected[40] over this range for a typical choice of parameters, starting with Eq.(4) at $\mu^2 = 10$. The result is certainly compatible with the extracted $G(\eta)$ shape at $\mu^2 = 100$, but also suggests that the true shape here is not a simple power formula like Eq.(8).

Open $c\bar{c}$ hadroproduction. Now we come to the bad news. The important new ingredient here is the hadronization, which is empirical and quite outside perturbative QCD, so we start by looking at the data. Instantly we meet two problems.

First problem: the model predictions for total charm production - assumed to be independent of hadronization details - are

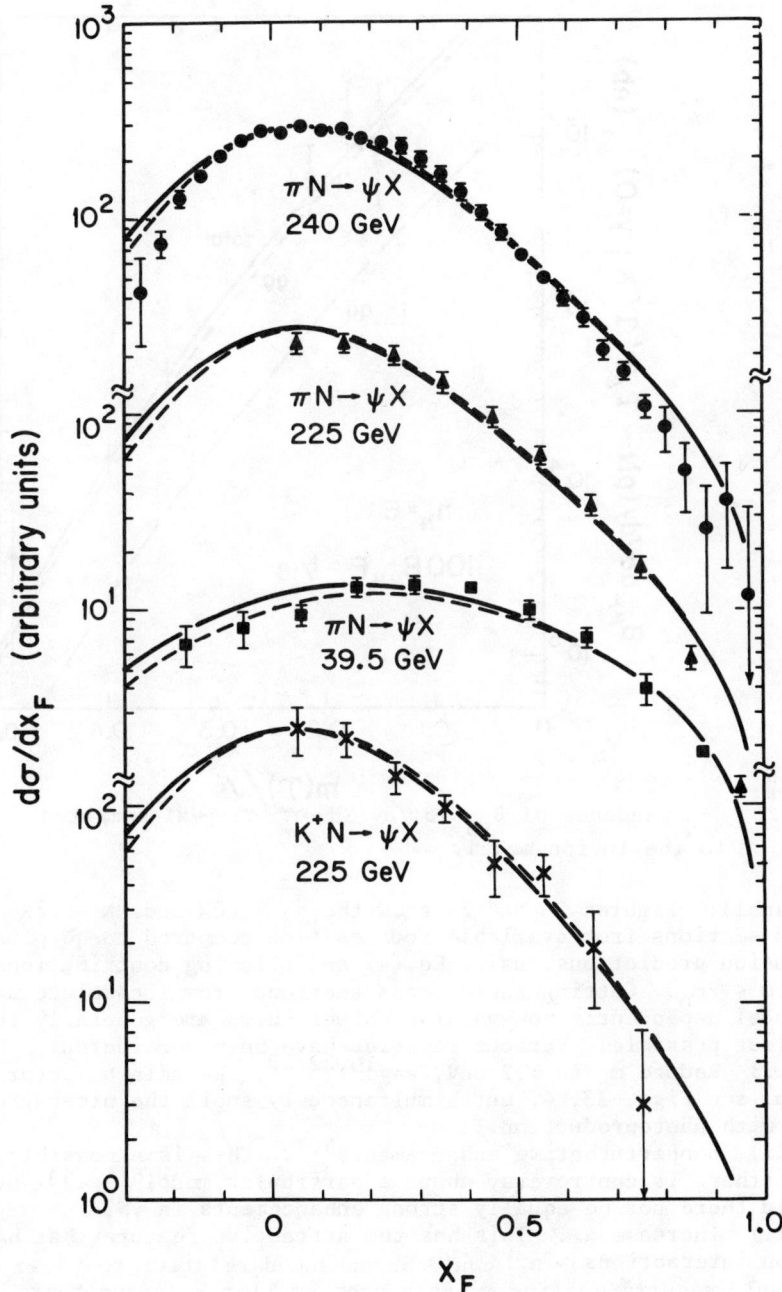

Fig. 20 $x_F$-dependence of $\pi N \to \psi X$, $K^+ N \to \psi X$ in the fusion model[20]

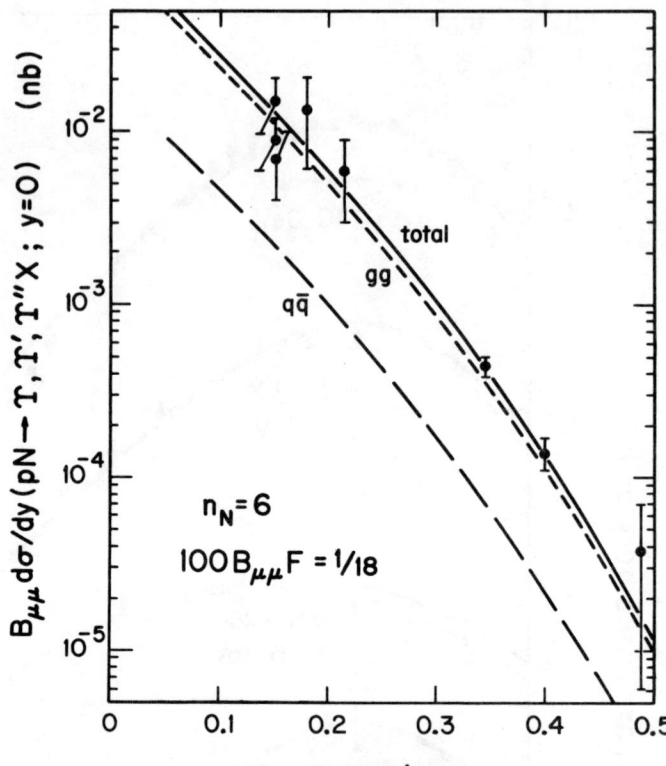

Fig. 21  s-dependence of $B_{\mu\mu} d\sigma/dy(NN \to \Upsilon + \Upsilon' + \Upsilon''x)$ compared to the fusion model.

too small. Figures 23 and 24 show the $\pi N \to c\bar{c}X$ and $NN \to c\bar{c}X$ cross sections from available sources[41-51] compared to $q\bar{q}$ plus gg fusion predictions, using Eq.(4) and allowing contributions from $m < 2m_D$. Getting these cross sections from incomplete data is model dependent; however the values shown are generally the smallest possible. Various remedies have been considered.

  i) Reduce $m_c$ to 1.2 GeV, say[10,22,23]. We gain a factor 4 or so, see Figs. 23,24, but simultaneously spoil the nice agreement with photoproduction.

  ii) Nonperturbative enhancements[6,7]. This is a possible escape (there is controversy about a particular model[6,52,53]) but why should there not be equally strong enhancements in $\gamma N$?

  iii) Increase $\alpha_s$. This has the attractive feature that hadron-hadron interactions $(\sim \alpha_s^2)$ could be enhanced relative to $\gamma N (\sim \alpha_s)$. But it would require scaling $\alpha_s$ at a much smaller $\mu^2/\Lambda^2$ and there is only limited scope in this direction.

  iv) Maybe higher-order corrections are much bigger for hadron-hadron than for $\gamma N$? Perhaps, but no solid argument or calculation has yet appeared.

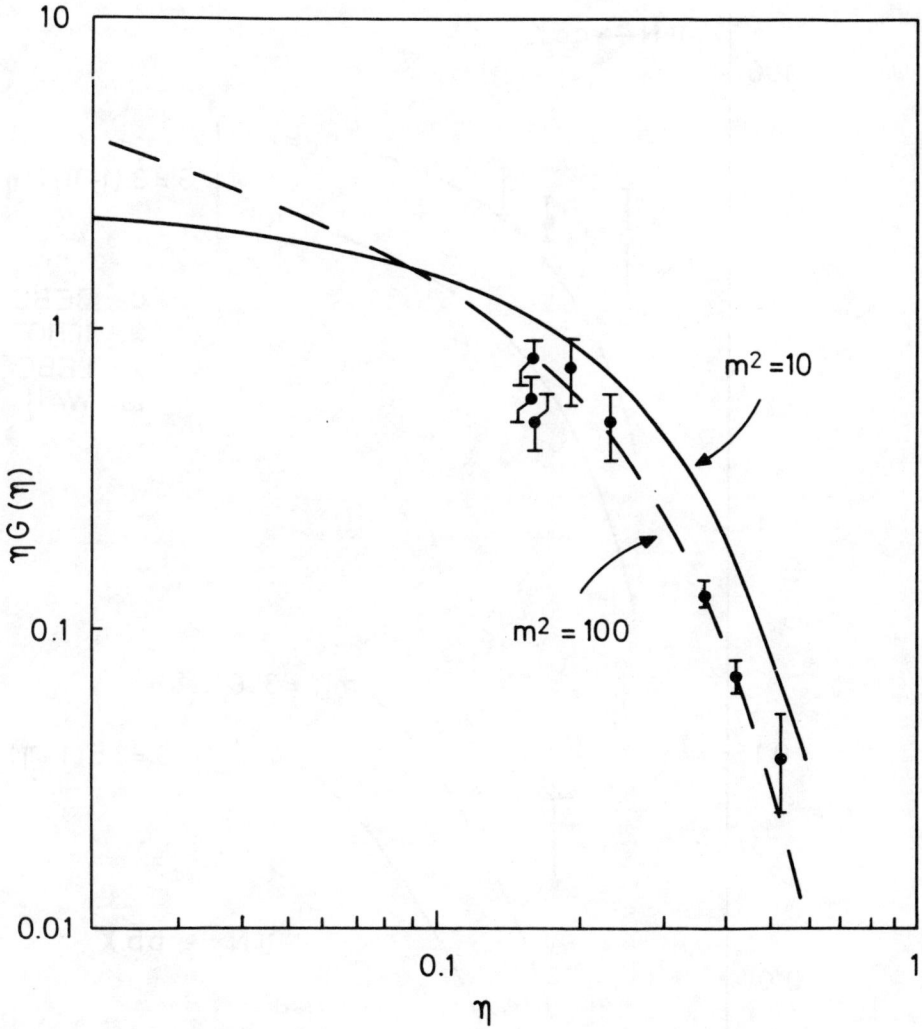

Fig. 22  $\eta G(\eta)$ at $\mu^2 = m_\Upsilon^2$ is compatible with QCD evolution.

To summarize, no single clear solution has yet emerged.

Second problem: the model gives rather central $c\bar{c}$ production (just right for the $\psi$ case) whereas open charm production occurs copiously at large $x_F$. Figure 25 shows the $x_F$-dependence of $pp \to \Lambda_c^+ X$ production measured by the Lampshade Magnet[45], UCLA-Saclay[46] and Split-Field Magnet[47] groups at the CERN ISR, compared with ordinary $\Lambda^0$ production[54] and with the $q\bar{q}$ + gg fusion model predictions for c-quark production (using $m_c$ = 1.5 plus Eq.(4) and including a contribution from $m < 2m_D$). Notice the fierce log

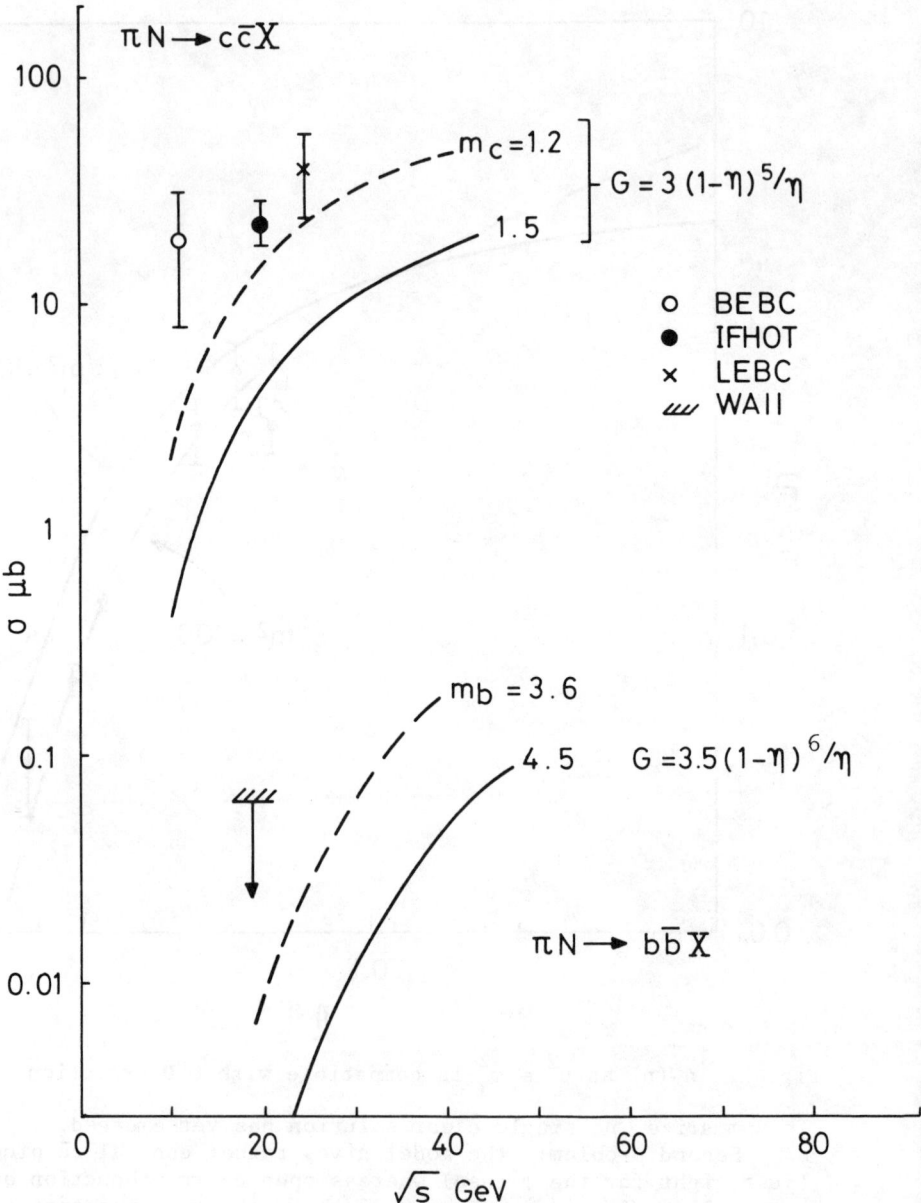

Fig. 23 $\sigma(\pi N \to c\bar{c}X)$ open charm production versus $\sqrt{s}$, from refs. 41-43. The $b\bar{b}$ upper limit is from ref.44.

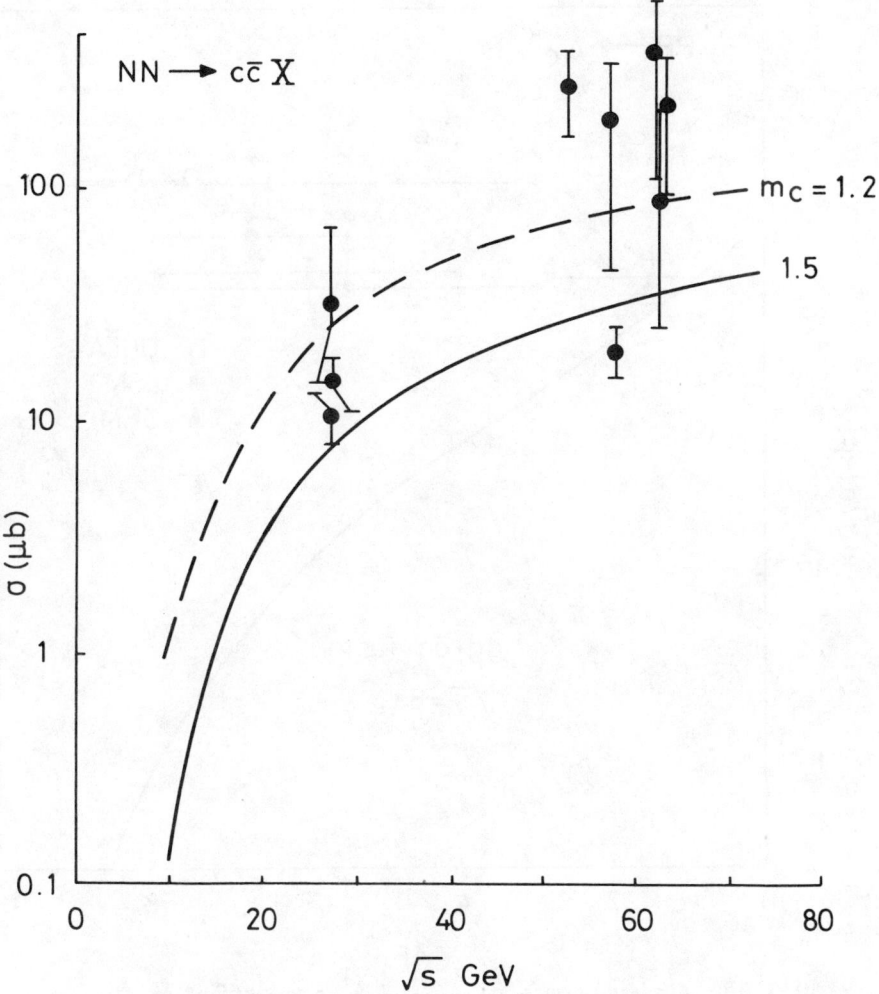

Fig. 24 $\sigma(NN \to c\bar{c}X)$ open charm production versus $\sqrt{s}$.

scale on the ordinate; at $x \sim 0.6$ the bare quark production is three or more orders of magnitude below the $\Lambda_c^+$ data. Remedies?

i) Maybe hadronization puts it right? We can imagine a slow c-quark being picked up by fast valence u+d quarks in one of the incident protons to make a fast final $\Lambda_c^+$. This is the Recombination Model[55] and it has the right features to explain the leading $\Lambda^0$ production all right. But for $\Lambda_c^+$ production it incorporates a recombination function

$$R(x_c/x, x_u/x, x_d/x)$$

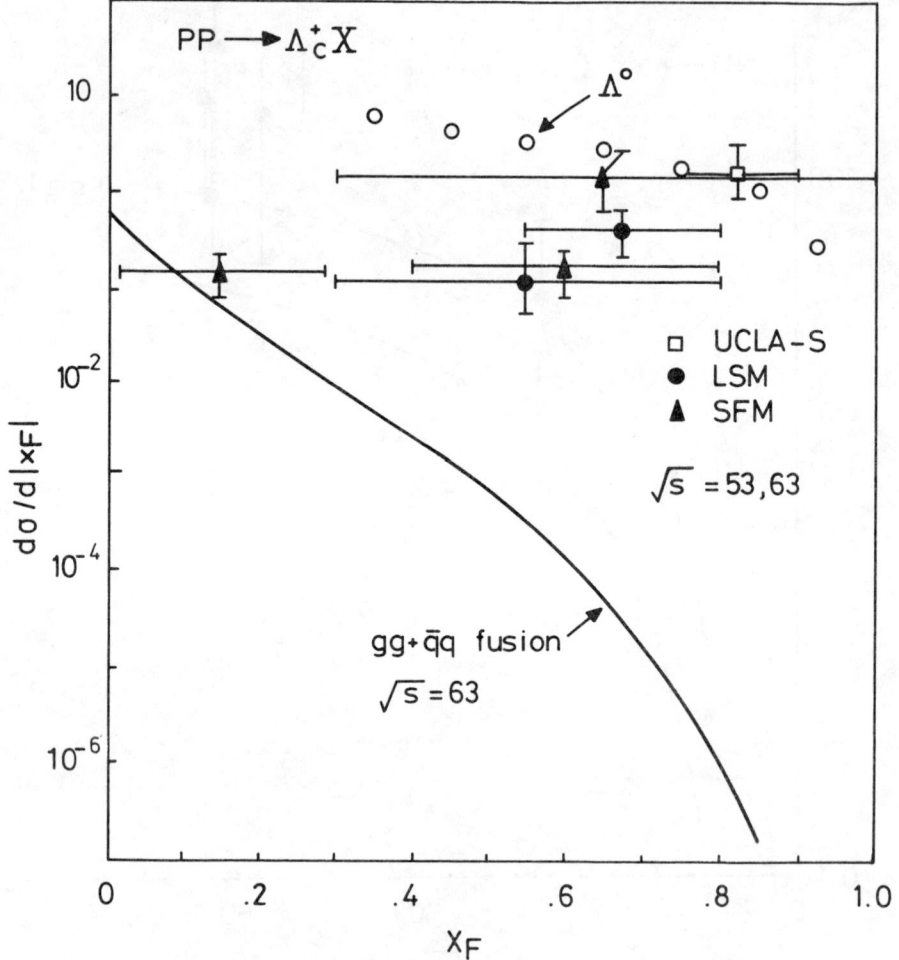

Fig. 25  $x_F$-dependence of $\Lambda^0$ and $\Lambda_c^+$ production at ISR

that should reflect the composition of $\Lambda_c^+$, in which the valence c-quark is expected to carry a large fraction of the momentum[56]. Thus R should favour substantial $x_c/x$, disfavour small $x_c/x$, making it hard to convert a central c-quark into a leading baryon; the $\Lambda_c^+$ data however do not look suppressed! Furthermore fast $D^+(c\bar{d})$ mesons are seen[47], whereas $\bar{d}$ is not a fast valence quark.

Other soft hadronization mechanisms are not expected to accelerate the original c and $\bar{c}$, nor to form fast $c\bar{c}$ pairs from the vacuum.

ii) Postulate a ($uudc\bar{c}$) five-quark Fock component in the incident proton wave function[57]. Such a component would be relatively long-lived. Hence, because of their large masses, the c and $\bar{c}$ would be expected [56] to carry large momentum fractions - 2/7 each in the

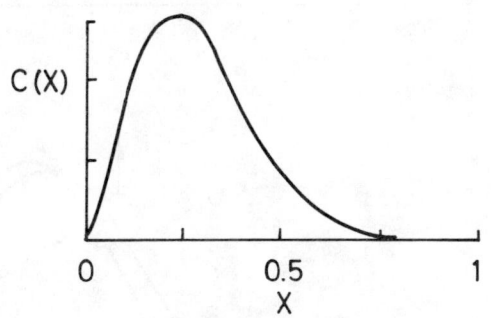

Fig. 26 c-quark probability distribution from ref.57

model of ref. 57; Fig. 26 shows the probability distribution c(x) for this model, where the integrated probability is about 0.01 per nucleon. The idea is that in pp collisions a (udc) or (uuc) subsystem could be shaken off to form a leading charmed baryon - nothing to do with hard QCD at all.

But how do we simultaneously forbid leading $\psi$ production that is not seen? It may be a difficulty.

And what about deep inelastic muon scattering data, that measures $F_2$(charm) $\simeq (8/9)xc(x)$ in the parton model? There is already borderline conflict between BFP data[25] and the model above[57], shown in Fig. 27. The data we met already; the dashed lines are the naive parton prediction for the model using standard $x = Q^2/(2m_N\nu)$; the solid lines are model predictions using the slow rescaling variable $x' = (Q^2 + 4m_c^2)/(2m_N\nu)$ instead. The model predictions are presumably additive to the QCD contributions previously shown in Fig.10.

The model almost escapes confrontation with experiment here, because c and $\bar{c}$ are at large x and hence large $Q^2$, where the cross section is suppressed. (Going to smaller $\nu$ helps, but eventually there is the threshold). However, the experiment has acceptance well beyond the published data points[34], so the confrontation can probably be sharpened. The EMC results should also be analyzed in these terms.

## CONCLUSIONS

The models considered are parton models with low-order QCD interactions, not complete theories. The hadronization stage, for example, is empirical. However they have fewer parameters than constraints; there are many correlations between $\nu$- and $Q^2$-dependences, between s- and $x_F$-dependences, between one reaction and another. Their successes therefore look significant.

There are problems in understanding hadroproduction of open charm, but these may be outside perturbative QCD - to do rather with final state interactions, wave functions, etc. We do not yet know enough experimentally to pinpoint the problems.

These parton-QCD models are helpful and physically appealing. It seems worthwhile to push them as far as they will go.

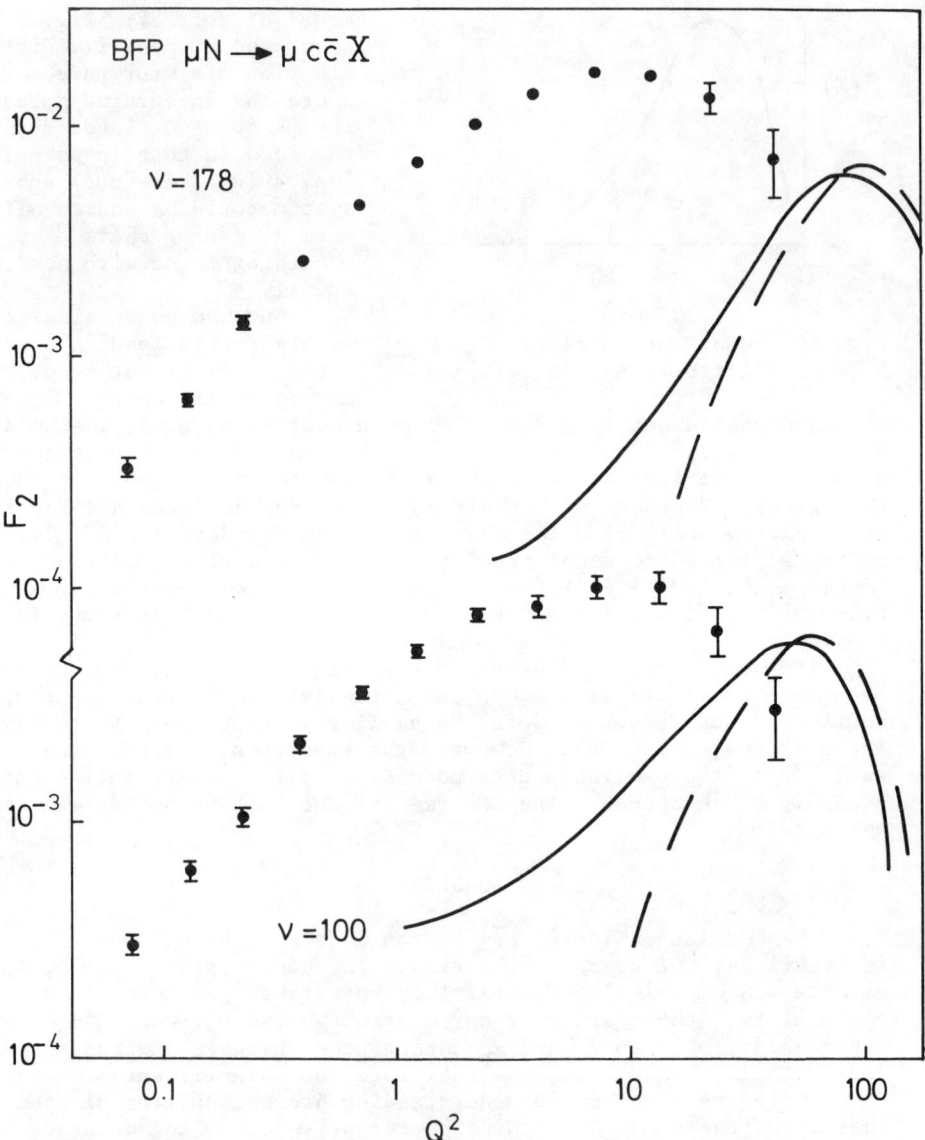

Fig. 27  $F_2$(charm) from BFP compared to the model of ref.57

## ACKNOWLEDGEMENT

I am grateful to many colleagues at Rutherford, Madison and elsewhere for helpful discussions and for early information about their results. Special thanks are due to Vernon Barger and Wai Yee Keung.

## REFERENCES

1. H. Fritzsch, Phys. Lett. $\underline{67B}$, 217 (1977).
2. F. Halzen, Phys. Lett. $\underline{69B}$, 105 (1977).
3. L.M. Jones and H.W. Wyld, Phys. Rev. $\underline{D17}$, 759, 1782, 2332 (1978).
4. M. Gluck and E. Reya, Phys. Lett. $\underline{79B}$, 453 (1978), $\underline{83B}$,98(1979).
5. M. Gluck, J.F. Owens and E. Reya, Phys. Rev. $\underline{D17}$, 2324, (1978).
6. H. Fritzsch and K.H. Streng, Phys. Lett. $\underline{72B}$,$\overline{385}$,78B,447(1978).
7. F. Halzen and D.M. Scott, Phys. Lett. $\underline{72B}$, 404 (1978)
8. M.A. Shifman, A.I. Vainshtein and V.I. Zacharov, Nucl. Phys. $\underline{B136}$, 157 (1978)
9. J. Babcock, D. Sivers and S. Wolfram, Phys. Rev. $\underline{D18}$, 162 (1978).
10. C.E. Carlson and R. Suaya, Phys. Rev. $\underline{D18}$, 760 (1978); Phys. Lett. $\underline{81B}$, 329 (1979).
11. H. Georgi et al, Ann. Phys. $\underline{114}$, 273 (1978).
12. J. Babcock and D. Sivers, Phys. Rev. $\underline{D18}$,2301 (1978).
13. R.J. De Witt, D.E. Willen and H.W. Wyld, Illinois report ILL-(TH)-78-45 (1978).
14. K. Hagiwara and T. Yoshino, Phys. Lett. $\underline{80B}$, 282 (1979).
15. J.P. Leveille and T. Weiler, Nucl. Phys. $\underline{B147}$, 147 (1979).
16. B.L. Combridge, Nucl. Phys. $\underline{B151}$, 429 (1979).
17. J.P. Leveille, Proc. of Topical Workshop on New Particles (eds. V. Barger, F. Halzen), Univ. of Wisconsin 1979
18. J.H. Kühn, Phys. Lett. $\underline{89B}$, 385 (1980); J.H. Kühn and R. Rückl, MPI-PAE/pTH 7/80.
19. T. Weiler, Phys. Rev. Lett. $\underline{44}$, 304 (1980).
20. V. Barger, W.Y. Keung and R.J.N. Phillips, Phys. Lett. $\underline{91B}$, 253, $\underline{92B}$, 179 (1980); Z. Phys. $\underline{C}$ to be published.
21. D.W. Duke and J.F. Owens, Phys. Rev. Lett. $\underline{44}$, 1173 (1980).
22. Y. Afek, C. Leroy and B. Margolis, Phys. Rev. $\underline{D22}$, 86, 93 (1980).
23. M. Gluck, E. Hoffmann and E. Reya, DO-TH 80/13.
24. D.W. Duke and J.F. Owens, FSU-HEP-800709.
25. A.R. Clark et al, Phys. Rev. Lett. $\underline{43}$, 187 (1979); LBL-10747, LBL-10879, LBL-11009, and report to this conference.
26. J.J. Aubert et al, Phys. Lett. $\underline{89B}$, 267; $\underline{94B}$, 96, 101 (1980); CERN-EP/80-84; R.Mount, report at this conference.
27. U. Camerini et al, Phys. Rev. Lett. $\underline{35}$, 483 (1975); B. Gittelman et al, Phys. Rev. Lett. $\underline{35}$, 1616 (1976).
28. B. Knapp et al, Phys. Rev. Lett. $\underline{34}$, 1040 (1975); T.Nash et al, Phys. Rev. Lett. $\underline{36}$, 1233 (1976).
29. R.P. Feynman, R.D. Field and G.C. Fox, Phys.Rev.$\underline{D18}$,3320(1978).
30. I. Gaines, report at this conference.
31. D. Aston et al, Phys. Lett. $\underline{94B}$, 113 (1980); B. D'Almagne, report at this conference.
32. G. Diambrini-Palazzi, report at this conference.
33. A.J. Buras and K.J.F. Gaemers, Nucl. Phys. $\underline{B132}$, 249 (1978).
34. M. Strovink, private communication.
35. T. Weiler, report to this conference.
36. D.P. Roy, report to this conference.
37. J. Knobloch, report to this conference.
38. T. Gottschalk, ANL-HEP-PR-80-35, report to this conference.

39. M. Teper, Phys. Lett. 78B, 148 (1978); M. Teper and D.W. Duke, Nucl. Phys. B166, 84 (1980).
40. I owe this calculation to D.W. Duke.
41. W. Allison et al, Phys. Lett. 93B, 509 (1980).
42. R. Barloutaud et al, Rutherford RL80/003.
43. L.J. Koester, report to this conference.
44. J.G. McEwen, report to this conference.
45. K.L. Giboni et al, Phys. Lett. 85B, 437 (1979);
    F.Muller, report to this conference.
46. W. Lockman, et al, Phys. Lett. 85B, 443 (1979).
47. D. Drijard et al, Phys. Lett. 81B, 250, 85B, 452 (1979);
    G. Sajot, report to this conference.
48. A. Chilingarov et al, Phys. Lett. 83B, 136 (1979).
49. P. Alibran et al, Phys. Lett. 74B, 134 (1978);
    T. Hansl et al, Phys. Lett. 74B, 139 (1978);
    P.C. Bosetti et al, Phys. Lett. 74B, 143 (1978);
    H. Wachsmuth, Proc. Lepton-Photon Conference, Batavia (1979).
50. A. Diamant-Berger et al, Phys. Rev. Lett. 43, 1774 (1979).
51. A. Kernan, Proc. Lepton-Photon Conference, Batavia (1979).
52. M. Gluck and E. Reya, Phys. Lett. 94B, 84 (1980).
53. H. Fritzsch, Bern report (1980).
54. S. Erhan et al, Phys. Lett. 85B, 447 (1979).
55. S. Pokorski and L. Van Hove, Acta Phys. Polon. B5, 229 (1974);
    K.P. Das and R.C. Hwa, Phys. Lett. 68B, 459 (1977);
    L. Van Hove, CERN TH. 2628 (1979).
56. M. Suzuki, Phys. Lett. 71B, 139 (1977).
    J.D. Bjorken, Phys. Rev. D17, 171 (1978).
57. S.J. Brodsky et al, Nordita-80/18.
58. H. Wahl, Proc. Bergen Neutrino Conf. 1979.

# NEW FLAVOR SPECTROSCOPY

K. Berkelman, Cornell, Speaker

E. Goldwasser, Illinois-Urbana-Champaign, Chairman

R. Loveless, Wisconsin-Madison,

R. Imlay, Louisiana,

D. Scott, Wisconsin-Madison,

Scientific Secretaries

# NEW FLAVOR SPECTROSCOPY

Karl Berkelman
Laboratory of Nuclear Studies, Cornell University, Ithaca, NY 14853

## ABSTRACT

This is a summary talk on the spectroscopy of the particles containing charm and bottom quarks. I review mainly the experimental results from the SPEAR, DORIS, and CESR $e^+e^-$ rings and the relevant theory, as presented in sessions C5, C6, and C7. Time and space limitations do not permit a comprehensive review of the field, nor even a complete coverage of the results obtained in the past year. Some recent reviews are contained in references 1 to 4.

## $c\bar{c}$ AND $b\bar{b}$ BOUND STATE SPECTROSCOPY

### Triplet State Masses and Spinless Potential Models

The $^3S\ 1^{--}$ states of a heavy quark-antiquark system (Fig. 1) can be produced as resonances in the total cross section for $e^+e^- \to$ hadrons (Figs. 2 and 3). Other states, such as $^1S\ 0^{-+}$ and $^3P\ J^{++}$, are reached only by radiative or hadronic transitions from the $^3S$ states. Even before the discovery of the $\psi$[7,8] it was suggested[9] that one could treat heavy quark binding nonrelativistically, in much the same way as positronium. Attempts to understand the $^3S$ masses and the centers of gravity of the $^3P$ and $^3D$ multiplets in the $\psi$ and $\Upsilon$ systems with flavor and spin independent potentials fall into two classes.

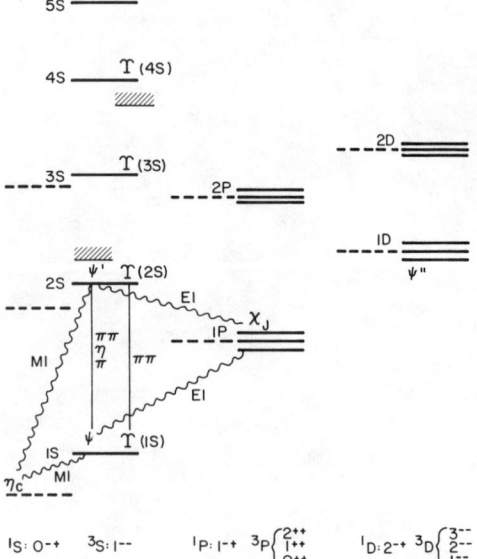

Fig. 1: The energy levels of the heavy quark antiquark system, showing the observed $c\bar{c}$ and $b\bar{b}$ states and transitions.

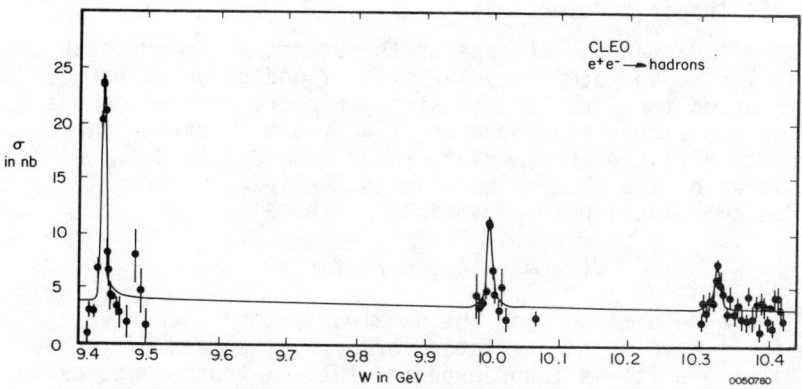

Fig. 2: Total cross section for $e^+e^- \to$ hadrons in the region of the first three upsilon states, as measured by CLEO (ref. 5).

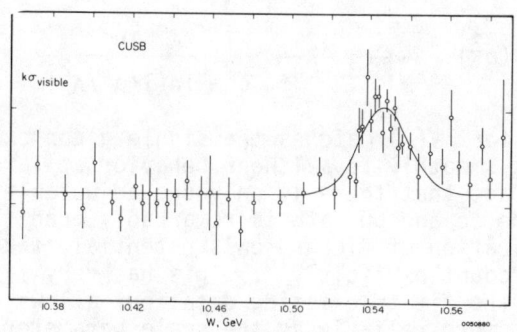

Fig. 3: Total cross section for $e^+e^- \to$ hadrons in the vicinity of the $\Upsilon(4S)$ resonance, as measured by CUSB (ref. 6).

a) Empirical potentials

Here one tries to get information about the potential from the data with a minimum of theoretical prejudice. For instance, the ordering of the levels implies constraints on the shape of the potential.[10] The similarity of the spacings in the $c\bar{c}$ and $b\bar{b}$ levels tells us that $V(r)$ is approximately logarithmic where the wave function is large.[11,12] A completely ad hoc power law potential $V(r) = A + Br^\nu$ with parameters fit to several lower levels is sufficient to make successful predictions[12,13] of level spacings (Table I and Fig. 4). A rather elegant procedure is the use of the inverse scattering formalism[14] to derive approximate potentials from the measured spectra, thereby giving an unbiased demonstration of the flavor independence of the heavy quark potential.

b) QCD-Inspired Potentials

At small distances one expects the potential to approach the Coulombic 1/r dependence characteristic of one gluon exchange. At large distances the flux tube or string picture of confinement leads us to expect a linear r dependence. The $\psi$ and $\Upsilon$ levels however are most sensitive to the intermediate range from 0.1 to 3 fm (Fig. 4), where neither of the theoretical limits applies. Nevertheless, the simple Coulomb-plus-linear potential[15] (Fig. 4)

$$V(r) = - (4/3)\alpha_s/r + r/a^2$$

with $\alpha_s$ and a determined from the data was rather successful in fitting the $\psi$ levels and predicted correctly the mass of the $^3D$ $\Upsilon''$ (3770) before it was found experimentally. Another successful variation was a piecewise combination of Coulomb for r<0.12 fm, linear for r>0.87 fm, and logarithmic in between[16], later modified by making the Coulomb coefficient $\alpha_s$ a function of the quark mass[17] (Fig. 4).

Richardson[18] has made a more serious attempt to incorporate first order QCD in the potential. In momentum space he writes

$$V(q^2) = - \frac{4}{3} \frac{12\pi}{33-2n_f} \frac{1}{q^2} \frac{1}{\ln(1+q^2/\Lambda^2)} .$$

The Fourier transform V(r) matches the single gluon exchange form at small r and joins smoothly to a linear behavior at large r (Fig. 4). In spite of the fact that there is only one adjustable parameter $\Lambda$, the fit to all the $c\bar{c}$ and $b\bar{b}$ data is remarkably good (Table I). We now have a modification of Richardson's potential, taking second order QCD into account explicitly[19]. This has very little effect on the shape and on the fit to existing data, but allows one to interpret the constant more reliably as the scale parameter of QCD. The fitting of the one adjustable parameter in the potential gives $\Lambda_{\overline{ms}} = 0.508$ GeV, in good agreement with other determinations. The slope of V(r) at large distances yields the accepted value for the Regge trajectory slope, $\alpha'=1.05$ GeV$^{-2}$. People have also used the MIT bag model, with several adjustable parameters, to give good fits to the $\psi$ and $\Upsilon$ spectra[20].

The potential models can predict mass differences in each of the two heavy quark systems, using one or several mass differences as input to fix the parameters of the potential. The actual mass of the lowest state is set by picking the value of the quark mass used in the Schrödinger equation. For the bound states below the continuum threshold the agreement with experiment is good (Table I), even too good, remembering that there must be corrections for relativistic, higher order QCD, and spin dependent effects. In the continuum ($\psi''$, $\Upsilon(4S)$, and higher states) the bound state model does not apply; it tends to overestimate the masses.

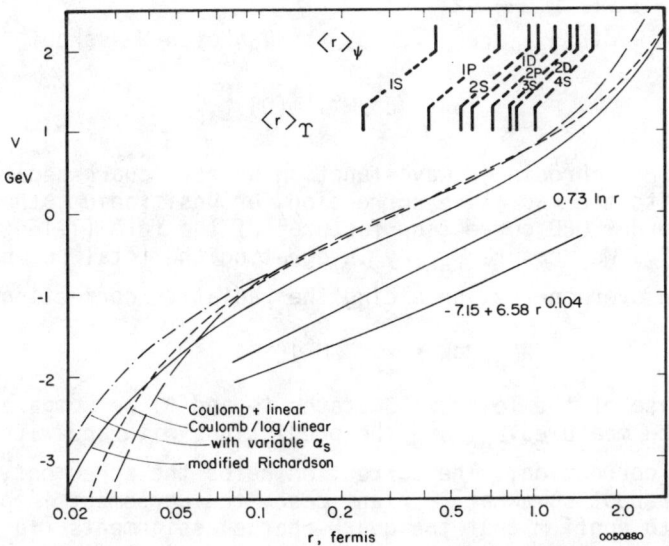

Fig. 4: Potentials of refs. 11, 13, 15, 16, 17, and 19. Mean radii for several states in the $\psi$ and $\Upsilon$ systems.

Table I Measured masses and mass differences in MeV, compared with predications of two representative potential models

|     |             | Expt. | Ref. | Martin[13] | Mod. Richardson[19] |
|-----|-------------|-------|------|------------|---------------------|
| M:  | $\psi$ (or J) | 3096.93±0.09 | 21 | – | – |
| $\Delta$M: | $\psi'$ | 589.07±0.10 | 21 | 589 | 589 |
|     | $\psi''$ | 671±2 | Table VIII | 705[a] | 715[b] |
|     | $\chi$ (c.of g.) | 426±4 | Table IV | 425 | 425 |
|     | $\eta_c$ | -118±8 | Table V | – | – |
| M:  | $\Upsilon$ (1S) | 9462±10 | 22 | – | – |
|     |             | 9433±28 | 5,6 |  |  |
| $\Delta$M: | $\Upsilon$ (2S) | 553±10 | 22 | (560) | (560) |
|     |             | 560±3 | 5,6 |  |  |
|     | $\Upsilon$ (3S) | 889±4 | 5,6 | (890) | 890 |
|     | $\Upsilon$ (4S) | 1114±5 | 5,6 | 1120 | 1160 |

[a] For $^3D_{1,2,3}$ center of gravity; observed $\psi''$ is $^3D_1$

## Leptonic Widths of $^3S$ States

Leptonic widths $\Gamma_{ee}$ are given by the VanRoyen-Weisskopf[23] formula,

$$\Gamma_{ee} = 16\pi(\alpha^2 q^2/M^2)|\psi(0)|^2,$$

in terms of the Schrödinger wave function at zero quark separation. In analogy with the radiative correction for positronium, there may be a higher order QCD correction factor[24] of the form $(1-16\alpha_s/3\pi)$. Experimentally, we measure $\Gamma_{ee}$ by integrating the total resonance cross section over energy, unfolding the radiative correction:

$$\int \sigma_{res}\, dW = (6\pi^2/M^2)\Gamma_{ee}.$$

In the case of the lowest $1^3S$ states ($\psi$ and $\Upsilon$) we compare (Table II) the measured $\Gamma_{ee}$ and the predictions with and without the higher order correction. The correction helps the agreement, but seems to overshoot somewhat. In any case, the agreement is probably good enough to confirm that the quark charge assignments ($|q_c| = 2/3$, $|q_b| = 1/3$) are correct.

The uncertainty in the higher order corrections should not affect the predictions for leptonic widths of higher $n^3S$ states provided they are normalized to $\Gamma_{ee}$ for the $1^3S$. These ratio values are much closer to the experimental results than are the straight $\Gamma_{ee}$ values. The QCD-inspired potential[19] does extremely well, while the power law potential[13] is clearly inferior. This is presumably because the leptonic widths are more sensitive than the mass levels to the small distance part of the potential, which is poorly given by the power law.

Table II  Measured and predicted leptonic widths (in keV) and ratios

|  |  | Expt. | Ref. | Martin[13] | Mod. Richardson[19] |
|---|---|---|---|---|---|
| $\Gamma_{ee}$: | $\psi$ | 4.8 ±0.6 | 25 | - | 8.0, 3.5[a] |
| $\Gamma_{ee}/\Gamma_{ee}(\psi)$: | $\psi'$ | 0.44±0.06 | 25 | 0.35 | 0.45 |
| $\Gamma_{ee}$: | $\Upsilon$ (1S) | 1.29±0.09±0.13<br>1.02±0.07±0.15 | 22<br>5,6 | - | 1.7, 1.1[a] |
| $\Gamma_{ee}/\Gamma_{ee}(\Upsilon)$: | $\Upsilon$ (2S) | 0.45±0.06±0.02<br>0.45±0.03±0.04 | 22<br>5,6 | 0.43 | 0.45 |
|  | $\Upsilon$ (3S) | 0.32±0.03±0.03 | 5,6 | 0.28 | 0.32 |
|  | $\Upsilon$ (4S) | 0.24±0.02±0.03 | 5,6 | 0.20 | 0.26 |

[a]Using higher order QCD correction

Table III Present status of unconfirmed levels

| M, MeV | Original Measurement | Ref. | Crystal Ball 90% c.l. limit | Ref. |
|---|---|---|---|---|
| 2820 | $B(\psi\to\gamma\chi)\cdot B(\chi\to\gamma\gamma)$ $(1.4\pm0.4)10^{-4}$ | 26 | $<0.3\times10^{-4}$ | 29 |
| 3455 | $B(\psi'\to\gamma\chi)\cdot B(\chi\to\gamma\psi)$ $0.8\pm0.4\%$ | 27 | $<0.04\%$ | 30 |
| 3591 | $B(\psi'\to\gamma\chi)\cdot B(\chi\to\gamma\psi)$ $0.18\pm0.06\%$ | 28 | $<0.04\%$ | 30 |

Table IV Crystal ball data on $\chi$ states

| State | $J^{PC}$ | Mass, MeV[a] | $B(\psi'\to\gamma\chi)$[b] | $B(\psi'\to\gamma\chi)\cdot B(\chi\to\gamma\psi)$[a] |
|---|---|---|---|---|
| $\chi_2$ | $2^{++}$ | $3553.9\pm0.5\pm4$ | $7.0\pm2.3\%$ | $1.26\pm0.09\pm0.20\%$ |
| $\chi_1$ | $1^{++}$ | $3508.4\pm0.4\pm4$ | $7.1\pm2.0\%$ | $2.38\pm0.12\pm0.38\%$ |
| $\chi_0$ | $0^{++}$ | $3413\pm5$ | $7.2\pm2.0\%$ | $0.059\pm0.015\pm0.009\%$ |

[a] ref. 30
[b] ref. 29

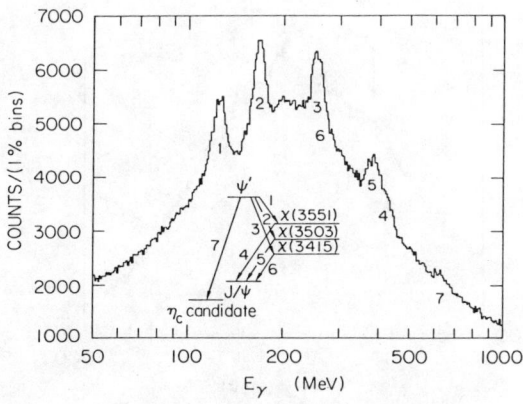

Fig. 5: Inclusive photon spectrum from the $\psi'$ measured by the Crystal Ball (ref. 31).

States Reached by Radiative Transitions:

The Crystal Ball detector at SPEAR is optimized for high resolution photon detection. Other experiments with limited statistics have seen indications of radiative transitions to states at 2820[26], 3455[27], 3591[28] MeV. With much greater sensitivity the Crystal Ball sees no such transitions[29,30] (Table III). Theorists who were discomforted by the possible existence of these levels can now be assured that they are most likely not there.

Fig. 5 shows the inclusive single photon spectrum observed in the Crystal Ball running at the $\psi'$ resonance[31]. The three most prominent peaks, marked 1,2,3, correspond to the transitions to the $^3P$ states $\chi(3413)$, $\chi(3508)$, and $\chi(3554)$. Three more Doppler broadened peaks, labeled 4,5,6, are the transitions $\chi \to \gamma \psi$. Table IV shows the latest results on the masses and radiative widths[29,30]. The Crystal Ball collaboration has done an extensive analysis[30] of the angular distributions of the two photons from $\psi' \to \gamma \chi$, $\chi \to \gamma \psi$ for the $\chi(3508)$ and $\chi(3554)$. They find that the spin assignments, $1^{++}$ and $2^{++}$ previously deduced from indirect arguments[32] are confirmed; other hypotheses are excluded at the 99% confidence level or better. As expected, the (four) transitions are electric dipole with less than 3% quadrupole intensity admixture.

Fig. 5 also shows a less prominent photon peak labeled 7, interpreted as the radiative decay of the $\psi'$ to an $\eta_c$ candidate. One also sees a peak at the corresponding energy, about 120 MeV, in the spectrum of single photons[31] from the $\psi$ (Fig. 6). A combined fit to both spectra, shown in more detail in Fig. 7, yields M=2981±15 MeV and $\Gamma_{tot} = 20^{+16}_{-11}$ MeV for the $\eta_c$, and branching ratios 0.43±0.08±0.18% and about 1% for $\psi' \to \gamma \eta$ and $\psi \to \gamma \eta_c$, respectively.[31]

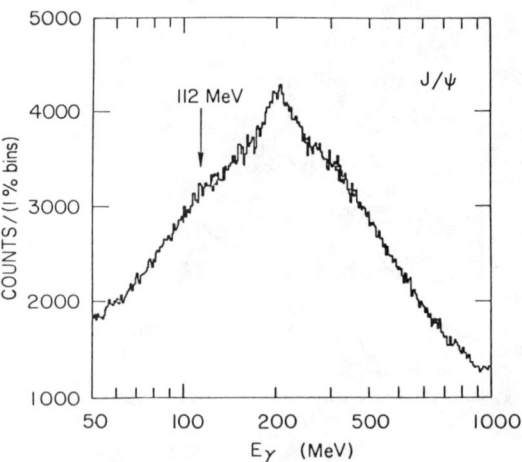

Fig. 6: Inclusive photon spectrum from the $\psi$ measured by the Crystal Ball (ref. 31).

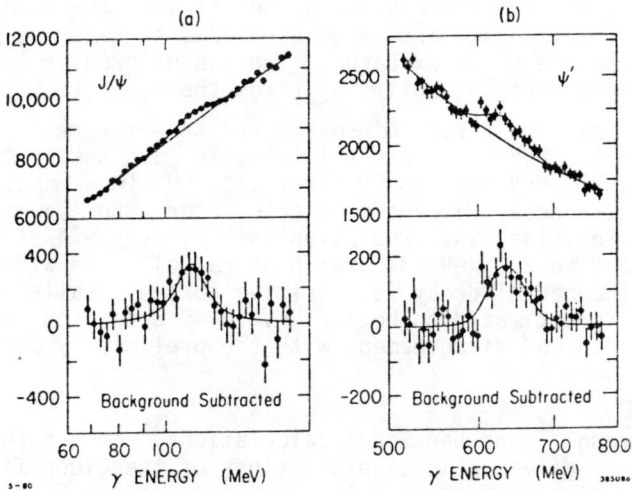

Fig. 7: Details of the inclusive photon spectra from the $\psi$ and $\psi'$ measured by the Crystal Ball (ref. 31).

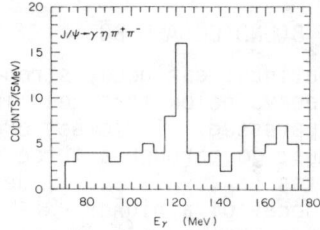

Fig. 8: Mass distribution of $\eta\pi^+\pi^-$ from $\psi \to \gamma\eta\pi^+\pi^-$ in the Crystal Ball (ref. 31).

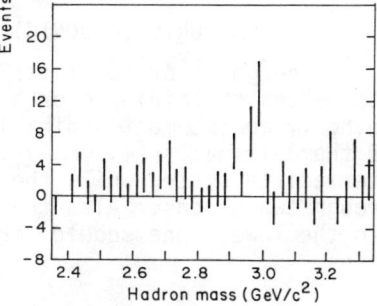

Fig. 9: Mass distribution of charged final states X (see text) in $\psi' \to \gamma X$ from Mark 2 (ref. 33).

We now have confirmation by observation of exclusive final states: $\eta_c \to \eta\pi^+\pi^-$ from the Crystal Ball[31] (Fig. 8), and $\eta_c \to K^\pm K^0_s \pi^\mp$ and a number of other charged modes from the Mark 2 group[33] (Fig. 9). These results, although involving a rather small number of events, actually provide more accurate mass and width measurements. The existing data on the $\eta_c$ are summarized in Table V. Although the spin has yet to be measured, we can probably assume that this is the pseudoscalar $1^1S_0$ ground state of the $c\bar{c}$ system.

First order QCD makes a prediction[4,32] for the spin dependence of the heavy quark-antiquark binding, but it is naive to expect the potential at intermediate and large distances to have the spin structure of single gluon exchange. One cannot predict either the hyperfine singlet-triplet splitting (from the $\vec{\sigma}_1 \cdot \vec{\sigma}_2$ term in V) or the $^3P$ fine structure (from spin-orbit and tensor terms). Nevertheless, the new 118 MeV $\psi$-$\eta_c$ mass splitting is much easier to accommodate than the 280 MeV implied by older data.[26] Once masses are known, radiative decay rates can be calculated. For $\psi \to \gamma \eta_c$, a favored M1 transition, the prediction is[15] $\Gamma(\psi \to \gamma \eta_c) = (4/3)(\alpha^2 q^2/M_q^2) k^3$, which works out to 2.0 keV (3% branching ratio). The width for the $\psi' \to \gamma \eta_c$ hindered M1 decay is sensitive to the details of the wavefunction and is estimated to be 1.0 keV (0.4% branching ratio). These are not in bad disagreement with the preliminary data.

## Other States

String model[34] and bag model calculations[35] have both suggested the possibility of vibrational excitations of the gluon field in heavy quark-antiquark systems. The lowest lying one in the $c\bar{c}$ system is expected to be in the continuum where it may be difficult to detect, but in the $b\bar{b}$ system the first such level may show up as a $^3S$ bound state between the normal $\Upsilon(3S)$ and $\Upsilon(4S)$. Although a low $\Gamma_{ee}$ may make a new peak hard to find, its discovery would be a beautiful demonstration of the reality of the gluon field and a nice confirmation of QCD.

## HADRONIC (GLUONIC) DECAYS OF BOUND $C\bar{C}$ AND $B\bar{B}$

Although $c\bar{c}$ or $b\bar{b}$ states in the continuum can decay strongly into mesons containing c or b quarks, decays below threshold in which the heavy quarks must annihilate are suppressed. In lowest order of QCD the J=1 states ($\psi$, $\psi'$, $\chi_1$, $\Upsilon$,...) must go through a three gluon intermediate state, while the J=0,2 states ($\eta_c$, $\chi_0$, $\chi_2$,...) decay through two gluons. Also the hadronic decay of a higher $^3S_1^{--}$ state into the lowest one should take place through the emission of two gluons.

## $\psi$ and $\Upsilon$ Decays into Three Gluons

For the decay $(^3S) \to ggg$ QCD predicts[36]

$$\Gamma_{ggg} = (160/81)(\pi^2-9)(\alpha_s^3/M^2)|\psi(0)|^2.$$

Reexpressing the wave function at the origin in terms of $\Gamma_{ee}$ gives

$$\Gamma_{ggg} = (10/81\pi)(\pi^2-9)(\alpha_s^3/\alpha^2 q^2)\Gamma_{ee}.$$

Experimentally, we first determine $\Gamma_{ee}$ from $\int \sigma_{res} dW$, then measure a leptonic branching ratio, say $B_{\mu\mu}$ for $^3S \to \mu^+\mu^-$. This then gives us the total width, $\Gamma_{tot} = \Gamma_{ee}/B_{\mu\mu}$. For the lowest $^3S$ state ($\psi$ or $\Upsilon$)

$$\Gamma_{ggg} = \Gamma_{tot} = (3+R)\Gamma_{ee},$$

where the subtraction accounts for all final states coming from a virtual photon producing a lepton pair or a quark pair.

We now have new data from DORIS on $B_{\mu\mu}$ for the $T(1S)$. The average value[22] from the PLUTO[37], DASP[38], and LENA[39] detectors is $B_{\mu\mu} = 3.0\pm0.8\%$, implying $\Gamma_{tot} = 43^{+20}_{-11}$ keV, to be compared with $67\pm12$ keV in the case of the $\psi$. Computing $\Gamma_{ggg}$ and solving for $\alpha_s$ yields $\alpha_s(T) = 0.17\pm0.02$, rather similar to $\alpha_s(\psi) = 0.19\pm0.02$ obtained by the same method. Since higher order QCD corrections have not been included, the actual values of $\alpha_s$ we have obtained are not significant, but the fact that they are reasonable and that the small variation from $\psi$ to $T$ is what one expects of the QCD running constant gives support to the three gluon decay hypothesis.

The corresponding prediction for the two gluon decay of a $^1S$ state is[36] $\Gamma_{gg} = (32\pi/3)(\alpha_s^2/M^2)|\psi(0)|^2$. If we substitute the $\alpha_s$ and $\psi(0)$ determined from the $\psi$ decay, we get $\Gamma_{gg} = 5$ MeV for the $\eta_c$, compatible with the experimental limit (Table V).

If the $\psi$ and $T$ decay predominantly through three gluons, we might see a tendency for the final state hadrons to concentrate in three jets, provided the typical angular separation between jets is larger than the angular spreading in the fragmentation process. In the case of the $\psi$ the energy is too low and the jets do not collimate enough to be recognizable. Even for the $T$ it is difficult to distinguish the decay hadron distributions from isotropic phase space.

There is not much to add to H. Meyer's review of the data from DORIS on $T$ decay distributions.[3] The PLUTO group has recalculated the distributions in thrust, triplicity, and other shape variables for their 1978 sample of $T$ decays with continuum and single-photon-mediated decays subtracted.[40] This time they give the distributions for all decay particles instead of only charged secondaries. The conclusions are the same, though. The experimental distributions (Figs. 10 and 11) contrast markedly with those in the two-jet continuum, but are less easily distinguishable from distributions from an isotropic phase space Monte Carlo. A three-gluon Monte Carlo, in which gluons fragment like quarks, matches the data rather well.

The CLEO group at CESR has obtained similar distributions with similar conclusions.[5] In the search for a shape variable which would more clearly distinguish the three-jet topology from phase space, they have come up with one they call triple sphericity. Three axes are found which minimize the sum of squares of track momentum components transverse to the nearest axis. The triple sphericity is this sum divided by the sum of squares of the momenta. The observed distribution for $T$ decays (Fig. 12) shows a fairly good contrast relative to a phase space Monte Carlo distribution.

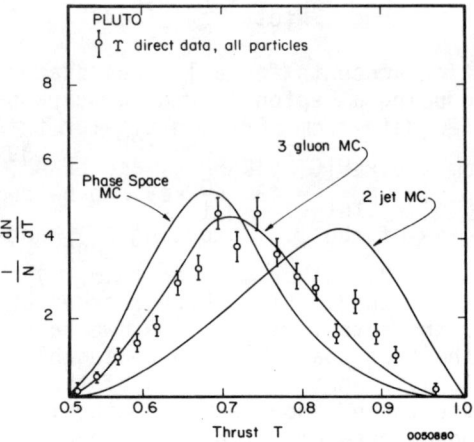

Fig. 10: Thrust distribution of $\Upsilon$ decays with continuum and photon-mediated decays subtracted, from PLUTO (ref. 40).

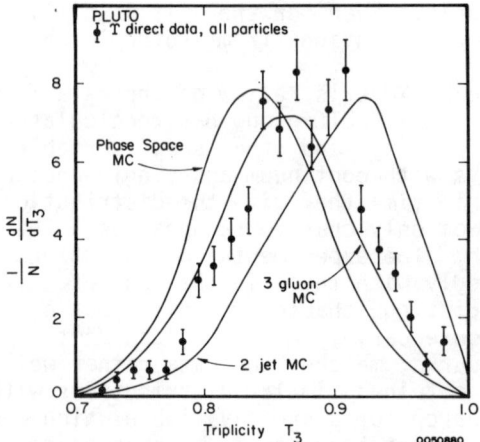

Fig. 11: Triplicity distribution of $\Upsilon$ decays with continuum and photon-mediated decays subtracted, from PLUTO (ref. 40).

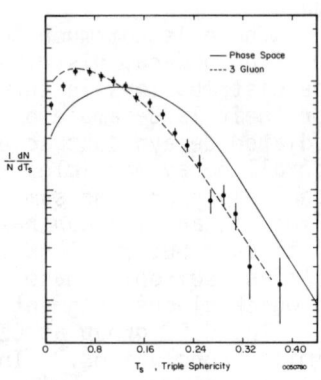

Fig. 12: Triple-sphericity distribution of $\Upsilon$ decays with continuum subtracted, from CLEO (ref. 5).

Table V Crystal Ball and Mark 2 Data on the $\eta_c$

| Property | Measured Value | Ref. |
|---|---|---|
| M | 2979±8 MeV | 31,33 (av.) |
| $\Gamma_{tot}$ | <20 MeV (90% c.l.) | 31 |
| Branching ratios: | | |
| $\psi \to \gamma \eta_c$ | ~1% | 31 |
| $\psi' \to \gamma \eta_c$ | 0.43±0.08±0.18% | 31 |
| $\eta_c \to K\bar{K}\pi$ | $10^{+7}_{-6}$% | 33[a] |
| $\eta_c \to \eta \pi^+ \pi^-$ | ~3% | 31[b] |
| $\eta_c \to 2\pi^+ 2\pi^-$ | $1.3^{+1.5}_{-1.0}$% | 33[a] |
| $\eta_c \to \pi^+ \pi^- K^+ K^-$ | $0.9^{+1.5}_{-0.7}$% | 33[a] |
| $\eta_c \to p\bar{p}$ | $0.2^{+0.2}_{-0.1}$% | 33[a] |
| $\eta_c \to \pi^+ \pi^- p\bar{p}$ | <1.1% (90% c.l.) | 33[a] |

[a] Assumes B = 0.43±0.20% for $\psi' \to \gamma \eta_c$ (ref. 31).
[b] Assumes B~1% for $\psi \to \gamma \eta_c$ (ref. 31)

Once we are convinced of the three jet nature of the $\Upsilon$ decay we can use the angular distributions of the jets to test the theory. For instance, the polar angle distribution of the thrust axis relative to the beam line (Fig. 13) distinguishes between scalar and vector gluons. The data[40] favor the distribution $1+0.39 \cos^2\theta$ for spin 1 gluons over the $1-\cos^2\theta$ for spin 0.

## Hadronic Cascades from Higher $\psi$ and $\Upsilon$ States

The observed hadronic transitions from the $\psi'$ to the $\psi$ are listed in Table VI. There are new data from the Crystal Ball[30] and Mark 2 [42] at SPEAR on the rare modes $\psi' \to \eta\psi$ and $\pi^0\psi$. The $\pi^0$ reaction is especially interesting as an example of isospin violation in strong interactions.[43]

We now have the first direct evidence of the fact that the $\Upsilon'$ and the $\Upsilon$ are closely related. The LENA group[44] at DORIS and the CLEO[5] and CUSB[6] groups at CESR have each seen the decay $\Upsilon' \to \Upsilon \pi^+ \pi^-$, identifying the $\Upsilon$ by either its $e^+e^-$ or $\mu^+\mu^-$ decay. LENA reports 7 candidates (5 $e^+e^-$ and 2 $\mu^+\mu^-$) with a background of about 1 event as determined by running on the $T$. CLEO reports two reconstructed events (one of each) and CUSB has 3 (all $e^+e^-$). Using the DORIS value[22] for the branching ratio for $T \to \mu^+\mu^-$, CLEO calculates a branching ratio for $T' \to T\pi^+\pi^-$ of 22±16±11% from the two events; CUSB quotes a branching ratio at the level of 10 to 20%. Figs. 14 and 15 show two of these events.

CLEO[5] has made a more accurate determination by comparing appropriately normalized inclusive charged particle momentum spectra of $T$ and $T'$ decays (Fig. 16). The excess at small momenta in the $T'$ spectrum comes presumably from the cascade pions. A one-parameter fit yields a branching ratio of 21.6±3.5%. LENA[44] does a similar analysis comparing the charged multiplicity distributions for $T'$ and $T$, giving a branching ratio of 27±9%.

Suppose we try to predict this branching ratio from theory and other measurements. First assume that for the $T'$ $\Gamma_{tot} = \Gamma_{ggg} + \Gamma_{T\pi\pi} + \Gamma_{em} + \Gamma_{other}$. The virtual-photon-mediated decay width is given in terms of the measured leptonic width by $\Gamma_{em} = (3+R) \Gamma_{ee} = 6\pm1$ keV (Table II). The three gluon width relative to that of the $T$ (which is known[22]) is in the ratio of $|\psi(0)|^2$ (or $\Gamma_{ee}$) values. This leads to $\Gamma_{ggg} = 16^{+9}_{-5}$ keV. The QCD prediction for $\Gamma_{T\pi\pi}$ based on a multipole expansion of the gluon field[45] is given in terms of corresponding $T'$ width and the average radii:

$$\frac{\Gamma(T' \to T\pi\pi)}{\Gamma(\psi' \to \psi\pi\pi)} = \left(\frac{\langle r_T^2 \rangle}{\langle r_\psi^2 \rangle}\right)^2 = \frac{1}{10}.$$

Table VI Cascade Decays $T' \to \psi X$

| Mode | Branching Ratio | Ref. |
|---|---|---|
| $\psi' \to \pi^+\pi^-\psi$ | 33±3% | 26, 41 (av.) |
| $\psi' \to \pi^0\pi^0\psi$ | 17±2% | 26, 41 (av.) |
| $\psi' \eta\psi$ | 3.9±0.3% | 26, 28, 41 (av.) |
| | 2.18±0.14% | 30 |
| | 2.5±0.6% | 42 |
| $\psi' \to \pi^0\psi$ | 0.09±0.02% | 30 |
| | 0.15±0.06% | 42 |

Thus we expect $\Gamma_{T\pi\pi} = 11\pm2$ keV. Assuming that $\Gamma_{other}$ ($\gamma\eta_b$, $\gamma\chi_b$, etc.) is small, we then have $\Gamma_{tot} = 33^{+9}_{-6}$ keV, and (2/3) $\Gamma_{T\pi\pi}/\Gamma_{tot} = 22\pm6\%$ for the branching ratio $T' \to \pi^+\pi^-T$. The factor 2/3 accounts for the

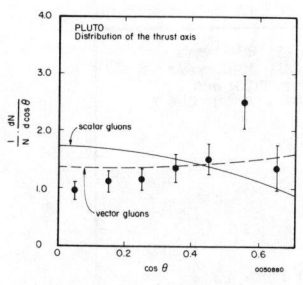

Fig. 13: Polar angle distribution of thrust axis relative to beam direction for T decays with continuum and photon-mediated decays subtracted, from PLUTO (ref.40)

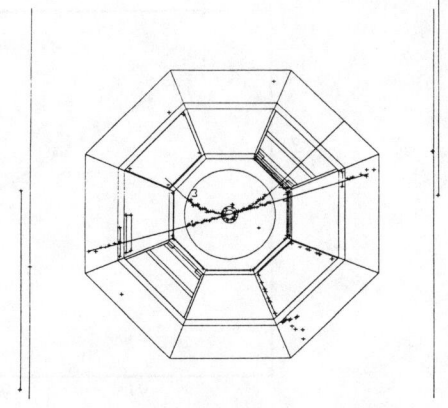

Fig. 14: An end-view of the CLEO detector (ref. 5) showing an event $T' \to T\pi^+\pi^-$, $T \to \mu^+\mu^-$

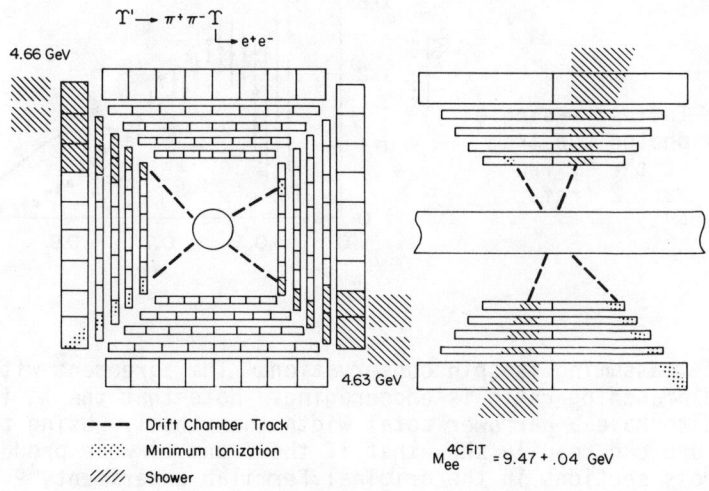

Fig. 15: A display of an event $T' \to T\pi^+\pi^-$, $T \to e^+e^-$ in the CUSB detector (ref. 6).

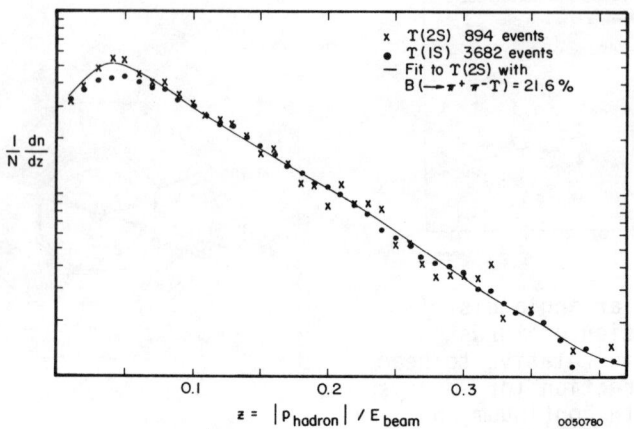

Fig. 16: Inclusive momentum spectra of charged particles from the T' and the T with continuum subtracted, from CLEO (ref. 5).

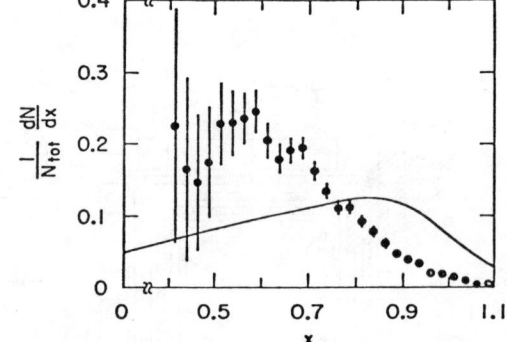

Fig. 17: Inclusive single photon spectrum at the T, from Mark 2 (ref. 48).

$\pi^\circ\pi^\circ$ mode, assuming isospin conservation. The agreement with the measured branching ratio is encouraging. Note that the T' is expected to have a narrower total width than the T. Using these numbers one can readily show that if the T and T' were produced with equal cross sections in the original Fermilab experiment,[46] the ratio of rates of muon pairs with invariant masses at the T' and at the T would be about 0.4, which is not too far from the measured ratio 0.31±0.03.

Decays into Gluons plus a Photon
If in the three gluon decays of a $^3S$ state such as the $\psi$ we replace one gluon by a photon, lowest order QCD predicts[47]

$$\Gamma_{\gamma gg}/\Gamma_{ggg} = (36/5)(\alpha q^2/\alpha_s) \sim 0.13.$$

Since $\Gamma_{ggg} = 0.64\ \Gamma_{tot}$, we then expect the branching ratio for $\psi \to \gamma$+hadrons to be about 8%.

At the upper end of the $\psi$ single photon energy spectrum observed in the Crystal Ball[31] (Fig. 6) we note several exclusive final state peaks ($\eta$, $\eta'$, and E(1420)) and a large continuum. Fig. 17 shows the Mark 2 data[48] with poorer energy resolution, but with a careful $\pi^\circ$ background subtraction and rate normalization. The lower energy range is not accessible because of the increasing $\pi^\circ$ background. However, one can integrate the measured photon spectrum from 60 to 100% of the end point to get a partial branching ratio of 4.1±0.8%, which is compatible with the QCD gluon-gluon-photon prediction. Qualitatively similar data have been obtained by the Lead Glass Wall collaboration.[49] The curve in Fig. 17 is a guess at the spectrum shape assuming massless gluons. The finite mass of the final state hadrons surely suppresses the high energy photon yield, so perhaps the failure of the naive prediction is not surprising.

Table VII summarizes the recent Crystal Ball[50], Mark 2,[51] and other data on the radiative decays of the $\psi$ to the $\pi^\circ$, $\eta$, $\eta'$, f(1270), and E(1420). It is surprising that the E is so prominent in $\psi$ decay as compared to hadronic experiments,[55] where E production is small and usually accompanied by D(1285) production. To explain this the Mark 2 experimenters[51] suggest that the E may be a bound gluonium state--a "glueball". We need to know more about the spin and decays of the E before we can decide whether it is really a glueball, or just an ordinary $q\bar{q}$ member of a 1+ nonet (along with the D(1285), $A_1$, and $Q_A$), or perhaps a radial excitation of the and $\eta'$.[56]

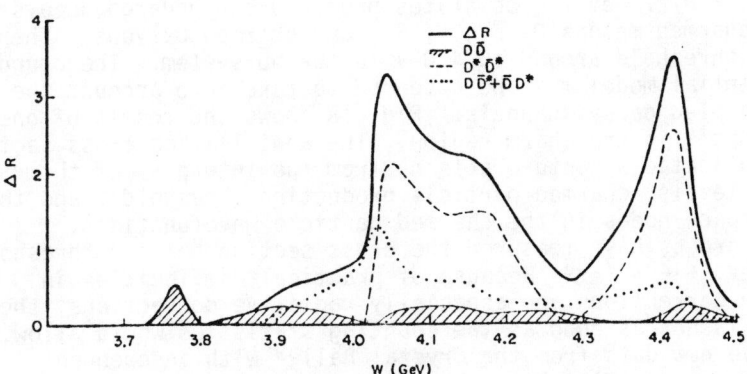

Fig. 18: Predicted total cross section for $e^+e^-$ production of charmed mesons (ref. 15).

Table VII  Radiative decays of the $\psi$: new data from Crystal Ball and Mark 2 compared with best previous measurements.

| Decay | Mode seen | Branching ratio | Ref. | Theory |
|---|---|---|---|---|
| $\psi \to \gamma \pi^0$ | $\gamma\gamma$ | $(7\pm5)\times10^{-5}$ | 26 | $2\times10^{-5}$ a |
| $\psi \to \gamma \eta$ | $\gamma\gamma$ | $(8\pm2)\times10^{-4}$ | 26 | $9\times10^{-4}$ b |
| | $\gamma\gamma$ | $(12\pm2)\times10^{-4}$ | 50 | |
| | $\gamma\gamma$ | $(9\pm4)\times10^{-4}$ | 51 | |
| $\psi \to \gamma \eta'$ | $\rho^0\gamma$ | $(24\pm7)\times10^{-4}$ | 52 | $33\times10^{-4}$ b |
| | $\gamma\gamma$ | $(69\pm17)\times10^{-4}$ | 50 | |
| | $\rho^0\gamma$ | $(34\pm7)\times10^{-4}$ | 51 | |
| $\psi \to \gamma f(1270)$ | $\pi^+\pi^-$ | $(20\pm3)\times10^{-4}$ | 53 | |
| | $\pi^+\pi^-$ | $(13\pm3)\times10^{-4}$ | 51 | |
| $\psi \to \gamma E(1420)$ | $\eta\pi\pi$ | $(23\pm15)\times10^{-4}$ c | 50 | |
| | $K\bar{K}\pi^0$ | $(34\pm24)\times10^{-4}$ c | 50 | |
| | $K\bar{K}\pi$ | $(36\pm14)\times10^{-4}$ c | 51 | |

a Vector meson dominance
b Ref. 54
c Product of branching ratios $B(\psi \to \gamma E)B(E \to \text{mode seen})$

## CONTINUUM $c\bar{c}$ AND $b\bar{b}$ PRODUCTION. D, F, AND B MESONS

### The $c\bar{c}$ Continuum and D* Decays

When the total energy in $e^+e^-$ annihilation exceeds the charm threshold at 3727 MeV the $c\bar{c}$ states produced can undergo decays into pairs of charmed mesons D, D*, F, F*, and charmed baryons. There is a similar threshold around 10.4 GeV in the $b\bar{b}$ system. The bound state potential model must be extended to take into account the several coupled decay channels. Fig. 18 shows the result of one such attempt[15] in the charm region. The annihilation cross section has a complicated structure arising from the interplay of the $^3S$ $c\bar{c}$ resonance levels, charmed particle production thresholds, and the momentum space nodes in the charmed particle wavefunctions.

Many groups have measured the cross section between threshold and 4.5 GeV.[2,26,57,58] Because of practical difficulties in acceptance corrections and especially radiative corrections, the agreement is not as good as the counting statistics would allow. We now have new data from the Crystal Ball[59] with independent systematic errors (Fig. 19). The radiative corrections, which enhance fluctuations, have not been applied to the data. Nevertheless, it is interesting to note how closely the Crystal Ball results resemble the theoretical cross section (Fig. 18). The cross section measurements on the $^3D$ $1^{--}$ $\psi''(3770)$ resonance just above threshold provide new values for its mass, $\Gamma_{ee}$, and total width (Table VIII).

In order to study the production and decay of the D* the Crystal Ball group made an extensive run at the W=4.028 GeV peak in the cross section, which according to the theory[15] (Fig. 18) should be prominantly $D\bar{D}^*$ or $D^*\bar{D}$, and $D^*\bar{D}^*$ (just above threshold). The inclusive photon spectrum[59] (Fig. 20) shows a broad peak centered at 70 MeV from $D^* \to D\pi^\circ$, $\pi^\circ \to \gamma\gamma$, and a narrower peak at about 140 MeV from $D^* \to D\gamma$. Each peak has two components, one from D* produced at rest in the process $e^+e^- \to D^*\bar{D}^*$ and a Doppler broadened one from D* produced with D. Events with multiple detected photons were analyzed to give the inclusive spectrum of $\pi^\circ$ (Fig. 21), which shows a sharp peak at about 140 MeV from $e^+e^- \to D^*\bar{D}^*$, $D^* \to D\pi^\circ$ and a broader distribution from about 135 to 160 MeV from $e^+e^- \to D^*\bar{D}$ or $D\bar{D}^*$, $D^* \to D\pi^\circ$. The two distributions (Figs. 20 and 21) are fitted to give the D and D* masses and combinations of cross sections and branching ratios (Table IX and Ref. 59). The results are generally compatible with previous measurements in the Mark 1 detector[60], but more accurate in some cases.

Evidence for the F Meson
----

The original DASP discovery of the F and F*[26] was based on two pieces of evidence.

a) The observation of six reconstructed events fitting the hypothesis $e^+e^- \to FF^* \to \gamma F^{\pm} F^{\mp}$, $F^{\pm} \to \eta\pi^{\pm}$, $\eta \to \gamma\gamma$. The measured masses

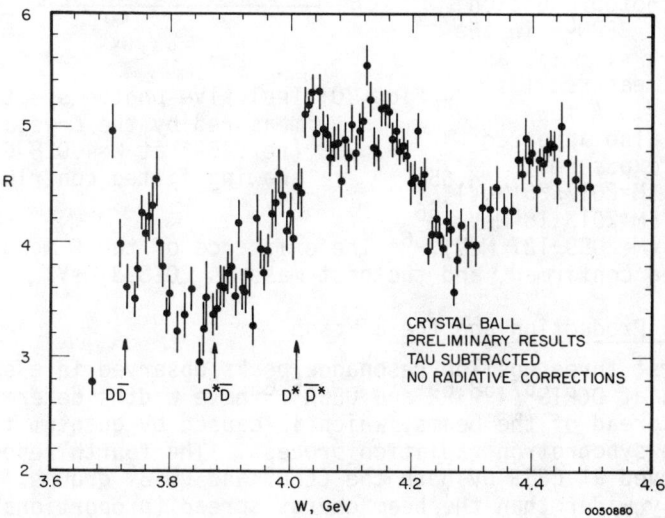

Fig. 19: Ratio of hadronic and mu pair cross sections for $e^+e^-$ annihilation, measured by the Crystal Ball (ref. 59).

Table VIII Properties of the $\psi''(3770)$

| Property | LGW | DELCO | Mark 2 | Cr.Ball | Av. |
|---|---|---|---|---|---|
| M, MeV | 3772±3±4 | 3770±2±4 | 3764±2±2 | 3768±2±4 | 3768±5 |
| $\gamma_{ee}$, eV | 345±85 | 180±60 | 276±50 | 308±58 | 270±60 |
| $\Gamma$, MeV | 28±5 | 24±5 | 24±5 | 36±8 | 26±5 |
| Ref. | 60 | 61 | 62 | 59 | |

were M = 2030±60 MeV, M* = 2140±60 MeV, and M*-M=110±46 MeV.

b) An apparent threshold in the inclusive process $e^+e^- \to \eta$ +2 charged tracks + anything, at about 4.1 GeV.

The Crystal Ball with better statistics and resolution has now measured the W-dependence of inclusive $\eta$ production.[63] They see plenty of production well below 4.1 GeV in contradiction to (b). However the F has now been seen in a high statistics photoproduction experiment at CERN[64] in the decay modes $\eta\pi$, $\eta 3\pi$, and $\eta 5\pi$ with a measured mass M=2020±10 MeV. A few events have also appeared in emulsion exposures:
$F^+ \to \pi^+\pi^+\pi^-\pi^0$ (M=2017±25 MeV),[65]
$F^- \to \pi^-\pi^-\pi^+\pi^0$ (M=2013±18 MeV),[66]
$F^+ \to K^+K^0\pi^+\pi^-$ (M=2089±121 MeV).[66]

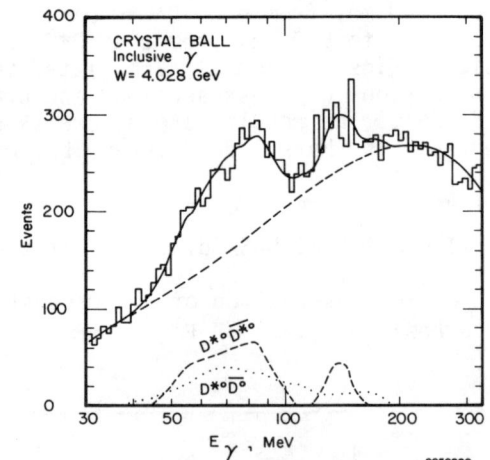

Fig. 20: Inclusive photon spectrum measured by the Crystal Ball (ref. 59) at W=4.028 GeV, showing fitted contributions.

The existence of the F meson can be considered confirmed, and the best mass is 2018±8 MeV.

## Continuum $b\bar{b}$ Production and the B Meson

The first three upsilon resonance peaks observed in $e^+e^-$ annihilation at DORIS[67,68,69] and CESR[5,6] have widths determined by the energy spread of the beams, which is caused by quantum fluctuations in the synchrotron radiation process. The fourth resonance $\Upsilon$(4S), observed at CESR by both the CLEO[5] and CUSB[6] groups, is significantly wider than the beam energy spread (proportional to $W^2$) extrapolated from the first three resonances (Fig. 22). After unfolding the radiative correction and the beam energy spread the CLEO data imply a total width $\Gamma_{4S}$ = 19±3 MeV[5]; the CUSB data give

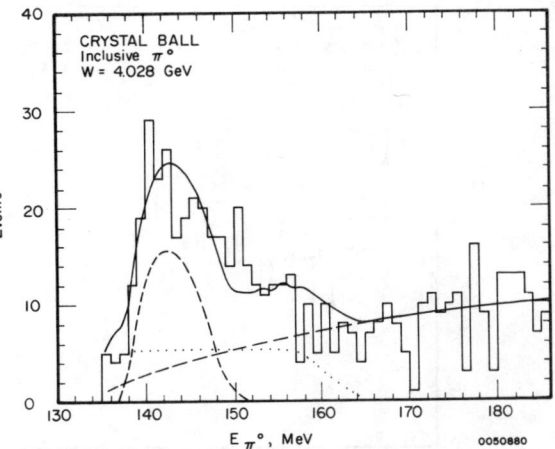

Fig. 21: Inclusive $\pi^0$ spectrum measured by the Crystal Ball (ref. 59) at W=4.028 GeV, showing fitted contributions.

$\Gamma_{4S} = 12.6 \pm 6$ MeV.[6]

The measured mass and $\Gamma_{ee}$ agree rather well with the potential model predictions for the $4^3S\ 1^{--}$ state of the upsilon family (Table I). The fact that the total width is much greater than for the lower 3S states indicates that the $\Upsilon(4S)$ lies just above the threshold for a strong decay channel. If the $\Upsilon(4S)$ decayed prominantly into a pair of slow heavy mesons, which then decay essentially isotropically into ordinary hadrons, one would expect much higher average sphericity for the $\Upsilon(4S)$ decays, compared with the nonresonant two-jet continuum. This is illustrated by Fig. 23. Suppose we take it as a working hypothesis that the $\Upsilon(4S)$ decays into a $B\bar{B}$ pair, where for now B denotes any heavy meson of mass near 5.25 GeV. We will then use the experimentally observed characteristics of the $\Upsilon(4S)$ decays to tell us about the nature of the B.

But first, what can we say about the mass of the B? One can estimate it theoretically from other masses:[15]

$$M_B = M_D + m_b - m_c + (3/4)(1-m_c/m_b)(M_{D^*}-M_D) = 5.26 \text{ GeV}$$

To quote experimental limits on $M_B$, we have to be more precise about the energy scale. Since the DORIS ring energy calibration is more accurate for the time being, I will rescale CESR mass measurements so that the $\Upsilon(1S)$ mass agrees with the DORIS value; that is;
$E = (M_{TDORIS}/M_{TCESR}) E_{CESR} = 1.029\ E_{CESR}$. We can set a conservative lower limit on $M_B$ by noting that the fact that the $\Upsilon(3S)$ is narrow means that it is below the threshold at $2M_B$. Thus, $M_B > 1/2\ M_{3S} = 5.18$ GeV. An upper limit comes from the assumption that the threshold at $2M_B$ must be below the mass (minus a half width) of the $\Upsilon(4S)$: $M_B > 1/2(M_{4S} - \Gamma/2) = 5.28$ GeV.

We can get a number to better than ±1% accuracy by accepting some model dependence. The decay width of the $\Upsilon(4S)$ is obviously an increasing function of how far it is above the $B\bar{B}$ threshold. The same kind of calculation used to predict the strucutre of the charm

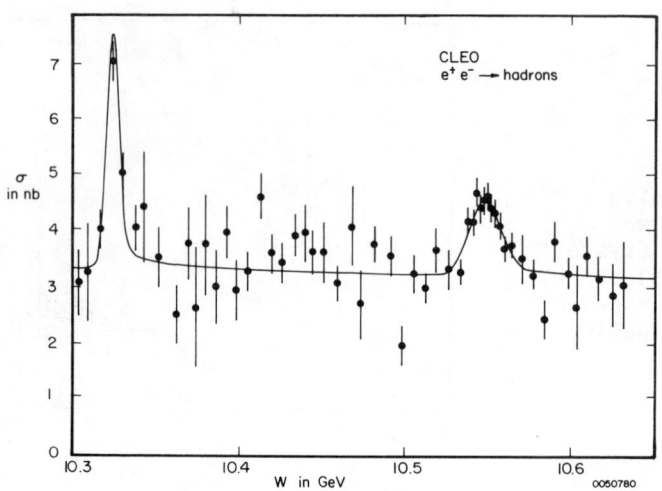

Fig. 22: Total cross section for $e^+e^-$ hadrons showing the $\Upsilon(3S)$ and $\Upsilon(4S)$ resonances, measured by CLEO (ref. 5).

cross section above threshold[15] has been extended to the $b\bar{b}$ case.[70] Fig. 24 shows the predicted shape of the $\Upsilon(4S)$ resonance for several choices of the energy above threshold $\Delta = M_{4S} - 2M_B$. Note that because of the successive opening up of the $B\bar{B}$, $B\bar{B}^*$ or $B^*\bar{B}$, and $B^*\bar{B}^*$ channels, the resonance shape is complicated and the dependence of width on $\Delta$ is not monotonic. The $B^*$-$B$ mass difference was assumed to be given by $M_{B^*} - M_B = (m_c/m_b)(M_{D^*} - M_D)$ ~50 MeV.[15] The observed $\Upsilon(4S)$ width, between 10 and 20 MeV, is consistent with either $\Delta = 40 \pm 10$ MeV, implying $M_B = 5.27 \pm 0.01$ GeV, or with $\Delta = 110 \pm 10$ MeV, in which case $M_B = 5.23 \pm 0.01$ GeV. If the latter mass were correct, the $\Upsilon(4S)$ would be above the threshold for $B\bar{B}^*$ or $B^*\bar{B}$, and perhaps even for $B^*\bar{B}^*$. One should therefore expect to see photons of about 50 MeV energy from $B^* \to \gamma B$. The sodium iodide detector of the CUSB

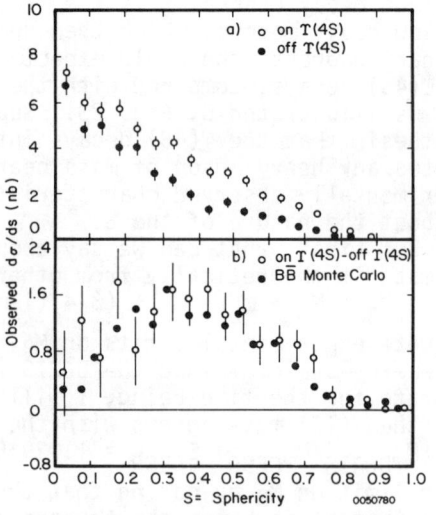

Fig. 23: Uncorrected sphericity distributions measured by CLEO (ref. 5) at the $\Upsilon(4S)$ and in the nearby continuum.

Table IX Crystal Ball (ref. 59) and Lead Glass Wall (ref. 60) results on D and D* masses compared (in MeV).

| Quantity | Crystal Ball | Lead Glass Wall |
|---|---|---|
| $M(D^{*\circ})-M(D^\circ)$ | 142.2±0.5±1.5 | 142.7±1.7 |
| $M(D^{*\circ})$ | 2006±2±1.5 | 2006±1.5 |
| $M(D^\circ)$ | 1864±2±1.5 | 1863±0.9 |

Table X. Properties of the B Meson (ref. 5)

| Property | Experiment | Standard Theory |
|---|---|---|
| M | 5.18 to 5.28 GeV (5.27±0.01 or 5.23±0.01 favored) | 5.26 GeV |
| $\tau$ | $< 3 \times 10^{-11}$ sec [a] | $\sim 10^{-13}$ sec |
| $<n_{ch}>$ in decay | 5.9±0.5 | high |
| $B_e + B_\mu$ | 24±6% | ~34% |
| $<K^{\pm\circ}>$ in decay | 2.4±0.5±0.5 | ~1.5 |

[a] Ref. 75

group is well suited to look for lines in the inclusive photon spectrum. The data[6] (Fig. 25) however show no sign of a photon peak between 40 and 60 MeV. The higher choice for the B mass (5.27±0.01 GeV) is therefore the more likely one.

Standard Model Predictions for B Decays

In order to anticipate what B decays might look like we make a few assumptions, which constitute the "standard model".[71,72] We assume that there are six kinds of quarks u, d, s, c, b, t and that the B mesons are the following combinations: $B^+ = \bar{b}u$, $B^- = b\bar{u}$, $B^\circ = \bar{b}d$, and $\bar{B}^\circ = b\bar{d}$. We assume that the B's are the lightest particles containing b quarks, that the b quantum number is conserved in strong and electromagnetic interactions, and hence that the B must decay weakly. We assume no flavor changing neutral currents, so the only allowed couplings are $W^- \to e^- \bar{\nu}_e$, $\mu^- \bar{\nu}_\mu$, $\tau^- \bar{\nu}_\tau$, $d'\bar{u}$, $s'\bar{c}$, and $b'\bar{t}$ (and charge conjugates). The primes refer to the Cabibbo-rotated weak eigenstates defined in terms of three mixing angles $\theta_1, \theta_2, \theta_3$ by the Kobayashi-Maskawa mixing matrix relating (d' s' b') to (d s b).[72] Since in B decay b→t is energetically forbidden, the b quark decay must proceed through the mixing in d' or s' to u or $\bar{c}$. That is, there are two possibilities with amplitudes proportional to the following K-M mixing coefficients.

$$b \rightarrow W^- c \sim c_1 c_2 c_3 - s_2 c_3 e^{i\delta},$$

$$b \rightarrow W^- u \sim s_1 s_3,$$

where $s_i = \sin\theta_i$, $c_i = \cos\theta_i$. It is likely that all of the angles $\theta_i$ are rather small,[73] so that unless there is a fortuitous cancellation of the two terms in the $b \rightarrow W^- c$ amplitude, it is likely to dominate over $b \rightarrow W^- u$.

We assume furthermore that for the B meson decay the spectator diagram (Fig. 26) is dominant; annihilation, exchange, and penguin diagrams do not contribute significantly. The various lepton pairs from the $W^-$ vertex should occur

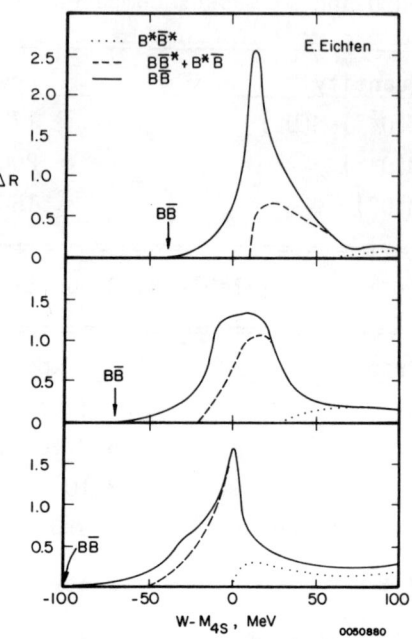

Fig. 24: Predicted total cross section for $e^+e^-$ production of b-flavored mesons in the vicinity of the $\Upsilon(4S)$ for several assumptions of the relative location of the $B\bar{B}$ threshold (ref. 70).

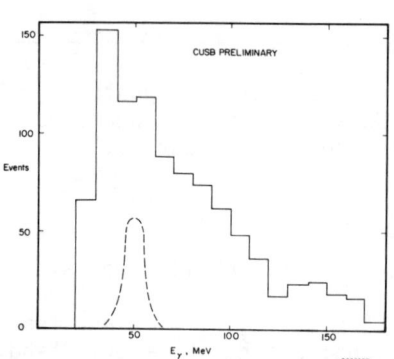

Fig. 25: Inclusive photon spectrum at the $\Upsilon(4S)$, measured by CUSB (ref. 6).

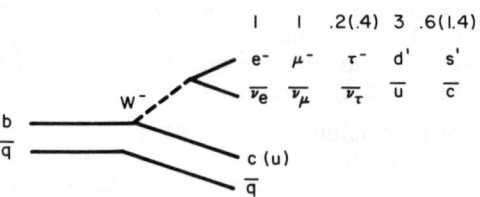

Fig. 26: The spectator model B decay, showing the relative weights for the various possible lepton pairs at the W vertex.

with equal probabilities except for the color factor (x3 for quark pairs) and a phase space suppression factor for the heavy particles t and c (0.2 if b→W$^-$c, about 0.4 if b→W$^-$u).

The semileptonic branching ratio (for B→e$\nu$X and for B→$\mu\nu$X) is then 17% if b→W$^-$c and 15% if b→W$^-$u. If each c or s quark always leads to one K meson in the final state, the average number of K's per B decay will be 1.23 if b→W$^-$c and 0.43 if b→W$^-$u. One should probably add a few tenths to each of these last numbers to account for s$\bar{s}$ pairs materialized from the vacuum in the fragmentation of the first generation quarks. The B lifetime prediction is $\tau_B$ = $4 \times 10^{-15}$sec/$(s_2^2 + s_3^2 + 2s_2 s_3 \cos\delta)$,[71] which could be anywhere from $10^{-10}$ to $10^{-14}$sec, but is most likely around $10^{-13}$ sec.

## Data on B Decays

To demonstrate the existence of a new particle, one's first thought might be to look for a peak in a mass plot for some simple final state that can be accurately reconstructed with low background, the classical method.[74] This has not been done in the CESR experiments, not only because of the practical difficulties--high multiplicities, no copious simple modes, etc.--but because the best signature for a new flavor is the presence of leptons. In leptonic decays the unseen neutrino always makes mass reconstruction very difficult.

Both of the CESR experimental groups have measured the yield of inclusive electrons as a function of beam energy. In the CUSB nonmagnetic detector[6] electrons are recognized as showering charged particles. In the CLEO magnetic detector[5] the additional requirements of a signal in a gas Cerenkov counter and an approximate match between the measured momentum and shower counter pulse height are imposed. Both experiments accept only electrons above 1 GeV. About 2/3 of the electrons from B→e$\nu$X should be above this cut, while most of the electrons from D decays (from the continuum or later stages of the B decay cascade) are eliminated. Figs. 27 and 28 show the raw measured inclusive electron cross sections in the vicinity of the $\Upsilon$(4S) resonance. In both experiments the electron rate increases at the $\Upsilon$(4S) by a much larger factor than does the total cross section. The low continuum electron yield is in fact consistent with the expected rate from charm and tau decays, QED processes, and misidentified hadrons. The high electron rate at the $\Upsilon$(4S) implies that the B meson has an appreciable semileptonic branching ratio, and is the best evidence that the B meson has indeed been found. Taking the acceptance and electron identification efficiency into account the two groups report the following branching ratios for B→e$\nu$X: 10 to 20% (CUSB[6]) and 16±4±7% (CLEO[5]).

The CLEO group has made a similar search for muons, identifying them with drift chambers outside of a 60cm thick iron shield surrounding the detector. The minimum momentum which can penetrate the shield is 1.0 to 1.6 GeV/c depending on the direction. The beam energy dependence of the observed rate (Fig. 29) shows a behavior similar to the electron rate. The implied branching ratio for $B \to \mu\nu X$ is 7.5±3.1 %.[5] Within the rather wide errors, all three semileptonic branching ratio measurements are compatible. The sum of the e and $\mu$ branching ratios is 24±8%, which is not far from the 34% expected in the standard model.

The CLEO time-of-flight scintillators identify charged K mesons with momenta between 0.6 and 1.0 GeV/c. The observed $K/\pi$ ratio in this momentum range (Fig. 30) shows a four standard deviation increase at the beam energy corresponding to the $\Upsilon(4S)$. The number of charged K's per event in the accepted momentum range is as follows:[5]

| | |
|---|---|
| observed from $B\bar{B}$ | 0.40±0.09±0.02 |
| from Monte Carlo, assuming $b \to W^- c$ | 0.22±0.05 |
| from Monte Carlo, assuming $b \to W^- u$ | 0.06±0.01 |
| observed from continuum | 0.06±0.01±0.01 |
| from Monte Carlo continuum | 0.08±0.02. |

The observed number of charged kaons per event disagrees with the $b \to W^- u$ hypothesis by about four standard deviations. It is marginally compatible with the theoretically favored $b \to W^- c$ hypothesis. The above numbers corrected for momentum acceptance using the $b \to W^- c$ Monte Carlo imply an average of 2.4±0.5±0.5 kaons (charged and neutral) per B decay.

Table X summarizes the experimental data on the B meson, as well as the predictions of the standard model. The agreement is impressive. There are many nonstandard models[76] in which the b has no t partner. The present B data are sufficient to rule out most of these. The most natural inference is that the standard model is essentially correct and there probably is a t quark.

## CONCLUSIONS

1. The QCD-inspired nonrelativistic potential theory is an excellent model of heavy quark binding, reproducing correctly mass differences and $\Gamma_{ee}$ ratios as well as yielding the accepted values for the QCD scale energy $\Lambda$ and Regge slope $\alpha'$. This gives strong support to the input assumptions: nonrelativistic kinematics, flavor independence, asymptotic freedom, one and two gluon exchange at small distances, linear confinement at large distances.

2. The lowest charmonium pseudosealar $\eta_c$ exists at a reasonable mass and with reasonable transition rates.

3. Lowest order QCD is qualitatively successful in understanding hadronic decay rates of heavy quarkonium. Theorists should get busy on the higher order corrections.

4. The B meson exists at the expected mass.

5. The b quark decays mainly via the c quark, and the decays to e, $\mu$, and s are consistent with the spectator model and simple lepton counting.

6. The chances for the existence of a t quark have improved.

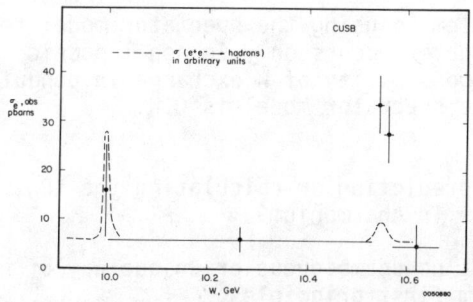

Fig. 27: CUSB data (ref. 6) for the visible inclusive electron cross section near the $\Upsilon(4S)$. The dashed line shows the behavior of the total hadronic cross section on an arbitrary scale.

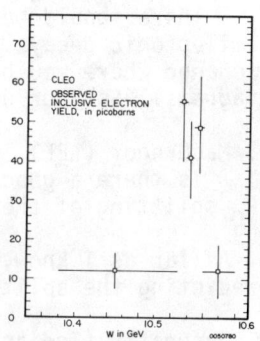

Fig. 28: CLEO data (ref. 5) for the visible inclusive electron cross section near the $\Upsilon(4S)$.

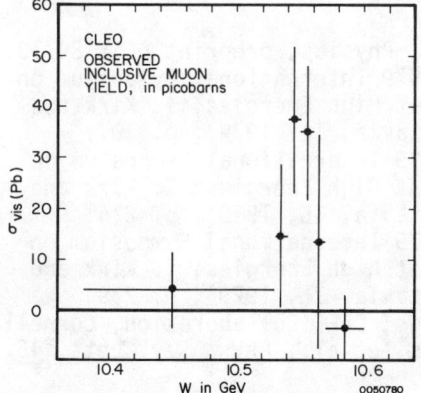

Fig. 29: CLEO data (ref. 5) for the visible inclusive muon cross section near the $\Upsilon(4S)$.

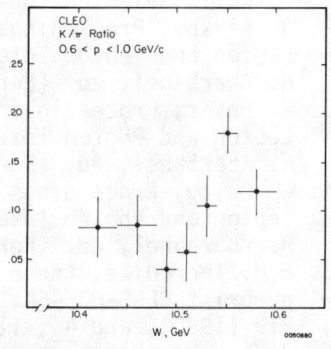

Fig. 30: CLEO data (ref. 5) for the charged $K/\pi$ ratio in the vicinity of the $\Upsilon(4S)$.

## DISCUSSION

Q. Tomozawa (Mich)
   In 1969 Muraki in Japan proposed a logarithmic potential to explain the spectrum of the observed particles. This was before the discovery of J/ψ. In 1976 Machacek and I proposed a logarithmic potential and power law potentials with a small power.

A. Thank you. I apologize that I cannot give credit to everyone.

Q. Deshpande (Oregon)
   There should be no problem in using the spectator model for semileptonic decays. The problem occurs only in non-leptonic exchange where you have the possibility of W exchange in penguin diagrams. So your use of the spectator model is O.K.

Q. MacGregor (LBL)
   Is there a good way of predicting or calculating the $^3P_2$, $^3P_1$, $^3P_0$ splitting of the χ levels in charmonium?

A. As far as I know, there is no unambiguous or unique way of predicting the splitting from first principles.

Q. Pasupathy (Indian Institute of Science)
   How soon do you expect to have results on the radiative decays of the upsilon states?

A. It is difficult to say. Although I make no promises, it may be possible within a year. It depends on how large the branching fractions are.

## REFERENCES

1. G. Wolf, Selected Topics on $e^+e^-$ Physics, preprint DESY 80/13.
2. J. Kirkby, Proceedings of the 1979 International Symposium on Lepton and Photon Interactions at High Energies, T. Kirk and H. Abarbanel, ed. (Fermilab, Batavia, IL, 1979), p. 107.
3. H. Meyer, Proceedings of the 1979 International Symposium on Lepton and Photon Interactions at High Energies, T. Kirk and H. Abarbanel, ed. (Fermilab, Batavia, IL, 1979), p. 214.
4. C. Quigg, Proceedings of the 1979 International Symposium on Lepton and Photon Interactions at High Energies, T. Kirk and H. Abarbanel, ed. (Fermilab, Batavia, IL, 1979), p. 239.
5. E.H. Thorndike, these proceedings; CLEO Collaboration, Cornell preprint CLNS-80/464, D. Andrews, et al., Phys. Rev. Lett. <u>45</u>, 219 (1980), and <u>44</u>, 1108 (1980).
6. J.K. Yoh, these proceedings and paper #933 submitted to this conference; G. Finocchiaro et al., Phys. Rev. Lett. <u>45</u>, 222 (1980); T. Böhringer, et al., Phys. Rev. Lett. <u>44</u>, 1111 (1980).

7. J.J. Aubert et al., Phys. Rev. Lett. 33, 1404 (1974).
8. J. Augustin et al., Phys. Rev. Lett. 33, 1406 (1974).
9. T. Appelquist and H.D. Politzer, Phys. Rev. Lett. 34, 43 (1975).
10. R.A. Bertlmann and A. Martin, Nucl. Phys. B in press; A. Khare, paper #539 submitted to this conference.
11. C. Quigg and J.L. Rosner, Phys. Lett. 71B, 153 (1977).
12. M. Machacek and Y. Tomozawa, Prog. Theor. Phys. 58, 1890 (1977), and Ann. Phys. (N.Y.) 110, 407 (1978).
13. A. Martin, these proceedings and CERN preprint Ref. TH. 2843.
14. J.L. Rosner, these proceedings; H.B. Thacker, C. Quigg, and J.L. Rosner, Phys. Rev. D18, 274 and 287 (1978).
15. E. Eichten, et al., Phys. Rev. D17, 3090 (1978) and D21, 203 (1980).
16. G. Bhanot and S. Rudaz, Phys. Lett. 78B, 119 (1978).
17. H. Krasemann and S. Ono, Nucl. Phys. B154, 283 (1979).
18. J.L. Richardson, Phys. Lett. 82B, 272 (1979).
19. W. Buchmüller, G. Grunberg, and S.-H. H. Tye, Phys. Rev. Lett. 45, 103 (1980).
20. J. Kuti, these proceedings; W.C. Haxton and L. Heller, paper #789 submitted to this conference.
21. A. Shrinsky, these proceedings.
22. W. Schmidt-Parzefall, these proceedings.
23. R. VanRoyen and V.F. Weisskopf, Nuovo Cimento 50A, 617 (1967).
24. R. Barbieri et al., Phys. Lett. 57B, 455 (1975); W. Celmaster, Phys. Rev. D19, 1517 (1979).
25. A.M. Boyarski et al., Phys. Rev. Lett. 34, 1357 (1975: V. Lüth et al., Phys. Rev. Lett. 35, 1124 (1975).
26. R. Brandelik et al., Z. Phys. C1, 233 (1979) and references therein.
27. J.S. Whitaker et al., Phys. Rev. Lett. 37, 1596 (1976).
28. W. Bartel, et al., Phys. Lett. 79B, 492 (1978).
29. C. Biddick et al., Phys. Rev. Lett. 38, 1324 (1977); R. Partridge, et al., Phys. Rev. Lett. 44, 712 (1980).
30. T.H. Burnett, these proceedings; M. Oreglia, Proceedings of the XV Rencontre de Moriond, Tran Than Van ed. (Les Arcs, 1980) and preprint SLAC-PUB-2529.
31. R. Partridge et al., Phys. Rev. Lett. in press; K. Königsmann, these proceedings.
32. J.D. Jackson, Proceedings of the 1977 European Conference on Particle Physics, L. Jenik and I. Montvay ed. (Budapest, 1977), p. 605.
33. T.M. Himel et al., Phys. Rev. Lett. in press; G.H. Trilling, these proceedings.
34. R.C. Giles and S.-H.H. Tye, Phys. Rev. Lett. 37, 1175 (1976), and Phys. Rev. D16, 1079 (1977); W. Buchmüller and S.-H.H. Tye, Phys. Rev. Lett. 44, 850 (1980).
35. P. Hasenfratz, et al., CERN preprint Ref. TH.2837.
36. T. Appelquist and H.D. Politzer, Phys. Rev. D12, 1404 (1975); R. Barbieri, R. Gatto, and R. Kögerler, Phys. Lett 60B, 183 (1976); R. Barbieri, R. Gatto, and E. Remiddi, Phys. Rev. Lett. 61B, 465 (1976).

37. C. Berger et al., Z. Phys. C1, 343 (1979), and DESY preprint 80/15.
38. H. Albrecht et al., DESY preprint 80/30.
39. B. Niczyporuk, et al., paper #383 submitted to this conference.
40. C. Grupen, these proceedings.
41. G.S. Abrams, et al., Phys. Rev. Lett. 34, 1181 (1975); W. Tanenbaum et al., Phys. Rev. Lett. 36, 402 (1976).
42. T.M. Himel, et al., Phys. Rev. Lett. 44, 920 (1980).
43. P. Langacker, Phys. Lett. 90B, 447 (1980); T.N. Pham, paper #732 submitted to this conference.
44. F. Messing, these proceedings.
45. T.-M. Yan, Cornell preprint CLNS 80/451 and other references therein.
46. S.W. Herb, et al., Phys. Rev. Lett. 39, 252 (1977); K. Ueno, et. al., Phys. Rev. Lett. 42, 486 (1979).
47. M. Chanowitz, Phys. Rev. D12, 918 (1975).
48. G.S. Abrams, et al., Phys. Rev. Lett. 44, 114 (1980).
49. M.T. Ronan, et al., Phys. Rev. Lett. 44, 367 (1980).
50. D.G. Aschman, Proceedings of the XV Rencontre de Moriond, Tran Than Van ed. (Les Arcs, 1980) and preprint SLAC-PUB-2550.
51. D.L. Scharre, these proceedings and preprint SLAC-PUB-2519.
52. W. Bartel, et al., Phys. Lett. 64B, 483 (1976), and 66B, 489 (1977).
53. G. Alexander, et al., Phys. Lett. 72B, 493 (1978).
54. H. Fritsch and J.D. Jackson, Phys. Lett. 66B, 365 (1977).
55. C. Dionisi, et al., Nucl. Phys. B in press (preprint CERN/EP 80-1 (1980)); C. Bromberg, et al., Caltech preprint CALT-68-747 (1980).
56. L. Montanet, J. Donoghue, and S. Meshkov, these proceedings.
57. J. Siegrist, et al., Phys. Rev. Lett. 36, 700 (1976).
58. J. Burmester, et al., Phys. Lett. 66B, 395 (1977).
59. H. Sadrozinski, these proceedings.
60. P.A. Rapidis, et al., Phys. Rev. Lett. 39, 526 (1977).
61. W. Bacino, et al., Phys. Rev. Lett. 40, 671 (1978).
62. V. Lüth, Proceedings of the 1979 International Symposium on Lepton and Photon Interactions at High Energies, T. Kirk and H. Abarbanel ed. (Fermilab, Batavia, IL, 1979), p. 78.
63. F. Porter, these proceedings.
64. B. d'Almagne, these proceedings.
65. R. Ammar, these proceedings.
66. K. Niu, these proceedings.
67. C. Berger, et al., Phys. Lett. 76B, 243 (1978).
68. C.W. Darden, et al., Phys. Lett. 76B, 246 (1978), and 76B, 364 (1978), and 80B, 419 (1979).
69. J.K. Bienlein, et al., Phys. Lett. 78B, 360 (1978).
70. E. Eichten, Harvard preprint HUTP-80/A027 (1980); S. Ono, Aachen preprint PITHA 80/2 (1980).
71. J. Ellis, M.K. Gaillard, D.V. Namopoulos, and S. Rudaz, Nucl. Phys. B131, 285 (1977).
72. M. Kobayashi and T. Maskawa, Prog. Theor. Phys. 49, 652 (1973).
73. R. Shrock, S.B. Treiman, and L.L. Wang, Phys. Rev. Lett. 42,

1589 (1979); V. Barger, W.F. Long, and S. Pakvasa, Phys. Rev. Lett. $\underline{42}$, 1585 (1979).
74. According to J.G. McEwen, these proceedings, earlier indications of a peak in $\psi K$ and $\psi K\pi$ have disappeared with four times the original data sample.
75. S. Yamada, these proceedings.
76. S. Pakvasa, these proceedings; G. Kane, paper #671 submitted to this conference.

Chapter 14

New Theoretical Developments

# RECENT PROGRESSES IN GAUGE THEORIES

G. Parisi, INFN-Frascati, Speaker

M.A.B. Beg, Rockefeller, Chairman

A. Jevicki, Brown,

S. Gottlieb, Argonne,

L. Dolan, Rockefeller,

Scientific Secretaries

# RECENT PROGRESSES IN GAUGE THEORIES

G. Parisi
INFN, Laboratori Nazionali di Frascati, 00044 Frascati, Italy

## 1. - INTRODUCTION

In the last years we have seen many developments in our understanding of gauge theories, expecially toward the construction of new tools for doing reliable computations in the non-perturbative region. The motivations are clear: we believe that quantum cromodynamics (QCD) is the true fundamental theory for strong interactions; this interaction is characterized by an effective running coupling constant $a(q^2)$ which goes to zero when $q^2 \to \infty$, i.e. at short distances (this property of the coupling constant is called asymptotic freedom)[1]. More precisely, if the number of quark flavours is four, we find:

$$a(q^2) \to \frac{12}{25 \ln(q^2/\Lambda^2)} \qquad q^2 \to \infty \ . \qquad (1.1)$$

The parameter $\Lambda^2$ is experimentally measured (e. g. in deep inelastic scattering) and it should be in the range 0.2-0.5 GeV$^2$. QCD is a complete theory of strong interactions: using $\Lambda^2$ and the quark masses as input, we should be able to compute all the physical quantities, in particular the mass spectrum of hadrons. However perturbation theory can be used only to compute hard processes (the coupling constant being small) and non-perturbative techniques are badly needed in the soft region.

Many interesting results have been obtained; for lack of time it is impossible to mention all of them: in this talk I will speak only those ideas which are more familiar to me.

For the time being most of the theoretical effort has been concentrated on the study of pure gauge theories without fermions, where only double coloured gluons interact; fermion should be included pertubatively at a later stage. In such a simplyfied theory $\Lambda$ is the only parameter: we want to comput the static potential between quarks (V(r)) and the glueball spectrum. Other quantities, like the energy dependance of the total cross section for glueball scattering, are more difficult to obtain[5]. In this talk I will try to give you a rough idea of how to carry on these non-perturbative calculations. In Section 2, I present the formalism of Euclidean quantum field theory, which is essential to master most of the new developments. In Section 3 I discuss

0094-243X/81/681531-38$1.50 Copyright 1981 American Institute of Physics

some qualitative ideas on confinement. Most of the efforts to obtain quantitative results can be divided into two categories: "brute force" computations and analytic computation. Brute force computations are usually very long and the results can be obtained only after spending a lot of human or computer time. They are mainly based on lattice gauge theories (Section 4) and can be divided into two large groups: computer simulations (Section 5) and high temperature expansions (Section 6). Some of the difficulties to deal with the high temperature expansion are connected with the possible existence of an elusive roughening transition (Section 7).

The most interesting analytic approach is based on the idea of writing equations for $W(C)$, the vacuum expectation value of the Wilson loop:

$$\overline{W}(C) = \langle W(C) \rangle ; \quad W(C) = \mathrm{Tr}\left\{P\left[\exp(i\oint_C A_\mu(x)\,dx_\mu)\right]\right\}. \qquad (1.2)$$

One obtains equations which are rather difficult to be solved, however it is known that rather impressive simplifications are present in the limit in which the number of colours become infinite. This fact enable us to write simple closed equations for an $SU(N)$ in the limit $N \to \infty$ (Section 8). Although these equations seem formidable there is some hope that they can be solved, at least approximatively.

## 2. - EUCLIDEAN FIELD THEORY

After the first works by Schwinger[6], it has strongly enphasized by Symanzik[7] that there are deep similarities between quantum field theory and classical statistical mechanics: indeed a field theory defined on a D-dimensional Minkowski space (D-1 space directions, one time direction) is connected to the corresponding field theory on a D dimensional Euclidean space by an analytic continuation (Wick rotation): the Minkowski metric $x^2 = \sum_{i=1}^{3} x_i^2 - x_o^2$ become the Euclidean metric $x^2 = \sum_{i=1}^{4} x_i^2$ if we set $x_4 = ix_o$.

The equivalence of the two theories has very deep consequences: if we quantize an Euclidean field theory using the Feynman path integral formulation, we obtain a special kind of classical statistical mechanics. Now classical statistical mechanics is a much older discipline than relativistic quantum field theory and we have a much better physical intuition of it: statistical mechanics deals with probabilities, not with amplitudes as quantum mechanics and the variety of statistical systems is large also in everyday life.

Only the beginning of the seventies Symanzik ideas became popular and they started to be applied in rather different fields: after the key works of Nelson[8] and of Osterwalder and Schrader[9] and the beautiful results of Guerra, Rosen and Simon[10] Euclidean field theory became an essential tool in the rigorous approach to the construction of an interacting quantum field theory[11].

In a different contest it was enphasized by Migdal and Poliakov[13] that the problem of computing the critical exponents for second order phase transitions is connected to the control of infrared divergencies in a theory with masless particles. Somewhat later in a beautiful series of paper Wilson[14] succeeded to compute the critical exponents; his approach was a combination of the block spin picture of Kadanoff[14] and the renormalization group which was used by Gellman and Low[15] to study the high energy limit of QED[16].

In these last years we have seen a very fruitful cross-fertilization[18] of statistical mechanics and quantum field theory[19]; from a conceptual point of view quantum field theory has started to be adsorbed in the general framework of classical statistical mechanics: this process is arrived to such a stage that it could be said (although it is not quite true) that quantum field theory is an high specialized branch of statistical mechanics.

To help the reader to orient himself in these recent developments I have inserted a table showing the relations between the main quantities in quantum field theory and the corresponding quantities in statistical mechanics.

In Table I the bracket indicates as usually the statistical expectation value, e.g.

$$\langle \varphi(x) \varphi(0) \rangle = \int d[\varphi]\, \varphi(x)\varphi(0) \exp(-\beta H) / \int d[\varphi] \exp(-\beta H). \quad (2.1)$$

Let us see an example in details: we suppose that the field creates from the vacuum an infinite number of particles of mass $m_n$. In momentum space we can write:

$$G(p) = \Sigma_n\, C_n / (p^2 + m_n^2). \quad (2.2)$$

In position space we obtain (to avoid Bessel functions let us consider the case D=3):

$$C(x) = \langle \varphi(x) \varphi(0) \rangle = \Sigma_n\, C_n \exp(-m_n |x|)/|x|. \quad (2.3)$$

If we know the function $C(x)$ analytically, it is trivial to compute all the $m_n$, however if $C(x)$ is only known numerically with some er-

## TABLE I

| Quantum field theory | Classical statistical mechanics |
|---|---|
| Minkowski space | Euclidean space |
| $\hbar$ | $\beta^{-1} = kT$ |
| $\mathscr{L}$ | $H$ |
| Feynman factor for amplitudes: $\exp(-i\mathscr{L}/h)$ | Boltzman factor for probabilities: $\exp(-\beta H)$ |
| Sum of all vacuum to vacuum diagrams: $\int d[\varphi] \exp(-i\mathscr{L}/h)$ | Partition function: $\int d[\varphi] \exp(-\beta H)$ |
| Vacuum energy | Free energy |
| Vacuum expectation value $\langle 0|A|0 \rangle$ | Statistical expectation value $\langle A \rangle$ |
| Quantum fluctuations | Statistical fluctuations |
| Time ordered products | Simple products |
| Existence of a mass gap | Exponential decrease of correlations |
| Mass | (Correlation length)$^{-1}$ |
| Green functions: $\langle 0|T[\varphi(x)\varphi(0)]|0\rangle \sim$ $\sim \exp i m|t|$ | Correlation functions: $\langle \varphi(x)\varphi(0) \rangle \sim \exp - m|x|$ |
| Changement of vacuum | Phase transition |
| Goldstone bosons | Spin waves |
| Decrease to zero the mass | Approach a second order phase transition |
| Free scalar bosons | Random walk (Free curves) |
| Scalar bosons with repulsive interaction | Self avoiding walk (Interacting curves) |
| Gauge theories | Interacting surfaces ? |
| Wightman axioms | Osterwalder and Schrader axioms |
| Cutoff | (Lattice spacing)$^{-1}$ |
| Hamiltonian | Transfer matrix |
| Instantons | Defects (vertices, dislocations) |

rors, it may be not so simple to extract $m_1$ and the numerical evaluation of $m_2$ may easyly present serious difficulties; however this is a pratical problem in numerical computations and it does not involve questions of principles.

The Euclidean formulation of gauge theories does not present any special difficulty. In the SU(N) case (N = 3 is the physical one) the gluon field is a doubly coloured vector:

$$A_\mu^{a,b}, \qquad a, b = 1, N$$

and b are the colour indeces.

The lagrangian which describes a pure gluonic wordl is:

$$\mathcal{L} = \frac{1}{g^2} \int d^D x F_{\mu\nu}^2 \qquad (2.4)$$

where g is the chromatic charge ($\alpha = g^2/4\pi$) and $F_{\mu\nu} = \partial_\mu A_\nu - \partial_\nu A_\mu + [A_\mu, A_\nu]$ [21]. The presence of the commutator in the definition of $F_{\mu\nu}$ is the origine of the interaction among gluons. The factor $1/g^2$ seems to be unusual: using the same convenction in electromagnetism, one would find:

$$\text{div } E/e^2 = \varrho . \qquad (2.5)$$

$\varrho$ being the electron density. In other words we have set the electron charge equal to one and we have rescaled the electric field; the conventional electron field $E_c$ can be easily obtained:

$$E_c = E/e . \qquad (2.6)$$

In order to define the theory in 4 dimension we must renormalize the coupling constant to avoid ultraviolet divergences; one finally obtains an effective running coupling constant which in the large $q^2$ region behaves as:

$$\alpha(q^2) \sim 12\pi/(11 N \ln q^2/\Lambda^2) . \qquad (2.7)$$

The only parameter of the theory is $\Lambda$; dimensionless quantities are therefore fixed and cannot be changed by changing the coupling constant.

Another important feature of the theory is gauge invariance: in the electromagnetic case only the fileds E and H and not the potentials A are well defined; the value of the potential in one point is arbitrary: it can be changed by a gauge transformation:

$$A_\mu^{(x)} \rightarrow A_\mu^{(x)} + \partial_\mu \lambda(x) \ . \tag{2.8}$$

Although in perturbation theory one usually remove this ambiguity by fixing the gauge[22], only gauge invariant quantities have a clear physical meaning.

All these features of the theory in the usual Minkowski formulation are also true in the Euclidean version of the theory.

As we said in the introduction our aim is to compute the glueball $m_G$ mass and the static potential between quarks. It is easy to prove that in Euclidean space:

$$m_G = -\lim_{r \to \infty} \frac{1}{r} \ln \langle F^2(r) F^2(0) \rangle \ , $$

$$V(L) = -\lim_{T \to \infty} \frac{1}{T} \ln \langle W(C) \rangle \ , \quad C = T \times L \tag{2.9}$$

where $C = T \times L$ indicates that the circuit $C$ is a rectangle of sizes $L$ and $T$.

As we see it is rather clear how to extract the most physically interesting information from the Euclidean version of QCD.

## 3. - CONFINEMENT

It is clear that the Wilson loop $W(C)$ plays a very important role in the study of gauge theories: it is the most natural gauge invariant observable. Indeed in the abelian case $\Phi(C) = i \ln W(C)$ (i.e. $W(C) = \exp -i\Phi(C)$) is the flux concatenated with the circuit $C$.

The quantity $A(C) = \ln \langle W(C) \rangle$ [24] is the contribution of the chromodynamic field to the action of a coloured particle having $C$ as trajectory. $A(C)$ is well defined in perturbation theory[25] apart from a linear divergence proportional to the length of the circuit $C$, which corresponds to the classical self energy of the electron.

At the first order in perturbation theory (or in the abelian case), i.e. neglecting gluons self couplings, $A(C)$ is the standard electromagnetic self induction of the circuit $C$:

$$A(C) = e^2 \oint_C dx_\mu \oint_C dy_\mu \, 1/(x-y)^2 \ . \tag{3.1}$$

Perturbation theory will obviously break down at large distances. If the potential $V(L)$ increase at infinity like $\sigma L$ (confinement), we find that:

$$\overline{W}(C) = \langle W(C)\rangle \sim \exp{-\sigma L \cdot T} = \exp{-\sigma S}. \tag{3.2}$$

The decrease of the expectation value of the Wilson loop like the surface may be considered as a criterion for confinement. In such a situation we expect the formation of a physical string between the two quarks where the energy is concentrated. In the time evolution the string will describe a surface: in the Euclidean space we expect that there will be a region of space on which the increase in action will be concentrated. We can introduce the parameter $a(x, C)$ defined by[26-28]:

$$a(x,C) = \langle F^2(x) W(C)\rangle / \langle W(C)\rangle. \tag{3.3}$$

Intuitively $a(x,C)$ has the meaning of the increase of the action at the point $x$ as effect of the Wilson loop: more precisely:

$$d/d(1/g^2) W(C) = \int d^4x\, W(C) \cdot a(x, C). \tag{3.4}$$

If the theory confines we expect that there will be a region in which $a(x,C)$ is substantially different from zero this region become a surface of some thickness[29] having $C$ as boundary in the limit in which the loop $C$ becomes very large.

Perturbation theory tell us that at short distances $V(L)$ defined in eq. (2.9) behaves as $\alpha(1/L^2)/L$ and it does give us no informations on the large distance behaviour.

"Has anybody proved confinement in QCD?" is the standard question to the expert. However this is not the most important question; there are general arguments showing that or QCD confines, or gluons take mass like in the Higgs mechanism, or there are long range forces[30]. This statement is very similar in spirit to the sentence "a material is solid or liquid or gas"; we need an explicit computation to find if Helium is liquid at zero temperature and it would very difficult to decide the issue using general theorems. Here the situation is the same: what we need in QCD, is efficient way to do computations in the low energy region: the output should be the whole mass spectrum.

Let me present for compleatness a simplified argument which shows how confined may be realized. I will neglect for semplicity the non abelian character of the theory. We consider a flat surface of area $S$. We divide it in $N$ smaller surfaces of area $S/N$. We obtain:

$$W(C) = \exp i\Phi(C) = \exp(i\sum_{i=1}^{N}\Phi_i) = \prod_{i=1}^{N} \exp(i\Phi_i), \tag{3.5}$$

where $\Phi_i$ is the flux going through each of the smaller surfaces. We have to compute the statistical average of W(C). Let us assume that the $\Phi_i^2$ are statistically independent (so happens in two dimensional theories):

$$\langle \Phi_i \rangle = 0, \quad \langle \Phi_i^2 \rangle = f, \quad \langle \Phi_i \Phi_j \rangle = 0, \quad i \neq j \quad (3.6)$$

and let us pospone the discussion on the origine of (3.6).

Now confinement is trivial; indeed:

$$\langle \exp - i \Phi_i \rangle \simeq \exp(-f/2),$$

$$\langle W(C) \rangle = \prod_{1}^{N} {}_i \langle \exp i \Phi_i \rangle \simeq \exp - \frac{Nf}{2}. \quad (3.7)$$

In other words the total flux $\Phi(C)$ is the sum of N statistically uncorrelated variables of zero mean and fixed variance; the central limit theorem tell us that the probability distribution of $\Phi$ is a gaussian with variance proportional to S:

$$P(\Phi) = \frac{1}{(2\pi f)^{1/2}} \exp\left(-\frac{N\Phi^2}{2f}\right). \quad (3.8)$$

We finally get:

$$\langle W(C) \rangle = \int d\Phi \, P(\Phi) \exp i\Phi = \exp - \frac{fN}{2}. \quad (3.9)$$

Confinement is a simple consequence of the large fluctuations of the flux concatenated to large circuits and the statistical independence hypothesis naturally lead to (3.8-9). Of course (3.6) is not true in perturbation theory, where the conservation law for $\nabla_\mu F_{\mu\nu} = 0$ gives strong costraints; beyond perturbation theory everything is possible, as it has been advocated by many authors[31, 32], the main difference being in mechanism producing eq. (3.6) chromomagnetic; monopoles, dense instantons, merons, condensations of flux tubes have been suggested; all these approaches share have one common point: the practically impossibility of using them to obtain reliable quantitative answers[33].

In the next Section we shall see other approach which should be able to give quantitative predictions.

## 4. - LATTICE GAUGE THEORIES

In the standard formulation of the theory ultraviolet divergences are present; although these ultraviolet divergences can be removed in perturbation theory, in order to give a non perturbative definition of the theory it is better not to introduce them from the beginning. This can be easily done by using a cutoff M and send M to infinity only at the end (momenta greater than M are disregarded). This can be done by discretizing the Euclidean space introducing a lattice, the fields will be defined only on the points or links of the lattice. In any computation the momenta will be bounded inside the first Brilloin zone; if we consider an hypercubic lattice of spacing a, each component p will belong to the interval

$$\left[-\frac{\pi}{a}, \frac{\pi}{a}\right], \qquad \text{M being equal to } \frac{\pi}{a}.$$

In principle it is also possible to work in the real Minkowski space and to discretize only the space and not the time. This approach has been suggested long time ago, but only recently it has been strongly developed. The introduction of a lattice is a device for dealing only with a finite number of degrees of freedom: it is very similar to the introduction of a mesh of points for solving differential equations[34].

There are many ways in which one can write a field theory on the lattice, however it is better to conserve the symmetries of the original problem as far as possible. The symmetry we want to preserve here is gauge invariance. As a first step we must define the gauge fields $A_\mu$ and the Wilson loop W(C). We associate to each link (i, k) of the lattice[35] a variable $U_{ik}$ belonging to the group[36,37]: it is the lattice equivalent of $\exp(i \int_i^k A_\mu dx_\mu)$[38]. The Wilson loop is simple given by:

$$W(C) = \text{Tr} \prod_{(i,k) \in C} U_{ik}, \qquad (4.1)$$

where the product runs over all the links belonging to the path C.

We notice that for a small loop $C_{\mu\nu}$ of area $a^2$ laying in the $\mu\nu$ plane we have in the continuum case:

$$W(C_{\mu\nu}) = 1 + i a^2 \text{Tr} F_{\mu\nu} - \frac{a^4}{2} \text{Tr} F_{\mu\nu}^2 + O(a^6). \qquad (4.2)$$

We find:

$$F^2_{\mu\nu} \simeq \frac{1}{a^4} \left[ 1 - W(C_{\mu\nu}) + \text{h. c.} \right],$$

(4.3)

$$\int (\sum_{\mu,\nu} F^2_{\mu\nu}) d^D x = a^{D-4} \sum_P \left[ 1 - W(P) + \text{h. c.} \right],$$

where the sum runs over all the plaquettes P (faces of the cubes) of the lattice (W(P) is the Wilson loop associated to the plaquette P).

The final expression for the Partition function and the Wilson loop are:

$$Z = \int dU \exp(-H/g^2_B),$$

$$\langle W(C) \rangle = \int dU \exp(-H/g^2_B) W(C)/Z,$$

(4.4)

$$H = a^{D-4} \sum_P \left[ 1 - W(P) + \text{h. c.} \right].$$

Formally $g^2_B$ and $1/g^2_B$ play the role of the temperature and $\beta$ respectively. The low coupling expansion (i.e. in powers of $g_B$) can be done using the saddle point method: Feynman rules can be derived[40, 41]; they are more complex than the usual ones: the interaction is non polynomial but no ultraviolet divergences are present, all momenta being bounded. In the limit a going to zero, one recovers the standard Feynman rules.

The bare coupling constant ($\alpha_B = g^2_B 4\pi$) is approximately the running coupling constant evaluated at $q^2 = \pi^2/a^2$. More precisely one obtains for the SU(2) group

$$\alpha(M^2) = \alpha(\pi^2/a^2) = \alpha_B + H\alpha^2_B + O(\alpha^3_B),$$

(4.5)

$$\frac{11}{6\pi} \alpha(q^2) = \frac{1}{\ln q^2/\Lambda^2 + \frac{102}{121} \ln \frac{11}{6\pi} \alpha(q^2)},$$

where the running coupling constant is defined in the momentum subtraction scheeme and the constant H has been computed by Hasenfratz and Hasenfratz[40]; (H $\simeq$ 3.39).

From eq. (4.5) we trivially get:

$$\Lambda = M(\frac{11}{6\pi} \alpha(M^2))^{-\frac{51}{121}} \exp(-\frac{3\pi}{11 \alpha(M^2)}) \simeq$$

$$\simeq M(\frac{11}{6\pi} a_B)^{-\frac{51}{121}} \exp\left[-\frac{3\pi}{11}(\frac{1}{a_B} - H)\right]. \tag{4.6}$$

It is clear that the continuum approximation can be good only in the region $a_B H \ll 1$. We need of such a small value of $a_B$ in order to apply perturbation theory: indeed a simple computation[42] shows that the mean value of the plaquette (U) (normalized to 1)[43] is equal to

$$U = 1 - \frac{3\pi}{4} a_B. \tag{4.7}$$

For $a_B$ as small as 0.2, U is equal to only 60% of its free value. The presence of terms proportional to $\pi a_B$ is typical of lattice gauge theories: the relevant expansion parameter is $g_B^2$ and not $a_B/\pi$.

H is a fundamental constant in the comparison of the results of the lattice theory with the continuum version; let us present a rough qualitative computation of H. We first notice that the renormalized charge is different from the free one also in the pure electromagnetic case, the origine of this difference is the non linearity of the lattice action. If $a$ is not zero, the thermal fluctuations renormalize the charge; let us try to estimate this effect. In order to compute the renormalized charge, we must know the variation of the action with respect to an external perturbation; let us decompose the field A as $A_f + A_e$: $A_f$ is the fluctuating part and $A_e$ is the external field; $F_{\mu\nu}$ is essentially given by:

$$\frac{1}{g_B^2} \cos A = \frac{1}{g_B^2} (\cos A_f \cos A_e - \sin A_f \sin A_e). \tag{4.8}$$

If we do the mean over the fluctuating field $A_f$ we get

$$\frac{1}{g_B^2} \cos A \sim \frac{\langle \cos A_f \rangle}{g_B^2} \cos A_e = \frac{\cos A_e}{\tilde{g}_B^2}, \tag{4.9}$$

$$\tilde{g}_B^2 = g_B^2/U.$$

In other words the more appropriate expansion parameter should be:

$$\tilde{a}_B = a_B/U = a_B + 0.75\, a_B^2 + \cdots. \tag{4.10}$$

This elementary computation give an estimate of H which is 70% of the correct value. Using the new variable $\tilde{a}_B$ one gets:

$$a(M^2) = \tilde{a}_B + 1.03\tilde{a}_B^2 + O(\tilde{a}_B^3),$$

$$\Lambda = 2.4 M \left(\frac{6\pi}{11\tilde{a}_B}\right)^{-\frac{51}{121}} \exp\left(-\frac{3\pi}{11 a_B}\right). \tag{4.11}$$

A good choice of the expansion parameter is very important: the two expression for $\Lambda$ in eq. (4.6); which are equivalent in the limit $a_B \to 0$, differ by a factor 10 for $a_B \simeq 0.15$. Although we need a formidable two loop computation to have reliable results, it may be useful to investigate the problem in an abelian theory where two loops computations are much simpler than in the non abelian case (in other words we should compute those diagrams giving contributions proportional to powers of $\frac{N^2-1}{N} a_B$)[43].

## 5. - COMPUTER SIMULATION

In the last years the most spectacular results have been obtained doing computer simulations using the Montecarlo technique. Although the Montecarlo technique is time honoured in the framework of statistical mechanics[44], only recently Wilson[45] suggested to apply it to the study of gauge theories; let me spend some time to give a physical picture of the method.

Suppose that we consider a finite piece of the lattice: a cube of size d (d >> $\Lambda$). In this situation we have pratically reached the thermodynamic limit $d \to \infty$; if periodic boundary conditions are used the corrections to the thermodynamic limit should be small as $\exp(-d/\Lambda)$[46].

Using the integral representation (4.4) the expectation value of the Wilson loop can be written as the integral over all the configurations of the fields in the cube. The number of fields N is of order $(d/a)^4$, so that also for small values of d the evaluation of the integral using the Simpson rule is practically impossible (i.e. the computer time needed is greater than the age of the universe). We need a method such that the time computer increase like N (or a small power of N) and not as $\exp(N)$.

In order to find the method we must go back in time and undo what Boltzmann and Maxwell did. Equilibrium statistical mechanics was introduced to study the large time behaviour of the system. Let us consider a classical example: we study the time behaviour of N particle in a box, whose trajectories $x_i(t)$ (i = 1 - N) satisfy the Newton law:

$$m \ddot{x}_i = -\frac{\partial U}{\partial x_i} = F_i \qquad (5.1)$$

where $U(x_1, \ldots, x_N) = U[x]$ is the interparticle potential.

Standard arguments based on ergodic theorems, tell us that in most of the cases after enough time the system will reach equilibrium; for large N the microcanonical distribution will be equivalent to the canonical distribution, given by the Boltzman factor. We finally find the highly non trivial result:

$$\lim_{\tau \to \infty} \frac{1}{\tau} \int_0^\tau f[x(t)] \, dt = \overline{f[x]} = \frac{\int dx_i \exp - \beta U[x] f[x]}{\int dx_i \exp - \beta [U]} \qquad (5.2)$$

The temperature T ($\beta = 1/kT$) can be computed as function of the energy E (which is a conserved quantity) using a thermometer; in this case the momenta themselves may play the role of the thermometer; indeed we know that:

$$\lim_{\tau \to \infty} \frac{1}{\tau} \int_0^\tau \frac{p_i^2(t)}{2m} \, dt = \overline{\frac{p_i^2}{2m}} = \frac{3}{2} kT . \qquad (5.3)$$

Now the point of view of a computer is the opposite of Boltzman; the number of steps needed to solve the coupled Newton equations is of order N x T, i.e. it increase linearly with N: the right hand side of eq. (5.2) is much easier to compute then the right hand side[48].

What happens for finite times? Also at equilibrium random thermodynamic fluctuations are present; we finally obtain:

$$\frac{1}{T} \int_0^\tau f(x) \, dt = \langle f \rangle + O(1/\tau^{1/2}) . \qquad (5.4)$$

The $\tau^{-1/2}$ law come from the mean of independent random fluctuations (the practically random behaviour of the deterministic system (5.1) is the basis of thermodynamics).

Of course we must understand how the time $\tau_e$ for which equilibrium is reached (i.e. the time for which $\tau^{-1/2}$ corrections are small), depends on N. The physical intuation tell us that if N is increased at fixed density and the potential is not pathological, equlibrium is reached locally in a time which is independent from N[49].

Near a second order phase transition (when long range correlations, i.e. zero mass particles, are present) the $\tau^{-1/2}$ law is no more valid and the pace of the approach to equlibrium is much lower, this phenomenum being called critical slowing down[50]. Large times are also needed to reach the equilibrium near a first order transition; the system may be locked into a metastable state untill a fluctuation grater than a critical size, is formed and becomes the germ of the condensation[51]; this difficulty may be avoided by fixing the initial conditions in such a way that in half of the box there is one phase (gas) and in the other half there is the other phase (liquid) and studying the movement of the interphase boundary[52].

In other words, instead of using the Boltzmann integral representation, it is more convenient to introduce a fictitious time t, to write appropriate equations of motion and to study the large time behaviour of the system.

The equation of motion can be freely chosen, provided that thermodynamic equilibrium is asymptotically reached. For example we can consider the same particles an in eq. (5.1) moving in highly viscous liquid at temperature T. One finds the Langevin equation[50, 53]:

$$\eta \dot{x}_i(t) = F_i[x(t)] + b_i(t) \tag{5.6}$$

$$\langle b_i(t) \rangle = 0, \quad \langle b_i'(t) b_j'(t) \rangle = \delta_{ik} \delta(t-t') B,$$

$$B = 2KT\eta,$$

where $\eta$ is the viscosity, $b_i(t)$ are random gaussian variables, uncorrelated in time and represent the effect of the Brownian motion. The relation among the viscosity, the temperature and B dates back to Einstein[54].

It is intuitive that at large time the particles must go to thermal equilibrium, they are in contact with the liquid that plays the role of heat reservoir. The formal proof of this statement can be done using the Fokker Plank equation[53, 55].

It is possible to do computer simulations based on the Langevin equation, however for practical purpose we must discretize the time interval and some errors are introduced. It is also possible to write random equation of motion for discreate times whose solution goes to equilibrium at large times: in this way we obtain the Montecarlo technique; in the limit of small steps the Montecarlo technique reduces to the Langevin equation: it can be considered as a discretized form (discretization in time) of the Langevin equation, such to preserve the asymptotic limit. From the physical point of view there is no substantial difference between the Langevin equation and the Montecarlo procedure[57]: My impression is that the Montecarlo method seems to be faster while the Langevin equation have the advantage of having a simpler analytic form[59].

Now in the last year many computer simulations have been done and many very interesting results have been obtained; we start to have a good understanding of the physics with finite groups and of the problem of approximating a continuous group with a discreate subgroup[61]. For lack of time I will not discuss here this very interesting problematics, and I will present the results for the continuous groups (mainly SU(2) and U(1)).

Let me describe a typical computer simulation: we start with a lattice having cubic shape[62], the length of the cube range from 4 to 16 lattice spacings in most of the computations.

The first thing to do is to look for phase transitions by studying the temperature dependence of the internal energy (the expectation value of the Wilson loop around one plaquette) on the temperature[63]. Phase transitions can be divided into two groups; first order: U is discontinuous but it is infinitely differentiable from both sides; it is believed that U can be approximatively analitically continued from each side beyond the transition point[64], giving the results for the metastable phase[64]. If a second order phase transition the internal energy is continuous, there is no metastability, but only a "critical slowing down", and the internal energy has a singularity proportional to $|T - T_c|^{-\alpha+1}$ $\alpha = 2 - D\nu$ [65], the energy connected correlation function (in the case of gauge theories the connected correlation functions of two plaquettes) decrease to zero like $r^{-2D+2/\nu}$. If $\alpha$ is negative it is very easy to see the transition by plotting U against T. For $\alpha$ strongly positive it is not so easy; it is also difficult to distinguish an higher power decrease of the correlation function[66] from an exponential decrease; it is therefore possible to miss a second order transition.

Let see some of the results: in Fig. 1 we have the internal energy of 5 dimensional SU(2) theory[42], some of the points are obtained decreasing the temperature, other by increasing it, so that we see an hysteresis loop which can be interpreted as a first order phase transition[67]. The high temperature expansion (described in the next Section) tell us that the theory is confined at

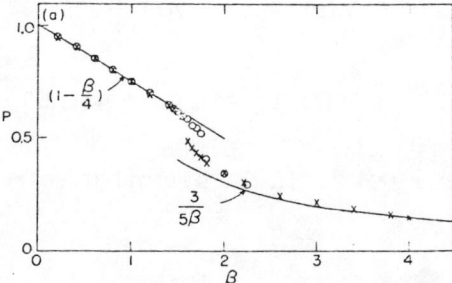

Fig. 1. The expectation value of a single plaquette P (defined in the text as 1-U) as function of $\beta$ in the 5-dimensional SU(2) gauge theory[42]. Crosses heating; circles cooling.

high temperature, the absence of infrared divergences in the low perturbative expansion (the standard perturbative expansion) suggest us that we see the transition from the confined and the unconfined phase.

In Fig. 2 we have shown the same plot for the 4 dimensional U(1) theory: we still have a transition between the confined and the unconfined phase; a carefull analysis shows that the hysteresis loop is not due to metastability, but to critical slowing down: the transition is a second order one $\nu \sim 1/3^{(69, 70)}$.

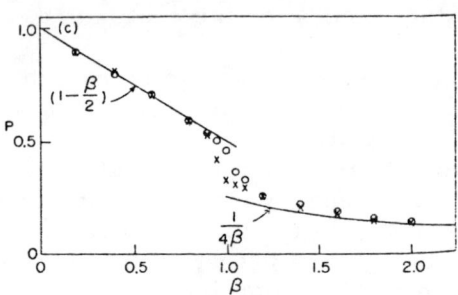

Fig. 2. The expectation value of a single plaquette P (defined in the text as 1-U) as function of $\beta$ in the 4-dimensional U(1) theory$^{(42)}$. Crosses heating; circles cooling.

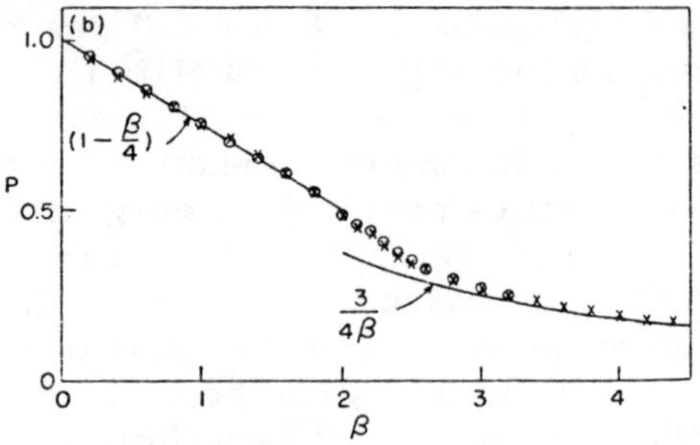

Fig. 3. The expectation value of a single plaquette P (defined in the text as 1-U) as function of $\beta$ in the 4-dimensional SU(2) theory$^{(42)}$. Crosses heating; circles cooling; no evident hysteresis loop is appreciable.

In Fig. 3 we see the internal energy for the more interesting case: the 4 dimensional SU(2) theory; there is no evidence for a phase transition, although the glitch around $\beta = 2$ ($a = 1/2$) may suggest a second order transition with a strong negative value of $\alpha$.

Let us see what happens to the Wilson loop in order to decide if the theory is confined also for $\beta > 2$. Let me recall what we expect in the confined phases for large r: for the static potential V(r) and for the expectation value of a Wilson loop of size L x T

$$V(r) \simeq \sigma r + \mu + \frac{\lambda}{r} , \tag{5.7}$$

$$W(L, T) \simeq LT\sigma + \mu(L+T) + \ln(L+T) + f_c(L/T) + \cdots$$

for $r \gg \Lambda^{-1}$. In the perturbative region a $r \ll \Lambda^{-1}$ we get:

$$V(r) \sim \mu + \frac{a}{r},$$

$$W(L,T) \sim \mu(L+T) + a f_p(L/T).$$
(5.8)

The presence of the last term in the confined case has been suggested in a beautiful fundamental paper of Lüscher, Symanzik and Weitsz[71] and it is connected to the presence of oscillations of the string (as we shall discuss in the Section 7).

A careful analysis is needed to separate the term proportional to the surface from the other ones[72]. This has been done by Rebbi[39] in the case of a discrete subgroup of SU(2) and by Creutz for the group SU(2)[73]. According to our policy of not discussing the results for finite groups for lack of space (although the analysis in more accurate) we present only the results for the SU(2) group.

Let us consider the quantity:

$$R(L,a) = \ln\left[\overline{W}(L,L)W(L-1,L-1)/W^2(L,L-1)\right]^2.$$
(5.9)

The contribution linear with L has been eliminated we expect that

$$R(L,a) \sim (a^2\sigma + \lambda^2/L^2), \quad L \gg \Lambda^{-1}(a),$$

$$R(L,a) \sim a(\pi^2/L^2), \quad L \ll \Lambda^{-1}(a).$$
(5.10)

The results are shown in Fig. 4. The straight line is a rough approximation to $\Lambda$. The value of $\sigma$ can be reasonably estimated in the region $\beta < 2.3$.

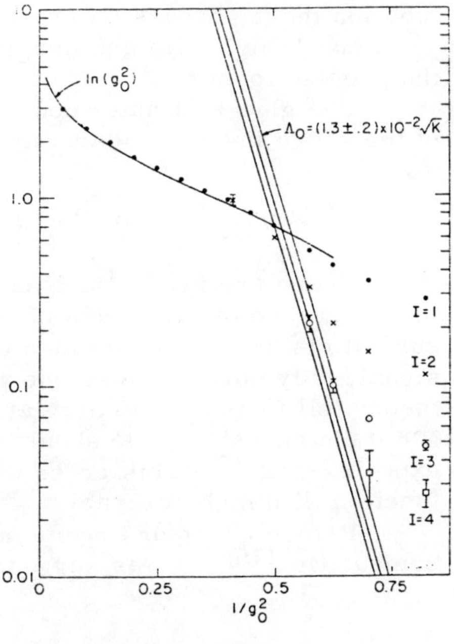

Fig. 4. The quantity $R(I,a)$ defined in the text agains $1/g_B^2 = \beta/4$[73] for the 4-dimensional SU(2) theory.

We notice that in the region of small $\alpha_B = 1/\beta\pi$ and large L, the function $R(L, \alpha)$ should depend only on a renormalization invariant quantity[27]:

$$\frac{L^2}{\alpha} R(L, \alpha) = f(z),$$

$$z = \frac{L^2}{\pi^2} (\frac{6\pi}{11} \alpha(M^2)) \exp(\frac{6\pi}{11\alpha(M^2)})$$

(5.11)

Standard perturbation theory allow us to compute the coefficients of the expansion in power of $\log(z)$ around $z = 0$, while if eq. holds, in the large z region we have:

$$f(z) \to z + \text{Const}.$$

(5.12)

The function $f(z)$ controll the cross over from perturbation theory to confinement: it would be rather interesting to extract it from the Montecarlo data[74] and to see if Const $\neq 0$.

It is rather unfortunately that the smallness of the lattice ($8^4$) prevents us from studying Wilson loop greater than 4 without feeling the effects of the periodic boundary conditions[75]. However from these relatively small lattices we can definitely say that the SU(2) theory confines also in the small coupling region as it was predicted by the Migdal Kadanof recurrence equation[76]: the exponential decrease of $\sigma$ with a slope similar to the one suggested by the renormalization group is a greater succes of the theory. The drastic changement in the behaviour around $\beta = 2$ ($\alpha = 1/2\pi$) can be easyly understood: the quadratic terms dominates over the linear ones in eq. (4.5) and pertubation theory breaks down. The ratio $\sigma/\Lambda^2$ is found to be of order 1, although higher values are not excluded, given our ignorance on the precise form of $\Lambda$.

The glue-ball mass could be extracted by studying the decrease of the gauge invariant plaquette-plaquette connected correlation function

$$C(r) = \langle W(P(0)) W(P(r)) \rangle - \langle W(P) \rangle^2.$$

(5.13)

There are two difficulties. The connected term is defined as the difference of two terms of order 1 and it is very difficult to measure at distances greater than one[77]; at high distances one sees no signal, only noise, unless one waits a very large time. Perturbation theory tell us that this correlation function decrease like $\alpha^2/r^8$ at small distances[78]; one should see a cross over from $1/r^8$ to $\exp(-M_G r)/r^{3/2}$. This cross over is invisible unless one knows the function with high accuracy.

Before changing argument let me mention a very interesting computation[79]. It was suggested[80,81] that QCD in Minkowski space has a phase transition from a confined to an unconfined phase when the temperature become greater than $T_c$ (Here the temperature is really the temperature, measured in Kelvin). In the interpretation of ref. (80) this phenomenum is connected to the exponential increase of the hadronic mass spectrum and the Hagerdon limiting temperature should be interpreted as a phase transition[82].

Finite temperature field theories can also simulated on the computer using the same Montecarlo approach. This was done in ref. (79) where some extimate of $T_c$ are presented; at this preliminary stage it is unclear if the transition is first or second: a more detailed analysis and more computer time is needed; the results are however very promising.

## 6. - THE HIGH TEMPERATURE EXPANSION

The high temperature expansion is a familiar technique is statistical mechanics[83]: at infinite temperature the theory is trivial: entropy dominates over energy and no correlation is present between different variables, i.e. all connected correlation functions are zero. If the lattice hamiltonian is enough simple (nearest neighnour interaction is the ideal) is easy do develop any physical quantity (e. g. $f(\beta)$) in powers of $\beta$.

$$f(\beta) = \sum_{0}^{\infty} {}_k f_k \beta^k . \tag{6.1}$$

Under general conditions it is possible to prove that this expansion has a finite radius of convergence (that is not true with the usual perturbative expansion), i.e. the function $f(\beta)$ is analytic around $\beta = 0$[84]. A well known theorem tell us that the knowledge of the function $f(\beta)$ around $\beta = 0$ (i.e. all the $f_k$) is sufficient to fix the function in the whole analyticity domain. The boundaries of the domain of analyticity are phase transitions for real temperatures and other physical uninteresting singularities for complex temperatures. If no phase transitions are present for real positive temperature[85], the behaviour at low temperatures can be extracted from the high temperature expansion![86].

So long with the theorems; let us come back with the reality. One could think that it would be enough to compute some of the $f_k$, construct the Padé approximations to $f(\beta)$ and look to the computer output. Although this procedure is convergent it is highly inefficient; it is a general rule that it is difficult to extract "quantitative" informations from the high temperature expansion without a "qualitative" input[88]. Let me give classical examples: we consider the three dimensional Ising model; the magnetic susceptibility $\chi(\beta)$ is believed to have a singularity of the form $(\beta_c-\beta)^{-\gamma}$. It is pure nonsense to use the first 10 or 20 $\chi_k$ known to construct Padé approximant to $\chi$ and to fit the output in order to find $\beta_c$ and $\gamma$. A Padé approximant has only simple poles and it does not approximate well a cut near the tip of the cut. It is much wiser to construct the logarithmic derivative

$$\mathrm{ld}(\beta) = \frac{d}{d\beta} \ln \chi(\beta). \qquad (6.2)$$

$\mathrm{ld}(\beta)$ has a simple pole at $\beta_c$ (plus a subdominant cut) and the residuum is the critical exponent $\gamma$.

This example shows that it is much better to use approximants[89] which have automatically the correct singularity structure. The informations on the nature of the singularities should be deduced from physical arguments. If we do not have this informations, we should try to extract it from the k dependence of the $f_k$. The following Appel comparison theorem is usual very useful.

If the nearest singularity to the origine has the form $(\beta_c-\beta)^{-\gamma}$, the asymptotic behaviour of the $f_k$ is given by:

$$f_k \propto k^{\gamma-1} |\beta_c|^{-k} \exp{-in\theta}, \qquad \beta_c = |\beta_c| \exp i\theta. \qquad (6.3)$$

If the function is real and $\beta_c$ is complex we must have a pair of complex conjugate singularities and $\exp-in\theta$ is substituted by $\cos(n\theta + \varphi_o)$. If all terms are positive the nearest singularity is on the positive real axis. We can therefore use the ratio test:

$$\beta_c = \lim_{k \to \infty} R_k, \qquad \gamma = \lim_{k \to \infty} \left[1 + k\left(\frac{R_k}{\beta_c} - 1\right)\right],$$

$$R_k = \frac{f_k}{f_{k-1}}. \qquad (6.4)$$

We must now extrapolate the values for $R_k$ and $\gamma_k$ we have computed, up to $k = \infty$. If the sequence is smooth a fit with inverse powers of k normally give the correct result[90], if the sequence is not smooth, you are in trouble.

Without a preliminary estimation of the effective radius of convergence of the series there is the danger of using it in the region where they are not convergent or slowly convergent. Unfortunately in many paper I have seen on the high temperature expansion for lattice gauge theories this elementary precaution has not been taken and unreliable results have been obtained.

Life is not always so easy. Sometimes the nearest singularity is on the negative axis: however a simple conformal mapping like $\beta = \dfrac{z}{b+z}$ (91) may map far away the singularity on the negative axis and the physical singularity on the real positive axis become the nearest to the origine.

We must also add a word of caution; first order transitions are normally invisible in the high temperature expansion, one obtains automatically the analytic continuation of the free energy in the metastable region. Of course the combined use of the high and the low temperature expansion is very efficient to locate first order transitions.

Let us come back to gauge theories. At infinite temperature there is no correlation among variables at different points. The mass of the glue-ball (the inverse of the correlation length) is infinite in this limit. Condition (3.6) is clearly satisfacted and the theory is confined. The construction of the high temperature expansion is straightforward: in order to compute the term proportional to $\beta^k$ in the free energy, we must count how many closed surfaces of k plaquette can be imbedded on the lattice and weight each surphace with a group theoretical factor depending on the topology of the surphace[92].

Before analyzing real interesting series for 4 dimensional models it is wise to study what happens on an hypercubic lattice when the dimension of the space (D) goes to infinity. In this situation the high temperature expansion can be exactly summed up[93]; in the SU(2) case (and also in the Z2 case) one finds[94]

$$U(\beta) = \beta \left[ u_0(\beta^2/D^{1/2}) + \frac{1}{D^{1/2}} u_1(\beta^2/D^{1/2}) + \cdots \right] \quad (6.5)$$

The functions $u_0$, $u_1$, etc. can be explicitely computed. When D goes to infinity one finds two singularity $\beta^2 = \pm 1/D^{1/2}$; the 1/D corrections show that the negative $\beta^2$ singularity is the nearest to the origine. As explained in details in ref. (93), near the transition point, the surface, on which is concentrated the energy in presence of a large Wilson loop, is no more flat, there are many tree like deformations, which form a branched polymer of cubes. At the transition point, the lenght of this tubes arrives to infinity, their thickness remaining fixed; we can call this transition a "local roughening" transition, because the deformation can start from a small size of the surface. The critical exponent $\alpha$ is 1/2.

A more careful analysis shows that there is a first order phase transition at $\beta = 1/D$ which separate the confined and the unconfined phases[39, 93]. The second order transition is in the metastable phase and it is a virtual transition not directly relevant for the physics.

Descreasing the dimensions the two transitions become rather near; the analysis is very clear in the 4 dimensional Z2 theory: there is a first order transition at the self dual point(37, 95, 96) and there are two virtual second order transition very near: about $\beta_{II} \simeq$ $\simeq 0.49$ and $0.40$ respectively, as can be easily seen using the ratio test and conformal mapping(97).

I have analyzed the 4-dimensional high temperature expansion for $U(\beta)$ in SU(2) up to $\beta^{15}$(39). A first sight one finds that a transition at $\beta \simeq 2.15$ with $\alpha \simeq 1/2$ is strongly suggested; however this result is in strong contrast with Montecarlo: if we resume the high temperature expansion according to this hypothesis the expression for $U(\beta)$ violently disagree with the Montecarlo results in the region near the critical temperature. In order to reach compatibility of the high temperature expansion with computer simulations we must decrease to 2 and to $-1/2$(98). This drastic solutions still compatible with the first fifteen orders of ref.(37), however a preliminary analysis of the 21 orders of Wilson shows that the higher orders have the tendency of preferring higher still values of $\beta$ and $\alpha$(100). Indeed a simple minded analysis would suggest $\beta_c \sim 3$ and $\alpha = 2$ which is a pure nonsense. The only deceiving conclusion is that it is impossible to use the high temperature expansion to extrapolate beyond $\beta \simeq 2.1$ unless 50 orders are computed. Similar conclusions can be extracted from the high temperature expansion for $\sigma$(101) and from $a(x)$(28) defined in eq. (3.3).

The reason for this debacle is clear. In the mean field theory the deformations of the surface have Hausdorf dimensions 4(93, 102), as ordinary branched polimers, in other words there radius increase like $N^{1/4}$, N being the number of steps(103). This means that these polimers of cubes, if self-repulsion is neglected, would be strongly overlapping. The self-repulsion, due to the non linear superposition of fluxes, decrease the phase space allowed to these deformation and probably forbids the transition(93). In the high temperature expansion this effect comes from excluded volume effects: however to construct a diagram of a tube which bends back whith the two ending point touching we must go to the 17$^{th}$ order for $U(\beta)$. In other words the effect that is potentially able to stop the transition appears only at rather high orders and low order computations do not contain enough informations on what happens beyond the would be transition point.

It may be possible that better results are obtained if we use all the informations we have on the high temperature expansion at arbitrary non integer dimensions; e.g. we set $D = 2 + \dfrac{2\beta^2}{1+\beta^2}$ and we extrapolate at $D = 4$ to the point $\beta^2 = \infty$ by changing $D$ together with $\beta^2$. This procedure avoid us to pass near the region $D = 4$, $\beta \sim 2.1$ which is not smooth. We can also try more fancy parametrizations like $D = 4 - \dfrac{8\beta^2}{16+\beta^4}$ to outflank the non existing local roughening transition.

The impossibility of extracting informations for the behaviour in the low coupling region from the high temperature expansion, should dissuate from using shorter series in the same region where longer series for $U(\beta)$ do not give the correct result.

As usually happens, it is possible that using unjustified procedures one gets the correct order of magnitude, only because it is very difficult to make errors in the order of magnitude; however I want to recall the attention of the reader on ref. (104, 105): in these two papers the same high temperature expansion is used to extrapolate at zero temperature and the results for a quantity like $\sigma/\Lambda^2$ differ of a few orders of magnitude. A blind use of the matching conditions may give rather serious errors.

## 7. - THE ROUGHENING TRANSITION

In the strong coupling limit the surface associated to the Wilson loop is very rigid, the surface tension being very high ($\sigma \sim -\ln\beta$); it coincides with the minimal area surface and no fluctuations are present. If we consider a Wilson loop on the $x_1$, $x_2$ plane the parameter $a(x)$ introduced in eq. (3.3) will be different from zero only on this plane:

$$a(x_1, x_2) = \delta(x_\perp). \qquad (7.1)$$

By decreasing the temperature, the surface tension decreases and the surface is no more flat. The simplest defect consistes of shifting one lattice element in one of the 2(D-2) directions and adding four more plaquettes. The probability for this deformation is proportional to $2(D-2)\beta^4 \equiv \lambda$. When the parameter $\lambda$ becomes of order 1 the surface full of defects; each of them corresponding to placing a three dimensional cube on the surface in one of the 2(D-2) allowed directions[106].

If two near by cubes are parallel they gain a factor $\beta^2$ in energy but they loose a factor 2(D-2) in entropy[96]. The probability of being parallel and vanish when the dimension go to infinity[107] at fixed $\lambda$, for lower dimensions it is possible that the cube organize themselves and their directions start to be correlated on large scale[108]. In this case we can speak of "global roughening" while if the direction of the cubes are not correlated we have a "local roughening".

A local roughening transition happens when adding cubes on the top of cubes the deformation may arrive to infinity (see Fig. 5). At the transition point one would find for Wilson loops of any size:

$$a(x) = \delta(x_\perp) + \tilde{a}(x) , \qquad (7.2)$$

$\tilde{a}(x)$ going slowing to zero at infinity.

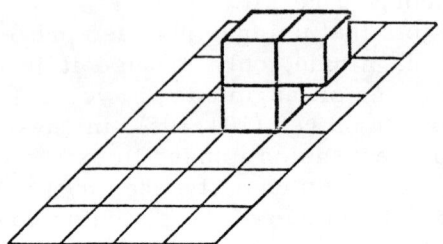

Fig. 5. A typical deformation of a surface (hangover) near a local roughening transition.

For a local roughening transition all thermodynamic quantities, in particular the expectation value of Wilson loops of any size, are singular.

A global roughening transition[28, 30] can be defined only in the limit of infinite size of the Wilson loop: the surface tension is singular but no singularity is present in the expectation value of the Wilson loop of finite size. In the case of a Wilson loop of size $L \times L$, the function $a(x)$ should be substantially different from zero inside a region having transversal dimensions $R(L)$, $R(L)$ going to infinity with $L$.

Both transitions have their correspective in dual models. The local roughening transition correspond to the tachion of the conventional dual models[93] and the global roughening transition is responsible of the last term in eq. (5.7) which was found in ref. (71) using a new solution of the equations of motions of the string, which should not suffer of the drawbacks of the conventional solution; according to this analysis the function $R(L)$ should behave like $\ln L$[30].

In 4-dimensional SU(2) theories the high temperature expansion suggest the presence of a local roughening transition at $\beta \simeq 2.15$, however as we said in the previous Section, this indication cannot be taken seriously. It is also possible that the self-repulsion effect transform an incipient local roughening transition in global one[110]. This issue may decided only if good quality compute simulations for the quantity $a(x)$ are available.

In 3-dimensional SU(2) theories the situation is much clearer. No singularity is seen in the high temperature expansion for the internal energy[111], although the extrapolation at zero temperature is problematic, while the surface tension clearly shows a singularity at about $\beta = 1.4$[112]. Summarizing, we have a good evidence for a global roughening transition in $D = 3$, a similar transition is likely present in the 4-dimensional case, but there is no serious evidence point.

## 8. - THE LARGE N EXPANSION

Many progresses have been done using the large N expansion. This technique has been introduced long time ago[113] in statistical mechanics : the main idea is that if the number of colours (N) goes to infinity, it is possible to use statistical theorems also in colour space. In the simplest situation fields have only one colour index and the problem can be exactly solved[114] : field theory can be reduced to an integral equation (the Hartree-Fock approximation is correct). This large N expansion is a wonderful laboratory to study the formal properties of field theory such as infrared finiteness[115], existence of non renormalizable interactions[116], analiticity in the Borel plane[117], etc.

It was remarked by t'Hooft[118] if the field is double coloured (as the gluon)[119], the theory can still be defined in the infinite N limit : remarkable semplifications are present, only planar diagrams survive[120]. Although in this case the theory cannot be solved (exception done for some notable cases), it has been argued that in this limit only zero with resonances are present[121]. In other words if we write formally :

$$SU(N) = SU(\infty) + A_1/N^2 + A_2/N^4$$

the mass spectrum is contained in the leading term ($SU(\infty)$) the whith of the resonances in the first correction ($A_1/N^2$) and the third term ($A_2/N^2$) gives (for N = 3) 1% corrections to the mass spectrum, which hopefully can be neglected[122].

If we want to solve the $SU(\infty)$ theory, we must fint something better than to sum all the planar diagrams ; we can take advantage of the fact that statistical (quantum) fluctuations are strongly depressed in the limit $N \to \infty$. Indeed if we consider two quantities having a finite expectation value when N goes to infinity, we find the factorization property[125] :

$$\langle AB \rangle = \langle A \rangle \langle B \rangle + O(\frac{1}{N^2}) \ . \tag{8.1}$$

No fluctuations are present (in other words the commutator A,B in the standard operatorial approach can be neglected). Although we are inclined to think that the saddle point method (or the classical equations of motion) must give the correct result, the situation is more complex. Let me spend some time to present the results for the quantum mechanics which, together with 2-dimensional QCD, is the better understood model.

In the first case we consider the hamiltonian

$$H = \frac{1}{2}P^2 + V(X^2), \quad P^2 = \frac{1}{N}\Sigma_i p_i^2, \quad X^2 = \frac{1}{N}\Sigma_i x_i^2 \qquad (8.2)$$

i.e. central potential for a particle moving in N dimensions.

As can be checked in the harmonic oscillator, the variables are well defined in the limit $N \to \infty$ however the commutator $P^2, X^2$ vanishes when N goes to infinity.

The ground state energy is given by[123]:

$$E_o = \min \left(\frac{1}{x^2} + V(x^2)\right), \qquad (8.3)$$

i.e. the classical result, plus the centrifugal term for a N dimensional s wave.

In the second case the variables X and P are NxN matrices: the Hamiltonian is:

$$H = \frac{1}{N}\text{Tr}\frac{P^2}{2} + \text{Tr}\left[V(X)\right]. \qquad (8.4)$$

The ground state energy is given by

$$E = \frac{1}{2\pi}\int h(p,x)\,\theta(\varepsilon - h(p,x)), \quad h(p,x) = \frac{p^2}{2} + V(x), \qquad (8.5)$$

where $\varepsilon$ is fixed by the condition:

$$\frac{1}{2\pi}\int \theta(\varepsilon - h(p,x)) = 1. \qquad (8.6)$$

Eqs. (8.5) and (8.6) can be considered as a variation of the conventional WKB-Thomas-Fermi approximation.

Quantum effects do not desappear in the limit $N \to \infty$ but the can be easyly computed. A better understanding of the physical origine of eqs. (8.5) and (8.6) (it is not self evident that they describe the sum of all the planar diagrams), would be very useful, expecially for extending these results to real scalar field theories in higher dimensions[126].

Fortunately the simple geometrical interpretation of gauge theories allow us to write directly useful equation in the limit $N \to \infty$. The fundamental variable are the expectation values of the Wilson loops associated to an arbitrary path C. It is convenient to introduce the functional derivative $\frac{\delta}{\delta\sigma_{\mu\nu}(x)}W(C)$, which quantifies the variation of W(C) under an infinitesimal deformation of the path C around the point $x$[127]. It was shown by Mandelstam[131] that:

$$\frac{\delta W}{\delta \sigma_{\mu\nu}(x)} = \text{Tr}\left[P(F_{\mu\nu}(x) \exp i \oint_x^x A_\mu(z) dz_\mu)\right]. \qquad (8.7)$$

We can now transcribe the "Maxwell" equations ($D_\mu F_{\mu\nu} = J_\nu$, $D_\mu \tilde{F}_{\mu\nu} = 0$) in functional equations for $W(C)$[132]; we find in the limit $N \to \infty$[128]:

$$\partial_\mu \frac{\delta}{\delta\sigma_{\mu\nu}(x)} \overline{W}(C) = \int dy_\nu \delta(x-y) W(C_{xy}) W(C_{yx})$$

$$\partial_\varrho \frac{\delta W}{\delta\sigma_{\mu\nu}} \varepsilon_{\varrho\mu\nu\lambda} = 0.$$
(8.8)

The $\delta$-function implies that the points $x$ and $y$ coincides: the loop $C$ looks like an eight, $C_{xy}$ and $C_{yx}$ are the two smaller loops into which the eight may be decomposed.

It is very important to understand the physical meaning of eq. (8.8).

A serious step in this direction has been done by Foester and by Migdal and Makeenko. These last two authors have been shown that an approximate solution of the eqs. (8.8) satisfy the integral equation[134]:

$$\frac{\delta}{\delta\sigma_{\mu\nu}(x)} \ln \overline{W}(C) = \int dy_\nu \, dt \, dP_{xy}^t[\omega] \, \omega_\mu(t) \cdot$$

$$\cdot \frac{\overline{W}(C_{xy\omega}) \overline{W}(C_{\omega xy})}{\overline{W}(C)} - (\mu \leftrightarrow \nu),$$
(8.9)

where $dP_{xy}^t[\omega]$ stands for the sum over all the paths $\omega(t')$ going from $x$ to $y$ in "time" $t$ (Wiener measure); the point $x$ and $y$ divide the closed path $C$ into two open paths $C_1$ and $C_2$, $C_{xy\omega}$ and $C_{\omega xy}$ denote the two closed path obtained by adding $\omega$ to $C_1$ and $C_2$.

The meaning of eq. (8.9) is clear[135]; $\delta \ln \overline{W}/\delta\sigma_{\mu\nu}$ is roughly speaking the field induced at the point $x$ by the Wilson loop: gluons coming from any point ($y$) of the loop contribute to it; in the abelian case, the factor $\overline{W}(C_{xy\omega})\overline{W}(C_{\omega xy})/\overline{W}(C)$ is substituted by 1 and the photon trajectories are free one, in the non abelian case the trajectories of the gluons are strongly influenced by the presence of the Wilson loop.

The most interesting result is contained in the last paper of Migdal[136]: he shows that the solution of eq. (8.8) can be written as a two dimensional field theory involving both bosons and fermions[137]. For lack of space I cannot enter in the details and the reader is strongly recomended to read the original literature on the subject.

I believe that this field is still in its infancy and that many very spectacular results are waiting us in the next future.

## DISCUSSION

Q1: Frampton, Harvard: In the relationship between the roughening transition in 4-dimensional QCD and the condensation of tachyons in dual models, how explicit can this relationship be because the dimension in dual models is fixed and cannot be varied continously, unless there is something completely new in dual models?

A1: Well, personally I an inclined to think that dualist ought to find something completely new in dual model, the impossibility of varying the dimensions being a spourious effect (Anyhow the high temperature expansion shows that gauge theories have a local roughening transition also at higher dimensions, at least in the metastable phases).
In my talk I wanted to underline the fact that also in conventionally defined dual models the surface of the string is not smooth, but locally rough.

Q2: Dolan, Rockfeller: How relevant is the existence of the roughening transition for the continuum limit?

A2: This has no effect on the continuum limit. It is only relevant to know if the strong coupling calculations on the lattice should be pursued, or how to modify them. For example, let us suppose that we have rather long series and we discover that there is a global roughening transition. In this case we should study not $\sigma(\beta)$, which has a singularity, but $W(C)$ for finite loop and extrapolate at infinitely large loops only when $\beta$ goes to infinity.

Q3: Kawamoto, Amsterdam: Could you comment about the introduction of fermions in the Montecarlo method?

A3: Work is in progress on this subject. I believe it is technically possible and in a year we will have some results.

A3: Nahm, CERN: Fermions are definitely very important. They dominate the high energy behavior of scattering (see the work of A. White). They may also suppress a roughening transition, if the analogy with tachyonic strings works, as tachyons are absent in the supersymmetric string model in 10 dimensions.

Q4: Brodsky, SLAC: Does the infrared singularity described by Frankel and Taylor for inelastic scattering in perturbation theory in non-Abelian gauge theories give any clue as to which theories confine?

A4: Frankel, Pennsylvania: We have to wait and see if results to all orders in perturbation theory support the existence of this singularity.

A4: I don't believe confinement can be seen in perturbation theory.

Q5: Pasupathy, Bangalore: How relevant are nontrivial topological configurations for confinement?

A5: Physicists are divided on this point. I believe since instantons are not apparent in the large N equations and since the large N limit appears to confine, instantons are not relevant, neither for confinement, nor to solve the $U(1)$ problem: the use of topological classification of configurations seems to be rather doubtful in an asymtotic free field theory, where fields strongly fluctuate at large distances and are very far from a pure gauge.

## FOOTNOTES AND REFERENCES

(1) - Long time ago, Landau[2], using general arguments, suggested that strong interaction should be asymptotically free: however the only asymptotically free theory known at that time was the $g\phi^4$ with negative coupling (Landau's arguments were too advanced for the time and they have been forgotten; in modern times the first example of an asymptotic free theory was given by Symanzik[3]: he also stressed the importance of being asymtotically free). A modern discussione of Landau philosophy can be found in ref. (4).

(2) - L. D. Landau, A. A. Abrikosv and I. M. Khalatnikov, Dokl. Akad. Nauk USSR 95, 773, 1177 (1954); 96, 261 (1954); L. D. Landau and I. Pomeranchuk, Dokl. Akad. Nauk USSR 102, 489 (1955); L. D. Landau, Niels Bohr and the Development of Physics, ed. by W. Pauli (Pergamon, 1955).

(3) - K. Symanzik, Lett. Nuovo Cimento 6, 420 (1973).

(4) - G. Parisi, Phys. Rev. D11, 909 (1975); L. Maiani, G. Parisi and R. Petronzio, Nuclear Phys. B136, 115 (1978); N. Cabibbo, L. Maiani, G. Parisi and R. Petronzio, Nuclear Phys. B158, 295 (1979).

(5) - V. Alessandrini and A. Krzywicki, Orsay Preprint LPTHE 80/24 (1980).

(6) - J. Schwinger, Proc. Nat. Acad. Sci. (U.S.) 44, 956 (1958).

(7) - K. Symanzik, J. Math. Phys. 7, 510 (1966); Rendiconti della Scuola Internazionale di Fisica "E. Fermi", Corso 45, Varenna 1968, ed. by R. Jost (Academic Press, 1969), pag. 152.

(8) - E. Nelson, in "ConstructiveQuantum Field Theory", ed. by K. Velo and A. S. Wighman (Springer, 1973); See also: F. Guerra, Phys. Rev. Letters 28, 1213 (1972), and F. Guerra and P. Ruggero, Phys. Rev. Letters 31, 1022 (1973).

(9) - K. Osterwalder and J. Schrader, Comm. Math. Phys. 31, 83 (1973).

(10) - F. Guerra, L. Rosen and B. Simon, Comm. Math. Phys. 27, 10 (1972).

(11) - Technically speaking the fact that probabilities (not amplitudes) are positive definite allow us to write very powerful inequalities (the use of inequalities is a very common procedure in statistical mechanics), e.g. the intuitive absence of a two particle bound state in presence of a repulsive interaction turns out to be a rigorous consequences of the Lebowitz inequalities[12] for Ising spin system.

(12) - For a review see: J. Glimm and A. Faffe, Cargese Summer School 1976 (Plenum Press, in press).

(13) - A. M. Poliakov, Soviet Phys. -JEPT 28, 533 (1969); A. A. Migdal, Soviet Phys. -JEPT 22, 1036 (1969).

(14) - K. G. Wilson and M. E. Fisher, Phys. Rev. Letters $\underline{28}$, 234 (1972); K. G. Wilson, Phys. Rev. Letters $\underline{28}$, 548 (1972); K. G. Wilson and J. Kogut, Phys. Rep. $\underline{12C}$, 77 (1974); L. P. Kadanoff, W. Götze, D. Hamblen, R. Hecht, E. A. S. Lewis, V. V. Palcianskas, M. Rayl, J. Swift, D. Aspnes and J. Kane, Rev. Mod. Phys. $\underline{39}$, 395 (1967).
(15) - M. Gellman and F. Low, Phys. Rev. $\underline{95}$, 1300 (1954).
(16) - The relevance of the renormalization group in the study of second order phase transitions was also stressed by Di Castro and Jona-Lasinio[17].
(17) - C. Di Castro and G. Jona-Lasinio, Phys. Letters $\underline{29A}$, 332 (1969); F. de Pasquale, C. Di Castro and G. Jona, Rendiconti della Scuola Internazionale di Fisica "E. Fermi", Corso 51, Varenna 1970, ed. by M. S. Green (Academic Press, 1971), pag. 123.
(18) - D. Amit, The role of statistical mechanics in contemporary physics, presented at Camerino (1979), unpublished.
(19) - For a more detailed discussions of the relations between statistical mechanics and quantum field theory at the beginning of the seventies see refs. (18, 20).
(20) - A. Baracca, G. Parisi, L. Peliti, M. Rasetti e M. Valdacchino, Le transizioni di fase e i problemi attuali della fisica delle particelle elementari, in: A. Baracca: Manuale Critico di Meccanica Statistica (CULP, Catania), in press.
(21) - I will use a matrix notation and I will not write the colour indices in most of the cases; here $[\ ,\ ]$ indicates the commutator.
(22) - The Langevin equation formulation of field theory (see Sect. 5) may be used to construct a perturbative diagrammatic approach to gauge theories in which no gauge fixing is needed[23].
(23) - G. Parisi and Wu Yong-shi, Preprint of the Institute of Theoretical Physics of the Academia Sinica ASITP-80-004 (1980); Scientia Sinica, to be published.
(24) - Please notice that $\ln \langle W(C) \rangle$ is very different from $\langle \ln W(C) \rangle$.
(25) - J. L. Gervais and A. Neveau, Nuclear Phys. $\underline{B163}$, 189 (1980); Cargese Summer School 1979 (Plenum Press, in press); Phys. Rep., to be published.
(26) - J. Groenveld, J. Jurkiewicz and C. P. Korthals Altes, Phys. Letters $\underline{92B}$, 312 (1980).
(27) - G. Mack and E. Pietarinen, Phys. Letters $\underline{94B}$, 397 (1980).
(28) - A. Hasenfratz, E. Hasenfratz and P. Hasenfratz, CERN Preprint TH 2890 (1980); C. Itzykson, M. Peskin and J. B. Zuber, Saclay Preprint, to be published; M. Lüsher, G. Münster and P. Weisz, DESY Preprint 80/63 (1980); G. Münster and P. Weisz, DESY Preprint 80/74 (1980).
(29) - As we shall see in the Sect. 7 we have two options: the thickness of the surphace may go to a limit when the diameter L of the circuit goes to infinity, or it may go to infinity with L.

(30) - G. 't Hooft, Nuclear Phys. B138, 1 (1978); B153, 141 (1979).
(31) - C. Callan, R. Dashen and D. Gross, Phys. Rev. D17, 2717 (1978); A. Mandelstam, Phys. Rep. 23C, 237 (1977); A. M. Poliakov, Nuclear Phys. B120, 429 (1977); T. Banks, R. Myerson and J. Kogut, Nuclear Phys. B129, 493 (1977); G. Parisi, Frascati Preprint LNF-76/15 (1976); B. Glimm and A. Jaffe, Phys. Letters 73B, 167 (1978); A. Jaffe, Introduction to Gauge Theories, presented at Helsinki Conference on Mathematics on Physics, 1978; Harward Preprint, unpublished; M. Stone and P. R. Thomas, Phys. Rev. Letters 41, 351 (1979); D. Foerster, Phys. Letters 76B, 597 (1978); T. Yoneya, Nuclear Phys. B144, 195 (1978).
(32) - N. K. Nielsen and P. Olsen, Nuclear Phys. B144, 376 (1978); Phys. Letters 79B, 304 (1978); J. Ambjørn, N. K. Nielsen and P. Olesen, Nuclear Phys. B152, 75 (1979); H. B. Nielsen and N. Ninomiya, Nuclear Phys. B156, 1 (1979); H. B. Nielsen and P. Olesen, Nuclear Phys. B160, 380 (1979); J. Ambjørn and P. Olesen, Nuclear Phys. B170 (FS1), 60 (1980).
(33) - The most serious effort to transform this qualitative model into a quantitative one has been done by the Copenhagen School (see refs. (32)).
(34) - In the study of partial differential equations one approximates the derivative with a finite difference operator, however in many cases the most efficient way for solving differential equations is the finite element method; it would be nice to see if and how this method can be transfered to the field theory framework.
(35) - i and k are two next neighbour points of the lattice.
(36) - K. G. Wilson, Phys. Rev. D10, 2445 (1974).
(37) - R. Balian, J. M. Drouffe and C. Itzykson, Phys. Rev. D10, 3376 (1974); D11, 2098 (1975); D11, 2104 (1975); D19, 2514 (1979).
(38) - The variables A belong to the Lie algebra, they are the generators of the Lie group; their exponent belongs to the group. Notice that in the lattice formulation of the gauge theory all variable belong to the group: it is possible to construct theories based on discreate groups for which no continuum formulation is possible (e. g. the group of rotations of a cube)[39]. Although in many cases no theory is obtained in the continuum limit (the mass gap is always proportional to the cutoff), a notable exception is the three dimensional $Z_2$ theory[37].
(39) - C. Rebbi, Brookhaven Preprint BNL-27203 (1980); Phys. Rep., to be published; D. Petcher and D. H. Weingarten, Indiana Preprint (1980).
(40) - A. Hasenfratz and P. Hasenfratz, CERN Preprint TH 2827 (1980).
(41) - V. F. Müller and W. Rühl, Kaiserlautern Preprint (1980).
(42) - M. Creutz, Phys. Rev. Letters 43, 553 (1979).

(43) - We notice that for general N, $H = \frac{N^2-1}{2N} \pi + 0.51 N$ and the prefactor 2.4 in eq. (4.11) is N independent.

(44) - N. Metropolis, A. W. Rosenbluth, A. H. Teller and E. Teller, J. Chem. Phys. <u>21</u>, 1087 (1953); for recent applications see for example: K. Binder, in Phse Transition and Critical Phenomena, ed. by C. Domb and M. Green (Academic Press, 1976), vol. 5B; Montecarlo Methods, ed. by K. Binder (Springer, 1979); S. Kirpratick, Les Houches 1978, ed. by R. Balian, J. Meynard and G. Toulouse (North Holland, 1980).

(45) - K. Wilson, talk given at the Crete Summer School 1977.

(46) - More precisely they should be of order $\exp(-d/M_G)$ where $M_G$ is the glueball mass. If we use the experimental information that $M_G$ is at least 2-3 $\Lambda$ (the value $M_G = 1500$ MeV was suggested in ref. (47), finite volume effects should be rather small for $d = \Lambda$).

(47) - G. Parisi and R. Petronzio, Phys. Letters <u>94B</u>, 51 (1980).

(48) - This procedure is currently used and it is called molecular dynamics.

(49) - That is true if we do not store energy in coherent motion of the particles (i. e. macroscopic motion): it would take some time to dissipate it into heat via turbolence.

(50) - L. Van Hove, Phys. Rev. <u>93</u>, 1374 (1954); See for a review: B. Halperin and P. C. Hoemberg, Rev. Mod. Phys. <u>49</u>, 435 (1977).

(51) - J. S. Langer, Ann. of Phys. <u>41</u>, 108 (1967).

(52) - Notice that the speed of the interphase boundary goes to zero, near the critical temperature $T_c$, like a power of $|T - T_c|$.

(53) - P. Langevin, Comptes Rendus <u>146</u>, 530 (1908); A. D. Fokker, Ann. d. Physik <u>43</u>, 812 (1914); M. Planck, Sitz. der Preuss. Akad. 324 (1917); For more recent references see for example: Noise and Stochastic Processes, ed. by N. Wax (Dover, 1954); R. F. Fox, Phys. Rep. <u>48C</u>, 179 (1978).

(54) - A. Einstein, Ann. d. Physik <u>17</u>, 549 (1905); <u>19</u>, 371 (1906).

(55) - I want to profit of my position to suggest to the constructivists that stochastic differential equations (see in this respect ref. (56)) may be an alternative technique to construct an interacting field theory in a mathematically rigorous way.

(56) - G. Parisi and N. Sorlas, Phys. Rev. Letters <u>43</u>, 744 (1979).

(57) - A detailed description of the Montecarlo procedure technique can be found in refs. (44, 58).

(58) - K. Wilson, Cargese Summer School 1979 (Plenum Press, in press).

(59) - It has been shown in ref. (60) that the computation of correlation functions cen be done with much higher accuracy using the Langevin equation.

(60) - G. Parisi, Correlation functions and computer simulations, Frascati Preprint LNF-80/54 (1980).

(61) - For example if we substitute to the group of rotation in the space O(3) the 60 elements group of rotations of the icosaedrum, there will be a first order transition at a temperature $T_c$, i. e. at a coupling $\alpha = \alpha_i$ ; at high temperature ($\alpha > \alpha_c$) the two theories will be very similar, while at low temperature ($\alpha > \alpha_c$) the two theories will behave very differently[37].

(62) - However for special problems it would be more convenient to use rectangular lattices.

(63) - We will use the following definitions $T = g_B^2/4 = \pi \alpha_B$; $\beta = T^{-1}$.

(64) - Some care must be used in doing the analytic continuation ; in reality the transition point is an essential singularity and the analytic continuation of U has an exponentially small immaginary part, proportional to the inverse of the mean life of the metastable state[51].

(65) - According to the original Erenfest classification second order transition are characterized by $\alpha \geq 0$, if $0 > \alpha \geq 1$ the transition is third order, the general rule if that the transition is of order k is the k'th derivative of the free energy is discontinuous, the $(k-1)^{th}$ being continous. However one often use the words "second order phase transitions" to indicate any transition of order higher than the first.

(66) - $\nu$ must satisfly the bounds $1/D < \nu \leq \infty$

(67) - Let me explain this misterious sentence. In order to fasten the approach to equilibrium in Montecarlo simulations it is usual to start at an high temperature, reach the eqlibrium, change, the temperature and restart the Montecarlo simulation using as starting point an equlibrium configuration of the previous temperature. It is also possible to go in the opposite direction, i. e. to start from low temperature and gradually increasing the temperature. In special cases it is convenient to have some oscillations in the temperature in order to annehale the defects[68] (a well known procedure in metallurgy).

(68) - S. Kirpatrick and D. Sherrington, Phys. Rev. 17B, 4385 (1978).

(69) - B. Lautrup and M. Nauenberger, CERN Preprint TH 2873 (1978).

(70) - Although Figs. 1 and 2 looks similar, longer computer simulations show a difference between the two cases[42,69].

(71) - M. Lüscher, K. Symanzik and P. Weisz, DESY Preprint 80/31 (1980). It would be interesting to compare the results of this paper with D. J. Wallace and R. K. Zia, Phys. Rev. Letters 43, 808 (1971) and M. J. Lowe and D. Wallace, Edinburgh Preprint 80/110 (1980).

(72) - For small $\alpha$, only $\sigma$ is exponentially small.

(73) - M. Creutz, Phys. Rev. Letters 45, 313 (1980).

(74) - It is also rather interesting to compute at least the first terms in the development of f(z) in powers of ln z and to compare it with the Montecarlo data.

(75) - I am convinced that asymmetric lattices like $6 \times 6 \times 12 \times 12$ or $4 \times 4 \times 12 \times 12$ are more appropriate to obtain informations on large Wilson loops. Moreover if one finds an observable dependence of the Wilson loop on the perpendicular boundary conditions, it would be rather interesting to analyze it.

(76) - A. A. Migdal, Z. Eksper. Theoret. Fiz. $\underline{69}$, 810 (1975); $\underline{69}$, 1457 (1975); L. P. Kadanoff, Ann. Phys. $\underline{100}$, 359 (1976); S. Caracciolo and P. Menotti, Ann. Phys. $\underline{122}$, 74 (1979); CERN Preprint TH 2899 (1980); G. Martinelli and G. Parisi, CERN Preprint TH 2882 (1980).

(77) - K. Wilson, C. Rebbi and M. Creutz, private communication.

(78) - It would be rather interesting to compute the perturbative expansion also for this function.

(79) - J. Polonyi, K. Szlachanyi and J. Kuti, Talk presented at this Conference.

(80) - N. Cabibbo and G. Parisi, Phys. Letters $\underline{59B}$, 67 (1975).

(81) - A. Poliakov, ICTP Lectures Notes taken by E. Gava, Report IC/78/4 (1978); W. Fischler, J. Kogut and L. Susskind, Phys. Rev. $\underline{D19}$, 1188 (1979); T. Banks and E. Rabinovici, Nuclear Phys. $\underline{B160}$, 349 (1979); E. Gava, R. Jengo and C. Omero, ICTP Preprint IC/80/77 (1980).

(82) - To study the developments of the limiting temperature hypothesis would be a nice subject for an historical reconstruction of the attitude toward physical laws which was shared by many physicists at that time.

(83) - For a review see: Phase Transitions and Critical Phenomena, ed. by C. Domb and M. Green (Academic Press, 1972), vol. 3.

(84) - See for example: D. Ruelle, Statistical Mechanics (Benjamin, 1972).

(85) - Negative temperature are well defined for bounded hamiltonian when the "field" variables are bounded: they are interesting because they correspond to antiordering.

(86) - This approach has been advocated by ref. (87).

(87) - L. I. Shiff, Phys. Rev. $\underline{92}$, 766 (1952); J. B. Kogut and L. Susskind, Phys. Rev. $\underline{D11}$, 395 (1975); C. J. Hamer, J. B. Kogut and L. Susskind, Phys. Rev. Letters $\underline{41}$, 1337 (1978); T. Banks, S. Rabi, L. Susskind, J. Kogut, D. R. T. Jones, P. N. Sharback and D. H. Sinclair, Phys. Rev. $\underline{D15}$, 1111 (1977).

(88) - The Lord, whose oracle is at Delphes, neither says nor hides, but hints (Heracleitus).

(89) - An approximant of order k ($f_k(\beta)$) is the only function, inside a given class, such that $f_k(\beta) - f(\beta) = 0(\beta^{k+1})$. If the class of function is given by the polynomial of order k, we have the standard Taylor expansion; if it is given by rational functions, we have the Padé approximants.

(90) - For a simple and efficient procedure to extrapolate see: J. Zinn Justin, Journ. de Physique $\underline{40}$, 969 (1979).

(91) - The value of b can be chosen such to fasten the convergence. The choice of the appropriate conformal mapping is very important.
(92) - J. M. Drouffe, Phys. Rev. D18, 1174 (1978).
(93) - J. M. Drouffe, G. Parisi and N. Sourlas, Nuclear Phys. B161, 397 (1979); G. Parisi, in Proc. Third J. Hopkins Workshop on Current Problems in High Energy Particle Theory, Florence 1979, ed. by R. Casalbuoni et al. (J. Hopkins Univ., 1979), pag. 179; Cargese Summer School 1979 (Plenum Press, in press); J. M. Drouffe, Saclay Preprint (1980).
(94) - In any dimensions $U(\beta)$ is an odd function of $\beta$ for these two groups.
(95) - M. Creutz, L. Jacobs and C. Rebbi, Phys. Rev. Letters 42, 1390 (1979); Phys. Rev. D20, 1915 (1979).
(96) - J. M. Drouffe, Nuclear Phys. B170 (FS1), 79 (1980).
(97) - I have done this computations using a pocket calculator. Using Padé approximants the higher singularity has been extimated at in the same region $\beta_{II} \simeq 0.55^{(96)}$ and $\beta_{II} \simeq 0.46^{(98)}$.
(98) - N. Kimura, Hokkaido Univ. Preprint (1980).
(99) - As clearly shown by eq. (6.4) the values of $\beta_c$ and $\alpha$ are strongly correlated.
(100) - I am grateful to K. Wilson for having comunicated to me his results prior to publication. I believe that he has reached the same conclusions as me; however the responsability of any errors is only mine.
(101) - G. Münster, DESY Preprint 80/44 (1980).
(102) - P. G. De Gennes, Academie des Sciences (Paris) Preprint (1980); G. Parisi, Phys. Letters 81B, 327 (1979).
(103) - B. Zimm and W. Stockmayer, J. Chem. Phys. 17, 301 (1949); T. Lubensky and J. Isaacson, Phys. Rev. Letters 41, 829 (1978).
(104) - D. S. Fisher and D. R. Nelson, Phys. Rev. B16, 2300 (1977).
(105) - G. Parisi, Phys. Letters 90B, 111 (1980).
(106) - We can also add a cube on the top of an other cube, but let us neglect this possibility for the time being.
(107) - In high dimensions this discussion is purely academic because $\lambda$ is of order 1 only in the metastable phase.
(108) - The argument on the infinite dimensional case should be more refined. The system of cubes is equivalent to a 2 dimensional Pott model with $q = 2(D-2)$ states and it is known that for q greater than 4 a first order transition is present[109].
(109) - R. G. Baxter, J. Phys. C6, L445 (1973); J. B. Kogut, Illinois Preprint (1980), to appear in Phys. Rev.; P. Ginsparg, Y. Y. Goldschmidt and J. B. Zuber, Saclay Preprint DPh-T/63/80 (1980).
(110) - The analogy with the Potts model suggests that the spontaneous orientation of cube-like deformations of the surface is a first order transition for $D > 4$ and it is a second order transition for $D \leq 4$.

(111) - A. Duncan and H. Waidya, Phys. Rev. D20, 903 (1979).
(112) - In this case after conformal mapping the ratio test gives good results in agreement with Padé approximants.
(113) - A. Stanley, Phys. Rev. 176, 718 (1968); R. Abe, Progr. Theor. Phys. 48, 1414 (1972); G. Parisi and L. Peliti, Phys. Letters 41A, 331 (1972); M. Suzuki, Phys. Letters 45A, 5 (1972); E. Brezin and D. J. Wallace, Phys. Rev. B7, 1967 (1973); R. A. Ferrel and D. J. Scalapino, Phys. Rev. Letters 29, 413 (1972); S. Ma, Phys. Rev. Letters 29, 1361 (1972); K. Wilson, Phys. Rev. D4, 2911 (1973).
(114) - The field transform as the fundamental representation of the O(N) or SU(N) group.
(115) - A. Jevick, Phys. Letters 71B, 327 (1977).
(116) - G. Parisi, Nuclear Phys. B100, 368 (1975); K. Symanzik, DESY Preprint 77/05 (1977); Y. Araf'eva, Theor. Math. Phys. 31, 279 (1977).
(117) - G. Parisi, Phys. Letters 76B, 65 (1978); Phys. Rep. 49C, 215 (1978); Nuclear Phys. B150, 153 (1979).
(118) - G. t'Hooft, Nuclear Phys. B72, 461 (1974).
(119) - The field transform as the adjoint representation of the O(N) or SU(N) group.
(120) - J. Koplizk, A. Neveau and S. Nussinov, Nuclear Phys. B123, 109 (1977).
(121) - G. Veneziano, Nuclear Phys. B117, 719 (1976); A. A. Migdal, Ann. of Phys. 109, 365 (1977).
(122) - In the quantum mechanics case studied in ref. (123), one find that the expansion parameter is about $1/N^2 \pi^2$ (124). The first two terms give the value of the ground state energy with a relative error of $10^{-4}$.
(123) - E. Brezin, C. Itzykson, G. Parisi and J. Zuber, Comm. Math. Phys. 59, 35 (1978).
(124) - E. Brezin, Phys. Rep. 49C, 221 (1978).
(125) - E. Witten, Nuclear Phys. B160, 57 (1979).
(126) - A. Jevicki and B. Sakita, Nuclear Phys. B165, 511 (1980); A. Jevicki and H. Levine, Brown Preprints HET-418 and HET-419 (1980); T. Yoneya, Tokyo Preprint (1980).
(127) - The precise definition of $\delta W/\delta\sigma_{\mu\nu}(x)$ can be found in refs. (128-130): the whole analysis become simpler on a lattice.
(128) - Yu. M. Makeenko and A. A. Migdal, Phys. Letters 88B, 135 (1979); A. A. Migdal, Chernolovka Preprint 27 (1979); Yu. M. Mekeenko, ITEP Preprint 141 (1979); Yu. M. Makeenko and A. A. Migdal, ITEP Preprint 23 (1980).
(129) - D. Foerster, Phys. Letters 87B, 87 (1979); T. Eguchi, Phys. Letters 87B, 91 (1979).
(130) - V. Volterra, Rend. Lincei 4a, III, 274 (1887), and 4a, VI, 127 (1890); A. T. Ogielski, Brookhaven Preprint (1980).
(131) - S. Mandelstam, Phys. Rev. 175, 1580 (1968).
(132) - This program was started in ref. (133).

(133) - G. De Angelis, D. De Falco and F. Guerra, Lett. Nuovo Cimento 19, 55 (1977); Y. Nambu, Phys. Letters 80B, 372 (1978); A. M. Poliakov, Phys. Letters 82B, 247 (1978); J. L. Gervais and A. Neveau, Phys. Letters 80B, 255 (1979); E. Corrigan and H. Hasslacher, Phys. Letters 81B, 181 (1979).
(134) - Yu. M. Makeenko and A. A. Migdal, ITEP Preprint 23, Sections 6-8 (1980), private comunication from A. A. Migdal.
(135) - I am trying to reproduce here what Migdal kindly explained to me more than one year ago at the Alustha Conference (1979).
(136) - A. A. Migdal, The Elf af the String, to be submitted to Phys. Letters, private comunication.
(137) - The need of introducing Fermionic degrees of freedom in the description of the string in QCD was stressed by D. Foerster, Nuclear Phys. 170B, 107 (1980).

# NEW IDEAS AND SPECULATIONS

L. Susskind, Stanford, Speaker

R. Marshak, VPISU, Chairman

M. Sokoloff, Lawrence Berkeley Lab,

P. Stevenson, Wisconsin-Madison,

Scientific Secretaries

(written version not available)

## SPECIAL SESSION:
## THE FUTURE OF HIGH ENERGY PHYSICS

ORGANIZED BY B. DURAND, WISCONSIN

M. GELL-MANN,  CALTECH
L. LEDERMAN,   FERMILAB
L. OKUN,       ITEP
A. SALAM,      ICTP, TRIESTE
Y. NE'EMAN,    TEL AVIV, CHM.

THERE IS NO WRITTEN VERSION OF THE SPECIAL SESSION HOWEVER THERE IS A VIDEOTAPE AVAILABLE.

## LOCAL ORGANIZERS

Vernon Barger
Adam Bincer
Ugo Camerini
Chen Hongfang
David Cline
Bernice Durand
Loyal Durand
Marvin Ebel
Albert Erwin
William F. Fry
Charles Goebel
Francis Halzen
Jacques Leveille
Richard Loveless

Robert March
Robert Morse
Martin Olsson
Lee Pondrom
Richard Prepost
Don Reeder
Marleigh Sheaff
David Scott
Paul Stevenson
Murray Thompson
Sau Lan Wu
Cosmas Zachos
Zhu Dongpei

## ADVISORY COMMITTEE, U.S. AND CANADA

| | |
|---|---|
| M.A.B. Bég | Rockefeller University |
| Peter Carruthers | Los Alamos Laboratory |
| Victor Cook | University of Washington |
| James Cronin | University of Chicago |
| Robert Diebold | Argonne National Laboratory |
| Stephen Gasiorowicz | University of Minnesota |
| Frederick Gilman | SLAC |
| Roman Jackiw | MIT |
| Lawrence Jones | University of Michigan |
| Juliet Lee-Franzini | Cornell University |
| Vera Luth | SLAC |
| James Prentice | University of Toronto |
| Chris Quigg | Fermilab |
| Jack Sandweiss | Yale University |
| James Sanford | Brookhaven National Laboratory |
| Frank Sciulli | Caltech |
| Samuel Treiman | Princeton University |
| George Trilling | Lawrence Berkeley Laboratory |
| Stanley Wojcicki | SLAC |
| Lincoln Wolfenstein | Carnegie-Mellon University |
| Taiji Yamanouchi | Fermilab |
| Gaurang Yodh | University of Maryland |
| Fredrik Zachariasen | Caltech |

## INTERNATIONAL ADVISORY COMMITTEE

| | |
|---|---|
| Marcello Conversi | Italy |
| Paul Falk-Vairant | Switzerland |
| Mary Kay Gaillard | Switzerland |
| Edwin Goldwasser | University of Illinois |
| Haim Harari | Stanford |
| Leon Lederman | Fermilab |
| Francis Low | MIT |
| Tetsuji Nishikawa | Japan |
| Wolfgang Panofsky | Stanford |
| W. Paul | Federal Republic of Germany |
| Abdus Salam | London |
| Herwig Schopper | DESY |
| A.N. Skrinsky | U.S.S.R. |
| L.D. Soloviev | U.S.S.R. |
| Godfrey Stafford | England |
| A.N. Tavkhelidze | U.S.S.R. |
| Leon van Hove | Switzerland |
| G. von Dardel | Sweden |
| Robert Wilson | Fermilab |
| Andrzej Wroblewski | Poland |
| Yoshio Yamaguchi | Japan |

## ASSISTANT SCIENTIFIC SECRETARIES

Gerry Bauer
Robert Benada
Robert Bosch
Eric Braaten
Max Chaves
Michael Cherney
David Christenson
Penelope Constanta
Max Dechantsreiter
Manuel Delfino
Daniel Dubbel
Mark Duffy
Guy Fogleman
James Green
Brian Heltsley
Richard Hollenhorst
Wai-Yee Keung
Thomas Kirkman
Jerome Krebs
Clara Kuehn
George Kuhr
Byron Lundberg
James Matthews
Patrick McCabe
Eduardo Mendel
Kevin Miller
Daniel Minette
Mohammad Mohammadi
Andrew More
Timothy Murphy
Manyee Ngai
Kenneth Nelson
Lyn J. Nichisch
Philip Nyman
Michael Procario
David Schumann
Elton Smith
Paul Smolensky
Daniel Solomon
Andrew Szentgyorgyi
Thomas Trinko
Robert Tucci
Kerry Whisnant
James Whitenton

CONFERENCE STAFF

R. Bonniwell
B. Cameron
S. Christenson
C. Cochrane
J. Danielson
M. Elliott
D. Goff
N. Goff
M. Haberland
W. Hughes
D. Jenkins
J. Kolonko
W. Long
M. Miller
S. O' onahue
G. Riedasch
S. Robinson
L. Schmidt
J. Smith
K. True
B. Wade
C. Wollensak

## CONFERENCE DELEGATES

| | | |
|---|---|---|
| Abrams, Gerald S. | Lawrence Berkeley Lab | USA |
| Abrams, Robert J. | Illinois-Chicago Circle | USA |
| Acharya, R.A. | Arizona State | USA |
| Achiman, Yoav | Wuppertal | WEST GERMANY |
| Adair, Robert K. | Yale | USA |
| Adjei, Samuel | Texas-Austin | USA |
| Adler, Stephen L. | Institute for Advanced Study | USA |
| Ahmad, Kamaluddin | Quaid-i-Azam | PAKISTAN |
| Ahmed, Tounsi | Paris VII | FRANCE |
| Akhababyan, N. | JINR | USSR |
| Albright, Carl H. | Northern Illinois | USA |
| Alessandrini, Victor | Orsay | FRANCE |
| Alexander, Gideon | Tel Aviv | ISRAEL |
| Ali, Ahmed | DESY | WEST GERMANY |
| Allison, J. | Manchester | GREAT BRITAIN |
| Alverson, George | Illinois-Urbana-Champaign | USA |
| Aly, Hadi H. | Baghdad | IRAQ |
| Amaldi, Ugo | CERN | CERN |
| Amatuni, A.Z. | Erevan | USSR |
| Amer, Ahmad | Cairo | EGYPT |
| Ammar, R. | Kansas | USA |
| Anderson, E. Walter | Iowa State | USA |
| Andrei, Natan | New York University | USA |
| Ansorge, R.E. | Cambridge | GREAT BRITAIN |
| Appel, Jeffrey A. | Fermilab | USA |
| Aragone, Carlos | Simon Bolivar | VENEZUELA |
| Arbuzov, B.A. | Serpukhov | USSR |
| Arenton, Michael W. | Argonne National Lab | USA |
| Armenteros, Rafael | CERN | CERN |
| Armitage, J.C. | Netherlands- Nikhef | USA |
| Arnowitt, Richard | Northeastern | USA |
| Aronson, Samuel | Brookhaven National Lab | USA |
| Ascoli, Giulio | Illinois-Urbana-Champaign | USA |
| Ash, W. | SLAC | USA |
| Atherton, Harry | CERN | GREAT BRITAIN |
| Auvil, Paul | Northwestern | USA |
| Ayres, David S. | Argonne National Lab | USA |
| Bacon, T.C. | Imperial College | GREAT BRITAIN |
| Baggett, Neil | Department of Energy | USA |
| Bailin, D. | Sussex | GREAT BRITAIN |
| Baillon, Paul | CERN | CERN |
| Baker, Winslow F. | Fermilab | USA |
| Bakich, A. M. | Sydney | AUSTRALIA |
| Baldini, R. | Frascati | ITALY |
| Baldo-Ceolin, Milla | INFN-Padua | ITALY |
| Ball, J.S. | Utah | USA |
| Ball, Robert | Michigan | USA |
| Baluni, Varouzhan | Institute for Advanced Study | USA |
| Barate, Robert | CERN | FRANCE |
| Bardeen, William A. | Fermilab | USA |
| Barger, Vernon | Wisconsin-Madison | USA |
| Barnes, Virgil E. | Purdue | USA |
| Barnett, Bruce A. | Johns Hopkins | USA |
| Barnett, Michael | SLAC | USA |
| Baroni, G. | INFN-Rome | ITALY |
| Barr, Stephen | Pennsylvania | USA |
| Bars, Itzhak | Yale | USA |
| Bartels, J. | Hamburg | WEST GERMANY |
| Bartley, J.H. | University College London | GREAT BRITAIN |
| Baubillier, Michel | Pierre et Marie Curie | FRANCE |
| Beg, M. A. B. | Rockefeller | USA |
| Bellini, Gianpaolo | INFN-Milan | ITALY |
| Bellotti, E. | INFN-Milan | ITALY |
| Bensinger, James | Brandeis | USA |
| Benvenuti, Alberto | CERN | CERN |
| Beretvas, Andrew | Rutgers | USA |
| Berger, Edmond L. | Argonne National Lab | USA |
| Bergstrom, L. | Royal Institute of Tech | SWEDEN |
| Berkelman, Karl | Cornell | USA |
| Bernard, Claude | California-Los Angeles | USA |
| Bernstein, Robert H. | Chicago | USA |

| | | |
|---|---|---|
| Bertanza, L. | INFN-Pisa | ITALY |
| Besch, Hans-Jurgen | CERN | CERN |
| Besset, Didier H. | Stanford | USA |
| Bettini, A. | INFN-Padua | ITALY |
| Bhamathi, G. | Madras | INDIA |
| Bijtebier, J. | VU-Brussels | BELGIUM |
| Bincer, Adam | Wisconsin-Madison | USA |
| Bingham, Harry H. | California-Berkeley | USA |
| Binkley, Morris | Fermilab | USA |
| Binon, F. | CERN | CERN |
| Biritz, Helmut | Georgia Tech | USA |
| Biswas, Nripen N. | Notre Dame | USA |
| Bizot, J.C. | Orsay | FRANCE |
| Bleser, Edward J. | Brookhaven National Lab | USA |
| Blondel, Alain | Lawrence Berkeley Lab | USA |
| Blumenfeld, Barry | Johns Hopkins | USA |
| Bock, Greg | Fermilab | USA |
| Bodek, Arie | Rochester | USA |
| Bodmer, Arnold | Illinois-Chicago Circle | USA |
| Bohm, Albrecht | RWTH Aachen | WEST GERMANY |
| Bohm, Arno | Texas-Austin | USA |
| Bohm, J. | JINR | USSR |
| Bohringer, Terence | Columbia | SWITZERLAND |
| Bose, Samir K. | Notre Dame | USA |
| Bosetti, Peter | CERN | CERN |
| Boulware, David G. | Washington | USA |
| Bozzo, Marco | CERN | CERN |
| Brabson, B. | Indiana | USA |
| Braccini, P. | INFN-Pisa | ITALY |
| Bradamante, F. | INFN-Trieste | ITALY |
| Bramon, Albert | Autonoma-Barcelona | SPAIN |
| Brandenburg, George W. | MIT | USA |
| Branson, J. | MIT | USA |
| Brasse, F.W. | CERN | CERN |
| Brau, James E. | SLAC | USA |
| Breakstone, Alan | Iowa State | USA |
| Brenner, A.E. | Fermilab | USA |
| Brenner, Richard | Caltech | USA |
| Breykin, V.A. | State Committee | USSR |
| Brick, David | Brown | USA |
| Bridges, David L. | LeMoyne College | USA |
| Brodsky, Stanley J. | SLAC | USA |
| Bronzan, J. | Rutgers | USA |
| Brown, Charles N. | Fermilab | USA |
| Brown, Lowell S. | Institute for Advanced Study | USA |
| Brown, Stanley G. | Physical Review | USA |
| Broyles, Arthur A. | Florida | USA |
| Brucker, E.B. | Stevens Institute of Tech | USA |
| Bruckner, Walter | CERN | WEST GERMANY |
| Brunet, Jean Michel | College de France | FRANCE |
| Buchholz, David A. | Northwestern | USA |
| Buchmuller, Wilfried | Bonn | WEST GERMANY |
| Bugg, William M. | Tennessee | USA |
| Buon, J. | Orsay | FRANCE |
| Buran, Torleiv | Oslo | NORWAY |
| Buras, Andrzej | Fermilab | USA |
| Burnett, Thompson H. | Washington | USA |
| Burns, Alan J. | CERN | CERN |
| Burnstein, R.A. | Illinois Institute of Tech | USA |
| Bussey, PJ | Glasgow | GREAT BRITAIN |
| Butterworth | Imperial College | GREAT BRITAIN |
| Byers, Nina | California-Los Angeles | USA |
| Cabibbo, N. | Rome | ITALY |
| Caldi, Daniel G. | Lawrence Berkeley Lab | USA |
| Camerini, Ugo | Wisconsin-Madison | USA |
| Cammarata, J. Barry | Virginia Polytechnic | USA |
| Campbell, Bruce | Toronto | CANADA |
| Campbell, William B. | Nebraska-Lincoln | USA |
| Candlin, D.J. | Edinburgh | GREAT BRITAIN |
| Caneschi, L. | INFN-Pisa | ITALY |
| Capps, Richard H. | Purdue | USA |

| | | |
|---|---|---|
| Caprasse, Hubert | Liege | BELGIUM |
| Carlitz, Robert | Pittsburgh | USA |
| Carlson, Carl Edwin | William and Mary | USA |
| Carrigan, Richard A. | Fermilab | USA |
| Cartacci, A.M. | INFN-Florence | ITALY |
| Carter, A.A. | CERN | GREAT BRITAIN |
| Carter, Ashton | Rockefeller | USA |
| Casalbuoni, Roberto | INFN-Florence | ITALY |
| Casella, Russell | National Bureau of Standards | USA |
| Caso, Carlo | INFN-Genoa | ITALY |
| Cason, Neal M. | Notre Dame | USA |
| Cassiday, George L. | Utah | USA |
| Castillejo, Leonardo | University College London | GREAT BRITAIN |
| Castoldi, P. | UL-Brussels | BELGIUM |
| Caswell, William E. | Maryland | USA |
| Catz, Philippe | Annecy-le-Vieux | FRANCE |
| Celeghini, Enrico | INFN-Florence | ITALY |
| Celmaster, William | Argonne National Lab | USA |
| Cerny, Joseph | Lawrence Berkeley Lab | USA |
| Cester, Rosanna | Turin | ITALY |
| Chadwick, George B. | SLAC | USA |
| Chahine, Charles | College de France | FRANCE |
| Chaichian, Masud | Helsinki | FINLAND |
| Chan, Lai Him | Louisiana State | USA |
| Chang, Chung Y. | Maryland | USA |
| Chang, Lay Nam | Virginia Polytechnic | USA |
| Chang, Lee | Qing Hua/Caltech | CHINA |
| Chang, Ngee-Pong | City College of New York | USA |
| Chao, Kuang Ta | Beijing Univ | CHINA |
| Chase, MK | Oxford | GREAT BRITAIN |
| Chela-Flores, Julian | Simon Bolivar | VENEZUELA |
| Chen, Chih Kwan | Purdue | USA |
| Chen, Hongfang | China Univ Sci & Tech/Wisconsi | CHINA |
| Chen, K. Wendell | Michigan State/CERN | USA |
| Chen, Min | DESY | WEST GERMANY |
| Cheng, Ta-Pei | Missouri-St. Louis | USA |
| Cheng, W. K. | Delaware | USA |
| Chertok, Benson | American | USA |
| Chevallier, Michel | Villeurbanne | FRANCE |
| Chew, Denyse | Lawrence Berkeley Lab | USA |
| Chiang, C.C. | Texas-Austin | TAIWAN, CHINA |
| Chiefari, Giovanni | INFN-Naples | ITALY |
| Chiu, Charles B. | Texas-Austin | USA |
| Chon, Yoo-Bang | Wisconsin-Madison | USA |
| Chou, T. T. | Georgia | USA |
| Chou, Yuehua | IHEP-Beijing/Rockefeller | CHINA |
| Christ, Norman | Columbia | USA |
| Christian, David | Johns Hopkins | USA |
| Chung, Suh-Urk | Brookhaven National Lab | USA |
| Ciafaloni, Marcello | Scuola Normale Superiore | ITALY |
| Clark, Alan R. | Lawrence Berkeley Lab | USA |
| Clark, Allan G. | CERN | CERN |
| Clark, Robert Beck | Texas A&M | USA |
| Clavelli, Louis J. | Bonn | WEST GERMANY |
| Cleymans, J. | Bielefeld | WEST GERMANY |
| Cline, David | Wisconsin-Madison | USA |
| Coffin, C. Tristram | Michigan | USA |
| Coleman, R. | Rochester | USA |
| Collins, John | Princeton | USA |
| Collins, PDB | Durham | GREAT BRITAIN |
| Conforto, Bianca | INFN-Florence | ITALY |
| Conforto, Gianni | INFN-Florence | ITALY |
| Contin, Andrea | CERN | CERN |
| Contogouris, Andreas P. | McGill | CANADA |
| Conversi, M. | INFN-Rome | ITALY |
| Cook, Victor | Washington | USA |
| Cool, Rodney L. | Rockefeller | USA |
| Cooper, John W. | Pennsylvania | USA |
| Corcoran, Marjorie D. | Rice | USA |
| Cords, Dieter | DESY | WEST GERMANY |
| Cormell, Laird R. | Pennsylvania | USA |

| | | |
|---|---|---|
| Costa, Giuseppe | INFN-Milan | ITALY |
| Cottingham, W.N. | Bristol | GREAT BRITAIN |
| Courant, Hans | Minnesota | USA |
| Cox, Bradley | Fermilab | USA |
| Coyne, Donald | Princeton | USA |
| Crawley, H.B. | Iowa State | USA |
| Creutz, Michael J. | Brookhaven National Lab | USA |
| Crutchfield, William Y. | SUNY-Stony Brook | USA |
| Cui, Huachuan | IHEP-Beijing/Illinois-Urbana-C | CHINA |
| Cumalat, John | Fermilab | USA |
| Curtright, T. | Chicago | USA |
| D'almagne, Bernard | Orsay | FRANCE |
| Dado, Shlomo | Technion | ISRAEL |
| Damgaard, Gunnar | Niels Bohr Institute | DENMARK |
| Dankowych, John A. | McGill | CANADA |
| Dar, Arnon | Technion | ISRAEL |
| Dasgupta, P. | Kalyani | INDIA |
| Dash, Jan | Marseille | FRANCE |
| Davier, Michel | Orsay | FRANCE |
| Davies, A. T. | Glasgow | GREAT BRITAIN |
| Davies, J.K. | Oxford | GREAT BRITAIN |
| Davis, A. C. | Imperial College | GREAT BRITAIN |
| De Calan, Claude | Ecole Polytechnique | FRANCE |
| De Swart, J.J. | Nijmegen | NETHERLANDS |
| DeGrand, Thomas A. | California-Santa Barbara | USA |
| DeTar, Carleton | Utah | USA |
| Decamp, Daniel | Orsay | FRANCE |
| Declais, Yves | Annecy-le-Vieux | FRANCE |
| Del Papa, Carlo | Washington | USA |
| Delpierre, Pierre | College de France | FRANCE |
| Derevtchikov, A.A. | Serpukhov | USSR |
| Derman, Emanuel | Colorado | USA |
| Derrick, M. | Argonne National Lab | USA |
| Deshpande, N.G | Oregon | USA |
| Devenski, P.A. | Fermilab | USA |
| Devlin, Thomas J. | Rutgers | USA |
| Diambrini-Palazzi, Giordano | CERN | CERN |
| Dibon, H. | Austrian Academy of Sciences | AUSTRIA |
| Diebold, Robert E. | Argonne National Lab | USA |
| Diekmann, Berud | Bonn | WEST GERMANY |
| Din, A.M. | Annecy-le-Vieux | FRANCE |
| Ditzler, W.R. | Argonne National Lab | USA |
| Dolan, Louise | Rockefeller | USA |
| Dolfini, R. | INFN-Pavia | ITALY |
| Dominguez, Cesareo A. | Texas A&M | USA |
| Domokos, Gabor | Johns Hopkins | USA |
| Donoghue, John F. | MIT | USA |
| Dorfan, David E. | California-Santa Cruz | USA |
| Dorfan, Jonathan | SLAC | USA |
| Dornan, P.J. | Imperial College | GREAT BRITAIN |
| Dosch, Hans Gunter | Hiedelberg | WEST GERMANY |
| Dothan, Yossef | Tel Aviv | ISRAEL |
| Dowd, John P. | Southeastern Massachusetts | USA |
| Drijard, Daniel | CERN | CERN |
| Ducros, Yves | Saclay | FRANCE |
| Dudelzak, Boris | Orsay | FRANCE |
| Dullemond, Cornelis | Catholic University-Nijmegen | NETHERLANDS |
| Duncan, Alan L. | Colorado | USA |
| Duong-Van, Minh | Los Alamos Scientific Lab | USA |
| Durand, Bernice | Wisconsin-Madison | USA |
| Durand, Loyal | Wisconsin-Madison | USA |
| Ebel, Marvin | Wisconsin-Madison | USA |
| Ebert, D. | JINR | USSR |
| Ecker, Gerhard | Vienna | AUSTRIA |
| Ecklund, Stan | SLAC | USA |
| Edwards, Bonnie Jean | Argonne National Lab | USA |
| Edwards, Kenneth W. | Carleton | CANADA |
| Eilam, Gad | Technion | ISRAEL |
| Einsweiler, Kevin | SLAC | USA |
| Eisner, Alan M. | California-Santa Barbara | USA |
| Elias, Victor | Toronto | CANADA |

| | | |
|---|---|---|
| Elliott, James | Colorado | USA |
| Ellis, Stephen D. | Washington | USA |
| Ellison, Robert John | Manchester | GREAT BRITAIN |
| Ellsworth, R. W. | Maryland | USA |
| Ely Jr., Robert P. | Lawrence Berkeley Lab | USA |
| Engler, Joachim | Karlsruhe | WEST GERMANY |
| Erickson, Roger | SLAC | USA |
| Ernwein, Jean | Fermilab | USA |
| Erwin, Albert | Wisconsin-Madison | USA |
| Eskreys, Andrzej | DESY | POLAND |
| Esposito, Belisario | CERN | CERN |
| Estabrooks, Penny | British Columbia | CANADA |
| Everett, Allen E. | Tufts | USA |
| Fabjan, Christian W. | CERN | CERN |
| Fabrizio, Ralph | California-Santa Cruz | USA |
| Fackler, Orrin | Rockefeller | USA |
| Fairbank, William | Stanford | USA |
| Falk-Vairant, Paul | CERN | CERN |
| Fanchiotti, Huner | La Plata | ARGENTINA |
| Fang, John | Georgia | USA |
| Fanourakis, George | Wisconsin-Madison | USA |
| Farmelo, Graham | Open | GREAT BRITAIN |
| Fayyazuddin | Quaid-i-Azam | PAKISTAN |
| Feldman, G. | SLAC | USA |
| Fenker, Howard | Fermilab | USA |
| Fernandez, Enrique | Argonne National Lab | USA |
| Ferro-Luzzi, M. | CERN | CERN |
| Ficenec, John R. | Virginia Polytechnic | USA |
| Fickinger, William J. | Case Western Reserve | USA |
| Fields, T.H. | Argonne National Lab | USA |
| Figiel, J. | Cracow | POLAND |
| Filho, Jose de sa Borges | Rio de Janeiro | BRAZIL |
| Finkelstein, Robert | California-Los Angeles | USA |
| Fischbach, Ephraim | Purdue | USA |
| Fisher, C. M. | Rutherford Lab | GREAT BRITAIN |
| Fleishon, Neil | Washington | USA |
| Flume, Rainald | Bonn | WEST GERMANY |
| Fogleman, Guy | Indiana | USA |
| Ford, William T. | Colorado | USA |
| Fowler, Earle C. | Purdue | USA |
| Fowler, William | | USA |
| Frampton, Paul | Harvard | USA |
| Franklin, Jerrold | Temple | USA |
| Franzini, Paulo | Columbia | USA |
| Fraser, Gordon | CERN Courier | CERN |
| Frazer, William R. | California-San Diego | USA |
| Fredriksson, Sverker | Royal Institute of Tech | SWEDEN |
| Freedman, Barry A. | Brookhaven National Lab | USA |
| French, B. | CERN | CERN |
| Frenkel, Josif | Sao Paulo | BRAZIL |
| Freudenreich, Klaus | CERN | WEST GERMANY |
| Freund, Peter G.O. | Chicago | USA |
| Friedman, Jerome I. | MIT | USA |
| Friedman, Marvin | Northeastern | USA |
| Fries, Rene | SLAC | USA |
| Fry, William F. | Wisconsin-Madison | USA |
| Fuchs, Norman H. | Purdue | USA |
| Fujii, Hirofumi | Tokyo | JAPAN |
| Fujii, Kanji | Hokkaido | JAPAN |
| Fujii, Tadao | Tokyo | JAPAN |
| Fujikawa, Kazuo | Tokyo-Tanashi | JAPAN |
| Fujimoto, Yasushi | Dublin | IRELAND |
| Fujimura, Kimio | Tokuyama | JAPAN |
| Fujisaki, Masakazu | Michigan | USA |
| Fujiwara, Kunio | Tokyo-Komaba | JAPAN |
| Fukawa, Mineo | KEK | JAPAN |
| Fulco, Jose R. | California-Santa Barbara | USA |
| Fulton, Thomas | Johns Hopkins | USA |
| Furman, Miguel A. | Columbia | USA |
| Gaemers, K.J.F. | Amsterdam | NETHERLANDS |
| Gago, Jose | Inst Superior Tecnico-Lisbon | PORTUGAL |

| | | |
|---|---|---|
| Gaidos, James | Purdue | USA |
| Gaillard, Mary Kath | Annecy-le-Vieux | SWITZERLAND |
| Gaines, Irwin | Fermilab | USA |
| Gaisser, Thomas K. | Deleware | USA |
| Galbraith, W. | Sheffield | GREAT BRITAIN |
| Gao, Chong-Shou | Beijing Univ/SLAC | CHINA |
| Garbincius, Peter H. | Fermilab | USA |
| Garcia, Augusto | CIEA-IPN | MEXICO |
| Garcia-Canal, Carlos Alberto | La Plata | ARGENTINA |
| Garelick, David | Northeastern | USA |
| Garren, Lynn | Illinois-Urbana-Champaign | USA |
| Gasiorowicz, S. | Minnesota | USA |
| Gavillet, Philippe | CERN | CERN |
| Geich-Gimbel, C. | Bonn | WEST GERMANY |
| Gelfand, Norman M. | Fermilab | USA |
| Gell-Mann, Murray | Caltech | USA |
| George, Roger | Pierre et Marie Curie | FRANCE |
| Giacomelli, Giorgio | INFN-Bologna | ITALY |
| Gibbard, Bruce G. | Brookhaven National Lab | USA |
| Giles, Roscoe C. | MIT | USA |
| Gilmore, Robin | SLAC | USA |
| Gittelman, Bernard | Cornell | USA |
| Glass, George | Los Alamos Scientific Lab | USA |
| Glasser, Robert G. | Maryland | USA |
| Glaubman, Michael | Northeastern | USA |
| Gluck, Moshe | Dortmund | WEST GERMANY |
| Goebel, Charles | Wisconsin-Madison | USA |
| Goldberg, Marvin | Syracuse | USA |
| Goldhaber, Maurice | Brookhaven National Lab | USA |
| Goldman, Joseph I. | American | USA |
| Goldman, Terry | Los Alamos Scientific Lab | USA |
| Goldschmidt-Clermont, Y. | CERN | CERN |
| Golowich, Eugene | Massachusetts | USA |
| Gorbics, Mark | Iowa State | USA |
| Gordon, Howard A. | Brookhaven National Lab | USA |
| Goshaw, Alfred T. | Duke | USA |
| Gottlieb, Steven | Argonne National Lab | USA |
| Gottschalk, Thomas D. | Argonne National Lab | USA |
| Grannis, Paul | SUNY-Stony Brook | USA |
| Grard, F. | l'Etat | BELGIUM |
| Grassler, Herbert | RWTH Aachen | WEST GERMANY |
| Green, M.B. | Queen Mary College | GREAT BRITAIN |
| Green, MG | Westfield College | GREAT BRITAIN |
| Greenberg, O. W. | Maryland | USA |
| Grisaru, Mark | Brandeis | USA |
| Gross, Franz L. | William and Mary | USA |
| Grote, Hans | CERN | CERN |
| Grupen, Claus | Siegen | WEST GERMANY |
| Gu, Yifan | IHEP-Beijing/SLAC | CHINA |
| Guerin, Francoise | Brown | USA |
| Gunderson, Bruce N. | Max Planck Institute | WEST GERMANY |
| Guryn, Wlodzimierz | SLAC | USA |
| Gustafson, H. Richard | Michigan | USA |
| Haacke, E. Mark | Case Western Reserve | USA |
| Haber, Howard | Lawrence Berkeley Lab | USA |
| Hafen, Elizabeth S. | MIT | USA |
| Hagberg, Eskil | Gustaf Werner's Institute | SWEDEN |
| Hagen, C. | Rochester | USA |
| Haggerty, Herman | Fermilab | USA |
| Hagiwara, T. | Brandeis | USA |
| Hagopian, Sharon | Florida State | USA |
| Hagopian, Vasken | Florida State | USA |
| Hahn, B. | Bern | WEST GERMANY |
| Haidt, D. | DESY | WEST GERMANY |
| Haller, Kurt | Connecticut | USA |
| Halliwell, Clive | Illinois-Chicago Circle | USA |
| Halprin, Arthur | Deleware | USA |
| Halzen, Francis | Wisconsin-Madison | USA |
| Han, M.Y. | Duke | USA |
| Handler, Robert | Wisconsin-Madison | USA |
| Hanke, Paul | CERN | WEST GERMANY |

| | | |
|---|---|---|
| Hansen, Jorn Dines | Niels Bohr Institute | DENMARK |
| Hanson, Gail | SLAC | USA |
| Harari, Haim | Weizmann Institute | ISRAEL |
| Harris, Frederick | Hawaii | USA |
| Hayashi, Mitsuo | Tokai | JAPAN |
| Haymaker, Richard | Louisiana State | USA |
| Hayot, Fernand | Saclay | FRANCE |
| Heisterberg, Richard H. | Virginia Polytechnic | USA |
| Heller, Ken | Minnesota | USA |
| Hemingway, R.J. | Carleton | CANADA |
| Hemmi, Yasuo | Kyoto | JAPAN |
| Hendrick, R.E. | St. Bonaventure | USA |
| Henri, Victor P. | l'Etat | BELGIUM |
| Henzi, R. | McGill | CANADA |
| Hepp, Volker | Hamburg | WEST GERMANY |
| Herb, Stephen | Columbia | USA |
| Herczeg, Peter | Los Alamos Scientific Lab | USA |
| Herquet, Philippe | l'Etat | BELGIUM |
| Higashijima, Kiyoshi | Tokyo | JAPAN |
| Hilger, Erwin | Bonn | WEST GERMANY |
| Hinchliffe, I. | Lawrence Berkeley Lab | USA |
| Hoek, J. | Amsterdam | NETHERLANDS |
| Hofer, H. | SIN | SWITZERLAND |
| Hoff, Gloria | Illinois-Chicago Circle | USA |
| Hoffmann, Lothar | CERN | CERN |
| Hofstadter, Robert | Stanford | USA |
| Hogaasen, Hallstein | Oslo | NORWAY |
| Holmes, Stephen | Columbia | USA |
| Holmgren, S.O. | Stockholm | SWEDEN |
| Homer, R.J. | Birmingham | GREAT BRITAIN |
| Honda, Keijiro | Hokkaido | JAPAN |
| Horwitz, Nahmin | Syracuse | USA |
| Hosoya, Akio | Osaka | JAPAN |
| Hou, Bo-Yu | Northwest (CHINA) | CHINA |
| Houlden, M.A. | Liverpool | GREAT BRITAIN |
| Hoyer, Paul | NORDITA | DENMARK |
| Hu, Bambi | Houston | USA |
| Huang, Justin C. | Missouri | USA |
| Huang, Tao | IHEP-Beijing/SLAC | CHINA |
| Hugentobler, E. | Bern | SWITZERLAND |
| Hughes, Ian | Glasgow | GREAT BRITAIN |
| Hughes, Vernon | Yale | USA |
| Hulth, P.O. | CERN | CERN |
| Hwa, Rudolph C. | Oregon | USA |
| Hylen, Jim | Michigan State | USA |
| Hyman, Lloyd G. | Argonne National Lab | USA |
| Igi, Keiji | Tokyo | JAPAN |
| Imlay, Richard L. | Louisiana State | USA |
| Imrie, Derek C. | University College London | GREAT BRITAIN |
| Irving, Alan C | Liverpool | GREAT BRITAIN |
| Irwin, George | SLAC | USA |
| Isgur, Nathan | Toronto | CANADA |
| Ishihara, Nobuhiro | KEK | JAPAN |
| Ishikawa, Kenzo | California-Los Angeles | USA |
| Islam, Muhammad M. | Connecticut | USA |
| Iso, Chikashi | Tokyo Institute of Tech | JAPAN |
| Iwao, Syurei | Kanazawa | JAPAN |
| Izatt, Dale | Utah | USA |
| Jackiw, Roman | MIT | USA |
| Jacobs, Laurence | UNA-Mexico | MEXICO |
| Jacquet, Francois | Ecole Polytechnique | FRANCE |
| Jaffe, Robert L. | MIT | USA |
| Jaffre, Michel | Orsay | FRANCE |
| Jagannathan, K. | Rochester | USA |
| Jain, P.L. | SUNY-Buffalo | USA |
| Jancso, G. | Hungarian Academy of Sciences | HUNGARY |
| Jarlskog, Goran | Lund | SWEDEN |
| Jaroszewicz, T. | Cracow | POLAND |
| Jenkins, Edgar W. | Arizona | USA |
| Jensen, Douglas A. | Massachusetts | USA |
| Jersak, J. | Aachen Hochschule | WEST GERMANY |

| | | |
|---|---|---|
| Jevicki, Antal | Brown | USA |
| Johnson, Denis P. | CERN | BELGIUM |
| Johnson, James | Wisconsin-Madison | USA |
| Johnson, Porter W. | Illinois Institute of Tech | USA |
| Johnson, Randy A. | Brookhaven National Lab | USA |
| Johnson, Rolland P. | Fermilab | USA |
| Johnson, William B. | SLAC | USA |
| Jonckheere, Alan M. | Fermilab | USA |
| Jones, Daniel Lee | Argonne National Lab | USA |
| Jones, H.F. | Imperial College | GREAT BRITAIN |
| Jones, Keith | Edit. Office of Nuclear Physic | DENMARK |
| Jones, Lawrence W. | Michigan | USA |
| Jones, Lorella | Illinois-Urbana-Champaign | USA |
| Joshi, G.C. | Melbourne | AUSTRALIA |
| Jostlein, Hans | Fermilab | USA |
| Jousset, Jacques | Clermont | FRANCE |
| Kabe, Seiji | KEK | JAPAN |
| Kafka, Tomas | SUNY-Stony Brook | USA |
| Kahn, Stephen | Brookhaven National Lab | USA |
| Kajikawa, Ryoichi | Nagoya | JAPAN |
| Kaku, Michio | City College of New York | USA |
| Kalbfleisch, George R. | Oklahoma | USA |
| Kamal, A.N. | Alberta | CANADA |
| Kane, Gordon L. | Michigan | USA |
| Kang, J.S. | Korea Standards Research Inst. | SOUTH KOREA |
| Kanofsky, Alvin | Lehigh | USA |
| Karczmarczuk, J. | Jagellonian | POLAND |
| Karshon, Uri | Weizmann Institute | ISRAEL |
| Kasha, Henry | Yale | USA |
| Kawamoto, N. | Amsterdam | NETHERLANDS |
| Kayser, Boris | National Science Foundation | USA |
| Kazama, Yoichi | Fermilab | USA |
| Kazes, Emil | Pennsylvania State | USA |
| Kendall, Henry W. | MIT | USA |
| Kennedy, A. D. | Maryland | USA |
| Kennett, Rosemary G. | Caltech | USA |
| Kenney, V. Paul | Notre Dame | USA |
| Kephart, Robert D. | Fermilab | USA |
| Khan, Irshadullah | Riyadh | SAUDI ARABIA |
| Khanna, Mohinder P. | Panjab | INDIA |
| Kichimi, Hiromichi | KEK | JAPAN |
| Kielanowski, Piotr | Warsaw | POLAND |
| Kikkawa, Keiji | Hiroshima | JAPAN |
| Kim, Chong Oh | Korea | SOUTH KOREA |
| Kim, D.Y. | Regina | CANADA |
| Kim, Jae Kwan | Korea Advanced Institute of Sc | SOUTH KOREA |
| Kim, Y. S. | Maryland | USA |
| Kinoshita, Toichiro | Cornell | USA |
| Kinson, J.B. | Birmingham | GREAT BRITAIN |
| Kirk, Thomas B.W. | Fermilab | USA |
| Kirkbride, Ian | Stanford | USA |
| Kiskis, Joe | Los Alamos Scientific Lab | USA |
| Kistiakowsky, Vera | MIT | USA |
| Kitagaki, Toshio | Tohoku | JAPAN |
| Kitazoe, Tetsuro | Kobe | JAPAN |
| Klein, M. | JINR | USSR |
| Kleinknecht, Konrad | Dortmund | WEST GERMANY |
| Kluge, E.E. | Heidelberg | WEST GERMANY |
| Knobloch, Jurgen R. | CERN | WEST GERMANY |
| Ko, Winston | California-Davis | USA |
| Kobayashi, Makoto | KEK | JAPAN |
| Kobrak, Hans | California-San Diego | USA |
| Koester Jr., Louis J. | Illinois-Urbana-Champaign | USA |
| Koh, In-Gyu | Sogang | SOUTH KOREA |
| Koller, E. L. | Stevens Institute of Tech | USA |
| Kolyvodiakos, Nick J. | Ministry of Education | GREECE |
| Kondo, Takahiko | Fermilab | USA |
| Konigsmann, Kay | Stanford | USA |
| Koshiba, Masatoshi | Tokyo | JAPAN |
| Kourkoumelis, Christine | Athens | GREECE |
| Kovacs, Eve | Melbourne | AUSTRALIA |

| | | |
|---|---|---|
| Kovesi-Domokos, Susan | Johns Hopkins | USA |
| Kraemer, Robert W. | Carnegie Mellon | USA |
| Kreisler, Michael N. | Massachusetts | USA |
| Kretzschmar, Martin | Johannes-Gutenberg | WEST GERMANY |
| Kroll, Norman M. | California-San Diego | USA |
| Kuang, Yu-Ping | Lanzhou | CHINA |
| Kubantsev, M.A. | ITEP | USSR |
| Kugler, Moshe | Weizmann Institute | ISRAEL |
| Kuhn, Johann H. | Max Planck Institute | WEST GERMANY |
| Kuo, T. K. | Purdue | USA |
| Kurtz, Nicole | Strasbourg | FRANCE |
| Kuti, J. | Hungarian Academy of Sciences | HUNGARY |
| Kwak, Nowhan | Kansas | USA |
| Kwiecinski, J. | Cracow | POLAND |
| Kycia, T. F. | Brookhaven National Lab | USA |
| Laasanen, Alvin T. | Purdue | USA |
| Lach, Joseph | Fermilab | USA |
| Lackner, Klaus | Caltech | USA |
| Lam, Harry C.S. | McGill | CANADA |
| Lander, Richard L. | California-Davis | USA |
| Langacker, Paul | Pennsylvania | USA |
| Lanius, Karl | Zeuthen | EAST GERMANY |
| Lankford, Andrew J. | Lawrence Berkeley Lab | USA |
| Lassila, K. | Iowa State | USA |
| Laursen, Morten | Oklahoma State | USA |
| Layas, M | | unknown |
| Layssac, Jacques | Montpellier | FRANCE |
| Layter, John G. | California-Riverside | USA |
| LeBritton, Joseph A. | Rochester | USA |
| Leacock, R.A. | Iowa State | USA |
| Leader, Elliot | Westfield College | GREAT BRITAIN |
| Lee, Choonkyu | Michigan | USA |
| Lee, H.C. | Chalk River Nuclear Lab | CANADA |
| Lee, James Roy | Caltech | USA |
| Lee, Richard Y.Y. | Tsing Hua | TAIWAN, CHINA |
| Lee-Franzini, Juliet | SUNY-Stony Brook | USA |
| Lefrancois, Jacques | Orsay | FRANCE |
| Lehman, Elliot | Cincinnati | USA |
| Leibbrandt, George | Guelph | CANADA |
| Leipuner, L. | Brookhaven National Lab | USA |
| Lemonne, J. | VU-Brussels | BELGIUM |
| Lennox, Arlene | Fermilab | USA |
| Leung, Pat | SLAC | USA |
| Leveille, Jacques | Wisconsin-Madison | USA |
| Levy, Aharon | Tel Aviv | ISRAEL |
| Levy, Ahron | MIT | USA |
| Li, Bing-An | IHEP-Beijing/SLAC | CHINA |
| Li, Ling-Fong | Carnegie Mellon | USA |
| Li, Xiaoyuan | ITP-Beijing | CHINA |
| Licht, Lewis | Illinois-Chicago Circle | USA |
| Lim, Y.K. | Singapore | SINGAPORE |
| Lindenbaum, S. J. | City College of New York | USA |
| Ling, Ta-Yung | Ohio State | USA |
| Linn, Stephan L. | Michigan | USA |
| Lipkin, H.J. | Fermilab | ISRAEL |
| Lipton, Ronald | Northwestern | USA |
| Lissauer, David | Tel Aviv | ISRAEL |
| Litke, Alan | SLAC | USA |
| Llewellyn-Smith, C.H. | Oxford | GREAT BRITAIN |
| LoSecco, John M. | Michigan | USA |
| Loh, Eugene | SLAC | USA |
| Loken, Stewart C. | Lawrence Berkeley Lab | USA |
| Long, Bill | Conference Staff | USA |
| Longacre, Ronald S. | Brookhaven National Lab | USA |
| Loomis, William A. | Harvard | USA |
| Lopez, Cayetano | Madrid | SPAIN |
| Love, Sherwin | Purdue | USA |
| Loveless, Richard | Wisconsin-Madison | USA |
| Loverre, Pier Ferruccio | CERN | ITALY |
| Low, Francis E. | MIT | USA |
| Lowenstein, John H. | New York University | USA |

| | | |
|---|---|---|
| Lubatti, H.J. | Washington | USA |
| Lucas, Peter W. | Duke | USA |
| Ludlam, Thomas W. | Brookhaven National Lab | USA |
| Lukierski, J. | Wrockaw | POLAND |
| Luste, G.J. | British Columbia | CANADA |
| Luth, Vera | SLAC | USA |
| Lys, Jeremy | Lawrence Berkeley Lab | USA |
| Ma, Ernest | Hawaii | USA |
| MacGregor, Malcolm H. | Lawrence Livermore Lab | USA |
| Madaras, Ronald | Lawrence Berkeley Lab | USA |
| Mahanthappa, Kalyana T. | Colorado | USA |
| Mainland, G. Bruce | Ohio State | USA |
| Majerotto, Walter | Austrian Academy of Sciences | AUSTRIA |
| Maki, Akihiro | Fermilab | USA |
| Maki, Ziro | Kyoto | JAPAN |
| Malamud, Ernest | Fermilab | USA |
| Malhotra, P.K. | Tata Institute | INDIA |
| Malos, J. | Bristol | GREAT BRITAIN |
| Mann, W. Anthony | Tufts | USA |
| Mansouri, Freydoon | Yale | USA |
| March, Robert | Wisconsin-Madison | USA |
| Margolis, B. | McGill | CANADA |
| Margulies, Seymour | Illinois-Chicago Circle | USA |
| Markees, Alf | Dortmund | WEST GERMANY |
| Markytan, Manfred | Austrian Academy of Sciences | AUSTRIA |
| Marnelius, Robert | Goteborg | SWEDEN |
| Marshak, Robert E. | Virginia Polytechnic | USA |
| Martellotti, G. | INFN-Rome | ITALY |
| Martin, Andre | CERN | CERN |
| Martin, Francois | CERN | CERN |
| Martin, John | Toronto | CANADA |
| Martin, Philip S. | Lawrence Berkeley Lab | USA |
| Marx, Michael D. | Brookhaven National Lab | USA |
| Mason, G. C. | Melbourne | AUSTRALIA |
| Massaro, G.G.G. | DESY | NETHERLANDS |
| Masuda, Naohiko | Yamagata | JAPAN |
| Mathur, V. | Rochester | USA |
| Matsuoka, Takeo | Nagoya | JAPAN |
| Matteuzzi, Clara | CERN | CERN |
| Matveev, V.A. | Academy of Sciences | USSR |
| May, Edward N. | Argonne National Lab | USA |
| Mazur, Peter O. | Fermilab | USA |
| McCal, Dennis | Wisconsin-Madison | USA |
| McCliment, Edward R. | Iowa | USA |
| McCusker, Brian | Sydney | AUSTRALIA |
| McDaniel, Boyce | Cornell | USA |
| McEwen, J.G. | Southampton | GREAT BRITAIN |
| McGlinn, William D. | Notre Dame | USA |
| McIntyre, Peter | Fermilab | USA |
| McLeod, Donald | Illinois-Chicago Circle | USA |
| Mendel, Roberto | MIT | USA |
| Merkel, B. | Orsay | FRANCE |
| Mermod, Ronald | Geneva | SWITZERLAND |
| Merritt, F. | Chicago | USA |
| Meshkov, Sydney | National Bureau of Standards | USA |
| Mess, Karl H. | CERN | CERN |
| Messing, Fredric | Carnegie Mellon | USA |
| Metcalf, William J. | Louisiana State | USA |
| Metzger, Wesley J. | Nijmegen | NETHERLANDS |
| Michelotti, Leo | Fermilab | USA |
| Miczaika, Thomas S. | Bonn | WEST GERMANY |
| Miettinen, Hannu E. | Rice | USA |
| Miettinen, Hannu I. | Helsinki | FINLAND |
| Migneron, R. | Western Ontario | CANADA |
| Mikaelian, Karnig O. | Oklahoma State | USA |
| Milburn, Richard H. | Tufts | USA |
| Miller, David H. | Purdue | USA |
| Milton, Kimball A. | California-Los Angeles | USA |
| Minakata, H. | Fermilab | USA |
| Mine, Philippe | Ecole Polytechnique | FRANCE |
| Mischke, Richard E. | Los Alamos Scientific Lab | USA |

| | | |
|---|---|---|
| Nishina, Masanori | Fermilab | USA |
| Mita, Katsunori | Oklahoma State | USA |
| Miyachi, Takashi | Tokyo-Tanashi | JAPAN |
| Miyake, Kozo | Kyoto | JAPAN |
| Mohapatra, Rabindra N. | City College of New York | USA |
| Moller, Rasmus | Niels Bohr Institute | DENMARK |
| Montanet, L. | CERN | CERN |
| Moore, Craig | Fermilab | USA |
| Morel, Benoit F. | Harvard | USA |
| Morgan, Thomas A. | Nebraska-Lincoln | USA |
| Moriyasu, Keihachiro | Washington | USA |
| Morpurgo, G. | INFN-Genoa | ITALY |
| Morrison, Rollin J. | California-Santa Barbara | USA |
| Morse, Robert | Wisconsin-Madison | USA |
| Mosca, Luigi | Saclay | FRANCE |
| Mount, R. | Oxford | GREAT BRITAIN |
| Mukhopadhyay, P. | Jadavpur | INDIA |
| Mulders, Peter J. | Nijmegen | NETHERLANDS |
| Muller, F. | CERN | CERN |
| Mulvey, John | Oxford | GREAT BRITAIN |
| Munczek, Herman | Kansas | USA |
| Murphy, C. Thornton | Fermilab | USA |
| Muta, Taizo | Kyoto | JAPAN |
| Myrheim, Jan | Niels Bohr Institute | DENMARK |
| Nagashima, Yorikiyo | KEK | JAPAN |
| Nahm, Werner | CERN | CERN |
| Nakagawa, N. | Purdue | USA |
| Nakamura, Seitaro | Tokai | JAPAN |
| Nakazawa, Nobuya | Kogakuin | JAPAN |
| Nambu, Yoichiro | Chicago | USA |
| Nandi, Satyanarayan | Ohio State | USA |
| Nanopoulos, Demetres | CERN | CERN |
| Narayan, D.S. | Tata Institute | INDIA |
| Narayana-Swamy, P. | Southern Illinois | USA |
| Nash, Thomas | Fermilab | USA |
| Nassalski, Jan P. | Warsaw | POLAND |
| Natali, Sergio | INFN-Bari | ITALY |
| Nath, Pran | Northeastern | INDIA |
| Navarria, F | INFN-Bologna | ITALY |
| Navelet, Henri | Saclay | FRANCE |
| Ne'eman, Yuval | Tel Aviv | ISRAEL |
| Nelson Jr., Charles A. | Fermilab | USA |
| Neuffer, David | Fermilab | USA |
| Newman, Harvey B. | DESY | WEST GERMANY |
| Nicole, Denis Alan | Pittsburgh | USA |
| Niebergall, Friedrich | Hamburg | WEST GERMANY |
| Nigro, M. | INFN-Padua | ITALY |
| Nikitin, Yu P. | State Committee | USSR |
| Nilsson, Sigward | Stockholm | SWEDEN |
| Ninomiya, Kansuke | Nihon Fukushi Daigaku | JAPAN |
| Nishikawa, Tetsuji | KEK | JAPAN |
| Niu, Kiyoshi | Nagoya | JAPAN |
| Nodulman, Larry | Argonne National Lab | USA |
| Nordberg, Emery | Cornell | USA |
| Norton, P.R. | Rutherford Lab | GREAT BRITAIN |
| Novikoff, David | Caltech | USA |
| Nuzzo, Salvatore | INFN-Bari | ITALY |
| O'Donnell, Patrick J. | Toronto | CANADA |
| O'Neill, Gerard K. | Princeton | USA |
| Oades, G.C. | Aarhus | DENMARK |
| Oberlack, Horst | Max Planck Institute | WEST GERMANY |
| Odorico, R. | INFN-Bologna | ITALY |
| Ogawa, Arthur | SLAC | USA |
| Ogielski, A. | Wrocław | POLAND |
| Oh, B.Y. | Michigan State | USA |
| Oh, Choo-Hiap | Malaysia | MALAYSIA |
| Ojima, Izumi | Kyoto | JAPAN |
| Okabayashi, M. | Shiga Daigaku | JAPAN |
| Oliver, William P. | Tufts | USA |
| Olsen, J.M. | Bergen | NORWAY |
| Olsen, Stephen L. | Rochester | USA |

| | | |
|---|---|---|
| Olsson, Martin | Wisconsin-Madison | USA |
| Oneda, Sadao | Maryland | USA |
| Ono, Masaaki | Fermilab | USA |
| Ono, Seiji | RWTH Aachen | WEST GERMANY |
| Orito, Shuji | Tokyo | JAPAN |
| Orr, J. Rich | Fermilab | USA |
| Overseth, Oliver E. | Michigan | USA |
| Ovrut, Burt | Brandeis | USA |
| Owen, Dan | Fermilab | USA |
| Owens, Joseph F. | Florida State | USA |
| Pac, Pong Youl | Seoul National University | SOUTH KOREA |
| Pagels, Heinz | Rockefeller | USA |
| Paige Jr., Frank E. | Brookhaven National Lab | USA |
| Paikeday, Joseph | SE Missouri State | USA |
| Pakvasa, Sandip | Hawaii | USA |
| Palmonari, Federico | INFN-Bologna | ITALY |
| Pancheri-Srivastava, Giulia | Northeastern | ITALY |
| Pandoulas, D. | Imperial College | WEST GERMANY |
| Parisi, G. | Frascati | ITALY |
| Park, Don | Chicago | USA |
| Parker, Don | Iowa State | USA |
| Pasierb, Elaine | California-Irvine | USA |
| Pasti, P. | INFN-Padua | ITALY |
| Pasupathy, J. | Bangalore | INDIA |
| Patel, P.M. | McGill | CANADA |
| Pati, Jogesh | ICTP-Trieste | ITALY |
| Patricelli, Sergio | INFN-Naples | ITALY |
| Paul, W. | Bonn | WEST GERMANY |
| Pauss, Felicitas | Cornell | USA |
| Paver, Nello | ICTP-Trieste | ITALY |
| Payne, F. | Liverpool | GREAT BRITAIN |
| Pearson, Robert | California-Santa Barbara | USA |
| Peaslee, David | Department of Energy | USA |
| Penev, Vladimir | IAIAE | BULGARIA |
| Peoples, John | Fermilab | USA |
| Pessard, Henri | Annecy-le-Vieux | FRANCE |
| Peters, Michael W. | Hawaii | USA |
| Petersen, Jens Lyng | Niels Bohr Institute | DENMARK |
| Peterson, Earl A. | Minnesota | USA |
| Peterson, Vincent Z. | Hawaii | USA |
| Petersson, Bengt | Bielefeld | WEST GERMANY |
| Petroff, Pierre | Orsay | FRANCE |
| Petronzio, R. | CERN | CERN |
| Peyaud, Bernard | Saclay | FRANCE |
| Peyrou, Charles | CERN | CERN |
| Pfeil, Walter | Toronto | CANADA |
| Pham, Trinang | Ecole Polytechnique | FRANCE |
| Pham, Xuan Yem | Pierre et Marie Curie | FRANCE |
| Phillips, James | Stanford | USA |
| Phillips, Roger | Rutherford Lab | GREAT BRITAIN |
| Pi, So-Young | SLAC | USA |
| Picasso, Emilo | CERN | CERN |
| Piccioni, Oreste | California-San Diego | USA |
| Pietschmann, H. | Vienna | AUSTRIA |
| Pinsky, Stephen S. | Ohio State | USA |
| Piroue, Pierre | Princeton | USA |
| Pisarski, Rob | Yale | USA |
| Pitthan, Rainer | SLAC | USA |
| Poirier, John A. | Notre Dame | USA |
| Pollard, David | Stanford | USA |
| Pondrom, Lee | Wisconsin-Madison | USA |
| Porter, Frank | Caltech | USA |
| Posa, F. | INFN-Bari | ITALY |
| Potter, Douglas N. | Rutgers | USA |
| Poulet, Maurice | Annecy-le-Vieux | FRANCE |
| Prentice, J.D. | Toronto | CANADA |
| Prepost, Richard | Wisconsin-Madison | USA |
| Pretzl, Klaus | Max Planck Institute | WEST GERMANY |
| Price, Lawrence E. | Argonne National Lab | USA |
| Pringle, Thomas H. | Oregon | USA |
| Pripstein, Morris | Lawrence Berkeley Lab | USA |

| | | |
|---|---|---|
| Prosperi, Giovanni | INFN-Milan | ITALY |
| Protopopescu, Serban | Brookhaven National Lab | USA |
| Quigg, Chris | Fermilab | USA |
| Radeka, V. | Brookhaven National Lab | USA |
| Rahm, D.C. | CERN | CERN |
| Raja, Rajendran | Fermilab | USA |
| Ram, Budh | New Mexico State | USA |
| Ramachandran, R. | Indian Inst of Tech-Kanpur | INDIA |
| Ramond, P. | Caltech | USA |
| Randa, James | Colorado | USA |
| Range, WH | Liverpool | GREAT BRITAIN |
| Rapidis, Petros A. | Fermilab | USA |
| Rapp, Patrick | Fermilab | USA |
| Ratti, S. | INFN-Pavia | ITALY |
| Ray, Asim K. | Visva Bharati | INDIA |
| Reay, Neville W. | Fermilab | USA |
| Recami, Erasmo | Catania | ITALY |
| Reeder, Don | Wisconsin-Madison | USA |
| Reibel, Kurt | Ohio State | USA |
| Reines, Frederick | California-Irvine | USA |
| Reinharz, Max | CERN | CERN |
| Rek, Zbigniew | Iowa State | USA |
| Repko, Wayne W. | Michigan State | USA |
| Resvanis, Leo | Athens | GREECE |
| Reucroft, Steve | CERN | GREAT BRITAIN |
| Revel, Daniel | Weizmann Institute | ISRAEL |
| Reya, E. | Dortmund | WEST GERMANY |
| Rhoades, Jean | Wisconsin-Madison | USA |
| Rich, James | Saclay | FRANCE |
| Ritson, David | SLAC | USA |
| Rivoal, Monique | Pierre et Marie Curie | FRANCE |
| Roberto, Vito | ICTP-Trieste | ITALY |
| Roberts, R.G. | Rutherford Lab | GREAT BRITAIN |
| Roberts, Thomas J. | Michigan | USA |
| Robinson, Barry | Pennsylvania | USA |
| Robinson, D. Keith | Case Western Reserve | USA |
| Rock, Stephen | American | USA |
| Roe, Byron P. | Michigan | USA |
| Roehrig, J. | SLAC | USA |
| Rohrlich, Fritz | Syracuse | USA |
| Romanowski, Thomas | Ohio State | USA |
| Ronat, Elhanan E. | Weizmann Institute | ISRAEL |
| Roos, Charles E. | Vanderbilt | USA |
| Rosen, Jerome L. | Northwestern | USA |
| Rosen, S. Peter | Purdue | USA |
| Rosenberg, Eli | CERN | USA |
| Rosenbladt, Bob | Washington | USA |
| Rosenfeld, C. | Rochester | USA |
| Rosner, Jonathan L. | Minnesota | USA |
| Ross, Richard T. | CERN | GREAT BRITAIN |
| Rosselet, Philippe | Lausanne | SWITZERLAND |
| Rotelli, Pietro | INFN-Bari | ITALY |
| Rothenberg, Allan | CERN | CERN |
| Roussarie, Andre C. | Saclay | USA |
| Rousset, Andre | DGRST | FRANCE |
| Roy, D.P. | Tata Institute | INDIA |
| Roy, S.M. | Tata Institute | INDIA |
| Ruan, Tunan | China Univ Sci & Tech | CHINA |
| Rubbia, C. | CERN | CERN |
| Rubinstein, Roy | Fermilab | USA |
| Rudolph, G. | Wisconsin-Madison | USA |
| Ruhl, Werner | Kaiserslautern | WEST GERMANY |
| Russ, James | Carnegie Mellon | USA |
| Sachs, Robert G. | Chicago | USA |
| Sadoulet, Bernard | CERN | CERN |
| Sadrozinski, Hartmut F.W. | Princeton | USA |
| Sajot, Gerard | College de France | FRANCE |
| Sakai, Norisuke | Tohoku | JAPAN |
| Sakita, Bunji | City College of New York | USA |
| Salam, Adbus | ICTP-Trieste | ITALY |
| Salomonson, Per | Goteborg | SWEDEN |

| | | |
|---|---|---|
| Salvini, Giorgio | CERN | ITALY |
| Samuel, Mark A. | Oklahoma State | USA |
| Sanders, Gary H. | Los Alamos Scientific Lab | USA |
| Sandweiss, Jack | Yale | USA |
| Sanford, J. | Brookhaven National Lab | USA |
| Santroni, Alberto | INFN-Genoa | ITALY |
| Sard, Robert D. | Illinois-Urbana-Champaign | USA |
| Sarmadi, Mohammad H. | Pittsburgh | USA |
| Sartorelli, Gabriella | CERN | CERN |
| Sasaki, Ken | Yokohama | JAPAN |
| Saulys, Alfred | Brookhaven National Lab | USA |
| Sayegh, Samir | Notre Dame | USA |
| Sazdjian, Hagop | Orsay | FRANCE |
| Scadron, Michael D. | Arizona | USA |
| Schachinger, L. | Chicago | USA |
| Schaeffer, Michel | Strasbourg | FRANCE |
| Schalk, Terry | California-Santa Cruz | USA |
| Schamberger, Robert Dean | SUNY-Stony Brook | USA |
| Scharre, Daniel L. | SLAC | USA |
| Schechter, Joe | Syracuse | USA |
| Schiavon, P. | INFN-Trieste | ITALY |
| Schiff, Dominique | Orsay | FRANCE |
| Schiller, Diethard | Siegen | WEST GERMANY |
| Schlatter, Dieter | SLAC | USA |
| Schluter, Robert A. | Northwestern | USA |
| Schmidt, Michael P. | Yale | USA |
| Schmidt-Parzefall, W. | DESY | WEST GERMANY |
| Schnell, W. | CERN | CERN |
| Schneps, Jack | Tufts | USA |
| Schnitzer, Howard J. | Brandeis | USA |
| Schomblond, CH | UL-brussels | BELGIUM |
| Schopper, Herwig | DESY | WEST GERMANY |
| Schuler, Peter | Yale | WEST GERMANY |
| Schultz, Jonas | California-Irvine | USA |
| Sciulli, Frank | Caltech | USA |
| Sciuto, S. | INFN-Turin | ITALY |
| Scott, David | Wisconsin-Madison | USA |
| Segler, Samuel L. | Fermilab | USA |
| Seiden, Abraham | California-Santa Cruz | USA |
| Selove, Walter | Pennsylvania | USA |
| Seltesse, Joel | Saclay | FRANCE |
| Senjanovic, Goran | Maryland | USA |
| Sepunaru, Dani | Duke | USA |
| Shaevitz, Michael H. | Caltech | USA |
| Shafi, Qaisar | CERN | GREAT BRITAIN |
| Shambroom, W. David | SLAC | USA |
| Shapiro, Gilbert | California-berkeley | USA |
| Shaw, Gordon | California-Irvine | USA |
| Shaw, Graham | Manchester | GREAT BRITAIN |
| Sheaff, Marleigh | Wisconsin-Madison | USA |
| Shen, Benjamin | California-Riverside | USA |
| Shepard, Paul F. | Pittsburgh | USA |
| Shephard, William D. | Notre Dame | USA |
| Sher, Marc | Colorado | USA |
| Sherry, Thomas N. | Galway | IRELAND |
| Shima, Kazunari | Saitama Institute of Tech | JAPAN |
| Shima, Tomi | Michigan | USA |
| Shimada, Tokuzo | Lawrence Berkeley Lab | USA |
| Shivpuri, R.K. | Delhi | INDIA |
| Shochet, M. | Chicago | USA |
| Shoemaker, Frank C. | Princeton | USA |
| Shore, Graham M. | Harvard | USA |
| Shrauner, J. Ely | Washington University | USA |
| Siebert, H.W. | CERN | WEST GERMANY |
| Siemann, Robert | Cornell | USA |
| Sijacki, Djordje | Boris Kidric Institute | YUGOSLAVIA |
| Sikivie, Pierre | Rockefeller | USA |
| Silverman, Dennis | California-Irvine | USA |
| Sinclair, Donald Keith | Rockefeller | USA |
| Singer, Paul | Technion | ISRAEL |
| Singh, V. | Tata Institute | INDIA |

| | | |
|---|---|---|
| Sirotenko, V.I. | Serpukhov | USSR |
| Sixel, P. | RWTH Aachen | WEST GERMANY |
| Skrinsky, A.N. | Novosibirsk | USSR |
| Skubic, Patrick | Rutgers | USA |
| Slansky, Richard | Los Alamos Scientific Lab | USA |
| Sleeman, John C. | Washington | USA |
| Smadja, Gerard | Saclay | FRANCE |
| Smith, Alasdair | CERN | GREAT BRITAIN |
| Smith, Dennis B. | California-Santa Cruz | USA |
| Smith, Gerald A. | Michigan State | USA |
| Smith, James | Colorado | USA |
| Snellman, Hakan | Royal Institute of Tech | SWEDEN |
| Sobel, H. | California-Irvine | USA |
| Soergel, Volker | CERN | CERN |
| Sokoloff, Michael | California-Berkeley | USA |
| Solomon, Julius | Illinois-Chicago Circle | USA |
| Soloviev, L.D. | Serpukhov | USSR |
| Song, Hi Sung | Seoul National University | SOUTH KOREA |
| Soni, Amarjit | California-Irvine | USA |
| Soper, Davison E. | Oregon | USA |
| Sorensen, Cristian | Argonne National Lab | USA |
| Sorokin, P.V. | Kharkov | USSR |
| Souder, Paul A. | Yale | USA |
| Sourlas, Nicolas | Ecole Normale Superieure | FRANCE |
| Spillantini, P. | Frascati | ITALY |
| Spinka, Harold | Argonne National Lab | USA |
| Spitzer, H. | Hamburg | WEST GERMANY |
| Stacey, Richard | Rutherford Lab | GREAT BRITAIN |
| Stafford, G.H. | Rutherford Lab | GREAT BRITAIN |
| Stanev, Todor | Sofia | BULGARIA |
| Stanton, Noel R. | Ohio State | USA |
| Stefanski, Raymond J. | Fermilab | USA |
| Stein, Peter | Cornell | USA |
| Stenger, Victor | Hawaii | USA |
| Stephenson, Gerard | Los Alamos Scientific Lab | USA |
| Sterman, G. | SUNY-Stony Brook | USA |
| Stern, Jan | Orsay | FRANCE |
| Stevens, A. J. | Brookhaven National Lab | USA |
| Stevenson, Paul | Wisconsin-Madison | USA |
| Sticker, Harry | Rockefeller | USA |
| Stiewe, Jurgen | Heidelberg | WEST GERMANY |
| Stirling, W. James | Washington | USA |
| Strait, James | Lawrence Berkeley Lab | USA |
| Strovink, Mark | California-Berkeley | USA |
| Stroynowski, Ryszard | SLAC | USA |
| Strub, Roger | Strasbourg | FRANCE |
| Stump, D. | Indiana | USA |
| Stutte, Linda | Fermilab | USA |
| Subramanian, A. | Tata Institute | INDIA |
| Sucher, Joseph | Maryland | USA |
| Sudarshan, E. C. G. | Texas-Austin | USA |
| Sugano, Katsuhito | Fermilab | USA |
| Sugar, Robert L. | California-Santa Barbara | USA |
| Sugawara, H. | KEK | JAPAN |
| Sukhatme, U. | Orsay | FRANCE |
| Sulak, Lawrence R. | Michigan | USA |
| Sumner, R. | Princeton | USA |
| Sun, Chih-Ree | SUNY-Albany | USA |
| Sundaresan, M.K. | Carleton | CANADA |
| Susskind, Leonard | Stanford | USA |
| Suura, Hiroshi | Minnesota | USA |
| Suzuki, Mahiko | Lawrence Berkeley Lab | USA |
| Suzuki, Yoichiro | Brown | USA |
| Svensson, B.E.Y. | Lund | SWEDEN |
| Swanson, Robert A. | California-San Diego | USA |
| Swider, Greg | Washington | USA |
| Szego, K. | Hungarian Academy of Sciences | HUNGARY |
| Szymacha, A. | Warsaw | POLAND |
| Tadic, Dubravko | Zagreb | YUGOSLAVIA |
| Taft, Horace D. | Yale | USA |
| Talaga, Richard | Los Alamos Scientific Lab | USA |

| | | |
|---|---|---|
| Tanaka, Katsumi | Ohio State | USA |
| Tarnopolsky, Giora J. | SLAC | ISRAEL |
| Tarrach, Rolf | Barcelona | SPAIN |
| Tatur, Stanislaw | Polish Academy of Sciences | POLAND |
| Tavernier, Stefaan | VU-Brussels | BELGIUM |
| Taylor, Frank E. | Northern Illinois | USA |
| Telegdi, V. | Swiss Federal Institute-Zurich | SWITZERLAND |
| Teli, M.T. | Marathwada | INDIA |
| Tenner, A.G. | NIKHEF-H Amsterdam | NETHERLANDS |
| Teplitz, Vigdor L. | Virginia State | USA |
| Terrano, Anthony | Caltech | USA |
| Terwilliger, Kent M. | Michigan | USA |
| Theodosiou, George | Argonne National Lab | USA |
| Theriot, Dennis | Fermilab | USA |
| Thews, Robert L. | Arizona | USA |
| Thompson, Julia A. | Pittsburgh | USA |
| Thompson, Murray | Wisconsin-Madison | USA |
| Thompson, P. A. | Rochester | USA |
| Thomson, Gordon | Wisconsin-Madison | USA |
| Thorndike, Alan | Brookhaven National Lab | USA |
| Thorndike, Edward H. | Rochester | USA |
| Thornton, Ronald K. | Tufts | USA |
| Thresher, J.J. | Rutherford Lab | GREAT BRITAIN |
| Ticho, Harold K. | California-Los Angeles | USA |
| Tiecke, H.G.J.M. | CERN | NETHERLANDS |
| Toki, Walter | SLAC | USA |
| Tollestrup, Alvin | Fermilab | USA |
| Tolun, Perihan | Middle East Tech | TURKEY |
| Tomozawa, Yukio | Michigan | USA |
| Tompkins, John C. | Stanford | USA |
| Tornqvist, Nils | Helsinki | FINLAND |
| Tosa, Yasunari | Rochester | USA |
| Touboul, Marie Claude | Pierre et Marie Curie | FRANCE |
| Toussaint, Doug | California-Santa Barbara | USA |
| Treille, Daniel | CERN | CERN |
| Trilling, George | Lawrence Berkeley Lab | USA |
| Tsao, Hung-Sheng | Rockefeller | USA |
| Tsuru, Tsuneaki | KEK | JAPAN |
| Tu, Tung-sheng | IHEP-Beijing | CHINA |
| Tuli, S.K. | Banaras Hindu | INDIA |
| Tung, Wu-Ki | Illinois Institute of Tech | USA |
| Tupper, Gary B. | Oklahoma State | USA |
| Turkot, Frank | Fermilab | USA |
| Turnau, J. | Cracow | POLAND |
| Tuts, Philip Michael | SUNY-Stony Brook | USA |
| Tze, Hsiung Chia | Yale | USA |
| Ueno, Koji | Fermilab | USA |
| Ugaz, Eduardo | Central de Venezuela | VENEZUELA |
| Underwood, David G. | Argonne National Lab | USA |
| Unger, David G. | Michigan | USA |
| Unno, Yoshinobu | SLAC | USA |
| Uschersohn, Julian | Nebraska-Lincoln | USA |
| Valenti, Giovanni | CERN | CERN |
| Van Dalen, Gordon | California-Riverside | USA |
| Van Dyck, Olin B. | Los Alamos Scientific Lab | USA |
| Van Hove, L. | CERN | CERN |
| Van der Spuy, E. | Atomic Energy Board-Pretoria | SOUTH AFRICA |
| Vaughn, Michael T. | Northeastern | USA |
| Vayonakis, Constantine | CERN | GREECE |
| Venturi, G. | INFN-Bologna | ITALY |
| Verma, R.C. | Alberta | CANADA |
| Vernon, Wayne | California-San Diego | USA |
| Villani, M. | INFN-Bari | ITALY |
| Violini, Galileo | INFN-Rome | ITALY |
| Visnjic, Vladimir | Chicago | YUGOSLAVIA |
| Viswanathan, K.S. | Simon Fraser | CANADA |
| Von Krogh, J. | Heidelberg | WEST GERMANY |
| Vorobiev, A.A. | Leningrad | USSR |
| Votano, L. | Frascati | ITALY |
| Voyvodic, Louis | Fermilab | USA |
| Wacker, Klaus | Harvard | USA |

| | | |
|---|---|---|
| Wada, S. | Cambridge | GREAT BRITAIN |
| Wada, Walter W. | Ohio State | USA |
| Wagner, Stephen R. | Johns Hopkins | USA |
| Wagner, Walter | DESY | WEST GERMANY |
| Wahl, Horst | Austrian Academy of Sciences | AUSTRIA |
| Wahlen, Helmut | Wuppertal | WEST GERMANY |
| Wali, Kameshwar C. | Syracuse | USA |
| Walker, William D. | Duke | USA |
| Wallenmeyer, William | Department of Energy | USA |
| Wallraff, Wolfgang | RWTH Aachen | WEST GERMANY |
| Wanderer, Peter J. | Brookhaven National Lab | USA |
| Wang, Ling-Lie | Brookhaven National Lab | USA |
| Watanabe, Tsunetoshi | Asia University | JAPAN |
| Watts, Terence | Rutgers | USA |
| Webb, Robert C. | Princeton | USA |
| Weber, Gustav | DESY | WEST GERMANY |
| Weerts, Harry | RWTH Aachen | WEST GERMANY |
| Weiler, Thomas J. | Northeastern | USA |
| Weiner, Richard | Marburg | WEST GERMANY |
| Weingarten, D. | Indiana | USA |
| Weisberg, Howard | Brookhaven National Lab | USA |
| Weisberger, William I. | SUNY-Stony Brook | USA |
| Weiss, Jeffrey M. | SLAC | USA |
| Weldon, Arthur | Pennsylvania | USA |
| Whitaker, Scott | MIT | USA |
| White, Alan R. | CERN | CERN |
| White, Herman | Fermilab | USA |
| White, Sebastian | Rockefeller | USA |
| Wicklund, A.B. | Argonne National Lab | USA |
| Widgoff, Mildred | Brown | USA |
| Wiik, Bjorn | DESY | WEST GERMANY |
| Wilczynski, Henry | Fermilab | USA |
| Wilkes, Richard Jeffrey | Washington | USA |
| Williams, P.K. | Department of Energy | USA |
| Williams, Robert W. | Washington | USA |
| Willis, Suzanne | Fermilab | USA |
| Wilquet, G. | UL-Brussels | BELGIUM |
| Wilson, Richard | Harvard | USA |
| Winstein, B. | Chicago | USA |
| Winston, Roland | Chicago | USA |
| Winter, Klaus | CERN | CERN |
| Wise, Mark B. | SLAC | USA |
| Wiser, David | Wisconsin-Madison | USA |
| Witherell, Michael S. | Princeton | USA |
| Wojcicki, Stanley | Stanford | USA |
| Wolf, Gunter | DESY | WEST GERMANY |
| Wosiek, Jacek | Jagellonian | POLAND |
| Wouthuysen, S. | Amsterdam | NETHERLANDS |
| Wright, S. C. | Chicago | USA |
| Wu, Alfred C.T. | Michigan | USA |
| Wu, Sau Lan | Wisconsin-Madison | USA |
| Wuthrick, Jean-Pierre | Ecole Polytechnique | FRANCE |
| Xian, Dingchang | IHEP-Beijing | CHINA |
| Yamada, Kazuo | Fermilab | USA |
| Yamada, Ryuji | Fermilab | USA |
| Yamada, Sakue | Tokyo | JAPAN |
| Yamagishi, Kenji | Tokuyama | JAPAN |
| Yamaguchi, Yoshio | Tokyo | JAPAN |
| Yamawaki, Koichi | Nagoya | JAPAN |
| Yamdagni, N. | Stockholm | SWEDEN |
| Yan, Wuguang K. | IHEP-Beijing/SLAC | CHINA |
| Yang, W | Fermilab | USA |
| Ye, Minghan H. | IHEP-Beijing/Princeton | CHINA |
| Yellin, Steven J. | California-Santa Barbara | USA |
| Yen, Edward | Tsing Hua | TAIWAN, CHINA |
| Yeung, Patrick S. | MIT | USA |
| Yock, P.C.M. | Auckland | NEW ZEALAND |
| Yodh, Gaurang B. | Maryland | USA |
| Yoh, John | Columbia | USA |
| Yokosawa, Akihiko | Argonne National Lab | USA |
| Yoon, T.S. | British Columbia | CANADA |

| | | |
|---|---|---|
| Young, Kenneth K. | Washington | USA |
| Yun, Suk Koo | Saginaw Valley State College | USA |
| Yuta, Haruo | Tohoku | JAPAN |
| Yvert, Michel | Annecy-le-Vieux | FRANCE |
| Zachos, Cosmas | Wisconsin-Madison | USA |
| Zagury, Nicim | Pontificia Universidade Catoli | BRAZIL |
| Zaimidoroga, O.A. | JINR | USSR |
| Zakharov, V. | ITEP | USSR |
| Zakrzewski, Wojciech J. | Durham | GREAT BRITAIN |
| Zanon, Daniela | Milsu | ITALY |
| Zepeda, Arnulfo D. | CIEA-IPN | MEXICO |
| Zheng, Lingsheng | IHEP-Beijing | CHINA |
| Zhu, Dongpei | China Univ Sci & Tech/Wisconsi | USA |
| Zingl, Harald F.K. | Graz | AUSTRIA |
| Zlatev, I. | JINR | USSR |
| Zmuidzinas, Jonas S. | Southern California | USA |
| Zobernig, Georg | Wisconsin-Madison | USA |
| Zumino, B. | CERN | CERN |
| Zweig, George | Caltech | USA |
| de Groot, J.G.H. | CERN | NETHERLANDS |
| de Notaristefani, Francesco | INFN-Rome | ITALY |
| van Neerven, Willy | CERN | CERN |
| von Dardel, Guy | Lund | SWEDEN |

## PARTICIPATION BY COUNTRY

| | |
|---|---|
| USA | 679 |
| ARGENTINA | 2 |
| AUSTRALIA | 5 |
| AUSTRIA | 7 |
| BELGIUM | 11 |
| BRAZIL | 3 |
| BULGARIA | 2 |
| CANADA | 26 |
| CHINA | 18 |
| DENMARK | 8 |
| EGYPT | 1 |
| FEDERAL REPUBLIC OF GERMANY | 60 |
| FINLAND | 3 |
| FRANCE | 54 |
| GERMAN DEMOCRATIC REPUBLIC | 1 |
| GREAT BRITAIN | 52 |
| GREECE | 4 |
| HUNGARY | 3 |
| INDIA | 17 |
| IRAQ | 1 |
| IRELAND | 2 |
| ISRAEL | 16 |
| ITALY | 57 |
| JAPAN | 50 |
| MALAYSIA | 1 |
| MEXICO | 3 |
| NETHERLANDS | 12 |
| NEW ZEALAND | 1 |
| NORWAY | 3 |
| PAKISTAN | 2 |
| POLAND | 13 |
| PORTUGAL | 1 |
| SINGAPORE | 1 |
| SOUTH AFRICA | 1 |
| SOUTH KOREA | 6 |
| SPAIN | 3 |
| SWEDEN | 12 |
| SWITZERLAND | 7 |
| TAIWAN, CHINA | 3 |
| TURKEY | 1 |
| USSR | 19 |
| VENEZUELA | 3 |
| YUGOSLAVIA | 3 |
| SAUDI ARABIA | 1 |
| CERN | 51 |

## CONTRIBUTED PAPERS

1. Quark-Like Composite Model of Quarks and Leptons, Mestres, L. Gonzalez, Paris-Sud/CERN
2. Comments on the High Temperature Yang-Mills Gas, Shepard, H. K., New Hampshire
3. Masses and Mixing Angles in SU(5) Gauge Model, Nandi, S. and Tanaka, K., Ohio State
4. Anisotropic Superfluidity of Hadronic Matter, Chela-Flores, J. and DeLisa, F., Simon Bolivar
5. Full Estimate of the Higher Order Corrections to the Drell-Yan Process in Quantum Chromodynamics, Harada, K. and Muta, T., Tohoku/Kyoto
6. Problem of Fermion Generations in Grand Unified Theories, Chakrabarti, J., et. al., City College/CUNY
7. Neutrino Mass and Spontaneous Parity Violation, Mohapatra, H. N. and Senjanovic, G., City College-NY/Maryland
8. Is Light Velocity in Vacuum Really a Constant? Is "Locali-zation really an Absolute Concept? A Way to Regularize..., Fujiwara, K., Tokyo/Komaba
9. Spatial Volume Integral of Tr F F for Dyons, Kim, Y., et. al., Sogang/Seoul
10. Dyon with New Charge Distributions, Koh, I-G. and Yongduk, K., Sogang/Seoul
11. D*F for the SU(2) Wu-Yang Solutions, Koh, I-G. and Yongduk, K., Sogang/Seoul
12. New Mass Relations in Mixing Angles in a SU(5) Model of the Electroweakstrong Interaction, Yun, S. K., Saginaw Valley State
13. Broken Color Symmetry and Gluon Masses, Yun, S. K., Saginaw Valley State
14. Particle Production and Search for Long Lived Particles in 200-240 GeV/c Proton-Nucleon Collisions, Bussiere, A., et. al., LAPP/INFN-Bologna/DPhPE CEN-Saclay
15. Comment on the Cascade Model for Quark Fragmentation, Ugaz, E., Univ Central de Venezuela
16. Solutions of the Sine-Gordon Equation in Higher Dimensions, Leibbrandt, G., et. al., Harvard/Guelph
17. Isospin-Violating Mixing in Meson Nonets, Isgur, N., et. al., Toronto/Max-Planck/Weizmann/Argonne
18. Isospin-Violating Mass Differences and Mixing Angles: The Role of Quark Masses, Isgur, N., Toronto
19. Baryon Decays in a Quark Model with Chromodynamics, Koniuk, R. and Isgur, N., Toronto
20. Ground State Baryon Magnetic Moments, Isgur, N. and Karl, G., Toronto/Guelph
21. Point Splitting Regularization in a Supersymmetric Gauge Field Theory, Hagiwara, T., et. al., Brandeis/SLAC/Rockefeller
22. Regularizations and Superconformal Anomalies, Hagiwara, T., et. al., Brandeis/MIT/Rockefeller
23. Width and Decay Modes of Charmed Tensor Mesons, Privman, V. and Singer P., Technion
24. Perturbative Effect of Heavy Particles in Effecive Langrangian Approach, Hagiwara, T. and Nakazawa, N., Brandeis/Harvard
25. Charmonium Model, Pseudo-Dimension Rule and the Suppres-sions of Some OZI-Allowed Decays of the Psi(4.03,4.15,4.42), Mukhopadhyay, P., Jadavpur
26. Problem of R in e+e- Annihilation, Barnett, R. M., et. al., SLAC
27A. Simpler, Cheaper and Reliable Pi-N Partial-Wave Analysis, Chew, D. M., LBL
28A. Pion Multiplicities in Inelastic Nucleus-Nucleus Interaction and in Central Nucleus-Nucleus Collisions at 4.5 GeV/c/N..., SKM-200 Collaboration, Alma-Ata-Dubna-Tbilisi-Warsaw Collaboration
29A. Rapidity Distributions for Pions Produced in Inelastic Nucleus-Nucleus Interactions at 4.5 GeV/c/N Incident Momentu, SKM-200 Collaboration, Collaboration: Alma-Ata-Dubna-Tbilisi-Warsaw
30A. Validity Test of the Coherent Tube Model CTM for Nucleus-Nucleus Collisions in GeV/Nucleon Energy Region, SKM-200 Collaboration, Collaboration: Alma-Ata-Dubna-Tbilisi-Warsaw
31A. Cross-sections for Inelastic (4)He and (12)C-Nucleus Collisions at 4.5 GeV/c/N Incident Momentum, SKM-200 Collaboration, Collaboration: Alma-Ata-Dubna-Tbilisi-Warsaw

| | |
|---|---|
| 32A | Alpha-Particle Scattering on C, Al and Cu Nuclei at 17.9 GeV/c, Ableev, V. G., et. al., JINR/Moscow/INR/PTI |
| 33A | Deutron Wave Function at Small Distances Defined from Measurements of the d-p Fragmentation Spectrum at 8.9 ..., Ableev, V. G., et. al., JINR/NPI/EASE/INR/ITP/HCTI |
| 34A | Forward Deep Inelastic Scattering dC-dX at 8.9 GeV/c, Ableev, V. G., et. al., JINR/NPI/EASE/INR/IPP/HCTI |
| 36 | Sum Rule for Light Quark Masses, Olsson, M. G. and Miller, K. J., LASL/Wisconsin |
| 37 | Heavy Quark Masses and Duality, Miller, K. J. and Olsson, M. G., Wisconsin |
| 38 | Nonlinear Superposition for Liouville's Equation in Three Spatial Dimensions, Leibbrandt, G., Guelph |
| 39 | Direct Production of Electrons in 70 GeV/c Pi-p Interactions, Barloutaud, R., et. al., Bologna-Glasgow-Rutherford-Saclay-Torino |
| 40 | New Wave Equation for the Relativistic Two-Body Problem, Bakri, M. M. and Mansour, H. H. M., Cairo/ICTP-Trieste |
| 41A | Experimental Study of Pi(o)-Meson Cumulative Generation in the Reaction Pi(-)-C-Pi(o) (180)X at a Momentum of ..., Astvatsaturov, R. G., et. al., JINR |
| 42A | Scaling in the Mean and Associative Multiplicities for the Inclusive Reactions p+p-K(o)+x and p+p-x at 22.4 GeV/c, Batyunya, B. V., et. al., Dubna-Alma-Ata-Helsinki-Moscow-Prague-Tbilisi |
| 43A | Multiquark Resonant States, Shahbazian, B. A., et. al., JINR |
| 44 | Gamma and Pi(o) Production in p-bar-p Interactions at 22.4 GeV/c, Batyunya, B. V., et. al., Dubna-Alma-Ata-Helsinki-Moscow-Prague-Tbilisi |
| 45A | Evidence for the Baryon Resonance with Isotopic Spin 5/2 in np Interactions at Energies of 4 + 5 GeV, Abdivaliev, A., et. al., JINR |
| 46 | Stationary Quantum States in Yang-Mills Theory as the Unitary Representation of Homotopy Group, Pervushin, V. N., et. al., JINR |
| 47 | General Scheme for Bidemensional Models with Associated Linear Set, D'Auria, R., et. al., IFT/INFN-Turin |
| 48 | Exterior Calculus and Two-Dimensional Supersymmetric Models, Sciuto, S., IFT/INFN-Turin |
| 49 | Group Theoretical Construction of Two-Dimensional Models with Infinite Set of Conservation Laws, D'Auria, R., et. al., IFT/INFN-Turin |
| 51 | Subquark Model of Leptons and Quarks, Terazawa, H., Tokyo |
| 52 | Electroweak Contributions to Quark Masses, Kiskis, J. and West, G. B., LASL |
| 53 | Possible Origin of Four Quark Flavours and their Quantum Number Assignments, Van der Spuy, E., Atomic Energy Bd, S Africa |
| 54 | Toward an Understanding of the Leptoproduction R Ratio, Haack, E. M., Case Western Res |
| 55 | Composite Quarks and Leptons and the Higgs Boson Mass, Derman, E., Colorado |
| 56 | Can Quark Models Explain Electroproduction of the L=1 Supermultiplet?, Alcock, J. W., et. al., Bristol |
| 57 | Q(2) Evolution of Flavour Singlet Wavefunctions in QCD, Chase, M. K., Oxford |
| 58 | Implications of Higgs Bosons on Second Class Currents n Neutrino Reactions, Rodenberg, R. and Stamm, Chr., III Phys Inst |
| 59 | Why the Productions of the Charmed Particles F, F*, Sigma(c) are Suppressed, Mukhopadhyay, P., Jadavpur |
| 60 | Lifetimes of Charmed Mesons, Jagannathan, K. and Mathur, V. S., Rochester |
| 61 | Quark Recombination Amplitudes and the Decays of Charmed Mesons, Jagannathan, K. and Mathur, V. S., Rochester |
| 62 | Fluctuons and Tubes in Multiparticle Production from Nuclear Targets, Fredriksson, S., Royal Inst Tech |
| 63 | Treatment of Hard Processes Sensitive to the Infra-red Structure of QCD, Amati, D., et. al., CERN/Paris-Sud/INFN-Pisa/INFN-Parma |
| 64 | Energy Dependence of Spin-spin Effects in p-p Elastic Scattering at $90(o)^{cm}$, Crosbie, E. A., et. al., Argonne/ANL/Michigan/Kiel/Miami/Oxford |
| 65 | Spin-spin Forces in GeV/c Neutron-Proton Elastic Scattering, Crabb, D. G., et. al., Michigan/ANL/Argonne/Abadan/Bell/Miami |

66  Phenomenological Model for Charmed Meson Decays, Jagannathan, K. and Mathur, V. S., Rochester
67  Inclusive Production of Xi+ and Xi- in K+p Interactions at 32 GeV/c, Ajinenko, I. V., et. al., IHEP-Serpukhov
68  Semileptonic Decay of Charmed Particles and Weak Form Factors, Yamada, K., FERMILAB
70A  Scattering of 17.9 GeV/c Alpha-particles from hydrogen and helium Nuclei, Ableev, V. G., et. al., JINR
71  Neutrino Mass Problem and Gauge hierarchy, Magg, M. and Wetterich, Ch., CERN/Freiburg
72  Nonlinear Approach to Electrodynamics, Righi, R. and Venturi, G., INFN-Bologna
73  Conformal Invariance and the Fine Structure Constant in a Nonlinear Approach to Electrodynamics, Righi, R. and Venturi, G., INFN-Bologna
74  Periodic Solutions in SU(2) Gauge Theory, Abdel-Rahman, A. M. M., Riyadh
75  Group Theoretical Construction of Two-Dimensional Supersymmetric Models, D'Auria, R., et. al., INFN/Geneve
76  Quark Loop Description of the Radiative Neutral Meson Decays (Pi,eta, eta', $K^L$)-2 gamma, (eta, eta', $K^L$)-Pi+Pi-gamm, Ebert, D. and Volkov, M. K., JINR
77  Clustering Hadronization of Quarks: A Treatment of the Low-P(T) Problem, Hwa, R. C., Oregon
79  Lectures on the Relativistic String, Rohrlich, F., Syracuse
80  Relativistic Hamiltonian Dynamics, Rohrlich, F., Syracuse
81  Relativistic Hamiltonian Dynamics I: Classical Mechanics, Rohrlich, F., Syracuse
82  Relativistic Particle Systems with Confining Interactions, Rohrlich, F., Syracuse
84  Covariant Hamiltonian Systems Leading to Confinement, Rohrlich, F., Syracuse
85  Non-Point-Like Parton Model with Asymptotic Scaling and with Scaling Violation at Moderate Q(2) Values, Chen, C. K., Purdue
86  Chi (3.51) Revisited, Mukhopadhyay, P., Jadavpur
87  Exotic New Quarks and Dynamical Symmetry Breaking, Marciano, W. J., Rockefeller
88  Neutrino Oscillations and the Modulation of Neutrino- Electron Scattering, Rosen, S. P. and Kayser, B., NSF/Purdue
89  Bound State Solutions of the Dirac Equation with Linear and Quadratic Scalar Potentials, Ram, B. and Arafah, M., New Mexico State
90  Inclusive Pi(o) and eta production from Kaon, Proton and Antiproton Beams in the Triple Regge Region, Kennett, R. G., et. al., Cal Tech/LBL
91  Pion Decay Constant, Electromagnetic Form Factor and Quark Electromagnetic Self Energy in QCD, Pagels, H. and Stokar, S., Rockefeller
92  Magnitude of the Light Current Quark Masses, Pagels, H. and Stokar, S., Rockefeller
93  Models of Dynamically Broken Gauge Theories, Pagels, H., Rockefeller
94  Possible Structures of Gauge Hierarchies in the SU(L) Unification, Umemura, I. and Yamamoto, K., Kyoto
95  Data Photographs from the Auckland Cosmic Ray Telescope, Yock, P. C. M., Auckland
96  Supersymmetric Ward-Takahashi Identities and the 'T Hooft- Veltman Dimensional Regulator, Hagiwara, T. and Majumdar, P., Brandeis
97  Proton Model with Transverse Momentum, Caprasse, H., Liege
98  Mixing Among Fermion Generations, Ma, E., Hawaii-Manoa
99  Quark Jets in Pion Diffractive Dissociation, Randa, J., Colorado
100  Comment on Higgs Photoproduction, Randa, J., Colorado
101  Color Singlet Jets and Large Rapidity Gaps in e+e- Annihilation, Randa, J., et. al., Colorado
102  Counter-Example to Non-Abelian Bloch-Nordsieck Theorem, Doria, R., et. al., Oxford/Sao Paulo
103  W Boson Pair Production by Electrons and Positrons of Same Helicity, Gourdin, M. and Pham, X. Y., Pierre et Marie Curie
104  Selection Rules for Charmed Meson Decays and how to Find the F Meson, Bace, M. and Pham, X. Y., Pierre et Marie Curie

105  Subcomponent Models of Quarks and Leptons, Casalbuoni, R. and Gatto, R., INFN/Geneve
106  Unified Theories for Quarks and Leptons based on Clifford Algebras, Casalbuoni, R. and Gatto, R., CERN/Geneve
107  Unified Description of Quarks and Leptons, Casalbuoni, R. and Gatto, R., CERN/Geneve
108  Indefinite-Metric Quantum Field Theory of General Relativity IX: "Coral" of Symmetries, Nakanishi, N., Kyoto
109  Indefinite-Metric Quantum Field Theory of General Relativity X: Sixteen-Dimensional Superspace, Nakanishi, N., Kyoto
110  Superalgebras of Non-Abelian Gauge Theories in the Manifestly-Covariant Canonical Formalism, Nakanishi, N. and Ojima, I., Kyoto
111  Local Covariant Operator Formalism of Non-Abelian Gauge Theories and Quark Confinement Problem, Kugo, T. and Ojima, I., Kyoto
112  Delta-T = 1 Constraints and Final State Interactions in Two-body Charm Decay, Rosen, S. P., Purdue
113  W-Exchange Dominance in Neutral B-Meson Decay, Rosen, S. P., Purdue
114  Evidence for Point-Like Structure in Neutral Pseudoscalar Mesons from Form Factor Studies, Bergstrom, L. and Snellman, H., RIT
115  Decays of Intermediate Vector Bosons, Radiative Corrections and QCD Jets, Albert, D., et. al., Rockefeller/NJIT
116  Decay of Psi and Upsilon into a Lepton Pair and Gluons, Leveille, J. P. and Scott, D. M., Wisconsin
117  Higgs Boson Decay and the Running Mass, Braaten, E. and Leveille, J. P., Wisconsin
118  Estimating Inclusive Psi and Psi-eta(C) Production in e+e- Annihilation, Kane, G. L., et. al., SLAC/Michigan/Wisconsin
119  Tau-nu Decay Signature for Detecting F+/- and B+/- Mesons in e+e- Collisions, Barger, V., et. al., Wisconsin/Rutherford
120  Can One Measure Strong Corrections to Weak Decays?, Eilam, G. and Leveille, J. P., Wisconsin
121  Magnetic Monopoles in Grand Unified Theories, Langacker, P. and Pi, S-Y., IAS/Pennsylvania/SLAC
122  Possible Indications of Neutrino Oscillations, Barger, V., et. al., Wisconsin/Fermilab/Rutherford
123  Flux Independent Tests of Neutrino Oscillations, Barger, V., et. al., Wisconsin/Fermilab/Rutherford
124  Gauge Model with Light W and Z Bosons, Barger, V., et. al., Wisconsin/Hawaii-Manoa
125  Sequential W and Z Bosons, Barger, V., et. al., Wisconsin/Hawaii-Manoa
126  Doubling of Weak Gauge Bosons in an Extension of the Standard Model, Barger, V., et. al., Wisconsin/Hawaii-Manoa
127  Inclusive Production of K* (892) and Sigma (1385) in pp Interactions at 3.6 GeV/c and a Test of Factorization ..., Banerjee, S., et. al., Tata
128  Non-Spectator Quark Interactins and the Lambda(+)$^C$ Lifetime, Barger, V., et. al., Wisconsin
129  Gluon Enhancements in Charmed Meson Decays, Barger, V., et. al., Wisconsin
130  Photon-Gluon Fusion: A Review, Leveille, J. P., Wisconsin
131  Higgs Boson Production in e+e- Annihilation, Leveille, J. P., Wisconsin
132  Diffractive and Non-Diffractive Psi Leptoproduction in QCD, Leveille, J. P. and Weiler, T., Wisconsin/Northeastern
133  On the Recent Observation of the Hadroproduction of Prompt Photons, Halzen, F. and Scott, D. M., Wisconsin
134  Properties and Signatures of Heavy Quarks, Pakvasa, S., et. al., Hawaii/Wisconsin
135  Structure of Direct Photon Events, Halzen, F., et. al., Wisconsin/Hawaii
136  Double Drell-Yan Annihilations in Hadron Collisions: Novel Tests of the Constituent Picture, Goebel, C., et. al., Wisconsin
137  Chromodynamics and Jet-Acollinearity in e+e- Annihilation: Determining the Quark-Gluon Coupling, Halzen, F. and Scott, D. M., Hawaii-Manoa/Wisconsin
138  Hadroproduction of Heavy Flavors at Collider Energies, Scott, D. M., et. al., Wisconsin

139  Do-Do-bar Mixing from Flavor-Changing Higgs Bosons, Barger, V. and Ma, E., Wisconsin/Hawaii
140  Neutrino Productin of Psi Via Neutral Currents, Barger, V., et. al., Wisconsin/Rutherford
141  On Psi and Upsilon Production via Gluons, Barger, V., et. al., Wisconsin/Rutherford
142  Infra-red and Mass Regularization in AF Field Theories II: QCD, Humpert, B. and Van Neerven, W. L., CERN
143  Inclusive Gamma-Production and a Search for Direct Gamma-Productin in p-bar-p Interaction at 0.76 GeV/c and 2.0 GeV/c, Banerjee, S., et. al., Tata
144  Study of p-bar-p Interactins at Low Energies in Terms of Principal Axis Variables, Banerjee, S., et. al., Tata
145  Infra-red and Mass Regularization in AF Field Theories: Theta $(3)^b$, Humpert, B. and Van Neerven, W. L., CERN
146  $SU(5) \times S(4)$ Model in Grand Unified Theories, Hayashi, H., et. al., Shizuoka/Tokai/Ecole Polytechnique
147  Kaon-Nucleus and Kaon-Nucleon Forward Scattering Amplitudes, Diu, B. and Ferraz de Camargo, A., LPT
148  New Regge Singularity, Diu, B. and Ferraz de Camargo, A., LPT
149  Exponentiating Soft Emission QCD, Ciafaloni, M., INFN-Pisa/Scuola Normale Superior-Pisa
150  Second Order QCD Effect: Quark-Quark Bremsstrahlung Contribution to Transverse Momentum of Lepton Pairs, Chaichian, M., et. al., Helsinki/Saitama
151  Local B-L Symmetry of Electroweak Interaction, Majorana Neutrinos and Neutron Oscillation, Mohapatra, R. N. and Marshak, R. E., CUNY/VPI-Blacksburg
152  Phenomenology of Neutron Oscillations, Mohapatra, R. N. and Marshak, R. E., CUNY/VPI-Blacksburg
153  Effective Gauge Theories, Hu, B., Houston
154  Preliminary Large p-perp Cross Sections Measured with a 2 Calorimeter Trigger, Favuzzi, C., et. al., Bari-k-L-M-N Coll
155  New Results in the Search of Quarks in Matter by the Magnetic Levitation Electrometer, Marinelli, M. and Morpurgo, G., Ist. Fisica Genova,Italy/Ist. Fisica Genova.
156  Confirmation of the Electric Neutrality of Matter at the Milligram., Marinelli, M. and Morpurgo, G., Genova/INFN
157  Measurement of Differential Cross Sections of Pi+P and Pi-P Elastic Scattering in the Region of Low-Lying Pion Nucleon, Gordeev, V.A., et. al., LNPI Leningrad USSR
158  Consequences of Majorana and Dirac Mass Mixing for Neutrino Oscillations, Barger, V., et. al., Wisconsin
159  Calculation of the Width of the First Excitations of Charmed Mesons D and D*, Amer, A. A. and El-Bialy, S., Cairo
160  Loop Space Representation and the Large N Behaviour of the One-Plaquette Kogut-Susskind Hamiltonian, Jevicki, A. and Sakita, B., Brown/CUNY
161  Semiclassical Approach to Planar Diagrams, Jevicki, A. and Levine, H., Brown/Harvard
162  Large N Classical Equations and their Quantum Significance, Jevicki, A. and Levine, H., Brown/Harvard
163  Dynamical Model for the Two-Component Pomeron, Barut, A. O. and Hacinliyan, A., Colorado/Bogazici
164  Narrow Resonances as an Eigenvalue Problem and Applications to High Energy Magnetic Resonances: An Exactly Soluble Model, Barut, A. O., et. al., Colorado/Mexico
165  $Q(2)$-Evolution Equations for a Multihadron Fragmentation Functions, Sukhatme, U. P., et. al., Iowa
166  Bose Form of Two Dimensional Quantum Chromodynamics, Baluni, V., IAS/SLAC
167  SU(N) Grand Unificatin with Several Quark-Lepton Generations, Frampton, P. H., Ohio
168  SU(9) Grand Unification of Flavor with Three Generations, Frampton, P. and Nandi, S., Harvard/Ohio
169  Unification of Flavor, Frampton, P. H., Harvard/Ohio
170  No-Go Theorem for SU(N) Unification of Extra-Strong Interactions, Frampton, P. H., Harvard/Ohio
171  Low-p(T) Production in Deuteron Fragmentation, Dasgupta, P., et. al., Kalyani/Uluberia/Burdwan

| | |
|---|---|
| 172 | Some Comments on Selection Rules for Nonleptonic Decays of Charmed Particles, Ziaoyuan, L., Chinese Academy of Sci |
| 173 | Alternative Way for the Relativistic Electron Rings Generation Using Gyrac Mechanism, Golovanivsky, K. S., Patrice Lumumba |
| 174 | Relativistic Q-bar-Q (U,D,S,C) Physics from Bethe-Salpeter Premises, Mitra, A. N. and Santhanam, I., Delhi |
| 175 | Relativistic QQQ Spectra from Bethe-Salpeter Premises, Mitra, A. N. and Santhanam, I., Delhi |
| 176 | Unstable Heavy Particles, Frampton, P. H. and Glashow, S. L., Harvard |
| 177 | Proton Diffraction Excitation Function and Scaling Contri- butions at Large x in Inclusive Inelastic Proton Collisions, Lee-Franzini, J., et. al., Stony Brook/Columbia |
| 178 | Lambda Production in K-p Interactions at 32 GeV/c, FSUCSU Collaboratin, France-Soviet Union/CERN-Soviet Coll |
| 179 | Dominant Euclidean Configurations for all N, Bardakci, K., et. al., LBL |
| 180 | Width of the S Meson From Duality, Igi, K., Tokyo |
| 181 | Orbital Excitation in Modified Yukawa Potential, Iwao, S., Kanazawa |
| 182 | Parton Description of Soft pp Annihilation, Sukhatme, U. P., Iowa |
| 183 | Nonleptonic Weak Decay Rate of Explicitly Flavored Heavy Mesons, Suzuki, M., LBL |
| 184 | Dynamical Symmetry Breakdown in Weak Interactions, Acharya, R., et. al., Arizona State/Southern Illinois |
| 185 | Zero-Point Energy of Confined Fermions, Milton, K. A., Ohio |
| 186 | Pseudoscalar Decay Constants and Nu-3Pi in Chiral And 1/N Perturbation Theory, Milton, K.A., et. al., Ohio/Calif. |
| 187 | Can Gluon Enhancement Explain the A-Dependence of Fragmentation of Jets Produced off Nuclei?, Takagi, F., Tohoku |
| 188 | Collective Quark Tubes in Hadron-Nucleus Collisions at high Energies, Takagi, F., Tohoku |
| 189 | Collective Quark Tube Model For Nucleus-Nucleus Collisions At high Energies, Takagi, F., Tohoku |
| 190 | Regge Behaviour of Symmetric Octet Exchange Amplitude For Gluon-Gluon Scattering, Kwiecinski, J. and Praszalowicz, M., INP/Jagellonian |
| 191 | Three Gluon Integral Equation and Odd C Singlet Regge Singularities in QCD, Kwiecinski, J. and Praszalwicz, M., INP/Jagellonian |
| 192A | On Adler's Quaternionic Chromostatics, Sequnaru, D., et. al., Duke/Tel Aviv |
| 193 | Transformation of a Non-Abelian Gauge Theory into an Abelian Gauge Theory, Khan, I., Riyad |
| 194 | QCD Corrections at the Z(o), Humpert, B. and Van Neerven, W. L., CERN |
| 195 | Exactly Solvable Eigenvalue Problem with Hypergometric Eigenfunctions, King, M. and Rohrlich, F., Syracuse |
| 196 | Stability Properties of Classical Solutions to Non-Linear Sigma Models, Din, A. M. and Zakrzewski, W. J., CERN |
| 197 | Embeddings of Classical Solutions of o(2p+1) Non-Linear Sigma Models in CP(n-1) Models, Din, A. M. and Zakrzewski, W. J., CERN |
| 199 | Relativistic Hamiltonian Dynamics II: Momentum Dependent Interactions, Confinement and Quantization, King, M. and Rohrlich, F., Syracuse |
| 200 | Dynamic Confinement from Velocity-Dependent Interactions, King, M. and Rohrlich, F., Syracuse |
| 201 | Bag Model Calculation of the Nucleon Lifetime in Grand Unified Theories, Din, A. M., et. al., CERN/LAPP |
| 202T | Dynamics of Low Energy Hadrons, Zakharov, V. E., ITEP-Moscow |
| 203T | Accelerator Possibilities of High Energy Physics, Skrinsky, A. N., NINP-Novosibirsk |
| 204T | Measurement of Real Part of Pi Minus P, PP-Scattering Amplitude and of the Diffraction Cone Slope Parameter at..., Vorobiev, A A., NPI-Leningrad |
| 205T | Study of PI Plus-Minus Helium Scattering and Small Angle Scattering at 50-300 GEV, Vorobiev, A. A., NPI-Leningrad |

| | |
|---|---|
| 207T | Production of Charmed Particles in NU NU n Interactions and C-Quark Fragmentation Functions, Nikitin, Y. P., IEP-Moscow |
| 208T | Gluonium and Theory of Light Hadronic States in Bag Model with Partially Degenerated Condensate, Kobzarev, I. Y., ITEP-Moscow |
| 209T | Possible Relation Between Processes at Small and Large Transferred Momenta, Kaydalov, A. B., ITEP-Moscow |
| 210 | Electromagnetic Formfactors of Hadrons at Large $Q^2$ and Confinement Effects, Kaydalov, A. B., ITEP-Moscow |
| 211T | New Arguments in Favour of the Existence Exotic Resonances, Kaydalov, A. B., ITEP-Moscow |
| 212T | Search of CP-Invariance Violation in K Zero-Decays by Means of Xenon Bubble Chamber, Chuvilo, I. V., ITEP-Moscow |
| 213 | Pi/p and Ch/n Ratios at 300-3000 GeV at Mountain Altitudes, Antonyan, K. G., et. al., Yerevan Physics Institute |
| 214 | Measurement of Cross Section Asymmetry for Polarized Photons and Models of Eta-Meson Photoproduction in the Energy ...., Vartapetyan, H. A. and Piloposyan, S. E., Yerevan Physics Institute |
| 215T | Polarization in Pi Minus P-Pi Zero N Charge-Exchange Reaction in the Low Momentum Transfer Region at 40 GeV/c, Derevtchikov, A. A., IHEP-Serpukhov |
| 216T | Hadron Production with High Transverse Momenta, Sulyaev, R. M., IHEP-Serpukhov |
| 217 | Group Representation Theory and Integration of Two-Dimen-sional Equations of Mathematical Physics, Leznov, A. N. and Saveliev, M. V., IHEP-Serpukhov |
| 218 | Cylindrically-Symmetric Configurations of Gauge Fields for an Arbitrary Lie Group, Leznov, A. N. and Saveliev, M. V., IHEP-Serpukhov |
| 219T | On Scaling Solutions in Gauge Theories, Arbuzov, B. A., IHEP-Serpukhov |
| 220T | Institute for High Energy Physics Accelerator Storage Complex (unk), Soloviev, L. D., IHEP-Serpukhov |
| 221T | Study of Asymmetry of the Crossection of Deuteron Photodesintegration by Polarized Photons, Sorokin, P. V., PI-Kharkov |
| 222 | Positive Pion Productin from Polarized Protons by Linearly Polarized Photons in the Energy Range 280-420 MeV, Get'man, V. A., et. al., Kharkov/PI-Moscow |
| 223T | Multiple Production of Charged Particles in Collision of Pi and K-Mesons with Nuclei at 40 GeV/c in the Spectrometer...., Boehm, J., JINR-Dubna |
| 224T | Description of Radiation Decays of Neutral Mesons with the Help of Quark Loops, Ebert, D., JINR-Dubna |
| 225 | Interference of Identical Secondary Particles and Sizes of their Production Volume in Pi-p and Pi-C Collisions at 40GEV, Akhababian, N., JINR-Dubna |
| 226T | Processes with High Momentum Transfer in Quantum Chromo-dynamics, Efremov, A. V., JINR-Dubna |
| 227T | Study of Diffraction Scattering of Mesons by Nuclei at 40 GeV, Zaimdoroga, O. A, JINR-Dubna |
| 228 | Relativistic Nuclear Collisions, Soloviev, M. I., JINR-Dubna |
| 229T | Pion Formfactor and its Asymptotic Behaviour, Mescheryakov, V. A., JINR-Dubna |
| 230T | Quark-Parton Distributions in Nuclei, Baldin, A. M., JINR-Dubna |
| 231T | Study of Photoproduction Processes at Energies of 15-38 GeV, Belousov, A. S., PI-Moscow |
| 232 | Charged Current Events with Neutral Strange Particles in High Energy Antineutrino Interactions, Ammosov, V. V., et. al., IHEP-Serpukhov/Fermilab/ITEP-Moscow/Michigan |
| 233 | Number of Higgs Scalars and m(b)/m(Tau) Ratio in the SU(5) Model, Komatsu, H., Kyushu |
| 234 | Possible Spin-One Resonances in the Strongly Coupled Higgs Sector, Kobayashi, M. and Matsuki, T., NLHEP-Ibaraki ken |
| 235 | Low Energy Parameters in Grand Unified Theories, Yoshimura, M., NLHEP-Ibaraki ken |
| 236 | Decoupling of Superheavy Particles in Grand Unified Theories, Yoshimura, M., NLHEP |
| 237 | Relation between Weinberg's and Georgi-Politzer's Renormalization Group Equations, Matsuki, T. and Yamamoto, N., NLHEP-Ibaraki/Osaka |

238 Effects of Superheavy Quarks and Leptons in Low-Energy Weak Processes $K(L)\to MUMU$, $K+\to PINUNU$ and $K(0)\to K-(0)$, Inami, T. and Lim, C. S., Tokyo-Kumaba
239 Problem of ETA in the Large N Limit--Effective Lagrangian Approach, Kawarabayashi, K. and Ohta, N., Tokyo
240 Radiative Corrections for $\mu$-e bar-Nu Decay in the Weinberg-Salam Model with Arbitrary Number of Generations, Inoue, K., et. al., Kinki
241 Hydrodynamical Study of Nuclear-Size Dependence of Particle Production at Large Transverse Momentum, Masuda, N., Yamagata
242 Method for Solving Some Classical Yang Mills Equations, Castillejo, L. and Kugler, M., University College/Weizmann Institute
243 Path Ordered Operator Formalism of Gauge Theories in the Two-Dimensional Space-Time, Kikkawa, K., Hiroshima
244 Photon Structure Function as Calculated Using Perturbative Quantum Chromodynamics, Duke, D.W., et. al., FSU
245 High $p(T)$ Production of Direct Photons and Jets in Quantum Chromodynamics, Cormell, L. and Owens, J. F., Pennsylvania/Florida State
246 Measurement of the Continuum of Dimuons Produced in High-Energy Proton-Nucleus Collisions, Ito, A.S., et. al., SUNY/Columbia/Fermilab
247 Observed Scaling of Particle Lifetimes, MacGregor, M. H., LLL
248 Multijet Structure in Quantum Chromodynamics, Lai, C.H., et. al., Niels Bohr/Nordita/DES
249 Renormalization Group Approach to Summing QCD Corrections In Hadronic Decays of the Higgs Boson, Inami, T. and Kubota, T., Tokyo-Komaba
250 Departure From Weinberg-Salam Model and Grand Unification, Deshpande, N. G., Oregon
251 New Look at the $A(1)$, $E$, $Q(1)(1400)$-Meson System: Derivation of Mass Formulas and Selection Rules, Oneda, S. and Kno, J. S., Maryland/R.Walters-Cincinnati
252 Why are the Isoscalar Neutral Current Axial-Vector Couplings and Isoscalar Nucleon Anomalous Moments Small?, Oneda, S., et. al., Maryland/LASL
253 Pseudoscalars and Vector Mesons in a Unitary and Self-Consistently Broken $SU(6)w$ Scheme, Roos, M. and Tornqvist, N., Helsinki
254 Baryon Production from Pion Fragmentation in the Quark Recombination Model, Migneron, R. and Robinson, J. L., Western Ontario
255 Triple Pomeron Coupling and Pion Inclusive Scattering, Migneron, R., et. al., U of W. Ontario
256 $K*$-Dominance of $K(e3)$ Formfactor and Sirlin's Relation, Yamada, K., Rochester/Fermilab
257 Possibility of Detecting Triple Gluon Coupling and Adler-Bell-Jackiw Anomaly in Polarized Deep Inelastic Scattering, Lam, C. S. and Li, B.A., SLAC/McGill/Academia Sincia-Beijing
259 String-Like Wave Functional in a Model of Colour Flux Collimation, Hosoya, A., Osaka
260 Supersymmetric Sigma Models and Composite Yang-Mills Theory, Lukierski, J., ICTP-Trieste
261 Pathological Lattice Field Theory for Interacting Strings, Weingarten, D., Indiana
262 Non-planar Diagrams in the Large N Limit of $U(N)$ and $SU(N)$ Lattice Gauge Theories, Weingarten, D., Indiana
263 String Equations for Lattice Gauge Theories with Quarks, Weingarten, D., Indiana
264 Monte Carlo Calculations and a Model of the Phase Structure for Gauge Theories on Discrete Subgroups of $SU(2)$, Petcher, D., et. al., Indiana
265 Observation of Charmed F+ Meson Produced by Neutrino Interaction in Emulsion, Ammar, R., et. al., Kansas/Fermi/IHEP/ITEP/INP/JINR/Wash/Gos
266 Continuum Limit of $QED(2)$ on a Lattice, Weingarten, D., Indiana
267 Continuum Limit of $QED(2)$ on a Lattice II, Weingarten, D., Indiana

268 Supersymmetric Dynamics on Pre-QCD Level with Elementary Quarks and Composite Gluons, Lukierski, J. and Milewski, B., Wroclaw
269 Analysis of Large-$p^t$ Single Particle Inclusive Cross Sections for p-p Reactions, Lim, Y. K. and Phua, K. K., Singapore/Nanyang
271 P-State Charmed Mesons--Especially the Charmed 1+ Mesons, Eno, J. S. and Oneda, S., Cincinnati/Maryland
272 Relativistic Equal-Time Equation, Tu-nan, R., et. al., CUST/ITP-Peking/IHEP-Peking
273 General Form of Effective Langrangian for Path Integral Quantizatin Formalism in Curved Space, Tu-nan, R., et. al., CUST
274 Quantum Field Theory of Composite Particles, Tu-nan, R. and Tsu-hsiu, H., CUST/ITP-Peking
275 General Form of the Path Integral Quantization Formalism, Hung-chun, Y., et. al., CUST/IHEP-Peking/ITP-Peking
276 Functional Integration Techniques of Negative Metric, Tu-nan, R., et. al., CUST
277 Asymmetry Parameter as a Function of Transverse Momentum in the Production of Dilepton with Polarized Beam and Target, Mani, H. S. and Noman, M., IIT-Kanpur
278 Extracting the Quark and Gluon Distribution in Mesons from Dilepton Data, Mani, H. S. and Noman, M., IIT-Kanpur
279 Study on the 'Compound Multiplicity' in P-Emulsion Interaction at 400 GeV/c, Ghosh, D., et. al., Jadavpur
280 Unusual Emulsion Stars Observed at High Energy Proton Interactions, Ghosh, D., et. al., Jadavpur
281 Study of the Cluster Characteristic in Hadron-Nucleus Interactions at Ultra-High Cosmic-Ray Energies, Ghosh, D., et. al., Jadavpur
282 Universal Mass Formula and its Particularization for Mesons, Baryons and Some of their Resonances, Petrescu, S. and Petrescu, V., IP-Bucharest
283 Exclusive Calculations for QCD Jets in a Monte Carlo Approach, Odorico, R., Bologna/INFN-Bologna
284 Grand Unification with the Exception Group E(8), Bars, I., Yale
285 Dynamical Theory of Subconstituents based on Ternary Algebras, Bars, I. and Gunaydin, M., Yale/Bonn
286 U(N) Integral for Generating Functional in Lattice Gauge Theory, Bars, I., Yale
287 Simple Lagrangian for a Superspace Affine Theory, Friedman, M. H., Northeastern
288 Mixing of Dim 5 and 6 Operators Induced by Higgs Scalars, Tosa, Y., Rochester
289 Color, Flavor, Duality, and the Pomeron, Dash, J. W., CNRS
290 Large P(T) Photoproduction and Tests of QCD, Fontannaz, M., et. al., Paris-Sud/EP-Palaiseau
291 How to Extract the Gluon Fragmentation Function from Large P Photoproduction, Fontannaz, M., et. al., LPTHE,Paris-Sud/EP-Palaiseau
292 Photoproduction with Polarized Photons as a Source of Polarized Gluon Jets, Petersson, B. and Pire, B., Bielefeld/Ecole Polytechnique
293 Energy and x Dependence of Inclusive Particle Production in Annihilation and Non-annihilation Reactions in the Quark...., Muirhead, H., et. al., Liverpool-Stockholm-Vienna Collaboration
294 Diffractive Proton Fragmentation Containing s-bar-s Quarks From 32 110 GeV/c K-p Interactions and Extrapolation to..., MacNaughton, J., et. al., Aachen-Berlin-Cracow-CERN-London-Vienna-Warsaw/SU.
296 On-Shell QCD Quark Form Factor and its Determination from Two-Particle Correlations in e+e- Annihilation, Marquardt, W. and Steiner, F., Hamburg
297 Asymptotoic Dynamics of QCD, Coherent States and the Quark Form Factor, Dahmen, H. D. and Steiner, F., DESY/Siegen/Hamburg
298 Global and Pole Duality Applied to Charm and Bottom Quarks, Miller, K. J. and Olsson, M. G., Wisconsin
299 Evidence For "Total Number" Conservation of Quarks in the D-Meson Decay?, Matsuda, M., et. al., Aichi/Meijo/Nagoya
300 Mechanism of "Delta I = 1/2 Enhancement" in the Relativistic Quark Model, Abe, Y., et. al., Hokkaido/Kyoto
301 Spectroscopy of Atomlike Mesons Q bar-q in the Dirac Equation with Logarithmic Confining Potential, Kaburagi, M. and Kawaguchi, M., Kobe

302 Correlations with Large Transverse Momentum Photons and the Gluon Structure Function, Baier, R., et. al., Bielefeld
303 Small Coupling (Low Temperature) Expansions of Gauge Field Models on a Lattice-Part II: Expansions For the Gauge..., Muller, V.F., et. al., Kaiserslautern
304 Production of Charged Hadrons, Pi(o) and Electron Pairs in 70 GeV/c Pi-p Interactions, Barloutaud, R., et. al., Bologna-Glasgow-Rutherford-Saclay...Collaboration
305 On the Nonleptonic Decay of Hadron and the Total Quark Number Conservation, Matsuda, M., et. al., Aichi/Meijo/Ngoya
306 Jet Mass Spectra in Perturbative QCD, Clavelli, L. and Nilles, H-P., Bonn
307 Jet Angular Momentum and Quantum Chromodynamics, Clavelli, L., Bonn
308 Poincare and Symplectic Properties of Realativistic Extended Hadrons, Kim, Y. S. and Noz, M. E., Maryland/New York
309 Extended Quark Recombination Model With Valence Quark Annihilation Compared to Inclusive Pi+/Pi- Ratios, Buschbeck, B., et. al., IHOAW-Vienna
310 Gauge Theory of the Integral Form: Lattice-Like Gauge Theory Invariant Under Translation and Rotation, Matsumoto, T., Tokyo
311 Quark Confinement within a Unified Theory of Strong and Gravitational Interactions, Recami, E., Catania
312 On the Questions of Phase Transition and Coupling Constant Renormalization in the Interacting Instanton Gas, Ilgenfritz, E-M. and Muller-Preussker, M., JINR
313 Recombination Model and Low-Transverse-Momentum Distribution, Gupt, C., et. al., Delhi/Kurukshetra/JINR
315 Flavor Unifying Schemes with a Single Fermionic Representation, Davidson, A., et. al., Syracuse
316 Behaviour of Sigma(e+e- → W+W-) Near Threshold Including Finite Width Effects, Layssac, J., Languedoc
317 Asymmetry in Dilepton Production with Polarised Beam and Target in Perturbative QCD, Bajpai, R.P., et. al., Himachal Pradesh/IIT-Kanpur
318 Order - R Vacum Action Functional In Scalar-Free Unified Theories with Spontaneous Scale Breaking, Adler, S. L., IAS
319 Formula for the Induced Gravitational Constant, Adler, S. L., IAS
320 Delta I = 1/2 Rule and the Relativistic Quark Model, Abe, Y., et. al., Hokkaido
321 Composite Model of Quarks and Leptons Related to Composite Structures of SO(10), Yasue, M., Nagoya
323 Symmetry Breaking of O(10) and Constraints on Higgs Potential (I), Yasue, M., Tokyo
324 Symmetry Restoration of the Electroweak Interactions, Midorikawa, S., Tokyo-Tanashi
325 Nuclear Limiting Fragmentation and Hot Spots, Stelte, N., et. al., Marburg
326 Coherence and Non-Linear Effects in High Energy Reactions, Fowler, G. N. and Weiner, R. M., Exeter/Marburg
326A QCD Approach to Reggeon Calculus, Cohen-Tannoudji, G., et. al., unkown (France)
329 Direct Electron Pair Production in Pi(-)p Interactions at 16 GeV/c, Stroynowski, R., SLAC-John Hopkins-Caltech Collaboration
332 Measurement of the Virtual Photon Total Hadronic Cross Section on Nuclei, Goodman, M.S., et. al., Harvard/Tufts/Chicago/Fermilab/MIT/Michigan/Oxford
333 QCD Predictions for Deep-Inelastic Photon-Photon Scatterings, Sasaki, K., Yokohama
334 Horizontal Symmetry and Masses of Neutrinos, Yanagida, T., Tohoku
335 Inclusive Muon Production at Petra Energies, Berger, Ch., et. al., PLUTO
336 One-Loop Corrections To Nu-e Scattering in Weinberg-Salam Theory--Neutral Current Processes, Aoki, K-I., et. al. and, Kyoto
337 Proton Decay Rate, Tomozawa, Y., Michigan
338 Neutrino Mass in the SO(10) Grand Unified Gauge Model, Tomozawa, Y., Michigan

339 Generalization of Quantum Mechanics for High Energies And Quark Physics, Saavedra, I. and Treras, C., Chile
340 Muon-Electron Mass Ratio in a Semi Classical Model, Alfaro, J., et. al., Chile
341 Cosmic Rays in the Atmosphere and Characteristics of High Energy Interactions, Yodh, G. B., et. al., Maryland/Delaware
342 J/Psi Production by Pi(-) at Momenta from 150 to 193 GeV/c, WA11 Collaboration, Saclay-Imperial-Southampton-Indiana
343 Second Order QCD Contributions to Thrust and Spherocity, Chandramohan, T., et. al., Bonn
344 Leading Particle and Nuclear-Mass Effects on Multiparticle Production in 50 GeV/c Pi(-) -Em Collision, Varma, S.C., et. al., Kurukshetra
346A Inclusive Production of Charmed Particles at the ISR, Irion, J., et. al., Aachen-CERN-Harvard-Munich... Coll.
347 Higher Twist Effects in Asymptotically Free Gauge Theories: The Anomalous Dimensions of Four-Quark Operators, Okawa, M., Tokyo
348A Search for Short-Lived Particles Produced on Nuclei with a Heavy Liquid Mini Bubble Chamber, Badertscher, A., et. al., Bern-Munich (MPI) Collaboration
349 Evidence for Higher Order Asymptotic Freedom Corrections To Deep Inelastic Scattering, Duke, D. W. and Roberts, D. W., Florida State/Rutherford
350 Dense Detector for Baryon Decay, Courant, H., et. al., Minnesota
351 Depolarization Parameter in pp Inclusive Scattering at 6 GeV/c, Courant, H., et. al., Minnesota/Houston/ANL
352 Why Most Flavor Dependence Predictions for Nonleptonic Charm Decays are Wrong, Lipkin, H. J. and Lipkin, H. J., Fermilab/ANL
353 Why Masses and Magnetic Moments Satisfy Naive Quark Model Predictions, Cohen, I. and Lipkin, H. J., Weizmann/Fermilab/ANL
354 SU(5) Without SU(5): Conservations Laws in Unified Models of Quarks and Leptons, Lipkin, H. J., Fermilab/ANL
355 BL Parity, A New Conserved Quantity in Weak Interactions of Quarks and Leptons, Lipkn, J. L., Fermilab/ANL
356 Magnetic Moments of Quarks, Leptons and Hadrons: A Serious Difficulty for Composite Models, Lipkin, H. J., Fermilab/ANL/Weizmann
357 Do Bound Color-Octet States of Liberated Quarks Exist?, Lipkin, H. J., et. al., ANL/Fermilab/Weizmann
358 Nonperturbative Gluoproduction vs. Perturbative QCD Fusion Models for Heavy Quark Production, Gluck, M. and Reya, E., Dortmund
359 Signature for QCD in Gluon and Quark Jet Broadenings, Gluck, M and Reya, E., Dortmund
360 Energy-Momentum Constraints on Higher-Twist Effects in Deep Inelastic Scattering, Gluck, M. and Reya, E., Dortmund
361 Direct Evidence for the Three-Gluon Coupling, Gluck, M. and Reya, E., Dortmund
362 Dip and Kink Structures in Hadron-Nucleus and Hadron- Hadron Diffraction Dissociation, Chou, T. T. and Yang, C. N., Georgia/SUNY-Stony Brook
363 Proton Lifetime and Fermion Masses in an SO(10) Model, Lazarides, G. and Wetterich, D., CERN/Freiburg
364 Space-Time Structure of Jet Hadronization, Minakata, H., Fermilab
365 Exact Classical Solution of Equation of Motivation for Gravitino and Topology of Flat Space in Supergravity, Shima, K. and Kasuya, M., Saitama/Tokyo
366 Dynamical Theory of Spontaneous Breakdown for Chiral- Invariant QCD: Resolution of the U(1) Problem, Patrascioiu, A.N., et. al., Arizona
367 Higher Spin Vierbein Gauge Fermions and Hypergravities, Aragone, C. and Deser, S., Brandeis
368 Equivalence Principle for Sourced and Higgsed Gauge Fields, Aragone, C., Simon Bolivar
369 Rates of Nuclear Double Beta Decay and Lepton Number Nonconservation, Haxton, W. C., et. al., LASL
370 Parity Violation in Proton Nucleon Scattering at 6 GeV/c, Lockyer, N., et. al., Ohio/LASL/Elmhurst/Enrico Fermi/Illinois/ANL
371 Test of a Liquid Argon Multistrip Ionization Chamber with 8.5 um RMS Resolution, Deiters, K., et. al., IHAW-Berlin/CERN/Munich

| | |
|---|---|
| 372 | New Method for Identifying Events of Coherent Multiple Production in Nuclear Emulsion, Kim, C. O., et. al., Korea |
| 373 | Nonlinear Realization of the Chiral Symmetry in the MIT Bag, Szymacha, A. and atur, S., Warsaw/Copernicus Astronomical Ctr |
| 374 | Strange Particle Production in a Neutron Beam with a Momentum of 40 GeV/c, Aleev, A. N., et. al., Berlin-Dubna-Moscow...Collaboration |
| 375 | Measurement of the Polarization of Lambda(o)'s Produced Inclusively by Neutrons at an Average Neutron Energy..., Aleev, A. N., et. al., Berlin-Dubna-Moscow-...Collaboration |
| 376 | Dynamical Properties of Many-Dimensional U(1) Solitons, Makhankov, V.G., et. al., JINR |
| 377 | On a Possible Mass-Generating Scheme of Spin(2) Mesons, Lee, Y. Y., California-Los Angeles |
| 378 | Symmetry of New Particles, Nakamura, S., et. al., Tokai/Harvard/Rikkyo |
| 379A | Probing Hadronic Structure with high Cross Sections, Cohen-Tannoudji, G., et al., unknown (France) |
| 380 | Some Features of a Six Quark Model, Ahmad, M. and Jallu, M. Shafi, Kashmir |
| 381 | Direct Comparison of the Pi- and K- Form Factors, Dally, E. B., et. al., UCLA/JINR/Fermilab/Notre Dame/Pittsburgh |
| 382 | Goldstone Pion with Bag Confinement, Goldman, T. and Haymaker, R. W., Caltech/LASL/Louisiana State |
| 383 | Total Width and Leptonic Branch Ratio of the T (9.46), Niczyporuk, B., et. al., LENA collaboration |
| 384 | Lagrangian Formalism for Superfields with S(N) Internal Symmetry, Changkeun, J., et. al., Texas |
| 385 | Unification through a Supergroup, Ne'eman, Y. and Sternberg, S., Tel Aviv |
| 386 | Geometrical Gauge Theory of Ghost and Goldstone Fields and of Ghost Symmetries, Ne'eman, Y. and Thierry-Mieg, J., Tel Aviv/Texas/Caltech/Meudon |
| 387 | Extended Geometric Supergravity on Group Manifolds with Spontaneous Fibration, Thierry-Mieg, J. and Ne'eman, Y., Caltech/Tel-Aviv/Texas |
| 388 | Soft-Group-Manifold-Becchi-Rouet-Stora Transformations and Unitarity for Gravity, Supergravity, and Extensions, Ne'eman, Y., et. al., Tel Aviv/Texas/Harvard/Meudon |
| 389 | Unified Affine Gauge Theory of Gravity and Strong Interactions with Finite and Infinite $GL(4,R)$ Spinor Fields, Ne'eman, Y. and Sijacki, Dj., Tel Aviv/Boris Kidric |
| 390 | Quark Recombination Model and Half-Flavored Inclusive Distributions, Su, Y. S. and Yen, E., Tsing Hua |
| 391 | Nucleon Time-Like Form Factors and New Experimental Data On e+e- - p-bar-p, Bardek, V. and Zovko, N., Rudjer Boskovic |
| 392 | Goldberger-Treiman Constants of Dynamical Near-Goldstone Modes, Beg, M. A. B., Rockefeller |
| 393 | Duality Invariant Renormalization Group Study of the Four- State Potts Model, Hu, B., Houston/SLAC |
| 394 | Reggeization of Elementary Fermions in Arbitrary Renormalizable Gauge Theories, Grisaru, M. T. and Schnitzer,H. J., Brandeis |
| 395 | Supersymmetry Ward Identity for the Supersymmetric Non- Abellian Gauge Theory, Majumdar, P., et. al., Brandeis/MIT |
| 397A | Supersymmetric Regulators and Super Current Anomalies, Majumdar, P., et. al., Brandeis/MIT |
| 398 | New Approach to Effective Field Theories, Ovrut, B. and Schnitzer, H. J., Brandeis |
| 399A | Decoupling Theories For Effective Field Theories, Ovrut, B. A. and Schnitzer, H. J., Brandeis |
| 400 | Large-$p(T)$ Direct Photon Production and Opposite-Side Photon-Hadron Correlations in QCD, Contogouris, A. P., et. al., McGill |
| 401 | Large $O(Alpha(s)^2)$ Corrections Near the Kinematic Boundaries and the Definition of the Gluon Density, Contogouris, A. P., et. al., McGill |
| 402 | QCD Corrections Due to Quark Bremsstrahlung of Virtual and Real Photons, Contogouris, A. P., et. al., McGill |
| 403 | Multi-Quark States: Their Classification, Production and Possible Presence in Nuclei, Tsai, S. Y., Nihon |

404 Are the Upsiron Resonances Colored Particles? , Tsai, S. Y., Nihon
405 Bilocal Field Model of Mesons and its Spectroscopy , Mita, K. and Laursen, M. L., Oklahoma
406 Slavnov-Taylor Identities in Weinberg's Renormalization Scheme, Pascual, P. and Tarrach, R., Barcelona
407 Strong Corrections to Quarkonium Annihilation in Weinberg's Renormalization Scheme, Pascual, P. and Tarrach, R., Barcelona
408 Projectile-Independent Scaling in Hadron-Nucleus Collision , Kumar, V., et. al., Kurukshetra
409 Further Measurement of Nucleon Structure Functions in High Energy Muon-Iron Interactions, Ball, R. C., et. al., Michigan/Fermilab
410 Quark Form Factors and Leading Double Logarithms in QCD , Ellis, S. D. and Stirling, W. J., Washington
411 Non-Perturbative Gauge Vaioable Separation with Nonlinear , Chang, S-S., Northeastern
412A Limits on Associated Charm Muoproduction at Threshold , Bledsoe, H. L., et. al., California-Santa Cruz/SLAC
413A Inclusive Production of Pi(o) and Eta(o) Mesons in High Energy Photon-Proton Collisions, Breakstone, A. M., et. al., California-Santa Cruz
414A Diffractive Production of Omega(o) Mesons by High Energy Photons, Breakstone, A. M., et. al., California-Santa Cruz
415 Elastic Compton Scattering in the 50-130 GeV Range , Breakstone, A. M., et. al., UC-Santa Cruz
416 High-p(T) Pion Production in Pi-Pi Scattering , Biswas, N. N., et. al., Notre Dame/Michigan/Duke/Iowa
417A Results of a Beam Dump Experiment at the CERN SPS , CDHS Collaboratio, CERN-Dortmund-Heidelberg...Collab
418A Evidence For Hard Gluon Bremsstrahlung in Neutrino-Induced Charm Production, CDHS Collaboration, CERN-Dortmund-Heidelberg...Coll
419A Measurement of Nucleon Structure Functions in Inclusive Neutrino Nucleon Scattering, CDHS Collaboration, CERN/Dortmund/Heidelberg/Saclay
420 Pi- - Neon Collisions at 200 GEV , Band, H. R., et. al., Duke/SUNY
421 Present Status of the Pion-Nucleon Sigma Term , Dominguez, C. A. and Langacker, P., Texas A /Pennsylvania
422 Study of the Gauge Group $SU(2)L \times (T3)(R) \times U(v)(1)$ , Deshpande, N. G. and Iskandar, D., Oregon
423 Mass and Width of a Gluonic-bound State in Psi Decays , Ishikawa, K., California-Los Angeles
424 Heavy Flavor Production in Proton-Proton Interaction , Afek, Y., et. al., McGill
425 Photon-Proton Reactions , Afek, Y., et. al., McGill
426 Determination of the Pi and K Meson Structure Functions From Massive Dimuons Produced at 150 and 200 GeV, Badier, J., et. al., CEN-CERN-France-Palaiseau-LAL
427A Massive Dimuon Transverse Momentum and Angular Distribution From Pi- Induced Reaction at 150 to 280 GeV, Badier, J., et. al., CEN-CERN-France-Palaiseau-LAL
428 High Statistics Study ($10^6$ Events) of J/Psi Production In The Energy Range 150 to 280 GeV by Pi+/-, K+/-,p+/-..., Badier, J., et. al., CEN-CERN-France-Palaiseau-LAL
429 Experimental Determination of the Antiproton Structure Function by the Drell-Yan Mechanism, Badier, J., et. al., CEN-CERN-France-Palaiseau-LAL
430 Connection between the Goldstone Theorem, Vacuum Symmetry and Neutral PCAC, Fuchs, N. H. and Scadron, M. D., Purdue/Arizona
431 Resolving the On- and Off-Shell Ambiguity of Perturbative QCD Using Sum Rules, Ambjorn, J. and Sakai, N., Niels Bohr/Nordita
432 Effect of the Nonperturbative Vacuum in e+e- Annihilation , Ptkos, A. and Sakai, N., Niels Bohr/Nordita
433 Simulation of Centauros, Ellsworth, R.W., et. al., Maryland/Bartol
434 Branching of the Running Coupling Constant in Grand Unified Theories, Kubo, J. and Sakaibara, S., Dortmund
435 Affine Gauge Theory of Gravitational Interactions of Elementary Particles, Sijacki, Dj., Boris Kidric

| | |
|---|---|
| 436 | Bag Model Superalgebra, Sijakci, Dj., Boris Idric |
| 437 | Radiative Corrections to the Neutral Current Interactions in the Weinberg-Salem Model, Sakakibara, S., Dortmund |
| 438 | Test of QED in the Reactions e+e- -e+e- And e+e- -Mu+Mu- At CMS Energies From 9.4 to 31.6 GeV, Berger, Ch., et. al., PLUTO |
| 439 | Determination of the Electronic Branching Ratio of the Upsilon(9.46) and an Upper Limit for its Total Width, Berger, Ch., et. al., PLUTO |
| 440 | Lepton and Hadron Pair Production in Two-Photon Reactions, Berger, Ch., et. al., PLUTO |
| 441 | Measurement of the Reaction e+e- - Gamma-Gamma at CMS Energies from 9.4 to 31.6 GeV., Berger, Ch., et. al., PLUTO |
| 442 | Kho' (1600) Meson, O'Donnell, P. J., Toronto |
| 443 | Testing the Spin of the Gluon in Large Transverse Momentum Lepton Pair Production, Johnson, P. W. and Tung, W. K., Illinois Inst. of Tech./Fermilab |
| 444 | Elastic Proton-Proton Diffraction Scattering and its Energy Dependence, Henzi, R. and Valin, P., McGill/Harvard |
| 445 | A-Dependence of Inclusive Hadron Scattering at 100 GeV, Brenner, A. E., et. al., Fermilab/Brown/INFN-Bari/MIT/INR |
| 446 | Multiplicity Measurements in Hadron Reactions at 100 and 175 GeV, Aitkenhead, W., et. al., MIT/INFN/Fermilab/CERN/Brown |
| 447A | Study of K-P Interactions at 11 GeV/c: New Results on Strange Meson System, Aston, D., et. al., SLAC/Carleton |
| 448 | Narrowness of High-Spin Beautiful Mesons, Hikasa, K. and Igi, K., Tokyo U |
| 449 | Measurement of the Polarization Parameter in pp Elastic Scattering at 200 GeV/c, Fidecaro, G., et. al., CERN/INFN-Padova*Trieste/Vienna |
| 450 | Inclusive Gamma and Pi(o) Production in p-bar-p Interactions at 32 GeV/c, Poiret, C., et. al., Frnce-Sov. Un.Cern-Sov. Un./France |
| 451 | Evidence for a Dip at t = -1.4 (GeV/c)2 in p-bar-p Differential Elastic Scattering Cross-Section at 50 GeV/c, Asa'd, Z., et. al., LAPP-CERN-Niels Bohr-Genova-Oslo-London |
| 452 | Jet Properties and the Parton Interpretation of the Leading Logarithmic Approximation Scheme, Kirschner, R., Karl Marx U |
| 453 | Two-Parton Mass and Longitudinal Spectra in QCD Jets, Kirschner, R., Karl Marx U |
| 454 | Scale Breaking Polarized Quark and Gluon Distributions, Perlt, H. and Ranft, G., Karl Marx Univ. |
| 455 | Interference of QCD Interactions and the Detection of Vector Bosons by Hadronic Jets+ PP and bar-PP..., Ranft, G. and Ranft, J., Karl Marx U |
| 456 | Simple Space-Time Description of High Energy Hadron Nucleus Collisions, Wei-qin, C., et. al., Inst. HEP-China/Texas/Inst. TP-China |
| 457 | Nu-bar-Universality in High Energy Hadron-Nucleus Collisions, Chiu, C. B. and Tow, D. M., Texas |
| 458 | Constraint of Color Separation of Perturbative QCD, Chiu, C. B., Max Planck/Texas |
| 459 | Small p(T)-Component of Hadronic Jets Based on the Chromoelectric Flux Tube Model, Chiu, C. B., Max Planck/Texas |
| 460 | Diagrammatic Approach to Pair Production in Slowly Varying and Constant Fields, Chiu, C. B. and Nussinov, S., Max Planck/Texas/Tel Aviv |
| 461 | Polarization at High Transverse Momentum Through Weak Interaction and its Interference with Strong Interaction, Ranft, G. and Ranft, J., Karl Marx |
| 462 | Parton Distribution and Fragmentation Functions for Processes Involving Real Photons and Electrons, Schiller, A., et. al., Karl Marx Univ.-DDR |
| 463 | Gluon Jet Production in Deep Inelastic Lepton Hadron Collisions: Monte Carlo Study of Jet Fragmentation Effects, Ritter, S., et. al., Karl Marx |
| 464 | O(14) Super Unification of Color, Flavor and Generations, Sato, H., Tokyo |
| 465 | Electron Proton Collisions at a Centre of Mass Energy of 200 GeV: Proposal for a Study for a Canadian Electron Ring ...., CHEER, CHEER |

466A Contributions from Higher Twist Effects to the Quark Fragmentation Functions in Neutrino Data, Hagunauer, M., et. al., CERN-Milano
467 Higher-Order Corrections in the Cut Vertex Theory and the Reciprocity Relation in (Phi3)6 Field Theory, Kubota, T., Tokyo
468 New Treatment of the Scaling Violation for Flavor Singlet Distribution Functions in QCD, Kato, K. and Shimizu, Y., Tokyo
469 Real Photon to Neutral Pion Ratios and Photon-Jet Correlations in Quantum Chromodynamics, Kato, K. and Yamamoto, H., Tokyo
470 Energy-Momentum Distribution in e+e- Annihilation, Pancheri-Srivastava, G. and Srivastava, Y., INFN-Frascati/Northeastern
471 Quantum-Chromodynamic Radiation and Mean Scaling for Hadronic and Current Processes, Pancheri-Srivastava, G. and Srivastava, Y., INFN-Frascati/Northeastern
472 Superspace Actions for Conformal and Extended Conformal Supergravity Theories, Mansouri, F., Yale
473 General Features of the Reactions Pi+p - Delta++Pi(o)Pi(o) at 8 GeV/c, Cason, N. M., et. al., Notre Dame/ANL
474 Evidence for l=1 (A1) and l=0 (H) Axial-Vector Resonances in Pi-p - Pi+Pi-Pi(o) n at 8.45 GeV/c, Dankowych, J. A., et. al., Toronto/BNL/Carleton/Unio/McGill
475 Real Part of the Forward Elastic Nuclear Amplitude for pp, p-bar-p, Pi+p, K+p, and K-p Scattering between 70 and 200 Ge, Fajardo, L. A., et. al., Yale/Fermilab
476 High Statistics Study of Pi+p, Pi-p, and pp Elastic Scattering at 200 GeV/c, Schiz, A., et. al., Yale/Fermilab
477 Measurement of $D^*$ Production by Pions on Nucleons at $s^{**}1/2 = 19$ GeV, Fitch, V. L., et. al., Princeton/Cen Saclay/Torino/ENL
478 Theoretical and Experimental Review of the Weak Neutral Current: A Determination of its Structure and Limits ...., Kim, J. E., et. al., Pennsylvania
479 Quadratic Mass Relations in Topological Bootstrap Theory, Jones, C. E. and Uschersohn, J., Behlen
480 Neutral Pion Production and Diffraction Dissociation in High Energy Pi-Nucleon Collisions, Band, H. R., et. al., Duke
481 Quark Matter Diagnostics, Domokos, G. and Goldman, J. I., Johns Hopkins
482 Magnetic Moments of Composite Quarks and Leptons--Further Difficulties, Lipkin, H. J., ANL/Fermilab
483 Stochastic Model for Sequential Production Processes, Blazek, M., Bratislava
484A Search for Narrow bar-pp States in the Reaction Pi+p- Delta(++)$^f$ bar-pp, Bionta, R.M., et. al., BNL-Carnegie-Mellon-Fermilab-S.E. Mass U
485A Search for Narrow bar-pp States in Pi+p Collisions, Bionta, R.M., et. al., BNL/Carnegie Mellon/Fermilab/Southeastern
486 On the Direct Emission Component of K(o){L} - Pi+Pi-Gamma Decay, Ozaki, K., Osaka
487 Natural Composite Model for Quarks and Leptons Visnjic-Triantafillou, V., Fermilab
488 Consequences of Flavor as a Dynamical Quantum Number, Visnjic-Triantafillou, V., Fermilab
489 Upper Limits on Phi-Phi Production in 350 GeV/c Proton - Beryllium Collisions, Yamonouchi, T., et. al., Fermilab/SUNY
490A Determination of the Ratio of the Axial Vector to Vector Coupling Constant in Lambda(o)-p + e bar-Nu, Wise, J., et. al., Massachusetts/BNL
491 Measurement of Lambda(o) Polarization in Inclusive Production at 28.5 GeV/c, Lomanno, F., et. al., Massachusetts/BNL
492 Measurement of Lambda (-o) Polarization in Inclusive Production at 28.5 GeV/c and Comparison of Lamba (-0)/..., Lomanno, F., et. al., Massachusetts/BNL
493 Precise Measurement of Gamma (Lambda(o) - p+e- + NU-bar)/ Gamma (Lambda(o) - p + Pi-), Wise, J., et. al., Massachusetts/BNL
494 Exclusive Processes in Perturbative Quantum Chromodynamics, Lepage, G. P. and Brodsky, S. J., Cornell/SLAC
496 Inclusive Non-Strange Meson Production and Particle Ratios in Pi-p Interactions at 6 GeV/c, Ochiai, F., et. al., KEK
497 Primordial Motion Studies from p(T) Distributions in pp Interactions at 405 GeV/c, Suzuki, A., et. al., KEK/Osaka City/Kinki

| | |
|---|---|
| 498 | Searches for Effects of Flavor-Changing Neutral Currents, Kane, G. L. and Thun, R., Michigan |
| 499 | Apparent Violation of Unitarity in Elastic Pi-p Scattering, Roy, S. M., Tata |
| 500 | Natural Quark Field Confinement, Van der Spuy, E., AEB-Pelindaba |
| 501 | Quark Mass Ratios and Generalized Cabibbo Angles in an SU(5) x S(4) Model, Hayashi, H., et. al., Shizuoka/Tokai/L'Ecole Polytechnique |
| 502 | Hidden Symmetry in Yang Mills: A Path Dependent Gauge Transformation, Dolan, L., Rockefeller |
| 503 | Nonlocal Currents as Noether Currents, Dolan, L. and Roos, A., Rockefeller |
| 504 | Three Neutrino Oscillations and Present Experimental Data, Barger, V., et. al., Wisconsin/Rutherford |
| 505 | Hadroproduction of Psi and Upsilon, Barger, V., et. al., Wisconsin/Rutherford |
| 508 | Further Measurements of Spin Dependent Structure Functions of the Proton in Deep Inelastic and Resonance Region..., Baum, G., et. al., Berne/Bielefeld/KEK/Kyoto/Peking/SACLAY/... |
| 509 | Monte Carlo Program for Quark and Gluon Jet Generation, Sjostrand, T., Lund |
| 510 | How to Find the Gluon Jets in e+e- Annihilation, Andersson, B., et. al., Lund |
| 511 | Model for the Reaction Mechanism and the Baryon Fragmentation Distributions in Low P-PERP Hadronic..., Andersson, B., et. al., Lund |
| 512 | Cross Sections for Diffractive Charm Production at the CERN ISR, Eickmeyer, J., et. al., Aachen/CERN/Harvard/Munich/Northwestern/Riverside |
| 513 | Diquark Fragmentation Studies at the CERN ISR, Hanna, D., et. al., Aachen/CERN/Harvard/Munich/Northwestern/Riverside |
| 514 | Search for Narrow Baryon-Antibaryon States in the Reaction K+p-K+bar-ppp At 13 GeV/c, Frame, D., et. al., Glasgow/Birmingham/CERN |
| 515 | Search for a Narrow State with S=+1 and Q=+2 Produced in K+p Interactions at 13 GeV/c, Frame, D., et. al., Glasgow/Birmingham/CERN |
| 516 | Search for Single e+/- Production in p-bar-p Interactions at 70 GeV/c, Dumont, J.J., et. al., IIHE/Helsinki/Liverpool/Mons/Stockholm |
| 517 | Why is B-L Conserved and Baryon Number Not in Unified Models of Quarks and Leptons, Lipkin, H. J., ANL/Fermilab |
| 518 | Superconductor Model win Baryon-Type Collective Modes, Matsumoto, T., Tokyo |
| 519 | Inclusive Vector Meson Production in Nu(Mu)D Charged Current Interactions, Chang, C.C., et. al., Tufts/IIT-Chicago/Maryland/SUNY/Tohoku |
| 520 | Some Remarks on Leptonic and Hadronic Cross Sections at High Energy, Bandyopadhyay, P., Indian Statistical Institute |
| 521 | Hyperon and Omega- Nonleptonic Weak Decays, Tadic, D. and Trampetic, J., Zagreb/Rudjer Boskovic |
| 522 | Non-Leptonic Hyperon Decays and Harmonic Oscillator Quark Model for Baryons, Tadic, D. and Trampetic, J., Zagreb/Rudjer Boskovic |
| 523 | Sign of 1/2- Resonance Contributions to Parity-Violating NuNuRho Couplings, Picek, I., et. al., Rudjer Boskovic/Zabreb |
| 524 | Production of Hadrons with Transverse Momentum in the Range 0.5 to 2.2 GeV/c in pd Collision at 70 GeV, Abramov, V.V., et. al., IHEP-Serpukhov |
| 525 | Scaling Violation in the Pion Structure Function, Thews, R.L., et. al., Arizona/BNL |
| 526 | Gas Ionization Sampling Electromagnetic Shower Detector, Atac, M., et. al., Fermilab/Caltech/ANL |
| 527 | Pion Exchange in Muon-Pair Production by Pions, Sarma, K. V. L., Tata |
| 528 | Inclusion of Generations in SO(14), Ida, M., et. al., Kobe |
| 529 | Inclusive Spectra and Two Particle Correlations in Small p(T) Hadron Jets from Quark Cascade Model Wtih ..., Fukuda, H., et. al., Tokyo Inst. of Tech |
| 530 | Search for Light Charged Scalar Bosons, Kim, D. Y., Regina |
| 531 | Quark Moment Contributions to Baryon Magnetic Moments, Franklin, J., Temple |

532  Cosmological and Astrophysical Implications of Heavy Majorana Particles, Yanagida, T. and Yoshimura, M., Tohoku/NLHEP-Ibaraki ken

533  Search for Baryonium States in the Reaction pp - ppp-bar-p at 11.75 GeV/c, Kooijman, S., et. al., ANL/Elmhurst/Enrico Fermi

534  How to Search for Dibaryon Resonances Using Deuteron Targets, Kanai, K., et. al., Tokyo/Kogoshima/Waseda/Saga//JAPAN

535  Search for Narrow States Produced in the Reaction Pi-p n + Neutrals at 13 GeV/c, Cniang, I-H., et. al., BNL/Illinois/Princeton

536  Transverse Momentum of Hadrons Produced in NU and NU-bar Interactions on an Isoscalar Target in BEBC, Deden, H., et. al., Aachen-Bonn-CERN-Demokritos-IC-Oxford-Saclay

537  How Well Can a Phenomenological Quark-Quark Interaction Approximate QCD?, Rohrlich, F. and Rohrlich, F., Syracuse

538  Quark-Nucleon Phase Diagram and Quantum Chromodynamics, Kuti, J., et. al., CERN/Central Research Inst. for Physics-Hungary

539  Rigorous Inequalities in Nonrelativistic Quantum Mechanics with Applications to Quarkonium Spectroscopy, Khare, A., Manchester

540  Study of the Interference Correlations Between Pions in K-p Interactions at 32 GeV/c, Babintsev, V.V., et. al., IHEP-Serpukhov/Berlin/CEN/Vienna/PITH

541  Search for Resonant States in the (Phi$^\circ$pi-) System in K-p Interactions at 32 GeV/c, Arestov, Yu. I., et. al., IHEP-Serpukhov/Berlin/CEN/Vienna/PITH

542  Inclusive Production of Non-Strange Resonances in K-p Interactions at 32 GeV/c, Arestov, Yu., et. al., France-USSR CERN-USSR

543  Jet-Like Properties of Multiparticle Systems Produced in K+p Interactions at 70 GeV/c, Barth, M., et. al., Brussels-CERN-Genova-Mons-Nijmegen-Serpukhov

544  On Inelastic Screening in K(S) Regeneration on Nuclei, Kopeliovich, B. Z and Nikolaev, N. N., JINR/CERN

545  Many Particle Correlations in the Multiple Production on Nuclei, Levin, E. M., et. al., Leningrad/CERN

546  Experimental Results on J/Psi Production by Pi+/-, K+/-, p and p-bar Beams at 39.5 GeV/c, Corden, M. J., et. al., Birmingham/CERN/Ecole Polytech.-France

547  Higher Order Calculation of the Radiatively Induced Higgs Meson Mass, Mahanthappa, K. T. and Sher, M. A., Colorado

548  Consistent Way of Treating Arbitrary Spin Polarization, Kim, J., et. al., Seoul

549  Moments of Structure Functions as a Test of QCD, Kaushal, R. S., Ramjas

550  Strong-Coupling Expansion for Fermion Field Theories and the Large-N Limit of the Gross-Neveu Model, Guerin, F. and Kenway, R. D., Brown

551  Photon Structure Functions: Are They Measureable?, Irving, A. C. and Newland, D. B., Liverpool

552  How to Deal with Renormalisation Scheme-Dependent Perturbation Theory, Stevenson, P. M., Wisconsin

553  Measurement of the Charm Structure Function and its Role in Scale-Noninvariance, Clark, A. R., et. al., LBL/Fermilab/Princeton

554  Limit on Upsilon Muoproduction at 209 GeV, Clark, A. R., et. al., LBL/Fermilab/Princeton

555  Cross Section Measurements for Charm Production by Muons and Photons, Clark, A. R., et. al., LBL/Fermilab/Princeton

556  Short Range Correlations and Baryon Decay, Karl, G. and Lipkin, H. J., Guelph/Fermilab/ANL

557  Differential and Total Proton Cross Sections, Particle Production and the Parton Model, Afek, Y., et. al., McGill

558  Embedding of Non-Simple Lie Groups, Coupling Constant Relations and Non-Uniqueness of Models of Unification, Pasupathy, J. and Sudarshan, E. C. G., Indian Inst. of Science

559  Production of Charmed Particles in Nu(Mu) Nu-bar(Mu) N-Interactions and the Fragmentations Function of C-Quark, Konoplich, R. V., et. al., Moscow Phys. Engr. Inst.

560  Heavy Leptons in Neutrino Interaction: Production and Decay, Nikitin, Y. P. and Rubin, S. G., Moscow Phys. Engr. Inst.

561 Quark-Parton Cascade in Nuclei , Nikitin, Yu. P., et. al., Moscow Phys. Engr. Inst.
562 Hyperon Excitation in the Nuclear Coulomb Field , Nikitin, Yu. P., et. al., Moscow Phys. Engr. Inst.
563 High Mass Meson Resonance Production in pp Interactions at 405 GeV/c, Suzuki, A., et. al., KEK/Osaka/Kinki
564 Rapidity Distributions and Charmed Multiplicity in High Energy pp Interactions, Suzuki, A., et. al., KEK/Osaka/Kinki
565 Inclusive Study of Single- and Double-Pion Production in pp Interactions at 40 GeV/c, Suzuki, S., et. al., KEK/Osaka/Kinki
568 PP Spin Correlations at High $p(T)$ , Auer, I. P., et. al., ANL
569 Nuclear Effects in Photoproduction and Leptoproduction: Where are They? Are They Negligible?, Nikolaev, N. N., CERN
570 Interaction of High-Energy Particles with Nuclei , Nikolaev, N. N., Landau
571 If Intermediate Weinberg-Salam Boson Will Not Be Found.... , Berezinsky, V. S. and Smirnov, A. Yu., INR
572 Practically Stable Proton in $Su(5)$ Model , Berezinsky, V. S. and Smirnov, A. Yu., INR
573 Elastic Scattering of p+/-, Pi+/- and K+/- on Protons at High Energies and Small Momentum Transfer, Cool, R. L., et. al., Rockefeller
574 Diffractive Hadron Dissociation at 100 and 200 GeV , Cool, R. L., et. al., Rockefeller
575 Baryonization of Primordial Free Quarks in the Very Early Universe, Yun, S. K., Saginaw Valley-MI
576 Anomalous Low Mass e+e- Pair Production in 17 GeV/c Pi- p Collisions, Abshire, G., et. al., BNL/Pennsylvania/SUNY
577 Octet Enhancement as a Nonplanar Effect in Hadronic Weak Decays, Igarashi, Y., et. al., Bielefeld/Wupperal
578 Soft Hadron Interaction Effects in Charmed Meson Decays , Igarashi, Y., et. al., Bielefeld
579 Measurement of the Elastic Electron-Neutron Cross Section at High $Q(2)$, Rock, S., et. al., American University
580 Hadronic Interactions around 50 TeV , Ellsworth, R. W., et. al., Maryland/Bartol
581 New Experimental Data and Present Status of Dinucleons , Auer, I. P., et. al., ANL
582 Evidence for a Stable Sigma-Hyperdeuteron , Strobele, H., et. al., Darmstadt/Heidelberg
583 Surprising Nonleptonic Decays , Bernreuther, W., et. al., Heidelberg
584 Exceptional Groups for Grand Unification , Stech, B. and Dosch, H. G., Heidelberg
585 Baryon Spectrum in Low Order QCD with Migdal Regularization , Dosch, H.G., et. al., Heidelberg
586 Study of Double Tagged Two-Photon Reactions at Spear , Biddick, C. J., et. al., California-San Diego
587 Experimental Study of Low-$p(t)$ Hadron Fragmentation , Cutts, D., et. al., Brown/CERN/Fermilab/INFN-Bari/MIT
588 Determination of the Pion and Kaon Structure Functions , Aitkenhead, W., et. al., MIT/INFN/Fermilab/CERN/Brown
589 Angular Dependence of High-$p(T)$ Pi(o) Production , Owen, D. L., et. al., SUNY/INFN-Pisa
590 Effects of QCD Multi-Jet Contributions in Electron-Positron Annihilation, Mazzanti, P. and Odorico, R., IFU/INFN-Bologna
591 Structure of Eta-Like Mesons and Violation of the Vacuum Symmetry under Strong Interaction (QCD) Anomaly, Aizawa, N., et. al., Kyoto
592 Measurement of e-D Inclusive Scattering at Large $Q(2)$ Near $x(D)=1$, Arnold, R. G., et. al., American Univrsity/Bonn
593 Dynamical Symmetry of the Magnetic Monopole , Jackiw, R., California-Santa Barbara
594 Delta*-Delta(o)-Mixing as the Threshold Phenomenon , Achasov, N.N., et. al., Novosibirsk
595 Nature of Scalar Resonances , Aschasov, N.N., et. al., Novosibirsk
596 Spontaneous CP Nonconservation and Natural Flavor Conservation: A Minimal Model, Branco, G. C., Carnegie-Mellon

597  Theoretical Pattern for Neutrino Oscillations , Wolfenstein, L., Carnegie-Mellon
599  New Notions About Space-Time And Gravitation , Logunov, A.A., et. al., IHEP-Serpukhov
600  Study of Eta - Mu+Me- Decay , Dzhelyadin, R. I., et. al., IHEP-Serpukhov
601  Investigation of Eta Electromagnetic Structure in Eta - Mu+mu-Gamma Decay, Dzhelyadin, R. I., et. al., IHEP-Serpukhov
602  Further Study of Eta' - Mu+Mu-Gamma Decay , Dzhelyadin, R. I., et. al., IHEP-Serpukhov
603  Production of Charged Hadrons with Large Transverse Momenta in pp Collisions at 70 GeV, Abramov, V. V., et. al., IHEP-Serpukhov
604  Unique Left-Right Symmetric Model from General Horizontal Symmetry, Ecker, G., et. al., Wien
605  Observation of KK-bar Enhancement at 1.85 GeV in the Reaction K-p KK-bar-Lambda At 8.25 GeV/c, Amirzadeh, J., et. al., CERN/Birmingham/Glasgow/Michigan/Paris LPNHE...
606  Fragmentation Spectra in K-p Interactions at 110 GeV/c , Gottgens, R., et. al., Aachen-Berlin/CERN/Cracow/London/Vienna...
607  General Classical Solutions in the CP(n-1) Model , Din, A. M. and Zakrzewski, W. J., LAPP/Durham
608  Properties of General Classical CP(n-1) Solutions , Din, A. M. and Zakrzewski, W. J., LAPP/Durham
609  Measurement of Lepton and Pion Pair Production in Photon- Photon Collisions at DCI
       Courau, A., et. al., Clermont II/Paris-Sud
610  Beautiful Signature F+ - pn-bar as an Unambiguous Proof of the Annihilation Mechanism, Pham, X. Y., Pierre et Marie Curie
611  Search for Narrow States in the Lambda-bar-Nucleon System Produced in K+p Interactions at 13 GeV/c, Frame, D., et. al., Glasgow/Birmingham/CERN
612  Chemistry of Free Quarks , Lackner, K. S. and Zweig, G., Caltech
613  Measurement of Exclusive Hypercharge-Exchange Reactions at 35 to 140 GeV/c, May, E.N., et. al., ANL/Fermilab/SLAC
614  Polarization of Baryons and Vector Mesons in Quark Jets in e+e- Annihilation, Bartl, A., et. al., Wien/Wurzburg/IHOAW-Wien
615  Determination of Electromagnetic and Speudoscalar Form Factors of Hadrons from Inverse Pion Electroproduction, Alizade, V.V., et. al., JINR
616  Elliptic Relations Between Degrees of Polarization of Scattered Spin 1/2 Particles, Teli, M. T., Marathwada
617  On the Symmetries Between Neutrino and Antineurino-Nucleon Quasi-Elastic Scattering, Teli, M. T., Marathwada
618  Diffractive Hadronic Production of D Mesons , Koester, L.J., et. al., Illinois/Fermilab/Harvard/Oxford/Tufts
619  Fit of Upsilon and Charmonium Spectra , Martin, A., CERN
620  Condensation of Color-Triplet Diquarks and Manifestation of Color Gauge Theory in Hadron Physics, Matsumoto, T., Tokyo
621  Jet-Chamber of the Jade Detector at Petra , Bartel, W., et. al., JADE Collaboration
622  Additional Evidence for Fractional Charge of 1/3 e on Matter , LaRue, G.S., et. al., Stanford
623  Particle Distribution in the Central Region of Parton Jets , Wada, S., Cambridge
624  Second Order QCD Effect in J/Psi-Photoproduction , Tajima, T. and Watanabe, T., SLAC
625  Observation and Quantum Numbers Determination of the E(1420) Meson in Pi-p Interactions at 3.95 GeV/c, Dionisi, C., et. al., CERN/France/Madrid/Stockholm
626  On the Possible Existence of Stable Four-Quark Scalar Mesons with Charm and Strangeness, Isgur, N. and Lpkin, H. J., Toronto/ANL/Fermilab/Weizmann
627  Finite Temperature Approach to Confinement , Gava, E., et. al., IFT/INFN/ICTP-Trieste
628  CP Noninvariance in the Decays of Heavy Charged Quark Systems, Bander, M., et. al., California-Irvine
629  Jets as a Source for the Observed Increase in the Gamma-P Cross Section, Gotsman, E., et. al., California-Irvine
630  Jets in Photon Collisions and Tests for a Pointlike Coupling of the Photon, Soni, A., SLAC/California-Irvine

631 Limits on Nonconservation of Baryon Number , Learned, J., et. al., California-Irvine
632 Mechanism for the Difference in Lifetimes of Charged and Neutral D Mesons, Bander, M., et. al., California-Irvine
633 Can Superheavy Stable Particles be Detected in Current Spectral Data, Thomas, G. H. and Jones, D., ANL
634 Non-Abelian Traveling Wave Solutions for Massless QCD(2) , Jones, D. and Thomas, G. H., ANL
635 Parton Distribution and Fragmentation Functions for Processes Involving Real Photons and Electrons, Schiller, A., Karl Marx
636 On Charmed Particle Hadronic Production , Kartvelsihvili, V.G., et al., IHEP-Serpukhov
637 On Maximum Rise of the Average Associated Multiplicity in Processes with Large Momentum Transfers, Logunov, A.A., et. al., IHEP-Serpukhov
638 Two and Three Kaon Systems in the Reaction K+p - K-K+k+p at 13 GeV/c, Frame, D., et. al., Glasgow/Birmingham/CERN
639 Pseudoscalar Mesons and Mixing Models , Gault, F. D. and Rimmer, A. B., Durham
641 Hadronic Wave Function in Quantum Chromodynamics , Brodsky, S. J., et. al., SLAC/Cornell
642 Mass Formula for Mesons and Baryons , Sakharov, A. D., none listed
643 Estimate of the Interaction Constant between Quarks and Gluon Field, Sakharov, A.,
644 Study of the $K(o)^S K(o)^S$ System Produced in the Reaction Pi-p - $K(o)^S K(o)^S$n At 3.95 GeV/c, Loverre, P.F., et. al., CERN/France/Madrid/Stockholm
645 Study of the Reactions Pi-p - K+Sigma- (1385) At 3.95 GeV/c , Aguilar-Benitez, M., et. al., Madrid/Stockholm/CERN/College de France
646A Study of Events with a Particle at High Transverse Momentum , ACCDHW Collaboration, Annecy-CERN-Coll France-Dortmund-Heidelberg-Warsaw
647 Study of the Reactions Pi-p - K(s)(890) Lambda,K(o)(890) Sigma(o) and K(o)(890) Sigma(o)(1385) at 3.95 Gev/c, Aguilar-Benitez, M., et. al., CERN/College de France/Madrid/Stockhold/Collab.
648 Six-Quark Dibaryon Resonances , Mulders, P. J. and DeSwart, J. J., Nijmegen
649 Dilepton Production in Charged Current Neutrino Interactions , Ballagh, H. C., et. al., Calif/Fermilab/Hawaii/Washington/Wisconsin
650 Chiral Lepto-Hadrons , Subramanian, A., CERN
651 Note on the Decays Pi(o) - NuNu' , Herczeg, P. and Hoffman, C. M., LASL
652 Cosmic Ray Interaction of Energy Greater than 130 TeV , Koss, T. A., et. al., Washington
654 Towards the Phase Structure of Euclidean Lattice Gauge Theories with Fermions, Kawamoto, N., Amsterdam
655A New Approach to Analysis of Hadron Jets by the Lobachevsky Velocity Space Method, Bubelev, E. G., JINR
656A Elastic Hadronic Diffraction off Nuclei in the Constituent Quark Model, Kopeliovich, B.Z., et. al., JINR
657 Backlund Transformation, Local and Non-Local Conservation Laws for Super-Chiral Fields, Popwicz, Z. and Wang, L-L. C, SUNY/BNL
658 Quark-Diagram Classification of Charm Decays , Rizzo, T. G. and Wang, L-L. C., BNL
659 Regge Slope and the Lambda Parameter in QCD: An Empirical Approach Via Quarkonia, Buchmuller, W., et. al., Cornell
660 Effects of Coloured Glue in the QCD Motivated Bag of Heavy Quark-Antiquark Systems, Hasenfratz, P., et. al., CERN
661 Heavy Baryon Spectroscopy in the QCD Bag Model , Hasenfratz, P., et. al., CERN
663 Are High-P(T) Events Dominated by Non-Jet Processes? , Selove, W., Pennsylvania
664 Patterns of Mass Degeneracy in the Baryonium , Anderson, R. and Joshi, G.C., Melbourne
665 Electromagnetic Decay Widths for L = 1 J(PC) = 1-- T- Baryonia, Ellis, R.G., et. al., Melbourne/Oxford

666 Baryonium and Non-Exotic Hadron Trajectories from a Color Dependent Potential, Anderson, R. and Joshi, G. C., Melbourne
667 Four-Body Potential in Multiquark States, Warner, R. C. and Joshi, G. C., Melbourne
668 Interaction Potentials for Multiquark States From Instantons and Other Background Gauge Field Configurations, Warner, R. C. and Joshi, G. C., Melbourne
669 Factorisation of Regge Slopes for Multiquark Hadrons, Anderson, R. and Joshi, G. C, Melbourne
670 Polarization in Large Angle Proton-Neutron Elastic Scattering, Makdisi, Y., et. al., Minnesota/Rice/ANL
671 B-Quark Decays Can Decide if there is a T-Quark, Kane, G. L., Michigan
672 Su(4) Clebsch-Gordan Coefficients for the Formation of Baryonium and Exotic Baryons, Anderson, R. and Joshi, G C., Melbourne
673 Hadronic Transitions for Heavy Quarkonia in the O-Meson Model, Belyea, C. l. and Joshi, G. C., Melbourne
674 Search for Fractional Charge and Heavy Stable Particles at Petra, Bartel, W., et. al., JADE Collaboration
675 Chromodynamic Corrections to Neutrino Production of Heavy Quarks, Gottschalk, T., ANL
676 Fast-Meson Production and the Recombination Model, Takasugi, E., et. al., Texas/Max Planck/IIS-Bangalore
677 Recombination Model and Baryon Production by pp and Pi p Collisions, Takasugi, E. and Tata, X., Oregon/Texas
678 Recombination Model and Multimeson Production, Takasagi, E. and Tata, λ., Texas
679 Groups and Supergroups in Exceptional Gravity, Morel, B. and Thierry-Mieg, J., Harvard
680 Inclusive Production of Rho(o) and Phi Mesons in K+p Interactions at 32 GeV/c, Chliapnikov, P.V., et. al., France-Soviet Union CERN-Soviet Union Collab.
681 Charged Pion Production in 70 GeV/c K+p Interactions, Barth, M., et. al., IIHE/CERN/INFN-ISF/Mons/Nijmegen/IHEP-Serpukhov
682 Parton and Valon Distributions in the Nucleon, Hwa, R. C. and Zair L. Sajjad, Oregon
686 Inclusive $K^*+(892)$ Production in $K^*(o)(892)$ Production in K+p Interactions at 32 GeV/c, Ajinenko, I.V., et. al., Cern/Soviet Union and France-Soviet Union Collab
687 Chromodynamics and the Transverse Momentum of Jets and Hadrons in e+e- Annihilation, Halzen, F. and Scott, D. M., Hawaii/Wisconsin
688 Large Orders in the 1/N Expansion of the Non-Linear Sigma Model and Hidden Symmetries, DeVega, H., Pierre et Marie Curie
689 Excitation of the Level $^{12}C(15.1)$ by 17 Gev Protons and the Quark Model, Piccioni, O., et. al., California-San Diego
690 High Spin Mesons and Zweig-Superconvergence Relations, Bramon, A. and Masso, E., Barcelona
691 Angular Distributions and the Physics of Charmed Meson Production at the 4.028 GeV Resonance, Cahn. R. N. and Kayser, B., LBL/NSF
692 Self-Consistent Solutions for Fermions in Constant SU(2) Gauge Potentials, Akhoury, R. and Weisberger, W., SUNY
693 Fermionic Solitons in Gauge Theories and their Quantum O Corrections, Akhoury, R. and Weisberger, W. I., SUNY
694 Phase Transition in the Su(5) Model at High Temperatures, Daniel, M. and Vayonakis, C. E., CERN
695A Intrinsic Transverse Momentum II: Gauge Theories, Collins, J. C. and Soper, D. E., Princeton/Oregon
696 Search For Narrow p-bar-p States in the Reaction Pi-p-p Pi p-bar-p at 16 GeV/c, Chung, S. U., et. al., BNL/Brandeis/City College-NY/Southeastern/Mass
697 Search for Narrow p-bar-p States in Reaction p-bar-p - p-bar-pPi(o) (or Rho$^o$) at 5 GeV/c, Chung, S. U., et. al., BNL/Brandeis/Cincinnati/Florida/Southeastern
698 Search for Narrow States in the Lambda p+/- System, Bar Yam, Z., et. al., BNL/Brandeis/City College-NY/Southeastern/Mass
699 Magnetic Moment of a Massive Neutrino and Neutrino Spin Rotation, Fujikawa, K. and Shrock, R., SUNY
700 Current and Target Jets Produced in High Energy Neutrino-Deuterium Interactions, Kitagaki, T., et. al., Tohoku/IIT/Maryland/CUNY/Tufts

701 Charmed Baryon Production in high Energy Neutrino- Deuterium Interactions, Kitagaki, T., et. al., Tohoku/IIT/Maryland/SUNY/Tufts
702 First Results on Muon Production in Multiparticle Events from Jade, Bartel, W., et. al., JADE Collaboration
703 Quark-Tadpole Origin of the Delta-I + 1/2 Rule, Scadron, M. D., Arizona
704 Properties of the Vacuum Medium in Non-Abelian Gauge Theory, Hosotani, Y. and Yamagishi, H., Chicago/Princeton
705 Measurement and Analysis of $F^2$(Nu-bar) $(x,Q^2)$ and $xF^3$(Nu-bar) $(x,Q^2)$ In the Q(2)-Region 0.5-30 GeV(2), Weerts, H.J.M., et. al., GARGAMELLE SPS
706 First Experiences with a Fastbus System at Brookhaven, Leipuner, L.B., et. al., BNL/Yale
709A Experimental Constraints on Models of CP Violation, Schmidt, M.P., et. al., Yale/BNL
710 Comment on Chiral and Conformal Anomalies, Fujikawa, K., SUNY/Tokyo
711 Development of High Gain Multigap Avalance Detectors for Cherenkov Ring Imaging, Gilmore, R. S., et. al., SLAC
712 On the Quark Content of Delta(980) and Other Scalar Mesons, Bramon, A. and Masso, E., Barcelona
713 Study of Double-Scattering Effects in PiD and Pd Interactions between 15 and 400 Gev, Lubatti, H. J., et. al., Washington/Warsaw
714 Production of Hadrons in Pi+/-Ne Interactions, Burnett, T.H., et. al., Washington/L. Pasteur/Warsaw
715 Inclusive Pion Production in Pi-Pi Collisions at high Energies, Ko, W., et. al., UC-Davis/Warsaw/INR/AMMINP-Krakow/Washington
716 Homestake Mine Nucleon Decay Search, Deakyne, M., et. al., Pennsylvania
717 Pseudoscalar Glueball Production in Psi - GammaX, T - GammaX, e+e- - e+e-λ, and pp - λ, Goldberg, H., Northeastern
718 Magnetic Superconductivity and a Linear Response Theory for Non-Abelian Gauge Fields, Hosotani, Y., Enrico Fermi
719 Calculation of the Quark-Quark Contribution to the Drell-Yan Process II: Identical Particle Effects, Schellekens, A. N. and Van Neerven, W. L., ITP-Nijmegen/CERN
720 Comment on the Decay (Omega Omega-bar)$^{0-+}$ Hadrons in QCD Perturbation Theory, Celmaster, W. and Sivers, D., ANL
722 Some Remarks about Possible Sensitive Tests of the Influence of Higher Order Weak and Electromagnetic Radiative ..., Rodenberg, R., Aachen
723 Forward and Backward Multiplicities in K-p Interactions at 110 GeV/c, Gottgens, R., et. al., Aachen/Berlin/CERN/Cracow/London(IC)/Vienna..
724 Inclusive K(o) and K*+(890) Production in 70 GeV/c K+p Interactions, Barth, M., et. al., IIHE/CERN/INFN-ISF/Mons/Nijmegen/IHEP-Serpukhov
725 What is Happening to Multiquark Baryons?, Hogaasen, H. and Sorba, P., Oslo/LAPP
726 Bag Model Calculation of the Nucleon Lifetime In Grand Unified Theories, Din, A. M., et. al., CERN/LAPP
727 Integrability Condition in Loop Space, Eguchi, T. and Hosotani, Y., Enrico Fermi
728 Rescattering as the Origin of "Breaks" in Elastic Proton- Proton Diffraction Scattering, Henzi, R. and Valin, P., McGill
729 Two-Dimensional Quantum Chromodynamics with Massless Fermions, Sorense, C. and Thomas, G. H., ANL
730 Slope Parameter for the Differential Cross-Section for the Reaction p+d X+d in the Region of Small Momentum Transfer.., Akimov, Yu. K., et. al., JINR/Sofia/INR/Warsaw/Fermilab/Rochester/Az.
731 Anomalous Magnetic Coupling of Quarks and the Relative Radiative Decay Rates Gamma(Eta' -Rho(o)Gamma)/Gamma(Eta'.., Pham, T. N., Ecole Polytechnique
732 Isospin Violating Psi' - Pi(o) and Eta - 3Pi Decays, Pham, T. N., Ecole Polytechnique
733 Z(o) Asymmetries in Jets in e+e- Annihilation as a Test of Quark Fragmentation Mo......, Punuala, M.J., et. al., Ames

735  Study of e+e- Annihilation into Hadrons in the 1400-2200 MeV Energy Range with the Magnetic Detector DM1 at DCI: Observ.., Bixot, J. C., et. al. and Bixot, J. C., et. al., LAL/USTL
736  Intrinsic Charm of the Proton, Brodsky, S.J., et. al., SLAC/Nordita
737  Study of the Reactions Pi-p - K(o)Lambda and Pi-p -K(o) Sigma(o) at 3.95 GeV/c, Loverre, P.F., et. al., CERN/France/Madrid/Stockholm
738  Characteristics of Charged Particle Multiplicity Distributions in High Energy Nu(n) and Nu(p) Interactions, Kitagaki, T., et. al., Tohoku/IIT/Maryland/SUNY/Tufts
739  Neutral Current Interactions in Nu(n) and Nu(p) Collisions, Sommars, S., et. al., SUNY/IIT/Maryland/Tohoku/Tufts
740  Study of Dimuon Events in 280 GeV Muon Interactions, Aubert, J.J., et. al., European Muon Collaboration
741  Measurement of J/Psi Production in 280 GeV/c Mu+ Iron Interactions, Aubert, J.J., et. al., European Muon Collaboration
742  Study of Trimuon Events in 280 GeV Muon Interactions, Aubert, J.J., et. al., European Muon Collaboration
743  Inelastic J/Psi Production in 280 GeV Muon-Iron Interactions, Aubert, J.J., et. al., The European Muon Collaboration
744  Cabibbo Angle and Quark Masses in a Left-Right Symmetric Electroweak Gauge Model with Horzontal Symmetry, Ray, A. K. and Kundu, M. M., Visva-Bharati
745  New Approach to Evaluation of Multiloop Feynman Integrals, Chetyrkin, K. G. and Tkachov, F. V., INR-Moscow
746  Higher Order Corrections to Sigma(tot) (e+e- -hadrons) in Quantum Chromodynamics, Chetyrkin, K.G., et. al., INR-Moscow
747  Inelastic Thresholds and Dibaryon Resonances, Edwards, B. J. and Thomas, G. H., ANL
748  Properties of the Omega p Resonance at 1.8 GeV Observed in pp - p Omega p at 11.75 GeV/c, Arenton, M.W., et. al., ANL
749  Sensitive Search for the Time Evolution of an Electron Neutrino Beam, Cortez, B., et. al., Harvard/Michigan
750  Triple Gluon Coupling, Adler-Bell-Jackiw Anomaly, and Polarized Deep Inelastic Scattering, Lam, C. S. and Li, B. A., SLAC
751  Reformulation of the Salam-Weinberg Unified Theory of Weak and Electromagnetic Interactions, Khan, I., Riyad
752  $CP^{N-1}$ Model With Unconstrained Variables, Haber, H. E., et. al., LBL
753  Analyticity, Cuts and Branch Points In Wisconsin, Lipkin, H. J., none listed
754  Evidence for Gluon Radiation in High Energy Neutrino Interactions, Ballagh, H. C., et. al., UC-Berkeley/Fermilab/Hawaii-Manoa/Washington/Wisc
755  Observation of Charm Events Produced by High-Energy Photons in Nuclear Emulsions Coupled with a Magnetic Spectrometer, Photon-emulsion/Omega-photon, Photon-emulsion/Omega-photon Collaborations
756  Measurements of the Total Pi-He and p-He Cross-Sections and the Slopes of the Forward Diffraction Peak at Energies......, Burq, J. P., et. al., NA8 (Lyon/LNPI/CERN/Gustaf Werner/Clermont)
757  Measurement of Pi-p Elastic Scattering in the Coulomb Interference Region at an Incident Momentum of 345 GeV/c, Burq, J. P., et. al., NA8 (Lyon/LNPI/CERN/Gustaf Werner/Clermont)
758  Scattering Processes in Lattice Gauge Theories, Alessandrini, V. and Krzywicki, A., LPTHE-Orsay
759  Operator Analysis of New Physics, Weldon, H. A. and Zee, A., Pennsylvania
760  Determination of the Quark Ratio d(x)/u(x) in the Proton Using ep and Nu p Data, Allen, P., et. a., Aachen-Bonn-CERN-Munich-Oxford Collaboration
761  Observations of the D and E Mesons and Possible Three Kaon Enhancements in Pi-p - KoK+/-Pi+/-X, KoK+K-Kat 50 and 100, Bromberg, C., et. al., Caltech/Fermilab/Illinois-Chicago Circle/Indiana
762  Cosmological Models of the Universe with Rotation of Time's Arrow, Sakharov, A.D.,
763  Resolution of the Renormalisation-Scheme Ambiguity in Perturbative QCD, Stevenson, P. M., Wisconsin

764  QCD and Jets in Leptoproduction , Stevenson, P. M., Imperial
765  QCD and Final-State Jet Measures in Leptoproduction , Stevenson, P. M., Imperial
766  Hadron Electromagnetic Mass Differences and a Prediction of B+ - Bo, Chan, L.H., Louisiana State
767A Measurement of Direct Electron Production from Hadronic Bremsstrahlung, Goshaw, A. T., et. al., Duke/SLAC/Imperial
768  Direct Photon Production from Pi+p Interactions at 10.5 GeV/c, Goshaw, A. T., et. a., Duke/SUNY-Albany
769  Unitary Description of Scalar Mesons in PiPi and K-bar-K , Wicklund, A. B., et. al., ANL
770  Perturbative Calculation of Jet Structure in e+e- Annihilation, Ellis, R. K., et. al., Caltech
771  Operator Ordering and Feynman Rules in Gauge Theories , Christ, N. H. and Lee, T. D., Columbia
772  On the Hyperon Magnetic Moments and the 1-- - 0-+ + Gamma Decays, Oneda, S., et. al., Maryland/LASL
773  Bottom Excitation by Antineutrinos , Canal, C.A.G., et. al., La Plata/Louis Pasteur
774  Grand Unified Model with a Stable Proton and No Axion Problem, Deshpande, N. G., Oregon
775  Helical Motion of the Particles of Light , Kolyvodiakos, N. J., none
776  Mass and Mixing Scales of Neutrino Oscillations , Barger, V., et. al., Wisconsin/Fermilab/Rutherford
777  Observation of the Eta$^c$(2980) Produced in the Radiative Decay of the Psi'(3684), Himel, T. M., et. al., SLAC/LBL/California-Berkeley
779  Diffractve Production of Omega(o) Mesons by high Energy Photons, Breakstone, A. M., et. al., California-Santa Cruz
780  Neutrino-Antineutrino Oscillation , Wu, D. D., Harvard/CAS-Beijing
781  Neutrino Mass in the SO(10) Model without Right-handed Heavy Neutral Lepton, Wu, D. D., SLAC/Harvard
782A Muon Storage Ring for Neutrino Oscillations Experiments , Cline, D. and Neuffer, D., Fermilab
783  Neutrino Mixing in SO(10) , Hama, S., et. al., Ohio State
784  On the Pomeranchuk Singularity in Massless Vector Theories , Bartels, J., Tel-Aviv/Hamburg
785  Generation Problem and SO(14) Grand Unified Theories , Ma, Z. Q., et. al., IHEP-Peking
786  Generation Problem and S(7) Grand Unified Theories , Ma, Z. Q., et. al., IHEP-Peking
787  Possible SU(7) Grand Unified Theory with Four Generations of Light Fermions, Ma, Z. Q., et. al., IHEP-Peking
788  Is There a "Signature" of the Delta(980)-Meson Four-Quark Nature?, Achasov, N. N., et. al., IM-Novosibirsk
789  Heavy Quark-Antiquark Potential in the MIT Bag Model , Haxton, W. C. and Heller, L., LASL
790  Comparison of Hadron Production in K-p Interactions with Quark Jet Behaviour in Lepton Interactions, Gottgens, R., et. al., Aachen-Berlin-CERN-Cracow-London-Vienna-Warsaw
791  Potential Energy of Three Heavy Quarks in the MIT Bag Model , Aerts, A. T. and Heller, L., LASL
792  Deep Inelastic Muon-Nucleon Scattering at High Q(2) , Bollini, D., et. al., IFU INFN-Bologna/CERN/JINR/Munchen/CEN-Saclay
793  Diffractive Photoproduction of Strangeness and Charm--Over- Lap of QCD and Regge Models, Roy, D. P., ANL
794  On Perturbative QCD of Hard and Soft Processes , Efremov, A. V. and Radyushkin, A. V., JINR
795  Live Target for Measuring the Lifetime of Charmed Particles , Albini, E., et. al., INFN-Milano
796  Multiplicity Distributions in Neutrino-Hydrogen Interactions , Allen, P., et. al., Aachen-Bonn-CERN-Munich-Oxford Collaboration
797  Quantum Numbers of Quark Jets , Grossmann, P., Oxford
798A Pade Approximants and Solution of the Dispersion Relations for the Pion Photoproduction Amplitudes, Lebedev, A. I. and Mangazeev, B.V., Lebedev
800  Three(Pi)-Nucleon Collision in Coherent Production on Nuclei at 40GeV/c, Bellini, G., et. al., Dubna-Milano Collaboration

801A Measurements of Hyperon Magnetic Moments, Handler, R., et. al., Michigan/Rutgers/Wisconsin/Minnesota/BNL
802 Evidence of the Same Multiparticle Production Mechanism in p-p Collisions as in e+e-Annihilation, Basile, M., et. al., CERN/IFU INFN-Bologna/INFN-Frascati
803 Fractional Momentum Distribution in p-p Collisions Compared with e+e- Annihilation, Basile, M., et. al., IFU INFN-Bologna/CERN/INFN-Frascati/IFU-Bari
804 Energy Dependence of Charged Particle Multiplicity in p-p Interactions, Basile, M., et. al., INFN-Bologna/CERN/INFN-Frascati/INFN-Bari
805 Planar Jets in Proton-Proton Collisions, Basile, M., et. al., IF INFN-Bologna/CERN/INFN-Frascati/INFN-Bari
806 Monte Carlo Program for QCD Event Simulation in e+e- Annihilation at LEP Energies, Mazzanti, P. and Odorico, R., IFU INFN-Bologna
807 Contribution of Chi Decay to J/Psi Production in Pi- Be Collisions at 175 GeV/c, WA11 Collaboration, Saclay-Imperial-Southampton-Indiana
809 Xi(320) and Xi(1530) Production in K-p Interactins at 4.2 GeV, Mazzucato, M., et. al., Amsterdam-CERN-Nijmegen-Oxford Collaboration
810 Charge Distributions in K-p Interactions and Separation of Beam and Target Fragments with Energy, ABCLVW Collaboratin, Aachen-Berlin-CERN-London-Vienna-Warsaw
812 Study of Direct Single Photons and Correlated Particles in Proton-Proton Collisions at s(1/2) = 62.4 GeV, Angelis, A. L. S., et. al., CERN-Columbia-Oxford-Rockefeller Collaboration
813 Quark-Parton Structure Functions of Nuclei, Baldin, A. M., JINR
814 New Arguments in Favour of Exotic Resonances, Grigoryan, A. A. and Kaidalov, A. B., Yerevan
815 Antiquark Distributions in Pion and Nucleon, Arakelian, G. G. et. al., ITEP-Moscow
816 Hadronic Mass-Relations from Topological Expansion and String Model, Kaidalov, A. B., ITEP-Moscow
817 On the Possible Connection between Hard and Soft Processes, Kaidalov, A. B., ITEP-Moscow
819 Baryon Magnetic Moments in broken SU(6), Bohm, M., et. al., Wurzburg/CIEAIPN-Mexico
820 Grand Unification and Proton Stability Based on a Chiral SU(8) Theory, Deshpande, N. G. and Mannheim, P. D., Oregon/Connecticut
822 Inclusive and Semi-Inclusive Production of Positive Pions and Protons in p-bar-p Interactions at 22.4 GeV/c, Boos, E. G., et. al., Alma-Ata-Dubna-Helsinki-Moscow-Prague-Tbilisi
823 High Statistics Study of Dimuon Production by 400 GeV/c Protons, Gustafson, H. R., et. al., Michigan/Washington/Northeastern/Tufts
824 Study of Inclusive Production of Boson Resonances in p-bar-p Interactions at 32 GeV/c, Chekulaev, S. V., et al., IHEP-Serpukhov/INR/Alma-Ata/Mons/CEN/ULB-VUB
825A Lambda+c and Do Production at the ISR at s(1/2) = 63 GeV, ACCDHW Collaboration, Annecy-CERN-Coll France-Dortmund-Heidelberg-Warsaw
826 Masses of Lowest Lying Heavy Mesons in QCD, Reinders, L. J., et. al., Rutherford
827 QCD Contributions to Vacuum Polarization, Reinders, L. J., et. al., Rutherford
900 Radiative Corrections in the SU(2)L x U(1) Theory: A Simple Renormalization Framework, Sirlin, A., Institute for Advanced Study
901 Isolation of Definite Naturality Exchanges: The General Solution, Stacey, R., London
902 Generation of F(Pi) not = 0 in a New Chromodynamics, Stacey, R., University College
903 U(1): A New Approach, Stacey, R., U College London
904 Three Dimensional Model for Quark and Gluon Jets, Andersson, B., et. al., Lund
905 Polarized Target Asymmetry of Gamma d - pn in the Energy Range between 300 and 650 MeV, Awaji, N., et. al., Nagoya/NUCMT/Tokyo/Osaka

906 Measurement of Polarized Target Asymmetry on Gamma n - Pi(-)p Around the Second Resonance Region, Fujii, K., et. al., Nagoya/NUCMT/Osaka
907 Measurement of Polarized Target Asymmetry on Gamma p - Pi(+)n in Third Resonance Region, Fujii, K., et. al., Nagoya/NUCMT/Osaka
908A Recent Results from the CFRR Collaboration at Fermilab, CFRR Collaboration, Caltech/Fermilab/Rochester/Rockefeller
909 Large QCD Corrections to Hadron Calorimeter Reactions, Furman, M.A., Columbia
910A Jet Production in High Energy Hadron Collisions, Jain, P. L. and Das, G., SUNY-Buffalo
911 Effects of the Charged Gluon Jets in the Transversely Polarized e(+)e(-) Annihilation, Ji, C. R. and Kim, J. K., KAIS
912 Large P(Gamma) Reactions in Broken Color Gauge Theory, Ji, C. R., et. al., KAIS-Seoul
913 How to Sum the Planar Diagrams-A Reformulation of U(N) Lattice Gauge Theory, For N -oo, in Terms of a Statistical, Foerster, D., CEN SACLAY
914 Dynamical Stability of Local Gauge Symmetry - Creation of Light from Chaos, Foerster, D. et. al., CEN-Saclay/Niels Bohr/NORDITA
915 Observation of Charmed Baryon Production in Nu p Interactions, Allen, P., et. al., Aachen-Bonn-CERN-Munich(MPI)-Oxford
916 Measurement of the Transverse Momenta of Partons, and of Jet Fragmentation as a Function of s(1/2) in p-p Collisions, Angelis, A. L. S., et. al., CERN/Columbia/Oxford/Rockefeller
917 Scaled Energy Distributions of Single Hadrons Observed in Muon Proton Scattering, Auber, J. J., et. al., European Muon Collaboration
918 Transverse Momentum of Produced Hadrons in Deep Inelastic Muon Scattering, Aubert, J. J., et. al., European Muon Collaboration
919 Two Russian-American Experiments Fail to be a Creator of the True Pion Charge Radius, Dubnicka, S., et. al., SAS-Bratislava/Comenius
920 J/Psi and Psi' Particle Production in Pi(-) Cu-Interaction at 50 GeV/C, Antipov, Yu. M., et. al., IHEP=Serpukhov
921 Study of the Reaction p-bar-p - p + X at 22.4 GeV/c, Boos, E. G., et. al., Alma-Dubna-Helsinki-Kosice-Moscow-Prage-Tbilisi
922 Invariant Differential Cross Section of Reaction at 5 GeV, Antos, J., et. al., JINR/Kosice/Minsk
923 Polarization of Lambda-Hyperons, Produced in Pi(-)-Propane Interactins at P(Pi-) = 40 GeV/c, Dshemuchadse, S. W., et. al., Tbilisi/Moscow
924 Investigation of the Pi(-) p - Pi(o)Pi(o)N Reaction Near the Threshold, Eeljkov, A. A., et. al., JINR
925 Total Cross Section of PiN - PiPiN Reaction near the Threshold in the Theory of Broken Chiral Symmetry, Belkov, A. A., et. al., JINR
926 High Energy-Proton-Proton Scattering and Geometrical Scaling in a Wide Momentum Transfer Region, Goloskokov, S. V., et. al., JINR
927 Amplitude Analysis of PiN-Scattering in Impact Parameter Representation, Eremyan, Sh. S and Khachatryan, G. N., Epebah
928 Generalized Effective Potential and its Motion Equations, Ananikyan, N. S. and Savvidy, G. K., Epebah
929 Neutrino and t-Quark Masses in the SO(10) Grand Unification Scheme, Asatryan, H. M. and Matinyan, S. G., Epebah
930 Precise Measurement of the Ratio of the Axial Vector Coupling to Vector Coupling in Lambda(o)-p+e(-)+Nu-bar, Jensen, D. A., et. al., Massachusetts/BNL
931 Direct Electron Production at the T(4S), Andrews, D., et. al., CLEO Collaboration
932 Observation of Three Upsilon States, Andrews, D., et. al., CLEO Collaboratin
933 Results on the Gamma Resonances from the CUSB Group at CESR, Yoh, J. K., Columbia
934 Observation of a Fourth Upsilon State in e+e- Annihilations, Andrews, D., et. al., CLEO Collaboration
935 Progress Report on Results from the CLEO Detector, CLEO Collaboration, CLEO Collaboration
936 Measurements of the Total Cross Section of Sigma(-) and Xi(-) on Protons and Deuterons between 75-135 GeV/c at ..., Carter, A. A., et. al., WA42 Collaboration

937 CP Violation in Cascade Decays of b Mesons, Carter, A and Sanda, A. I., Rockefeller
939 Origin of Cancellation of Infrared Divergences in Coherent State Approach: Forward Process qq - qq + Gluon, Nelson, C.A., Fermilab
940 Transverse Momentum Distribution of Drell-Yan Pairs in Perturbation Theory, Jones, H. F. and Wyndham, J., Imperial College
941 Flavour Mixing and Proton Instability in Grand Unification Schemes, Asatryan, G.M. and Matinyan, S.G., Yerevan
942 Does the Standard Quark Model of D-Meson Nonleptonic Decays Survive?, Dulyan, L. S. and Khodjamirian, A. Yu., Epeban
943 Dispersion Sum Rules and Residues of Isoscalar Reggeons, Grigoryan, A. A., et. al., Yerevan
944 Radiative Transitions in Quarkonium and Quantum Chromodynamics, Khodjamirian, A. Yu., Yerevan
945 On Some Possibilities of Extraction of Local Ionization on the Background of Continuously Distributed One in ...., Astabatyan, R. A., et. al., Yerevan
946A Measurement of the Proton Structure Function F(2) in Muon-Hydrogen Interactions at 280 and 120 GeV, Aubert, J. J., et. al., European Muon Collaboration
947 Measurement of the Nucleon Structure Function F(2) by Muon Iron Interactions at 280, 250 and 120 GeV, Aubert, J. J., et. al., European Muon Collaboration
948 Scaling Violation in MuN Scattering and a Possible Interpretation, Aubert, J. J., et. al., European Muon Collaboration
949A Hyperon Polarization at Fermilab, Heller, K., Michigan-Minnesota-Rutgers-Wisconsin
950 Measurement of D(+), F(+), and Lambda(+)$^c$ Charmed Particle Lifetimes, Ushida, N., et. al., Aichi/Fermi/Kobe/Korea/McGill/Nagoya/Ohio/Okayama
951 Measurement of the D(o) Lifetime, Ushida, N., et. al., Aichi/Fermilab/Kobe/Korea/McGill/Nagoya/Ohio...
953 Anomalous Dimension Quark Counting at Large Transferred Momenta in QCD, Matveev, V. A., et. al., JINR
954 Integer Charge Quarks and Spontaneously Broken Color Symmetry, Chetyrkin, K. G., et. al., INR-Moscow
955 Is the Electric Charge Conserved?, Ignatiev, A. Yu., et. al., INR-Moscow
956 Electron Stability and Charge Fragmentation in Gauge Theories, Ignatiev, A. Yu., et. al., INR-Moscow
957 Neutron-Antineutron Oscillations, Chetyrkin, K. G., et. al., INR-Moscow
958 Baryon Asymmetry of the Universe versus Left-Right Symmetry, Kuzmin, V. A. and Shaposhnikov, M. E., INR-Moscow
959 Hadron Production in Pi(+)p, K(+)p and pp Collisions at 147 GeV/c and Properties of Jet-Like Multiparticle Systems, Brick, D., et. al., International Hybrid Spectrometer Consortium
960 Pi(-)-p Differential Elastic Cross-Section at 50 GeV/c in the Intermediate t Region, Asa'd, Z., et. al., LAPP/CERN/Copenhagen/Genova/Oslo/London
961 Production of Muon Pairs in the Continuum Region by 39.5 GeV/c Pi(-), K(+/-), p and p Beams Incident on a ...., Corden, M. J., et. al., Birmingham/CERN/Palaiseau
962A Use of 168/E Micro-Processors in Offline Analysis at SLAC, Aston, D., et. al., SLAC
963 Diffractive Photoproduction of D-D-bar Pairs on Silicon Target for Measuring Charmed Meson Lifetime, Baldini, R., et. al., NA-1 (FRAMM) Collaboration
964 Elastic Scattering Determination of the Negative Pion Form Factor, Dally, E. B., et. al., UCLA/JINR/Fermilab/Notre Dame/Pittsburgh
966 Experimental Study of Neutral and Charged Current Cross Sections and Y-Distributions for (Anti) Neutrinos, Jonker, M., et. al., CHARM Collaboration
967 , ,
969 Experimental Study of the K(o) - 3Pi(o) Decay, Barmin, V. V., et. al., ITEP-Moscow/IFUDS-Padova/INFN-Padova
971 Measurement of High Mass Gamma-Gamma and Pi(o)Pi(o) Production in 400 GeV/c p be Interactions; A Search for Eta(c), Baltrusaitis, R. M., et. al., Fermilab/Johns Hopkins

972   Two New Triggerable Track Chamber-Targets, Potter, D. M., Rutgers
973   Search for Baryonium Production in p-bar-p Interactions, Armstrong, T., et. al., CERN/Oliver Lodge-Liverpool
974   Search for Narrow Resonances in pp - pPi(+)n at 2 GeV/c, Sauer, J. R., et. al., ANL/Rice/Elmhurst/Enrico Fermi
975   Limits on Bottom Meson Production in Pi(-)N Interactions at 165 GeV/c, Barate, R., et. al., Saclay/Imperial/Southampton/CERN/Indiana
976   Inclusive Neutral Strange Particle Production in Nu p Interactions, Allen, P., et. al., Aachen-Bonn-CERN-Munchen(MPI)-Oxford
977   Discrimination between the Standard Theory of Electroweak Interactions and its Extensions, Costa, G. and Kabir, P., Virginia
978   , ,
979   Test of Nu Stability Using a 200 GeV Narrowband Neutrino Beam at EEBC, ABCDLOS Collaboration, Aachen-Bonn-CERN-Demokritos-London-Oxford-Saclay
980   Search for Neutrino Oscillations, Allred, J. C., et. al., LASL-Houston-Rice
981   On the Loop-Space Formulation of Gauge Theories, Gu, C. and Wang, L-L. C., SUNY-Stony Brook/IHES-France/BNL
982   Chi Spectroscopy in Pi(-) Be Collisions at 185 GeV/c, Lemoigne, Y., et. al., CEN-Saclay/Imperial/Southampton/Indiana/CERN
983   Pion Polarizability in a Chiral Quark Model, Ebert, D. and Volkov, M. K., JINR
985   Measurement of the Ratio of the Total Charged-Current Interaction Cross Sections of high Energy Neutrinos on Neutrons.., Armenise, N., et. al., BEBC TST Neutrino Collaboration
986   Search for Narrow Diquonium States in Pi(-)p Interactions at 10 GeV/c, Evangelista, C., et. al., Bari-Bonn-CERN-Daresbury-Glagow-Liverpool....
989A  Numerical Investigation of n = 2 and 3 Monopole Solutions, Adler, S. L. and Piran, T., IAS
990   Observation of Charmed Particles Produced by High-Energy Photons in Nuclear Emulsions Coupled with a Magnetic ..., Adamovich, M. I., et. al., Photon-Emulsion Collaboration
991   Full Operator Structure of the Nonleptonic Delta-S = 1 Weak Hamiltonian, Hill, C. T. and Ross, G. G., Fermilab/Oxford
992   Study of Dimuon Transverse Momentum, Scaling and a Dependence from Pi(-) Induced Reaction at 150-280 GeV/c, Badier, J., et. al., Saclay-CERN-College de France-Palaiseau-Orsay
993   Polarization in Charge Exchange Reaction Pi(-)p - Pi(o) n in Small Momentum Transfer Range at 40 GeV/c, Avvakumov, I. A., et. al., IHEP-Serpukhov
994   Further Study of the Narrow Hyperon State of Mass 3.17 GeV, Amirzadeh, J., et. al., Birmingham-CERN-Glasgow-Michigan-LPNHE/Cambridge
995   Study of Strangeness -3 Systems in K(-)p Interactions at 8.25 GeV/c, Al-Harran, S., et. al., Birmingham-CERN-Glasgow-Michigan-LPNHE
996   Further Study of the Xi*(2370) Produced in 8.25 GeV/c K(-)p Interactions, Amirzadeh, J., et. al., Birmingham-CERN-Glasgow-Michign-LPNHE
997   High Energy Antineutrino Interactions in Deuterium, Alassia, D., et. al., Amsterdam/IFU INFN-Bologna,Padova,Pisa,Torino/CEN
1000  Average Multiplicities of Secondary Particles in Hadron- Hadron Collisions and Quark Combinatorics, Anisovich, V. V., et. al., JINR-Dubna
1001  Does the Standard Quark Model of D-Meson Nonleptonic Decays Survive?, Dulyan, L. S. and Khodjamirian, A. Yu., Yerevan Physics Institute
1002  Massive Lepton Pair Production in Pip and pp Collisions Arakelian, G. G. and Boreskov, K. G., ITEP-Moscow
1005  Unique Solution for the Weak Neutral Current Coupling Constants in Purely Leptonic Interactions, Mark J Collaboration, PITHA-Mark J Coll
1006  Anomalies in Supersymmetric Gauge Theories, Grigoryan, G. V. and Grigoryan, R. P., Yerevan Physics Institute

1007 Photoproduction of Pi(-)-Mesons on Neutrons with Polarized Photons in the Energy Range $E^{Gamma}$=0.9 - 1.65 GeV, Abrahamian, L. O., et. al., Yerevan Physics Institute

1008 Measurement of the Ratio of the Total Charged-Current Interaction Cross Sections of High Energy Neutrinos on ...., Armenise, N., et. al., BEBC TST Neutrino Collaboration

1009 Some Comments on the Theory and the Experiments on Direct Photon Production, Kienzle-Focacci, M. N., Geneve

1010 Observation of a Spin 4 K* in the Reaction $K(+-)p - K(o)^S Pi(+-)p$ at 50 GeV/c, Cleland, W. E., et. al., Pittsburg/Geneva/Lausanne

1011 Observation of a Spin 6 Isopsin 1 Boson Resonance in the Charged KK-bar System, Cleland, W. E., et. al., Pittsburgh/Geneva/Lausanne

1012 Model for Parton Showers in QCD, Fox, G. C. and Wolfram, S., Caltech

1013 Parton Evolution as a branching Process, Cvitanovic, P., et. al., Nordita

1014 Model for the A Dependence of Inclusive Hadronic Cross Sections at Large Y and Small p- Perp, Durand, B. and Krebs, J., UW-Madison

1015 Classical Solutions of Non-Linear Sigma-Models and their Quantum Fluctuations, Din, A. M., LAPP Annecy

1016 Phase Diagram of SU(2) Lattice Gauge Theory in Four Dimensions, Moriarty, K. J. M., DESY

1017 Constructive Approach to the Polynomial Interaction in Four-Dimensional Space-Time, Osipov, E. P., IN-Novosibirsk

AIP Conference Proceedings

|  |  | L.C. Number | ISBN |
|---|---|---|---|
| No.1 | Feedback and Dynamic Control of Plasmas | 70-141596 | 0-88318-100-2 |
| No.2 | Particles and Fields - 1971 (Rochester) | 71-184662 | 0-88318-101-0 |
| No.3 | Thermal Expansion - 1971 (Corning) | 72-76970 | 0-88318-102-9 |
| No.4 | Superconductivity in d-and f-Band Metals (Rochester, 1971) | 74-18879 | 0-88318-103-7 |
| No.5 | Magnetism and Magnetic Materials - 1971 (2 parts) (Chicago) | 59-2468 | 0-88318-104-5 |
| No.6 | Particle Physics (Irvine, 1971) | 72-81239 | 0-88318-105-3 |
| No.7 | Exploring the History of Nuclear Physics | 72-81883 | 0-88318-106-1 |
| No.8 | Experimental Meson Spectroscopy - 1972 | 72-88226 | 0-88318-107-X |
| No.9 | Cyclotrons - 1972 (Vancouver) | 72-92798 | 0-88318-108-8 |
| No.10 | Magnetism and Magnetic Materials - 1972 | 72-623469 | 0-88318-109-6 |
| No.11 | Transport Phenomena - 1973 (Brown University Conference) | 73-80682 | 0-88318-110-X |
| No.12 | Experiments on High Energy Particle Collisions - 1973 (Vanderbilt Conference) | 73-81705 | 0-88318-111-8 |
| No.13 | π-π Scattering - 1973 (Tallahassee Conference) | 73-81704 | 0-88318-112-6 |
| No.14 | Particles and Fields - 1973 (APS/DPF Berkeley) | 73-91923 | 0-88318-113-4 |
| No.15 | High Energy Collisions - 1973 (Stony Brook) | 73-92324 | 0-88318-114-2 |
| No.16 | Causality and Physical Theories (Wayne State University, 1973) | 73-93420 | 0-88318-115-0 |
| No.17 | Thermal Expansion - 1973 (lake of the Ozarks) | 73-94415 | 0-88318-116-9 |
| No.18 | Magnetism and Magnetic Materials - 1973 (2 parts) (Boston) | 59-2468 | 0-88318-117-7 |
| No.19 | Physics and the Energy Problem - 1974 (APS Chicago) | 73-94416 | 0-88318-118-5 |
| No.20 | Tetrahedrally Bonded Amorphous Semiconductors (Yorktown Heights, 1974) | 74-80145 | 0-88318-119-3 |
| No.21 | Experimental Meson Spectroscopy - 1974 (Boston) | 74-82628 | 0-88318-120-7 |
| No.22 | Neutrinos - 1974 (Philadelphia) | 74-82413 | 0-88318-121-5 |
| No.23 | Particles and Fields - 1974 (APS/DPF Williamsburg) | 74-27575 | 0-88318-122-3 |

| No. | Title | LCCN | ISBN |
|---|---|---|---|
| No.24 | Magnetism and Magnetic Materials - 1974 (20th Annual Conference, San Francisco) | 75-2647 | 0-88318-123-1 |
| No.25 | Efficient Use of Energy (The APS Studies on the Technical Aspects of the More Efficient Use of Energy) | 75-18227 | 0-88318-124-X |
| No.26 | High-Energy Physics and Nuclear Structure - 1975 (Santa Fe and Los Alamos) | 75-26411 | 0-88318-125-8 |
| No.27 | Topics in Statistical Mechanics and Biophysics: A Memorial to Julius L. Jackson (Wayne State University, 1975) | 75-36309 | 0-88318-126-6 |
| No.28 | Physics and Our World: A Symposium in Honor of Victor F. Weisskopf (M.I.T., 1974) | 76-7207 | 0-88318-127-4 |
| No.29 | Magnetism and Magnetic Materials - 1975 (21st Annual Conference, Philadelphia) | 76-10931 | 0-88318-128-2 |
| No.30 | Particle Searches and Discoveries - 1976 (Vanderbilt Conference) | 76-19949 | 0-88318-129-0 |
| No.31 | Structure and Excitations of Amorphous Solids (Williamsburg, VA., 1976) | 76-22279 | 0-88318-130-4 |
| No.32 | Materials Technology - 1975 (APS New York Meeting) | 76-27967 | 0-88318-131-2 |
| No.33 | Meson-Nuclear Physics - 1976 (Carnegie-Mellon Conference) | 76-26811 | 0-88318-132-0 |
| No.34 | Magnetism and Magnetic Materials - 1976 (Joint MMM-Intermag Conference, Pittsburgh) | 76-47106 | 0-88318-133-9 |
| No.35 | High Energy Physics with Polarized Beams and Targets (Argonne, 1976) | 76-50181 | 0-88318-134-7 |
| No.36 | Momentum Wave Functions - 1976 (Indiana University) | 77-82145 | 0-88318-135-5 |
| No.37 | Weak Interaction Physics - 1977 (Indiana University) | 77-83344 | 0-88318-136-3 |
| No.38 | Workshop on New Directions in Mossbauer Spectroscopy (Argonne, 1977) | 77-90635 | 0-88318-137-1 |
| No.39 | Physics Careers, Employment and Education (Penn State, 1977) | 77-94053 | 0-88318-138-X |
| No.40 | Electrical Transport and Optical Properties of Inhomogeneous Media (Ohio State University, 1977) | 78-54319 | 0-88318-139-8 |
| No.41 | Nucleon-Nucleon Interactions - 1977 (Vancouver) | 78-54249 | 0-88318-140-1 |
| No.42 | Higher Energy Polarized Proton Beams (Ann Arbor, 1977) | 78-55682 | 0-88318-141-X |
| No.43 | Particles and Fields - 1977 (APS/DPF, Argonne) | 78-55683 | 0-88318-142-8 |
| No.44 | Future Trends in Superconductive Electronics (Charlottesville, 1978) | 77-9240 | 0-88318-143-6 |

| No. | Title | | |
|---|---|---|---|
| No. 45 | New Results in High Energy Physics - 1978 (Vanderbilt Conference) | 78-67196 | 0-88318-144-4 |
| No. 46 | Topics in Nonlinear Dynamics (La Jolla Institute) | 78-057870 | 0-88318-145-2 |
| No. 47 | Clustering Aspects of Nuclear Structure and Nuclear Reactions (Winnepeg, 1978) | 78-64942 | 0-88318-146-0 |
| No. 48 | Current Trends in the Theory of Fields (Tallahassee, 1978) | 78-72948 | 0-88318-147-9 |
| No. 49 | Cosmic Rays and Particle Physics - 1978 (Bartol Conference) | 79-50489 | 0-88318-148-7 |
| No. 50 | Laser-Solid Interactions and Laser Processing - 1978 (Boston) | 79-51564 | 0-88318-149-5 |
| No. 51 | High Energy Physics with Polarized Beams and Polarized Targets (Argonne, 1978) | 79-64565 | 0-88318-150-9 |
| No. 52 | Long-Distance Neutrino Detection - 1978 (C.L. Cowan Memorial Symposium) | 79-52078 | 0-88318-151-7 |
| No. 53 | Modulated Structures - 1979 (Kailua Kona, Hawaii) | 79-53846 | 0-88318-152-5 |
| No. 54 | Meson-Nuclear Physics - 1979 (Houston) | 79-53978 | 0-88318-153-3 |
| No. 55 | Quantum Chromodynamics (La Jolla, 1978) | 79-54969 | 0-88318-154-1 |
| No. 56 | Particle Acceleration Mechanisms in Astrophysics (La Jolla, 1979) | 79-55844 | 0-88318-155-X |
| No. 57 | Nonlinear Dynamics and the Beam-Beam Interaction (Brookhaven, 1979) | 79-57341 | 0-88318-156-8 |
| No. 58 | Inhomogeneous Superconductors - 1979 (Berkeley Springs, W.V.) | 79-57620 | 0-88318-157-6 |
| No. 59 | Particles and Fields - 1979 (APS/DPF Montreal) | 80-66631 | 0-88318-158-4 |
| No. 60 | History of the ZGS (Argonne, 1979) | 80-67694 | 0-88318-159-2 |
| No. 61 | Aspects of the Kinetics and Dynamics of Surface Reactions (La Jolla Institute, 1979) | 80-68004 | 0-88318-160-6 |
| No. 62 | High Energy $e^+e^-$ Interactions (Vanderbilt, 1980) | 80-53377 | 0-88318-161-4 |
| No. 63 | Supernovae Spectra (La Jolla, 1980) | 80-70019 | 0-88318-162-2 |
| No. 64 | Laboratory EXAFS Facilities - 1980 (Univ. of Washington) | 80-70579 | 0-88318-163-0 |
| No. 65 | Optics in Four Dimensions - 1980 (ICO, Ensenada) | 80-70771 | 0-88318-164-9 |
| No. 66 | Physics in the Automotive Industry - 1980 (APS/AAPT Topical Conference) | 80-70987 | 0-88318-165-7 |
| No. 67 | Experimental Meson Spectroscopy - 1980 (Sixth International Conference, Brookhaven) | 80-71123 | 0-88318-166-5 |
| No. 68 | High Energy Physics - 1980 (XX International Conference, Madison) | 81-65032 | 0-88318-167-3 |
| No. 69 | Polarization Phenomena in Nuclear Physics - 1980 (Fifth International Symposium, Santa Fe) | 81-65107 | 0-88318-168-1 |
| No. 70 | Chemistry and Physics of Coal Utilization - 1980 (APS, Morgantown) | 81-65106 | 0-88318-169-X |

**RAYMOND H. FOGLER LIBRARY**
**DATE DUE**

BOOKS ARE SUBJECT TO
RECALL AFTER TWO WEEKS